Guidebook to
Molecular Chaperones and Protein-Folding Catalysts

Other books by Sambrook & Tooze Publications

Guidebook to the Cytoskeletal and Motor Proteins
Edited by Thomas Kreis and Ronald Vale

Guidebook to the Extracellular Matrix and Adhesion Proteins
Edited by Thomas Kreis and Ronald Vale

Guidebook to the Homeobox Genes
Edited by Denis Duboule

Guidebook to Cytokines and Their Receptors
Edited by Nicos A. Nicola

Guidebook to the Secretory Pathway
Edited by Jonathan Rothblatt, Peter Novick, and Tom H. Stevens

Guidebook to the Small GTPases
Edited by Marino Zerial and Lukas A. Huber

Guidebook to the Calcium-Binding Proteins
Edited by Marco Celio, with Thomas Pauls and Beat Schwaller

Guidebook to Protein Toxins and Their Use in Cell Biology
Edited by Rino Rappuoli and Cesare Montecucco

Guidebook to
Molecular Chaperones and Protein-Folding Catalysts

Edited by

Mary-Jane Gething
Department of Biochemistry and Molecular Biology,
University of Melbourne
Australia

A SAMBROOK & TOOZE PUBLICATION
AT OXFORD UNIVERSITY PRESS
1997

Oxford University Press, Great Clarendon Street, Oxford OX2 6DP
Oxford   New York

Athens Auckland Bangkok Bogota Bombay Buenos Aires
Calcutta Cape Town Dar es Salaam Delhi Florence Hong Kong
Istanbul Karachi Kuala Lumpur Madras Madrid Melbourne
Mexico City Nairobi Paris Singapore Taipei Tokyo Warsaw

and associated companies in
Berlin   Ibadan

Oxford is a trade mark of Oxford University Press

Published in the United States
by Oxford University Press Inc., New York

© Sambrook and Tooze Publishing Partnership, 1997

All rights reserved. No part of this publication may be
reproduced, stored in a retrieval system, or transmitted, in any
form or by any means, without the prior permission in writing of Oxford
University Press. Within the UK, exceptions are allowed in respect of any
fair dealing for the purpose of research or private study, or criticism or
review, as permitted under the Copyright, Designs and Patents Act, 1988, or
in the case of reprographic reproduction in accordance with the terms of
licences issued by the Copyright Licensing Agency. Enquiries concerning
reproduction outside those terms and in other countries should be sent to
the Rights Department, Oxford University Press, at the address above.

This book is sold subject to the condition that it shall not,
by way of trade or otherwise, be lent, re-sold, hired out, or otherwise
circulated without the publisher's prior consent in any form of binding
or cover other than that in which it is published and without a similar
condition including this condition being imposed
on the subsequent purchaser.

A catalogue record for this book is available from the British Library

Library of Congress Cataloging in Publication Data
(Data available)

ISBN 0 19 859949 8 (Hbk)
ISBN 0 19 859948 X (Pbk)

Typeset by
EXPO Holdings, Malaysia
Printed in Great Britain by
The Bath Press

Preface

The past ten to fifteen years have been enormously exciting for scientists interested in the mechanistic and biological aspects of protein folding. We have entered this field from many directions, bringing a great diversity of systems and a myriad of skills. Many of us would not have suspected initially that problems of protein folding lay at the core of our observations. Some of us would even have been reluctant to admit that polypeptides need any assistance in attaining their correct three-dimensional structures. But we all now know that molecular chaperones and protein folding catalysts lie at the heart of biology, presiding over the birth, life, and death of most (if not all) cellular proteins, and over the functioning and regulation of many cellular processes and responses.

Because each of us usually focuses on only a single protein or a single biological process, even the 'experts' in the field cannot but be helped by this compendium of current knowledge about the multitude of chaperones and folding catalysts, and about the protein families to which most belong. For the newcomer to the field, I believe this guidebook will be indispensible. To the 260 experts who contributed to the book I give heartfelt thanks on behalf of myself and future readers. In particular, I thank those who wrote the overviews for the protein families and also helped me plan the appropriate contributors for the individual entries in each section. Above all I thank Zoe Ord, whose organizational skills and cheerfulness continually saved me as each deluge of discs, faxes, and emails (and then the dreaded proofs) threatened to engulf my office and my sanity.

Melbourne M. J. G.
June 1997

Footnote: a note about nomenclature

Many different conventions are used in naming the proteins and protein families described in this guidebook. To bring some temporary order to the chaos, I have chosen to use fully capitalized names to denote the protein family (thus HSP70, CPN60, HSP90 etc.) and an initial capital letter for specific family members (thus Hsc70, Cpn60, Hsp83 etc.). Italicized names denote genes, with some variation in the use of upper or lower case letters depending on the species.

A computer system will be available from December 1997 to accompany Guidebook to Molecular Chaperones and Protein-Folding Catalysts

Due to the rapid pace of biological research, the editor and publishers of this book believe it is important that its readers are kept informed of recent developments on these proteins. For this purpose, we have established a computer database that can be accessed through the worldwide web. This database will not include the full entries shown in this book; instead the authors have been asked to add, periodically, any new information on their protein that has been published since they wrote their original entry. Authors will be asked to deposit their updates from December 1997.

The update system can be accessed using any of the standard tools for browsing the worldwide web, such as Netscape or Mosaic. The URL for information relating to this book is *http://www.oup.co.uk/guidebooks/chaperon*. For information on other Sambrook & Tooze Guidebooks, start from the Oxford University Press home page at *http://www.oup.co.uk/*. and follow the links to the Guidebooks series.

Contents

List of contributors — xv
List of abbreviations — xxv

Part 1. HSP70 Proteins

The Hsp70 family — an overview
B. Miao, J. Davis and E.A. Craig — 3

Three-dimensional structure of Hsc70
S. Wilbanks — 13

Three-dimensional structure of the peptide-binding domain of DnaK
M.-J. Gething — 18

***Escherichia coli* DnaK**
A. Buchberger and B. Bukau — 22

***Escherichia coli* Hsc66**
T.H. Kawula — 25

***Saccharomyces cerevisiae* Ssa proteins**
T. Ziegelhoffer and E.A. Craig — 26

***Saccharomyces cerevisiae* Ssb proteins**
T. Ziegelhoffer, Christine Pfund and E.A. Craig — 29

***Saccharomyces cerevisiae* Ssc1**
J. Davis, B. Miao and E.A. Craig — 30

***Saccharomyces cerevisiae* Ssh1p**
E.A. Craig — 32

***Saccharomyces cerevisiae* Kar2p**
J.L. Brodsky and M.D. Rose — 33

***Saccharomyces cerevisiae* Ssi1p**
B.K. Baxter and E.A. Craig — 36

***Schizosaccharomyces pombe* BiP**
J. Armstrong — 37

Plant BiP proteins
J.W. Gillikin and R.S. Boston — 38

***Trypanosoma brucei* BiP**
J.D. Bangs — 41

***Drosophila melanogaster* Hsc4p**
A.D. Mehta, D.M. Rubin, F. Elefant and K.B. Palter — 42

***Drosophila melanogaster* Hsc3p**
D. Rubin and K. Palter — 45

Fish HSP70 proteins
L.E. Hightower, J.A. Ryan and C.E. Norris — 47

The *hsp70* genes of mice and men
M. Tavaria, I. Kola and R.L. Anderson — 49

Mammalian Hsc70 and Hsp70 proteins
L.E. Hightower and S.-M. Leung — 53

Mammalian Prp73
J.F. Dice — 58

Mammalian BiP
M.-J. Gething — 59

Mammalian Pbp74
S.K. Pierce and J. Dahlseid — 65

Mammalian mitochondrial Hsp70
N.J. Hoogenraad, P.B. Høj, D. Naylor and T. Webster — 67

STCH
F.J. Kaye, G.A. Otterson and G.C. Flynn — 69

Part 2. HSP110/SSE Proteins

The HSP110/SSE stress proteins — an overview
D.P. Easton and J.R. Subjeck — 73

***Saccharomyces cerevisiae* Sse1 and Sse2 proteins**
H. Mukai, H. Shuntoh and T. Kuno — 76

***Neurospora crassa* Hsp88**
N. Plesofsky-Vig and R. Brambl — 77

The sea urchin egg receptor for sperm
R.L. Stears, M.L. Just, and W.J. Lennarz — 78

Mammalian Hsp110
D.P. Easton and J.R. Subjeck — 81

Mammalian Hsp70h
K.D. Dyer and H.F. Rosenberg — 82

Mammalian OSP94
S.R. Gullans and R. Kojima   84

Mammalian Grp170
D.P. Easton and and J.R. Subjeck   85

Part 3. HSP40 (DnaJ-related) Proteins

HSP40 (DnaJ-related) proteins — an overview
D.M. Cyr   89

***Escherichia coli* DnaJ**
A. Wawrzynów and M. Zylicz   95

***Escherichia coli* CbpA**
T. Mizuno and C. Ueguchi   98

***Escherichia coli* RcsG**
W.L. Kelley and C. Georgopoulos   100

***Saccharomyces cerevisiae* Ydj1**
D.M. Cyr and A.J. Caplan   102

***Saccharomyces cerevisiae* Sis1**
K.T. Arndt   104

***Saccharomyces cerevisiae* Zuotin**
D.M. Cyr   105

***Saccharomyces cerevisiae* Mdj1p**
E. Schwarz   106

***Saccharomyces cerevisiae* Sec63p**
J.L. Brodsky   108

***Saccharomyces cerevisiae* Scj1p**
P.A. Silver   110

Plant Anj1
J.-K. Zhu   112

DnaJ-like proteins from leek: Ldj1 and Ldj2
J.-J. Bessoule   113

Cysteine-string proteins
K.E. Zinsmaier   115

***Drosophila melanogaster* Tid56**
U. Kurzik-Dumke, A. Debes and D. Gundacker   117

Mammalian Hsp40
K. Ohtsuka   121

Human Hdj2
A. Chellaiah, A. Davis and T. Mohanakumar   123

Auxillin
E. Eisenberg and L.E. Greene   124

Human neurone-specific Hsj1 proteins
M.E. Cheetham and M. Tighe   126

Mammalian Mtj1
M.W. Lässle, B.R. Zetter, S.E. Brightman and G.L. Blatch   129

Part 4. GrpE-Like Proteins

The GrpE family of proteins — an overview
O. Deloche, D. Ang and C. Georgopoulos   133

***Escherichia coli* GrpE**
A. Buchberger and B. Bukau   137

***Saccharomyces cerevisiae* Mge1**
S. Laloraya and E.A. Craig   139

***Drosophila melanogaster* Droe1p**
A.D. Mehta, J.Y. Lee and K.B. Palter   141

Mammalian mitochondrial GrpE
D.J. Naylor, N.J. Hoogenraad and P.B. Høj   142

Part 5. HSP90 Proteins

The HSP90 family — an overview
T. Scheibel and J. Buchner   147

***Escherichia coli* HtpG**
U. Jakob and J.C.A. Bardwell   151

***Saccharomyces cerevisiae* Hsp90**
Y. Kimura and S. Lindquist   152

***Drosophila melanogaster* Hsp83**
D.F. Nathan and S. Lindquist   154

Plant Hsp90
P. Krishna   156

Mammalian Hsp90
U. Jakob and J. Buchner   158

Mammalian Grp94
J. Melnick and Y. Argon   161

Part 6. CPN60 and CPN10 Proteins

The CPN60 and CPN10 families — an overview
F.U. Hartl and M. Mayhew — 167

***Escherichia coli* GroEL, structure and function**
J.S. Weissman, W.A. Fenton, K. Braig,
P.D. Adams and A. Horwich — 173

The structure of *Escherichia coli* GroES
J.F. Hunt and J. Deisenhofer — 179

***Escherichia coli* GroES**
A. Taher and S.J. Landry — 183

Bacteriophage T4 Gp31 protein
S.M. van der Vies — 185

***Thermus thermophilus* GroEL and GroES**
H. Taguchi and M. Yoshida — 187

***Saccharomyces cerevisiae* Hsp60**
T. Langer and W. Neupert — 189

***Saccharomyces cerevisiae* Hsp10**
J. Höhfeld — 191

Plastid Cpn60
R.J. Ellis — 192

Plastid Cpn21
A.A. Gatenby and H. Chen — 194

Mammalian Cpn60
P.B. Høj, N.J. Hoogenraad and D. Hartman — 197

Mammalian Cpn10
P.V. Viitanen — 199

The neuroendocrine polypeptide 7B2
G.J. Martens — 201

Part 7. Cytosolic Chaperonins

Cytosolic chaperonins — an overview
H. Kubota and K. Willison — 207

Thermosome
T. Waldmann, A. Lupas and W. Baumeister — 211

CCTα
H. Kubota and K. Willison — 215

CCTβ
H. Kubota and K. Willison — 217

CCTγ
H. Kubota and K. Willison — 218

CCTδ
H. Kubota and K. Willison — 220

CCTε
H. Kubota and K. Willison — 221

CCTζ
H. Kubota and K. Willison — 222

CCTη
H. Kubota and K. Willison — 224

CCTϑ
H. Kubota and K. Willison — 225

Cofactors in the facilitated folding of α- and β-tubulin
N.J. Cowan — 226

Part 8. HSP100 Proteins (Clps)

The HSP100 family — an overview
E.C. Schirmer and S. Lindquist — 231

***Escherichia coli* ClpA**
S. Gottesman, S. Wickner and M.R. Maurizi — 236

***Escherichia coli* ClpB**
C.L. Squires and C. Squires — 238

***Escherichia coli* ClpX**
A. Wawrzynów and M. Zylicz — 240

***Escherichia coli* ClpY**
S. Gottesman, W.-F. Wu and D. Missiakas — 242

***Bacillus subtilis* ClpC**
E. Krüger and M. Hecker — 243

***Synechococcus* sp. PCC 7942 ClpB**
A.K. Clarke and M.-J. Eriksson — 246

***Synechococcus* sp. PCC 7942 ClpC**
A.K. Clarke — 247

Saccharomyces cerevisiae Hsp104
E.C. Schirmer and S. Lindquist   249

Saccharomyces cerevisiae Hsp78
M. Schmitt and T. Langer   251

Plant Hsp101/ClpB
E. Vierling   253

Chloroplast-localized Clp proteins
E. Vierling   255

Leishmania Hsp100
J. Clos, A. Hübel, S. Brandau, A. Dresel and
A. Hörauf   259

Malaria plasmid ClpC
I. Wilson   261

Mammalian HSP100
J.R. Glover   264

Part 9. Small HSPs

The small heat shock proteins — an overview
J. Buchner, M. Gaestel and E. Vierling   269

Escherichia coli small heat shock proteins IbpA and IbpB
M. Gaestel and J. Buchner   273

Saccharomyces cerevisiae Hsp26
M. Tuite, D.R.J. Rahman and N.J. Bentley   274

Plant small heat shock proteins (sHSPs)
E. Vierling and G.J. Lee   277

Drosophila melanogaster small heat shock proteins
S. Michaud, D.R. Joanisse and R.M. Tanguay   280

Mammalian Hsp27
J. Landry   283

Small heat shock protein Hsp25 from mouse and rat
M. Gaestel and J. Buchner   285

Mammalian α-Crystallins
W.W. de Jong and W.C. Boelens   288

Part 10. Calnexin and Calreticulin

Calnexin and calreticulin proteins — an overview
D.B. Williams   293

Saccharomyces cerevisiae Cne1
F. Parlati, D.Y. Thomas and J. Bergeron   296

Schizosaccharomyces pombe Cnx1
F. Parlati, D.Y. Thomas and J. Bergeron   298

Mammalian calnexin
S. Pind and D.B. Williams   299

Mammalian calreticulin
J.R. Peterson and A. Helenius   304

Part 11. PDI and Thioredoxin-Related Proteins

PDI and thioredoxin-related proteins — an overview
J.C.A. Bardwell   311

Escherichia coli thioredoxin
H. Nakamura, M. Björnstedt and A. Holmgren   314

Escherichia coli glutaredoxin
H. Nakamura, F. Åslund and A. Holmgren   316

Escherichia coli DsbA
J.C.A. Bardwell   318

Structure of DsbA
L.W. Guddat and J.L. Martin   320

Escherichia coli DsbB
J.C.A. Bardwell   322

Escherichia coli DsbC
D. Missiakis, A. Zapun, T.E. Creighton and S. Raina   324

Escherichia coli DsbD
D. Missiakis and S. Raina   326

Escherichia coli DsbE
D. Missiakis and S. Raina   327

Synechococcus TxlA
A. Grossman and J. Collier   329

Bacillus brevis Bdb
S. Udaka   330

Rhodobacter capsulatus HelX
R.G. Kranz   331

Bradyrhizobium japonicum TlpA
H. Loferer   332

Saccharomyces cerevisiae protein disulfide isomerase (PDI)
M.F. Tuite and A. Dunn   335

Saccharomyces cerevisiae Eug1
T.H. Stevens   339

Saccharomyces cerevisiae Mpd1p and Mpd2p
H. Tachikawa and T. Mizunaga   341

BS2: Trypanosoma brucei protein disulfide isomerase homologue
J.D. Bangs   342

Mammalian thioredoxin
H. Nakamura, M. Björnstedt and A. Holmgren   343

Mammalian glutaredoxin
H. Nakamura, F. Åslund and A. Holmgren   346

Mammalian protein disulfide isomerases
R. B. Freedman   348

Mammalian Erp61
N. Marcus and M. Green   351

Mammalian ERp72
D. Ferrari and H.-D. Söling   353

Mammalian CaBP1
J. Fuellekrug and H.-D. Söling   354

Part 12. Peptidyl-Prolyl Isomerases (PPIases)

Peptidyl-prolyl isomerases — an overview of the cyclophilin, FKBP and parvulin families
K. Dolinski and J. Heitman   359

Part 12A. Cyclophilin PPIases

Escherichia coli cyclophilins
N. Takahashi   370

Saccharomyces cerevisiae cyclophilin A /Cpr1/Cyp1
B. Haendler   372

Saccharomyces cerevisiae cyclophilin B /Cpr2/Cyp2
R. Cafferkey, R. Stan, K. Freeman and G.P. Livi   373

Saccharomyces cerevisiae Cpr6
H.-C.J. Chang and S. Lindquist   376

Saccharomyces cerevisiae Cpr7
H.-C.J. Chang and S. Lindquist   377

Saccharomyces cerevisiae mitochondrial cyclophilin Cpr3
A. Matouschek   378

Saccharomyces cerevisiae CYPD/Cpr5
G. Frigerio   380

Saccharomyces cerevisiae Scc3/Cpr4
R. Contreras and J. Demolder   381

Neurospora crassa cyclophilins
M. Tropschug   382

Drosophila melanogaster ninaA
C.S. Zuker and S.L. Rutherford   384

Mammalian cyclophilin A
B. Haendler   386

Vertebrate cyclophilin B
E.R. Price and F. McKeon   388

Mammalian cyclophilin C
M. Trahey and I.L. Weissman   390

Mammalian cyclophilin D
G.P. Livi and D.J. Bergsma   392

Mammalian cyclophilin 40 (Cyp40)
K. Hoffman and R.E. Handschumacher   394

Part 12B. FK506-Binding Proteins (FKBPs)

Three-dimensional structure of FKBPs
J. Liang and J. Clardy — 397

Mip
G. Fischer and J.-U. Rahfeld — 399

***Escherichia coli* FkpA**
S. Raina and D. Missiakas — 402

***Escherichia coli* trigger factor**
R. Moerschell, O. Kandror, A.L. Goldberg,
G. Fischer and J.-U. Rahfeld — 404

***Saccharomyces cerevisiae* FKBP12**
K. Dolinski and J. Heitman — 408

***Saccharomyces cerevisiae* FKBP13**
S.P. Parent and J.B. Neilsen — 411

***Saccharomyces cerevisiae* Npi46p**
X. Shan, Z. Xue and T. Melese — 414

***Neurospora crassa* FKBPs**
M. Tropschug — 416

Plant FKBP73
O. Blecher and A. Breiman — 417

***Drosophila* FKBP39**
U. Theopold and D. Hultmark — 419

Mammalian FKBP12
J.J. Siekierka — 420

Mammalian FKBP12.6
G.J. Wiederrecht — 422

Mammalian FKBP25
A. Galat — 424

Mammalian FKBP51
D.F. Smith — 428

Mammalian FKBP52
D.O. Toft and L.E. Faber — 430

Mammalian FKBP13
S.K. Nigam and S.J. Burakoff — 432

Part 12C. Parvulin PPIases

***Escherichia coli* parvulin**
G. Fischer and J.-U. Rahfeld — 434

***Escherichia coli* SurA**
C. Gross, L. Connolly and P. Rouvière — 436

***Saccharomyces cerevisiae* Ess1**
S.D. Hanes — 438

***Drosophila melanogaster* dodo**
R. Maleszka and G.L. Gabor Miklos — 440

Mammalian Pin1
K.P. Lu and T. Hunter — 443

Part 13. Individual Chaperones

***Escherichia coli* SecB**
L.L. Randall — 449

***Escherichia coli* FtsH**
Y. Akiyama, T. Ogura and K. Ito — 451

Mammalian MSF
K. Mihara and M. Sakaguchi — 453

Mammalian Hip
J. Höhfeld — 455

Vertebrate p23
J.L. Johnson — 457

Vertebrate p60
J.L. Johnson — 458

***Saccharomyces cerevisiae* Sti1**
J.L. Johnson — 459

Part 14. Protein-Specific Chaperones

Escherichia coli PapD
M.-J. Lombardo, D.G. Thanassi and S.J. Hultgren   463

Vertebrate Hsp47
K. Nagata   465

Part 15. Intramolecular Chaperones

Subtilisin
U. Shinde and M. Inouye   471

Carboxypeptidase Y and its propeptide
J.R. Winther   472

Proteinase A and its propeptide
J.R. Winther and H.B. van den Hazel   475

Part 16. Molecular Chaperone Machines

The role of molecular chaperones in DNA replication
A. Wawrzynów and M. Zylicz   481

Ribosome-associated chaperones and protein synthesis: molecular machines catalysing protein targeting, folding and assembly
D. McColl and F.U. Hartl   489

Functions of molecular chaperone proteins in the biogenesis of mitochondria
T. Langer, W. Neupert and E. Schwarz   499

Protein translocation into the endoplasmic reticulum
S.K. Lyman and R. Schekman   506

Quality control in the endoplasmic reticulum
J.F. Simons and A. Helenius   515

The pathway of assembly of the progesterone receptor
D.F. Smith   518

Part 17. Cellular Regulation of Chaperone Activity

The heat shock response in *Escherichia coli*
B. Bukau   525

The periplasmic unfolded protein response in *Escherichia coli*
S. Raina and D. Missiakas   529

Transcriptional regulation of eukaryotic heat shock genes
R.I. Morimoto   534

Signalling of the unfolded protein response from the endoplasmic reticulum to the nucleus
C. McNees and M.-J. Gething   541

Index   549

Contributors

Paul D. Adams Yale University, Department of Molecular Biophysics and Biochemistry, 295 Congress Avenue, New Haven, CT 06510, USA. Tel. 1 203 737 4431, Fax. 1 203 737 1761, E-mail: horwich@hhvms8.csb.yale.edu

Yoshinori Akiyama Kyoto University Institute for Virus Research, Department of Cell Biology, Sakyo-ku, Kyoto 606-01, Japan. Tel. 81 75 751 4015, Fax. 81 75 761 5699 or 5626, E-mail: yakiyama@virus.kyoto-u.ac.jp

Robin L. Anderson Peter MacCallum Cancer Institute, East Melbourne, Vic 3002, Australia. Tel. 61 3 9656 1284, Fax. 61 3 9656 1411, E-mail: anderson@petermac.unimelb.edu.au

Debbie Ang University of Geneva, Biochemie Medicale CMU, 1, rue Michel–Servet, 1211 Geneva 4, Switzerland. Tel. 41 22 702 5512, Fax. 41 22 347 3334, E-mail: georgopo@cmu.unige.ch

Yair Argon The University of Chicago, Dept. of Pathology, Cmtee. on Immunology, 5841 S. Maryland Ave., MC1089, Chicago, IL 60637, USA. Tel. 1 773 702 6388, Fax. 1 773 702 3701, E-mail: yargon@midway.uchicago.edu

John Armstrong University of Sussex, School of Biological Sciences, Falmer BN1 9QG, UK. Tel. 44 1273 678576, Fax. 44 1273 678433, E-mail: J.Armstrong@sussex.ac.uk

Kim T. Arndt Cold Spring Harbor Laboratory, P.O. Box 100, Delbruck Lab, Cold Spring Harbor, NY 11724-2212, USA. Tel. 1 516 367 8836 (lab), Tel. 1 516 367 8835 (off), Fax. 1 516 367 8369, E-mail: arndt@cshl.org

Fredrik Åslund Medical Nobel Institute for Biochemistry, Department of Medical Biochemistry and Biophysics, Karolinska Institute, S-171 77 Stockholm, Sweden. Tel. 46 8 728 7686, Fax. 46 8 728 4716, E-mail: Arne.Holmgren@mbb.ki.se

James Bangs University of Wisconsin, Department of Medical Microbiology and Immunology, 1300 University Avenue, Madison, WI 53706, USA. Tel. 1 608 262 3110, Fax. 1 608 262 8418, E-mail: bangs@macc.wisc.edu

James Bardwell University of Michigan, Department of Biology, Ann Arbor, MI 48109, USA. Tel. 1 313 764 8028, Fax. 1 313 647 0884, E-mail: JBardwell@biologie.lsa.umich.edu

Wolfgang Baumeister Max Planck Institut, Structural Biology, Am Klopferspitz 18A, D-82152 Martinsried, Germany. Tel. 49 89 8578 2652, Fax. 49 89 8578 2641, E-mail: baumeist@alf.biochem.mpg.de

Bonnie K. Baxter University of Wisconsin-Madison, Department of Biomolecular Chemistry, 1300 University Avenue, Madison, WI 53706, USA. Tel. 1 608 263 7105, Fax. 1 608 262 5253, E-mail: bcraig@facstaff.wisc.edu

Nicola J. Bentley University of Kent, Research School of Biosciences, Canterbury CT2 7NJ, UK. Tel. 44 1227 823699, Fax. 44 1227 763912, E-mail: M.F.Tuite@ukc.ac.uk

John J.M. Bergeron Department of Anatomy and Cell Biology, McGill University, 3640 University Street, Montreal, PQ H3A 2B2, Canada. Tel. 1 514 398 6351, Fax. 1 514 398 5049

Derk Bergsma SmithKline Beecham Pharmaceuticals, Department of Molecular Genetics, UE0548, P.O. Box 1539, King of Prussia, PA 19406, USA. Tel. 1 610 270 7610, Fax. 1 610 270 7962, E-mail: Derk_J_Bergsma@sbphrd.com

Jean-Jacques Bessoule UMR-CNRS 5544, Universite Victor Ségalen, Bordeaux II, Case 92, Laboratoire de Biogenese Membranaire, Boite 92-146, Rue Leo Saignat, 33076 Bordeaux Cedex, France. Tel. 33 557 571 274, Fax. 33 556 518 361, E-mail: Jean-Jacques.Bessoule@biomemb.u-bordeaux2.fr

Mikael Björnstedt Medical Nobel Institute for Biochemistry, Department of Medical Biochemistry and Biophysics, Karolinska Institute, S-171 77 Stockholm, Sweden. Tel. 46 8 728 7686, Fax. 46 8 728 4716, E-mail: Arne.Holmgren@mbb.ki.se

Gregory Blatch Department of Biochemistry, WITS University, P.O. WITS, Johannesburg, 2050, South Africa. Tel. 11 716 2204, Fax. 11 716 4479, E-mail: 089BLG@cosmaos.wits.as.za

Oshra Blecher Tel Aviv University, Department of Botany, George S. Wise Faculty of Life Sciences, 69978 Ramat Aviv, Israel. Tel. 972 3 640 9377, Fax. 972 3 640 9380, E-mail: adina@ccsg.tau.ac.il

Wilbert C. Boelens University of Nijmegen, Department of Biochemistry, PO Box 9101, 6500 HB Nijmegan, Netherlands. Tel. 31 24 3616848, Fax. 31 24 3540525, E-mail: w.Boelens@bioch.kun.nl

Rebecca S. Boston North Carolina State University, Department of Botany, Box 7612, Raleigh, NC 27695-7612, USA. Tel. 1 919 515 3390, Fax. 1 919 515 3436 E-mail: cornbip@unity.ncsu.edu

Kerstin Braig Yale University, Department of Genetics, BCMM 154, 295 Congress Avenue, New Haven, CT 06510 USA. Tel. 1 203 737 4431, Fax. 1 203 737 1761, E-mail: horwich@hhvms8.csb.yale.edu

Robert Brambl The University of Minnesota, Department of Plant Biology, 220 Biol. Sci. Center/1445 Gortner Ave, St. Paul, Minnesota 55108, USA. Tel. 1 612 624 5375, Fax. 1 612 625 1738, E-mail: nora@biosci.cbs.umn.edu

Sven Brandau Bernhard Nocht Inst. für Tropenmedizin, Bernhard Nocht Strasse 74, 20359 Hamburg, Germany. Tel. 49 40 31182 485, Fax. 49 40 31182 400

Adina Breiman Tel Aviv University, Department of Botany, Geprge S. Wise Faculty of Life Sciences, 69978

Ramat Aviv, Israel. Tel. 972 3 640 9377, Fax. 972 3 640 9380, E-mail: adina@ccsg.tau.ac.il

Shannon Brightman Department of Biology, Sacred Heart University, 5151 Park Avenue, Fairfield, CT 06432, USA. Tel. 1 203 365 7597, Fax. 1 203 371 7888, E-mail: Brightman@sacredheart.edu

Jeffrey Brodsky University of Pittsburgh, Department of Biological Sciences, A234 Langley Hall, Pittsburgh, PA 15260, USA. Tel. 1 412 624 4831, Fax. 1 412 624 4759, E-mail: jbrodsky+@pitt.edu

Alexander Buchberger University of Heidelberg, Zentrum für Molekulare Biologie, Im Neuenheimer Feld 282, Heidelberg D-69120, Germany. Tel. 49 6221 546865, Fax. 49 6221 565892

Johannes Buchner Universität Regensburg, Biochemie 2, Universitätsstrasse 31, Regensburg D-93040, Germany. Tel. 49 941 943 3039, Fax. 49 941 943 2813, E-mail: Johannes.Buchner@biologie.uni-regensburg.de

Bernd Bukau Institut für Biochemie und Molekularbiologie, Universität Freiburg, Hermann Herder Str. 7, D-79104 Freiburg, Germany. Tel: 49 761 203 5221, Fax. 49 761 203 5257, E-mail: bukau@sun2.ruf.uni-freiburg.de

Steven Burakoff Sidney Farber Cancer Institute, Division of Pediatric Oncology, 44 Binney Street, Boston, MA 02115-6084, USA. Tel. 1 617 632 3564, Fax. 1 617 735 8647

Robert Cafferkey SmithKline Beecham Pharmaceuticals, Department of Molecular Diagnostics, UE0548, P.O. Box 1539, King of Prussia, PA 19406, USA. Tel. 1 610 270 6614, Fax. 1 610 270 5093, E-mail: Robert_Cafferkey-1@sbphrd.com

Avrom J. Caplan Mount Sinai School of Medicine, Department of Cell Biology & Anatomy, one Gustave L. Levy Place, New York, NY 10029, USA. Tel. 1 212 241 6563, Fax. 1 212 860 1174, E-mail: caplan@msvax.mssm.edu

Hui-Chen Jane Chang University of Chicago, Department of Molecular and Cell Biology, 5841 South Maryland Avenue, MC 1028, Chicago, IL 60637, USA. Tel. 1 773 702 8049, Fax. 1 773 702 7254, E-mail: jhchang@midway.uchicago.edu

Michael Cheetham University College London, Department of Pathology, Institute of Opthamology, Bath Street, London EC1V 9EL, UK. Tel. 44 171 608 6944, Fax. 44 171 608 6862, E-mail: michael.cheetham@ucl.ac.uk

Arasu Chellaiah Department of Molecular Biology and Pharmacology, Campus Box 8103, Washington University School of Medicine, St. Louis, MO 63110, USA. Tel. (314) 362-5074, E-mail: arasu@pharmdec.wustl.edu

Huanfeng Chen El DuPont and Co., Central R&D, Experimental Station, P.O. Box 80402, Wilmington, DE 19880-0402, USA. Tel. 1 302 695 7437, Fax. 1 302 695 1374, E-mail: gatenbaa@a1.esvax.umc.dupont.com

Jon Clardy Cornell University, Department of Chemistry, Baker Laboratory, Ithaca, NY 14853-1301, USA. Tel. 1 607 255 7685, Fax. 1 607 255 1253, E-mail: jcc12@cornell.edu

Adrian Clarke University of Umea, Department of Plant Physiology, S-901 87 Umeå E5, Sweden. Tel. 46 90 7865209, Fax. 46 90 7866676, E-mail: Adrian.Clarke@plantphys.umu.se

Joachim Clos Bernhard Nocht Inst. für Tropenmedizin, Bernhard Nocht Strasse 74, 20359 Hamburg, Germany. Tel. 49 40 31182 485, Fax. 49 40 31182 400, E-mail: joachim-clos@MagicVillage.de

Jackie Collier Rensselaer Polytechnic Institute, Biology 304 Materials Research Center, 110 8th Street, Troy, NY 12180-3590. Tel. 1 518 276 2178, Fax. 1 518 276 2344, E-mail: collij3@rpi.edu

Lynn Connolly University of California San Francisco, Department of Biochemistry and Biophysics, 513 Parnassus Ave., San Francisco, CA 94143, USA. Tel. 1 415 476 1493, Fax. 1 415 476 4204, E-mail: lcon@itsa.ucsf.edu

Roland Contreras State University of Gent, Laboratory of Molecular Biology, K.L. Ledeganckstraat 35, B-9000 Gent, Belgium. Tel. 32 9 264 5136, Fax. 32 9 264 53 48, E-mail: Roland@lmb1.rug.ac.be

Nicholas Cowan New York University Medical Center, Department of Biochemistry, 550 First Avenue, New York, NY 10016, USA. Tel. 1 212 263 5809, Fax. 1 212 263 8166, E-mail: Cowann01@MCRCR6.MED.NYU.EDU

Elizabeth Craig University of Wisconsin-Madison, Department of Biomolecular Chemistry, 1300 University Avenue, Madison, WI 53706, USA. Tel. 1 608 263 7105, Fax. 1 608 262 5253, E-mail: bcraig@facstaff.wisc.edu

Tom E. Creighton European Molecular Biology Laboratory, Meyerhofstrasse 1, D-69012 Heidelberg, Germany.

Douglas Cyr University of Alabama at Birmingham, Department of Cell Biology, 1918 University Boulevard, Birmingham, AL 35294-0005, USA. Tel. 1 205 975 4892, Fax. 1 205 934 0950, E-mail: DCYR@cellbio.bhs.uab.edu

Jeffrey Dahlseid R.M. Bock Laboratories, Laboratory of Molecular Biology, 1525 Linden Avenue, Madison, WI 53706, USA. Tel: 1 608 262 4381, Fax. 1 608 262 4570. E-mail: dahlseid@facstaff.wisc.edu

Angela Davis Washington University School of Medicine, Department of Surgery; Campus box 8109, 4939 Children's Place, St. Louis, MO 63110, USA

Julie Davis University of Wisconsin-Madison, Department of Biomolecular Chemistry, 1300 University Avenue, Madison, WI 53706, USA. Tel. 1 608 263 7105, Fax. 1 608 262 5253

Wilfried De Jong University of Nijmegen, Department of Biochemistry, PO Box 9101, 6500 HB Nijmegan, Netherlands. Tel. 31 24 3616848, Fax. 31 24 3540525, E-mail: w.dejong@bioch.kun.nl

Johann Deisenhofer UT Southwestern Medical Center, Howard Hughes Medical Institute, 5323 Harry Hines Boulevard, Dallas, TX 75235-9050, USA. Tel. 1 214 648 5089, Fax. 1 214 648 5095, E-mail: jd@howie.swmed.edu

Olivier Deloche University of Geneva, Biochemie Medicale CMU, 1, rue Michel–Servet, 1211 Geneva 4,

Switzerland. Tel. 41 22 702 5512, Fax. 41 22 347 3334, E-mail: georgopo@cmu.unige.ch

Jan Demolder Flanders Interuniversity Institute for Biotechnology, Rijvisschestraat 118, box 1, B-9052 Zwijnaarde, Belgium. Tel. 32 9 244 6611, Fax. 32 9 244 6610, E-mail: Jan.Demolder@vib.be

J. Fred Dice Tufts University School of Medicine, Department of Physiology, 136 Harrison Avenue, Boston, MA 02111-1800, USA. Tel. 1 617 636 6707, Fax. 1 617 636 0445, E-mail: jdice@opal.tufts.edu

Kara Dolinski Department of Genetics, Howard Hughes Medical Institute, Duke University Medical Center, 322 CARL Bldg, Research Drive, Durham, NC 27710 USA. Tel. 1 919 684 2809/2824, Fax. 1 919 684 5458, E-mail: kd2@acpub.duke.edu

Annette Dresel Bernhard Nocht Inst. für Tropenmedizin, Bernhard Nocht Strasse 74, 20359 Hamburg, Germany. Tel. 49 40 31182 485, Fax. 49 40 31182 400

Angela M. Dunn University of Kent, Research School of Biosciences, Canterbury CT2 7NJ, UK. Tel. 44 1227 823699, Fax. 44 1227 763912

Kimberly Dyer National Institutes of Health, Laboratory of Host Defences, NIAID, Bldg. 10, Room 11N104, Bethesda, MD 20892-1886, USA. Tel. 1 301 496 2877, Fax. 1 301 402 4369, E-mail: kd26h@nih.gov

Doug Easton State University College at Buffalo, 1300 Elmwood Ave, Buffalo, NY 14222, USA. Tel. 1 716 878 3050, Fax. 1 716 878 4028, E-mail: eastondp@snybufaa.cs.snybuf.edu

Evan Eisenberg National Institutes of Health, Laboratory of Cell Biology, Bethesda, Maryland 20890-0301, USA. Tel. 1 301 496 2846, Fax. 1 301 402 1519

Felice Elefant Temple University, Department of Biology, 12th St. and Norris St., Philadelphia, PA 19122, USA. Tel. 1 215 204 8845, Fax. 1 215 204 6646

R. John Ellis University of Warwick, Department of Biological Sciences, Coventry CV4 7AL, UK. Tel. 44 1203 523509, Fax. 44 1203 523568 or 701, E-mail: je@dna.bio.warwick.ac.uk

Mats-Jerry Eriksson University of Umea, Department of Plant Physiology, S-901 87 Umeå E5, Sweden. Tel. 46 90 7865209, Fax. 46 90 7866676, E-mail: mats-jerry.eriksson@plantphys.umu.se

Lee. E. Faber Medical College of Ohio, Department of Physiology, Toledo, OH 43699, USA. Tel. 419 381 4584, Fax. 419 381 3124, E-mail: lfaber@gemini.mco.edu

Wayne A. Fenton Yale University, Department of Genetics, BCMM 154, 295 Congress Avenue, New Haven, CT 06510, USA. Tel. 1 203 737 4431, Fax. 1 203 737 1761, E-mail: horwich@hhvms8.csb.yale.edu

David Ferrari Universität Göttingen, Abteilung Klinische Biochemie, Zentrum Innere Medizine, D-37070 Göttingen, Germany. Tel. 49 551 39 6389, Fax. 49 551 39 2953, E-mail: dferrari@med.uni-goettingen.de

Gunter Fischer Max Planck Research, Unit-Enzymology of Protein Folding, Kurt-Mothes-Str. 3, D-06120 Halle, Germany. Tel. 49 345 552 2801, Fax. 49 345 551 1972, E-mail: fischer@cis. biochemtech.uni-halle.de

Gregory Flynn Institute of Molecular Biology, University of Oregon, Eugene, OR 97403, USA. Tel. 1 503 346 1534/1535, Fax. 1 503 346 5891, E-mail: gflynn@molbio. uoregon.edu

Robert B. Freedman University of Kent, Dept. of Biosciences, Canterbury CT2 7NJ, UK. Tel. 44 1 227 823226, Fax. 44 1 227 763912, E-mail: r.b.freedman@ukc.ac.uk

Katie Freeman SmithKline Beecham Pharmaceuticals, Dept. of Comparative Genetics, UE0548, P.O. Box 1539, King of Prussia, PA 19406, USA. Tel. 1 610 270 7535, Fax. 1 610 270 5093, E-mail: Katie_B_Freeman@sbphrd.com

Gabriella Frigerio Sanger Center, Hinxton Hall, Cambridge, CB10 1RQ, UK. Tel. 44 1223 494954, Fax. 44 1223 494919, E-mail: gcf@sanger.ac.uk

Joachim Fuellekrug European Molecular Biology Laboratory, Cell Biology Programme, Meyerhofstr. 1, D-69112 Heidelberg, Germany. Tel. 49 6221 387408, Fax. 49 6221 387512, E-mail: fuellekr@embl-heidelberg.de

Matthais Gaestel AG Stressproteine, Max-Delbrück-Center for Mol. Medicine, R.-Rössle-Str. 10, 13122 Berlin, Germany. Tel. 49 30 9406 3785, Fax. 49 30 9406 3798, E-mail: gaestel@mdc-berlin.de

Andrzej Galat DIEP/DSV, Bat. 152, CE-Saclay, 91191 Gif-sur-Yvette, France. Tel. 33 1 690 830 40, Fax. 33 1 690 891 37

Anthony A. Gatenby El DuPont and Co., Central R&D, Experimental Station, P.O. Box 80402, Wilmington, DE 19880-0402, USA. Tel. 1 302 695 7437, Fax. 1 302 695 1374, E-mail: gatenbaa@a1.esvax.umc.dupont.com

Costa Georgopoulos University of Geneva, Biochemie Medicale CMU, 1, rue Michel–Servet, 1211 Geneva 4, Switzerland. Tel. 41 22 702 5512, Fax. 41 22 347 3334, E-mail: georgopo@cmu.unige.ch

Mary-Jane Gething University of Melbourne, Department of Biochemistry and Molecular Biology, Parkville, Melbourne, Victoria 3052, Australia. Tel. 61 3 9344 5948, Fax. 61 3 9347 9109, E-mail: gething@ariel.ucs.unimelb.edu.au

Jeffrey W. Gillikin North Carolina State University, Department of Botany, Box 7612, Raleigh, NC 27695-7612, USA. Tel. 1 919 515 3570, Home. 1 919 515 2727, Fax. 1 919 515 3436, E-mail: cornbip@unity.ncsu.edu

John R. Glover University of Chicago, Department of Molecular Genetics and Cell Biology, 5841 South Maryland Avenue, MC1028, Chicago, IL 60637, USA. Tel. 1 773 702 8048, Fax. 1 773 702 7254, E-mail: jrglover@uchicago.edu

Alfred L. Goldberg Harvard Medical School, Department of Cell Biology, 240 Longwood Ave., Boston, MA 02115, USA. Tel. 1 617 432 1855, Fax. 1 617 232 0173, E-mail: agoldber@bcmp.med.harvard.edu

Susan Gottesman National Institutes of Health, NCI, Department of Molecular Biology, Building 37, Room 2E18, Bethesda, MD 20892, USA. Tel. 1 301 496 3524, Fax. 1 301 496 3875, E-mail: susang@helix.nih.gov

Michael Green St. Louis University School of Medicine, Dept. of Mol. Microbiology & Immunology, 1402 S. Grand Boulevard, St. Louis, MO 63104, USA. Tel. 1 314 577 8445, Fax. 1 314 773 3403, E-mail: greenmi@wpogate.slu.edu

Lois E. Greene National Institutes of Health, Laboratory of Cell Biology, Bethesda, Maryland 20890, USA. Tel. 1 301 496 2846, Fax. 1 301 402 1519, E-mail: greenel@helix.nih.gov

Carol Gross University of California–San Francisco, Departments of Stomatology and Microbiology, 513 Parnassus Ave., San Francisco, CA 94143-0512, USA. Tel. 1 415 476 4161, Fax. 1 415 476 4204, E-mail: cgross@cgl.ucsf.edu

Arthur R. Grossman Carnegie Institution of Washington, Department of Plant Biology, 290 Panama St., Stanford, CA 94305, USA. Tel. 1 415 325 1521, Fax. 1 415 325 6857, E-mail: arthur@andrew.stanford.edu

Luke W. Guddat University of Queensland, Centre for Drug Design and Development, St. Lucia, Queensland 4072, Australia. Tel. 61 7 3365 4942, Fax. 61 7 3365 1990, E-mail: J.Martin@mailbox.uq.oz.au

Steven R. Gullans Harvard Institutes of Medicine, Room 554, 77 Avenue Louis Pasteur, Boston, MA 02115 USA. Tel. 617 525 5712, Fax. 617 525 5711, E-mail: sgullans@rics.bwh.harvard.edu

Bernard Haendler Schering AG, Institute of Cellular and Molecular Biology, D-13342 Berlin, Germany. Tel. 49 30 468 12669, Fax. 49 30 468 16707, E-mail: Bernard.Haendler@Schering.DE

Robert Handschumacher Yale Univ. School of Medicine, Department of Pharmacology, 333 Cedar Street, New Haven, CT 06520-8066, USA. Tel. 1 203 785 4385, Fax. 1 203 785 7670, E-mail: HandschuRE@maspo2.mas.yale.edu

Steven D. Hanes SUNY Albany, Wadsworth Center, NY State Dept. Health, Dept. of Biomedical Sciences, Albany, NY 12203, USA. Tel. 1 518 473 4213, Fax. 1 518 474 3181, E-mail: steven.hanes@wadsworth.org

F. Ulrich Hartl Max-Planck-Institut für Biochemie, Department of Cellular Biochemistry, Am Klopferspitz 18a, D-82152, Martinsried bei München, Germany. Tel. 49 89 8578 2233/2244, Fax. 49 89 8578 2211, E-mail: uhartl@biochem.mpg.de

Dadna Hartman Victorian Institute of Animal Science, 475-485 Mickleham Rd., Attwood, Victoria 3049, Australia. Tel. 61 3 9217 4349, Fax. 61 3 9217 4299, E-mail: hartmand@woody.as-vic.gov.au

Michael Hecker Inst. für Mikrobiol. und Molekularbiol., Ernst-Moritz-Arndt-UniversitätGreifswald, Jahnstr. 15, D-17487 Greifswald, Germany. Tel. 49 3834 864200, Fax. 49 3834 864202, E-mail: Hecker@microbio7.biologie.uni-greifswald.de

Joseph Heitman Howard Hughes Medical Institute, Department of Genetics, 321 Carl Building, Box 3546, Durham, NC 27710, USA. Tel. 1 919 684 2824, Fax. 1 919 684 5458, E-mail: heitm001@mc.duke.edu

Ari Helenius Yale University School of Medicine, Department of Cell Biology, 333 Cedar Street, P.O. Box 3333, New Haven, CT 06520-8002, USA. Tel. 1 203 785 4313, Fax. 1 203 785 7226, E-mail: ARI_HELENIUS@qm.yale.edu

Lawrence Hightower University of Connecticut, Department of Molecular & Cell Biology, 75 N. Eagleville Road, Storrs, CT 06269-3044, USA. Tel. 1 860 486 4257, Fax. 1 860 486 1784, E-mail: HIGHTOWER@biotek.mcb.uconn.edu

Kai Hoffman Yale Univ. School of Medicine, Department of Pharmacology, 333 Cedar Street, New Haven, CT 06520-8066, USA. Tel. 1 203 785 4385, Fax. 1 203 785 7670, E-mail: HandschuRE@ maspo2.mas.yale. edu

Jörg Höhfeld Universität Heidelberg, Zentrum für Molekulare Biologie, ZMBH, Im Neuenheimer Feld 282, D-69120 Heidelberg, Germany. Tel. 49 6221 54 6738, Fax. 49 6221 54 5891, E-mail: j-hoehfeld@sun0.urz.uni-heidelberg.de

Peter Høj Waite Campus, Department of Viticulture, PMBI, Glen Osmond, South Australia 5064, Australia. Tel. 61 8 8303 7248, Fax. 61 8 8303 7117, E-mail: phoj@waite.adelaide.edu.au

Arne Holmgren Medical Nobel Institute for Biochemistry, Department of Medical Biochemistry and Biophysics, Karolinska Institute, S-171 77 Stockholm, Sweden. Tel. 46 8 728 7686, Fax. 46 8 728 4716, E-mail: Arne.Holmgren@mbb.ki.se

Nick Hoogenraad La Trobe University, Department of Biochemistry, Bundoora, Victoria 3083, Australia. Tel. 61 3 9479 2175, Fax. 61 3 9479 2467, E-mail: N.Hoogenraad@Latrobe.edu. au

Achim Hörauf Bernhard Nocht Institute für Tropenmedizin, Bernhard Nocht Strasse 74, 20359, Germany. Tel. 49 40 31182 485, Fax. 49 40 31182 400

Arthur Horwich Yale University, HHMI, Department of Molecular Genetics, BCMM 154, 295 Congress Avenue, New Haven, CT 06510, USA. Tel. 1 203 737 4431, Fax. 1 203 737 1761, E-mail: horwich@hhvms8.csb.yale.edu

Andreas Hübel Bernhard Nocht Inst. für Tropenmedizin, Bernhard Nocht Strasse 74, 20359 Hamburg, Germany. Tel. 49 40 31182 485, Fax. 49 40 31182 400

Scott J. Hultgren Washington Univ. School of Medicine, Department of Molecular Microbiology, St. Louis, MO 63110-1093, USA. Tel. 1 314 362 6788, Fax. 1 314 362 1998, E-mail: hultgren@borcim.wustl.edu

Dan Hultmark Stockholm University, Department of Developmental Biology, S-106 91 Stockholm, Sweden. Tel. 46 8 164 153, Fax. 46 8 152 350, E-mail: dan@molbio.su.se

John F. Hunt UT Southwestern Medical Center, Howard Hughes Medical Institute, 5323 Harry Hines Boulevard, Dallas, TX 75235-9050, USA. Tel. 1 214 648 5089, Fax. 1 214 648 5095, E-mail: hunt@howie.swmed.edu

Tony Hunter The Salk Institute, Molecular Biology and Virology Lab, 10010 North Torrey Pines Road, La Jolla, CA 92037, USA. Tel. 1 619 453 4100 x1387, Fax. 1 619 457 4765, E-mail: Hunter@salk.edu

Masayori Inouye UMDNJ-RW Johnson Medical School, Department of Biochemistry, 675 Hoes Lane, Piscataway, NJ 08854-5635, USA. Tel. 1 908 235 4115, Fax. 1 908 235 4559, E-mail: inouye@rwja.umdnj.edu

Koreaki Ito Kyoto University Institute for Virus Research, Department of Cell Biology, Sakyo-ku, Kyoto 606-01, Japan. Tel. 81 75 751 4015, Fax. 81 75 761 5626, E-mail: kito@virus.kyoto-u.ac.jp

Ursula Jakob University of Michigan, Department of Biology, Ann Arbor, MI 48109-1048, USA. Tel. 1 313 764 8028, Fax. 1 313 647 0884, E-mail: ujakob@biology.lsa.umich.edu

Denis R. Joanisse Laboratoire de génétique cellulaire et développementale, RSVS, Pav. CE Marchand, Université Laval, Ste-Foy, Québec G1K 7P4, Canada. Tel. 1 418 656 3339, Fax. 1 418 656 7176

Jill L. Johnson University of Wisconsin-Madison, Department of Biomolecular Chemistry, 1300 University Avenue, Madison, WI 53706, USA. Tel. 1 608 262 1358, Fax. 1 608 262 5253, E-mail: jjohns39@facstaff.wisc.edu

Margaret L. Just SUNY Stony Brook, Department of Biochemistry and Cell Biology, Stony Brook, NY 11794, USA. Tel. 1 516 632 8560, Fax. 1 516 632 8575, E-mail: wlennarz@ccmail.sunysb.edu

Olga Kandror Harvard Medical School, Department of Cell Biology, 240 Longwood Ave., Boston, MA 02115, USA. Tel. 1 617 432 1855, Fax. 1 617 232 0173

Thomas H. Kawula University of North Carolina, Department of Microbiology and Immunology, CB#7290, 804 Mary Ellen Jones Bldg., Chapel Hill, N. Carolina 27599-7290, USA. Tel. 1 919 966 2637, Fax. 1 919 962 8103, E-mail: kawula@med.unc.edu

Frederic J. Kaye NCI-Navy Medical Oncology Branch, Building 8, Room 5101, Naval Hospital, Bethesda, MD 20889, USA. Tel. 1 301 496 0916, Fax. 1 301 496 0047, E-mail: fkaye@helix.nih.gov

William L. Kelley University of Geneva–CMU, Department of Medical Biochemistry, 1, rue Michel–Servet, 1211 Geneva 4, Switzerland. Tel. 41 22 70 25514, Fax. 41 22 34 73334, E-mail: william.kelley@medecine.unige.ch

Yoko Kimura Tokyo Metropolitan Inst. Med. Sci, Department of Tumor Biology, 3-18-22, Honkomagome, Bunkyo-ku, Tokyo, 113, Japan. Fax. 81 3 5685 2932, E-mail: ykimura@rinshoken.or.jp

Ryoji Kojima Meijo University, Dept. of Pharmacology, Faculty of Parmacy, Nagoya, 468, Japan. Tel. 81 52 832 1781, Fax. 81 52 834 8780, E-mail: kojima@meiju-u.ac.jp

Ismail Kola Monash University, Molecular Genetics and Development Group, Institute of Reproduction and Development, Clayton, Melbourne, Victoria 3168, Australia. Tel. 61 3 9550 5480, Fax. 61 3 9550 5568, E-mail: ismailko@silas.cc.monash.edu.au

R.G. Kranz Washington University, Department of Biology, Campus Box 1132, 1 Brookings Dr., St. Louis, MO 63130, USA. Tel. 1 314 935 4278, Fax. 1 314 935 4432, E-mail: Kranz@wustlb.wustl.edu

Priti Krishna The University of Western Ontario, Dept. of Plant Sciences, 1151 Richmond St. N., London, ON, N6A 5B7, Canada. Tel. 1 519 679 2111 ext. 6406, Fax. 1 519 661 3935, E-mail: pkrishna@julian.uwo.ca

Elke Krüger Institut für Mikrobiol. und Molekularbiol., Ernst-Moritz-Arndt-UniversitätGreifswald, Jahnstr. 15, D-17487 Greifswald, Germany. Tel. 49 0 3834 864200, Fax. 49 0 3834 864202, E-mail: ekrueger@microbio4.biologie.uni-greifswald.de

Hiroshi Kubota HSP Research Insitute, Kyoto Research Park, 17 Chudoji-minamimachi, Shimogyo-ku, Kyoto, 600, Japan. Tel. 81 75 315 8656, Fax. 81 75 315 8659, E-mail: kubota@hsp.co.jp

Takayashi Kuno Kobe University School of Medicine, Department of Pharmacology, 7-5-1 Kusunoki-cho, Chuo-ku, Kobe 650, Japan. Tel. 81 78 341 7451, Fax. 81 78 351 6531, E-mail: tkuno@kobe-u.ac.jp

Ursula Kurzik-Dumke Johannes Gutenberg Universität, Institut für Genetik, Saarstrasse 21, 55099 Mainz, Germany. Tel. 49 6131 395223, Fax. 49 6131 395845, E-mail: Kurzik@mzdmza.zdv.uni-mainz.de

Shikha Laloraya University of Wisconsin-Madison, Department of Biomolecular Chemistry, 1300 University Avenue, Madison, WI 53706, USA. Tel. 1 608 263 7105, Fax. 1 608 262 5253, E-mail: bcraig@macc.wisc.edu

Jacques Landry University of Laval, Centre de Recherche en Cancérologie, 1, rue de l'Arsenal, Quebec, G1R 2J6, Canada. Tel. 1 418 691 5281, Fax. 1 418 691 5439, E-mail: jacques.landry@med.ulaval.ca

Samuel J. Landry Tulane University School of Medicine, Department of Biochemistry, SL43, 1430 Tulane Avenue, New Orleans, LA 70112-2699, USA. Tel. 1 504 586 3990, Fax. 1 504 584 2739, E-mail: landry@mailhost.tcs.tulane.edu

Thomas Langer Ludwig-Maximilians-Universität, Institut für Physiologische Chemie, Goethestrasse 33, D-80336 München, Germany. Tel. 49 89 5996 283, Fax. 49 89 5996 270, E-mail: Langer@bio.med.uni-muenchen.de

Michael Lässle Harvard Medical School, Children's Hosp., Ender's Building, Room 1077, 300 Longwood Avenue, Boston, MA 02115, USA. Tel. 1 617 355 6767, Fax. 1 617 355 7043, E-mail: lassle@mit.edu

John Y. Lee Temple University, Department of Biology, 12th Street and Norris Street, Philadelphia, PA 19122, USA. Tel. 1 215 787 8845, Fax. 1 215 787 6646, E-mail: palter@astro.ocis.temple.edu

Garrett J. Lee University of Arizona, Department of Biochemistry, Life Sciences S Building, Tucson, AZ 85721-0001, USA. Tel. 1 602 621 1601, Fax. 1 602 621 3709, E-mail: eliz@biosci.arizona.edu

William Lennarz SUNY Stony Brook, Department of Biochemistry and Cell Biology, Stony Brook, NY 11794, USA. Tel. 1 516 632 8560, Fax. 1 516 632 8575, E-mail: wlennarz@ccmail.sunysb.edu

Sau-Mei Leung University of Connecticut, Department of Molecular and Cell Biology, 75 N. Eagleville Road, Storrs, CT 06269-3044, USA. Tel. 1 860 486 4257, Fax. 1 860 486 1784, E-mail: HIGHTOWER@biotek.mcb.uconn.edu

Jun Liang Cornell University, Department of Chemistry, Baker Laboratory, Ithaca, NY 14853-1301, USA. Tel. 1 607 255 7685, Fax. 1 607 255 1253, E-mail: jl40@cornell.edu

Susan Lindquist University of Chicago, Howard Hughes Medical Institute, 5841 South Maryland Avenue, Box 391, Chicago, IL 60637, USA. Tel. 1 773 702 8049, Fax. 1 773 702 7254, E-mail: S-Lindquist@uchicago.edu

George Livi SmithKline Beecham Pharmaceuticals, Department of Comparative Genetics, UE0548, P.O. Box 1539, King of Prussia, PA 19406, USA. Tel. 1 610 270 7717, Fax. 1 610 270 7962, E-mail: George_P_Livi@sbphrd.com

Hannes Loferer Geneva Biomedical Research Institute, Glaxo Wellcome Research and Development S.A., 14, chemin des Aulx, CH-1228 Plan-les-Quates, Geneva, Switzerland. Tel. 41 22 706 9608, Fax. 41 22 794 6965, E-mail: jhl39437@ggr.co.uk

Mary-Jane Lombardo Washington University School of Medicine, Department of Molecular Microbiology, St. Louis, MO 63110-1093, USA. Tel. 1 314 362 6788, Fax. 1 314 362 1998, E-mail: hultgren@borcim.wustl.edu

Damian McColl Department of Biochemistry and Biophysics, University of California San Francisco, 513 Parnassus Avenue, San Francisco, CA 94143-0448, USA.

Carolyn McNees Department of Biochemistry and Molecular Biology, University of Melbourne, Parkville, Victoria 3052, Australia. Tel. 61 3 9344 8172, Fax. 61 3 9347 9109

Andrei Lupas Max Planck Institut, Structural Biology, Am Klopferspitz 18A, D-82152 Martinsried, Germany. Tel. 49 89 8578 2652, Fax. 49 89 8578 2641, E-mail: lupas@genmic.biochem.mpg.de

Susan Lyman HHMI/University of California, Berkeley, Department of Molecular and Cell Biology, 401 Barker Hall, Berkeley, CA 94720-0001, USA. Tel. 1 510 642 5756, Fax. 1 510 642 7846

Richard Maleszka The Australian National University, Visual Sciences, Research School of Biological Sciences, Canberra, ACT 0200, Australia. Tel. 61 6 249 0451, Fax. 61 6 249 3784, E-mail: maleszka@rsbs-central.anu.edu.au

Nancy Marcus St. Louis University School of Medicine, Department of Molecular Microbiology and Immunology, 1402 S. Grand Boulevard, St. Louis, MO 63104, USA. Tel. 1 314 577 8445, Fax. 1 314 773 3403

Gerard J. M. Martens University of Nijmegen, Department of Animal Physiology, Toernooiveld, Nijmegen, 6525 ED, The Netherlands. Tel. 31 24 3652601, Fax. 31 24 3652714, E-mail: gmart@sci.kun.nl

Jennifer F. Martin University of Queensland, Centre for Drug Design and Development, St. Lucia, Qld 4072, Australia. Tel. 61 7 3365 4942, Fax. 61 7 3365 1990, E-mail: J.Martin@mailbox.uq.oz.au

Andreas Matouschek Northwestern University, Dept. of Biochem., Mol. Biol. & Cell Biol., 2153 Sheridan Rd, Evanston, IL 60208-3500, USA. Tel. 1 847 467 3570, Fax. 1 847 467 1380, E-mail: matousche@nwu.edu

Michael Maurizi NCI/NIH, Department of Cell Biology, Building 37, Room 1B07, Bethesda, MD 20892, USA. Tel. 1 301 496 7961, Fax. 1 301 402 0450

Mark Mayhew Memorial Sloan-Kettering Cancer Centre, HHMI Cellular Biochemistry and Biophysics, 1275 York Avenue, Box 519, New York, New York 10021, USA. Tel. 1 212 639 2300, Fax. 1 212 717 3604, E-mail: m-mayhew@ski.mskcc.org

Frank McKeon Harvard Medical School, Dept. of Cell Biology, 25 Shattuck Street, Boston, MA 02115, USA. Tel. 1 617 432 0994 or 0327, Fax. 1 617 432 1144

Ashwin Mehta North Carolina State University, Department of Botany, Raleigh, NC 27695, USA. Tel. 1 919 515 3570, Fax. 1 919 515 3436, E-mail: aswin@unity.ncsu.edu

Teri Melese Columbia University, Department of Biological Sciences, 702 Fairchild, New York, NY 10027, USA. Tel. 1 212 854 5443, Fax. 1 212 865 8246, E-mail: teri@cubsps.bio.columbia.edu

Jeffrey Melnick Department of Pathology, Box 8818, Washington University Medical Center, 660 S. Euclid Ave., St. Louis, MO 63110, Tel: 314 836 5458, Fax: 314 836 8950, E-mail: melnick@path.wustl.edu

Bingjie Miao University of Wisconsin-Madison, Department of Biomolecular Chemistry, 1300 University Avenue, Madison, WI 53706, USA. Tel. 1 608 263 7105, Fax. 1 608 262 5253

Sébastien Michaud Université Laval, Laboratoire de génétique cellulaire et développementale, RSVS, Pav. CE Marchand, Ste-Foy, Québec G1K 7P4, Canada. Tel. 1 418 656 3339, Fax. 1 418 656 7176

Katsuyoshi Mihara Kyushu University, Graduate School of Medical Science, Department of Molecular Biology, Maidashi, Higashi-ku, Fukuoka 812-12, Japan. Tel. 81 92 642 6176, Fax. 81 92 642 6183, E-mail: mihara@cell.med.kyushu-u.ac.jp

G.L. Gabor Miklos The Neurosciences Institute, 10640 John Jay Hopkins Drive, San Diego, CA 92121, USA. Tel. 1 619 626 2000, Fax. 1 619 626 2099, E-mail: miklos@nsi.edu

Dominique Missiakas CNRS UPR9027, 31, Chemin J. Aiguier, 13402 Marseille Cedex 20, France. Tel. 33 4 91 76 03 59, Fax. 33 4 91 71 21 24, E-mail: missiaka@ibsm.cnrs-mrs.fr

Takemitsu Mizunaga Keisen College, 1436 Sannomiya, Isehara City, Kanagawa 259-11, Japan. Tel. 81 463 95 1010, Fax. 81 463 96 6219

Takeshi Mizuno Nagoya University, Laboratory of Molecular Microbiology, School of Agriculture, Japan. Tel. 81 52 789 4089, Fax. 81 52 789 4091, E-mail: i45455a@nucc.cc.nagoya-u.ac.jp

Richard Moerschell Harvard Medical School, Department of Cell Biology, 240 Longwood Ave., Boston, MA 02115, USA. Tel. 1 617 432 1855, Fax. 1 617 232 0173

T. Mohanakumar Washington University School of Medicine, Department of Surgery; Campus box 8109, 4939 Children's Place, St. Louis, MO 63110, USA.

Richard Morimoto Northwestern University, Department of Biochemistry, Molecular and Cell Biology, 2153 Sheridan Road, Evanston, IL 60201, USA. Tel. 1 847 491 3340, Fax. 1 847 491 4461, E-mail: morimoto@casbah.acns.nwu.edu

Hideyuki Mukai Kobe University School of Medicine, Department of Pharmacology, 7-5-1 Kusunoki-cho, Chuo-ku, Kobe 650, Japan. Tel. 81 78 341 7451, Fax. 81 78 351 6531, E-mail: tkuno@kobe-u.ac.jp

Kazuhiro Nagata Kyoto University, Department of Cell Biology, Chest Disease Research Institute, Sakyo-ku, Kyoto 606-01, Japan. Tel. 81 75 751 3848, Fax. 81 75 751 4645 or 752 9017, E-mail: nagata@chest.kyoto-u.ac.jp

Hajime Nakamura Medical Nobel Institute for Biochemistry, Department of Medical Biochemistry and Biophysics, Karolinska Institute, S-171 77 Stockholm, Sweden. Tel. 46 8 728 7686, Fax. 46 8 728 4716

Debra F. Nathan University of Chicago, Howard Hughes Medical Institute, 5841 South Maryland Avenue, Box 391, Chicago, IL 60637, USA. Tel. 1 773 702 8049, Fax. 1 773 702 7254, E-mail: dnathan@midway.uchicago.edu

Dean Naylor Waite Campus, Department of Viticulture, PMBI, Glen Osmond, SA 5064, Australia. Tel. 61 8 8303 7248, Fax. 61 8 8303 7117, E-mail: dnaylor@waite.adelaide.edu.au

Walter Neupert University of Munich, Physiologische Chemie, Goethestrasse 33, München, Germany. Tel. 49 89 599 6313, Fax. 49 89 599 6270, E-mail: neupert@bio.med.uni-muenchen.d

Ailsa Nielsen Northwestern University, Dept. of Biochemistry, Mol. & Cell Biol., 2153 North Campus Drive, Evanston, IL 60208-3500, USA. Tel. 1 708 491 7147, Fax. 1 708 467 1610, E-mail: skp843@lulu.acns.nwu.edu

Jennifer B. Nielsen Merck Research Laboratories, P.O. Box 2000, RY 80Y-210, Rahway, NJ 07065, USA. Tel. 1 908 594 6799, Fax. 1 908 594 1399

Sanjay Nigam Brigham and Women's Hospital, Department of Medicine, Renal Division, Harvard Medical School, 75 Francis St. Boston, MA 02115, USA. Tel. 1 617 525 8880, Fax. 1 617 525 8881, E-mail: sknigam@bics.bwh.harvard.edu

Carol E. Norris University of Connecticut, Department of Molecular & Cell Biology, 75 N. Eagleville Road, Storrs, CT 06269-3044, USA. Tel. 1 860 486 4257, Fax. 1 860 486 1784, E-mail: HIGHTOWER@biotek.mcb.uconn.edu

Teru Ogura Kumamoto University School of Medicine, Dept. Molecular Cell Biology, Institute of Molecular Embryology & Genetics, Kumamoto, 862, Japan. Tel: 81 96 373 5336, Fax. 81 96 371 2408, E-mail: ogura@gpo.kumamoto-u.ac.jp

Kenzo Ohtsuka Aichi Cancer Center Research Institute, Laboratory of Experimental Radiology, 1-1 Kanokoden, Chikusa-ku, Nagoya 464, Japan. Tel. 81 52 762 6111, Ext. 8813, Fax. 81 52 763 5233, E-mail: kohtsuka@aichi-cc.pref.aichi.jp

Gregory Otterson NCI Navy Medical Oncology Branch, Building 8, Room 5101, Naval Hospital, Bethesda, MD 20889, USA. Tel. 1 301 496 0916, Fax. 1 301 496 0047

Karen Palter Temple University, Department of Biology, 12th Street and Norris Street, Philadelphia, PA 19122, USA. Tel. 1 215 204 8845, Fax. 1 215 204 6646, E-mail: palter@astro.ocis.temple.edu

Stephen Parent Merck Research Laboratories, P.O. Box 2000, RY 80Y-300, Rahway, NJ 07065-0900, USA. Tel. 1 908 594 5787, Fax. 1 908 594 5878, E-mail: steve_parent@merck.com

Frank Parlatti McGill University, Dept. of Biology, Stewart Biology Building, 1205 Dr. Penfield Av., Montreal, Quebec, H3A 1B1, Canada. Tel. 1 514 496 6155, Fax. 1 514 496 6213, E-mail: parlati@biotech.lan.nrc.ca

Jeffrey R. Peterson Yale University School of Medicine, Department of Cell Biology, 333 Cedar Street, P.O. Box 3333, New Haven, CT 06510-3219, USA. Tel. 1 203 785 4313, Fax. 1 203 785 7226, E-mail: JPETERSO@biomed.med.yale.edu

Christine Pfund University of Wisconsin-Madison, Department of Biomolecular Chemistry, 1300 University Avenue, Madison, WI 53706, USA. Tel. 1 608 263 7105, Fax. 1 608 262 5253

Susan K. Pierce Northwestern University, Dept. of Biochemistry, Mol. & Cell Biol., 2153 North Campus Drive, Evanston, IL 60208-3500, USA. Tel. 1 847 491 5089, Fax. 1 847 467 1610, E-mail: skpierce@merle.acns.nwu.edu

Steven Pind University of Manitoba, Faculty of Medicine, Dept. of Biochem. & Mol. Biol., Winnipeg, Manitoba, R3E 0W3, Canada. Tel. 1 20 4 789 3603, Fax. 1 204 783 0864, E-mail: spind@cc.umanitoba.ca

Nora Plesofsky-Vig The University of Minnesota, Dept. of Genetics and Cell Biology, 220 Biol. Sci. Center/1445 Gortner Ave, St. Paul, Minnesota 55108, USA. Tel. 1 612 624 5375, Fax. 1 612 625 1738, E-mail: nora@biosci.cbs.umn.edu

E. Roydon Price Harvard Medical School, Dept. of Cell Biology, 25 Shattuck Street, Boston, MA 02115, USA. Tel. 1 617 432 0994 or 0327, Fax. 1 617 432 1144, E-mail: eprice@warren.med.harvard.edu

Jens-U. Rahfeld Max-Planck-Research, Unit, Enzymology of Protein Folding, Kurt-Mothes-Str. 3, D-06120 Halle, Germany. Tel: 49 345 552 2801, Fax. 49 345 551 1972

Daisy R.J. Rahman University of Kent, Research School of Biosciences, Canterbury CT2 7NJ, UK. Tel. 44 1227 823699, Fax. 44 1227 763912, E-mail: M.F.Tuite@ukc.ac.uk

Satish Raina University of Geneva, Dept. de Biochemie Medicale CMU, 1, rue Michel–Servet, CH-1211, Geneva 4, Switzerland. Tel. 41 22 702 5511, Fax. 41 22 702 5502, E-mail: Satish.Raina@medecine.unige.ch

Linda L. Randall Washington State University, Department of Biochemistry and Biophysics, Pullman, WA 99164-4660, USA. Tel. 1 509 335 6398, Fax. 1 509 335 9688

Mark Rose Princeton University, Department of Molecular Biology, Lewis Thomas Laboratory, Princeton, NJ 08544-1014, USA. Tel. 1 609 258 2804, Fax. 1 609 259 6175, E-mail: mrose@molecular.princeton.edu

Helene Rosenberg National Institutes of Health, Laboratory of Host Defences, NIAID, Bldg. 10, Room 11N104, Bethesda, MD 20892, USA. Tel. 1 301 402 9131, Fax. 1 301 402 4369, E-mail: hr2k@nih.gov

Pierre Rouvière Du Pont Company, Environmental Biotechnology, Central Research and Development,

Wilmington, DE 19880-0328, USA. Tel. 1 302 695 1782, Fax. 1 302 695 1829, E-mail: rouviepe@al.esvax.umc.dupont.com

David Rubin Harvard Medical School, Department of Cell Biology, 240 Longwood Ave., Boston, MA 02115, USA. Tel. 1 617 432 1519, Fax. 1 617 432 1144, E-mail: luckyleo@warren.med.harvard.edu

Suzanne Rutherford HHMI- Uni of Chicago, 5841 S. Maryland Avenue, MC1028, AMB-101, Chicago, IL 60637, USA. Tel. 1 773 702 0868, Fax. 1 773 702 7254, E-mail: srutherf@midway.unchicago.edu

John A. Ryan University of Connecticut, Department of Molecular & Cell Biology, 75 N. Eagleville Road, Storrs, CT 06269-3044, USA. Tel. 1 860 486 4257, Fax. 1 860 486 1784, E-mail: HIGHTOWER@biotek.mcb.uconn.edu

Masao Sakaguchi Kyushu University, Graduate School of Medical Science, Department of Molecular Biology, Maidashi, Higashi-ku, Fukuoka 812-12, Japan. Tel. 81 92 642 6176, Fax. 81 92 642 6183, E-mail: sakag@cell.med.kyushu-u.ac.jp

Thomas Scheibel Universität Regensburg, Biochemie 2, Universitätsstrasse 31, Regensburg D-93040, Germany. Tel. 49 941 943 3039, Fax. 49 941 943 2813, E-mail: Johannes.Buchner@biologie.uni-regensburg-de

Randy W. Schekman HHMI/University of California, Berkeley, Department of Molecular and Cell Biology, 401 Barker Hall, Berkeley, CA 94720-0001, USA. Tel. 1 510 642 5686, Fax. 1 510 642 7846, E-mail: schekman@mendel.Berkeley.edu

Eric C. Schirmer University of Chicago, Howard Hughes Medical Institute, 5841 South Maryland Avenue, Box 391, Chicago, IL 60637, USA. Tel. 1 773 702 8049, Fax. 1 773 702 7254, E-mail: er93@midway.uchicago.edu

Matthias Schmitt Ludwig-Maximilians-Universität, Institut für Physiologische Chemie, Goethestrasse 33, D-80336 München, Germany. Tel. 49 89 5996 283, Fax. 49 89 5996 270, E-mail: Langer@bio.med.uni-muenchen.de

Elisabeth Schwarz Institut für Physiologische, Martin-Luther-Universität, Halle-Wittenberg, Kurt-Mothes-Str. 3, D-06120 Halle, Germany. Tel. 49 345 5524 856, Fax. 49 345 5527 013, E-mail: Elisabeth.Schwarz@biochemtech.uni-halle.de

Xiaoyin Shan Columbia University, Department of Biological Sciences, 702 Fairchild, New York, NY 10027, USA. Tel. 1 212 854 5443, Fax. 1 212 865 8246, E-mail: shan@cubsps.bio.columbia.edu

Ujwal Shinde UMDNJ-RW Johnson Medical School, Department of Biochemistry, 675 Hoes Lane, Piscataway, NJ 08854-5635, USA. Tel. 1 908 235 4115, Fax. 1 908 235 4559, E-mail: Shinde@rwja.umdnj.edu

Hisato Shuntoh Kobe University School of Medicine, Department of Pharmacology, 7-5-1 Kusunoki-cho, Chuo-ku, Kobe 650, Japan. Tel. 81 78 341 7451, Home. Ext. 3271, Fax. 81 78 351 6531, E-mail: tkuno@icluna.kobe-u.ac.jp

John J. Siekierka The R.W. Johnson Pharmaceutical Research Institute, Route 202 South, Raritan, NJ 08869, USA. Tel. 1 908 704 4599, Fax. 1 908 526 7118, E-mail: jsiekieka@prius.jnj.com

Pamela Silver Dana Farber Cancer Institute, Biological Chemistry & Molecular Pharmacology, 44 Binney St., Mayer 849, Boston, MA 02115, USA. Tel. 1 617 632 5102, Fax. 1 617 632 5103, E-mail: pamela_silver@macmailgw.dfci.harvard.edu

Jan Fredrik Simons Yale University School of Medicine, Department of Cell Biology, 333 Cedar Street, P.O. Box 3333, New Haven, CT 06520-8002, USA. Tel. 1 203 785 4303, Fax. 1 203 737 7446, E-mail: Ari.Helenius@ yale.edu

David F. Smith University of Nebraska Medical Center, Department of Pharmacology, 600 South 42nd Street, Omaha, NE 68198, USA. Tel. 1 402 559 4044, Fax. 1 402 559 7495, E-mail: dfsmith@mail.unmc.edu

Hans-Dieter Söling Universität Göttingen, Abteilung Klinische Biochemie, Zentrum Innere Medizine, D-37070 Göttingen, Germany. Tel. 49 551 39 6389, Fax. 49 551 39 2953, E-mail: hsoelin@gwdg.de

Catherine L. Squires Tufts University School of Medicine, Department of Mol. Biol. and Microbiol., 136 Harrison Avenue, Boston, MA 02111, USA. Tel. 1 617 636 6947, Fax. 1 617 636 0337, E-mail: c.squires_rib@opal.tufts.edu

Craig Squires Tufts University School of Medicine, Department of Mol. Biol. and Microbiol., 136 Harrison Avenue, Boston, MA 02111, USA. Tel. 1 617 636 6947, Fax. 1 617 636 0337, E-mail: c.squires_rib@opal.tufts. edu

Rodica Stan SmithKline Beecham Pharmaceuticals, Department of Comparative Genetics, UE0548, P.O. Box 1539, King of Prussia, PA 19406, USA. Tel. 1 610 270 7717, Fax. 1 610 270 7962, E-mail: Rodica_Stan 1.@sbphrd.com

Robin Stears SUNY Stony Brook, Department of Biochemistry and Cell Biology, Stony Brook, NY 11794, USA. Tel. 1 516 632 8560, Fax. 1 516 632 8575, E-mail: wlennarz@ccmail.sunysb.edu

Tom H. Stevens Institute of Molecular Biology, University of Oregon, Eugene, OR 97403-1229, USA. Tel. 1 541 346 5884, Fax. 1 541 346 4854, E-mail stevens@molbio.uoregon.edu

John R. Subjeck Roswell Park Cancer Institute, Department of Radiation Biology, P.O. Box 84563, Elm and Carlton Streets, Buffalo, NY 14263-0001, USA. Tel. 1 716 845 4425, E-mail: subjeck@sc3010.med. buffalo.edu

Hiroyuki Tachikawa Tokyo University, Dept. of Applied Biological Science, 3-5-8 Saiwaicho, Fuchu-shi, Tokyo, 183, Japan. Tel. 81 423 67 5703, Fax. 81 423 67 5703, E-mail: tachi@cc.tuat.ac.jp

Hideki Taguchi Tokyo Institute of Technology, Research Lab. of Resources Utilization, Yokohama, Japan. Tel. 81 45 924 5233, Fax. 81 45 924 5277, E-mail: myoshida@res.titech.ac.jp

Abida Taher Tulane University School of Medicine, Department of Biochemistry, SL43, 1430 Tulane Avenue, New Orleans, LA 70112-2699, USA. Tel. 1 504 586 3990, Fax. 1 504 584 2739, E-mail: landry@mailhost.tsc.tulane.edu

Nobuhiro Takahashi Tonen K. K., Corporate Research & Development Lab., 1-3-1 Nishi-tsurugaoka, Ohimachi, Iruma-gun, Saitama 354, Japan. Tel. 81 492 66

8371, Fax. 81 492 66 8359, E-mail: LDM03454@niftyserve. or.jp

Robert M. Tanguay Laboratoire de génétique cellulaire et développementale, RSVS, Pav. CE Marchand, Université Laval, Ste-Foy, Québec G1K 7P4, Canada. Tel. 1 418 656 3339, Fax. 1 418 656 7176, E-mail: Robert.Tanguay@rsvs.ulaval.ca

Michael Tavaria Monash University, Molecular Genetics & Development Group, Institute of Reproduction & Development, Clayton, Melbourne, Vic 3168, Australia. Tel. 61 3 9550 5480, Fax. 61 3 9550 5568, E-mail: ismailko@silas.cc.monash.edu.au

David G. Thanassi Washington University Medical School, Department of Molecular Microbiology, Box 8230, 660 South Euclid Ave., St. Louis, MO 63110-9109, USA. Tel. 1 314 362 1998, Fax. 1 314 362 6788, E-mail: thanassi@borcim.wustl.edu

Ulrich Theopold University of Adelaide, Department of Crop Protection, Glen Osmond, SA 5064, Australia. Tel. 61 8 8303 6565, Fax. 61 8 8379 4095, E-mail: utheopold@waite.adelaide. edu.au

David Thomas Biotechnology Research Institute, Eukaryotic Genetics Group, 6100 Avenue Royalmount, Montreal, Quebec, H4P 2R2, Canada. Tel. 1 514 496 6156, Fax. 1 514 496 6213, E-mail: dave.thomas@ bri.ncc.ca

Miranda Tighe University College London, Department of Pathology, Institute of Opthamology, Bath Street, London EC1V 9EL. UK. Tel. 44 171 608 6944, Fax. 44 171 608 6862

David O. Toft Mayo Clinic, Dept. of Biochem. & Mol. Biol., 200 1st St., SW, Rochester, MN 55905-0001, USA. Tel. 1 507 284 8401, Fax. 1 507 284 2053, E-mail: toft@mayo.edu

Mary Trahey Stanford University, Department of Pathology, B257 Beckman Center, Stanford, CA 94305-5324, USA. Tel. 1 415 723 7389, Fax. 1 415 498 6255, E-mail: traheym@selway.umt.edu

Maximilian Tropschug Institut für Biochemie und Molekularbiologie, Hermann-Herder-Strasse 7, D-79104 Freiburg im Breisgau, Germany. Tel. 49 761 203 5244, Fax. 49 761 203 5253, E-mail: tropschu@sun2.ruf.uni-freiburg.de

Mick F. Tuite University of Kent, Research School of Biosciences, Canterbury CT2 7NJ, UK. Tel. 44 1227 823699, Fax. 44 1227 763912, E-mail: M.F.Tuite@ukc.ac.uk

Shigeza Udaka Tokyo University of Agriculture, Dept. Brewing and Fermentation, 1-1 Sakuragaoka, Setagaya, Tokyo, 156, Japan. Tel. 81 3 5477 2385, Fax. 81 3 5477 2622

Chiharu Ueguchi Nagoya University, Laboratory of Molecular Microbiology, School of Agriculture, Japan. Tel. 81 52 789 4089, Fax. 81 52 789 4091, E-mail: i45455a@nucc.cc.nagoya-u.ac.jp

H. Bart van den Hazel Department of Yeast Genetics, Carlsberg Laboratory, Gamle Carlsberg Vej 10, DK-2500 Copenhagen Valby, Denmark. Tel. 45 3327 5282, Fax. 45 3327 4765, E-mail: carllab@biobase.dk

Saskia M. van der Vies University of Geneva, Department of Medical Biochemistry, CMU, 1, rue Michel-Servet, 1211 Geneva-4, Switzerland. Tel. 41 22 702 5515, Fax. 41 22 702 5502, E-mail: Saskia.VanDerVies@medecine.unige.ch

Elizabeth Vierling University of Arizona, Department of Biochemistry, Life Sciences S Building, Tucson, AZ 85721-0001, USA. Tel. 1 520 621 1601, Fax. 1 520 621 3709, E-mail: eliz@biosci.arizona.edu

Paul Viitanen El DuPont and Co., Central R&D, Experimental Station, P.O. Box 80402, Wilmington, DE 19880-0402, USA. Tel. 1 302 695 7032, Fax. 1 302 695 4509, E-mail: viitanPV@esvax.dnet.dupont.com

Thomas Waldmann Max Planck Institut, Structural Biology, Am Klopferspitz 18A, D-82152 Martinsried, Germany. Tel. 49 89 8578 2652, Fax. 49 89 8578 2641

Alicja Wawrzynow University of Gdansk, Department of Molecular and Cellular Biology, Kladki 24, 80 822 Gdansk, Poland. Tel. 48 58 31 92 22, Fax. 48 58 31 00 72, E-mail: zylicz@biotech.univ.gda.pl

Tracie Webster Victorian Institute of Animal Science, 475 Mickleham Rd., Attwood, Melbourne, Victoria 3049, Australia. Tel. 61 3 9217 4273, Fax. 61 3 9217 4299, E-mail: webstert@woody.agvic.gov.au

Irving L. Weissman Stanford University School of Medicine, Department of Pathology, B257 Beckman Center, Stanford, CA 94305-5324, USA. Tel. 1 415 723 7389, Fax 1 415 498 6255

Jonathan S. Weissman Yale University, HHMI, Department of Molecular Genetics, BCMM 154, 295 Congress Avenue, New Haven, CT 06510, USA. Tel. 1 203 737 4431, Fax. 1 203 737 1761, E-mail: horwich@hhvms8.csb.yale.edu

Sue Wickner National Institutes of Health, NCI, Department of Molecular Biology, Building 37, Room 2E18, Bethesda, MD 20892, USA. Tel. 1 301 496 3524, Fax. 1 301 496 3875, E-mail: susang@helix.nih.gov

Gregory Wiederrecht Merck Research Laboratories, P.O. Box 2000, Mail Code R80M-260B, Rahway, NJ 07065-0900, USA. Tel. 1 908 594 6576, Fax. 1 908 594 7140, E-mail: greg_wiederrecht@merck.com

Sigurd Wilbanks Stanford University, Department of Structural Biology, Stanford School of Medicine, Stanford, CA 94305-5400, USA. Tel. 1 415 723 7095, Fax. 1 415 723 8464, E-mail: WILBANKS@cellbio.stanford.edu

David B. Williams University of Toronto, Department of Biochemistry, Medical Science Bldg, Toronto, Ontario M5S 1A8, Canada. Tel. 1 416 978 2546, Fax. 1 416 978 8548, E-mail: david.williams@utoronto.ca

Keith Willison Institute of Cancer Research, Chester Beatty Laboratories, 237 Fulham Road, London SW3 6JB, UK. Tel. 44 171 352 8133, Fax. 44 171 351 3325, E-mail: willison@icr.uk.ac

Iain Wilson National Institute for Medical Research, The Ridgeway, Mill Hill, London, NW7 1AA, UK. Tel. 44 181 959 3666 ext. 2177, Fax. 44 181 913 8593, E-mail: r-wilson@nimr.mrc.ac.uk

Jakob Winther Carlsberg Laboratory, Department of Yeast Genetics, Gamle Carlsberg Vej 10, DK-2500

Copenhagen Valby, Denmark. Tel. 45 33 27 5282, Home. 45 33 27 5391, Fax. 45 33 27 4765 or 4708, E-mail: JRW@CRC.DK

Whi-Fin Wu   National Institutes of Health, NCI, Department of Molecular Biology, Building 37, Room 2E18, Bethesda, MD 20892, USA. Tel. 1 301 496 3524, Fax. 1 301 496 3875, E-mail: susang@helix.nih.gov

Zhixiong Xue   DuPont Central Research and Development, P.O. Box 80402, Wilmington, DE 19880-0402, USA. Tel. 1 302 695 9465, Fax. 1 302 695 8480, E-mail: xuez@esvac.dnet.dupont.com

Masasuke Yoshida   Tokyo Institute of Technology, Research Lab. of Resources Utilization, Yokohama, Japan. Tel. 81 45 924 5233, Fax. 81 45 924 5277, E-mail: myoshida@res.titech.ac.jp

André Zapun   European Molecular Biology Laboratory, Meyerhofstrasse 1, D-69012 Heidelberg, Germany

Bruce R. Zetter   Harvard Medical School, Children's Hosp., Ender's Building, Room 1077, 300 Longwood Avenue, Boston, MA 02115, USA. Tel. 1 617 355 6767, Fax. 1 617 355 7043, E-mail: zetter@a1.tch.harvard.edu

Jian-Kang Zhu   University of Arizona, Department of Plant Sciences, Tucson, AZ 85721, USA. Tel. 1 520 621 9567, Fax. 1 520 621 7186, E-mail: jkzhu@ag. arizona. edu

Thomas Ziegelhoffer   University of Wisconsin-Madison, Department of Biomolecular Chemistry, 1300 University Avenue, Madison, WI 53706, USA. Tel. 1 608 263 7105, Fax. 1 608 262 5253

Konrad Zinsmaier   University of Penn School of Medicine, Department of Neuroscience, 232a Stemmler Hall, Philadelphia, PA 19104-6074, USA. Tel. 1 215 898 2576, Fax. 1 215 573 2015, E-mail: zinsmaie@mail.med. upenn.edu

Charles Zuker   HHMI, University of California, San Diego, Departments of Biology and Neuroscience, 9500 Gilman Drive, CMM 355, Department 0649, La Jolla, CA 92093-0649, USA. Tel. 1 619 534 5528, Fax. 1 619 534 8510, E-mail: charles@flyeye.ucsd.edu

Maciej Zylicz   University of Gdansk, Department of Molecular and Cellular Biology, Kladki 24, 80 822 Gdansk, Poland. Tel. 48 58 31 92 22, Fax. 48 58 31 00 72, E-mail: zylicz@biotech.univ.gda.pl

Abbreviations

20-HOE	20-hydroxyecdysone	IMC	intramolecular chaperone
aa	amino acid	IP3R	inositol-1,4,5-triphosphate receptor
ADF	adult T cell leukaemia-derived factor	LPS	lipopolysaccharide
ANS	anilino-naphthalene-8-sulphonate	MAP	microtubule-associated protein
BPTI	bovine pancreatic trypsin inhibitor	MDR	multidrug resistance
BRK	baby rat kidney	MHC	major histocompatibility complex
CaN	calcineurin	Mip	macrophage infectivity potentiator
CAP	cyclophilin-associated protein	mMDH	mitochondrial malate dehydrogenase
CBP	calcium-binding protein	MSF	mitochondrial import stimulation factor
CCT	cytosolic chaperonins	NAC	nascent polypeptide-associated complex
CD	circular dichroism	NADP	nicotinamide-adenine dinucleotide phosphate
CIRCE	controlling inverted repeat of chaperone expression	NEM	N-ethylmaleimide
CPT	carnitine medium/long chain acyltransferase	NID	NIMA protein interaction domain
CPY	carboxypeptidase Y	NLS	nuclear targeting (localization) signal
CRC	Ca^{2+} release channel	NTP	nucleoside triphosphate
CRE	cAMP response element	OMP	outer membrane protein
CSA	C3H strain-specific antigen	ORF	open reading frame
CsA	cyclosporin A	OSP	osmotic stress protein
csp	capsular polysaccharide	OTC	ornithine transcarbamylase
csp	cysteine-string protein	PAPS	3′-phosphoadenosine 5′-phosphosulfate
CypA	cyclophilin A	PBF	presequence binding factor
CypCAP	cyclophilin C-associated protein	PHA	phytohemagglutinin
DHA	dihydroascorbic acid	PHF	paired helical filaments
DHFR	(mouse) dihydrofolate reductase	PIPLC	phosphatidylinositol-specific phospholipase C
DTT	dithiothreitol	ppaF	pre-pro-alpha factor
EAT	Ehrlich ascites tumour cells	PR	progesterone receptor
EBV	Epstein–Barr virus	RACE	rapid amplification of DNA ends
EcRE	ecdysterone response element	RAP	rapamycin
EM	electron microscopy	RCMLA	reduced and carboxy-methylated lactalbumin
EMS	ethylmethane sulphonate		
Endo H	endoglycosidase H	RL	rabbit reticulocyte lysate
EPF	early pregnancy factor	RR	ribonucleotide reductase
ER	endoplasmic reticulum	RT–PCR	reverse transcriptase–polymerase chain reaction
ERBC	oestrogen receptor-binding cyclophilin		
Erps	ER-specific proteins	rubisco	ribulose biphosphate carboxylase oxygenase
GIT	glutathione–insulin transhydrogenase	RyR	ryanodine receptor
GPI	glycosylphosphatidylinositol	SDS–PAGE	sodium dodecyl sulphate–polyacrylamide gel electrophoresis
GR	glucocorticoid receptor		
GRP	glucose-regulated protein	SEC	size exclusion chromatography
GSBP	glycosylation site-binding protein	serpin	serine protease inhibitor family
GSH	glutathione	SGNE-1	secretory granule neuroendocrine protein-1
GST	glutathione S-transferase	SHR	steroid receptor family proteins
Hip	Hsc70-interacting protein	SR	SRP receptor
HIP	HSP70-interacting protein	SRP	signal recognition particle
HSE	heat shock element	STRE	stress response element
HSF	heat shock factor	TCP-1	t-complex polypeptide 1
HSG	heat shock granule	TCR	T cell receptor
HSP	heat shock protein	TF	trigger factor
HUVE	human umbilical vein endothelial cDNA library	TF55	thermophilic factor of 55 kDa
		TPR	tetratricopeptide repeat
IAA	indoleacetic acid	TRG	thioredoxin-related glycoprotein
Ibp	inclusion body proteins	ts	temperature sensitive
Ig	immunoglobulin	UPRE	unfolded protein response element

UTR	untranslated region
VSG	variant surface glycoprotein
WT	wild-type

HSP70 Proteins

The HSP70 family — an overview

The HSP70 proteins are a highly conserved ubiquitous group of proteins found in all species examined. Many species have multiple types of HSP70s which are located in cellular compartments including mitochondria, the endoplasmic reticulum, lysosomes, and chloroplasts, as well as the cytosol and nucleus. Through their ability to bind short stretches of polypeptide HSP70s are involved in prevention of aggregation of unfolded polypeptides and the disassembly of multimeric protein complexes. HSP70 function encompasses a wide variety of cellular processes, including protein trafficking between cellular compartments, protein folding, and regulation of the heat shock response.

■ Distribution of HSP70s and relatedness

hsp70 genes were first visualized as puffs on polytene chromosomes isolated from salivary glands upon exposure of *Drosophila* larvae to high temperature or chemicals (Ritossa, 1962). More than a decade later it was determined that the major gene being transcribed and translated upon exposure to stress encoded a 70 kDa protein (reviewed in Ashburner and Bonner, 1979; Craig, 1985)). Proteins related to the *Drosophila* HSP70s have now been identified in all organisms examined (Fig. 1). Many HSP70s are very abundant proteins accounting for as much as 1–2% of cellular protein (Herendeen *et al.*, 1979). Comparison of HSP70s from 24 divergent eukaryotic and prokaryotic species demonstrates their extreme conservation (Boorstein *et al.*, 1994) (Figs. 2 and 3). Proteins from the most distantly related species are at least 45% identical in sequence. Phylogenetic trees indicate that ancient events gave rise to at least four groups of HSP70s with each group having a common cellular localization: cytosol, endoplasmic reticulum (ER), mitochondria, and chloroplasts (Fig. 1). HSP70s of the mitochondria and plastids are most similar to the HSP70s of purple bacteria and cyanobacteria, respectively, which is consistent with the proposed prokaryotic origin of these organelles. For years it has been assumed that prokaryotes have a single HSP70; however, a second has been found recently (Kawula and Lelivelt, 1994; Seaton and Vickery, 1994). The cellular function of this newly identified HSP70 remains to be defined.

Recent completion of the sequence of the genome of the budding yeast *Saccharomyces cerevisiae* allows analysis of a complete set of HSP70s within an organism. Since the sequence was very recently completed, the analysis is preliminary, but nevertheless revealing. Fourteen HSP70-related proteins have been identified. The cytosol, ER, and mitochondria all contain more than one HSP70. The ER has two (Baxter *et al.*, 1996; Craven *et al.*, 1996; Hamilton and Flynn, 1996), mitochondria two or three (Schilke *et al.*, 1996; S. Laloraya and E. Craig, unpublished results), and the cytosol at least six (Boorstein *et al.*, 1994). A crude family tree resulting from the comparison of the newly identified members with the well-studied family members is shown in Fig. 4. While some of the HSP70 proteins are highly conserved, some are as little as 24% identical with other family members.

The degree of conservation of HSP70 proteins varies throughout their length. The N-terminal ATPase domain is the most highly conserved region, with the residues in the ATP-binding cleft being particularly so. The degree of conservation diminishes towards the C-termini with no conservation detected in the last 10 kDa between the subcellular groups. However, the extreme C-termini are highly conserved within subfamily groups in some cases (Fig. 3). Proteins making up the large cytosolic subclass end in GP(T/K)(V/I)EEVD. The ER-localized HSP70s have the well-characterized ER retrieval sequence (KDEL, HDEL, etc.; Pelham, 1989) at their extreme C-termini.

■ Structure of HSP70 proteins

HSP70 proteins transiently bind short polypeptide segments and possess a weak ATPase activity which is stimulated by the binding of peptides or polypeptides. Our understanding of the basic domain structure of HSP70s was deduced from studies of truncated forms of the protein (Chappell *et al.*, 1987; Wang *et al.*, 1993). The N-terminal 44 kDa segment which contains the adenosine nucleotide binding site has a weak ATPase activity, in the range of 0.03–0.27 min^{-1} (McKay *et al.*, 1994). The structure of this domain is similar to that of hexokinase and actin (Flaherty *et al.*, 1990, 1991; Bork *et al.*, 1992, see entry p. 13). The adjacent 18 kDa segment has been shown to bind peptide (Wang *et al.*, 1993). Based on computer modeling studies this domain was initially proposed to be similar to MHC class I and II peptide-binding domains (Rippmann *et al.*, 1991; Flajnik *et al.*, 1991). However, structural studies using NMR (Morshauser *et al.*, 1995) showed that the connections between strands within β-sheets were not compatible with a MHC class I- or class II-like structure. Recently the structure of a crystal of the binding domain of DnaK complexed with peptide has been defined by X-ray crystallographic studies (Zhu *et al.*, 1996, see entry p. 18) revealing that the peptide-binding region does not resemble any previously solved

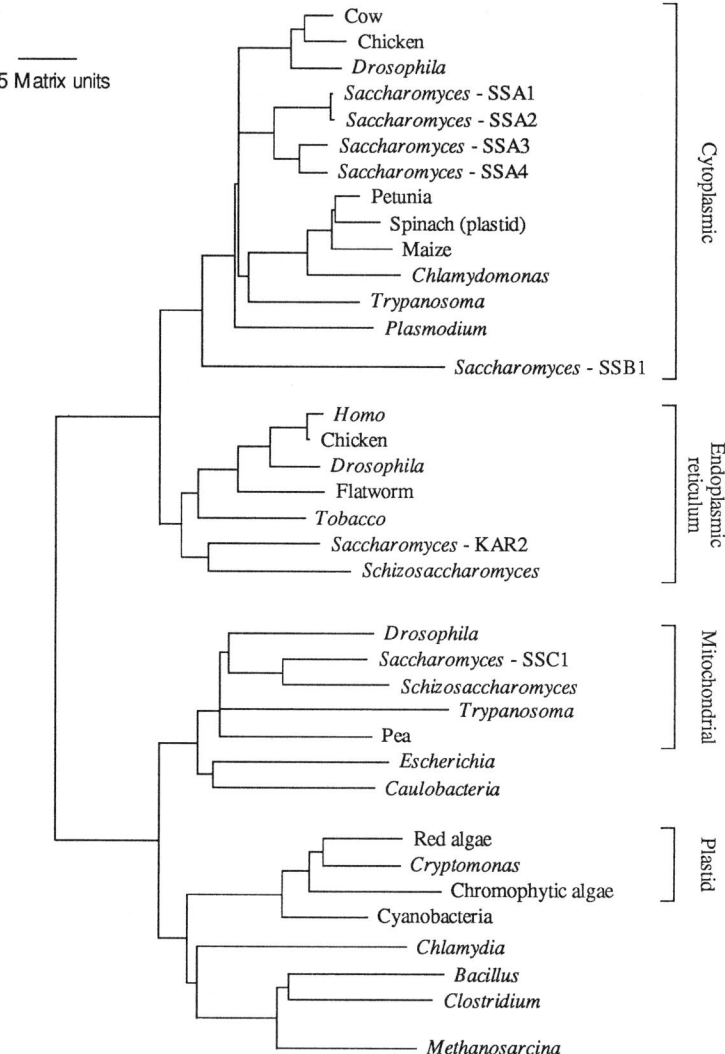

Figure 1. Distance-matrix based phylogenetic tree depicting relationships among 36 HSP70 proteins from 25 different species. The tree was constructed from a common progressive alignment of sequences as described in Boorstein *et al.* (1994). Intracellular localizations of the eukaryotic HSP70 proteins analysed are indicated by the square brackets at the right of the figures. Species names and sequence accession numbers are presented in Table 1. Taken from Boorstein *et al.* (1994) with permission.

structure, and that the peptide is bound in a channel formed by loops protruding from one end of a β-sandwich structure. The peptide-binding channel is capped by an α helix which is predominantly encompassed in the 18 kDa fragment. The C-terminal 10 kDa segment, which is α-helical in nature, does not interact with bound peptide. It is essential, but its function remains to be determined.

While the great majority of HSP70 proteins conform to the general domain structure described above, several exceptions have been found. STCH is a 471 amino acid protein related to HSP70 proteins but lacking the peptide-binding domain (Otterson *et al.*, 1994; see entry p. 69). Its function is not known. A small group of HSP70-related proteins, which in this volume are described in Part 2 'HSP110/SSE proteins', differ in lacking a highly conserved region of the HSP70 peptide-binding domain and in containing a non-homologous insertion in the C-terminal domain. For example, Hsp110 is an 858 amino acid mammalian stress-induced protein which is 33% identical to mammalian Hsp70. Most of the identity is in the ATPase domain; a highly conserved region of the HSP70 peptide-binding domain is lacking in Hsp110. The larger size of Hsp110 is mostly a result of a 100 amino acid non-homologous insertion near amino acid 500 (Lee-Yoon et al., 1995, see entry p. 81).

HSP70 proteins undergo autophosphorylation *in vitro* (Zylicz et al., 1983; Freiden et al., 1992). Autophosphorylation activity is stimulated by the presence of Ca^{2+}, while the ATPase activity, which requires Mg^{2+}, is inhibited in the presence of Ca^{2+}. While HSP70s are found in phosphorylated forms *in vivo*, the sites of phosphorylation differ from those found *in vitro* as the result of autophosphorylation (Gaut and Hendershot, 1993). The function of the phosphorylation found *in vivo* remains obscure.

Under certain conditions purified HSP70 proteins are known to exist as oligomers, which are converted to monomers in the presence of peptides or ATP (Palleros et al., 1991; Blond-Elguindi et al., 1993b). Genetic data obtained from the studies of dominant *hsp70* mutations and intergenic complementation between *hsp70* mutants suggest that oligomers also form *in vivo* (Nicolet and Craig, 1991; Wild et al., 1992).

■ Interacting proteins

Initial studies identified two proteins, GrpE and DnaJ, which interact with DnaK, the major HSP70 of *E. coli* (reviewed in Georgopoulos, 1992). Both proteins bind DnaK and together increase the steady state ATPase activity up to 100-fold. DnaJ stimulates the rate of ATP hydrolysis, while GrpE acts as a nucleotide release factor (Liberek et al., 1991; Buchberger et al., 1994; McCarty et al., 1995). The site of interaction of DnaJ with DnaK is thought to be in the C-terminal peptide-binding domain, while a conserved loop in the ATPase domain of the DnaK chaperone has been shown to be essential for stable binding of GrpE (Buchberger et al., 1994). Numerous DnaJ-related proteins have been identified in eukaryotes in several cellular compartments; biochemical and genetic data indicate that the eukaryotic proteins act in a manner similar to prokaryotic DnaJ (for review see Cyr et al., 1994, and the overview of the HSP40 (DnaJ-related) proteins, p. 89). GrpE-related proteins have also been found in eukaryotes, but only in mitochondria thus far in eukaryotic cells (see overview p. 133). Differences in the biochemical properties between cytosolic and mitochondrial or prokaryotic HSP70s may obviate the need for a eukaryotic cytosolic nucleotide release factor (Hohfeld et al., 1995; Ziegelhoffer et al., 1995).

More recently other HSP70-interacting proteins have been identified. Hip interacts with the ATPase domain of cytosolic Hsc70 and appears to stabilize the ADP-bound form of the protein (Hohfeld et al., 1995). Hip also appears to interact with HSP40 (DnaJ) and HSP90 and has been identified as a component of the complex involved in hormone receptor activation in vertebrates (Prapapanich et al., 1996). Hsc70, HSP90, Hip, and other proteins, including various immunophilins, have all been found in inactive steroid hormone receptor complexes, while Hsc70 and Hip are not found in mature receptor complexes that are competent for binding hormone (Smith et al., 1995, see review p. 518).

Finally, Tim44p (also called Mim44p, ISP45, Mpi1p), a protein of the mitochondrial inner membrane, interacts with Ssc1p, an abundant mitochondrial HSP70 required for translocation of many nuclear-encoded proteins into mitochondria (for review see Pfanner et al., 1994). This interaction, which appears to be critical for the translocation process, is disrupted by ATP.

■ Interaction of HSP70 proteins with substrate polypeptides

HSP70 proteins bind peptides and proteins via their C-terminal peptide-binding domains. A minimum peptide length of seven residues is required for optimal binding to BiP, the major ER-located HSP70 (Flynn et al., 1991), and a peptide bound to DnaK has been shown to be in an extended conformation (Landry et al., 1992), consistent with the role of HSP70 proteins in binding unfolded polypeptides. In cases where HSP70s bind to folded proteins such as clathrin, it has been suggested that they recognize conformationally flexible structures such as loops and turns (DeLuca-Flaherty et al., 1990). In general, HSP70s preferentially bind hydrophobic peptides, while the presence of acidic residues is unfavorable. Basic residues are tolerated, although there seem to be some position-specific tolerance differences among different HSP70 family members (Fourie et al., 1994; Gragerov et al., 1994). It has been proposed that BiP preferentially binds peptides containing hydrophobic residues in alternating positions (Blond-Elguindi et al., 1993a).

The nucleotide content of HSP70 proteins has a great effect on their affinity for peptides. When ADP is bound, HSP70s have a higher affinity for peptides, but the kinetics of peptide binding and release is relatively slow. In contrast, when ATP is bound, HSP70s binds peptides very rapidly, but the release of peptides is also fast, resulting in a lower overall affinity for peptides (Schmid et al., 1994). At least two models have been proposed concerning the reaction cycle of HSP70 peptide binding and release. In one model (Palleros et al., 1994), the ADP-bound form of HSP70 is responsible for peptide binding, subsequent exchange of ATP for ADP causes the release of bound peptide, and ATP hydrolysis returns HSP70 to the ADP-bound form. In an another model (Greene et al., 1995; McCarty, et al., 1995), the ATP-bound form of HSP70 is the active peptide-binding form owing to its fast peptide-binding properties. Subsequent ATP hydrolysis converts the unstable HSP70–ATP–peptide complex into the stable HSP70–ADP–peptide complex. ATP can then replace ADP in the complex and cause peptide release. In both models DnaJ homologs facilitate the ATP hydrolysis step while GrpE homologs facilitate nucleotide exchange. While further biochemical analysis is needed to distinguish between these and other models, it is generally accepted that transient interaction with short peptide sequences regulated by interaction with nucleotide is critical for the ability of HSP70 proteins to function in essential biological processes.

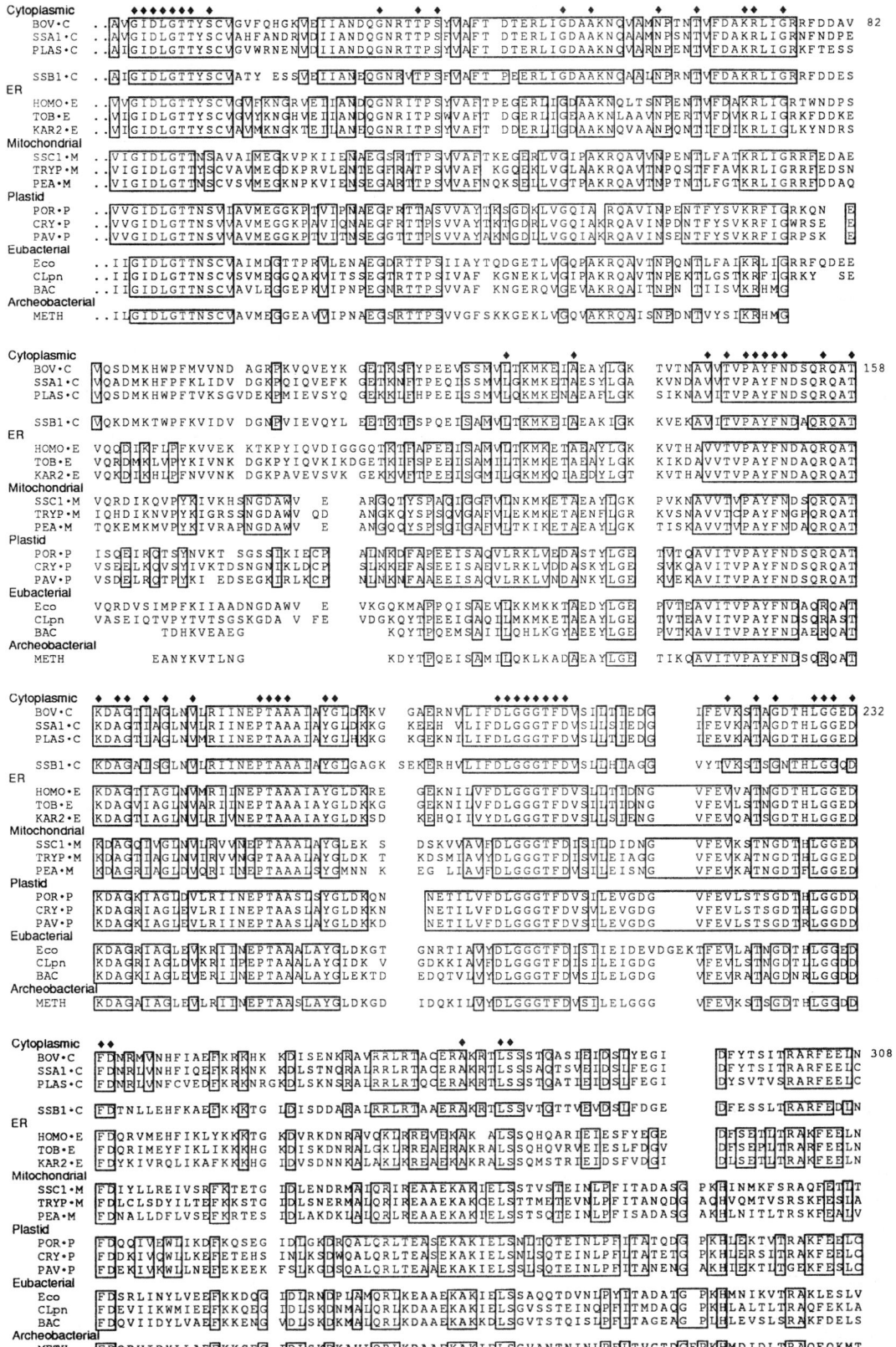

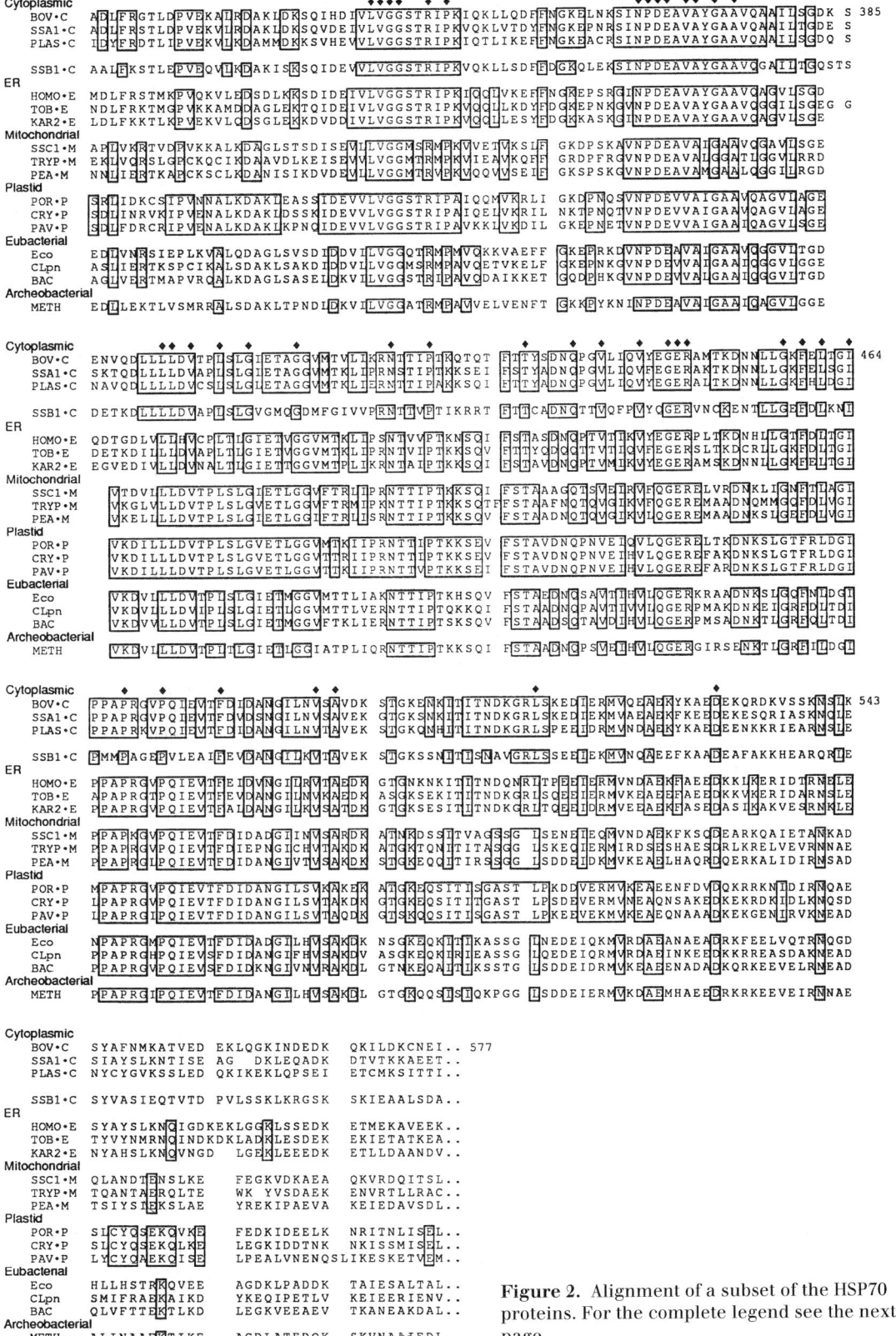

Figure 2. Alignment of a subset of the HSP70 proteins. For the complete legend see the next page.

Figure 2. (Previous pages) Alignment of a subset of the HSP70 proteins. The sequences shown here are truncated at amino- and carboxy-termini to exclude regions that do not appear to be conserved at the level of primary sequence composition. The amino acids of the bovine cytoplasmic Hsc70 are numbered with respect to the initiating methionine as +1. Spaces indicate positions where gaps were inserted in the alignment. Three of the most distantly related members of the clusters from the phylogenetic trees of Fig. 1 were selected for inclusion here. The progressive alignment of the 17 sequences shown here was performed together with 19 additional sequences (all 36 HSP70s from Fig. 1); therefore the subset presented may not appear optimally aligned at some positions and there are several positions in which gaps appear in all sequences. Diamonds above the sequence denote positions at which the same amino acid is present in all 36 HSP70s. Boxes are drawn around amino acids that are identical in all members of a cluster analysed in this study; this includes 12 cytoplasmic, seven ER, five mitochondrial, three plastid and six eubacterial HSP70s. Because both the number of sequences included in each group and their phylogenetic diversity differ widely, the significance of the boxed residues differs between groups; this explains the large number of boxes around the plastid sequences relative to the other groups. Because the *Saccharomyces cerevisiae* Ssb proteins are so distantly related to other eukaryotic cytoplasmic HSP70s, and this difference is substantiated by genetic and biochemical experiments, Ssb1p was not included in the cytoplasmic group. Instead, Ssb1p amino acids are boxed if identical to residues conserved among all of the other cytoplasmic HSP70s. The *Methanosarcina* HSP70 (abbreviated METH) amino acids are boxed if identical to conserved eubacterial residues. The nuclear encoded, chloroplast localized spinach gene was excluded from the information used to box conserved residues as it does not clearly fit into any group. A key to HSP70 abbreviations and accession numbers are presented in Table 1. Taken with permission from Boorstein *et al.* (1994).

```
Cytoplasmic
  BOV·C    ..INWLDKN  QTAEKEEFEHQQKELEKVCNPIITKLYQ         S AGGMPGGMPGGM         PGGFPG  GGAP  635
  SSA1·C   ..ISWLDSN  TTASKEEFDDKLKELQDIANPIMSKLYQ         A GGAPGGAAGGA          PGGFPG  GA
  PLAS·C   ..LEWLEKN  QLASKEEYESKQKELAESVCAPIMSKIYQDVGGAAGGMPGGMPGGM               PGGMPGGMPGGG
  SSB1·C   ..LAALQI   EDPSADELRKAEVGLKRVVTKAMSSR·
ER
  HOMO·E   ..IEWLESH  QDADIEDFKAKKKELEEIVQFIISKLY          GSAG   P PP             TGE
  TOB·E    ..LEWLDDN  QSAEKEDYEEKLKEVEAVCNPIITAVY          QKSGGAPGGE             SGA
  KAR2·E   ..LEWLDDNFETAIAEDFDEKFESLSKVAYPITSKLYG          GADGSGAADYDD           EDE
Mitochondrial
  SSC1·M   ..KELVARVQ GGEE VNAEELKTKTEELQTSSMKLFEQLYKN                              DSNN       NNN
  TRYP·M   ..RKSMENPN VTKDEL SAATDKLQKAVMECGRTEYQQAAAG              NSSSS          SGNT       DSS
  PEA·M    ..RTAMAGEN ADDIKAKLDAANKAVSKI  GQHMSGGSSGG              PSEG            GSQG        G
Plastid
  POR·P    ..RSNLEKEE LDSIEANSEKLQNALMEI   GKNA   TSAEKD            TQ             N    A     SND
  CRY·P    ..RNAINNEN YDEMRDLNSKLQTALMDL   GKSVYEKTSKEQT            ST             SSPT       NSN
  PAV·P    ..LKENIKKED YDKIKENLKKLQEKLMEI  GQKAYAKKEPLKD            ED             SNKA       GSQ
Eubacterial
  Eco      ..ETALKGED KAAIEAKMQELAQVSQKLMEIAQQQHAQQQTA              GADA            SANN        AKD
  CLpn     ..RNALKDDAPIEKIKEVTEDLSKHMQKI   GESMQSQSASAAASSAAANAKGGPNINTEDLKKHSFSTKPPSNNGSSE
  BAC      ..KAAIEKND LEEIKAKKDELQEIVQAL   TVKLYEAQQQAQQA           GE             QGA         QN
Archeobacterial
  METH     ..KKALEGKD AEDIKAKTEALQESVYPI   STAMYQKAQQAQQAA          GGE            GGAAGTDARGPD

Cytoplasmic
  BOV·C    PSGG                AS SGPT IEEVD*   650
  SSA1·C   PPAP                EAH GPT VEEVD*
  PLAS·C   MPGGMNFPGGMPGGGMPGGAPAGS GPT VEEVD*
  SSB1·C
ER
  HOMO·E   EDTA                EK      DEL*
  TOB·E    SEDD                DH      DEL*
  KAR2·E   DDDG                DYFEH   DEL*
Mitochondrial
  SSC1·M   NNGN                NAESGETKQ*
  TRYP·M   QGEQ                QQQGDQQKQ*
  PEA·M    EQAP                EAEYEEVKK*
Plastid
  POR·P    DTVI                DIDE    SEAK*
  CRY·P    DSVI                DADF    SETK*
  PAV·P    DDFI                DADF    TESK*
Eubacterial
  Eco      DDVV                DAEFEEVKDKK*
  CLpn     DHIE                EADVEIIDNDDK*
  BAC      DDVV                DAEFEEVNDDKK*
Archeobacterial
  METH     ETVV                DADYEVVDDEKRK*
```

Figure 3. Carboxy-terminal sequences of HSP70 proteins illustrating limited identity among members of closely related groups and divergence between different groups. The sequences are contiguous with those shown in Fig. 2. These 17 sequences were progressively aligned together with the full set of 36 HSP70s in the same order used in the alignment presented in Fig. 2. Conserved amino acids are boxed as described in the legend to Fig. 2. The position of the termination codons are indicated by asterisks. Taken from Boorstein *et al.* (1994) with permission.

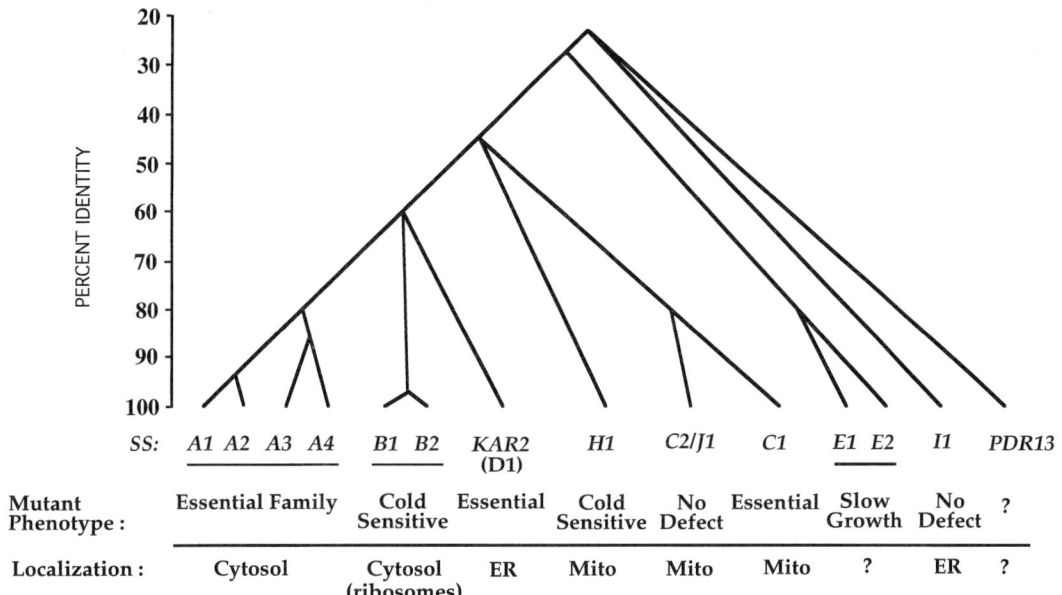

Figure 4. The *S. cerevisiae* HSP70 proteins. The tree was constructed simply by comparing the newly identified sequences of Sseps, Ssh1p, Ssi1p, Ssj1p (Ssc2p) and Pdr13p (S. Moye-Rowley, personal communication) to the well-characterized Ssaps, Kar2p, Ssbps and Ssc1p sequences (Boorstein *et al.*, 1994). The cellular location of Sse1p, Sse2p and Pdr13 have not been determined; Ssc2/J1 has a putative mitochondrial leader sequence but its location has not been determined.

■ Cellular functions of HSP70 proteins

HSP70s, with other interacting components such as DnaJ and its homologs, form a central chaperone system of cells. Through their interactions with short peptide sequences they are involved in a wide array of biological processes. These processes can be divided into two general types: (1) those involving specific interactions with a protein having a native conformation, for example during initiation of λ DNA replication and uncoating of clathrin-coated vesicles; and (2) those involving interactions with a variety of unfolded or partially unfolded polypeptides, for example translocation of proteins across membranes and prevention of protein aggregation.

The first functional identification of an HSP70 was the isolation of mutations in *dnaK* that rendered cells resistant to bacteriophage λ lytic infection. DnaK's main function in infection is the release of the λP protein from the DnaB helicase, allowing λP to proceed along the DNA and unwind the DNA template (Georgopoulos *et al.*, 1990). Early biochemical information about HSP70s came from the characterization of the protein that facilitates the uncoating of clathrin-coated vesicles, which mediate selective endocytosis and the sorting of proteins in the *trans*-Golgi network. Clathrin triskelions, the basic structural component of the clathrin coat, are composed of three heavy chains and three light chains. Hsc70 binding to a particular peptide sequence of the light chain, which is proposed to be exposed as a result of a Ca^{2+} flux, facilitates disassembly of the cage (Brodsky *et al.*, 1991).

Apparently, peptide substrates bind the same site on HSP70 as clathrin, since peptide substrates can compete for clathrin binding (Greene *et al.*, 1995).

More general chaperone functions of HSP70s include roles in protein translocation, folding, and proteolysis (for reviews see Brodsky and Schekman, 1994; Pfanner *et al.*, 1994; Stuart *et al.*, 1994; Pfanner and Meijer, 1995; Hayes and Dice, 1996; and the various entries on HSP70 proteins that follow this overview). HSP70 proteins of the mitochondrial matrix (Ssc1p/Pbp74/mtHsp70) and the lumen of the ER (Kar2p/BiP) are required for translocation of precursor proteins from the cytosol to the interior of these organelles through direct interactions with the translocating polypeptide. Whether these HSP70s act by exerting a 'pulling' motion on the polypeptide, or whether they act by preventing backsliding is the subject of debate (Stuart *et al.*, 1994; Glick, 1995; Pfanner and Meijer, 1995).

Roles in protein folding and the suppression of aggregation have been studied most extensively in purified systems. For example, HSP70 with its cohort proteins, such as DnaJ and GrpE or their homologs, are sufficient to facilitate the refolding of denatured luciferase *in vitro* (Schroder *et al.*, 1993). Compelling *in vivo* evidence implicates BiP in the refolding of a vacuolar protein after traversing the ER membrane (Simons *et al.*, 1995).

HSP70 proteins are also involved in proteolysis. Abnormal proteins are degraded more slowly than normal in *dnaK* mutants (Straus *et al.*, 1988) and DnaK has been found in a complex with the protease La and an abnormal protein (Sherman and Goldberg, 1992).

Similarly, Pim1 protease, the yeast mitochondrial homolog of protease La, requires the mitochondrial HSP70, Ssc1, for efficient degradation of two abnormal proteins (Wagner et al., 1994). Mammalian cytosolic HSP70 (Hsc70 or Prp73) stimulates lysosomal uptake of polypeptides destined for degradation, and thus stimulates lysosomal degradation (Chiang et al., 1989). The role of HSP70s in proteolysis may be to prevent aggregation and maintain the accessibility of substrate proteins to the proteolytic system rather than to actively facilitate protein degradation.

■ Regulation of synthesis of HSP70 proteins

The majority of HSP70 proteins are synthesized constitutively and abundantly under normal growth conditions, but their expression may be increased following a change in temperature, or a variety of other environmental stresses. Other HSP70 family members are only synthesized following a specific stress, such as heat shock. In mammalian cells changes in the levels of expression of HSP70 proteins are also observed during the cell cycle, and during differentiation and development, as well as in response to abnormal physiological states such as fever, inflammation, and wound responses (reviewed in Morimoto et al., 1993).

In E. coli, expression of heat shock genes, including dnaK, is dependent on the heat shock gene promoter-specific sigma subunit (σ^{32}) of RNA polymerase, while many eukaryotic HSP70 proteins are induced following heat stress by a transcriptional mechanism involving the well-defined positively acting transcription factor, HSF, which binds to DNA elements called HSEs (see entry p. 534). Other HSP70s are regulated in additional or completely different ways. For example, the ER-localized HSP70s, commonly called BiPs, are regulated in response to increased levels of unfolded proteins in the ER lumen. Therefore, inhibitors of glycosylation such as tunicamycin and certain secretory mutants, which result in build-up of unfolded proteins in the lumen, cause induction. Such induction is mediated through a defined DNA element, called a UPRE, which is distinct from an HSE (Mori et al., 1992).

Certain HSP70s are not induced by heat shock, while expression of others shuts off upon a heat shock, or increases upon a temperature downshift. For example, the most ubiquitous cytosolic HSP70 class contains members that are heat inducible as well as members whose expression does not change upon temperature upshift (Fig. 5). Expression of a class of cytosolic HSP70s, the Ssb proteins, is turned off upon a temperature upshift (Werner-Washburne et al., 1989). Finally, the second identified HSP70 of E. coli, Hsc66, is induced about 11-fold after a cold shock (Lelivelt and Kawula, 1995).

The heat shock response is autoregulated. The fact that the lowering of the amount of the major cytosolic form of HSP70 (Hsc/Hsp70 of mammals, Ssa proteins of yeast, and DnaK of E. coli) results in the induction of the heat shock response, and that overexpression results in diminution of the response led to the hypothesis that HSP70s play a critical role in this regulation both in prokaryotes and eukaryotes (Craig and Gross, 1991; Morimoto et al., 1993). In E. coli DnaK interacts directly with σ^{32}, thereby preventing σ^{32} interaction with core RNA polymerase and favoring the formation of holoenzyme containing the major sigma factor, σ^{70} (Blaszczak et al., 1995). According to current thinking a heat shock increases the concentration of partially unfolded substrate proteins that bind DnaK, thus freeing σ^{32} for interaction with core polymerase. The mechanism of the involvement of eukaryotic HSP70 in the regulation of the heat shock response is less well defined, but a similar model has been proposed. While direct interaction between HSF and HSP70 has been observed (Abravaya et al., 1992) compelling data to implicate the direct interaction between the two proteins as the key regulatory mechanism in vivo has yet to be forthcoming.

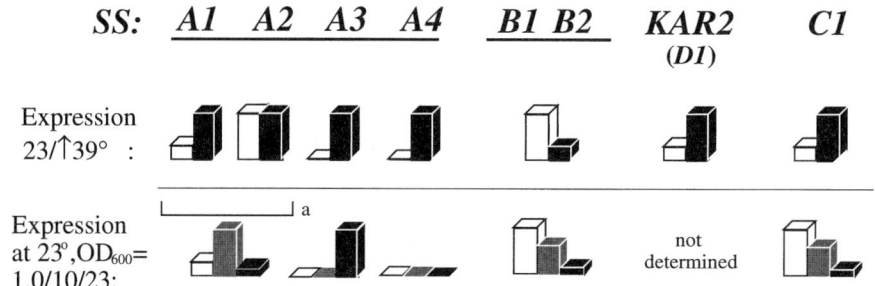

Figure 5. Expression of selected HSP70 genes of S. cerevisiae. 'Expression 23/↑39°:' depicts the relative level of expression of each gene or gene subfamily before (white columns) and 20 min after (black columns) a shift from 23 to 39°C, which induces the heat shock response. "Expression at 23°, OD_{600}=1.0/10/23:" depicts the relative level of expression of each gene or gene subfamily as cells progress from rapid exponential growth (OD_{600}=1.0, white columns) through diauxic shift (OD_{600}=10, grey columns) to stationary phase (OD_{600}=23, black columns). a: the probe used in these experiments was incapable of distinguishing between the nearly identical SSA1 and SSA2 genes.

References

Abravaya, K., Meyers, M.P., Murphy, S.P. and Morimoto, R.I. (1992). The human heat shock protein hsp70 interacts with HSF, the transcription factor that regulates heat shock gene transcription. Genes Dev. **6**, 1153–1164.

Ashburner, M. and Bonner, J.J. (1979). The induction of gene activity in Drosophila by heat shock. Cell **17**, 241–254.

Baxter, B.K., James, P., Evans, T. and Craig, E.A. *SSI1*, which encodes a novel Hsp70 homolog of the *S. cerevisiae* endoplasmic reticulum, exhibits genetic interactions with *KAR2*. Mol. Cell. Biol. **16**, 6444–6456.

Blaszczak, A., Zylicz, M., Georgopoulos, C. and Liberek, K. (1995). Both ambient temperature and the DnaK chaperone machine modulate the heat shock response by regulating the switch between sigma 70 and sigma 32 factors assembled with RNA polymerase. EMBO J. **14**, 5085–5093.

Blond-Elguindi, S., Cwirla, S., Dower, W., Lipshutz, R., Sprang, S., Sambrook, J. and Gething, M.-J. (1993a). Affinity panning of a library of peptides displayed on bacteriophages reveals the binding specificity of BiP. Cell **75**, 717–728.

Blond-Elguindi, S., Fourie, A.M., Sambrook, J.F. and Gething, M.-J.H. (1993b). Peptide-dependent stimulation of the ATPase activity of the molecular chaperone BiP is the result of conversion of oligomers to active monomers. J. Biol. Chem. **268**, 12730–12735.

Boorstein, W.R., Ziegelhoffer, T. and Craig, E.A. (1994). Molecular evolution of the HSP70 multigene family. J. Mol. Evol. **38**, 1–17.

Bork, P., Sander, C. and Valencia, A. (1992). An ATPase domain common to procaryotic cell cycle proteins, sugar kinases, actin and hsp70 heat shock proteins. Proc. Natl. Acad. Sci. USA **89**, 7290–7294.

Brodsky, J. and Schekman, R. (1994). Heat shock cognate proteins and polypeptide translocation across the endoplasmic reticulum membrane. In the biology of heat shock proteins and molecular chaperones (R.I. Morimoto, A. Tissiéres and C. Georgopoulos, eds). Cold Spring Harbor Laboratory Press, New York, pp. 85–110.

Brodsky, F., Hill, B.L., Acton, S.L., Nathke, I., Wong, D., Ponnambalam, S. and Parham, P. (1991). Clathrin light chains: arrays of protein motifs that regulate coated-vesicle dynamics. Trends Biochem. Sci. **16**, 208–213.

Table 1. Species names and sequence accession numbers of the HSP70 proteins analysed in Figure 1.

Code[a]	Common name	Scientific name	Accession #[b]
BAC		*Bacillus megaterium*	P05646
BOV•C	Cow	*Bos taurus*	P19120
CAUL		*Caulobacteria crescentus*	P20442
CHIK•C	Chicken	*Gallus gallus*	P08106
CHIK•E	Chicken	*Gallus gallus*	M27260
CLOS		*Clostridium perfringens*	X62915
CLpn		*Chlamydia pneumoniae*	X60083
CLre•C		*Chlamydomonas reinhardtii*	M76725
CRYP•P		*Cryptomonas phi*	M76547
CYAN	Cyanobacteria	*Synechocystis* sp. (strain PCC 6803)	P22358
DRO•C	Fruit fly	*Drosophila melanogaster*	L01500
DRO•E	Fruit fly	*Drosophila melanogaster*	L01498
DRO•M	Fruit fly	*Drosophila melanogaster*	L01502, L01503
ECH•E	Flatworm	*Echinococcus multilocularis*	M63604
Eco		*Escherichia coli*	P04475
HOMO•E	Human	*Homo sapiens*	P11021
KAR2•C	Baker's yeast	*Saccharomyces cerevisiae*	M25064, M31006
METH	Methanogen	*Methanosarcina mazei*	X60265
PAV•P	Chromophytic algae	*Pavlova lutherii*	X59555
PEA•M	Pea	*Pisum sativum*	X54739
PET•C	Petunia	*Petunia hybrida*	P09189
PLAS•C		*Plasmodium cynomolgi* (strain Berok)	M90978
POR•P	Red algae	*Porphyra umbilicalis*	X62240
SPIN•P	Spinach	*Spinacia oleracea*	X61491
Spom•E	Fission yeast	*Schizosaccharomyces pombe*	X64416
Spom•M	Fission yeast	*Schizosaccharomyces pombe*	P22774
SSA1•C	Baker's yeast	*Saccharomyces cerevisiae*	X12926
SSA2•C	Baker's yeast	*Saccharomyces cerevisiae*	X12927
SSA3•C	Baker's yeast	*Saccharomyces cerevisiae*	M97225
SSA4•C	Baker's yeast	*Saccharomyces cerevisiae*	J05637
SSB1•C	Baker's yeast	*Saccharomyces cerevisiae*	X13713
SSB2•C	Baker's yeast	*Saccharomyces cerevisiae*	L11262
SSC1•M	Baker's yeast	*Saccharomyces cerevisiae*	M27229
TOB•E	Tobacco	*Nicotiana tabacum*	X60057
TRYP•C		*Trypanosoma cruzi*	M26595
TRYP•M		*Trypanosoma cruzi*	M73627
ZEA•M	Maize	*Zea mays*	P11143

[a] Key to eukaryotic localization abbreviations: •E = endoplasmic reticulum; •M = mitochondrial; •C = cytoplasmic; •P = plastid.
[b] Numbers preceded by a "P" are SwissProt database entries; all others are GenBank entries.

Buchberger, A., Schroder, H., Buttner, M., Valencia, A. and Bukau, B. (1994). A conserved loop in the ATPase domain of the DnaK chaperone is essential for stable binding of GrpE. Nature Structural Biol. **1**, 95–101.

Chappell, T.G., Konforti, B.B., Schmid, S.L. and Rothman, J.E. (1987). The ATPase core of a clathrin uncoating protein. J. Biol. Chem. **262**, 746–751.

Chiang, H.L., Terlecky, S., Plant, C. and Dice, J.F. (1989). A role for a 70-kilodalton heat shock protein in lysosomal degradation of intracellular proteins. Science **246**, 382–385.

Craig, E.A. (1985). The heat shock response. CRC Crit. Revs. in Biochem. **18**, 239–280.

Craig, E.A. and Gross, C.A. (1991). Is hsp70 the cellular thermometer? Trends Biochem. Sci. **16**, 135–140.

Craven, R.A., Egerton, M., and Stirling, C.J. (1996). A novel Hsp70 of the yeast ER lumen is required for the efficient translocation of a number of protein precursors. EMBO J. **15**, 2640–2650.

Cyr, D.M., Langer, T. and Douglas, M.G. (1994). DnaJ-like proteins: molecular chaperones and specific regulators of Hsp70. Trends Biochem. Sci. **19**, 176–181.

DeLuca-Flaherty, C., McKay, B.B., Parham, P. and Hill, B.L. (1990). Uncoating protein (hsc70) binds a conformationally labile domain of clathrin light chain LCa to stimulate ATP hydrolysis. Cell **62**, 875–887.

Flaherty, K.M., DeLuca-Flaherty, C. and McKay, D.B. (1990). Three dimensional structure of the ATPase fragment of a 70K Heat-shock Cognate Protein. Nature **346**, 623–628.

Flaherty, K.M., McKay, D.B., Kabash, W. and Holmes, K. (1991). Similarity of the three-dimensional structure of actin and the ATPase fragment of a 70 kDa heat shock cognate protein. Proc. Natl. Acad. Sci. USA **88**, 5041–5045.

Flajnik, M.F., Canel, C., Kramer, J. and Kasahara, M. (1991). Which came first, MHC class I or class II? Immunogenetics **33**, 295–300.

Flynn, G., Pohl, J., Flocco, M. and Rothman, J. (1991). Peptide-binding specificity of the molecular chaperone BiP. Nature **353**, 726–730.

Fourie, A.M., Sambrook, J.F. and Gething, M.-J.H. (1994). Common and divergent peptide binding specificities of hsp70 molecular chaperones. J. Biol. Chem. **269**, 30470–30478.

Freeman, B., Myers, M., Schumacher, R. and Morimoto, R. (1995). Identification of a regulatory motif in Hsp70 that affects ATPase activity, substrate binding and interaction with HDJ-1. EMBO J. **14**, 2281–2292.

Freiden, P.J., Gaut, J.R. and Hendershot, L.M. (1992). Interconversion of three differentially modified and assembled forms of BiP. EMBO J. **11**, 63–70.

Gaut, J.R. and Hendershot, L.M. (1993). The immunoglobulin-binding protein *in vitro* autophosphorylation site maps to a threonine within the ATP binding cleft but is not a detectable site of *in vivo* phosphorylation. J. Biol. Chem. **268**, 12691–12698.

Georgopoulos, C. (1992). The emergence of the chaperone machines. Trends Biochem. Sci. **17**, 294–299.

Georgopoulos, C., Ang, D., Liberek, K. and Zylicz, M. (1990). Properties of the *Escherichia coli* heat shock proteins and their role in bacteriophage lambda growth. In Stress proteins in biology and medicine (R.I. Morimoto, A. Tissieres and C. Georgopoulos, eds). Cold Spring Harbor Laboratory Press, pp. 191–222.

Glick, B.S. (1995). Can Hsp70 proteins act as force-generating motors? Cell **80**, 11–14.

Gragerov, A., Zeng, L., Zhao, X., Burkholder, W. and Gottesman, M.E. (1994). Specificity of DnaK-peptide binding. J. Mol. Biol. **235**, 848–854.

Greene, L., Zinner, R., Naficy, S. and Eisenberg, E. (1995). Effect of nucleotide on the binding of peptides to 70-kDa heat shock proteins. J. Biol. Chem. **270**, 2967–2973.

Hamilton, T.G. and Flynn, G.C. (1996). Cer1p, a novel hsp70-related protein required for posttranslational endoplasmic reticulum translocation in yeast. J. Biol. Chem. **271**, 30610–30613.

Hayes, S.A. and Dice, J.F. (1996). Roles of molecular chaperones in protein degradation. J. Cell Biol. **132**, 255–258.

Herendeen, S.L., Van Bogelen, R.A. and Neidhardt, F.C. (1979). Levels of major proteins of *Escherichia coli* during growth at different temperatures. J. Bacteriol. **139**, 185–194.

Hohfeld, J., Minami, Y. and Hartl, F.-U. (1995). Hip, a novel cochaperone involved in the eukaryotic Hsc70/Hsp40 reaction cycle. Cell **83**, 589–598.

Kawula, T. and Lelivelt, M. (1994). Mutations in a gene encoding a new Hsp70 suppress rapid DNA inversion and *bgl* activation, but not proU derepression in *hns-1* mutant *Escherichia coli*. J. Bact. **176**, 610–619.

Landry, S.J., Jordan, R., McMacken, R. and Gierasch, L.M. (1992). Different conformations for the same polypeptide bound to chaperones DnaK and GroEL. Nature **355**, 455–457.

Lee-Yoon, D., Easton, D., Murawski, M., Burd, R. and Subjeck, J.R. (1995). Identification of a major subfamily of large hsp70-like proteins through the cloning of the mammalian 110-kDa heat shock protein. J. Biol. Chem. **270**, 1 5725–15733.

Lelivelt, M. and Kawula, T. (1995). Hsc66, an Hsp70 homolog in *Escherichia coli*, is induced by cold shock but not by heat shock. J. Bact. **177**, 4900–4907.

Liberek, K., Marszalek, J., Ang, D. and Georgopoulos, C. (1991). *Escherichia coli* DnaJ and GrpE heat shock proteins jointly stimulate ATPase activity of DnaK. Proc. Natl. Acad. Sci. USA **88**, 2874–2878.

McCarty, J., Buchberger, A., Reinstein, J. and Bukau, B. (1995). The role of ATP in the functional cycle of the DnaK chaperone system. J. Mol. Biol. **249**, 126–137.

McKay, D.B., Wilbanks, S.M., Flaherty, K.M., Ha, J.-H., O.Brien, M.C. and Shirvanee, L. (1994). Stress–70 proteins and their interactions with nucleotides. in The biology of heat shock proteins and molecular chaperones. Cold Spring Harbor Laboratory Press, pp. 153–177.

Mori, K., Sant, A., Kohno, K., Normington, K., Gething, M.J. and Sambrook, J. (1992). A 22-bp cis-acting element is necessary and sufficient for the induction the yeast *KAR2* (BiP) gene by unfolded proteins. EMBO J. **7**, 2583–2593.

Morimoto, R.I., Sarge, K.D. and Abravaya, K. (1993). Transcriptional regulation of heat shock genes. J. Biol. Chem. **267**, 21987–21990.

Morshauser, R.C., Wang, H., Flynn, G. and Zuiderweg, E.R.P. (1995). The peptide-binding domain of the chaperone protein Hsc70 has an unusual structure topology. Biochem. **34**, 6261–6266.

Nicolet, C. and Craig, E. (1991). Functional analysis of a conserved amino terminal region of HSP70 by site-directed mutagenesis. Yeast **7**, 699–716.

Otterson, G.A., Flynn, G.C., Kratzke, R.A., Coxon, A., Johnston, P.G. and Kaye, F.J. (1994). Stch encodes the 'ATPase core' of a microsomal stress 70 protein. EMBO J. **13**, 1216–1225.

Palleros, D.R., Welch, W.J. and Fink, A.L. (1991). Interaction of hsp70 with unfolded proteins: effects of temperature and nucleotides on the kinetics of binding. Proc. Natl. Acad. Sci. USA **88**, 5719–5723.

Palleros, D.R., Shi, L., Reid, K.L. and Fink, A. (1994). hsp70-protein complexes. J. Biol. Chem. **269**, 13107–13114.

Pelham, H.R.B. (1989). Control of protein exit from the endoplasmic reticulum. Annu. Rev. Cell Biol. **5**, 1–23.

Pfanner, N. and Meijer, M. (1995). Protein sorting: pulling in the proteins. Curr. Biol. **5**, 132–135.

Pfanner, N., Craig, E.A. and Meijer, M. (1994). The protein import machinery of the mitochondrial inner membrane. Trends Biochem. Sci. **19**, 368–372.

Prapapanich, V., Chen, S., Nair, S., Rimerman, R.A. and Smith, D.F. (1996). Molecular cloning of p48, a transient component of progesterone receptor complexes and an Hsp70-binding protein. Mol. Endocrinol. **10**, 420–431.

Rippmann, F., Taylor, W.R., Rothbard, J.B., and Green, N.M. (1991). A hypothetical model for the peptide binding domain of hsp70 based on the peptide binding domain of HLA. EMBO J. **10**, 1053–1059.

Ritossa, F. (1962). A new puffing pattern induced by temperature shock and DNP in *Drosophila*. Experientia **18**, 571–573.

Schilke, B., Forster, J., Davis, J., James, P., Walter, W., Laloraya, S., Johnson, J., Miao, B. and Craig, E. (1996). Cold-sensitivity of a *S. cerevisiae* mutant lacking a newly identified mitochondrial Hsp70 is suppressed by loss of mitochondrial DNA. J. Cell Biol. **134**, 603–613.

Schmid, D., Baici, A., Gehring, H. and Christen, P. (1994). Kinetics of molecular chaperone action. Science **263**, 971–973.

Schroder, H., Langer, T., Hartl, F.-U. and Bukau, B. (1993). DnaK, DnaJ and GrpE form a cellular chaperone machinery capable of repairing heat-induced protein damage. EMBO J. **12**, 4137–4144.

Seaton, B.L. and Vickery, L.E. (1994). A gene encoding a DnaK/hsp70 homolog in *Escherichia coli*. Proc. Natl. Acad. Sci. USA **91**, 2066–2070.

Sherman, M. and Goldberg, A. (1992). Involvement of the chaperonin DnaK in the rapid degradation of a mutant protein in *Escherichia coli*. EMBO J. **11**, 71–77.

Simons, J.F., Ferro-Novick, S., Rose, M.D. and Helenius, A. (1995). BiP/Kar2p serves as a molecular chaperone during carboxypeptidase Y folding in yeast. J. Cell Biol. **130**, 41–49.

Smith, D.F., Whitesell, L., Nair, S., Chen, S., Prapapanich, V. and Rimerman, R. A. (1995). Progesterone receptor structure and function altered by geldanamycin, an hsp90-binding agent. Mol. Cell. Biol. **15**, 6804–6812.

Straus, D., Walter, W. and Gross, C. (1988). *Escherichia coli* heat shock gene mutants are defective in proteolysis. Genes Dev. **2**, 1851–1858.

Stuart, R., Cyr, D., Craig, E.A and Neupert, W. (1994). Mitochondrial molecular chaperones: their role in protein translocation. Trends Biochem. Sci. **19**, 87–92.

Wagner, I., Arlt, H., van Dyck, L., Langer, T. and Neupert, T. (1994). Molecular chaperones cooperate with PIM1 protease in the degradation of misfolded protein in mitochondria. EMBO J. **13**, 5135–45.

Wang, T.-F., Chang, J.-H. and Wang, C. (1993). Identification of the peptide binding domain of hsc70. J. Biol. Chem. **268**, 26049–26051.

Werner-Washburne, M., Becker, J., Kosics-Smithers, J. and Craig, E. A. (1989). Yeast Hsp70 RNA levels change in response to the physiological status of the cell. J. Bacteriol. **171**, 2680–2688.

Wild, J., Kamath-Loeb, A., Ziegelhoffer, E., Lonetto, M., Kawasaki, Y. and Gross, C. A. (1992). Partial loss of function mutations in DnaK, the *Escherichia coli* homologue of the 70-kDa heat shock proteins, affect highly conserved amino acids implicated in ATP binding and hydrolysis. Proc. Natl. Acad. Sci. USA **89**, 7139–7143.

Zhu, X., Zhao, X., Burkholder, W.F., Gragerov, A., Ogata, C.M., Gottesman, M.E. and Hendrickson, W. A. (1996). Structural analysis of substrate binding by the molecular chaperone DnaK. Science **272**, 1606–1614.

Ziegelhoffer, T., Lopez-Buesa, P. and Craig, E.A. (1995). The dissociation of ATP from hsp70 of *Saccharomyces cerevisiae* is stimulated by both Ydj1p and peptide substrates. J. Biol. Chem. **270**, 10412–10419.

Zylicz, M., LeBowitz, J.H., McMacken, R. and Georgopoulos, C. (1983). The *dnaK* gene of *Escherichia coli* possesses an ATPase and autophosphorylating activity and is essential in an *in vitro* DNA replication system. Proc. Natl. Acad. Sci. USA **80**, 6431–6435.

■ *Bingjie Miao, Julie Davis, and Elizabeth Craig:*
Department of Biomolecular Chemistry
University of Wisconsin–Madison
1300 University Avenue
Madison, WI 53706, USA
Tel. 1 608 263 7105
Fax. 1 608 262 5253
E-mail: ecraig@facstaff.wisc.edu

Three-dimensional structure of Hsc70

The HSP70 proteins are composed of multiple domains. An atomic resolution structure (from X-ray crystallography) is available for the isolated ATPase domain of bovine Hsc70, which is bi-lobed, with the nucleotide bound at the base of the cleft between the two lobes. The secondary structure of the peptide-binding domain of Hsc70, solved by NMR, is largely a β sheet and matches the three-dimensional structure of the DnaK peptide-binding domain of DnaK.

■ Sequence and homology

The X-ray crystal structure of the isolated ATPase domain (residues 1–384) of bovine Hsc70 (Swiss Protein Accession # Hs7c_Bovin) has been solved (Flaherty et al., 1990). The secondary structural elements of a recombinant peptide-binding fragment (residues 385–543) of rat Hsc70 (Swiss Protein Accession # Hs7c_Rat) have been identified by

NMR (Morshauser et al., 1995), and the X-ray crystal structure of a peptide complex with the substrate-binding unit of DnaK has been reported (Zhu et al., 1996; see entry p. 18), confirming that this domain of HSP70 proteins has a completely novel topology. Recent publication of the X-ray crystal structure of the DnaK ATPase domain complexed with GrpE confirmed that features seen in the structures discussed here are present in other members of the HSP70 family (Harrison et al. 1997). The bovine and rat versions are >99% identical and have no insertions relative to one another within the regions discussed here. The residue numbers used in this entry are those of the bovine Hsc70 sequence.

■ Domain structure and topology of Hsc70

The domain structure of Hsc70 was deduced by characterization of truncated forms produced in proteolysis experiments (Chappell et al., 1987; Wang et al., 1993). The most evident domain boundaries thus defined are near residues 385, 550 and 610. Residues 1–384 bind ATP and retain ATPase activity (Chappell et al., 1987). Residues 385-543 bind to protein substrates of Hsc70 (Wang et al., 1993). There is scant structural information for the C-terminal residues 550–650, nor have the residues involved in the interface between the ATPase and peptide-binding domains been identified.

Figure 1 shows the secondary elements and topology of residues 1–543, inferred from either the X-ray structure (1–384) or NMR (385–543). The ATPase domain is composed of four subdomains arranged in two lobes (I and II). Subdomains IA and IIA of the ATPase domain are very similar to each other in secondary structure and topology. Strands 1–3 and 10 and helices E and F of subdomain IA correspond to strands 12–14 and 17 and helices K and N of subdomain IIA. The coil between helices N and O occupies the homologous position to that of strand 11 and makes some β sheet-like contacts. Subdomains IA and IIA make up a unit which is homologous to the nucleotide-binding sites of hexokinase and actin (Flaherty et al., 1991). In those proteins residues 30–37 are absent, β strand 3 is uninterrupted and subdomains IB and IIB are different from those of Hsc70. Subdomains IB and IIB can be viewed as insertions within IA and IIA, respectively.

Estimates of the abundance of secondary structure features in residues 385–650 and predictions of secondary structure were reviewed recently (Hightower et al., 1994). The secondary structure of the recombinant peptide-binding domain (in 50 mM sodium phosphate/pH 7/10% acetonitrile) was determined by NMR (Morshauser et al., 1995) and is shown in Fig. 1. The topology of the peptide-binding domain does not match that of any known protein. In particular, the identified secondary elements and topology are inconsistent with the hypothesis that the HSP70 peptide-binding domain is homologous to the human leukocyte antigen peptide-presenting protein of the class I major histocompatibility complex (Rippmann et al., 1991; Flajnik et al., 1991). The three-dimensional structure of this domain is not yet available for Hsc70, but has recently been reported for DnaK (Zhu et al., 1996; see entry p. 18).

■ Three-dimensional structure of the ATPase domain of Hsc70

The crystal structure of the ATPase domain of wild-type bovine Hsc70 has been refined under several different conditions (Flaherty et al., 1990; Wilbanks and McKay 1995) and the structures of the ATPase domains of several mutant Hsc70 proteins have also been defined (O'Brien and McKay 1993; Flaherty et al., 1994). The structures can be divided into those with ATP bound and those with ADP bound; within each group the structures are practically superimposable. The differences between the ADP- and ATP-bound states are limited to minor rearrangements of solvent molecules and protein side chains near the scissile bond. A schematic representation of the protein is shown in Fig. 2.

The ATPase fragment is made up of four subdomains, each of which has a β sheet core flanked by α helices. As noted above, subdomains IA and IIA are homologous to the nucleotide-binding subdomains of hexokinase and actin. Subdomains IA and IB appear tightly bound to one another and make up one lobe of the ATPase domain; IIA and IIB make up the other lobe. The two lobes are distinct, with only two covalent connections: helices G and O between subdomains IA and IIA.

Domains IB and IIB do not interact closely, and many ordered water molecules are trapped in the interface. It seems likely that this interface shifts during conformational change. The structure of DnaK bound to the nucleotide exchange factor GrpE suggests that lobe IIB can rotate outward, facilitating nucleotide exchange (Harrison et al., 1997).

■ The nucleotide-binding site of Hsc70

The adenine base of the nucleotide is bound in a hydrophobic pocket, made in part by the aliphatic methylene segments of lysine 271, arginine 272 and arginine 342, of which K271 and R272 reach around the base and hydrogen bond to oxygens of the ribose moiety. Atoms N6 and N7 (and to a lesser extent C5 and C6) of the base are the only solvent-exposed atoms of the nucleotide.

Three short glycine and threonine rich stretches (13–15, 202–204 and 339–340) provide a majority of the protein interactions with the phosphate moieties. These three loops are positioned so that every main chain amide is oriented toward the phosphate oxygens. Figure 3 shows an ion cluster in the area around the scissile bond. Several acidic residues, along with a few main chain carbonyl oxygens form the anchor points for a network made up of several water molecules, two potassium ions and a magnesium ion. The positioning of the magnesium ion allows ATP to grasp it in a bidentate complex, which is observed in the structure of Hsc70 mutant K71M

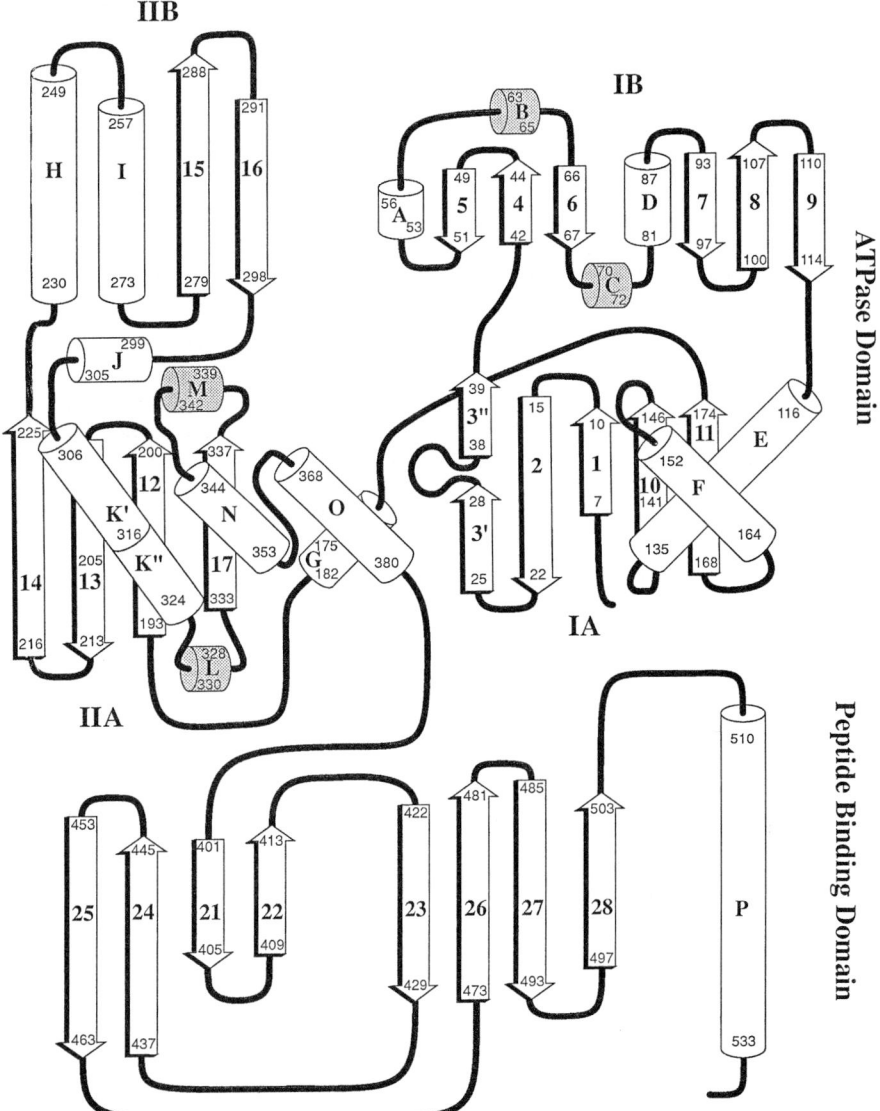

Figure 1. The secondary structure and topology of Hsc70. The β strands are shown as sequentially numbered arrows while helices are shown as cylinders identified by letters; 3_{10} helices are shaded. The limits of each structural element are noted on the element. Strand 3 is interrupted by a loop of eight residues. Helix K is divided in two at proline-316. The elements of the ATPase domain are identified from the X-ray crystal structure of bovine Hsc70 and arranged to reflect that structure. The elements of the peptide-binding domain are identified by NMR and arranged to indicate the observed organization into two β sheets. The strands of the peptide-binding domain are numbered 21–28 to correspond to the numbering (1–8) in Morshauser et al. (1995).

(O'Brien et al., 1996). . The two potassium ions flank the magnesium. Both magnesium and potassium are required in Hsc70 for full activity.

The side chains shown in Fig. 3 were individually mutated and expressed in the ATPase domain (O'Brien and McKay, 1993; Wilbanks, et al., 1994). Changes in side chains D10 and D199 markedly reduced the steady state ATPase rate. Changes to T204 and D206 increased the K_m. Changes to E175 had both effects. Changes to K71 (not shown) abolished activity (O'Brien et al., 1996). The crystal stuctures of these mutant proteins are listed in Table 1.

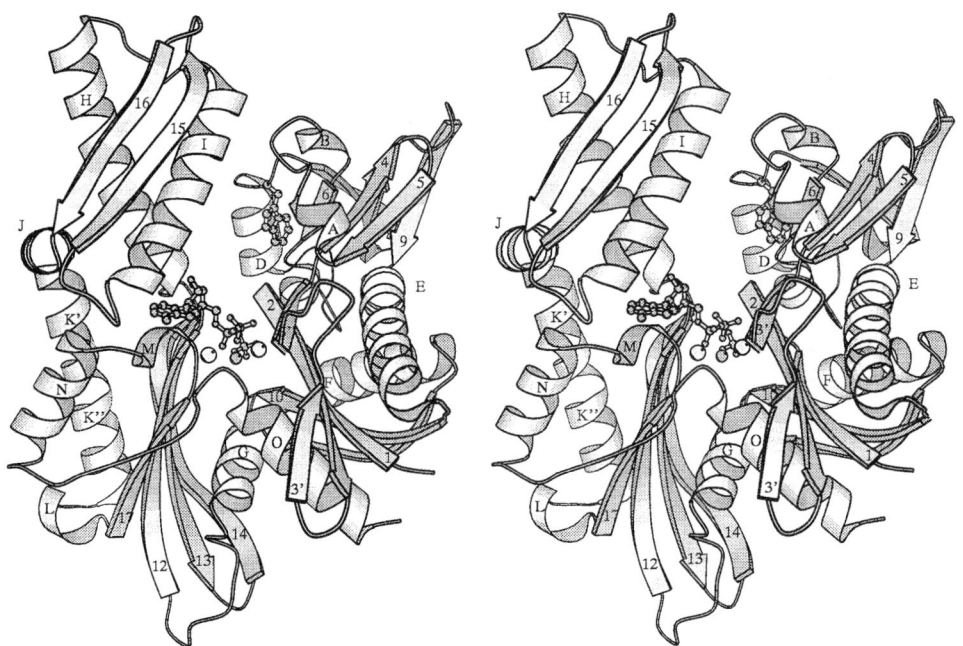

Figure 2. Ribbons diagram of the ATPase domain. Most secondary structural elements are identified by the letters and numbers used in Fig. 1. Tryptophan-90 (the only tryptophan in the ATPase domain) is shown as balls and sticks. Nucleotide (ADP), inorganic phosphate, two potassium ions (larger, lightly shaded spheres) and the magnesium ion are shown in the active site. The diagram is in stereo.

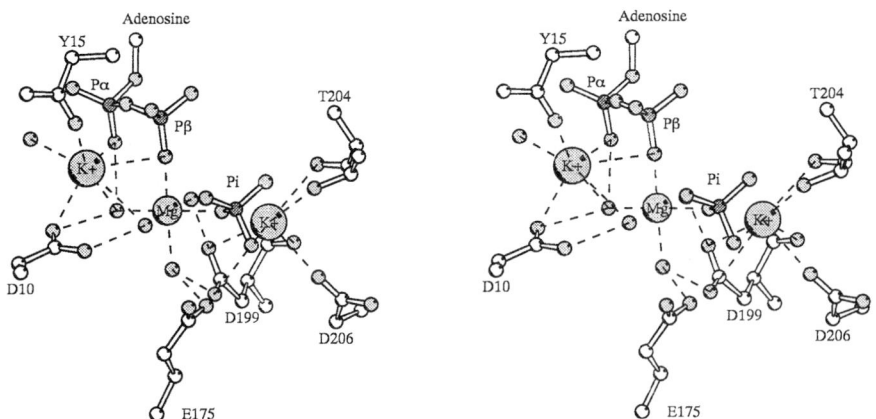

Figure 3. The ion cluster near the scissile bond. The active site is shown with ADP and Pi bound. Only the phosphates of ADP are shown, the position of the adenosine moiety is indicated. Three cations (labeled) are shown along with a number of water molecules (small spheres). In addition, several residues which interact with these solvent molecules are shown (only main chain atoms of Tyr-15 are shown). Hydrogen bonds and ionic interactions are shown with dashed lines. The diagram is in stereo.

Table 1. Atomic coordinates for the ATPase fragment of wild-type and mutant forms of bovine Hsc70. The following coordinate sets for the ATPase fragment of bovine Hsc70 have been deposited with the Protein Data Bank. In general the coordinates include residues 3–381

PDB#	Mutant	Nucleotide bound	Divalent cation	Monovalent cation	Resolution (Å)	Reference
1HSC	w.t.*	ADP+Pi	Mg	Na (n.s.)	2.3	Flaherty et al., 1990
2HSC	w.t.	ADP+Pi	Mg	Na (n.s.)	1.9	Flaherty et al., 1994
3HSC	w.t.	ADP+Pi	Mg	Two Na	1.9	Wilbanks and McKay, 1995
1KAX,Y,Z	K71M,A,E	ATP	Mg	Three K	1.7	O'Brien et al., 1996
1ATR	T204V	ADP+Pi	Mg	Na (n.s.)	2.0	O'Brien and McKay, 1993
1ATS	T204E	ADP	Mg	Na (n.s.)	2.0	O'Brien and McKay, 1993
1NGA	E175S	ADP	Mg	Na (n.s.)	2.4	Flaherty et al., 1994
1NGB	E175Q	ADP+Pi	Mg	Na (n.s.)	2.4	Flaherty et al., 1994
1NGC	D206S	ADP+Pi	Mg	Na (n.s.)	2.5	Flaherty et al., 1994
1NGD	D206N	ADP+Pi	Mg	Na (n.s.)	2.4	Flaherty et al., 1994
1NGE	D199S	ATP	Mg	Na (n.s.)	2.4	Flaherty et al., 1994
1NGF	D199N	ATP	Mg	Na (n.s.)	2.4	Flaherty et al., 1994
1NGG	D10S	ATP	Mg	Na (n.s.)	2.4	Flaherty et al., 1994
1NGH	D10N	ATP	Mg	Na (n.s.)	2.5	Flaherty et al., 1994
1NGI	w.t.	AMPPNP	Ca	Na (n.s.)	2.4	Flaherty et al., 1994
1NGJ	w.t	AMPPNP	Mg	Na (n.s.)	2.4	Flaherty et al., 1994
1HPM	w.t	ADP+Pi	Mg	Two K	1.7	Wilbanks and McKay, 1995

* Protein purified from bovine brain. All the other structures were solved using recombinant protein.
n.s. = ion is present in crystals but not identified in deposited coordinates.

■ Surface features of the Hsc70 ATPase domain

The ATPase domain of HSP70 proteins is known to interact with the more C-terminal part of the protein and is believed to interact with protein cofactors. Of these, only the site of interaction of DnaK with GrpE has been mapped (Buchberger et al., 1994). The loop inserted in β strand 3 (equivalent to residues 30–37 in Hsc70) is required for GrpE binding. Within eukaryotic cytosolic Hsp70 proteins this loop and all of strand 3 are particularly highly conserved. Together with strands 4 and 5 from subdomain IB and portions of helices I and M of subdomain IIB it forms a highly conserved patch on the protein surface. This patch covers the center of the front of the ATPase domain as viewed in Fig. 2 and spans the nucleotide-binding cleft.

A second highly conserved surface patch is low in the center of the back of the ATPase domain (as viewed in Fig. 2) and includes portions of helices E, F, G and H. The loops at the base of domain IIA contain the least well ordered main chain in the crystal structure, suggesting an interaction not preserved in the isolated, truncated protein.

■ References

Buchberger, A., Schroeder, H., Buttner, M., Valencia, A., and Bukau, B. (1994). A conserved loop in the ATPase domain of the DnaK chaperone is essential for stable binding of GrpE. Struct. Biol. **1**, 95–101.

Chappell, T.G., Konforti, B.B., Schmid, S.L., and Rothman, J.E. (1987). The ATPase core of a clathrin uncoating protein. J. Biol. Chem. **262**, 746–751.

Flaherty, K.M., DeLuca-Flaherty, C., and McKay, D.B. (1990). Three-dimensional structure of the ATPase fragment of a 70 kilodalton heat shock cognate protein. Nature **346**, 623–628.

Flaherty, K.M., McKay, D.B., Kabsch, W., and Holmes, K.C. (1991). Similarity of the three-dimensional structures of actin and the ATPase fragment of a 70-kDa heat shock cognate protein. Proc. Natl. Acad. Sci. USA. **88**, 5041–5045.

Flaherty, K.M., Wilbanks, S.M., DeLuca-Flaherty, C., and McKay, D.B. (1994). Structural basis of the 70-kilodalton heat shock cognate protein ATP hydrolytic activity. II. Structure of the active site with ADP or ATP bound to wild-type and mutant ATPase fragment. J. Biol. Chem. **269**, 12899–12907.

Flajnik, M.F., Canel, C., Kramer, J., and Kasahara, M. (1991). Which came first, MHC class I or class II? Immunogenetics **33**, 295–300.

Harrison, C.J., Hayer-Hartl, M., Di Liberto, M., Hartl, F.-U. and Kuriyan, J. (1997). Crystal structure of the nucleotide exchange factor GrpE bound to the ATPase domain of the molecular chaperone DnaK. Science **276**, 431–435.

Hightower, L.E., Sadis, S. E., and Takenaka, I.M. (1994). Interactions of vertebrate hsc70 and hsp70 with unfolded proteins and peptides. In The biology of heat shock proteins and molecular chaperones (Morimoto, R. I., Tissieres, A., and Georgopoulos, C., eds). Cold Spring Harbor Laboratory Press, pp. 179–20.

Morshauser, R.C., Wang, H., Flynn, G.C., and Zuiderweg, E.R.P. (1995). The peptide-binding domain of the chaperone protein hsc70 has an unusual secondary structure topology. Biochemistry **34**, 6261–6266.

O'Brien, M.C. and McKay, D.B. (1993). Threonine 204 of the chaperone protein hsc70 influences the structure of the active site but is not essential for ATP hydrolysis. J. Biol. Chem. **268**, 24323–24329.

O'Brien, M.C., Flaherty, K.M. and McKay, D.B. (1996). Lysine 71 of the chaperone protein Hsc70 is essential for ATP hydrolysis. J. Biol. Chem. **271**, 15874–15878.

Rippmann, F., Taylor, W.R., Rothbard, J.B., and Green, N.M. (1991). A hypothetical model for the peptide binding domain of hsp70 based on the peptide binding domain of HLA. EMBO J. **10**, 1053–1059.

Wang, T., Chang, J., and Wang, C. (1993). Identification of the peptide binding domain of hsc70. J. Biol. Chem. **268**, 26049–26051.

Wilbanks, S.M. and McKay, D.B. (1995). How potassium affects the activity of the molecular chaperone Hsc70: II. Potassium binds specifically in the ATPase active site. J. Biol. Chem. **270**, 2251–2257.

Wilbanks, S.M., DeLuca-Flaherty, C., and McKay, D.B. (1994). Structural basis of the 70-kilodalton heat shock cognate protein ATP hydrolytic activity I. Kinetic analyses of active site mutants. J. Biol. Chem. **269**, 12893–12898.

Zhu, X., Zhao, X., Burkholder, W.F., Gragerov, A., Ogata, C.M., Gottesman, M.E., and Hendrickson, W.A. (1996). Structural analysis of substrate binding by the molecular chaperone DnaK. Science **272**, 1606–1614.

■ Sigurd Wilbanks:
Department of Structural Biology
Stanford School of Medicine
Stanford, CA 94305-5400, USA
Tel. 1 415 723 7095
Fax. 1 415 723 8464
E-mail: WILBANKS@cellbio.standford.edu

The 3-dimensional structure of the substrate binding domain of *E. coli* DnaK

The determination of an X-ray crystallographic structure, at 2.0 Å resolution, of a C-terminal fragment of DnaK with peptide bound revealed that the substrate binding domain consists of a compact β sandwich followed by helical elements. The peptide is bound in an extended conformation in a channel formed by loops that connect strands of the sandwich. The helical elements are not directly involved in binding to the peptide, but rather stabilize the β domain and act as a lid to encapsulate the peptide.

HSP70 proteins contain two major domains — an N-terminal domain that contains the ATPase catalytic site and a C-terminal substrate binding domain — that communicate to regulate the affinity and duration of (poly)peptide binding (reviewed by McKay et al., 1993). The 3-dimensional structure of the isolated ATPase domain of the constitutively-expressed bovine cytosolic Hsc70 was reported some time ago (Flaherty et al., 1990), but the determination of the structure of the C-terminal domain of an HSP70 protein remained elusive. Intriguingly, two totally indepent approaches using structure prediction algorithms suggested that the peptide binding domain might resemble that of MHC class I antigens (Flajnik et al., 1991; Rippmann et al., 1991). However, the secondary structure assignments derived for the C-terminal domain of bovine hsc70 using NMR analysis indicated a totally novel fold (Morshauser et al., 1995). The recent publication of an X-ray crystallographic structure, at 2.0 Å resolution, of a C-terminal fragment of DnaK with peptide bound (Zhu et al., 1996) revealed that the substrate binding domain consists of a compact β sandwich followed by helical elements (Fig. 1a). The β sandwich is arranged in two sheets with four antiparallel β strands in each, and the peptide is bound in an extended conformation in a channel formed by loops that connect strands of the sandwich. The particular topology of this β sandwich has not been observed previously in other protein structures. The five helical elements are not directly involved in binding to the peptide, but rather stabilize the β domain and act as a lid to encapsulate the peptide. Interestingly, this five helix unit is similar in topology and general structure to a portion of a helical domain which entraps the bound guanine nucleotide in the α subunits of G proteins (Noel et al., 1993; Coleman et al., 1994).

■ Interactions between the peptide and DnaK

The bound peptide, which has the sequence NRLLLTG, was originally identified in a bacteriophage display screen for peptides that bind DnaK with high affinity (Gragerov et al., 1994a). Binding interactions between this peptide and DnaK are centred and concentrated on Leu[4], which is completely buried in a deep pocket, and involve almost exclusively the middle five residues of the heptapeptide (Zhu et al., 1996; Fig. 1b). Leu[3] is also highly buried, and together with Leu[4] contributes the majority of the side-chain contacts with DnaK. Arg[2], Leu[5], and Thr[6] are moderately buried and contribute the remainder of the side-chain contacts. The terminal residues, Asn[1] and Gly[7], make no significant side-chain contacts with DnaK, although the question of whether a residue other than glycine at position 7 could provide stabilizing side-chain interactions remains open. Consistent with predictions

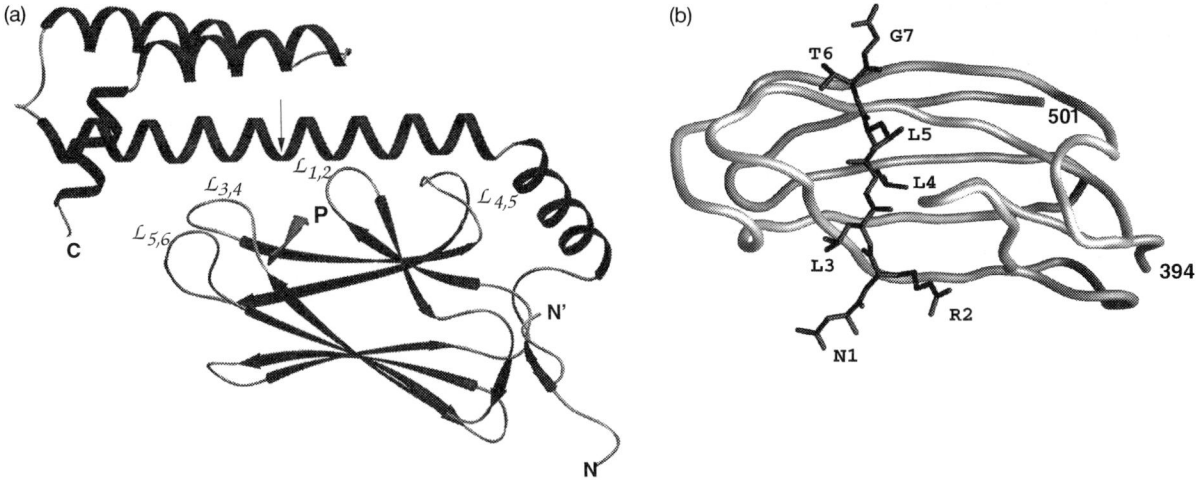

Figure 1. Three dimensional structure of the substrate binding domain of DnaK. (a) Ribbon diagram of a C_α trace of the structure of the DnaK substrate binding domain with peptide (P) bound in a channel that is formed by loops off the β sandwich and capped by an α-helical domain. Loops $L_{1,2}$ and $L_{3,4}$ directly form the channel for substrate binding, while loops $L_{4,5}$ and $L_{5,6}$ stabilize the channel by buttressing loops $L_{1,2}$ and $L_{3,4}$, respectively through main chain and side chain hydrogen bonds and hydrophobic interactions. The arrow shows the point at which the long αβ-helix kinks upward in the type 2 crystal structure. The N-terminus of the substrate binding domain, which in the intact molecule would be connected to the ATPase domain, lies almost at the opposite end of the structure from the substrate channel and the α-helical lid. the N-terminal residues can take up two alternative conformations, the second being shown as N'. (b) Binding of peptide NRLLLTG to the β subdomain. The protein and the peptide backbones are shown in light grey, and dark grey respectively. The side chains of the peptide are shown. These images were generously prepared and provided by Wayne Hendrickson and Xiaotian Zhu.

from peptide binding studies, the side-chain van der Waals contacts between the peptide and DnaK are dominantly hydrophobic. The peptide main-chain contributes seven hydrogen bonds, five of which are made with DnaK backbone groups. The pocket that binds Leu⁴, designated central site '0' by Zhu et al., appears to be the crucial determinant of peptide binding. Contact analysis indicates that leucine is the ideal occupant for this site, while methionine and isoleucine could fit, albeit very snugly (Zhu et al., 1996). Adjustments of the protein or peptide backbone might allow the pocket to accept phenylalanine, but probably could not accomodate tyrosine or tryptophan. Smaller side chains would fit, paying an energetic penalty for the residual cavity, which in the cases of threonine or serine might be compensated for by formation of a hydrogen bond to the side-chain O of Thr^{437} in the β4 strand of the sandwich. Geometric constraints do not appear to provide significant restrictions on the identity of residues at positions in the peptide other than the anchor residue at site 0, although the observed peptide φ angles appear to exclude proline. A general hydrophobic character at sites −1 and +1 is generated by a hydrophobic arch over the binding channel that is constructed by contacts between the two loops that form the channel (see Fig. 1b). Sites outside the central triad are quite open, and the electrostatic potential of the surface of DnaK at these positions is largely negative, accounting for the observed exclusion of acidic residues and favouring of basic residues at the N-and C-termini of hexa- and heptapeptides that bind DnaK (Gragerov et al., 1994a). Rüdiger et al. (1997) have recently identified DnaK binding sites within 4360 overlapping 13-mer peptides that scan 37 protein sequences. The binding motif that emerged from this study consists of a 4–5 residue hydrophobic core, particularly enriched in leucine, flanked by short basic segments. This motif is consistent with earlier observations of Gragerov et al. (1994a) using peptides displayed by bacteriophages, and is completely compatible with the structure of the DnaK substrate binding site (Zhu et al., 1996).

■ How do polypeptide substrates enter and exit the binding site?

Zhu et al. (1996) point out that it would be extremely difficult for even a short peptide to thread its way into or out of the binding channel. A threading mechanism of entry or exit would be impossible for a natural substrate, in which the binding site is part of a continuous (and often partly folded) polypeptide chain. It has long been recognized from protease sensitivity and spectroscopic studies (recently reviewed by Fung et al., 1996) that large conformational changes take place in the substrate binding domain of HSP70 proteins, potentiated first when ATP binds to the N-terminal domain (promoting release of bound polypeptide) and then when ADP is generated following hydrolysis of the ATP (which induces a high affinity state for substrate binding). While an X-ray structure of the complex between the peptide and the isolated substrate binding domain can present only a stationary snapshot of one stage of the 'HSP70 reaction cycle' (Hartl, 1996) of substrate binding and release, analysis of two different crystal forms of the complex provide important clues as to the mechanism of this process (Zhu et al., 1996). Such clues include observations that (i) contacts between the α-helical lid that caps the binding channel and the loops that form the channel involve only a relatively small buried surface area, and (ii) the structures of the complex in the two different crystal forms differ only in that these contacts are disrupted and the 'lid' is partially lifted by a hinge-like transformation around a kink half way along the capping helix (Fig. 1b). The peptide remains bound but is less well ordered, and Zhu et al. consider the conformation in the type 2 crystal as being incipient to the full-scale change required for peptide exchange. It thus appears that on binding of ATP to the N-terminal domain, the helical domain hinges up and away, uncapping the peptide channel. Even once this uncapping is achieved, substantial movement of the channel-forming loops must occur before a bound polypeptide can be released from their grasp, or a new polypeptide can enter the binding site. Binding of the new substrate would itself cause a conformational change that, once propagated to the N-terminal domain would stimulate ATP hydrolysis, and the presence of ADP in the nucleotide binding site would in turn cause a conformational change in the ATPase domain that, when in turn propagated back to the C-terminal domain, would stabilize the binding of the substrate. Replacement of the bound ADP by ATP, a rate-limiting step for DnaK that is accelerated by the co-chaperone GrpE (Hartl, 1996), would restart the cycle of polypeptide release and rebinding. The second co-chaperone, DnaJ, which potentiates the substrate-induced stimulation of ATP hydrolysis (Hartl, 1996), is thought to bind to the region of DnaK corresponding to the α-helical domain (Wawrzynow and Zylicz, 1995). Zhu et al. propose that binding of peptide or DnaJ could stabilize the closed state of the α-helical lid, thus favouring the ADP bound conformation and consequently stimulating ATP hydrolysis.

■ How do the substrate binding and ATPase domains interact?

While no structure is yet available for an intact HSP70 molecule, the structure of the isolated substrate binding domain provides a significant insight into how the N-terminal and C-terminal domains interact to facilitate coupling of the conformational changes that occur upon the binding of polypeptide and nucleotides. The N-terminus of the substrate binding domain, which in the intact molecule would be connected to the ATPase domain, lies almost at the opposite end of the structure from the substrate channel and the α-helical lid (Zhu et al. 1996; see Fig. 1a). Thus conformational changes in the ATPase domain cannot directly affect the structure of the lid. Inspection of two alternative dispositions of the N-terminal residues of the substrate binding domain that occur in the type 1 crystal indicate that these residues can be either extended or folded into a hydrophobic pocket near the juncture between the C-terminus of the β sandwich domain and the beginning of the α-helical domain. Zhu et al. surmise that conformational changes in the ATPase domain could displace the N-terminal sequence from the hydrophobic pocket, causing changes in the positioning of the α-helical domain relative to the β sandwich and affecting the equilibria between the open and closed states of the lid.

■ What lessons does the DnaK structure provide for other HSP70 family members?

Zhu et al. (1996) report that the pattern of sequence conservation within the family is such that the backbone conformation in the β subdomain would be virtually identical in all HSP70 proteins. Furthermore, all of the backbone hydrogen bonds and one of the two side-chain hydrogen bonds between DnaK and the peptide backbone should be maintained in all members of the family. Apart from the central pocket, which is very highly conserved, residues that in DnaK contact the side-chains of the peptide can vary in the different family members, and the electrostatic potential of the surfaces surrounding the substrate binding channel is also likely to vary in different HSP70 proteins. Such similarities and differences between the peptide binding channels of HSP70 proteins will account for the observation that DnaK, Hsc70, and BiP display both overlapping and distinct peptide binding specificities (Fourie et al., 1994; Gragerov et al., 1994b). While the three chaperones generally display similar high affinities for highly hydrophobic peptides, significant differences in binding affinities are quite frequently observed. How differential binding specificities between DnaK and other HSP70 proteins are reflected in differences in the detailed topography of their substrate binding sites must await the definition of additional structures for other family members.

■ Acknowledgement

This entry is based closely on a *Dispatch* entitled 'Molecular Chaperones: Clasping the Prize' by M.-J. Gething published in *Current Biology* **6**, 1573–1576 (1996).

References

Becker, J. and Craig, E.A. (1994). Heat-shock proteins as molecular chaperones. Eur. J. Biochem. **219**, 11–23.

Blond-Elguindi, S., Cwirla, S.E., Dower, W.J., Lipshutz, R.J., Sprang, S.R., Sambrook, J.F., and Gething, M.J. (1993). Affinity panning of a library of peptides displayed on bacteriophages reveals the binding specificity of BiP. Cell **75**, 717–728.

Boorstein, W.R., Ziegelhoffer, T., and Craig, E.A. (1994). Molecular evolution of the HSP70 multigene family. J. Mol. Evol. **38**, 1–17.

Coleman, D., Berghuis, A.M., Lee, E., Linder, M.E., Gilman, A.G., and Sprang, S.R. (1994). Structures of active conformations of $G_{i\alpha 1}$ and the mechanism of GTP hydrolysis. Science **265**, 1405–1412.

Flaherty, K.M., DeLuca-Flaherty, C., and McKay, D.B. (1990). Three-dimensional structure of the ATPase fragment of a 70K heat-shock cognate protein. Nature **346**, 623–628.

Flajnik, M.F., Canel, C., Kramer, J., and Kasahara, M. (1991). Which came first, MHC class I or class II? Immunogenetics **33**, 295–300.

Flynn, G.C., Pohl, J., Flocco, M.T., and Rothman, J.E. (1991). Peptide-binding specificity of the molecular chaperone BiP. Nature **353**, 726–730.

Fourie, A.M., Sambrook, J.F., and Gething, M.-J.H. (1994). Common and divergent peptide binding specificities of hsp70 molecular chaperones. J. Biol. Chem. **269**, 30470–30478.

Fung, K.L., Hilgenberg, L., Wang, N., and Chirico, W.J. (1996). Conformations of the nucleotide and polypeptide binding domains of cytosolic Hsp70 molecular chaperones are coupled. J. Biol. Chem. **271**, 21559–21565.

Gething, M.J. and Sambrook, J.F. (1992). Protein folding in the cell. Nature **355**, 33–45.

Gragerov, A. and Gottesman, M.E. (1994a). Different peptide binding specificities of hsp70 family members. J. Mol. Biol. **241**, 133–135.

Gragerov, A., Zeng, L., Zhao, X., Burkholder, W., and Gottesman, M.E. (1994b). Specificity of DnaK-peptide binding. J. Mol. Biol. **235**, 848–854.

Hartl, F.U. (1996). Molecular chaperones in cellular protein folding. Nature **381**, 571–580.

Knarr, G., Gething, M.-J., Modrow, S., and Buchner, J. (1995). BiP-binding sequences in antibodies. J. Biol. Chem. **270**, 27589–27594.

McKay, D.B. (1993). Structure and mechanism of 70-kDa heat-shock-related proteins. Advances in Protein Chemistry **44**, 67–80.

Morshauser, R.C., Wang, H., Flynn, G.C., and Zuiderweg, E.R.P. (1995). The peptide-binding domain of the chaperone protein Hsc70 has an unusual secondary structure topology. Biochemistry **34**, 6261–6266.

Noel, J.P., Hamm, H. E., and Sigler, P.B. (1993). The 2.2 Å crystal structure of transducin-α complexed with GTPγS. Nature **366**, 654–663.

Rippmann, F., Taylor, W.R., Rothbard, J. B., and Green, N.M. (1991). A hypothetical model for the peptide binding domain of hsp70 based on the peptide binding domain of HLA. EMBO J. **10**, 1053–1059.

Rüdiger, S., Germeroth, L., Schneider-Mergener, J., and Bukau, B. (1997). Substrate specificity of the DnaK chaperone identified by screening of cellulose-bound peptide libraries. EMBO J. **16**, 1501–1507 (1997).

Takenaka, I.M., Leung, S.-M., McAndrew, S.J., Brown, J.P., and Hightower, L.E. (1995). Hsc70-binding peptides selected from a phage display peptide library that resemble organellar targeting sequences. J. Biol. Chem. **270**, 19839–19844.

Wawrzynów, A. and Zylicz, M. (1995). Divergent effects of ATP on the binding of the DnaK and DnaJ chaperones to each other, or to their various native and denatured protein substates. J. Biol. Chem. **270**, 19300–19306.

Zhu, X., Zhao, X., Burkholder, W.F., Gragerov, A., Ogata, C.M., Gottesman, M.E., and Hendrickson, W.A. (1996). Structural analysis of substrate binding by the molecular chaperone DnaK. Science **272**, 1606–1614.

Mary-Jane Gething
Department of Biochemistry and Molecular Biology
University of Melbourne
Parkville
Victoria 3052
Australia
Tel. 61 3 9344 5948
Fax. 61 3 9347 9109
Email: gething@ariel.ucs.unimelb.edu.au

Escherichia coli DnaK

DnaK of Escherichia coli *is a stress-inducible member of the HSP70 protein family. It cooperates with the DnaJ and GrpE cochaperones to form an ATP-dependent chaperone system involved in central cellular processes including the prevention of aggregation and refolding of misfolded proteins, degradation of unstable proteins, translocation of proteins, synthesis of DNA and RNA and regulation of the heat shock response.*

■ Alternative names

GroPC (Georgopoulos, 1977).

■ Isolation of *dnaK*

The *dnaK* gene was identified by mutations rendering *E. coli* resistant to lytic infection by bacteriophage λ (Georgopoulos, 1977; Saito and Uchida, 1977).

■ *dnaK* gene and sequence

The *dnaK* gene is the promoter proximal gene of the *dnaK dnaJ* operon located at 0.3 minutes on the *E. coli* map. The nucleotide sequence of *E. coli dnaK* (GenBank accession number K01298 for strain K12; DDBJ accession number D10765 for strain B) encodes a 69-kDa protein of 638 residues (Bardwell and Craig, 1984). The N-terminal formylmethionine is cleaved off *in vivo*.

■ DnaK protein

DnaK belongs to the HSP70 family and shares approx. 50% identical amino acids with eucaryotic members (Bardwell and Craig, 1984). Conservation is highest for the N-terminal 540 residues of DnaK, while the C-terminal 100 residues are less well conserved. Sequence alignments group DnaK into a branch of prokaryotic and organellar HSP70 family members, the next relatives being DnaK from *Brucella bovis* and *Caulobacter crescentus* (67% amino acid identity).

DnaK is an abundant cytosolic protein of *E. coli* accounting for approx. 1 and 2% of total cellular protein at 30 and 42°C, respectively (Herendeen *et al.*, 1979). Its purification using standard protocols involves ATP-agarose affinity chromatography (Cegielska and Georgopoulos 1989; Buchberger *et al.*, 1994b). DnaK-specific antisera are commercially available (StressGen Biotechnologies Corp., Victoria, Canada). Like other HSP70 family members, DnaK is structurally organized in at least three domains (Fig. 1). Residues 1–386 form the ATPase domain that is structurally highly similar to the actin-like three-dimensional structure of bovine brain Hsc70 (Flaherty *et al.*, 1990; Harrison *et al.*, 1997). Residues 387 to 537 form the substrate binding domain (Zhu *et al.*, 1996) which consists of a β-sandwich subdomain with four loops defining the peptide binding

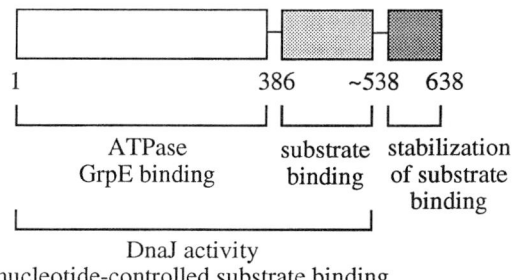

Figure 1. Assignment of functions to domains of DnaK. A model of DnaK with three domains is shown. The numbers indicate the residues defining domain borders, and functions correlated to particular domains are indicated below.

channel, followed by two α-helical segments constituting a lid for that channel (see Fig. 1 of previous entry, p. 19). Residues 539 to 639 form a predominantly α-helical domain (Zhu *et al.*, 1996) that plays a role in the stabilization of substrate binding (H. Schröder and B. Bukau, unpublished results).

DnaK exists in an equilibrium of monomeric, dimeric and higher oligomeric species that is influenced by DnaK concentration, temperature, ATP and substrates (Schönfeld *et al.*, 1995). Oligomerization is mediated through the C-terminal residues 387–638 (H. Schröder and B. Bukau, unpublished results). The biological significance of oligomerization is unclear.

■ Biological activities of DnaK

DnaK is the main *E. coli* member of the HSP70 family and together with DnaJ and GrpE forms a central chaperone system of the cell. It is involved in many metabolic processes of stressed as well as unstressed cells (for reviews see Georgopoulos *et al.*, 1990, 1994; Gross *et al.*, 1990; Frydman and Hartl, 1994; and references therein), including: (i) prevention of aggregation and refolding of misfolded proteins; (ii) mediation of degradation of unstable proteins by proteases; (iii) modulation of the heat shock response; (iv) initiation of bacteriophage λ DNA replication; (v) activation of DnaA, RepA and RepE for initiation of replication of the *E. coli* chromosome and plasmids derived from P1 and F, respectively; (vi) protein translocation; and (vii) flagellar synthesis. Roles for DnaK in the folding of nascent polypeptide chains (Hendrick et al., 1993) and in cell division (Bukau and Walker, 1989) are proposed.

Biological regulation of DnaK

The expression of the *dnaK* gene is controlled by the heat shock promoter-specific σ^{32} subunit of RNA polymerase (see entry on regulation of the *E. coli* heat shock response, p. 525). Expression is induced by various forms of stress, including temperature upshift, as part of the heat shock response.

The activity of the DnaK protein in chaperone functions is controlled by nucleotides (Schmid et al., 1994) and the DnaJ and GrpE co-chaperones (Fig. 2). DnaK interacts with protein substrates by binding to short peptide segments containing a hydrophobic core of 4–5 residues and two flanking regions enriched in basic residues (Rüdiger et al. 1997). DnaK·ADP is the high affinity form characterized by low on and off rates for DnaK–substrate complexes. DnaK·ATP is the low affinity form which, although rapid in association with substrates, exhibits high off rates for DnaK–substrate complexes resulting in complex destabilization. This coupling is essential for DnaK chaperone function (Buchberger et al., 1994b) and involves physical interactions between the ATPase and the substrate-binding domains (Liberek et al., 1991b; Buchberger et al., 1995). DnaK has a low intrinsic ATPase activity (0.02–0.05 min^{-1} at 30°C [Buchberger et al., 1994b]) which is stimulated by DnaJ and GrpE (Liberek et al., 1991a) over 100-fold (McCarty et al., 1995). DnaJ, through interaction with the ATPase and substrate-binding domains (H. Schröder and B. Bukau, unpublished results), efficiently stimulates hydrolysis of DnaK-bound ATP (Liberek et al., 1991a) by acting at the rate-limiting λ phosphate cleavage step of the DnaK ATPase (McCarty et al., 1995; R. McMacken, J. Reinstein, personal communications). GrpE, through binding to the ATPase domain (Buchberger et al., 1994a), acts as a nucleotide-dissociation factor (Liberek et al., 1991a). The tight regulation of the ATPase cycle of DnaK by DnaJ and GrpE is thought to be important for the productive interaction of this chaperone system with substrates (Fig. 2).

Mutagenesis studies with *dnaK*

The first *dnaK* mutant alleles isolated were obtained upon selection of *E. coli* mutants resistant to lytic infection by bacteriophage λ (Georgopoulos, 1977). These mutants, as well as a Δ*dnaK* deletion mutant, had additional phenotypes including temperature-sensitive growth and defects in cell division, heat shock gene regulation, proteolysis and replication of chromosomal and plasmid DNA (Georgopoulos et al., 1990). Point mutations in *dnaK* obtained by genetic selection for mutants deficient in heat shock gene regulation (Wild

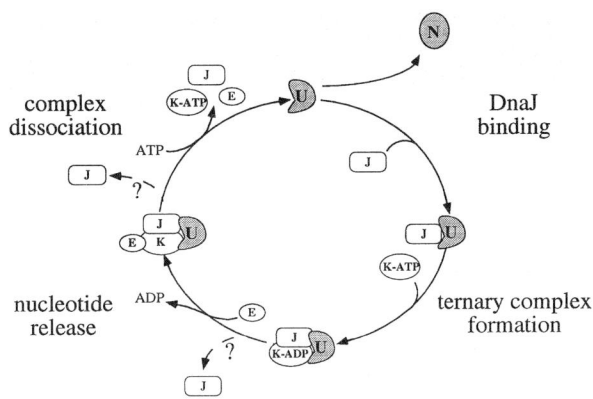

Figure 2. Model of the functional cycle of the DnaK chaperone system. This cycle relies on models proposed earlier (Gamer et al., 1996; Szabo et al., 1994). It starts with the rapid association of DnaJ (J) with a non-native protein substrate (U). DnaK (K), in its ATP-bound form and in absence of DnaJ, interacts only transiently with the substrate. DnaJ associated with the substrate interacts with DnaK·ATP and, through rapid stimulation of ATP hydrolysis by DnaK, mediates efficient formation of DnaK–DnaJ–substrate complexes containing ADP. DnaJ might dissociate from these complexes independently of DnaK and thus be capable of acting in this cycle in substoichiometric amounts. GrpE (E) can associate with DnaK complexed to the substrate, thereby triggering nucleotide release and forming quaternary GrpE–DnaK–DnaJ–substrate complexes lacking nucleotide. Spontaneous nucleotide release in the absence of GrpE might also occur but is considered less important given that GrpE is essential for most chaperone functions of the DnaK system. Binding of ATP induces dissociation of the substrate from DnaK. Then, in a process controlled by kinetic partitioning, the dissociated substrate will either undergo further folding (N) or reassociate with the DnaJ and DnaK components of this chaperone system.

et al., 1992) or by site-specific mutagenesis (Buchberger et al., 1994b; McCarty and Walker, 1994) display subsets of these phenotypes *in vivo*.

In vitro, certain substitutions of residue E171 of DnaK cause an uncoupling of substrate binding from the ATPase activity, probably by interfering with ATP-induced conformational changes in the ATPase domain of DnaK, and thus result in a loss of chaperone function (Buchberger et al., 1994b; Kamath-Loeb et al., 1995). A D201N alteration causes very similar defects. An A174T alteration specifically affects the stimulation of the ATPase activity by the combined action of GrpE and DnaJ (Kamath-Loeb et al., 1995). Alterations in residue T199 abolish autophosphorylation and, depending on the mutant side chain, decrease or extinguish the ATPase activity of DnaK, resulting in the loss of chaperone function (McCarty and Walker, 1991). *dnaK756* was the first dnaK mutant allele mapped (Yochem et al., 1978) and was later shown to contain three point mutations resulting in glycine to aspartate alterations at positions 32, 455 and 468 (Miyazaki et al., 1992; Buchberger et al., 1994a). The reduced affinity of GrpE for the DnaK756 mutant protein (Johnson et al., 1989) was demonstrated to be caused by the G32D alteration residing in an exposed, conserved loop of the DnaK ATPase domain (Buchberger et al., 1994a). Deletion of the entire loop (Δ28–33) resulted in destabilization of GrpE binding, demonstrating that this loop is a major interaction site of DnaK and GrpE. Determination of the crystal structure of the DnaK ATPase domain bound to GrpE (Harrison et al., 1997) confirmed that the loop is one of six surface regions on the ATPase domain that contact GrpE.

References

Bardwell, J.C.A. and Craig, E.A. (1984). Major heat shock gene of *Drosophila* and the *Escherichia coli* heat-inducible *dnaK* gene are homologous. Proc. Natl. Acad. Sci. USA **81**, 848–852.

Buchberger, A., Schröder, H., Büttner, M., Valencia, A., and Bukau. (1994a). A conserved loop in the ATPase domain of the DnaK chaperone is essential for the stable binding of GrpE. Nature Struct. Biol. **1**, 95–101.

Buchberger, A., Valencia, A., McMacken, R., Sander, C., and Bukau, B. (1994b). The chaperone function of DnaK requires the coupling of ATPase activity with substrate binding through residue E171. EMBO J. **13**, 1687–1695.

Buchberger, A., Theyssen, H., Schröder, H., McCarty, J.S., Virgallita, G., Milkereit, P., Reinstein, J., and Bukau, B. (1995). Nucleotide-induced conformational changes in the ATPase and substrate binding domains of the DnaK chaperone provide evidence for interdomain communication. J. Biol. Chem. **270**, 16903–16910.

Bukau, B., and Walker, G.C. (1989). Cellular defects caused by deletion of the *Escherichia coli dnaK* gene indicate roles for heat shock protein in normal metabolism. J. Bacteriol. **171**, 2337–2346.

Cegielska, A. and Georgopoulos, C. (1989). Functional domains of the *Escherichia coli* dnaK heat shock protein as revealed by mutational analysis. J. Biol. Chem. **264**, 21122–21130.

Flaherty, K.M., Deluca-Flaherty, C., and McKay, D.B. (1990). Three-dimensional structure of the ATPase fragment of a 70K heat-shock cognate protein. Nature (London) **346**, 623–628.

Frydman, J. and Hartl, F.-U. (1994). Molecular chaperone functions of hsp70 and hsp60 in protein folding. In The biology of heat shock proteins and molecular chaperones (R.I. Morimoto, A. Tissiéres and C. Georgopoulos, eds). Cold Spring Harbor Laboratory Press, New York, pp. 251–283.

Gamer, J., Multhaup, G., Tomoyasu, T., McCarty, J.S., Rüdiger, S., Schönfeld, H.-J., Schirra, C., Bujard, H., and Bukau, B. (1996). A cycle of binding and release of the DnaK, DnaJ and GrpE chaperones regulate activity of the *E. coli* heat shock transcription factor σ^{32}. EMBO J. **15**, 607–617.

Georgopoulos, C.P. (1977). A new bacterial gene (*groPC*) which affects lambda DNA replication. Molec. Gen. Genet. **151**, 35–39.

Georgopoulos, C., Ang, D., Liberek, K., and Zylicz, M. (1990). Properties of the *Escherichia coli* heat shock proteins and their role in bacteriophage λ growth. In Stress proteins in biology and medicine (R.I. Morimoto, A. Tissieres and C. Georgopoulos, eds). Cold Spring Harbor Laboratory Press, New York, pp. 191–222.

Georgopoulos, C., Liberek, K., Zylicz, M., and Ang, D. (1994). Properties of the heat shock proteins of *Escherichia coli* and the autoregulation of the heat shock response. In The biology of heat shock proteins and Molecular Chaperones (R.I. Morimoto, A. Tissiéres and C. Georgopoulos, eds). Cold Spring Harbor Laboratory Press, New York, pp. 209–249.

Gross, C.A., Straus, D.B., and Erickson, J.W. (1990). The function and regulation of heat shock proteins in *Escherichia coli*. In Stress proteins in biology and medicine (R.I. Morimoto, A. Tissieres and C. Georgopoulos, eds). Cold Spring Harbor Laboratory Press, New York, pp. 167–190.

Harrison, C.J., Hayer-Hartl, M., Di Liberto, M., Hartl, F.-U., and Kuriyan, J. (1997). Crystal structure of the nucleotide exchange factor GrpE bound to the ATPase domain of the molecular chaperone DnaK. Science **276**, 431–435.

Hendrick, J.P., Langer, T., Davis, T.A., Hartl, F.-U., and Wiedman, M. (1993). Control of folding and membrane translocation by binding of the chaperone DnaJ to nascent polypeptides. Proc. Natl. Acad. Sci. USA **90**, 10216–10220.

Herendeen, S.L., VanBogelen, R.A., and Neidhardt, F.C. (1979). Levels of major proteins of *Escherichia coli* during growth at different temperatures. J. Bacteriol. **139**, 185–194.

Johnson, C., Chandrasekhar, G.N., and C. Georgopoulos (1989). DnaK and GrpE heat shock proteins interact both in vivo and in vitro. J. Bacteriol. **171**, 1590–1596.

Kamath-Loeb, A.S., Lu, C.Z., Suh, W.-C., Lonetto, M.A., and Gross, C.A. (1995). Analysis of three DnaK mutant proteins suggests that progression through the ATPase cycle requires conformational changes. J. Biol. Chem. **270**, 30051–30059.

Liberek, K., Marszalek, J., Ang, D., Georgopoulos, C., and Zylicz, M. (1991a). *Escherichia coli* DnaJ and GrpE heat shock proteins jointly stimulate ATPase activity of DnaK. Proc. Natl. Acad. Sci. USA **88**, 2874–2878.

Liberek, K., Skowyra, D., Zylicz, M., Johnson, C., and Georgopoulos, C. (1991b). The *Escherichia coli* DnaK chaperone, the 70-kDa heat shock protein eukaryotic equivalent, changes conformation upon ATP hydrolysis, thus triggering its dissociation from a bound target protein. J. Biol. Chem. **266**, 14491–14496.

McCarty, J.S. and Walker, G.C. (1991). DnaK as a thermometer: threonine-199 is site of autophosphorylation and is critical for ATPase activity. Proc. Natl. Acad. Sci. USA **88**, 9513–9517.

McCarty, J.S. and Walker, G.C. (1994). DnaK mutants defective in ATPase activity are defective in negative regulation of the heat shock response: Expression of mutant DnaK protein results in filamentation. J. Bacteriol. **176**, 764–780.

McCarty, J.S., Buchberger, A., Reinstein, J., and Bukau, B. (1995). The Role of ATP in the functional cycle of the DnaK chaperone system. J. Mol. Biol. **249**, 126–137.

Miyazaki, T., Tanaka, S., Fujita, H., and Itikawa, H. (1992). DNA sequence analysis of the dnaK gene of *Escherichia coli* B and two dnaK genes carrying the temperature-sensitive mutations dnaK(Ts) and dnaK756(Ts). J. Bacteriol. **174**, 3715–3722.

Saito, H. and Uchida, H. (1977). Initiation of the DNA replication of bacteriophage lambda in *Escherichia coli* K12. J. Mol. Biol. **113**, 1–25.

Schmid, D., Baici, A., Gehring, H., and Christen, P. (1994). Kinetics of molecular chaperone action. Science **263**, 971–973.

Schönfeld, H.-J., Schmidt, D., Schröder, H., and Bukau, B. (1995). The DnaK chaperone system of *Escherichia coli*: quarternary structures and interactions of the DnaK and GrpE components. J. Biol. Chem. **270**, 2183–2189.

Szabo, A., Langer, T., Schröder, H., Flanagan, J., Bukau, B., and Hartl, F.U. (1994). The ATP hydrolysis-dependent reaction cycle of the *Escherichia coli* Hsp70 system-DnaK, DnaJ, and GrpE. Proc. Natl. Acad. Sci. USA **91**, 10345–10349.

Wild, J., Kamath, L.A., Ziegelhoffer, E., Lonetto, M., Kawasaki, Y., and Gross, C.A. (1992). Partial loss of function mutations in DnaK, the *Escherichia coli* homologue of the 70-kDa heat shock proteins, affect highly conserved amino acids implicated in ATP binding and hydrolysis. Proc Natl Acad Sci USA **89**, 7139–7143.

Yochem, J., Uchida, H., Sunshien, M., Saito, H., Georgopoulos, C.P., and Feiss, M. (1978). Genetic analysis of two genes, *dnaJ* and *dnaK*, necessary for *Escherichia coli* and bacteriophage lambda DNA replication. Molec. Gen. Genet. **164**, 9–14.

Zhu, X., Zhao, X., Burkholder, W.F., Gragerov, A., Ogata, C.M., Gottesman, M.E., and Hendrickson, W.A. (1996). Structural analysis of substrate binding by the molecular chaperone DnaK. Science **272**, 1606–1614.

■ *Alexander Buchberger and Bernd Bukau:*
Institut für Biochemie und Molekularbiologie
Universität Freiburg
Hermann Herder Str. 7
D-79104 Freiburg
Germany
Tel. 49 761 203 5221
Fax. 49 761 203 5257
E-mail: bukau@sun2.ruf.uni-freiburg.de

Escherichia coli Hsc66

Hsc66 is a newly described member of the HSP70 protein family in *Escherichia coli*. Unlike other HSP70 proteins, Hsc66 expression is not induced following heat shock, but is induced following cold shock. The biochemical properties of the Hsc66 protein are yet to be determined; however, mutations in the gene encoding this protein profoundly affect the cold shock-induced protein profile in E. coli.

■ Hsc66 sequence

The *hscA* gene sequence (*E. coli* K-12 accession number U01827, *E. coli* B U05338) encoding Hsc66 revealed that this protein is composed of 615 amino acids totalling 65 913 Da with an isoelectric point of 4.93. Hsc66 classification as an HSP70 protein is based on the fact that the amino acid sequence of this protein is 42% identical and 62% similar to the classic *E. coli* HSP70 protein DnaK (Kawula and Lelivelt, 1994; Seaton and Vickery, 1994), and contains perfect matches to two amino acid sequences conserved among all of the known HSP70 proteins. These sequences most likely comprise the ATP-binding domains of HSP70 proteins and are the octapeptides IDLGTTNS and DLGGGTFD beginning at positions 29 and 212 of Hsc66, respectively.

■ Biological activities of Hsc66

Hsc66 is a 66-kDa protein that is a member of the HSP70 protein family, yet it lacks some of the fundamental characteristics of HSP70 proteins. Unlike all other described HSP70 proteins, Hsc66 expression is not increased as a result of environmental shifts that induce protein misfolding and aggregation (Lelivelt and Kawula, 1995). Also, even though Hsc66 contains perfect homology to the ATP-binding domain of other described HSP70 proteins, Hsc66 does not bind to ATP-agarose columns under the same conditions that allow ATP binding by other HSP70 proteins (our unpublished results).

Though direct functions of Hsc66 have yet to be described, it is evident that this protein has biological activities in *E. coli* that affect gene expression. Mutations in the *hscA* gene encoding Hsc66 can compensate for some expression defects associated with mutations in the gene encoding the nucleoid-associated DNA-binding protein H-NS (Kawula and Lelivelt, 1994). Also, strains devoid of Hsc66 expression exhibit drastically altered cold shock protein induction profiles (Lelivelt and Kawula, 1995).

■ Gene regulation of *hscA*

Hsc66 expression is not induced upon heat shock, but is induced approximately 11-fold following cold shock (Lelivelt and Kawula, 1995). Primer extension analysis defined the transcription initiation site for the *hscA* gene encoding Hsc66. The DNA adjacent to this start site did not contain any of the known regulatory sequences associated with other cold-inducible genes (Lelivelt and Kawula, 1995). *hscA* is cotranscribed with two other genes, *hscB* and *fdx* (Seaton and Vickery, 1994; Lelivelt and Kawula, 1995). The *hscB* gene encodes a DnaJ-like protein (Kawula and Lelivelt, 1994), and the *fdx* gene encodes a ferredoxin-binding protein (Ta and Vickery, 1992). While it is well documented that HSP70 and DnaJ-like proteins often function in concert (Hendrick and Hartl, 1993), the biological ramifications of these HSP proteins being coexpressed with a ferredoxin-binding protein are unclear.

References

Hendrick, J.P. and Hartl, F.-U. (1993). Molecular chaperone functions of heat-shock proteins. Annu. Rev. Biochem. **62**, 349–384.

Kawula, T.H. and Lelivelt, M.J. (1994). Mutations in a gene encoding a new Hsp70 suppress rapid DNA inversion and *bgl* activation, but not *proU* derepression, in *hns-1* mutant *Escherichia coli*. J. Bacteriol. **176**, 610–619.

Lelivelt, M.J. and Kawula, T.H. (1995). Hsc66, an Hsp70 homolog in *Escherichia coli*, is induced by cold shock but not by heat shock. J. Bacteriol. **177**, 4900–4907.

Seaton, B.L. and Vickery, L.E. (1994). A gene encoding a DnaK/hsp70 homolog in *Escherichia coli*. Proc. Natl. Acad. Sci. USA **91**, 2066–2070.

Ta, T.D. and Vickery, L.E. (1992). Cloning, sequencing, and overexpression of a [2Fe-2S] ferridoxin gene from *Escherichia coli*. J . Biol. Chem. **267**, 11120–11125.

■ Tom H. Kawula:
Department of Microbiology and Immunology
CB#7290, 804 Mary Ellen Jones Bldg.
University of North Carolina
Chapel Hill, NC 27599-7290, USA
Tel. 1 919 966 2637
Fax. 1 919 962 8103
E-mail: kawula@med.unc.edu

Saccharomyces cerevisiae Ssa proteins

The Ssa proteins are the most abundant HSP70 molecules of the Saccharomyces cerevisiae cytosol and are encoded by a family of four highly homologous genes. Although the Ssa proteins appear to be functionally redundant, each of the four genes has a unique pattern of expression. The Ssa proteins are essential for viability. Strains which are defective in Ssa protein function exhibit perturbations in protein localization and heat shock regulation.

■ Isolation

The *SSA1–4* genes (initially known as *YG100*, *YG102*, *YG106* and *YG107*, respectively) were isolated by virtue of their high degree of homology to *Drosophila HSP70* genes (Ingolia *et al.*, 1982).

■ *SSA* genes and sequence

The *SSA* genes are highly homologous to other *HSP70* genes. Like other eukaryotic heat shock genes, *SSA1*, *3* and *4* possess promoter elements known as HSEs (heat shock elements) in their 5′ untranslated regions. The GenBank accession numbers for *SSA1–4* are X12926, X12927, M97225 and J05637, respectively. *SSA1*, *SSA3* and *SSA4* are on chromosomes 1, 2 and 5, respectively. The chromosomal location of *SSA2* has not been determined.

■ Ssa proteins

The *SSA* genes 1–4 encode proteins of c.70 000 Da which are more than 80% identical at the amino acid level (Fig. 1). The Ssa1p and Ssa2p proteins have been purified from yeast using ATP-agarose affinity chromatography (Chirico *et al.*, 1988; Gao *et al.*, 1991), analogous to schemes developed for *E. coli* DnaK and bovine brain Hsc70. The intrinsic ATPase activities reported for Ssa1p and Ssa2p are in the range of 2–5 nmole/min mg (Cyr *et al.*, 1992; Ziegelhoffer *et al.*, 1995).

■ Biological regulation of *SSA* genes

Initial analysis of the *SSA* genes established the complex regulation of this gene family, as depicted in Fig. 1. *SSA1* is a classical *HSP70* gene whose expression rises several fold from a significant basal level upon induction of the heat shock response (Werner-Washburne *et al.*, 1989). In contrast, *SSA2*, although 99% identical in amino acid sequence, is more akin to mammalian Hsc70 in that its expression is essentially constitutive. Both *SSA3* and *SSA4* are not expressed at significant levels under normal logarithmic growth but are highly heat inducible. *SSA3* expression is also increased upon approach to stationary phase (the diauxic shift) (Boorstein and Craig, 1990a). HSEs 5′ of the transcriptional initiation site in the *SSA1,3* and *4* genes have been shown to be responsible for heat induction (Park and Craig, 1989; Boorstein and Craig, 1990b, c).

SSA3 and *SSA4* expression is constitutively induced in *ssa1 ssa2* cells at permissive temperatures, suggesting that *SSA* gene products also negatively regulate their own synthesis (Craig and Jacobsen, 1984). Consistent with this idea, overexpression of Ssa1 or Ssa4 protein suppresses induction of an HSE-driven promoter upon a heat shock (Stone and Craig, 1990). This increase in expression

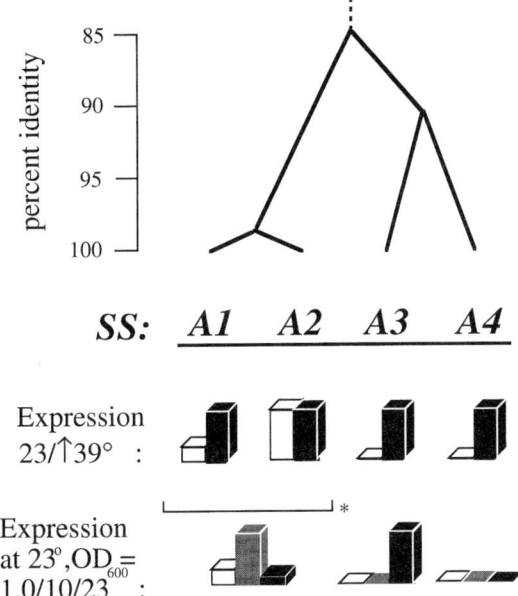

Figure 1. *Saccharomyces cerevisiae* SSA multigene family. The top portion of the figure illustrates the degree of evolutionary relatedness among the members of the *SSA* family, with shared branch points indicating the per cent amino acid identity between any two members. 'Expression 23/↑39°:' depicts the relative level of expression of each gene or gene subfamily before (white columns) and after (black columns) a shift from 23 to 39°C, which induces the heat shock response. 'Expression at 23°, OD_{600} = 1.0/10/23:' depicts the relative level of expression of each gene or gene subfamily as cells progress from rapid exponential growth (OD_{600} = 1.0, white columns) through diauxic shift (OD_{600} = 10, grey columns) to stationary phase (OD_{600} = 23, black columns). *The probe used in these experiments was incapable of distinguishing between the highly homologous *SSA1* and *SSA2* genes.

from heat shock promoters appears to be the result of increased HSF activity since an HSE is sufficient for increased expression (Boorstein and Craig, 1990c). This autoregulation extends to the more general heat shock response, since Hsp90 and Hsp104 expression is also induced in an *ssa1 ssa2* mutant strain (Craig and Jacobsen, 1984; Halladay and Craig, 1995).

■ Mutagenesis studies with *SSA* genes

The *SSA* gene family is essential for viability. Most genetic studies have been carried out using disruption alleles of the four *SSA* genes. From early experiments the similarity in the function of the *SSA* proteins was clear. For example, the disruption of the more closely related *SSA1* and *SSA2* genes results in slow growth at most temperatures compared with wild-type cells, but inability to grow at 35°C or higher (Craig and Jacobsen, 1984). However, an *ssa1 ssa2 ssa4* strain is non-viable, indicating that the presence of Ssa4p permits viability (Werner-Washburne *et al.*, 1987). In addition, the expression of the related *SSA3* gene can rescue a *ssa1 ssa2 ssa4* strain when driven by the *SSA2* promoter.

However, the question remains whether all four proteins, even though their expression is regulated very differently, are functionally identical rather than merely similar. While an *ssa1 ssa2* strain cannot form colonies at 37°C, the presence of either *SSA3* or *SSA4* on a multicopy plasmid allows cell growth at 37°C in this genetic background (Craig *et al.*, 1995). An *SSA1* mutant has wild-type growth properties on rich media; however on non-fermentable carbon sources the *SSA1* mutant strain does not grow at 37°C, while *ssa2* cells grow normally under these conditions. At first glance this phenotype of *ssa1* cells suggests a specialized function of Ssa1p. However, increasing the copy number of *SSA2* in a *SSA1* mutant strain obviates the requirement for Ssa1p for growth on non-fermentable carbon sources. In these experiments, even though the *SSA2* gene was present on a high copy number plasmid the level of these *SSA* proteins was increased by less than 50% owing to the tight regulation of the expression of *SSA* genes. In conclusion, although strains containing different combinations of *SSA* mutants grew poorly under certain conditions, in all cases tested a small increase in the concentration of another member of the *SSA* subfamily compensated for the growth defects. Therefore there is no genetic evidence at this time for differences in function among the *SSA* family members.

■ Biological activities of Ssa proteins

Experiments in which Ssa protein levels were regulated via the *GAL10* promoter showed that depletion of Ssa protein results in the accumulation of precursors to both alpha factor (a secreted mating pheromone peptide) and the β subunit of the mitochondrial F_1ATPase (Deshaies *et al.*, 1988). Similar perturbations in precursor maturation were also observed in strains carrying a temperature sensitive allele of the *SSA1* gene, *ssa1–45* (J. Becker and E.A. Craig, unpublished results). Furthermore, addition of SSA proteins to an *in vitro* translocation system increased the efficiency of import (Chirico *et al.*, 1988). Given the general role of HSP70 as a molecular chaperone, it is likely that Ssa proteins serve to stabilize precursor proteins in an unfolded state prior to their interaction with the translocation apparatus.

The Ssa1p and Ssa2p isoenzymes have also been shown to function as clathrin-uncoating ATPases *in vitro* (Gao *et al.*, 1991), demonstrating their functional relatedness to mammalian Hsc70.

The Ssa proteins also play a role in thermotolerance. Overexpression of Ssa1 protein in an *hsp104* mutant partially restores the thermotolerance defect of these mutants. Strains lacking various *SSA* genes show no alteration in thermotolerance. However, in combination with an *hsp104* mutation, *ssa2 ssa3 ssa4* mutations cause significantly decreased levels of thermotolerance compared with strains carrying only the *hsp104* mutation (Sanchez et al., 1993).

As described above in the biological regulation section, Ssa proteins also function in the regulation of the heat shock response.

■ Biological and genetic interactions

The Ssa1p and Ssa2p isoenzymes interact with the yeast DnaJ homologue, Ydj1p, both *in vivo* and *in vitro* (Cyr et al., 1992; Ziegelhoffer et al., 1995). The basal ATPase activity of Ssa protein is stimulated by as much as 20-fold in the presence of Ydj1p. Ssa1p, together with Ydj1p, has been shown to assist in the *in vitro* renaturation of unfolded firefly luciferase, demonstrating a chaperone function for this pair of proteins (Levy et al., 1995).

Ssa proteins have also been shown to be in a complex with Hsp82, the HSP90 protein of *S. cerevisiae* (Chang and Lindquist, 1994). This result is analogous to the observations made in mammalian cells in which HSP90 and HSP70 proteins are part of a multiprotein complex that functions in regulating certain steroid receptors.

Analysis of extragenic suppressors of the temperature sensitive and slow growth phenotype of *ssa1 ssa2* mutants showed that reduced activity of the heat shock transcription factor, HSF1, partially suppresses these phenotypes (Nelson et al., 1992; Halladay and Craig 1995). Apparently, the overexpression of certain heat-inducible genes is toxic to cells and reduction in transcription of heat shock proteins (HSPs) partially relieves the growth defect.

UBP3 which encodes a ubiquitin-processing protease, partially suppresses the temperature-sensitive growth of an *ssa1 ssa2* strain (Craig et al., 1994). Although the reason for this suppression is not understood, the suppression establishes a link between cytosolic HSP70 and the ubiquitin proteolytic system.

■ References

Boorstein, W.R. and Craig, E.A. (1990a). Regulation of a yeast HSP70 gene by cAMP responsive transcriptional control element. EMBO J. **9**, 2543–3267.

Boorstein, W.R. and Craig, E.A. (1990b). Transcriptional regulation of *SSA3*, an HSP70 gene from *Saccharomyces cerevisiae*. Mol. Cell. Biol. **10**, 3262–2553.

Boorstein, W.R. and Craig, E.A. (1990c). Structure and regulation of the *SSA4* HSP70 gene of *Saccharomyces cerevisiae*. J. Biol. Chem. **265**, 18912–18921.

Chang, H.-C.J. and Linquist, S. (1994). Conservation of Hsp90 macromolecular complexes in *Saccharomyces cerevisiae*. J. Biol. Chem. **269**, 24983–24988.

Chirico, W., Waters, M.G., and Blobel, G. (1988). 70K heat shock related proteins stimulate protein translocation into microsomes. Nature **332**, 805–810.

Craig, E.A. and Jacobsen, K. (1984). Mutations of the heat-inducible 70 kilodalton genes of yeast confer temperature-sensitive growth. Cell **38**, 841–849.

Craig, E.A., Baxter, B.K., Becker, J., Halladay, J., and Ziegelhoffer, T. (1994). Cytosolic hsp70s of *Sacchromyces cerevisiae*: roles in protein synthesis, protein translocation, proteolysis, and regulation. In *The biology of heat shock proteins and molecular chaperones*., R. Morimoto, A. Tissieres and C. Georgopoulos (eds). Cold Spring Harbor Laboratory Press, New York, p. 31–52

Craig, E.A., Ziegelhoffer, T., Nelson, J., Laloraya, S., and Halladay, J. (1995). Complex multigene family of functionally distinct Hsp70s of yeast. *Cold Spring Harbor Symposia on Quantitative Biology: protein kinesis: the dynamics of protein trafficking and stability*, Vol. LX, 441–449.

Cyr, D., Lu, X., and Douglas, M. (1992). Regulation of Hsp70 function by a eukaryotic DnaJ homolog. J. Biol. Chem. **267**, 20927–20931.

Deshaies, R., Koch, B., Werner-Washburne, M., Craig, E., and Schekman, R. (1988). A subfamily of stress proteins faciliates translocation of secretory and mitochondrial precursor polypeptides. Nature **332**, 800–805.

Gao, B., Biosca, J., Craig, E.A., Greene, L.E., and Eisenberg, E. (1991). Uncoating of coated vesicles by yeast hsp70 proteins. J. Biol. Chem. **266**, 19565–19571.

Halladay, J. and Craig, E. (1995). A heat shock transcription factor with reduced activity suppresses a yeast HSP70 mutant. Mol. Cell. Biol. **15**, 4890–4897.

Ingolia, T.D., Slater, M.R., and Craig, E.A. (1982). *Saccharomyces cerevisiae* contains a complex multigene family related to the major heat shock inducible gene of *Drosophila*. Mol. Cell. Biol. **2**, 1388–1398.

Levy, E., McCarty, J., Bukau, B., and Chirico, W. (1995). Conserved ATPase and luciferase refolding activities between bacteria and yeast Hsp70 chaperones and modulators. FEBS Lett. **368**, 435–440.

Nelson, R.J., Heschl, M., and Craig, E.A. (1992). Isolation and characterization of extragenic suppressors of mutations in the *SSA* hsp70 genes of *Saccharomyces cerevisiae*. Genetics **131**, 277–285.

Park, H.-O. and Craig, E.A. (1989). Positive and negative regulation of basal expression of a yeast hsp70 gene. Mol. Cell. Biol. **9**, 2025–2033.

Sanchez, Y., Parells, D.A., Taulien, J., Vogel, J.L., Craig, E.A., and Lindquist, S. (1993). Genetic evidence for a functional relationship between Hsp104 and Hsp70. J. Bacteriol. **175**, 6484–6491.

Stone, D.E. and Craig, E.A. (1990). Self regulation of 70 kilodalton heat shock proteins in *Saccharomyces cerevisiae*. Mol. Cell. Biol. **10**, 1622–1632.

Werner-Washburne, M., Stone, D.E., and Craig, E.A. (1987). Complex interactions among members of an essential subfamily of hsp70 genes in *Saccharomyces cerevisiae*. Mol. Cell. Biol. **7**, 2568–2577.

Werner-Washburne, M., Becker, J., Kosics-Smithers, J., and Craig, E.A. (1989). Yeast Hsp70 RNA levels change in response to the physiological status of the cell. J. Bacteriol. **171**, 2680–2688.

Ziegelhoffer, T., Lopez-Buesa, P., and Craig, E.A. (1995). The dissociation of ATP from hsp70 of *Saccharomyces cerevisiae* is stimulated by both Ydj1p and peptide substrates. J. Biol. Chem. **270**, 10412–10419.

■ *Thomas Ziegelhoffer and Elizabeth A. Craig*
Department of Biomolecular Chemistry
University of Wisconsin–Madison
1300 University Avenue
Madison, WI 53706, USA.
Tel. 1 608 263 7105
Fax. 1 608 262 5253
E-mail: ecraig@facstaff.wisc.edu

Saccharomyces cerevisiae Ssb proteins

The Ssb proteins of Saccharomyces cerevisiae *represent a novel class of HSP70 homologues which facilitate translation. Under logarithmic growth conditions, a majority of the Ssb protein in the cell is associated with ribosomes, most likely through direct interaction with the nascent polypeptide chain.*

■ Isolation

The *SSB1* and *SSB2* genes (initially known as *YG101* and *YG103*) were isolated by virtue of their high degree of similarity to *Drosophila HSP70* genes (Ingolia *et al.*, 1982; Craig and Jacobsen, 1985).

■ *SSB* genes and sequences

The cytosolic Ssb proteins of *S. cerevisiae* are encoded by two nearly identical genes (Boorstein *et al.*, 1994), *SSB1* and *SSB2*., which differ at only 47 positions within their protein coding regions, although their promoters share no obvious homology. The 66 594 dalton Ssb proteins are identical at all but 4 of their 613 amino acids (Boorstein *et al.*, 1994). The GenBank accession numbers are X13713 for *SSB1* and L11262 for *SSB2*. *SSB1* and *SSB2* are on chromosomes 4 and 14, respectively (unpublished data).

■ Ssb proteins

Although sharing more than 50% identity with the cytosolic Ssa proteins (as well as with human HSP70 and DnaK of *E. coli*), the Ssb family is genetically distinct and appears to be a specialized type of HSP70 homologue whose function is to facilitate the translation of mRNA into protein (Craig and Jacobsen, 1985). The Ssb proteins have been shown to cosediment with polysomes upon sucrose gradient fractionation of crude cell extracts (Nelson *et al.*, 1992). This interaction appears to be specific for translating ribosomes and is disrupted in the presence of puromycin and high salt, a treatment which results in the release of nascent chains from ribosomes. Given the propensity of HSP70 homologues to bind to polypeptides which are in an extended conformation, it has been proposed that Ssb proteins bind to nascent chains as they emerge from the large ribosomal subunit, thereby preventing their premature folding.

The Ssb1/2 proteins have been purified from yeast using ATP-agarose affinity chromatography and have been found to have a basal ATPase activity similar to that of Ssa1 protein (Cyr and Douglas, 1994). However, the K_m for ATP of the Ssb1 protein is at least 100x higher than that of the Ssa1 protein (T. Ziegelhoffer, unpublished results). Unlike Ssa1 or Ssa2 proteins, the Ssb proteins fail to depolymerize clathrin baskets *in vitro*, further supporting the idea that these two classes of proteins are different (Gao *et al.*, 1991).

■ Biological regulation of *SSB* genes

The expression of *SSB* genes decreases upon heat shock (Craig and Jacobsen, 1985; Werner-Washburne *et al.*, 1987).

■ Mutagenesis studies with *SSB* genes

Strains lacking functional *SSB* genes grow poorly at all temperatures, with the growth defect being more pronounced with decreasing temperature (Craig and Jacobsen, 1985). Such strains are also hypersensitive to aminoglycoside antibiotics and verrucarin A, both of which are inhibitors of translation (Nelson *et al.*, 1992). Fusions between domains of the Ssb1 protein and the Ssa1 protein have been analyzed (James *et al.*, 1997). The Ssa1 18 kDa peptide binding domain can substitute for the Ssb1 peptide binding domain. The 44 kDa ATPase domain of Ssb1 is required for rescue of the cold-sensitive phenotype of an *ssb1 ssb2* mutant. Two of the three domains (44 kDa ATPase, 18 kDa peptide binding and 10 kDa C-terminal) of the Ssb1 protein are required for rescue of antibiotic sensitivity, but any two will suffice.

■ Biological or genetic interactions

A genetic screen for multicopy suppressors of the slow growth phenotype of *ssb1 ssb2* strains resulted in the isolation of a novel homologue of the GTP-binding protein, EF1α, which catalyses the binding of aminoacyl-transfer RNAs to the ribosome. This gene was termed *HBS1* (high copy *ssb* suppressor) (Nelson *et al.*, 1992). The function of the Hbs1 protein is unknown, although its similarity to EF1α is consistent with the role of Ssb protein in translation.

SSB1 itself has been isolated as a multicopy suppressor of two very different cellular defects. Increased copy number of *SSB1* suppresses a temperature-sensitive mutation in the Y7 subunit of the multicatalytic proteosome, which functions to degrade both ubiquitin-conjugated and ubiquitin-free substrates (Ohba, 1994). In addition, *SSB1* has been identified as a multicopy suppressor of the temperature-sensitive *dbf3-1* allele (Shea *et al.*, 1994) which was originally characterized as having a cell cycle phenotype due to a defect in the initiation of DNA synthesis. More recently, *DBF3* was discovered to be identical to *PRP8* which encodes the U5 snRNP, an important component of the yeast spliceosome. It is unclear if these latter genetic interactions represent specific functions of the Ssb proteins or are simply an indirect consequence of their role in translation.

References

Boorstein, W.R., Ziegelhoffer, T., and Craig, E.A. (1994). Molecular evolution of the HSP70 multigene family. J. Mol. Evol. **38**, 1–17.

Craig, E.A. and Jacobsen, K. (1985). Mutations in cognate gene of *Saccharomyces cerevisiae* HSP70 result in reduced growth rates at low temperatures. Mol. Cell. Biol. **5**, 3517–3524.

Cyr, D.M. and Douglas, M.G. (1994). Differential regulation of Hsp70 subfamilies by the eukaryotic DnaJ homologue YDJ1. J. Biol. Chem. **269**, 9798–9804.

Gao, B., Biosca, J., Craig, E.A., Greene, L.E., and Eisenberg, E. (1991). Uncoating of coated vesicles by yeast hsp70 proteins. J. Biol. Chem. **266**, 19565–19571.

Ingolia, T.D., Slater, M.R., and Craig, E.A. (1982). Saccharomyces cerevisiae contains a complex multigene family related to the major heat shock inducible gene of Drosophila. Mol. Cell. Biol. **2**, 1388–1398.

James, P., Pfund, C. and Craig, E.A. (1997). Functional specificity among Hsp70 molecular chaperones. Science **275**, 387–389.

Nelson, R.J., Ziegelhoffer, T., Nicolet, C., Werner-Washburne, M., and Craig, E.A. (1992). The translation machinery and seventy kilodalton heat shock protein cooperate in protein synthesis. Cell **71**, 97–105.

Ohba, M. (1994). A 70-kDa heat shock cognate protein suppresses the dfects caused by a proteasome mutation in *Saccharomyces cerevisiae*. FEBS Lett. **351**, 263–266.

Shea, J.E., Toyn, J.H., and Johnston, L.H. (1994). The budding yeast U5 snRNP Prp8 is a highly conserved protein which links RNA splicing with cell cycle progression. Nucl. Acid Res. **22**, 5555–5564.

Werner-Washburne, M., Stone, D.E., and Craig, E.A. (1987). Complex interactions among members of an essential subfamily of hsp70 genes in *Saccharomyces cerevisiae*. Mol. Cell. Biol. **7**, 2568–2577.

■ *Thomas Ziegelhoffer, Christine Pfund and Elizabeth A. Craig:*
Department of Biomolecular Chemistry
University of Wisconsin–Madison
1300 University Avenue
Madison, WI 53706, USA.
Tel. 1 608 263 7105
Fax. 1 608 262 5253
E-mail: ecraig@facstaff.wisc.edu

Saccharomyces cerevisiae Ssc1

The Ssc1 protein is the HSP70 family member located in the matrix compartment of mitochondria of Saccharomyces cerevisiae. Ssc1 is an abundant, essential protein required for translocation of many proteins from the cytosol into the matrix, and also appears to play roles in protein folding and proteolysis.

■ Alternative names

Mitochondrial hsp70 (mt-hsp70), Ens1p (Morishima *et al.*, 1990).

■ Isolation

SSC1 was originally isolated by hybridization based on its sequence similarity to the *SSA1* gene of *S. cerevisiae* (Craig *et al.*, 1987).

■ Ssc1 protein

The Ssc1 protein (Ssc1p) is an abundant protein of the matrix of mitochondria (Craig *et al.*, 1987, 1989). The nucleotide sequence of the *SSC1* gene (GenBank accession number M27229) encodes a 654 amino acid polypeptide that includes a leader sequence of 23 amino acids, which is removed upon import into mitochondria. The calculated molecular weight of the precursor and mature forms of the protein are 70 634 and 68 101 daltons respectively. Ssc1p is more closely related to the *E. coli* DnaK protein than any other family of HSP70 proteins, having an identity of 62% with DnaK compared with 54% with the cytosolic HSP70 proteins (Ssa1p and Ssa2p) from *S. cerevisiae*. On two-dimensional gels the Ssc1 protein runs as three isoforms owing to phosphorylation (Craig *et al.*, 1989; E. Craig, unpublished data).

■ Function of Ssc1p

SSC1 is an essential gene (Craig *et al.*, 1987). Ssc1p is required for the translocation of a number of cytosolic precursor proteins into mitochondria (Kang *et al.*, 1990). Ssc1p directly interacts with incoming precursor proteins as they cross the mitochondrial membranes (Scherer *et al.*, 1990), binding to both the presequence and the mature portions of the protein. Two basic models exist for the role of Ssc1p in the translocation of proteins into the matrix (Pfanner and Meijer, 1995): The 'Brownian ratchet' model proposes that the energy for translocation is driven by Brownian motion and Ssc1p promotes directionality by preventing backsliding (retrograde movement) of the precursor into the cytosol. In the 'pulling' model it is proposed that Ssc1p provides the force required via a conformational change requiring ATP binding and/or hydrolysis.

Analysis of *ssc1* mutants has also provided evidence for the role of Ssc1p in additional mitochondrial processes including acting as a chaperone for at least some proteins encoded by the mitochondrial genome (Hermann *et al.*, 1994) and folding of proteins translocated from the cytosol. Degradation of matrix proteins by the Pim1 protease is also hampered in *ssc1* mutants (Wagner *et al.*, 1994).

Ssc1p has been isolated as the non-catalytic subunit of the mitochondrial site-specific endonuclease Endo.*Sce*I (Morishima *et al.*, 1990). Ssc1p is required for significant catalytic activity and is present in a 1:1 stoichiometry in purified enzyme.

Mutagenesis studies with *SSC1*

Two temperature-sensitive mutant alleles of *SSC1*, *ssc1–3* and *ssc1–2*, which were used extensively in the functional studies described above, have been isolated and sequenced (Kang *et al.*, 1990; Gambill *et al.*, 1993; Voos *et al.*, 1993). Ssc1-3p, containing a G to S substitution at amino acid 65 in the ATPase domain, appears to interact only weakly if at all with polypeptides being translocated into mitochondria. Ssc1-2p, containing a P to S substitution at amino acid 419 of the peptide-binding domain, has been shown to bind translocation intermediates as well as protein completely imported into mitochondria. Ssc1-2p appears to be defective in release of bound polypeptides.

Biological interactions of Ssc1p

Ssc1 protein interacts with other proteins during the import process, including Tim44 (also called Isp45 and Mim44) (Kronidou *et al.*, 1994; Rassow *et al.*, 1994; Schneider *et al.*, 1994), a component of the protein import machinery which is localized on the matrix side of the mitochondrial inner membrane. Tim44 is found in a complex with Ssc1p and precursor protein during an early step of the translocation process. It is thought that Tim44 recruits Ssc1p to the import machinery. The interaction between Tim44 and Ssc1p is disrupted in the presence of ATP.

Ssc1p also interacts with Mge1p (Bolliger *et al.*, 1994; Ikeda *et al.*, 1994; Laloraya *et al.*, 1994), an essential protein of the mitochondrial matrix which is related to bacterial GrpE. Analogous to GrpE, Mge1p interacts with Ssc1p and acts as a nucleotide release factor (Miao *et al.*, 1997). The Mge1p–Ssc1p interaction is disrupted by ATP (Voos *et al.*, 1994). Two types of Ssc1:Mge1 complexes have been identified in mitochondria: a Ssc1:Mge1:Tim44 and a Ssc1:Mge1:Mdj1 complex (Horst *et al*, 1997).

Regulation of *SSC1*

SSC1 is constitutively expressed under normal growth conditions. The expression of *SSC1* is induced about twofold upon a heat shock (Craig *et al.* 1987). The level of Ssc1 protein is several fold higher in cells grown on non-fermentable carbon sources compared with cells grown on glucose-based medium (P.J. Kang and E. Craig, unpublished results).

References

Bolliger, L., Deloche, O., Glick, B., Georgopoulos, C., Jeno, P., Kronidou, N., Horst, M., Morishima, N., and Schatz, G. (1994). A mitochondrial homolog of bacterial GrpE interacts with mitochondrial hsp70 and is essential for viability. EMBO J. **13**, 1998–2006.

Craig, E.A., Kramer, J., and Kosic-Smithers, J. (1987). SSC1, a member of the 70-kDa heat shock protein multigene family of *Saccharomyces cerevisiae*, is essential for growth. Proc. Natl. Acad. Sci. USA **84**, 4156–4160.

Craig, E.A., Kramer, J., Shilling, J., Werner-Washburne, M., Holmes, S., Kosic-Smithers, J., and Nicolet, C.M. (1989). *SSC1*, an essential member of the *S. cerevisiae* HSP70 multigene family, encodes a mitochondrial protein. Mol. Cell. Biol. **9**, 3000–3008.

Gambill, B.D., Voos, W., Kang, P.J., Miao, B., Langer, T., Craig, E.A., and Pfanner, N. (1993). A dual role for mitochondrial Heat shock protein 70 in membrane translocation of preproteins. J. Cell Biol. **123**, 109–117.

Hermann, J., Stuart, R., Craig, E., and Neupert, W. (1994). Mitochondrial heat shock protein 70, a molecular chaperone for proteins encoded by mitochondrial DNA. J. Cell Biol. **127**, 893–902.

Horst, M., Oppliger, W., Rospert, S., Schonfeld, H.-J., Schatz, G., and Azem, A. (1997). Sequential action of two Hsp70 complexes during protein import into mitochondria. EMBO J. **16**, 1842–1849.

Ikeda, E., Yoshida, S., Mitsuzawa, H., Uno, I. , and Toh-e, A. (1994). YGE1 is a yeast homologue of Escherichia coli *grpE* and is required for maintenance of mitochondrial function. FEBS Lett. **339**, 265–268.

Kang, P.J., Ostermann, J., Shilling, J., Neupert, W., Craig, E.A., and Pfanner, N. (1990). Hsp70 in the mitochondrial matrix is required for translocation and folding of precursor proteins. Nature **348**, 137–143.

Kronidou, N.G., Opplinger, W., Bollinger, L., Hannavy, K., Glick, B., Schatz, G., and Horst, M. (1994). Dynamic interaction between Isp45 and mitochondrial hsp70 in the protein import system of yeast mitochondrial inner membrane. Proc. Natl. Acad. Sci. USA **91**, 12818–12822.

Laloraya, S., Gambill, B.D., and Craig, E.A. (1994). A role for a eukaryotic GrpE-related protein, Mge1p, in protein translocation. Proc. Natl. Acad. Sci. USA **91**, 6481–6485.

Miao, B., Davis, J. and Craig, E.A. (1997). Mge1 functions as a nucleotide release factor for Ssc1, a mitochondrial Hsp70 of *Saccharomyces cerevisiae*. J. Mol. Biol. **265**, 541–552.

Morishima, N., Nakagawa, K., Yamamoto, E., and Shibata, T. (1990). A subunit of yeast site-specific endonuclease SceI is a mitochondrial version of the 70-kDa heat shock protein. J. Biol. Chem. **265**, 15189–15197.

Pfanner, N. and Meijer, M. (1995). Protein folding: pulling in the proteins. Curr. Biol. **5**, 132–135.

Rassow, J., Maarse, A., Krainer, E., Kubrich, M., Muller, H., Meijer, M., Craig, E., and Pfanner, N. (1994). Mitochondrial protein import: biochemical and genetic evidence for interaction of matrix hsp70 and the inner membrane protein MIM44. J. Cell Biol. **127**, 1547–1556.

Scherer, P., Krieg, U., Hwang, S., Vestweber, D., and Schatz, G. (1990). A precursor protein partially translocated into yeast mitochondria is bound to a 70kd mitochondrial stress protein. EMBO J. **9**, 4315–4322.

Schneider, H.-C., Berthold, J., Bauer, M.F., Dietmeier, K., Guiard, B., Brunner, M., and Neupert, W. (1994). Mitochondrial Hsp70/MIM44 complex facilitates protein import. Nature **371**, 768–774.

Voos, W., Gambill, B.D., Guiard, B., Pfanner, N., and Craig, E.A. (1993). Presequence and mature parts of preproteins strongly influence the dependence of mitochondrial protein import on heat shock protein 70 in the matrix. J. Cell Biol. **129**, 119–123.

Voos, W., Gambill, B.D., Laloraya, S., Ang, D., Craig, E.A., and Pfanner, N. (1994). Mitochondrial GrpE is present in a complex with hsp70 and preproteins in transit across membranes. Mol. Cell. Biol. **14**, 6627–6634.

Wagner, I., Arlt, H., van Dyck, L., Langer, T., and Neupert, T. (1994). Molecular chaperones cooperate with PIM1 protease in the degradation of misfolded protein in mitochondria. EMBO J. **13**, 5135–51345.

■ *Julie Davis, Bingjie Miao, and Elizabeth Craig:*
Department of Biomolecular Chemistry
University of Wisconsin–Madison
1300 University Avenue
Madison, WI 53706, USA.
Tel. 1 608 263 7105
Fax. 1 608 262 5253
E-mail: ecraig@facstaff.wisc.edu

Saccharomyces cerevisiae Ssh1p

The Ssh1 protein is the second member of the HSP70 family identified in the matrix compartment of mitochondria of Saccharomyces cerevisiae. The cold-sensitive phenotype of ssh1 mutants is suppressed by loss of mitochondrial DNA, suggesting a role for this HSP70 in mitochondrial DNA replication or folding and/or assembly of mitochondrially encoded proteins.

■ Isolation

SSH1 was identified during a search of the yeast genome database using the yeast Ssa1 protein as a query (Schilke et al., 1996).

■ Ssh1 protein

The Ssh1 protein is encoded by an open reading frame on chromosome 12 (GenBank YSCL8039, accession number U19103) predicted to encode a 657 amino acid protein. The predicted sequence includes an N-terminal sequence with similarity to mitochondrial presequences. Ssh1p is imported into the mitochondrial matrix and is cleaved in the process. Ssh1p shares most identity (53%) with Ssc1p. It is 46 and 43% identical to Ssa1p and Kar2p of the cytosol and ER lumen, respectively (Schilke et al., 1996).

■ Function of Ssh1p

The function of Ssh1p is not known. Initial analysis revealed no defects in the translocation of preproteins from the cytosol nor in the synthesis of mitochondrially encoded proteins. However, the characterization of suppressor mutants has provided some clues (Schilke et al., 1996). Spontaneous suppressors of the cold-sensitive phenotype of an *SSH1* null mutation were obtained at high frequency. All were found to be respiratory deficient; 15/16 lacked mitochondrial DNA and the sixteenth had reduced amounts of DNA. Ssh1p may be involved in mitochondrial DNA replication or folding and assembly of mitochondrially encoded proteins.

■ Biological interactions

Overexpression of *SSC1* can suppress the cold-sensitive phenotype of an *ssh1* deletion mutant, suggesting that Ssc1p can carry out the functions of Ssh1p, which is normally expressed at much lower levels than Ssc1p. However, a fusion that expresses Ssh1p under control of the *SSC1* promoter does not suppress the phenotypes of an *ssc1* temperature-sensitive mutant (Schilke et al., 1996).

■ Regulation of *SSH1*

SSH1 appears to be expressed at very low levels, as indicated by analysis of codon usage. In addition, an *SSC1:lacZ* fusion is expressed at more than 100 times the level of an *SSH1:lacZ* fusion, as measured by β-galactosidase activity. β-Galactosidase activity is about twofold higher in cells grown on galactose compared with an easily fermented carbon source, such as glucose.

■ References

Schilke, B., Forster, J., Davis, J., James, P., Walter, W., Laloraya, S., Johnson, J., Miao, B. and Craig, E. (1996). Cold-sensitivity of a S. cerevisiae mutant lacking a newly identified mitochondrial Hsp70 is suppressed by loss of mitochondrial DNA. J. Cell Biol. **134**, 603–613.

■ Elizabeth Craig
Department of Biomolecular Chemistry
University of Wisconsin–Madison
1300 University Avenue
Madison, WI 53706, USA
Tel. 1 608 263 7105
Fax. 1 608 262 5253
E-mail: ecraig@facstaff.wisc.edu

Saccharomyces cerevisiae Kar2p

The KAR2 gene encodes the Saccharomyces cerevisiae homologue of mammalian BiP/Grp78, the major ER-resident member of the HSP70 family. In yeast, the gene product of KAR2 (Kar2p) is required for the translocation of proteins into the lumen of the ER, for facilitating protein folding, and for efficient nuclear fusion during mating.

■ Alternative names

Yeast BiP and *SSD1* (Craig et al., 1987; Normington et al., 1989; Rose et al., 1989; Nicholson et al., 1990).

■ *KAR2* isolation

The *KAR2* gene was isolated initially in a selection for karyogamy (nuclear fusion) defective mutants in yeast (Polaina and Conde, 1982; Rose et al., 1989). Subsequent to the original genetic identification, *KAR2* was identified independently by several laboratories using DNA probes homologous to HSP70 (Craig et al., 1987; Normington et al., 1989; Nicholson et al., 1990).

■ *KAR2* gene and sequence

The DNA sequence of *KAR2* (GenBank accession numbers M25064 and M25394; Normington et al., 1989; Rose et al., 1989) predicts a protein product of 682 amino acids (Normington et al., 1989; Rose et al., 1989; Nicholson et al., 1990). The amino acid sequence shows 50% identity to the *Escherichia coli* HSP70, DnaK, 63% identity to the *S. cerevisiae* cytosolic HSP70, Ssa1p, 25% identity to the recently identified second HSP70 family member in the yeast ER, Ssi1p (Baxter et al., 1996; Craven et al., 1996; see entry p. 36), and 67% identity to rat BiP (Normington et al., 1989; Rose et al., 1989; Nicholson et al., 1990). The N-terminus of Kar2p contains a signal sequence that is cleaved after residue 42 (Tokunaga et al., 1992) and a yeast ER retention signal (HDEL) is present at the extreme C-terminus (Hardwick et al., 1990). Kar2p lacks potential N-linked glycosylation sites. The 5′ flanking regions of both the *KAR2* and the cytoplasmic *SSA* genes contain multiple conserved heat shock elements (GAAnnTTC), while the gene encoding mammalian BiP lacks this motif (Normington et al., 1989; Rose et al., 1989).

■ *KAR2* regulation

Unlike its mammalian counterpart, *KAR2* is induced by two regulatory systems. Firstly, *KAR2* is induced by treatments that lead to the accumulation of proteins in the lumen of the ER (Normington et al., 1989; Rose et al., 1989). These include glycosylation inhibitors such as tunicamycin and specific secretory mutants (e.g. *sec53*, *sec18*, and *sec11*). Secondly, consistent with the presence of the conserved heat shock elements, *KAR2* mRNA is also independently induced by heat shock (Craig et al., 1987; Normington et al., 1989; Rose et al., 1989; Nicholson et al., 1990). In addition, *kar2* mutations lead to induction of *KAR2* mRNA, suggesting that the gene is regulated by the level of functional Kar2 protein (Scidmore et al., 1993). Two additional upstream regulatory regions have been defined that are required for the expression of *KAR2*. One is a small GC-rich region necessary for basal level expression and the second is a 22 bp sequence (the unfolded protein response element, UPRE) required for induction in response to protein accumulation in the ER (Mori et al., 1992; Kohno et al., 1993). A genetic screen for mutants in which this response is defective has yielded an ER membrane protein kinase, Ern1p/Ire1p, that might transduce a signal for misfolded protein accumulation (Cox et al., 1993; Mori et al., 1993).

■ Kar2 protein

Like other HSP70 proteins, Kar2p is a two domain protein containing an N-terminal ATPase domain and a C-terminal peptide-binding domain (see entry for mammalian BiP p. 59; Normington et al., 1989; Rose et al., 1989; Nicholson et al., 1990). Kar2p has been purified and demonstrates a low level of ATPase activity (Tokunaga et al., 1992; Brodsky et al., 1995). Polyclonal antisera raised against peptides and a C-terminal TrpE fusion protein detect a polypeptide of c. 78 kDa (Normington et al., 1989; Rose et al., 1989). In translocation-defective secretory mutants, a cytoplasmic precursor of 82 kDa accumulates, consistent with the presence of a cleaved signal sequence (Rose et al., 1989).

ER localization of Kar2p has been established by a variety of biochemical, cytological, and genetic techniques. Firstly, the wild-type protein is imported into a membranous compartment both *in vitro* and *in vivo* and *KAR2–lacZ* hybrid proteins are imported into the ER and glycosylated *in vivo* (Normington et al., 1989; Rose et al., 1989). Secondly, the protein exists in a perinuclear membranous compartment as determined by immunofluorescent microscopy (Rose et al., 1989), consistent with its localization in the yeast ER, and immunoelectron microscopy has confirmed that Kar2p is present in the same compartment as secretory precursor proteins (Preuss et al., 1991). Finally, the maturation of Kar2p is sensitive to mutations that block protein translocation (Rose et al., 1989).

Mutagenesis studies with *KAR2*

A temperature-sensitive mutant allele of *KAR2*, *kar2-1*, was first isolated in a selection for mutants defective for karyogamy (Polaina and Conde, 1982). Construction of deletion alleles established that *KAR2* is an essential gene (Normington et al., 1989; Rose et al., 1989; Nicholson et al., 1990). Several additional recessive temperature-sensitive alleles were then isolated by the plasmid shuffling technique using *in vitro* mutagenized *KAR2* plasmid (Vogel et al., 1990). Three of these recessive gene products were purified from yeast; one, Kar2-113p, exhibited wild-type levels of ATPase activity, one (Kar2-159p) was unable to bind to ATP-agarose, and a third (Kar2-203p) was unstable in yeast and present at only c. 28% of steady-state wild-type levels (Brodsky et al., 1995). Vogel and Rose have also isolated four dominant mutant alleles of *KAR2*, and we have now discovered that the purified gene products are severely defective for ATPase activity, and for the ability to bind to Sec63p *in vitro* (see below; A. McClellan, J. Vogel, M. Rose and J. Brodsky, unpublished data).

Mutant phenotypes

Temperature-sensitive alleles of *KAR2* exhibit defects in nuclear fusion *in vivo* and in homotypic ER fusion *in vitro* (Vogel et al., 1990; Latterich and Schekman, 1994), demonstrating that Kar2p is directly required for membrane fusion. Analysis of nuclear fusion in *kar2* mutants by electron microscopy reveals that while the nuclei are juxtaposed, nuclear membrane fusion does not occur (Kurihara et al., 1994). *kar2* mutants are placed into two classes based on their viability and effects on secretion at the restrictive temperature. Class I alleles (e.g. *kar2-159*) block translocation of secretory proteins into the lumen of the ER and exhibit a corresponding rapid loss in cell viability at the restrictive temperature (Vogel et al., 1990). Class II alleles (e.g. *kar2-1*) do not block secretion but instead show a variety of post-translocational defects and remain viable at the restrictive temperature (J. Vogel and M.D. Rose, unpublished data). Expression of the wild-type gene on a high copy number plasmid causes slow growth and accumulation of the Kar2p precursor (Rose et al., 1989). Deletion of the ER retention signal results in the secretion of Kar2p and a compensatory increase in *KAR2* transcription (Hardwick et al., 1990). Indeed, the viability of this mutant is dependent upon increased Kar2p expression (Beh and Rose, 1995). Deletion of larger segments of the carboxy-terminus can be tolerated (as many as 75 residues) but cells bearing these deletions grow normally only when the protein is overexpressed (Rose et al., 1989).

Role of Kar2p in translocation and protein folding

Kar2p has been shown by a variety of techniques to be required for the import of proteins into the ER. Cells depleted for Kar2p by repression of an inducible construct (*pGAL1–KAR2*) or the use of recessive, temperature-sensitive mutants (e.g. *kar2-159*) accumulate untranslocated precursor proteins (Vogel et al., 1990; Nguyen et al., 1991). Microsomes prepared from the class I mutants are defective for both the co- and post-translational import of secretory precursors (Sanders et al., 1992; Brodsky et al., 1995). Reconstitution of the defective microsomes in the presence of wild-type Kar2p can restore translocation competence (Brodsky et al., 1993), and addition of Kar2p to reconstituted vesicles in the absence of other luminal components stimulates translocation fourfold (Panzner et al., 1995). Finally, Kar2p can be cross-linked to a partially imported nascent chain during translocation (Sanders et al., 1992), suggesting that Kar2p may drive protein import by successive rounds of protein binding and release (Simon et al., 1992).

Karp2 may also play an additional role in an earlier stage in the translocation of post-translationally imported proteins (Sanders et al., 1992; Lyman and Schekman, 1995; Lyman and Schekman, 1997). For these secretory proteins, Kar2p is required for the ATP-dependent transfer of the precursors from a complex comprised of Sec62p, Sec71p and Sec72p to the translocation pore protein Sec61p (Lyman and Schekman, 1997).

Because of its involvement in the assembly and targeting of heterologous and mutant proteins (Schonberger et al., 1991; Te Heesen and Aebi, 1994), Kar2p has been implicated in the process of protein folding in the ER. Confirmation of this hypothesis has come from studies on the folding of the vacuolar protease, carboxypeptidase Y (CPY), in different *kar2* mutants (Simons et al., 1995). Wild-type CPY aggregates and remains in the ER in strains containing class I *kar2* alleles upon a temperature shift to 34°C.

Biological and genetic interactions

Both biochemical and genetic data indicate that Kar2p interacts specifically with the Sec63 protein. Kar2p copurifies with a solubilized ER membrane fraction that contains Sec63p, Sec71p, and Sec72p, but is released from the complex in the presence of ATPγS, or if the complex is prepared from a strain containing the temperature-sensitive *sec63-1* allele (Brodsky and Schekman, 1993). Furthermore, the *kar2-159* mutation is synthetically lethal with the *sec63-1* allele and several dominant alleles of *KAR2* suppress the *sec63-1* mutation (Scidmore et al., 1993). Unlike the other translocation-defective mutations, *sec63-1* causes *KAR2* mRNA to be induced to high levels at the non-permissive temperature (Scidmore et al., 1993). Together, these results are particularly intriguing because Kar2p is c. 50% identical to the bacterial DnaK protein and a luminal segment of Sec63p, with which Kar2p interacts, is 43% identical to the bacterial DnaJ protein (Sadler et al., 1989; Feldheim et al., 1992); DnaK and DnaJ act as a cochaperones in *E. coli* during multiple cellular processes (reviewed by Frydman and Hartl, 1994). Thus, the Sec63p–BiP complex may be part of a translocation engine that drives protein import (see Sec63p entry, p.108 and Brodsky, 1996).

The DnaJ domain from Sec63p may be replaced with the corresponding domain from another luminal DnaJ homologue, Scj1p, and a *scj1* disruption and the *kar2-159* mutation are synthetically lethal (Schlenstedt et al., 1995). This result suggests that Kar2p may also normally interact with Scj1p in the ER lumen.

References

Baxter, B.K., James, P., Evans, T., and Craig, E.A. *SSI1*, which encodes a novel Hsp70 homolog of the *S. cerevisiae* endoplasmic reticulum, exhibits genetic interactions with *KAR2*. submitted.

Beh, C.T., and Rose, M.D. (1995). Two redundant systems maintain levels of resident proteins within the yeast endoplasmic reticulum. Proc. Natl. Acad. Sci. USA **92**, 9820–9823.

Brosky, J.L. (1996). Post-translational protein translocation: not all hsc70s are created equal. Trends Biochem. Sci. **21**, 122–126.

Brodsky, J.L. and Schekman, R. (1993). A Sec63p-BiP complex from yeast is required for protein translocation in a reconstituted proteoliposome. J. Cell Biol. **123**, 1355–1363.

Brodsky, J.L., Hamamoto, S., Feldheim, D., and Schekman, R. (1993). Reconstitution of protein translocation from solubilized yeast membranes reveals topologically distinct roles for BiP and cytosolic hsc70. J. Cell Biol. **120**, 95–102.

Brodsky, J.L., Goeckeler, J., and Schekman, R. (1995). Sec63p and BiP are required for both co- and post-translational protein translocation into yeast microsomes. Proc. Natl. Acad. Sci. USA **92**, 9643–9646.

Cox, J.S., Shamu, C.E., and Walter, P. (1993). Transcriptional induction of genes encoding endoplasmic reticulum resident proteins requires a transmembrane protein kinase. Cell **73**, 1197–1206.

Craig, E.A., Kramer, J., and Kosic-Smithers, J. (1987). *SSC1*, a member of the 70-kDa heat shock protein multigene family of *Saccharomyces cerevisiae*, is essential for growth. Proc. Natl. Acad. Sci. USA **84**, 4156–4160.

Craven, R.A., Egerton, M., and Stirling, C.J. (1996). A novel Hsp70 of the yeast ER lumen is required for the efficient translocation of a number of protein precursors. EMBO J. **15**, 2640–2650.

Feldheim, D., Rothblatt, J., and Schekman, R. (1992). Topology and functional domains of Sec63p, an endoplasmic reticulum membrane protein required for secretory protein translocation. Mol. Cell. Biol. **12**, 3288–3296.

Frydman, J. and Hartl, F.-U. (1994). Molecular chaperone functions of hsp70 and hsp60 in protein folding, in The biology of heat shock proteins and molecular chaperones (Morimoto, R.I., Tessieres, A. and Georgopoulos, C., eds). Cold Spring Harbor Laboratory Press, New York, pp. 251–284.

Hardwick, K.G., Lewis, M.J., Semenza, J., Dean, N., and Pelhman, H.R.B. (1990). *ERD1*, a yeast gene required for the retention of luminal endoplasmic reticulum proteins, affects glycoprotein processing at the Golgi apparatus. EMBO J. **9**, 623–630.

Kohno, K., Normington, K., Sambrook, J., Gething, M.J., and Mori, K. (1993). The promoter region of the yeast *KAR2* (BiP) gene contains a regulatory domain that responds to the presence of unfolded proteins in the endoplasmic reticulum. Mol. Cell. Biol. **13**, 877–890.

Kurihara, L.J., Beh, C.T., Latterich, M., Schekman, R., and Rose, M.D. (1994). Nuclear congression and membrane fusion: two distinct pathways in the yeast karyogamy pathway. J. Cell Biol. **126**, 911–923.

Latterich, M. and Schekman, R. (1994). The karyogamy gene *KAR2* and novel proteins are required for ER-membrane fusion. Cell **78**, 87–98.

Lyman, S.K. and Schekman, R. (1995). Interaction between BiP and Sec63p is required for the completion of protein translocation into the ER of *Saccharomyces cerevisiae*. J. Cell Biol. **131**, 1163–1171.

Lyman, S.K. and Schekman, R. (1997). Binding of secretory precursor polypeptides to a translocon subcomplex is regulated by BiP. Cell **88**, 85–96.

Mori, K., Sant, A., Kohno, K., Normington, K., Gething, M.J., and Sambrook, J. (1992). A 22 bp cis-acting element is necessary and sufficient for the induction of the yeast *KAR2* (BiP) gene by unfolded proteins. EMBO J. **7**, 2583–2593.

Mori, K., Ma, W., Gething, M.J., and Sambrook, J. (1993). A transmembrane protein with a cdc2+/CDC28-related kinase activity is required for signaling from the ER to the nucleus. Cell **74**, 743–756.

Nguyen, T.H., Law, D.T.S., and Williams, D.S. (1991). Binding protein BiP is required for translocation of secretory proteins into the endoplasmic reticulum in *Saccharomyces cerevisiae*. Proc. Natl. Acad. Sci. USA **88**, 1565–1569.

Nicholson, R.C., Williams, D.B., and Moran, L.A. (1990). An essential member of the hsp70 family of *Saccharomyces cerevisiae* is homologous to immunoglobulin heavy chain binding protein. Proc. Natl. Acad. Sci. USA **86**, 1159–1163.

Normington, K., Kohno, K., Kozutsumi, Y., Gething, M.J., and Sambrook, J. (1989). *S. cerevisiae* encodes an essential protein in sequence and function to mammalian BiP. Cell **57**, 1223–1236.

Panzner, S., Dreier, L., Hartmann, E., Kostka, S., and Rapoport, T.A. (1995). Posttranslational protein translocation in yeast reconstituted with a purified complex of Sec proteins and Kar2p. Cell **81**, 561–570.

Polaina, J. and Conde, J. (1982). Genes involved in the control of nuclear fusion during the sexual cycle of *Saccharomyces cerevisiae*. Mol. Gen. Genet. **186**, 253–258.

Preuss, D., Mulholland, J., Kaiser, C.A., Orlean, P., Albright, C., Rose, M.D., Robbins, P.W., and Botstein, D. (1991). Structure of the yeast endoplasmic reticulum: localization of ER proteins using immunofluorescence and immunoelectron microscopy. Yeast **7**, 891–911.

Rose, M.D., Misra, L.M., and Vogel, J.P. (1989). *KAR2*, a karyogamy gene, is the yeast homolog of the mammalian BiP/GRP78 gene. Cell **57**, 1211–1221.

Sadler, I., Chiang, A., Kurihara, T., Rothblatt, J., Way, J., and Silver, P. (1989). A yeast gene important for protein assembly into the endoplasmic reticulum and the nucleus has homology to dnaJ, an *Escherichia coli* heat shock protein. J. Cell Biol. **109**, 2665–2675.

Sanders, S.L., Whitfield, K., Vogel, J.P., Rose, M.D., and Schekman, R.W. (1992). Sec61p and BiP directly facilitate polypeptide translocation into the ER. Cell **69**, 353–365.

Schlenstedt, G., Harris, S., Risse, B., Lill, R., and Silver, P. (1995). A yeast homolog, Scj1p, can function in the endoplasmic reticulum with BiP/Kar2p via a conserved domain that specifies interactions with hsp70s. J. Cell Biol. **129**, 979–988.

Schonberger, O., Hirst, T.R., and Pines, O. (1991). Targeting and assembly of an oligomeric bacterial enterotoxin in the endoplasmic reticulum of *Saccharomyces cerevisiae*. Mol. Microbiol. **5**, 2663–2671.

Scidmore, M., Okamura, H.H., and Rose, M.D. (1993). Genetic interactions between *KAR2* and *SEC63*, encoding eukaryotic homologues of DnaK and DnaJ in the endoplasmic reticulum. Mol. Biol. Cell **4**, 1145–1159.

Simon, S.M., Peskin, C.S., and Oster, G.F. (1992). What drives the translocation of proteins? Proc. Natl. Acad. Sci. USA **89**, 3770–3774.

Simons, J.F., Ferro-Novick, S., Rose, M.D., and Helenius, A. (1995). BiP/Kar2p serves as a molecular chaperone during carboxypeptidase Y folding in yeast. J. Cell Biol. **130**, 41–49.

Te Heesen, S. and Aebi, M. (1994). The genetic interaction of *kar2* and *wbp1* mutations. Eur. J. Biochem. **222**, 631–637.

Tokunaga, M., Kawamura, A., and Kohno, K. (1992). Purification and characterization of BiP/Kar2 protein from *Saccharomyces cerevisiae*. J. Biol. Chem. **267**, 17553–17559.

Vogel, J.P., Misra, L.M., and Rose, M.D. (1990). Loss of BiP/GRP78 function blocks translocation of secretory proteins in yeast. J. Cell Biol. **110**, 1885–1895.

■ Jeffrey L. Brodsky:
Department of Biological Sciences
University of Pittsburgh
Pittsburgh, PA 15260, USA
Tel. 1 412 624 4831
Fax. 1 412 624 4759
E-mail: jbrodsky+@pitt.edu

■ Mark D. Rose:
Department of Molecular Biology
Princeton University
Princeton, NJ 08544-1014, USA
Tel. 1 609 258 2804
Fax. 1 609 258 6175
E-mail: mrose@molecular.princeton.edu

Saccharomyces cerevisiae Ssi1p

SSI1 encodes a 100-kDa homolog of the HSP70 proteins which localizes to the endoplasmic reticulum. While not essential for growth under normal conditions, Ssi1p is required for normal translocation of some cytoplasmic precursors into the ER and is essential for viability in the presence of certain mutant alleles of KAR2.

■ Alternative name

Lhs1p (Craven *et al.*, 1996); Cer1p (Hamilton and Flynn, 1996).

■ Isolation

SSI1 was sequenced as part of the yeast genome sequencing project, and identified by its homology of its deduced gene product to known HSP70 proteins (Rasmussen, 1994; Craven *et al.*, 1996; Baxter *et al.*, 1996).

■ *SSI1* gene and sequence

The *SSI1* gene (GenBank accession number Z28073) has an open reading frame of 2643 nucleotides, potentially encoding a protein of 881 amino acids with a predicted size of 99.6 kDa. The putative Ssi1p sequence includes the C-terminal motif HDEL, the ER-retention signal of *S. cerevisiae*; an N-terminal signal sequence with a predicted ER cleavage site after the first 20 amino acids; and eight potential sites of N-linked ER glycosylation. Analysis of the codon usage of *SSI1* gives a codon adaptation index of 0.147 on a scale of 0 to 1, suggesting a low level of expression.

Ssi1p shares similar levels of identity (approximately 25%) with yeast HSP70 proteins of various families, including Ssa1p, Ssb1p, Ssc1p, Sse1p, and Kar2p. Alignment of Ssi1p and Kar2p requires the introduction of six gaps in the Kar2p ATPase domain of four amino acids or more; these gaps all correspond to locations within one residue of turn or loop regions in the three-dimensional structure of bovine Hsc70 (Flaherty *et al.*, 1990). This suggests structural conservation of the ATPase domain between Kar2p and Ssi1p (Baxter *et al.*, 1996).

■ Ssi1 protein

Ssi1p is localized to the endoplasmic reticulum, as evidenced by protease protection and cell fractionation experiments. Ssi1p receives N-linked glycosylation, and its maturation is inhibited in strains deficient in ER translocation. (Craven *et al.*, 1996; Baxter *et al.*, 1996).

■ Mutagenesis studies with *SSI1*

Strains carrying a deletion of *SSI1* are viable and grow well under normal growth conditions, although slight cold sensitivity and an enhanced resistance to manganese have been observed (Baxter *et al.*, 1996).

■ Genetic interactions

SSI1 shows a complex set of interactions with *KAR2*, demonstrating divergent phenotypes which are separable by *kar2* mutant class (Brodsky and Rose, this volume p. 33; M.D. Rose, J.P. Vogel, and M.A. Scidmore, personal communication). Deletion of *SSI1* is synthetically lethal in combination with class I alleles of *KAR2*: *kar2-159*, or *kar2-113* (Craven *et al.*, 1996; Baxter *et al.*, submitted). On the other hand, growth of strains carrying a class III allele of *KAR2*, *kar2-191*, is improved by deletion of *SSI1*. Class II *KAR2* alleles show an intermediate phenotype:

growth of kar2-1 or kar2-133 strains under normal growth conditions is furthered impaired by deletion of SSI1 (Baxter et al., 1996).

■ Biological activities of Ssi1p

Strains carrying a deletion of SSI1 show impaired translocation of some cytosolic precursors into the endoplasmic reticulum, including the mating pheromone alpha factor (Craven et al., 1996; Baxter et al., 1996). This defect can be suppressed by overexpression of SCj1p (Hamilton and Flynn, 1996). In addition, Ssi1p has been implicated in repair of heat-denatured proteins. The yeast endoplasmic reticulum harbors an ATP-dependent heat-resistant machinery, which is able to refold proteins denatured by thermal insult to the cells (Jamsa et al., 1995). In the absence of Ssi1p, an in vivo heat-denatured and aggregated reporter enzyme failed to be reactivated and solubilized, as it is in wild-type cells, and was slowly degraded (Vakula et al., 1996).

■ References

Baxter, B.K., James, P., Evans, T., and Craig, E.A. (1996). SSI1 encodes a novel Hsp70 of the Saccharomyces cerevisiae endoplasmic reticulum. Mol. Cell Biol. **16**, 6444–6456.

Craven, R.A., Egerton, M., and Stirling, C.J. (1996). A novel Hsp70 of the yeast ER lumen is required for the efficient translocation of a number of protein precursors. EMBO J. **15**, 2640–2650.

Flaherty, K.M., DeLuca-Flaherty, C., and McKay, D.B. (1990). Three dimensional structure of the ATPase fragment of a 70K heat-shock cognate protein. Nature **346**, 623–628.

Hamilton, T.G., and Flynn, G.C. (1996). Cer1p, a novel hsp70-related protein required for post-translational endoplasmic reticulum translocation in yeast. J. Biol. Chem. **271**, 30610–30613.

Jamsa, E., Vakula, N., Arffman, A., Kilpelainen, I., and Makarow, M. (1995). In vivo reactivation of heat-denatured protein in the endoplasmic reticulum of yeast. EMBO J. **14**, 6028–6033.

Rasmussen, S.W. (1994). Sequence of a 20.7 kb region of yeast chromosome XI includes the NUP100 gene, an open reading frame (ORF) possibly representing a nucleoside diphosphate kinase gene, tRNAs for His, Val and Trp in addition to seven ORFs with weak or no significant similarity to known proteins. Yeast **10**, S69–S74.

Vakula, N., Holkeri, H., and Makarow, M. (1996). Repair of heat-denatured proteins in the yeast endoplasmic reticulum requires a novel hsp70 homolog. In Molecular chaperones and the heat shock response. Cold Spring Harbor Laboratory, New York, p. 229.

■ *Bonnie K. Baxter and Elizabeth A. Craig:*
Department of Biomolecular Chemistry
University of Wisconsin–Madison
1300 University Avenue
Madison, WI 53706, USA.
Tel. 1 608 263 7105
Fax. 1 608 262 5253
E-mail: ecraig@facstaff.wisc.edu

Schizosaccharomyces pombe BiP

BiP from the fission yeast Schizosaccharomyces pombe is approximately equally diverged in sequence from the BiP proteins of mammalian cells and of the budding yeast Saccharomyces cerevisiae. The protein is localized to the endoplasmic reticulum and contains the C-terminal ER retention sequence ADEL. Unusually among BiP proteins, approximately 10% of the protein is cotranslationally N-glycosylated. The functional significance of this modification is unknown. No direct studies of the protein's role as a chaperone have yet been reported; however, gene disruption shows it to be essential for viability.

■ Isolation

A homologue of the BiP/Kar2p proteins of mammalian cells and budding yeast was identified in the fission yeast Schizosaccharomyces pombe (Pidoux and Armstrong, 1992). A portion of the gene was recovered by polymerase chain reaction using primers designed from sequences conserved between mammalian and S. cerevisiae BiP. This fragment was used to isolate the complete gene by hybridization. A further S. pombe HSP70 homologue, which is probably located in the endoplasmic reticulum, has been identified from genome sequencing (protein accession number e212000) but no studies of this gene have been reported.

■ *Schizosaccharomyces pombe bip* gene and sequence

The gene encodes, without introns, a protein whose predicted molecular weight is 73 kDa. The protein is 67% identical to S. cerevisiae BiP/Kar2p and 66% identical to rat BiP (Pidoux and Armstrong, 1992). It has a predicted site for signal peptide cleavage between residues 24 and 25, a predicted site of N-glycosylation at residue 29, and a C-terminal sequence ADEL. Two sequences with homology to a heat shock element, TTCTGGAA and TTCTGGTA, lie upstream of the coding sequence. The gene sequence is available from Genbank/EMBL with accession number X64416.

From sequencing of the *S. pombe* genome, a second gene has been discovered. The encoded protein has a predicted signal peptide and a C-terminal sequence of SDEL; however the remainder of the protein shows only approximately 10–12% identity to members of the HSP70 family, and no specific homologue exists in the *S. cerevisiae* genome. Thus far no studies of this unusual gene have been reported. The gene sequence is available as part of cosmid SPAC1F5, Genbank/EMBL accession number Z68136. The protein sequence is available as Swissprot Q10061.

■ *S. pombe* BiP protein

The *S. pombe* BiP protein has been studied both as an epitope-tagged form (Pidoux and Armstrong, 1992) and using rabbit antibodies raised to a C-terminal portion of the protein expressed in bacteria (Pidoux and Armstrong, 1993), giving consistent results. By immunofluorescence, the protein is localized to the nuclear envelope and filaments connected to a tubular reticulum, which together constitute the endoplasmic reticulum (ER) of *S. pombe*. Three-dimensional reconstruction from confocal sections (M. Shipman, N. Bone and J. Armstrong, unpublished) shows that the reticulum underlies the plasma membrane but is separate from it. Electrophoretic analysis showed that the protein migrates as a doublet of species with apparent molecular masses of 75 and 80 kDa, the latter representing approximately 10% of the total. This larger band is N-glycosylated, an unusual finding among BiP proteins. The proportion of the larger species does not change with time, suggesting that glycosylation probably occurs cotranslationally. The oligosaccharide side chain is attached at a site near the cleaved N-terminus (Pidoux and Armstrong, 1993). It is not known if glycosylation has any function, or even if the glycosylated protein is functional.

The C-terminal sequence ADEL, when attached to another secretory protein, causes it to be retained in the ER (Pidoux and Armstrong, 1992).

■ Biological regulation

The *bip* gene from *S. pombe* is induced by similar stresses as in other yeasts (Pidoux and Armstrong, 1992). Temperature shift from 30 to 39°C, or incubation with tunicamycin, 2-deoxyglucose or the calcium ionophore A23187 all result in the appearance of an additional mRNA which is smaller than the constitutive mRNA but of unknown relationship to it. All of these treatments may cause the accumulation of misfolded or insoluble proteins in the ER.

■ Mutagenesis studies with *S. pombe bip*

The *bip* gene was disrupted in a diploid *S. pombe* by homologous integration of DNA in which most of the coding sequence was replaced with a selectable marker gene (Pidoux and Armstrong, 1992). After sporulation, no haploids were recovered that contained the marker gene. Therefore the BiP gene appears to be essential for viability.

■ References

Pidoux, A. and Armstrong, J. (1992) Analysis of the BiP gene and identification of an ER retention signal in *Schizosaccharomyces pombe*. EMBO J. **11**, 1583–1591.

Pidoux, A. and Armstrong, J. (1993) The BiP protein and the endoplasmic reticulum of *Schizosaccharomyces pombe*: fate of the nuclear envelope during cell division. J. Cell Sci. **105**, 1115–1120.

■ John Armstrong:
School of Biological Sciences
University of Sussex
Falmer, UK
Tel. 44 1273 678576
Fax. 44 1273 678433
E-mail :J.Armstrong@sussex.ac.uk

Plant BiP Proteins

Plant BiP is the ER-localized member of the HSP70 family of proteins and is encoded by a multigene family in some, but not all plant species. BiP is involved in such vital functions as protein translocation, assembly of proteins into oligomeric structures and suppression of protein aggregation.

■ Alternative names

b-70 (Galante *et al.*, 1983), luminal binding protein (Denecke *et al.*, 1991).

■ Isolation

Plant BiP was originally purified from maize kernels using procedures that employed ATP-agarose affinity chromatography (Fontes *et al.*, 1991; Marocco *et al.*, 1991). Analysis of affinity purified protein by two-dimensional gel electrophoresis revealed the presence of three polypeptides (Fontes *et al.*, 1991). However, only two of these cross-react with anti-HDEL antibody (Boston *et al.*, 1991). The N-terminal amino acid sequences of the two maize BiP isoforms are identical and show a high degree of similarity with mammalian Grp78/BiP (Marocco *et al.*, 1991). Antibodies produced against the maize polypeptides were used to isolate a corresponding cDNA clone (Fontes et al., 1991).

■ Plant *bip* gene and sequence

bip cDNAs containing the entire coding region have been cloned from a variety of plant species including *Arabidopsis* (DDBJ accession number D84414), maize (GenBank accession numbers M59449, U58208 and U58209), soybean (accession numbers U08384 and U08383), spinach (L23551), tobacco (X60057 and X60058), and tomato (L08830). One unique feature of plants is that in some species, BiP is encoded by a small multigene family (Denecke et al., 1991; Kalinski et al., 1995; R.L. Wrobel and R.S. Boston, unpublished results). However, little is known regarding the expression or the function(s) of different family members.

Intraspecies comparisons of amino acid sequences deduced from the nucleotide sequences of the cDNAs yielded identities of 99% (maize), 96.7% (tobacco), and 92.7% (soybean), while interspecies comparisons revealed amino acid identities between 87 and 93%. Comparisons of plant BiP with either Kar2p (the yeast BiP homolog) or Chinese hamster BiP show amino acid identities of 62 and 70%, respectively. All plant BiPs possess an N-terminal signal sequence of variable length and the C-terminal tetrapeptide (His–Asp–Glu–Leu or HDEL) that is apparently required for retrieval of BiP that escapes the ER (Pelham, 1989).

■ Plant BiP protein

Plant BiP proteins are members of the HSP70 (stress-70) family of proteins having apparent molecular masses between 75 and 80 kDa (Fontes et al., 1991; Denecke et al., 1991; Giorini and Galili, 1991; D'Amico et al., 1992; Anderson et al., 1994b). Labeling of maize proteins *in vivo* and *in vitro* clearly shows that only one of the two isoforms is capable of being phosphorylated and ADP-ribosylated (Fontes et al., 1991). The role these modifications play in protein function has not yet been determined. Plant BiPs, like mammalian BiP (Freiden et al., 1992; Blond-Elguindi et al., 1993), can be found in higher order structures. Anderson et al. (1994b) reported that spinach BiP purified by ATP-agarose affinity chromatography exists in several higher order complexes. Furthermore, they found that the addition of ATP alters the migration of dimeric and large oligomeric forms during rechromatography on a gel filtration column (Anderson et al., 1994b). Miernyk and Hayman (1996) also observed monomeric, dimeric, and oligomeric forms of BiP. Higher order complexes of BiP and prolamine storage proteins are observed in rice (Li et al., 1993), but it is not clear whether these complexes represent true oligomeric BiP complexes, multiple monomeric forms of BiP bound to a single prolamine polypeptide, BiP associated with other ER-localized chaperones and a prolamine polypeptide, or a combination of these possibilities.

BiP has been shown to be localized to the ER or ER-derived structures by a variety of subcellular fractionation techniques (Galante et al., 1983; Denecke et al., 1991; Fontes et al., 1991; Giorini and Galili, 1991; Marocco et al., 1991; D'Amico et al., 1992; Li et al., 1993). Immunocytochemical studies show that BiP is localized to the nuclear envelope and to the ER of germinating tobacco seeds (Denecke et al., 1995). In addition to being immunolocalized to the ER, BiP is observed in association with ER-derived protein bodies present in the developing endosperm of maize and rice (Zhang and Boston, 1992; X. Li, T. Okita, and R.S.Boston, unpublished results) and a subset of protein bodies in wheat (Rubin et al., 1992).

■ Biological activities of plant BiPs

Plant BiPs, like their mammalian counterparts (Gething and Sambrook, 1992), have an ATPase activity that can be stimulated by the addition of short peptides, but the degree of stimulation is peptide dependent (Zhou and Miernyk, 1997; Miernyk 1997). Plant BiP interacts transiently with normal proteins entering the ER and enters into a more stable interaction with proteins that cannot attain their proper tertiary or quaternary structures (Gillikin et al., 1995; Li et al., 1993; Pedrazzini et al., 1994; Vitale et al., 1995). The interaction of BiP with these polypeptides is typically disrupted with ATP, but not the non-hydrolysable analog ATPγS or GTP (Denecke and Vitale, 1995; Gillikin et al., 1995; D'Amico et al., 1992; Li et al., 1993). The inability of ATPγS or GTP to disrupt BiP–protein interactions implies a role for ATP hydrolysis in the dissociation of BiP from its bound polypeptide. Interestingly, in a rice protein body fraction, BiP apparently exists in three distinct pools that differ with respect to their location within the developing protein body and their sensitivity to ATP (Li et al., 1993). BiP is observed in ATP-sensitive complexes with mature prolamines at the protein body surface and lumen and in ATP-insensitive complexes with polysomal nascent polypeptide chains (Li et al., 1993). The presence of ATP-insensitive complexes may imply role(s) for accessory proteins that are necessary for ATP-dependent dissociation of complexes.

■ Biological regulation of plant BiPs

Plant BiPs are expressed constitutively in plant tissues, but the level of expression varies depending upon tissue type (Anderson et al., 1994a; Denecke et al., 1995). Plant BiP is induced by tunicamycin and the proline analog, L-azetidine-2-carboxylic acid, pharmacological agents that are thought to affect protein folding and assembly in the ER (Denecke et al., 1991; Fontes et al., 1991; D'Amico et al., 1992, R.L. Wrobel and R.S. Boston, unpublished results). Several maize endosperm mutants express elevated levels of BiP (Galante et al., 1983; Boston et al., 1991; Fontes et al., 1991; Marocco et al., 1991). The elevated levels of BiP observed in these mutants are thought to result from the exposure of non-native polypeptide regions in the aqueous environment of the

ER lumen. Plant BiP genes are responsive to temporal control and environmental stimuli such as cold acclimatization, heat shock, and wounding (Anderson et al., 1994a; Denecke et al., 1995; Kalinski et al., 1995; Koizumi, 1996). Heat shock treatments resulting in increased, decreased, and unchanged levels of BiP mRNA have all been reported (Denecke et al., 1991, 1995; Anderson et al., 1994a). These conflicting reports may reflect differences in heat shock treatments or in the steady state level of BiP mRNAs in different tissues at the time of treatment. Although cold acclimatization, heat shock, and wounding appear to affect RNA levels, the protein levels are not dramatically affected by these treatments (Anderson et al., 1994a).

■ Biological or genetic interactions

Plant BiP has been implicated in a variety of vital functions that include protein translocation, prevention of proteins from entering an aggregation pathway, and assembly of proteins into oligomeric structures (Boston et al., 1996; Pedrazzini and Vitale, 1996). Recent work clearly demonstrated the interaction of BiP with nascent prolamine polypeptides in rice as well as with completely synthesized phaseolin in common bean and β-conglycinin in soybean (Gillikin et al., 1995; Li et al., 1993; Vitale et al., 1995). BiP is also found in association with underglycosylated proteins and proteins that cannot obtain their proper quaternary structures (D'Amico et al., 1992; Pedrazzini et al., 1994). Recently, Vitale et al. (1995) reported that BiP only interacts with the monomeric form of phaseolin, but does not interact with the mature trimeric form of the protein. These recent findings suggest that plant BiP performs many of the same functions as its mammalian counterpart (Gething and Sambrook, 1992), but leaves open questions regarding the significance of the BiP multigene families observed in some plant species.

■ References

Anderson, J.V., Li, Q.-B., Haskell, D.W., and Guy, C.L. (1994a). Structural organization of the spinach endoplasmic reticulum-luminal 70-kilodalton heat-shock cognate gene and expression of 70-kilodalton heat-shock genes during cold acclimation. Plant Physiol. **104**, 1359–1370.

Anderson, J.V., Haskell, D.W., and Guy, C.L. (1994b). Differential influence of ATP on native spinach 70-kilodalton heat-shock cognates. Plant Physiol. **104**, 1371–1380.

Blond-Elguindi, S., Fourie, A.M., Sambrook, J.F., and Gething, M.J.H. (1993). Peptide-dependent stimulation of the ATPase activity of the molecular chaperone BiP is the result of conversion of oligomers to active monomers. J. Biol. Chem. **268**, 12730–12735.

Boston, R.S., Fontes, E.B.P., Shank, B.B., and Wrobel, R.L. (1991). Increased expression of the maize immunoglobulin binding protein homolog b-70 in three zein regulatory mutants. Plant Cell **3**, 497–505.

Boston, R.S., Viitanen, P.V., and Vierling, E. (1996). Molecular chaperones and protein folding in plants. Plant Mol. Biol. **32**, 191–222.

D'Amico, L., Valsasina, B., Daminati, M.G., Fabbrini, M.S., Nitti, G., Bollini, R., Ceriotti, A., and Vitale, A. (1992). Bean homologs of the mammalian glucose-regulated proteins: induction by tunicamycin and interaction with newly synthesized seed storage proteins in the endoplasmic reticulum. Plant J. **2**, 443–455.

Denecke, J. and Vitale, A. (1995). The use of protoplasts to study protein synthesis and transport by the plant endomembrane system. In "Methods in Cell Biology," Vol. 50B, (D.W. Galbraith, D.P. Bourque, and H.J. Bohnert), Academic Press, San Diego, pp. 335–348.

Denecke, J., Goldman, M.H.S., Demolder, J., Seurinck, J., and Botterman, J. (1991). The tobacco luminal binding protein is encoded by a multigene family. Plant Cell **3**, 1025–1035.

Denecke, J., Carlsson, L.E., Vidal, S., Höglund, A.-S., Ek, B., van Zeijl, M.J., Sinjorgo, K.M.C., and Palva, E.T. (1995). The tobacco homolog of mammalian calreticulin is present in protein complexes in vivo. Plant Cell **7**, 391–406.

Fontes, E.B.P., Shank, B.B., Wrobel, R.L., Moose, S.P., OBrian, G.R., Wurtzel, E.T., and Boston, R.S. (1991). Characterization of an immunoglobulin binding protein homolog in the maize *floury-2* endosperm mutant. Plant Cell **3**, 483–496.

Freiden, P.J., Gaut, J.R., and Hendershot, L.M. (1992). Interconversion of three differentially modified and assembled forms of BiP. EMBO J. **11**, 63–70.

Galante, E., Vitale, A., Manzocchi, L., Soave, C., and Salamini, F. (1983). Genetic control of a membrane component and zein deposition in maize endosperm. Mol. Gen. Genet. **192**, 316–321.

Gething, M.J. and Sambrook, J. (1992). Protein folding in the cell. Nature **355**, 33–44.

Gillikin, J.W., Fontes, E.P.B. and Boston, R.S. (1995). Protein-protein interactions within the endoplasmic reticulum. In "Methods in Cell Biology," Vol. 50B, (D.W. Galbraith, D.P. Bourque, and H.J. Bohnert), Academic Press, San Diego, pp. 309–323.

Giorini, S. and Galili, G. (1991). Characterization of HSP-70 cognate proteins from wheat. Theor. Appl. Genet. **82**, 615–620.

Kalinski, A., Rowley, D.L., Loer, D.S., Foley, C., Buta, G., and Herman, E.M. (1995). Binding-protein expression is subject to temporal, developmental and stress-induced regulation in terminally differentiated soybean organs. Planta **195**, 611–621.

Koizumi, N. (1996). Isolation and responses to stress of a gene that encodes a luminal binding protein in *Arabidopsis thaliana*. Plant Cell Physiol. **37**, 862–865.

Li, X., Wu, Y., Zhang, D.Z., Gillikin, J.W., Boston, R.S., Franceschi, V.R., and Okita, T.W. (1993). Rice prolamine protein body biogenesis: a BiP-mediated process. Science **262**, 1054–1056.

Marocco, A., Santucci, A., Cerioli, S., Motto, M., DiFonzo, N., Thompson, R., and Salamini, F. (1991). Three high-lysine mutations control the level of ATP-binding hsp70-like proteins in the maize endosperm. Plant Cell **3**, 507–515.

Miernyk, J.A. (1997). The 70 kDa stress-related proteins as molecular chaperones. Trends in Plant Science **2**, 180–187.

Miernyk, J.A. and Hayman, G.T. (1996). ATPase activity and molecular chaperone function of the stress70 proteins. Plant Physiol. **110**, 419–424..

Pedrazzini, E. and Vitale, A. (1996). The binding protein, BiP, and the synthesis of secretory proteins. Plant Physiol. Biochem. **34**, 207–216.

Pedrazzini, E., Giovinazzo, G., Bollini, R., Ceriotti, A., and Vitale, A. (1994). Binding of BiP to an assembly-defective protein in plant cells. Plant J. **5**, 103–110.

Pelham, H.R.B. (1989). Control of protein exit from the endoplasmic reticulum. Annu. Rev. Cell Biol. **5**, 1–23.

Rubin, R., Levanony, H., and Galili, G. (1992). Evidence for the presence of two different types of protein bodies in wheat endosperm. Plant Physiol. **99**, 718–724.

Vitale, A., Bielli, A., and Ceriotti, A. (1995). The binding protein associates with monomeric phaseolin. Plant Physiol. **107**, 1411–1418.

Zhang, F. and Boston, R.S. (1992). Increases in binding protein (BiP) accompany changes in protein body morphology in three high-lysine mutants of maize. Protoplasma **171**, 142–152.

Zhou, R. and Miernyk, J.A. (1997). ATPase activities of the maize stress70 molecular chaperone proteins. J. Biol. Chem. in press.

■ Jeffrey W. Gillikin and Rebecca S. Boston:
Department of Botany
North Carolina State University
Raleigh, NC 27695-7612, USA.
Tel. 1 919 515 3570 (J. W. G.)
Tel. 1 919 515 3390 or 2727 (R. S. B.)
Fax. 1 919 515 3436
E-mail: cornbip@unity.ncsu.edu and boston@unity.ncsu.edu

Trypanosoma brucei BiP

TbBiP is the Trypanosoma brucei homologue of the eukaryotic HSP70 protein BiP, and has been thoroughly characterized as a soluble luminal ER resident in this parasitic protozoan. Trypanosomes are the causative agent of African Sleeping Sickness and are among the most ancient of all eukaryotic lineages. TbBiP transiently associates with newly synthesized variant surface glycoprotein (VSG), the major surface antigen of the bloodstream stage of the parasite. The C-terminal tetrapeptide sequence, MDDL, functions as an ER localization signal.

■ *Trypanosoma brucei*

African trypanosomes of the *T. brucei* species (Order Kinetoplastida) are parasitic protozoa that are the causative agents of African Sleeping Sickness in humans and Nagana in cattle (Vickerman et al., 1993). Analyses of 18S rRNA sequence homologies indicate that these organisms comprise one of the most ancient of all eukaryotic lineages, being as distant phylogenetically from yeast as they are from humans (Sogin et al., 1989). They are digenetic parasites living alternately in the bloodstream of the mammalian host and the midgut and mouthparts of the insect vector, the tsetse fly. Complex differentiation processes occur at each stage of the life cycle and survival in the mammalian host is dependent on the sequential expression of different variant surface glycoproteins (VSG) in a process called antigenic variation. VSG is the major cell surface secretory product of bloodstream trypanosomes comprising 10–20% of total cell protein.

■ *TbBiP* gene and sequence

The cloned *T. brucei* BiP (TbBiP) gene sequence (GenBank accession number L14477), derived from a genomic library, is a single contiguous open reading frame of 1959 nucleotides (Bangs et al., 1993). Like all known trypanosomal genes it has no introns. The deduced amino acid sequence indicates a protein of 653 residues (71.5 kDa) including a putative N-terminal signal sequence (c. 30 amino acids), no canonical N-linked glycosylation sites, and the C-terminal tetrapeptide Met–Asp–Asp–Leu (MDDL). It has overall amino acid identities of 64 and 62%, respectively, with rat BiP (see entry p. 59) and yeast Kar2p/BiP (see entry p. 33).

■ TbBiP protein

Antibody to recombinant TbBiP has been used to characterize the native protein (Bangs et al., 1993). The patterns seen in indirect immunofluorescence assays and immunoelectron microscopy (EM) are consistent with an ER localization. Staining is tubular, reticular, often perinuclear, and extends both posterior and anterior to the centrally located nucleus (trypanosomes are elongated cells of c. 2.5 × 25 microns). Subcellular fractionation, in conjunction with protease protection and alkaline carbonate experiments, indicates it to be a soluble luminal resident of crude microsomes vesicles. Proteolysis of the native protein results in the generation of discrete fragments of c. 45 and 30 kDa corresponding, respectively, to the conserved N-terminal ATPase and C-terminal peptide-binding domains of all HSP70 proteins (Chappell et al., 1987).

Deletion of the C-terminal tetrapeptide MDDL allows secretion of the TbBiP protein and addition of MDDL to heterologous reporters results in intracellular retention (Bangs et al., 1996). Thus MDDL, although somewhat divergent from ER localization signals in other eukaryotes (e.g. KDEL, mammals; HDEL, *Saccharomyces cerevisiae*), does serve this function in trypanosomes. Other C-terminal tetrapeptides that can function for retention in trypanosomes include KDEL and KQDL. The latter is the native C-terminal sequence of BS2, a putative trypanosomal protein disulfide isomerase homologue (see BS2 entry p. 342).

Interestingly, deletion of MDDL alone does not lead to the efficient secretion of TbBiP. Maximal export requires further deletion of the entire 30 kDa C-terminal peptide-binding domain (Bangs et al., 1996). Apparently in the absence of a *bona fide* retrieval signal this domain can retard export by virtue of its binding functionality. More direct evidence for the secretory molecular chaperone function of TbBiP comes from coprecipitation of newly synthesized VSG polypeptides with anti-TbBiP antibody (Bangs et al., 1996). This non-covalent association with TbBiP is transient and rapidly disappears as VSG is exported from the ER to the cell surface, as would be expected for a chaperone interaction.

■ Gene regulation of TbBiP

TbBiP protein is apparently expressed at a modest 2–3-fold higher level per cell in bloodstream vs. insect stage trypanosomes (Bangs et al., 1993). This may reflect increased demand on the secretory machinery by the synthesis of large amounts of VSG. Elevated synthesis of TbBiP itself is not induced (J.D. Bangs, unpublished results) by at least two treatments that are known to induce BiP in other systems, tunicamycin (Lee, 1987) and brefeldin A (Liu et al., 1992). The latter result is not surprising since brefeldin A has no effect on secretion in trypanosomes (J.D. Bangs, unpublished results). Owing to the peculiarities of gene expression in trypanosomes (transcription is polycistronic and no RNA polymerase II promoters have been identified), searches for upstream stress-induced regulatory sequences have not been performed.

■ References

Bangs, J.D., Uyetake, L., Brickman, M.J., Balber, A.E., and Boothroyd, J.C. (1993). Molecular cloning and cellular localization of a BiP homologue in *Trypanosoma brucei*: divergent ER retention signals in a lower eukaryote. J. Cell Sci. **105**, 1101–1113.

Bangs, J.D., Brouch, E.M., Ransom, D.M., and Roggy, J.L. (1996). A soluble secretory reporter system in *Trypanosoma brucei*: studies on endoplasmic reticulum targeting. J. Biol. Chem. **271**, 18387–18393.

Chappell, T.G., Konforti, B.B., Schmid, S.L., and Rothman, J.E. (1987). The ATPase core of a clathrin uncoating protein. J. Biol. Chem. **262**, 746–751.

Lee, A.S. (1987). Coordinated regulation of a set of genes induced by glucose and calcium ionophores in mammalian cells. Trends Biochem. Sci. **12**, 20–23.

Liu, E.S., Ou, J., and Lee, A.S. (1992). Brefeldin A as a regulator of grp78 gene expression in mammalian cells. J. Biol. Chem. **267**, 7128–7133.

Sogin, M.L., Gunderson, J.H., Elwood, H.J., Alonoso, R.A., and Peattie, D.A. (1989). Phylogenetic meaning of the kingdom concept: an unusual ribosomal RNA from *Giardia lamblia*. Science **243**, 75–77.

Vickerman, K., Myler, P.J., and Stuart, K.D. (1993) African trypanosomiasis. In Immunology and molecular biology of parasitic infections (K.S. Warren, ed.). Blackwell Scientific Publications, Boston, pp. 170–212.

■ *James D. Bangs:*
Department of Medical Microbiology and Immunology
University of Wisconsin–Madison
1300 University Avenue
Madison, WI 53706, USA
Tel. 1 608 262 3110
Fax. 1 608 262 8418
E-mail: bangs@facstaff.wisc.edu

Drosophila melanogaster Hsc4p

The Hsc4 protein of *Drosophila melanogaster is a constitutively expressed member of the HSP70 (stress-70) protein family. Hsc4p is a very abundant cytoplasmic protein present in all* Drosophila *tissues. Across species, Hsc70 homologs have been implicated in nascent protein folding, translocation of proteins across intracellular membranes, dissolution of protein aggregates and uncoating of clathrin-coated vesicles.*

■ Alternative names

hsc70 (Palter et al., 1986).

■ *HSC4* sequence

The nucleotide sequence for Drosophila *HSC4* was determined from a genomic clone [GenBank accession numbers M36114 (Perkins et al., 1990) and L01500 (Rubin et al., 1993)]. Comparison of the genomic sequence with that of a cDNA clone revealed the presence of a 1.6-kb intron located 5′ of the initiating ATG (Perkins et al., 1990). The DNA sequence of *HSC4* revealed an open reading frame of 1953 bp that would encode a protein of 651 amino acids with a molecular mass of 71 130 daltons. Hsc4p shares 74% amino acid identity with the heat-inducible Drosophila Hsp70 protein, 82% with the minor Drosophila cytoplasmic Hsc70 proteins, Hsc1p and Hsc2p, and approximately 60% identity with non-cytoplasmic HSP70-related proteins within and across species. As is true for all family members, the N-terminal region of Hsc4p is highly conserved and contains an ATP-binding domain.

Hsc4 protein

Hsc4p is synthesized constitutively and comprises approximately 1% of the total protein in adult flies and 5% in embryos (Palter et al., 1986). Although Hsc4p is present in all tissues examined, *HSC4* transcripts are enriched in cells active in endocytosis and those undergoing rapid growth (Perkins et al., 1990). *HSC4* is expressed at a level 30–70 times greater than *HSC1* and *HSC2* which also encode cytoplasmic Hsc70 proteins (Craig et al., 1983). Although the level of *HSC4* RNA doubles after heat shock, the protein level does not increase significantly. Under normal growth conditions, Hsc4p appears to be associated with cytoplasmic fibers (Palter et al., 1986). Upon stress induction, Hsc4p localizes to the nucleus, similar to Hsp70 (K.B. Palter and E. Craig, unpublished data). Hsc4p has been purified from *Drosophila* embryos and a bacterial expression system using ATP-affinity chromatography (A.D. Mehta, X. Chen and K.B. Palter, unpublished data). Bacterially expressed Hsc4p is isolated from inclusion bodies, but can be efficiently refolded *in vitro*. Purified Hsc4p is monomeric, with a sedimentation coefficient of 4.2 s and a pI of 5.75 (A.D. Mehta, X. Chen and K.B. Palter, unpublished data). A polyclonal rabbit antiserum has been raised to a non-conserved peptide of Hsc4p (anti-Hsc4) that is specific for *Drosophila* Hsc70 (F. Elefant and K.B. Palter, unpublished data). A polyclonal rabbit antiserum raised to a highly conserved N-terminal peptide of mammalian Hsc70 (Chappell et al., 1986) and a rat monoclonal antibody (7.10) raised to *Drosophila* Hsp70 (Velazquez et al., 1983) will also recognize *Drosophila* Hsc70.

Biological activities of Hsc4p

HSP70 proteins act as 'molecular chaperones' in that they bind to the non-native conformations of other proteins and then facilitate their correct folding (Hendrick and Hartl, 1993; Hartl, 1996). Across species, HSP70 proteins have been implicated in nascent protein folding, translocation of proteins across intracellular membranes, dissolution of protein aggregates and uncoating of clathrin-coated vesicles (reviewed in Gething and Sambrook, 1992). The ATPase domain of HSP70 homologs is coupled with the substrate-binding domain such that HSP70 cycles between conformations that have either high or low affinity for substrates depending on the bound nucleotide (reviewed in Hendrick and Hartl, 1993; Hartl, 1996). Similar to other HSP70 homologs, Hsc4p interacts with a variety of substrates, and the interactions are weakened in the presence of Mg^{2+}/ATP (D.M. Rubin and K.B. Palter, unpublished data). Both bacterially expressed and native *Drosophila* Hsc70 have similar activity to bovine Hsc70 in an *in vitro* clathrin-uncoating assay (A.D. Mehta and K.B. Palter, unpublished data). Only a single cycle of uncoating was detected in this assay and therefore accessory proteins may be required for efficient Hsc70 turnover (A.D. Mehta and K.B. Palter, unpublished data).

Studies with the *E.coli* HSP70, DnaK, have demonstrated that for most biological activities, DnaK works in conjunction with two regulatory proteins, DnaJ and GrpE (Georgopoulos, 1992). DnaJ stimulates the ATPase activity of DnaK and may also play a role in substrate recognition. The GrpE protein accelerates the rate of nucleotide exchange (for both ADP and ATP) from DnaK. Eukaryotic homologs of DnaJ and GrpE have been found in organisms ranging from yeast to plants to humans (Cyr et al., 1994). Thus far, four *Drosophila* DnaJ homologs have been identified; cysteine-string protein (csp) (Zinsmaier et al., 1990; see entry p. 115), l(2)tid, a tumor suppressor protein (Kurzik-Dumke et al., 1995; see entry p. 117) and Droj1p and Droj5p (J.Y. Lee and K.B. Palter, unpublished data). Although a *Drosophila* mitochondrial protein homolog of GrpE has been identified (see entry for Droe1p, p. 141), no eukaryotic cytoplasmic homolog has yet been reported.

Mutagenesis studies with *HSC4*

HSC4 has been localized by *in situ* hybridization to band 88E5,6-E13 of polytene chromosomes (Craig et al., 1983, K.B. Palter, unpublished data). An EMS induced dominant negative mutation of *HSC4* (A148V) was obtained as a modifier of the *Notch* locus (H. Hing, X. Sun and S. Artavanis-Tsakonis, pers. commun.). Heterozygotes for this mutation display eye and wing defects, whereas homozygotes die as embryos. A P-element insertion (lethal 3350, A. Spradling pers. commun.) disrupts the *HSC4* promoter and a deficiency, Df (3) PG4 from 88E3, 4-88E8,9 includes the *HSC4* locus (H. Hing, X. Sun and S. Artavanis-Tsakonis, pers. commun.).

Site-directed mutagenesis has been used to introduce substitutions of conserved residues into the ATPase domain of *Drosophila* Hsc4p (D.M. Rubin, Q. Wells, and K.B. Palter, unpublished results). Mutations D10S, D206S and K71S all caused loss of Hsc70 biological activity in the *in vitro* clathrin-uncoating assay. Furthermore, mutation K71S also prevented protein retention on ATP-affinity columns, while the other substitutions did not. Hsc4p D206S dominantly inhibited the dissociation of coated vesicles by wild-type Hsc70 *in vitro*. When mutant Hsc4p D206S and Hsc4p K71S proteins were expressed *in vivo* using a GAL4-targeted gene expression system (Brand and Perrimon, 1993), both proteins caused a dominant loss of viability at varying stages of development depending on the level of mutant protein produced (F. Elefant and K.B. Palter, unpublished data). Earlier developmental lethality was found to correlate with a higher level of mutant protein production for different insertion sites of *HSC4* in independent fly lines. A global accumulation of misfolded proteins may be responsible for the lethality. When mutant protein was targeted to either the peripheral nervous system or mesoderm (muscles), the phenotypes produced resembled those of known mutations affecting these tissues. Larvae expressing Hsc4p D206S or Hsc4p K71S induced a stress response indicated by the presence of the heat-inducible Hsp70 protein.

An X-ray crystallographic analysis of the ATPase fragment of the wild-type bovine Hsc70 protein and eight mutant proteins has been reported (Wilbanks et al., 1994; Flaherty et al., 1994). The analysis suggests that Asp-10 and Asp-199 are involved in positioning a Mg^{2+} atom (via water molecules) that is essential for the hydrolytic activity. Based on the wild-type structure, Lys-71 may bind the H_2O molecule (or OH^- ion) that attacks the γ-phosphate. The exact role of Asp-206 is not clear and it has minimal effect on the catalytic activity (Wilbanks et al., 1994). Asp-206 could be important for the subdomain movements that couple the ATPase domain with the substrate domain as has been shown for Glu-171 of DnaK (Buchberger et al., 1994). The dominant phenotypes produced by mutant Hsc4p in Drosophila may result from competition with wild-type Hsc70 for either substrates or regulatory proteins, thereby interfering with the normal function of Hsc70. A mutant Hsc4p that failed to release substrate upon binding ATP might be especially toxic to the cell. Alternatively, mutant Hsc4p might poison oligomeric Hsc70 complexes. Dominant phenotypes have also been reported for mutations of the ATPase domain of bacterial DnaK (Wild et al., 1992; Buchberger et al., 1994) and mammalian BiP (Hendershot et al., 1995).

■ References

Brand, A.H. and Perrimon, N. (1993). Targeted gene expression as a means of altering cell fates and generating dominant phenotypes. Development **118**, 401–415.

Buchberger, A., Valencia, A., McMacken, R., Sander, C., and Bukau, B. (1994). The chaperone function of DnaK requires the coupling of ATPase activity with substrate binding through residue E171. EMBO J. **13**, 1687–1695.

Chappell, T.G., Welch, W.J., Schlossman, D.M., Palter, K.B., Schlesinger, M.J., and Rothman, J.E. (1986). Uncoating ATPase is a member of the 70 kilodalton family of stress proteins. Cell **45**, 3–13.

Craig, E.A., Ingolia, T.D., and Manseau, L.J. (1983). Expression of Drosophila heat-shock cognate genes during heat shock and development. Develop. Biol. **99**, 418–426.

Cyr, D.M., Langer, T., and Douglas, M.G. (1994). DnaJ-like proteins: molecular chaperones and specific regulators of Hsp70. Trends Biochem. Sci. **19**, 176–181.

Flaherty, K.M., Wilbanks, W.M., DeLuca-Flaherty, C., and McKay, D.B. (1994). Structural basis of the 70-kilodalton heat shock cognate protein ATP hydrolytic activity. J. Biol. Chem. **17**, 12899–12907.

Georgopoulos, C. (1992). The emergence of the chaperone machines. Trends Biochem. Sci. **17**, 295–299.

Gething, M.-J. and Sambrook, J. (1992). Protein folding in the cell. Nature **355**, 33–45.

Hartl, F.U. (1996). Molecular chaperones in cellular protein folding. Nature **381**, 571–580.

Hendershot, L.M., Wei, J.-Y., Gaut, J.R., Lawson, B., Freiden, P.J., and Murti, K.G. (1995). In vivo expression of mammalian BiP ATPase mutants cause disruption of the endoplasmic reticulum. Mol. Biol. Cell **6**, 283–296.

Hendrick, J.P. and Hartl, F.-U. (1993). Molecular chaperone functions of heat-shock proteins. Annu. Rev. Biochem. **62**, 349–384.

Kurzik-Dumke, U., Gundacker, D., Rendrop, M., and Gateff, E. (1995). Tumor suppression in Drosophila is causually related to the function of the lethal(2)tumorous imaginal discs gene, a DnaJ homolog. Develop. Genet. **16**, 64–76.

Lindsley, D.L. and Sandler, L. (1972). Segmental aneuploidy and the gross structure of the Drosophila genome. Genetics **71**, 157–184.

Lindsley, D.L. and Zimm, G.G. (eds) (1992). The genome of Drosophila melanogaster. Academic Press, Inc., San Diego.

Palter, K.B., Watanabe, M., Stinson, L., Mahowald, A.P., and Craig, E.A. (1986). Expression and localization of Drosophila melanogaster hsp70 cognate proteins. Mol. Cell. Biol. **6**, 1187–1203.

Perkins, A.L., Doctor, J.S., Zhang, K., Stinson, L., Perrimon, N., and Craig, E.A. (1990). Molecular and developmental characterization of the heat shock cognate 4 gene of Drosophila melanogaster. Mol. Cell. Biol. **10**, 3232–3238.

Rubin, D.R., Metha, A.D., Zhu, J., Shoham, S., Chen, X., Wells, Q.R., and Palter, K.B. (1993). Genomic structure and sequence analysis of Drosophila melanogaster HSC70 genes. Gene **128**, 155–163.

Velazquez, J.M., Sonoda, S., Bugaisky, G., and Lindquist, S. (1983). Is the major Drosophila heat shock protein present in cells that have not been heat shocked? J. Cell Biol. **96**, 286–290.

Wilbanks, S.M., DeLuca-Flaherty, C., and McKay, D.B. (1994). Structural basis of the 70-kilodalton heat shock cognate protein ATP hydrolytic activity. J. Biol. Chem. **17**, 12893–12898.

Wild, J., Kamath-Loeb, A., Ziegelhoffer, E., Lonetto, M., Kawasaki, Y., and Gross, C.A. (1992). Partial loss of function mutations in DnaK, the Escherichia coli homologue of the 70-kDa heat shock proteins, affect highly conserved amino acids implicated in ATP binding and hydrolysis. Proc. Natl. Acad. Sci. USA **89**, 7139–7143.

Zinsmaier, K.E., Hofbauer, A., Heimbeck, G., Pflugelder, G.O., Buchner, S., and Buchner, E. (1990). A cysteine-string protein is expressed in retina and brain of Drosophila. J. Neurogenet. **7**, 15–29.

■ *Ashwin D. Mehta:*
North Carolina State University
Department of Botany
Raleigh, NC 27695, USA
Tel. 1 919 515 3570
Fax. 1 919 515 3436
E-mail: ashwin@unity.ncsu.edu

■ *David M. Rubin:*
Department of Cell Biology
Harvard Medical School
240 Longwood Ave
Boston, MA 02115, USA
Tel. 1 617 432 1519
Fax. 1 617 432 1144
E-mail: luckyleo@warren.med.harvard.edu

Felice Elefant and Karen B. Palter:
Department of Biology
Temple University
12th St. and Norris St.
Philadelphia, PA 19122, USA
Tel. 1 215 204 8845
Fax. 1 215 204 6646
E-mail: palter@astro.ocis.temple.edu

Drosophila melanogaster Hsc3p

The Hsc3 protein of Drosophila melanogaster *is a homolog of mammalian BiP and yeast Kar2p/BiP. Hsc3p is the sole member of the* Drosophila *HSP70 (stress-70) protein family localized to the lumen of the endoplasmic reticulum (ER). Hsc3p is believed to play a role in translocation of secretory proteins into the ER and then to facilitate their correct folding and oligomerization.*

■ Alternative names

hsc72 (Palter *et al.*, 1986).

■ *HSC3* gene and sequence

The nucleotide sequence for *Drosophila HSC3* (Rubin *et al.*, 1993) was determined from both genomic and larval cDNA clones (GenBank accession numbers L01498 and L01499). The genomic clone was shown to contain introns of 1.98 and 0.37 kb. The cDNA has a single open reading frame (ORF) of 1968 bp that would encode a protein of 656 amino acids, including an 18-residue N-terminal hydrophobic signal sequence and a C-terminal retrieval tetrapeptide, KDEL (Rubin *et al.*, 1993). The 3′ untranslated region of *HSC3* is unusually long, 904 nt from the stop codon to the first base of the poly(A) tail (Rubin *et al.*, 1993). The signal sequence of Hsc3p is cleaved upon entry into the ER to produce a mature protein that is 83% identical to rat BiP, 65% identical to yeast Kar2p/BiP, and approximately 60% identical to non-ER HSP70-related proteins both within and across species (Rubin *et al.*, 1993). The *Drosophila* heat-inducible *HSP70* promoter contains four heat shock regulatory elements (HSEs) consisting of tandemly arranged inverted repeats of the modular sequence, nGAAn (Pelham, 1982; Xiao *et al.*, 1991). A minimum of three modules is required to bind heat shock factor (HSF) and form a functional HSE. The *HSC3* promoter contains a single pair of inverted repeats, which is believed to be insufficient for heat-induced transcription (Rubin *et al.*, 1993). Consistent with this observation, *HSC3* message levels do not increase in response to heat (K.B. Palter, unpublished data). A sequence similar to a 28-base pair conserved sequence shared by the promoters of genes regulated by glucose starvation and calcium ionophores, including rat BiP/GRP78 and human GRP94 (Resendez *et al.*, 1988; Chang *et al.*, 1989) can be found within the *HSC3* promoter (Rubin *et al.*, 1993). A related sequence located within the *KAR2* promoter is required for the 'unfolded protein response' of that gene (Mori *et al.*, 1992; Kohno *et al.*, 1993).

■ Hsc3 protein

Hsc3p migrates on SDS–PAGE gels with an apparent molecular weight of 72 kDa and as a doublet of pI 5.6 and 5.5 on two-dimensional isoelectric focusing gels (Palter *et al.*, 1986). The acidic isoform of Hsc3p is post-translationally modifed and can be labeled with both inorganic phosphate and adenosine (Q. Wells, D.M. Rubin, and K.B. Palter, unpublished data), as is true for mammalian BiP (Carlsson and Lazarides, 1983). The levels of both forms of Hsc3p are developmentally regulated. The level of Hsc3p in embryos is about 10% that of Hsc70 (see entry for Hsc4, p. 42), is barely detectable in first and second instar larvae, is comparable to Hsc70 in third instar larvae and pupae and is about 30% the level of Hsc70 in adult flies (Q. Wells, D.M. Rubin and K.B. Palter, unpublished data). The level of modified Hsc3p is also highest in late larval and pupal stages. *Drosophila* imaginal discs actively secrete the pupal cuticle in response to the steriod hormone, 20-hyroxyecdysone (20-HOE) (Silvert *et al.*, 1984). The 3.1 kb *HSC3* transcript is strongly induced when mass isolated imaginal discs are cultured in the presence of 20-HOE, consistent with Hsc3p playing a role in secretion (J. Doctor, J. Natzle, E. Craig and K.B. Palter, unpublished data). Hsc3p purified from a bacterial expression system using ATP-affinity chromatography is monomeric with a sedimentation coefficient of 4.2 *s* (A. Mehta and K.B. Palter, unpublished data). Both isoforms of Hsc3p purified from adult flies bind ATP-agarose (D.M. Rubin and K.B. Palter, unpublished data). Three rat monoclonal antibodies (5/24.1, 5/80.5 and 5/46.5) raised to flight muscle troponin H of water bugs (made by Belinda Bullard, Heidelberg, Germany) cross-react with Hsc3p. Troponin H is only present in flight muscles and the antibodies do not recognize other HSP70 family members (D.M. Rubin and K.B. Palter, unpublished data).

■ Biological regulation of Hsc3p

Mammalian BiP is reported to be post-translationally modified by both phosphorylation and ADP-ribosylation (Carlsson and Lazarides, 1983; Leno and Ledford, 1990; Freiden et al., 1992). Labeling of Drosophila Hsc3p by both inorganic phosphate and adenosine is inhibited by nicotinamide, a competitive inhibitor of NAD (Q. Wells and K.B. Palter, unpublished data) suggesting that Hsc3p is post-translationally modified by ADP-ribosylation. The observation that the modified form of mammalian BiP is not bound to proteins led to the proposal that it may represent an inactive form of BiP and that BiP modification regulates the amount of active BiP in the ER (Freiden et al., 1992). However, in vivo studies of Hsc3p have shown that when the requirement for active Hsc3p should be high, such as late larval and early pupal stages when proteins are secreted to form the pupal cuticle and during recovery from heat shock, the levels of both total Hsc3p and modified Hsc3p are highest (Q. Wells, D.M. Rubin and K.B. Palter, unpublished data). Conversely, treatments that cause an accumulation of misfolded proteins in the ER and retard their secretion, such as exposure to tunicamycin and heat shock, result in complete loss of the modified form of Hsc3p. Expression of dominant Hsc3p mutants that may cause an accumulation of misfolded proteins in the ER, also inhibits the modification of wild-type Hsc3p (D.M. Rubin and K.B. Palter, unpublished data). These observations have led to a proposal that modification marks Hsc3p molecules that have exited the ER and reached the pre-Golgi compartment (Q. Wells, D.M. Rubin and K.B. Palter, unpublished data). In this model, increased traffic of secretory proteins and Hsc3p from the ER would be predicted to increase the pool of modified Hsc3p, as is observed. The modification may prevent Hsc3p in the pre-Golgi compartment from rebinding proteins before being returned via the KDEL receptor to the ER. The level of BiP in the ER is believed to be transcriptionally regulated by the 'unfolded protein-response pathway' (reviewed in Shamu et al., 1994). The total level of Hsc3p is augmented in response to tunicamycin, suggesting that Hsc3p is similarly regulated.

■ Mutagenesis studies with *HSC3*

HSC3 has been localized by in situ hybridization to band 10E3,4 of polytene chromosomes (K.B. Palter and E. Craig, unpublished data). Thus far, no lethal mutations within that interval have been rescued by a wild-type HSC3 gene (K. B. Palter, unpublished data).

Site-directed mutagenesis has been used to introduce substitutions of conserved residues into the ATPase domain of Drosophila Hsc3p (D.M. Rubin, Q. Wells, F. Elefant and K.B. Palter, unpublished data). In vivo expression in a wild-type background of Hsc3p K97S and Hsc3p D231S from either a heat shock promoter or using a GAL4-targeted gene expression system (Brand and Perrimon, 1993) caused a dominant loss of viability. Expression of Hsc3p E201S from a heat shock promoter also caused a dominant loss of viability whereas expression of Hsc3p D35S produced a wild-type phenotype. Residues D35, K97, E201 and D231 in Hsc3p correspond respectively to residues D10, K71, E175 and D206 in both bovine and Drosophila Hsc70. Hsc3p D35S, E201S and D231S bind ATP-agarose similarly to the wild type protein. The mutant Hsc3p K97S may have reduced affinity for ATP as the bacterially expressed protein binds ATP-agarose but the modest amount of mutant protein present in fly extracts does not. Both Hsc3p K97S and Hsc3p D231S fail to release substrates upon binding ATP. Expression of dominant-acting mutant Hsc3p probably causes an accumulation of misfolded proteins in the ER that eventually results in lethality. Expression of mutant Hsc3p K97S and Hsc3p D231S resulted in a loss of modified wild-type Hsc3p, perhaps owing to the retention of the wild-type protein in the ER.

■ References

Brand, A.H. and Perrimon, N. (1993). Targeted gene expression as a means of altering cell fates an generating dominant phenotypes. Development **118**, 401–415.

Carlsson, L. and Lazarides, E. (1983). ADP-ribosylation of the Mr 83 000 stress-inducible and glucose-regulated protein avian and mammalian cells: modulation by heat shock and glucose starvation. Proc. Natl. Acad. Sci. USA **80**, 4664–4668.

Chang, S.C., Erwin, A.E., and Lee, A.S. (1989). Glucose-regulated protein (GRP94 and GRP78) genes share common regulatory domains and are coordinately regulated by common trans-acting factors. Mol. Cell. Biol. **9**, 2153–2162.

Freiden, P.J., Gaut, J.R., and Hendershot, L.M. (1992). Interconversion of three differentially modified and assembled forms of BiP. EMBO J. **11**, 63–70.

Kohno, K., Normington, K., Sambrook, J., Gething, M.-J., and Mori, K. (1993). The promoter region of the yeast KAR2 (BiP) gene contains a regulatory domain that responds to the presence of unfolded proteins in the endoplasmic reticulum. Mol. Cell. Biol. **13**, 877–890.

Leno, G.H. and Ledford, B.E. (1990). Reversible ADP-ribsoylation of the 78-kDa glucose-regulated protein. FEBS Lett. **276**, 29–33.

Mori, K., Sant, A., Kohno, K., Normington, K., Gething, M.-J., and Sambrook, J.F. (1992). A 22 bp cis-acting element is necessary and sufficient for the induction of the yeast KAR2 (BiP) gene by unfolded proteins. EMBO J. **11**, 2583–2593.

Palter, K.B., Watanabe, M., Stinson, L., Mahowald, A.P., and Craig, E.A. (1986). Expression and localization of Drosophila melanogaster hsp70 cognate proteins. Mol. Cell. Biol. **6**, 1187–1203.

Pelham, H.R.B. (1982). A regulatory upstream promoter element in the Drosophila hsp70 gene. Cell **30**, 517–528.

Resendez Jr., E., Wooden, S.K., and Lee, A.S. (1988). Identification of the highly conserved regulatory domains and protein-binding sites in the promoters of the rat and human genes encoding the stress-inducible 78-kilodalton glucose-regulated protein. Mol. Cell. Biol. **8**, 4579–4584.

Rubin, D.R., Metha, A.D., Zhu, J., Shoham, S., Chen, X., Wells, Q.R., and Palter, K.B. (1993). Genomic structure and sequence analysis of Drosophila melanogaster HSC70 genes. Gene **128**, 155–163.

Shamu, C.E., Cox, J.S., and Walter, P. (1994). The unfolded-protein-response pathway in yeast. Trends Cell Biol. **4**, 56–60.

Silvert, D.J., Doctor, J., Quesada, L., and Fristrom, J.W. (1984). Pupal and larval cuticle proteins of Drosophila melanogaster. Biochemistry **23**, 5767–5774.

Xiao, H., Perisic, O., and Lis, J.T. (1991). Cooperative binding of Drosophila heat shock factor to arrays of a conserved 5 bp unit. Cell **64**, 585–593.

■ David M. Rubin:
Department of Cell Biology
Harvard Medical School
240 Longwood Ave.
Boston, MA 02115, USA
Tel. 1 617 432 1519
Fax. 1 617 432 1144
E-mail: luckyleo@warren.med.harvard.edu

■ Karen B. Palter:
Department of Biology
Temple University
12th St. and Norris St.
Philadelphia, PA 19122, USA
Tel. 1 215 204 8845
Fax. 1 215 204 6646
E-mail: palter@astro.ocis.temple.edu

Fish HSP70 proteins

Studies of the HSP70 family in fish have emphasized their roles in environmental biology and toxicology. Inducible Hsp70 holds promise as a molecular biomarker for sublethal effects of chemical toxicants in both intact animals and cultured cells. Biochemical variants of Hsp70 but not Hsc70 and Grp78 exist among closely related species of desert fish, indicating that Hsp70 is evolutionarily less constrained and likely to be different in function or regulation than Hsc70.

■ Alternative names

The designation stress-70 is frequently used in the environmental toxicology literature (Sanders, 1993).

■ Isolation

Fish Hsp70s were originally identified as proteins induced in cultured fish cells by exposure to heat, heavy metal ions or sodium arsenite. Initial biochemical characterizations were done using denaturing polyacrylamide gel electrophoresis (Heikkila et al., 1982; Kothary and Candido, 1982).

■ Fish *hsp70* genes and sequences

Few fish *hsp* genes have been cloned and sequenced. GenBank contains an 837 bp cDNA that encodes part of the inducible Hsp70 of rainbow trout (*Salmo gairdneri*) (accession number K02549, Kothary et al., 1984) corresponding to amino acids 128–406 of *Drosophila* Hsp70 to which it is 72% identical. Genomic and cDNA clones of a complete *hsc70* gene from rainbow trout (accession number S85730 available through protein sequence accession number 246719) have been isolated and characterized (Zafarullah et al., 1992). This fish *hsc70* gene is very similar to a human *hsc70* gene (94% identical amino acid sequence) and contains eight introns. Its corresponding mRNA is constitutively abundant in several trout tissues and salmonid cell lines. The coding region of a complete chinook salmon *hsp70* cDNA sequence (accession number U35064) shares 83% amino acid seqence identity with trout Hsc70 with most of the differences occurring at the extreme C-terminus (Lee Weber, personal communication).

■ Fish HSP70 proteins

Hsc70 was purified from livers of unstressed channel catfish (*Ictalurus punctatus*) and a mixture of Hsp70 and Hsc70 was purified from heat shocked fish by the same methods used to isolate mammalian Hsp70 (Abukhalaf et al., 1994). As expected for an evolutionarily highly conserved protein, polyclonal antibodies raised to catfish liver Hsp70s cross-react with Hsp70 in a variety of catfish tissues and to Hsp70 from four different genera of fish. Antisera raised against gel-purified Hsp70 was used to localize Hsc70/Hsp70 in heat shocked carp (*Cyprinus carpio*) fibroblasts in culture (Ku *et al.*, 1994). Like mammalian Hsp70, fish Hsp70 localized to nuclei and nucleoli in response to heat shock and then redistributed to the cytoplasm during recovery. In general, antibodies from mammalian species that recognize both Hsc70 and Hsp70 cross-react with fish Hsp70 and Hsc70, whereas antibodies specific for a mammalian-inducible Hsp70 do not cross-react. Fish Hsp70s have been studied most frequently using one- and two-dimensional polyacrylamide gel electrophoresis and immunoblotting. The overall patterns of fish Hsp70 families on two-dimensional gels are similar to those of mammalian species. The most acidic member corresponds to BiP (Grp78), usually represented by both a phosphorylated and unphosphorylated isoform, followed by a more basic form of Hsc70 (one or two isoforms), and the inducible Hsp70s, the most basic, are usually represented by multiple isoforms. Biochemical variation in Hsp70 isoforms exists among closely related species of the desert fish *Poeciliopsis* but only one Hsc70 isoform was found, indicating that evolutionary constraints on these two nucleocytoplasmic members of the HSP70 family are different (White *et al.*, 1994). These data suggest that stress-inducible Hsp70 may differ from constitutive Hsc70 in function or in regulation by accessory proteins. Biochemical variation in Hsp70 exists even within species of tropical and desert *Poeciliopsis* (Norris *et al.*, 1995). A single tropical species contains almost all of the Hsp70 isoforms found in six desert species, suggesting that tropical species may represent a reservoir of stress protein variation.

■ Biological activities of fish HSP70 proteins

It is generally assumed that the functions of fish HSP70s are essentially the same as their mammalian counterparts, except that there may be some differences in their regulation by accessory proteins owing to the poikilothermic lifestyle of fishes. A more novel feature of the study of fish HSP70s is that they have been considered as molecular biomarkers in environmental toxicology (Sanders, 1993). Molecular biomarkers rely primarily on changes in DNA and proteins to indicate that an organism has been either exposed to or has responded to the effects of toxic chemicals. HSPs are considered to be biomarkers of *effect*, since there is considerable evidence that they are induced in response to protein damage to cells, i.e. proteotoxicity (Hightower, 1991). In a combined stress protein and cytotoxicity assay (neutral red uptake) using cultured fish cells, there was a direct concentration-dependent relationship between sublethal cytotoxic effects of heavy metal ions and increases in Hsp70 levels (Ryan and Hightower, 1994). In biomarker assays, HSP levels are usually assayed by ELISA, immunoblots, or by silver staining proteins on polyacrylamide gels. The predictive value of the induction of Hsp70 in intact animals as an early indicator of population-level changes remains to be determined.

■ Biological regulation of fish *hsp70* genes

Detailed information on the mechanisms of heat shock gene regulation have come from studies of yeast, *Drosophila*, and cultured mammalian cells; however, studies of intact fish and their cells have provided our most advanced understanding of the effects of the natural environment on regulation of the response (Bols *et al.*, 1992). Threshold temperatures for induction of Hsp70 are closely tied to optimum growth temperatures of fish. For example, Hsp70 is induced 2–3°C above the optimum temperature of the desert fish *Poeciliopsis* (P.J. diIorio, unpublished observations). Therefore, inducible Hsp70 is routinely present in the upper part of the *normal* temperature tolerance range; this may be true of most poikilotherms. When four species of marine fishes from different thermal environments were acclimatized to the same temperature, large variations among species in induction temperatures persisted, indicating that there is a genetically determined component to this set point (Dietz and Somero, 1993). Seasonal variation is superimposed on this fixed component of the threshold induction temperature (Dietz, 1994). Large differences in constitutive levels of Hsc70 occur in different tissues within species and the levels vary dramatically for the same tissue from different species, even after acclimatization to the same temperature. Recent studies aimed at establishing correlations between levels of Hsp70, Hsc70, and survival at elevated temperatures have shown that individuals from natural populations of a tropical species of *Poeciliopsis* vary considerably in survival of a challenge heat stress following acquisition of thermotolerance. Survival time increases roughly with increasing levels of Hsp70 in gill tissue. Interestingly, abundance of Hsp70 is directly related to abundance of Hsc70, suggesting that fish with higher levels of Hsc70 produce more Hsp70 and are more thermoresistant (Norris *et al.*, 1995). Hemiclonal hybrids of *Poeciliopsis* vary in their abilities to survive heat stress. Survival is highest when rates of temperature increase are slowest, allowing maximal acquisition of thermotolerance, and when the hybrids contain male and female genomes from the same local, i.e. sympatric genomes (diIorio *et al.*, 1996).

Fish HSP70s are induced by the same stressors that induce their mammalian counterparts. Because of the interest in fish HSPs as molecular biomarkers, chemical stressors, particularly heavy metal ions and sodium arsenite, have received considerable attention and the list of chemical stressors has been expanded considerably by the inclusion of organic chemicals of importance in environmental toxicology (Sanders, 1993). As shown for heat shock, both qualitative and quantitative differences in the stress responses of particular tissues to sodium arsenite also occur in intact fish (Dyer *et al.*, 1993).

■ References

Abukhalaf, I.K., Covington, S., Zimmerman, E.G., Dickson, K.L., Masaracchia, R.A., and Donahue, M.J. (1994). Purification of the 70-kDa heat-shock protein from catfish liver:

immunological comparison of the protein in different fish species and its potential use as a stress indicator. Environ. Toxicol. Chem. **13**, 1251–1257.

Bols, N.C., Mosser, D.D., and Steels, G.B. (1992). Temperature studies and recent advances with fish cells *in vitro*. Comp. Biochem. Physiol. **103A**, 1–14.

Dietz, T.J. (1994). Acclimation of the threshold induction temperatures for 70-kDa and 90-kDa heat shock proteins in the fish *Gillichthys mirabilis*. J. Exp. Biol. **188**, 333–338.

Dietz, T.J. and Somero, G.N. (1993). Species- and tissue-specific synthesis patterns for heat-shock proteins hsp70 and hsp90 in several marine teleost fishes. Physiol. Zool. **66**, 863–880.

diIorio, P.J., Holsinger, K., Schultz, R.J., and Hightower, L.E. (1996). Quantitative evidence that both hsc70 and hsp70 contribute to thermal adaptation in hybrids of the livebearing fishes *Poeciliopsis*. Cell Stress Chap. **1**, 139–148.

Dyer, S.D., Brooks, G.L., Dickson, K.L., Sanders, B.M., and Zimmerman, E.G. (1993). Synthesis and accumulation of stress proteins in tissues of arsenite-exposed fathead minnows (*Pimephales Promelas*). Environ. Toxicol. Chem. **12**, 913–924.

Heikkila, J., Schultz, G., Iatrou, K., and Gedamu, L. (1982). Expression of a set of fish genes following heat or metal ion exposure. J. Biol. Chem. **337**, 12000–12005.

Hightower, L.E. (1991). Heat shock, stress proteins, chaperones, and proteotoxicity. Cell **66**, 191–197.

Kothary, R.K. and Candido, E.P.M. (1982). Induction of a novel set of polypeptides by heat shock or sodium arsenite in cultured cells of rainbow trout, *Salmo gairdnerii*. Can. J. Biochem. **60**, 347–355.

Kothary, R.K., Jones, D., and Candido, E.P.M. (1984). 70-kilodalton heat shock polypeptides from rainbow trout: characterization of cDNA sequences. Mol. Cell. Biol. **4**, 1785–1791.

Ku, C.C., Lu, C.H., Kou, G.H., and Chen, S.N. (1994). Purification and immunological characterization of color carp (*Cyprinus carpio*) fibroblast heat-shock proteins. Comp. Biochem. Physiol. **107B**, 147–159.

Norris, C.E., diIorio, P.J., Schultz, R.J., and Hightower, L.E. (1995). Variation in heat shock proteins within tropical and desert species of poeciliid fishes. Mol. Biol. Evol. **12**, 1048–1062.

Ryan, J.A. and Hightower, L.E. (1994). Evaluation of heavy metal ion toxicity in fish cells using a combined stress protein and cytotoxicity assay. Environ. Toxicol. Chem. **13**, 1231–1240.

Sanders, B.M. (1993). Stress proteins in aquatic organisms — an environmental perspective. Crit. Rev. Toxicol. **23**, 49–75.

White, C.N., Hightower, L.E., and Schultz, R.J. (1994). Variation in heat-shock proteins among species of desert fishes (Poeciliidae, *Poeciliopsis*). Mol. Biol. Evol. **11**, 106–109.

Zafarullah, M., Wisniewski, J., Shworak, N.W., Schieman, S., Misra, S., and Gedamu, L. (1992). Molecular cloning and characterization of a constitutively expressed heat-shock cognate hsc71 gene from rainbow trout. Eur. J. Biochem. **204**, 893–900.

Lawrence E. Hightower, John A. Ryan and Carol E. Norris:
Deptartment of Molecular and Cell Biology
University of Connecticut
Storrs, CT 06269-3044, USA.
Tel. 1 860 486 4257
Fax. 1 860 486 1784
E-mail: hightower@biotek.mcb.uconn.edu

The *hsp70* genes of mice and men

Genes encoding proteins of the HSP70 family have been identified and cloned from numerous species, ranging from bacteria to humans. Phylogenetic analysis reveals that these genes have been highly conserved during evolution (Rensing and Maier, 1994). Despite being first identified as stress-inducible proteins, many of these isoforms are expressed in the absence of stress. The constitutively expressed isoforms are expressed throughout the cell and are concentrated at sites of protein synthesis and translocation. These HSP70 proteins act as molecular chaperones which, in association with other molecular chaperones, oversee the folding process of newly synthesized and transported proteins (see reviews by Gething and Sambrook, 1992; Hartl and Martin 1995; this volume, overview of HSP70, p. 3). The role of the stress-inducible HSP70 isoforms is less well understood, although increasing evidence supports the long-held belief that these isoforms are involved in the protection and/or repair of stress-induced protein damage.

■ Chromosomal locations of human *hsp70* genes

The various HSP70 isoforms are encoded by a multigene family consisting of at least eleven distinct genes in humans (Table 1, see review by Tavaria *et al.*, 1996). Three *hsp70* genes are contained within the major histocompatibility class III (MHC class III) region on chromosome 6 (Harrison *et al.*, 1987; Goate *et al.*, 1987; Sargent *et al.*, 1989). Two of these genes encode an identical protein, the major cytoplasmic-inducible HSP70 (Hsp70-1/Hsp70-2), whilst the third gene encodes a less well characterized protein (Hsp70-Hom). Two inducible HSP70 proteins (Hsp70-6 and Hsp70-7) are encoded by genes located close together on the long arm of chromosome 1 (Voellmy *et al.*, 1985; Leung *et al.*, 1990, 1992). Other *hsp70* genes have been localized to chromosomes 5 (Fathallah *et al.*, 1993), 9 (Hendershot *et al.*, 1994), 11 (Tavaria *et al.*, 1995) and 14 (Bonnycastle *et al.*, 1994;

Table 1. Nomenclature for the gene loci encoding human and mouse HSP70 proteins. *hsp70* gene locus symbols were obtained from the Genome Database. Alternative names and intracellular localizations were obtained from published literature. Acc. no. is the GDB/NCBI database accession number (maximum of two shown where more exist). Column A is the intracellular expression localization. Nu/Cyto = nucleus/cytoplasm, ER = endoplasmic reticulum, Mito = mitochondria. Column B indicates whether expression is detectable in unstressed cells

Human							Mouse			
Locus symbol	Protein	Acc. no.	Locus site	A	B	Ref.	Mouse homolog	Acc. no.	Locus site	Ref.
HSPA1A	Hsp70,Hsx70,Hsp72, Hsp70i, Hsp70-1	M59828 M34267	6p21.3	Nu/Cyto	Yes	1,2	hsp70A1, hsp70.3	M76613	17	16,17
HSPA1B	Hsp70-2	M59830 M34269	6p21.3	Nu/Cyto	Yes	2	hsp70A2, hsp70.1	M35021	17	18
HSPA1L	Hsp70-Hom, Hsp70t	M59829 M34268	6p21.3	?	Yes	2	hsc70t	M32218	17	19
HSPA2A	Hsp70-3	U10149	14q22	?	Yes	3				
HSPA2B	Hsp70-3	L26336	14q24.1	?	Yes	4	hsp70.2, hsc70B	M20567	12	20
HSPA3	—	?	21(?)	?	?	5				
HSPA4	Hsp70RY, Hsp70H	L12723	5q31.1	?	Yes	6	hsr.1	U08215	2	21
HSPA5	BiP, Grp78	M19645 X87949	9q34	ER	Yes	7,8	BiP, Grp78	U16277		
HSPA6	Hsp70-6, Hsp70B'	X51757	1q	?	No	9,10				
HSPA7	Hsp70-7, Hsp70B	M11236	1q	?	No	10,11				
HSPA8	Hsc70, Hsp73	Y00371	11q24	Nu/Cyto	Yes	12,13	hsc70A, hsc73	M19141 U27129		22
HSPA9	Grp75, Pbp74, mtHSP75	L15189	?	Mito	Yes	14,15	Pbp74, Grp75, CSA, p66^{mot-1}	D17556 D11089	18	14,23,24

References: 1. Wu et al., 1985; 2. Milner and Campbell, 1990; 3. Roux et al., 1994; 4. Bonnycastle et al., 1994; 5. Harrison et al., 1987; 6. Fathallah et al., 1993; 7. Ting and Lee, 1988; 8. Hendershot et al., 1994; 9. Leung et al., 1990; 10. Leung et al., 1992; 11. Voellmy et al., 1985; 12. Dwornizcak and Miraut, 1987; 13. Tavaria et al., 1995; 14. Domanico et al., 1993; 15. Bhattacharyya et al., 1995; 16. Lowe and Moran, 1986; 17. Perry et al., 1994; 18. Hunt and Calderwood, 1990; 19. Matsumoto and Fujimoto, 1990; 20. Zakeri et al., 1988; 21. Gaskins et al., 1992; 22. Giebel et al., 1988; 23. Michikawa et al., 1993; 24. Kaul et al., 1995.

Roux et al., 1994). In addition to these genes, it has been proposed that an inducible HSP70 protein is encoded by a gene on chromosome 21 (Harrison et al., 1987), although analysis of chromosome 21-containing hybrids, Southern blot analysis and cloning attempts have not identified the gene involved (Goate et al., 1987; Gabriele et al., 1996).

■ Mouse *hsp70* genes

Homologues of many of the human *hsp70* genes have been identified in other species, including the mouse, and are highly conserved both in primary sequence and genomic organization (Table 1). As with the human MHC locus, the mouse MHC locus on chromosome 17 contains a similar arrangement of three *hsp70* genes (Snoek et al., 1993; Hunt et al., 1993), two of which encode a heat-inducible isoform. The third gene is expressed only in male germ cells (Matsumoto and Fujimoto, 1990) leading to suggestions that the human homologue, Hsp70-Hom, may be a testis-specific gene. The mouse HSP70 family includes another member, Hsp70.2 or Hsc70B, which is expressed at high levels in the testes during meiosis (Allen et al., 1988, 1996; Zakeri et al., 1988). This gene is located on mouse chromosome 12 in a region of synteny with human chromosome 14. Two *hsp70* genes are located in this region of human chromosome 14, one of which is highly expressed in the testis (Bonnycastle et al., 1994). The mouse and human genes encoding the endoplasmic reticulum HSP70 isoform, BiP or Grp78, are also located on syntenic chromosomes: chromosome 9 in humans (Hendershot et al., 1994), chromosome 2 in mouse (Gaskins et al., 1992; Hunt et al., 1993; Pilz et al., 1994).

The only other mouse *hsp70* gene to be localized is that of the mitochondrial HSP70 isoform, variously known as Pbp74, Grp75, CSA and p66^{mot-1} (mortalin), to chromosome 18 (Kaul et al., 1995) which is syntenic with human chromosomes 18 and 5. The chromosomal localization of the human mitochondrial HSP70 isoform has not yet been determined. Interestingly, *in situ* hybridization using the mouse mortalin sequence identifies a locus on human chromosome 5q31.1 (Kaul et al., 1995). An HSP70-related gene, Hsp70RY or Hsp70H (see entry p. 82), which is only distantly related to other HSP70 family members, including the mitochondrial HSP70 isoform, has already been localized to human 5q31.1 (Fathallah et al., 1993). The *hsp70RY* gene encodes a protein that is most highly related to the mouse hsr.1 protein (Yang, 1994), the sea urchin sperm receptor (Foltz et al., 1993), yeast Sse1 and Sse2 (Mukai et al., 1993) and to recently cloned members of the mouse (Yasuda et al., 1995) and Chinese hamster (Lee-Yoon et al., 1995) HSP110 family. These proteins therefore represent a novel class of HSP70-related proteins.

■ Nomenclature for human and mouse *hsp70* genes

In recent years there has been a rapid expansion in the characterization of genes encoding stress proteins and an increasing understanding of the role of stress proteins in the physiology of the normal and stressed cell. This progress has, however, been complicated by the use of conflicting nomenclature and confusion over the identity of some isoforms. We have recently described a nomenclature which we have found useful in identifying/distinguishing the different human HSP70 isoforms (Tavaria et al., 1996) and which we hope will be the catalyst for further rationalization of the HSP70 nomenclature. Indeed, such a process has recently been undertaken and has led to the standardization of nomenclature for the mitochondrial import proteins (Pfanner et al., 1996). In this review, we have extended our survey of the HSP70 family to include murine homologues where sequence and/or functional information are available. The functional relationship between individual HSP70 isoforms should become clearer with further genetic and functional characterization of the HSP70 family in these and other species.

■ References

Allen, R.L., O'Brien, D.A., Jones, C.C., Rockette, D.L., and Eddy, E.M. (1988). The expression of heat shock proteins by isolated mouse spermatogenic cells. Mol. Cell. Biol. **8**, 3260–3266.

Allen, J.W., Dix, D.J., Collins, B.W., Merrick, B.A., He, C., Selkir, J.K., Poorman-Allen, P., Dresser, M.E., and Eddy, E.M. (1996). HSP70-2 is part of the synaptosomal complex in mouse and hamster spermatocytes. Chromosoma **104**, 414–421.

Bhattacharyya, T., Karnezis, A.N., Murphy, S.P., Hoang, T., Freeman, B.C., Phillips, B., and Morimoto, R.I. (1995). Cloning and subcellular localization of human mitochondrial hsp70. J. Biol. Chem. **270**, 1705–1710.

Bonnycastle, L.L., Yu, C-E., Hunt, C.R., Trask, B.J., Clancy, K.P., Weber, J.L., Patterson, D., and Schellenberg, G.D. (1994). Cloning, sequencing, and mapping of the human chromosome 14 heat shock protein gene (HSPA2). Genomics **23**, 85–93.

Domanico, S.Z., DeNagel, D.C., Dahlseid, J.N., Greene, A.E., and Mullivor, R.A. (1993). Cloning of the gene encoding peptide-binding protein 74 shows that it is a new member of the heat shock protein 70 family. Mol. Cell. Biol. **13**, 3598–3610.

Dwornizcak, B. and Mirault, M.E. (1987). Structure and expression of a human gene coding for a 71 kd heat shock 'cognate' protein. Nucleic Acids Res. **15**, 5181–5197.

Fathallah, D.M., Cherif, D., Dellagi, K., and Arnaout, M.A. (1993). Molecular cloning of a novel human Hsp70 from a B cell line and its assignment to chromosome 5. J. Immunol. **151**, 810–813.

Foltz, K.R., Partin, J.S., and Lennarz, W.J. (1993). Sea urchin egg receptor for sperm: sequence similarity of binding domain and hsp70. Science **259**, 1421–1425.

Gabriele, T., Tavaria, M., Anderson, R.L., and Kola, I. (1996). Analysis of heat shock protein 70 in human chromosome 21 containing hybrids. Int. J. Biochem. Cell Biol. **28**, 905–910.

Gaskins, H.R., Prochazka, M., and Leiter, E.H. (1992). Chromosomal localization of the murine stress gene encoding glucose-regulated protein 78 (BiP). Mamm. Genome **3**, 836–837.

Gething, M-J. and Sambrook, J. (1992). Protein folding in the cell. Nature **355**, 33–45.

Giebel, L.B., Dworniczak, B.P., and Bautz, E.K. (1988). Developmental regulation of a constitutively expressed mouse mRNA encoding a 72-kDa heat shock-like protein. Develop. Biol. **125**, 200–207.

Goate, A.M., Cooper, D.N., Hall, C., Leung, T.K., Solomon, E., and Lim, L. (1987). Localization of a human heat shock Hsp70 gene sequence to chromosome 6 and detection of two other loci by somatic-cell hybrid and restriction fragment length polymorphism analysis. Human. Genet. **75**, 123–128.

Harrison, G.S., Drabkin, H.A., Kao, F.T., Hartz, J., Hart, I.M., Chu, E.H., Wu, B.J., and Morimito, R.I. (1987). Chromosomal location of human genes encoding major heat shock protein Hsp70. Somat. Cell Mol. Genet. **13**, 119–130.

Hartl, F.U. and Martin, J. (1995). Molecular chaperones in cellular protein folding. Curr. Opin. Struct. Biol. **5**, 92–102.

Hendershot, L.M., Valentine, V.A., Lee, A.S., Morris, S.W., and Shapiro, D.N. (1994). Localization of the gene encoding BiP/GRP78, the endoplasmic reticulum cognate of the Hsp70 family, to chromsome 9q34. Genomics **20**, 281–284.

Hunt, C. and Calderwood, S. (1990). Characterization and sequence of a mouse hsp70 gene and its expression in mouse cell lines. Gene **87**, 199–204.

Hunt, C.R., Gasser, D., Chaplin, D.D., Pierce, J.C., and Kozak, C.A. (1993). Chromosomal localization of five murine HSP70 gene family members: *Hsp70-1*, *Hsp70-2*, *Hsp70-3*, *Hsc70t*, and *Grp78*. Genomics **16**, 193–198.

Kaul, S.C., Wadhwa, R., Matsuda, Y., Hensler, P.J., Pereira-Smith, O.M., Komatsu, Y., and Mitsui, Y. (1995). Mouse and human chromosomal assignments of mortalin, a novel member of the hsp70 family of proteins. FEBS Lett. **361**, 269–272.

Lee-Yoon, D., Easton, D., Murawski, M., Burd, R., and Subjeck, J.R. (1995). Identification of a major subfamily of large hsp70-like proteins through the cloning of the mammalian 110-kDa heat shock protein. J. Biol. Chem. **270**, 15725–15733.

Leung, T.K., Rajendran, M., Monfries, C., Hall, C., and Lim, L. (1990). The human heat shock protein family. Expression of a novel heat-inducible Hsp70 (Hsp70B′) and isolation of its cDNA and genomic DNA. Biochem. J. **267**, 125–132.

Leung, T.K., Hall, C., Rajendran, M., Spurr, N.K., and Lim, L. (1992). The human heat shock genes HSPA6 and HSPA7 are both expressed and localize to chromosome 1. Genomics **12**, 74–79.

Lowe, D. and Moran, L.A. (1986). Molecular cloning and analysis of DNA complementary to three mouse M_r = 68,000 heat shock protein mRNAs. J. Biol. Chem. **261**, 2102–2112.

Matsumoto, M. and Fujimoto, H. (1990). Cloning of a hsp70-related gene expressed in mouse spermatids. Biochem. Biophys. Res. Commun. **166**, 43–49.

Michikawa, Y., Baba, T., Arai, Y., Sakakura, T., and Kusakabe, M. (1993). Structure and organization of the gene encoding a mouse mitochondrial stress-70 protein. FEBS Lett. **336**, 27–33.

Milner, C.M. and Campbell, R.D. (1990). Structure and expression of the three MHC-linked Hsp70 genes. Immunogenetics **32**, 242–251.

Mukai, H., Kuno, T., Tanaka, H., Hirata, D., Miyakawa, T., and Tanaka, C. (1993). Isolation and characterization of SSE1 and SSE2, new members of the yeast HSP70 multigene family. Gene **132**, 57–66.

Perry, M.D., Aujame, L., Shtang, S., and Moran, L. (1994). Structure and expression of an inducible HSP70-encoding gene from *Mus musculus*. Gene **146**, 273–278.

Pfanner, N., Douglas, M.G., Endo, T., Hoogenraad, N.J., Jensen, R.E., Meijer, M., Neupert, W., Schatz, G., Schmitz, U.K., and Shore, G.C. (1996). Uniform nomenclature for the protein transport machinery of the mitochondrial membranes. Trends Biochem. Sci. **21**, 51–52.

Pilz, A., Prohaska, R., Peters, J., and Abbott, C. (1994). Genetic linkage analysis of the *Ak1*, *Col5a1*, *Epb7.2*, *Fpgs*, *Grp78*, *Pbx3*, and *Notch1* genes in the region of mouse chromosome 2 homologous to human chromosome 9q. Genomics **21**, 104–109.

Rensing, S.A. and Maier, U-G. (1994). Phylogenetic analysis of the stress 70 protein family. J. Mol. Evol. **39**, 80–86.

Roux, A-F., Nguyen, V.T.T., Squire, J.A., and Cox, D.W. (1994). A heat shock gene at 14q22: mapping and expression. Hum. Mol. Genet. **3**, 1819–1822.

Sargent, C.A., Dunham, I., Trowsdale, J. and Campbell, R.D. (1989). Human major histocompatibility complex contains genes for the major heat shock protein Hsp70. Proc. Natl. Acad. Sci. USA **86**, 1968–1972.

Snoek, M., Jansen, M., Olavesen, M.G., Campbell, R.D., Teuscher, C., and van Vugt, H. (1993). Three *Hsp70* genes are located in the *C4-H-2D* region: possible candidates for the *Orch-1* locus. Genomics **15**, 350–356.

Tavaria, M., Gabriele, T., Anderson, R.L., Mirault, M-E., Baker, E., Sutherland, G., and Kola, I. (1995). Localization of the gene encoding the human heat shock cognate protein, HSP73, to chromosome 11. Genomics **29**, 266–268.

Tavaria, M., Gabriele, T., Kola, I., and Anderson, R.L. (1996). A hitchhikers guide to the human Hsp70 family. Cell Stress Chaperones **1**, 23–28.

Ting, J. and Lee, A.S. (1988). Human gene encoding the 78,000-dalton glucose-regulated protein and its pseudogene: structure, conservation and regulation. DNA **7**, 275–286.

Voellmy, R., Ahmed, A., Schiller, P., Bromley, P., and Rungger, D. (1985). Isolation and functional analysis of a human 70,000-dalton heat shock protein gene segment. Proc. Natl. Acad. Sci. USA **82**, 4949–4953.

Wu, B., Hunt, C., and Morimoto, R.I. (1985). Structure and expression of the human gene encoding the major heat shock protein Hsp70. Mol. Cell. Biol. **5**, 330–341.

Yang, C. (1994). The molecular cloning and characterization of NST-1 (hsr.1), a novel member of the Hsp70 superfamily. Ph.D. Thesis. The Rockefeller University, New York.

Yasuda, K., Nakai, A., Hatayama, T., and Nagata, K. (1995). Cloning and expression of murine high molecular mass heat shock proteins, HSP105. J. Biol. Chem. **270**, 29718–29723.

Zakeri, Z.F., Wolgemuth, D.J., and Hunt, C.R. (1988). Identification and sequence analysis of a new member of the mouse *HSP70* gene family and characterization of its unique cellular and developmental pattern of expression in the male germ line. Mol. Cell. Biol. **8**, 2925–2932.

■ *Michael Tavaria and Ismail Kola:*
Molecular Genetics and Development Group
Institute of Reproduction and Development
Monash University
Clayton
Victoria 3168, Australia
Tel. 61 3 9550 5480
Fax. 61 3 9550 5568
E-mail: ismailko@silas.cc.monash.edu.au

■ *Robin L. Anderson:*
Peter MacCallum Cancer Institute
East Melbourne
Victoria 3002, Australia
Tel. 61 3 9656 1284
Fax. 61 3 9656 1411
E-mail: anderson@petermac.unimelb.edu.au

Mammalian Hsc70 and Hsp70 proteins

Constitutively expressed Hsc70 and stress-inducible Hsp70 proteins are located in the nucleocytoplasmic compartment of mammalian cells. Both are peptide/unfolded protein-stimulated ATPases. Hsc70 is the best characterized biochemically; it can distinguish between native and unfolded forms of the same protein in vitro, and binds with high affinity both hydrophobic/aromatic sequences like FYQLALT and hydrophobic/basic sequences like NIVRKKK. Hsc70 is part of a large chaperoning complex containing nascent polypeptide chains and another containing steroid hormone receptors, it facilitates protein renaturation, and it chaperones proteins for membrane translocation. The role of Hsp70 in stressed cells is not clearly defined but it has been implicated in thermotolerance. This family of chaperones works with other chaperones and accessory proteins, notably Hsp90 and DnaJ homologs.

■ Alternative names

Hsc70 is also known as clathrin-uncoating ATPase (Ungewickell, 1985), Hsc73 (Welch, 1992), Prp73 (Terlecky et al., 1992, see entry p. 58). Hsp70 is also named Hsp72 (Welch, 1992) and may generally be designated a stress-70 protein (Gething and Sambrook, 1992).

■ Isolation

Mammalian Hsp70 was originally described as a heat-inducible protein characterized on denaturing polyacrylamide gels (Kelley and Schlesinger, 1978). Mammalian Hsc70 was originally identified in rat cells as a normally abundant protein with a peptide map very similar to stress-inducible Hsp70 (Hightower and White, 1981).

■ *hsp70* genes and sequences

Human *hsp70* genes belong to multigene families whose individual members differ in levels of basal expression and induction. They are located on several different chromosomes (Harrison et al., 1987; see also previous entry, p. 49). GenBank contains complete nucleotide sequences for *hsp70* and *hsc70* genes from human, rat and mouse along with additional bovine and hamster *hsc70* sequences. A human *hsc70* gene (accession number Y00371) contains eight introns and encodes a protein (predicted M_r = 70 899 Da) which is 87% identical to human-inducible Hsp70 (predicted M_r = 69 800, *hsp70* accession number M11717). A small nucleolar RNA (U14 snoRNA) essential for pre-rRNA processing is encoded by introns 5, 6 and 8 of mouse, hamster, rat, human, frog and trout *hsc70* genes (Leverette et al., 1992). Amino acid sequence comparisons suggest that vertebrate Hsp70s are evolving faster than vertebrate Hsc70s. For example, mouse and human Hsc70 differ in only two amino acids whereas mouse and human homologous Hsp70 proteins are different at 34 positions.

■ Hsc70 and Hsp70 proteins

Hsc70 is usually purified as a mixture of monomers and oligomers from bovine brain, and Hsp70 from heated or chemically stressed cultured cells, e.g. HeLa cells. The application of ATP-affinity column chromatography was a major improvement in purification (Welch and Feramisco, 1985). Physical studies have focused on Hsc70 (reviewed in Hightower et al., 1994; McKay et al., 1994): the extinction coefficients at 205 and 280 nm at concentrations of 1 mg/ml are 29.2 and 0.54, respectively; circular dichroism spectroscopy and computer algorithms predict a secondary structure content of 41% α helix, 24% β strand, 24% β turn and 11% aperiodic structure; purified Hsc70 undergoes a thermal transition into an irreversibly aggregating molten globule with a transition midpoint of 43°C, which is shifted to 57°C by bound ATP/ADP; binding studies provide evidence of single nucleotide and peptide-binding sites (Greene et al., 1995). However, Hsc70 undergoes a conformational change upon binding hydrophobic/aromatic peptides which may mask a second peptide-binding site (Takenaka et al., 1995).

Structural studies have focused on Hsc70 (Fig. 1 and see entry on The three-dimensional structure of Hsc70, p. 13). The meandering β sheet topology of the peptide-binding domain defined by NMR (Morshauser et al., 1995) rules out previous computer models based on MHC class I and II peptide-binding domains. Using structural and protein sequence information on the ATPase domain, a common evolutionary origin has been proposed for the HSP70 family, actin, sugar kinases and prokaryotic cell cycle proteins (Bork et al., 1992). Computerized secondary structure algorithms for Hsc70 are still useful in predicting beyond the known helix α512, after which two strong α helix predictions are made (Fig. 1). The known and predicted helices contain a repeated heptad substructure, raising the possibility that the helices could be coiled or bundled. Hydropathic plots suggest that the β strands in the peptide-binding domain are the only regions in the C-terminal domain sufficiently hydrophobic to bind peptides like FYQLALT (see below).

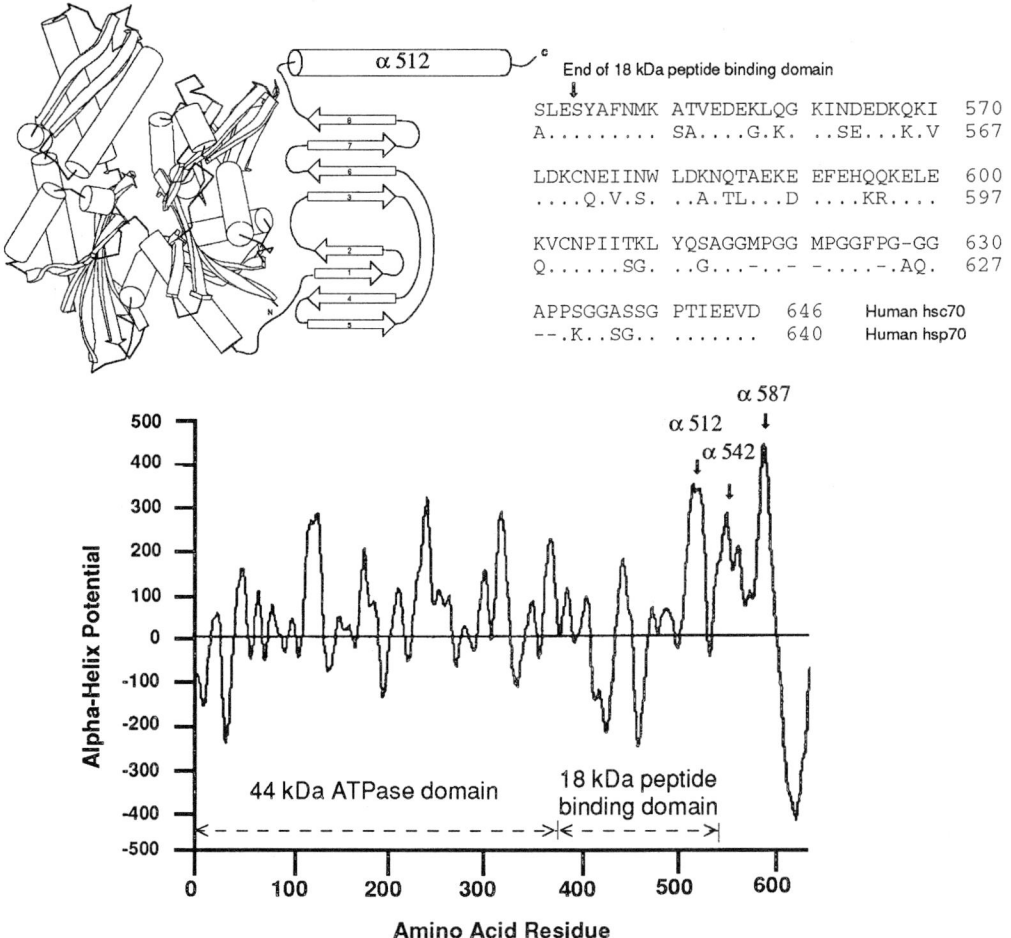

Figure 1. Molecular structure of Hsc70. The 44 kDa ATPase domain (residues 1–386) was determined by X-ray crystallography and consists of four domains forming two lobes with a deep cleft between them in which nucleotide binds (McKay et al., 1994; see also entry p. 13). The 18 kDa peptide-binding domain (residues 384–543) consists of two four-stranded antiparallel β sheets and a single α helix, detemined using multidimensional NMR (Morshauser et al., 1995). The C-terminal 10 kDa (residues 542–646) of Hsc70 is predicted using the Garnier algorithm to be primarily α helix followed by a Gly/Pro-rich aperiodic segment next to the highly conserved EEVD terminal sequence (Hightower et al., 1994). An amino acid sequence comparison of the least conserved C-terminal region of a human Hsp70 and Hsc70 is shown.

Hsc70 in unstressed cells is distributed in the cytoplasm and nucleus. In heat shocked cells, the amounts of Hsc70 along with newly synthesized Hsp70 increase in the nucleus (and particularly in nucleoli) and then redistribute to the cytoplasm during recovery (reviewed in Welch, 1992). Hsc70 and Hsp70 have been isolated with microtubules and microfilaments and have been identified in centrosomes. Hsc70 and Hsp70 are concentrated in regions active in protein synthesis in cells and may bind directly to nascent polypeptide chains (Georgopoulos and Welch, 1993). Hsp70 associates with ribosomes in polysomes of thermotolerant cells, suggesting a direct role in rescuing translation (Beck and Demaio, 1994). Portions of both Hsc70 and Hsp70 are easily releasable from cells by a non-ER–Golgi pathway, and Hsp70 binds to cell surfaces (Multhoff and Hightower, 1996).

Purified bovine Hsc70 and antibodies against both Hsp70 and Hsc70 are available commercially (StressGen Biotechnologies, Affinity Bioreagents, Sigma Chemical Co., Amersham). Because of the high sequence similarity between Hsc70 and Hsp70, most antibodies raised against either protein bind both proteins; those that bind only Hsp70 are quite species specific.

Biological activities of Hsc70 and Hsp70

Hsc70 and Hsp70 can be described functionally as peptide/unfolded protein-stimulated ATPases (Hightower et al., 1994; McKay et al., 1994). One of the major unanswered questions is whether Hsp70 and Hsc70 are functionally the same or not. As discussed above, Hsc70 sequences are more highly conserved among mammals than Hsp70 sequences, indicating that the evolutionary constraints on these two proteins are different, particularly near the C-terminus (Fig. 1), a region that may be involved in interdomain communication and interactions with accessory factors. This suggests that the two proteins differ somewhat in function and/or regulation.

Hsc70 binds denatured but not native forms of the same protein (Palleros et al., 1991; Sadis and Hightower, 1992). Hsc70 binds two different peptide sequence motifs, based on screening phage display peptide libraries. One motif, e.g. FYQLALT, containing large hydrophobic and aromatic residues is a relatively good stimulator of ATPase activity and probably represents sites used by Hsc70 to distinguish native from unfolded conformations (Fourie et al., 1994). The second motif, e.g. NIVRKKK, containing a combination of hydrophobic and basic residues is a poor stimulator of ATPase activity and may be used for chaperoning proteins to organelles or to bind certain folded proteins such as itself, p53 and DnaJ homologs (Takenaka et al., 1995). Remarkably, peptides naturally associated with Hsc70 purified from tumors can induce tumor-specific immunity (Udono and Srivastava, 1994).

A fluorescently labeled version of FYQLALT has been used to measure the kinetics of peptide binding and release to recombinant Hsc70 (Takeda and McKay, 1996). Peptide binding is a two-step process in the presence of MgADP. A low affinity peptide–Hsc70 complex is created in the first step (K_d = 14 mM) and the peptide is locked into a higher affinity complex (K_d = 4.3 mM) during the second step. In contrast, peptide binding is a one-step process in the presence of MgATP (K_d = 40–50 mM). A model was suggested for domain interactions in which the ATPase domain with bound ATP constrains the peptide-binding domain to a lower affinity conformation. After ATP hydrolysis, the peptide-binding domain is proposed to be less constrained by the ATPase domain, allowing a higher affinity interaction with the bound peptide.

In vitro, assembled clathrin baskets preferentially bind Hsc70–ATP and hydrolysis generates a relatively stable Hsc70–ADP–clathrin complex during disassembly (Greene et al., 1995). ATP hydrolysis is not required for release of substrates; exchange of ADP for ATP is sufficient (Palleros et al., 1993). Potassium ions bind specifically in the ATPase active site and are required for substrate binding/release and for optimum ATPase activity (Wilbanks and McKay, 1995). Kinetic analyses of recombinant bovine Hsc70 in the absence of peptide/unfolded protein substrate show that Pi release and chemical hydrolysis of MgATP have similar rates which are slower than ADP release (McKay et al., 1994).

Clathrin–Hsc70 complexes formed in vivo can be recovered from rat neurons by non-denaturing immunoprecipitation (Black et al., 1991), and microinjection into cells of a monoclonal antibody against the Hsc70 peptide-binding domain blocks receptor-mediated endocytosis (Honing et al., 1994). Microinjection of cells with anti-Hsc70 antibodies also inhibits import of peroxisomal proteins (Walton et al., 1994). In an in vitro study of synthesis and import of ornithine transcarbamylase precursor into mitochondria, the major requirement for Hsc70 is during synthesis, presumably to hold newly made precursor in an import-competent form (Terada et al., 1995). Both Hsp70 and Hsc70 contain a sequence beginning at residue 250 (FKRKHKKDISQNKRAVRR) that can target normally cytoplasmic proteins to the nucleus and nucleolus (Dang and Lee, 1989). This sequence may be responsible for entry of Hsp70 and Hsc70 into these compartments.

In cells microinjected with anti-Hsp70 monoclonal antibodies, heat-induced nuclear translocation of Hsp70 is blocked and the cells are unable to survive a heat shock (Georgopoulos and Welch, 1993). Increases in cellular levels of Hsp70 following a triggering heat shock parallel the increased protection against nuclear protein aggregation (Stege et al., 1995). A number of studies have shown that constitutive expression of inducible Hsp70 in transfected cells confers thermoresistance, increased survival and rapid recovery of translation (Liu et al., 1992; Kim et al., 1995). Heat-resistant mutants of a mouse fibrosarcoma contain a new Hsc70-like isoform, suggesting that Hsc70 is involved in the adaptation (Anderson et al., 1994). However, human/hamster hybrid cell lines, which vary broadly in intrinsic thermal resistance, all have similar levels of Hsc70, suggesting that the amounts of the constitutive form are not the only determinant (Anderson et al., 1994).

Biological interactions of Hsc70

HSP70 family proteins do not chaperone alone. In a mammalian translation assay using firefly luciferase as the model substrate (Frydman et al., 1994), the nascent polypeptide chain is in a large complex containing Hsc70, Hsp40 (the mammalian DnaJ homolog, see entry p. 121) and the chaperonin TRiC (see entry p. 207). Hsc70 also participates along with other chaperones and accessory proteins in firefly luciferase refolding in reticulocyte lysates (Frydman et al., 1994). Human DnaJ-like proteins bind Hsc70, stimulating ATP hydrolysis fivefold and lowering the affinity of Hsc70 for unfolded proteins, perhaps mimicking the effects of K^+ (Cheetham et al., 1994). Purified Hsc70 protects proteins from thermal denaturation and reactivates bacterial and mammalian DNA polymerases, and a mammalian DnaJ homolog lowers the amount of Hsc70 needed for reactivation tenfold (Ziemienowicz et al., 1995). Mutant forms of the p53 tumor suppressor gene product are immunopurified from cell extracts as complexes with Hsc70 and Hsp40 (Sugito et al., 1995). Hsc70 and Hsp90 are present in immunopurified complexes of a variety of steroid hormone receptors (Johnson and Toft, 1994). Three new accessory proteins have been found: Hip (Hsc70-interacting protein), Hop (Hsp70/Hsp90-organizing protein), and Hup (Hsc70-unbinding protein), an NDP kinase which releases Hsc70 from substrates (Frydman and Höhfeld, 1997; Leung and Hightower, 1997).

Biological regulation of Hsp70

Hsp70 is induced in mammalian cells by environmental stressors such as heat, during normal processes such as the cell cycle, development and differentiation, and during pathophysiological states such as fever, inflammation and wound responses (reviewed by Morimoto et al., 1994). The latter are complex responses involving both stressors and response modifiers like hormones and cytokines, so it can be anticipated that regulation of hsp70 gene expression in animals is considerably more complex than in cell culture models using single stressors. For example, pretreatment of cells with arachidonate, a mediator of inflammatory responses, lowers the temperature threshold for induction of heat shock gene transcription from 42 to 39°C (Morimoto et al., 1994). Hsp70 induction is down-regulated in certain differentiated cells, e.g. neurons, and during aging in cell culture and in animals. At least two heat shock transcription factors (HSF) are present in mammals; HSF1 is stress-inducible whereas HSF2 is used during differentiation and development. Evidence from cultured cells indicates that Hsp70 and Hsc70 are involved in autoregulation. Current models propose that HSF1 is kept in an inactive monomeric state by transient interactions with Hsc70 (Morimoto et al., 1994). During stress, Hsc70 would bind to damaged proteins, lowering free pools, and allowing HSF1 monomer to change conformations and form trimers. Efforts to induce Hsp70 by reducing Hsc70 levels by pseudogenetic approaches have been complicated by the fact that Hsc70 antisense RNA is much more effective in targeting newly synthesized transcripts than existing mRNA and because cells have compensatory regulatory mechanisms for maintaining Hsc70 levels (Li and Hightower, 1995). During recovery, increased amounts of Hsp70 in cells are proposed to cause active DNA-bound trimers of HSF1 to diassemble into inactive monomers again. Hsp70 may accelerate the recovery of heat shocked mammalian cells through alterations in HSF1 phosphorylation and HSE binding (Kim et al., 1995). These models all carry the prediction that Hsc70–HSF1 and Hsp70–HSF1 complexes should exist in cells. Activated HSF1 purified from heat shocked HeLa cells binds purified mixtures of Hsp70 and Hsc70 (Voellmy, 1995), and antibody shift experiments using anti-Hsp70 antisera indicate that a portion of HSF1 in extracts of heat shocked cells is bound to Hsp70 (Morimoto et al., 1994).

Biological regulation of Hsc70

Differential scanning calorimetry has been used to determine the thermal transitions of bovine Hsc70 (Leung et al., 1996). A novel finding was an exothermic transition, reflecting a conformational change in Hsc70, beginning at about 30°C with a transition midpoint of 41°C. Chaperoning activity, measured by formation of complexes between Hsc70 and apocytochrome c in vitro increases over the same temperature range as the exothermic conformational change. Thus, Hsc70 may be a thermal sensor that matches the supply of chaperoning activity with demand over the physiological temperature range of mammalian cells. Thermal activation of Hsc70 may also have a role in acquired thermotolerance.

Mutagenesis studies

Active site mutants of the Hsc70 ATPase domain have been characterized: substitutions at Glu-175 and Asp-206 reduce k_{cat} 100-fold; changes at Glu175 and Asp-203 reduce k_{cat} 10-fold; changes at Asp-199 and Asp-206 have little effect on K_m; and changes at Asp-10 and Glu-175 increase K_m 10–100-fold (reviewed by McKay et al., 1994). No single Asp or Glu residue in the active site dominates ATP hydrolysis. The bound Mg^{2+} and its local environment have major effects on catalysis. Based on structural and kinetic studies, it is proposed that the γ-phosphate of bound ATP reorients to form a β, γ-bidentate phosphate complex with Mg^{2+} which allows direct nucleophilic attack on the γ-phosphate by a water molecule or OH– followed by Pi release (McKay et al., 1994).

In recombinant Hsc70 produced in E. coli, substitution of Asp-10 with Asn also reduces ATPase activity; the 18 kDa fragment immediately following the ATPase domain is sufficient for peptide binding; and removal of the C-terminal 10 kDa eliminates clathrin-uncoating activity but not peptide-stimulated ATPase activity (Tsai and Wang, 1994). Deletion of the highly conserved C-terminal EEVD sequence of Hsp70 increases ATPase activity, reduces unfolded protein binding and inhibits interactions with the DnaJ homolog HDJ-1 (Freeman et al., 1995). C-terminal deletions of Hsc70 (67.5, 59.5 and 56.5 kDa fragments) are impaired in nuclear transport (Mandell and Feldherr, 1992). CHO cell mutants resistant to microtubule inhibitors have altered isoforms of Hsc70, suggesting a role in microtubule assembly or function (Ahmad et al., 1990). In tumor cell variants with either enhanced or reduced expression of Hsp70, tumor immunogenicity cosegregates with expression of Hsp70 but not Hsc70 in rat colon carcinomas (Menoret et al., 1995).

References

Ahmad, S., Ahuja, R., Venner, T.J., and Gupta, R.S. (1990). Identification of a protein altered in mutants resistant to microtubule inhibitors as a member of the major heat shock protein (hsp70) family. Mol. Cell. Biol. **10**, 5160–5165.

Anderson, R.L., Naylor, D.J., Gabriele, T., Tetaz, T., and Hoj, P.B. (1994). Characterization of novel hsp70 in mammalian cells. Int. J. Hyperthermia **10**, 419–428.

Beck, S.C. and Demaio, A. (1994). Stabilization of protein synthesis in thermotolerant cells during heat shock — association of heat shock protein-72 with ribosomal subunits of polysomes. J. Biol. Chem. **269**, 21803–21811.

Black, M.M., Chestnut, M.H., Pleasure, I.T., and Keen, J.H. (1991). Stable clathrin:uncoating protein (hsc70) complexes in intact neurons and their axonal transport. J. Neurosci. **11**, 1163–1172.

Bork, P., Sander, C., and Valencia, A. (1992). An ATPase domain common to prokaryotic cell cycle proteins, sugar kinases, actin, and hsp70 heat shock proteins. Proc. Natl. Acad. Sci. USA **89**, 7290–7294.

Cheetham, M.E., Jackson, A.P., and Anderton, B.H. (1994). Regulation of 70-kDa heat-shock-protein ATPase activity and substrate binding by human DnaJ-like proteins, HSJ1a and HSJ1b. Eur. J. Biochem. **226**, 99–107.

Dang, C. and Lee, W. (1989). Nuclear and nucleolar targeting sequences of c-erb-A, c-myc, N-myc, p53, HSP70, and HIV tat proteins. J. Biol. Chem. **264**, 18019–18023.

Fourie, A.M., Sambrook, J.F., and Gething, M.J.H. (1994). Common and divergent peptide binding specificities of hsp70 molecular chaperones. J. Biol. Chem. **269**, 30470–30478.

Freeman, B.C., Myers, M.P., Schumacher, R., and Morimoto, R.I. (1995). Identification of a regulatory motif in Hsp70 that affects ATPase activity, substrate binding and interaction with HDJ-1. EMBO J. **14**, 2281–2292.

Frydman, J., and Höhfeld, J. (1997). Chaperones get in touch: the Hip-Hop connection. Trends. Biochem. Sci. **22**, 89–92.

Frydman, J., Nimmesgern, E., Ohtsuka, K., and Hartl, F.U. (1994). Folding of nascent polypeptide chains in a high molecular mass assembly with molecular chaperones. Nature **370**, 111–117.

Georgopoulos, C. and Welch, W.J. (1993). Role of the major heat shock proteins as molecular chaperones. Annu. Rev. Cell Biol. **9**, 601–634.

Gething, M. and Sambrook, J. (1992). Protein folding in the cell. Nature **355**, 33–45.

Greene, L.E., Zinner, R., Naficy, S., and Eisenberg, E. (1995). Effect of nucleotide on the binding of peptides to 70-kDa heat shock protein. J. Biol. Chem. **270**, 2967–2973.

Harrison, G.S., Drabkin, H.A., Kao, F.T., Hartz, J., Hart, I.M., Chu, E.H., Wu, B.J., and Morimoto, R.I. (1987). Chromosomal location of human genes encoding major heat-shock protein hsp70. Somat. Cell Mol. Genet. **13**, 119–30.

Hightower, L.E. and White, F.P. (1981). Cellular responses to stress: comparison of a family of 71-73 kilodalton proteins rapidly synthesized in rat tissue slices and canavanine-treated cells in culture. J. Cell. Physiol. **108**, 261–275.

Hightower, L.E., Sadis, S.E., and Takenaka, I.M. (1994). Interactions of vertebrate hsc70 and hsp70 with unfolded proteins and peptides. In *The Biology of Heat Shock Proteins and Molecular Chaperones*, (R.I. Morimoto, A. Tissieres and C. Georgopoulos, eds). Col Spring Harbor Laboratory Press, New York, pp. 179–208.

Honing, S., Kreimer, G., Robenek, H., and Jockusch, B.M. (1994). Receptor-mediated endocytosis is sensitive to antibodies against the uncoating ATPase (hsc70). J. Cell Sci. **107**, 1185–1196.

Johnson, J.L. and Toft, D.O. (1994). A novel chaperone complex for steriod receptors involving heat shock proteins, immunophilins, and p23. J. Biol. Chem. **269**, 24989–24993.

Kelley, P.M. and Schlesinger, M.J. (1978). The effect of amino acid anologues and heat shock on gene expression in chicken embryo fibroblasts. Cell **15**, 1277–1286.

Kim, D.H., Ouyang, H., and Li, G.C. (1995). Heat shock protein hsp70 accelerates the recovery of heat-shocked mammalian cells through its modulation of heat shock transcription factor HSF1. Proc. Natl. Acad. Sci. USA **92**, 2126–2130.

Leverette, R.D., Andrews, M.T., and Maxwell, E.S. (1992). Mouse U14 snRNA is a processed intron of the cognate hsc70 heat shock premessenger RNA. Cell **71**, 1215–1221.

Leung, S.-M. and Hightower, L.E. (1997). A 16-kDa protein functions as a new regulatory protein for Hsc70 molecular chaperone and is identified as a member of the Nm23/nucleoside diphosphate kinase family. J. Biol. Chem. **272**, 2607–2614.

Leung, S.-M., Senisterra, G., Ritchie, K.P., Sadis, S.E., Lepock, J.R., and Hightower, L.E. (1996). Thermal activation of the bovine Hsc70 molecular chaperone at physiological temperatures: physical evidence of a molecular thermometer. Cell Stress Chap. **1**, 78–89.

Li, T. and Hightower, L.E. (1995). Effects of dexamethasone, heat shock, and serum responses on the inhibition of hsc70 synthesis by antisense RNA in NIH 3T3 cells. J. Cell. Physiol. **164**, 344–355.

Liu, R.Y., Li, X.G., and Li, G.C. (1992). Expression of human hsp70 in rat fibroblasts enhances cell survival and facilitates recovery from translational and transcriptional inhibition following heat shock. Cancer Res. **52**, 3997–3673.

Mandell, R. and Feldherr, C. (1992). The effect of carboxyl-terminal deletions on the nuclear transport rate of rat hsc70. Exp. Cell Res. **198**, 164–169.

McKay, D.B., Wilbanks, S.M., Flaherty, K.M., Ha, J.-H., O'Brien, M.C., and Shirvanee, L.L. (1994). Stress-70 proteins and their interaction with nucleotides. In *The Biology of Heat Shock Proteins and Molecular Chaperones*, (R.I. Morimoto, A. Tissieres and C. Georgopoulos, eds). Cold Spring Harbor Laboratory Press, New York, pp. 153–178.

Menoret, A., Patry, Y., Burg, C., and Lependu, J. (1995). Co-segregation of tumor immunogenicity with expression of inducible but not constitutive hsp70 in rat colon carcinomas. J. Immunol. **155**, 740–747.

Morimoto, R.I., Jurivich, D.A., Kroeger, P.E., Mathur, S.K., Murphy, S.P., Nakai, A., Sarge, K., Abravaya, K., and Sistonen, L.T. (1994). Regulation of heat shock gene transcription by a family of heat shock factors. In *The Biology of Heat Shock Proteins and Molecular Chaperones*, (R.I. Morimoto, A. Tissieres and C. Georgopoulos, eds). Cold Spring Harbor Laboratory Press, New York, pp. 417–456.

Morshauser, R.C., Wang, H., Flynn, G.C., and Zuiderweg, E.R.P. (1995). The peptide-binding domain of the chaperone protein hsc70 has an unusual secondary structure topology. Biochemistry **34**, 6261–6266.

Multhoff, G., and Hightower, L.E. (1996). Cell surface expression of heat shock proteins and the immune response. Cell Stress Chap. **1**, 167–176.

Palleros, D., Welch, W., and Fink, A. (1991). Interaction of Hsp70 with unfolded proteins: effects of temperature and nucleotides on the kinetics of binding. Proc. Natl. Acad. Sci. USA **88**, 5719–5723.

Palleros, D., Reid, K., Li, S., Welch, W., and Fink, A. (1993). ATP- induced protein-hsp70 complex dissociation requires K$^+$ and does not involve ATP hydrolysis. Nature **365**, 664–666.

Sadis, K. and Hightower, L.E. (1992). Unfolded proteins stimulate molecular chaperone Hsc70 ATPase by accelerating ADP/ATP exchange. Biochemistry **31**, 9406–9412.

Stege, G.J.J., Brunsting, J.F., Kampinga, H.H., and Konings, A.W.T. (1995). Thermotolerance and nuclear protein aggregation: protection against initial damage or better recovery? J. Cell. Physiol. **164**, 579–586.

Sugito, K., Yamane, M., Hattori, H., Hayashi, Y., Tohnai, I., Ueda, M., Tsuchinda, N., and Ohtsuka, K. (1995). Interaction between hsp70 and hsp40, eukaryotic homologues of DnaK and DnaJ, in human cells expressing mutant-type p53. FEBS Lett. **358**, 161–164.

Takeda, S. and Mckay, D.B. (1996). Kinetics of peptide binding to the bovine 70 kDa heat shock cognate protein, a molecular chaperone. Biochemistry **35**, 4636–4644.

Takenaka, I.M., Leung, S.-M., McAndrew, S.J., Brown, J.P., and Hightower, L.E. (1995). Hsc70-binding peptides selected from a phage display peptide library that resemble organellar targeting sequences. J. Biol. Chem. **270**, 19839–19844.

Terada, K., Ohtsuka, K., Imamoto, N., Yoneda, Y., and Mori, M. (1995). Role of heat shock cognate 70 protein in import of ornithine transcarbamylase precursor into mammalian mitochondria. Mol. Cell. Biol. **15**, 3708–3713.

Terlecky, S., Chiang, H., Olson, T., and Dice, J.F. (1992). Protein and peptide binding and stimulation of *in vitro* lysosomal proteolysis by the 73-kDa heat shock cognate proteins. J. Biol. Chem. **267**, 9202–9209.

Tsai, M.Y. and Wang, C. (1994). Uncoupling of peptide-stimulated ATPase and clathrin-uncoating activity in deletion mutant of Hsc70. J. Biol. Chem. **269**, 5958–5962.

Udono, H. and Srivastava, P.K. (1994). Comparison of tumor-specific immunogenicities of stress-induced proteins gp96, hsp90, and hsp70. J. Immunol. **152**, 5398–5403.

Ungewickell, E. (1985). The 70-kd mammalian heat shock proteins are structurally and functionally related to the uncoating protein that releases clathrin triskelion from coated vesicles. EMBO J. **4**, 3385–3391.

Voellmy, R. (1995). Transduction of the stress signal, and mechanisms of transcriptional regulation of heat shock/stress protein gene expression in higher eukaryotes. Crit. Rev. Eukaryotic Gene Expr. **4**, 357–401.

Walton, P.A., Wendland, M., Subramani, S., Rachubinski, R.A., and Welch, W.J. (1994). Involvement of 70-kD heat-shock proteins in peroxisomal import. J. Cell Biol. **125**, 1037–1046.

Welch, W.J. (1992). Mammalian stress response: cell physiology, structure/function of stress proteins and implications for medicine and disease. Physiol. Rev. **72**, 1063–81.

Welch, W.J. and Feramisco, J.R. (1985). Rapid purification of mammalian 70,000 dalton stress proteins: affinity of the proteins for nucleotides. Mol. Cell. Biol. **5**, 1229–1237.

Wilbanks, S.M. and McKay, D.B. (1995). How potassium affects the activity of the molecular chaperone hsc70. II. Potassium binds specifically in the ATPase active site. J. Biol. Chem. **270**, 2251–2257.

Ziemienowicz, A., Zylicz, M., Floth, C., and Hubscher, U. (1995). Calf thymus hsc70 protein protects and reactivates prokaryotic and eukaryotic enzymes. J. Biol. Chem. **270**, 15479–15484.

■ *Lawrence E. Hightower and Sau-Mei Leung:*
Deptartment of Molecular and Cell Biology
University of Connecticut
Storrs, CT 06269-3044, USA.
Tel. 1 860 486 4257
Fax. 1 860 486 1784
E-mail: hightower@biotek.mcb.uconn.edu

Mammalian Prp73

Prp73, the RNase S-peptide recognition protein of 73 kDa, is a member of the HSP70 family. This protein stimulates the uptake of substrate polypeptides by lysosomes and thereby facilitates this selective pathway of lysosomal proteolysis. Prp73 has been identified by sequencing and functional analysis to be the major constitutively expressed HSP70 in mammalian cells, Hsc70.

■ Alternative names

Hsc70; Hsc73 (Terlecky et al., 1992); clathrin uncoating ATPase (Deluca-Flaherty et al., 1990).

■ Isolation

Prp73 was identified as a cytosolic protein able to bind specifically to RNase S-peptide (residues 1–20 of RNase A), a known substrate for a selective pathway of lysosomal proteolysis (Chiang et al., 1989; Dice, 1990). Prp73 binds to RNase S-peptide–Sepharose with a K_d of approximately 5 μM (Terlecky et al., 1992) and was able to stimulate uptake and degradation of RNase S-peptide and RNase A by lysosomes (Chiang et al., 1989; Terlecky et al., 1992). Prp73 is now known to be the major constitutively expressed HSP70, Hsc70. This conclusion is based on limited sequence analysis of Prp73, which shows 100% identity with Hsc70 (Chiang et al., 1989) as well as comparisons of Prp73, isolated by RNase S-peptide affinity chromatography, and Hsc70, isolated by ATP-affinity chromatography. Both proteins bind to RNase S-peptide, RNase A, aspartate aminotransferase, and pyruvate kinase, proteins that contain KFERQ peptide motifs and are substrates of the selective lysosomal degradation pathway; and not to β-galactosidase, lysozyme, ovalbumin, or ubiquitin, proteins that do not contain KFERQ peptide motifs and are not substrates of this lysosomal degradation pathway (Terlecky et al., 1992). Furthermore, in an *in vitro* assay utilizing purified lysosomes, uptake and degradation of RNase S-peptide was stimulated equivalently by added Prp73 or Hsc70 while three other HSP70 family members were inactive (Terlecky et al., 1992). More recently we have shown that recombinant human Hsc70 produced in *E. coli* is active in stimulating this selective pathway of lysosomal proteolysis using RNase S-peptide or glyceraldehyde-3-phosphate dehydrogenase (Cuervo et al., 1994) as substrates (A. Cuervo, E. Knecht, and J.F. Dice, unpublished data).

For a detailed description of the structure, function, and biological activities of Hsc70, see the entries on pp. 13 and 53.

■ Biological regulation

We reported that levels of intracellular Prp73 were increased in confluent fibroblast monolayers in response to serum withdrawal, the same stimulus that activates the selective lysosomal pathway of proteolysis (Chiang et al., 1989). We have not been able to reproduce this finding, and we believe the original results were erroneous because of incomplete recovery of Hsc70 from cells grown in the presence of serum (Dice et al., 1994). Although most intracellular Hsc70 is cytosolic, a small proportion of Hsc70 is associated with lysosomes (Terlecky and Dice, 1993; Cuervo et al., 1995), and this proportion reproducibly increases in response to serum withdrawal (Terlecky and Dice, 1993). This lysosomal Hsc70 is important for the import of substrate proteins into lysosomes (Dice et al., 1994), so the amount of Hsc70 within lysosomes may be a critical factor in regulating this selective pathway of lysosomal proteolysis. The lysosomal membrane receptor (Cuervo and Dice, 1996) also appears to regulate the activity of this proteolytic pathway.

■ References

Agarraberes, F.A., Terlecky, S.R., and Dice, J.F. (1997). An intralysosomal hsp70 is required for a selective pathway of lysosomal protein degradation. J. Cell Biol., in press.

Chiang, H.-L., Terlecky, S.R., Plant, C.P., and Dice, J.F. (1989). A role for a 70-kilodalton heat shock protein in lysosomal proteolysis of intracellular proteins. Science **246**, 282–285.

Cuervo, A.M., Terlecky, S.R., Dice, J.F., and Knecht, E. (1994). Selective binding and uptake of ribonuclease A and glyceraldehyde-3-phosphate dehydrogenase by isolated rat liver lysosomes. J. Biol. Chem. **269**, 26374–26380.

Cuervo, A.M., Terlecky, S.R., Knecht, E., and Dice, J.F. (1995). Activation of a selective pathway of lysosomal proteolysis in rat liver by prolonged starvation. Am. J. Physiol. **69**, C1200–C1208.

Cuervo, A.M. and Dice, J.F. (1996). A receptor for the selective uptake and degradation of proteins by lysosomes. Science **273**, 501–503.

Cuervo, A.M., Dice, J.F., and Knecht, E. (1997). A population of rat liver lysosomes responsible for the selective uptake and degradation of cytosolic proteins. J. Biol. Chem. **272**, 5606–5615.

Deluca-Flaherty, C., McKay, D.B., Parnum, P., and Hill, B.L. (1990). Uncoating protein (hsc70) binds a conformationally labile domain of clathrin chain LCa to stimulate ATP hydrolysis. Cell **62**, 875–887

Dice, J.F. (1990). Peptide sequences that target cytosolic proteins for lysosomal proteolysis. Trends Biochem. Sci. **15**, 305–309.

Dice, J.F., Agarraberes, F., Kirven-Brooks, M., and Terlecky, S.R. (1994). Heat shock 70-kD proteins and lysosomal proteolysis. In The Biology of Heat Shock Proteins and Molecular Chaperones, (R.I. Morimoto, A Tissieres, and C. Georgopoulos, eds). Cold Sping Harbor Laboratory Press, New York, pp. 137–151.

Terlecky, S.R. and Dice, J.F. (1993). Polypeptide import and degradation by isolated lysosomes. J. Biol. Chem. **268**, 23490–23495.

Terlecky, S.R., Chiang, H.-L., Olson, T.S., and Dice, J.F. (1992). Protein and peptide binding and stimulation of in vitro lysosomal proteolysis by the 73 kDa heat shock cognate protein. J. Biol. Chem. **267**, 9202–9209.

■ J. Fred Dice:
Department of Physiology
Tufts University School of Medicine
136 Harrison Avenue
Boston, MA 02111, USA
Tel. 1 617 636 6707
Fax. 1 617 636 0445
E-mail: dice@opal.tufts.edu

Mammalian BiP

The BiP protein of mammalian cells is a member of the HSP70 protein family. BiP, which is located in the lumen of the endoplasmic reticulum (ER), was originally identified independently as the immunoglobulin heavy chain binding protein and the glucose-regulated protein, Grp78. The presumed role of BiP is to bind newly synthesized secretory proteins as they are translocated into the ER and to maintain them in a state competent for subsequent folding and oligomerization. BiP is an abundant protein under all growth conditions, but its synthesis is markedly induced under conditions that lead to the accumulation of unfolded polypeptides in the ER.

■ Alternative names

Grp78 (Pouyssegur et al., 1977); immunoglobulin heavy chain binding protein (Haas and Wabl, 1983).

■ BiP gene and sequence

The nucleotide sequences of BiP cDNAs cloned from mammalian species (Munro and Pelham, 1986; Ting et al., 1987; Ting and Lee, 1988; Kozutsumi et al., 1989) (GenBank accession numbers M19645 and M17169) predict a protein of 635 amino acids that includes an 18-residue N-terminal signal sequence, and a C-terminal tetrapeptide (Lys–Asp–Glu–-Leu or KDEL) required for retrieval of BiP molecules that leave the ER (Pelham, 1989). Clusters of acidic amino acids, thought to be involved in the low affinity, high capacity Ca^{2+}-binding activity of the molecule (Macer and Koch, 1988), are found at both the N- and C-termini of the molecule. BiP proteins are highly conserved, there being 99% amino acid sequence identity between the mature proteins from human and rodent cells (Kozutsumi et al., 1989) and 67% identity between mouse BiP and the Saccharomyces cerevisiae BiP protein, Kar2p (Normington et al., 1989; see entry, p. 33).

The gene encoding murine BiP maps to mouse chromosome 2 (Haas et al., 1992), while the human gene has been localized to chromosome 9q34 (Hendershot et al., 1994).

■ BiP protein

The c. 78 kDa BiP protein is the only ER-located member of the HSP70 protein family identified to date in mammalian cells. The wild-type protein has been purified from dog pancreas microsomes (Kassenbrock and Kelly, 1989) and from bovine liver (Flynn et al., 1989; Rowling et al., 1994), using protocols involving affinity chromatography on ATP agarose. Wild-type and mutant forms of recombinant BiP have been purified following expression in E. coli (Blond-Elguindi et al., 1993b; Wei and Hendershot, 1995), and recombinant hamster BiP is available from StressGen Biotechnologies Corp. Polyclonal antibodies that recognize rodent BiP proteins but not human BiP are available from Affinity BioReagents, Inc. and StressGen Biotechnologies Corp.

Like other HSP70 proteins, BiP binds ATP with high affinity (Kassenbrock and Kelly, 1989; Wei and Hendershot, 1995) and displays a weak ATPase activity (Kassenbrock and Kelly, 1989; Flynn et al., 1989) that is stimulated by binding of unfolded proteins and some, but not all, synthetic peptides (Flynn et al., 1989; Blond-Elguindi et al., 1993a, b; Fourie et al., 1994; Knarr et al., 1995). Binding of adenine nucleotides causes conformational changes in the protein that result in altered sensitivity to proteases (Kassenbrock and Kelly, 1989; Wei and Hendershot, 1995): ATP protects an c. 60 kDa fragment while ADP protects an c. 44–48 kDa fragment. This effect of ATP does not require hydrolysis since ATPγS can substitute. By contrast, addition of ATP, but not of non-hydrolysable analogues, to cell extracts causes dissolution of complexes between BiP and protein substrates such as immunoglobulin heavy chains (Munro and Pelham, 1986). The analysis of ATPase-defective mutants of BiP confirmed that hydrolysis is not necessary for ATP-induced peptide release (Wei et al., 1995).

Although the three-dimensional structure of BiP has not been determined, an N-terminal c. 45 kDa fragment of bovine Hsc70, which retains the ATPase activity of the molecule, has been crystallized and its structure determined to a resolution of 2.2 Å by X-ray crystallography (Flaherty et al., 1990; see entry p. 13). The amino acid sequence of this domain of BiP, which shares 66% identity and 81% similarity with that of the cytosolic Hsc70 protein, can be mapped without difficulty on to the backbone of the Hsc70 structure, suggesting that the two proteins have very similar conformations (D. McKay, personal communication). Recently the structure of the C-terminal polypeptide-binding domain of DnaK has also been determined by X-ray crystallography (Zhu et al., 1996; see entry p. 18), revealing a novel fold that is also likely to be very similar in all members of the HSP70 protein family.

■ Post-translational modification of BiP

In vivo mammalian BiP exists in interconvertible monomeric and oligomeric forms (Hendershot et al., 1988; Carlino et al., 1992; Freiden et al., 1992; Blond-Elguindi et al., 1993b), and can be post-translationally modified by phosphorylation on serine and threonine residues (Welch et al., 1983; Hendershot et al., 1988; Leustek et al., 1991) and by ADP ribosylation (Carlsson and Lazarides, 1983; Leno and Ledford, 1990; Ledford and Leno, 1994). These modifications may be important in regulating the synthesis and polypeptide-binding activity of the molecule (Freiden et al., 1992). In vitro BiP, like several other HSP70 family members, has a Ca^{2+}-dependent autophosphorylation activity (Leustek et al., 1991, 1992) whose substrate is a threonine residue located in the ATP-binding cleft (Gaut and Hendershot, 1993a). This threonine is apparently not a detectable site of phosphorylation in vivo (Gaut and Hendershot, 1993a) and its modification may occur as a side reaction occurring during ATP hydrolysis. To what extent the degree of post-translational modification affects the basal ATPase activity of the BiP molecules or their propensity to oligomerize is not known. However, conditions that increase the levels of unfolded polypeptides in the ER lumen cause a decrease in the extent of modification of BiP (Carlsson and Lazarides, 1983; Hendershot et al., 1988; Leustek et al., 1991; Leno and Ledford, 1990) and an increase in the proportion of monomeric species (Freiden et al., 1992). Only unmodified, monomeric BiP molecules are found in complexes with unfolded or unassembled polypeptides (Hendershot et al., 1988; Freiden et al., 1992).

■ Recognition of polypeptide substrates by BiP

BiP's role as a general chaperone during protein folding in the ER lumen depends on its ability to recognize a wide variety of nascent polypeptides that share no obvious sequence similarity, while accurately discriminating between properly folded and unfolded structures. In vitro, BiP can interact with short synthetic peptides whose binding stimulates its ATPase activity (Flynn et al., 1989; Blond-Elguindi et al., 1993a, b; Fourie et al., 1994; Knarr et al., 1995) and alters its oligomeric state (Blond-Elguindi et al., 1993b). Experiments using a small set of randomly chosen synthetic peptides showed that their affinities for BiP varied over a 1000-fold range (Flynn et al., 1989). Analysis of the BiP-binding capacity of a collection of synthetic peptides of defined length but random sequence (Flynn et al., 1991) demonstrated that the optimum length for binding was 7–8 residues and revealed a positive selection for amino acids with aliphatic side chains, particularly at positions 3 to 6. The characteristics of peptides that bind to BiP was also investigated using affinity screening of large, highly diverse libraries of peptides displayed by fusion to the N-termini of the 4–5 copies of the pIII protein present at one tip of bacteriophage fd particles (Blond-Elguindi et al., 1993a). The peptides displayed by BiP-binding bacteriophages selected by affinity panning showed extensive sequence diversity, consistent with the observed 'promiscuity' of BiP's interaction with a wide variety of unrelated nascent polypeptides, and usually exhibit marked hydrophobicity, consistent with the likelihood that BiP interacts with sequences normally located in the interior of a fully folded protein. Tryptophan, phenylalanine and leucine, and to a lesser extent methionine and isoleucine, were increased in abundance in the peptides displayed by BiP-binding bacteriophages. The analysis revealed a heptameric motif best described as Hy(W/X)HyXHyXHy, where Hy is a bulky aromatic or hydrophobic residue (most frequently tryptophan, phenylalanine or leucine, but also methionine and isoleucine), W is tryptophan and X is any amino acid.

A computer scoring system that codifies the observed positional preferences and calculates the scores for all possible heptapeptides in a polypeptide sequence can be used to predict BiP-binding sites in natural proteins (Blond-Elguindi et al., 1993a). Analysis of the ATPase stimulating activity of putative BiP-binding sequences identified in immunoglobulin heavy chains using this BiP-Score program demonstrated its predictive power to be > 50% for positively scoring peptides, rising to > 80% when peptides with scores >10 were considered, and to be close to 100% for negatively scoring peptides (Knarr et al., 1995). Interestingly, when the BiP-binding sequences were mapped on to the three-dimensional structure of the Fd antibody fragment, the majority involve residues that participate in contact sites between the heavy and light chains. This suggests that, in vivo, BiP chaperones the folding and assembly of antibody molecules by binding to hydrophobic surface regions on the isolated subunits that subsequently participate in interchain contacts.

Comparative studies of the binding of synthetic peptides to BiP and other HSP70 family members indicate that HSP70 proteins can exhibit common or exclusive binding specificities, depending on the peptide sequence (Fourie et al., 1994; Gragerov and Gottesman, 1994).

■ Function of BiP

BiP binds transiently to newly synthesized wild-type proteins in the ER and more permanently to misfolded, underglycosylated or unassembled proteins whose transport from the ER is blocked (reviewed in Gething et al., 1994). BiP does not interact with native polypeptides. Like all HSP70 proteins (reviewed by Gething and Sambrook, 1992; Hartl, 1996), BiP is thought to play a role in the folding and assembly of newly synthesized proteins by recognizing unfolded polypeptides and by inhibiting intra- or intermolecular aggregation, maintaining them in a state competent for subsequent folding and oligomerization. Experiments performed in yeast also indicate a role for BiP during protein folding and assembly in the ER as well as in translocation of newly synthesized polypeptides across the ER membrane (see the Kar2p entry, page 33).

■ Interaction of BiP with HSP40/DnaJ-like cochaperones

Studies in S. cerevisiae demonstrate that the yeast BiP protein, Kar2p, interacts with two different ER-located HSP40 proteins, Sec63p (see entry p. 108) and Scj1p (see entry p. 110). Sec63p is a multispanning transmembrane protein whose J-domain is located in the ER lumen (Feldheim et al., 1992). The association of Kar2p with components of the ER translocation machinery is mediated through a nucleotide-dependent interaction which is compromised by a single amino acid substitution of a conserved residue in the DnaJ-like domain of Sec63p (Brodsky and Schekman, 1993). A possible mammalian counterpart of Sec63p has been identified, Mtj1 (Brightman et al., 1995, see entry p. 129) but no evidence is available for a physical or functional interaction between Mtj1 and BiP in the mammalian ER. Scj1 is a luminal ER homologue of DnaJ that is thought to interact with Kar2p and facilitate its function in protein folding (Schlenstedt et al., 1995). No mammalian homologue of Scj1 has been identified.

■ Cooperation between BiP and other chaperones

Studies on the folding of newly synthesised polypeptides in the ER indicate that BiP and other chaperones, including Grp94 (see entry p. 161) and calnexin (see entry p. 299), cooperate to assist the folding and assembly of secretory proteins. BiP and Grp94 can be isolated in a ternary complex with newly synthesised immunoglobulin chains (Melnick et al., 1992). BiP preferentially binds an early disulfide intermediate of Ig light chain and dissociates within a few minutes, while Grp94 exclusively binds fully oxidized molecules and dissociates with a half-time of 50 min (Melnick et al., 1994). Similarly, following cross-linking, BiP, Grp94 and ERp72 (a PDI-related protein of the ER lumen, see entry p. 353) can be isolated in complexes with newly synthesized thyroglobulin (Kuznetsov et al., 1994). When nascent thyroglobulin is induced to aggregate in the ER by treatment of live cells with DTT, the protein binds calnexin (Kim and Arvan, 1995). Following washout of the reducing agent, calnexin dissociates and increasing amounts of the thyroglobulin are found associated with BiP. Subsequently BiP dissociates and ultimately the folded thyroglobulin is secreted from the cells. A reverse sequence of BiP binding followed by calnexin binding was observed during the folding of vesicular stomatitis virus G protein (Hammond and Helenius, 1994). BiP bound maximally to early folding intermediates of G-protein, whereas calnexin bound after a short lag to more folded molecules. Newly synthesized major histocompatibility complex class I and II chains also interact transiently, but not necessarily simultaneously, with both BiP and calnexin (Marks et al., 1995; Nossner and Parham, 1995). Taken together these studies indicate that different chaperones may act either simultaneously or sequentially during the conformational maturation of secretory proteins, with the nature and order of involvement of the various chaperones depending perhaps on the particular structural features displayed by the folding polypeptides at different stages of their maturation.

■ Mutagenesis studies on BiP

Hendershot and colleagues have mutated residues in hamster BiP that, based on the known structure of the bovine Hsc70 ATPase domain (Flaherty et al., 1990), were implicated in ATP binding or hydrolysis (Gaut and Hendershot, 1993a, b; Wei et al., 1995). Analysis of the phenotypes of bacterially produced recombinant BiP proteins bearing such mutations revealed that amino acid substitutions within the nucleotide-binding site either inhibit ATP binding (G226D, G227D), ATP hydrolysis (E201G, T229G), autophosphorylation (T37G, T229G) or the ATP-induced conformational change in the BiP molecule (T37G). The in vitro autophosphorylation site was mapped to T229, but this was not a detectable site of phosphorylation of BiP in cells (Gaut and Hendershot, 1993b). ATP binding and the ATPase-induced conformation change, but not ATP hydrolysis, was found to be necessary for release of immunoglobulin heavy chains from complexes with BiP (Wei et al., 1995). Finally, in vivo expression of ATPase mutants, or of the isolated C-terminal protein-binding domain, causes disruption of the ER, indicating that these mutants can act in a dominant negative fashion in the presence of the endogenous wild-type BiP protein (Hendershot et al., 1995).

■ Gene regulation

Under normal growth conditions BiP is synthesized constitutively and abundantly, comprising c. 5% of the luminal content of the ER. In mammalian cells, its synthesis can be further induced by the accumulation of mutant proteins in the ER, or by a variety of stress conditions, including glucose starvation and treatment with calcium ionophores, amino acid analogues or drugs that inhibit glycosylation (reviewed by Lee, 1987, 1992), whose common denominator is believed to be the accumulation in the ER of unfolded polypeptides (Kozutsumi et al., 1988; Nakaki et al., 1989). Studies of BiP induction in mammalian cells and in yeast (reviewed by Kohno et al., 1993) suggest that the signal for induction is a decrease in the concentration of free BiP in the ER, rather than an increase in the concentration of complexes between BiP and unfolded polypeptides. Studies using protein kinase inhibitors indicate that tyrosine- and serine/threonine kinases may be required in the transcriptional activation of the promoter of the rat grp78 (BiP) gene by thapsigargin, an inhibitor of the ER Ca^{2+}–ATPase (Cao et al., 1995).

The promoters of mammalian grp78 genes contain cis-acting regulatory elements required for high basal level expression and for induction by malfolded proteins, glycosylation block or the calcium ionophore A23187 (Wooden et al., 1991; Lee, 1992). These include: (i) a 36 bp 'grp core' element involved in stress inducibility that includes a 28 base pair sequence that is highly conserved in the promoters of mammalian and avian genes that encode Grp78 and another stress-responsive ER resident protein, Grp94 (Chang et al., 1989; Liu and Lee, 1991); (ii) a series of CCAAT- or CCAAT-like motifs, the most proximal of which is required to mediate the effects of the upstream regulatory elements and which binds the heteromeric transcription factor CBF whose in vitro binding activity is sensitive to calcium ion concentration (Roy and Lee, 1995; (iii) a second stress responsive element which provides partial stress inducibility when the grp core and proximal CCAAT motif are deleted (Wooden et al., 1991); and (iv) a cAMP response element (CRE) which functions as a major basal level regulatory element and is also necessary to maintain high promoter activity under stress-induced conditions (Alexandre et al., 1991).

Synthetic oligomers of the grp core element are capable of conferring partial stress inducibility to a heterologous promoter (Li et al., 1993) and can compete for nuclear factors binding to grp promoters in vitro and in vivo (Chang et al., 1989; Liu and Lee, 1991). Stable integration and amplification of the grp core in CHO cells coordinately down-regulate the expression of the endogenous grp78 and grp94 genes (Li and Lee, 1991). Specific changes in factor occupancy and DNA methylation patterns occur within a cluster of bases located in the 3′ half of the grp core, whereas other regulatory elements are constitutively occupied (Li et al., 1994). A 70-kDa binding factor, p70CORE, which forms a complex specific to the grp core has recently been shown to be identical to YY1, a member of the GLI zinc finger family of transcription factors (Li et al., 1997).

■ References

Alexandre, S., Nakaki, T., Vanhamme, L., and Lee, A.S. (1991). A binding site for the cyclic adenosine 3′,5′-monophosphate-response element-binding protein as a regulatory element in the grp78 promoter. Mol. Endocrin. **5**, 1862–1872.

Blond-Elguindi, S., Cwirla, S.E., Dower, W.J., Lipshutz, R.J., Sprang, S.R., Sambrook, J.F., and Gething, M.J. (1993a). Affinity panning of a library of peptides displayed on bacteriophages reveals the binding specificity of BiP. Cell **75**, 717–728.

Blond-Elguindi, S., Fourie, A.M., Sambrook, J.F., and Gething, M.J.H. (1993b). Peptide-dependent stimulation of the ATPase activity of the molecular chaperone BiP is the result of conversion of oligomers to active monomers. J. Biol. Chem. **268**, 12730–12735.

Brightman, S.E., Blatch, G.L., and Zetter, B.R. (1995). Isolation of a mouse cDNA encoding MTJ1, a new murine member of the DnaJ family of proteins. Gene **153**, 249–254.

Brodsky, J.L. and Schekman, R. (1993). A Sec63p–BiP complex from yeast is required for protein translocation in a reconstituted proteoliposome. J. Cell Biol. **123**, 1355–1363.

Cao, X., Zhou, Y., and Lee, A.S (1995). Requirement of tyrosine- and serine/threonine kinases in the transcriptional activation of the mammalian grp78/BiP promoter by thapsigargin. J. Biol. Chem. **270**, 494–502.

Carlino, A., Toledo, H., Skaleris, D., DeLisio, R., Weissbach, H., and Brot, N. (1992). Interactions of liver Grp78 and Escherichia coli recombinant Grp78 with ATP: multiple species and disaggregation. Proc. Natl. Acad. Sci. USA **89**, 2081–2085.

Carlsson, L. and Lazarides, E. (1983). ADP-ribosylation of the Mr 83,000 stress-inducible and glucose-regulated protein in avian and mammalian cells: modulation by heat shock and glucose starvation. Proc. Natl. Acad. Sci. USA **80**, 4664–4668.

Chang, S.C., Erwin, A.E., and Lee, A.S. (1989). Glucose-regulated protein (GRP94 and GRP78) genes share common regulatory domains and are coordinately regulated by common trans-acting factors. Mol. Cell. Biol. **9**, 2153–2162.

Feldheim, D., Rothblatt, J., and Schekman, R. (1992). Topology and functional domains of Sec63p, an endoplasmic reticulum membrane protein required for secretory protein translocation. Mol. Cell. Biol. **12**, 3288–3296.

Flaherty, K.M., DeLuca-Flaherty, C., and McKay, D.B. (1990). Three-dimensional structure of the ATPase fragment of a 70K heat-shock cognate protein. Nature **346**, 623–628.

Flynn, G.C., Chappell, T.G., and Rothman, J.E. (1989). Peptide binding and release by proteins implicated as catalysts of protein assembly. Science **245**, 385–390.

Flynn, G.C., Pohl, J., Flocco, M.T., and Rothman, J.E. (1991). Peptide-binding specificity of the molecular chaperone BiP. Nature **353**, 726–730.

Fourie, A.M., Sambrook, J.F., and Gething, M.-J.H. (1994). Common and divergent peptide binding specificities of hsp70 molecular chaperones. J. Biol. Chem. **269**, 30470–30478.

Freiden, P.J., Gaut, J.R., and Hendershot, L.M. (1992). Interconversion of three differently modified and assembled forms of BiP. EMBO J. **11**, 63–70.

Gaut, J.R. and Hendershot, L.M. (1993a). The immunoglobulin-binding protein in vitro autophosphorylation site maps to a threonine within the ATP binding cleft but is not a detectable site of in vivo phosphorylation. J. Biol. Chem. **268**, 12691–12698.

Gaut, J.R. and Hendershot, L.M. (1993b). Mutations within the nucleotide binding site of immunoglobulin-binding protein inhibit ATPase activity and interfere with release of immunoglobulin heavy chain. J. Biol. Chem. **268**, 7248–7255.

Gething, M.J. and Sambrook, J.F. (1992). Protein folding in the cell. Nature **355**, 33–45.

Gething, M.J., Blond-Elguindi, S., Mori, K., and Sambrook, J.F. (1994). Structure, function and regulation of the endoplasmic reticulum chaperone, BiP. In The Biology of Heat Shock Proteins and Molecular Chaperones, (R.I. Morimoto, A. Tissières, and C. Georgopoulos, eds). Cold Spring Harbor Laboratory Press, New York, pp. 111–135.

Gragerov, A. and Gottesman, M.E. (1994). Different peptide binding specificities of hsp70 family members. J. Mol. Biol. **241**, 133–135.

Haas, I.G. and Wabl, M. (1983). Immunoglobulin heavy chain binding protein. Nature **306**, 387–389.

Haas, I.G., Simon-Chazottes, D., and Guenet, J.-L. (1992). The gene coding for the immunoglobulin heavy chain binding protein BiP (Hsce-70) maps to mouse chromosome 2. Mammalian Genome **3**, 659–660.

Hammond, C. and Helenius, A. (1994). Folding of VSV G protein: sequential interaction with BiP and calnexin. Science **266**, 456–458.

Hartl, F.U. (1996). Molecular chaperones in cellular protein folding. Nature **381**, 571–580.

Hendershot, L.M., Ting, J., and Lee, A.S. (1988). Identity of the immunoglobulin heavy-chain-binding protein with the 78,000-dalton glucose-regulated protein and the role of posttranslational modifications in its binding function. Mol. Cell. Biol. **8**, 4250–4256.

Hendershot, L.M., Valentine, V.A., Lee, A.S., Morris, S.W., and Shapiro, D.N. (1994). Localization of the gene encoding human BiP/GRP78, the endoplasmic reticulum cognate of the HSP70 family, to chromosome 9q34. Genomics **20**, 281–284.

Hendershot, L.M., Wei, J-Y., Gaut, J.R., Lawson, B., Freiden, P.J., and Murti, K.G. (1995). In vivo expression of mammalian BiP ATPase mutants causes disruption of the endoplasmic reticulum. Mol. Biol. Cell **6**, 283–296.

Kassenbrock, C.K. and Kelly, R.B. (1989). Interaction of heavy chain binding protein (BIP/GRP78) with adenine nucleotides. EMBO J. **8**, 1461–1467.

Kim, P.S. and Arvan, P. (1995). Calnexin and BiP act as sequential molecular chaperones during thyroglobulin folding in the endoplasmic reticulum. J. Cell Biol. **128**, 29–38.

Knarr, G., Gething, M.-J., Modrow, S., and Buchner, J. (1995). BiP-binding sequences in antibodies. J. Biol. Chem. **270**, 27589–27594.

Kohno, K., Normington, K., Sambrook, J., Gething, M.J., and Mori, K. (1993). The promoter region of the yeast KAR2 (BiP) gene contains a regulatory domain that responds to the presence of unfolded proteins in the endoplasmic reticulum. Mol. Cell. Biol. **13**, 877–890.

Kozutsumi, Y., Segal, M., Normington, K., Gething, M.J., and Sambrook, J. (1988). The presence of malfolded proteins in the endoplasmic reticulum signals the induction of glucose regulated proteins. Nature **332**, 462–464.

Kozutsumi, Y., Normington, K., Press, E., Slaughter, C., Sambrook, J., and Gething, M.J. (1989). Identification of immunoglobulin heavy chain binding protein as glucose regulated protein 78 on the basis of amino acid sequence, immunological crossreactivity and functional activity. J. Cell Sci. **11**, 115–137.

Kuznetsov, G., Chen, L.B., and Nigam, S.K. (1994). Several endoplasmic reticulum stress proteins, including ERp72, interact with thyroglobulin during its maturation. J. Biol. Chem. **269**, 22990–22995.

Ledford, B.E. and Leno, G.H. (1994). ADP-ribosylation of the molecular chaperone GRP78/BiP. Mol. Cell. Biochem. **138**, 141–148.

Lee, A.S. (1987). Coordinated regulation of a set of genes by glucose and calcium ionophores in mammalian cells. Trends Biochem. Sci. **12**, 20–23.

Lee, A.S. (1992). Mammalian stress response: induction of the glucose-regulated protein family. Curr. Opin. Cell Biol. **4**, 267–273.

Leno, G.H. and Ledford, B.E. (1990). Reversible ADP-ribosylation of the 78 kDa glucose-regulated protein. FEBS **276**, 29–33.

Leustek, T., Toledo, H., Brot, N., and Weissbach, H. (1991). Calcium-dependent autophosphorylation of the glucose-regulated protein, grp78. Arch. Biochem. Biophys. **289**, 256–261.

Leustek, T., Amir-Shapira, D., Toledo, H., Brot, N., and Weissbach, H. (1992). Autophosphorylation of 70 kDa heat shock proteins. Cell. Mol. Biol. **38**, 1–10.

Li, W.W., Alexandre, S., Cao, X., and Lee, A.S. (1993). Transactivation of the grp78 promoter by Ca^{2+} depletion. J. Biol. Chem. **268**, 12003–12009.

Li, W.W., Sistonen, L., Morimoto, R.I., and Lee, A.S. (1994). Stress induction of the mammalian GRP78/BiP protein gene: in vivo genomic footprinting and identification of p70CORE from human nuclear extract as a DNA-binding component specific to the stress regulatory element. Mol. Cell. Biol. **14**, 5533–5546.

Li, X. and Lee, A.S. (1991). Competitive inhibition of a set of endoplasmic reticulum protein genes (GRP78, GRP94, and

ERp72) retards cell growth and lowers viability after ionophore treatment. Mol. Cell. Biol. **11**, 3446–3453.

Li, W.W., Hsiung, Y., Zhou, Y., Roy, B., and Lee, A.S. (1997). Induction of the mammalian GRP78/BiP gene by Ca^{2+} depletion and formation of aberrant proteins: Activation of the conserved stress-inducible *grp* core promoter element by the human nuclear factor YY1. Mol. Cell. Biol. **17**, 54–60.

Liu, E.S. and Lee, A.S. (1991). Common sets of nuclear factors binding to the conserved promoter sequence motif of two coordinately regulated ER protein genes, GRP78 and GRP94. Nucl. Acids Res. **19**, 5425–5431.

Macer, D.R.J. and Koch, G.L.E. (1988). Identification of a set of calcium-binding proteins in reticuloplasm, the luminal content of the endoplasmic reticulum. J. Cell Sci. **91**, 61–70.

Marks, M.S., Germain, R.N., and Bonifacino, J.S. (1995). Transient aggregation of major histocompatibility complex class II chains during assembly in normal spleen cells. J. Biol. Chem. **270**, 10475–10481.

Melnick, J., Aviel, S., and Argon, Y. (1992). The endoplasmic reticulum stress protein GRP94, in addition to BiP, associates with unassembled immunoglobulin chains. J. Biol. Chem. **267**, 21303–21306.

Melnick, J., Dul, J.L., and Argon, Y. (1994). Sequential interaction of the chaperones BiP and GRP94 with immunoglobulin chains in the endoplasmic reticulum. Nature **370**, 373–375.

Munro, S. and Pelham, H.R.B. (1986). An Hsp70-like protein in the ER: identity with the 78 kd glucose-regulated protein and immunoglobulin heavy chain binding protein. Cell **46**, 291–300.

Nakaki, T., Deans, R.J., and Lee, A.S. (1989). Enhanced transcription of the 78,000-dalton glucose-regulated protein (GRP78) gene and association of GRP78 with immunoglobulin light chains in a nonsecreting B-cell myeloma line (NS-1). Mol. Cell. Biol. **9**, 2233–2238.

Normington, K., Kohno, K., Kozutsumi, Y., Gething, M.J., and Sambrook, J. (1989). *S. cerevisiae* encodes an essential protein homologous in sequence and function to mammalian BiP. Cell **57**, 1223–1236.

Nossner, E. and Parham, P. (1995). Species-specific differences in chaperone interaction of human and mouse major histocompatibility complex class I molecules. J. Exp. Med. **181**, 327–337.

Pelham, H.R.B. (1989). Control of protein exit from the endoplasmic reticulum. Annu. Rev. Cell Biol. **5**, 1–23.

Pouyssegur, J., Shiu, R.P.C., and Pastan, I. (1977). Induction of two transformation-sensitive membrane polypeptides in normal fibroblasts by a block in glycoprotein synthesis or glucose deprivation. Cell **11**, 941–947.

Rowling, P.J., McLaughlin, S.H., Pollock, G.S., and Freedman, R.B. (1994). A single purification procedure for the major resident proteins of the ER lumen: endoplasmin, BiP, calreticulin and protein disulfide isomerase. Protein Expr. Purif. **5**, 331–336.

Roy, B. and Lee, A.S. (1995). Transduction of calcium stress through interaction of the human transcription factor CBF with the proximal CCAAT regulatory element of the *grp78*/BiP promoter. Mol. Cell. Biol. **15**, 2263–2274.

Schlenstedt, G., Harris, S., Risse, B., Lill, R., and Silver, P.A. (1995). A yeast DnaJ homologue, Scj1p, can function in the endoplasmic reticulum with BiP/Kar2p via a conserved domain that specifies interactions with Hsp70s. J. Cell Biol. **129**, 979–988.

Ting, J. and Lee, A.S. (1988). Human gene encoding the 78,000-dalton glucose-regulated protein and its pseudogene: structure, conservation, and regulation. DNA **7**, 275–286.

Ting, J., Wooden, S.K., Kriz, R., Kelleher, K., Kaufman, R.J., and Lee, A.S. (1987). The nucleotide sequence encoding the hamster 78-kDa glucose-regulated protein (GRP78) and its conservation between hamster and rat. Gene **55**, 147–152.

Wei, J. and Hendershot, L.M. (1995). Characterization of the nucleotide binding properties and ATPase activity of recombinant hamster BiP purified from bacteria. J. Biol. Chem. **270**, 26670–26676.

Wei, J., Gaut, J.R., and Hendershot, L.M. (1995). *In vitro* dissociation of BiP–peptide complexes requires a conformational change in BiP after ATP binding but does not require ATP hydrolysis. J. Biol. Chem. **270**, 26677–26682.

Welch, W.J., Garrels, J.I., Thomas, G.P., Lin, J.J.-C., and Feramisco, J.R. (1983). Biochemical characterization of the mammalian stress proteins and identification of two stress proteins as glucose- and Ca^{2+}-ionophore-regulated proteins. J. Biol. Chem. **258**, 7102–7111.

Wooden, S.K., Li, L.J., Navarro, D., Qadri, I., Pereira, L., and Lee, A.S. (1991). Transactivation of the grp78 promoter by malfolded proteins, glycosylation block, and calcium ionophore is mediated through a proximal region containing a CCAAT motif which interacts with CTF/N-I. Mol. Cell. Biol. **11**, 5612–5623.

Zhu, X., Zhao, X., Burkholder, W.F., Gragerov, A., Ogata, C.M., Gottesman, M.E., and Hendrickson, W.A. (1996). Structural analysis of substrate binding by the molecular chaperone DnaK. Science **272**, 1606–1614.

■ *Mary-Jane Gething:*
Department of Biochemistry and Molecular Biology
University of Melbourne
Parkville
Victoria 3052, Australia
Tel. 61 3 9344 5948
Fax. 61 3 9347 9109
E-mail: gething@ariel.ucs.unimelb.edu.au

Mammalian Pbp74

Peptide-binding protein 74 (Pbp74) is a member of the mammalian HSP70 protein family. Pbp74, which is located in the matrix space of mitochondria, was originally identified as one of a mixture of peptide-binding proteins referred to as Pbp72/74. Under the pseudonym Grp75, Pbp74 was shown to associate with newly synthesized mitochondrial proteins and the mammalian mitochondrial Hsp60. Based on this data and analyses of the well-characterized Pbp74 yeast homolog, Ssc1p, Pbp74 presumably functions to aid translocation of precursor proteins into the mitochondrial matrix and to assist in their relocalization, folding and/or assembly.

■ Alternative names

Grp75 (Mizzen et al., 1989); P71 (Leustek et al., 1989); C3H strain-specific antigen (CSA) (Michikawa et al., 1993); mortalin (Wadhwa et al., 1993a); mtHsp70 (Webster et al., 1994); mthsp75 (Bhattacharyya et al., 1995).

■ Isolation and cloning

Pbp74 was originally isolated as one of a mixture of peptide-binding proteins termed Pbp72/74 that was purified by peptide affinity chromatography from mouse spleen cells (Lakey et al., 1987). Antisera to this mixture of peptide-binding proteins blocked the immune cell function of antigen processing (Lakey et al., 1987) and in immunoelectron microscopy stained subcellular endosomal compartments in which antigen processing occurs (Van Buskirk et al., 1991). Subsequent serological and biochemical characterization of the peptide-binding proteins showed that some were members of the HSP70 family, including BiP, Pbp74 and Hsc70 (Van Buskirk et al., 1989; Domanico et al., 1993).

Affinity purified Pbp74, separated from the other peptide-binding proteins, was subjected to N-terminal amino acid sequence analysis and 16 N-terminal residues were identified. An additional 89 residues of internal amino acid sequence was determined from five peptides derived from Pbp74 (Domanico et al., 1993). The amino acid sequence was used to design degenerate oligonucleotide primers for use in a coupled reverse-transcription PCR. This strategy led to the isolation of cDNA clones representing the entire mouse *pbp74* sequence and subsequently cDNA clones representing the human *pbp74* sequence (Domanico et al., 1993). Further characterization of Pbp74 (see below) definitively showed that it is the mammalian mitochondrial Hsp70 and therefore Pbp74 is no longer believed to play a role in antigen processing (Dahlseid et al., 1994).

■ *pbp74* gene and sequence

The nucleotide sequence of *pbp74* cDNAs cloned from mouse, rat, and human (GenBank accession numbers L06896, L11066, D17556, D11089, L17189, S75280) predicts a protein of 679 amino acids including a 46 amino acid presequence deduced from N-terminal amino acid sequencing of the mature protein (Domanico et al., 1993). The predicted protein sequences for mouse and human are nearly 99% identical, with only six residue changes each in the presequence and in the mature protein (Domanico et al., 1993). Nucleotide sequences for *pbp74* were independently reported (GenBank numbers included above) by four additional laboratories (Michikawa et al., 1993; Wadhwa et al., 1993a; Webster et al., 1994; Bhattacharyya et al., 1995). Among the mouse sequences reported, there is confirmed polymorphism at amino acid positions 572 and 578 of the mature protein sequence (Michikawa et al., 1993; Wadhwa et al., 1993b). These sequences, in order and pairwise, are either Met/Gly or Val/Arg.

The N-terminus (residues 8–16) of the mature Pbp74 protein contains the highly conserved sequence GIDLG found in the ATP-binding domain of all known HSP70 members (Flaherty et al., 1990). Comparison of the Pbp74 sequences with those of other HSP70 family members shows a high degree of identity, and includes greater sequence conservation through the N-terminal two-thirds than the C-terminal third of the proteins, which is characteristic for this protein family (Gething and Sambrook, 1992). The Pbp74 sequence shares the highest degree of identity with the mitochondrial HSP70 protein homologs from *Drosophilia melanogaster* (74% identical) and *Saccharomyces cerevisiae* (65% identical) (Craig et al., 1989; Rubin et al., 1993). Pbp74 is presumed to be identical to the mitochondrial HSP70 protein, Grp75, identified by a biochemical means by Mizzen et al. (1989).

■ Pbp74 protein

Pbp74 is a 74-kDa member of the HSP70 protein family. The original Pbp74 purification was based on its ability to bind peptide, specifically that of pigeon cytochrome *c* residues 81–104 (Van Buskirk et al., 1989). Moreover, peptide-bound Pbp74 can be released with ATP, but not with ADP or AMP (Van Buskirk et al., 1989). Pbp74 also binds ATP and can be purified by this criteria (Leustek et al., 1989; Mizzen et al., 1989; Van Buskirk et al., 1989; Kaul et al., 1993; Webster et al., 1994). It has been reported that Pbp74 (Grp75, P71, mtHsp70) can undergo autophosphorylation *in vitro*, in the presence of calcium (20–50 mM) and low pH (5.5–6.0) (Leustek *et al.*, 1989; Mizzen *et al.*, 1989; Webster *et al.*, 1994).

The cloning of the cDNA encoding Pbp74 allowed for a definitive description of its subcellular location. Using a variety of immunological and biochemical approaches both *in vitro* and *in vivo*, Pbp74 was demonstrated to reside in the mitochondria of cells (Dahlseid *et al.*, 1994). An influenza virus hemagglutinin epitope tag was introduced into the *pbp74* cDNA. The epitope-tagged Pbp74 protein transiently expressed in L cells localized to the mitochondria. Deletion of the N-terminal 46 amino acid presequence resulted in a cytosolic localization of the epitope-tagged protein. Cell fractionation studies demonstrated Pbp74 in purified mitochondria in a protease-protected location. Using a well-characterized protein import system developed with isolated yeast mitochondria, the precursor of the Pbp74 protein was shown to be transported into yeast mitochondria where it is proteolytically processed to the mature protein. Pbp74 was found to localize to the matrix space. Import of Pbp74 into mitochondria is time- and temperature-dependent, requires matrix ATP and is abolished upon depletion of the membrane potential across the mitochondrial inner membrane. In mammalian cells, Pbp74 is synthesized as a precursor that requires membrane potential-dependent import into mitochondria for its maturation (Dahlseid *et al.*, 1994). Monoclonal antibodies have now been generated which recognize Pbp74 in immunoprecipitation, in western blotting and in immunofluorescence (Green *et al.*, 1995).

■ Function of Pbp74

Pbp74 (Grp75) has been shown to interact transiently with newly synthesized mitochondrial proteins (Mizzen *et al.*, 1991). This association becomes stable if cells are incubated in the presence of amino acid analogs to cause newly synthesized proteins to be malfolded (Mizzen *et al.*, 1991). Pbp74 has also been shown to coassociate with the mammalian mitochondrial HSP60 (Hsp58) in an ATP-dependent manner (Mizzen *et al.*, 1991). The present evidence indicates that Pbp74 is the mammalian homolog of the well-characterized *Saccharomyces cerevisiae* mitochondrial HSP70, Ssc1p (Craig *et al.*, 1989, see entry p. 30) and identical to mtHsp70 (see next entry p. 67). Ssc1p associates with partially translocated precursor proteins and is required in the mitochondrial matrix for proper translocation and folding of precursor proteins (Kang *et al.*, 1990; Ostermann *et al.*, 1990). Studies of temperature-sensitive alleles of *SSC1* led to the proposal that the mitochondrial Ssc1p plays a dual role in preprotein translocation by: (i) facilitating the unfolding of the polypeptide chain for translocation across mitochondrial membranes and (ii) binding the polypeptide chain to help drive the transport of matrix proteins to completion (Gambill *et al.*, 1993; Voos *et al.*, 1993). This translocation requires a functional ATPase domain and ATP.

■ References

Bhattacharyya, T., Karnezis, A.N., Murphy, S.P., Hoang, T., Freeman, B.C., Phillips, B., and Morimoto, R.I. (1995). Cloning and subcellular localization of human mitochondrial hsp70. J. Biol. Chem. **270**, 1705–1710.

Craig, E.A., Kramer, J., Shilling, J., Werner-Washburne, M., Holmes, S., Kosic-Smithers, J., and Nicolet, C.M. (1989). SSC1, an essential member of the yeast HSP70 multigene family, encodes a mitochondrial protein. Mol. Cell. Biol. **9**, 3000–3008.

Dahlseid, J.N., Lill, R., Green, J.M., Xu, X., Qiu, Y., and Pierce, S.K. (1994). PBP74, a new member of the mammalian 70 kDa heat shock protein family, is a mitochondrial protein. Mol. Biol. Cell **5**, 1265–1275.

Domanico, S.Z., DeNagel, D.C., Dahlseid, J.N., Green, J.M., and Pierce, S.K. (1993). Cloning of the gene encoding peptide-binding protein 74 shows that it is a new member of the heat shock protein 70 family. Mol. Cell. Biol. **13**, 3598–3610.

Flaherty, K.M., DeLuca-Flaherty, C., and McKay, D.M. (1990). Three-dimensional structure of the ATPase fragment of a 70 kDa heat-shock cognate protein. Nature **346**, 623–628.

Gambill, B.D., Voos, W., Kang, P.J., Miao, B., Langer, T., Craig, E.A., and Pfanner, N. (1993). A dual role for mitochondrial heat shock protein 70 in membrane translocation of preproteins. J. Cell Biol. **123**, 109–117.

Gething, M-J. and Sambrook, J. (1992). Protein folding in the cell. Nature **355**, 33–45.

Green, J.M., Gu, L., Iskovits, C., Kaumaya, P.T.P., Conrad, S., and Pierce, S.K. (1995). Generation and characterization of monoclonal antibodies specific for members of the mamalian 70kD heat shock protein family. Hybridoma **14**, 347.

Kang, P.J., Ostermann, J., Shilling, J., Neupert, W., Craig, E.A., and Pfanner, N. (1990). Requirement for hsp70 in the mitochondrial matrix for translocation and folding of precursor proteins. Nature **348**, 137–143.

Kaul, S.C., Wadhwa, R., Komatsu, Y., Sugimoto, Y., and Mitsui, Y. (1993). On the cytosolic and perinuclear mortalin: an insight by heat shock. Biochem. Biophys. Res. Commun. **193**, 348–355.

Lakey, E.K., Margoliash, E., and Pierce, S.K. (1987). Identification of a peptide binding protein which plays a role in antigen presentation. Proc. Natl. Acad. Sci. USA. **84**, 1659–1663.

Leustek, T., Dalie, B., Amir-Shapira, D., Brot, N., and Weissbach, H. (1989). A member of the hsp70 family is localized in mitochondria and resembles *Escherichia coli* DnaK. Proc. Natl. Acad. Sci. USA. **86**, 7805–7808.

Michikawa, Y., Baba, T., Arai, Y., Sakakura, T., Tanaka, M., and Kusakabe, M. (1993). Antigenic protein specific for C3H strain mouse is a mitochondrial stress-70 protein. Biochem. Biophys. Res. Commun. **196**, 223–232.

Mizzen, L.A., Chang, C., Garrels, J.I., and Welch, W.J. (1989). Identification, characterization, and purification of two mammalian stress proteins present in mitochondria, grp 75, a member of the hsp 70 family and hsp 58, a homolog of the bacterial groEL protein. J. Biol. Chem. **264**, 20664–20675.

Mizzen, L.A., Kabiling, A.N., and Welch, W.J. (1991). The two mammalian mitochondrial stress proteins, grp 75 and hsp 58, transiently interact with newly synthesized mitochondrial proteins. Cell Regul. **2**, 165–179.

Ostermann, J., Voos, W., Kang, P.J., Craig, E.A., Neupert, W., and Pfanner, N. (1990). Precursor proteins in transit through the mitochondrial contact sites interact with hsp70 in the matrix. FEBS Lett. **277**, 281–284.

Rubin, D.M., Mehta, A.D., Zhu, J., Shoham, S., Chen, X., Wells, Q.R., and Palter, K.B. (1993). Genomic structure and sequence analysis of *Drosophila melanogaster* HSC70 genes. Gene **128**, 155–163.

Van Buskirk, A., Crump, B.L., Margoliash, E., and Pierce, S.K. (1989). A peptide binding protein having a role in antigen presentation is a member of the HSP70 heat shock family. J. Exp. Med. **170**, 1799–1809.

Van Buskirk, A., DeNagel, D.C., Guagliardi, L.E., Brodsky, F.M., and Pierce, S.K. (1991). Cellular and subcellular distribution of PBP72/74, a peptide-binding protein that plays a role in antigen processing. J. Immunol. **146**, 500–506.

Voos, W., Gambill, B.D., Guiard, B., Pfanner, N., and Craig, E.A. (1993). Presequence and mature part of preproteins strongly influence the dependence of mitochondrial protein import on heat shock protein 70 in the matrix. J. Cell Biol. **123**, 119–126.

Wadhwa, R., Kaul, S.C., Ikawa, Y., and Sugimoto, Y. (1993a). Identification of a novel member of mouse hsp 70 family: its association with cellular mortal phenotype. J. Biol. Chem. **268**, 6615–6621.

Wadhwa, R., Kaul, S.C., Sugimoto, Y., and Mitsui, Y. (1993b). Induction of cellular senescence by transfection of cytosolic mortalin cDNA in NIH 3T3 cells. J. Biol. Chem. **268**, 22239–22242.

Webster, T.J., Naylor, D.J., Hartman, D.J., Hoj, P.B. and Hoogenraad, N.J. (1994). cDNA cloning and efficient mitochondrial import of pre-mtHSP70 from rat liver. DNA Cell Biol. **12**, 1213–1220.

■ Susan K. Pierce:
Department of Biochemistry, Molecular Biology and Cell Biology
Northwestern University
2153 North Campus Drive, Hogan 3-120
Evanston, IL 60208, USA
Tel. 1 847 491 5089
Fax. 1 847 467 1610
E-mail: skpierce@merle.acns.nwu.edu

■ Jeffrey Dahlseid:
Laboratory of Molecular Biology
R.M. Bock Laboratories
University of Wisconsin-Madison
1525 Linden Avenue
Madison, WI 53706, USA
Tel: 1 608 262 4381
Fax. 1 608 262 4570
E-mail: dahlseid@facstaff.wisc.edu

Mammalian mitochondrial Hsp70

Mammalian mtHsp70 is a mitochondrial matrix protein which is a member of the HSP70 superfamily. It is a protein of 74 kDa with a large, 46 amino acid, cleavable signal sequence. It shows a high degree of homology with other members of the HSP70 family but displays a distinct autophosphorylation pattern.

■ Alternative names

Grp 75 (Mizzen et al., 1989); P71 (Leustek et al., 1989); Pbp74 (Domanico et al., 1993); C3H strain specific antigen (CSA) (Michikawa et al., 1993); mortalin (Wadhwa et al., 1993); mthsp75 (Bhattacharyya et al., 1995).

■ Isolation

Rat mtHsp70 was purified from rat liver mitochondria by ATP-agarose affinity chromatography through elution with 5 mM ATP plus 500 mM NaCl, after a prior wash of the column with 0.5 M KCl and 1 mM GTP (Mizzen et al., 1989; Webster et al., 1994). A total of 800 μg mtHsp70 was obtained from 120 g liver.

■ mtHsp70 sequence

Based on protein sequence obtained from endo Lys-C peptide fragments of the isolated protein, a rat cDNA sequence of 2.1kb (S75280) encoding a 679 amino acid precursor protein (p-mtHsp70) was isolated (Webster et al., 1994). Comparison of the cDNA sequence with the N-terminal sequence of the isolated protein shows that the precursor protein contains a 46 residue, cleavable presequence. Based on the sequence, p-mtHsp70 has a mass of 74 kDa and mature mtHsp70 a mass of 69 kDa (Webster et al., 1994).

mtHsp70 shows a high degree of sequence identity with other members of the HSP70 family such as E. coli DnaK (51%; Bardwell and Craig, 1984), S. cerevisiae Ssc1p (65%; Craig et al., 1989), constitutive cytosolic Hsc70 from rat (46%; O'Malley et al., 1985) and the endoplasmic reticulum isoform, BiP from rat (49%; Munro and Pelham, 1986). Rat mtHsp70 differs by only 12 residues from the mouse protein Pbp74 (Domanico et al., 1993, see preceeding entry p. 65) and nine residues from the human mtHsp75 (Bhattacharyya et al., 1995; there are an additional seven differences in the 46 amino acid signal peptide).

■ mtHsp70 protein

The mitochondrial isoform of HSP70 has a unique autophosphorylation pattern and can be distinguished from other mammalian isoforms by its ability to auto-phosphorylate in a Ca^{2+}-dependent manner (Leustek et al., 1989). This pattern is the same as that observed for DnaK from E. coli. In contrast, the cytosolic forms are not phosphorylatable in the presence of divalent cations (Hendershot et al., 1988).

Although mtHsp70 shows a high degree of sequence identity with other isoforms of HSP70, the sequences diverge considerably within the C-terminal peptide-binding domain of the protein. No section of the 46 amino acid targeting sequence is predicted to form a basic amphiphilic helix with a high hydrophobic moment characteristic of mitochondrial signal peptides (Lemire et al., 1989). Yet p-mtHsp70 is imported into mitochondria and processed from the 74 kDa precursor to a mature 69 kDa protein (Webster et al., 1994). Antibodies to the purified protein, made in chickens, show the protein to be localized to the mitochondrial matrix in rat liver. However, a number of reports suggest that the protein is also found in extramitochondrial locations (Domanico et al., 1993, but see Pbp74 entry p. 65; Wadhwa et al., 1993).

■ Biological regulation of mtHsp70

mtHsp70 is constitutively expressed under normal cell growth conditions. The gene has been isolated and the promoter contains two heat shock elements (Michikawa et al., 1993b). However, it does not appear to be induced by heat stress although it is induced by the amino acid analogue L-azetidine-2-carboxylic acid. The protein has been suggested to be induced by glucose deprivation (Mizzen et al., 1989).

■ Function of mtHsp70

Whereas Drosophila mt-Hsp70 has shown to be localized near the inner mitochondrial membrane (Carbajal et al., 1993) and, additionally in yeast, to play a role in import of proteins into the mitochondria, via a 'Brownian ratchet' like mechanism (Glick, 1995; Pfanner and Meijer, 1995), and in their subsequent folding in the matrix (Martinus et al., 1995), the role of the mammalian homologue awaits resolution. Similarly a role for auto-phosphorylation has not been established.

■ References

Bardwell, J. and Craig, E.A. (1984). Major heat shock gene of Drosphila and the Escherichia coli heat inducible DnaK gene are homologous. Proc. Natl. Acad. Sci. USA **81**, 848–852.

Bhattacharyya, T., Karnezis, A.N., Murphy, S.P., Hoang, T., Freeman, B.C., Phillips, B., and Morimoto, R.I. (1995). Cloning and subcellular localization of human mitochondrial hsp70. J. Biol. Chem. **270**, 1705–1710.

Carbajal, E., Beaulieu, J.-F., Nicole, L.M., and Tanguay, R.M. (1993). Intramitochondrial localization of the main 70-kDa heat shock cognate protein in Drosophila cells. Exp. Cell. Res. **207**, 300–309.

Craig, E.A., Kramer, J., Shilling, J., Werner-Washburne, M., Holmes, S., Kosic-Smithers, J., and Nicolet, C.M. (1989). SSC1, an essential member of the yeast HSP70 multigene family, encodes a mitochondrial protein. Mol. Cell. Biol. **9**, 3000–3008.

Domanico, S.Z., DeNagel, D.C., Dahlseid, J.N., Green, J.M., and Pierce, S.K. (1993). Cloning of the gene encoding peptide-binding protein 74 shows that it is a new member of the heat shock protein 70 family. Mol. Cell. Biol. **13**, 3598–3610.

Glick, B.S. (1995). Can Hsp70 proteins act as force-generating motors? Cell **80**, 11–14.

Hendershot, L.M., Ting, J., and Lee, A.S. (1988). Identity of the immunoglobulin heavy-chain binding protein with the 78,000-dalton glucose-regulated protein and the role of translational modifications in its binding function. Mol. Cell. Biol. **8**, 4250–4256.

Lemire, B.D., Fankhauser, C., Baker, A., and Schatz, G. (1989). The mitochondrial targeting function of randomly generated peptide sequences correlates with predicted helical amphiphilicity. J. Biol. Chem. **264**, 20206–20215.

Leustek, T., Dalie, B., Amir-Shapira, D., Brot, N., and Weissbach, H. (1989). A member of the hsp70 family is localized in the mitochondria and resembles Escherichia coli DnaK. Proc. Natl. Acad. Sci. USA **86**, 7805–7808.

Martinus, R.M., Ryan, M.T., Naylor, D.J., Herd, S.M., Hoogenraad, N.J., and Høj, P.B. (1995). Role of chaperones in the biogenesis and maintenance of the mitochondrion. FASEB J. **9**, 371–378.

Michikawa, Y., Baba, T., Arai, Y., Sakakura, T., Tanaka, M., and Kusakabe, M. (1993a). Antigenic protein specific for C3H strain mouse is a mitochondrial stress-70 protein. Biochem. Biophys. Res. Commun. **196**, 223–232.

Michikawa, Y., Baba, T., Arai, Y., Sakakura, T., and Kusakabe, M. (1993b). Structure and organization of the gene encoding mouse mitochondrial stress 70 protein. FEBS Lett. **336**, 27–33.

Mizzen, L.A., Chang, C., Garrels, J.I., and Welch, W.J. (1989). Identification, characterization and purification of two mammalian stress proteins present in mitochondria, grp75, a member of the hsp70 family and hsp58, a homolog of the bacterial GroEL protein. J. Biol. Chem. **264**, 20664–20675.

Munro, S. and Pelham, H.R.B. (1986). Rat BiP: an Hsp70-like protein in the ER: identity with the 78 kd glucose-regulated protein and immunoglobulin heavy chain binding protein. Cell **46**, 291–300.

O'Malley, K., Mauron, A., Barchas, J.D., and Kedes, L. (1985). Constitutive cytosolic Hsp70 from rat: constitutively expressed rat mRNA encoding a 70-kilodalton heat-shock-like protein. Mol. Cell. Biol. **5**, 3476–3483.

Pfanner, N. and Meijer, M. (1995). Pulling in the proteins. Curr. Biol. **5**, 132–135.

Wadhwa, R., Kaul, S.C, Ikawa, Y., and Sugimoto, Y. (1993). Identification of a novel member of the mouse hsp 70 family. J. Biol. Chem. **268**, 6615–6621.

Webster, T.J., Naylor, D.J., Hartman, D.J., Høj, P.B., and Hoogenraad, N.J. (1994). cDNA cloning and efficient mitochondrial import of pre-mtHSP70 from rat liver. DNA Cell Biol. **12**, 1213–1220.

■ Nicholas J Hoogenraad;
School of Biochemistry
La Trobe University
Bundoora
Victoria 3083, Australia
Tel. 61 3 9479 2175
Fax. 61 3 9479 2467
E-mail: N.Hoogenraad@Latrobe.edu.au

Peter B Høj and Dean Naylor:
Department of Horticulture, Viticulture and Oenology
University of Adelaide
Waite Campus PMB1
Glen Osmond
South Australia 5064, Australia
Tel. 61 8 8303 6663
Fax. 61 8 8303 7117
E-mail: phoj@waite.adelaide.edu.au
E-mail: dnaylor@waite.adelaide.edu.au

Tracie Webster:
Victorian Institute of Animal Science
475 Mickleham Road
Attwood
Victoria 3049, Australia
Tel. 61 3 9 217 4273
Fax. 61 3 9 217 4299
E-mail: webstert@woody.agvic.gov.au

STCH

STCH is a constitutively expressed member of the HSP70 protein chaperone family. The predicted STCH product is a 471 amino acid protein that has significant homology to the N-terminal ATPase domain of HSP70 proteins, but is truncated prior to the peptide/substrate-binding domain. The STCH protein product is localized to a crude microsomal fraction of cells and migrates on SDS–PAGE as a 60 kDa species.

■ *Stch* gene and sequence

Stch was isolated from a human cDNA library derived from the K562 erythroleukemia cell line and its sequence (GenBank/EMBL accession number U04735) predicts a peptide of 471 amino acids (Otterson *et al.*, 1994). The STCH product contains a unique hydrophobic leader sequence, shares homology within the N-terminal domains of the HSP70 gene family (33% identity and 43% homology to both human Hsp70 and BiP ATPase domains), has a unique 45-residue insertion within the ATP-binding domains, and truncates the C-terminal peptide-binding region. Of interest, STCH terminates two residues downstream from the putative cleavage site for the 44 kDa N-terminal fragment of Hsc70 (Chappell *et al.*, 1987). In addition, we have recently isolated the cDNA for the rat homolog of *Stch* and have found that the predicted rat and human STCH peptides have conserved each of these unique structural features. We have also isolated a partial human genomic *Stch* clone and have localized the *Stch* gene to human chromosome 21q11.1 (Brodsky *et al.*, 1995).

■ STCH protein

Specific polyclonal antisera were raised against recombinant STCH peptides that represent two internal domains with reduced homology to Hsc70/BiP (Otterson *et al.*, 1994). SDS–PAGE analysis recognizes a predominant 60 kDa protein that is absent in nuclear preparations and is strongly enriched in a crude microsomal fraction. In addition, a second species that migrated at approximately 72 kDa (intermediate between the migration patterns of Hsc70 and BiP) was weakly detected. Immunofluorescence confirmed a cytoplasmic/reticular localization for STCH and demonstrated that it resembled, but was not identical to the BiP localization. Protease digestion experiments were consistent with an intraluminal localization within the microsomal preparations. A purified GST–STCH recombinant protein had basal levels of ATPase activity which were not stimulated by the presence of exogenously added synthetic peptides (Otterson *et al.*, 1994). The absence of peptide-stimulated ATPase activity is consistent with the truncation of the STCH product prior to the peptide-binding domains found in other HSP70 members.

■ Biological regulation of *stch*

Stch mRNA is constitutively expressed as 2.4 and 4.4 kilobase bands in all tissues tested (although the relative amounts of the 2.4 and 4.4 kb mRNA species vary). The molecular basis for the two distinct transcripts is not known. *Stch* mRNA is not induced in response to heat shock, but modest induction is observed following a 24-hour exposure to the calcium ionophore A23187. Although STCH has basal ATPase activity, the biological function of the STCH protein is presently unknown. Understanding how STCH is retained in its intraluminal compartment, how it regulates its ATPase activity, and whether it can associate reversibly with other members of the protein chaperone family are important issues to address.

■ References

Brodsky, G., Otterson, G.A., Parry, B.B., Hart, I., Patterson, D., and Kaye, F.J. (1995). Localization of *Stch* to human chromosome 21q11.1. Genomics **30**, 627–628.

Chappell, T.., Konforti, B.B., Schmid, S.L., and Rothman, J.E. (1987). The ATPase core of a clathrin uncoating protein. J. Biol. Chem. **262**, 746–751.

Otterson, G.A., Flynn, G.C., Kratzke, R.A., Coxon, A., Johnston, P. G., and Kaye, F.J. (1994). *Stch* encodes the 'ATPase core' of a microsomal stress70 protein. EMBO J. **13**, 1216–1225.

■ *Frederic J. Kaye and Gregory A. Otterson:*
NCI–Navy Medical Oncology Branch
Building 8, Room 5101
National Naval Medical Center
Bethesda, MD 20889, USA
Tel. 1 301 496 0916
Fax. 1 301 496 0047
E-mail: fkaye@helix.nih.gov

■ *Gregory C. Flynn;*
Institute of Molecular Biology
University of Oregon
Eugene, OR 97403-1229, USA
Tel. 1 503 346 1534/1535
Fax. 1 503 346 5891
E-mail: gflynn@molbio.uoregon.edu

2

HSP110/SSE Proteins

The HSP110/SSE stress proteins — an overview

The HSP110-like proteins make up a family of unusual high molecular weight HSP70-related proteins which share closer sequence similarity to one another than they do to the HSP70 family. They are most similar to HSP70 proteins (and to each other) in the N-terminal 'ATPase domain'. They are diverged from HSP70 proteins, and from one another, in their C-terminal domain. A number of specific conserved sequences are interspersed, among more diverged sequences, throughout the C-terminal half of the HSP110-like proteins. These conserved sequences, and their linear arrangement in the C-terminal domain, define the HSP110/SSE subfamily of HSP70-related proteins.

■ Common structural features of the family

The proteins so far identified as members of the HSP110/SSE family (Lee-Yoon et al., 1995) are: yeast stress seventy proteins Sse1 and Sse2 (D13908 and D13909, Mukai et al., 1993; Shirayama et al., 1993), C30c11.4, Caenorhabditis elegans hypothetical protein (GenBank accession number Q05036, Wilson et al., 1994); sea urchin egg receptor for sperm (L04969, D67016, Foltz et al., 1993; but see Just et al., 1997), human B lymphocyte HSP70-like protein, Hsp70RY (L127123, Fatallah et al., 1993); mouse Hsp110 (L40406, Morozov et al., 1995; Yasuda et al., 1995), mouse osmotic stress protein Osp94 (U23921, Kojima et al., 1996) also known as Apg-1 (D49482, Kaneko et al., 1997), mouse Apg-2 (D85904) and Chinese hamster Hsp110 (Z47807, Lee-Yoon et al., 1995). Grp170, an endoplasmic reticulum-retained, glucose-regulated stress protein (Lin et al., 1993; Chen et al., 1996), also shares some conserved features common to this family. We have made pairwise comparisons of the sequences for many of these proteins using the Clustal method, and constructed a 'phylogenetic' tree expressing their mutual relatedness and their relationship to the stress-70 proteins (Fig. 1). These proteins are clearly divided into three groups by this approach: (i) the stress-70 group (including E. coli DnaK, bovine Hsp70, yeast Ssa2p, BiP/Grp78); (ii) the HSP110/SSE group (Hsp110, Hsp70RY, sea urchin sperm receptor, Sse1 and Sse2); and (iii) Grp170. The HSP110/SSE group appears to be no more closely related to the stress-70 group than to Grp170. However, the overall organization of sequence elements in Grp170 appears to be more similar to that of the HSP110/SSE subfamily than to that of the stress-70 group (Fig. 2).

HSP110-related proteins contain a stretch of approximately 100 residues (beginning at about residue 500 in Hsp110) in which little similarity can be found among the members of the family. However, the structure-forming properties of the amino acids in this region are remarkably similar in all family members; they are highly hydrophilic, promote flexibility and are likely to form turns. The remainder of the C-terminal half of these proteins consists of short stretches of conserved sequence interspersed with more diverged sequences. Figure 2 details the general relationships between these structural elements in four members of the family and in Grp170. Although Grp170 is clearly a large HSP70-like protein and shares some features with the HSP110s, it is not strictly a member of the HSP110/SSE family as indicated by a computer phylogenetic analysis (Fig. 1). Moreover, additional large HSP70-like sequences have recently appeared in the databases which have in common with Grp170, apparent C-terminal endoplasmic reticulum retention signals.

■ Biological functions of HSP110/SSE family members

Hsp110 is a heat shock protein whose induction correlates with thermal tolerance expression in mammalian cells (Subjeck and Sciandra, 1982). The yeast proteins, Sse1p and Sse2p, are stress proteins that can be induced by heat shock (Mukai et al., 1993). Sse1p is required for growth at high temperature, indicating that it too may be involved in thermotolerance. Grp170 appears to be a molecular chaperone analogous to Grp78/BiP (Lin et al., 1993). However, its sequence in the region of the protein believed to be responsible for peptide binding is more

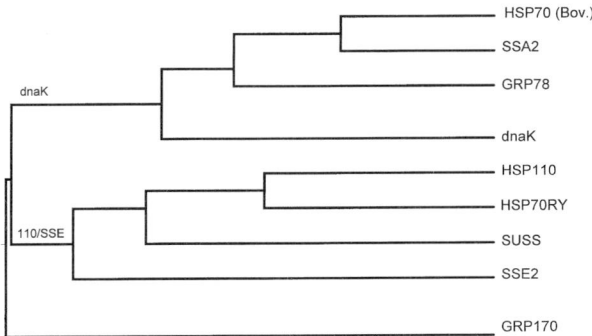

Figure 1. Family tree of HSP70-related proteins. The tree was determined by pairwise alignment using the Clustal method as implemented in DNASTAR's lazergene. The sequences used (except Grp170) are the same as in Lee-Yoon et al. (1995).

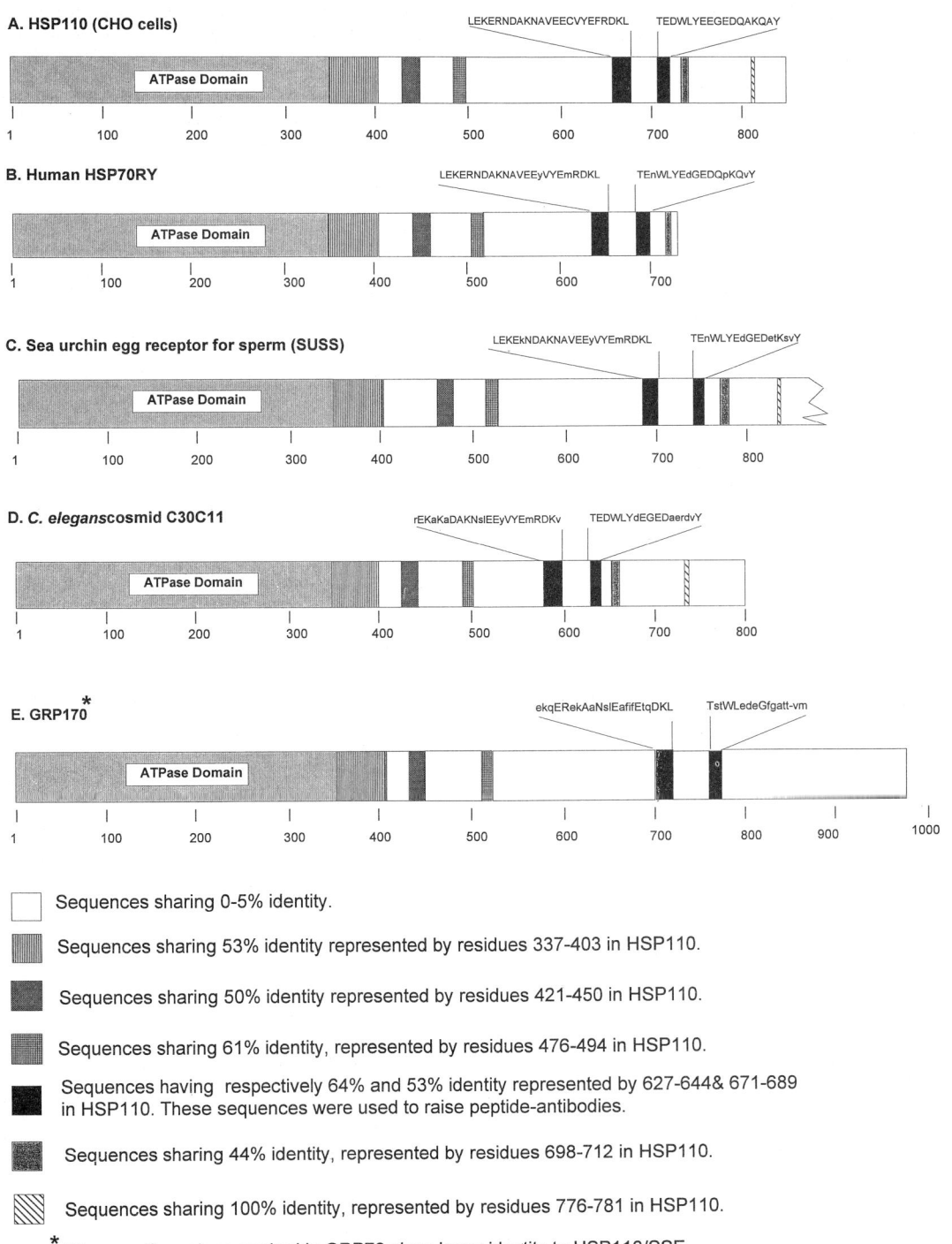

Figure 2. Schematic representation of selected members of the HSP110/SSE family and Grp170. The variously hatched regions represent positions of high identity amongst the family members, as determined by multiple alignment (Lee-Yoon *et al.*, 1995). Regions of Grp170 aligning with and showing the highest levels of identity with the HSP110/SSE family members are indicated in E. The levels of identity in these regions are not as high as this between members of the HSP110/SSE family. The designation 'ATPase domain' does not indicate ATPase function but rather sequence similarity between the designated region and the ATPase domain of HSP70.

similar to Hsp110 than to Grp78 or Hsp70, suggesting that the function of Grp170 may be similar to that of Hsp110 rather than Grp78. Grp170 has been shown to be coinduced with Grp78 and Grp94, with similar kinetics and under the same stress conditions that induce the other Grps (Chang et al., 1989). The sea urchin receptor for sperm is unique in having a cell surface location (Giusti et al., 1997). The function of this protein is to bind spermatozoa to the surface of the egg.

Mouse Hsp110 shares 94% amino acid sequence identity with hamster Hsp110 (Morozov et al., 1995), while Hsp70RY displays 65% identity with Hsp110. Since mammals contain both an Hsp110 and the 78-kDa Hsp70RY protein (Dyer and Rosenberg, 1994), multiple HSP110/SSE subfamily members can coexist in an individual species (i.e. similar to what has long been known for members of the HSP70 protein family). The mouse Apg-2 protein includes a sequence that is nearly identical to Hsp70RY; however the *apg-2* cDNA (GenBank accession number D85904) encodes a protein much larger than Hsp70RY, being approximately the same size as Hsp110 and Apg-1 (J. Fujita, personal communication). In addition, mouse Apg-1 and Osp94 represent identical proteins that differ from Hsp110, Apg-1 and Hsp70RY (see entry on Osp94, p. 84). This suggests that at least three members of the HSP110/SSE family exist in mammals (Hsp110, Apg-1/Osp94, and Apg-2). The common architecture of the HSP110/SSE proteins would suggest common functions, and all of these proteins may function as molecular chaperones. The commonalities of the group should serve as a guide in our efforts to determine the structure/function relationships for individual family members.

■ References

Chang, S.C., Erwin, A.E., and Lee, A.S. (1989). Glucose-regulated protein (GRP94 and GRP78) genes share common regulatory domains and are coordinately regulated by common trans-acting factors. Mol. Cell. Biol. **9**, 2153–2162.

Chen, X., Easton, D.P., Lee-Yoon, D., Liu, X., and Subjeck, J.R. (1996). The 170 kDa glucose regulated protein is a large hsp70/Hsp110-like protein of the endoplasmic reticulum. FEBS Lett. **380**, 68–72.

Dyer, K.D. and Rosenberg, H.F. (1994). Hsp70RY: further characterization of a novel member of the hsp70 protein family. Biochem. Biophys. Res. Commun. **203**, 577–581.

Fatallah, D.M., Cherif, D., Dellagi, K., and Arnaout, M. A. (1993). Molecular cloning of a novel human hsp70 from a B cell line and its assignment to chromosome 5. J. Immunol. **151**, 810–813.

Foltz, K.R., Partin, J.S., and Lennarz, W.J. (1993). Sea urchin egg receptor for sperm: sequence similarity of binding domain and hsp70. Science **259**, 1421–1425.

Giusti, A.F., Hoang, K.M., and Foltz, K.R. (1997). Surface localization of the sea urchin egg receptor for sperm. Develop. Biol. **184**, 10–24.

Hatayama, T., Nishiyama, E., and Yasuda, K. (1994). Cellular localization of high-molecular-mass heat shock proteins in murine cells. Biochem. Biophys. Res. Commun. **200**, 1367–1373.

Just, M.L., and Lennarz, W.J. (1997). Re-examination of the sequence of the sea urchin egg receptor for sperm: implications with respect to its properties. Develop Biol. **184**, 25–30.

Kaneko, Y., Nishiyama, H., Nonoguchi, K., Higashituji, H., Nakayama, H. and Frujita, J. (1997). A novel hsp110-related gene, apg-1, that is abundantly expressed in the testis responds to a low temperature heat shock rather than the traditional elevated temperatures. J. Biol. Chem. **272**, 2640–2645.

Kojima, R., Randall, J., Brenner, B.M. and Gullans, S.R. (1996). Osmotic stress protein 94 (Osp94): A new member of the Hsp110/SSE gene subfamily. J. Biol. Chem. **271**, 12327–12332.

Lee-Yoon, D., Easton, D.P., Murawski, M., Burd, R., and Subjeck, J.R. (1995). Identification of a major subfamiy of large hsp70-like proteins through the cloning of the mammalian 110 kDa heat shock protein. J. Biol. Chem. **270**, 15725–15733.

Lin, H.Y., Masso-Welch, P., Di, Y. P., Cai, J. W., Shen, J. W., and Subjeck, J.R. (1993). The 170 kDa glucose regulated protein is an endoplasmic reticulum proten that binds immunoglobulin. Mol. Biol. Cell **4**, 1109–1119.

Morozov, A., Subjeck, J., and Raychaudhuri, P. (1995). Molecular cloning of a cDNA fro a 110 kDa heat shock protein related polypeptide as an HPV16 E7-inducible gene. FEBS Lett. **371**, 214–218.

Mukai, H., Kuno, T., Tanaka, H., Hirata, D., Miyakawa, T., and Tanaka, C. (1993). Isolation and characterization of SSE1 and SSE2, new members of the yeast hsp70 multigene family. Gene **132**, 57–66.

Shirayama, M., Kawakami, K., Matsui, Y., Tanaka, K., and Toh-E, A. (1993). MSI3, a multicopy suppressor of mutants hyperactivated in the RAS–cAMP pathway, encodes a novel Hsp70 protein of *Saccharomyces cerevisiae*. Mol. Gen. Genet. **240**, 323–332.

Subjeck, J. R. and Sciandra (1982). In *Heat Shock: From Bacteria to Man*, (M. Schlesinger, M. Ashburner and A. Tissieres, eds). Cold Spring Harbor Press, New York, pp. 405–411.

Wilson, R., Ainscough, R., Anderson, K., Baynes, C., Berks, M., Bonfield, J., Burton, J., Connel, M., Copsey, T., Cooper, J., Coulson, A., Craxton, M., Dear, S., Du, Z., Durbin, R., Favello, A., Fraser, A., Fulton, L., Gardner, A., Green, P., Hawkins, T., Hillier, L., Jier, M., Johnston, L., Jones, M., Kershaw, J., Kirsten, J., Laisster, N., Latreille, P., Lightning, J., Lloyd, C., Mortimore, B., O'Callaghan, M., Parsons, J., Percy, C., Rifken, L., Roopra, A., Saunders, D., Shownkeen, R., Sims, M., Smaldon, N., Smith, A., Smith, M., Sonnhammer, E., Staden, R., Sulston, J., Thierry-Mieg, J., Thomas, K., Vaudin, M., Vaughan, K., Waterston, R., Watson, A., Weinstock, L., Wilkinson-Sproat, J., and Wohldman, P. (1994) 2.2 Mb of contiguous nucleotide sequence from chromosome III of C .elegans. Nature **368**, 32–38.

Yasuda, K., Naki, A., Hatayama, T., and Nagata, K. (1995). Cloning and expression of a murine high-molecular weight heat shock protein, Hsp105. J. Biol. Chem. **270**, 29718–29723.

■ *Douglas P. Easton:*
Department of Biology
State University College at Buffalo
1300 Elmwood Ave.
Buffalo, NY 14222, USA
Tel. 1 716 878 3050
Fax. 1 716 878 4028
E-mail: eastondp@snybufaa.cs.snybuf.edu

■ *John R. Subjeck:*
Dept. of Molecular and Cellular Biology
Roswell Park Cancer Institute
Buffalo, NY 14263, USA
Tel. 1 716 845 4425
Fax. 1 716 845 5639
E-mail: subjeck@sc3101.med.buffalo.edu

Saccharomyces cerevisiae Sse1 and Sse2 proteins

Sse1 and Sse2 proteins are highly diverged members of the HSP70 protein family, originally identified as the major calmodulin-binding proteins of the yeast Saccharomyces cerevisiae. These proteins display 38–40% amino acid identities over their entire lengths to the mammalian Chinese hamster ovary Hsp110 protein. Haploid yeast cells carrying a disrupted SSE1 gene grow at slower rates than the parental, wild-type strain.

■ Alternative names

Msi3 (Sse1) (Shirayama et al., 1993).

■ Isolation

Complementary DNAs encoding Sse1 and Sse2 proteins were isolated from a S. cerevisiae expression library using antisera made against a yeast calmodulin-binding fraction (Mukai et al., 1993). The SSE1 gene was also isolated as a multicopy suppressor of the heat shock-sensitive phenotype of the ira1 mutation, which causes hyperactivation of the RAS–cAMP pathway (Shirayama et al., 1993).

■ SSE genes and sequences

The nucleotide sequences of SSE1 cDNA and genomic DNA (GenBank accession numbers D13908 and D13910) predict a protein of 693 amino acids with calculated molecular weight of 77 408 (Mukai et al., 1993). The nucleotide sequences of SSE2 cDNA and genomic DNA (GenBank accession numbers D13909 and D13911) predict a protein of 693 amino acids with a calculated molecular weight of 77 619 (Mukai et al., 1993). The Sse1 and Sse2 proteins are 75.5% identical to each other, and are 27.1 and 26.8% identical, respectively, to the Ssa1 protein. Ssa1p, a cytosolic HSP70 protein of S. cerevisiae (see entry p. 26) displays the highest homology to Sse proteins among other HSP70 family members. The N-terminal 382 amino acids of Sse1p and Sse2p, which are thought to be involved in ATP binding, are 35.6 and 36.1% identical with Ssa1p, while their C-terminal 311 amino acids are only 14.8 and 14.7% identical, respectively, with Ssa1p. These yeast proteins display 38–40% identities over their entire lengths to the mammalian Chinese hamster ovary Hsp110 protein (Lee-Yoon et al., 1995). The two yeast proteins also share basic organizational characteristics with other Hsp110-related proteins such as human Hsp70RY (Fathallah et al., 1993), sea urchin egg receptor for sperm (Foltz et al., 1993), Caenorhabdidtis elegans cosmid C30c11 (Wilson et al., 1994), e.g. the alignment of highly conserved C-terminal regions, the absence of a section of a sequence conserved in HSP70 proteins (Rippmann et al., 1991), and the presence of a large central non-homologous region.

■ Sse1 and 2 proteins, function, and mutagenesis studies

Sse1 and Sse2 proteins are c. 78-kDa members of the stress-70 protein family, belonging to the HSP110/SSE subfamily. Sse1p and Sse2p bind calmodulin, but the functional significance of this is unknown (Mukai et al., 1993). Overexpression of Sse1p suppresses the heat shock-sensitive phenotype of S. cerevisiae ira1 and bcy1 mutants (Shirayama et al., 1993). Haploid cells carrying a disrupted SSE1 gene grow at slower rates than wild-type cells at 20, 25, or 37°C, while haploid cells carrying a disrupted SSE2 gene are indistinguishable from wild-type cells; double-mutant cells are indistinguishable from the haploid strain containing only the disrupted SSE1 gene (Mukai et al., 1993). The slow growth phenotype resulting from the SSE1 gene disruption could be suppressed by introduction of the SSE2 gene on a multicopy plasmid.

■ SSE gene regulation

SSE1 mRNAs are moderately abundant during steady-state growth at 23°C but increase a few-fold after upshift to 37°C. In the case of SSE2 mRNA, less abundant transcripts present at 23°C increase dramatically upon temperature upshift to 37°C. The potential heat shock elements, tGAAggTTCt and aTTCtaGAAgaTTCc, were found at positions −315 and −187 respectively, in the SSE1 gene, with numbering based on the assignment of +1 to A in the initiating ATG (Mukai et al., 1993). In the case of the SSE2 gene, the potential heat shock elements, cTTCagGAAa and gTTCtaGAAc, were found at position −364 and −238, with numbering based on the same rule.

■ References

Fatallah, D.M., Cherif, D., Dellagi, K., and Arnaout, M.A. (1993). Molecular cloning of a novel human hsp70 from a B cell line and its assignment to chromosome 5. J. Immunol. **151**, 810–813.

Foltz, K.R., Partin, J.S., and Lennarz, W.J. (1993). Sea urchin egg receptor for sperm: sequence similarity of binding domain and hsp70. Science **259**, 1421–1425.

Lee-Yoon, D., Easton, D.P., Murawski, M., Burd, R., and Subjeck, J.R. (1995). Identification of a major subfamiy of large hsp70-like proteins through the cloning of the mammalian 110 kDa heat shock protein. J. Biol. Chem. **270**, 15725–15733.

Mukai, H., Kuno, T., Tanaka, H., Hirata, D., Miyakawa, T., and Tanaka, C. (1993). Isolation and characterization of SSE1 and SSE2, new members of the yeast hsp70 multigene family. Gene **132**, 57–66.

Rippmann, F., Taylor, W.R., Rothbard, J.B., and Green, N.M. (1991). A hypothetical model for the peptide binding domain of hsp70 based on the peptide binding domain of HLA. EMBO J. **10**, 1053–1059.

Shirayama, M., Kawakami, K., Matsui, Y., Tanaka, K., and Toh-E, A. (1993). MSI3, a multicopy suppressor of mutants hyperactivated in the RAS–cAMP pathway, encodes a novel Hsp70 protein of Saccharomyces cerevisiae. Mol. Gen. Genet. **240**, 323–332.

Wilson, R., Ainscough, R., Anderson, K., Baynes, C., Berks, M., Bonfield, J., Burton, J., Connel, M., Copsey, T., Cooper, J., Coulson, A., Craxton, M., Dear, S., Du, Z., Durbin, R., Favello, A., Fraser, A., Fulton, L., Gardner, A., Green, P., Hawkins, T., Hillier, L., Jier, M., Johnston, L., Jones, M., Kershaw, J., Kirsten, J., Laisster, N., Latreille, P., Lightning, J., Lloyd, C., Mortimore, B., O'Callaghan, M., Parsons, J., Percy, C., Rifken, L., Roopra, A., Saunders, D., Shownkeen, R., Sims, M., Smaldon, N., Smith, A., Smith, M., Sonnhammer, E., Staden, R., Sulston, J., Thierry-Mieg, J., Thomas, K., Vaudin, M., Vaughan, K., Waterston, R., Watson, A., Weinstock, L., Wilkinson-Sproat, J., and Wohldman, P. (1994). 2.2 Mb of contiguous nucleotide sequence from chromosome III of C .elegans. Nature **368**, 32–38.

■ *Hideyuki Mukai, Hisato Shuntoh, and Takayoshi Kuno:*
Department of Pharmacology
Kobe University School of Medicine
7-5-1 Kusunoki-cho, Chuo-ku
Kobe 650, Japan
Tel. 81 78 341 7451 ext. 3270
Fax. 81 78 351 6531
E-mail: tkuno@kobe-u.ac.jp

Neurospora crassa Hsp88

Hsp88 of Neurospora crassa *was identified by its binding to Hsp30, the α-crystallin-related heat shock protein of N. crassa; this property led to the isolation of Hsp88 by affinity chromatography. Hsp88 is closely related to proteins that have been characterized from mammals, yeast, and nematodes; all these proteins are members of the HSP110/SSE family and are distantly related to HSP70 proteins.*

■ Isolation

We first identified Hsp88 in our search for proteins from heat shocked cells that interact with the small HSP of *N. crassa*. Two detectable proteins, Hsp70 and Hsp88, were found to bind specifically to Hsp30 (N. Plesovsky-Vig and R. Brambl, unpublished results). Hsp88 was isolated by affinity chromatography and gel electrophoresis. After digestion of Hsp88 by trypsin, several peptides were N-terminally sequenced, leading to the design of conceptually coding oligonucleotides. An amplified gene fragment, produced by the polymerase chain reaction, was used to probe a *N. crassa* cDNA library, and a 2.6 kb cDNA was identified and sequenced.

■ *hsp88* gene and sequence

The cDNA sequence for Hsp88 predicts a protein of 707 amino acids that is 30% identical to HSP70. It shows greater similarity to a newly characterized class of HSP110/SSE proteins, showing 48% identity to Hsp79 (Sse2p) of *Saccharomyces cerevisiae* (Mukai et al., 1993) and 45 and 40% identity, respectively, to the human Hsp70RY (Fathallah et al., 1993) and hamster Hsp110 (Lee-Yoon et al., 1995) within a 500–600 amino acid overlap. The putative ATPase domains of these proteins are more conserved. In this N-terminal, 385 amino acid region, Hsp88 has 56% identity to yeast Hsp79 and 49% identity to Hsp70RY. The conserved amino acids within the ATP-binding site of HSP70 (Bork et al., 1992) are present in Hsp88 and most related proteins. Hsp110, however, deviates at one residue in the phosphate 2 region of this site.

■ Hsp88 localization

One known activity of Hsp88 is its *in vitro* binding to recombinant Hsp30. However, these two Hsps are located in different cellular compartments during heat shock. Hsp30 is found predominantly associated with mitochondria (Plesofsky-Vig and Brambl, 1990), whereas Hsp88 is mainly a soluble cytosolic protein. Nevertheless, a small amount of Hsp88 cosediments with mitochondria, and

Hsp30 itself becomes soluble in cells that have been shifted to a normal temperature.

Function of Hsp88

Interaction of ATP-dependent chaperones, such as Hsp70 and most likely Hsp88, with Hsp30 may assist this small Hsp in its putative role of preventing protein aggregation, or it may regulate Hsp30 chaperone activity. Hsp88 and Hsp70 may also facilitate the dissociation of Hsp30 from mitochondria.

Regulation of *hsp88*

Neurospora crassa cells express the gene for Hsp88 during growth at normal temperature, as indicated by its RNA level in a Northern blot. However, cellular exposure to heat shock strongly stimulates the accumulation of *hsp88* RNA at least tenfold (N. Plesovsky-Vig and R. Brambl, unpublished results).

Biological interactions of Hsp88

Both Hsp88 and Hsp70 bind to Hsp30 in the presence of cellular components. However, there is no obvious stoichiometry between Hsp88 and Hsp70, since their relative binding in two cell fractions, cytosol and mitochondrial lysate, reflected their concentration in these fractions. This suggests that Hsp88 and/or Hsp70 may bind independently to Hsp30.

References

Bork, P., Sander, C., and Valencia, A. (1992). An ATPase domain common to prokaryotic cell cycle proteins, sugar kinases, actin, and hsp70 heat shock proteins. Proc. Natl. Acad. Sci. USA **89**, 7290–7294.

Fathallah, D.M., Cherif, D., Dallagi, K., and Arnaout, M.A. (1993). Molecular cloning of a novel human hsp70 from a B cell line and its assignment to chromosome 5. J. Immunol. **151**, 810–813.

Lee-Yoon, D., Easton, D., Murawski, M., Burd, R., and Subjeck, J.R. (1995). Identification of a major subfamily of large hsp70-like proteins through the cloning of the mammalian 110-kDa heat shock protein. J. Biol. Chem. **270**, 15725–15733.

Mukai, H., Kuno, T., Tanaka, H., Hirata, D., Miyakawa, T., and Tanaka, C. (1993). Isolation and characterization of SSE1 and SSE2, new members of the yeast HSP70 multigene family. Gene **132**, 57–66.

Plesofsky-Vig, N. and Brambl, R. (1990). Gene sequence and analysis of hsp30, a small heat shock protein of Neurospora crassa which associates with mitochondria. J. Biol. Chem. **265**, 15432–15440.

■ Nora Plesofsky-Vig and Robert Brambl:
The University of Minnesota
Department of Genetics and Cell Biology
Department of Plant Biology
220 Biological Sciences Center
1445 Gortner Avenue
St. Paul, MN 55108, USA
Tel. 1 612 624 5375
Fax. 1 612 625 1738
E-mail: nora@biosci.cbs.umn.edu

The sea urchin egg receptor for sperm

The egg receptor for sperm is a unique cell surface glycoprotein that mediates egg–sperm interaction in fertilization in sea urchins. Much of the egg receptor for sperm contains high sequence identity to the HSP70 subfamily termed HSP110/SSE.

Isolation

A 70 kDa glycoprotein fragment of the egg receptor for sperm was isolated by endolysyl proteinase C digestion of dejellied *Strongylocentrotus purpuratus* eggs, followed by DEAE and gel exclusion chromatography. The receptor glycoprotein fragment was detected by its ability to inhibit fertilization in competitive fertilization bioassays (Foltz and Lennarz, 1990). A polyclonal antibody generated against the 70 kDa fragment was used to screen an ovary/oocyte expression cDNA library in order to clone the receptor (Foltz *et al.*, 1993). The intact egg receptor for sperm was isolated by detergent solubilization of egg plasma membrane/vitelline layer preparations, followed by affinity purification with WGA chromatography then DEAE anion exchange chromatography (Ohlendiek *et al.*, 1993).

Gene and sequence

Sequence changes from those reported for the original cDNA clones (Foltz *et al.*, 1993) now predict that the egg receptor for sperm is 889 amino acids (c. 99 kDa) and contains a cysteine rich region at the N-terminus and 17 O-linked and 5 N-linked putative sites for glycosylation (Just and Lennarz, 1997). The new sequence is more than 44% identical to Hsp110 (Lee-Yoon et al., 1995) across the entire molecule, with the amino terminus being more highly conserved than the C-terminus (Fig.1).

■ Egg receptor for sperm protein

The egg receptor for sperm exhibits novel properties. Unlike all other members of the HSP70 and HSP110 families characterized so far, the egg receptor for sperm associates with the cell surface and the extracellular environment (Foltz and Lennarz, 1992; Ohlendieck et al., 1994a; Giusti et al., 1997). Furthermore glycosylation of the receptor comprises approximately 70% of the mass weighting the 99 kDa peptide backbone to approximately 350 kDa in its native state (Ohlendieck et al., 1993; Kitazume-Kawaguchi et al., 1997). Currently no other HSP110 family members have been shown to be glycosylated. The majority of the HSP110 members are localized in the cytoplasm or nucleus. One interesting feature of this new family of heat shock proteins is its involvement and localization in gonadal tissue. Hsp110 was identified in Chinese hamster ovary cells (Lee-Yoon et al. 1995), and Apg-1 was cloned from mouse testis cells (Kaneko et al., 1997). The role for heat shock proteins in gamete production is not so difficult to understand given the need to regulate body temperature in testis and to package the egg with sufficient protein to sustain the early developing embryo.

■ Biological activities

The egg receptor for sperm functions in fertilization by mediating genus-specific gamete interactions, i.e. binding sperm. Although both the carbohydrate elements and the polypeptide backbone appear to participate in sperm binding (Foltz et al., 1993; Dhume and Lennarz, 1995), it is the unglycosylated receptor that mediates selectivity with respect to the sperm (Stears and Lennarz, 1997). The sulfated O-linked carbohydrate chains also bind sperm, although less actively than the polypeptide backbone (Dhume and Lennarz, 1995). These biologically active carbohydrate chains inhibit fertilization with an IC_{50} 20-fold lower than the recombinant proteins or intact receptor, (Dhume et al., 1996). Their role appears to be in stabilizing sperm binding.

■ Gene regulation

Developmental northern blot analysis reveals that the receptor transcript is present at low levels during oocyte development (R. Stears, unpublished observations) with an increase in the message levels during egg maturation. Following fertilization the transcript quickly disappears. This finding is in good agreement with the changes in the protein levels observed during and post insemination (Ohlendieck et al., 1994b). Recent work indicates that expression occurs again during development in the gastrula stage (M.L. Just and W.J. Lennarz, unpublished observation). It is not certain if this form of the transcript becomes glycosylated and localized to the cell surface. Furthermore, its role during this stage of development is unknown. It is interesting to note that in a different sea urchin species, Hsp70 expression is triggered at the time of gastrulation (Sconzo et al., 1992).

■ Biological interactions

Two sperm-binding domains have been identified in a portion of the polypeptide chain. One domain near the amino terminus binds sperm without specificity; it can bind several different species and genuses of sperm. The carboxy terminal domain binds sperm in a more discriminate manner; it prefers sperm of the same genus and excludes binding to more distantly related sperm. A small region has been identified in the carboxy terminal domain containing 32 amino acids which accounts for the sperm binding selectivity (Fig. 2). A model for sperm interaction with the egg receptor for sperm suggests that the sperm receptor interacts with sperm on three levels: the amino terminus of the receptor binds to sperm with little species/genus selectivity and the carbohydrate elements stabilize this binding. It is proposed that the receptor undergoes a conformational change upon initial sperm binding allowing for the 32 amino acid genus specific domain to interact with the sperm (Stears and Lennarz, 1997). This binding allows the sperm to continue along the pathway to successful fertilization.

Figure 1. Representation of comparison between the sequence of the sea urchin egg receptor for sperm and hsp110. Slashed boxes represent greater than 50% identity while the dotted boxes represent 30–49% identity. Unshaded boxes indicate regions of no identity to hsp110.

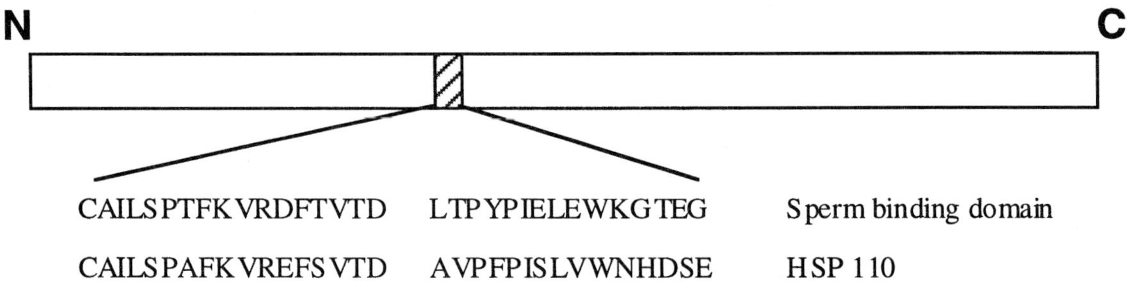

```
        N                                                    C
    ┌─────────────────────────┬──┬──────────────────────────┐
    │                         │▨▨│                          │
    └─────────────────────────┴──┴──────────────────────────┘
                         ╱         ╲
                        ╱           ╲
```

CAILSPTFKVRDFTVTD   LTPYPIELEWKGTEG   Sperm binding domain

CAILSPAFKVREFSVTD   AVPFPISLVWNHDSE   HSP 110

Figure 2. HSP homologies within the sperm binding domains. The identified sperm binding domains were compared to the HSP protein sequences by the Blast-p program and GCG sequence alignments. Modified from Stears and Lennarz (1997).

■ References

Dhume, S.T. and Lennarz, W.J. (1995). The involvement of O-linked oligosaccharide chains of the sea urchin egg receptor for sperm in fertilization. Glycobiology **5**, 11–17.

Dhume, S., Stears, R.L., and Lennarz, W.J. (1996). Sea urchin egg receptor for sperm: the oligosaccharide chains stabilize sperm binding. Glycobiology **6**, 59–64.

Foltz, K. and Lennarz, W.J. (1990). Purification and characterization of an extracellular fragment of the sea urchin egg receptor for sperm. J. Cell Biol. **111**, 2931–2959.

Foltz, K.R. and Lennarz, W.J. (1992). Identification of the sea urchin egg receptor for sperm using an antiserum raised against a fragment of its extracellular domain. J. Cell Biol. **116**, 647–658.

Foltz, K., Partin, J., and Lennarz, W.J. (1993). Sea urchin egg receptor for sperm: sequence similarity of binding domain and hsp70. Science **229**, 1421–1425.

Giusti, A.F., Hoang, K.M., and Foltz, K.R. (1997). Surface localization of the sea urchin egg receptor for sperm. Develop. Biol. **184**, 10–24.

Just, M.L., and Lennarz, W.J. (1997). Re-examination of the sequence of the sea urchin egg receptor for sperm: implications with respect to its properties. Develop. Biol. **184**, 25–30.

Kaneko, Y., Nishiyama, H., Nonoguchi, K., Higashituji, H., Kishishita, M., and Fujita, J. (1997). A novel hsp 110-related gene, apg-1, that is abundantly expressed in the testis responds to a low temperature heat shock rather than the traditional elevated temperatures. J. Biol. Chem. **272**, 2640–2645.

Kitazume-Kawaguchi, S., Inoue, S., Inoue, Y., and Lennarz, W.J. (1997). Identification of sulfated oligosialic acid units in the O-linked glycan of the sea urchin egg receptor for sperm. Proc. Natl. Acad. Sci. USA **94**, 3650–3655.

Lee-Yoon, D., Easton, D., Murawski, M., Burd, R., and Subjeck, J. (1995). Identification of a major subfamily of large hsp70 like proteins through the cloning of the mammalian 110-kDa heat shock protein. J. Cell Biol. **270**, 15725–15733.

Ohlendiek, K., Dhume, S., Partin, J., and Lennarz, W.J. (1993). The sea urchin egg receptor for sperm: isolation and characterization of the intact biologially active receptor. J. Cell Biol. **122**, 887–895.

Ohlendiek, K., Partin, J., and Lennarz, W.J. (1994a). The biologically active form of the sea urchin egg receptor for sperm is a disulfide-bonded homo-multimer. J. Cell Biol. **125**, 817–824.

Ohlendiek, K., Partin, J.S., Stears, R.L., and Lennarz, W.J. (1994b). Developmental expression of the sea urchin egg receptor for sperm. Develop. Biol. **165**, 53–62.

Sconzo, G., Scardina, G., and Ferraro, M.G. (1992). Characterization of a new member of the sea urchin *Paracentrotus lividus* hsp70 gene family and its expression. Gene **121**, 353–358.

Stears, R.L. and Lennarz, W.J. (1997). Mapping the sperm binding domains on the sea urchin egg receptor for sperm. Develop. Biol. In press.

■ *Robin L. Stears, Margaret L. Just, and*
William J. Lennarz:
Dept. of Biochemistry and Cell Biology
State University of New York at Stony Brook
Stony Brook, NY 11794, USA
Tel. 1 516 632 8560
Fax. 1 516 632 8575
E-mail: wlennarz@life.bio.sunysb.edu

Mammalian Hsp110

Hsp110 is a major mammalian heat shock protein which has been long observed but has only recently been cloned and sequenced. This protein is synthesized constitutively and its synthesis may be strongly enhanced by the same treatments/agents that induce HSP70 proteins. Hsp110 displays 30–35% amino acid sequence identity with HSP70 proteins; most of this sequence conservation is found in the N-terminal ATPase domain. Hsp110 protein shares specific structural organizational features with other large and unusual HSP70-like proteins. Induction of Hsp110 strongly correlates with thermotolerance.

■ Alternative names

Although the Hsp110 designation is now highly prevalent, the protein has also been referred to as Hsp107 (Landry et al., 1982), Hsp112 (Tomasovic et al., 1983) and Hsp105 (Hatayama et al., 1994).

■ Sequence

The hamster *hsp110* cDNA has recently been cloned and sequenced (EMBL/GenBank™/DDBL accession number Z47807, Lee-Yoon et al., 1995). The nucleotide sequence predicts that Hsp110 is an 858 amino acid protein with a calculated molecular weight of 96 159 Da. The sequence has a 'nuclear localization' signal KKPK starting at position 586. BLAST searches of the GenBank™ databases show that Hsp110, although similar to the major family of HSP70 proteins, is much more closely related to a number of unusual high molecular weight HSP70-like proteins, now designated the HSP110/SSE subfamily: sea urchin egg receptor for sperm (L04964), human Hsp70RY (P34932), mouse Hsp110 (L40406, D67016), mouse APG-1/Osp94 (D49482, U23921), *Caenorhabdidtis elegans* C30c11 hypothetical protein (Q05036) and the yeast Sse1 and Sse2 proteins (JN0842 and JN0843). All these proteins are most similar to each other, and to bovine Hsp70, in their N-terminal half. Their sequences are most diverged in the region that is equivalent to residues 500–600 in Hsp110. The remainder of the C-terminal end, equivalent to residues 600–858 in Hsp110, contains regions of high similarity to the other members of the group (excluding Hsp70) interspersed with regions of lower similarity (see Fig. 2 in the overview of the HSP110/SSE sub-family, p. 73) (Lee-Yoon et al., 1995). The residues in Hsp110 that align with the active site motifs in HSP70 proteins meet the specifications of Bork et al. (1992) for an ATP-binding site.

Hatayama et al. (1994) sequenced a peptide of an endopeptidase-digested 105 kDa heat shock protein of mouse FM3A cells. Recently, the complete cDNA for Hsp105 has been obtained, and determined to be 96% identical to Hsp110 (Yasuda et al., 1995).

■ Hsp110 protein

Early studies of the mammalian heat shock proteins identified, in addition to the familiar HSP70 proteins, heat shock proteins of 28, 90 and 110 kDa. Of these proteins, Hsp110 is the least well studied. Hsp110 is a constitutive protein of mammalian cells, whose cellular levels are augmented by heat shock (Subjeck et al., 1982; Welch et al., 1983). Its induction correlates well with expression of thermotolerance (Subjeck et al., 1982). Hsp110 is localized throughout the nucleoplasm and cytoplasm and can appear in a halo-like formation around nucleoli (Lee-Yoon et al., 1995; Yasuda et al., 1995). The general similarity in sequence between Hsp70 and Hsp110 suggests that these proteins may perform somewhat similar chaperoning functions in cells.

■ Gene regulation of *hsp110*

Hsp110 is a constitutive protein in mammalian cells whose synthesis can be induced by the same conditions/agents and with the same kinetics as Hsp70 (Welch et al., 1983). The regulation of the promoter of the *hsp110* gene is assumed to be the same as that for *hsp70*., but isolation of genomic clones and sequence analysis of the promoter region is required to confirm this hypothesis. Recently, Morozov et al. (1995) reported that human papillomavirus protein E7 can specifically induce transcription of *hsp110* mRNA. The mechanism and physiological consequences of this induction remain to be determined. In mouse FM3A cells, two forms of Hsp105 have been observed, Hsp105 itself and a smaller 42°C Hsp which is induced only at 42°C (Yasuda et al., 1995). The 42°C Hsp is shortened by the omission of amino acids 530–573. These residues lie within the diverged region of HSP110 proteins.

■ Function of the Hsp110 protein

Data on the role of Hsp110 in cellular physiology are sparse. Besides information on heat shock induction and correlation with thermotolerance, very little additional information is available. Hsp110 has been identified as a 'glial transfer protein'. The glial transfer proteins, which include Hsp70 as well as Hsp110 and other unidentified proteins, are transferred from glial cells to the axons of squid giant motor neurones where they are hypothesized to cooperate in the maintenance and repair of the neurones (Tytell et al., 1986). Hsp110 constitutes 0.05–0.075% of the total protein in mouse brain, suggesting that it may be important in the function of mammalian neurones.

■ References

Bork, P., Sander, C., and Valencia, A. (1992). An ATPase domain common to prokaryotic cell cycle proteins, sugar kinases, actin and hsp70 heat shock proteins. Proc. Natl. Acad. Sci. USA **89**, 7290–7294.

Hatayama, T., Yasuda, K., and Nishiyama, E. (1994). Characterization of high molecular mass heat shock proteins and 42°C specific heat shock proteins of murine cells. Biophys. Res. Commun. **204**: 357–365.

Landry, J., Bernier, D., Chretien, P., Nicole, L. M., and Tanguay, R. M. (1982). Synthesis and degradation of heat shock proteins during development and decay of thermotolerance. Cancer Res. **61**, 428–437.

Lee-Yoon, D., Easton, D.P., Murawski, M., Burd, R., and Subjeck, J.R. (1995). Identification of a major subfamiy of large hsp70-like proteins through the cloning of the mammalian 110 kDa heat shock protein. J. Biol. Chem. **270**, 15725–15733.

Morozov, A., Subjeck, J., and Raychaudhuri, P. (1995). Molecular cloning of a cDNA from a 110 kDa heat shock protein related polypeptide as an HPV16 E7-inducible gene. FEBS Lett. **371**, 214–218.

Subjeck, J.R. and Sciandra J.J. (1982). In *Heat Shock: From Bacteria to Man*, (M. Schlesinger, M. Ashburner and A. Tissieres, eds). Cold Spring Harbor Press, New York, pp. 105–111.

Subjeck, J.R., Sciandra, J.J., and Johnson, J.R. (1982). Heat shock proteins and thermotolerance: a comparison of induction kinetics. Br. J. Radiol. **55**, 127–131.

Tomasovic, S.P., Steck, P.A., and Heitzman, D. (1983). Heat-stress proteins and thermal resistance in rat mammary tumor cells. Radiat. Res. **95**, 399–413.

Tytell, M, Greenburg, S.G., and Lasek, R.J. (1986). Heat shock-like protein is transferred from glia to axon. Brain Res. **363**, 1161–1164.

Welch, W.J., Garrels, J.I., Thomas, G.P., Lin, J.J.-C., and Feramisco, J.R. (1983). Biochemical characterization of the mammalian stress proteins and identification of two stress proteins as glucose- and Ca^{2+}-ionophore regulated proteins. J. Biol. Chem. **258**, 7102–7111.

Yasuda, K., Nakai, A., Hatayama, T., and Nagata, K. (1995). Cloning and expression of murine high-molecular-weight heat shock proteins, HSP105. J. Biol. Chem. **270**, 29718–29723.

■ *Douglas P. Easton:*
Department of Biology
State University College at Buffalo
1300 Elmwood Ave.
Buffalo, NY 14222, USA
Tel. 1 716 878 3050
Fax. 1 716 878 4028
E-mail: eastondp@snybufaa.cs.snybuf.edu

■ *John R. Subjeck:*
Dept. of Molecular and Cellular Biology
Roswell Park Cancer Institute
Buffalo, NY 14263, USA
Tel. 1 716 845 4425
Fax. 1 716 845 5639
E-mail: subjeck@sc3101.med.buffalo.edu

Mammalian Hsp70h

Hsp70h has been identified as a distant member of the HSP70 protein family on the basis of cDNA sequence homology. The mRNA encoding Hsp70h is expressed in many cell lines and the gene has been localized in humans to the long arm of chromosome 5.

■ Alternative names

hsp70RY (Fathallah et al., 1993, Dyer and Rosenberg, 1994, Lee-Yoon et al., 1995)

■ Isolation

The cDNA encoding Hsp70h was isolated from a human EBV-transformed B cell cDNA library using a CD18-derived cDNA probe (Fathallah et al., 1993). The open reading frame was later confirmed and further characterized by Dyer and Rosenberg (1994).

■ *hsp70h* gene and sequence

The *hsp70h* gene was determined to be a single copy gene localized to chromosome 5q31.1-5q31.2 (Fathallah et al., 1993). Southern blot analysis indicates the presence of a murine homolog of *hsp70h* and suggests a multi-exonic genomic structure (Dyer and Rosenberg, 1994). The *hsp70h* cDNA sequence (GenBank accession number L12723) contains the entire Hsp70h coding region in addition to 133 nucleotides of 5′ and 152 nucleotides of 3′ non-coding sequences (Fathallah et al., 1993). The cDNA encodes a 701 amino acid protein with a predicted molecular weight of c. 77 kDa. Figure 1 shows the predicted amino acid sequence of Hsp70h. The shaded boxes highlight the predicted ATP-binding pocket (residues 3–9, 271–274, 338–346, and 369) (Fathallah et al., 1993). A novel 100 amino acid motif (residues 500–600) is highlighted by the open box. This sequence is rich in glutamatic acid (E) and glutamine (Q) residues, and is more acidic than the remainder of the protein. The calculated pI of this region is 4.07 compared with a pI of 8.02 for amino acids 1–500 (Dyer and Rosenberg, 1994). This motif is not shared by other members of the HSP70 family and has no striking homology to any other protein listed in the Swiss Protein Database.

■ *hsp70h* expression

The mRNA (2.8 kb) encoding Hsp70h is expressed in a variety of human cell lines of both ectodermal (MeWo and HeLa) and mesodermal (K-562, HL-60, Jurkat, and EBV-transformed B cell line) origin (Dyer and Rosenberg, 1994). In experiments in which various hematopoietic cell lines were induced to differentiate towards different myeloid lineages, no change in the level of expression of message for Hsp70h was detected, indicating that Hsp70h is expressed constitutively rather than developmentally in these cell lines (Dyer and Rosenberg, 1994).

```
  1  MSVVGIDLGF QSCYVAVARA GGIETIANEY SDRCTPACIS FGPKNRSIGA
 51  AAKSQVISNA KNTVQGFKRF HGRAFSDPFV EAEKSNLAYD IVQWPTGLTG
101  IKVTYMEEER NFTTEQVTAM LLSKLKETAE SVLKKPVVDC VVSVPCFYTD
151  AERRSVMDAT QIAGLNCLRL MNETTAVALA YGIYKQDLPR LEEKPRNVVF
201  VDMGHSAYQV SVCAFNRGKL KVLATAFDTT LGGRKFDEVL VNHFCEEFGK
251  KYKLDIKSKI RALLRLSQEC EKLKKLMSAN ASDLPLSIEC FMNDVDVSGT
301  MNRGKFLEMC NDLLARVEPP LRSVLEQTKL KKEDIYAVEI VGGATRIPAV
351  KEKISKFFGK ELSTTLNADE AVTRGCALQC AILSPAFKVR EFSITDVVPY
401  PISLRWNSPA EEGSSDCEVF SKNHAAPFSK VLTFYRKEPF TLEAYYSSPQ
451  DLPYPDPAIA QFSVQKVTPQ SDGSSSKVKV KVRVNVHGIF SVSSASLVEV
501  HKSEENEEPM ETDQNAKEEE KMQVDQEEPH VEEQQQQTPA ENKAESEEME
551  TSQAGSKDKK MDQPPQCQEG KSEDQYCGPA NRESAIWQID REMLNLYIEN
601  EGKMIMQDKL EKERNDAKNA VEEYVYEMRD KLSGEYEKFV SEDDRNSFTL
651  KLEDTENWLY EDGEDQPKQV YVDKLAELKN LGQPIKIRFQ ESEERPNYLK
701  N
```

Figure 1. Predicted amino acid sequence of hsp70h. The shaded boxes show the predicted ATP binding pocket and the open box highlights a unique motif which contains an unusual concentration of glutamate (E) and glutamine (Q) residues which impart a pI of 4.07 to this portion of the protein.

Cellular distribution of Hsp70h

Analysis of the cDNA encoding Hsp70h suggests a cytoplasmic distribution for the encoded protein based on the lack of an endoplasmic retention signal or a leader sequence (Fathallah et al., 1993).

References

Dyer, K.D. and Rosenberg, H.F. HSP70RY: further characterization of a novel member of the hsp70 protein family. Biochem. Biophys. Res. Commun. **203**, 577–581 (1994).

Fathallah, D.M., Cherif, D., Dellagi, K., and Arnaout, M.A. Molecular cloning of a novel human hsp70 from a B cell line and its assignment to chromosome 5. J. Immunol. **151**, 810–813 (1993).

Lee-Yoon, D., Easton, D., Murawski, M., Burd, R., and Subjeck, J.R. Identification of a major subfamily of large hsp70-like proteins through the cloning of the mammalian 110-kDa heat shock protein. J. Biol. Chem. **270**, 15725–15733 (1995).

- *Kimberly D. Dyer and Helene F. Rosenberg:*
  Laboratory of Host Defenses, NIAID, NIH
  Building 10 Room 11N104
  10 Center Drive MSC 1886
  Bethesda, MD 20892-1886, USA
  Tel. 1 301 496 2877
  Fax. 1 301 402 4369
  E-mail: kd26h@nih.gov

Mammalian Osp94

Osp94 (osmotic stress protein, 94 kDa) is a recently discovered mammalian heat shock protein that is a member of the HSP110/SSE gene subfamily, which includes Hsp110, Hsp70RY, yeast Sse1 and Sse2, and the sea urchin sperm receptor. Osp94 displays 62–65% amino acid sequence identity with Hsp110 and Hsp70RY but only 30–35% identity with members of the HSP70 protein family. The N-terminal half of the protein contains an ATP-binding domain and the C-terminal half contains a putative peptide-binding domain. Osp94 expression is increased with heat shock, hyperosmotic stress, and a variety of other stresses. In vivo it is highly expressed in renal medulla and testis.

Alternative names

Osp94 is also known as APG-1 (Kaneko et al., 1997).

Isolation

osp94 was identified during an analysis of mRNAs that are up-regulated by hyperosmotic stress in mouse inner medullary collecting duct (mIMCD3) cells (Kojima et al., 1996). APG-1 was identified during a search for germ cell-specific mRNAs in testes (Kaneko et al., 1997).

osp94 gene and sequence

The *osp94* cDNA was cloned and sequenced from a mouse kidney cell line cDNA library (EMBL/GenBank/DDBL accession number U23921, Kojima et al., 1996). The cDNA is 2898 nucleotides in length, not including the 38 nucleotide poly(A) tail. Northern analysis indicated the presence of multiple *osp94* transcripts with sizes of 2.9, 4.1, and 11.2 kb (Kojima et al., 1996). 3′ RACE showed the 4.1 transcript contains a longer 3′ UTR; there is no evidence to suggest it encodes a different protein.

The putative ORF of *osp94* is 2514 nucleotides long beginning at nucleotide 183 and ending at nucleotide 2698. The predicted protein contains 838 amino acids (SwissProt accession number P34932) with a calculated molecular mass of 94 278 Da. Sequence alignment with members of the HSP110/SSE gene subfamily, which includes Hsp110 (accession numbers Z47807, L40406) and Hsp70RY (P34932), shows highest sequence homology in the N-terminal ATP-binding half of the molecule. The C-terminal domain has two regions of low sequence similarity: Val^{497}–Gln^{594} and Lys^{707}–Asp^{838}. Independent BLASTP searches of GenBank using these divergent regions of Osp94 failed to find proteins with significant homology.

apg-1 was cloned from mouse testis germ cells (accession number D49482, Kaneko et al., 1997). *apg-1* is a 2770-bp cDNA which is 99.4% identical to *osp94* and appears to be the same gene. Compared to *osp94*, the *apg-1* cDNA is truncated at the 5′ and 3′ ends. Alignment with *osp94* shows the *apg-1* cDNA begins at nucleotide 124 and ends at nucleotide 2893. *apg-1* and *osp94* encode the same protein.

A mouse genomic clone containing the 5′ flanking region of the *apg-1/osp94* gene was recently isolated and sequenced (accession number D70845). This 720-bp genomic sequence contains a consensus heat shock promoter element which can bind HSF1 (Kaneko et al., 1997).

Osp94 protein

There is virtually no information available regarding Osp94 protein expression. *In vitro* translation of the *osp94* cDNA identified a major protein product with an apparent molecular mass of 105–110 kDa on SDS–PAGE (Kojima et al., 1996). This molecular mass of Osp94 is greater than the calculated mass of 94 kDa, an observation similar to that seen for Hsp110 (Lee-Yoon et al., 1995). The intracellular localization of Osp94 is not known. It has no apparent N-terminal signal sequence, no endoplasmic reticulum retention signal, nor any apparent transmembrane domains. It is likely to be localized within the cytosol and/or nucleus.

Biological activities of Osp94

To date there have been no studies of Osp94 function. The deduced structure of the protein suggests it acts as a molecular chaperone and has ATPase activity.

Biological regulation of *osp94*

osp94 mRNA appears to be ubiquitously expressed at low levels in all tissues. Highest levels of expression are apparent in kidney inner medulla and testis (Kaneko et al., 1997; Kojima et al., 1996). *osp94* mRNA levels are greatly increased with heat shock and hyperosmotic stress but not with tunicamycin. In mouse kidney, there is a corticomedullary gradient of Osp94 expression which parallels the increasing osmolarity gradient. Moreover, with water restriction, renal medullary expression of Osp94 is increased in concert with increased renal medullary osmolality.

References

Kaneko, Y., Nishiyama, H., Nonoguchi, K., Higashituji, H., Kishishita, M. and Frujita, J. (1997). A novel hsp110-related gene, apg-1, that is abundantly expressed in the testis responds to a low temperature heat shock rather than the traditional elevated temperatures. J. Biol. Chem. **272**, 2640–2645.

Kojima, R., Randall, J., Brenner, B.M. and Gullans, S.R. (1996). Osmotic stress protein 94 (Osp94): a new member of the Hsp110/SSE gene subfamily. J. Biol. Chem. **271**, 12327–12332.

Lee-Yoon, D., Easton, D., Murawski, M., Burd, R. and Subjeck, J.R. (1995). Identification of a major subfamily of large hsp70-like proteins through the cloning of the mammalian 110-kDa heat shock protein. J. Biol. Chem. **270**, 15727–15733.

■ Steven R. Gullans:
Renal Division
Brigham and Women's Hospital
Harvard Institutes of Medicine
77 Avenue Louis Pasteur
Boston, MA 02115, USA
Tel. 1 617 525 5712
Fax. 1 617 525 5711
E-mail: gullans@bustoff.bwh.harvard.edu

■ Ryoji Kojima:
Department of Pharmacology
Faculty of Pharmacy
Meijo University
Nagoya 468, Japan
Tel. 81 52 832 1781 ext 318
Fax. 1 81 52 834 8780
E-mail: kojima@meijo-u.ac.jp

Mammalian Grp170

Grp170 is a luminal glycoprotein of the endoplasmic reticulum of mammalian cells. This protein can be induced by the same conditions that induce Grp78 and Grp94. Grp170 is a highly diverged relative of the HSP70 family as are several members of the HSP110/SSE protein subfamily. In vivo, Grp170 binds to newly synthesized proteins, including immunoglobulins. Grp170 may play a cooperative role with Grp78 and Grp94 in the folding and/or assembly of proteins in the ER lumen.

Alternative names

CBP140 (Naved et al., 1995).

Grp170 sequence

Chinese hamster Grp170 is encoded as a 999 amino acid precursor with a 34 residue signal peptide (GenBank accession number U34206, Chen et al., 1996). The calculated molecular weight of the processed protein is 107 461 Da. Grp170 is a glycoprotein with a calculated M_r of 164 kDa, determined by SDS–gel electrophoresis. The peptide terminates in NDEL which serves as an ER retention signal. The Grp170 sequence shows 35% identity to the N-terminal 350 residues of the HSP70 proteins. In this region of the molecule, the amino acids that align with residues of the ATP-binding motifs identified in HSP70 proteins by Bork et al. (1992), have the properties determined by these authors to define an ATP-binding pocket. Recent work shows that Grp170 binds ATP, suggesting that Grp170 may have an ATPase activity similar to that of the HSP70 proteins (Chen et al., 1996). The C-terminal half of the molecule, which is only 15% identical to HSP70 proteins, contains several regions which display sequence features conserved in members of the HSP110/SSE protein family (see Lee Yoon et al., 1995 and the overview of this protein family, p. 73). Recent submissions in GenBank deposited by the *Caenorhabdidtis elegans* and yeast genome programs identify additional sequences having homology to Grp170 (accession numbers Z67884 and U28940 from *C. elegans*; Z68136 from *S. pombe*; see Chen et al., 1996). Phylogenetic analysis (see overview of HSP110/SSE proteins, p. 73) suggests that Grp170 and a few other sequences constitute a group of large HSP70-related proteins which are distinct from HSP110/SSE family members. *C. elegans* has two such molecules each having an apparent endoplasmic reticulum retention sequence.

■ Grp170 protein, localization and function

Grp170 is found in all murine tissues that have been examined. Brain and liver are particularly rich in this protein, while muscle contains little of the protein. Modest amounts of Grp170 are found in immune system tissues such as spleen, lymph nodes and Pyer's patches (Lin et al., 1993). Immunofluorescence studies show a perinuclear reticular localization for Grp170. With calcium stress, the intensity of fluorescence increases but the overall localization of Grp170 does not change. Grp170 is found in microsomal membranes and is resistant to protease K treatment, consistent with it being resident in the ER lumen. Furthermore, Grp170 remains endoglycosidase sensitive for 4 h after synthesis, indicating that it is not transported into the *trans*-Golgi but is retained in the ER (Lin et al., 1993). When detergent lysates of immunoglobulin-producing cells are probed with antibody raised against immunoglobulin chains, or with antibodies raised against Grp78/BiP or Grp94, all three GRPs coprecipitate in the immune complexes (Lin et al., 1993). This suggests that the three GRPs and their substrate proteins form a processive complex *in vivo*., in which the GRP proteins work cooperatively to fold and/or assemble the substrate for secretion (Lin et al., 1993). A recent report (Naved et al., 1995) describes the cloning of a partial sequence for a protein described as a calcium-binding protein of 140 kDa (CBP140). This sequence and the protein's proposed localization in the ER strongly suggests that CBP140 is the mouse equivalent of Grp170.

grp170 gene regulation

Grp170 is synthesized constitutively and accumulates to high levels in some tissues, e.g. brain. Induction of enhanced Grp170 synthesis can be effected by treatment of cells with the same agents/conditions that induce synthesis of Grp78 and Grp94, i.e. treatments that result in the accumulation of unfolded proteins in the ER lumen or alter the functional requirements of the ER (Shen et al., 1987; Lin et al., 1993). The regulation of the promoter of the *grp170* gene is assumed to be the same as that for other *grp* genes (Chang et al., 1989), but isolation of genomic clones and sequence analysis of the promoter region is required to confirm this hypothesis. Anoxia, a general inducer of GRPs, increases the *grp170* mRNA level in CHO cells several-fold. Moreover, Grp170 and other GRPs have been found to be induced during growth of a murine tumor (Cai et al., 1993), presumably as a result of the ischemic environment characteristic of larger tumors.

■ References

Bork, P., Sander, C., and Valencia, A. (1992). An ATPase domain common to prokaryotic cell cycle proteins, sugar kinases, actin and hsp70 heat shock proteins. Proc. Natl. Acad. Sci. USA **89**, 7290–7294.

Cai, J.W., Henderson, B.W., Shen, J.W., and Subjeck, J.R. (1993). Induction of glucose regulated proteins during growth of a murine tumor. J. Cell. Physiol. **154**, 229–237.

Chang, S.C., Erwin, A.E., and Lee, A.S. (1989). Glucose-regulated protein (GRP94 and GRP78) genes share common regulatory domains and are coordinately regulated by common trans-acting factors. Mol. Cell. Biol. **9**, 2153–2162.

Chen, X., Easton, D.P. Lee-Yoon, D., Liu, X. and Subjeck, J.R. (1996). The 170 kDa glucose regulated protein is a large hsp70/110-like protein of the endoplasmic reticulum. FEBS Lett. **380**, 68–72.

Lee-Yoon, D., Easton, D.P., Murawski, M., Burd, R., and Subjeck, J.R. (1995). Identification of a major subfamily of large hsp70-like proteins through the cloning of the 110 kDa heat shock protein. J. Biol. Chem. **270**, 15725–15733.

Lin, H.Y., Masso-Welch, P., Di, Y.P., Cai, J.W., Shen, J.W., and Subjeck, J.R. (1993). The 170 kDa glucose regulated protein is an endoplasmic reticulum protein that binds immunoglobulin. Mol. Biol. Cell **4**: 1109–1119.

Naved, A.F., Ozawa, M., Yu, S., Miyauchi, T., Maramatsu, H., and Mauramatsu, T. (1995). CBP-140, a novel endoplasmic reticulum resident Ca^{2+}-binding protein with a carboxyl terminal NDEL sequence showed partial homology with 70 kDa heat shock protein (hsp70). Cell Struct. Funct. **20**, 133–141.

Shen, J., Hughes, C., Chao, C., Cai, J., Bartels, C., Gessner, T., and Subjeck, J. (1987). Coinduction of glucose regulated proteins and doxorubicin resistance in Chinese hamster cells. Proc. Natl. Acad. Sci. USA **84**, 3278–3282.

■ *Douglas P. Easton:*
Department of Biology
State University College at Buffalo
1300 Elmwood Ave.
Buffalo, NY 14222, USA
Tel. 1 716 878 3050
Fax. 1 716 878 4028
E-mail: eastondp@snybufaa.cs.snybuf.edu

■ *John R. Subjeck:*
Dept. of Molecular and Cellular Biology
Roswell Park Cancer Institute
Buffalo, NY 14263, USA
Tel. 1 716 845 4425
Fax. 1 716 845 5639
E-mail: subjeck@sc3101.med.buffalo.edu

HSP40 (DnaJ-related Proteins

HSP40 (DnaJ-related) proteins — an overview

HSP40 proteins are a family of cochaperones that regulate the activity of HSP70 proteins. The function of several HSP40 family members is essential for cell viability, whereas those of others are dispensable. The most prominent feature of HSP40 family members is the ability they possess to stimulate hydrolysis of ATP by specific HSP70 partners. A subset of HSP40 family members also recognize and bind non-native conformations of polypeptides and can be classified as molecular chaperones. Other HSP40 proteins interact with specific nucleic acid or protein substrates and appear to direct their partner HSP70 to interact with these molecules. The prototype of the HSP40 family is Escherischia coli DnaJ. The presence of different combinations of four conserved regions that are present in DnaJ define the HSP40 subfamilies. HSP40 family members exhibit different modes of transcriptional regulation; some are constitutively expressed whereas others are transcribed conditionally. Different HSP40 proteins are colocalized with HSP70 proteins to subcellular compartments that include the cytosol, nucleus, mitochondria, and endoplasmic reticulum. Some HSP40 proteins contain transmembrane domains or are modified post-translationally with lipids and are concentrated to membrane surfaces. Reactions facilitated by HSP40 proteins include protein synthesis, protein folding, protein translocation, signal transduction, assembly of macromolecular complexes, renaturation of misfolded proteins, and proteolysis.

■ Alternative names

DnaJ-like proteins; DnaJ-homologs, DnaJ.

■ Overview

The purpose of this overview is to provide an introduction to the key features of the HSP40 or DnaJ family of proteins. General aspects of HSP40 function will be discussed and special features of individual family members will be highlighted. To limit overlap, the details of the functions of specific HSP40 proteins will be discussed in the entries on individual family members. At present there are a total of 219 entries in GenBank, some redundant, for proteins with homology to the progenitor of the HSP40 family, *E. coli* DnaJ. A limited number of HSP40 proteins from both *E. coli* and eukaryotic organisms whose sequence and function are known have been selected for detailed discussion below. A number of recent reviews with material relevant to the HSP40 family are also available (Ang et al., 1991; Bork et al., 1992; Gething and Sambrook, 1992; Caplan et al., 1993; Silver and Way, 1993; Cyr et al., 1994; Hartl et al., 1994; Hartl, 1996).

■ HSP40/DnaJ sequences

The HSP40 family consists of a diverse group of proteins that are related to the DnaJ protein of *E. coli* (Bardwell et al., 1986). Eukaryotic genomes encode multiple homologs of DnaJ per cell; for example, yeast encodes at least 18 HSP40 proteins (Cyr et al., 1994; Mukai et al., 1994; F. Volkert, personnel communication). DnaJ homologs are now referred to as the HSP40 family because a number of them are approximately 40 kDa in size and transcriptionally regulated by heat stress. However, there are a number of exceptions to these criteria (Caplan et al., 1993; Silver and Way, 1993). HSP40 proteins typically share 35–50% overall sequence identity with DnaJ (Caplan et al., 1993). All HSP40 proteins contain a highly conserved domain (termed the J-domain) corresponding to the N-terminal 70 amino acids of DnaJ (Fig. 1). The presence of a J-domain on a protein warrants its classification as an HSP40 family member (Fig. 2). The J-domain confers the ability of HSP40 proteins to regulate the ATPase activity of HSP70. The J-domain contains a tripeptide, HPD, that is found in all HSP40 proteins (Zhang et al., 1992; Wall et al., 1994; Tsai and Douglas, 1996). Analysis of DNA sequences that encode a number of HSP40 proteins has revealed a consensus sequence of KYHPDK that is found in the majority of HSP40 proteins (Zhang et al., 1992). A functional J-domain is essential for the function of HSP40 proteins since its alteration or deletion inactivates HSP40 proteins (Feldheim et al., 1992; Wall et al., 1994; Dey et al., 1996; Tsai and Douglas 1996). Fragments of HSP40 proteins that include a J-domain are capable of stimulating the ATPase activity of purified HSP70 to levels observed with full length proteins (Wall et al., 1994). Expression *in vivo* of truncated forms of DnaJ that contain its J-domain partially complement defects in cell physiology that result from its deletion (Wall et al., 1994). The NMR solution structure of the J-domain from DnaJ has been solved and indicates that this motif folds into three regions of helical structure (Szyperski et al., 1994; Hill et al., 1995). Two of these helices flank a turn formed by the HPD motif. Mutations in the KYHPDK motif, or in the faces of the helices that

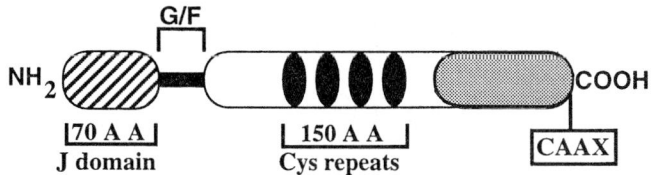

A. Domains and motifs found in hsp40 family members

B. Concensus sequences found within the conserved domains of hsp40 proteins

J-domain-KYHPDK

G/F rich region-GG.GG....GGFGGGGG..D.F.DIF.SFFGGG

Cys- repeats-CXXCXGXG

Figure 1. Model of an HSP40 family member depicting the conserved domains from DnaJ characteristic of this protein family. A. The domains shown correspond to regions in DnaJ (Bardwell et al., 1986) that are conserved throughout the HSP40 family of proteins. The CAAX motif is found only in eukaryotic family members and is responsible for the post-translational modification of HSP40 proteins with farnesyl (Caplan et al., 1992a, b). References that contain the sequence alignments that reveal the consensus motifs shown in panel B are as follows: J-domain (Zhang et al., 1992); G/F rich region (Wall et al., 1995); cysteine-rich region (Szabo et al., 1996).

flank it, give rise to severe growth defects (Feldheim et al., 1992; Dey et al., 1996; Tsai and Douglas, 1996; Wall et al., 1994). However, the KYHPDK peptide alone is not sufficient for regulation of HSP70 ATPase activity (Tsai and Douglas, 1996).

Open reading frames recently identified by genome sequencing projects for various eukaryotic organisms have been grouped as members of the HSP40 family on the basis that they contain an HDP peptide in their sequence (see the yeast protein data base for examples, http://quest7.proteome.com/YPDhome.html). Direct tests are now required to determine whether all these proteins interact with HSP70 family members and are thus true HSP40 family members.

Three additional regions of DnaJ are conserved in various members of the HSP40 family (Bork et al., 1992). The first is the glycine/phenylalanine-rich region (Fig. 1), a 35 amino acid stretch that is thought to be a flexible spacer that separates the J-domain from other parts of HSP40 proteins (Silver and Way, 1993; Szyperski et al., 1994). Deletion of the glycine/phenylalanine-rich region limits the ability of DnaJ to assist DnaK (the major HSP70 family member in E. coli) in binding protein substrates (Wall et al., 1995). The second additional conserved domain is the cysteine-rich region that corresponds to a 150 amino acid stretch in the middle of the protein that has the four repeated sequence motifs, CXXGXGXG, where X denotes a charged or polar amino acid (Fig. 1). This motif resembles a zinc finger (Caplan and Douglas, 1991) and was recently shown to bind zinc and help specify the substrates with which HSP40 proteins interacts (Szabo et al., 1996). The C-terminal third of DnaJ does not contain any consensus motifs, but is conserved in subfamilies of HSP40 proteins (Fig. 2). This region appears to contain the binding site that recognizes non-native proteins (Cyr et al., 1994; Wall et al., 1994). The glycine/phenylalanine-rich region and the cysteine-rich region are present in different combinations in about 50% of HSP40 family members. The C-terminal third of DnaJ is present in only 30% of HSP40 family members (Fig. 2).

HSP40 proteins also contain a number of specialized structural features that are unique to different subfamilies and to specific individual family members. Eukaryotic HSP40 proteins with high sequence identity to Ydj1p of the yeast Saccharomyces cerevisiae contain a C-terminal CaaX motif (a is an aliphatic amino acid and X is any amino acid) that directs the post-translational addition of farnesyl to these chaperone proteins (Caplan et al., 1992a; Zhu et al., 1993b; Pressig-Muller et al., 1994). The addition of a farnesyl-moiety to HSP40 proteins is essential for growth of cells at elevated temperatures (Caplan et al., 1992a,b). Other HSP40 proteins contain motifs such as leucine zippers (Mukai et al., 1994), nucleic acid binding motifs (Zhang et al., 1992), clathrin-binding sites (Ungewickell et al., 1995), and transmembrane domains (Silver and Way, 1993; Kelley and Georgopoulos, 1996). In summary, the HSP40 family is structurally diverse with all members containing a J-domain that allows them to interact with HSP70 proteins.

| Hsp40 Subfamilies | Representative members |

(figure showing domain structures)

- DnaJ, Xdj1
- &Scj1, Mdj1, Tid56
- Ydj1, Hdj2, Anj1, Csj1 (with CAAX)
- CpA, Sis1, Hsp40, Hsj1, HLJ1*
- Csp32, Caj1
- Auxillin, RsgA#
- Sec63#, Mtj1#, Zuotin

Figure 2. Subfamilies of HSP40 proteins. Representative proteins were grouped into subfamilies based on the number and configuration of the conserved domains from DnaJ that they contain. This list is not exhaustive: HSP40 proteins discussed in the text of this chapter and individual sections on respective proteins are included in the list. Search the string DnaJ in Genbank to obtain information on additional HSP40 family members. & Denotes that this subfamily of HSP40 proteins are synthesized as preproteins with N-terminal protein targeting signals. # Denotes that these polypeptides contain transmembrane domains. *The sequence for HLJ1 is not yet published, but has a GenBank accession number of Z49705x10 and was identified by Dr Dennis Volkert in a screen for proteins that are lethal when overexpressed. For original references on the individual HSP40 family members listed see the individual entries on these proteins in this section. CAJ1, whose function is unknown, was identified by Mukai and co-workers (Mukai et al., 1994).

■ Transcriptional regulation of *hsp40* genes

Prokaryotic and eukaryotic organisms encode multiple HSP40 family members whose expression is controlled by different modes of transcriptional regulation. The *E. coli* genome encodes HSP40 proteins that include DnaJ, CpbA, and RcsG. *dnaJ* is transcribed from the *dnaK/dnaJ* operon which is exclusively under the control of the transcription factor σ^{32} and subject to autoregulation by its gene products (Ang et al., 1991). Rates of DnaJ synthesis increase approximately 10 fold following heat shock (Ang et al., 1991). On the other hand, *cbpA*, which encodes a multicopy suppressor of mutations in DnaJ, is not heat inducible and its expression is under the control of σ^{S} (Yamashino et al., 1994). CbpA is expressed at much lower levels than DnaJ and appears to be required to facilitate specialized reactions that take place during the onset of the stationary phase and metabolite starvation (Ueguchi et al., 1994, 1995). RcsGp is expressed constitutively and induced by cold shock (W. Kelley and C. Georgopoulos, unpublished results). In eukaryotes such as *S. cerevisiae*, expression of HSP40 proteins is also regulated by different mechanisms. For example, the promoter regions of Ydj1p and Sis1p, proteins that are both localized to the cytosol, have different degrees of sequence identity with the consensus heat shock element and are expressed to different levels (Caplan and Douglas, 1991; Luke et al., 1991). The *SIS1* gene is also subject to negative control by Sis1p (Zhong et al., 1996). Expression of Scj1p, an ER cognate of DnaJ in yeast, is not regulated by heat shock, but is induced around fourfold over normal constitutive levels by tunicamycin (Blumberg and Silver, 1991; Schlenstedt et al., 1995). Such an observation is consistent with transcription of *SCJ1* being under control of the unfolded protein response element that controls the expression of its partner protein BiP (the

ER cognate of HSP70; Mori et al., 1992). *ANJ1*, a plant *hsp40* gene, is expressed differentially during growth and development (Zhu et al., 1993a). Anj1p is induced at high levels shortly after heat shock and also under salt stress. Interestingly, expression of the Anj1p partner protein, Hsp70, is not induced during salt stress (Zhu et al., 1993a). This observation suggests that HSP40 and HSP70 expression are not always coupled and that situations exist where HSP40 proteins function independently of HSP70 proteins. In summary, a multitude of mechanisms for transcriptional regulation of individual HSP40 family members has evolved. These regulatory mechanisms direct the expression of specialized HSP40 proteins that facilitate specific biochemical reactions in a temporal manner.

■ Localization of HSP40 proteins

Localization of HSP40 proteins to subcompartments in eukaryotic cells parallels that of HSP70 proteins (Caplan et al., 1993; Cyr et al., 1994). HSP40 proteins are colocalized to the cytosol, nucleus, the lumen of the ER, and the matrix space of mitochondria with HSP70 proteins. Localization of HSP40 proteins to organelles occurs by virtue of specific family members being encoded as preproteins that contain targeting signals (Fig. 2). HSP40 proteins are also localized to a number of different macromolecular protein complexes and membrane surfaces in the cell. Concentration of HSP40 proteins to distinct locations within the cell helps direct HSP70 partner proteins to catalyse specific cellular reactions at these sites. For example, auxilin is a soluble HSP40 protein that binds to clathrin on coated vesicles (Ungewickell et al., 1995). The C-terminal J-domain (Fig. 2) on this polypeptide helps target Hsc70 to bind clathrin and uncoat vesicles (Ungewickell et al., 1995). If the J-domain is deleted, auxilin can still bind clathrin-coated vesicles, but Hsc70 can no longer uncoat them with high efficiency (Ungewickell et al., 1995). Csp32 also has a C-terminal J-domain, but it is localized to the surface of synaptic vesicles where it plays a critical role in neurotransmitter release (Zinsmaier et al., 1994). Binding of Csp32 to secretory vesicles is not due to protein-protein interactions. Instead, binding results from lipid-lipid interactions between the palmitoyl groups that are present on Csp32 and components of the vesicle bilayer (Zinsmaier et al., 1994). Sec63p of yeast is an integral membrane HSP40 protein that exposes a J-domain in the lumen of the ER (Feldheim et al., 1992). The J-domain of Sec63p forms a complex with BiP that is important for a number of steps in the pathway of protein translocation into the ER (Brodsky, 1996). Tim44p, a peripheral mitochondrial inner membrane protein which has limited homology to DnaJ, forms a complex with the mitochondrial form of HSP70. Formation of the Tim44p–mtHsp70 complex concentrates mtHsp70 to the import import channel where it drives protein translocation into the matrix (Rassow et al., 1994; Schneider et al., 1994; Ungermann et al., 1994, 1996; Cyr and Neupert, 1996; Schatz and Doberstein, 1996). Ydj1p is soluble in the cytosol where it plays a role in protein translocation and trafficking reactions (Atencio and Yaffe, 1992; Caplan et al., 1992a,b, 1995; Kimura et al., 1995). A portion of Ydj1p is also concentrated (by virtue of its C-terminus being modified post-translationally with farnesyl) to the surface of the ER membrane, where it facilitates reactions that are essential for cell viability at elevated temperature (Caplan et al., 1992a,b). Thus, HSP40 proteins are localized to a variety of locations within the cell where they present their J-domains to specific HSP70 proteins. Recognition of the J-domain on HSP40 proteins helps concentrate the HSP70 family members to discrete cellular locations and thereby specifies their function.

■ Characteristics of purified HSP40 proteins

A number of HSP40 proteins have been overexpressed, purified, and characterized biochemically (Liberek et al., 1991; Cyr et al., 1992; Langer et al., 1992; Cheetham et al., 1994; Cyr and Douglas, 1994; Jordan and McMacken, 1995). All purified HSP40 proteins tested have the ability to stimulate the intrinsic ATPase activity of HSP70 proteins by about one order of magnitude. The binding of HSP40 appears to induce a conformational change in HSP70 that increases the maximal velocity of the ATPase reaction, without altering the Km for ATP. Maximal stimulation of ATPase activity occurs at a 1:1 molar ratio of HSP70 to HSP40. Some HSP40 proteins also form complexes with non-native polypeptides to prevent their aggregation independently of HSP70 and are therefore classified as molecular chaperone proteins (Wickner et al., 1991; Langer et al., 1992; Cyr, 1995). HSP70 and HSP40 proteins bind similar substrates with widely varying affinities (Langer et al., 1992; Cyr et al., 1994; Wawrzynow and Zylicz, 1995). However, neither the substrate specificity of HSP40 proteins, nor the mechanism for the binding and release of polypeptides has been determined.

Purified HSP40 proteins that contain only a J-domain (Fig. 2) regulate the ATPase activity of HSP70 and also have activities that are not related to binding non-native polypeptides. For instance, Zuotin contains a J-domain, but instead of having a putative polypeptide-binding domain it contains a region of homology to histone H1, which confers its ability to bind Z-DNA and tRNA molecules (Zhang et al., 1992; Wilhelm et al., 1994).

A different twist to HSP40 protein function comes from the recent observation that purified DnaJ is capable of catalysing *cis–trans* prolyl isomerization reactions (Crouy-Chanel et al., 1995). It is propsed that the cysteine-rich region in HSP40 members plays an important role in catalysing this reaction (Crouy-Chanel et al., 1995). It will be interesting to determine if the same mechanism that allows the cysteine-rich region to specify the substrates of HSP40 (Szabo et al., 1996) is also involved in facilitating *cis–trans* prolyl isomerization reactions (Crouy-Chanel et al., 1995). Whether additional HSP40 family members that contain the cysteine-rich region (Fig. 2) also catalyse *cis–trans* prolyl isomerization reactions remains to be tested.

Specificity of interactions between HSP40 and HSP70 family members

How do HSP70 proteins recognize HSP40 proteins? Is there specificity in the interactions between different pairs of HSP70 and HSP40 proteins which are localized to the same subcellular space? It appears that HSP70 proteins contain a regulatory site that is independent of the polypeptide-binding groove at which an unknown region of the J-domain is recognized (Cyr and Douglas, 1994; Tsai and Douglas, 1996). Binding of the J-domain to this regulatory site appears to transmit intramolecular signals to subdomains of HSP70 that regulate its chaperone activity (Cyr and Douglas, 1994; Freeman et al., 1995). A number of observations suggest that J-domains presented by different HSP40 proteins are specifically recognized by individual HSP70 family members (Brodsky and Schekman, 1993; Cyr and Douglas, 1994; Schlenstedt et al., 1995). The HSP70 Ssa and Ssb subfamilies are localized to the cytosol with Ydj1p (Craig et al., 1994). Purified Ydj1p interacts with Ssa1p and Ssa2p to regulate ATP hydrolysis and prevent protein aggregation, but similar interactions with the HSP70 Ssb proteins are not observed (Cyr et al., 1992; Cyr and Douglas, 1994; Cyr, 1995). Furthermore, Ssa1p is not capable of substituting for BiP in assays that reconstitute Sec63p-dependent protein translocation into the ER (Brodsky and Schekman, 1993). Finally, domain swap experiments demonstrate that the J-domain of Scj1p (Blumberg and Silver, 1991), which is colocalized to the ER lumen with Sec63p, is capable of substituting for that of Sec63p, but similar domains from mitochondrial and cytosolic HSP40 proteins cannot (Schlenstedt et al., 1995). Identification of the regions in the J-domain and HSP70 that interact is required in order to uncover the molecular basis for the specificity of interactions observed between these cochaperone proteins.

Mechanism of action of HSP40 proteins

A major role for HSP40 proteins in protein metabolism is to drive conversion of the ATP-bound form of HSP70 to the ADP-bound form. This is an important function because the ATP form of HSP70 is thought to initiate interactions with substrate proteins (Palleros et al., 1993; Szabo et al., 1994; Ungermann et al., 1996). Hydrolysis of ATP and generation of the ADP form of this chaperone increases the affinity of HSP70 for polypeptides (Palleros et al., 1991). HSP40 proteins are therefore proposed to help stabilize HSP70–polypeptide complexes (Langer et al., 1992). However, interactions between HSP40 and HSP70 proteins have also been observed to destabilize HSP70–polypeptide complexes in a manner that requires ATP hydrolysis (Cheetham et al., 1994; Cyr et al., 1992). The differences in these results appear to arise from differences in the stability of complexes between ADP and the purified HSP70 family members that were used in these different studies (Hohfeld et al., 1995; Ziegelhoffer et al., 1995). HSP70 proteins such as DnaK have a high affinity for the ADP formed upon ATP hydrolysis and ADP–DnaK–polypeptide complexes are very stable (Liberek et al., 1991; Palleros et al., 1991). Dissociation of DnaK–polypeptide complexes is triggered by GrpE-dependent release of ADP followed by the rebinding of ATP to HSP70 (Liberek et al., 1991; Palleros et al., 1993; Szabo et al., 1994). In contrast, HSP70 family members such as Ssa1 and Hsc70 appear to have much lower affinities for ADP than DnaK and do not require a GrpE-like protein to promote dissociation of ADP or polypeptides (Cyr et al., 1992, 1994; Cheetham et al., 1994; Hohfeld et al., 1995; Ziegelhoffer et al., 1995). In these situations it appears that HSP40 proteins drive the formation of an unstable ADP–Hsc70–polypeptide complex (Cyr et al., 1992; Cheetham et al., 1994). Hartl and co-workers have recently identified a new cochaperone protein termed HIP (HSP70 interacting protein) that interacts with HSP70 protein in an HSP40- and ATP-dependent manner (Hohfeld et al., 1995). HIP proteins facilitate the accumulation of ADP–Hsc70 complexes by blocking the dissociation of ADP (Hohfeld et al., 1995). Thus, multiple mechanisms by which cochaperones regulate the activity of HSP70 have evolved. The mechanism that applies appears to be specific for the HSP70 family member under study and may also be dependent upon the type of reaction that is being catalysed.

References

Ang, D., Liberek, K., Skowyra, D., Zylicz, M., and Georgopoulos, C. (1991). Biological role and regulation of the universally conserved heat shock proteins. J. Biol. Chem. **266**, 24233–24236.

Atencio, D.P. and Yaffe M.P. (1992). Mas5, a yeast homolog of DnaJ involved in mitochondrial protein import. Mol. Cell. Biol. **12**, 283–291.

Bardwell, J.C., Tilly, K., Craig, E., King, J., Zylicz, M., and Georgopoulos, C. (1986). The nucleotide sequence of the Escherichia coli K12 dnaJ+ gene. A gene that encodes a heat shock protein. J. Biol. Chem. **261**, 1782–1785.

Blumberg, H. and Silver, P.A. (1991). A homologue of the bacterial heat-shock gene DnaJ that alters protein sorting in yeast. Nature **349**, 627–630.

Bork, P., Sander, C., Valencia, A., and Bukau, B. (1992). A module of the DnaJ heat shock proteins found in malaria parasites. Trends Biochem. Sci. **17**, 129.

Brodsky, J.L. (1996). Post-translocational protein translocation: not all hsc70s are created equal. Trends Biochem. Sci. **21**, 122–126.

Brodsky, J.L. and Schekman, R. (1993). A Sec63p–BiP complex from yeast is required for protein translocation in a reconstituted proteoliposome. J. Cell Biol. **123**, 1355–1363.

Caplan, A.J. and Douglas, M.G. (1991). Characterization of YDJ1: a yeast homolog of the E. coli dnaJ gene. J. Cell Biol. **114**, 609–622.

Caplan, A.J., Tsai, J., Casey, P., and Douglas M.G. (1992a). Farnesylation of YDJ1p is required for function at elevated growth temperatures in Saccharomyces cerevisiae. J. Biol. Chem. **267**, 18890–18895.

Caplan, A.J., Cyr, D.M., and Douglas, M.G. (1992b). YDJ1 facilitates polypeptide translocation across different intercellular membranes by a conserved mechanism. Cell **71**, 1143–1155.

Caplan, A.J., Cyr, D.M., and Douglas, M.G. (1993). Eukaryotic homologs of E. coli DnaJ: a diverse protein family that functions with the Hsp70 stress proteins. Mol. Cell. Biol. **4**, 555–563.

Caplan, A.J., Langley, E., Wilson, E.M., and Vidal, J. (1995). Hormone-dependent transactivation by the human androgen receptor is regulated by a dnaJ protein. J. Biol. Chem. **270**, 5251–5257.

Cheetham, M.E., Jackson, A.P., and Anderton, B.H. (1994). Regulation of 70-kDa heat-shock-protein ATPase activity and substrate binding by human DnaJ-like protiens, HSJ1a and HSJ1b. Eur. J. Biochem. **226**, 99–107.

Craig, E.A., Baxter, B.K., Becker, J., Halladay, J., and Zeigelhoffer, T. (1994). Cytosolic HSP70s of Saccharomyces cerevisiae: roles in protein synthesis, protein translocation, proteolysis and regulation. In *The Biology of Heat Shock Proteins and Molecular Chaperones*, (eds R.I. Morimoto, A. Tissieres, and C. Georgopoulos). Cold Spring Harbor Laboratory Press, pp. 31–52.

Crouy-Chanel, A. de., Kohiyama, M., and Richarme, G. (1995). A novel function of Escherichia coli chaperone. DnaJ protein-disulfide isomerase. J. Biol. Chem. **270**, 22669–22672.

Cyr, D.M. (1995). Cooperation of the molecular chaperone Ydj1 with specific Hsp70 homologs to suppress protein aggregation. FEBS Lett. **359**, 129–132.

Cyr, D.M. and Douglas, M.G. (1994). Differential regulation of hsp70 subfamilies by the eukaryotic DnaJ homolog YDJ1. J. Biol. Chem. **269**, 9798–9804.

Cyr, D.M. and Neupert, W. (1996). Roles for Hsp70 proteins in intracellular protein, transport. In *Stress-Inducible Cellular Responses*, (ed. U. Feige, R. Morimoto, I. Yahara and B. Polla. Birkhauser/Springer Verlag, pp. 25–40.

Cyr, D.M., Lu, X., and Douglas, M.G. (1992). Regulation of eukaryotic hsp70 function by a DnaJ homolog. J. Biol. Chem. **267**, 20927–20931.

Cyr, D.M., Langer, T., and Douglas, D.M. (1994). DnaJ-like proteins: molecular chaperones and specific regulators of hsp70. Trends Biochem. Sci. **19**, 176–181.

Dey, B., Caplan, A.J., and Boschelli, F. (1996). The Ydj1 molecular chaperone facilitates formation of active p60v-src in yeast. Mol. Biol. Cell **7**, 91–100.

Feldheim, D., Rothblatt, J., and Schekman, R. (1992). Topology and functional domains of Sec63p, an endoplasmic reticulum membrane protein required for secretory protein translocation. Mol. Cell. Biol. **12**, 3288–3296.

Freeman, B.C., Myers, M.P., Schumacher, R., and Morimoto, R.I. (1995) Identification of a regulatory motif in Hsp70 that affects ATPase activity, substrate binding and interaction with HDJ-1. EMBO J. **14**, 2281–2292.

Frydman, J., Nimmesgern, E., Ohtsuka, K., and Hartl F-U. (1994). Folding of nascent polypeptide chains in a high molecular mass assembly with molecular chaperones. Nature (London) **370**, 111–117.

Gething, M.-J. and Sambrook, J. (1992). Protein folding in the cell. Nature (London) **355**, 33–45.

Hartl, F.-U. (1996). Molecular chaperones in cellular protein folding. Nature **381**, 571–580.

Hartl, F.-U., Holdan, R., and Langer, T. (1994). Molecular chaperones in protein folding: the art of avoiding sticky situations. Trends Biochem. Sci. **19**, 20–25.

Hill, R.B., Flanagan, J.M., and Prestegard, J.H. (1995). 1H and 15N magnetic resonance assignments, secondary structure, and tertiary fold of Escherichia coli. Biochemistry **34**, 5587–5596.

Hohfeld, J., Minami, Y., and Hartl, F.U. (1995). Hip, a novel cochaperone involved in the eukaryotic Hsc70/Hsp40 reaction cycle. Cell **83**, 589–598.

Jordan, R. and McMacken, R. (1995). Modulation of the ATPase activity of the molecular chaperone DnaK by peptides and the DnaJ and GrpE heat shock proteins. J. Biol. Chem. **270**, 4563–4569.

Kimura, Y., Yahara, I., and Lindquist, S. (1995). Role of the protein chaperone YDJ1 in establishing Hsp90-mediated signal transduction pathways. Science **268**, 1362–1365.

Langer, T., Lu, C., Echols, H., Flanagan, J., Hayer, M.K., and Hartl, F.U. (1992). Successive action of DnaK, DnaJ and GroEL along the pathway of chaperone-mediated protein folding. Nature **356**, 683–689.

Liberek, K., Marszalek, J., Ang, D., Georgopoulos, C., and Zylicz, M. (1991). Escherichia coli DnaJ and GrpE heat shock proteins jointly stimulate ATPase activity of DnaK. Proc. Natl. Acad. Sci. USA **88**, 2874–2878.

Luke, M.M., Sutton, A., and Arndt, K.T. (1991). Characterization of SIS1, a Saccharomyces cerevisiae homologue of bacterial dnaJ proteins. J. Cell Biol. **114**, 623–638.

Mori, K., Sant, A., Kohno, K., Normington, K., Gething, M.J., and Sambrook, J.F. (1992). A 22 bp cis-acting element is necessary and sufficient for the induction of the yeast KAR2 (BiP) gene by unfolded proteins. EMBO J. **11**, 2583–2593.

Mukai, H., Shuntoh, H., Chang, C.D., Asami, M., Ueno, M., Suzuki, K., and Kuno, T. (1994). Isolation and characterization of CAJ1, a novel yeast homolog of dnaJ. Gene **145**, 125–127.

Palleros, D.R., Welch, W.J., and Fink, A.L. (1991). Interaction of hsp70 with unfolded proteins: effects of nature and nucleotides on the kinetics of binding. Proc. Natl. Acad. Sci. USA **88**, 5719–5723.

Palleros, D.R., Reid, K.L., Shi, L., Welch, W.J., and Fink, A.L. (1993). ATP-induced protein-Hsp70 complex dissociation requires K+ but not ATP hydrolysis. Nature **365**, 664–666.

Preisig-Muller, R., Muster, G., and Kindl, H. (1994). Heat shock enhances the amount of prenylated Dnaj protein at membranes of glyoxysomes. Eur. J. Biochem. **219**, 57–63.

Rassow, J., Maarse, A.C., Krainer, F., Kubrich, M., Muller., Meijer, M., Craig, E.A., and Pfanner, N. (1994). Mitochondrial protein import: biochemical and genetic evidence for the interaction of matrix hsp70 and the inner membrane protein Mim44. J. Cell Biol. **127**, 1547–1556.

Rowley, N., Prip-Buus, C., Westermann, B., Brown, C., Schwarz, E., Barrel, B., and Neupert, W. (1994). Mdj1, a novel chaperone of the DnaJ family, is involved in mitochondrial biogenesis and protein folding. Cell **77**, 249–259.

Schatz, G. and Dobberstein, B. (1996). Common principles of protein translocation across membranes. Science **271**, 1519–1526.

Schlenstedt, G., Harris, S., Risse, B., Lill, R., and Silver, P.A. (1995). A yeast DnaJ homologue, Scj1p, can function in the endoplasmic reticulum with BiP/Kar2p via a conserved domain that specifies interactions with Hsp70s. EMBO J. **14**, 5367–5378.

Schneider, H.C., Berthold, J., Bauer, M.F., Dietmeier, K., Guiard, B., Brunner, M., and Neupert ,W. (1994) Mitochondrial Hsp70/MIM44 complex facilitates protein import. Nature **371**, 768–774.

Silver, P.A. and Way, J.C. (1993). Eukaryotic DnaJ homologs and the specificity of Hsp70 activity. Cell **74**, 5–6.

Szabo, A., Langer, T., H. Schroder, Flanagan, J., Bukau, B., and Hartl, F.U. (1994). The ATP hydrolysis-dependent reaction cycle of the Escherichia coli Hsp70 system DnaK, DnaJ, and GrpE. Proc. Natl. Acad. Sci. USA **91**, 10345–10349.

Szabo, A., Korszun, R., Hartl, F-U., and Flanagan, J. (1996). A zinc finger-like domain of the molecular chaperone DnaJ is involved in the binding of denatured substrates. EMBO J. **15**, 408–417.

Szyperski, T., Pellecchia, M., Wall, D., Georgopoulos, C., and Wuthrich, K. (1994). NMR structure determination of the Escherichia coli DnaJ molecular chaperone: secondary structure and backbone fold of the N-terminal region (residues 2–108) containing the highly conserved J domain. Proc. Natl. Acad. Sci. USA **91**, 11343–11347.

Tsai, J. and Douglas, M.G. (1996). A conserved HPD sequence of the J-domain is necessary for YDJ1 stimulation of Hsp70 ATPase activity at a site distinct from substratebinding. J. Biol. Chem. **271**, 9347–9354

Ueguchi, C., Kakeda, M., Yamada, H., and Mizuno, T. (1994). An analogue of the DnaJ molecular chaperone in Escherichia coli. Proc. Natl. Acad. Sci. USA **91**, 1054–1058.

Ueguchi, C., Shiozawa, T., Kakeda, M., Yamada, H., and Mizuno, T. (1995). A study of the double mutation of dnaJ and cbpA, whose gene products function as molecular chaperones in Escherichia coli. J. Bacteriol. **177**, 3894–3896.

Ungermann, C., Neupert, W., and Cyr, D.M. (1994). The role of hsp70 in conferring unidirectionality on protein translocation into mitochondria. Science **266**, 1250–1253.

Ungermann, C., Guiard, B., Neupert, W., and Cyr, D.M. (1996). The DY and Hsp70/MIM44 dependent reaction cycle driving early steps of protein translocation into mitochondria. EMBO J. **15**, 735–744.

Ungewickell, E., Ungewickell, H., Holstein, S. E., Lindner, R., Prasad, K., Barouch, W., Martin, B., Greene, L. E., and Eisenberg, E. (1995). Role of auxilin in uncoating clathrin-coated vesicles. Nature **378**, 632–635.

Wall, D., Zylicz, M., and Georgopoulos, C. (1994). The NH2-terminal 108 amino acids of the Escherichia coli DnaJ protein stimulate the ATPase activity of DnaK and are sufficient for lambda replication. J. Biol. Chem. **269**, 5446–5451.

Wall, D., Zylicz, M., and Georgopoulos, C. (1995). The conserved G/F motif of the DnaJ chaperone is necessary for the activation of the substrate binding properties of the DnaK chaperone. J. Biol. Chem. **270**, 2139–2144.

Wawrzynow, A. and Zylicz, M. (1995). Divergent effects of ATP on the binding of the DnaK and DnaJ chaperones to each other, or to their various native and denatured protein substrates. J. Biol. Chem. **270**, 19300–19306.

Wickner, S., Hoskins, J., and McKenney, K. (1991). Function of DnaJ and DnaK as chaperones in origin-specific DNA binding by RepA. Nature **350**, 165–167.

Wilhelm, M.L., Reinbolt, J., Gangloff, J., Dirheimer, D., and Wilhelm, F.X. (1994). Transfer RNA binding protein in the nucleus of Saccharomyces cerevisiae. FEBS Lett. **349**, 260–264.

Yamashino, T., Kakeda, M., Ueguchi, C., and Mizuno, J. (1994). An analogue of the DnaJ molecular chaperone whose expression is controlled by sigma s during the stationary phase and phosphate starvation in Escherichia coli. Mol. Microbiol. **13**, 475–483.

Zhang, S., Lockshin, C., Herbert, A., Winter, E., and Rich, A. (1992). Zuotin, a putative Z-DNA binding protein in Saccharomyces cerevisiae. EMBO J. **11**, 3787–3796.

Zhong, T., Luke, M.M., and Arndt, K.T. (1996). Transcriptional regulation of the yeast DnaJ homologue SIS1. J. Biol. Chem. **271**, 1349–1356.

Zhu, J.K., Shi, J., Bressan, R.A., and Hasegawa, P.M. (1993a). Expression of an Atriplex nummularia gene encoding a protein homologous to the bacterial molecular chaperone DnaJ. Plant Cell **5**, 341–349.

Zhu, J.K., Bressan, R.A., and Hasegawa, P.M. (1993b). Isoprenylation of the plant molecular chaperone ANJ1 facilitates membrane association and function at high temperature. Proc. Natl. Acad. Sci. USA **90**, 8557–8561.

Ziegelhoffer, T., Lopez-Buesa, P., and Craig, E.A. (1995). The dissociation of ATP from hsp70 of Saccharomyces cerevisiae is stimulated by both Ydj1p and peptide substrates. J. Biol. Chem. **270**, 10412–10419.

Zinsmaier, K.E., Eberle, K.K., Buchner, E., Walter, N., and Benzer, S. (1994). Paralysis and early death in cysteine string protein mutants of Drosophila. Science **263**, 977–980.

■ *Douglas M. Cyr:*
Department of Cell Biology
652 Basic Health Sciences Building
School of Medicine
University of Alabama at Birmingham
Birmingham, AL 35294-0005, USA
Tel. 1 205 975 4892
Fax. 1 205 934 0950
E-mail: DCYR@cellbio.bhs.uab.edu

Escherichia coli DnaJ

The Escherichia coli *heat shock protein DnaJ was originally discovered through the isolation of a mutant which blocked λDNA replication. DnaJ forms part of the DnaK/DnaJ/GrpE chaperone machinery involved in protein folding, translocation, proteolysis, DNA replication and autoregulation of heat shock gene expression. The presumed role of DnaJ is to bind to DnaK chaperone and stabilize the DnaK–substrate complex. Additionally, in cases where DnaJ itself interacts with a protein substrate, it may target the substrate for binding to DnaK.*

■ DnaJ sequence and structure

The *dnaJ* gene is located in an operon with *dnaK* at 0.3 minutes on the *E. coli* genetic map (Georgopoulos et al., 1994). The nucleotide sequence of the *dnaJ* gene (GenBank accession number M12565) predicts a protein of 375 amino acids (Bardwell et al., 1986; Ohki et al., 1986), which is highly homologous to DnaJ-like proteins from the cytosol, endoplasmic reticulum and mitochondria of eukaryotic cells. Comparison of the sequences of over 30 members of the DnaJ-like protein family reveals that they possess a modular architecture (Cyr et al., 1994) (see Fig.1). The N-terminal domain of the DnaJ protein constitutes the most highly conserved domain (at least 35% identity between different family members) and is used as the signature sequence of members of the DnaJ-like protein family. This so-called J-domain is important for the functions of DnaJ, including its interaction with DnaK chaperone protein (Wall et al., 1994; Szabo et al., 1996; Karzai and McMacken, 1996). Immediately adjacent to the C-terminus of the J-domain is a 35 amino acid region that is rich in both glycine and phenylalanine (the G/F module), which with the J domain is required for DnaJ function in activation of DnaK-mediated ATP hydrolysis and initiation of λ DNA replication (Wall et al., 1995; Karzai and McMacken, 1996). The structure of the truncated DnaJ12 protein, which contains only the J-domain and the G/F module, has been determined using NMR studies (Szyperski et al., 1994). Downstream of the G/F module, bacterial DnaJ and some of its eukaryotic homologues possess a cysteine-rich region which contains two Zn^{2+} binding sites (Szabo et al., 1996; Banecki et al., 1996). The C-terminal portion of DnaJ is the least conserved among family members and is most to be likely involved in substrate binding. It is not yet known which domain of the DnaJ protein is involved in its dimerization.

■ DnaJ protein

Bacterial DnaJ is a 41 kDa member of the HSP40 stress protein family. It is a basic and hydrophobic protein, which in its purified form has been identified as a dimer (Zylicz et al., 1985). It aggregates spontaneously at low salt concentrations and after cell lysis can be found in membrane fractions. The protocol for purification of DnaJ is based on the relatively high affinity of DnaJ for hydroxylapatite and phosphocellulose columns (Zylicz et al., 1985).

DnaJ interacts with the E. coli HSP70 family member DnaK, inducing conformational changes in DnaK that stabilize the complex between the DnaK chaperone and its protein substrate (Wawrzynów and Zylicz, 1995; Banecki and Zylicz, 1996). This DnaJ-dependent activation of a DnaK conformation that binds to both native and denatured protein substrates requires ATP hydrolysis (Szabo et al., 1994; Wawrzynów et al., 1995). The presence of both DnaJ and GrpE stimulates the ATPase activity of DnaK by up to 50-fold (Liberek et al., 1991). DnaJ alone stimulates DnaK's ATPase activity to a lesser extent (Jordan and McMacken, 1991; McCarty et al., 1995). DnaJ protein interacts directly with several native or denatured proteins (σ^{32}, RepA, denatured luciferase, rhodanase), subsequently targetting DnaK to those substrates (Wawrzynów and Zylicz, 1995; Szabo et al., 1996).

■ DnaJ function

The detailed mode of action of DnaJ has been investigated using several in vitro systems: activation of the prepriming complex during initiation of λDNA replication (Zylicz, 1993); activation of RepA for binding to the P1 ori sequence during initiation of P1 replication (Wickner et al., 1992); and protein folding (Hendrick et al., 1993). In these systems DnaJ acts synergistically with DnaK and their cochaperone GrpE (for review see Gross et al., 1990; Hendrick and Hartl, 1993; Georgopoulos et al., 1994). The DnaK/DnaJ/GrpE chaperone machinery suppresses aggregation of both prokaryotic and eukaryotic polypeptides, thus promoting protein folding or protein translocation and secretion (for review see Hendrick and Hartl, 1993). This chaperone machinery can also protect polypeptides from aggregation or even reactivate some partially aggregated enzymes, such as E. coli RNAP, DNA polymerase III and luciferase. In the absence of DnaK, DnaJ alone can also protect luciferase or rhodanese from aggregation (Schroder et al., 1993; Ziemienowicz et al., 1993, 1995; Szabo et al., 1996). DnaJ protein is involved in abnormal protein degradation, presumably by helping to maintain their soluble form (Jubete et al., 1996).

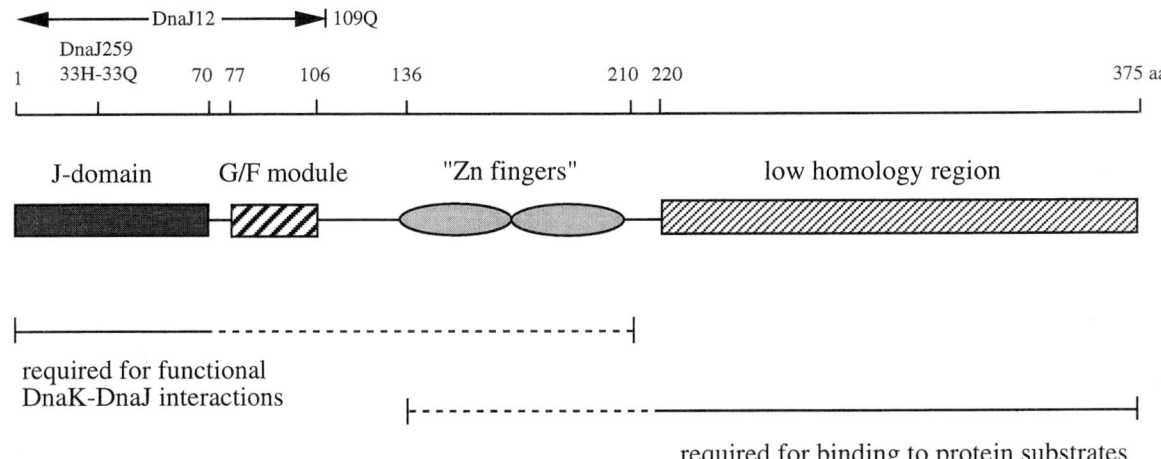

Figure 1. Structure of the *Escherichia coli* DnaJ protein.

Mutagenesis studies with DnaJ

The *dnaJ* gene is not absolutely essential for *E. coli* viability, since it can be deleted in all genetic backgrounds tested at temperatures up to 42–43°C. At this temperature *dnaJ* mutant strains grow more slowly and rapidly accumulate extragenic suppressors (Georgopoulos et al., 1994). This 'weak' phenotype of *dnaJ*-deletion mutants has been explained by the presence in *E. coli* of another *dnaJ*-homolog, CbpA (Ueguchi et al., 1994, see entry, p. 98). Double *dnaJ*/*cbpA* mutants are unable to grow at 37°C (Ueguchi et al., 1994).

dnaJ mutants exhibit a plethora of phenotypes including: (i) extensive filamentation even at permissive temperatures and arrest of the cell cycle at defined stages; (ii) a block of both host DNA and RNA synthesis and replication of λ, P1 and mini-FDNA; and (iii) overproduction of heat shock proteins (for review see Gross et al., 1990, Georgopoulos et al., 1994).

A point mutation in the J-domain (e.g. *dnaJ259*) specifically interferes with the interaction of DnaJ with DnaK but not with substrate (Wall et al., 1994). Deletion of the 35 residue G/F module (*dnaJD77–107*) drastically interferes with DnaJ-dependent stabilization of the DnaK–substrate complex (Wall et al., 1995). The DnaJ12 mutant, which contains only the J-domain and the G/F module, is partially active in λDNA replication, stimulates the ATPase activity of DnaK and activates σ^{32} for binding to DnaK protein without any apparent complex formation between σ^{32} and DnaJ12 (Wall et al., 1994; Georgopoulos et al., 1994; Karzai and McMacken 1996). An internal 57 amino acid deletion of the cysteine-rich region (*dnaJD144–200*) reduces the affinity of DnaJ for some protein substrates, interferes with the DnaJ-induced conformational changes in DnaK and creates a temperature-sensitive phenotype for bacteriophage λ growth. (Banecki et al., 1996). Binding experiments and crosslinking studies performed with several truncated DnaJ proteins suggest that the cysteine-rich region is required for binding of the DnaJ molecular chaperone to some denatured proteins (Szabo et al., 1996).

Gene regulation of *dnaJ*

The *E. coli* *dnaK*–*dnaJ* operon is exclusively under σ^{32} regulation (for review see Gross et al., 1990). Although the *dnaJ* and *dnaK* genes are both transcribed from the same strong heat shock promoter, the levels of the gene products within the cell are very different. During steady-state growth at 37°C the level of DnaJ is about 500 copies/cell, while the level of DnaK is 10-fold higher (Bardwell et al., 1986). It is not clear if this phenomenon is a result of premature termination of transcription, RNA processing or different efficiencies of translation. The rate of synthesis of DnaJ following heat shock is at least 10-fold higher than that at 30°C (Bardwell et al., 1986).

Mutations in *dnaK*, *dnaJ* and *grpE* are negative modulators of the heat shock response (for review see Gross et al., 1990). It appears that during recovery from heat shock conditions the cooperative action of DnaK, DnaJ and GrpE heat shock proteins switches transcription back to the level characteristic for constitutively expressed genes. The DnaK/DnaJ/GrpE machine simultaneously reactivates heat aggregated σ^{70} and sequesters the σ^{32} from RNA polymerase (Blaszczak et al., 1995). Sequestered σ^{32} is efficiently degraded by FtsH protease (Tomayasu et al., 1995). These observations explain previous results of genetic experiments which showed that the half-life of σ^{32} is increased approximately fivefold in *dnaK*, *dnaJ*, *grpE* mutant backgrounds (Gross et al., 1990).

References

Banecki, B. and Zylicz, M. (1996). Real time kinetics of the DnaK/DnaJ/GrpE molecular chaperone machine action. J. Biol. Chem. **271**, 6137–6143.

Banecki, B., Liberek, K., Wall, D., Wawrzynow, A., Georgopoulos, C., Bertoli, E., Tanfani, F. and Zylicz, M. (1996). Structure-functional analysis of the zinc finger region of the DnaJ molecular chaperone. J. Biol. Chem. **271**, 14840–14848.

Bardwell, J.C.A., Tilly, K., Craig, E., King, J., Zylicz, M., and Georgopoulos, C. (1986). The nucleotide sequence of the *Escherichia coli* K12*dnaJ*+ gene. J. Biol. Chem. **261**, 1782–1785.

Blaszczak, A., Zylicz, M., Georgopoulos, C., and Liberek, K. (1995). Both ambient temperature and the DnaK chaperone machine modulate the heat shock response in *Escherichia coli* by regulating the switch between σ^{70} and σ^{32} factors assembled with RNA polymerase. EMBO J. **14**, 5085–5093.

Cyr, D.M., Langer, T., and Douglas, M., (1994). DnaJ-like proteins: molecular chaperones and specific regulators of Hsp70. Trends Biochem. Sci. **19**, 176–181.

Georgopoulos, C., Liberek, K., Zylicz, M. and Ang, D. (1994). Properties of heat shock proteins of *Escherichia coli* and autoregulation of heat shock response. In *The Biology of Heat Shock Proteins and Molecular Chaperones* (eds R.I. Morimoto, A. Tissieres and C. Georgopoulos). Cold Spring Harbor Laboratory Press, pp. 209–249.

Gross, C.A., Straus, D.B., Erickson, J.W., and Yura, T. (1990). The function and regulation of heat shock proteins in *Escherichia coli*. In *Stress Proteins in Biology and Medicine* (eds R.I. Morimoto, A. Tissieres and C. Georgopoulos). Cold Spring Harbor Laboratory Press, pp.167–190.

Hendrick, J.P. and Hartl, F.-H. (1993). Molecular chaperone function of heat-shock proteins. Annu. Rev. Biochem. **62**, 349–384.

Hendrick, J.P., Langer, T., Davis, T.A., Hartl, F.-U., and Wiedmann, M. (1993). Control of folding and membrane translocation by binding of the chaperone DnaJ to nascent polypeptides. Proc. Natl. Acad. Sci. USA **90**, 10216–10220.

Jordan, R. and McMacken, R. (1995). Modulation of the ATPase activity of the molecular chaperone DnaK by peptides and DnaJ and GrpE heat shock proteins. J. Biol. Chem. **270**, 4563–4569.

Jubete, Y., Maurizi, M.R. and Gottesman, S. (1996). Role of the heat shock protein DnaJ in the Lon-dependent degradation of naturally unstable proteins. J. Biol. Chem. **271**, 30798–30803.

Karzai, A.W. and McMacken, R. (1996). A bipartite signaling mechanism involved in DnaJ-mediated activation of the *Escherichia coli* DnaK protein. J. Biol. Chem. **271**, 11236–11246.

Liberek, K., Marszalek, J., Ang, D., Georgopoulos, C., and Zylicz, M. (1991). *Escherichia coli* DnaJ and GrpE heat shock proteins jointly stimulate ATPase activity of DnaK. Proc. Natl. Acad. Sci. USA **88**, 2874–2878.

McCarty, J.S., Buchberger, A., Reinstein, J. and Bukau, B. (1995). The role of ATP in the functional cycle of the DnaK chaperone system. J. Mol. Biol. **249**, 126–137.

Ohki, M., Tamura, F., Nishimura, S., and Uchida, H. (1986). Nucleotide sequence of *Escherichia coli dnaJ* gene and purification of the gene product. J. Biol. Chem. **261**, 1778–1781.

Schroder, H., Langer, T., Hartl, F.-U., and Bukau, B. (1993). DnaK, DnaJ, GrpE form a cellular chaperone machinery capable of repairing heat-induced protein damage. EMBO J. **12**, 4137–4144.

Szabo, A., Langer, T., Schroder, H., Flanagan, J., Bukau, B., and Hartl, F.-U. (1994). The ATP hydrolysis-dependent reaction cycle of the *Escherichia coli* Hsp70 system-DnaK, DnaJ and GrpE. Proc. Natl. Acad. Sci. USA **91**, 10345–10349.

Szabo, A., Korszun, R., Hartl, F.-U., and Flanagan, J. (1996). A zinc finger-like domain of the molecular chaperone DnaJ is involved in binding to denatured protein substrates. EMBO J. **15**, 408–417.

Szyperski, T., Pellecchia, M., Wall, D., Georgopoulos, C., and Wuthrich, K. (1994). NMR structure determination of the *Escherichia coli* DnaJ molecular chaperone: secondary structure and backbone fold of the N-terminal region (residues 2–208) containing the highly conserved J domain. Proc. Natl. Acad. Sci. USA **91**, 11343–11347.

Tomayasu, T., Gamer, J., Bukau, B., Kanemori, M., Mori, H., Rutman, A.J., Oppenheim, A.B., Yura, T., Yamanaka, K., Niki, H., Hiraga, S., and Ogura T. (1995). *Escherichia coli* FtsH is a membrane-bound, ATP-dependent protease which degrades the heat-shock transcription factor σ^{32}. EMBO J. **14**, 2551–2560.

Wall, D., Zylicz, M., and Georgopoulos, C. (1994). The NH_2-terminal 108 amino acids of the *Escherichia coli* DnaJ protein stimulate the ATPase activity of DnaK and are sufficient for λ replication. J. Biol. Chem. **269**, 5446–5451.

Wall, D., Zylicz, M., and Georgopoulos, C. (1995). The conserved G/F motif of the DnaJ chaperone is necessary for the activation of the substrate binding properties of the DnaK chaperone. J. Biol. Chem. **270**, 2139–2144.

Wawrzynów, A. and Zylicz, M. (1995). Divergent effect of ATP on the binding of the DnaK and DnaJ chaperones to each other, or to their various native and denatured protein substrates. J. Biol. Chem. **270**, 19300–19306.

Wawrzynów, A., Banecki, B., Wall, D., Liberek, K., Georgopoulos, C., and Zylicz, M. (1995). ATP hydrolysis is required for the DnaJ-dependent activation of DnaK chaperone for binding to both native and denatured protein substrates. J. Biol. Chem. **270**, 19307–19311.

Wickner, S., Skowyra, D., Hoskins, J., and McKenney, K. (1992). DnaJ, DnaK, and GrpE heat shock proteins are required in oriP1 DNA replication solely at the RepA monomerization step. Proc. Natl. Acad. Sci. USA **89**, 10345–10349.

Ueguchi, C., Kakeda, M., Yamada, H., and Mizuno, T. (1994). An analogue of DnaJ molecular chaperone in *Escherichia coli*. Proc. Natl. Acad. Sci. USA **91**, 1054–1058.

Ziemienowicz, A., Skowyra, D., Zeilstra-Ryals, J., Fayet, O., Georgopoulos, C., and Zylicz, M. (1993). Both the *Escherichia coli* chaperone systems, GroEL/GroES and DnaK/DnaJ/GrpE, can reactivate heat-treated RNA polymerase. J. Biol. Chem. **268**, 25425–25431.

Ziemienowicz, A., Zylicz, M., Floth, C. and Hubscher, U. (1995). Calf thymus Hsc70 protein protects and reactivates prokaryotic and eukaryotic enzymes. J. Biol. Chem. **270**, 15479–15484.

Zylicz, M. (1993). The *Escherichia coli* chaperones involved in DNA replication. Phil. Trans. R. Soc. Lond. **339**, 271–278.

Zylicz, M., Yamamoto, T., McKittrick N., Sell, S., and Georgopoulos, C. (1985). Purification and properties of the dnaJ replication protein of *Escherichia coli*. J. Biol. Chem. **260**, 7591–7598.

■ Alicja Wawrzynów and Maciej Zylicz:
Department of Molecular and Cellular Biology
University of Gdansk
Kladki 24
80-822 Gdansk, Poland
Tel. 40 50 319222
Fax. 48 58 310072
E-mail: zylicz@biotech.univ.gda.pl

Escherichia coli CbpA

CbpA from *Escherichia coli appears to be an analogue of the well-characterized cochaperone, DnaJ, as judged from not only its structure but also its function. Characteristic lesions of* dnaJ *mutants (e.g. temperature sensitivity for growth and defects in* λ *phage and mini-F plasmid replication) are restored upon introduction of the* cbpA *gene on a multicopy plasmid. While the expression of* dnaJ *is dependent on the heat shock sigma factor (σ^{32}), that of* cbpA *is under the control of the stationary phase-specific sigma factor (σ^{38} or σ^{5}).*

■ Isolation of CbpA

CbpA was first identified and purified as an *E. coli* DNA-binding protein that could preferentially bind to a synthetic curved DNA sequence by means of a DNA-binding gel shift assay in the presence of an excess amount of a non-curved competitor sequence (Ueguchi *et al.*, 1994). The curved DNA sequence, used as the probe, contains (dA)5 stretches which were designed so as to be phased at 10 bp intervals in the DNA helix (c. 200 bp).

■ *cbpA* gene and sequence

The amino acid sequence deduced from the nucleotide sequence of the *cbpA* gene (DDBJ accession number D16500) reveals an open reading frame of 297 amino acids and a calculated molecular weight of 33 407 daltons (Ueguchi *et al.*, 1994). Optimal alignment of the *E. coli* CbpA and DnaJ sequences indicates 39% amino acid identity, plus 17% conserved substitutions. Similar degrees of identity are also seen with *Bacillus subtilis* and

Mycobacterium tuberculosis DnaJ homologues (37 and 31% identity, respectively). CbpA also exhibits significant sequence similarity to members of eukaryotic DnaJ homologues (e.g. Ydj1p of *Saccharomyces cerevisiae*), particularly in its N-terminal portion of 70 amino acids, which is called the canonical J-region. It is worth mentioning that CbpA lacks the cysteine-rich repeats (CXXCXGXG) that are found in some DnaJ homologues.

■ CbpA protein

CbpA has the ability to bind to DNA. Although this protein was first isolated as a cytoplasmic soluble protein, the overproduced protein was mainly found in the membrane fraction. Thus, CbpA may bind peripherally to the cytoplasmic membrane, as reported in the case of DnaJ. CbpA seems to be a relatively minor protein in *E. coli* cells, although its cellular level changes depending on growth conditions (see below). An expression vector for CbpA is available, by which a large amount of CbpA can be purified easily (Ueguchi *et al.*, 1994).

■ Biological activities of CbpA

Several lines of genetic evidence suggest that CbpA can function as an analogue of DnaJ in *E. coli* (Ueguchi *et al.*, 1994, 1995). For example, the *cbpA* gene functions as a multicopy suppressor of *dnaJ* mutations. Mutational lesions characteristic of *dnaJ* null mutants, namely, temperature sensitivity for growth and defects in λ phage and mini-F plasmid DNA replication, are restored upon introduction of the *cbpA* gene on a multicopy plasmid. This is compatible with the idea that the function(s) of CbpA in *E. coli* is closely related to that of DnaJ.

■ Biological regulation of CbpA

The characteristics of the expression of *cbpA* are as follows (Yamashino *et al.*, 1994). *cbpA* expression is not enhanced upon abrupt heat shock. However, it is markedly induced under certain growth conditions (onset of stationary phase or upon phosphate starvation). Such conditional expression of *cbpA* is largely dependent on the stationary phase-specific sigma factor, σ^{38}. The regulation of *cbpA* is thus in sharp contrast to that of *dnaJ*, which is induced upon heat shock, and is dependent on the heat shock sigma factor, σ^{32}. This suggests that CbpA and DnaJ display similar (or overlapping) activities, but function preferentially under different circumstances.

■ Mutagenesis studies with CbpA

An insertional inactivation mutant of *cbpA* was constructed in which the coding region on the chromosome was replaced by the kanamycin resistance gene (*kan*) (Ueguchi *et al.*, 1994). This *cbpA::kan* mutant strain was not noticeably changed in phenotype under the laboratory conditions tested. However, a *cbpA–dnaJ* double-null mutant was revealed to exhibit severe defects in cell growth; namely, a very narrow temperature range for growth, a defect in cell division and a susceptibility to killing by carbon starvation (Ueguchi *et al.*, 1995). This strain is sensitive for growth not only at high temperatures ($\geq 37°C$) but also at low temperatures ($\leq 16°C$). It produces filamentous cells without septa very vigorously. These phenotypes are very similar to those reported for *dnaK* null mutants but not to those of *dnaJ* null mutants. These observations are also compatible with the idea that the function(s) of CbpA is closely related to that of DnaJ.

■ CbpA biological interactions

Escherichia coli DnaJ is known to function *in vivo* and *in vitro* as a typical cochaperone in intimate coordination (interaction) with DnaK (HSP70) and GrpE in a variety of cellular processes (Georgopoulos, 1992). The current model for DnaJ function is that this protein, together with GrpE, stimulates the ATPase activity of DnaK. However, it is not known whether or not CbpA interacts directly with DnaK and/or GrpE.

■ References

Georgopoulos, C. (1992). The emergence of the chaperone machines. Trends Biochem. Sci. **17**, 295–299.

Ueguchi, C., Kakeda, M., Yamada, H., and Mizuno, T. (1994). An analogue of the DnaJ molecular chaperone in *Escherichia coli*. Proc. Natl. Acd. Sci. USA **91**, 1054–1058.

Ueguchi, C., Shiozawa, T., Kakeda, M., Yamada, H., and Mizuno, T. (1995). A study of the double mutation of *dnaJ* and *cbpA*, whose gene products function as molecular chaperones in *Escherichia coli*. J. Bacteriol. **177**, 3894–3896.

Yamashino, T., Kakeda, M., Ueguchi, C., and Mizuno, T. (1994). An analogue of the DnaJ molecular chaperone whose expression is controlled by σ^s during the stationary phase and phosphate starvation in *Escherichia coli*. Mol. Microbiol. **13**, 475–483.

■ *Takeshi Mizuno and Chiharu Ueguchi:*
Laboratory of Molecular Microbiology
School of Agriculture, Nagoya University
Chikusa-ku
Nagoya 464, Japan
Tel. 81 52 789 4089
Fax. 81 52 789 4091
E-mail: i45455a@nucc.cc.nagoya-u.ac.jp

Escherichia coli RcsG

The RcsG protein represents the third member of the DnaJ-domain family of Escherichia coli *that includes DnaJ and CbpA. RcsG possesses a J-domain at its extreme C-terminus, but shares no additional homology with DnaJ. Genetic analysis suggests that RcsG acts in concert with the RcsA/B/C signal transduction system to induce the cps (capsular polysaccharide) operon. rcsG is not an essential gene under all conditions tested, nor is it essential for mucoid capsule biosynthesis. The positive activation of the cps operon by RcsG requires DnaK and GrpE, suggesting that there is a role for these chaperones in the regulation of a bacterial signal transduction pathway. The RcsG J-domain is essential for the observed stimulation of this pathway. Evidence suggests that steady-state rcsG message is induced or stabilized in response to cold shock.*

■ Alternative names and homologs

RcsG has also been isolated as DjlA (Clarke *et al.*, 1966), a multicopy suppressor conferring resistance to the anticalmodulin drug, W7. Two bacterial genes having significant homology to *rcsG* have also been described. The gene *mucZ* (GenBank accession numbers L42518, P44607) of *Coxiella burnetti* shows 42% amino acid identity to *rcsG*, and, when present in a multicopy vector, it confers mucoidy upon the *E. coli* host. Preliminary characterization of *C. burnetti mucZ* in *E. coli* suggests that it may indeed encode a functional homolog of RcsG as it can positively stimulate the *cps* operon and requires DnaK and GrpE for this activity (Zuber *et al.*, 1995). A second *rcsG* homologue exists in *Haemophilus influenzae* as an uncharacterized reading frame which upon conceptual translation reveals 57% identity to RcsG (accession numbers U32713, HI0271; Fleishmann *et al.*, 1995).

■ *rcsG* gene and sequence

On the physical map, *rcsG* is located at 1.23 minutes on the *E. coli* chromosome. The gene was originally described as a hypothetical open reading frame in the *folA–hepA* intergenic region that was homolgous to *Saccharomyces cerevisiae SEC63* by virtue of its J-domain only (accession numbers D10483, P31680; Yura *et al.*, 1992). Our studies (W.L. Kelley and C. Georgopolous, manuscript submitted) of *rcsG* derive from subclones obtained from Kohara phage λ8D2 (106), or by PCR amplification from MC4100 genomic DNA. Sequence analysis of both strands confirmed the coding sequence deposited in the database D10483. The naming of *rcsG* follows the convention of regulator of capsule synthesis (Gottesman *et al.*, 1985).

■ RcsG protein

RcsG is a 30.6 kDa protein of 271 amino acid residues that contains a C-terminal DnaJ-domain motif and a hydrophobic N-terminal membrane anchor sequence (Fig. 1). RcsG shows no significant homology to anything currently in the database. Subcellular fractionation experiments show that RcsG is membrane associated. Inspection of the hydrophobic anchor sequence and the apparent absence of a canonical signal sequence suggests that RcsG is cytoplasmically oriented. The dependence of RcsG upon DnaK and GrpE for its activity, and the involvement of the J-domain in modulation of the RcsA/B/C signal transduction system described below, supports the argument that the RcsG J-domain is likely to be cytoplasmically oriented.

■ Regulation of *rcsG*

The expression of *rcsG* appears to be tightly regulated. Genomic subclones that contain *rcsG* and its putative promoter region are unstable or unclonable in high copy number vectors in *E. coli*. Northern analysis using strand-specific riboprobes shows that *rcsG* is most likely transcribed from a 1.3 kb monocistronic message in the clockwise sense on the genetic map. Steady-state mRNA was detected only under cold shock conditions and was

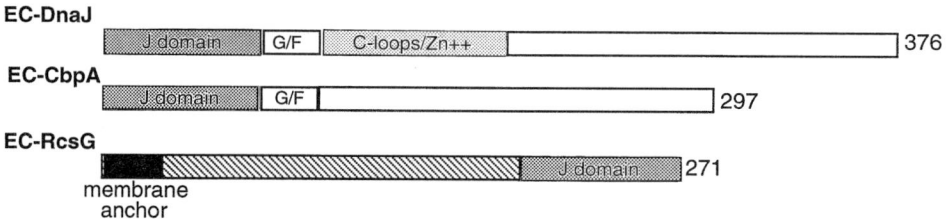

Figure 1. Comparison of *E. coli* DnaJ, CbpA and RcsG proteins.

not observed under a wide range of other conditions tested. Kinetic analysis of *rcsG* mRNA revealed detectable message within 15 minutes after shifting mid-log phase cultures from 30 to 16°C (cold shock), and a peak concentration of message following 1–3 h at 16°C. Details of the mechanism of transduction of environmental signals that provoke colanic acid mucoid capsule synthesis in *E. coli* have remained elusive.

■ Function of RcsG

Genomic subclones of *rcsG* in multicopy give a mucoid phenotype that suggested a potential involvment of the gene in the regulation of the *cps* operon responsible for capsule biosynthesis. To further dissect the mechanism of *cps* gene induction, we amplified the *rcsG* gene by PCR and subcloned it into the arabinose-inducible pBAD22A vector system (Guzman, *et al.*, 1995) that permitted us to exploit conditional induction of RcsG and RcsG mutants by the simple addition of L-arabinose as an inducer to the medium. Using appropriate fusion reporter strains, we have been able to dissect the effect of RcsG induction upon the RcsA/B/C signalling system (Gottesman *et al.*, 1985; Brill, *et al.*, 1988; Stout and Gottesmann, 1990). *cps::lacZ* induction mediated by RcsG requires the *rcsA/B/C* genes, since null mutations in any of the three abolishes *cps* induction by RcsG. RcsG does not apparently stimulate transcription from any of the described positive activators of *cps* : *rcsA*, *rcsB* or *rcsF* , as judged by the appropriate *lacZ* promoter fusions (Gottesman *et al.*, 1985; Stout *et al.*, 1991; Gervais and Drapeau, 1992). Collectively, these results suggest mechanisms whereby RcsG may be (i) acting to stabilize one or both of the RcsA/B transcriptional acitivators, (ii) interacting with the RcsC membrane sensor, or (iii) perturbing cellular physiology indirectly so as to activate an RcsA/B/C sensory response for *cps* gene expression.

The *cpsB10::lacZ* reporter fusion is sensitive to temperature since β-galactosidase activity is drastically diminished if cells to be assayed are grown at 37–39°C instead of 30°C in a *lon*– background (Gottesman *et al.*, 1985, cited as unpublished observations; Gottesman and Stout, 1991; Painbeni *et al.*, 1993). This has been hypothesized to be owing to an instability in a the promoter–RcsA/B transcription complex (Gottesman and Stout, 1991). A mutant, *rcsC137*, has been described that appears to be a constitutive activator since *cpsB10::lacZ* reporter activity is no longer temperature sensitive in the presence of this mutation in *lon*+ strains. In the absence of the Lon protease, expression of RcsG results in a hyperinduction of the *cps* operon. In addition, induction of RcsG abolishes some of the temperature sensitivity observed in the RcsC/B/A system, since there is only slight diminution of *lacZ* reporter gene activity when cells are grown at 37°C instead of at 30°C (in *lon*– strains). Thus, strikingly, induction of RcsG results in certain phenotypes ascribed to the *rcsC137* mutation. This result suggested the possibility that RcsG may act directly upon RcsC, perhaps converting it to a form similar to that expressed by *rcsC137*, or alternatively, RcsG could act upon a substrate of RcsC (perhaps RcsB) thereby altering the phosphorylation state or concomitantly the stability of the RcsB/A activator pair. Our present genetic analysis cannot distinguish between these models.

RcsG cannot substitute for DnaJ despite the presence of a DnaJ domain as judged by (i) the failure to complement the Ts or cold-sensitive phenotype of *dnaJ*–*cbpA* double-null strains (Ueguchi *et al.*, 1995) at temperatures above 37°C or below 16°C, or (ii) the inability to support the growth of bacteriophage λ. These results suggest that although RcsG represents a third DnaJ domain-containing protein in *E. coli*, RcsG alone cannot subsititute for the semi-overlapping chaperone activites of DnaJ and CbpA.

■ Mutagenesis studies with RcsG

Mutational analysis of RcsG reveals that its function as an inducer of *cps* depends critically upon the presence of the DnaJ domain since either deletion of the domain, or a point mutation, H233Q, equivalent to the canonical *dnaJ259* H33Q mutation (Wall *et al.*, 1994), abolishes RcsG activity. Furthermore, the presence of the membrane anchor sequence is essential for RcsG function, as deletion of this segment also abolishes its activity. Control experiments show that RcsG protein levels of the described mutants are comparable to wild-type RcsG levels in all cases.

The fact that RcsG-mediated induction of *cps* required the RcsG J-domain suggested that RcsG may require additional chaperones in order to function. To test this hypothesis, we constructed *cps::lacZ* reporter strains carrying null mutations in one or more chaperone genes. Our results revealed that, indeed, RcsG-mediated induction of *cps* required DnaK and GrpE. Interestingly, null mutations in *dnaJ* actually augmented RcsG-mediated *cps* induction, suggesting that DnaJ may play a negative role in this activation process. Induction of *cps* by RcsG was not blocked by mutations in *htpG* (HSP90), *cbpA* (DnaJ-like) or *hscA* (HSP70 family) genes.

rcsG is not an essential gene as judged by null mutations constructed in three different strain backgrounds and under all laboratory conditions so far tested. Furthermore, RcsG *per se* is not essential for the elucidation of the mucoid response, indicating that it is not essential for the proper functioning of the RcsA/B/C system. The only phenotype of the *rcsG*::Ωspc null mutation reproducibly noted was a slight retardation of the onset of mucoidy in strain W3110 Δ*rcsG*::Ωspc growing at 16°C. It remains to be shown in detail at what level(s) the *rcsG* gene is regulated and how, once expressed, RcsG acts together with the components of the RcsA/B/C signalling system to induce capsule synthesis.

■ Perspectives

As the RcsA/B/C system possesses extensive homology to other well-characterized members of the bacterial histidine kinase two-component signalling systems, our results suggest that the ensemble of the RcsG/J-domain cochaperone and the DnaK/GrpE chaperone machine could, together, modulate a phosphotransfer signal cascade in *E. coli*. The further elucidation of the mechanism of action of this RcsG/J-domain cochaperone may shed considerable light on the manner of chaperone regulation of signalling sytems, and in addition, provide insights into the potential mechanistic roles of proteins that contain only J-domains.

References

Brill, J.A., Quinlan-Walshe, C., and Gottesman, S. (1988). Fine-structure mapping and identification of two regulators of capsule synthesis in *Escherichia coli* K-12. J. Bacteriol. **170**, 2599–2611.

Clarke, D.J., Jacq, A., and Holland I.B. (1996). A novel DnaJ-like protein in *Escherichia coli* inserts into the cytoplasmic membrane with type III topology. Mol. Micro. **20**, 1273–1286.

Fleishmann, R.D., *et al.* (1995). Whole-genome random sequencing and assembly of *Haemophilus influenzae* Rd. Science **269**, 496–512.

Gervais, F.G. and Drapeau, G.R. (1992). Identification, cloning, and characterization of *rcsF*, a new regulator gene for exopolysaccharide synthesis that suppresses the division mutation *ftsZ84* in *Escherichia coli* K-12. J. Bacteriol. **174**, 8016–8022.

Gottesman, S. and Stout, V. (1991). Regulation of capsular polysaccharide synthesis in *Escherichia coli* K-12. Mol. Microbiol. **5**, 1599–1606.

Gottesman, S. Trisler, P., and Torres-Cabassa, A. (1985). Regulation of capsular polysaccharide synthesis in *Escherichia coli* K-12: characterization of three regulatory genes. J. Bacteriol. **162**, 1111–1119.

Guzman, L.-M., Belin, D., Carson, M.J., and Beckwith, J. (1995). Tight regulation, modulation, and high-level expression by vectors containing the arabinose pBAD promoter. J. Bacteriol. **177**, 4121–4130.

Kelley, W.L. and Georgopoulos, C. (1996). Characterization of rcsG of *Escherichia coli*: a positive activator of colanic acid capsule synthesis that is a member of the DnaJ-domain family of molecular chaperones. Submitted.

Painbeni, E., Mouray, E., Gottesman, S., and Rouviere-Yaniv, J. (1993). An imbalance of HU synthesis induces mucoidy in *Esherichia coli*. J. Mol. Biol. **234**, 1021–1037.

Stout, V. and Gottesman, S. (1990). RcsB and RcsC: a two-component regulator of capsule synthesis in Escherichia coli. J. Bacteriol. **172**, 659–669.

Stout, V., Torres-Cabassa, A., Maurizi, M.R., Gutnick, D., and Gottesman, S (1991). RcsA, an unstable positive regulator of capsular polysaccharide synthesis. J. Bacteriol. **173**, 1738–1747.

Ueguchi, C., Shiozawa, T., Kakeda, M., Yamada, H., and Mizuno, T. (1995). A study of the double mutation of *dnaJ* and *cbpA*, whose gene products function as molecular chaperones in *Escherichia coli*. J. Bacteriol. **177**, 3894–3896.

Wall, D., Zylicz, M., and Georgopoulos, C. (1994). The NH_2 108 amino acids of the *Escherichia coli* DnaJ protein stimulate the ATPase activity of DnaK and are sufficient for λ replication. J. Biol. Chem. **269**, 5446–5451.

Yura, T., Mori, H., Nagai, H., Nagata, T., Ishihama, A., Fujitan, N., Isono, K., Mizobuchi, K., and Nakata, A. (1992). Systematic sequencing of the *Escherichia coli* genome: analysis of the 0–2.4 min region. Nucl. Acids. Res. **20**, 3305–3308.

Zuber, M., Hoover, T.A., and Court, D.L. (1995). Analysis of a *Coxiella burnetti* gene product that activates capsule synthesis in *Escherichia coli*: requirement for the heat shock chaperone DnaK and the two-component regulator RcsC. J. Bacteriol. **177**, 4238–4244.

■ *William L. Kelley and Costa Georgopoulos:*
Département de Biochemie Médical
Centre Médicale Universitaire
Université de Genève
1, rue Michel-Servet
1211 Genève 4, Switzerland
Tel. 41 22 702 5514
Fax. 41 22 347 3334
E-mail: william.kelly@medecine.unige.ch

Saccharomyces cerevisiae Ydj1

Yeast DnaJ 1 gene (YDJ1) encodes a DnaJ homolog that is localized in the cytosol of Saccharomyces cerevisiae with the HSP70 Ssa and Ssb proteins. The roles that Ydj1 protein plays in cellular physiology include facilitation of protein translocation across membranes, trafficking of hormone receptors to the nucleus, and degradation of proteins via the ubiqutin/proteasome-dependent pathway. Ydj1 regulates the ATP hydrolytic cycle of Ssa proteins and acts as a molecular chaperone to catalyse these biochemical processes.

■ Alternative names

Mas5 (Atencio and Yaffe, 1992).

■ *YDJ1* gene and sequence

Ydj1 was identified independently by two laboratories. Caplan and Douglas (1991) identified Ydj1 as a component of a protein extract generated from the nuclear matrix of *Saccharomyces cerevisiae*. Atencio and Yaffe (1992) identified a protein identical to Ydj1, Mas5p, in a search for proteins that facilitate the transport of proteins into mitochondria. *YDJ1* (GenBank accession number X56560) is located on chromosome XIV and has a predicted open reading frame of 409 amino acids encoding a 44.6 kDa protein that shares 32% sequence identity with E. coli DnaJ. Expression of Ydj1 is induced about twofold upon heat shock (Atencio and Yaffe, 1992).

Regions of Ydj1p that exhibit a high degree of sequence identity with DnaJ include an N-terminal J-domain, a glycine- and phenylalanine-rich region, a cysteine-rich region, and a conserved C-terminus (for details see Fig. 1 of the overview on the HSP40 family, p. 90). Ydj1 has the additional feature of having at its C-terminus a CaaX motif (C is a cysteine and X is an aliphatic amino acid) that directs its post-translational modification with the isoprenoid farnesyl (Caplan et al., 1992a).

■ Ydj1 protein

Ydj1 is required for normal cell growth at 23°C and essential for growth at 37°C (Caplan and Douglas, 1991; Atencio and Yaffe, 1992). Immunofluorescence microscopy and subcellular fractionation experiments indicate that Ydj1 is present in the cytosol and is also concentrated around the cell nucleus on the surface of ER membranes (Caplan and Douglas, 1991). The farnesyl moiety on Ydj1 helps to facilitate the specific interaction between Ydj1 and the ER membrane (Caplan et al., 1992a). Inactivation of the CaaX motif on Ydj1 hinders an essential function of Ydj1 at 37°C and confers a temperature-sensitive growth phenotype on cells (Caplan et al., 1992a). The function of Ydj1 is conserved because E. coli DnaJ can complement the growth defects observed in ydj1 mutant strains (Caplan et al., 1992b). Purified Ydj1 regulates the affinity of Ssa1 for ATP (Ziegelhoffer et al., 1995). Ydj1 also strongly stimulates the ATPase activity of the Ssa proteins (Cyr et al., 1992), but not the Ssb proteins (Cyr and Douglas, 1994) that are colocalized with it in the yeast cytosol. Purified Ydj1 can also form complexes with protein folding intermediates to suppress their aggregation (Cyr, 1995), and cooperate with Ssa proteins to refold denatured proteins (Levy et al., 1995). Ydj1 can also regulate the oligomeric state of Hsc70 via a catalytic mechanism (King et al., 1995). Overall, Ydj1 and Ssa proteins appear to function as a specific cochaperone team in the yeast cytosol (Caplan et al., 1993; Cyr et al., 1994).

■ Function of Ydj1

Ydj1 catalyses several discrete reactions in cellular protein metabolism. Cells harboring mutant forms of Ydj1 exhibit defects in the translocation of nascent preproteins across membranes of the ER and mitochondria (Caplan et al., 1992b). Defects in protein transport correlate with the inability of Ydj1 mutants to stimulate the ATPase activity of Ssa1 (Caplan et al., 1992b). Ydj1 also shares partial functional overlap with another cytosolic yeast DnaJ protein, Sis1 (Caplan and Douglas, 1991) and plays a role in mediating signal transduction events that are dependent on the molecular chaperone protein Hsp90 (Caplan et al., 1995; Kimura et al., 1995; Dey et al., 1996). Ydj1 appears to interact physically with substrates of Hsp90, such as the glucocorticoid receptor. This interaction facilitates their subsequent binding by Hsp90. Genetic evidence also suggests that Ydj1 interacts directly with Hsp90 (Kimura et al., 1995). Biochemical studies with purified Hsp90 and Ydj1 are now required to support such an interpretation. A third function ascribed to Ydj1 is the delivery of damaged proteins to the ubiquitin/proteasome-dependent pathway for degradation of polypeptides (Lee et al., 1996). Inactivation of Ydj1 leads to a large decrease in the concentration of ubquitinated forms of proteins that have short half-lives in cells, which is accompanied by a drop in the rate of their degradation. Ssa proteins also appear to participate in the process by which protein substrates are conjugated to ubiquitin (Craig et al., 1994). How Ydj1 and other DnaJ homologs recognize polypeptide substrates and are recognized by Ssa proteins to facilitate protein metabolism is the subject of current study.

■ References

Atencio, D.P. and Yaffe, M.P. (1992). MAS5, a yeast homolog of DnaJ involved in mitochondrial protein import. Mol. Cell. Biol. **12**, 283–291.

Caplan, A.J. and Douglas, M.G. (1991). Characterization of YDJ1: a yeast homologue of the bacterial dnaJ protein. J. Cell. Biol. **114**, 609–621.

Caplan, A.J., Tsai, J., Casey, P., and Douglas M.G. (1992a). Farnesylation of Ydj1 is required for function at elevated growth temperatures in Saccharomyces cerevisiae. J. Biol. Chem. **267**, 18890–18895.

Caplan, A.J., Cyr, D.M., and Douglas, M.G. (1992b). Ydj1 facilitates polypeptide translocation across different intracellular membranes by a conserved mechanism. Cell **71**, 1143–1155.

Caplan, A.J., Cyr, D.M., and Douglas, M.G. (1993). Eukaryotic homologs of E. coli DnaJ: a diverse protein family that functions with the Hsp70 stress proteins. Mol. Cell. Biol. **4**, 555–563.

Caplan, A.J., Langley, E., Wilson, E.M., and Vidal, J. (1995). Hormone-dependent transactivation by the human androgen receptor is regulated by a dnaJ protein. J. Biol. Chem. **270**, 5251–5257.

Craig, E.A., Baxter, B.K., Becker, J., Halladay, J., and Ziegelhoffer, T. (1994) Cytosolic hsp70s of Saccharomyces cerevisiae: a role in protein synthesis, protein translocation, proteolysis, and regulation. In *The Biology of Heat Shock Proteins and Molecular Chaperones*, (R.I. Morimoto, A. Tissieres, and C. Georgopoulos eds). Cold Spring Harbor Laboratory Press, New York, pp. 31–52.

Cyr, D.M. (1995). Cooperation of the molecular chaperone Ydj1 with specific Hsp70 homologs to suppress protein aggregation. FEBS Lett. **359**, 129–132.

Cyr, D.M. and Douglas, M.G. (1994). Differential regulation of Hsp70 subfamilies by the eukaryotic DnaJ homologue YDJ1. J. Biol. Chem. **269**, 9798–9804.

Cyr, D.M., Lu, X., and Douglas, M.G. (1992). Regulation of Hsp70 function by a eukaryotic DnaJ homolog. J. Biol. Chem. **267**, 20927–20931.

Cyr, D.M., Langer, T., and Douglas, M.G. (1994). DnaJ-like proteins: molecular chaperones and specific regulators of Hsp70. Trends Biochem. Sci. **19**, 176–181.

Dey, B., Caplan, A.J., and Boschelli, F. (1996). The YDJ1 molecular chaperone facilitates formation of active p60v-src in yeast. Mol. Biol. Cell **7**, 91–100.

Lee, D.H., Sherman, M.Y. and Goldberg, A.L. (1996). Involvement of the molecular chaperone Ydj1 in the ubiquitin-dependent degradation of short-lived and abnormal proteins in *Saccharomyces cerevisiae*. Mol. Cell Biol. **16**, 4773–4781.

Levy, E.J., McCarty, J., Bukau, B., and Chirico, W.J. (1995) Conserved ATPase and luciferase refolding activities between bacterial and yeast Hsp70 chaperones and modulators. FEBS Lett. **368**, 435–440.

Kimura, Y., Yahara, I., and Lindquist, S. (1995) Role of the protein chaperone YDJ1 in establishing Hsp90-mediated signal transduction pathways. Science **268**, 1362–1365.

King, C., Eisenberg, E., and Greene, L. (1995) Polymerization of 70-kDa heat shock protein by yeast DnaJ in ATP. J. Biol. Chem. **270**, 22535–22540.

Ziegelhoffer, T., Lopez-Buesa, P., and Craig, E.A. (1995) The dissociation of ATP from hsp70 of Saccharomyces cerevisiae is stimulated by both Ydj1 and peptide substrates. J. Biol. Chem. **270**, 10412–10419.

■ Douglas M. Cyr:
Department of Cell Biology
652 Basic Health Sciences Building
School of Medicine
University of Alabama at Birmingham
Birmingham, AL 35294-0005, USA
Tel. 1 205 975 4892
Fax. 1 205 934 0950
E-mail: DCYR@cellbio.bhs.uab.edu

■ Avrom J. Caplan:
Department of Cell Biology and Anatomy
Mount Sinai School of Medicine
One Gustave L. Levy Place
New York, NY 10029, USA
Tel. 1 212 241 6563
Fax. 1 212 860 1174
E-mail: caplan@msvax.mssm.edu

Saccharomyces cerevisiae Sis1

Sis1 is an essential protein in the yeast *Saccharomyces cerevisiae* that is related to *Escherichia coli* DnaJ in the N-terminal third and the C-terminal third of the proteins. Sis1 may have multiple cellular functions, at least one of which is the normal initiation of translation.

■ *SIS1* gene and sequence

The DNA sequence of *SIS1* (GenBank accession number X58460) predicts a protein of 37.6 kDa with a pI of 9.15 and a codon bias of 0.163 (Luke *et al.*, 1991). The N-terminal third and the C-terminal third of Sis1p are related to the corresponding regions of bacterial DnaJ. The middle third of Sis1p is unrelated to DnaJ and contains a glycine/methionine-rich region.

■ Sis1 protein and biological interactions

Sis1p is localized throughout the cell, but is more concentrated in the region of the nucleus (Luke *et al.*, 1991). Sis1p coimmunoprecipitates with a 40 kDa protein (p40 in Luke *et al.*, 1991). Gel filtration shows that only a very small percentage (less than 5%) of Sis1p is bound to p40 and that the SIS1–p40 complex has a very large size (about $> 2 \times 10^6$). The function of p40 is not known.

■ Function of Sis1p

The *SIS1* gene was originally identified by its ability, when present on a high copy number plasmid, to suppress partially the slow growth rate of *sit4* mutants (Luke *et al.*, 1991). *SIT4* encodes the catalytic subunit of a type 2A-related protein phosphatase that is required for G1 cyclin expression and for bud initiation (Sutton *et al.*, 1991; Fernandez-Sarabia *et al.*, 1992). The mechanism (possibly relating to proteolysis or translation initiation, see below) by which overexpression of Sis1p stimulates the growth rate of *sit4* mutants is not known at present. *SIS1* is an essential gene (Luke *et al.*, 1991). To understand the biological function of Sis1p, extragenic suppressors of a temperature-sensitive *sis1* mutant were isolated (Zhong and Arndt, 1993). This analysis led to the finding that mutations in genes encoding 60S ribosomal subunit proteins suppressed the temperature-sensitive phenotype of the *sis1* mutant. The same alterations in 60S ribosomal subunits can also suppress the normally lethal effect caused by deletion of *PAB1*, which encodes the poly(A) binding protein that is required for the normal initiation of translation (Sachs and Davis, 1989). When temperature-sensitive *sis1* mutants are shifted to the non-permissive temperature, they have defects in polysome and ribosome levels that are indicative of a defect in the initiation of translation (Zhong and Arndt, 1993). In addition, a substantial fraction of Sis1 protein cosediments with ribosomes and polysomes. These findings indicate that Sis1p is required for the normal initiation of translation.

Transcriptional regulation of *SIS1*

The promoter of the *SIS1* gene contains a heat shock consensus element and the steady state levels of Sis1 protein increase two- to threefold upon heat shock (Luke et al., 1991). The *SIS1* RNA levels increase transiently after heat shock: after 20 minutes the RNA levels are induced about fivefold and return to near normal levels by 90 minutes (Zhong et al., 1996). The heat shock induction of the *SIS1* promoter requires only the heat shock element. In additiion, transcription from the *SIS1* promoter is negatively regulated by Sis1p function: defects in *sis1* itself highly activate the *SIS1* promoter and overexpression of Sis1p represses the *SIS1* promoter (Zhong et al., 1996). The Sis1p regulation of the *SIS1* promoter requires the intact heat shock element plus an additional promoter element that is adjacent to the heat shock element. Transcription of *SIS1* is also induced twofold in *cim3* and *cim5* mutants, which have defects in proteasome-mediated proteolysis.

References

Fernandez-Sarabia, M.J., Sutton, A., Zhong, T., and Arndt, K.T. (1992). SIT4 protein phosphatase is required for the normal accumulation of *SWI4*, *CLN1*, *CLN2*, and *HCS26* RNAs during late G1. Genes Dev. **6**, 2417–2428.

Luke, M.M., Sutton, A., and Arndt, K.T. (1991). Characterization of SIS1, a Saccharomyces cerevisiae homologue of bacterial dnaJ proteins. J. Cell Biol. **114**, 623–38.

Sachs, A.B. and Davis, R.W. (1989). The poly(A) binding protein is required for poly(A) shortening and 60S ribosomal subunit-dependent translation initiation. Cell **58**, 857–867.

Sutton, A., Immanuel, D., and Arndt, K.T. (1991). The SIT4 protein phosphatase functions in late G1 for progression into S phase. Mol. Cell. Biol. **11**, 2133–2148.

Zhong, T. and Arndt, K.T. (1993). The yeast SIS1 protein, a DnaJ homolog, is required for the initiation of translation. Cell **73**, 1175–1186.

Zhong, T., Luke, M.M., and Arndt, K.T. (1996). Transcriptional regulation of the yeast DnaJ homologue SIS1. J. Biol. Chem. **271**, 1349–1356.

Kim T. Arndt:
*Cold Spring Harbor Laboratory
1 Bungtown Road, Delbruck Building
Cold Spring Harbor, NY 11724-2212, USA
Tel. 1 516 367 8835/8836 (office/lab)
Fax. 1 516 367 8369
E-mail: arndt@cshl.org*

Saccharomyces cerevisiae Zuotin

Zuotin is a DnaJ homolog in the yeast Saccharomyces cerevisiae *that was identified via a biochemical screen for proteins that bind the Z-conformation of DNA. Zuotin also binds tRNA in nuclear extracts and may play a role in the processing or transport of these molecules. Zuotin and its mammalian homolog, MIDA1, are required for normal cell growth. These data suggest that zuotin plays an important role in the maintenance of cellular homeostasis via a mechanism that is conserved throughout evolution.*

Isolation of Zuotin

Zuotin was purified from yeast nuclear matrix based on its ability to bind a Z-DNA affinity column (Zhang et al., 1992).

ZUO1 gene and sequence

The *ZUO1* gene (GenBank accession number X63612) is located on yeast chromosome VII near *ADE3* and encodes a 433 residue protein that has predicted molecular weight of 49 kDa (Zhang et al., 1992). Zuotin is a basic protein with a pI of 8.8 and, based on sequence analysis, has a number of interesting features (Zhang et al., 1992). Zuotin contains an internal J-domain located between amino acids 111 and 165 and a histone H1 homology region between amino acids 300 and 363. The J-domains of zuotin and *E. coli* DnaJ share 46% sequence identity. The histone H1-like region shares 46% identity with sea urchin H1. Zuotin also contains two potential CDC28 phosphorylation sites and a bipartite nuclear targeting signal located between amino acids 340 and 356.

Zuotin protein

Deletion of the *ZUO1* gene results in a slow growth phenotype of null yeast strains (Zhang et al., 1992). A similar phenotype was observed when the mammalian zuotin homolog, MIDA1, a protein identified based on its ability to associate with helix–loop–helix proteins, was inactivated in cultured murine erythroleukemia cells (Shoji et al., 1995). Zuotin binds Z-DNA in a manner that is not competed by B-DNA competitor molecules (Zhang et al., 1992). Zuotin also binds tRNA molecules in RNA mobility shift assays (Wilhelm et al., 1994). The presence of a J-domain in zuotin suggests that an HSP70 protein may participate with it in certain aspects of its function (Cyr et al., 1994), but this remains to be tested.

■ References

Cyr, D.M., Langer, T., and Douglas, M.G. (1994). DnaJ-like proteins: molecular chaperones and specific regulators of Hsp70. Trends Biochem. Sci. **19**, 176–181.

Shoji, W., Inoue, T., Yamamoto, T., and Obinata, M. (1995). MIDA1, a protein associated with Id, regulates cell growth. J. Biol. Chem. **270**, 24818–24825.

Wilhelm, M.L., Reinbolt, J., Gangloff, J., Dirheimer, D., and Wilhelm, F.X. (1994). Transfer RNA binding protein in the nucleus of Saccharomyces cerevisiae. FEBS Lett. **349**, 260–264.

Zhang, S., Lockshin, C., Herbert, A., Winter, E. and Rich, A. (1992), Zuotin, a putative Z-DNA binding protein in Saccharomyces cerevisiae. EMBO J. **11**, 3787–3796.

■ *Douglas M. Cyr:*
Department of Cell Biology
652 Basic Health Sciences Building
School of Medicine
University of Alabama at Birmingham
Birmingham, AL 35294-0005, USA
Tel. 1 205 975 4892
Fax. 1 205 934 0950
E-mail: DCYR@cellbio.bhs.uab.edu

Saccharomyces cerevisiae Mdj1p

Mdj1p, a mitochondrial DnaJ homologue from Saccharomyces cerevisiae, is encoded by a nuclear gene and synthesized in the cytosol as a preprotein. Gene disruption of MDJ1 results in temperature-sensitive growth and loss of mitochondrial DNA. Mdj1p is involved in the folding of imported tester proteins and transiently protects against heat-induced protein aggregation.

■ *MDJ1* gene and sequence

The nuclear gene for mitochondrial DnaJ (*MDJ1*) was identified during the sequence analysis of chromosome VI (GenBank accession number D50617x53). The *MDJ1* open reading frame encodes a polypeptide of 511 residues (Rowley et al., 1994) that possesses all the features characteristic of a DnaJ protein: (i) an N-terminal J-domain of about 70 amino acid residues; (ii) a 60-residue segment rich in glycine and phenylalanine; (iii) four cysteine-containing repeats with the sequence motif: CXXCXGXG (with X being predominantly a polar or charged residue); and (iv) a C-terminal domain with little sequence conservation to other DnaJ proteins (for detailed reviews about the HSP40/DnaJ protein family see Caplan et al., 1993; Cyr et al., 1994; and the introductory overview to this section, p. 89).

■ Mdj1 protein and cellular location

Mdj1p is a mitochondrial homologue in *S. cerevisiae* of the still expanding HSP40/DnaJ protein family. The protein is encoded by a nuclear gene and synthesized in the cytoplasm as a precursor form. Upon import into the mitochondrial matrix the protein becomes processed to the mature form with a molecular mass of 49 kDa (Rowley et al., 1994). After *in vitro* synthesis the Mdj1 precursor protein can be imported into isolated mitochondria. Import into the mitochondrial matrix depends on the presence of a potential across the mitochondrial inner membrane. Endogenous Mdj1p appears to be associated with the inner leaflet of the inner membrane. Alkaline carbonate extraction releases Mdj1p from the membrane demonstrating that Mdj1p is not an integral membrane protein (Rowley et al., 1994). The intracellular and submitochondrial location of Mdj1p can be monitored immunologically using polyclonal antibodies directed against the full length protein and the C-terminal half (Rowley et al., 1994).

■ Function of Mdj1p

Analysis of mitochondria from the *mdj1* deletion strain demonstrated that Mdj1p is involved in the folding of newly imported tester proteins, such as DHFR (mouse dihydrofolate reductase) and firefly luciferase (Rowley et al., 1994). In addition, mitochondria lacking Mdj1p show enhanced protein aggregation of thermolabile tester proteins after shift to elevated temperature. It is assumed that Mdj1p interacts with mitochondrial Hsp70 (Ssc1p, see entry p. 30) during folding of newly imported proteins and heat stress. Lack of Mdj1p results in a reduced ability of Ssc1p to associate with matrix-located substrate proteins (Prip-Buus et al., 1996). This reduced ability of Ssc1p to associate with substrate proteins is probably also the reason why, in the absence of Mdj1p, Pim1p protease-mediated degradation of misfolded tester proteins is impaired (Wagner et al., 1994). Nevertheless, Mdj1p is not essential for mitochondrial protein import, a process that strictly relies on Ssc1p (Kang et al., 1990; Rowley et al., 1994). Apparently, Ssc1p is able to facilitate precursor import independently of Mdj1p.

■ Mutagenesis studies with *MDJ1*

Several point mutations and deletions have been introduced into *MDJ1* by both random and site-directed mutagenesis (Westerman et al., 1996). A mutation in the conserved J-domain (H_{34} to Q) results in a knock-out phenotype. The corresponding mutation in *E. coli* DnaJ causes a temperature-sensitive phenotype. Substitution of T_{53} by A yields a temperature-sensitive phenotype, as does the corresponding mutation in the J-like Sec63 protein (Feldheim et al., 1992). Deletion of one of the cysteine-containing repeats results in temperature-sensitive growth, as do several deletions in the C-terminal domain.

Mitochondria containing mutant Mdj1p show enhanced aggregation of the mitochondrially encoded Var1 protein, indicating that Mdj1p is also involved in the biogenesis of organelle-encoded gene products (Westerman et al., 1996).

Cells containing mutant Mdj1p lose their mitochondrial DNA while being propagated at restrictive temperature (Westerman et al., 1996). Whether Mdj1p is directly or indirectly involved in mitochondrial DNA maintenance remains to be resolved.

■ Biological regulation of *MDJ1*

The open reading frame in the *MDJ1* gene is preceded by several weak matches to the heat shock element nGAAnnTTCn (Amin et al., 1988; Young and Craig, 1993). Furthermore, five copies of the motif CCCCT (STRE for stress response element) were identified in the promoter region, which have been shown to confer a heat shock factor-independent gene induction (Kobayashi and McEntee, 1993). Upon shift of *S. cerevisiae* cultures from 24°C to 37°C a transient two- to threefold increase of transcript abundance classifies Mdj1p as a heat shock protein (Rowley et al., 1994).

■ References

Amin, J., Ananthan, J., and Voellmy, R. (1988). Key features of heat shock regulatory elements. Mol. Cell. Biol. **8**, 3761–3769.

Caplan, A.J., Cyr, D.M., and Douglas, M.G. (1993). Eukaryotic homologues of Escherichia coli dnaJ: a diverse protein family that functions with HSP70 stress proteins. Mol. Biol. Cell **4**, 555–563.

Cyr, D.M., Langer, T., and Douglas, M.G. (1994). DnaJ-like proteins: molecular chaperones and specific regulators of Hsp70. Trends Biochem. Sci. **19**, 176–181.

Feldheim, D., Rothblatt, J., and Schekman, R. (1992). Topology and functional domains of Sec63p, an endoplasmic reticulum membrane protein required for secretory protein translocation. Mol. Cell. Biol. **12**, 3288–3296.

Kang, P.-J., Ostermann, J., Shilling, J., Neupert, W., Craig, E.A., and Pfanner, N. (1990). Requirement for hsp70 in the mitochondrial matrix for translocation and folding of precursor proteins. Nature **348**, 137–143.

Kobayashi, N. and McEntee, K. (1993). Identification of cis and trans components of a novel heat shock stress regulatory pathway in Saccharomyces cerevisiae. Mol. Cell. Biol. **13**, 248–256.

Prip-Buus, C., Westerman, B., Schmitt, M., Langer, T., Neupert, W. and Schwarz, E. (1996). Role of the mitochondrial DnaJ homologue, Mdj1p, in the prevention of heat-induced protein aggregation. FEBS Lett. **380**, 142–146.

Rowley, N., Prip-Buus, C., Westermann, B., Brown, C., Schwarz, E., Barrell, B., and Neupert, W. (1994). Mdj1p, a novel chaperone of the DnaJ family, is involved in mitochondrial biogenesis and protein folding. Cell **77**, 249–259.

Wagner, I., Arlt, H., van Dyck, L., Langer, T., and Neupert, W. (1994). Molecular chaperones cooperate with PIM1 protease in the degradation of misfolded proteins in mitochondria. EMBO J. **13**, 5135–5145.

Westerman, B., Gaume, B., Herrmann, J.M., Neupert, W. and Schwarz, E. (1996). Role of the mitochondrial DnaJ homolog Mdj1p as a chaperone for mitochondrially synthesized and imported proteins. Mol. Cell. Biol. **16**, 7063–7071.

Young, M.R. and Craig, E.A. (1993). Saccharomyces cerevisiae HSP70 heat shock elements are functionally distinct. Mol. Cell. Biol. **13**, 5637–5646.

■ *Elisabeth Schwarz:*
Institut für Biotechnologie
Martin-Luther-Universität Halle-Wittenberg
Kurt-Mothes-Str. 3
D-06120 Halle
Germany
Tel. 49 345 5524 856
Fax. 49 345 5527 013
E-mail: Elisabeth.Schwartz@biochemtech.uni-halle.de

Saccharomyces cerevisiae Sec63p

The yeast SEC63 gene encodes an integral membrane protein that is required for nuclear protein localization and for both co- and post-translational protein translocation into the endoplasmic reticulum (ER). A luminal domain of Sec63p, which is 43% identical to the N-terminus of the bacterial DnaJ protein, mediates the interaction of Sec63p with Kar2p/BiP, a luminal HSP70 that may drive protein import. Sec63p also interacts with other factors required for protein translocation, suggesting that Sec63p is a central component of the translocation machine in the ER membrane.

■ Alternative names

NPL1 and *PTL1* (Toyn *et al.*, 1988; Rothblatt *et al.*, 1989; Sadler *et al.*, 1989).

■ Isolation

The *SEC63* gene was isolated initially in genetic selections for temperature-sensitive mutants in yeast that accumulate an untranslocated hybrid protein when shifted to the non-permissive temperature (Toyn *et al.*, 1988; Rothblatt *et al.*, 1989). The gene was also obtained in a selection for temperature-sensitive mutants that mislocalize a nuclear-targeted fusion protein at the non-permissive temperature (Sadler *et al.*, 1989; Bossie *et al.*, 1992). Disruption of *SEC63* indicates that the gene product is essential for spore germination (Sadler *et al.*, 1989).

■ *SEC63* gene and sequence

The nucleotide sequence of *SEC63* (GenBank accession number X16388) predicts a 663 amino acid polypeptide (73 kDa) with three transmembrane regions (Sadler *et al.*, 1989). Between the second and third transmembrane segments are 73 amino acids that share 43% sequence identity to the N-terminus of the *E. coli* DnaJ protein. The C-terminus of *SEC63* is highly acidic, containing 52 amino acids, of which 27 are aspartate or glutamate.

■ Sec63 protein

Antisera raised against Sec63–protein-A or lacZ fusion proteins detect an unglycosylated polypeptide of c. 70 kDa (Feldheim *et al.*, 1992). Indirect immunofluorescence staining shows that Sec63p is in the nuclear envelope–ER membrane network. Sec63p is not extracted from membranes by carbonate, demonstrating that it is an integral membrane protein. The topology of Sec63p in the ER membrane is shown in Fig. 1 and was determined using a series of Sec63p–invertase fusion proteins that become glycosylated when the invertase moiety resides inside the ER (Feldheim *et al.*, 1992).

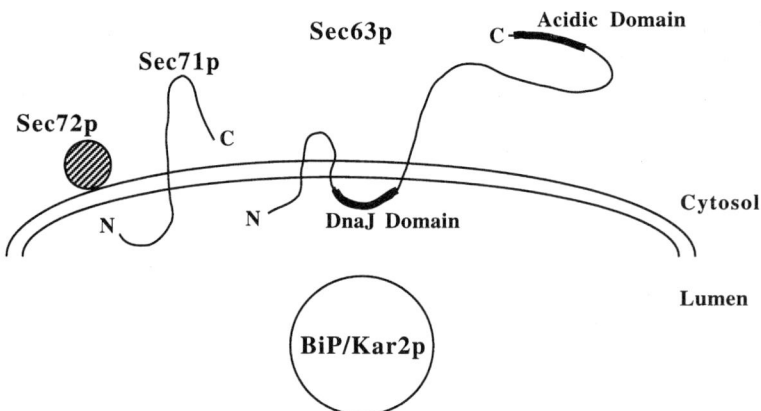

Figure 1. The Sec63p complex. Proteins that associate with Sec63p upon purification from detergent-solubilized microsomes are shown, and addition of this complex to either *sec63-1*, *sec71*, or *sec72* reconstituted proteoliposomes restores translocation activity (Brodsky and Schekman, 1993). *SEC63*, *KAR2*, and *SCE71* also interact genetically (Kurihara and Silver, 1993; Scidmore *et al.*, 1993). The complex may also be required for nuclear/ER membrane fusion (Ng and Walter, 1996).

Requirement for Sec63p during protein translocation

Strains containing specific temperature-sensitive mutations in *sec63* accumulate protein precursors in the cytosol at the non-permissive temperature (Toyn et al., 1988; Rothblatt et al., 1989; Ng et al., 1996). Microsomes prepared from the *sec63-1* mutant are unable to translocate efficiently either co- or post-translationally translocated substrates (Toyn et al., 1988; Rothblatt et al., 1989; Brodsky et al., 1995). Although reconstituted vesicles prepared from *sec63-1* microsomes are defective for protein translocation, transport is restored when the purified Sec63p complex is added (see below), showing that the *sec63-1* translocation defect is not a result of secondary effects (Brodsky and Schekman, 1993). In addition, the targeting of specific nuclear proteins is defective in *sec63* strains when incubated at the non-permissive temperature (Sadler et al., 1989; Bossie et al., 1992).

Mutagenesis studies with *SEC63p*

The *sec63-1* allele encodes a mutant protein containing a threonine at position 179 in the DnaJ domain in place of a conserved alanine (Nelson et al., 1993). Furthermore, plasmids encoding mutants with site-directed substitutions in either one of two other conserved residues in the DnaJ domain of *SEC63* fail to complement the temperature sensitivity of the *sec63-1* strain (Feldheim et al., 1992). The acidic domain of Sec63p is also vital because a plasmid encoding Sec63p lacking the C-terminal 28 amino acids fails to restore growth to the *sec63-1* strain at the non-permissive temperature (Feldheim et al., 1992).

Other mutations in *sec63* were isolated for their inability to direct a fusion protein to the nucleus at the non-permissive temperature (Sadler et al., 1989; Bossie et al., 1992). These mutants map to the cytosolic region, after the third transmembrane segment of Sec63p but preceding the acidic region (Nelson et al., 1993).

The DnaJ domain of Sec63p can be replaced by the corresponding domain from Scj1p, another ER luminal DnaJ homolog (Schlenstedt et al., 1995). However, the corresponding regions from two other DnaJ homologs in yeast are unable to substitute functionally for the DnaJ domain of Sec63p, suggesting that specific protein–protein contacts are mediated by this domain (see below).

Biological interactions and the mechanism of Sec63p function

Sec63p forms multiple contacts in the ER membrane. Triton X-100-solubilized yeast microsomes treated with anti-Sec63p antibody precipitate a complex containing Sec61p, Sec62p, Sec63p, Sec71p, and Sec72p (Deshaies et al., 1991). Each of these components are required for protein translocation into the yeast ER (reviewed by Brodsky and Schekman, 1994).

A complex containing Sec63p, Kar2p/BiP, Sec71p, and Sec72p can be purified from octylglucoside-solubilized yeast microsomes (see Fig. 1) and restores translocation to reconstituted vesicles prepared from the *sec63-1* strain (Brodsky and Schekman, 1993). The ER luminal HSP70 protein Kar2p/BiP dissociates from the complex in the presence of ATPγS or if the complex is isolated from *sec63-1* cells, demonstrating that the Sec63p–Kar2p/BiP interaction is regulated by ATP and is mediated by the DnaJ domain of Sec63p (see Kar2p entry, p. 33; Brodsky and Schekman, 1993). Panzner et al. (1995) recently purified a detergent-solubilized complex from the ER that contains Sec62p, Sec63p, Sec71p, Sec72p, and three proteins that form the Sec61p complex. Reconstituted vesicles prepared with this complex in the presence of Kar2p/BiP are translocation competent (Panzner et al., 1995).

There is a clear correlation between the biochemical and genetic interactions between Sec63p and other factors: the *sec63-1* and *kar2-159* mutations are synthetically lethal, and dominant suppressors of *sec63-1* temperature sensitivity map to the *KAR2* gene (Scidmore et al., 1993). In addition, multiple copies of the *SEC71* gene suppress the temperature-sensitive defect of a *sec63* mutant strain (Kurihara and Silver, 1993). Finally, *sec63* is synthetically lethal with both *sec62* and *sec61*, and *SEC62* expressed from a multicopy plasmid rescues the temperature sensitivity of a *sec63* strain (Rothblatt et al., 1989; Deshaies and Schekman, 1990).

The association between Sec63p and Kar2p/BiP was anticipated because Kar2p/BiP is c. 50% identical to the bacterial DnaK protein and DnaK and DnaJ interact during protein folding, protein secretion, and phage λ DNA replication in *E. coli* (Sadler et al., 1989; Frydman and Hartl, 1994). Furthermore, the ATPase activity of DnaK can be stimulated by DnaJ (Liberek et al., 1991). Lyman and Schekman (1996, 1997) have recently shown that the interaction between Kar2p/BiP and Sec63p is required both to facilitate the transfer of a preprotein to the translocation pore and to complete protein import into the ER lumen.

Sec63p might also be a component of the protein import machinery in the nucleus. Strains containing mutations in *sec63* mislocalize nuclear-targeted fusion proteins at the non-permissive temperature (Sadler et al., 1989; Bossie et al., 1992). The *SON1* gene, which encodes a nuclear protein, was isolated as a suppressor of a *sec63* temperature-sensitive mutation (Nelson et al., 1993). Although *SON1* is not essential, a strain deleted for *SON1* displays a partial defect in nuclear protein localization (Nelson et al., 1993).

Together, these results suggest that multiprotein import machines exist in which Sec63p is a central figure. During protein translocation into the ER, Sec63p may anchor Kar2p/BiP adjacent to the translocation pore (Brodsky and Schekman, 1994), permitting the HSP70 to drive protein import concomitant with ATP hydrolysis (Lyman and Schekman, 1996; see Kar2p entry, p. 33). The mechanism of Sec63p action during nuclear protein import remains obscure.

References

Bossie, M.A., Dehoratius, C., Barcelo, G., and Silver, P. (1992). A mutant nuclear protein with similarity to RNA binding proteins interferes with nuclear import in yeast. Mol. Biol. Cell **3**, 875–893.

Brodsky, J.L. and Schekman, R. (1993). A Sec63p–BiP complex from yeast is required for protein translocation in a reconstituted proteoliposome. J. Cell Biol. **123**, 1355–1363.

Brodsky, J.L. and R. Schekman. (1994). Heat shock cognate proteins and polypeptide translocation across the endoplasmic reticulum membrane, in *The Biology of Heat Shock Proteins and Molecular Chaperones* (Morimoto, R.I., Tessieres, A. and Georgopoulos, C., eds. Cold Spring Harbor Laboratory Press, Cold Spring Harbor, New York, pp. 85–110.

Brodsky, J.L., Goeckeler, J., and Schekman, R. (1995). Sec63p and BiP are required for both co- and post-translational protein translocation into yeast microsomes. Proc. Natl. Acad. Sci. USA **92**, 9643–9646.

Deshaies, R.J. and Schekman, R. (1990). Structural and functional dissection of Sec62p, a membrane bound component of the yeast endoplasmic reticulum protein import machinery. Mol. Cell. Biol. **10**, 6024–6035.

Deshaies, R.J., Sanders, S., Feldheim, D.A., and Schekman, R. (1991). Assembly of yeast Sec proteins involved in translocation into the endoplasmic reticulum into a membrane-bound multisubunit complex. Nature **349**, 806–808.

Feldheim, D., Rothblatt, J., and Schekman, R. (1992). Topology and functional domains of Sec63p, an endoplasmic reticulum membrane protein required for secretory protein translocation. Mol. Cell. Biol. **12**, 3288–3296.

Frydman, J. and Hartl, F.-U. (1994). Molecular chaperone functions of hsp70 and hsp60 in protein folding, in *The Biology of Heat Shock Proteins and Molecular Chaperones* (Morimoto, R.I., Tessieres, A. and Georgopoulos, C., eds). Cold Spring Harbor Laboratory Press, Cold Spring Harbor, New York, pp. 251–284.

Kurihara, T. and Silver, P. (1993). Suppression of a sec63 mutation identifies a novel component of the yeast ER translocation apparatus. Mol. Biol. Cell **4**, 919–930.

Liberek, K., Marszalek, J., Ang, D., Georgopoulos, C., and Zylicz, M. (1991). *Escherichia coli* dnaJ and grpE heat shock proteins jointly stimulate ATPase activity of dnaK. Proc. Natl. Acad. Sci. USA **88**, 2874–2878.

Lyman, S.K. and Schekman, R. (1996). Interaction between BiP and Sec63p is required for the completion of protein translocation into the ER of *Saccharomyces cerevisiae*. J. Cell Biol. **131**, 1163–1171.

Lyman, S.K. and Schekman, R. (1997). Binding of secretory precursor polypeptides to a translocon subcomplex is regulated by BiP. Cell **88**, 85–96.

Nelson, M.K., Kurihara, T., and Silver, P.A. (1993). Extragenic suppressors of mutations in the cytoplasmic C terminus of SEC63 define five genes in *Saccharomyces cerevisiae*. Genetics **134**, 159–173.

Ng, D.T.W and Walter, P. (1996). ER membrane protein complex required for nuclear fusion. J. Cell Biol. **132**, 499–509.

Ng, D.T.W., Brown, J.D., and Walter, P. (1996). Signal sequences specify the targeting route to the endoplasmic reticulum membrane. J. Cell Biol. **134**, 269–278.

Panzner, S., Dreier, L., Hartmann, E., Kostka, S., and Rapoport, T.A. (1995). Posttranslational protein translocation in yeast reconstituted with a purified complex of Sec proteins and Kar2p. Cell **81**, 561–570.

Rothblatt, J.A., Deshaies, R.J., Sanders, S.L., Daum, G., and Schekman, R. (1989). Multiple genes are required for proper insertion of secretory proteins into the endoplasmic reticulum in yeast. J. Cell Biol. **109**, 2641–2652.

Sadler, I., Chiang, A., Kurihara, T., Rothblatt, J., Way, J., and Silver, P. (1989). A yeast gene important for protein assembly into the endoplasmic reticulum and the nucleus has homology to dnaJ, an *Escherichia coli* heat shock protein. J. Cell Biol. **109**, 2665–2675.

Schlenstedt, G., Harris, S., Risse, B., Lill, R., and Silver, P. (1995). A yeast homolog, Scj1p, can function in the endoplasmic reticulum with BiP/Kar2p via a conserved domain that specifies interactions with hsp70s. J. Cell Biol. **129**, 979–988.

Scidmore, M., Okamura, H.H., and Rose, M.D. (1993). Genetic interactions between *KAR2* and *SEC63*, encoding eukaryotic homologues of DnaK and DnaJ in the endoplasmic reticulum. Mol. Biol. Cell **4**, 1145–1159.

Toyn, J., Hibbs, A.R., Sanz, P., Crowe, J., and Meyer, D.I. (1988). *In vivo* and *in vitro* analysis of ptl1, a yeast ts mutant with a membrane-associated defect in protein translocation. EMBO J. **7**, 4347–4353.

■ Jeffrey L. Brodsky:
Department of Biological Sciences
University of Pittsburgh
Pittsburgh, PA 15260, USA
Tel. 1 412 624 4831
Fax. 1 412 624 4759
E-mail: jbrodsky+@pitt.edu

Saccharomyces cerevisiae Scj1p

The *SCJ1* gene of the yeast *Saccharomyces cerevisiae* encodes one of several eukaryotic DnaJ homologues. Scj1p is located mainly in the endoplasmic reticulum where it may interact and perform charperone functions with the HSP70 family member Kar2p/BiP.

■ Alternative name

$\underline{S}$accharomyces $\underline{c}$erevisiae Dna$\underline{J}$ homologue (Blumberg and Silver, 1991).

■ Identification of *SCJ1*

SCJ1 was originally identified by a genetic screen in yeast in which overexpression of candidate genes could result in mis-sorting of a nuclear-targeted reporter protein (Blumberg and Silver, 1991).

■ SCJ1 gene and sequence

The *SCJ1* gene was isolated from a yeast genomic library. Low stringency Southern blot analysis of yeast genomic DNA using *SCJ1* DNA as probe revealed four to five cross-hybridizing sequences in addition to *SCJ1* itself, which most likely correspond to the genes encoding several other DnaJ homologues in yeast (Blumberg and Silver, 1991). The deduced nucleotide sequence of *SCJ1* (GenBank accession number X58679) could encode a 404 amino acid protein homologous to the bacterial heat shock protein DnaJ (Blumberg and Silver, 1991). The Scj1 protein is 37% identical to *Escherichia coli* DnaJ over its entire sequence, but the highest similarity lies in the region between residues 48 and 115 (numbering starting at the first predicted methionine), which correspond to the highly conserved J-domain (Silver and Way, 1993). A glycine-rich region common to many DnaJ-like proteins is found between residues 122 and 159 and has been proposed to separate functional domains (Silver and Way, 1993) or to be necessary for activation of DnaJ substrate-binding properties (Wall *et al.*, 1995). Towards the C-terminus Scj1p contains two sets of four cysteine residues that could form a structure resembling a zinc finger.

Relative to *E. coli* DnaJ, *SCJ1* could encode a protein with an N-terminal extension of at least 45 amino acids preceeding the DnaJ region (Blumberg and Silver, 1991). Amino acids 1–25 are rich in basic and hydroxylated residues that could form an amphiphilic α helix similar to that found in mitochondrial targeting sequences. A second form of Scj1p could also be encoded by translation from a second in frame ATG. This protein would contain an N-terminal extension relative to *E. coli* DnaJ with a basic residue at the fourth amino acid followed by 16 hydrophobic amino acids, making it similar to a signal sequence for transit into the ER. The potential ER-retention sequence KDEL appears at the C-terminus of Scj1p (Blumberg and Silver, 1991), although HDEL is preferred in *S. cerevisiae* (Semenza *et al.*, 1990).

■ Scj1 protein

Scj1p is a 40 kDa member of the growing HSP40/DnaJ protein family. Immunoblot analysis with anti-Scj1p antibody originally revealed one major and one minor cross-reacting protein species, both close to the molecular mass predicted from the *SCJ1* sequence (Blumberg and Silver, 1991). Yeast cells lacking the *SCJ1* gene are viable and show no marked growth defects indicating that *SCJ1* is not an essential gene for normal cell growth.

In initial cell fractionation experiments, Scj1p remained in a membrane pellet, cofractionated to some extent with mitochondria and was resistant to protease digestion unless the membranes were further permeabilized with detergent (Blumberg and Silver, 1991). Subsequent, more extensive fractionation studies indicated that the primary location for Scj1p is in the ER (Schlenstedt *et al.*, 1995). Further evidence for an ER location of Scj1p comes from the following results (Schlenstedt *et al.*, 1995): (i) Scj1p lacks sites for N-linked glycosylation. However, analysis of a mutated version that contains a consensus site for glycosylation indicated that most, if not all, of Scj1p must gain access to the ER lumen. (ii) Scj1p produced by translation *in vitro* migrates at a slightly higher size than wild-type Scj1p. In the presence of dog pancreas microsomes, the *in vitro* synthesized Scj1p is processed and incorporated into the microsomes. (iii) A higher molecular weight form of Scj1p accumulates in yeast *kar2* and *sec61* mutants that are thermosensitive for protein translocation across the ER.

■ SCJ1 gene regulation

Under normal growth conditions, Scj1p is produced constitutively and corresponds to about 0.1% of total cellular protein. There is no evidence for increased synthesis in response to heat shock. However, tunicamycin treatment stimulates the production of Scj1p approximately four-fold (Schlenstedt *et al.*, 1995). This is consistent with a sequence in the *SCJ1* promoter that has similarity to an unfolded protein response element (UPRE) found in the *KAR2* (BiP) promoter region (Mori *et al.*, 1992).

■ Function of Scj1p

Clues to the function of Scj1p come primarily from studies of other DnaJ homologues (reviewed in Kurihara and Silver, 1992 and in the introductory overview of this section, p. 89). In general, DnaJ and DnaJ-like proteins function with a corresponding DnaK/HSP70 protein in ATP-dependent protein folding and assembly reactions. DnaJ is thought to be necessary to stimulate the ATPase activity of DnaK and/or to target DnaK to its proper substrate. The location of Scj1p in the ER lumen suggests that it may function with Kar2p/BiP, the ER luminal HSP70 (Normington *et al.*, 1989; Rose *et al.*, 1989). Support for this proposal comes thus far only from genetic interactions (Schlenstedt *et al.*, 1995). The region common to all DnaJ homologues (the J-domain) from Scj1p can be swapped for a similar region in Sec63p, an ER transmembrane protein which is known to interact with Kar2p (Brodsky and Schekman, 1993; Scidmore *et al.*, 1993) suggesting that Kar2p interacts with two different DnaJ-like proteins. If so, the function of Scj1p may be to catalyse repeated rounds of ATP hydrolysis by BiP as it interacts with unfolded proteins emerging from the ER translocation machinery.

■ References

Blumberg, H. and Silver, P. (1991). SCJ1, a DNAJ homologue that alters protein sorting in yeast. Nature **349**, 627–630.

Brodsky, J.L. and Schekman, R. (1993). A sec63–BiP complex from yeast is required for protein translocation in a reconstituted proteoliposome. J. Cell Biol. **123**, 1355–1363.

Kurihara, T. and Silver, P. (1992). DnaJ homologues and protein transport. In *Membrane Biogenesis and Protein Targeting*, (W. Neupert and R. Lill, eds). Elsevier Science Publishers, Amsterdam, pp. 309–327.

Mori, K., Sant, A., Kohno, K., Normington, K., Gething, M.J., and Sambrook, J.F. (1992). A 22 bp cis-acting element is necessary and sufficient for the induction of the yeast KAR2 (BiP) gene by unfolded proteins. EMBO J. **11**, 2583–2593.

Normington, K., Kohno, K., Kozutsumi, Y., Gething, M.J., and Sambrook, J. (1989). S. cerevisiae encodes an essential protein homologous in sequence and function to mammalian BiP. Cell **57**, 1223–1236.

Rose, M.D., Misra, L.M., and Vogel, J.P. (1989). KAR2, a karyogamy gene, is the yeast homologue of the mammalian BiP/GRP78 gene. Cell **57**, 1211–1221.

Schlenstedt, G., Harris, S., Risse, B., Lill, R., and Silver, P.A. (1995). A yeast DnaJ homologue, Scj1p, can function in the endoplasmic reticulum with BiP/Kar2p via a conserved domain that specifies interactions with Hsp70s. J. Cell Biol. **129**, 979–988.

Scidmore, M.A., Okamura, H.H., and Rose, M.D. (1993). Genetic interactions between KAR2 and SEC63, encoding eukaryotic homologues of DnaK and DnaJ in the endoplasmic reticulum. Mol. Biol. Cell **4**, 1145–1159.

Semenza, J.C., Hardwick, K.G., Dean, N., and Pelham, H.R. (1990). ERD2, a yeast gene required for receptor-mediated retrieval of luminal ER proteins from the secretory pathway. Cell **61**, 1349–1357.

Silver, P.A. and Way, J.C. (1993). Eukaryotic DnaJ homologues and the specificity of Hsp70 activity. Cell **74**, 5–6.

Wall, D., Zylicz, M., and Georgopoulos, C. (1995). The conserved G/F motif of the DnaJ chaperone is necessary for the activation of the substrate binding properties of the DnaK chaperone. J. Biol. Chem. **270**, 2139–2144.

■ Pamela A. Silver:
Department of Biological Chemistry and Molecular Biology
Harvard Medical School and Dana Farber Cancer Institute
44 Binney Street
Boston, MA 02115, USA
Tel. 1 617 632 5102
Fax. 1 617 632 5103
E-mail: pamela_silver@macmailgw.dfci.harvard.edu

Plant Anj1

Anj1 is a member of the DnaJ family of molecular chaperones in higher plants. It is a functional homolog of the yeast cytoplasmic DnaJ-related protein, Ydj1. Anj1 is a membrane-associated protein due to farnesylation near its C-terminus. Anj1 mRNA is strongly induced by heat and salt stresses.

■ ANJ1 gene and sequence

The nucleotide sequence of *ANJ1* cDNA cloned from the halophytic higher plant *Atriplex nummularia* (GenBank accession number L09124) predicts a polypeptide of 417 amino acids without signal sequence or transmembrane domain (Zhu *et al.*, 1993a). At the N-terminus is a J-domain of approximately 70 amino acids which is conserved in all DnaJ-related proteins from both prokaryotes and eukaryotes (Caplan *et al.*, 1993; Zhu *et al.*, 1993a). J-domains are known to interact with HSP70 homologs (Schlenstedt *et al.*, 1995). A glycine-rich motif (or G/F domain) (Wall *et al.*, 1995) connects the J-domain to a putative 'zinc finger' which consists of four CXXCXGXG repeats (Caplan *et al.*, 1993; Zhu *et al.*, 1993a). The C-terminal half of Anj1 has less homology with other DnaJ family members which may reflect its potential role as a substrate-binding domain (Zhu *et al.*, 1993a). The C-terminal CAQQ sequence conforms to the CXXX motif for isoprenylation (Zhu *et al.*, 1993a). Homologs of Anj1 from other higher plant species are more than 80% identical to Anj1 (Bessoule, 1993).

■ Anj1 protein

Although the predicted molecular mass of Anj1 is 46.6 kDa, based on the length of its open reading frame, it has an apparent molecular mass of 51 kDa on SDS–PAGE (Zhu *et al.*, 1993b). Anj1 is associated with cellular membrane(s), however, the specific membrane(s) it associates with is not known (Zhu *et al.*, 1993b). Its membrane association is mediated through isoprenylation at cysteine-414 near the C-terminus (Zhu *et al.*, 1993b). *In vitro*, farnesylation dominates the post-translational modification of Anj1, but a low level of geranylgeranylation can also occur (Zhu *et al.*, 1993b). The ratio between farnesylation and geranylgeranylation is not known *in vivo*. The isoprenylation of Anj1 is required for its function (Zhu *et al.*, 1993b). Nothing is known about the interaction between Anj1 and its potential substrates or other molecular chaperones such as HSP70 proteins.

■ *ANJ1* gene regulation

ANJ1 mRNA is expressed constitutively in the absence of heat stress (Zhu *et al*., 1993a). The expression changes during the cell cycle and plant development, which may suggest a role for Anj1 in plant growth and development (Zhu *et al*., 1993a). Upon heat stress, *ANJ1* mRNA is dramatically induced as early as 10 minutes after heat exposure (Zhu *et al*., 1993a). Heat induction of *ANJ1* mRNA parallels that of *HSP70* (Zhu *et al*., 1993a). *ANJ1* mRNA is also induced by salt stress which does not affect *HSP70* mRNA accumulation (Zhu *et al*., 1993a). Conditions such as heat shock that induce *ANJ1* mRNA may not induce the Anj1 protein (Zhu *et al*., 1993b).

■ Function of Anj1

Anj1 appears to function as a molecular chaperone. It can suppress the temperature-sensitive phenotype of a *mas5/ydj1* mutant of yeast (Atencio and Yaffe, 1992; Caplan *et al*., 1992) which is defective in protein translocations into mitochondria and endoplasmic reticula (Zhu *et al*., 1993a). The C-terminal isoprenylation of Anj1 is essential for its function (Zhu *et al*., 1993b).

■ References

Atencio, D.P. and Yaffe, M.P. (1992). MAS5, a yeast homolg of dnaJ involved in mitochondrial protein import. Mol. Cell. Biol. **12**, 283–291.

Bessoule, J.J. (1993). Occurrence and sequence of a DnaJ protein in plant (Allium porrum) epidermal cells. FEBS Lett. **323**, 51–54.

Caplan, A.J., Cyr, D.M., and Douglas, M.G. (1992). YDJ1p facilitates polypeptide translocation across different intracellular membranes by a conserved mechanism. Cell **71**, 1143–1155.

Caplan, A.J., Cyr, D.M. and Douglas, M.G. (1993). Eukaryotic homologues of Escherichia coli dnaJ: a diverse protein family that functions with HSP70 stress proteins. Mol. Biol. Cell **4**, 555–563.

Schlenstedt, G., Harris, S., Risse, B., Lill, R., and Silver, P.A. (1995). A yeast DnaJ homologue, Scj1p, can function in the endoplasmic reticulum with BiP/Kar2p via a conserved domain that specifies interactions with Hsp70s. J. Cell Biol. **129**, 979–988.

Wall, D., Zylicz, M., and Georgopoulos, C. (1995). The conserved G/F motif of the DnaJ chaperone is necessary for the activation of the substrate binding properties of the DnaK chaperone. J. Biol. Chem. **270**, 2139–2144.

Zhu, J.-K., Shi, J., Bressan, R.A., and Hasegawa, P.M. (1993a). Expression of an Atriplex nummularia gene encoding a protein homologous to the bacterial molecular chaperone DnaJ. Plant Cell **5**, 341–349.

Zhu, J.-K., Bressan, R.A. and Hasegawa, P.M. (1993b). Isoprenylation of the plant molecular chaperone ANJ1 facilitates membrane association and function at high temperature. Proc. Natl. Acad. Sci. USA **90**, 8557–8561.

■ *Jian-Kang Zhu:*
Department of Plant Sciences
University of Arizona
Tucson, AZ 85721, USA
Tel. 1 520 621 9567
Fax. 1 520 621 7186
E-mail: jkzhu@ag.arizona.edu

DnaJ-like proteins from leek: Ldj1 and Ldj2

DnaJ proteins are involved with DnaK, GrpE and GroEL and their homologs in the folding and the assembly of proteins in the cell. Ldj are DnaJ proteins from leek epidermal cells. Except in this tissue, no report, to our knowledge, in the literature has shown the presence of two isoforms of a Dna J protein in the same plant cell.

■ Ldj sequences

Four different Dna J proteins have been reported in the yeast *Saccharomyces cerevisiae* (for review see Silver and Way, 1993); in plant cells, only proteins homologous to one of them (Ydj1) have been described: Anj1 (see previous entry, p. 112), LdJ1, Ldj2, Csdna J1 and Atj2 from *Atriplex nummularia*, *Allium porrum*, *Cucumis sativus* and *Arabidopsis thaliana*, respectively (Bessoule, 1993; Preisig-Muller and Kindl, 1993; Zhu *et al*., 1993a; Bessoule *et al*., 1994; Zhou *et al*., 1995). The presence of two isoforms of a Dna J protein in the same plant cell has only been reported for leek epidermal cells.

The amino acid sequence deduced from the nucleotide sequence (EMBL data accession number X69436) of a cDNA isolated from leek epidermal cells predicts a protein (Ldj1) with the N-terminal 52 residues sharing homology with the consensus sequence ('J-region') of the previously sequenced DnaJ proteins (Bessoule, 1993). Between residues 77 and 87 lies a Gly/Phe-rich region that has also been described in several DnaJ proteins. Ldj1 also contains four Cys–X–X–Cys–X–Gly–X–Gly units organized into two series of direct repeats of the sequence Cys–X–X–Cys–X–Gly–X–Gly–(X)$_8$–Cys–X–X–Cys–X–Gly–X–Gly. Furthermore, the C-terminal sequence of Ldj1 is VQCAQQ, which is very similar to the C-terminal sequence of Ydj1p: VQCASQ. Such a sequence corresponds to a farnesylation sites in Ydj1p and Anj1 (Caplan et al., 1992; Zhu et al., 1993b).

The amino acid sequence of Ldj2 (EMBL data accession number X77632) does not differ greatly from the sequence of Ldj1. The isolated cDNA contains 1511 bp and encodes 418 amino acids (Bessoule et al., 1994). The Cys–X–X–Cys–X–Gly–X–Gly units and the potential farnesylation site (CAQQ) remain unchanged. The sequence of the 'J-region' of Ldj2 (first 73 amino acids) only differs from that of Ldj1 by five amino acids, and four of them are found in positions where there is no identity between the sequences of Ydj1, Ldj1, Anj1 and CsdJ1. The four substitutions are conservative between Ldj1 and Ldj2. Only one of the five amino acids, a serine found 14 amino acids before the end of the 'J-region' of Ldj1 was also present in AnJ1, CsdJ1 and YdJ1, but is replaced by an asparagine in Ldj2. An asparagine is also found at this position in Sis1p, another yeast DnaJ protein (Luke at al., 1991, see entry p. 104). Ldj2 also contains four Cys–X–X–Cys–X–Gly–X–Gly units organized in two series of direct repeats of the sequence Cys–X–X–Cys–X–Gly–X–Gly–(X)$_8$–Cys–X–X–Cys–X–Gly–X–Gly) and the C-terminal sequence is VQCAQQ.

■ Ldj proteins

The Cys–X–X–Cys–X–Gly–X–Gly units present in Ldj proteins are similar to those described in the family of zinc finger proteins (Caplan and Douglas, 1991). However, as noted by Preisig-Muller and Kindl (1993) 'they do not conform to the exact consensus sequence commonly found in zinc finger DNA-binding proteins', which also contains histidine (Berg, 1990). Nevertheless, it has been shown that a Cys–X–X–Cys–(X)$_{13}$–Cys–X–X–Cys unit present in a glucocorticoid receptor (similar to the Cys–X–X–Cys–(X)$_{12}$–Cys–X–X–Cys units found in the DnaJ proteins) is unambiguously a zinc finger domain (Hard et al., 1990).

Prenylation, which occurs on the C-terminal cysteine residue, is required for Ydj1p to function at elevated temperature and to promote the attachment of proteins to the membranes (Caplan et al., 1992). Since the C-terminal sequences of Ldj1 and Ldj2 are very similar to the C-terminal sequence of Ydj1, and since the C-terminal amino acid is a glutamine residue, it is possible that Ldj proteins are prenylated and that the latter prenyl group is a farnesyl isoprenoid, rather than a geranylgeranyl group. In any case, Ldj1 and Ldj2 could be membrane-bound proteins despite the absence of large hydrophobic regions.

■ Biological interactions of Ldj proteins

As for other DnaJ-like proteins, Ldj could be inferred to function with homologs of DnaK, GrpE and GroEL in the folding and assembly of proteins of the cell. It has been proposed that DnaJ and its homologs 'protect nascent polypeptide chains against aggregation, and in cooperation with DnaK or other partner HSP70 proteins, control their productive folding' (Hendrick et al., 1993, and see the introductory overview to this section, p. 89). DnaJ interacts directly with DnaK stimulating its rate of ATP hydrolysis (Liberek et al., 1991; Langer et al., 1992). This phenomenon correlates with an increase in the affinity of DnaK for some proteins as luciferase, rhodanese (see Wall et al., 1995). In addition, DnaJ is also able to activate DnaK to bind a transcription factor involved in the heat shock response (Liberek et al., 1995). No partner HSP70 protein has yet been identified for Ldj1 or Ldj2.

■ References

Berg, J.M. (1990). Zinc fingers and other metal-binding domains. J. Biol. Chem., **265**, 6513–6516.

Bessoule, J-J. (1993). Occurrence and sequence of a DnaJ protein in plant (Allium porrum) epidermal cells. FEBS Lett., **323**, 51–54.

Bessoule, J.-J., Testet, E., and Cassagne, C. (1994). Cloning of a new isoform of a DnaJ protein from *Allium porrum* epidermal cells. Plant Physiol. Biochem., **32**, 723–727.

Caplan, A.V. and Douglas, M.G. (1991). Characterization of YDJ1: a yeast homologue of the bacterial DnaJ protein. J. Cell Biol., **114**, 609–621

Caplan, A.V., Tsai, J., Casey, P.J., and Douglas, M.G. (1992). Farnesylation of YDJ1P is required for function at elevated growth temperature in *Saccharomyces. cerevisiae*. J. Biol. Chem., **267**, 18890–18895.

Hard, T., Kellenbach, E., Boelens, R., Maler, B.A., Dahlman, K., Freedman, L.P., Carlstedt-Duke, J., Yamamoto, K.R., Gustafsson, J.-A., and Kaptein, R. (1990). Solution structure of the glucocorticoid receptor DNA-binding domain. Science, **249**, 157–160.

Hendrick, J.P., Langer, T., Davis, T.A., Hartl, F.U., and Wiedmann, M. (1993). Control of folding and membrane translocation by binding of the chaperone Dna J to nascent polypeptides. Proc. Natl. Acad. Sci. USA, **90**, 10216–10220.

Langer, T., Lu, C., Echols, H., Flanagan, J., Hayer, M.K., and Hartl, F.U. (1992). Successive action of DnaK, DnaJ and GroEL along the pathway of chaperone-mediated protein folding. Nature, **356**, 683–689.

Liberek, K., Marzalek, J., Ang, D., Georgopoulos, C., and Zylicz, M. (1991). Escherishia Coli DnaJ and GrpE heat shosk proteins jointly stimulate ATPase activity of DnaK. Proc. Natl. Acad. Sci. USA, **88**, 2874–2878.

Liberek, K, Wall, D., and Georgopoulos, C. (1995). The DnaJ chaperone catalytically activates the DnaK chaperone to preferentially bind the σ^{32} heat shock transcriptional regulator. Proc. Natl. Acad. Sci. USA, **92**, 6224–6228.

Luke, M.M., Sutton, A. and Arndt, K.T. (1991). Characterization of SIS1: a *Saccharomyces cerevisiae* homologue of bacterial dnaJ proteins. J. Cell Biol., **114**, 623–638.

Preisig-Muller, R. and Kindl, H. (1993). Plant dnaJ homologue: molecular cloning, bacterial expression, and expression analysis in tissues of Cucumber seedlings. Arch. Biochem. Biophys., **305**, 30–37.

Silver, P.A. and Way, J.C. (1993). Eucaryotic DnaJ homologs and the specificity of Hsp70 activity. Cell, **74**, 5–6

Wall, D., Zylicz, M., and Georgopoulos, C. (1995). The conserved G/F motif of the DnaJ chaperone is necessary for the activation of the substrate binding properties of the DnaK chaperone. J. Biol. Chem., **270**, 2139–2144.

Zhou, R., Kroczynska, B., Hayman, G.T., and Miernyk, J.A. (1995). Atj2, an Arabiodopsis homolog of *Escherichia coli* dnaj. Plant Physiol., **108**, 821–822.

Zhu, J.K., Shi, J., Bressan, R.A., and Hasegawa, P.M. (1993a). Expression of an Atriplex nummularia gene encoding a protein homologous to the bacterial molecular chaperone DnaJ. Plant Cell, **5**, 341–349.

Zhu, J.K., Bressan, R.A. and Hasegawa, P.M. (1993b). Isoprenylation of the plant molecular chaperone ANJ1 facilitates membrane association and function at high temperature. Proc. Natl. Acad. Sci. USA, **90**, 8557–8561.

■ *J.-J. Bessoule:*
UMR-CNRS 5544,
Université Victor Ségalen Bordeaux II, Case 92,
146 Rue Léo Saignat,
33076 Bordeaux Cédex, France
Tel. 33 557 571 274
Fax. 33 556 518 361
E-mail: Jean-Jacques.Bessoule@biomemb.
u-bordeaux2.fr

Cysteine-string proteins

Cysteine-string proteins (csps), originally identified as synaptic antigens in Drosophila, are small lipidated components of synaptic vesicles. The csp proteins have been identified from invertebrates to mammals, where they exhibit their characteristic cysteine-string domain and the 'J-domain' that defines them as distantly related paralogs to the DnaJ proteins. The phenotypic analysis of Dcsp mutants demonstrates a crucial role for csps in the process of evoked transmitter release. Furthermore, it has been suggested that csp proteins may link synaptic vesicles to presynaptic calcium channels and modulate channel activity.

■ Alternative names

Omega-conotoxin receptor or CCCS1 for candidate calcium channel subunit 1 (Mastrogiacomo et al., 1994b), now denoted Tcsp.

■ *csp* sequence

The *Drosophila csp* gene was isolated using the antibody MAb DCSP-1 (former Mab ab49) (Zinsmaier et al., 1990, 1994). The gene (Flybase gene id #Fbgn0004179) is located at 79E1-2 on the third chromosome and expresses three alternatively spliced RNA (2.3–2.9 kb) transcripts (Zinsmaier et al., 1994). The amino acid sequence deduced from the nucleotide sequence of two cDNA clones isolated from a *Drosophila* head cDNA library (GenBank accession numbers M63421 and M63008; Zinsmaier et al., 1990) reveals proteins of 249 and 223 amino acids (designated Dcsp-32 and Dcsp29a). A third cDNA (Zinsmaier et al., 1994) has been isolated with a deduced protein sequence of 228 amino acids (designated Dcsp29b; K.E. Zinsmaier, unpublished results). The *Drosophila* gene appears to be exclusively expressed in neurones (Zinsmaier et al., 1990, 1994). The predicted molecular weights of these proline-rich (8%) and cysteine-rich (6%) proteins ranges from 26 896 to 23 849, although the molecular masses deduced from their electrophoretic mobility during electrophoresis on SDS–PAGE of the native proteins are significantly higher (32–36 kDa). The N-terminal first third of the Dcsp proteins contains the highly conserved 'J-domain' (40–61%) of the DnaJ proteins (Kurihara and Silver, 1992). Just after the 'J-domain' the cysteine-string motif is located in the middle of the protein; it comprises a string of 11 cysteines flanked by a pair of cysteines on each side (C2X5C11X2C2). The C-terminal sequences of the Dcsp proteins contain two variable regions owing to alternative splicing. The Dcsp-32 protein shows a 21 amino acid insertion shortly after the cysteine-string, and Dcsp29a and Dcsp29b differ at the extreme C-terminal by seven amino acids which are identical between Dcsp32 and Dcsp29b.

Homologous genes have been identified in *Torpedo californica* (Gundersen and Umbach, 1992) and in *Rattus norvegicus* (Mastrogiacomo and Gundersen, 1995). Three Tcsp mRNAs of 7.4, 3, and 1.3 kb are reported from *Torpedo* electric lobe (Gundersen and Umbach, 1992). The cDNA corresponding to the largest mRNA has been sequenced and the deduced protein sequence codes for a 195 amino acid protein with a predicted molecular mass of 21 791 daltons (GenBank accession number M99327; NCBI gi: 213227). The Tcsp protein is, overall, 69–70% identical to the Dcsp proteins. The rat *csp* gene encoding Rcsp has been cloned using affinity purified antibodies against the C-terminal undecapeptide of Tcsp (Mastrogiacomo et al., 1994). The rat gene expresses two RNAs of 5 and 1.3 kb in adult brain. A 2.7 kb cDNA encoding Rcsp has been cloned (GenBank accession number S77549) and the predicted amino acid sequence contains 198 residues yielding a mass of 22 131 dalton. Overall Rcsp is 82% identical to the fish Tcsp protein and 54% identical to the fly Dcsp proteins. In general, sequence conservation is highest in the N-terminal portion, specifically the J-domain and the cysteine-string motif, and lowest at the C-terminus.

Csp proteins

The csp proteins form a novel class of vesicle-associated proteins that are highly conserved from flies to mammals. The *Dcsp* gene of *Drosophila* expresses three to five protein isoforms, all of which are exclusively localized to synaptic terminals of neurones including neuromuscular junctions (Zinsmaier et al., 1994). Dcsp protein expression starts before synaptogenesis in 11–13 hour old embryos and is evident in all neural tissues throughout all life stages. Further on in development, Dcsp proteins are differentially expressed at distinct synaptic terminals of the *Drosophila* brain. In contrast, only one 34 kDa csp protein has been identified in the electric organ of *Torpedo*, and no immunoreactivity was observed in liver (Mastrogiacomo et al., 1994a). In rat, two csp immuno-positive proteins have been identified; a 35 kDa protein appears to be the major form, whereas a 72 kDa species shows only weak immunoreactivity (Kohan et al., 1995; Mastrogiacomo and Gundersen, 1995). Immunohistochemistry showed that Rcsp is ubiquitously distributed in the rat nervous system and is abundant in synapse-rich regions, similar to *Drosophila* csp. Rcsp-expressing tissues include those of the retina, main olfactory bulb, hippocampal formation, and cerebellum. Immunoreactivity was also detected in the adrenal gland of rat suggesting an additional non-neuronal function of csp (Kohan et al., 1995).

The csp proteins are ultrastrucurally localized to synaptic vesicle membranes. In *Torpedo* and rat, csp proteins copurify with membrane-associated synaptic vesicle proteins, like SV2 (Kohan et al., 1995; Mastrogiacomo and Gundersen, 1995). Similar observations were made in *Drosophila* where Dcsp is present in slowly sedimenting membrane fractions (Littleton et al., 1993; Zinsmaier et al., 1994; Schulze et al., 1995). The identity of synaptic vesicles in these fractions has been confirmed and the Dcsp protein association with synaptic vesicles has been legitimized (Van de Goor et al., 1995).

A pervasive question has concerned the role of the highly conserved cysteine-string domain. The discovery that a large majority of the cysteine residues are fatty acylated has significant implications for csp protein structure and function (Gundersen et al., 1994). At least 11 of the 13 cysteine residues of the Tcsp are modified mostly by palmitoyl moieties. In the case of Rcsp and Dcsp, evidence for a similar extensive lipidation of cysteine residues has been observed; deacylation of the 35 kDa Rcsp protein reduces the electrophoretic mobility by 7–8 kDa (Mastrogiacomo and Gundersen, 1995). A similar reduction of the electrophoretic mobility of Dcsp after deacylation is observed in *Drosophila* (K. Eberle and K.E. Zinsmaier, unpublished work).

Function of csp proteins

The csp proteins appear to function as synaptic vesicle-associated proteins. The membrane association is caused by an extensive lipidation of the cysteine-string motif exposing both the N- and C-termini into the cytoplasm (C. Gundersen, unpublished data; K.E. Zinsmaier, unpublished data). Although the true function of the proteins still remains a mystery, some insights were gained from Tcsp RNA expression in frog oocytes and the phenotypical analysis of Dcsp mutants. The inhibition of N-type calcium channel activity by coinjection of Tcsp anti-sense mRNA with native *Torpedo* RNA into frog oocytes suggested a role for the Tcsp protein as a modulatory subunit of N-type calcium channels (Gundersen and Umbach, 1992). The phenotypic analysis of Dcsp gene deletions revealed temperature-sensitive (ts) paralysis and premature death (Zinsmaier et al., 1994), a very surprising phenotype for deletions. Lethality is first observed in structurally normally developed early larvae, which fail to hatch from the embryonic shell. At lower temperature some mutant flies reach adulthood, but early death is preceded by increasingly sluggish behavior and uncoordinated locomotion. The ts behavior of Dcsp mutants correlates with defects of synaptic transmission (Umbach et al., 1994). At permissive temperatures, evoked synaptic transmission is impaired; at high temperatures it is completely blocked. However, spontaneous release events persist. The electrophysiological analysis indicates that mutations of the *csp* gene impair synaptic vesicle exocytosis, probably by disrupting the coupling of nerve impulse activity and vesicle fusion (Gundersen and Umbach, 1992; Umbach et al., 1994).

As already described, the Dcsp proteins contain a 'J-domain' which has been shown to mediate modulatory DnaJ–HSP70 protein interactions (Kurihara and Silver, 1992; Caplan et al., 1993; Silver and Way, 1993). The high conservation of the 'J-domain' suggests HSP70-like proteins as potential ligands of the csp proteins. In this context, it is noteworthy that the entire deletion of *dnaJ* in *E. coli*, a distantly related paralog of *csp* (Caplan et al., 1993), exhibits a ts phenotype (Ohki et al., 1992). Recently, it has been shown that csp specifically activates the Hsc70 ATPase from bovine brain (Brain et al., 1996) suggesting that csp resembles a cofactor for a not yet identified Hsc70 activity in evoked exocytosis.

References

Braun, J.E.A., Wilbanks, S.M., and Scheller, R.H. (1996). The cysteine string secretory vesicle protein activates Hsc70 ATPase. J. Biol. Chem. **271**, 25989–25993.

Caplan, A.J., Cyr, D.M., and Douglas, M.G. (1993). Eukaryotic homologues of Escherichia coli dnaJ: a diverse protein family that functions with HSP70 stress proteins. Mol. Biol. Cell **4**, 555–563.

Gundersen, C.B. and Umbach, J.A. (1992). Suppression cloning of the cDNA for a candidate subunit of a presynaptic calcium channel. Neuron **9**, 527–537.

Gundersen, C.B., Mastrogiacomo, A., Faull, K., and Umbach, J.A. (1994). Extensive lipidation of a torpedo cysteine-string protein. J. Biol. Chem. **269**, 19197–19199.

Kohan, S.A., Pescatori, M., Brecha, N.C., Mastrogiacomo, A., Umbach, J.A., and Gundersen, C.B. (1995). Cysteine string protein immunoreactivity in the nervous system and adrenal gland of rat. J. Neurosci. **15**, 6230–6238.

Kurihara, T. and Silver, P. (1992). DnaJ homologs and protein transport, in Membrane Biogenesis and Protein Targeting, (W.N.a.R. Lill, ed.). Elsevier, New York, pp. 309–327.

Littleton, J.T., Bellen, H.J., and Perin, M.S. (1993). Expression of synaptotagmin in Drosophila reveals transport and localization of synaptic vesicles to the synapse. Development **118**, 1077–1088.

Mastrogiacomo, A. and Gundersen, C.B. (1995). The nucleotide and deduced amino acid sequence of a rat cysteine string protein. Mol. Brain Res. **28**, 12–18.

Mastrogiacomo, A., Evans, C.E., and Gundersen, C.B (1994a). Antipeptide antibodies against a Torpedo cysteine-string protein. J. Neurochem. **62**, 873–880.

Mastrogiacomo, A., Parsons, S.M., Zampighi, G.A., Jenden, D.J., Umbach, J.A., and Gundersen, C.B. (1994b). Cysteine string proteins — a potential link between synaptic vesicles and presynaptic Ca^{2+} channels. Science **263**, 981–982.

Ohki, R., Kawamata, T., Katoh, Y., Hosoda, F., and Ohki, M. (1992). Escherichia coli DnaJ deletion mutation results in loss of stability of a positive regulator, CRP. J. Biol. Chem. **267**, 13180–13184.

Schulze, K.L., Broadie, K., Perin, M.S., and Bellen, H. (1995). Genetic and electrophysiological studies of Drosophila syntaxin-1A demonstrate its role in nonneuronal secretion and neurotransmission. Cell **80**, 311–320.

Silver, P.A. and Way, J.W. (1993). Eukaryotic DnaJ homologs and the specificity of Hsp70 activity. Cell **74**, 5–6.

Umbach, J.A., Zinsmaier, K.E., Eberle, K.K., Buchner, E., Benzer, S., and Gundersen, C.B. (1994). Presynaptic dysfunction in Drosophila csp mutants. Neuron **13**, 899–907.

Van de Goor, J., Ramaswami, M., and Kelly, R. (1995). Redistribution of synaptic vesicles and their proteins in temperature-sensitive shibirets1 mutant Drosophila. Proc. Natl. Acad. Sci. USA **92**, 5739–5743.

Zinsmaier, K.E., Hofbauer, A., Heimbeck, G., Pflugfelder, G.O., Buchner, S., and Buchner, E. (1990). A cysteine-string protein is expressed in retina and brain of Drosophila. J. Neurogenet. **7**, 15–29.

Zinsmaier, K.E., Eberle, K.K., Buchner, E., Walter, N., and Benzer, S. (1994). Paralysis and early death in cysteine-string protein mutants of Drosophila. Science **263**, 977–980.

■ Konrad E. Zinsmaier:
University of Pennsylvania School of Medicine
Department of Neuroscience
232a Stemmler Hall
Pennsylvania, PA 19104-6074, USA
Tel. 1 215 898 2576
Fax. 1 215 573 2015
E-mail: zinsmaie@mail.med.upenn.edu

Drosophila melanogaster Tid56

Tid56 is the precursor of the mitochondrial protein Tid50 encoded by the Drosophila melanogaster *tumour suppressor gene* lethal(2) tumorous imaginal discs [l(2)tid]. *It shows significant homology to all known DnaJ-related proteins from bacteria, yeast and humans.*

■ Isolation

The starting material for the identification of the *Drosophila melangaster* tumour suppressor gene *lethal(2) tumorous imaginal discs* [*l(2)tid*] was a mutant, *l(2)tid¹*, generated via EMS mutagenesis and characterized by neoplastic growth of imaginal discs (Kurzik-Dumke *et al.*, 1992). The *l(2)tid* gene was mapped genetically to the position 104 ± 1 cM on the right arm of the second chromosome (Kurzik-Dumke *et al.*, 1992). The cytogenetic locus was further determined by deletion mapping and an extensive chromosome walk encompassing the bands 59D–60A (Kurzik-Dumke *et al.*, 1992; Gundacker *et al.*, 1993) to a 20 kb genomic region (Kurzik-Dumke *et al.*, 1995) harbouring at least four nested genes (Kurzik-Dumke *et al.*, 1996). The identification of the *l(2)tid* gene was performed using molecular techniques such as: (i) genomic Southern analysis of wild-type and hetero- and homozygous animals from *l(2)tid* allelic strains; (ii) temporal transcript analysis of wild-type and *l(2)tid* allelic strains; (iii) *in situ* determination of *l(2)tid* transcripts in normal and tumorous tissues; and (iv) germ line transformation experiments (Kurzik-Dumke *et al.*, 1995; Kurzik-Dumke, 1996). *l(2)tid* tumorous tissue shows alterations in expression of *l(2)tid*-encoded RNA and protein (Kurzik-Dumke, 1996; U. Kurzik-Dumke, A. Debes, M. Kaymes and P. Dienes, manuscript submitted).

■ Mutagenesis studies

Recessive mutations at the *l(2)tid* locus cause different developmental abnormalities including neoplastic transformation of imaginal discs (Table 1). The first *l(2)tid* allele, ethylmethane sulfonate (EMS)-induced, is characterized by neoplastic growth of imaginal discs (Kurzik-Dumke *et al.*, 1992, 1995) which are the adult integumental primordia (Poodry, 1980). In contrast to the wild-type imaginal disc differentiating finally, during metamorphosis at the pupal stage, into their respective adult structures of the head, thorax, appendages, and genitalia, the mutant imaginal discs are incapable of differentiation and grow during the prolonged larval life into lethal tumours that finally kill the host. In addition

Table 1. Phenotype and mode of induction of eight *l(2)tid* alleles

Designation of allele	Mode of induction	Lethal phase	Reference
l(2)tid¹	EMS	LP	Kurzik Dumke et al., 1992; 1995; Kurzik-Dumke, 1996
l(2)tid²	X-ray	LP	—
l(2)tid³	spontaneous	LP	—
l(2)tid⁴	X-ray	PA	—
l(2)Az	EMS	E	Reed, 1992; Kurzik-Dumke, 1996
l(2)Ag-2	EMS	E	—
l(2)Ao-3	EMS	E	—
l(2)Ax-4	EMS	E	—

LP: imaginal discs tumours causing lethality during transition from larval to pupal; PA: pharate adults; E: embryonic lethality.

to *l(2)tid¹*, seven further allelic strains, generated either via EMS or X-ray mutagenesis and exhibiting different phenotype, were isolated (Table 1). The malignant state of the *l(2)tid* tumorous tissues was determined using an in vivo implantation technique (Kurzik-Dumke et al., 1992).

■ Tid56 gene and sequence

The amino acid sequence deduced from a cDNA clone isolated from a *Drosophila* embryonic λ-gt10 cDNA library and from the sequence of the corresponding genomic region (GenBank accession number X877822) reveals a putative protein of 518 amino acids with a high content of glycine (64) and alanine residues (49) (Kurzik-Dumke et al., 1995). The calculated molecular weight of the deduced Tid56 protein, 56 kDa, corresponds well with the size of the precursor protein detected in western analysis (U. Kurzik-Dumke, unpublished result). Tid56 is highly charged and basic with a theoretical isoelectric point of 9.5. It shows significant amino acid identity with all DnaJ homologues known to date (Kurzik-Dumke et al., 1995). Figure 1 shows the amino acid identity between the Tid56 protein and a few representative members of the DnaJ protein family established with the help of the FASTA-program (EMBL Databases, HUSAR software package). Further characteristics of the Tid56 protein are structural elements such as: (i) the N-terminal 'J-domain' defining it as a member of the DnaJ-like protein family (Caplan et al., 1993(ii) the glycine-rich spacer; (iii) four CXXCXGXG repeats; and (iv) the less conserved C-terminal part (Kurzik-Dume et al., 1995). The 'J-domain' of the Tid56 protein is preceded by a 60 amino acid leader resembling a mitochondrial presequence. (Cyr et al., 1994).

■ Intracellular localization of Tid56

To determine the cellular localization of the Tid56 protein, *Drosophilia* Schneider cells and embryonic cells derived from the wild-type strain *Oregon R* were separated by differential centrifugation and by continuous and discontinuous sucrose gradients ultracentrifugation, according to standard procedures. Each cellular fraction was characterized by biochemical methods using enzymes specific for different cellular compartments and investigated by Western blotting. This analysis revealed a nonsoluble mitochondrial protein of 50 kDa, Tid50, and a soluble protein of 56 kDa. Because of that and the fact mentioned above that the putative Tid56 protein is characterized by a mitochondrial leader sequence we assume that Tid50 represents the mature protein whereas Tid56 corresponds to the precursor (U. Kurzik-Dumke, A. Debes, M. Kaymer, and P. Dienes, submitted).

■ *l(2)tid* gene regulation

The *l(2)tid* gene is localized in an overlapping gene cluster consisting of three genes, the *l(2)tid* gene and the genes *lethal(2) neighbour of tid* [*l(2)not*] (Kurzik-Dumke et al., 1995, 1997; EMBL Data Library accession number X77820) and *lethal(2) relative of tid* [*l(2)rot*] (Kurzik-Dumke et al., 1997; EMBL Data Library accession number X77821) spanning a 7047 base-pair genomic HindIII fragment (Kurzik-Dumke et al., 1997; EMBL Data Library accession number X95241) (Fig. 2). Rescue of the *l(2)tid* mutant phenotype, performed via the germ line transformation experiment, could be achieved only with a construct including the three genes intact (Kurzik-Dumke et al., 1995; Kurzik-Dumke, 1996). This fact suggests a functional relationship between these genes. The present state of investigations does not allow us to speculate on the possible function of the putative *l(2)not* gene product, the Not56 protein, in the context of the functional relationship of the three nested genes. In the case of the *l(2)rot* gene, however, the idea of its function as a regulator of the *l(2)tid* expression via antisense RNA, established on the basis of the fact that *l(2)tid*- and *l(2)rot*-RNA are complementary over the genomic part corresponding to exon 1 of the *l(2)tid* gene, was confirmed by detection of stable *l(2)tid*–*l(2)rot*-RNA hybrids during early embryonic development. Furthermore, Western blot analysis of the total number of extracts

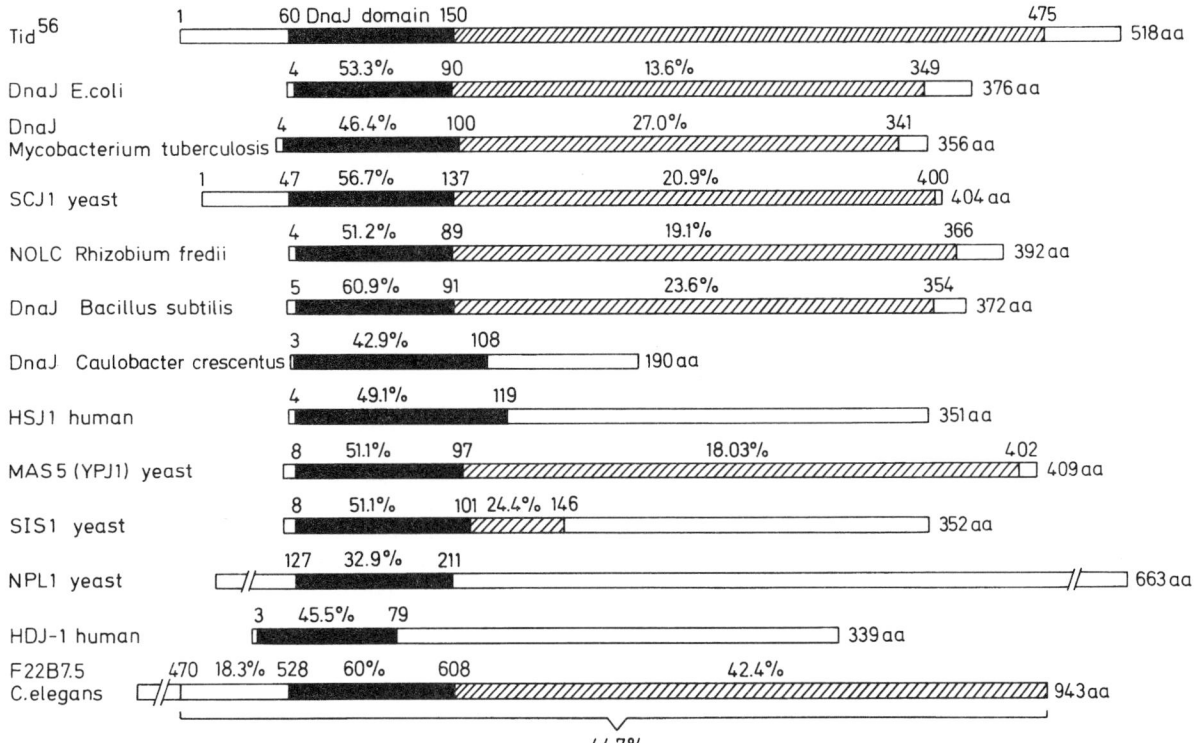

Figure 1. Comparison of twelve prokaryotic and eukaryotic DnaJ homologs in relation to the putative *Drosophila* Tid56 protein. The putative *Drosophila* Tid56 protein is compared with twelve prokaryotic (Bardwell *et al.*, 1986; Ohki *et al.*, 1986; Lathigra *et al.*, 1988; Rothblatt *et al.*, 1989; Gomez *et al.*, 1990; Wetzstein *et al.*, 1990; Krishnan and Pueppke, 1991) and eukaryotic (Sadler *et al.*, 1989; Blumberg and Silver, 1991; Caplan and Douglas, 1991; Luke *et al.*, 1991; Raabe and Manley, 1991; Atencio and Yaffe, 1992; Cheetham *et al.*, 1992; Wilson *et al.*, 1994) DnaJ homologues. The black bars represent the DnaJ-domain. Striped bars illustrate regions of lower homology and white bars non-homologous domains. The amino acid identities are indicated in percentages. The numbers above the regions represent the number of the amino acid in the sequence.

derived from all developmental stages, as well as staining of *Drosophila* Schneider cells with polyclonal anti-Rot antibody, revealed no detection of expression of the putative *l(2)rot* gene product, the Rot57 protein (U. Kurzik-Dumke *et al.*, 1997). This result implies either complete lack of translation of the *l(2)rot* gene or specific spartiotemporal expression that cannot be detected with the methods used.

DnaJ proteins are components of the HSP70 chaperone machinery and either belong to the heat shock proteins or are closely related to them (Silver and Way, 1993). The expression of the *l(2)tid* gene is definitely not heat inducible (Kurzik-Dumke *et al.*, 1996).

■ References

Atencio, D.P. and Yaffe, M.P. (1992). Mas5, a yeast homolog of DnaJ involved in mitochondrial protein import. Mol. Cell. Biol. **122**, 283–291.

Bardwell, J.C.A., Tily, K., Craig, E., King, J., Zylicz, M., and Georgopoulos, C. (1986). The nucleotide sequence of the *Escherichia coli* K12 *dnaJ+* gene: a gene that encodes a heat shock protein. J. Biol. Chem. **261**, 1782–1785.

Blumberg, H. and Silver, P.A. (1991). A homologue of the bacterial heat shock gene DnaJ that alters protein sorting in yeast. Nature **349**, 627–630.

Caplan, A.J. and Douglas, M.G. (1991). Characterization of YDJ1: a yeast homologue of the bacterial dnaJ protein. J. Cell Biol. **114**, 609–621.

Caplan, A.J., Cyr, D.M., and Douglas, M.C. (1993). Eukaryotic homologues of the *Escherichia coli* dnaJ: a diverse protein family that functions with HSP70 stress proteins. Mol. Biol. Cell **4**, 555–563.

Cyr, D.M., Langer, T., and Douglas, M.G. 1994. DnaJ-like proteins: molecular chaperones and specific regulators of Hsp70. TIBS **19**, 176–181.

Cheetham, M.E., Brion, J.-P., and Anderton, B.H. (1992). Human homologues of the bacterial heat-shock protein DnaJ are preferentially expressed in neurons. Biochem. J. **284**, 469–476.

Fleischer, S. and Kervina, M. (1994). Multiple fractions from a single tissue. In *Methods of enzymology*, (S.P. Colowick and N.O. Kaplan, eds). Academic Press, Inc., New York, vol. 31, pp. 1–41.

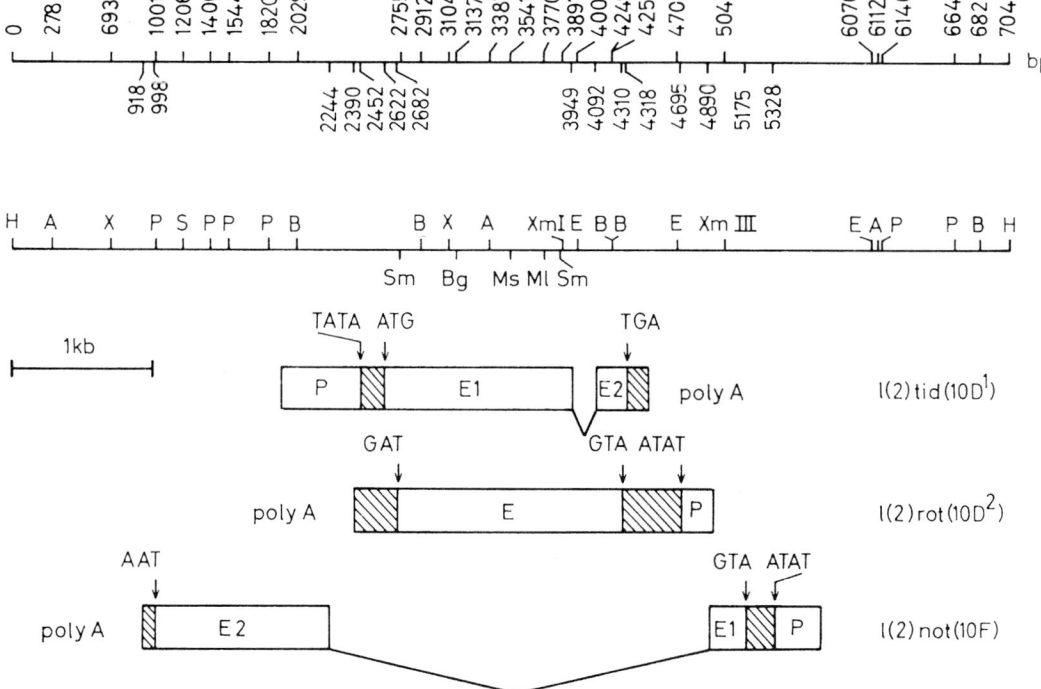

Figure 2. 'Gene within a gene' arrangement of the genes *l(2)tid*, *l(2)not*, and *l(2)rot.*. The genes *l(2)tid*, *l(2)not*, and *l(2)rot* are located in a 7047 base-pair genomic *Hin*dIII fragment. (B: *Bam*HI; Bg: *Bgl*II; E: *Eco*RI; H: *Hin*dIII; Ml: *Mlu*I; Ms: *Mst*I; P: *Pst*I; S: *Sal*I; Sm: *Sma*I; X: *Xho*I; XmI: *Xma*I; XmIII: *Xma*III).

Gomez, S.L., Goter, J.W., and Shapiro, L. (1990). Expression of the Caulobacter heat shock gene dnaK is developmentally controlled during growth at normal temperatures. J. Bacteriol. **172**, 3051–3059.

Gundacker, D., Phannavong, B., Vef, O., Gateff, E., and Kurzik-Dumke, U. (1993). Genetic and molecular characterization of breakpoints of five deficiencies in the genomic region 59F-60A. *Drosophila* Inf. Service **72**, 129–131.

Krishnan, H.B. and Pueppke, S.G. (1991). nolc, a *Rhizobium fredii* gene involved in cultivar specific nodulation of soybean, shares homology with a heat-shock gene. Mol. Microbiol. **5**, 737–745.

Kurzik-Dumke, U. (1996). Nested gene arrangement of the *Drosophila melanogaster* tumor suppressor locus *lethal(2)tumorous imaginal discs (l(2)tid)* consisting of at least three genes. In *Control mechanisms of carcinogenesis*, (J.G. Hengstler and F. Oesch, eds). Thieme Verlag, Meissen, pp. 224–244.

Kurzik-Dumke, U., Phannavong, B., Gundacker, D., and Gateff, E. (1992). Genetic, cytogenetic and developmental analysis of the *Drosophila melanogaster* tumor suppressor gene *lethal(2)tumorous imaginal discs (l(2)tid)*. Differentiation **51**, 91–104.

Kurzik-Dumke, U., Gundacker, D., Rentrop, M., and Gateff, E. (1995). Tumor suppression in *Drosophila* is causally related to the function of the *lethal(2)tumorous imaginal discs* gene, a dnaJ homolog. Develop. Genet. **16**, 64–76.

Kurzik-Dumke, U., Neubauerer, M., and Debes, A. (1996). Identification of a novel *Drosophila melanogaster* heat shock gene *lethal(2)denticless (l(2)dtl)*, coding for an 83-kDa protein. Gene **171**, 163–170.

Kurzik-Dumke, U., Gundacker, D., Kaymer, M., Debes, A., and Labitzke, K. 1977. Gene within gene configuration and expression pattern of the *Drosophila melanogaster* genes *lethal (2) neighbour of tid (l(2)not)* and *lethal(2)relative of tid (l(2)rot)*. Gene, in press.

Lathigra, R.B., Young, D.B., Sweetser, D., and Young, R.A. (1988). A gene from *Mycobacterium tuberculosis* which is homologous to the DnaJ heat shock protein of *E. coli*. Nucl. Acids Res. **16**, 1636.

Luke, M.M., Sutton, A., and Arndt, K.T. (1991). Characterization of SIS1, a *Saccharomyces cerevisiae* homolgue of bacterial dnaJ proteins. J. Cell Biol. **114**, 623–638.

Ohki, M., Tamura, F., Nishimura, S., and Uchida, H. (1986). Nucleotide sequence of the *Escherichia coli dnaJ* gene and purification of the gene product. J. Biol. Chem. **261**, 1778–1781.

Poodry, C.A. (1980). Imaginal discs: morphology and development. In *The genetics and biology of Drosophila*, (M. Ashburner and T. Wright, eds). Academic Press, New York, vol. 2d, pp. 407–432.

Raabe, T. and Manley, J.L. (1991). A human homologue of the *Escherichia coli* DnaJ heat-shock protein. Nucl. Acids Res. **19**, 6645.

Reed, B. (1992). The genetic analysis of endoreduplication in *Drosophila melanogaster*. Ph.D Thesis, University of Cambridge, Cambridge.

Rothblatt, J.A., Deshaies, R.J., Sanders, S.L., Daum, G., and Schekman, R. (1989). Multiple genes are required for proper insertion of secretory proteins. J.Cell Biol. **109**, 2641–2652.

Sadler, I., Chaing, A., Kurihara, T., Rothblatt, J.A., Way, J., and Silver, P. (1989). A yeast gene important for protein assembly into the endoplasmic reticulum and the nucleus has homology to DnaJ, an *Escherichia coli* heat shock protein. J. Cell Biol. **109**, 2665–2675.

Silver, P.A. and Way, J.C. (1993). Eukaryotic DnaJ homologs and the specifity of Hsp70 activity. Cell **74**, 5–6.

Wetzstein, M., Dedio, J., and Schumann, W. (1990). Complete nucleotide sequence of the *Bacillus subtilis dnaK* gene. Nucl. Acids Res. **18**, 2172.

Wilson, R., Ainscough, R., Anderson, K., *et al.* (1994). 2.2 Mb of contiguous nucleotide sequence from chromosome III of *C. elegans*. Nature **368**, 32–38.

■ Ursula Kurzik-Dumke
Institute of Genetics
Johannes Gutenberg-University
55099 Mainz, Germany
Tel. 49 6131 395223
Fax. 49 6131 395845
E-mail: Kurzik@mzdmza.zdv.uni-mainz.de

Mammalian Hsp40

Hsp40, originally identified as a 40-kDa heat- and stress-inducible protein in HeLa cells, is a member of the HSP40/DnaJ protein family. Hsp40, which is located in the cytoplasm at normal growth temperature, translocates into the nuclei and nucleoli upon heat shock and colocalizes with Hsp70 in the same nucleoli in the same cell. Mammalian Hsp40 and Hsc70 associate with nascent polypeptide chains emerging from ribosomes and facilitate their folding.

■ Alternative names

The *hsp40* gene was independently isolated as *HDJ1* by cDNA cloning from a human placenta cDNA library (Raabe and Manley, 1991).

■ *hsp40* sequence

The nucleotide sequence of human *hsp40* cDNA (GenBank accession numbers D17749 and D49547; Ohtsuka, 1993) is almost identical (98%) to that of independently isolated *HDJ1* cDNA (GenBank accession number X62421; Raabe and Manley, 1991). The *hsp40* cDNA encodes a protein of 340 amino acids (calculated molecular mass of 38 042 Da and pI of 8.92). The N-terminal 48 amino acids of purified Hsp40 (Hattori *et al.*, 1992) are completely identical to those deduced from the nucleotide sequence of the cDNA (the initiating methionine residue is cleaved *in vivo*). Hsp40 is 34–40 % identical, over 340–400 amino acid residues, to bacterial DnaJ and its homologues in yeast, plants and mammals. The region of highest homology among DnaJ proteins corresponds to the N-terminal 70 residues of DnaJ. This 'J-domain' is highly conserved throughout evolution. Some members of the DnaJ protein family (DnaJ, Scj1, Ydj1, Mdj1, Anj1 and Hdj2/HsdJ) also have a glycine- and phenylalanine-rich region (G/F domain) distal to the J-domain, and a cysteine-rich region (C-domain) in the middle of the protein (Bork *et al.*, 1992). Mammalian Hsp40 contains the J- and G/F-domains but not the C-domain. The J-domain of Hsp40 is thought to interact with Hsc70/Hsp70. A protein motif search revealed that Hsp40 has potential glycosylation, phosphorylation and myristylation sites (K. Ohtsuka, unpublished results).

■ Hsp40 protein

Hsp40 was originally identified as a 40 kDa protein induced by heat shock and other stresses in human, rat, mouse and chicken cells, as determined by O'Farrell's two-dimensional gel electrophoresis (NEPHGE/SDS–PAGE) (Ohtsuka *et al.*, 1990). Hsp40 was purified from HeLa cells by modified two-dimensional gel electrophoresis, then a partial amino acid sequence of the N-terminal 48 residues was determined and a polyclonal antibody against Hsp40 was prepared (Hattori *et al.*, 1992). The majority of Hsp40 is localized in the cytoplasm at normal growth temperature (Ohtsuka *et al.*, 1993). Upon heat shock, Hsp40 translocates into the nuclei, especially into the nucleoli; subsequently it returns again to the cytoplasm during the recovery period. The translocation kinetics of Hsp40 are very similar to those of Hsp70/Hsc70. Double immuno fluorescence staining showed the colocalization of Hsp40 and Hsc70 in the nucleoli of heat shocked mammalian cells (Hattori *et al.*, 1993; Yamane *et al.*, 1995). Hsp40 and Hsc70 are coimmunoprecipitated with antibodies against each other (Sugito *et al.*, 1995). Furthermore, expression of not only Hsp70 but also Hsp40 is well correlated with the development of thermotolerance in murine cells (Kaneko *et al.*, 1995). The three-dimensional structure of the N-terminal 77-residue polypeptide (J-domain) of a human Hsp40 has been determined by nuclear magnetic resonance spectroscopy in solution (Qian *et al.*, 1996).

Hsp40 function

The Hsp40/Hdj1 protein interacts with unfolded proteins and stimulates the ATPase activity of Hsp70 (Freeman et al., 1995; Minami et al., 1996). Mammalian Hsp40 and Hsp70 associate with nascent polypeptide chains emerging from ribosomes and mediate their correct folding (Frydman et al., 1994). These results strongly suggest the existence of an HSP70(DnaK)/HSP40(DnaJ) chaperone system in mammalian cells (Caplan et al., 1993; Silver and Way, 1993).

hsp40 gene and regulation

The *hsp40* gene was isolated from a human genomic DNA library (GenBank accession number D85429, Hata et al., 1996). The gene consists of three exons divided by two introns. At least two cis-acting heat shock elements (HSE) are present in the promoter region. The HSE is shown to be activated by heat shock as determined by a promoter activity assay using luciferase as a reporter gene. The human *hsp40* gene maps to chromosome location 19p13.2 by FISH method (Hata et al., 1996).

References

Bork, P., Sander, C., Valencia, A., and Bukau, B. (1992). A module of the DnaJ heat shock proteins found in malaria parasites. Trends Biochem. Sci. **17**, 129.

Caplan, A.J., Cyr, D.M., and Douglas, M.G. (1993). Eukaryotic homologues of Escherichia coli dnaJ: a diverse protein family that functions with HSP70 stress proteins. Mol. Biol. Cell **4**, 555–563.

Freeman, B.C., Myers, M.P., Schumacher, R., and Morimoto, R. (1995). Identification of a reguratory motif in Hsp70 that affects ATPase activity, substrate binding and interaction with HDJ-1. EMBO J. **14**, 2281–2292.

Frydman, J., Nimmesgern, E., Ohtsuka, K., and Hartl, F.U. (1994). Folding of nascent polypeptide chains in a high molecular mass assembly with molecular chaperons. Nature **370**, 111–117.

Hata, M., Okumura, K., Seto, M., and Ohtsuka, K. (1996). Genomic cloning of a human heat shock protein 40 (Hsp40) gene (HSPF1) and its chromosomal localization to 19p13.2. Genomics **38**, 446–449.

Hattori, H., Liu, Y.-C., Tohnai, I., Ueda, M., Kaneda, T., Kobayashi, K., Tanabe, K., and Ohtsuka, K. (1992). Intracelluar localization and partial amino acid sequence of a stress-inducible 40-kDa protein in HeLa cells. Cell Struct. Funct. **17**, 77–86.

Hattori, H., Kaneda, T., Lokeshwar, B., Laszlo, A., and Ohtsuka, K. (1993). A stress-inducible 40-kDa protein (hsp40): purification by modified two-dimensional gel electrophoresis and co-localization with hsc70(p73) in heat-shocked HeLa cells. J. Cell Sci. **104**, 629–638.

Kaneko, R., Hattori, H., Hayashi, Y., Tohnai, I., Ueda, M., and Ohtsuka, K. (1995). Heat-shock protein 40, a novel predictor of thermotolerance in murine cells. Radiat. Res. **142**, 91–97.

Minami, Y., Höhfeld, J. Ohtsuka, K. and Hartl, F.-U. (1996). Regulation of the heat-shock protein 70 reaction cycle by the mammalian DnaJ homolog, Hsp40. J. Biol. Chem. **271**, 19617–19624.

Ohtsuka, K. (1993). Cloning of a cDNA for heat-shock protein hsp40, a human homologue of bacterial DnaJ. Biochem. Biophys. Res. Commun. **197**, 235–240.

Ohtsuka, K., Masuda, A., Nakai, A., and Nagata, K. (1990). A novel 40-kDa protein induced by heat shock and other stresses in mammalian and avian cells. Biochem. Biophys. Res. Commun. **166**, 642–647.

Ohtsuka, K., Utsumi, K.R., Kaneda, T., and Hattori, H. (1993). Effect of ATP on the release of hsp 70 and hsp 40 from the nucleus in heat-shocked HeLa cells. Exptl. Cell Res. **209**, 357–366

Qian, Y.Q., Patel, D., Hartl, F.-U., and McColl, D. (1996). Nuclear magnetic resonance solution structure of the human Hsp40 (HDJ1) J-domain. J. Mol. Biol. **260**, 224–235.

Raabe, T. and Manley, J.L. (1991). A human homologue of the Escherichia coli DnaJ heat-shock protein. Nucl. Acids Res. **19**, 6645.

Silver, P.A. and Way, J.C. (1993). Eukaryotic DnaJ homologs and the specificity of Hsp70 activity. Cell **74**, 5–6.

Sugito, K., Yamane, M., Hattori, H., Hayashi, Y., Tohnai, I., Ueda, M., Tsuchida, N., and Ohtsuka, K. (1995). Interaction between hsp70 and hsp40, eukaryotic homologues of DnaK and DnaJ, in human cells expressing mutant-type p53. FEBS Lett. **358**, 161–164.

Yamane, M., Sugito, K., Hattori, H., Hayashi, Y., Tohnai, I., Ueda, M., Nishizawa, K., and Ohtsuka, K. (1995). Cotranslocation and colocalization of hsp40 (DnaJ) with hsp70 (DnaK) in mammalian cells. Cell Struct. Funct. **20**, 157–166.

■ *Kenzo Ohtsuka:*
Laboratory of Experimental Radiology
Aichi Cancer Center Research Institute
1-1 Kanokoden, Chikusa-ku
Nagoya 464, Japan
Tel. 81 52 762 6111, ext. 8843
Fax. 81 52 763 5233
E-mail: kohtsuka@aichi-cc.pref.aichi.jp

Human Hdj2

The human hdj2 cDNA, isolated from a human endothelial monocyte cDNA library, encodes a homologue of the bacterial DnaJ heat shock protein. Immunofluorescence experiments using antibodies raised against the Hdj2 polypeptide showed that it is normally a cytoplasmic protein. Following heat stress Hdj2 migrates to the Golgi and the nuclear envelope and nucleolus, suggesting an unique cellular localization and potential membrane association for the protein. Hdj2 may be a chaperone protein for translocation to the Golgi and the nucleus.

Alternative name

HSDJ (Oh et al., 1993).

hdj2 isolation and sequence

An hdj2 cDNA clone was originally isolated from a λ-gt11 human umbilical vein endothelial cDNA library (HUVE) using a monoclonal antibody, SK2H10, which reacts specifically to human endothelial cells and monocytes (Schook et al., 1987; Wood et al., 1988). The cDNA clone consists of 1275 nucleotides (GenBank accession number L08063) with an open reading frame (ORF) of 1191 nucleotides (Chellaiah et al., 1993). BLAST comparison of nucleic sequences showed that Hdj2 is 99% homologous to another human cDNA, isolated from a human fibrosarcoma HT-1080 cDNA library, that encodes a homologue (HSDJ) of the DnaJ protein (Oh et al., 1993). Considering the utilization by the yeast Saccharomyces cerevisiae of multiple DnaJ-like proteins (see earlier entries in this section), it is not surprising to find three DnaJ homologues in humans (Cheetham et al., 1992; Chellaiah et al., 1993; Oh et al., 1993).

Hdj2 protein sequence

The open reading frame of hdj2 predicts a 397 amino acid protein with a molecular weight of 45 kDa (Chellaiah et al., 1993). The polypeptide does not contain a recognizable signal sequence or transmembrane domain. It is homologous to many of the HSP40/DnaJ-like proteins. Hdj2 shows higher homology (49% identical overall) to the yeast DnaJ homologue Ydj1p (see entry p. 102) than to any of the other yeast DnaJ homologues such as Scj1p, Sis1p and Sec63p (Caplan and Douglas, 1991; see entries pp. 110, 104 and 108). The majority of the sequence homology to DnaJ occurs within the N-terminal half of the Hdj2 protein. Like the well-characterized Ydj1 protein, Hdj2 has a putative CaaX box at its C-terminus (CQTS) with a serine residue at the extreme C-terminus. Hence, it is possible that, like Ydj1p (Caplan et al., 1992), Hdj2 is farnesylated, increasing its hydrophobicity and facilitating its association with membranes.

Hdj2 biological interactions and function

Immunofluorencence studies to detect the location of Hdj2 within cells during normal and heat shock conditions were carried out in U-937, MRS-5 (a fibroblast cell line) and GG3348 (an EBV-transformed B cell line) cells. The cells were grown at 37, 42 or 45°C for different time periods and then fixed and stained. In cells grown at 37°C the fluorescence was mainly located in the cytoplasm, with faint staining at the nuclear membrane. Heat shock of the cells at 45°C caused intense staining of the Golgi complex, nuclear membrane and nucleolus, with less intense fluorescence in the cytoplasm (A. Davis, T. Mohanakumar and A. Chellaiah, unpublished data). Change in a protein location upon heat stress is a common occurrence among several DNAJ homologues. The human DnaJ homologue Hsp40 changes from a cytoplasmic localization to a nucleolar localization upon heat stress (Hattori et al., 1992), while Ydj1p goes from 20% membrane to 50% membrane association upon heat stress (Caplan et al., 1992). Since Hdj2 is highly similar to Ydj1p and also has a potential farnesylation site, it is likely they have similar functions. In the case of Ydj1, it is known that this function is to assist cytoplasmic HSP70 proteins (Ssa1/2, see entry p. 26) in the translocation of proteins across the endoplasmic reticulum and mitochondrial membranes (Caplan et al., 1993).

References

Caplan, A. and Douglas, M. (1991). Characterization of YDJ1: a yeast homologue of the bacterial DNAJ protein. J. Cell Biol. **114**, 609–621.

Caplan, A., Tsai, I., Casey, P., and Douglas, M. (1992). Farnesylation of YDJ1p is required for function at elevated growth temperatures in Saccharomyccs cerevisiae. J. Biol. Chem. **267**, 18890–18895.

Caplan, A.J., Cyr, D.M., and Douglas, M.G. (1993). Eukaryotic homologues of Escherichia coli DNAJ: a diverse protein family that functions with hsp 70 stress proteins. Mol. Biol. Cell **4**, 555–563.

Cheetham, M., Brion, J., and Anderton, B. (1992). Human homologues of the bacterial heat-shock protein DNAJ are preferentially expressed in neurons. Biochem. J. **284**, 467–476.

Chellaiah, A., Davis, A., and Mohanakumar, T. (1993). Cloning of a unique homologue of the Escherichia coli DNAJ heat shock protein. Biochim. Biophys. Acta **1174**, 111–113.

Hattori, H., Liu, Y., Tohnai, I., Ueda, M., Kaneda, T., Kobayashi, T., Tanabe, K., and Ohtsuka, K. (1992). Intracellular localization and partial amino acid sequence of a stress-inducible 40 kDa protein in HeLa cells. Cell Struct. Funct. **17**, 77–86.

Oh, S., Iwahori, A., and Kato, S. (1993). Human cDNA encoding DNAJ protein homologue. Biochim. Biophys. Acta **1174**, 114–116.

Schook, L., Wood, N., and Mohanakumar, T. (1987). Identification of human vascular endothelial cell-monocyte antigenic system using monoclonal antibodies. Transplantation **44**, 412–416.

Wood, N., Schook, L., Studer, E., and Mohanakumar, T. (1988). Biochemical characterization of human vascular endothelial cell-monocyte antigens defined by monoclonal antibodies. Transplantation **45**, 787–792.

■ *Arasu Chellaiah:*
Department of Molecular Biology and Pharmacology
Campus Box 8103
Washington University School of Medicine
660 S. Euclid
St. Louis, MO 63110, USA
Tel. 1 314 362 5074
Fax. 1 314 362 7058
E-mail: arasu@pharmdec.wustl.edu

■ *Angela Davis and T. Mohanakumar:*
Deptartment of Surgery
Washington University School of Medicine
Campus Box 8109
4939 Children's Place
St. Louis, MO 63110, USA

Auxilin

Auxilin is a neuronal specific 100 kDa protein isolated from bovine brain clathrin-coated vesicles. Its major function is to act as a necessary cofactor for the Hsc70-dependent uncoating of clathrin-coated vesicles. In agreement with this function, the sequence of the C-terminal domain of auxilin is homologous to the J-domain of DnaJ-like proteins, suggesting that auxilin may be a DnaJ homolog.

■ Isolation

Auxilin was first isolated from bovine brain clathrin-coated vesicles by Ahle and Ungewickell (1990). It was purified from a Tris extract of clathrin-coated vesicles by first separating the assembly protein fraction from clathrin by gel chromatography and then purifying the auxilin by hydroxyapatite chromatography followed by Superose 6 gel chromatography. Clathrin-coated vesicles, formed during receptor-mediated endocytosis, are composed primarily of clathrin and assembly proteins (or adaptins). The latter proteins promote the assembly of clathrin into clathrin coats. Auxilin was originally identified as a minor assembly protein (Ahle and Ungewickell, 1990; Lindner and Ungewickell, 1992), but more recently it has been identified as a cofactor necessary for the Hsc70-dependent dissociation of clathrin from clathrin cages (Ungewickell *et al.*, 1995).

■ Auxilin gene and sequence

From a bovine cDNA library, Schroder *et al.* (1995) found an open reading frame of 4531 nucleotides encoding auxilin (GenBank accession number U09237). There appears to be only one gene coding for auxilin in bovine genomic DNA. The primary sequence of 910 amino acids predicts that auxilin contains three domains with differing secondary structures: an N-terminal β sheet region, a C-terminal α helical domain, and an open middle region dominated by β turns (Schroder *et al.*, 1995). Auxilin does not show sequence similarity to any of the other assembly proteins. Although auxilin shares sequence similarity with the 200 kDa actin-binding protein, tensin, it does not appear to bind to actin (Schroder *et al.*, 1995). Rather, its most interesting sequence similarity occurs in the C-terminal region where there is limited, but significant homology between the C-terminal residues 846 to 910 and the J-domain of DnaJ-like proteins (Caplan *et al.*, 1993). Overall the similarity between the C-terminal segment of auxilin and known J-domains is about 55% (Ungewickell *et al.*, 1995).

■ Auxilin protein

Auxilin, with a molecular mass of 99 504 Da, has an isoelectric point of about 7, which probably explains its low solubility at or below pH 7 (Schroder *et al.*, 1995). Its sensitivity to proteolytic digestion may be explained by the open structure of the middle region between the N-terminal β sheet region and the C-terminal α helical region (Schroder *et al.*, 1995).

Immunoblotting studies with auxilin-directed mAb 100/4 suggest that the expression of auxilin is limited to neuronal cells (Ahle and Ungewickell, 1990) which is in agreement with the observation that auxilin mRNA is only expressed in significant amounts in brain (Schroder et al., 1995). Thus, like the assembly protein AP180, which is also neuronal specific (Ahle and Ungewickell, 1986; Kohtz and Puszkin, 1988, Morris et al., 1993), auxilin may play a specific role in endocytosis in nerve cells, i.e. it may play a role in the retrieval of synaptic vesicle membrane after neurotransmitter release.

In addition to isolation of auxilin from brain coated vesicles by Tris extraction (Ahle and Unglewickell, 1990; Prasad et al., 1993), auxilin has also been purified using immunoaffinity chromatography with the auxilin-directed mAb 100/4 (Ahle and Ungewickell, 1990). Recently, auxilin has been expressed with a six-histidine tag on its N-terminal end in E. coli and can be purified as a recombinant protein (Schroder et al., 1995).

■ Function of auxilin

The first biological activity discovered for auxilin was its ability to act as an assembly protein (Ahle and Ungewickell, 1990), that is to catalyse the assembly of clathrin triskelions into artificial clathrin baskets which have the same structure as the polymerized clathrin present in clathrin-coated vesicles (Pearse, 1987). In these baskets, auxilin was found to bind to clathrin at a ratio of three auxilins per clathrin triskelion, i.e. one auxilin per clathrin heavy chain (Ahle and Ungewickell, 1990). However, since auxilin was present in relatively low amounts compared with other assembly proteins (Lindner and Ungewickell, 1992), it was never clear whether this was the primary physiological function of auxilin.

More recently a much more specific function was found for auxilin. It had previously been shown that Hsc70 uncoats clathrin-coated vesicles in vitro in an ATP-dependent process. However, when artificial clathrin baskets were prepared from highly purified clathrin and assembly proteins, it was found that Hsc70 was unable to uncoat them, i.e. depolymerize the clathrin baskets, unless a cofactor was present (Prasad et al., 1993). This cofactor was later identified as auxilin (Ungewickell et al., 1995). Therefore, auxilin appears to be required for Hsc70 to uncoat clathrin-coated vesicles.

This effect of auxilin appears to be a general one since it occurs with clathrin baskets prepared with a variety of assembly proteins (Prasad et al., 1993; Barouch et al., 1994). Auxilin acts by increasing both the rate and magnitude of the initial burst of uncoating (Fig. 1). Apparently it works catalytically (Fig. 1) since maximum uncoating occurs at about a 1:10 molar ratio of auxilin to both clathrin baskets and Hsc70 (Prasad et al., 1993). In addition to being required for uncoating, the auxilin is also required for the increase in Hsc70 ATPase activity which is associated with uncoating (Barouch et al., 1994).

A possible mechanism for auxilin action was revealed by experiments at pH 6 or 6.5 where uncoating by Hsc70

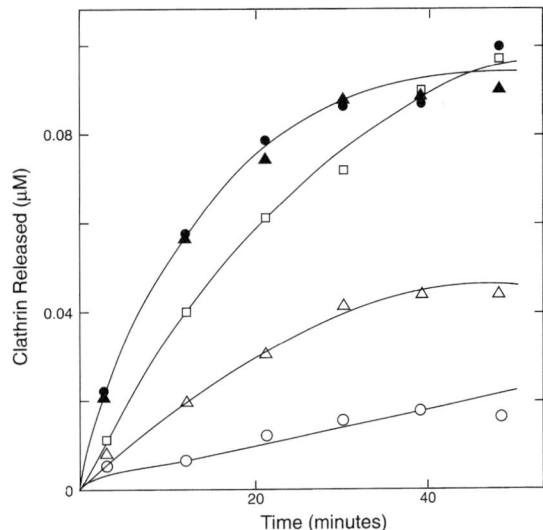

Figure 1. Time course of uncoating of AP_2–clathrin baskets (0.2 µM) by Hsc70 (0.3 mM) at several concentrations of auxilin. ○, no auxilin; △, 0.01 µM auxilin; □, 0.02 µM auxilin; ▲, 0.05 mM auxilin; ●, 0.15 mM auxilin (from Prasad et al., 1993).

is inhibited (Braell et al., 1984). Here it was found that auxilin was required for the binding of Hsc70 to clathrin (Ungewickell et al., 1995). This binding required ATP and, in fact, was reversed when the ATP was replaced by ADP. Furthermore, this binding was accompanied by almost 100-fold activation of the Hsc70 ATPase activity by the clathrin baskets at saturating levels of auxilin (Barouch et al., 1997). Therefore, auxilin may act by catalytically inducing the binding of Hsc70 to clathrin baskets; at pH 7 this leads to uncoating of the baskets while at low pH it simply leads to an increase in the Hsc70 ATPase activity. This effect of auxilin in inducing the binding of Hsc70 to clathrin baskets resembles the effect of other members of the HSP40/DnaJ family of proteins which are required to present substrates to Hsc70. However, the effect of auxilin is specific; other members of the HSP40/DnaJ family do not substitute for auxilin in supporting uncoating by Hsc70 (King et al., 1997). Therefore, specific members of the HSP40/DnaJ family of proteins may present specific substrates to Hsc70.

■ Mutagenesis studies with auxilin

To test whether the J-domain was required for the action of auxilin, this region of the protein was deleted from the recombinant protein and the resulting truncated auxilin tested for its biological activity in vitro. As expected the truncated auxilin still interacted with clathrin but could not support Hsc70 in uncoating clathrin baskets (Ungewickell et al., 1995; Holstein, et al., 1996).

Biological regulation of auxilin

In vitro, auxilin can be phosphorylated at one or more serine residues by casein kinase II (Schroder *et al.*, 1995). Sequence analysis has shown 14 potential casein kinase II phosphorylation sites, a possible tyrosine kinase phosphorylation site, a p34 cdc2 kinase site, several potential protein kinase C sites, a cAMP- or cGMP-dependent protein kinase site, and finally an SH3 domain (Schroder *et al.*, 1995). Whether any of these sites is involved in regulation of auxilin activity *in vivo* is not yet known.

References

Ahle, S. and Ungewickell, E. (1986). Purification and properties of a new clathrin assembly protein. EMBO J. **5**, 3143–3149.

Ahle, S. and Ungewickell, E. (1990). Auxilin, a newly identified clathrin-associated protein in coated vesicles from bovine brain. J. Cell Biol. **111**, 19–29.

Barouch, W., Prasad, K., Greene, L., and Eisenberg, E. (1994). ATPase activity associated with the uncoating of clathrin baskets by Hsp70. J. Biol. Chem. **269**, 28563–28568.

Barouch, W., Prasad, K., Greene, L., and Eisenberg, E. (1997). Auxilin-induced interaction of molecular chaperone Hsc70 with clathrin baskets. Biochemistry **36**, 4303–4308.

Braell, W., Schlossman, D., Schmid, S., and Rothman, J. (1984). Dissociation of clathrin coats coupled to the hyrolysis of ATP: role of uncoating ATPase. J. Cell Biol. **99**, 734–741.

Caplan, A., Cyr, D., and Douglas, M. (1993). Eukaryotic homologues of *Escherichia coli* dnaJ: a diverse protein family that functions with HSP70 stress proteins. Mol. Biol. Cell **4**, 555–563.

Holstein, E., Ungewickell, H. and Ungewickell, E. (1996). Mechanism of clathrin basket dissociation: separate functions of protein domains of the DnaJ homologue auxilin. J. Cell Biol. **135**, 925–937.

King, C., Eisenberg, E. and Greene, L., (1997). Effect of yeast and human DnaJ homologs on clathrin uncoating by 70 kDa heat shock protein. Biochemistry **26**, 4062.

Kohtz, D.S. and Puszkin, S. (1988). A neuronal protein (NP185) associated with clathrin-coated vesicles: characterization of NP185 with monoclonal antibodies. J. Biol. Chem. **263**, 7418–7425.

Lindner, R. and Ungewickell, E. (1992). Clathrin-associated proteins of bovine brain coated vesicles: an analysis of their number and assembly-promoting activity. J. Biol. Chem. **267**, 16567–16573.

Morris, S.A., Schroder, S., Plessmann, U., Weber K., and Ungewickell, E. (1993). Clathrin assembly protein AP180 : primary structure, domain organization and identification of a clathrin binding site. EMBO J. **12**, 667–675.

Pearse, B.M.F. (1987). Clathrin and coated vesicles. EMBO J. **6**, 2507–2512.

Prasad, K., Barouch, W., Greene, L., and Eisenberg, E. (1993). A protein cofactor is required for uncoating of clathrin baskets by uncoating ATPase. J. Biol. Chem. **268**, 23758–23761.

Schroder, S., Morris, S.A, Knorr, R., Plessmann, U., Weber, K., Vinh, N.G., and Ungewickell, E. (1995). Primary struture of the neuronal clathrin-associated protein auxilin and its expression in bacteria. Eur. J. Biochem. **228**, 297–304.

Ungewickell, E., Ungewickell, H., Holstein, S.E.H., Lindner, R., Prasad, K., Barouch, W., Martin, B., Greene, L., and Eisenberg, E., (1995). Role of auxilin in uncoating clathrin-coated vesicles. Nature **378**, 632–635.

■ *Evan Eisenberg and Lois E. Greene:*
National Institutes of Health
NHLBI, MSC0301
9000 Rockville Pike, 3 Center Drive
Bethesda, MD 20892-0301, USA
Tel. 1 301 496 2846
Fax. 1 301 402 1519
E mail: greene@helix.nih.gov

Human neurone-specific Hsj1 proteins

Hsj1 proteins are human neurone-specific members of the DnaJ-family of molecular chaperones. There are two isoforms of Hsj1 with different C-termini (Hsj1a and Hsj1b), which are generated by alternative splicing from a single copy gene. Although the similarity to the DnaJ family is limited to a 70 amino acid J-domain at the N-terminus, Hsj1 proteins show the conservation of the interaction with HSP70 proteins that is characteristic of the HSP40/DnaJ family. They can partially complement mutations in the yeast DnaJ-like protein Ydj1, and can regulate the activity of HSP70 in several different systems by stimulating HSP70 ATP hydrolysis and altering substrate binding.

Alternative name

Human neurone-specific DnaJ-like proteins.

Isolation

Hsj1 proteins were initially identified by immunoscreening an Alzheimer's disease brain expression cDNA library using a polyclonal antiserum raised to a preparation of paired helical filaments (PHF), one of the pathological hallmarks of Alzheimer's disease. However, to date we have been unable to confirm the association of Hsj1 proteins with PHF (Cheetham *et al.*, 1992).

hsj1 gene and sequence

Hsj1 proteins are encoded by a single copy gene found on chromosome 2 (D.A. Collier and M.E.

Cheetham, unpublished data; GenBank accession numbers X63368.gb_pr, X63368.em_pr, S37374, S37375). The gene transcript undergoes alternative splicing to produce two mRNA species of approximately 2 and 3 kb. These mRNA species differ by the presence of a 1.1 kb insert that predicts an alternative C-terminus in the larger transcript. Both transcripts have a short 5′ untranslated leader and long 3′ untranslated regions of undefined function. The predicted amino acid sequences of the N-termini of the Hsj1 proteins include a J-domain of approximately 70 residues but no leader or signal peptide sequence. The amino acid identity to other members of the HSP40/DnaJ family in this region is approximately 50% and many changes are semiconservative. Adjacent to the J-domain is a short region that is rich in glycine residues (9 of 24 residues), followed by a high density of phenylalanine. This 'G/F' domain is present in DnaJ and many other DnaJ-like proteins and separates the J-domain from the rest of the polypeptide. Unlike DnaJ, the phenylalanine-rich region in the Hsj1 proteins extends to residue 174 and is interspersed with serine residues forming repeated motifs of FSF (where the number of serines varies between one and four). This serine-rich region shows similarity to a great number of polypeptides that have a high serine content which probably has little functional significance (Cheetham et al., 1994a). However, this region does have a limited similarity to other DnaJ-like proteins, including Sis1p and Hsp40, that do not possess the cysteine repeat found in DnaJ (see Fig. 1).

The extra 1.1 kb sequence in the larger transcript encodes a C-terminal amino acid sequence that is 74 residues longer than the shorter form. We have designated these isoforms HSJ1a and HSJ1b.

Hsj1 protein

The *hsj1* mRNA transcripts are translated to form two protein products as identified by polyclonal sera raised to recombinant protein and synthetic peptides (Cheetham et al., 1992). Initially, there was some concern that the larger transcript could represent an unspliced hnRNA as the 1.1 kb insert showed some similarity to the highly conserved intron splice consensus sequence and splice donor and acceptor sites. However, another polyclonal sera raised against a peptide corresponding to the predicted C-terminus of HSJ1b recognizes the same slower migrating polypeptide as the previous antisera (Cheetham et al., 1994a). Therefore, Hsj1a and Hsj1b correspond to the translated products of the two mRNA species. Although Hsj1a has a predicted molecular weight of 32 kDa, its apparent molecular mass measured by acrylamide gel electrophoresis is 36 kDa. HSJ1b has a predicted molecular weight of 38 kDa, but an apparent molecular mass of 42 kDa.

Western blot analysis has shown that the expression of Hsj1a and Hsj1b is restricted to brain. However, even in brain the abundance of these proteins is very low, constituting less than 0.0025% of total brain protein. Subcellular fractionation studies indicate that Hsj1a and Hsj1b are not found exclusively in one organelle but are found in both the cytoplasm and the nucleus, although it would appear that they are not enriched in membrane fractions and are unlikely to be resident in the ER or mitochondria (M.E. Cheetham and M. Tighe, unpublished data).

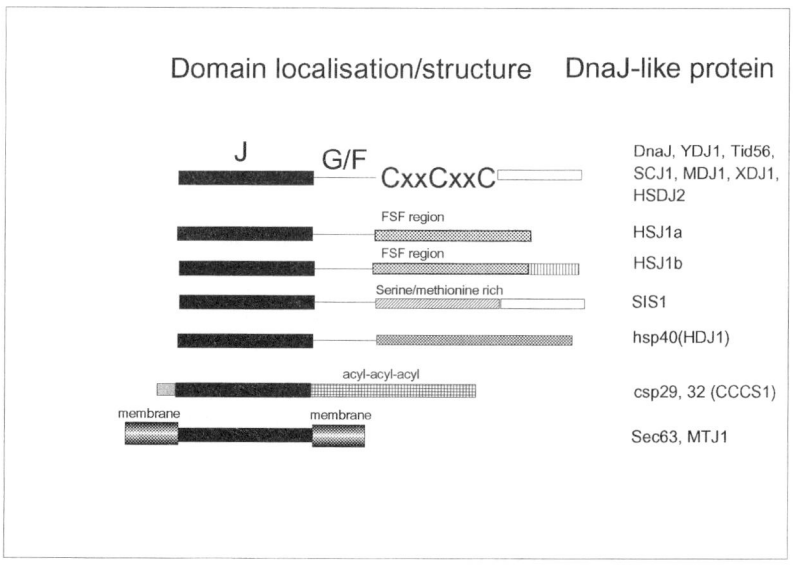

Figure 1. Schematic representation of HSP40/DnaJ family domain structure. The archetypal protein DnaJ has four domains, J, G/F, a cysteine repeat motif (CXXCXXC) and a C-terminal domain (many other DnaJ-like proteins share this structure but may have, in addition, leader sequences or C-terminal tags). Hsj1 proteins, Sis1p and Hsp40 share the J- and G/F-domains with DnaJ but are not similar throughout the rest of their sequences. Csp proteins lack the G/F-domain but have a cysteine string that is acylated allowing membrane attachment. Sec63p and Mtj1 are large proteins with a J-domain located between two membrane-spanning domains.

■ Hsj1 activities and biological interactions

Studies using recombinant Hsj1a and Hsj1b proteins purified from *E. coli* have shown that, like DnaJ, they can stimulate the ATPase activity of HSP70 proteins, in particular the constitutively expressed brain Hsc70, although they have no intrinsic ATPase activity of their own (Cheetham et al., 1994b). This enhancement of the intrinsic ATPase activity of Hsc70 is mediated via an increase in the rate of hydrolysis of bound ATP; the rate of nucleotide exchange is unaffected. For Hsc70 the stimulation of ATP hydrolysis is over fivefold and brings the rate of hydrolysis close to the rate of exchange, so that in the presence of Hsj1 proteins nucleotide exchange probably becomes the rate-limiting step. The stimulation of Hsc70 ATPase activity is maximal in the presence of stoichiometric amounts of Hsj1 proteins (Cheetham et al., 1994b).

Hsj1 proteins also affect Hsc70 substrate binding. The effect of Hsj1 proteins on the binding of Hsc70 to the permanently unfolded substrate reduced and carboxymethylated lactalbumin (RCMLA) was analysed by both gel-shift analysis and size exclusion chromatograhy (SEC) (Cheetham et al., 1994b). In the presence of ATP, Hsj1 proteins reduced the affinity of Hsc70 for RCMLA. However, when non-hydrolysable analogues of ATP were present the affinity of Hsc70 for RCMLA was increased. Additionally, the K^+ requirement for release of substrate from HSP70 (Palleros et al., 1993) can be bypassed in the presence of Hsj1 proteins. Therefore, Hsj1 proteins can regulate both the ATPase activity and substrate binding of Hsc70, suggesting that the *in vivo* function of Hsj1 proteins is to modulate HSP70 substrate binding and present putative substrates to HSP70 for chaperone action. The interaction of Hsj1 with Hsc70 appears to be transient, as the two proteins do not appear to form a stable complex, as judged by their failure to associate when analysed by immunoprecipitation, gel-shift analysis or size exclusion chromatography. It is not yet known whether Hsj1 proteins self-associate to form dimers or oligomers, as do some other molecular chaperones, or form stable complexes with other neuronal proteins.

The first documented function of Hsc70 was the ability to uncoat clathrin-coated vesicles. Therefore, we analysed the effect of Hsj1 proteins on this physiologically important process *in vitro* (Cheetham et al., 1996). Hsj1 proteins inhibit the Hsc70-mediated uncoating of structurally intact coated vesicles by over 40%. This appeared to be caused by the non-productive stabilization of Hsc70–clathrin complex in the first round of uncoating. Hsj1-like proteins are not enriched in the coated vesicle fraction of P12 cells, suggesting that Hsj1 proteins are inhibiting uncoating by interfering with an endogenous HSP40 protein in the coated vesicle, most probably auxilin (see entry p. 124). This new data further highlights the specialization of function within the HSP40/DnaJ family and the ability of these proteins to target normally promiscuous HSP70 proteins to particular substrates or even particular sites on a substrate.

The physiological chaperone partners and substrates of Hsj1 proteins remain to be defined, but it would appear that, although they show no affinity for RCMLA or native lactalbumin (Cheetham et al., 1994b), they can bind denatured lactalbumin (P. Csermely and M.E. Cheetham, unpublished data). Therefore, Hsj1 proteins do not appear to be merely HSP70 cofactors but are potentially molecular chaperones in their own right.

■ Mutagenesis studies

The ability of Hsj1 proteins to complement partially for null mutations in the yeast DnaJ-like protein Ydj1 (Caplan et al., 1992) has facilitated mutagenesis studies. Deletion mutagenesis of Hsj1a and Hsj1b from the C-terminus has shown that the first 100 residues of Hsj1 are sufficient to mediate this complementation (M.E. Cheetham and A.J. Caplan, unpublished data). As these 100 residues contain only the J-domain and some of the 'G/F' region it would appear that these domains alone are sufficient to replace some of the functions of Ydj1p. This is most probably mediated by the N-terminus of Hsj1 interacting with the yeast HSP70 protein, Ssa1p, as mutagenesis studies of DnaJ have shown that the first 100 amino acids of DnaJ are sufficient to mediate the interaction with DnaK (Wall et al., 1994).

■ Biological regulation of *hsj1*

Analysis of *hsj1* mRNA expression by Northern blotting and RNase protection assays suggests that expression is limited almost exclusively to the brain, and, more specifically, that the level of expression is determined by the area of brain analysed. *hsj1* expression is evident in the hippocampus and cerebellum, but the area with the strongest signal is located in the brain frontal cortex. Liver, kidney and testis failed to give any signal, and only an extremely weak signal was detectable in spleen and skeletal muscle (Cheetham et al., 1992). Preliminary studies have hinted that *hsj1* expression in neuroblastoma cells is up-regulated in response to stress and suggest that Hsj1 will behave as a neuronal heat shock protein.

In situ hybridization studies have further delineated *hsj1* expression to neuronal layers within the brain; however the level of expression is so low even in brain that single cell localization has not been possible. Notwithstanding this, the relative intensity of the neuronal layers strongly suggests that Hsj1a and Hsj1b proteins are expressed at much higher levels in some CNS neurones than in any other cell type (Cheetham et al., 1992, 1994a).

■ References

Caplan, A.J., Cyr, D.M., and Douglas, M.G. (1992). YDJ1p facilitates polypeptide translocation across different intracellular membranes by a conserved mechanism. Cell **71**, 1143–1155.

Cheetham, M.E., Brion, J.P., and Anderton, B.H. (1992). Human homologues of the bacterial heat-shock protein DnaJ are preferentially expressed in neurons. Biochem. J. **284**, 469–476.

Cheetham, M.E., Brion, J.P., and Anderton, B.H. (1994a). Neuronal homologues of the bacterial heat shock protein DnaJ. In *Heat-shock proteins in the nervous system*, (R.J. Mayer and I.R. Brown, eds), pp. 169–190. Academic Press, London.

Cheetham, M.E., Jackson, A.P., and Anderton, B.H. (1994b). Regulation of 70-kDa heat-shock-protein ATPase activity and substrate binding by human DnaJ-like proteins, HSJ1a and HSJ1b. Eur. J. Biochem. **226**, 99–107.

Cheetham, M.E., Anderton, B.H., and Jackson, A.P. (1996). Inhibition of hsc70 catalysed clathrin uncoating by HSJ1 proteins. Biochem. J. **319**, 103–108.

Palleros, D.R., Reid, K.L., Shi, L., Welch, W.J., and Fink, A.L. (1993). ATP-induced protein–Hsp70 complex dissociation requires K^+ but not ATP hydrolysis. Nature **365**, 664–666.

Wall, D., Zylicz, M., and Georgopoulos, C. (1995). The conserved G/F motif of the DnaJ chaperone is necessary for the activation of the substrate binding properties of the DnaK chaperone. J. Biol. Chem. **270**, 2139–2144.

■ Michael E. Cheetham and Miranda Tighe:
Department of Pathology
Institute of Ophthalmology
University College London
Bath Street
London EC1V 9EL, UK
Tel. 44 171 608 6944
Fax. 44 171 608 6862
E-mail: michael.cheetham@ucl.ac.uk

Mammalian Mtj1

Mtj1 is the first DnaJ-like protein isolated from a murine cell line. Of the structural domains characteristic of DnaJ-like proteins, the only one present in Mtj1 is the so-called J-domain. It is bordered by putative transmembrane regions in a way similar to the J-domain found in the yeast DnaJ-like protein Sec63p.

■ Isolation

The cDNA encoding Mtj1 was isolated from an M27 Lewis lung carcinoma expression library (Brightman *et al.*, 1995). Antibodies used for the screen were prepared against 50–60 kDa proteins that were retained on a hydrophobic peptide affinity column (YVGVAPG).

■ *mtj1* gene and sequence

The *mtj1* cDNA sequence (GenBank accession number L16953) includes an open reading frame of 1659 bp that predicts a protein of 63.9 kDa. Apart from a 3′ poly(A) tract and a potential polyadenylation signal 22 nucleotides upstream thereof, the isolated cDNA sequence contains several internal ATGs (Brightman *et al.*, 1995). The amino acid composition deduced from the open reading frame contains a large number of charged amino acids (16% D+E, 17% K+R). The calculated pI is 9.7. The Mtj1 amino acid sequence contains one potential site for N-linked glycosylation (Asn-475) and several potential sites for O-linked glycosylation or phosphorylation (Brightman *et al.*, 1995).

One classification of DnaJ-like proteins distinguishes functional chaperone homologues of *E. coli* DnaJ from more distantly related family members by the degree of sequence homology in different domains (Caplan *et al.*, 1993). Of the domains present in *E. coli* DnaJ only the J-domain is found in the Mtj1 open reading frame (amino acids 56–129). The glycine/phenylalanine-rich region is missing, as are the so-called cysteine-rich repeats. The list of DnaJ-like proteins that contain only the J-domain is still growing (Cyr *et al.*, 1994). Examples include the yeast proteins Sec63p (Rothblatt *et al.*, 1989; Sadler *et al.*, 1989; see entry p. 108), Zuotin (Zhang *et al.*, 1992, see entry p. 105) and YIB4w (Voss *et al.*, 1995). Mtj1 lacks significant sequence homology to any known protein outside its J-domain (Brightman *et al.*, 1995).

Potential transmembrane domains are located at amino acids 25–41 and 148–171 in the Mtj1 sequence (Brightman *et al.*, 1995). Their presence suggests that Mtj1 might be the second integral membrane protein of the DnaJ-like protein family. The arrangement of the two potential transmembrane regions on either side of the J-domain is reminiscent of the yeast protein Sec63p (Feldheim *et al.*, 1992). However, Mtj1 differs from Sec63p in not having a conventional signal sequence, a third N-terminal transmembrane domain or a C-terminal acidic region (Brightman *et al.*, 1995)

Northern analysis with an *mtj1*-specific probe and poly(A)$^+$ RNA from M27 cells revealed a major transcript of 3.2 kb and a minor one of 6.5 kb. Assuming that the cDNA contains a complete ORF, the presence of a 3.2 kb transcript indicates a lengthy 5′ untranslated region. The size of the 6.5 kb transcript may be owing to alternative splicing events (Brightman *et al.*, 1995). Tissue specificity analysis indicates that both transcripts are found in all adult mouse tissues examined, with brain and heart showing relatively lower levels (Brightman et al., 1995).

■ Mtj1 protein

Preliminary characterization of the Mtj1 protein suggests that it is an ER resident or nuclear protein (Brightman et al., 1995). Western analysis of different cell fractions identified major antigens of 62, 42 and 41 kDa. The 62 kDa protein, which corresponds well to the longest ORF of the cDNA clone, is found in both the nuclear and microsomal fraction. The 41 and 42 kDa antigens, which could represent translation products of an ORF beginning at nucleotide 440, would not contain the J-domain and are found in the nuclear or microsomal fraction, respectively (Brightman et al., 1995).

■ Biological activities of Mtj1

The function of Mtj1 is not known. The lack of homology between Mtj1 and E. coli DnaJ outside of the J-domain suggests that Mtj1 may not be a member of the chaperone subclass of DnaJ-like molecules (Cyr et al., 1994). Yet, through its J-domain it could still have regulatory effects on HSP70 family members (Silver and Way, 1993). The topological similarities and similar subcellular localizations of Mtj1 and the yeast translocation protein Sec63p are intriguing (Brightman et al., 1995). In the case of Sec63p, it has been demonstrated that the J-domain projects into the lumen of the ER (Feldheim et al., 1992). Perhaps the J-domain of Mtj1 projects into the same compartment. Here it could 'compete' with a yet to be found mammalian Scj1 homologue, i.e. a luminal DnaJ-like protein of the ER compartment, for interaction with the ER–HSP70 member BiP (Feldheim et al., 1992; Brodsky and Schekman, 1993; Schlenstedt et al., 1995).

■ References

Brightman, S.E., Blatch, G.L., and Zetter, B.R. (1995). Isolation of a mouse cDNA encoding MTJ1, a new murine member of the DnaJ family of proteins. Gene **153**, 249–254.

Brodsky, J.L. and Schekman, R. (1993). A sec63–BiP complex from yeast is required for protein translocation in a reconstituted proteoliposome. J. Cell Biol. **123**, 1355–1363.

Caplan, A.J., Cyr, D.M., and Douglas, M.G. (1993). Eukaryotic homologues of Escherichia coli dnaJ: a diverse protein family that functions with HSP70 stress proteins. Mol. Biol. Cell **4**, 555–563.

Cyr, D.M., Langer, T., and Douglas, M.G. (1994). DnaJ-like proteins: molecular chaperones and specific regulators of Hsp70. Trends Biochem. Sci. **19**, 176–181.

Feldheim, D., Rothblatt, J., and Schekman, R. (1992). Topology and functional domains of Sec63p, an endoplasmic reticulum membrane protein required for secretory protein translocation. Mol. Cell. Biol. **12**, 3288–3296.

Rothblatt, J.A., Deshaies, R.J., Sanders, S.L., Daum, G., and Schekman, R. (1989). Multiple genes are required for proper insertion of secretory proteins into the endoplasmic reticulum in yeast. J. Cell Biol. **109**, 2641–2652.

Sadler, I., Chiang, A., Kurihara, T., Rothblatt, J., Way, J., and Silver, P. (1989). A yeast gene important for protein assembly into the endoplasmic reticulum and the nucleus has homology to DnaJ, an Escherichia coli heat shock protein. J. Cell Biol. **109**, 2665–2675.

Schlenstedt, G., Harris, S., Risse, B., Lill, R., and Silver, P.A. (1995). A yeast DnaJ homologue, Scj1, can function in the endoplasmic reticulum with BiP/Kar2p via a conserved domain that specifies interactions with Hsp70s. J. Cell Biol. **129**, 979–988.

Silver, P.A. and Way, J.C. (1993). Eukaryotic DnaJ homologs and the specificity of Hsp70 activity. Cell **74**, 5–6.

Voss, H., Tamames, J., Teodoru, C., Valencia, A., Sensen, C., Wiemann, S., Schwager, C., Zimmermann, J., Sander, C., and Ansorge, W. (1995). Nucleotide sequence and analysis of the centromeric region of yeast chromosome IX. Yeast **11**, 61–78.

Zhang, S., Lockshin, C., Herbert, A., Winter, E., and A. Rich (1992). Zuotin, a putative Z-DNA binding protein in Saccharomyces cerevisiae. EMBO J. **11**, 3787–3796.

■ *Michael W. Lässle and Bruce R. Zetter:*
Department of Cell Biology and Surgery
Harvard Medical School
300 Longwood Avenue
Boston, MA 02115, USA
Tel. 1 617 355 6767
Fax. 1 617 355 7043
E-mail: lassle@mit.edu

■ *Shannon E. Brightman:*
Department of Biology
Sacred Heart University
5151 Park Avenue
Fairfield, CT 06432, USA
Tel. 1 203 365 7597
Fax. 1 203 371 7888
E-mail: Brightman@sacredheart.edu

■ *Gregory L. Blatch:*
WITS University / Department of Biochemistry
P.O. WITS
Johannesburg 2050, South Africa
Tel. 11 716 2204
Fax. 11 716 4479
E-mail: 089BLG@cosmaos.wits.ac.za

4

GrpE-Like Proteins

The GrpE family of proteins — an overview

The GrpE protein of Escherichia coli was first identified because a mutation in its corresponding gene completely blocked bacteriophage λ DNA replication. Subsequent analyses indicated that the GrpE protein is universally conserved and is an integral component of the DnaK/GrpE/DnaJ chaperone machine, which protects certain unfolded polypeptides from aggregation, and disaggregates certain protein aggregates as well. The GrpE protein, unlike DnaK and DnaJ, is not a molecular chaperone. Its sole biological role is to interact with the appropriate DnaK family member and act as an ADP–ATP exchange factor. This action aids in the acceleration of the DnaK–DnaJ–substrate release cycle, thus resulting in a more efficient DnaK/GrpE/DnaJ chaperone machine. In the absence of the GrpE family member, the residual DnaK(Hsp70) chaperone action is not enough to support growth in either E. coli or Saccharomyces cerevisiae.

■ Alternative names

GrpE (GroP-like protein) is the name usually reserved for the prokaryotic counterparts of this family. The independent cloning of the S. cerevisiae grpE-like gene resulted in its product being given three different names; namely, Yge1 (yeast GrpE; Ikeda et al., 1994), Mge1p (mitochondrial GrpE; Laloraya et al., 1994) and GrpEp (Bolliger et al., 1994).

■ Isolation of GrpE proteins

The E. coli grpE gene was first identified by Saito and Uchida (1977) in a massive mutant hunt, designed to identify genes whose products are essential for bacteriophage λ replication. The particular selection employed was bacterial survival at 42°C in a lysogenic strain that carried a thermolabile λcI857 repressor. The grpE280 mutation was thereby isolated, defining the grpE gene (Saito and Uchida, 1977). Other prokaryotic grpE-like genes were identified as being part of the dnaK operon (Wetzstein and Schumann, 1990). A grpE gene has also been detected in Methanosarcina mazei as being part of the dnaK operon (Conway de Macario et al., 1994).

Mammalian GrpE-like protein was purified from bovine, porcine and rat liver mitochondria as binding to an E. coli DnaK-affinity column and eluting with ATP (Naylor et al., 1995). Similarly, the Drosophila melanogaster GrpE-like protein (Droe1) was purified on an Hsc70 affinity column, and the S. cerevisiae GrpE-like protein was purified on an Ssc1p affinity column (Bolliger et al., 1994). The S. cerevisiae grpE-like gene (YGE1) was also identified by Ikeda et al. (1994) because, for unknown reasons, when present on a multicopy plasmid it conferred resistance to the antibiotic staurosporine.

■ GrpE family members

Swiss-Prot accession numbers for GrpE sequences are: P09372 for Escherichia coli; P48204 for Francisella tularensis; P15874 for Bacillus subtilis; P28609 for Borrelia burgdorferi; P48195 for Clostridium acetyobutylicum; P48604 for Drosophila melanogaster; P43732 for Haemophilus influenzae; P42369 for Lactococcus lactis; P45553 for Staphylococcus aureus; Q05562 for Streptomyces coelicolor; P38523 for Saccharomyces cerevisiae; the GenBank accession number for rat liver mt-GrpE is U62940.

■ Sequence and structure of GrpE proteins

The GrpE family of proteins is the least well conserved component of the DnaK/DnaJ/GrpE chaperone machine, with overall identity being as low as 19% between the characterized GrpE family members. An analysis of the known protein sequences suggests that various blocks are better conserved than others in the approximately 200 amino acid residue proteins (Fig. 1). According to the analysis of Naylor et al. (1996), only five amino acid residues are absolutely conserved between all family members (Fig. 1). Various mutations have been isolated and sequenced in the E. coli and S. cerevisiae genes. These include the canonical E. coli grpE280 mutation, which results in a Gly-122 to Asp substitution. Other mutations have been isolated and characterized by Wu et al. (1994), Laloraya et al. (1995) and Westermann et al. (1995). The detailed biochemical analysis of six mutant GrpE proteins resulted in the assignment of particular functions to various segments of the protein (Fig. 1; Wu et al., 1996).

■ Function of GrpE proteins

The original grpE280 mutation of E. coli was isolated at 42°C (Saito and Uchida, 1977). Subsequent analyses demonstrated that grpE280 bacteria do not grow at 42°C, but do at lower temperatures (Ang et al., 1986). At the non-permissive temperature of 42°C, E. coli grpE280 mutant bacteria exhibit a general block in all macromolecular processes, including DNA, RNA and protein syntheses.

Efforts to delete the grpE gene were unsuccessful under all temperature and growth conditions attempted

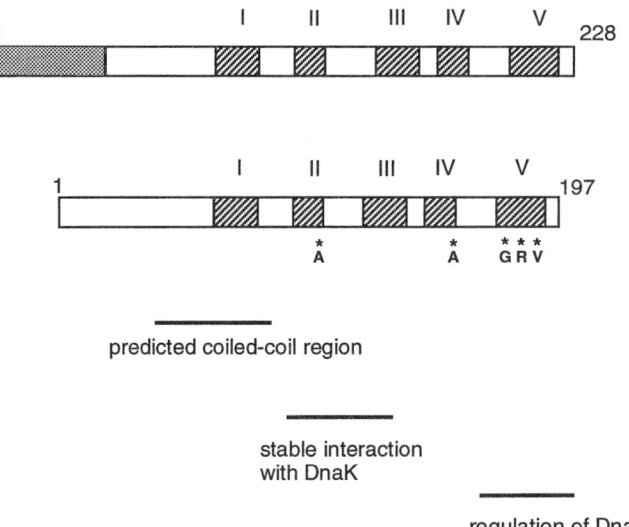

Figure 1. Structure/function correlation of two members of the GrpE family of proteins. Yeast Mge1p is shown at the top, the *E. coli* GrpE below. The comparison of the amino acid sequences of several GrpE family members has led to the demonstration of five highly conserved regions (hatched boxes; Wu *et al.*, 1994) (▨). The *S. cerevisiae* Mge1p contains a 43 amino acid precursor sequence (▨), which is cleaved upon entry into mitochondria. According to the analysis of Naylor *et al.* (1996), the five amino acids indicated underneath GrpE are the only universally conserved ones. The approximate locations of a predicted coiled-coil region, which may play a role in GrpE dimerization/oligomerization, a region thought to mediate stable interaction with DnaK, and a region thought to regulate the ATPase activity of DnaK at a step following binding are also shown [see Wu *et al.* (1996) for details].

(Ang and Georgopoulos, 1989). However, the *grpE* gene could be deleted in certain *E. coli* mutant backgrounds which had previously been selected to grow faster in the presence of certain debilitating *dnaK* gene mutations. These *dnaK* mutant strains accumulated various unmapped suppressor mutations which enabled their fast growth. This genetic approach clearly established that the sole biological role of the GrpE protein is to help DnaK in carrying out its biological functions efficiently.

This genetic conclusion was strongly reinforced by extensive biochemical studies on the GrpE protein. These showed that GrpE binds tightly to DnaK in a salt-resistant manner (up to 2M KCl), a fact that permitted its facile purification (Zylicz *et al.*, 1987). The GrpE–DnaK interaction is disrupted in the presence of ATP. Buchberger *et al.* (1994) showed that a conserved loop in the ATPase domain of DnaK, spanning amino acid residues 28–33, is important for GrpE protein binding. The evidence for this conclusion rests primarily on the fact that the deletion of this amino acid segment results in a DnaK (Δ28–33) mutant protein which has intact ATPase activity and is capable of binding to its protein substrates (Buchberger *et al.*, 1994). However, DnaK (Δ28–33) does not bind to GrpE and its ATPase is not further stimulated by GrpE–DnaJ. As a consequence, the DnaK (Δ28–33) mutant protein is 'biologically' dead; i.e. it cannot substitute for wild-type DnaK for bacterial growth. In addition, DnaK756, the product of the classical *dnaK*756 mutation, which was isolated on the basis of blocking bacteriophage λ growth (Georgopoulos, 1977), possesses a Gly-32 to Asp substitution in this conserved loop, and does not interact stably with GrpE, as judged by coimmunoprecipitation (Johnson *et al.*, 1989) or native gel electrophoresis (Buchberger *et al.*, 1994). In this respect it is interesting that amino acid residue 31 (DnaK's numbering system) is glutamic acid in all prokaryotic and mitochondrial DnaK homologues, but the corresponding amino acid in all of the other eukaryotic counterparts is a highly conserved glutamine. Perhaps this amino acid residue plays a key role in the stabilization of the GrpE–DnaK complex, since the GrpE-like proteins have been found only in bacteria and mitochondria so far (see below).

The purified GrpE protein behaves primarily as a dimer in solution, as judged by analytical ultracentrifugation, dynamic light scattering and native electrophoresis (Schönfeld *et al.*, 1995) as well as glutaraldehyde cross-linking (Ang, 1988; Osipiuk *et al.*, 1993). However, at high protein concentrations, GrpE dimers appear to oligomerize, suggesting the existence of a biologically active, putative hexameric form (Wu *et al.*, 1996). The crystal structure of GrpE bound to the ATPase domain of DnaK has recently been solved (Harrison *et al.*, 1997). GrpE is shown to be a dimer, whose binding to DnaK results in an opening of the nucleotide binding cleft of DnaK.

In a purified system, the GrpE protein was shown to release efficiently either ADP or ATP bound to DnaK (Liberek *et al.*, 1991). The joint presence of GrpE and DnaJ results in a massive acceleration in DnaK's weak

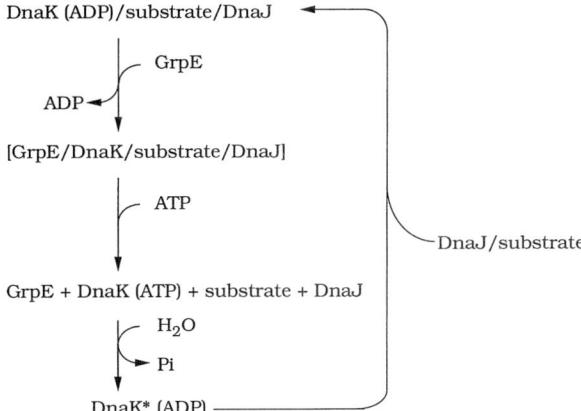

Figure 2. Mechanism of action of GrpE. *In vivo*, because of the large excess of ATP (3–5 mM) over ADP (0.3 mM), GrpE most likely plays the role of an ATP/ADP exchange factor for DnaK. The DnaK (ATP) moiety binds and releases substrate very rapidly. Following ATP hydrolysis, the DnaK* (ADP) form generated will bind DnaJ (either free or already bound to the substrate), thus restarting the DnaK/substrate/DnaJ cycle [see Banecki and Zylicz (1996) for details].

ATPase activity (Liberek et al., 1991). However, in vivo, because of the higher intracellular levels of ATP as opposed to ADP, coupled with the higher affinity of DnaK for ATP as opposed to ADP, the role of GrpE may be slightly different, e.g. the DnaK protein would be primarily found in the DnaK* (ADP) form, shown to bind and release substrates very fast, and to interact efficiently with DnaJ but not with GrpE (Fig. 2; Banecki and Zylicz, 1996). In this DnaK cycle, the GrpE protein is thought to interact with the DnaK (ADP)–substrate–DnaJ complex, thus giving rise to the intermediate GrpE–DnaK (ADP)–substrate–DnaJ form, reported by Gamer et al. (1996). The presence of GrpE will result in the release of ADP, thus exchanging for ATP, and may also contribute (sterically?) in the release of the DnaK-bound substrate (Fig. 2). Thus, the action of GrpE is essential for efficient recycling of the DnaK chaperone machine (see Banecki and Zylicz, 1996 for details).

In an *in vitro* reconstituted λ DNA replication system, consisting entirely of purified proteins, the presence of GrpE lowers the requirement for DnaK protein tenfold (Alfano and McMacken, 1989; Zylicz et al., 1989). This clearly exemplifies GrpE's biological role as necessary to efficiently 'recycle' the DnaK chaperone protein. GrpE also lowers the requirement for DnaK in the renaturation of heat-inactivated RNA polymerase (Zieminowicz et al., 1993), but its presence is absolutely required in the renaturation of firefly luciferase (Szabo et al., 1994). Skowyra and Wickner (1995) showed that GrpE not only lowers DnaK's affinity for ATP, but also increases DnaK's affinity for Mg^{2+} ions, thus weakening the interactions of Mg^{2+} with nucleotide prior to its release. In a purified system,

the GrpE protein was shown to 'displace' DnaK from its bound polypeptide substrates in the absence of nucleotide (Osipiuk et al., 1993). This action can be understood in terms of the formation of a stable DnaK–GrpE complex, thus sequestering DnaK away from its polypeptide substrate. In the same work, evidence was presented showing that the DnaK–GrpE complex binds preferentially to λP–DnaJ, as opposed to λP alone, suggesting the possibility that GrpE, under certain circumstances, can act as a DnaK 'specificity' factor.

The *S. cerevisiae* Mge1p protein is synthesized as a 28 000 Da precursor, whose leader sequence is cleaved upon entry into mitochondria (Laloraya et al., 1994). Deloche and Georgopoulos (1996) have presented data showing that the approximately 23 000 Da mature form of Mge1p is created following the removal of the 43 N-terminal amino acid residues of the precursor protein, and Bolliger et al. (1994) have shown that the N-terminal amino acid of the mature form of Mge1p is covalently modified. Mge1p is important for both the import of proteins into mitochondria, as well as their subsequent folding by Ssc1p, the Hsp70 family member resident in mitochondria (Laloraya et al., 1995; Westermann et al., 1995). The purified mature Mge1p was shown to behave, both qualitatively and quantitatively, in a fashion identical to that of GrpE of *E. coli*; namely, (i) its native size, as judged by its elution profile on a size column, (ii) its ability to stimulate DnaK's ATPase activity by acting as a nucleotide exchange factor, (iii) its binding to DnaK, as judged by glutaraldehyde cross-linking, which showed that Mge1p binds predominantly as a dimer, and (iv) its participation in the proper renaturation of guanidinium-inactivated firefly luciferase (Deloche and Georgopoulos, 1996). In this biochemical assay, mature Mge1p was capable of assisting the DnaK/DnaJ chaperone to the same extent as the GrpE protein of *E. coli*. Taken together, all these biochemical results demonstrate that mitochondrial mature Mge1p interacts with *E. coli* DnaK in a manner exactly analogous to that of GrpE, and serve to highlight the evolutionary conservation of the DnaK chaperone machine. In this respect, it is not surprising that the presence of the Mge1p protein allows the deletion of the otherwise essential *grpE* gene of *E. coli* (O. Deloche and C. Georgopoulos, unpublished results).

■ Do all HSP70 proteins need a GrpE-like cohort?

The fact that no GrpE homologue has been detected thus far in either the cytosol or the endoplasmic reticulum of eukaryotic cells begs the question of why not? An unlikely explanation is that such GrpE homologues do exist, but have gone undetected up to now because their complex with their corresponding HSP70 family member is not strong enough to be detected by affinity chromatography or coimmunoprecipitation. A more likely explanation is that the cytosolic and endoplasmic reticulum HSP70 resident members have evolved so that they do not require a GrpE-like protein for their activation cycle. In support of this, Ziegelhoffer et al. (1995) have provided a clue as to why the cytosolic Ssa1p of S. cerevisiae may function in the absence of a GrpE-like cytosolic homologue. These co-workers showed that although Ssa1p binds tightly to its ATP substrate, it releases ADP spontaneously, unlike the DnaK protein of E. coli which binds stably to ADP. As a consequence, the isolated Ssa1p (ATP) complex is devoid of ADP, in sharp contrast with DnaK, where approximately equimolar amounts of both the ATP- and ADP-bound forms are found. Thus, the rate-limiting step in the ATPase reaction cycle of Ssa1p appears to be the hydrolysis of the bound ATP, and hence the usefulness of the putative GrpE-like nucleotide exchange factor would be limited in the Ssa1p system. Consistent with this conclusion, Levy et al. (1995) and Freeman et al. (1995) have shown that cytosolic Hsc70 proteins, together with their corresponding cytosolic DnaJ homologues, can reactivate denatured proteins in a GrpE-independent reaction.

Another possibility is that additional regulatory factors exist in the cytosol or endoplasmic reticulum, which are not related in sequence to GrpE, but still modulate the biochemical properties of some of the cytosolic or endoplasmic reticulum HSP70 proteins. For example, using the S. cerevisiae two-hybrid system, Höhfeld et al. (1995) have discovered a gene that encodes a protein called Hip, which interacts tightly with the ATPase domain of rat cytosolic Hsc70 in an Hsp40- and ATP-dependent reaction. This interaction stabilizes the Hsc70 (ADP) form, thought to bind best to the polypeptide substrates. Finally, recently, a 66 000 M_r protein has been purified from rabbit reticulocytes on the basis of its association with cytosolic Hsc70 (Gross and Hessefort, 1996). This protein plays a recycling role for Hsc70, inasmuch as it catalyses the dissociation of the Hsc70-bound ADP in exchange for ATP. This process represents adenine nucleotide exchange, because ATP must be present for ADP's dissociation from Hsc70. Interestingly, the 66 000 M_r protein appears to be similar to the p60 protein component of the Hsp90 chaperone machine (Smith et al., 1993), and to the heat shock-inducible Sti1p of S. cerevisiae (Nicolet and Craig, 1989). Clearly, the Hip and 66 000 M_r proteins mentioned above must play an important regulatory role in modulating cytosolic Hsc70's biochemical activities, thus bypassing the necessity for the presence of a bona fide GrpE family member in the cytosol.

References

Alfano, C. and McMacken, R. (1989). Heat shock protein-mediated disassembly of nucleoprotein structures is required for the initiation of bacteriophage lambda DNA replication. Ph.D. Thesis, J. Biol. Chem. **264**, 10709–10718.

Ang, D. (1988). The role of the Escherichia coli heat shock protein, GrpE in Escherichia coli growth and λ DNA replication. Ph. D. Thesis, University of Utah, Salt Lake City.

Ang, D. and Georgopoulos, C. (1989). The GrpE heat shock protein is essential for Escherichia coli viability at all temperatures but is dispensable in certain mutant backgrounds. J. Bacteriol. **171**, 2748–2755.

Ang, D., Chandrasekhar, G.N., Zylicz, M., and Georgopoulos, C. (1986). Escherichia coli grpE gene codes for heat shock protein B25.3, essential for both λDNA replication at all temperatures and host growth at high temperature. J. Bacteriol. **167**, 25–29.

Banecki, B. and Zylicz, M. (1996). Real time kinetics of the DnaK/DnaJ/GrpE molecular chaperone action. J. Biol. Chem. **271**, 6137–6144.

Bolliger, L., Deloche, O., Glick, B.S., Georgopoulos, C., Jenö, P., Kronidou, N., Horst, M., Morishima, N., and Schatz, G. (1994). A mitochondrial homolog of bacterial GrpE interacts with mitochondrial Hsp70 and is essential for viability. EMBO J. **13**, 1998–2006.

Buchberger, A., Schröder, H., Büttner, M., Valencia, A., and Bukau, B. (1994). A conserved loop in the ATPase domain of the DnaK chaperone is essential for stable binding of GrpE. Struct. Biol. **1**, 95–101.

Conway de Macario, E., Dugan, C.B., and Macario, A.J.L. (1994). Identification of a grpE heat-shock gene homolog in the archaeon Methanosarcina mazei. J. Mol. Biol. **240**, 95–101.

Deloche, O. and Georgopoulos, C. (1996). Purification and biochemical properties of Saccharomyces cerevisiae's Mge1p, the mitochondrial cochaperone of Ssc1p. J. Biol. Chem. **271**, 23960–23966.

Freeman, B.C., Myers, M.P., Schumacher, R., and Morimoto, R.I. (1995). Identification of a regulatory motif in Hsp70 that affects ATPase activity, substrate binding and interaction with HDJ-1. EMBO J. **14**, 2281–2292.

Gamer, J., Multhaup, G., Tomoyasu, T., McCarty, J.S., Rüdiger, S., Schönfeld, H.J., Schirra, C., Bujard, H., and Bukau, B. (1996). A cycle of binding and release of the DnaK, DnaJ and GrpE chaperones regulates activity of the Escherichia coli heat shock transcription factor σ^{32}. EMBO J. **15**, 607–617.

Georgopoulos, C. (1977). A new bacterial gene (groPC) which affects λ DNA replication. Mol. Gen. Genet. **151**, 35–39.

Gross, M. and Hessefort, S. (1996). Purification and characterization of a 66-kDa protein from rabbit reticulocyte lysate which promotes the recycling of Hsp70. J. Biol. Chem. **271**, 16833–16841.

Harrison, C. J., Hayer-Hartl, M., Di Liberto, M., Hartl, F.-U., and Kuriyan, J. (1997). Crystal structure of the nucleotide exchange factor GrpE bound to the ATPase domain of the molecular chaperone DnaK. Science **276**, 431–435.

Höhfeld, J., Minami, Y., and Hartl, F.-U. (1995). Hip, a novel cochaperone involved in the eukaryotic Hsc70*Hsp40 reaction cycle. Cell **83**, 589–598.

Ikeda, E., Yoshida, S., Mitsuzawa, H., Uno, I., and Toh-e, A. (1994). YGE1 is a yeast homologue of Escherichia coli grpE and is required for maintenance of mitochondrial functions. FEBS Lett. **339**, 265–268.

Johnson, C., Chandrasekhar, G.N., and Georgopoulos, C. (1989). The DnaK and GrpE heat shock proteins of Escherichia coli interact both in vivo and in vitro. J. Bacteriol. **171**, 1590–1596.

Laloraya, S., Gambill, B.D., and Craig, E.A. (1994). A role for a eukaryotic GrpE-related protein, Mge1p, in protein translocation. Proc. Nat. Acad. Sci. USA **91**, 6481–6485.

Laloraya, S., Dekker, P.J.T., Voos, W., Craig, E.A., and Pfanner, N. (1995). Mitochondrial GrpE modulates the function of matrix Hsp70 in translocation and maturation of preproteins. Mol. Cell. Biol. **15**, 7098–7105.

Levy, E.J., McCarty, J., Bukau, B., and Chirico, W.J. (1995). Conserved ATPase and luciferase refolding activities between bacteria and yeast Hsp70 chaperones and modulators. FEBS Lett. **368**, 435–440.

Liberek, K., Marszalek, J., Ang, D., Georgopoulos, C., and Zylicz, M. (1991). The Escherichia coli DnaJ and GrpE heat shock proteins jointly stimulate DnaK's ATPase activity. Proc. Nat. Acad. Sci. USA **88**, 2874–2878.

Naylor, D.J., Ryan, M.T., Condron, R., Hoogenraad, N.J., and Høj, P.B. (1995). Affinity-purification and identification of GrpE homologues from mammalian mitochondria. Biochim. Biophys. Acta **1248**, 75–79.

Naylor, D.J., Hoogenraad, N.J., and Høj, P.B. (1996). Isolation and characterization of a cDNA encoding rat mitochondrial GrpE, a stress-inducible chaperone of ubiquitous appearance in mammalian organs. FEBS Lett., **396**, 181–188.

Nicolet, C.M. and Craig, E.A. (1989). Isolation and characterization of Sti1, a stress-inducible gene from *Saccharomyces cerevisiae*. Mol. Cell. Biol. **9**, 3638–3646.

Osipiuk, J., Georgopoulos, C., and Zylicz, M. (1993). Initiation of λ DNA replication: the *Escherichia coli* small heat-shock proteins, DnaJ and GrpE, increase DnaK's affinity for the λP protein. J. Biol. Chem. **268**, 4821–4827.

Saito, H. and Uchida, H. (1977). Initiation of the DNA replication of bacteriophage lambda in *Escherichia coli* K12. J. Mol. Biol. **113**, 1–25.

Schönfeld, H.-J., Schmidt, D., Schröder, H., and Bukau, B. (1995). The DnaK chaperone system of *Escherichia coli*: quaternary structures and interactions of the DnaK and GrpE components. J. Biol. Chem. **270**, 2183–2189.

Skowyra, D. and Wickner, S. (1995). GrpE alters the affinity of DnaK for ATP and Mg^{2+}. Implications for the mechanism of nucleotide exchange. J. Biol. Chem. **17**, 26282–26286.

Szabo, A., Langer, T., Schröder, H., Flanagan, J., Bukau, B., and Hartl, F.-U. (1994). The ATP hydrolysis-dependent reaction cycle of the *Escherichia coli* Hsp70 system-DnaK, DnaJ, and GrpE. Proc. Natl. Acad. Sci. USA **91**, 10345–10349.

Westermann, B., Prip-Buus, C., Neupert, W., and Schwarz, E. (1995). The role of the GrpE homologue, Mge1p, in mediating protein import and protein folding in mitochondria. EMBO J. **14**, 3452–3460.

Wetzstein, M. and Schumann, W. (1990). Nucleotide sequence of a *Bacillus subtilis* gene homologous to the *grpE* gene of *E. coli* located immediately upstream of the *dnaK* gene. Nucl. Acids Res. **18**, 1289.

Wu, B., Ang, D., Snavely, M., and Georgopoulos, C. (1994). Isolation and characterization of point mutations in the *Escherichia coli grpE* heat shock gene. J. Bacteriol. **176**, 6965–6973.

Wu, B., Wawrzynow, A., Zylicz, M., and Georgopoulos, C. (1996). Structure-function analysis of the *Escherichia coli* GrpE heat shock protein. EMBO J., **15**, 4806–4816

Ziegelhoffer, T., Lopez-Buesa, P., and Craig, E.A. (1995). The dissociation of ATP from Hsp70 of *Saccharomyces cerevisiae* is stimulated by both Ydj1p and peptide substrates. J. Biol. Chem. **270**, 10412–10419.

Ziemienowicz, A., Skowyra, D., Zeilstra-Ryalls, J., Fayet, O., Georgopoulos, C., and Zylicz, M. (1993). Either of the *Escherichia coli* GroEL/GroES and DnaK/DnaJ/GrpE chaperone machines can reactivate heat-treated RNA polymerase: different mechanisms for the same activity. J. Biol. Chem. **268**, 25425–25431.

Zylicz, M., Ang, D., and Georgopoulos, C. (1987). The GrpE protein of *Escherichia coli* : purification and properties. J. Biol. Chem. **262**, 17437–17442.

Zylicz, M., Ang, D., Liberek, K., and Georgopoulos, C. (1989). Initiation of λ DNA replication with purified host- and bacteriophage-encoded proteins: the role of the DnaK, DnaJ and GrpE heat shock proteins. EMBO J. **8**, 1601–1608.

■ *Olivier Deloche, Debbie Ang and Costa Georgopoulos:*
Département de Biochimie Médicale
Centre Médicale Universitaire
1, rue Michel-Servet
1211 Geneva 4, Switzerland
Tel. 41 22 702 5512
Fax. 41 22 702 5502
E-mail: georgopo@cmu.unige.ch

Escherichia coli GrpE

GrpE is an *Escherichia coli* heat shock protein that acts as cochaperone in the functional cycle of the DnaK chaperone system. It binds with high affinity to the ATPase domain of DnaK, thereby mediating dissociation of DnaK-bound nucleotides. This activity accelerates the nucleotide binding/release cycle of DnaK and, consequently, the ATP-mediated dissociation of substrates bound to DnaK. GrpE is essential for virtually all chaperone activities of the DnaK system.

■ Isolation

The first *grpE* mutation, *grpE280*, was obtained upon selection for mutants resistant to lytic infection by bacteriophage λ (Saito and Uchida, 1977). The *grpE* gene product was later shown to be identical to heat shock protein B25.3 (Ang et al., 1986).

■ *grpE* gene and sequence

The *grpE* gene is located at 57 minutes on the *E. coli* genetic map. Its sequence (EMBL bank accession number X07863) predicts a 197 amino acid residue protein of 21.8 kDa.

■ GrpE protein

GrpE is conserved among eubacteria (Wetzstein et al., 1992), archea (Conway deMacario et al., 1994) and mitochondria (Bolliger et al., 1994; Laloraya et al., 1994), with typically 30% identical and 45–60% similar amino acids. The N-terminal region of variable length (30–70 residues) exhibits no detectable sequence homology. GrpE is predicted to have a high content of α helices but to lack β sheets. GrpE purification protocols have been described (Zylicz et al., 1987; Schönfeld et al., 1995). GrpE forms stable dimers of elongated shape (Schönfeld et al., 1995; Harrison et al., 1997). The asymmetric dimer consists of two long parallel N-terminal α helices that lead into a small four-helix bundle, and of two small C-terminal β sheet domains (Harrison et al., 1997). The β sheet domains and the four-helix bundle form major areas of contact with the DnaK ATPase domain, whereas the long α helices are mainly involved in dimerization (Harrison et al., 1997).

■ Biological activities of GrpE

The biological functions of GrpE can be attributed to its role in the functional cycle of the DnaK chaperone system (Frydman and Hartl, 1994; Georgopoulos et al., 1994) (see Fig. 2 of the DnaK entry, p. 23). GrpE dimers form stable 1:1 complexes with DnaK that dissociate on addition of ATP (Zylicz et al., 1987; Schönfeld et al., 1995). The GrpE binding site lies within the ATPase domain of DnaK and involves an exposed, conserved loop structure (Buchberger et al., 1994; Harrison et al., 1997). The interaction of GrpE with DnaK accelerates dissociation of DnaK-bound nucleotide (Liberek et al., 1991; McCarty et al., 1995) and thus allows rapid rebinding of nucleotide. Together with the activity of the DnaJ cochaperone (see entry p. 95), which stimulates γ-phosphate hydrolysis, this results in a stimulation of over 100-fold of the steady-state ATPase activity of DnaK (Liberek et al., 1991; McCarty et al., 1995). The activity of GrpE to accelerate nucleotide exchange also accelerates the ATP-induced dissociation of substrates bound to DnaK (see Fig. 2 of DnaK entry, p. 23) (Szabo et al., 1994; McCarty et al., 1995; Gamer et al., 1996).

GrpE is essential for most, if not all, in vivo activities of the DnaK chaperone system. In vitro, GrpE is dispensable under certain conditions for DnaK- and DnaJ-dependent replication of bacteriophage λ and plasmid P1 DNA. However, GrpE greatly increases the apparent specific activity of DnaK in λ replication (Alfano and McMacken, 1989; Zylicz et al., 1989), thereby influencing the mode of replication (Wyman et al., 1993), and is essential under physiological buffer conditions for P1 replication (Skowyra and Wickner, 1993).

■ Regulation of *grpE* expression

The expression of the *grpE* gene is transcriptionally controlled by the heat shock promoter-specific σ^{32} subunit of RNA polymerase (see entry on regulation of the *E. coli* heat shock response, p. 525). Expression is induced as part of the heat shock response by various forms of stress, including temperature upshift.

■ Mutagenesis studies

The *grpE280* allele was isolated as a mutation blocking λ DNA replication (Saito and Uchida, 1977) and was subsequently shown to confer to the cells temperature-sensitive growth and defects in various cellular processes, including heat shock gene regulation (Gross et al., 1990; Georgopoulos et al., 1994), DNA and RNA synthesis (Ang et al., 1986) and protein degradation (Straus et al., 1988). The underlying molecular basis for these defects seems to be a reduced affinity of the GrpE280 mutant protein for DnaK (Johnson et al., 1989; Liberek et al., 1991) resulting in decreased chaperone activity of the DnaK system. Further studies revealed that *grpE* is essential at all temperatures in an otherwise wild-type strain, but is dispensable in the presence of an unknown extragenic suppressor of a *dnaK* mutation (Ang and Georgopoulos, 1989). A recent approach to isolate new *grpE* missense mutations yielded two groups of temperature sensitive mutants conferring bacteriophage λ resistance to the cells at both 30°C and 42°C (group I) or at 42°C but not 30°C (group II) (Wu et al., 1994). Subsequently, group I mutants including the re-isolated *grpE280* allele were demonstrated *in vitro* to be impaired in stable and productive interaction with DnaK, whereas group II mutants showed less severe defects resulting from altered properties with respect to dimerization and modulation of DnaK function (Wu et al., 1996).

■ References

Alfano, C. and McMacken, R. (1989). Heat shock protein-mediated disassembly of nucleoprotein structures is required for the initiation of bacteriophage lambda DNA replication. J. Biol. Chem. **264**, 10709–10718.

Ang, D. and Georgopoulos, C. (1989). The heat-shock-regulated *grpE* gene of Escherichia coli is required for bacterial growth at all temperatures but is dispensable in certain mutant backgrounds. J. Bacteriol. **171**, 2748–2755.

Ang, D., Chandrasekhar, G.N., Zylicz, M., and Georgopoulos, C. (1986). Escherichia coli grpE gene codes for heat shock protein B25.3, essential for both lambda DNA replication at all temperatures and host growth at high temperature. J. Bacteriol. **167**, 25–29.

Bolliger, L., Deloche, O., Glick, B. S., Georgoploulus, C., Jen, P., Kronidou, N., Horst, M., Morishima, N., and Schatz, G. (1994). A mitochondrial homolog of bacterial GrpE interacts with mitochondrial hsp70 and is essential for viability. EMBO J. **13**, 1998–2006.

Buchberger, A., Schröder, H., Büttner, M., Valencia, A., and Bukau, B. (1994). A conserved loop in the ATPase domain of the DnaK chaperone is essential for stable binding of GrpE. Nature Struct. Biol. **1**, 95–101.

Conway de Macario, E., Dugan, C.B., and Macario, A.J.L. (1994). Identification of a grpE heat-shock gene homolog in the archeon Methanosarcina mazei. J. Mol. Biol. **240**, 95–101.

Frydman, J. and Hartl, F.-U. (1994). Molecular chaperone functions of hsp70 and hsp60 in protein folding. In *The Biology of Heat Shock Proteins and Molecular Chaperones* (R.I. Morimoto, A. Tissières and C. Georgopoulos, eds). Cold Spring Harbor Laboratory Press, New York, pp. 251–283.

Gamer, J., Multhaup, G., Tomoyasu, T., McCarty, J.S., Rüdiger, S., Schönfeld, H.-J., Schirra, C., Bujard, H., and Bukau, B. (1996). A cycle of binding and release of the DnaK, DnaJ and GrpE chaperones regulate activity of the *E. coli* heat shock transcription factor σ^{32}. EMBO J. **15**, 607–617.

Georgopoulos, C., Liberek, K., Zylicz, M., and Ang, D. (1994). Properties of the heat shock proteins of *Escherichia coli* and the autoregulation of the heat shock response. In *The Biology of Heat Shock Proteins and Molecular Chaperones* (R.I.

Morimoto, A. Tissiéres and C. Georgopoulos, eds). Cold Spring Harbor Laboratory Press, New York, pp. 209–249.

Gross, C.A., Straus, D.B., and Erickson, J.W. (1990). The function and regulation of heat shock proteins in *Escherichia coli*. In *Stress Proteins in Biology and Medicine* (R.I. Morimoto, A. Tissieres and C. Georgopoulos, eds). Cold Spring Harbor Laboratory Press, New York, pp. 167–190.

Harrison, C. J., Hayer-Hartl, M., Di Liberto, M., Hartl, F.-U., and Kuriyan, J. (1997). Crystal structure of the nucleotide exchange factor GrpE bound to the ATPase domain of the molecular chaperone DnaK. Science **276**, 431–435.

Johnson, C., Chandrasekhar, G.N., and Georgopoulos, C. (1989). *Eschericia coli* dnaK and grpE heat shock genes interact both in vivo and in vitro. J. Bacteriol. **171**, 1590–1596.

Laloraya, S., Gambill, B.D., and Craig, E.A. (1994). A role for a eukaryotic GrpE-related protein, Mge1p, in protein translocation. Proc. Natl. Acad. Sci. USA **91**, 6481–6485.

Liberek, K., Marszalek, J., Ang, D., Georgopoulos, C., and Zylicz, M. (1991). *Escherichia coli* DnaJ and GrpE heat shock proteins jointly stimulate ATPase activity of DnaK. Proc. Natl. Acad. Sci. USA **88**, 2874–2878.

McCarty, J.S., Buchberger, A., Reinstein, J., and Bukau, B. (1995). The role of ATP in the functional cycle of the DnaK chaperone system. J. Mol. Biol. **249**, 126–137.

Saito, H. and Uchida, H. (1977). Initiation of the DNA replication of bacteriophage lambda in *Escherichia coli* K-12. J. Mol. Biol. **113**, 1–25.

Schönfeld, H.-J., Schmidt, D., Schröder, H., and Bukau, B. (1995). The DnaK chaperone system of *Escherichia coli*: quarternary structures and interactions of the DnaK and GrpE components. J. Biol. Chem. **270**, 2183–2189.

Skowyra, D. and Wickner, S. (1993). The interplay of the GrpE heat shock protein and Mg^{2+} in RepA monomerization by DnaJ and DnaK. J. Biol. Chem. **268**, 25296–25301.

Straus, D.B., Walter, W.A., and Gross, C.A. (1988). *Escherichia coli* heat shock gene mutants are defective in proteolysis. Genes Dev. **2**, 1851–1858.

Szabo, A., Langer, T., Schröder, H., Flanagan, J., Bukau, B., and Hartl, F. U. (1994). The ATP hydrolysis-dependent reaction cycle of the *Escherichia coli* Hsp70 system-DnaK, DnaJ, and GrpE. Proc. Natl. Acad. Sci. USA **91**, 10345–10349.

Wetzstein, M., Volker, U., Dedio, J., Lobau, S., Zuber, U., Schiesswohl, M., Herget, C., Hecker, M., and Schumann, W. (1992). Cloning, sequencing, and molecular analysis of the dnaK locus from *Bacillus sutilis*. J. Bacteriol. **174**, 3300–3310.

Wu, B., Ang, D., Snavely, M., and Georgopoupos, C. (1994). Isolation and characterization of point mutations in the *Escherichia coli grpE* heat shock gene. J. Bacteriol. **176**, 6965–6973.

Wu, B., Wawrzynow, A., Zylicz, M., and Georgopoulos, C. (1996). Structure-function analysis of the Escherichia coli GrpE heat shock protein. EMBO J. **15**, 4806–4816.

Wyman, C., Vasilikiotis, C., Ang, D., Georgopoulos, C., and Echols, H. (1993). Function of the GrpE heat shock protein in bidirectional unwinding and replication from the origin of phage l. J. Biol. Chem. **268**, 25192–25196.

Zylicz, M., Ang, D., and Georgopoulos, C. (1987). The grpE protein of Escherichia coli. Purification and properties. J. Biol. Chem. **262**, 17437–17442.

Zylicz, M., Ang, D., Liberek, K., and Georgopoulos, C. (1989). Initiation of λ DNA replication with purified host- and bacteriophage-encoded proteins: the role of the DnaK, DnaJ and GrpE heat shock proteins. EMBO J. **8**, 1601–1608.

■ *Alexander Buchberger and Bernd Bukau:*
Institut für Biochemie und Molekularbiologie
Universität Freiburg
Hermann Herder Str. 7
D-791104 Freiburg
Germany
Tel. 49 761 203 5221
Fax. 49 761 203 5257

Saccharomyces cerevisiae Mge1

The Mge1 protein of Saccharomyces cerevisiae is a mitochondrial homologue of the bacterial GrpE family of proteins. Mge1p is an abundant soluble protein of the mitochondrial matrix where it modulates the activity of mitochondrial Hsp70 (Ssc1 protein).

■ Alternative names

Yge1 (yeast GrpE, Ikeda *et al*., 1994; Nakai *et al*., 1994), GrpEp (Bolliger *et al*., 1994), mitochondrial GrpE (mt-GrpE).

■ Isolation

MGE1 was found in the GenBank database based on its similarity to *E. coli grpE* (Bolliger *et al*., 1994; Laloraya *et al*., 1994). It was also found as a gene (*YGE1*) that, when present in multiple copies, conferred resistance to staurosporine, a protein kinase C inhibitor (Ikeda *et al*., 1994).

■ *MGE1* gene and sequence

The nucleotide sequence of the *MGE1* gene (GenBank accession number U09565) from *Saccharomyces cerevisiae* (S288C) predicts a preprotein of 228 amino acids of 26 066 daltons which shares 34% identity and 57% similarity with the *E.coli* GrpE protein. The Mge1 protein has an N-terminal extension of 43 amino acids as compared to the GrpE of *E. coli*. This N-terminal region is rich in hydroxylated (S and T) and positively charged (K and R) residues and lacks negatively charged residues, features reminiscent of typical mitochondrial leader sequences.

Within the GrpE family of proteins the N-terminal one-third of the polypeptide is more divergent than the

C-terminal two-thirds. Comparison of the Mge1p sequence with GrpE-like protein sequences from seven different bacteria reveals that it is most closely related to E. coli GrpE and that there are only nine absolutely conserved (identical) residues in this family of proteins. Among higher eukaryotes, cDNAs encoding GrpE-like proteins have been identified from C. elegans and humans (Laloraya et al., 1994).

■ Mge1 protein

The MGE1 gene encodes a preprotein with a mobility of approximately 28 kDa (Laloraya et al., 1994). In wild-type cells the preprotein is efficiently processed to cleave off the leader and generate the mature Mge1 protein of approximately 23 kDa. The exact cleavage site is not known but has been estimated to be at approximately amino acid 43. Neurospora crassa mitochondria also have a 24 kDa protein which cross-reacts with antiserum against E.coli GrpE (Voos et al., 1994). Mge1p is a soluble mitochondrial matrix protein that colocalizes with mt-hsp70 (Ssc1p).

■ Biological interactions of Mge1p

Mge1p is associated with Ssc1p in an ATP-sensitive manner (Bolliger et al., 1994; Nakai et al., 1994; Voos et al., 1994). In mitochondria the molar ratio of Mge1 to Ssc1p is approximately 1:3 and virtually all of Mge1p has been found associated with Ssc1p in the absence of ATP. Mge1p has also been found associated with Ssc1–preprotein complexes (Voos et al., 1994).

■ Function of Mge1p

The biochemical function of Mge1p is likely to be similar to that of the GrpE protein of E. coli (i.e stimulation of release of bound nucleotides from its partner HSP70 protein). Mge1p can functionally substitute for GrpE as determined by complementation of the temperature-sensitive phenotype of the grpE280 mutant (Ikeda et al., 1994). In addition, using isolated components, Mge1 proteins stimulate the release of bound nucleotides from Ssc1p (Maio et al., 1997)

Cells with reduced levels of wild-type Mge1p or mutant alleles accumulate precursors of some mitochondrial preproteins (e.g. pre-hsp60) (Nakai et al., 1994; Laloraya et al., 1994). The function of Mge1p has also been investigated by studies of temperature-sensitive mutant alleles of MGE1 (Laloraya et al., 1995; Westerman et al., 1995). Mge1p modulates the function of Ssc1p in facilitating import of nuclear-encoded precursors of mitochondrial proteins from the cytosol into mitochondria. Mge1p function is required for binding of Ssc1p to preproteins and also for facilitating release of fully imported proteins from the Ssc1p. There is evidence that Mge1p function is required for maturation and folding of some preproteins following import. Mge1p has also been suggested to favour ATP-dependent dissociation of a complex of Ssc1p with Mim44, a protein attached to the matrix side of the inner mitochondrial membrane (Westerman et al., 1995).

Cells depleted of Mge1 protein also show altered distribution of mitochondria and produce daughter cells that may lack mitochondria. Cells with mutant alleles of MGE1 show a higher frequency of appearance of petite (respiratory deficient) colonies. Overexpression of Mge1p is lethal in some yeast strain backgrounds (Laloraya et al., 1994). For unknown reasons, introduction of the MGE1 gene on a multicopy vector confers resistance to staurosporine, an inhibitor of protein kinase C (Ikeda et al., 1994).

■ Mutagenesis studies

Several mge1 temperature-sensitive alleles have been isolated (Laloraya et al., 1995; Westerman et al., 1995). mge1-2 to 5 have the following amino acid changes: mge1-2, A_{81} to D; mge1-3, A_{134} to V; mge 1-4, E_{123} to K; and mge1-5, T_{102} to I. In addition, these mutants all have a T_{199} to A change that alone has no phenotypic effect. Of these alleles mge1-2 and mge1-5 have been used most extensively. mge1-2 causes a reduction in Mge1 protein levels at the non-permissive temperature, while mge1-5 has normal levels. An independently isolated allele, mge1-100, has also been extensively studied (Laloraya et al., 1995). mge1-100 has a substitution, R_{216} to K, of an arginine residue that is found in all known GrpE-related proteins.

■ References

Bolliger, L., Deloche, O., Glick, B., Georgopoulos, C., Jeno, P., Kronidou, N., Horst, M., Morishima, N., and Schatz, G. (1994). A mitochondrial homolog of bacterial GrpE interacts with mitochondrial hsp70 and is essential for viability. EMBO J. **13**, 1998–2006.

Ikeda, E., Yoshida, S., Mitsuzawa, H., Uno, I., and Toh-e, A. (1994). YGE1 is a yeast homologue of Escherichia coli grpE and is required for maintenance of mitochondrial function. FEBS Lett. **339**, 265–268.

Laloraya, S., Gambill, B.D., and Craig, E.A. (1994). A role for a eukaryotic GrpE-related protein, Mge1p, in protein translocation. Proc. Natl. Acad. Sci. USA **91**, 6481–6485.

Laloraya, S., Dekker, P., Voos, W., Craig, E., and Pfanner, N. (1995). Mitochondrial GrpE modulates the function of matrix Hsp70 in translocation and maturation of proteins. Mol Cell. Biol. **15**, 7098–7105.

Miao, B., Davis, J., and Craig, E.A. (1997). Mge1 functions as a nucleotide release factor for Ssc1, a mitochondrial Hsp70 of S. cerevisiae. J. Mol. Biol. **265**, 541–552.

Nakai, M., Kato, Y., Ikeda, E., Toh-e, A., and Endo, T. (1994). YGE1p, a eukaryotic Grpe homologue, is localized in the mitochondrial matrix aand interacts with mitochondrial hsp70. Biochem. Biophy. Res. Commun. **200**, 435–441.

Voos, W., Gambill, B.D., Laloraya, S., Ang, D., Craig, E.A., and Pfanner, N. (1994). Mitochondrial GrpE is present in a complex with hsp70 and preproteins in transit across membranes. Mol. Cell. Biol. **14**, 6627–6634.

Westerman, B., Prip-Buus, C., Neupert, W., and Schwarz, E. (1995). The role of the GrpE homologue, Mge1p, in mediating protein import and protein folding in mitochondria. EMBO J. **13**, 1998–2006.

Shikha Laloraya[1,2] and Elizabeth Craig:[1]
[1] Department of Biomolecular Chemistry
[2] Department of Oncology
University of Wisconsin–Madison
1300 University Avenue
Madison, WI 53706, USA
Tel. 1 608 263 7105
Fax. 1 608 262 5253
E-mail: ecraig@facstaff.wisc.edu

Drosophila melanogaster Droe1p

Drosophila *Droe1p* is a mitochondrial protein homolog of Escherichia coli *GrpE*. GrpE accelerates the rate of nucleotide release from the bacterial HSP70, DnaK. Droe1p is believed to function with the resident mitochondrial matrix Hsp70 and a DnaJ homolog to facilitate translocation of proteins into the mitochondria and then to help them attain their native conformation, as is the case for the related yeast chaperones. Expression of Droe1p has been shown to complement genetically bacterial grpE deletion strains demonstrating that Droe1p is a functional GrpE homolog.

■ Isolation

Droe1p was isolated from embryos using *Drosophila* Hsc4p-affinity chromatography (A.D. Mehta and K.B. Palter, unpublished data; Hsc4p is an Hsc70 homolog, see entry p. 42). The Hsc4p–Droe1p complex is resistant to dissociation by 2M KCl but dissociates in the presence of Mg^{2+}/ATP, as is true for bacterial GrpE (Zylicz *et al.*, 1987). Similar results were obtained using *Drosophila* Hsc3p-affinity chromatography (Hsc3p is the ER HSP70 protein, see entry p. 45). The sequences obtained from tryptic fragments of purified Droe1p were used to isolate a fragment of the gene using degenerate PCR. Full-length cDNA clones were obtained from both embryonic and larval cDNA libraries (J.Y. Lee and K.B. Palter, unpublished data). Droe1p has been shown to be a mitochondrial protein. It is significant that this purification protocol did not identify a cytoplasmic GrpE homolog. The GrpE homologs identified in yeast (Bolliger *et al.*, 1994; Ikeda *et al.*, 1994; Voos *et al.*, 1994) and mammalian cells (Naylor *et al.*, 1995) have all been mitochondrial proteins. These data suggest that a cytoplasmic GrpE homolog may have distinct biochemical properties from the mitochondrial and prokaryotic GrpE proteins, or may not exist.

■ *DROE1* sequence

The nucleotide sequence for *Drosophila DROE1* was determined from a larval cDNA clone (GenBank accession number U34903). The cDNA has a single open reading frame of 639 bp that would encode a protein of 213 amino acids with a molecular weight of 23 960 daltons. Droe1p shares 31% amino acid identity with *E. coli* GrpE and 38% identity with yeast GrpE. The GrpE proteins are conserved throughout their sequences, with the highest conservation near the C-terminus. The Droe1 protein contains an N-terminal extension characteristic of mitochondrial import sequences. *DROE1* produces two transcripts of 1.05 and 1.15 kb throughout development and neither is heat inducible.

■ Droe1 protein

The mature Droe1 protein migrates on SDS–PAGE gels with an apparent molecular weight of 22 kDa and has a pI of 6.9. *In vitro* synthesized Droe1p can be imported into isolated yeast mitochondria to a protease-inaccessible location while undergoing concomitant cleavage (J.Y. Lee, D. Gambill, K.B. Palter, unpublished data). Droe1p cofractionates with *Drosophila* Hsc5p, the mitochondrial HSP70. A polyclonal rabbit antiserum has been raised to a non-conserved peptide of Droe1p (anti-Droe1) (J.Y. Lee and K.B. Palter, unpublished data). The *DROE1* clone can complement *E. coli grpE* mutant strains defective for lambda infectivity and growth at 42°C (J.Y. Lee, K. Tilly and K.B. Palter, unpublished data).

■ References

Bolliger, L., Deloche, O., Glick, B.S., Georgopoulos, C., Jenoe, P., Kronidou, N., Horst, M., Morishima, N., and Schatz, G. (1994). A mitochondrial homolog of bacterial GrpE interacts with mitochondrial hsp70 and is essential for viability. EMBO J. **13**, 1998–2006.

Ikeda, E., Yoshida, S., Mitsuzawa, H., Uno, I., and Toh-e, A. (1994). YGE1 is a yeast homologue of *Escherichia coli grpE* and is required for maintenance of mitochondrial functions. FEBS Lett. **339**, 265–268.

Naylor, D.J., Ryan, M.T., Condron, R., Hoogenraad, N.J., and Høj, P.B. (1995). Affinity-purification and identification of GrpE homologues from mammalian mitochondria. Biochim. Biophys. Acta **1248**, 75–79.

Voos, W., Gambill, B.D., Laloraya, S., Ang, D., Craig, E.A., and Pfanner, N. (1994). Mitochondrial GrpE is present in a complex with hsp70 and preproteins in transit across membranes. Mol. Cell. Biol. **14**, 6627–6634.

Zylicz, M., Ang, D., and Georgopoulos, C. (1987). The *grpE* protein of *Escherichia coli*. J. Biol. Chem. **262**, 17437–17442.

■ *Ashwin D. Mehta:*
North Carolina State University
Department of Botany
Raleigh, NC 27695, USA

Tel. 1 919 515 3570
Fax. 1 919 515 3436
E-mail: ashwin@unity.ncsu.edu

■ *John Y. Lee and Karen B. Palter:*
Department of Biology
Temple University
12 St. and Norris St.
Philadelphia, PA 19122, USA
Tel. 1 215 204 8845
Fax. 1 215 204 6646
E-mail: palter@astro.ocis.temple.edu

Mammalian mitochondrial GrpE

mt-GrpE is a stress-inducible, mitochondrial protein of ubiquitous distribution in mammalian organs. It is a member of the GrpE family of molecular chaperones and its presumed role is to enhance ADP/ATP nucleotide exchange on mt-Hsp70 during mitochondrial protein import and subsequent folding in the matrix.

■ mt-GrpE sequence

The amino acid sequence deduced from the cDNA sequence of rat liver mt-GrpE (GenBank accession number U62940; Naylor *et al.* 1996) reveals a precursor protein of 217 amino acids (c. 24.3 kDa) with a 27 residue N-terminal targeting sequence, which upon mitochondrial protein import is proteolytically removed to generate the mature protein of 190 amino acids (c. 21.3 kDa). The nucleotide sequences of two partial cDNA clones from human white blood cells (TIGR EST#121800) and normal foreskin melanocyctes (Genbank accession number N28384), together encode a precursor protein of 217 amino acids. A partial mouse cDNA clone (GenBank accession number W08216) has also been isolated. The predicted isoelectric point (pI) of the mature rat mt-GrpE protein is 6.5, whilst the precursor due to its basic mitochondrial signal sequence has a predicted pI of 8.5 (Naylor *et al.* 1996).

The deduced amino acid sequences of rat and human mt-GrpE exhibit 90% positional identity, while positional identities with other GrpE family members are: 45% for mitochondrial Droe1p from *Drosophila melanogaster* (GenBank accession number U34903), 36% for *Caenorhabditis elegans* GrpE (EMBL accession number Z46996), 29% for mitochondrial Yge1p from *Saccharomyces cerevisiae* (Ikeda *et al.* 1994), 22% for *Escherichia coli* GrpE (Lipinska *et al.* 1988) and 20% for archaebacterial *Methanosracina mazei* GrpE (Conway de Marcario *et al.* 1994). Comparison of the deduced amino acid sequences of nineteen GrpE family members reveals that five residues are strictly conserved, four of which reside in the C-terminal third of GrpE (Naylor *et al.* 1996). A segment within the N-terminal third of bacterial GrpE has been predicted to form a coiled-coil secondary structure which has been postulated to facilitate protein–protein interaction (Wu *et al.* 1994). Genomic GrpE sequences have only been obtained from two eukaryotes. The *C. elegans* GrpE (Wilson *et al.* 1994) and *S. cerevisiae* Yge1p (Ikeda *et al.* 1994) genes contain four and no introns, respectively.

■ mt-GrpE protein

Mammalian mt-GrpE from bovine, porcine and rat liver mitochondria has been purified on DnaK affinity columns (Naylor *et al.* 1995). The association with the *E. coli* DnaK is resistant to 1 M KCl but is readily disrupted by 5 mM ATP/ 0.5 M KCl. Antibodies raised against mt-GrpE reveal an exclusive mitochondrial location for the 21 kDa protein. The protein is of low abundance constituting about 0.03% of the mitochondrial protein (Naylor *et al.* 1996), which can be compared with about 1% for mt-Hsp70 (Webster *et al.* 1994). As opposed to the yeast homologue (Bolliger *et al.* 1994), the mammalian homologue is not N-terminally blocked (Naylor *et al.* 1995). Recently, the crystal structure of *E. coli* GrpE has been solved (Harrison *et al.*, 1997).

■ mt-GrpE gene regulation

No genomic clones have been reported for mammalian mt-GrpE. The mt-GrpE transcripts are present in most if not all organs. Transcript levels in cultured cells rise slightly in response to growth in the presence of the proline analogue L-azetidine-2-carboxylic acid but appears not to be modulated by heat shock (Naylor *et al.* 1996). Consistent with this observation, the yeast *YGE1* gene is not induced by heat shock (Ikeda et al. 1994) and it appears that neither the S. cerevisiae nor the C. elegans grpE gene contain heat shock elements (Ikeda et al. 1994; Wilson et al. 1994).

■ Function of mt-GrpE

Apart from its tight association with HSP70 homologues, the biochemical function of mammalian mt-GrpE has not yet been elucidated. Consistent with the finding of mammalian mt-GrpE transcripts in every organ studied thus far, both the bacterial (Ang et al. 1986) and yeast (Bolliger et al. 1994; Ikeda et al. 1994) counterparts are essential for cell survival. By analogy to the proven function of GrpE in these organisms, mammalian mt-GrpE most likely acts as an ADP/ATP nucleotide exchange factor (Liberek et al. 1991) which modulates binding of substrate proteins to mt-Hsp70 during protein import (Laloraya et al. 1994; Nakai et al. 1994) and folding (Westermann et al. 1995) in the matrix (Martinus et al. 1995). In support of this suggestion, we have found mt-GrpE to be both free and associated with membranes in mitochondria (Naylor et al. 1996).

■ References

Ang, D.A., Chandrasekhar, G.N., Zylicz, M. and Georgopoulos, C. (1986) Escherichia coli grpE gene codes for heat shock protein B25.3, essential for both λ DNA replication at all temperatures and host growth at high temperature. J. Bacteriol. **167**, 25–29.

Bolliger, L., Deloche, O., Glick, B.S., Georgopoulos, C., Jenö, P., Kronidou, N., Horst, M., Morishima, N. and Schatz, G. (1994) A mitochondrial homolog of bacterial GrpE interacts with mitochondrial hsp70 and is essential for viability. EMBO J. **13**, 1998–2006.

Conway de Macario, E., Dugan, C.B. and Macario, A.J.L. (1994) Identification of a grpE heat-shock gene homolog in the archaeon Methanosarcina mazei. J. Mol. Biol. **240**, 95–101.

Ikeda, E., Yoshida, S., Mitsuzawa, H., Uno, I. and Toh-e A. (1994) YGE1 is a yeast homologue of Escherichia coli grpE and is required for maintaince of mitochondrial functions. FEBS Lett. **339**, 265–268.

Harrison, C.J., Hayer-Hartl, M., Di Liberto, M., Hartl, F.-U., and Kuriyan, J. (1997). Crystal structure of the nucleotide exchange factor GrpE bound to the ATPase domain of the molecular chaperone DnaK. Science **276**, 431–435.

Laloraya, S., Gambill, B.D. and Craig, E.A. (1994) A role for a eukaryotic GrpE-related protein, Mge1p, in protein translocation. Proc. Natl. Acad. Sci. USA **91**, 6481–6485.

Liberek, K., Marszalek, J., Ang, D., Georgopoulos, C. and Zylicz, M. (1991) Escherichia coli DnaJ and GrpE heat shock proteins jointly stimulate ATPase activity of DnaK. Proc. Natl. Acad. Sci. USA **88**, 2874–2878.

Lipinska, B., King, J., Ang, D. and Georgopoulos, C. (1988) Sequence analysis and transcriptional regulation of the Escherichia coli grpE gene, encoding a heat shock protein. Nucl. Acids Res. **16**, 7545–7562.

Martinus, R.D., Naylor, D.J., Herd, S.M., Hoogenraad, N.J. and Høj, P.B. (1995) Role of chaperones in the biogenesis and maintenance of the mitochondrion. FASEB J. **9**, 371–378.

Nakai, M., Kato, Y., Ikeda, E., Toh-e, A. and Endo, T. (1994) Yge1p, a eukaryotic Grp homolog, is localized in the mitochondrial matrix and interacts with mitochondrial hsp70. Biochem. Biophys. Res. Commun. **200**, 435–442.

Naylor, D.J., Ryan, M.T., Condron, R., Hoogenraad, N.J. and Høj, P.B. (1995) Affinity-purification and identification of GrpE homologs from mammalian mitochondria. Biochim. Biophys. Acta **1248**, 75–79.

Naylor, D.J., Hoogenraad, N.J., and Hoj, P.B. (1996). Isolation and characterisation of a cDNA encoding rat mitochondrial GrpE, a stress-inducible nucleotide exchange factor of ubiquitous appearance in mammalian organs. FEBS Lett., **396**, 181–188.

Webster, T.J., Naylor, D.J., Hartman D.J., Høj, P.B. and Hoogenraad, N.J. (1994) cDNA cloning and efficient mitochondrial import of pre-mt-Hsp70 from rat liver. DNA Cell Biol. **13**, 1213–1220.

Westermann, B., Prip-Buus, C., Neupert, W. and Schwarz, E. (1995) The role of the GrpE homologue, Mge1p, in mediating protein import and folding in mitochondria. EMBO J. **14**, 3452–3460.

Wilson, R., et al. (1994) 2.2 Mb of contiguous nucleotide sequence from chromosome III of C. elegans. Nature **368**, 32–38.

Wu, B., Ang, D., Snavely, M. and Georgopoulos, C. (1994) Isolation and characterisation of point mutations in the Escherichia coli grpE heat shock gene. J. Bacteriol. **176**, 6965–6973.

■ *Dean J. Naylor and Peter B. Høj:*
Department of Horticulture, Viticulture and Oenology
The University of Adelaide
PMB1, Glen Osmond
South Australia 5064, Australia
Tel. 61 8 8303 6670/6663 (D. J. N./ P. B. H.)
Fax. 61 8 8303 7117
E-mail dnaylor@waite.adelaide.edu.au and phoj@waite.adelaide.edu.au

■ *Nicholas J. Hoogenraad:*
School of Biochemistry
La Trobe University
Bundoora
Victoria 3083, Australia
Tel. 61 3 9479 2175
Fax. 61 3 9479 2467
E-mail: N. Hoogenraad@Latrobe.edu.au

5

HSP90 Proteins

The HSP90 family — an overview

HSP90 proteins constitute one of the most abundant and conserved HSP families. However, their structure and function is far from being understood in detail. Constitutively expressed and stress-induced members have been described and homologues have also been found in the endoplasmic reticulum (ER). While in vivo experiments suggest that eukaryotic Hsp90 is a specific chaperone involved in regulating signal transduction pathways by assisting structural changes of certain kinases and steroid receptors, results from in vitro studies highlight the general chaperone properties of HSP90 proteins. The existence of an ATPase activity associated with HSP90 remains under discussion although direct evidence for it is lacking. Like other chaperones, HSP90 proteins seems to perform at least part of their activity in association with partner proteins of so far unknown function.

■ Alternative names for family members

Hsp90, Hsp80, Hsp81, Hsp82, Hsc82, Hsp83, Hsp85, Hsp100, Grp94, endoplasmin, Gp96, HtpG, c62.5, Erp99, Hsp108.

■ Accession numbers for HSP90 sequences

Arabidopsis capsulatum, g-S21764; *Arabidopsis thaliana*, s-P27323; *Candida albicans*, g-X81025; *Canis familiaris*, (e) g-U01153; *Cricetulus griseus*, g-L33676; *Catharanthus roseus*, (e) g-L14594; *Drosophila melanogaster*, s-P02828; *Escherichia coli*, s-P10413; *Gallus gallus*, (a) s-P11501; *G. gallus*, (b) s-Q04619; *G. gallus*, (e) s-P08110; *Histoplasma capsulatum*, g-M55629; *Homo sapiens*, (a) s-P07900; *H. sapiens*, (b) s-P08238; *H. sapiens*, (e) s-P24625; *Hordeum vulgare*, (e) e-S31862; *Leishmania amazonensis*, s-P27741; *Lycopersicon esculentum*, g-M96549; *Mus musculus*, (a) s-P07901; *M. musculus*, (b) g-M36829; *M. musculus*, (e) s-P08113; *Oryza sativa*, s-P33126; *Plasmodium falciparum*, g-L34027; *Pharbitis nil*, g-M99431; *Rattus rattus*, (b) s-S45392; *Secale cereale*, (e) g-Z30243; *Saccharomyces cerevisiae*, (c) s-P15108; *Saccharomyces cerevisiae*, (h) s-P02829; *Schizosaccharomyces pombe*, g-L35550; *Sus scrofa* (e), g-X76301; *Trypanosoma brucei*, s-P12861; *Trypanosoma cruzi*, s-P06660; *Theileria parva*, g-M57386; *Zea mays*, g-S59780.

Note: The abbreviations s, g, and em indicate Swissprot, GenBank, and EMBL databases, respectively. The letters c, h, and e indicate cognate, heat-induced, and ER-resident forms of HSP90, respectively, and a and b stand for the α and β isoform of cytosolic HSP90s.

■ Sequences of HSP90 proteins

Members of the HSP90 family are found in the cytosol and in different cell compartments including the ER. All members of the family are well conserved from bacteria to humans. Common to all members of the HSP90 family are two highly conserved regions. A comparison of 33 HSP90 sequences in the database shows up to 73% identity from position number 1 to 205 (cf. Fig. 1) and up to 70% identity from position number 206 to 624 (cf. Fig. 1 and Table 1a, c). In all eukaryotic HSP90s a highly charged region with a length varying between 34 and 52 amino acids connects the two conserved portions of HSP90. This charged region shows no significant homology between the different members of the HSP90 family (cf. Table 1b, Fig. 1) and, interestingly, is lacking in the *E. coli* family member HtpG (Jakob and Buchner, 1994). This region is thought to be involved in targeting HSP90 to the nucleus (Kang et al., 1994). A second highly charged region near the C-terminus is well conserved between all HSP90 members identified. Furthermore, the C-terminal sequence EEVD is a common feature in all cytosolic members of the eukaryotic HSP90 family (Borkovich et al., 1989). In ER-resident members the sequence is EEXD followed by the retrieval signal; prokaryotic HSP90s lack this sequence (Table 1c).

The apparent molecular mass of eukaryotic members of the HSP90 family ranges from 80 to 108 kDa dependent on the size of the charged region, or on glycosylation in the case of ER-resident members. In addition, the ER-resident isoform (Grp94) contains an N-terminal signal sequence and a C-terminal retrieval signal (KDEL). Interestingly, a search of the yeast genome database did not reveal the existence of a Grp94 homologue in *Saccharomyces cerevisiae*.

Although not much is known about structure–function relationships of HSP90 proteins, some regions of functional importance have been described. The steroid receptor and peptide-binding sites are proposed to be located between residues 381 and 677 (Minami et al., 1993; Sullivan and Toft, 1993; Fig. 2) and the dimerization site comprises the 191 amino acids at the C-terminal end of HSP90, with the exception of the last 35 amino acids (Nemoto et al., 1995; Fig. 2). Unlike members of the HSP60, HSP70, or HSP100 protein families, the HSP90 proteins do not contain sequence motifs characteristic of an ATP-binding site (Jakob et al., 1996).

Regulation of HSP90 proteins

Although isoforms of HSP90 have interchangeable functions (Borkovich et al., 1989), the respective genes are differentially regulated. In most eukaryotic cells one of the two cytosolic members (Hsc82 in yeast, Hsp90β in

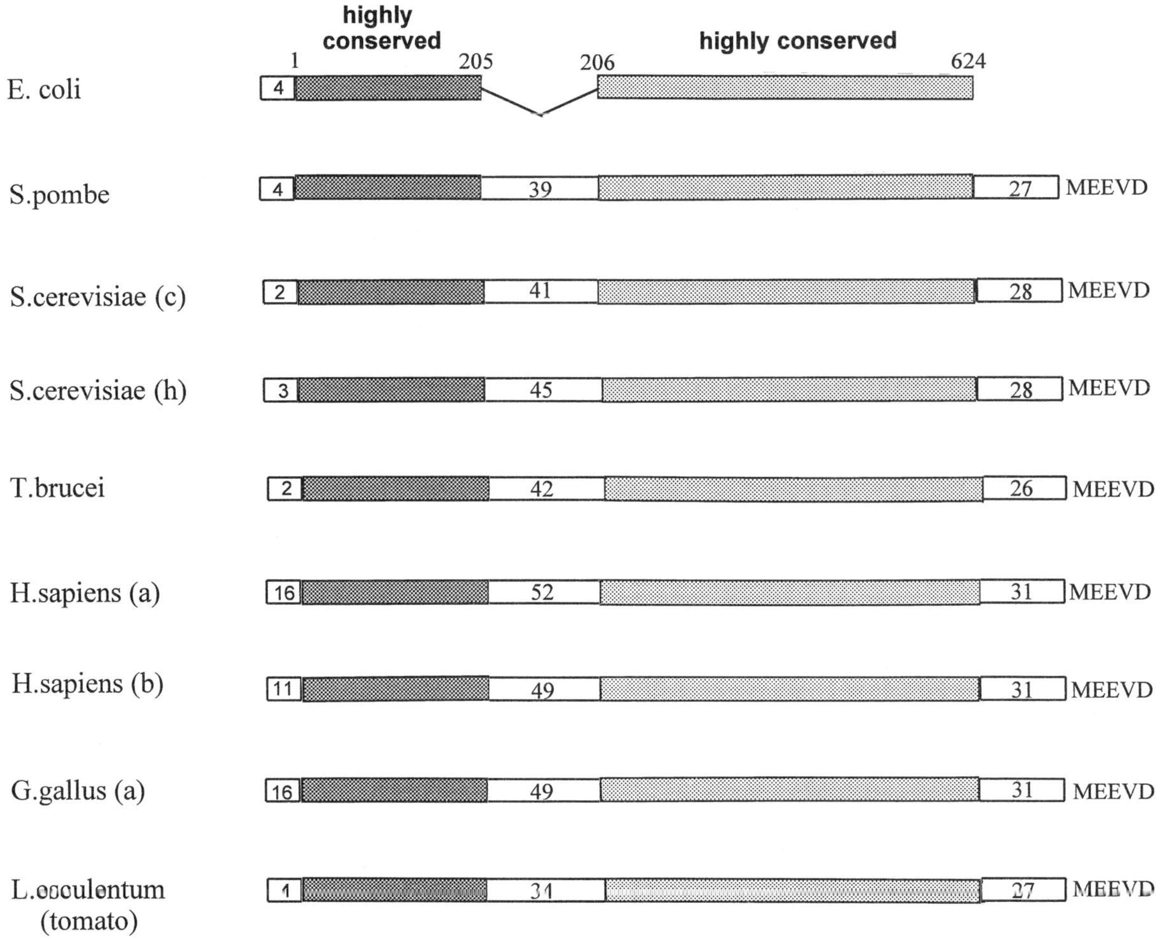

Figure 1. The multiple sequence alignment of cytosolic HSP90s was performed using Multialign, Malign and Clustal within HUSAR (GCG package). Two highly conserved regions (shaded areas) have been numbered 1–205 (N-terminal region) and 206–624 (C-terminal region). The open boxes are of variable length and the number of amino acids in this region is given for each fragment. The abbreviations a, b, c, and h are used as described for the accession numbers.

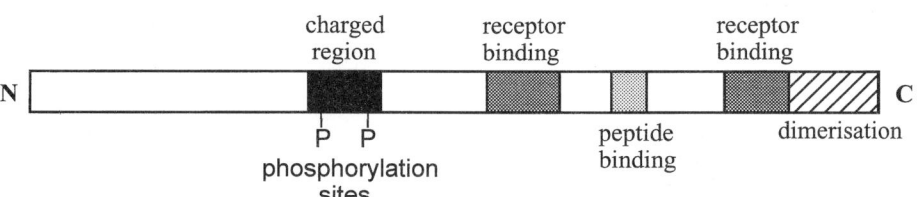

Figure 2. Structure–function relationship of HSP90. The charged region is only found in eukaryotic members of the HSP90 family (Jakob and Buchner, 1994). The phosphorylation sites were determined by Lees-Miller and Anderson (1989) and the peptide binding by Minami *et al.* (1993). Deletion of the two shaded regions abolishes receptor binding (Sullivan and Toft, 1993). The 191 C-terminal amino acids have been described to be involved in dimerization (Minami *et al.*, 1993; Nemoto *et al.*, 1995).

Table 1. Sequence alignments of HSP90

(a) N-terminal region: 73% indentity between position numbers 1–205 (as numbered in Fig. 1, highly conserved within 33 prokaryotic, eukaryotic, and ER-resident HSP90 sequences)

(b) Charged region:

```
ecoli    206  PVEI-------------------------------------------------------EKREEKDGETVISWEKI NKA
spombe   218  PIQLVVTREVEKEVPEEEETE-------EVKNEEDDKA-------PKIEEVDDESE-----K-KEKKTKKVKETTTETEEL NKT
sacchc   216  PIQLLVTKEVEKEVPIPEEEKKDEEKKDE-----DDKK-------PKLEEVDEEEE----K-KPK-TKKVKEEVQELEEL NKT
sacchh   203  PIQLVVTKEVEKEVPIPEEEKKDEEKKDEDDKK-------PKLEEVDEEE-----K-KPK-TKKVKEEVQEIEEL NKT
tbrucei  201  DIELMVENTTEKEVTDEDEDEEAAKKAEEGEEPK-------------VEEVKDGVDADAK----KKKTKKVKEVKQEFVVQ NKH
humanb   211  PITLYLEKEREKEISDDEAEEEKG---EKEEEDKDDEEKPK------IEDVGSDEEDDSG-KDKKKKTKKIKEKYIDQEEL NKT
chicka   216  PIRLFVEKERDKEVSDDEAEEKE---EEKEEKEEKTEDKPE-----IEDVGSDEEEE-KKDGDKKKKKKIKEKYIDEEEL NKT
humana   216  PITLFVEKERDKEVSDDEAEEKEDKEEEKEKEEKESEDKPE------IEDVGSDEEEE-KKDGDKKKKKKIKEKYIDQEEL NKT
tomato   208  PISLWVEKTIEKEIS-------------DDEEEEEKKD-EEGK---VEEVD-EE----KEKEEKKKKK-VKEVSNEWS LVNKQ

CONSENSUS      PI L   EKE EKEV     EEEE K     E              EEV     EEE       KKPKTKKVKE      EEL NKT
```

(c) C-teminal region: 70% identity between position numbers 206–624 (as numbered in Fig. 1)

(cytosolic)	625	I	D	E	D	X_{22-27}	M	E	E	V	D
(ER resident)	625	I	D	X	D/E	X_{22-27}	X	E	E	X	D	X_{4-17}	K D E L

vertebrates) is expressed constitutively at a high level at normal temperatures and induced only 2–3 times by heat shock (Borkovich et al., 1989; Krone and Sass, 1994). The second HSP90 gene (encoding Hsp82 in yeast, Hsp90α in vertebrates) is expressed at a low basal level at normal temperatures, but expression is enhanced strongly by heat treatment. In some plant species, members of the HSP90 family are encoded by a multigene family consisting of 5–6 proteins (Felsheim and Das, 1992; Krishna et al., 1995). Again, one or more members of the same gene family are strongly expressed at normal temperatures and show less heat inducibility, whereas others are strongly expressed under heat shock with a low basal level at normal growth conditions (Marrs et al., 1993; Yabe et al., 1994). In addition to heat, HSP90 mRNA levels in plants can be increased by shifting plants from dark to light (Felsheim and Das, 1992), during cold stress (Krishna et al., 1995), or by chemical treatments (Yabe et al., 1994).

In contrast, Drosophila contains only one known cytosolic HSP90, Hsp83, which is constitutively expressed under permissive conditions. Heat shock results in a 10–15-fold increase in Hsp83 mRNA levels, which is dependent on three tandem heat shock consensus elements (Xiao and Lis, 1989). The E. coli member of the HSP90 family, HtpG, which is controlled by the heat shock sigma factor σ^{32}, is expressed under normal growth conditions at up to 0.5% of the cellular mass (Herendeen, 1979) and shows an increased synthesis rate upon shift to higher temperatures (Neidhardt et al., 1984; Zhou et al., 1988) or other stress situations like carbon starvation (Jenkins et al., 1991) and acid shock (Heyde and Portalier, 1990).

The promotor of the grp94 gene shares a common regulatory domain with the promotor of the grp78 gene that encodes BiP, the ER member of the HSP70 family (Liu and Lee, 1991). The constitutive expression of each of these genes is up-regulated by a broad variety of stress conditions including glucose starvation, reducing reagents, and calcium ionophores, all of which are thought to increase the population of misfolded polypeptide species in the ER (Gething et al., 1994).

■ Localization of HSP90 proteins

Members of the HSP90 family have been found in the cytosol of prokaryotes and eukaryotes and in the ER of higher eukaryotes. In addition, cytosolic HSP90 protein(s) in eukaryotes is also localized to some extent to the nuclear compartment (Kang et al., 1994), in agreement with its function in the regulation of steroid receptors.

■ Structure and stability of HSP90 proteins

The three-dimensional structure of HSP90 is not known. The secondary structure of HSP90 proteins is composed of both α helices (about 60%) and β strands (about 15%) as deduced from CD spectroscopy (Jakob et al., 1995). Proteolysis experiments suggest that HSP90s are composed of several domains (T. Scheibel and J. Buchner, unpublished data). The structure of the N-terminal domain, which binds the anti-tumor antibiotic geldanamycin, has been resolved to a resolution of 1.9 Å. This domain has nine helices and an antiparallel β-sheet of eight strands that fold into an α–β sandwich (Prodromou et al., 1997; Stebbins et al., 1997). The functional form of HSP90 is the dimer (Jakob et al., 1995).

The stability against thermal and denaturant-induced unfolding has been determined for prokaryotic and eukaryotic HSP90 proteins. The midpoints for the thermal unfolding transitions were found to be between 50°C for human Hsp90 (Lanks et al., 1992) and 64°C for E. coli HtpG (Jakob *et al.*, 1995). Denaturant-induced unfolding showed biphasic transitions reflecting distinct steps during the unfolding process.

■ Function of HSP90 proteins

Several aspects of the function of HSP90 proteins are still discussed in a controversial manner, the most important being whether HSP90 proteins are general or specific chaperones. *In vivo* experiments performed so far mainly suggest that Hsp90 is a specific chaperone involved in regulating the structure and activity of proteins involved in signal transduction (Xu and Lindquist, 1993). There is compelling evidence for an essential function of Hsp90 in complex with specific partner proteins such as prolyl isomerases and p23 in steroid receptor activation and the structure formation of kinases regulating the cell cycle (see entry p. 518). Furthermore, for ER-resident members a specific function in the presentation of antigenic peptides has been suggested (Li and Srivastava, 1993). In contrast, Grp94 was also found to be involved in the folding of secretory proteins (Melnick *et al.*, 1994). This is in agreement with *in vitro* experiments demonstrating general chaperone properties of HSP90 proteins in protein folding under physiological and heat shock conditions (Jakob and Buchner, 1994; Buchner, 1996). Taken together, these experiments show that HSP90 is able to interact selectively with non-native polypeptides and to effectively prevent irreversible aggregation. The *in vitro* chaperone function of HSP90 was found to be ATP-independent in all cases investigated so far (Jakob and Buchner, 1994; Buchner, 1996) and there is no direct evidence for ATP binding to HSP90 (Jakob *et al.*, 1996). However, ATP binding may be involved in the association of partner proteins such as p23 to Hsp90.

In principle, HSP90 proteins are able to cooperate with the HSP70 system in protein folding (Freeman and Morimoto, 1996). This cooperation seems to be guarded and guided by Hop (also called p60 or Sti1, see entries pp. 458 and 459), a protein that seems to bind specifically to Hsp70 and Hsp90, thus bringing the two chaperone machineries into physical contact (Smith *et al.*, 1993). The role of HSP90 partner proteins in this context is unclear; however, some of them seem to exhibit properties of molecular chaperones themselves (B. Freeman, D. Toft and R. Morimoto, unpublished data; S. Bose, T. Weikl, H. Bügl and J. Buchner, unpublished data).

■ References

Borkovich, K.A., Farrelly, F.W., Finkelstein, D.B., Taulien, J., and Lindquist, S. (1989). Hsp82 is an essential protein that is required in higher concentrations for growth of cells at higher temperatures. Mol. Cell. Biol. **9**, 3919–3930

Buchner, J. (1996). Supervising the fold. Functional principles of molecular chaperones. FASEB J. **10**, 10–19

Felsheim, R.F. and Das, A. (1992). Structure and expression of a heat shock protein 83 gene of *Pharbitis nil*. Plant Physiol. **100**, 1764–1771

Freeman, B.C. and Morimoto, R.I. (1996). The human cytosolic molecular chaperones Hsp90, hsp70 (hsc70) and hdj-1 have distinct roles in recognition of a non-native protein and protein refolding. EMBO J. **15**, 2969–2979

Gething, M.-J., Blond-Elguindi, S., Mori, K., and Sambrook, S.F. (1994). Structure, function and regulation of the endoplasmic reticulum chaperone BiP. In *The biology of heat shock proteins and molecular chaperones*, (Morimoto, R.I., Tissieres, A. and Georgopoulos, C. eds). Cold Spring Harbor Laboratory Press, New York, pp. 111–135

Herendeen, S.L., vanBogelen, R.A., and Neidhardt, F.C. (1979) Levels of major proteins of Escherichia coli during growth at different temperatures. J. Bacteriol. **139**, 185–194

Heyde, M. and Portalier, R. (1990). Acid shock proteins of Escherichia coli. FEMS Micro. Lett. **57**, 19–26

Jakob, U. and Buchner, J. (1994). Assisting spontaneity — the role of Hsp90 and small Hsps as molecular chaperones. Trends Biochem. Sci. **19**, 205–211

Jakob, U., Meyer, I., Bügl, H., André, S., Bardwell, J.C.A., and Buchner, J. (1995). Structural organization of procaryotic and eucaryotic Hsp90. J. Biol. Chem. **270**, 14112–14119

Jakob, U., Scheibel, T., Bose, S., Reinstein, J., and Buchner, J. (1996). Assessment of the ATP binding properties of Hsp90. J. Biol. Chem. **271**, 10035–10041

Jenkins, D.E., Auger, E.A., and Matin, A. (1991). Role of RpoH, a heat shock regulator protein, in *Escherichia coli* carbon starvation protein synthesis and survival. J. Bacteriol. **173**, 1992–1996

Kang, K.I., Devin, J., Cadepond, F., Jibard, N., Guiochion-Mantel, A., Baulieu, E.-E., and Catelli, M.-G. (1994). *In vivo* functional protein–protein interaction: nuclear targeted hsp90 shifts cytoplasmic steroid receptor mutants into the nucleus. Proc. Natl. Acad. Sci. USA **91**, 340–344

Krishna, P., Sacco, M., Cherutti, J.F., and Hill, S. (1995). Cold induced accumulation of Hsp90 transcripts in *Brassica napus*. Plant Physiol. **107**, 915–923

Krone, P.H. and Sass, J.B. (1994). Hsp90 alpha and Hsp90 beta are present in the zebrafish and are differentially regulated in developing embryos. Biochem. Biophys. Res. Commun. **204**, 746–752

Lanks, K.W., London, E., and Dong, D.L. (1992), Hsp85 conformational change within the heat shock temperature range. Biochem. Biophys. Res. Commun. **184**, 394–399

Lees-Miller, S.P. and Anderson, C.W. (1989). The human double stranded DNA-activated protein kinase phosphorylates the 90-kDa heat shock protein Hsp90 alpha at two NH_2-terminal threonine residues. J. Biol. Chem. **264**, 17275–17280.

Li, Z. and Srivastava, P.K. (1993) Tumor rejection antigen gp96/grp94 is an ATPase: implications for protein folding and antigen presentation. EMBO J. **12**, 3143–3151

Liu, E.S. and Lee, A.S. (1991). Common sets of nuclear factors binding to the conserved promotor sequence motif of two coordinately regulated ER protein genes GRP78 and GRP94. Nucl. Acids Res. **19**, 5425–5431

Marrs, K.A., Casey, E.S., Capitant, S.A., Bouchard, R.A., Dietrich, R.S., Mettler, I.J., and Sinibaldi, R.M. (1993). Characterization of two maize Hsp90 heat shock genes: expression during heat shock, embryogenesis and pollen development. Develop. Genet. **14**, 27–41

Melnick, J., Dul, J.L., and Argon, Y. (1994). Sequential interaction of the chaperones BiP and GRP94 with immunoglobulin light chain in the endoplasmic reticulum. Nature **370**, 373–375

Minami, Y., Kawasaki, A., Susuki, K., and Yahara, I. (1993) The calmodulin binding domain of the mouse 90 kDa heat shock protein. J. Biol. Chem. **268**, 9604–9610

Neidhardt, F.C., van Bogelen, R.A., and Vaughn, V. (1984) The genetics and regulation of heat shock proteins. Annu. Rev. Genet. **18**, 295–329.

Nemoto, T., Ohara-Nemoto, Y., Ota, M., Takagi, T., and Yokoyama, K. (1995). Mechanism of dimer formation of the 90 kDa heat shock protein. Eur. J. Biochem. **233**, 1–8.

Prodromou, C., Roe, S.-M., Piper, P.W. and Pearl, L.H. (1997). A molecular clamp in the crystal structure of the yeast Hsp90 chaperone. Nature Struct. Biol. **4**, in press.

Smith, D. F., Sullivan, W.P., Marion, T.N., Zaitsu, K., Madden, B., McCormick, D.J., and Toft, D.O. (1993). Identification of a 60 kilodalton stress-related protein, p60, which interacts with hsp90 and hsp70. Mol. Cell. Biol. **13**, 869–876

Stebbins, C.E., Russo, A.A., Schneider, C., Rosen, N., Hartl, F.U. and Pavletich, N.P. (1997). Crystal structure of an Hsp90-geldanamycin complex: targeting of a protein chaperone by an anti-tumor agent. Cell **89**, 239–250.

Sullivan, W.P. and Toft, D.O. (1993). Mutational analysis of hsp90 binding to the progesterone receptor. J. Biol. Chem. **268**, 20373–20379

Xiao, H. and Lis, J.T. (1989). Heat shock and developmental regulation of the Drosophila melanogaster hsp83 gene. Mol. Cell. Biol. **9**, 1746–1753

Xu, Y. and Lindquist, S. (1993). Heat shock protein hsp90 governs the activity of pp60 v-src kinase. Proc. Natl. Acad. Sci. USA **90**, 7074–7078

Yabe, N., Takahashi, T., and Komeda, Y. (1994). Analysis of tissue-specific expression of Arabidopsis thaliana HSP90 family gene hsp81. Plant Cell. Physiol. **35**, 1207–1219

Zhou, Y.N., Kusukawa, N., Erikson, J.W., Gross, C.A., and Yura, T. (1988). Isolation and characterization of Escherichia coli mutants that lack the heat shock sigma factor sigma 32. J. Bacteriol. **170**, 3640–3649.

■ *Thomas Scheibel and Johannes Buchner:*
Universität Regensburg
Institut für Biophysik und Physikalische Biochemie
93040 Regensbur, Germany
Tel. 49 941 943 3039
Fax. 49 941 943 2813
E-mail: johannes.buchner@biologie.
uni-regensburg.de

Escherichia coli HtpG

The htpG *gene of* Escherichia coli *encodes a heat-inducible, cytoplasmic protein highly homologous to members of the HSP90 protein family of eukaryotes. Like its eukaryotic homologues it possesses protein chaperone activity in vitro.*

■ Alternative names

E. coli Hsp90 (Jakob et al., 1995a, b), C62.5 (Bardwell and Craig, 1987).

■ *htpG* gene and sequence

htpG was originally isolated based on hybridization to Drosophila hsp90 probes (Bardwell and Craig, 1987). htpG is located at 10.7 minutes on the E. coli chromosome. The nucleotide sequence of htpG (GenBank accession number M38777) predicts a protein that shares 40% amino acid identity with the eukaryotic members of the HSP90 family (Bardwell and Craig, 1987). Two 20 amino acid stretches near the C-terminus have > 90% identity. Major deletions have occurred in the htpG gene relative to its eukaryotic homologues; a highly charged region near the N-terminus is shortened by 50 amino acids and another deletion shortens the C-terminus by 35 amino acids. These deletions are unlikely to be of great functional significance since they lie in the most poorly conserved regions of HSP90 family members, but they result in a smaller protein with a predicted molecular mass of 71 429 Da.

■ HtpG protein

The HtpG protein has been purified from overexpression strains (Jakob et al., 1995b; Spence and Georgopoulos, 1989). It is active as a dimer but readily forms higher oligomers, especially in the presence of divalent cations (Jakob et al., 1995b). A small portion of HtpG is phosphorylated *in vivo* (Spence and Georgopoulos, 1989).

■ Role of HtpG in protein folding

Purified HtpG functions as an ATP-independent molecular chaperone *in vitro* (Jakob et al., 1995a, b). It recognizes and transiently binds non-native folding intermediates, reducing their free concentration in solution and thus preventing unspecific aggregation. These activities are very similar, both qualitatively and quantitatively, to bovine or yeast Hsp90 protein (Jakob et al., 1995a, b). Overproduction of HtpG compensates for decreased levels of FtsH, a membrane-bound protein that may have a chaperone-like function (Kihara et al., 1995; Shirai et al., 1996; see entry p. 451).

■ Mutagenesis studies

The htpG gene is non-essential (Bardwell and Craig, 1988). However, null mutations have a growth disadvantage at 37°C that increases with increasing temperature, becoming severe around 46°C.

■ HtpG regulation

The HtpG protein is an abundant cytoplasmic protein that makes up 0.45% of *E. coli*'s total protein mass at 37°C (Herendeen et al., 1979). On a molar basis, the HtpG dimer is about as abundant as the GroEL 14mer. Its transcription is controlled by the heat shock sigma factor, σ^{32}, and thus its synthesis rate is induced upon shift to high temperatures (Neidhardt et al., 1984; Cowing et al., 1985; Zhou et al., 1988). Upstream from the *htpG* gene are two overlapping promoter sequences that are transcribed by RNA polymerase containing σ^{32} *in vivo* and *in vitro* (Cowing et al., 1985). HtpG expression is also induced by carbon starvation (Jenkins et al., 1991), acid shock (Heyde and Portalier, 1990), 2,4-dinitrophenol (Gage and Neidhardt, 1993) and various other compounds. In all cases tested, this induction is under σ^{32} control.

■ References

Bardwell, J.C.A. and Craig, E.A. (1987). Eukaryotic Mr 83,000 heat shock protein has a homologue in *Escherichia coli*. Proc. Natl. Acad. Sci. USA **84**, 5177–5181.

Bardwell, J.C.A. and Craig, E.A. (1988). Ancient heat shock protein is dispensable. J. Bacteriol. **170**, 2977–2983.

Cowing, D.S., Bardwell, J.C.A., Craig, E.A., Woolford, C., Hendrix, R.W., and Gross, C.A. (1985). Consensus sequence for *Escherichia coli* heat shock gene promoters. Proc. Natl. Acad. Sci. USA **82**, 2679–2683.

Gage, D.J. and Neidhardt, F.C. (1993). Adaptation of *Escherichia coli* to the uncoupler of oxidative phosphorylation 2,4-dinitrophenol. J. Bacteriol. **175**, 7105–7108.

Herendeen, S.L., VanBogelen, R.A., and Neidhardt, F.C. (1979). Levels of major proteins of *Escherichia coli* during growth at different temperatures. J. Bacteriol. **139**, 185–194.

Heyde, M. and Portalier, R. (1990). Acid shock proteins of *Escherichia coli*. FEMS Micro. Lett. **69**, 19–26.

Jakob, U., Lilie, H., Meyer, I., and Buchner, J. (1995a). Transient interaction of Hsp90 with early unfolding intermediates of citrate synthase: Implications for heat shock *in vivo*. J. Biol. Chem. **270**, 7288–7294.

Jakob, U., Meyer, I., Bügl, H., André, S., Bardwell, J.C.A., and Buchner, J. (1995b). Structural organization of prokaryotic and eukaryotic Hsp90: influence of divalent cations on structure and function. J. Biol. Chem. **270**, 14412–14419.

Jenkins, D.E., Auger, E.A., and Matin, A. (1991). Role of RpoH, a heat shock regulator protein, in Escherichia coli carbon starvation protein synthesis and survival. J. Bacteriol. **173**, 1992–1996.

Kihara, A., Akiyama, Y., and Ito, K. (1995). FtsH is required for proteolytic elimination of uncomplexed forms of SecY, an essential protein translocation subunit. Proc. Natl. Acad. Sci. USA **92**, 4532–4536.

Neidhardt, F.C., VanBogelen, R.A., and Vaughn, V. (1984). The genetics and regulation of heat shock proteins. Annu. Rev. Genet. **18**, 295–329.

Shirai, Y., Akiyama, Y., and Ito, K. (1996). Suppression of ftsH mutant phenotypes by overproduction of molecular chaperones. J. Bacteriol. **178**, 1141–1145.

Spence, J. and Georgopoulos, C. (1989). Purification and properties of the *Escherichia coli* heat shock protein, HtpG. J. Biol. Chem. **264**, 4398–4403.

Zhou, Y.-N., Kusukawa, N., Erickson, J.W., Gross, C.A., and Yura, T. (1988). Isolation and characterization of *Escherichia coli* mutants that lack the heat shock sigma factor σ^{32}. J. Bacteriol. 170, 3640–3649.

■ Ursula Jakob and James C. A. Bardwell:
Department of Biology
University of Michigan
Ann Arbor, MI 48109-1048, USA
Tel. 1 313 764 8028
Fax. 1 313 647 0884
E-mail: Ujakob@biology.lsa.umich.edu and Jbardwel@biology.lsa.umich.edu

Saccharomyces cerevisiae Hsp90

Hsp90 is a protein chaperone that plays a role in diverse signal transduction pathways. In the yeast Saccharomyces cerevisiae, Hsp90 is encoded by two nearly identical but differentially regulated genes, HSC82 and HSP82. Yeast Hsp90 is structurally and functionally conserved with mammalian Hsp90. Mutational studies of yeast Hsp90 revealed that Hsp90 is required to promote and maintain various signal transduction pathways by direct interaction with transcription factors and kinases.

■ Alternative names

Hsc82/Hsp82.

■ *HSC82* and *HSP82* genes and sequences

The *HSP90* gene family of *Saccharomyces cerevisiae* consists of two members, *HSC82* (GenBank accession number M26044) and *HSP82* (GenBank accession number K01387) (Farrelly and Finkelstein, 1984; Borkovich et al., 1989). The genes encode products that are 97% identical to each other at the amino acid level and share 62% amino acid identity with human Hsp90. Although the genes are regulated differently, the Hsp82 and Hsc82 proteins appear to have interchangeable functions (Borkovich et al., 1989). The predicted gene products of *HSC82* and *HSP82* contain two highly charged regions and their four carboxy-terminal amino acids are EEVD, a feature shared with all eukaryotic members of cytosolic HSP90 and HSP70 families (Borkovich et al., 1989).

■ Gene regulation

HSP82 is constitutively expressed at a low level and strongly induced by stress. HSC82 is constitutively expressed at a high level and only slightly induced by stress (Borkovich et al., 1989). The promoters of the two genes contain heat shock elements (HSEs) as well as other potential regulatory elements (Erkine et al., 1995). The HSP82 upstream region contains one strong HSE and two weak HSEs. The presence of a nucleosome-free DNase I hypersensitive region is mediated by binding of heat shock factor (HSF) (Gross et al., 1993). HSF binds to HSE1 before heat shock. Increased binding to the HSEs occurs upon heat shock, and HSF dissociates from the HSEs during recovery (Giardina and Lis, 1995).

■ Yeast Hsp90 protein and function

The molecular weights of Hsp82 and Hsc82 are 81 419 and 80 885 daltons, respectively (Farrelly and Finkelstein, 1984; Borkovich et al., 1989). Hsp90 has been reported to have ATPase activity (Nadeau et al., 1993), but others have disputed this finding. Indeed, its ability to bind ATP is questionable (Jakob et al., 1996).

Yeast Hsp90 is essential for viability at normal temperatures and increasing concentrations of Hsp90 are required for survival at higher temperatures (Borkovich et al., 1989). Hsp90 plays an important role in the maturation of a wide variety of proteins that function in signal transduction (Picard et al., 1990; Bohen and Yamamoto, 1993, 1994; Xu and Lindquist, 1993; Carver et al., 1994; Nathan and Lindquist, 1995; Whitelaw et al., 1995). Based on the following results, the functions of yeast Hsp90 and mammalian Hsp90 are conserved. (1) Human Hsp90 complements null mutations of yeast Hsp90 (Picard et al., 1990; Minami et al., 1994). (2) Yeast Hsp90s form functional interactions with and promote the activation of several vertebrate target proteins, including steroid receptor family proteins (SHR) or $p60^{v\text{-}src}$ family proteins (Picard et al., 1990; Bohen and Yamamoto, 1993; Xu and Lindquist, 1993; Chang and Lindquist, 1994; Nathan and Lindquist, 1995). (3) Recombinant yeast Hsp90 activates casein kinase II in vitro, as does mammalian Hsp90 (Miyata and Yahara, 1992) . (4) Purified yeast Hsp90 effectively prevents aggregation of citrate synthase, as does mammalian Hsp90 (Jakob et al., 1995).

■ Mutagenesis studies

Mutagenesis studies were performed to create three kinds of mutants: (i) those that have low levels of Hsp90, (ii) those that show decreased activities for SHR activation, and (iii) temperature-sensitive mutants of Hsp90 (Picard et al., 1990; Bohen and Yamamoto, 1993; Xu and Lindquist, 1993; Kimura et al., 1994; Nathan and Lindquist, 1995). Point mutations of Hsp90 are broadly distributed throughout the protein (Bohen and Yamamoto, 1993; Kimura et al., 1994; Nathan and Lindquist, 1995). Studies in which heterologous target proteins were expressed in yeast Hsp90 mutants revealed that Hsp90 regulates the activities of its target proteins and promotes the acquisition and maintenance of their active conformations (Picard et al., 1990; Bohen and Yamamoto, 1993, 1994; Xu and Lindquist, 1993; Carver et al., 1994; Nathan and Lindquist, 1995; Whitelaw et al., 1995). In one study eight out of eight single amino acid subtitutions selected for temperature-sensitive growth and located in different regions of the protein affected the activity of both a transcription factor (the glucocorticoid receptor) and a tyrosine kinase ($pp60^{v\text{-}src}$), suggesting that Hsp90 promotes the activation of even very different target proteins through a common mechanism (Nathan and Lindquist, 1995). Hsp90 mutants also show a defect in DNA synthesis and a delayed cell cycle progression at non-permissive temperatures (Kimura et al., 1994).

■ Biological and genetic interactions

As observed for mammalian Hsp90, yeast Hsp90 is found in macromolecular complexes with some members of HSP70 family (Ssa, but not Ssb or BiP), p60 (Sti1p) and immunophilins including Cyp-40, FKBP59 and CypA (Nadeau et al., 1993; Chang and Lindquist, 1994). hsp82 mutations are synthetically lethal with a mutation in the gene encoding Ydj1p, a member of the DnaJ family (Kimura et al., 1995). Yeast Hsp90 binds to HSF (heat shock factor), suggesting it may be involved in the regulation of the heat shock response (Nadeau et al., 1993).

■ References

Bohen, S.P. and Yamamoto, K.R. (1993). Isolation of Hsp90 mutants by screening for decreased steroid receptor function. Proc. Natl. Acad. Sci. USA **90**, 11424–11428.

Bohen, S.P. and Yamamoto, K.R. (1994). Modulation of steroid receptor signal transduction by heat shock proteins. In The Biology of Heat Shock Proteins and Molecular Chaperones, (R.I. Morimoto, A. Tissiéres and C. Georgopoulos, eds). Cold Spring Harbor Laboratory Press, New York, pp. 313–334.

Borkovich, K.A., Farrelly, F.W., Finkelstein, D.B., Taulien, J., and Lindquist, S. (1989). Hsp82 is an essential protein that is required in higher concentrations for growth of cells at higher temperatures. Mol. Cell. Biol. **9**, 3919–3930.

Carver, L.A., Jackiw, V., and Bradfield, C.A. (1994). The 90-kDa heat shock protein is essential for Ah receptor signaling in a yeast expression system. J. Biol. Chem. **269**, 30109–30112.

Chang, H.-C.J. and Lindquist, S.L. (1994). Conservation of Hsp90 macromolecular complexes in Saccharomyces cerevisiae. J. Biol. Chem. **269**, 24983–24988.

Erkine, A.M., Szent-Gyorgyi, C., Simmons, S.F., and Gross, D. (1995). The upstream sequences of the HSP82 and HSC82 genes of Saccharomyces cerevisiae: regulatory elements and nucleosome positioning motifs. Yeast **11**, 573–580.

Farrelly, F.W. and Finkelstein, D.B. (1984). Complete sequence of the heat shock-inducible HSP90 gene of Saccharomyces cerevisiae. J. Biol. Chem. **259**, 5746–5751.

Giardina, C. and Lis, J.T. (1995). Dynamic protein–DNA architecture of a yeast heat shock promoter. Mol. Cell. Biol. **15**, 2737–2744.

Gross, D.S., Adams, C.C., Lee, S., and Stenz, B. (1993). A critical role for heat shock transcription factor in establishing a nucleosome-free region over the TATA-initiation site of the yeast HSP82 heat shock gene. EMBO J. **12**, 3931–3945.

Jakob, U., Meyer, I., Buggi, H., Andre, S., Bardwell, C.A., and Buchner, J. (1995). Structural organization of procaryotic and eucaryotic Hsp90. J. Biol. Chem. **270**, 14412–14419.

Jakob, U., Scheibel, T., Bose, S., Reinstein, J., and Buchner, J. (1996). Assessment of the ATP binding properties of Hsp90. J. Biol. Chem. **271**, 10035–10041.

Kimura, Y., Matsumoto, S., and Yahara, I. (1994). Temperature-sensitive mutants of *hsp82* of the budding yeast *Saccharomyces cerevisiae*. Mol. Gen. Genet. **242**, 517–527.

Kimura, Y., Yahara, I., and Lindquist, S. (1995). Role of the protein chaperone YDJ1 in establishing Hsp90-mediated signal transduction pathways. Science **268**, 1362–1365.

Minami, Y., Kimura, Y., Kawasaki, H., Suzuki, K., and Yahara, I. (1994). The carboxy-terminal region of mammalian HSP90 is required for its dimerization and function in vivo. Mol. Cell. Biol. **14**, 1459–1464.

Miyata, Y. and Yahara, I. (1992). The 90-kDa heat shock protein, HSP90, binds and protects casein kinase II from self-aggregation and enhances its kinase activity. J. Biol. Chem. **267**, 7042–7047.

Nadeau, K., Das, A., and Walsh, C.T. (1993). Hsp90 chaperonins possess ATPase activity and bind heat shock transcription factors and peptidyl prolyl isomerases. J. Biol. Chem. **268**, 1479–1487.

Nathan, D. and Lindquist, S. (1995). Mutational analysis of Hsp90 function: interactions with a steroid receptor and a protein kinase. Mol. Cell. Biol. **15**, 3917–3925.

Picard, D., Khursheed, B., Garabedian, M.J., Fortin, M.G., Lindquist, S., and Yamamoto, K. R. (1990). Reduced levels of hsp90 compromise steroid receptor action in vivo. Nature **348**, 166–168.

Whitelaw, M.L., McGuire, J., Picard, D., Gustafsson, J.A., and Poellinger, L. (1995). Heat shock protein hsp90 regulates dioxin receptor function in vivo. Proc. Natl. Acad. Sci. **92**, 4437–4441.

Xu, Y. and Lindquist, S.L. (1993). Heat-shock protein hsp90 governs the activity of pp60^{v-src} kinase. Proc. Natl. Acad. Sci. USA **90**, 7074–7078.

■ *Yoko Kimura:*
Department of Tumor Biology
The Tokyo Metropolitan Institute of Medical Science
3-18-22, Honkomagome, Bunkyo-ku
Tokyo 113, Japan
Fax. 81 3 5685 2932
E-mail: ykimura@rinshoken.or.jp

■ *Susan Lindquist:*
Department of Molecular Genetics and Cell Biology
and Howard Hughes Medical Institute
The University of Chicago
5841 S. Maryland Ave.
Room N339, MC1028
Chicago, IL 60637, USA
Tel. 1 773 702 8049
Fax. 1 773 702 7254
E-mail: S-Lindquist@uchicago.edu

Drosophila melanogaster Hsp83

Hsp83 is the sole member of the HSP90 family in Drosophila melanogaster. It is expressed constitutively in all cells and is induced by stress and during development. Drosophila Hsp83 interacts genetically with the receptor tyrosine kinases Sevenless and Torso and is believed to promote signaling through these molecules.

■ Alternative names

Hsp82 (Blackman and Meselson, 1986), Hsp90.

■ HSP83 gene and sequence

Drosophila melanogaster HSP83 is a single copy gene positioned at cytogenic locus 63B (GenBank accession numbers X03810 and K01685). Unlike the majority of *Drosophila HSP* genes, which do not contain intervening sequences, the *HSP83* gene contains an intron just upstream of the translation start site (Hackett and Lis, 1983; Blackman and Meselson, 1986). The second exon encodes a 718 amino acid protein that is highly homologous to other members of the HSP90 family. It shares greater than 50% amino acid identity with even the most distantly related eukaryotic HSP90s and 40% amino acid identity with the *Escherichia coli* protein, HtpG (Lindquist and Craig, 1988; see entry p. 151). Like all other eukaryotic HSP90 proteins, *Drosophila* Hsp83 contains two highly charged domains and terminates with the sequence EEVD (Hackett and Lis, 1983; Blackman and Meselson, 1986; Lindquist and Craig, 1988).

■ Hsp83 protein

Drosophila melanogaster Hsp83 exists as a dimer under non-denaturing conditions (Carbajal et al., 1990). In logarithmically growing *Drosophila* tissue culture cells, Hsp83 is predominantly cytoplasmic with some cells exhibiting weak nuclear staining. Immunogold labeling reveals an association with vacuolar structures in the cytoplasm and with perichromatin ribonucleoprotein fibrils in the nucleus. During heat shock, both cytoplasmic and nuclear staining increase and bright staining is seen near the plasma membrane. After a period of recovery, perinuculear staining is observed (Carbajal et al., 1990). Salivary glands of heat-stressed third instar larvae exhibit an even, reticulated pattern of cytoplasmic staining and specific chromosomal staining of the 93D, *hsrw* locus. This chromosomal staining parallels the heat-induced puffing pattern and does not occur when the locus is induced by other agents (Morcillo et al., 1993).

■ Biological regulation of *HSP83*

Drosophila Hsp83 is constitutively expressed in all cells. Heat shock results in a 10–15-fold increase in *HSP83* mRNA levels that is dependent on three tandem heat shock consensus elements located between nucleotides −88 and −49 (Xiao and Lis, 1989). Unlike most *Drosophila* heat shock proteins, which are maximally induced at 36.5°C, expression of the Hsp83 protein is maximally induced between 33 and 35°C. This is owing to a block in RNA splicing that occurs at higher temperatures (Yost and Lindquist, 1986).

In situ hybridization reveals that the levels of *HSP83* mRNA are regulated during oogenesis and embryogenesis. The *HSP83* message is present during all stages of oogenesis with two notable exceptions. It is not detected during the earliest stage and it is absent during stages 6–8. Following stage 8, the *HSP83* message is expressed in the nurse cells and during later stages it is transported from the nurse cells into the oocyte (Ding et al., 1993). In contrast, the Hsp83 protein is expressed at high levels throughout oogenesis (X. Yue and S. Lindquist, personal communication). During the first five nuclear divisions of embryogenesis, the maternal *HSP83* mRNA is uniformly distributed throughout the embryo. Through a combination of generalized mRNA degradation throughout the embryo and local protection at the posterior, the maternal *HSP83* message is localized to the posterior pole cells by the cellular blastoderm stage (Ding et al., 1993). Zygotic *HSP83* mRNA expression occurs at the syncytial blastoderm stage in the anterior third of the embryo and continues through gastrulation (Ding et al., 1993). Zygotic expression is dependent on promoter sequences upstream of the heat shock consensus elements between nucleotides −880 and −170 (Xiao and Lis, 1989) and appears to be regulated by the homeodomain protein bicoid (Ding et al., 1993) and the steroid hormone ecdysone (Thomas and Lengyel, 1986).

■ Mutagenesis studies with *HSP83*

Five alleles of *hsp83* were isolated in a screen for mutations that enhanced the phenotype of a barely functional allele of the receptor tyrosine kinase, Sevenless. Each of the *hsp83* mutations contained a single amino acid substitution (S38L, S574C, S655F, S592F, E317K), several of which are in close proximity to loss of function mutations isolated in the *Saccharomyces cerevisiae HSP82* gene (Cutforth and Rubin, 1994). *hsp83* mutations also suppressed a gain-of-function allele of the related receptor tyrosine kinase Torso (Doyle and Bishop, 1993).

■ References

Blackman, R.K. and Meselson, M. (1986). Interspecific nucleotide sequence comparisons used to identify regulatory and structural features of the *Drosophila hsp82* gene. J. Mol. Biol. **188**, 499–515.

Carbajal, M.E., Valet, J.-P., Charest, P.M., and Tanguay, R.M. (1990). Identification of Drosophila hsp 83 and immunoelectron microscopic localization. J. Cell Biol. **52**, 147–156.

Cutforth, T. and Rubin, G.M. (1994). Mutations in *Hsp83* and *cdc37* impair signaling by the Sevenless receptor tyrosine kinase in Drosophila. Cell **77**, 1027–1036.

Ding, D., Parkhurst, S.M., Halsell, S.R., and Lipshitz, H.D. (1993). Dynamic *Hsp83* RNA localization during *Drosophila* oogenesis and embryogenesis. Mol. Cell. Biol. **13**, 3778–3781.

Doyle, H.J. and Bishop, J.M. (1993). Torso, a receptor tyrosine kinase required for embryonic pattern formation, shares substrates with the Sevenless and EGF-R pathways in *Drosophila*. Genes Dev. **7**, 633–646.

Hackett, R.W. and Lis, J.T. (1983). Localization of the hsp83 transcript within a 3292 nucleotide sequence from the 63B heat shock locus of *D. melanogaster*. Nucl. Acids Res. **11**, 7011–7030.

Lindquist, S. and Craig, E.K. (1988). The heat-shock proteins. Annu. Rev. Genet. **22**, 631–677.

Morcillo, G., Diez, J.L., Carbajal, M.E., and Tanguay, R.M. (1993). HSP90 associates with specific heat shock puffs (hsrw) in polytene chromosomes of *Drosophila* and *Chironomus*. Chromosoma **102**, 648–659.

Thomas, S.R. and Lengyel, J.A. (1986). Ecdysteroid-regulated heat-shock gene expression during *Drosophila melanogaster* development. Develop. Biol. **115**, 434–438.

Xiao, H. and Lis, J.T. (1989). Heat shock and developmental regulation of the *Drosophila melanogaster hsp83* gene. Mol. Cell. Biol. **9**, 1746–1753.

Yost, H.J. and Lindquist, S. (1986). RNA splicing is interrupted by heat shock and is rescued by heat shock protein synthesis. Cell **45**, 185–193.

Zimmerman, J.L., Petri, W., and Meselson, M. (1983). Accumulation of a specific subset of D. melanogaster heat shock mRNAs in normal development without heat shock. Cell **32**, 1161–1170.

Debra F. Nathan and Susan Lindquist:
Department of Molecular Genetics and Cell Biology
The University of Chicago
5841 S. Maryland Ave.
Room N339, MC 1028
Chicago, IL 60637, USA
Tel. 1 773 702 8049
Fax. 1 773 702 7254
E-mail: dnathan@midway.uchicago.edu and S-Lindquist@uchicago.edu

Plant Hsp90

The 90 kDa heat shock protein, Hsp90, is known to function as a molecular chaperone that can regulate the activity of the protein to which it binds. In recent years HSP90 homologs have been identified in several plant species. Although a detailed understanding of HSP90 functions in plant cells is currently not available, studies of HSP90 expression suggest functions related to plant growth and development and to environmental stress.

■ Plant *hsp90* genes and sequences

Nucleotide sequence analysis of plant *hsp90* genes (GenBank accession numbers M36960, M96549 and M99431) predicts a molecular mass between 80 and 82 kDa, and pIs of 5.73 and 5.76 in two cases, for the encoded proteins (Conner *et al.*, 1990; Felsheim and Das, 1992; Koning *et al.*, 1992; Takahashi *et al.*, 1992; Marrs *et al.*, 1993; Yabe *et al.*, 1994). Comparison of the predicted amino acid sequences of plant HSP90s with those of HSP90 family members of yeast, *Drosophila*, human and chicken shows identities ranging from 63 to 71%, and between HSP90 proteins from different plant species, from 88 to 93%. The presence of two highly charged regions (amino acids 221-290 and 530-581) and the leucine heptad repeat region (between amino acids 530 and 581) in HSP90s from other organisms is conserved in plant HSP90 proteins (Koning *et al.*, 1992; Yabe *et al.*, 1994; P. Krishna and J.F. Cherutti, unpublished data). The five C-terminal amino acids, MEEVD, which are characteristic of eukaryotic HSP90 proteins, are also present in all plant HSP90 proteins identified so far. In addition to genes encoding the putative cytoplasm-localized Hsp90, cDNAs for the endoplasmic reticulum-localized Grp94 homolog (Schroder *et al.*, 1993; Walther-Larsen *et al.*, 1993) and a plastid-specific homolog (Schmitz *et al.*, 1996) have been isolated from plants.

■ *hsp90* gene regulation

Proteins of the HSP90 family in plants are encoded by a multigene family, consisting of 5-6 members in some plant species (Felsheim and Das, 1992; Krishna *et al.*, 1995). Analysis of the mRNA expression patterns indicates that one or more members of the same gene family are strongly expressed under normal growth conditions and are only mildly heat inducible, whereas others are strongly heat inducible but weakly expressed in the absence of heat shock (Marrs *et al.*, 1993; Yabe *et al.*, 1994). The expression of *hsp90* genes in plants is developmentally regulated, with high levels of mRNA being present in root and shoot apices, flower buds, flowers, embryos and developing pollen (Koning *et al.*, 1992; Marrs *et al.*, 1993; Yabe *et al.*, 1994; Krishna *et al.*, 1995). *hsp90* mRNA levels increased dramatically in cotyledons of *Pharbitis nil* when plants were shifted from dark to light (Felsheim and Das, 1992), and in leaves of *Brassica napus* when plants grown at 20°C were exposed to 5°C (Krishna *et al.*, 1995). The rapid increase in *hsp90* mRNA levels was accompanied by a gradual accumulation of the protein in the leaf tissue of cold-stressed plants (Krishna *et al.*, 1995). In rice, a putative Hsp90 protein was noted to accumulate in response to low temperature, salt and water stress (Pareek *et al.*, 1995). This suggests a role for Hsp90 in adaptation to cold temperature stress.

Exogenous treatment with either the phytohormone indoleacetic acid (IAA) or 0.1 M NaCl also led to increases in *hsp90* transcript levels (Yabe *et al.*, 1994). Although *hsp90* genes are induced by stresses other than heat shock and by chemical treatments, *cis* elements that may be important in regulating pathways other than the heat shock response have not as yet been identified in the promoter region of *hsp90* genes.

■ Hsp90 protein

A polyclonal antibody that specifically reacts with plant Hsp90 was raised in rabbits (Krishna *et al.*, 1995). Preliminary studies of the native forms of plant Hsp90 in crude extracts of spinach and *Brassica napus* indicate that the protein exists in monomeric, dimeric and high molecular mass forms (Krishna *et al.*, 1997). Recently a *B. napus* Hsp90, modified to contain six histidines at the C-terminus, was expressed and purified from insect cells (M. Park, C.Y. Kang and P. Krishna, unpublished data). This plant Hsp90 protein was observed to undergo autophosphorylation and to phosphorylate other proteins such as histones and casein in the presence of Mn^{2+}.

Protein–protein interactions

The interaction of vertebrate Hsp90 with steroid hormone receptors has been studied in the most detail (Pratt, 1993). To determine if plant Hsp90 can enter into a complex with the mammalian glucocorticoid receptor, the immunoadsorbed receptor was incubated with wheat germ lysate. Not only could the functional (steroid-binding) receptor–wheat Hsp90 heterocomplex form (Stancato et al., 1996), but the assembly of this heterocomplex could be stimulated by human p23 (Hutchison et al., 1995), a member of the mammalian steroid receptor–Hsp90 heterocomplex (Pratt, 1993). Using purified p23 and an immunoadsorbing antibody against this protein, a p23–wheat Hsp90 heterocomplex was isolated from wheat germ lysate. As with mammalian Hsp90 heterocomplexes, this complex was shown to contain one or more plant immunophilins of the FK506 binding class (Owens-Grillo et al., 1996). Together, these results attest to the conservation of the chaperone complex in plants.

References

Conner, T.W., LaFayette, P.R., Nagao, R.T., and Key, J.L. (1990). Sequence and expression of a HSP83 from *Arabidopsis thaliana*. Plant Physiol. **94**, 1689–1695.

Felsheim, R.F. and Das, A. (1992). Structure and expression of a heat-shock protein 83 gene of *Pharbitis nil*. Plant Physiol. **100**, 1764–1771.

Hutchison, K.A., Stancato, L.F., Owens-Grillo, J.K., Johnson, J.L., Krishna, P., Toft, D.O., and Pratt, W.B. (1995). The 23-kDa acidic protein in reticulocyte lysate is the weakly bound component of the hsp foldosome that is required for assembly of the glucocorticoid receptor into a functional heterocomplex with hsp90. J. Biol. Chem. **270**, 18841–18847.

Koning, A.J., Rose, R., and Comai, L. (1992). Developmental expression of tomato heat-shock cognate protein 80. Plant Physiol. **100**, 801–811.

Krishna, P., Sacco, M., Cherutti, J.F., and Hill, S. (1995). Cold-induced accumulation of hsp90 transcripts in *Brassica napus*. Plant Physiol. **107**, 915–923.

Krishna, P., Reddy, R.K., Sacco, M., Frappier, J.R.H. and Felsheim, R.F. (1997). Analysis of the native forms of the 90 kDa heat shock protein (hsp90) in plant cytosolic extracts. Plant Mol. Biol. **33**, 457–466.

Marrs, K.A., Casey, E.S., Capitant, S.A., Bouchard, R.A., Dietrich, P.S., Mettler, I.J., and Sinibaldi, R. (1993). Characterization of two maize HSP90 heat shock protein genes: expression during heat shock, embryogenesis and pollen development. Develop. Genet. **14**, 27–41.

Owens-Grillo, J.K., Stancato, L.F., Hoffmann, K., Pratt, W.B., and Krishna, P. (1996). Binding of immunophilins to the 90 kDa heat shock protein (hsp90) via a tetratricopeptide repeat domain is a conserved protein interaction in plants. Biochemistry **35**, 15249–15255.

Pareek, A., Singla, S.L., and Grover, A. (1995). Immunological evidence for accumulation of two high-molecular-weight (104 and 90 kDa) HSPs in response to different stresses in rice and in response to high temperature stress in diverse plant genera. Plant Mol. Biol. **29**, 293–301.

Pratt, W.B. (1993). The role of heat shock proteins in regulating the function, folding and trafficking of the glucocorticoid receptor. J. Biol. Chem. **268**, 21455–21458.

Schmitz, G., Schmidt, M. and Feiderabend, J. (1996). Characterization of a plastid-specific HSP90 homologue: identification of a cDNA sequence, phylogenetic descendence and analysis of its mRNA and protein expression. Plant Mol. Biol. **30**, 479–492.

Schroder, G., Beck, M., Eichel, J., Vetter, H.-P., and Schroder, J. (1993). Hsp90 homologue from Madagascar periwinkle (*Catharanthus roseus*): cDNA sequence, regulation of protein expression and location in the endoplasmic reticulum. Plant Mol. Biol. **23**, 583–594.

Stancato, L.F., Hutchison, K.A., Krishna, P., and Pratt, W.B. (1996) Animal and plant cell lysates share a conserved chaperone system that assembles the glucocorticoid receptor into a functional heterocomplex with hsp90. Biochemistry, **35**, 554–561.

Takahashi, T., Naito, S. and Komeda, Y. (1992). Isolation and analysis of the expression of two genes for the 81-kilodalton heat-shock proteins from *Arabidopsis*. Plant Physiol. **99**, 383–390.

Walther-Larsen, H., Brandt, J., Collinge, D.B., and Thordal-Christensen, H. (1993). A pathogen-induced gene of barley encodes a HSP90 homologue showing striking similarity to vertebrate forms resident in the endoplasmic reticulum. Plant Mol. Biol. **21**, 1097–1108.

Yabe, N., Takahashi, T., and Komeda, Y. (1994). Analysis of tissue-specific expression of *Arabidopsis thaliana* hsp90-family gene HSP81. Plant Cell Physiol. **35**, 1207–1219.

■ Priti Krishna:
Department of Plant Sciences
The University of Western Ontario
1151 Richmond Street North
London, Ontario, N6A 5B7, Canada
Tel. 1 519 679 2111 ext. 6406
Fax. 1 519 661 3935
E-mail: pkrishna@julian.uwo.ca

Mammalian Hsp90

Hsp90 is the most prominent heat shock protein in the cytosol of various mammalian cells. It is mainly found in stable association with untransformed steroid receptors and newly synthesized kinases. Interaction with Hsp90 in complexes with specific partner proteins (e.g. Hsp70, immunophilins and proteins of unknown function) keeps these protein substrates in an activatable state. More recently, it was demonstrated that Hsp90 exhibits properties of molecular chaperones.

■ Alternative names

The alternative names of members of the Hsp90 family correspond to their apparent molecular weight on SDS–PAGE and depend on the context in which they have been identified: Hsp84/Hsp86 (mouse; Moore et al., 1989), Hsp85 (L929-cells; Lanks, 1989), Hsp89/Hsp90 (HeLa cells; Welch and Feramisco, 1982).

■ *hsp90* genes and sequence

Mammalian cells express two highly related Hsp90 isoforms, αHsp90 and βHsp90, which are encoded by two separate genes. The products of human *hsp90α* and *hsp90β* are polypeptides of 732 and 724 amino acids, respectively, which are predicted to be identical at 630 out of the 724 possible residue matches (Rebbe et al., 1987). The nucleotide sequences reveal three highly charged regions, which are present in all eukaryotic HSP90s but are partly missing in the *Escherichia coli* HSP90 protein HtpG (reviewed in Jakob and Buchner, 1994; see also entry p. 151). Furthermore, members of the eukaryotic HSP90 family contain four highly conserved amino acids (EEVD) at the C-terminus, which are also found with slight variations in some members of the HSP70 family (Lindquist and Craig, 1988). The function of this motif in Hsp90 has yet to be elucidated.

■ Hsp90 protein

Hsp90 is localized in the cytosol and probably the nuclear compartment of eukaryotic cells (Lindquist and Craig, 1988). It was purified from the cytosol of HeLa cells and bovine tissue using standard purification techniques (Welch and Feramisco, 1982; Wiech et al., 1993). Complexes of Hsp90 with partner and/or substrate proteins were isolated by coimmunoprecipitation, using monoclonal antibodies (Perdew and Whitelaw, 1991; Smith and Toft, 1993; Pratt, 1993). Monoclonal antibodies against Hsp90 from different species are commercially available (Stressgen, Affinity Bioreagents).

The active form of Hsp90 appears to be the dimer (Welch and Feramisco, 1982; Minami et al., 1991). Studies on murine Hsp90 reveal that the two isoforms Hsp90α and Hsp90β are expressed in nearly equal amounts and are mainly present as α–α and β–β homodimers. Only a minor population of Hsp90β but not of Hsp90α exists also as monomers (Minami et al., 1991). Both isoforms undergo post-translational modifications and for human α- and βHsp90 at least two homologous serine residues could be identified as potential phosphorylation sites *in vivo* (Lees-Miller and Anderson, 1989). Divalent cations were shown to induce significant changes in the oligomeric state of Hsp90, which subsequently lead to a loss of chaperone function *in vitro* (Jakob et al., 1995a).

Comparative structural studies demonstrated that eukaryotic and prokaryotic Hsp90 have very similar physicochemical properties. Hsp90s are α helical proteins with a substantial amount of β sheet structure (Jakob et al., 1995a). The crystal structure of an N-terminal fragment of Hsp90 has recently been solved (Stebbins et al., 1997; Prodromou et al., 1997). The domain contains both alpha helices and β strands and binds the anti-tumor drug geldanamycin which blocks Hsp90 function (see below). Mammalian Hsp90, although synthesized and active under heat shock conditions, is only stable against thermal denaturation up to 50°C (Lanks et al., 1992).

■ Biological activities of Hsp90

Despite the abundance of Hsp90 in the cytosol of eukaryotic cells, the interactions observed so far involve only a limited number of 'substrate proteins'. Under physiological conditions, Hsp90 has been found in association with untransformed steroid receptors, several kinases, actin, tubulin and calmodulin (summarized in Bohen and Yamamoto, 1994). While not much is known about interactions with the cellular cytoskeleton and calmodulin, substantial progress has been made concerning the association with receptors and kinases. Complex formation between Hsp90 and specific partner proteins (see below) with hormone-free steroid receptors represents a major regulatory step in signal transduction (summarized in Bohen and Yamamoto, 1994). It is the necessary prerequisite for successful ligand binding and subsequent activation of transcription. Therefore, the interactions with Hsp90 keep receptors in an activatable state rather then in an inactive state. Similarly, complex formation of Hsp90 with non-receptor protein kinases also reveals regulatory character. Again, in association with specific partner proteins, Hsp90 interacts with newly synthesized protein kinases (e.g. p60[v-src]) until the protein becomes attached to the plasma membrane via myristylation (Brugge, 1986).

Under heat shock conditions, Hsp90 expression is increased several fold. A decrease in the level of Hsp90 expression results in an increased mortality of the cells at elevated temperatures (Bansal et al., 1991). This protective role of Hsp90 under conditions where unfolding and subsequent aggregation of polypeptides occurs is consistent with the observed chaperone activities of Hsp90 *in vitro* (Jakob et al., 1995b). Studies on the renaturation of completely unfolded proteins revealed that stochiometric concentrations of Hsp90 suppress unspecific side reactions (aggregation) and therefore promote functional refolding to the native state (Wiech et al., 1992). Qualitatively similar data were obtained when the helix–loop–helix transcription factor MyoD was used as an unfolded substrate. Hsp90 interacts transiently with non-native MyoD which subsequently leads to its increased DNA-binding activity (Shaknovich et al., 1992). Supporting evidence for the role of Hsp90 as a chaperone came also from *in vitro* studies on casein kinase II, an *in vivo* substrate of Hsp90. Interaction of Hsp90 with casein kinase II protected the protein from unfolding and subsequent aggregation under low salt conditions (Miyata and Yahara, 1992). Using an *in vitro* thermal unfolding assay, it was shown that Hsp90 is able to stabilize unfolding proteins by transient interactions with structured folding intermediates. These ATP-independent associations of Hsp90 with non-native protein substrates observed *in vitro*, are likely to reflect the mechanism of Hsp90 function *in vivo* (Jakob and Buchner, 1994; Jakob et al., 1995b).

For a long time, conflicting evidence existed concerning ATP hydrolysis and ATP binding of Hsp90 (summarized in Jakob and Buchner, 1994). Recently, it was calculated that only a few molecules of highly active kinases or phosphatases could be responsible for the observed ATPase activity (Jakob et al., 1996). In this context, the proposed interaction of Hsp90 and ATP was re-examined and it was shown that Hsp90 does not bind ATP tightly in the absence of potential modulators (partner proteins or transition state metals) (Jakob et al., 1996).

■ Interaction of Hsp90 with partner proteins

Coimmunoprecipitated complexes of Hsp90 with steroid receptors and kinases contain a set of specific partner proteins (for reviews see p. 518 and Pratt, 1993; Smith and Toft, 1993; Jakob and Buchner, 1994). While in the case of Hsp90–kinase complexes, only Hsp70 and p50 seem to be involved (Whitelaw et al., 1991), additional proteins are part of the Hsp90–steroid receptor complex (Smith and Toft, 1993). These partner proteins are either transiently (p60, p48, Hsp70) or permanently (Hsp56, cyp40, p23) associated with the Hsp90–steroid receptor complex. Evidence exists that p60 is needed only for complex formation (Smith and Toft, 1993). Interaction of Hsp90, Hsp70 and ATP was shown to be a necessary prerequisite for successful binding of unliganded glucocorticoid receptors (Hutchison et al., 1994). At present, the function of the prolyl isomerases (Hsp56, cyp40), as well as that of the non-heat shock proteins (p23, p60), in the complex is not well understood. Molybdate, containing a transition state metal, has been found to be an important tool for stabilizing Hsp90–steroid receptor complexes, thus facilitating their purification (Housley et al., 1985), and for studying the interaction between Hsp90 and p23 (Johnson and Toft, 1995). It was suggested that molybdate mimicks some endogenous 'modulators' of Hsp90 (Bodine et al., 1995). Recently, geldanamycin, a benzochinone ansamycin, was discovered to bind to Hsp90 (Whitesell et al., 1994). It seems to block the interaction of Hsp90 with p23. As a result, complex formation of Hsp90 with receptors is also affected (Smith et al., 1995).

The isolation of complexes of Hsp90 with partner proteins (Hsp70, Hsp56, p63, p60) in the absence of the respective substrates led to the suggestion that Hsp90 generally exerts its function together with a subset of other proteins, and raised the possibility of Hsp90 being part of a general cytosolic chaperone machine (reviewed in Smith, 1993; Jakob and Buchner, 1994). However, one has to take into account that in some cases the level of Hsp90 greatly exceeds that of the partner proteins. Therefore, there is good reason to assume that some of the important functions of Hsp90 are performed by Hsp90 free of partner proteins.

■ Mutagenesis studies with hsp90

Mutagenesis studies were mainly performed to identify regions of Hsp90 involved in receptor binding. *In vitro* translation of truncated and internal deletion mutants of chicken Hsp90 (highly homologous to mammalian Hsp90) in reticulocyte lysate and subsequent reconstitution of progesterone receptor complexes revealed that regions in the C-terminal half of Hsp90 are important for successful receptor binding (Sullivan and Toft, 1993). While the N-terminal 380 amino acids of Hsp90 can be removed without a significant loss of receptor binding activity, the regions between amino acid 381–441 and 601–677 seem to be crucial for interaction with the progesterone receptors and calmodulin. Similar regions in murine Hsp90 (amino acids 643–690) have recently been shown to chaperone the folding of MyoD (Shaknovich et al., 1992; Shue and Kohtz, 1994). Different results have been obtained by analysing complex formation of deletion mutants of chicken Hsp90 with various steroid receptors (Cadepond et al., 1993; Sullivan and Toft, 1993). The deletion of the highly charged region (amino acids 221–290) of Hsp90 precluded the interaction of Hsp90 with the receptor, while deletion of the second charged region (amino acids 530–581), as well as the leucine heptad repeat region (amino acids 392–419), led to the formation of abnormal Hsp90–receptor complexes, which were no longer able to bind hormones. In addition, the C-terminal region seems to be important for dimerization of human Hsp90α (Minami et al., 1994). Deletion of the N-terminal 117 amino acids had no influence on the dimerization of Hsp90, whereas the last 50 amino acids were essential for the formation of active Hsp90α dimers (Minami et al., 1994).

References

Bansal, G.S., Norton, P.M., and Latchman, D.S. (1991). The 90-kDa heat shock protein protects mammalian cells from thermal stress but not from viral infection. Exp. Cell Res. **195**, 303–306.

Bodine, P.V.N., Alnemri, E.S., and Litwack, G. (1995). Synthetic peptides derived from the steroid binding domain block modulator and molybdate action toward the rat glucocorticoid receptor. Receptor **5**, 117–122.

Bohen, S.P. and Yamamoto, R.I. (1994). Modulation of steroid receptor signal transduction by heat shock proteins. In *The Biology of Heat Shock Proteins and Molecular Chaperones* (Morimoto, R.I., Tissières, A. and Georgopoulos, C., eds. CSHL Press, Cold Spring Harbour, New York, pp. 313–334.

Brugge, J.S. (1986). Interaction of the Rous sarcoma virus protein pp60[src] with the cellular proteins pp50 and pp90. Curr. Top. Microbiol. Immunol. **123**, 1–22.

Cadepond, F., Binart, N., Chambraud, B., Jibard, N., Schweizer-Groyer, G., Segard-Maurel, I., and Baulieu, E.E. (1993). Interaction of glucocorticoid receptor and wild-type or mutated 90-kDa heatshock protein coexpressed in baculovirus-infected Sf9 cells. Proc. Natl. Acad. Sci. USA **90**, 10434–10438.

Housley, P.R., Sanchez, E.R., Westphal, H.M., Beato, M., and Pratt, W.B. (1985). The molybdate-stabilized L-cell glucocorticoid receptor isolated by affinity chromatography or with monoclonal antibodies is associated with a 90–92-kDa nonsteroid-binding phosphoprotein. J. Biol. Chem. **260**, 13810–13817.

Hutchison, K.A., Dittmar, K.D., Czar, M.J., and Pratt, W.B. (1994). Proof that hsp70 is required for assembly of the gluticord receptor into a complex with hsp90. J. Biol. Chem. **269**, 5043–5049.

Jakob, U. and Buchner, J. (1994). Assisting spontaneity: the role of Hsp90 and small Hsps as molecular chaperones. Trends Biochem. Sci. **19**, 205–211.

Jakob, U., Meyer, I., Bügl, H., Andrè, S., Bardwell, J.C.A., and Buchner, J. (1995a). Structural organzation of prokaryotic and eukaryotic Hsp90-Influence of divalent cations on structure and function. J. Biol. Chem. **270**, 14412–14419.

Jakob, U., Lilie, H., Meyer, I., and Buchner, J., (1995b). Transient interaction of Hsp90 with early unfolding intermediates of citrate synthase-Implications for heat shock *in vivo*. J. Biol. Chem. **270**, 7288–7294.

Jakob, U. Scheibel, T., Bose, S., Reinstein, J. and Buchner, J. (1996). Assessment of the ATP binding properties of Hsp90. J. Biol. Chem. **271**, 10035–10041.

Johnson, J.L. and Toft, D.O. (1995). Binding of p23 and Hsp90 during assembly with the progesterone receptor. Mol. Endocrinol. **9**, 670–678.

Lanks, K.W. (1989). Temperature-dependent oligomerization of Hsp85 *in vitro*. J. Cell Physiol. **140**, 601–607.

Lanks, K.W., London, E., and Dong, D.L. (1992). Hsp85 conformational change within the heat shock temperature range. Biochem. Biophys. Res. Commun. **184**, 394–399.

Lees-Miller, S.P. and Anderson, C.W. (1989). Two human 90-kDa heat shock proteins are phosphorylated *in vivo* at conserved serines that are phosphorylated *in vitro* by casein kinase II. J. Biol. Chem. **264**, 2431–2437.

Lindquist, S.C. and Craig, E.A. (1988). The heat shock proteins. Annu. Rev. Genet. **22**, 631–677.

Minami, Y., Kawasaki, H., Miyata, Y., Suzuki, K., and Yahara, I. (1991). Analysis of native forms and isoform compositions of the mouse 90-kDa heat shock protein, Hsp90. J. Biol. Chem. **266**, 10099–10103.

Minami, Y., Kimura, Y., Kawasaki, H., Suzuki, K., and Yahara, I. (1994). The carboxy-terminal region of mammalian Hsp90 is required for its dimerization and function *in vivo*. Mol. Cell. Biol. **14**, 1459–1464.

Miyata, Y. and Yahara, I. (1992). The 90-kDa heat shock protein, Hsp90, binds and protects casein kinase II from self-aggregation and enhances its kinase activity. J. Biol. Chem. **267**, 7042–7047.

Moore, S.K., Kozak, C., Robinson, E.A., Ullrich, S.J., and Appella, E. (1989). Murine 86- and 84-kDa heat shock proteins, cDNA sequences, chromosome assignments and evolutionary origins. J. Biol. Chem. **264**, 5343–5351.

Perdew, G.H. and Whitelaw, M.L. (1991). Evidence that the 90 kDa heat shock protein (Hsp90) exists in the cytosol in heteromeric complexes containing Hsp70 and three other proteins with Mr 63.000, 56.000 and 50.000. J. Biol. Chem. **266**, 6708–6713.

Pratt, W.B. (1993). The role of heat shock proteins in regulating the function, folding, and trafficking of the glucocorticoid receptor. J. Biol. Chem. **268**, 21455–21458.

Prodromou, C., Roe, S.-M., Piper, P.W. and Pearl, L.H. (1997). A molecular clamp in the crystal structure of the yeast Hsp90 chaperone. Nature Struct. Biol. **4**, in press.

Rebbe, N.F., Ware, J., Bertina, R.M., Modrich, P., and Stafford, D. (1987). Nucleotide sequence of a cDNA for a member of the human 90-kDa heat shock protein family. Gene **53**, 235–245.

Shaknovich, R., Shue, G. and Kohtz, S. (1992). Conformational activation of basic helix-loop-helix protein (MyoD) by the C-terminal region of murine Hsp90 (Hsp84). Mol. Cell. Biol. **12**, 5059–5068.

Shue, G. and Kohtz, D.S. (1994). Structural and functional aspects of basic helix-loop-helix protein folding by heat shock protein Hsp90. J. Biol. Chem. **269**, 2707–2711.

Smith, D.F. (1993). Dynamics of heat shock protein 90-progesterone receptor binding and the disactivation loop model for steroid receptor complexes. Mol. Endocrinol. **7**, 1418–1429.

Smith, D.F. and Toft, D.O. (1993). Steroid receptors and their associated proteins. Mol. Endocrinol. **7**, 4–11.

Smith, D.F., Whitesell, L., Nair, S.C., Chen, S., Prapapanich, V., and Rimerman, R.A. (1995). Geldanamycin, a specific Hsp90-blocking reagent, inhibits progesterone receptor function by blocking Hsp90-p23 interactions. Mol. Cell. Biol. **15**, 6804–6812.

Stebbins, C.E., Russo, A.A., Schneider, C., Rosen, N., Hartl, F.-U. and Pavletich, N.P. (1997). Crystal structure of an Hsp90-geldanamycin complex: targeting of a protein chaperone by an anti tumor agent. Cell **89**, 239–250.

Sullivan, W.P. and Toft, D.O. (1993). Mutational analysis of Hsp90 binding to the progesterone receptor. J. Biol. Chem. **268**, 20373–20379.

Welch, W.J. and Feramisco, J.R. (1982). Purification of major mammalian heat shock proteins. J. Biol. Chem. **257**, 14949–14959.

Whitelaw, M.L., Hutchison, K., and Perdew, G.H. (1991). A 50-kDa cytosolic protein complex with the 90-kDa heat shock protein (Hsp90) is the same protein complexed with pp60v-src Hsp90 in cells transformed by the Rous sarcoma virus. J. Biol. Chem. **266**, 16436–16440.

Whitesell, L., Mimnaugh, E.G., De Costa, B., Myers, C.E., and Neckers, L.M. (1994). Inhibition of heat shock protein Hsp90-pp60[v-src] heteroprotein complex formation by benzoquinone ansamycins: essential role for stress proteins in oncogenic transformation. Proc. Natl. Acad. Sci. USA **91**, 8324–8328.

Wiech, H., Buchner, J., Zimmermann, R., and Jakob, U. (1992). Hsp90 chaperones protein folding *in vitro*. Nature **358**, 169–170.

Wiech, H., Buchner, J., Zimmermann, M., Zimmermann, R., and Jakob, U. (1993). Hsc70, BiP and Hsp90 differ in their ability to stimulate transport of precursor proteins into mammalian microsomes. J. Biol. Chem. **268**, 7414–7421.

- Ursula Jakob:
  Department of Biology
  University of Michigan
  Ann Arbor, MI 48109-1048, USA
  Tel. 1 313 764 8028
  Fax. 1 313 647 0884
  E-mail: Ujakob@biology.lsa.umich.edu

- Johannes Buchner:
  Department of Biophysics und Physical Biochemistry
  Universität Regensburg
  93040 Regensburg, Germany
  Tel. 49 941 943 3039
  Fax. 49 941 943 2813
  E-mail: Johannes.Buchner@biologie.uni-regensburg.de

Mammalian Grp94

Grp94, a member of the HSP90 family of stress proteins, is an abundant ER protein. Recent studies suggest that Grp94 is a molecular chaperone involved in the folding and assembly of a limited spectrum of newly synthesized polypeptides. Grp94 is also an important immune modulator, involved in tumor rejection and possibly in antigen presentation.

Alternative names

Endoplasmin (Koch et al., 1986), ERp99 (Mazzarella and Green, 1987), hsp108 (Sargan et al., 1986), gp96 (Li and Srivastava, 1993).

Isolation

Grp94 was isolated by several independent approaches. Biochemically, it was identified as the major concanavalin A-binding protein in rough microsomes (Koch et al., 1986), as a component of the chicken progesterone receptor (Sargan et al., 1986), and as the major protein responding to either glucose starvation (Lee et al., 1983) or to differentiation signals (Mazzarella and Green, 1987). Immunologically, it was identified as a tumor-specific antigen (Li and Srivastava, 1993). grp94 genes were independently cloned by four of these groups (Lee et al., 1983; Li and Srivastava, 1993; Mazzarella and Green, 1987; Sargan et al., 1986).

grp94 gene and sequence

The gene encoding murine grp94 is located on chromosome 10, while the human homologue is on chromosome 12q24 (Maki et al., 1993). So far, only one functional gene has been characterized in the mouse (GenBank accession number J03297), rat (S69315), human (X15187), chicken (X04961), and barley (X67960) genomes. The encoded proteins are highly homologous; even as distant a pair as human and barley show 50% identity between their amino acid sequences. The grp94 genes also display a high level of homology over their entire length with genes encoding cytosolic Hsp90 proteins. For example, human Grp94 and Saccharomyces cerevisiae Hsp90 show 50% identity of amino acids. Like all other members of the HSP90 family, Grp94 contains the signature hexapeptide, NKEIFL, in its N-terminal domain.

Each grp94 gene encodes a protein of 802–809 amino acids. The calculated molecular weight of the mature murine protein is 90 096 Da. Grp94 has a 21 residue, N-terminal signal sequence and a C-terminal tetrapeptide, KDEL, which serves as the retention/retrieval signal for a number of ER luminal proteins (Munro and Pelham, 1987). Of the six predicted N-glycosylation sites, only one (Asn-196 in the mouse protein) seems to be used regularly, while other sites in the C-terminal half are glycosylated in a small fraction of molecules (Qu et al., 1994). Multiple potential phosphorylation sites for different types of protein kinases are scattered along the sequence. Hydropathy plots of the sequence predict a string of 17 hydrophobic residues starting at Leu-170, which may serve as a transmembrane domain. There are also two highly charged domains containing frequent runs of acidic residues: Thr-245 to Glu-347 and Asp-731 to the C-terminus. A predicted leucine zipper-like motif starting at Leu-479 is present in Grp94 and other HSP90 proteins.

grp94 gene regulation

Grp94 is transcriptionally coregulated with BiP, the ER member of the HSP70 family of stress proteins. The promoters of the two genes share a common regulatory domain and bind similar trans-acting factors (Liu and Lee, 1991). Constitutive expression of each is up-regulated by a variety of stress conditions, including glucose starvation (hence the name, glucose-regulated protein), glycosylation inhibitors, reducing agents, amino acid analogues, and calcium ionophores, all of which are thought to cause accumulation of improperly folded polypeptides in the ER (Gething and Sambrook, 1992; Lee, 1987). This coregulation suggests that BiP and Grp94 have related functions.

■ Grp94 protein

Grp94 is the only known HSP90 in the endoplasmic reticulum. The protein is found in virtually all mammalian tissues, with the possible exception of some embryonic cells. It has been identified in multiple species, including mouse, monkey, human, chicken, and mosquito, but not in *Dictyostelium*, *Drosophila*, or yeast. Conservation of the protein is evident not only by sequence analysis, but also serologically. For example, the monoclonal antibody 9G10 was raised against the chicken protein, but recognizes Grp94 from all five species.

Grp94 has been isolated as both monomers and dimers, which are primarily non-covalent (Li and Srivastava, 1993; Qu *et al.*, 1994). Its purification from microsomes, either by anion exchange and gel filtration or by lectin-affinity chromatography, does not necessitate detergents. Together with the C-terminal KDEL motif, this finding indicates that it is a soluble, luminal ER protein. However, other purification procedures and proteolytic digestion of microsomes have provided evidence for an additional, membrane-bound form of Grp94 (Koch *et al.*, 1986; Mazzarella and Green, 1987; Kang and Welch, 1991; Qu *et al.*, 1994). This is consistent with the hydropathy analysis and with the identification of Grp94 as a cell surface protein in a sarcoma cell line (Li and Srivastava, 1993).

Grp94 is phosphorylated on serine and threonine residues, possibly by a casein kinase-like enzyme. Only the non-phosphorylated form binds to immunoglobulin chains, but the relation of phosphorylation to activity remains to be established. Grp94 binds ATP, but possesses only a weak ATPase activity of unproven physiological relevance (Clairmont *et al.*, 1992; Li and Srivastava, 1993). Grp94 displays heterogeneous levels of N-glycosylation, with the main murine form possessing one oligosaccharide attached to Asn-196 (Qu *et al.*, 1994). Glycosylation of Grp94 is not essential for its immunoglobulin-binding activity. Grp94 is a major Ca^{2+}-binding protein (Koch *et al.*, 1986), with its acidic regions presumably mediating low affinity, high capacity Ca^{2+} binding.

■ Grp94 function

Two findings suggest that Grp94 has an important general function: its widespread species and tissue distribution, and the fact that it accounts for as much as 10% of the luminal ER protein. One important role, particularly in the sarcoplasmic reticulum, may involve Ca^{2+} storage. Recent discoveries have suggested two additional functions for Grp94.

Grp94/gp96 was identified as a tumor-specific rejection antigen, mediating the rejection of at least one sarcoma by specific T lymphocytes. Because there were no discernable differences between Grp94 present in the tumor line and in other cells, Grp94/gp96 was hypothesized to act by presenting peptide antigens specific for the sarcoma to T cells. The reported peptide-binding activities and association with MHC class I molecules, either in the ER or on the cell surface, are consistent with this function (Li and Srivastava, 1993). Recently, purified Grp94 was also shown to present a specific viral antigen to cytotoxic T cells, in a manner dependent on MHC class I proteins on macrophages that interact with these T cells (Suto and Srivastava, 1995). Thus, Grp94 may facilitate the loading of MHC class I with antigenic peptides.

Grp94 was suggested to be a molecular chaperone following observations of its association with a mutant viral glycoprotein, an unassembled MHC class II molecule, and immunoglobulin light and heavy chains (Navarro *et al.*, 1991; Melnick *et al.*, 1992, 1994; Schaiff *et al.*, 1992). In the case of wild-type, secretable immunoglobulin light chains, Grp94 association is transient (Melnick *et al.*, 1994). BiP also associates with light chains transiently, but Grp94 association is biochemically, kinetically, and structurally distinct. Grp94 associates with and dissociates from light chains subsequently to BiP, and binds only to relatively mature, fully oxidized molecules, in contrast to BiP's strong preference for earlier intermediates. Moreover, ternary complexes of Grp94–light chain–BiP have been isolated, suggesting a relay mechanism linking the two chaperones (Melnick and Argon, 1995; Melnick *et al.*, 1994).

The specificity of the Grp94–polypeptide interaction is not understood. Consistent with its implicated role in antigen presentation, Grp94 can bind short peptides. On the other hand, its interactions with relatively mature folding intermediates and the limited substrate spectrum indicate that, like cytosolic Hsp90, Grp94 recognizes more mature structural determinants (Melnick and Argon, 1995).

The consequences of Grp94–polypeptide interaction are likewise poorly understood. Enhancement of protein folding *in vitro* has been demonstrated for cytosolic Hsp90 (Wiech *et al.*, 1992), but not yet for Grp94. Nonetheless, it seems likely that Grp94 plays a role in the folding and/or assembly of polypeptides. Indeed, inhibition of Grp94 induction decreases the transport of an overexpressed lysosomal protein (Little and Lee, 1995). Additional functions, including ER to nucleus signalling and specific protection against Ca^{2+} depletion, are also under investigation (Li and Srivastava, 1993; Little and Lee, 1995).

■ References

Clairmont, C.A., De, M.A., and Hirschberg, C.B. (1992). Translocation of ATP into the lumen of rough endoplasmic reticulum-derived vesicles and its binding to luminal proteins including BiP (GRP 78) and GRP 94. J. Biol. Chem. **267**, 3983–3990.

Gething, M.J. and Sambrook, J. (1992). Protein folding in the cell. Nature **355**, 33–45.

Kang, H.S. and Welch, W.J. (1991). Characterization and purification of the 94-kDa glucose-regulated protein. J. Biol. Chem. **266**, 5643–5649.

Koch, G., Smith, M., Macer, D., Webster, P., and Mortara, R. (1986). Endoplasmic reticulum contains a common, abundant calcium-binding glycoprotein, endoplasmin. J. Cell Sci. **86**, 217–232.

Lee, A.S. (1987). Coordinated regulation of a set of genes by glucose and calcium ionophores in mammalian cells. Trends Biochem. Sci. **12**, 20–23.

Lee, A.S., Delegeane, A.M., Baker, V., and Chow, P.C. (1983). Transcriptional regulation of two genes specifically induced by glucose starvation in a hamster mutant fibroblast cell line. J. Biol. Chem. **258**, 597–603.

Li, Z. and Srivastava, P.K. (1993). Tumor rejection antigen gp96/grp94 is an ATPase: implications for protein folding and antigen presentation. EMBO J. **12**, 3143–3151.

Little, E. and Lee, A.S. (1995). Generation of a mammalian cell line deficient in glucose-regulated protein stress induction through targeted ribozyme driven by a stress-inducible promoter. J. Biol. Chem. **270**, 9526–9534.

Liu, E.S. and Lee, A.S. (1991). Common sets of nuclear factors binding to the conserved promoter sequence motif of two coordinately regulated ER protein genes, GRP78 and GRP94. Nucl. Acids Res. **19**, 5425–5431.

Maki, R.G., Eddy, R.L., Byers, M., Shows, T.B., and Srivastava, P.K. (1993). Mapping of the genes for human endoplasmic reticular heat shock protein gp96/grp94. Somat. Cell Molec. Genet. **19**, 73–81.

Mazzarella, R.A. and Green, M. (1987). ERp99, an abundant, conserved glycoprotein of the endoplasmic reticulum, is homologous to the 90-kDa heat shock protein (hsp90) and the 94-kDa glucose regulated protein (GRP94). J. Biol. Chem. **262**, 8875–8883.

Melnick, J. and Argon, Y. (1995). Molecular chaperones and the biosynthesis of antigen receptors. Immunol. Today **16**, 243–250.

Melnick, J., Aviel, S., and Argon, Y. (1992). The endoplasmic reticulum stress protein GRP94, in addition to BiP, associates with unassembled Immunoglobulin chains. J. Biol. Chem. **267**, 21303–21306.

Melnick, J., Dul, J.L., and Argon, Y. (1994). Sequential interaction of the chaperones BiP and GRP94 with immunoglobulin chains in the endoplasmic reticulum. Nature **370**, 373–375.

Munro, S. and Pelham, H.R.B. (1987). A C-terminal signal prevents secretion of luminal ER proteins. Cell **48**, 899–907.

Navarro, D., Qadri, I., and Pereira, L. (1991). A mutation in the ectodomain of herpes simplex virus 1 glycoprotein B causes defective processing and retention in the endoplasmic reticulum. Virology **184**, 253–264.

Qu, D., Mazzarella, R.A., and Green, M. (1994). Analysis of the structure and synthesis of GRP94, an abundant stress protein of the endoplasmic reticulum. DNA Cell Biol. **13**, 117–124.

Sargan, D.R., Tsai, M.J., and O'Malley, B.W. (1986). hsp108, a novel heat shock inducible protein of chicken. Biochemistry **25**, 6252–258.

Schaiff, W.T., Hruska, K.A., McCourt, D.W., Green, M., and Schwartz, B.D. (1992). HLA-DR associates with specific stress proteins and is retained in the endoplasmic reticulum in invariant chain negative cells. J. Exp. Med. **176**, 657–666.

Suto, R. and Srivastava, P.K. (1995). A mechanism for the specific immunogenicity of heat shock protein-chaperoned peptides. Science **269**, 1585–1588.

Wiech, H., Buchner, J., Zimmermann, R., and Jakob, U. (1992). Hsp90 chaperones protein folding in vitro. Nature **358**, 169–170.

■ *Jeffrey Melnick:*
Department of Pathology, Box 8818
Washington University Medical Center
660 S. Euclid Ave.
St. Louis, MO 63310, USA
Tel. 1 314 836 5458
Fax. 1 314 836 8950
E-mail: melnick@path.wustl.edu

■ *Yair Argon:*
Department of Pathology and Committee on Immunology
5841 S. Maryland Ave., MC 1089
University of Chicago
Chicago, IL 60637, USA
Tel. 1 773 702 6388
Fax. 1 773 702 3701
E-mail: yargon@midway.uchicago.edu

6

CPN60/CPN10 Proteins

The CPN60 and CPN10 families — an overview

The members of the CPN60 family of molecular chaperones play a fundamental role in mediating the folding of many proteins in an ATP-dependent manner (reviewed in Hendrick and Hartl, 1995; Hartl, 1996). While being ubiquitous components of all eubacteria, CPN60 family members are also found in the endosymbiotic organelles of eukaryotic cells (Hsp60 in mitochondria and rubisco (ribulose biphosphate carboxylase oxygenase) subunit binding protein in chloroplasts). Prokaryotic members of the CPN60 family fold newly synthesised proteins, while the eukaryotic forms fold newly imported proteins. Although CPN60 proteins are constitutively expressed, their expression is further induced under conditions of stress. Under such conditions they have the task of refolding destabilized proteins. The CPN60 molecular chaperones perform their function in association with the CPN10 family of proteins. The latter act as cofactors in the refolding reaction.

■ Alternative names

CPN60 (or chaperonin 60; Hemmingsen et al., 1988) family members are also called HSP60. CPN10 family members are also called chaperonin 10, HSP10 or cochaperonin.

■ CPN60 and CPN10 sequences

Sequence alignment of CPN60 family members reveals at least 70% similarity and 50–70% identity between any two members. Residues 86–91 of the CPN60 protein from *Escherichia coli*, GroEL, form part of the ATP-binding site that includes the consensus sequence GDGTT found in all chaperonins, which is similar to the sequence GXGXX(G) in the ATP-binding sites of many type I kinases (Kim et al., 1994). Finally, all CPN60 family members contain two or three GGM repeats at their extreme C-termini. These sequences can be proteolytically cleaved without any detrimental effect on chaperonin function (Langer et al., 1992). While the conservation between CPN10 family members is not as great as between CPN60 members, the most distantly related share at least 55% similarity and 35% identity, while closely related members share 75% similarity and 55% identity. A limited sequence homology also exists between *E. coli* GroES and GroEL which was the basis for the classification of GroES as a chaperonin (Hemmingsen et al., 1988).

■ Gene regulation

All CPN60 and CPN10 family members are constitutively expressed (Mizzen et al., 1989; Gross et al., 1990; Hartman et al., 1992; Monzini et al., 1994; Viitanen et al., 1995). Under non-stress conditions *E. coli* GroEL constitutes approximately 1% of total cellular protein, and after up-regulation caused by heat shock amounts to 10% of all cellular protein. The prokaryotic members of the CPN60 and CPN10 family are up-regulated when the sigma factor, σ^{32}, whose expression is induced by heat shock, forms a complex with RNA polymerase (Gross et al., 1990). Interestingly, unlike other prokaryotes, the nitrogen fixing bacteria possesses more than one CPN60/10-like operon. *Bradyrhizobium japonicum* codes for five distinct CPN60/CPN10 operons that are apparently regulated differentially (Fischer et al., 1993). Heat shock regulation has been detected in three of the five operons; however, in one of these the expression is regulated by an inverted repeat DNA structure called CIRCE (controlling inverted repeat of chaperone expression) rather than by the sigma factor σ^{32} (Babst et al., 1996). CIRCE has been found as a controlling element of many heat shock proteins (Narberhaus and Bahl, 1992 and references therein). The inverted repeat seems to act as a negative *cis* element which, under non-stress conditions, represses the expression of the genes that it preceeds. The mechanism by which CIRCE regulates expression is unknown but it seems clear that it functions in association with an as yet unidentified protein (Zuber and Schumann, 1994). Finally, at least one of the CPN60/10 operons from *B. japonicum* is not regulated by heat shock but rather by the nitrogen fixation regulatory protein NifA and transcribed by the sigma factor σ^{54} (Fischer et al., 1993). Transcription of genes encoding the eukaryotic members of the CPN60 and CPN10 families is up-regulated 2–3-fold following heat shock in a process dependent on a heat shock transcription factor, HSF (Johnson et al., 1989; Mizzen et al., 1989; Reading et al., 1989; Hartman et al., 1992; Monzini et al., 1994). Chloroplast CPN60 and CPN10 family members are only slightly up-regulated by heat stress (Viitanen et al., 1995).

■ Structure of CPN60 proteins

The unusual and characteristic structure of the CPN60 proteins was initially visualized by electron microscopy (Hendrix, 1979; Hohn et al., 1979). With the possible exception of mammalian mitochondrial Cpn60, all members of the CPN60 family have been shown to consist of a complex of 14 subunits arranged in two heptameric rings stacked back-to-back, which define a large central cavity (Saibil et al., 1993). It has been proposed that mammalian Cpn60 forms single ring structures (Viitanen et al., 1992b), however it is unclear whether this is the functionally active form. The recent solution of the crystal structure of GroEL has now provided a molecular framework for all members of the CPN60 family (Braig et al., 1994, see also entry p. 173). The GroEL tetradecamer is 137 Å in diameter and 146 Å in length. The central cavity measures 45 Å in diameter. Each subunit of GroEL consists of three domains, an apical domain which defines the entrance into the cavity and to which both substrate and GroES bind, an equatorial domain which contains the ATP-binding site and is responsible for all of the ring–ring interactions and many of the subunit–subunit interactions, and, finally, the intermediate domain which connects the apical and equatorial domains and possibly acts as a molecular hinge to transmit conformational changes. The crystal structure has also revealed that the N- and C-termini of GroEL project into the cavity at the level of the equatorial domain suggesting that the cavity is occluded at this level.

The binding of nucleotide to Cpn60 has been shown by electron microscopy to induce significant conformational changes (Saibil et al., 1993), characterized by a 5–10° rotation of the subunits, particularly at the level of the apical domains. These nucleotide-induced conformational changes were not as apparent in the crystal structure of nucleotide-bound GroEL (Boisvert et al., 1996). This may be owing to the fact that the crystal structure of this complex was solved with a mutant form of GroEL which has lost many of its allosteric attributes (Aharoni and Horovitz, 1996).

■ Structure of CPN10 proteins

Electron microscopy revealed the single ring structure of CPN10 proteins (Chandrasekhar et al., 1986; Lubben et al., 1990; Hartman et al., 1992). The crystal structures of two CPN10 molecules have now been solved, that of E. coli GroES (Hunt et al., 1996) and the Cpn10 from Mycobacterium leprae (Mande et al., 1996). Indeed, they form single rings of seven subunits that are assembled in a dome-like structure approximately 75 Å in diameter and 30 Å in height. A large loop region that extends at the lower rim of the molecule (residues 16–33 in GroES) is involved in CPN60 binding (Landry et al., 1993) and a second loop at the top of the molecule defines the roof of the assembled CPN10 heptamer. Of particular interest is the hydrophilic nature of the inner surface of the GroES dome and the significant flexibility of the structure as suggested by its deviation from strict sevenfold symmetry (Hunt et al., 1996). In contrast to all other CPN10 family members, the chloroplast Cpn24 subunit consists of a single polypeptide comprised of two distinct CPN10 homologous domains in tandem, presumably the result of a gene duplication event (Bertsch et al., 1992; Baneyx et al., 1995, see also entry p. 179). The functional implications of this structure remain elusive.

■ CPN60–CPN10 complexes

Under most conditions CPN60 and CPN10 proteins form nucleotide-dependent asymmetric complexes (one 14mer binds to one 7mer), which have been studied in detail with GroEL and GroES from E. coli (Langer et al., 1992). These complexes are particularly stable when formed with ADP (Martin et al., 1993; Todd et al., 1993). The binding of GroES to GroEL is mediated in part by the mobile loop located at the lower rim of GroES, which has been shown to form a defined β hairpin in the GroEL–GroES complex (Landry et al., 1993). Both residues at the top surface of GroEL and hydrophobic residues on the apical domains facing the central cavity are involved in the interaction with GroES (Fenton et al., 1994). Thus it is possible that there are two distinct interactive sites between GroEL and GroES. The structure of the nucleotide-bound complex between GroEL and GroES has been visualized by electron microscopy (Chen et al., 1994). The binding of GroES induces significant conformational changes in the apical domains of the GroEL ring to which it is bound. The domains move upwards and outwards, thereby significantly increasing the volume of the central cavity such that proteins up to at least 50 kDa could, in principle, be accomodated within the enclosed cavity underneath GroES.

At higher levels of Mg^{2+}–ATP, GroEL and GroES form symmetrical complexes with two GroES heptamers bound to GroEL, one at each end of the tetradecamer (Azem et al., 1994, Llorca et al., 1994, Schmidt et al., 1994b). These complexes have been proposed to represent an intermediate in the GroEL–GroES reaction cycle (Todd et al., 1994), but their physiological relevance remains to be clarified (Engel et al., 1995; Hayer-Hartl et al., 1995).

■ Substrate binding to CPN60

It seemed intuitive that the large central cavity in CPN60 complexes would be the site of substrate binding. This was indeed shown to be the case by negative stain electron microscopy of complexes of GroEL with two different substrate proteins, rhodanese and alcohol oxidase (Langer et al., 1992). This observation has since been confirmed by other electron microscopy studies both with GroEL (Braig et al., 1993) and Cpn60 from Rhodobacter sphaeroides (Saibil et al., 1993), and most recently by small-angle neutron scattering of GroEL–rhodanese complexes (Thiyagarajan et al., 1996). Polypeptide binds at the level of the apical domains of the GroEL subunits. Mutagenesis studies have identified a number of hydrophobic residues on the inner surface of the apical domains which, when replaced by polar residues, prevent substrate binding (Fenton et al., 1994). These same residues were also identified as the site of substrate binding by a cross-linking approach in which dihydrofolate reductase was covalently linked to GroEL after it had been bound as a substrate (Mayhew et al., 1996).

The structure of CPN60-bound polypeptides has been analysed with GroEL using a variety of biophysical methods. Initially tryptophan fluorescence and ANS-binding studies suggested that GroEL bound compact folding intermediates of the molten globule type (Martin et al., 1991). Further investigation using fluorescence spectroscopy and NMR and electrospray ionization mass spectrometry, both in conjunction with hydrogen exchange, have confirmed that the substrate bound by GroEL is characterized by a fluctuating hydrophobic core and transient secondary structure, but no tertiary structure, which are hallmarks of the molten globule state (Hayer-Hartl et al., 1994; Robinson et al., 1994; Zahn et al., 1994). While it is clear that GroEL interacts with substrate proteins via hydrophobic interactions (Fenton et al., 1994; Hayer-Hartl et al., 1994; Hlodan et al., 1995; Lin et al., 1995), it remains unclear whether GroEL also recognizes specific structural features in substrate proteins. Both α helical (Landry and Gierasch, 1991) and β stranded (Schmidt and Buchner, 1992) structures can bind to GroEL.

■ Mechanism of action of CPN60 and CPN10

CPN60 and CPN10 are encoded by essential genes in both prokaryotes and eukaryotes (Fayet et al., 1989; Johnson et al., 1989; Reading et al., 1989; Rospert et al., 1993; Höhfeld and Hartl, 1994). These chaperones constitute a protein folding machinery that is found ubiquitously in eubacteria and in the endosymbiotic organelles, mitochondria and chloroplasts. Indeed, it was in chloroplasts that the role of chaperones in protein assembly was originally recognized (Barraclough and Ellis, 1980; Hemmingsen et al., 1988). It then became clear that members of the CPN60 family were involved in the ATP-dependent folding of monomeric proteins or the subunits of protein assemblies (Ostermann et al., 1989; Horwich et al., 1993). Furthermore, mitochondrial Hsp60 has been shown to stabilize pre-existent proteins under conditions of heat stress and mediate their refolding upon return to normal growth conditions (Martin et al., 1992). Although it is unclear how many proteins require the CPN60/CPN10 machinery for their folding, or how many proteins actually use the machinery (Ellis and Hartl, 1996; Lorimer, 1996), it is clear that a large fracton of all soluble proteins, when unfolded, can bind to CPN60 family members (Viitanen et al., 1992a; Martin et al., 1992) and a defined number of proteins are dependent on CPN60/CPN10 for their folding (Horwich et al., 1993; Rospert et al., 1996).

Many years of study have provided a basic understanding of the mechanism by which CPN60 and CPN10 cooperate in the promotion of protein folding. The chaperonin system of E. coli has been analysed in most detail. A model of the reaction cycle of GroEL–GroES-dependent protein folding is shown in Fig. 1. GroEL has a weak K^+- and Mg^{2+}-dependent ATPase (Ishihama et al., 1976; Viitanen et al., 1990). The binding of GroES to GroEL is nucleotide dependent (Chandrasekhar et al., 1986; Langer et al., 1992). Complex formation with GroES increases the cooperativity of ATP hydrolysis by GroEL while lowering the rate (Gray and Fersht, 1991; Jackson et al., 1993; Todd et al., 1993). In the absence of GroES, a bound polypeptide will cycle on and off GroEL, dependent upon ATP hydrolysis (Martin et al., 1991). If the polypeptide has the ability to fold spontaneously, then this GroES-independent cycling is sufficient to promote folding (Martin et al., 1991). However, other proteins with a tendency to aggregate or become kinetically trapped, depend on a productive complex between GroEL and GroES in order to fold to the native state (Martin et al., 1991, 1993; Schmidt et al., 1994b). Under these conditions the ability of GroEL to iteratively bind and release substrate polypeptides still seems to be the key to efficient folding. However, these events occur productively only when GroES binds to the same ring of GroEL as the substrate, thereby enclosing the substrate polypeptide in the cavity (Martin et al., 1993; Weissman et al., 1995, 1996; Mayhew et al., 1996). Since GroES and substrate polypeptide have overlapping binding sites on GroEL (Fenton et al., 1994), it seems likely that the binding of GroES is sufficient to release the polypeptide into the GroEL cavity (Hartl, 1994; Mayhew and Hartl, 1996). This in turn allows the polypeptide to attempt to fold in a sequestered environment in which intermolecular aggregation is prevented. ATP hydrolysis in the opposite ring of the GroEL–substrate–GroES complex induces release of GroES (Todd et al., 1994; Burston et al., 1995; Hayer-Hartl et al., 1995) allowing folded protein to exit from the complex (Mayhew et al., 1996; Weissman et al., 1996). Any substrate that did not reach a sufficiently stable conformation or has been trapped as a kinetcally stable intermediate will rebind to GroEL, possibly without being released from the central cavity. Rebinding may allow unfolding and structural rearrangement of unproductive intermediates, preparing them for another folding attempt upon the rebinding of GroES.

In summary, three functional elements contribute to the high efficiency of chaperonin-assisted protein folding. (1) The prevention of aggregation by binding unfolded or kinetically trapped folding intermediates that expose hydrophobic surfaces. (2) The GroES-induced release of unfolded polypeptides into a sequestered environment that is permissive for folding. (3) A proofreading mechanism by virtue of rebinding and structurally rearranging polypeptides that failed to fold sufficiently upon release, thereby preparing them for another folding trial in the GroEL cavity.

■ References

Aharoni, A. and Horovitz, A. (1996). Inter-ring communication is disrupted in the GroEL mutant Arg13 → Gly; Ala126 → Val with known crystal structure. J. Mol. Biol. **258**, 732–735.

Azem, A., Kessel, M., and Goloubinoff, P. (1994). Characterization of a functional $GroEL_{14}(GroES_7)_2$ chaperonin hetero-oligomer. Science **265**, 653–656.

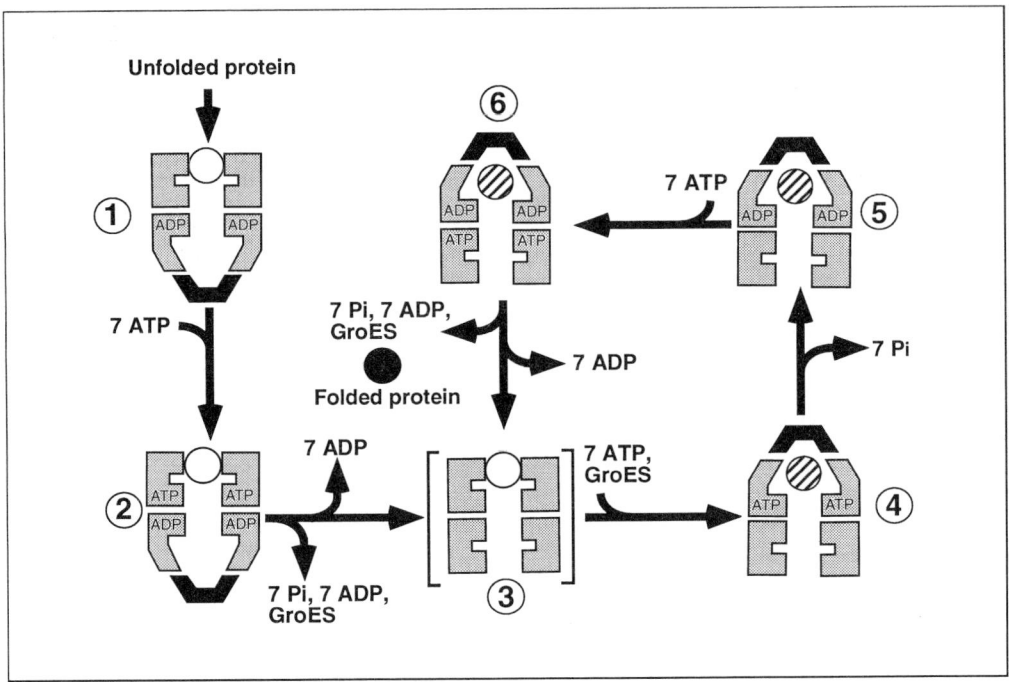

Figure 1. Model of the GroEL–GroES reaction cycle in folding (adapted from Hartl, 1996). The substrate polypeptide will initially bind to the unoccupied ring of an ADP-bound, asymmetrical, GroEL–GroES complex (1). Polypeptide binding to this ring of GroEL, as well as ATP binding and hydrolysis, results in the release of GroES and nucleotide from the opposite ring of GroEL (2, 3). The nucleotide-free state shown in (3) may only occur transiently and is shown for simplicity. Nucleotide and GroES will then rebind to either ring of GroEL, thereby enclosing 50% of the bound polypeptide in a cavity defined by GroEL and GroES (4). The binding of GroES to the substrate-bound ring of GroEL may in itself induce the release of substrate polypeptide from its binding sites on GroEL into the cavity, at which point it can attempt to fold (4). The hydrolysis of ATP in the substrate-bound ring ensures the release of substrate polypeptide into the cavity, possibly by inducing a tighter binding between GroEL and GroES (4, 5). The binding and hydrolysis of ATP in the opposite ring of GroEL triggers the release of GroES allowing folded protein to emerge from the cavity (6). Any incompletely folded substrate polypeptide rebinds to GroEL. Light and dark spheres represent unfolded and folded substrate protein, respectively, and the hatched sphere indicates the presence of a mixture of folded and unfolded substrate in the population of GroEL molecules. Unfolded protein can be retained in steps (6) to (3).

Babst, M., Hennecke, H., and Fischer, H.M. (1996). Two different mechanisms are involved in the heat-shock regulation of chaperonin gene expression in *Bradyrhizobium japonicum*. Mol. Microbiol. **19**, 827–839.

Baneyx, F., Bertsch, U., Kalbach, C.E., Vandervies, S.M., Soll, J., and Gatenby, A.A. (1995). Spinach chloroplast cpn21 co-chaperonin possesses two functional domains fused together in a toroidal structure and exhibits nucleotide-dependent binding to plastid chaperonin 60. J. Biol. Chem. **270**, 10695–10702.

Barraclough, R. and Ellis, R.J. (1980). Protein synthesis in chloroplasts. IX. Assembly of newly-synthesized large subunits into ribulose bisphosphate carboxylase in isolated intact pea chloroplasts. Biochim. Biophys. Acta **608**, 18–31.

Bertsch, U., Soll, J., Seetharam, R., and Viitanen, P. V. (1992). Identification, characterization, and DNA sequence of a functional double groES-like chaperonin from chloroplasts of higher plants. Proc. Natl. Acad. Sci. USA **89**, 8696–8700.

Boisvert, D.C., Wang, J.M., Otwinowski, Z., Horwich, A.L., and Sigler, P.B. (1996). The 2.4 angstrom crystal structure of the bacterial chaperonin GroEL complexed with ATP-g-S. Nature Struct. Biol. **3**, 170–177.

Braig, K., Simon, M., Furuya, F., Hainfeld, J.F., and Horwich, A.L. (1993). A polypeptide bound by the chaperonin groEL is localized within a central cavity. Proc. Natl. Acad. Sci. USA **90**, 3978–3982.

Braig, K., Otwinowski, Z., Hegde, R., Boisvert, D.C., Joachimiak, A., Horwich, A.L., and Sigler, P.B. (1994). The crystal structure of the bacterial chaperonin GroEL at 2.8 Å. Nature **371**, 578–586.

Burston, S.G., Ranson, N.A., and Clarke, A.R. (1995). The origins and consequences of asymmetry in the chaperonin reaction cycle. J. Mol. Biol. **249**, 138–152.

Chandrasekhar, G.N., Tilly, K., Woolford, C., Hendrix, R., and Georgopoulos, C. (1986). Purification and properties of the groES morphogenetic protein of *Escherichia coli*. J. Biol. Chem. **261**, 12414–12419.

Chen, S., Roseman, A.M., Hunter, A.S., Wood, S.P., Burston, S.G., Ranson, N.A., Clarke, A.R., and Saibil, H.R. (1994). Location of a folding protein and shape changes in GroEL–GroES complexes imaged by cryo-electron microscopy. Nature **371**, 261–264.

Ellis, R.J. and Hartl, F.U. (1996). Protein folding in the cell: competing models of chaperonin function. FASEB J. **10**, 20–26.

Engel, A., Hayer-Hartl, M.K., Goldie, K.N., Pfeifer, G., Hegerl, R., Muller, S., Dasilva, A.C.R., Baumeister, W., and Hartl, F.U. (1995). Functional significance of symmetrical versus asymmetrical GroEL–GroES chaperonin complexes. Science **269**, 832–836.

Fayet, O., Ziegelhoffer, T., and Georgopoulos, C. (1989). The groES and groEL heat shock gene products of *Escherichia coli* are essential for bacterial growth at all temperatures. J. Bacteriol. **171**, 1379–1385.

Fenton, W.A., Kashi, Y., Furtak, K., and Horwich, A.L. (1994). Residues in chaperonin GroEL required for polypeptide binding and release. Nature **371**, 614–619.

Fischer, H.M., Babst, M., Kaspar, T., Acuna, G., Arigoni, F., and Hennecke, H. (1993). One member of a gro-ESL-like chaperonin multigene family in *Bradyrhizobium japonicum* is co-regulated with symbiotic nitrogen fixation genes. EMBO J. **12**, 2901–2912.

Gray, T.E. and Fersht, A.R. (1991). Cooperativity in ATP hydrolysis by GroEL is increased by GroES. FEBS Lett. **292**, 254–258.

Gross, C.A., Straus, D.B., Erickson, J.W., and Yura, T. (1990). The function and regulation of heat shock proteins in *Escherichia coli*., In 'Stress Proteins in Biology and Medicine'. (R.I. Morimoto, A. Tissieres and C. Georgopoulos, eds). Cold Spring Harbor Laboratory Press, New York, pp. 167–189.

Hartl, F.U. (1994). Protein folding. Secrets of a double-doughnut. Nature **371**, 557–559.

Hartl, F.U. (1996). Molecular chaperones in cellular protein folding. Nature **381**, 571–580.

Hartman, D.J., Hoogenraad, N.J., Condron, R., and Hoj, P.B. (1992). Identification of a mammalian 10-kDa heat shock protein, a mitochondrial chaperonin 10 homologue essential for assisted folding of trimeric ornithine transcarbamoylase in vitro. Proc. Natl. Acad. Sci. USA **89**, 3394–3398.

Hayer-Hartl, M.K., Ewbank, J.J., Creighton, T.E., and Hartl, F.U. (1994). Conformational specificity of the chaperonin GroEL for the compact folding intermediates of alpha-lactalbumin. EMBO J. **13**, 3192–3202.

Hayer-Hartl, M.K., Martin, J., and Hartl, F.U. (1995). Asymmetrical interaction of GroEL and GroES in the ATPase cycle of assisted protein folding. Science **269**, 836–841.

Hemmingsen, S.M., Woolford, C., van der Vies, S.M., Tilly, K., Dennis, D.T., Georgopoulos, C.P., Hendrix, R.W., and Ellis, R.J. (1988). Homologous plant and bacterial proteins chaperone oligomeric protein assembly. Nature **333**, 330–334.

Hendrick, J.P. and Hartl, F.U. (1995). The role of molecular chaperones in protein folding. FASEB J. **9**, 1559–1569.

Hendrix, R.W. (1979). Purification and properties of GroE, a host protein involved in bacteriophage assembly. J. Mol. Biol. **129**, 375–392.

Hlodan, R., Tempst, P., and Hartl, F.U. (1995). Binding of defined regions of a polypeptide to GroEL and its implications for chaperonin-mediated protein folding. Nature Struct. Biol. **2**, 587–595.

Höhfeld, J. and Hartl, F.U. (1994). Role of the chaperonin cofactor Hsp10 in protein folding and sorting in yeast mitochondria. J. Cell. Biol. **126**, 305–315.

Hohn, T., Hohn, B., Engel, A., Wortz, M., and Smith, P.R. (1979). Isolation and characterisation of the host protein GroE involved in bacteriophage lambda assembly. J. Mol. Biol. **129**, 359–373.

Horwich, A.L., Low, K.B., Fenton, W.A., Hirshfield, I.N., and Furtak, K. (1993). Folding in vivo of bacterial cytoplasmic proteins: role of GroEL. Cell **74**, 909–917.

Hunt, J.F., Weaver, A.J., Landry, S.J., Gierasch, L., and Deisenhofer, J. (1996). The crystal structure of the GroES co-chaperonin at 2.8 angstrom resolution. Nature **379**, 37–45.

Ishihama, A., Ikeuchi, T., Matsumoto, A., and Yamamoto, S. (1976). A novel adenosine triphosphatase isolated from RNA polymerase preparations of *Escherichia coli*. II. Enzymatic properties and molecular structure. J. Biochem. (Tokyo) **79**, 927–937.

Jackson, G.S., Staniforth, R.A., Halsall, D.J., Atkinson, T., Holbrook, J.J., Clarke, A.R., and Burston, S.G. (1993). Binding and hydrolysis of nucleotides in the chaperonin catalytic cycle: implications for the mechanism of assisted protein folding. Biochemistry **32**, 2554–2563.

Johnson, R.B., Fearon, K., Mason, T., and Jindal, S. (1989). Cloning and characterization of the yeast chaperonin HSP60 gene. Gene **84**, 295–302.

Kim, S., Willison, K.R., and Horwich, A.L. (1994). Cystosolic chaperonin subunits have a conserved ATPase domain but diverged polypeptide-binding domains. Trends Biochem. Sci. **19**, 543–548.

Landry, S.J. and Gierasch, L.M. (1991). The chaperonin GroEL binds a polypeptide in an alpha-helical conformation. Biochemistry **30**, 7359–7362.

Landry, S.J., Zeilstra, R.J., Fayet, O., Georgopoulos, C., and Gierasch, L.M. (1993). Characterization of a functionally important mobile domain of GroES. Nature **364**, 255–258.

Langer, T., Pfeifer, G., Martin, J., Baumeister, W., and Hartl, F.U. (1992). Chaperonin-mediated protein folding: GroES binds to one end of the GroEL cylinder, which accommodates the protein substrate within its central cavity. EMBO J. **11**, 4757–4765.

Lin, Z., Schwartz, F.P., and Eisenstein, E. (1995). The hydrophobic nature of GroEL-substrate binding. J. Biol. Chem. **270**, 1011–1014.

Llorca, O., Marco, S., Carrascosa, J.L., and Valpuesta, J.M. (1994). The formation of symmetrical GroEL–GroES complexes in the presence of ATP. FEBS Lett. **345**, 181–186.

Lorimer, G.H. (1996). A quantitative assessment of the role of chaperonin proteins in protein folding in vivo. FASEB J. **10**, 5–9.

Lubben, T.H., Gatenby, A.A., Donaldson, G.K., Lorimer, G.H., and Viitanen, P.V. (1990). Identification of a groES-like chaperonin in mitochondria that facilitates protein folding. Proc. Natl. Acad. Sci. USA **87**, 7683–7687.

Mande, S.C., Mehra, V., Bloom, B.R., and Hol, W.G.J. (1996). Structure of the heat shock protein chaperonin-10 of *Mycobacterium leprae*. Science **271**, 203–207.

Martin, J., Langer, T., Boteva, R., Schramel, A., Horwich, A.L., and 'artl, F.U. (1991). Chaperonin-mediated protein folding at the surface of groEL through a 'molten globule'-like intermediate. Nature **352**, 36–42.

Martin, J., Horwich, A.L., and Hartl, F.U. (1992). Prevention of protein denaturation under heat stress by the chaperonin Hsp60. Science **258**, 995–998.

Martin, J., Mayhew, M., Langer, T., and Hartl, F.U. (1993). The reaction cycle of GroEL and GroES in chaperonin-assisted protein folding. Nature **366**, 228–233.

Mayhew, M. and Hartl, F.U. (1996). Lord of the rings: GroES structure. Science **271**, 161–162.

Mayhew, M., Da Silva, A.C.R., Martin, J., Erdjument-Bromage, H., Tempst, P., and Hartl, F.U. (1996). Protein folding in the central cavity of the GroEL–GroES chaperonin complex. Nature **379**, 420–426.

Mizzen, L.A., Chang, C., Garrels, J.I., and Welch, W.J. (1989). Identification, characterization, and purification of two mammalian stress proteins present in mitochondria, grp75, a member of the hsp70 family, and hsp58, a homolog of the bacterial groEL protein. J Biol. Chem. **264**, 20664–20675.

Monzini, N., Legname, G., Marcucci, F., Gromo, G., and Modena, D. (1994). Identification and cloning of human chaperonin 10 homologue. Biochim. Biophys. Acta **1218**, 478–480.

Narberhaus, F. and Bahl, H. (1992). Cloning, sequencing, and molecular analysis of the groESL operon of *Clostridium acetobutylicum*. J. Bacteriol. **174**, 3282–3289.

Ostermann, J., Horwich, A.L., Neupert, W., and Hartl, F.-U. (1989). Protein folding in mitochondria requires complex formation with hsp60 and ATP hydrolysis. Nature **341**, 125–130.

Reading, D.S., Hallberg, R.L., and Meyers, A.M. (1989). Characterization of the yeast *HSP60* gene coding for a mitochondrial assembly factor. Nature **337**, 655–659.

Robinson, C.V., Gross, M., Eyles, S.J., Ewbank, J.J., Mayhew, M., Hartl, F.U., Dobson, C.M., and Radford, S.E. (1994). Conformation of GroEL-bound a-lactalbumin probed by mass spectrometry. Nature **372**, 646–651.

Rospert, S., Junne, T., Glick, B.S., and Schatz, G. (1993). Cloning and disruption of the gene encoding yeast mitochondrial chaperonin 10, the homolog of *E. coli* groES. FEBS Lett. **335**, 358–360.

Rospert, S., Looser, R., Dubaquie, Y., Matouschek, A., Glick, B.S., and Schatz, G. (1996). Hsp60-independent protein folding in the matrix of yeast mitochondria. EMBO J. **15**, 764–774.

Saibil, H.R., Zheng, D., Roseman, A.M., Hunter, A.S., Watson, G.M.F., Chen, S., auf der Mauer, A., O'Hara, B.P., Wood, S.P., Mann, N.H., Barnett, L.K., and Ellis, R.J. (1993). ATP induces large quarternary rearrangements in a cage-like chaperonin structure. Curr. Biol. **3**, 265–273.

Schmidt, M. and Buchner, J. (1992). Interaction of GroE with an all-beta-protein. J. Biol. Chem. **267**, 16829–16833.

Schmidt, M., Buchner, J., Todd, M.J., Lorimer, G.H., and Viitanen, P.V. (1994a). On the role of groES in the chaperonin-assisted folding reaction. Three case studies. J. Biol. Chem. **269**, 10304–10311.

Schmidt, M., Rutkat, K., Rachel, R., Pfeifer, G., Jaenicke, R., Viitanen, P., Lorimer, G., and Buchner, J. (1994b). Symmetric complexes of GroE chaperonins as part of the functional cycle. Science 265, 656–659.

Thiyagarajan, P., Henderson, S.J., and Joachimiak, A. (1996). Solution structures of GroEL and its complex with rhodanese from small-angle neutron scattering. Structure **4**, 79–88.

Todd, M.J., Viitanen, P.V., and Lorimer, G.H. (1993). Hydrolysis of adenosine 5'-triphosphate by *Escherichia coli* GroEL: effects of GroES and potassium ion. Biochemistry **32**, 8560–8567.

Todd, M.J., Viitanen, P.V., and Lorimer, G.H. (1994). Dynamics of the chaperonin ATPase cycle: implications for facilitated protein folding. Science **265**, 659–666.

Viitanen, P.V., Lubben, T.H., Reed, J., Goloubinoff, P., O'Keefe, D.P., and Lorimer, G.H. (1990). Chaperonin-facilitated refolding of ribulosebisphosphate carboxylase and ATP hydrolysis by chaperonin 60 (groEL) are K^+ dependent. Biochemistry **29**, 5665–5671.

Viitanen, P.V., Gatenby, A.A., and Lorimer, G.H. (1992a). Purified chaperonin 60 (groEL) interacts with the nonnative states of a multitude of *Escherichia coli* proteins. Protein Sci. **1**, 363–369.

Viitanen, P.V., Lorimer, G.H., Seetharam, R., Gupta, R.S., Oppenheim, J., Thomas, J.O., and Cowan, N.J. (1992b). Mammalian mitochondrial chaperonin 60 functions as a single toroidal ring. J. Biol. Chem. **267**, 695–698.

Viitanen, P.V., Schmidt, M., Buchner, J., Suzuki, T., Vierling, E., Dickson, R., Lorimer, G.H., Gatenby, A., and Soll, J. (1995). Functional characterization of the higher plant chloroplast chaperonins. J. Biol. Chem. **270**, 18158–18164.

Weissman, J.S., Hohl, C.M., Kovalenko, O., Kashi, Y., Chen, S.X., Braig, K., Saibil, H.R., Fenton, W.A., and Horwich, A.L. (1995). Mechanism of GroEL action: productive release of polypeptide from a sequestered position under GroES. Cell **83**, 577–587.

Weissman, J.S., Rye, H.S., Fenton, W.A., Beechem, J.M., and Horwich, A.L. (1996). Characterization of the active intermediate of a GroEL–GroES-mediated protein folding reaction. Cell **84**, 481–490.

Zahn, R., Spitzfaden, C., Ottiger, M., Wuthrich, K., and Pluckthun, A. (1994). Destabilization of the complete protein secondary structure on binding to the chaperone GroEL. Nature **368**, 261–265.

Zuber, U. and Schumann, W. (1994). CIRCE, a novel heat shock element involved in regulation of heat shock operon dnaK of *Bacillus subtilis*. J. Bacteriol. **176**, 1359–1363.

■ *Mark Mayhew:*
Howard Hughes Medical Institute
Cellular Biochemistry and Biophysics Program
Memorial Sloan–Kettering Cancer Center
1275 York Avenue
New York, NY 10021, USA
Tel. 1 212 638 6505
Fax. 1 212 717 3604
E-mail: m-mayhew@ski.mskcc.org

■ *F. Ulrich Hartl:*
Max-Planck-Institut Für Biochemie
Am Klopferspitz 18a
D-82152 Martinsried
Germany
Tel. 49 89 8578 2211
Fax. 49 89 8578 2244/2233
E-mail: uhartl@biochem.mpg.de

Escherichia coli GroEL, structure and function

GroEL in the bacterial cytoplasm is an abundant and essential 800 kDa protein complex, composed of two seven-membered rings stacked back-to-back, that mediates ATP-dependent folding of proteins to their native form. GroEL binds many newly synthesized proteins, as well as proteins misfolding under stress, in its central channel as non-native folding intermediates. This binding appears to be mediated primarily through hydrophobic interactions. The single ring cochaperone GroES is able to bind to the same ring of GroEL as the polypeptide, thereby sequestering the peptide in the GroEL central cavity. GroES binding in the presence of adenine nucleotide is associated with upward and twisting movements of the apical domains, the most severe twisting, as observed by cryo-electron microscopy, occurring with ATP, and GroES appears to make contact with the mobilized peptide binding sites. The rearrangement of GroEL upon GroES binding results in an approximate doubling of the volume of the central channel on the cis side, and the peptide binding sites appear to be removed from facing the channel, potentially explaining rapid release of peptide with commencement of folding in the presence of ATP/GroES. Subsequent ATP action in the opposite GroEL ring promotes the release of GroES and polypeptide. Upon release, a fraction of polypeptide can have already reached native form, another fraction is committed to reaching native form in solution, and a third fraction remains non-native and undergoes kinetic partitioning between GroEL or other chaperones or proteases.

■ Alternative names

Chaperonin 60, cpn60 (Hemmingsen et al., 1988; Goloubinoff et al., 1989b); GroES as chaperonin 10, cpn10 (Goloubinoff et al., 1989a).

■ Isolation of GroEL

GroEL was originally identified through mutations of Escherichia coli that led to defective λ phage biogenesis (Georgopoulos et al., 1973; Sternberg, 1973). In such mutant cells, λ phage heads were found as aggregates ('monsters') unable to produce infectious phage particles, thus implicating GroEL in protein assembly. The phage biogenesis defect could be overcome in part by mutations affecting the E protein of λ, which abrogated its aggregation, hence the name GroE. Subsequently, GroEL was recognized as an abundant heat shock protein, whose already great abundance of 1% total protein is increased to as much as 10% under heat shock conditions (Fayet et al., 1989). The striking ring-shaped quaternary structure was revealed by electron microscopy (Hendrix, 1979; Hohn et al., 1979). Both GroEL and its cochaperone GroES, itself a single ring composed of seven radially arranged 10 kDa subunits, were shown to be essential at all temperatures (Fayet et al., 1989).

■ groEL gene and sequence

A cloned operon from 94.3 minutes on the E. coli chromosome rescued GroEL-deficient mutants, and its sequence (GenBank accession number X07850) contains: a 40 base promoter region bearing a regulatory site for the heat shock factor, σ^{32}, that is also recognizable by the constitutive σ^{70} factor; 97 codons encoding the 10 kDa GroES subunit; a 43 base intergenic region; and 548 codons encoding the 57 kDa GroEL subunit (Hemmingsen et al., 1988). When purified, both GroEL and GroES are devoid of their initiator methionines. The deduced primary sequence has 50–60% identity to the eukaryotic mitochondrial protein Hsp60 (Jindal et al., 1989; Reading et al., 1989), shown to be required for folding of a large number of imported mitochondrial proteins (Cheng et al., 1989), and to the chloroplast protein rubisco-binding protein, implicated in the assembly of hexadecameric rubisco (Barraclough and Ellis, 1980). These three so-called chaperonins (Ellis, 1987) are presumably related evolutionarily by endosymbiosis. The primary sequence of GroEL predicted residues likely to participate in nucleotide binding at positions 87–92 (Lewis et al., 1992).

■ GroEL protein

GroEL is an 800 kDa assembly, a tetradecamer of 57 kDa subunits arranged in two seven-membered rings arranged back-to-back (Fig. 1). As revealed by electron microscopy studies (Hutchinson et al., 1989; Zwickl et al. 1990; Saibil et al., 1993; Chen et al., 1994) and by the crystal structure at 2.8 Å (Braig et al., 1994), the overall dimensions of the cylinder-shaped assembly are 146 Å in height and 137 Å in diameter, with a central channel measuring c. 45 Å in diameter (see Fig. 1a). The channel is occluded at the equatorial level of both rings, as judged by microscopic analysis and neutron scattering (Thiyagarajan et al., 1996). The occluding structure is apparently the collective of COOH-termini of the GroEL subunits, residues 523–548, which are not resolved in crystallographic studies and may be non-ordered at least in part as the result of a

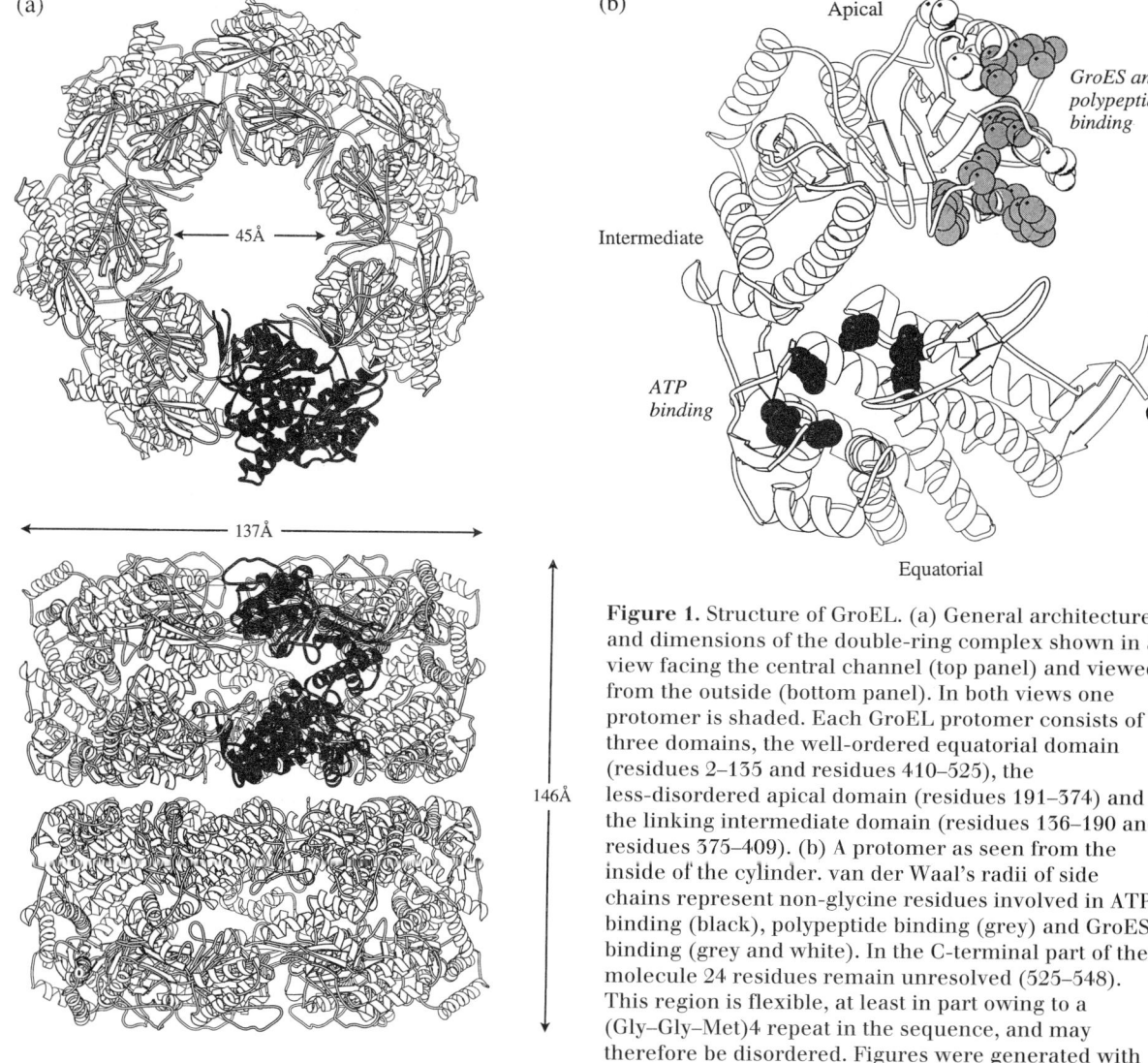

Figure 1. Structure of GroEL. (a) General architecture and dimensions of the double-ring complex shown in a view facing the central channel (top panel) and viewed from the outside (bottom panel). In both views one protomer is shaded. Each GroEL protomer consists of three domains, the well-ordered equatorial domain (residues 2–135 and residues 410–525), the less-disordered apical domain (residues 191–374) and the linking intermediate domain (residues 136–190 and residues 375–409). (b) A protomer as seen from the inside of the cylinder. van der Waal's radii of side chains represent non-glycine residues involved in ATP binding (black), polypeptide binding (grey) and GroES binding (grey and white). In the C-terminal part of the molecule 24 residues remain unresolved (525–548). This region is flexible, at least in part owing to a (Gly–Gly–Met)4 repeat in the sequence, and may therefore be disordered. Figures were generated with the program MOLSCRIPT (Kraulis, 1991).

terminal repeating Gly–Gly–Met motif. The single ring cochaperonin GroES measures 70–80 Å in outside diameter and is c. 30 Å in height, with a smaller central channel of c. 30 Å diameter (Hunt *et al.*, 1996; Mande *et al.*, 1996; see next entry p. 179). Abundance of GroES is similar to that of GroEL on a molar basis.

GroEL subunits are composed of three domains (Fig. 1b; Braig *et al.*, 1994). The apical domain, at the terminus of the chaperonin cylinder, is composed of a central orthogonal β sheet structure flanked at both central and outside aspects by α helices. This domain was less well resolved crystallographically than the other two, with a less defined electron density map and overall higher B factors (Braig *et al.*, 1995). This may reflect that this domain occupies a flexible position to accomodate binding of a large range of polypeptides. The equatorial domain, at the waist of the cylinder, was crystallographically well resolved. It is composed of long α helices and makes tight contacts both around and within the chaperonin rings. The intermediate domain is a smaller hinge-like domain, between the other two, composed of ascending/descending α helices under a β sheet roof structure. A cavity in the top of the equatorial domain c. 15 Å in diameter is the site of nucleotide binding (Boisvert *et al.*, 1996).

Biological activity of GroEL

In vivo GroEL appears to assist the folding of a considerable number of soluble cytoplasmic proteins, as suggested by effects on solubility and half-life of many proteins in a severe temperature-sensitive lethal allele (Horwich et al., 1993). Whether the class of proteins destined for secretion interact generally with GroEL remains unclear, but an effect on β-lactamase export in a GroEL mutant has been reported (Kusukawa et al., 1989). In vitro GroEL binds a wide variety of proteins: c. 50% of soluble species from E.coli bind to GroEL upon dilution from denaturant (Viitanen et al., 1992). There appears to be a preference to form stable binary complex with proteins that are prone to aggregation, suggesting that GroEL acts to assist folding of kinetically trapped or aggregation prone proteins (Buchner et al., 1991; Jackson et al., 1993; Todd et al., 1994; Weissman et al., 1994; Ranson et al., 1995). In general, smaller proteins may fold without entering such traps and as such may not form stable complexes with GroEL, although the folding of the 12 kDa protein, barnase, has been shown to be slowed by the presence of GroEL (Gray and Fersht, 1993; Corrales and Fersht, 1995; Zahn et al., 1996). Proteins greater than c. 60 kDa may be too large to be accommodated underneath GroES during a folding cycle and would need to bind and release from the GroEL ring opposite that bound by GroES. Whether such proteins can be assisted in vivo by GroEL–GroES is unclear at present.

GroEL binds ATP in the seven equatorial binding sites of one ring at a time, with c. 10 μM affinity (Jackson et al., 1993), and hydrolyses ATP in a reaction dependent on both Mg^{2+} and K^+ (Viitanen et al., 1990; Todd et al., 1993) at a rate of c. 35 per complex per minute in the absence of GroES at 25°C. GroES binding requires the presence of adenine nucleotide, and the cochaperone increases the cooperativity of both ATP binding and hydrolysis (Gray and Fersht, 1991; Bochkareva et al., 1992; Jackson et al., 1993). GroES binding is stabilized by the presence of ADP in the GroES-bound ring, and binding in turn stabilizes the binding of ADP (Jackson et al., 1993; Martin et al., 1993; Todd et al., 1993). Bound GroES is released by the hydrolysis of ATP in the opposite ring (Todd et al., 1994; Burston et al., 1995). Given such considerations, GroES would likely be in dynamic association with GroEL in vivo. Since GroEL is occupied most of the time by GroES, polypeptide is likely to be bound by an asymmetric complex containing GroES at one end (Saibil et al., 1991; Ishii et al., 1992; Langer et al., 1992; Todd et al., 1993). Polypeptide is bound in the central channel of the unoccupied GroEL ring through hydrophobic residues in the apical domain facing the channel. Mutation of these residues to charged or polar character abrogates binding (Fenton et al., 1994). Binding presumably occurs by interaction of exposed hydrophobic surface of the bound substrate protein (Landry and Gierasch, 1994; Zahn and Pluckthun, 1994; Zahn et al., 1994a; Itzhaki et al., 1995) with the hydrophobic apical side chains. The bound folding intermediate may have a collapsed but non-organized tertiary structure, because bound substrates are very susceptible to proteolysis. Bound substrate protein may exhibit a degree of native-like secondary structure, however, based on physical measurements. Studies of various oxidized 2,3, and 4-disulfide forms of α-lactalbumin reveal that species with 3-disulfides, which contain some helical secondary structure, can bind efficiently to GroEL. Intriguingly, however, the native-like molten globule form of α-lactalbumin is not recognized by GroEL (Hayer-Hartl et al., 1994; Okazaki et al., 1994). Moreover, deuterium exchange experiments with bound cyclophilin (Zahn et al., 1994b) and α-lactalbumin (Robinson et al., 1994) have found, at best, only modest protection of amide protons in bound substrate. More recently, deuterium exchange and NMR spectroscopy was carried out with human DHFR bound to GroEL, allowing assignment of protected protons (Goldberg et al., 1997). Modest protection was observed, as in other studies, but here they could be mapped, localizing to the regions that are the most protected in native DHFR, suggesting that there is already a native-like global fold, albeit unstable, in the bound substrate protein. Stopped flow studies of human DHFR showed that it could be rapidly bound by GroEL very early following dilution, in the dead time of mixing (corresponding to the intermediate in the NMR study), but also at later times, even minutes after initiation of refolding, suggesting that GroEL can possibly recognize different conformations of the non-native protein. GroEL may in some cases recognize relatively unstructured intermediates that represent earlier conformations than the molten globule; in other cases, it may recognize more structured intermediates.

While ATP alone has been found in vitro to promote the productive release of a number of GroEL-bound polypeptides, even in the absence of the cochaperonin GroES (e.g. Laminet et al., 1990; Buchner et al., 1991; Martin et al., 1991; Viitanen et al., 1991), several observations emphasize the physiological requirement for GroES and suggest that a GroEL-mediated folding reaction in the presence of GroES is qualitatively different from one in its absence. Firstly, GroES and its yeast mitochondrial homologue, Hsp10, are essential under all growth conditions (Fayet et al., 1989; Rospert et al., 1993; Hohfeld and Hartl, 1994), and conditional deficiency in Hsp10 leads to defective polypeptide folding (Hohfeld and Hartl, 1994). Secondly, overexpression of GroEL and GroES together, but not GroEL alone, has been found to increase the yield of the native state of a number of proteins in vivo (e.g. Fayet et al., 1986; Goloubinoff et al., 1989a; Gordon et al., 1994; Cole, 1996). Thirdly, the productive release of substrates from GroEL in vitro in the absence of GroES occurs only for folding conditions where substantial spontaneous folding can be observed (e.g. see Schmidt et al., 1994). Finally, in the presence of GroES, the rate of GroEL-mediated folding reactions is generally comparable to the rate of spontaneous folding, and in some cases is substantially faster. By contrast, a GroEL-mediated folding reaction in the absence of GroES is generally a slow process compared with the rate of spontaneous folding (e.g. Martin et al., 1991; Viitanen et al., 1991; Gray and Fersht, 1993; Staniforth et al., 1994).

A variety of structural and biochemical studies have begun to elucidate the mechanism by which GroES allows GroEL to support folding under conditions where spontaneous folding is not possible. GroES binding is found to induce an opening upward and outward of the GroEL apical domains, as observed by electron microscopy, greatly increasing the volume of the GroEL central cavity (Chen et al., 1994; Roseman et al., 1996). The same GroEL residues involved in polypeptide binding are required for binding of GroES (Fenton et al., 1994), which is presumed to occur through a mobile domain that itself contains hydrophobic residues, detected from NMR studies (Landry et al., 1993). Thus, polypeptide and GroES appear to compete in part for the same apical sites on GroEL. GroES has been shown, in order-of-addition experiments, to be capable of binding to the same GroEL ring as that occupied by polypeptide (cis), thereby sequestering the polypeptide in the central channel of GroEL a space c. 60 Å × 60 Å in dimension. While GroES can also bind to the opposite ring of GroEL (trans), single turnover experiments reveal that productive folding and release occur exclusively from the cis complexes (Weissman et al., 1995). Stopped-flow fluorescence anisotropy studies indicate that, in the presence of ATP (but not ADP), folding initiates in the active cis ternary complex. The preferred action of ATP in supporting cis folding can be correlated with recent cryoelectron microscopy studies with image averaging that indicate that while ADP, AMP-PNP, and ATP all support the same GroES-induced large conformational changes of the GroEL apical domains, ATP produces the furthest twisting movements (Roseman et al., 1996). This may be associated with a complete release of peptide, that correlates both with the ATP/GroES-triggered drop of anisotropy and with efficient refolding of rhodanese in ATP-cis ternary complexes (Weissman et al., 1996). By contrast the lesser movements in ADP and AMP-PNP correlate with observations that neither ADP/GroES nor AMP-PNP/GroES produce significant anisotropy changes, and with findings that the kinetics of rhodanese refolding in such complexes are substantially slower (Weissman et al., 1996; Hayer-Hartl et al., 1996). This suggests that cis complexes formed in ATP are the most efficient in promoting the chaperonin reaction. For some substrates such complexes may be essential to productive folding.

The folding-active cis ternary complex is dissociated by timed release of GroES, triggered by action of ATP (Todd et al., 1994). The set-point of this timer allows nearly all of certain substrate polypeptides, such as ornithine transcarbamylase (OTC), to fold to completion within this time (Weissman et al., 1996), while only a small fraction of others, such as rhodanese, achieve the native state (or a form committed to it) in a single round of ATP hydrolysis and GroES release (Mayhew et al., 1996; Weissman et al., 1996). Thus, after GroES release, substrate polypeptides can leave in any of several states, as determined by kinetic partitioning: native (Mayhew et al., 1996; Weissman et al., 1996); a form committed to reach the native state in solution (Weissman et al., 1995); or a non-native state that can rebind to GroEL and undergo additional rounds of release (Jackson et al., 1993; Todd et al., 1994; Weissman et al., 1994; Ranson et al., 1995; Smith and Fisher, 1995; Taguchi and Yoshida, 1995; Burston et al., 1996).

Despite the recent progress, important questions remain unresolved. For example, how is GroEL able to 'rescue' stable folding intermediates of proteins like mitochondrial malate dehydrogenase (mMDH) (Peralta et al., 1994; Ranson et al., 1995) and dimeric rubisco (Todd et al., 1994, 1996), which do not readily fold to the native state but also do not form irreversible aggregates? Conversely, how does GroEL–GroES actively facilitate folding for proteins such as OTC (Zheng et al., 1993), where enhancements in the rate of folding appear to occur, or mMDH (Ranson et al., 1995), where substoichiometric amounts of GroEL are sufficient to produce a maximum refolding rate? Finally, is there a role for the residues lining the central channel in the GroEL–GroES complex in monitoring initial folding and directing a protein down a particular pathway towards the native state?

■ References

Barraclough, R. and Ellis, R.J. (1980). Protein synthesis in chloroplasts. IX. Assembly of newly-synthesized large subunits into ribulose bisphosphate carboxylase in isolated intact pea chloroplasts. Biochim. Biophys. Acta **608**, 19–31.

Bochkareva, E.S., Lissin, N.M., Flynn, G.C., Rothman, J.E., and Girshovich, A. S. (1992). Positive cooperativity in the functioning of molecular chaperone GroEL. J. Biol. Chem. **267**, 6796–6800.

Boisvert, D.C., Wang, J., Otwinowski, Z., Horwich, A.L., and Sigler, P.B. (1996). The 2.4 Å crystal structure of the bacterial chaperonin GroEL complexed with ATPγS. Nature Struct. Biol. **3**, 170–177.

Braig, K., Otwinowski, Z., Hegde, R., Boisvert, D.C., Joachimiak, A., Horwich, A.L., and Sigler, P.B. (1994). The crystal structure of the bacterial chaperonin GroEL at 2.8 A. Nature **371**, 578–586.

Braig, K., Adams, P.D., and Brunger, A.T. (1995). Conformational variability in the refined structure of the chaperonin GroEL at 2.8 Å resolution. Nature Struct. Biol. **2**, 1083–1094.

Buchner, J., Schmidt, M., Fuchs, M., Jaenicke, R., Rudolph, R., Schmid, F.X., and Kiefhaber, T. (1991). GroE facilitates refolding of citrate synthase by suppressing aggregation. Biochemistry **30**, 1586–1591.

Burston, S., Ranson, N.A., and Clarke, A.R. (1995). The origins and consequences of asymmetry in the chaperonin reaction cycle. J. Mol. Biol. **249**, 138–152.

Burston, S.G., Weissman, J.S., Farr, G.W., Fenton, W.A., and Horwich, A.L. (1996). Release of both native and non-native proteins from a cis-only GroEL ternary complex. Nature **383**, 96–99.

Chen, S., Roseman, A.M., Hunter, A.S., Wood, S.P., Burston, S.G., Ranson, N.A., Clarke, A.R., and Saibil, H.R. (1994). Location of a folding protein and shape changes in GroEL-GroES complexes imaged by cryo-electron microscopy. Nature **371**, 261–264.

Cheng, M.Y., Hartl, F.U., Martin, J., Pollock, R.A., Kalousek, F., Neupert, W., Hallberg, E.M., Hallberg, R.L., and Horwich, A.L. (1989). Mitochondrial heat-shock protein hsp60 is essential for proteins imported into yeast mitochondria. Nature **337**, 620–625.

Cole, P.A. (1996). Chaperone-assisted protein expression. Structure **4**, 239–242.

Corrales, F.J. and Fersht, A.R. (1995). The folding of GroEL-bound barnase as a model for chaperonin-mediated protein folding. Proc. Natl. Acad. Sci. USA **92**, 5326–5330.

Ellis, R.J. (1987). Proteins as molecular chaperones. Nature **328**, 378–379.

Fayet, O., Louarn, J.M., and Georgopoulos, C. (1986). Suppression of the Escherichia coli dnaA46 mutation by amplification of the groES and groEL genes. Mol. Gen. Genet. **202**, 435–445.

Fayet, O., Ziegelhoffer, T., and Georgopoulos, C. (1989). The groES and groEL heat shock gene products of Escherichia coli are essential for bacterial growth at all temperatures. J. Bacteriol. **171**, 1379–1385.

Fenton, W.A., Kashi, Y., Furtak, K., and Horwich, A.L. (1994). Residues in chaperonin GroEL required for polypeptide binding and release. Nature **371**, 614–619.

Georgopoulos, C., Hendrix, R., Casjens, S., and Kaiser, A. (1973). Host participation in bacteriophage lambda head assembly. J. Mol. Biol. **76**, 45–60.

Goldberg, M.S., Zhang, J., Sondek, S., Matthews, C.R., Fox, R.O., and Horwich, A.L. (1997). Native-like structure of a protein-folding intermediate bound to the chaperonin GroEL. Proc. Natl. Acad. Sci. USA **94**, 1080–1085.

Goloubinoff, P., Christeller, J.T., Gatenby, A.A., and Lorimer, G.H. (1989a). Reconstitution of active dimeric ribulose bisphosphate carboxylase from an unfolded state depends on two chaperonin proteins and Mg-ATP. Nature **342**, 884–889.

Goloubinoff, P., Gatenby, A.A., and Lorimer, G.H. (1989b). GroE heat-shock proteins promote assembly of foreign prokaryotic ribulose bisphosphate carboxylase oligomers in *Escherichia coli*. Nature **337**, 44–47.

Gordon, C.L., Sather, S.K., Casjens, S., and King, J. (1994). Selective in vivo rescue by GroEL/ES of thermolabile folding intermediates to phage P22 structural proteins. J. Biol. Chem. **269**, 27941–24951.

Gray, T.E. and Fersht, A.R. (1991). Cooperativity in ATP hydrolysis by GroEL is increased by GroES. FEBS Lett. **292**, 254–258.

Gray, T.E. and Fersht, A.R. (1993). Refolding of barnase in the presence of GroE. J. Mol. Biol. **232**, 1197–1207.

Hayer-Hartl, M., Ewbank, J.J., Creighton, T.E., and Hartl, F.U. (1994). Conformational specificity of the chaperonin GroEL for the compact folding intermediates of alpha-lactalbumin. EMBO J. **13**, 3192–3202.

Hayer-Hartl, M.K., Weber, F., and Hartl, F.-U. (1996). Mechanism of chaperonin action: GroES binding and release can drive GroEL-mediated protein folding in the absence of ATP hydrolysis. EMBO J. **15**, 6111–6121.

Hemmingsen, S., Woolford, C., van der Vies, S., Tilly, K., Dennis, D., Georgopoulos, C., Hendrix, R., and Ellis, R. (1988). Homologous plant and bacterial proteins chaperone oligomeric protein assembly. Nature **333**, 330–334.

Hendrix, R.W. (1979). Purification and properties of groE, a host protein involved in bacteriophage assembly. J. Mol. Biol. **129**, 375–392.

Hohfeld, J. and Hartl, F.U. (1994). Role of the chaperonin cofactor Hsp10 in protein folding and sorting in yeast mitochondria. J. Cell Biol. **126**, 305–315.

Hohn, T., Hohn, B., Engel, A., and Wurtz, M. (1979). Isolation and characterization of the host protein groE involved in bacteriophage lambda assembly. J. Mol. Biol. **129**, 359–373.

Horwich, A.L., Low, K.B., Fenton, W.A., Hirshfield, I.N., and Furtak, K. (1993). Folding in vivo of bacterial cytoplasmic proteins: role of GroEL. Cell **74**, 909–917.

Hunt, J.F., Weaver, A.J., Landry, S.J., Gierasch, L., and Deisenhofer, J. (1996). The crystal structure of the GroES co-chaperonin at 2.8 Å resolution. Nature **379**, 37–45.

Hutchinson, E.G., Tichelaar, W., Hofhaus, G., Weiss, H., and Leonard, K.R. (1989). Identification and electron microscopic analysis of a chaperonin oligomer from Neurospora crassa mitochondria. EMBO J. **8**, 1485–1490.

Ishii, N., Taguchi, H., Sumi, M., and Yoshida, M. (1992). Structure of holo-chaperonin studied with electron microscopy. Oligomeric cpn10 on top of two layers of cpn60 rings with two stripes each. FEBS Lett. **299**, 169–174.

Itzhaki, L.S., Otzen, D.E., and Fersht, A.R. (1995). Nature and consequences of GroEL-protein interactions. Biochemistry **34**, 14581–14587.

Jackson, G.S., Staniforth, R.A., Halsall, D.J., Atkinson, T., Holbrook, J.J., Clarke, A.R., and Burston, S.G. (1993). Binding and hydrolysis of nucleotides in the chaperonin catalytic cycle: implications for the mechanism of assisted protein folding. Biochemistry **32**, 2554–2563.

Jindal, S., Dudani, A.K., Singh, B., Harley, C.B., and Gupta, R.S. (1989). Primary structure of a human mitochondrial protein homologous to the bacterial and plant chaperonins and to the 65-kilodalton mycobacterial antigen. Mol. Cell. Biol. **9**, 2279–2283.

Kraulis, P.J. (1991). MOLSCRIPT: a program to produce both detailed and schematic plots of protein structures. J. Appl. Cryst. **24**, 946–950.

Kusukawa, N., Yura, T., Ueguchi, C., Akiyama, Y., and Ito, K. (1989). Effects of mutations in heat-shock genes groES and groEL on protein export in Escherichia coli. EMBO J. **8**, 3517–3521.

Laminet, A.A., Ziegelhoffer, T., Georgopoulos, C., and Pluckthun, A. (1990). The Escherichia coli heat shock proteins GroEL and GroES modulate the folding of the beta-lactamase precursor. EMBO J. **9**, 2315–2319.

Landry, S.J. and Gierasch, L.M. (1994). Polypeptide interactions with molecular chaperones and their relationship to in vivo protein folding. Annu. Rev. Biophys. Biomol. Struct. **23**, 645–669.

Landry, S.J., Zeilstra, R.J., Fayet, O., Georgopoulos, C., and Gierasch, L.M. (1993). Characterization of a functionally important mobile domain of GroES. Nature **364**, 255–258.

Langer, T., Pfeifer, G., Martin, J., Baumeister, W., and Hartl, F.U. (1992). Chaperonin-mediated protein folding: GroES binds to one end of the GroEL cylinder, which accommodates the protein substrate within its central cavity. EMBO J. **11**, 4757–4765.

Lewis, V.A., Hynes, G.M., Zheng, D., Saibil, H., and Willison, K. (1992). T-complex polypeptide-1 is a subunit of a heteromeric particle in the eukaryotic cytosol. Nature **358**, 249–252.

Mande, S.C., Mehra, V., Bloom, B.R., and Hol, W.G.J. (1996). Structure of the heat shock protein chaperonin-10 of *Mycobacterium leprae*. Science **271**, 203–207.

Martin, J., Langer, T., Boteva, R., Schramel, A., Horwich, A.L., and Hartl, F.U. (1991). Chaperonin-mediated protein folding at the surface of groEL through a 'molten globule'-like intermediate. Nature **352**, 36–42.

Martin, J., Mayhew, M., Langer, T., and Hartl, F.U. (1993). The reaction cycle of GroEL and GroES in chaperonin-assisted protein folding . Nature **366**, 228–233.

Mayhew, M., da Silva, A.C., Martin, J., Erdjument-Bromage, H., Tempst, P., and Hartl, F.-U. (1996). Protein folding in the central cavity of the GroEL–GroES chaperonin complex. Nature **379**, 420–426.

Okazaki, A., Ikura, T., Nikaido, K., and Kuwajima, K. (1994). The chaperonin GroEL does not recognize apo–lactalbumin in the molten globule state. Nature Struct. Biol. **1**, 439–446.

Peralta, D., Hartman, D.J., Hoogenraad, N.J., and Hoj, P.B. (1994). Generation of a stable folding intermediate which can be rescued by the chaperonins GroEL and GroES. FEBS Lett. **339**, 45–49.

Ranson, N.A., Dunster, N.J., Burston, S.G., and Clarke, A.R. (1995). Chaperonins can catalyze the reveral of early aggregation steps when a protein misfolds. J. Mol. Biol. **250**, 581–586.

Reading, D.S., Hallberg, R.L., and Myers, A.M. (1989). Characterization of the yeast HSP60 gene coding for a mitochandrial assembly factor. Nature **337**, 655–659.

Robinson, C.V., Gross, M., Eyles, S.J., Ewbank, J.J., Mayhew, M., Hartl, F.U., Dobson, C.M., and Radford, S.E. (1994). Conformation of GroEL-bound alpha-lactalbumin probed by mass spectrometry. Nature **372**, 646–651.

Roseman, A.M., Chen, S., White, H., Braig, K., and Saibil, H.R. (1996). The chaperonin ATPase cycle: mechanism of allosteric switching and movements of substrate-binding domains in GroEL. Cell **87**, 241–251.

Rospert, S., Junne, T., Glick, B.S. and Schatz, G. (1993). Cloning and disruption of the gene encoding yeast mitochondrial chaperonin 10, the homolog of E. coli groES. FEBS Lett. **335**, 358–360.

Saibil, H., Dong, Z., Wood, S., auf der Mauer, A. (1991). Binding of chaperonins. Nature **353**, 25–26.

Saibil, H.R., Zheng, D., Roseman, A.M., Hunter, A.S., Watson, G.M.F., Chen, S., auf der Mauer, A., O'Hara, B.P., Wood, S.P., Mann, N.H., Barnett, L.K., and Ellis, R.J. (1993). ATP induces large quarternary rearrangements in a cage-like chaperonin structure. Curr. Biol. **3**, 265–273.

Schmidt, M., Buchner, J., Todd, M.J., Lorimer, G.H., and Viitanen, P.V. (1994). On the role of GroES in the chaperonin-assisted folding reaction. J. Biol. Chem. **14**, 10304–10311.

Smith, K.E. and Fisher, M.T. (1995). Interactions between the GroE chaperonins and rhodanese. Multiple intermediates and release and rebinding. J. Biol. Chem. **270**, 21517–21523.

Staniforth, R.A., Burston, S.G., Atkinson, T., and Clarke, A.R. (1994). Affinity of chaperonin-60 for a protein substrate and its modulation by nucleotides and chaperonin-10. Biochem. J. **300**, 651–658.

Sternberg, N. (1973). Properties of a mutant of Escherichia coli defective in bacteriophage lambda head formation (groE). II. The propagation of phage lambda. J. Mol. Biol. **76**, 25–44.

Taguchi, H. and Yoshida, M. (1995). Chaperonin releases the substrate protein in a form with tendency to aggregate and ability to rebind to chaperonin. FEBS Lett. **359**, 195–198.

Thiyagarajan, P., Henderson, S.J., and Joachimiak, A. (1996). Solution structures of GroEL and its complex with rhodanese from small-angle neutron scattering. Structure **4**, 79–88.

Todd, M.J., Viitanen, P.V., and Lorimer, G.H. (1993). Hydrolysis of adenosine 5'-triphosphate by Escherichia coli GroEL: effects of GroES and potassium ion. Biochemistry **32**, 8560–8567.

Todd, M.J., Viitanen, P.V., and Lorimer, G.H. (1994). Dynamics of the chaperonin ATPase cycle: implications for facilitated protein folding. Science **265**, 659–666.

Todd, M.J., Lorimer, G.H., and Thirumalai, D. (1996). Chaperonin-facilitated protein folding: Optimization of rate and yield by an iterative annealing mechanism. Proc. Natl. Acad. Sci. USA **93**, 4030–4035.

Viitanen, P., Lubben, T., Reed, J., Goloubinoff, P., O'Keefe, D., and Lorimer, G. (1990). Chaperonin-facilitated refolding of ribulosebisphosphate carboxylase and ATP hydrolysis by chaperonin 60 (groEL) are K^+ dependent. Biochemistry **29**, 5665–5671.

Viitanen, P.V., Donaldson, G.K., Lorimer, G.H., Lubben, T.H., and Gatenby, A.A. (1991). Complex interactions between the chaperonin 60 molecular chaperone and dihydrofolate reductase. Biochemistry **30**, 9716–9723.

Viitanen, P.V., Gatenby, A.A., and Lorimer, G.H. (1992). Purified chaperonin 60 (groEL) interacts with the nonnative states of a multitude of Escherichia coli proteins. Protein Sci. **1**, 363–369.

Weissman, J.S., Kashi, Y., Fenton, W.A., and Horwich, A.L. (1994). GroEL-mediated protein folding proceeds by multiple rounds of binding and release of nonnative forms. Cell **78**, 693–702.

Weissman, J.S., Hohl, C.M., Kovalenko, O., Kashi, Y., Chen, S., Braig, K., Saibil, H.R., Fenton, W.A., and Horwich, A.L. (1995). Mechanism of GroEL action: productive release of polypeptide from a sequestered position under GroES. Cell **83**, 577–588.

Weissman, J.S., Rye, H.S., Fenton, W.A., Beechem, J.M., and Horwich, A.L. (1996). Characterization of the active intermediate of a GroEL–GroES-mediated protein folding reaction. Cell **84**, 481–490.

Zahn, R. and Pluckthun, A. (1994). Thermodynamic partitioning model for hydrophobic binding of polypeptides by GroEL. II. GroEL recognizes thermally unfolded mature beta-lactamase. J. Mol. Biol. **242**, 165–174.

Zahn, R., Axmann, S.E., Rucknagel, K.P., Jaeger, E., Laminet, A.A., and Pluckthun, A. (1994a). Thermodynamic partitioning model for hydrophobic binding of polypeptides by GroEL. I. GroEL recognizes the signal sequences of beta-lactamase precursor. J. Mol. Biol. **242**, 150–164.

Zahn, R., Spitzfaden, C., Ottiger, M., Wuthrich, K., and Pluckthun, A. (1994b). Destabilization of the complete protein secondary structure on binding to the chaperone GroEL. Nature **368**, 261–265.

Zahn, R., Perrett, S. Stenberg, G., and Fersht, A.T. (1996). Catalysis of amide proton exchange by the molecular chaperones GroEL and SecB. Science **271**, 642–645.

Zheng, X., Rosenberg, L.E., Kalousek, F., and Fenton, W.A. (1993). GroEL, GroES, and ATP-dependent folding and spontaneous assembly of ornithine transcarbamylase. J. Biol. Chem. **268**, 7489–7493.

Zwickl, P., Pfeifer, G., Lottspeich, F., Kopp, F., Dahlmann, B., and Baumeister, W. (1990). Electron microscopy and image analysis reveal common principles of organization in two large protein complexes: groEL-type proteins and proteasomes. J. Struct. Biol. **103**, 197–203.

■ *Jonathan S. Weissman,*[1,3] *Wayne A. Fenton,*[1] *Kerstin Braig,*[1] *Paul D. Adams,*[2] *and Arthur Horwich:*[1,3]
Departments of Genetics[1] *and Molecular Biophysics and Biochemistry*[2]
and the Howard Hughes Medical Institute[3]
Yale University School of Medicine,
New Haven, CT 06510, USA
Tel. 1 203 737 4431
Fax. 1 203 737 1761
E-mail: horwich@hhvms8.csb.yale.edu

The structure of *Escherichia coli* GroES

Structural interaction between the chaperonin GroEL and the co-chaperonin GroES is required for chaperonin-assisted folding of substrate polypeptides under conditions that do not support spontaneous folding of the substrate in the absence of the chaperonin. The biochemistry of the GroEL–GroES interaction has been studied in detail, and the structures of both proteins have been determined using X-ray crystallography. Theories regarding the functional role of the co-chaperonin GroES in the assisted folding reaction are discussed in the context of its crystal structure.

■ The structure of the GroES dome

The crystal structure of GroES shows that the 10 kDa monomer forms a topologically irregular four-stranded β-barrel (Hunt et al., 1996; Mande et al., 1996). This structure is homologous to that of a β-barrel domain found in the enzymes alcohol dehydrogenase and quinone oxidoreductase (Murzin, 1996); the β-barrel motif in these proteins is structurally equivalent to that of the SH3 domain but differs in the pattern of connections between the constituent β-strands. In the GroES heptamer, the β-barrels are arranged in a ring with the axes of the barrels parallel to the 7-fold symmetry axis of the oligomer (Fig. 1). An antiparallel β-hairpin extends inward from the top of each β-barrel towards the 7-fold symmetry axis of the heptamer, forming a 'roof' over the top of the cavity within the ring. The resulting structure can be described as a dome approximately 30 Å high and 75 Å in diameter with an inner cavity approximately 20 Å high and 30 Å in diameter.

The 'mobile loop' of GroES is a 16 residue segment in each monomer that is disordered in the free heptamer but becomes immobilized upon formation of the complex with GroEL (Landry et al., 1993, 1996; see also the entries on *E. coli* GroES, p. 183, and *E. coli* GroEL, p. 173). In the crystal structure, the mobile loop is disordered in six of the subunits, but it adopts a well-ordered conformation in the final subunit, where it is sandwiched between two heptamers in a crystal packing contact. In this conformation, the mobile loop bridges the interface of two adjacent subunits in the heptamer to which it is covalently attached. The mobile loop extends outward from the monomer at the diametrically opposite point on the β-barrel from the roof β-hairpin, i.e. at the bottom outer rim of the GroES dome.

In the crystal structure, the packing of the adjacent subunits in the GroES dome is irregular, producing a substantial deviation from proper 7-fold symmetry in the heptamer. The core structure of the GroES β-barrel is highly conserved in all subunits, and the irregular packing is caused by differences in the structure of the interface between the adjacent subunits in different positions around the ring. The deformability of the interface is mediated by the variable conformation of the

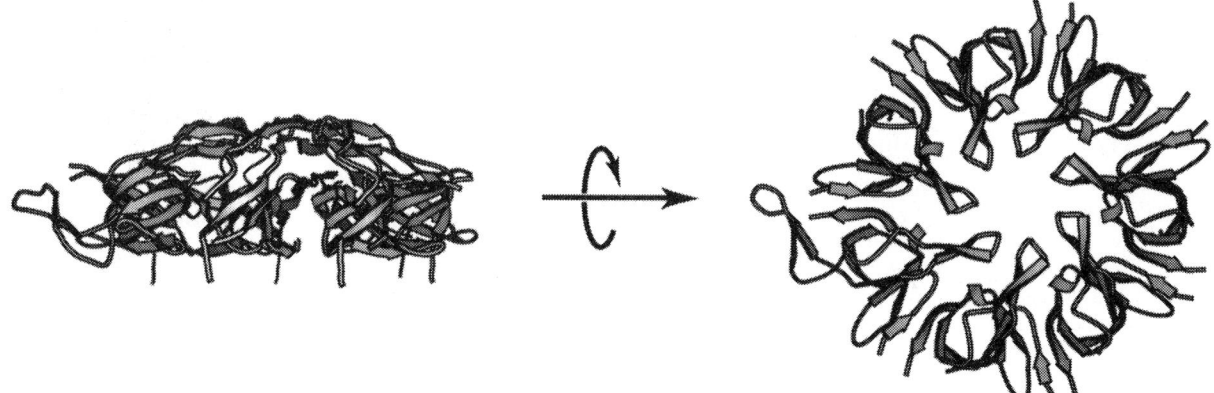

Figure 1. Two orthogonal views of a ribbon diagram of the GroES heptamer. The mobile loop (i.e. residues 17–32 in the monomer) is observed in the crystal structure in only one of the seven subunits; the single observed copy is visible on the left side of the heptamer in these images. The images were produced using the program MOLSCRIPT (Kraulis, 1991).

polypeptide backbone and the prevalence of long, flexible side chains in the sequences participating in the interface. It has been shown that the GroES heptamer dissociates to monomers below a subunit concentration of 0.7 mM (Zondlo et al., 1995), indicating that the binding energy of the subunit–subunit interface is modest (approximately −7 kcal/mole). The irregular packing of the subunits observed in the crystal structure is consistent with the relatively weak binding energy measured for the subunit–subunit interface is modest, and, taken together, these observations suggest that the GroES dome is a flexible structure.

■ The mobile loop and the interaction between GroES and GroEL

GroES and GroEL form a highly stable asymmetrical GroES$_7$–GroEL$_{14}$ complex in the presence of ADP (Lissin et al., 1990; Martin et al., 1993; Todd et al., 1994). The binding of GroES to GroEL in this complex is mediated, at least in part, by the mobile loop (Landry et al., 1993; Landry and Gierasch, 1994). Therefore, the observed location of the mobile loop in the GroES structure suggests that the bottom outer rim of the GroES dome is in contact with GroEL in the chaperonin complex. Given this constraint and the known structure of the GroEL double cylinder (Braig et al., 1994), GroES must be positioned like a cap over the top of one of the two hollow polypeptide binding chambers of GroEL in order to maintain sevenfold symmetry in the GroEL–GroES complex. Cryoelectron microscopy image reconstruction studies of the asymmetrical complex support this inference, showing the addition of a dumb bell shaped particle over the top of one of the binding chambers of GroEL after the addition of GroES and ADP (Chen et al., 1994; Langer et al., 1992; Schmidt et al., 1994b); the shape of the putative GroES density in these images matches that seen in projection in the ribbon diagram on the left side of Fig. 1. In this orientation, the cavity within the GroES dome will be continuous with the volume in the polypeptide binding chamber of GroEL (Fenton et al., 1994) in the chaperonin complex.

■ The roof of the GroES dome

The β-hairpins comprising the roof of the GroES dome form a pinwheel structure that is visible at the center of the molecule in the view shown on the right side of Fig. 1. The orifice in the center of the roof is approximately 8 Å in diameter. The protein sequences surrounding this orifice are highly acidic. The glutamic acid residue at position 50 in GroES is located in the β-turn at the tip of the roof β-hairpin; this residue is the closest one in the structure to the sevenfold symmetry axis. A second glutamic acid residue occurs immediately on the C-terminal side of the same β-turn at position 53 in the monomer. Although basic residues occur at flanking positions in the roof β-hairpin (i.e. Arg-47 and Lys-55), their presence is not adequate to shield the acidic residues near the symmetry axis, and a deeply negative electrostatic potential is observed in this region of the molecule. Moreover, the packing interactions within the roof appear to be weak. There are relatively few van der Waals contacts or hydrogen bonds between the adjacent roof β-hairpins, and the amount of buried solvent-accessible surface area in their interface is also small. The electrostatic repulsion associated with the concentration of negative charge near the symmetry axis combined with the weak packing interactions have led us to hypothesize that the roof of the GroES dome could be a metastable structure.

A phylogenetic alignment of GroES with other members of the CPN10 family shows stringent overall sequence conservation, including the absolute conservation of four glycine residues in the immediate vicinity of the roof β-hairpin (Hunt et al., 1996; Lorimer and Todd, 1996). However, significant divergence is observed in the sequence of the roof β-hairpin itself. In the human mitochondrial CPN10 homolog, this sequence is rich in glycine and lysine residues that will tend to disfavor the formation of an ordered β-hairpin structure. The crystal structure of this mitochondrial Cpn10 has been solved using molecular replacement (J.F. Hunt, S.J. Landry, and J. Deisenhofer, manuscript in preparation) and shows no evidence of well-defined electron density in the region of the roof β-hairpin, suggesting that it is disordered in this homologue in the crystal lattice. Direct NMR measurement of protein backbone dynamics confirm this inference, indicating that the roof β-hairpin in human Cpn10 is dynamically disordered in the heptamer free in solution (S.J. Landry, N.K. Steede, and K. Maskos, manuscript submitted).

Sequence alignment of the bacteriophage T4 Gp31 protein (Fig. 2) suggested that the roof of the dome may be deleted entirely in this GroES ortholog (Konnin and van der Vies, 1995; Lorimer and Todd, 1996). The structure of this protein has also been resolved using molecular replacement (J.F. Hunt, S.M. van der Vies, and J. Deisenhofer, manuscript submitted) showing that the roof β-hairpin is indeed deleted, leaving a hole of at least 16 Å diameter through the centre of the Grp31 heptamer. (See also the entry on bacteriophage T4 Grp31, p. 185).

■ What is the function of GroES in the chaperonin-assisted protein folding reaction?

Recent experiments on the mechanism of the GroEL–GroES-assisted protein folding reaction have shown that some proteins can fold to the native state while sequestered beneath GroES in the binding cavity of the GroEL–GroES complex, i.e. with GroES and the substrate in cis in the asymmetrical complex (Mayhew et al., 1996; Weissman et al., 1996; Hayer-Hartl et al., 1996). These results are consistent with the 'Anfinsen cage' hypothesis which proposes that the function of the chaperonin

		Mobile Loop	Roof β-Hairpin

```
E. coli    5-PLHDRVIVKRK    EVETKSAGGIVLTGSAAA KSTRGEVLAVGNGRILENGEVKPLDVKVGDIVIF-67

Human     10-PLFDRVLVERS    AAETVTKGGIMLPEKSQG KVLQATVVAVGSGSKGKGGEIQPVSVKVGDKVLL-72

Gp31      11-AVGEYVILVSEPAQAGDEEVTESGLIIGKRVQGEVPELCVVHSVGPDV          PEGFCEVGDLTSL-71
```

Figure 2. Sequence alignment of *E. coli* and human CPN10 sequences with that of the Gp31 protein from bacteriophage T4.

complex is to provide a chemical environment where folding to the native state can be promoted while protecting the aggregation-prone substrate polypeptide from interaction with the reservoir of other hydrophobic and aggregation-prone species in the cellular cytoplasm (Ellis, 1994). In the *in vitro* folding experiments, a complex is formed between GroEL and the substrate polypeptide, and folding of the substrate is observed after GroES and adenine nucleotides are added to the complex in a second step. Electron microscopy shows a dramatic expansion of the volume of the polypeptide binding cavity in the GroEL ring to which the co-chaperonin GroES binds (Chen *et al.*, 1994). On this basis, the binding of GroES to the *cis* ring relative to the substrate is believed to displace the substrate from its binding sites on GroEL and thereby allow the substrate to fold in the enclosed cavity.

It has been widely suggested that this displacement could be mediated by interaction of the hydrophobic cluster in the mobile loop of GroES with the hydrophobic residues on the inner wall of the GroEL ring that have been identified by mutagenesis (Fenton *et al.*, 1994) as comprising the binding site for non-native substrate polypeptides. The validity of this inference is one of the most important issues that will be resolved by the forthcoming crystal structure of the GroEL–GroES–ADP complex. In any event, one critical function of the GroES ring is to deliver the mobile loops simultaneously to the seven subunits of GroEL in order to coordinate release of the bound substrate polypeptide (Gray and Fersht, 1991; Todd *et al.*, 1994). The flexibility of the GroES ring could facilitate the accommodation of structurally diverse substrate polypeptides in the Anfinsen cage, and the metastability of the GroES roof could allow a further expansion in the volume of the cage when necessary. In this context, it has been suggested that the high density of electrostatic charge on the inner surface of the GroES dome could function to make the environment of the substrate extremely hydrophilic, thereby minimizing non-specific hydrophobic interactions with the non-native polypeptide in the Anfinsen cage (Mande *et al.*, 1996). However, there is a ring of phylogenetically conserved aromatic residues that lines the inner rim at the bottom of the GroES dome making the region of the co-chaperonin that will be in closest proximity to the substrate rather hydrophobic (Hunt *et al.*, 1996; Lorimer and Todd, 1996).

It is clear that ATP hydrolysis on the *trans* ring of the GroEL double cylinder relative to GroES–ADP induces efficient release of the co-chaperonin from the *cis* ring (Martin *et al.*, 1993; Todd *et al.*, 1994). Therefore, because folding can be observed in the Anfinsen cage without the hydrolysis of ATP, it has been proposed that the hydrolysis reaction functions primarily as a timing mechanism to periodically eject the lid from the Anfinsen cage in order to give the substrate polypeptide an opportunity to diffuse away if it has successfully folded to the native state (Mayhew *et al.*, 1996; Weissman *et al.*, 1996). In contrast, we have suggested that there may be a cryptic mechanical phase to the protein denaturation reaction in which the energy of ATP hydrolysis is directly coupled to the unfolding of the substrate (Hunt *et al.*, 1996). This speculative hypothesis is motivated by the observed structural properties of GroES. Specifically, the flexibility of the GroES ring and the potentially metastable conformation of its roof suggested to us that the co-chaperonin could function in a manner equivalent to a valve in a mechanical denaturation reaction. Transient forces generated in the binding cavity of GroEL during the hydrolysis reaction could do mechanical work on the substrate in order to promote unfolding of kinetically trapped intermediates (Hubbard and Sander, 1991; Todd *et al.*, 1994; Weissman *et al.*, 1994). If such a mechanistic phase were to exist, the elastic ring of GroES sitting over the top of the binding cavity would prevent the substrate from being pushed out into solution and thereby enhance the mechanical efficiency of the denaturation reaction. While the *in vitro* experimentation has shown that such a mechanism is not an obligate component of the chaperonin-assisted protein folding reaction, we believe that it is possible that the chaperonin could exploit mechanistic diversity (i.e. more than one detailed mechanism) as a strategy to deal with the chemical and physical complexity of the protein folding reaction (Schmidt *et al.*, 1994a).

References

Braig, K., Otwinowski, Z., Hegde, R., Boisvert, D.C., Joachimiak, A., Horwich, A.L., and Sigler, P.B. (1994). The crystal structure of the bacterial chaperonin GroEL at 2.8 Å. Nature **371**, 578–586.

Chen, S., Roseman, A.M., Hunter, A.S., Wood, S.P., Burston, S.G., Ranson, N.A., Clarke, A.R., and Saibil, H.R. (1994). Location of a folding protein and shape changes in GroEL–GroES complexes imaged by cryo-electron microscopy. Nature **371**, 261–264.

Ellis, R.J. (1994). Molecular chaperones. Opening and closing the Anfinsen cage. Curr. Biol. **4**, 633–635.

Fenton, W.A., Kashi, Y., Furtak, K., and Horwich, A.L. (1994). Residues in chaperonin GroEL required for polypeptide binding and release. Nature **371**, 614–619.

Gray, T.E. and Fersht, A.R. (1991). Cooperativity in ATP hydrolysis by GroEL is increased by GroES. FEBS Lett. **292**, 254–258.

Hayer-Hartl, M.K., Weber, F., and Hartl, F.U. (1996). Mechanism of chaperonin action: GroES binding and release can drive GroEL-mediated protein folding in the absence of ATP hydrolysis. EMBO J. **15**, 6111–6121.

Hubbard, T.J.P. and Sander, C. (1991). The role of heat-shock and chaperone proteins in protein folding: possible molecular mechanisms. Protein Eng. **4**, 711–717.

Hunt, J.F., Weaver, A.J., Landry, S.J., Gierasch, L., and Deisenhofer, J. (1996). The crystal structure of the GroES co-chaperonin at 2.8 Å resolution. Nature **379**, 37–45.

Koonin, E.V. and van der Vies, S.M. (1995). Conserved sequence motifs in bacterial and bacteriophage chaperonins. Trends Biochem. Sci. **20**, 14–15.

Kraulis, P.J. (1991). MOLSCRIPT: a program to produce both detailed and schematic plots of proteins. J. Appl. Cryst. **24**, 946–950.

Landry, S.J. and Gierasch, L.M. (1994). Polypeptide interactions with molecular chaperones and their relationship to in vivo protein folding. Annu. Rev. Biophys. Biomol. Struct. **23**, 645–669.

Landry, S.J., Zeilstra-Ryalls, J., Fayet, O., Georgopoulos, C., and Gierasch, L.M. (1993). Characterization of a functionally important mobile domain of GroES. Nature **364**, 255–258.

Landry, S.J. Taher, A., Georgopoulos, C., and van der Vies, S.M. (1996). Interplay of structure and disorder in co-chaperonin mobile loop. Proc. Natl. Acad. Sci. USA **93**, 11622–11627.

Langer, T., Pfeifer, G., Martin, J., Baumeister, W., and Hartl, F.U. (1992). Chaperonin-mediated protein folding: GroES binds to one end of the GroEL cylinder, which accommodates the protein substrate within its central cavity. EMBO J. **11**, 4757–4765.

Lissin, N.M., Venyaminov, S.Y., and Girshovich, A.S. (1990). (Mg-ATP)-dependent self-assembly of molecular chaperone GroEL. Nature **348**, 339–342.

Lorimer, G.H. and Todd, M.J. (1996). GroE structures galore. Nature Struct. Biol. **3**, 116–121.

Mande, C.S., Mehra, V., Bloom, B.R., and Hol, W.G.J. (1996). Structure of the heat shock protein chaperonin-10 of *Mycobacterium leprae*. Science **271**, 203–207.

Martin, J., Mayhew, M., Langer, T., and Hartl, F.U. (1993). The reaction cycle of GroEL and GroES in chaperonin-assisted protein folding. Nature **366**, 228–233.

Mayhew, M., da Silva, A.C., Martin, J., Erdjument-Bromage, H., Tempst, P., and Hartl, F.-U. (1996). Protein folding in the central cavity of the GroEL–GroES chaperonin complex. Nature **379**, 420–426.

Murzin, A.G. (1996). Structural classification of proteins: new superfamilies. Curr. Opin. Struct. Biol. **6**, 386–394.

Schmidt, M., Buchner, J., Todd, M.J., Lorimer, G.H., and Viitanen, P.V. (1994a). On the role of groES in the chaperonin-assisted folding reaction. Three case studies. J. Biol. Chem. **269**, 10304–10311.

Schmidt, M., Rutkat, K., Rachel, R., Pfeifer, G., Jaenicke, R., Viitanen, P.V., and Buchner, J. (1994b). Symmetric complexes of GroE chaperonins as part of the functional cycle. Science **265**, 656–659.

Todd, M.J., Viitanen, P.V., and Lorimer, G.H. (1994). Dynamics of the chaperonin ATPase cycle: implications for facilitated protein folding. Science **265**, 659–666.

Weissman, J.S., Kashi, Y., Fenton, W.A., and Horwich, A.L. (1994). GroEL-mediated protein folding proceeds by multiple rounds of binding and release of nonnative forms. Cell **78**, 693–702.

Weissman, J.S., Rye, H.S., Fenton,W.A., Beechem, J.M., and Horwich, A.L. (1996). Characterization of the active intermediate of a GroEL-GroES-mediated protein folding reaction. Cell **84**, 481–490.

Zondlo, J., Fisher, K.E., Lin, Z., Ducote, K.R., and Eisenstein, E. (1995). Reversible dissociation of the GroES chaperonin. Biochemistry **34**, 10334–10339.

John F. Hunt and Johann Deisenhofer:
HHMI and Department of Biochemistry
University of Texas Southwestern Medical Center
5323 Harry Hines Boulevard
Dallas, TX 75235-9050, USA
Tel. 1 214 648 5089
Fax. 1 214 648 5095
E-mail: hunt@howie.swmed.edu and jd@howie.swmed.edu

Escherichia coli GroES

GroES is the Escherichia coli *member of the CPN10 family of heat shock proteins and serves as cochaperonin for GroEL-assisted protein folding* in vivo *and* in vitro. *GroES increases the cooperativity of the GroEL ATPase activity and promotes ATP-dependent discharge and folding of GroEL-bound substrates.*

■ Alternative names

Chaperonin-10, cpn10 (Golubinoff et al., 1989); cochaperonin (Lubben et al., 1990); Hsp10 (Hartman et al., 1992).

■ *groES* gene and sequence

The *groE* locus maps at 94 minutes on the *E. coli* chromosome and encodes two polypeptides, GroES and GroEL, in that order (Genbank accession number X07850) (Hemmingsen et al., 1988). The 97 residue GroES polypeptide has a predicted M_r of 10 368 daltons and a pI of 4.92. There exists approximately 35% sequence identity between the GroES and bovine cpn10 amino acid sequences. Genetic interaction between GroES and GroEL is indicated by the existence of *groEL* mutant suppressors in *groES*, and vice versa (Zeilstra-Ryalls et al., 1994). Genetic, biochemical and biophysical experiments led to the identification of a 17 residue 'mobile' segment in the GroES polypeptide that plays an essential role in GroES–GroEL interaction (Landry et al., 1993). This segment of GroES is conserved in all known CPN10s (Koonin and van der Vies, 1995).

■ Regulation of *groES*

The *groE* cistron is under control of two promoters that assure expression at all temperatures (Georgopoulos et al., 1994). Under physiological conditions, transcription is principally controlled by the σ^{32}-directed promoter which blocks the σ^{70}-directed promoter. The presence of the σ^{70}-directed promoter probably guarantees expression under all conditions. Expression of *groES* and *groEL* increases as much as tenfold during heat shock or any other natural or experimental stress that depletes chaperone pools; e.g. bacteriophage infection, foreign protein expression, secretion block and mutations in other heat shock proteins such as DnaJ or DnaK.

■ GroES protein

GroES appears not to be post-translationally modified, and the measured pI is 5.2 (Tilly et al., 1981). In contrast to its eukaryotic homologues, the initiator Met is retained in the native GroES protein (Dickson et al., 1994). GroES is composed of seven subunits arranged with sevenfold symmetry, which is compatible with its observed binding to one end of the sevenfold symmetric, 14-subunit GroEL protein (Chen et al., 1994). The 'mobile loop' of GroES appears to mediate, at least partially, GroES binding to GroEL (Landry et al., 1993). Transferred nuclear Overhauser effects in NMR spectra of a synthetic mobile loop peptide indicate that it adopts a hairpin conformation while bound to GroEL (Landry et al., 1993, 1996). Mobility in the loop may facilitate cycles of GroES binding and release (Landry et al., 1996). Binding of GroES to one end of the GroEL double toroid in the presence of ATP or ADP causes a major conformational change in GroEL (Chen et al., 1994) and reduces the affinity for GroES on the opposite end of GroEL (Todd et al., 1994). However, in the presence of ATP symmetrical complexes of $(GroES)_2GroEL$ form and may represent important intermediates of the chaperonin cycle (Todd et al., 1994). Indeed, chaperonin-dependent refolding activity is most efficient in conditions where symmetrical complexes are highly populated (Azem et al., 1995).

■ GroES function

Both GroEL and GroES are essential proteins at all temperatures and have been implicated in a wide variety of cellular processes in the bacterial cell (Fayet et al., 1989). Absence of either gene product results in pleiotropic defects in *E. coli*, including defective DNA and RNA synthesis, inability to support bacteriophage growth, block in cell division and a decrease in overall proteolysis in the cell. The *groE* gene products are fundamental for bacteriophage morphogenesis, except in the case of the bacteriophage T4 (Zeilstra-Ryalls et al., 1991). T4 capsid assembly is blocked by mutations in *groEL* but not in *groES*. T4 specifically requires the product of T4 gene *31*, Gp31 (see entry p. 185), to serve as cochaperonin with GroEL. Overproduction of GroES and GroEL has been shown to suppress a number of temperature-sensitive mutants, presumably by stabilizing the defective gene product (Van Dyk et al., 1989). *In vitro* chaperonin-assisted refolding has been demonstrated for several substrate proteins, including ribulose bisphosphate carboxylase oxygenase (rubisco) (Goloubinoff et al., 1989), citrate synthase (Buchner et al., 1991; Zhi et al., 1992), rhodanese (Martin et al., 1991; Mendoza et al., 1991), ornithine transcarbamylase (Zheng et al., 1993) and malate dehydrogenase (Hartman et al., 1993). Under conditions that otherwise give poor refolding yield, efficient refolding of these proteins requires the complete chaperonin system composed of GroEL, GroES, Mg–ATP and K⁺ ion. Multiple cycles of substrate binding to GroEL followed by ATP- and GroES-dependent release appear to be necessary for enhanced refolding yield (Martin et al., 1993; Weissman et al., 1994). Under conditions where a substrate can fold unassisted, GroES accelerates nucleotide-dependent substrate release (Laminet and Plückthun, 1990; Fisher, 1994). GroES enhances the cooperativity of ATP binding and hydrolysis by GroEL (Gray and Fersht, 1991; Bochkareva et al., 1992), and it inhibits the ATPase activity of GroEL to the extent of 50% at high [K⁺] and up to 100% at low [K⁺] by increasing the affinity of GroEL for ADP (Todd et al., 1993). The net effect of GroES is 'quantized' nucleotide hydrolysis and substrate dissociation, wherein each cycle of GroES binding and release is accompanied by the binding and hydrolysis of seven molecules of ATP (Todd et al., 1994). Current data are consistent with a role for GroES in protein folding as coordinator of simultaneous discharge of the substrate protein from multiple GroEL subunits. However, observations of native protein entrapped within GroES–GroEL complexes (Weissman et al., 1995, 1996; Mayhew et al., 1996) allow the possibility that GroES promotes formation of the native state, or at least protects the folding protein from intermolecular interactions in an 'Anfinsen cage' fashion (Saibil et al., 1993). Uncapping of GroEL and substrate discharge into the bulk solution may be triggered by GroES and ATP binding to the opposite end of GroEL (Torok et al., 1996).

■ References

Azem, A., Diamant, S., Kessel, M., Weiss, C., and Goloubinoff, P. (1995). The protein-folding activity of chaperonins correlates with the symmetric GroEL-14 (GroES-7)-2 heterooligomer. Proc. Natl. Acad. Sci. U.S.A. **92**, 12021–12025.

Bochkareva, E.S., Lissin, N.M., Flynn, G.C., Rothman, J.E., and Girshovich, A.S. (1992). Positive cooperativity in the functioning of molecular chaperone GroEL. J. Biol. Chem. **267**, 6796–6800.

Buchner, J., Schmidt, M., Fuchs, M., Jaenicke, R., Rudolph, R., Schmid, F.X., and Kiefhaber, T. (1991). GroE facilitates refolding of citrate synthase by suppressing aggregation. Biochemistry **30**, 1586–1591.

Chen, S., Roseman, A.M., Hunter, A.S., Wood, S.P., Burston, S.G., Ranson, N.A., Clarke, A.R., and Saibil, H.R. (1994). Location of a folding protein and shape changes in GroEL–GroES complexes imaged by cryo-electron microscopy. Nature **371**, 261–264.

Dickson, R., Larsen, B., Viitanen, P.V., Tormey, M.B., Geske, J., Strange, R., and Bemis, L.T. (1994). Cloning, expression, and purification of a functional nonacetylated mammalian mitochondrial chaperonin 10. J. Biol. Chem. **269**, 26858–26864.

Fayet, O., Ziegelhoffer, T., and Georgopoulos, C. (1989). The groES and groEL heat shock gene products of Escherichia coli are essential for bacterial growth at all temperatures. J. Bacteriol. **171**, 1379–1385.

Fisher, M.T. (1994). The effect of GroES on the GroEL-dependent assembly of dodecameric glutamine synthetase in the presence of ATP and ADP. J. Biol. Chem. **269**, 13629–13636.

Georgopoulos, C., Liberek, K., Zylicz, M., and Ang, D. (1994). Properties of the heat shock proteins of Escherichia coli and the autoregulation of the heat shock response. In *The Biology of Heat Shock Proteins and Molecular Chaperones* (R.I. Morimoto, A. Tissiéres, and C. Georgopoulos, eds). Cold Spring Harbor Laboratory Press, New York, pp. 209–249.

Goloubinoff, P., Christeller, J.T., Gatenby, A.A., and Lorimer, G.H. (1989). Reconstitution of active dimeric ribulose bisphosphate carboxylase from an unfolded state depends on two chaperonin proteins and Mg–ATP. Nature **342**, 884–889.

Gray, T.E. and Fersht, A.R. (1991). Cooperativity in ATP hydrolysis by GroEL is increased by GroES. FEBS Lett. **292**, 254–258.

Hartman, D.J., Hoogenraad, N.J., Condron, R., and Høj, P.B. (1992). Identification of a mammalian 10-kDa heat shock protein, a mitochondrial chaperonin 10 homologue essential for assisted folding of trimeric ornithine transcarbamoylase in vitro. Proc. Natl. Acad. Sci. USA **89**, 3394–3398.

Hartman, D.J., Surin, B.P., Dixon, N.E., Hoogenraad, N.J., and Høj, P.B. (1993). Substoichiometric amounts of the molecular chaperones GroEL and GroES prevent thermal denaturation and aggregation of mammalian mitochondrial malate dehydrogenase invitro. Proc. Natl. Acad. Sci. USA **90**, 2276–2280.

Hemmingsen, S.M., Woolford, C., van der Vies, S.M., Tilly, K., Dennis, D.T., Georgopoulos, C.P., Hendrix, R.W., and Ellis, R.J. (1988). Homologous plant and bacterial proteins chaperone oligomeric protein assembly. Nature **333**, 330–334.

Koonin, E.V. and van der Vies, S.M. (1995). Conserved sequence motifs in bacterial and bacteriophage chaperonins. Trends Biochem. Sci. **20**, 14–15.

Laminet, A.A., Ziegelhoffer, T., Georgopoulos, C., and Plückthun, A. (1990). The *Escherichia coli* heat shock proteins GroEL and GroES modulate the folding of the beta-lactamase precursor. EMBO J. **9**, 2315–2319.

Landry, S.J., Zeilstra-Ryalls, J., Fayet, O., Georgopoulos, C., and Gierasch, L.M. (1993). Characterization of a functionally important mobile domain of GroES. Nature **364**, 255–258.

Landry, S.J., Taher, A., Georgopoulos, C., and van der Vies, S.M. (1996). Interplay of structural and disorder in co-chaperonin mobile loops. Proc. Natl. Acad. Sci. USA, **93**, 11622–11627.

Lubben, T.H., Gatenby, A.A., Donaldson, G.K., Lorimer, G.H., and Viitanen, P.V. (1990). Identification of a groES-like chaperonin in mitochondria that facilitates protein folding. Proc. Natl. Acad. Sci. USA **87**, 7683–7687.

Martin, J., Langer, T., Boteva, R., Schramel, A., Horwich, A.L., and Hartl, F.U. (1991). Chaperonin-mediated protein folding at the surface of groEL through a molten globule-like intermediate. Nature **352**, 36–42.

Martin, J., Mayhew, M., Langer, T., and Hartl, F.U. (1993). The reaction cycle of GroEL and GroES in chaperonin-assisted protein folding. Nature **366**, 228–233.

Mayhew, M., da Silva, A.C., Martin, J., Erdjument-Bromage, H., Tempst, P., and Hartl, F.-U. (1996). Protein folding in the central cavity of the GroEL–GroES chaperonin complex. Nature **379**, 420–426.

Mendoza, J.A., Rogers, E., Lorimer, G.H., and Horowitz, P.M. (1991). Chaperonins facilitate the in vitro folding of monomeric mitochondrial rhodanese. J. Biol. Chem. **266**, 13044–13049.

Saibil, H.R., Zheng, D., Roseman, A.M., Hunter, A.S., Watson, G.M.F., Chen, S., auf der Mauer, A., O'Hara, B.P., Wood, S.P., Mann, N.H., Barnett, L.K., and Ellis, R.J. (1993). ATP induces large quarternary rearrangements in a cage-like chaperonin structure. Curr. Biol. **3**, 265–273.

Tilly, K., Murialdo, H., and Georgopoulos, C. (1981). Identification of a second Escherichia coli groE gene whose product is necessary for bacteriophage morphogenesis. Proc. Natl. Acad. Sci. USA **78**, 1629–1633.

Todd, M.J., Viitanen, P.V., and Lorimer, G.H. (1993). Hydrolysis of adenosine 5′-triphosphate by Escherichia coli GroEL: effects of GroES and potassium ion. Biochemistry **32**, 8560–8567.

Todd, M.J., Viitanen, P.V., and Lorimer, G.H. (1994). Dynamics of the chaperonin ATPase cycle: implications for facilitated protein folding. Science **265**, 659–666.

Török, Z., Vigh, L., and Goloubinoff, P. (1996). Fluorescence detection of symmetric GroEL(14) (GroES(7)) (2) heterooligomers involved in protein release during the chaperonin cycle. J. Biol. Chem. **271**, 16180–16186.

Van Dyk, T.K., Gatenby, A.A., and LaRossa, R.A. (1989). Demonstration by genetic suppression of interaction of GroE products with many proteins. Nature **342**, 451–453.

Weissman, J.S., Kashi, Y., Fenton, W.A., and Horwich, A.L. (1994). GroEL-mediated protein folding proceeds by multiple rounds of binding and release of nonnative forms. Cell **78**, 693–702.

Weissman, J.S., Hohl, C.M., Kovalenko, O., Kashi, Y., Chen, S., Braig, K., Saibil, H.R., Fenton, W.A., and Horwich, A.L. (1995). Mechanism of GroEL action: productive release of polypeptide from a sequestered position under GroES. Cell **83**, 577–588.

Weissman, J.S., Rye, H.S., Fenton, W.A., Beechem, J.M., and Horwich, A.L. (1996). Characterization of the active intermediate of a GroEL-GroES-mediated protein folding reaction. Cell **84**, 481–490.

Zeilstra-Ryalls, J., Fayet, O., and Georgopoulos, C. (1991). The universally conserved GroE (Hsp60) chaperonins. Annu. Rev. Microbiol. **45**, 301–325.

Zeilstra-Ryalls, J., Fayet, O., and Georgopoulos, C. (1994). Two classes of extragenic suppressor mutations identify functionally distinct regions of the GroEL chaperone of Escherichia coli. J. Bacteriol. **176**, 6558–6565.

Zheng, X.X., Rosenberg, L.E., Kalousek, F., and Fenton, W.A. (1993). GroEL, GroES, and ATP-dependent folding and spontaneous assembly of ornithine transcarbamylase. J. Biol. Chem. **268**, 7489–7493.

Zhi, W., Landry, S.J., Gierasch, L.M., and Srere, P.A. (1992). Renaturation of citrate synthase: influence of denaturant and folding assistants. Protein Sci. **1**, 522–529.

■ *Abida Taher and Samuel J. Landry:*
Department of Biochemistry—SL43
Tulane University School of Medicine
1430 Tulane Avenue
New Orleans, LA 70112-2699, USA
Tel. 1 504 586 3990
Fax. 1 504 584 2739
E-mail: ataher@mailhost.tcs.tulane.edu and landry@mailhost.tcs.tulane.edu

Bacteriophage T4 Gp31 protein

The Gp31 protein is encoded by bacteriophage T4 and is a member of the CPN10 subfamily of chaperonins. Gp31 interacts with the Escherichia coli *host CPN60 (GroEL) to facilitate the correct assembly of the major capsid protein Gp23 during bacteriophage morphogenesis.*

■ Alternative names

Cochaperonin; cpn10 (van der Vies *et al.*, 1994).

■ Gp31 sequence

The nucleotide sequence of the bacteriophage-encoded gene *31* (GenBank accession numbers M34502, X17657, X54536 and M37882) predicts a polypeptide of 111 amino acids, with a calculated M_r of 12 064 daltons and a pI of 4.88 (Nivinskas and Black, 1988; Keppel *et al.*, 1990; Prilipov, *et al.*, 1990; Raudonikiene and Nivinskas, 1990). Initial database searches failed to detect statistically significant similarities between the Gp31 amino acid sequence and members of the CPN10 subfamily. However, the demonstration that Gp31 is a functional analogue of the *E. coli* CPN10, GroES, prompted a more detailed amino acid sequence analysis, which revealed the presence of three conserved motifs (Koonin and van der Vies, 1995). The second motif includes residues of GroES and Gp31 that have been identified as a mobile loop segment, probably representing the region that interacts with a unique site on the *E. coli* CPN60, GroEL (Landry *et al.* 1993, 1996).

Comparison of the Gp31 amino acid sequence with 38 CPN10 sequences from chloroplast, mitochondria and bacteria, reveals a striking lack of nine amino acids in the Gp31 protein (between amino acid 57 and 58, Gp31 numbering, Fig. 1). This region is visible in the crystal structures of the CPN10s of *E. coli* and *Mycobacterium leprae* (Hunt *et al.*, 1996; Mande *et al.*, 1996) where it forms the 'roof loop' of the CNP10 dome. The absence of this roof loop in Grp31 may reflect a feature that is essential for T4 bacteriophage capsid assembly.

■ Gp31 synthesis and its role in bacteriophage T4 morphogenesis

Gp31 acts transiently at an early stage of bacteriophage T4 prohead assembly, allowing the ordered assembly of the Gp23 capsid protein (Laemmli *et al.*, 1970). In the absence of a functional Gp31, the Gp23 capsid proteins accumulate as amorphous aggregates on the bacterial membrane. The same phenotype is observed in the absence of a functional GroEL (Georgopoulos *et al.*, 1972; Takano and Kakefuda, 1972; Revel *et al.*, 1980). In addition, mutations in *groEL* can be suppressed by mutations

```
Motif               I                   II
                                     mobile loop
                                     ──────────────────────────
Gp31    .....MSEVQQLPIR  AVGEYVILVS  EPAQAGDEEV  TESGLIIGKR  VQGEVPELCV

Ecoli   ..........MNIR   PLHDRVIVK-  ---RKEVETK  SAGGIVLTGS  AAA-KSTRGE
Mycle   .........AKVKIK  PLEDKILVQ-  ---AGEAETM  TPSGLVIPEN  AKE-KPQEGT
Legmi   .........MKIR    PLHDRVVVR-  ---RMEEERT  TAGGIVIPDS  ATE-KPTRGE
Ricts   ..........MKYQ   PLYDRVLVE-  ---PIQNDE-  AHGKILIPDT  AKE-KPTEGI
Rat     ......AGQAFRKFL  PLFDRVLVE-  ---RSAAETV  TKGGIMLPEK  SQG-KVLQAT
Yeast   ...MSTLLKSAKSIV  PLMDRVLVQ-  ---RIKAQAK  TASGLYLPEK  NVE-KLNQAE
Human   ......AGQAFRKFL  PLFDRVLVE-  ---RSAAETV  TKGGIMLPEK  SQG-KVLQAT
Sp12    VRAASITTSKYTSVK  PLGDRVLIK-  ---TKIVEEK  TTSGIFLPTA  AQK-KPQSGE
Sp11    ..........KDLK   PLNDRLLIK-  ---VAEVENK  TSGGLLLAES  SKE-KPSFGT


Motif                                       III
        roof loop

Gp31    VHSVGPD---  ------VPEG  FCEVGDLTSL  PVG--QIRNV  PHPFVALGLK  QPKEIKQKFV  TCHYKAIPCL  YK

Ecoli   VLAVGNGRIL  ENG-EVKPLD  VK-VGDIVIF  NDGYGVKSEK  ID-NEEVLIM  SESDILAIVE  A.........  ..
Mycle   VVAVGPGRWD  EDGAKRIPVD  VS-EGDIVIY  SKYGGTE-IK  YN-GEEYLIL  SARDLAVVSK  ..........  ..
Legmi   IIAVGPGKVL  ENG-DVRALA  VK-VGDVVLF  GKYSGTE-VK  IS-GQELVVM  REDDIMGVIE  K.........  ..
Ricts   VVMVGGGYRN  DKG-DITPLK  VK-KGDTIVY  TKWAGTE-IK  LE-SKDYVVI  KESDILLVKS  ..........  ..
Rat     VVAVGSGGKG  KGG-EIQPVS  VK-VGDKVLL  PEYGGTKVVL  D--DKDYFLF  RDGDILGKYV  D.........  ..
Yeast   VVAVGPGFTD  ANG-NKVVPQ  VK-VGDQVLI  PQFGGSTIKL  GN-DDEVILF  RDAEILAKIA  KD........  ..
Human   VVAVGSGSKG  KGG-EIQPVS  VK-VGDKVLL  PEYGGTKVVL  D--DKDYFLF  RDGDILGKYV  D.........  ..
Sp12    VVAIGSGKKV  --GDKKLPVA  VK-TGAEVVY  SKYTGTE-IE  VD-GSSHLIV  KEDDIIGILE  TDDV......  ..
Sp11    VVATGPGVLD  EEG-NRIPLP  VC-SGNTVLY  SKYAGNDFKG  VD-GSDYMVL  RVSDVMAVLS  ..........  ..
```

Figure 1. Conserved amino acid sequence motifs in bacteriophage and bacterial chaperonins.

in gene *31* (the so-called ε mutations) suggesting that the two proteins interact functionally during bacteriophage morphogenesis (Keppel et al., 1990). Interestingly, these mutations are all located in the mobile loop segment of Gp31, suggesting that this region may be involved in modulating the GroEL chaperonin activity.

Gp31 belongs to the group of bacteriophage T4 proteins that are synthesized between one and three minutes after infection and are made throughout infection, with a maximum level between three and eleven minutes (Cowan et al., 1994). This pattern of synthesis does not coincide with that of the Gp23 major capsid protein, whose synthesis starts about seven minutes after infection. This raises the interesting possibility that Gp31 may not only be required for capsid assembly but may have an additional regulatory role, for example in the initial take-over of the cell by the bacteriophage.

■ Gp31 functions as a molecular chaperone

In contrast to bacteriophage T4, several other bacteriophages such as λ and T5 use the host CPN10 to assemble their morphogenic structures (head and tail respectively). Gp31 can not only substitute for *E. coli* GroES for λ and T5, but can also replace GroES for general *E. coli* growth, since Gp31 suppresses the temperature-sensitive growth defect at 43°C of *E. coli groES* mutant cells (van der Vies et al., 1994).

The purified Gp31 protein is a homo-oligomer of seven identical subunits arranged in a single ring. Like GroES, Gp31 forms a stable complex with GroEL in the presence of Mg–ATP *in vitro*, thereby inhibiting GroEL's ATPase activity. Gp31 can also replace GroES in the chaperonin-dependent folding of the bacterial enzyme ribulose bisphosphate carboxylase (rubisco), both *in vivo* and *in vitro*. The absolute requirement of Gp31 for the correct assembly of the Gp23 capsid protein *in vivo* suggests that Gp31 may possess specific cochaperonin properties that are absent in GroES, or it may function similarly to GroEL but with improved efficiency.

■ References

Cowan, J., d'Acci, K., Guttman, B., and Kutter, E. (1994). Gel analysis of T4 prereplicative proteins. In *Molecular biology of bacteriophage T4* (J.D. Karam, ed.). ASM Press, Washington D.C., pp. 520–527.

Georgopoulos, C., Hendrix, R.W., Kaiser, A.D., and Wood, W.B. (1972). Role of the host cell in bacteriophage morphogenesis:

Effects of a bacterial mutation on T4 head assembly. Nature New Biol. *239*, 38–41.

Hunt, J.F., Weaver, A.J., Landry, S.J., Gierasch, L., and Deisenhofer, J. (1996). The crystal structure of the GroES co-chaperonin at 2.8 Å resolution. Nature **379**, 37–45.

Keppel, F., Lipinska, B., Ang, D., and Georgopoulos, C. (1990). Mutational analysis of the phage T4 morphogentic gene *31*, whose product interacts with the *Escherichia coli* GroEL protein. Gene **86**, 19–25.

Koonin, E.V. and van der Vies, S.M. (1995). Conserved sequence motifs in bacterial and bacteriophage chaperonins. Trends Biochem. Sci. **20**, 14–15.

Laemmli, U.K., Beguin, F., and Guyer-Kellenberger, G. (1970). A factor preventing the major head protein of bacteriophage T4 from random aggregation. J. Mol. Biol. **47**, 69–85.

Landry, S.J., Zeilstra-Ryalls, Fayet, O., Georgopoulos, C., and Gierasch, L.M. (1993). Characterization of a functionally important mobile domain of GroES. Nature **364**, 255–258.

Landry, S.J., Taher, A., Georgopoulos, C., and van der Vies, S.M. (1996). Interplay of structure and disorder in the chaperonin-10 mobile loops. Proc. Natl. Acad. Sci. USA **93**, 11622–11627.

Mande, C.S., Mehra, V., Bloom, B.R., and Hol, W.G.J. (1996) Structure of the heat shock protein chaperonin-10 of *Mycobacterium leprae*. Science **271**, 203–207.

Nivinskas, R. and Black, L.W. (1988). Cloning, sequencing and expression of the temperature-dependent phage T4 capsid assembly gene *31*. Gene **73**, 251–257.

Prilipov, A.P., Mesyanzhinov, V.V., Aebi, U., and Kellenberger, E. (1990). Cloning and sequencing of the bacteriophage T4 genes between map position 128.3-130.3. Nucl. Acids Res. **18**, 3635.

Raudonikiene, A. and Nivinskas, R. (1990). Nucleotide sequence of bacteriophage T4 gene *31* region. Nucl. Acids Res. **18**, 4280.

Revel, H.R., Stitt, B.L., Lielausis, I., and Wood, W.B. (1980). Role of the host cell in bacteriophage T4 development, I. Characterization of host mutants that block T4 assembly. J. Virol. **33**, 366–376.

Takano, T. and Kakefuda, T. (1972). Involvement of a bacterial factor in morphogenesis of bacteriophage capsid. Nature New Biol. **239**, 34–37.

van der Vies, S.M., Gatenby, A.A., and Georgopoulos, C. (1994). Bacteriophage T4 encodes a co-chaperonin that can substitute for *Escherichia coli* GroES in protein folding. Nature **368**, 654–656.

■ *Saskia M. van der Vies:*
Département de Biochimie Médicale
Centre Médical Universitaire
Université de Genève
1211 Genève 4, Switzerland
Tel. 4 22 702 5515
Fax. 41 22 702 5502
E-mail: Saskia.VanDerVies@medicine.unige.ch

Thermus thermophilus GroEL and GroES

A unique chaperonin system has been developed using the chaperonin from a thermophilic eubacteria, *Thermus thermophilus*. Unlike GroEL and GroES of *Escherichia coli*, the *Thermus* GroEL homologue (cpn60) is copurified with the GroES homologue (cpn10) as a large complex containing endogenous ADP (the holo-chaperonin or bullet-shaped chaperonin complex). The *Thermus* chaperonin is thermostable up to around 80°C facilitating studies of the chaperonin's function during protein folding at high temperatures and during heat denaturation of relatively heat labile enzymes.

■ Alternative names

Thermus chaperonin means the chaperonin from *T. thermophilus*, consisting of 14 copies of GroEL homologues (*T. th* cpn60) and seven copies of GroES homologues (*T. th* cpn10) (Taguchi *et al*., 1991; Amada *et al*., 1995).

■ Isolation of *Thermus* chaperonin

The chaperonin has been purified from *T. thermophilus* using protocols involving DEAE ion exchange chromatography and gel filtration (e.g. Sepharose CL4B) (Taguchi *et al*., 1991). Unlike *E. coli* chaperonin GroEL and GroES, the chaperonin of *T. thermophilus* is purified as a complex including both the cpn60 and the cpn10 subunits, and endogenously bound ADP (Taguchi *et al*., 1991; Ishii *et al*., 1992; Yoshida *et al*., 1993). The stoichiometry of the bound ADP differed from preparation to preparation, ranging from 0.3 to 2.8 mol ADP per mol of the *Thermus* chaperonin complex (Yoshida *et al*., 1993). *T. th* cpn60 and *T. th* cpn10 have been expressed separately in *E. coli* cells, purified, and characterized (Amada *et al*., 1995).

■ *Thermus* chaperonin sequences

Amino acid sequences deduced from the nucleotide sequence of *Thermus* chaperonin (GenBank accession number D45880) reveal two open reading frames corresponding to *T. th* cpn10 (101 amino acids) and *T. th* cpn60 (543 amino acids) (Amada *et al*., 1995). The calculated molecular weights of the cpn10 and cpn60 subunits are 10 996 and 57 888 daltons, respectively. The homology scores are as follows: *T. th* cpn60 is 52.0% identical with *E. coli* GroEL; *T. th* cpn10 is 54.3% identical with *E. coli* GroES. Although most CPN60 proteins, including GroEL, have a GGM repeat at the C-terminus, it is not found in *T. th* cpn60. Both *T. th* cpn10 and *T. th* cpn60 lack cysteine and tryptophan residues.

Function of *Thermus* chaperonin proteins

Although *Thermus* chaperonin is highly homologous with GroEL and GroES, it contains several distinct features.

1. Judging from the temperature dependency of ATPase activity and the decay of the CD (circular dichroism) spectrum, *Thermus* chaperonin is stable up to around 80°C (Taguchi et al., 1991). This thermostability facilitated investigation of the role of the chaperonin at high temperature (Taguchi et al., 1991; Taguchi and Yoshida, 1993). The effect of *Thermus* chaperonin on protein folding is different at higher and lower temperature ranges (Taguchi et al., 1991; Yoshida et al., 1993). At higher temperatures, where the native proteins are stable but spontaneous folding does not occur, the chaperonin induces productive folding in an ATP-dependent manner. At lower temperatures where spontaneous folding occurs, the chaperonin slows down the rate of folding without changing the final yield. Taking advantage of this thermostability, the effect of *Thermus* chaperonin on several relatively heat labile enzymes was also studied (Taguchi and Yoshida, 1993). When the proteins are incubated at their denaturing temperatures, the presence of *Thermus* chaperonin prevents the proteins from irreversible heat denaturation by capturing the denaturation intermediate. Once the heat denaturing (or folding) intermediate is captured by chaperonin, it retains the ability to resume productive folding even after exposure to the otherwise denaturing high temperatures. The heat stability of the folding intermediate seems to be limited solely by that of the chaperonin.

2. Monomeric *T. th* cpn60 and its proteolytic 50 kDa fragment, which lacks the N-terminal 78 amino acid residues, possess the ability to interact with non-native protein, to suppress aggregation, and to promote protein folding under certain conditions (Taguchi et al., 1994). However, different from tetradecameric cpn60, folding promoted by the monomeric cpn60 and the 50 kDa fragment does not require ATP or *T. th* cpn10.

3. When bullet-shaped *Thermus* chaperonin is incubated with ATP–Mg and c. 100 mM KCl, cone-shaped particles (*T. th* cpn60$_7$–*T. th* cpn10$_7$) appear (Ishii et al., 1995). However their functional significance is not yet clear.

Electron microscopic studies of *Thermus* chaperonin

Thermus chaperonin shows a bullet-like shape in the side view observed by electron microscopy, and antibody against *T. th* cpn10 binds only to the round side of the bullet (Ishii et al., 1992). Complexes of *Thermus* chaperonin and a folding intermediate have been observed by immunoelectron microscopy after decoration by labelling with IgG against substrate protein (Ishii et al., 1994). EM images show that the IgG is bound to the bottom end of the bullet-shaped *Thermus* chaperonin.

References

Amada, K., Yohda, M., Odaka, M., Endo, I., Ishii, N., Taguchi, H., and Yoshida, M. (1995). Molecular cloning, expression, and characterization of chaperonin-60 and chaperonin-10 from a thermophilic bacterium, *Thermus thermophilus* HB8. J. Biochem. **118**, 347–354.

Ishii, N., Taguchi, H., Sumi, M., and Yoshida, M. (1992). Structure of holo-chaperonin studied with electron microscopy: oligomeric cpn10 on top of two layers of cpn60 rings with two stripes each. FEBS Lett. **299**, 169–174.

Ishii, N., Taguchi, H., Sasabe, H., and Yoshida, M. (1994). Folding intermediate binds to the bottom of bullet-shaped holo-chaperonin and is readily accessible to antibody. J. Mol. Biol. **236**, 691–696.

Ishii, N., Taguchi, H., Sasabe, H., and Yoshida, M. (1995). Equatorial split of holo-chaperonin from Thermus thermophilus by ATP and K$^+$. FEBS Lett. **362**, 121–125.

Taguchi, H. and Yoshida, M. (1993). Chaperonin from Thermus thermophilus can protect several enzymes from irreversible heat denaturation by capturing denaturation intermediate. J. Biol. Chem. **268**, 5371–5375.

Taguchi, H., Konishi, J., Ishii, N., and Yoshida, M. (1991). A chaperonin from a thermophilic bacterium, Thermus thermophilus, that controls refolding of several thermophilic enzymes. J. Biol. Chem. **266**, 22411–22418.

Taguchi, H., Makino, Y., and Yoshida, M. (1994). Monomeric chaperonin-60 and its 50-kDa fragment possess the ability to interact with non-native proteins, to suppress aggregation, and to promote protein folding. J. Biol. Chem. **269**, 8529–8534.

Yoshida, M., Ishii, N., Muneyuki, E. and Taguchi, H. (1993). A chaperonin from a thermophilic bacterium, Thermus thermophilus. Phil. Trans. R. Soc. Lond. B **339**, 305–312.

■ *Hideki Taguchi and Masasuke Yoshida:*
Research Laboratory of Resources Utilization
Tokyo Institute of Technology
R-1, 4259 Nagatsuta, Midori-ku
Yokohama 226, Japan
Tel: 81 45 924 5233
Fax: 81 45 924 5277
E-mail: myoshida@res.titech.ac.jp

Saccharomyces cerevisiae Hsp60

Mitochondrial Hsp60 of Saccharomyces cerevisiae *belongs to the CPN60 family of molecular chaperones, termed chaperonins. It forms a barrel-like structure composed of 14 subunits of 60 kDa that are arranged in two layers with sevenfold symmetry. In cooperation with the cochaperonin Hsp10, Hsp60 mediates the ATP-dependent folding of nuclear-encoded proteins after import into mitochondria and prevents the heat denaturation of mitochondrial proteins under heat stress.*

∎ Alternative names

mt-Hsp60 (mitochondrial Hsp60), mt-cpn60 (mitochondrial chaperonin 60).

∎ Isolation of *HSP60*

The *HSP60* gene of *Saccharomyces cerevisiae* was identified in the conditional mutant *mif4* that lacked enzymatic activity of imported ornithine transcarbamylase (Cheng et al., 1989). Concurrently, the yeast *HSP60* gene was isolated by screening a λ-gt11 genomic library with a polyclonal antiserum raised against purified *Tetrahymena thermophila* Hsp60 (Reading et al., 1989) or monoclonal antibodies directed against a protein fraction enriched in mitochondrial ribosomal proteins, respectively (Johnson et al., 1989). Independently, Hsp60 was identified as a component of the stimulatory factor I of DNA polymerase ε (Smiley et al., 1992). The *HSP60* gene of *S. cerevisiae* is located on chromosome XII. Accession numbers: GenBank U17244x3; PIR JQ0157; SwissProt P19882.

∎ *HSP60* gene and sequence

The *HSP60* gene encodes a protein of 572 amino acids with a relative molecular mass of 60 830 Da that is 54% identical to *Escherichia coli* GroEL and 43% identical to the rubisco subunit-binding protein of chloroplasts (Reading et al., 1989). In comparison to GroEL, Hsp60 contains an N-terminal extension of 21 amino acids. This sequence is characteristic for mitochondrial targeting signals in that it is enriched in positively charged and hydroxylated amino acid resdues, but lacks acidic residues. The mitochondrial targeting sequence is cleaved on import into mitochondria (Cheng et al., 1990).

Amino acid residues conserved between Hsp60 and *E. coli* GroEL include Tyr-502 and Asp-102 of Hsp60. The corresponding residues of GroEL, Tyr-477 and Asp-78, have been identified to be involved in ATP binding and hydrolysis by mutational or biochemical analysis (Martin et al., 1993; Fenton et al., 1994). Substitution of Ala-47 by Val and of Val-77 by Ile in mitochondrial Hsp60 impairs the respiratory competence of yeast cells at elevated temperature (Sanyal et al., 1995). At the C-terminus of Hsp60, a conserved glycine- and methionine-rich sequence of unknown function is found.

∎ Hsp60 protein

Hsp60 is a soluble protein localized in the mitochondrial matrix space (McMullin and Hallberg, 1987). It assembles into a homo-oligomeric structure (McMullin and Hallberg, 1988). The Hsp60 homologues of *S. cerevisiae* and of *Tetrahymena thermophila* sediment in sucrose gradients as a 20S–25S complex (McMullin and Hallberg, 1987, 1988; Johnson et al., 1989). Electron microscopic analysis of the *Neurospora crassa* protein revealed a structure very similar to *E. coli* GroEL. Two heptameric rings are arranged in two layers, thereby forming a barrel-like structure with a diameter of about 12 nm (Hutchinson et al., 1989). In mitochondria of mammals, a heptameric structure of Hsp60 was also reported (Viitanen et al., 1992).

Hsp60 has been purified to homogeneity from yeast (Rospert et al., 1993). Mitochondrial Hsp60 exibits an ATPase activity of about 6 mol/min per mol protomer, that is stimulated by K^+ ions (Rospert et al., 1993). In contrast to bacterial chaperonins, however, the ATPase of Hsp60 is not inhibited upon binding of the cochaperonin Hsp10 (Rospert et al., 1993).

∎ Function of Hsp60

Hsp60 is a central component of the folding machinery of the mitochondrial matrix (Langer and Neupert, 1994). Its function is regulated by the cochaperonin Hsp10 which binds to Hsp60 in a nucleotide-dependent manner (Rospert et al., 1993; Höhfeld and Hartl, 1994). Assembly of a variety of mitochondrial proteins is impaired in a temperature-sensitive mutant form of mt-Hsp60 under restrictive conditions (Cheng et al., 1989) or after genetic depletion from the mitochondrial matrix of mt-Hsp60 (Hallberg et al., 1993). Mt-Hsp60 acts at the level of polypeptide chain folding, rather than affecting the assembly of multi-subunit complexes (Ostermann et al., 1989; Saijo and Tanaka, 1995). The folding of proteins encoded by the mitochondrial genome is also assisted by mt-Hsp60 (Prasad et al., 1990; Horwich et al., 1992). Nevertheless, some model proteins were recently found to fold independently of Hsp60 when imported into mitochondria (Rospert et al., 1996). At elevated temperatures, mt-Hsp60 is required to maintain thermolabile proteins in an enzymatically active conformation (Martin et al., 1992). In the presence of a conditional mutant form of mt-Hsp60, thermolabile proteins form aggregates under restrictive conditions (Martin et al., 1992).

■ HSP60 gene regulation

Hsp60 is constitutively expressed under normal growth conditions. The start point of transcription has been identified by S1 nuclease analysis to be 33 bp upstream of the beginning of the ATG start codon (Johnson et al., 1989). HSP60 expression is increased 2–3-fold above basal levels after heat shock (McMullin and Hallberg, 1987). Under these conditions, Hsp60 represents about 0.3% of total cellular protein. Heat shock regulatory elements, allowing the binding of the heat shock transcription factor, are present in the promotor region of HSP60 (Reading et al., 1989; Johnson et al., 1989).

■ References

Cheng, M.Y., Hartl, F.U., Martin, J., Pollock, R.A., Kalousek, F., Neupert, W., Hallberg, E.M., Hallberg, R.L., and Horwich, A.L. (1989). Mitochondrial heat-shock protein hsp60 is essential for assembly of proteins imported into yeast mitochondria. Nature **337**, 620–625.

Cheng, M.Y., Hartl, F.U., and Horwich, A.L. (1990). The mitochondrial chaperonin hsp60 is required for its own assembly. Nature **348**, 455–458.

Fenton, W.A., Kashi, Y., Furtak, K., and Horwich, A. (1994). Residues in chaperonin GroEL required for polypeptide binding and release. Nature **371**, 614–619.

Hallberg, E.M., Shu, Y., and Hallberg, R.L. (1993). Loss of mitochondrial hsp60 function: nonequivalent effects on matrix-targeted and intermembrane-targeted proteins. Mol. Cell. Biol. **13**, 3050–3057.

Höhfeld, J. and Hartl, F.-U. (1994). Requirement of the chaperonin cofactor HSP10 for protein folding and sorting in yeast mitochondria. J. Cell Biol.**126**, 305–315.

Horwich, A., Caplan, S., Wall, J.S., and Hartl, F.U. (1992). Chaperonin-mediated protein folding. In Membrane Biogenesis and Protein Targeting (Neupert, W. and Lill, R., eds). Elsevier, Amsterdam, pp. 329–337.

Hutchinson, E.G., Tichelaar, W., Hofhaus, G., Weiss, H., and Leonard, K.R. (1989). Identification and electron microscopic analysis of a chaperonin oligomer from Neurospora crassa mitochondria. EMBO J. **8**, 1485–1490.

Johnson, R.B., Fearon, K., Mason, T., and Jindal, S. (1989). Cloning and characterization of the yeast chaperonin HSP60 gene. Gene **4**, 295–302.

Langer, T. and Neupert, W. (1994). Chaperoning mitochondrial biogenesis. In The Biology of Heat Shock Proteins and Molecular Chaperones (Morimoto, R.I., Tissieres, A., Georgopoulos, C., eds). Cold Spring Harbor Laboratory Press, New York, pp. 53–83.

Martin, J., Horwich, A.L., and Hartl, F.U. (1992). Prevention of protein denaturation under heat stress by the chaperonin Hsp60. Science **258**, 995–998.

Martin, J., Geromanos, S., Tempst, P., and Hartl, F.-U. (1993). Identification of nucleotide-binding regions in the chaperonin proteins GroEL and GroES. Nature **366**, 279–282.

McMullin, T. W. and Hallberg, R.L. (1987). A normal mitochondrial protein is structurally related to the protein encoded by the Escherichia coli groEL gene. Mol. Cell. Biol. **7**, 4414–4423.

McMullin, T.W. and Hallberg, R.L. (1988). A highly evolutionarily conserved mitochondrial protein is structurally related to the protein encoded by the Escherichia coli groEL gene. Mol. Cell. Biol. **8**, 371–380.

Ostermann, J., Horwich, A.L., Neupert, W. and Hartl, F.U. (1989). Protein folding in mitochondria requires complex formation with hsp60 and ATP hydrolysis. Nature **341**, 125–130.

Prasad, T.K., Hack, E. and Hallberg, R.L. (1990). Function of the maize mitochondrial chaperonin hsp60: specific association between hsp60 and newly synthesized F1-ATPase alpha subunits. Mol. Cell. Biol. **10**, 3979–3986.

Reading, D.S., Hallberg, R.L. and Myers, A.M. (1989). Characterization of the yeast HSP60 gene coding for a mitochondrial assembly factor. Nature **337**, 655–659.

Rospert, S., Glick, B.S., Jeno, P., Schatz, G., Todd, M.J., Lorimer, G.H. and Viitanen, P.V. (1993). Identification and functional analysis of chaperonin 10, the groES homolog from yeast mitochondria. Proc. Natl. Acad. Sci. USA **90**, 10967–10971.

Rospert, S., Looser, R., Dubaquie, Matouschek, A., Glick, B.S. and Schatz, G. (1996). Hsp60-independent protein folding in the matrix of yeast mitochondria. EMBO J. **15**, 764–774.

Sajio, T. and Tanaka, K. (1995). Isoalloxazine ring of FAD is required for the formation of the core of the Hsp60-assisted folding of medium chain acyl-CoA dehydrogenase subunit in the assembly competent conformation in mitochondria. J. Biol. Chem. **270**, 1899–1907.

Sanyal, A., Harington, A. Herbert, C.J., .Groudinsky, O., Slonimski, P.P., Tung, B. and Getz, G.S. (1995). Heat shock protein HSP60 can alleviate the phenotype of mitochondrial RNA-deficient temperature-sensitive mna2 pet mutants. Mol. Gen. Genet. **6**, 56–64.

Smiley, J.K., Brown, W.C., and Campbell, J.L. (1992). The 66 kDa component of yeast SFI, stimulatory factorI, is hsp60. Nucl. Acid Res. **20**, 4913–4918.

Viitanen, P.V., Lorimer, G.H., Seetharam, R., Gupta, R.S., Oppenheim, J., Thomas, J.O., and Cowan, N.J. (1992). Mammalian mitochondrial chaperonin 60 functions as a single toroidal ring. J. Biol. Chem. **267**, 695–698.

Zheng, X., Rosenberg, L.E., Kalousek, F., and Fenton, W.A. (1993). GroEL, GroES, and ATP-dependent folding and spontaneous assembly of ornithine transcarbamylase. J. Biol. Chem. **268**, 7489–7493.

■ *Thomas Langer and Walter Neupert:*
Institut fur Physiologische Chemie
Goethestr. 33
80336 Munchen, Germany
Tel. 49 89 5996 283
Fax. 49 89 5996 270
E-mail: Langer@bio.med.uni-muenchen.de

Saccharomyces cerevisiae Hsp10

Hsp10, the eukaryotic homologue of Escherichia coli GroES, is an essential component of the mitochondrial protein-folding machinery (Rospert et al., 1993a; Höhfeld and Hartl, 1994). Localized in the matrix space it participates in the folding of newly imported polypeptide chains as a regulator of the chaperonin Hsp60. Hsp10 is also required for Hsp60-dependent intramitochondrial protein sorting, further revealing Hsp60/Hsp10 as a cooperating chaperonin team in mitochondria similar to the bacterial GroEL/GroES system (Hendrick and Hartl, 1993). By analogy to GroES, Hsp10 probably synchronizes the ATPase activity of the individual subunits within the Hsp60 double ring complex.

■ Alternative names

Chaperonin 10, cpn10 (Rospert et al., 1993a). Since this entry focuses on mitochondrial cpn10 which cooperates with Hsp60 in the matrix compartment, the term Hsp10 will be used throughout the text.

■ Hsp10 sequence and structure

The nuclear *HSP10* gene of S. cerevisiae (GenBank accession number X75754) encodes a polypeptide of 106 amino acids with a deduced molecular mass of 11 374 Da (Rospert et al., 1993a; Höhfeld and Hartl, 1994). Hsp10 displays a native mass of about 80 kDa reflecting the assembly of seven subunits to a homo-oligomeric ring structure characteristic of GroES and its homologues (Mande et al., 1996; Hunt et al., 1996). The amino acid sequence of yeast Hsp10 is 36.5% identical to that of E. coli GroES and 43.6% identical to rat liver Hsp10. Significantly, the eukaryotic proteins possess an N-terminal extension which forms an amphiphilic helix–turn–helix structure and functions as a non-cleavable mitochondrial targeting sequence (Rospert et al., 1993a; Jarvis et al., 1995; Ryan et al., 1994). Acetylation of the N-terminus may contribute to efficient import of Hsp10 into mitochondria. Structural conservation between the prokaryotic and eukaryotic homologues is found in a 'mobile loop' domain (residues 17–32 of GroES and 25–40 of yeast Hsp10), shown to represent the GroEL-binding region of GroES (Landry et al., 1993). The ability of mitochondrial Hsp10 to interact functionally with bacterial GroEL confirms the conservation of this binding region and actually provided the means by which yeast Hsp10 was isolated (Rospert et al., 1993b; Höhfeld and Hartl, 1994). Indeed, a temperature-sensitive mutation, which disrupts the interaction of Hsp10 with Hsp60, maps to the mobile loop domain (exchange of a proline residue at position 36 to serine, P36S) (Höhfeld and Hartl, 1994).

■ Function of Hsp10

HSP10 is an essential gene of S. cerevisiae demonstrating the vital role of the Hsp10 protein for mitochondrial biogenesis (Rospert et al., 1993a; Höhfeld and Hartl, 1994). Characterization of the conditional P36S mutant strain revealed a requirement for Hsp10 during protein folding and assembly in the mitochondrial matrix (Höhfeld and Hartl, 1994). In the absence of functional Hsp10, newly imported precursor of the α subunit of the matrix processing protease fails to reach its native state and instead aggregates. Similarly, rat ornithine transcarbamylase, imported into mutant yeast mitochondria, does not assemble into the enzymatically active trimer. Both proteins interact with the chaperonin Hsp60 following their translocation (Cheng et al., 1989; Manning-Krieg et al., 1991). The regulatory function of Hsp10 is however required to achieve an interaction productive for folding and assembly.

Hsp10 also cooperates with Hsp60 in the sorting of proteins from the mitochondrial matrix to the inner membrane and intermembrane space. For example, the Rieske iron–sulfur protein is translocated into the matrix, where it interacts with Hsp60, prior to export across the inner membrane (Hartl and Neupert, 1990). Presumably, Hsp60 maintains the protein in a loosely folded conformation competent for the second translocation step. In hsp10-mutant mitochondria, the iron–sulfur protein is unable to reach its final destination owing to aggregation in the matrix (Höhfeld and Hartl, 1994). Apparently, Hsp10 regulates the 'antifolding' activity of Hsp60 during intramitochondrial sorting.

Following heat shock Hsp10 expression is induced about threefold, and thus to comparable levels with Hsp60, indicating that the chaperonin relies on the regulatory cofactor for the prevention of protein denaturation under stress conditions (Martin et al., 1992; Höhfeld and Hartl, 1994).

As mitochondrial Hsp10 from both yeast and mammalian cells can substitute for bacterial GroES in GroEL-mediated folding reactions (Lubben et al., 1990; Hartman et al., 1992; Rospert et al., 1993b), insight into the molecular mechanism of Hsp10-mediated regulation is gained by studies on GroES function (Martin et al., 1993; Hartl et al., 1994). Binding of GroES results in cooperative ATP binding and hydrolysis by GroEL. The achieved synchronization of individual GroEL subunits within the chaperonin double ring cylinder provides an efficient and instantaneous release of bound substrate protein. It thus appears that Hsp60 goes through Hsp10-synchronized cycles of ATP binding and hydrolysis accompanied by binding and release of the substrate protein until folding (in the case of a matrix protein) or initiation of re-export (in the case of an intermembrane space protein) is achieved.

References

Cheng, M.Y., Hartl, F.-U., Martin, J., Pollock, R.A., Kalousek, F., Neupert, W., Hallberg, E.M., Hallberg, R.L., and Horwich A.L. (1989). Mitochondrial heat-shock protein hsp60 is essential for assembly of proteins imported into yeast mitochondria. Nature **337**, 620–625.

Hartl, F.-U. and Neupert, W. (1990). Protein sorting to mitochondria: evolutionary conservations of folding and assembly. Science **247**, 930–938.

Hartl, F.-U., Hlodan, R., and Langer T. (1994). Molecular chaperones in protein folding: the art of avoiding sticky situations. Trends Biochem. Sci. **19**, 20–25.

Hartman, D.J., Hoogenraad, N.J., Condron, R., and Høj, P.B. (1992). Identification of a mammalian 10-kDa heat shock protein, a mitochondrial chaperonin 10 homologue essential for assisted folding of trimeric ornithine transcarbamoylase in vitro. Proc. Natl. Acad. Sci. USA **89**, 3394–3398.

Hendrick, J.P. and Hartl, F.-U. (1993). Molecular chaperone functions of heat shock proteins. Annu. Rev. Biochem. **62**, 349–384.

Höhfeld, J. and Hartl., F.-U. (1994). Role of the chaperonin cofactor hsp10 in protein folding and sorting in yeast mitochondria. J. Cell Biol. **126**, 305–315.

Hunt, J.F., Weaver, A.J., Laundry, S.J., Gierasch, L., and Deisenhofer, J. (1996). The crystal structure of the GroES co-chaperonin at 2.8 Å resolution. Nature **379**, 37–45.

Jarvis, J.A., Ryan, M.T., Hoogenraad, N.J., Craik, D.J., and Høj, P.B. (1995). Solution structure of the acetylated and noncleavable mitochondrial targeting signal of rat chaperonin 10. J. Biol. Chem. **270**, 1323–1331.

Landry, S.J., Zeilstra-Ryalls, J., Fayet, O., Georgopoulos, C., and Gierasch, L.M. (1993). Characterization of a functionally important mobile domain of GroES. Nature **364**, 255–258.

Lubben, T.H., Gatenby, A.A., Donaldson, G.K., Lorimer, G.H., and Viitanen, P.V. (1990). Identification of a groES-like chaperonin in mitochondria that facilitates protein folding. Proc. Natl. Acad. Sci. USA **87**, 7683–7687.

Mande, S.C., Mehra, V., Bloom, B.R., and Hol, W.G.J. (1996). Structure of the heat shock protein chaperonin-10 of Mycobacterium leprae. Science **271**, 203–207.

Manning-Krieg, U.C., Scherer, P.E., and Schatz, G. (1991). Sequential action of mitochondrial chaperonines in protein import into the matrix. EMBO J. **10**, 3273–3280.

Martin, J., Horwich, A.L., and Hartl, F.-U. (1992). Prevention of protein denaturation under heat stress by the chaperonin hsp60. Science **258**, 995–998.

Martin, J., Mayhew, M., Langer, T., and Hartl, F.-U. (1993). The reaction cycle of GroEL and GroES in chaperonin-assisted refolding. Nature **366**, 228–233.

Rospert, S., Junne, T., Glick, B.S., and Schatz, G. (1993a). Cloning and disruption of the gene encoding yeast mitochondrial chaperonin 10, the homolog of E. coli groES. FEBS Lett. **335**, 358–360.

Rospert, S., Glick, B.S., Jenö, P., Schatz, G., Todd, M.J., Lorimer, G.H., and Viitanen, P.V. (1993b). Identification and functional analysis of chaperonin 10 in the groES homolog form yeast mitochondria. Proc. Natl. Acad. Sci. USA **90**, 10967–10971.

Ryan, M.T., Hoogenraad, N.J., and Høj, P.B. (1994). Isolation of a cDNA clone specifying rat chaperonin 10, a stress-inducible mitochondrial matrix protein synthesised without a cleavable presequence. FEBS Lett. **337**, 152–156.

■ Jörg Höhfeld:
Zentrum für Molekulare Biologie, ZMBH
Universität Heidelberg
Im Neuenheimer Feld 282
D-69120 Heidelberg, Germany
Tel.: 49 6221 54 6837
Fax. 49 6221 54 5891
E-mail: j-hoehfeld@sun0.urz.uni-heidelberg.de

Plastid Cpn60

Plastid chaperonin 60 (Cpn60) is the plastid homologue of the GroE chaperonin 60 found in eubacteria and mitochondria. It occurs at high concentrations in the chloroplast stroma and is implicated in the assembly of ribulose bisphosphate carboxylase oxygenase (rubisco) and other plastid proteins imported from the cytosol.

Alternative names

Rubisco large subunit binding protein (Ellis, 1981; Lennox and Ellis, 1986; Musgrove and Ellis, 1986: Musgrove et al., 1987); rubisco subunit binding protein (Ellis and van der Vies, 1988); chloroplast chaperonin 60 (Gutteridge and Gatenby, 1995); chl cpn60.

Plastid Cpn60 protein

The chloroplast protein is an abundant soluble oligomer of 14 subunits that resembles GroEL from *Escherichia coli* in being arranged in two stacked rings, each of seven subunits, surrounding a central cavity (Pushkin et al., 1982; Viitanen et al., 1995). Unlike GroEL, the purified chloroplast protein consists of equal amounts of two types of related subunit termed α and β (Hemmingsen and Ellis, 1986: Musgrove et al., 1987; Martel et al., 1990). The disposition of the α and β subunits in the oligomer is unknown, but the equal amounts of the two subunits supports an α_7/β_7 arrangement, with a calculated molecular weight of 801 108 for the oligomer from Pisum sativum. Sequence analyses of isolated subunits and cDNAs indicate that each type of subunit initially contains a typical chloroplast targeting presequence at the N-terminus that is removed in the chloroplast stroma to produce a highly conserved 56–58 kDa mature region (Musgrove et al., 1987; Martel et al., 1990; Zabaleta et al., 1992). The mature α and β subunits are about 50% identical to each other (Martel et al., 1990) and to GroEL (Hemmingsen et al., 1988) and the latter discovery led to the coining of the term 'chaperonin' to describe this family of proteins (Hemmingsen et al., 1988). The subunits from Pisum sativum chloroplasts run anomalously on SDS polyacrylamide gels; the slower α subunits have a slightly smaller mass (56 499 daltons) than the faster β subunits (57 945), and a ratio of acrylamide to bis-acrylamide of 100 is required to resolve them (Hemmingsen and Ellis, 1986). The α and β subunits are immunologically distinct (Musgrove et al., 1987), and have been expressed separately and together in E. coli (Cloney et al., 1992).

■ Plastid *CPN60* genes and sequences

Genomic clones have been isolated from Arabidopsis thaliana (Zabaleta et al., 1992) and Pisum sativum (S.P. Alldrick and R.J. Ellis, unpublished data). The β subunits are encoded by a small gene family, the sequenced members of which each contain 13 introns, while the only α gene sequenced from Pisum sativum contains no introns. Accession numbers for cDNAs: SWISSPROT, Triticum aestivum α, P08823; Arabidopsis thaliana β, P21240; Brassica napus α, P21239; Brassica napus β, P21241; Cyanidium caldarium, P28256. GenBank, Pisum sativum α, U21105; Pisum sativum β, U21139.

■ Biological activities of plastid Cpn60

This protein was discovered by its ability to bind noncovalently large subunits of rubisco newly synthesized by intact chloroplasts isolated from leaves of Pisum sativum (Barraclough and Ellis, 1980). These large subunits are subsequently transferred to the holoenzyme of rubisco, prompting the suggestion that this rubisco subunit binding protein is involved in the assembly of rubisco (Barraclough and Ellis, 1980), and can be regarded as the second example of a molecular chaperone (Musgrove and Ellis, 1986; Ellis, 1987). The higher plant rubisco, unlike that from bacteria, has resisted all attempts to either express it in E. coli or reassemble it from the denatured state (Gatenby and Ellis, 1990; Gutteridge and Gatenby, 1995), so there is no direct evidence that shows that the chloroplast Cpn60 is required for the assembly of plant rubisco. However, experiments with chloroplast extracts indicate that the transfer of newly synthesized rubisco large subunits from the binding protein to the holenzyme is inhibited by antisera to the binding protein (Cannon et al., 1986) and requires Mg–ATP (Milos and Roy, 1984).

The rubisco large subunit-binding protein also binds several chloroplast proteins, including rubisco small subunits, that are imported into isolated chloroplasts after synthesis by cytosolic ribosomes (Gatenby, 1996; Gutteridge and Gatenby, 1995), and occurs in non-photosynthetic plastids as well as in all chloroplasts studied (Hemmingsen and Ellis, 1986; Viitanen et al., 1995). Transgenic antisense plants expressing reduced levels of the β subunit show severe pleiotropic defects but still contain rubisco; the levels of α subunits in these plants were not reported (Zabaleta et al., 1994). The protein is abundant in chloroplasts, i.e. 12 μM in Pisum chloroplasts (Lennox and Ellis, 1986), and its amount increases only slightly after heat stress (Viitanen et al., 1995). By analogy with the Cpn60 of E. coli (GroEL), the rubisco subunit-binding protein probably functions to assist the correct folding of several plastid proteins, and should thus be regarded as a plastid chaperonin rather than one specific for rubisco. The significance of the divergent subunits is not clear, but may indicate a degree of specialization to assist the correct folding of the large subunit of chloroplast rubisco, which is not only very abundant but also unusually prone to self aggregation (Gatenby and Ellis, 1990).

Purified plastid Cpn60 forms a 1:1 complex with both GroES from Escherichia coli and the double-headed Cpn21 (see entry p. 194) from chloroplasts in the presence of either ADP or ATP. These binary complexes increase the yield of enzymatically active bacterial rubisco and mitochondrial malate dehydrogenase if present when these enzymes refold after removal from denaturing conditions (Viitanen et al., 1995). This assisted refolding requires ATP hydrolysis, as does that by chaperonins from eubacteria and mitochondria, but does not require K^+ ions. Another difference between the eubacterial and plastid Cpn60 proteins is that, in the presence of ATP, the latter dissociate in a reversible manner into monomers (Bloom et al., 1983; Hemmingsen and Ellis, 1986; Musgrove et al., 1987). This effect is dependent on the concentration of the protein, but is unlikely to have physiological relevance since it is observed only at concentrations well below those found inside chloroplasts (Lennox and Ellis, 1986; Lissin, 1995).

■ References

Barraclough, R. and Ellis, R.J. (1980). Protein synthesis in chloroplasts IX. Assembly of newly-synthesized large subunits into ribulose bisphosphate carboxylase in isolated intact pea chloroplasts. Biochim. Biophys. Acta **608**, 19–31.

Bloom, M.V., Milos, P. and Roy, H. (1983). Light-dependent assembly of ribulose-1,5-bisphosphate carboxylase. Proc. Natl. Acad. Sci. USA **80**, 1013–1017.

Cannon, S., Wang, P., and Roy, H. (1986). Inhibition of ribulose bisphosphate carboxylase assembly by antibody to a binding protein. J. Cell Biol. **103**, 1327–1335.

Cloney, L.P., Wu, H.B., and Hemmingsen, S.M. (1992). Expression of plant chaperonin-60 genes in *Escherichia coli*. J. Biol. Chem. **267**, 23327–23332.

Ellis, R.J. (1981). Chloroplast proteins: synthesis, transport and assembly. Annu. Rev. Plant Physiol. **32**, 111–137.

Ellis, R.J. (1987). Proteins as molecular chaperones. Nature **328**, 378–379.

Ellis, R.J. and van der Vies, S.M. (1988). The rubisco subunit binding protein. Photosyn.Res. **16**, 101–115.

Gatenby, A.A. (1996). The chaperonins of photosynthetic organisms. In *The Chaperonins* (Ellis, R.J., ed.), pp 65–90. Academic Press, Orlando.

Gatenby, A.A. and Ellis, R.J. (1990). Chaperone function: the assembly of ribulose bisphosphate carboxylase-oxygenase. Annu. Rev. Cell Biol. **6**, 125–149.

Gutteridge, S. and Gatenby, A.A. (1995). Rubisco synthesis, assembly, mechanism and regulation. Plant Cell **7**, 809–819.

Hemmingsen, S.M. and Ellis, R.J. (1986). Purification and properties of the ribulose bisphosphate carboxylase large subunit binding protein. Plant Physiol. **80**, 269–276.

Hemmingsen, S.M., Woolford, C., van der Vies, S.M., Tilly, K., Dennis, D.T., Georgopoulos, C.P., Hendrix, R.W., and Ellis, R.J. (1988). Homologous plant and bacterial proteins chaperone oligomeric protein assembly. Nature **333**, 330–334.

Lennox, C.R. and Ellis, R.J. (1986). The carboxylase large subunit binding protein: photoregulation and reversible dissociation. Biochem. Soc. Trans. **14**, 9–11.

Lissin, N.M. (1995). In vitro dissociation and self-assembly of three chaperonin 60s: the role of ATP. FEBS Letts. **361**, 55–60.

Martel, R., Cloney, L.P., Pelcher, L.E., and Hemmingsen, S.M. (1990). Unique composition of plastid chaperonin-60; α and β polypeptide-encoding genes are highly divergent. Gene **94**, 181–187.

Milos, P. and Roy, H. (1984). ATP-released large subunits participate in the assembly of RuBP carboxylase. J. Cell. Biochem. **24**, 153–162.

Musgrove, J.E. and Ellis, R.J. (1986). The rubisco large subunit binding protein. Phil. Trans. R. Soc. Lond. B **313**, 418–428.

Musgrove, J.E., Johnson, R.A., and Ellis, R.J. (1987). Dissociation of ribulose bisphosphate carboxylase large subunit binding protein into dissimilar subunits. Eur. J. Biochem. **163**, 529–534.

Pushkin, A.V., Tsuprun, V.L., Salovjeva, N.A., Shubin, V.V., Erstigneeva, Z.G., and Ktretovitch, W.L. (1982). High molecular weight pea leaf protein similar to the groE protein of *Escherichia coli*. Biochim. Biophys. Acta **704**, 379–384.

Viitanen, P.V., Schmidt, M., Buchner, J., Suzuki, T., Vierling, E., Dickson, R., Lorimer, G.H., Gatenby, A.A., and Soll, J. (1995). Functional characterisation of the higher plant chloroplast chaperonins. J. Biol. Chem. **270**, 18158–18164.

Zabaleta, E., Oropeza, A., Jimenez, B., Salerno, G., Crespi, M., and Herrera-Estrella, L. (1992). Isolation and characterisation of genes encoding chaperonin 60β in *Arabidopsis thaliana*. Gene **111**, 175–181.

Zabaleta, E., Oropeza, A., Assad, N., Mandel, A., Salerno, G., and Herrera-Estrella, L. (1994). Antisense expression of chaperonin 60β in transgenic tobacco plants leads to abnormal phenotypes and altered distribution of photoassimilates. Plant J. **6**, 425–432.

■ R. John Ellis:
Department of Biological Sciences
University of Warwick
Coventry CV4 7AL, UK
Tel. 44 1203 523509
Fax. 44 1203 523568
E-mail: je@dna.bio.warwick.ac.uk

Plastid Cpn21

Cpn21 is a nuclear encoded chloroplast cochaperonin with an unusual structural feature, consisting of two cpn10-like domains fused together in tandem. A transit peptide targets the polypeptide into chloroplasts where it can function in plastid protein folding. Cpn21 physically interacts with chloroplast Cpn60 or Escherichia coli GroEL to assist in the chaperonin-dependent refolding of denatured target proteins, and can also efficiently suppress groES *mutations in bacteria, despite its binary nature.*

■ Alternative names

Ch-cpn10 (Bertsch *et al.*, 1992), cpn24 (Chen and Jagendorf, 1994).

■ Cpn21 identification and isolation

Cpn21 was originally identified in pea (*Pisum sativum*) and spinach (*Spinacea oleracea*) chloroplasts (Bertsch et al., 1992). The pea protein was isolated as a factor that substitutes for the GroES cochaperonin in GroEL-dependent refolding of denatured ribulose-1,5-bisphosphate carboxylase (Goloubinoff et al., 1989), and the spinach Cpn21 identification was based on DNA sequence analysis and similarity of the deduced protein sequence with GroES and other co-chaperonins.

▬ CPN21 gene and sequence

The pea cpn21 DNA sequence (Bertsch et al., 1992) is deposited in the GenBank database (accession number M87646). Two important features were immediately apparent from analysis of the deduced amino acid sequence. The first is that an N-terminal extension of 53 amino acids is present that exhibits typical characteristics of a chloroplast transit peptide, suggesting that Cpn21 polypeptides are synthesized as higher molecular mass (26 872 Da) precursors in the plant cytosol, and that these are subsequently targeted to chloroplasts. This plastid localization was confirmed by import and processing of the Cpn21 precursor with isolated pea chloroplasts (Bertsch et al., 1992). The second unusual feature is that the 202 amino acid mature coding sequence is about twice the length of other CPN10 cochaperonins (Gatenby and Viitanen, 1994). Examination of deduced amino acid sequences indicate that the N or C halves of Cpn21 have 37 or 31% identity (58–60% similarity) to GroES, and the two halves exhibit 38% amino acid identity to each other (64% similarity). Cpn21 is highly polarized with a basic N-domain (pI 9.95) and an acidic C-domain (pI 4.43), compared with a pI of 4.99 for GroES (R. Dickson, unpublished data). Apparently, chloroplast Cpn21 evolved from some form of gene duplication and fusion event that brought two CPN10-like coding sequences together to form a two-domain polypeptide. This event occurred early in plant evolution, and has been highly conserved, since a Cpn21-type protein is present in a wide range of photosynthetic organisms including mosses, liverworts, ferns, gymnosperms and angiosperms (Baneyx et al., 1995). It is possible that smaller Cpn10 molecules may also coexist together with Cpn21 in chloroplasts (Hartman et al., 1992), but these have not been characterized in detail. Photosynthetic bacteria appear to posses the more orthodox Cpn10 cochaperonin (Gatenby, 1996).

Consistent with a domain fusion origin for Cpn21, a potential oligopeptide linker (TDDVKD) is located in the mature protein at T103 (Baneyx et al., 1995). The oligopeptide has a composition highly favorable for domain linkage, and is effectively positioned midway along the Cpn21 polypeptide to provide an internal symmetry axis around which a pseudosymmetric structure could form following collapse and folding of the two domains (Gatenby, 1996). Each of the two domains contain a sequence (V25–Q40; V123–K138) with similarity to the highly mobile polypeptide loop in GroES (Landry et al., 1993), and this, together with the location of eight residues found in most CPN10 proteins (Bertsch et al., 1992), suggests that each domain is functional. It is not clear why chloroplasts require a 'double-domain' cochaperonin. In contrast, plant mitochondria make use of the more usual CPN10-type of cochaperonin (Burt and Leaver, 1994; Gatenby, 1996). Cpn21 gene expression and protein accumulation do not appear to be significantly induced by heat stress and the protein is constitutively expressed in leaves (Viitanen et al., 1995).

▬ Cpn21 protein

Although purified mature Cpn21 analysed by SDS–PAGE has an apparent molecular mass of 24 kDa (Bertsch et al., 1992), the calculated molecular mass of the polypeptide from its DNA sequence is 21.4 kDa, resulting in its designation as Cpn21 (Baneyx et al., 1995; Gatenby, 1996). Although a partial purification of Cpn21 from pea chloroplasts is reported (Bertsch et al., 1992), purification to homogeneity and in large amounts required the construction of a spinach Cpn21 expression plasmid for use in E. coli (Baneyx et al., 1995; Viitanen et al., 1995). The initiator methionine of E. coli-expressed spinach Cpn21 is removed in vivo resulting in the production of a polypeptide identical to that found in chloroplasts. An active Cpn21 bearing an N-terminal polyhistidine sequence can also be purified by metal chelate affinity chromatography (Baneyx et al., 1995).

Despite the availability of significant amounts of purified spinach Cpn21, attempts to examine its structure have proven difficult. To keep the protein soluble at high concentrations (> 5 mg/ml) NaCl concentrations of 0.3 M or greater are needed, and the propensity for Cpn21 to dissociate or aggregate readily has made gel filtration sizing estimates inaccurate (Bertsch et al., 1992). The ease with which spinach Cpn21 substitutes for E. coli GroES in its interactions with GroEL suggests a high probability of sevenfold rotational symmetry and the potential to form toroidal structures, as observed for GroES (Chandrasekhar et al., 1986). Indeed, toroids of Cpn21 are observed using electron microscopy (Baneyx et al., 1995), although the resolution is not sufficient to provide information on subunit arrangements. Results from velocity sedimentation using histidine-tagged Cpn21 in an analytical ultracentrifuge give an S value of 6 (H. Chen and A.A. Gatenby, unpublished data).

▬ Cpn21 function

Cpn21 functions in vitro as a cochaperonin to facilitate successful discharge of chaperonin-bound denatured proteins and their partitioning to native states, with concomitant ATP hydrolysis (Bertsch et al., 1992; Baneyx et al., 1995; Viitanen et al., 1995). Cpn21 interacts physically with either GroEL from E. coli or Cpn60 from pea chloroplasts, with both α and β subunits of Cpn60 present in similar amounts when associated with Cpn21 (Baneyx et al., 1995). Asymmetric complexes form between Cpn21 and Cpn60 in the presence of ADP, to yield bullet-shaped particles (Viitanen et al., 1995), and these complexes can be isolated using gel filtration or affinity chromatography in the presence of adenine nucleotides (Bertsch et al., 1992; Baneyx et al., 1995). Although binding of Cpn21 to E. coli GroEL inhibits its ATPase activity (Baneyx et al., 1995; Bertsch and Soll, 1995), surprisingly the ATPase activity of chloroplast Cpn60 is not inhibited following its interaction with Cpn21 (H. Chen and A.A. Gatenby, unpublished data). Another unusual feature exhibited by the chloroplast chaperonins during assisted-folding reactions is that potassium ions are not required (Viitanen et al., 1995).

■ Mutagenesis studies with Cpn21

Three aspects of the Cpn21 protein have been examined by mutagenesis. To see if a single domain of the binary protein retains function, large deletions were used to remove either of the two domains (Baneyx et al., 1995). Site-specific mutations were introduced to test: (a) if highly conserved residues are functionally important (Bertsch and Soll, 1995), and (b) whether the two flexible loops in Cpn21 are required for activity (H. Chen and A. A. Gatenby, unpublished data). A useful feature of Cpn21 is that when expressed in E. coli cells that are groES defective the plant protein can substitute for GroES. Therefore, bacteriophages that are unable to assemble in groES mutant strains will do so if Cpn21 is expressed (Baneyx et al., 1995). This property allows the influence of mutations to be examined both in vivo for phage growth, and in vitro during refolding of target proteins, sometimes with conflicting results.

Baneyx et al. (1995) found that either the N- or C-domains alone are functional as cochaperones in vivo to allow phage λ growth in groES30 cells, but with different efficiencies. Native Cpn21 suppressed the mutation as effectively as GroES, but the individual N- or C-domains were, respectively, only 70 or 30% efficient in supporting λ growth, and failed to support λ or T5 growth in groES619 cells. This reduction is possibly the result of lower expression levels or instability of the heptamers. Purified individual domains do not function during in vitro folding reactions (Baneyx et al., 1995; Bertsch and Soll, 1995). Bertsch and Soll (1995) also tested a number of mutations in highly conserved residues and found that most single substitutions allowed efficient λ growth, but when a double mutation (G52I/G54I) was tested phage growth was reduced to less than 1%. Interestingly, all the single or double mutations were impaired during in vitro folding reactions and none can inhibit GroEL–ATPase activity. Recently, we have introduced substitutions into either or both of the flexible loops in Cpn21 (G32D, G130D) and find that all still permit λ growth in cells (H. Chen and A.A. Gatenby, unpublished data) but interfere to different extents with in vitro folding activity. In contrast to groES30, both λ and T5 fail to grow on groES30 cells if either one or both loops are mutated. The data indicate that λ morphogenesis in groES30 can tolerate significant disturbances to Cpn21 structure and still form viable phages, but that λ and T5 growth in groES619 and in vitro folding reactions are less forgiving to such perturbations.

■ References

Baneyx, F., Bertsch, U., Kalbach, C.E., van der Vies, S.M., Soll, J., and Gatenby, A.A. (1995). Spinach chloroplast cpn21 co-chaperonin possesses two functional domains fused together in a toroidal structure, and exhibits nucleotide-dependent binding to plastid chaperonin 60. J. Biol. Chem. **270**, 10695–10702.

Bertsch, U. and Soll, J. (1995). Functional analysis of isolated cpn10 domains and conserved amino acid residues in spinach chloroplast co-chaperonin by site-directed mutagenesis. Plant Mol. Biol. **29**, 1039–1055.

Bertsch, U., Soll, J., Seetharam, R., and Viitanen, P.V. (1992). Identification, characterization, and DNA sequence of a functional 'double' groES-like chaperonin from chloroplasts of higher plants. Proc. Natl. Acad. Sci. USA **89**, 8696–8700.

Burt, W.J.E. and Leaver, C.J. (1994). Identification of a chaperonin-10 homologue in plant mitochondria. FEBS Lett. **399**, 139–141.

Chandrasekhar, G.N., Tilly, K., Woolford, C., Hendrix, R., and Georgopoulos, C. (1986). Purification and properties of the groES morphogenetic protein of Escherichia coli. J. Biol. Chem. **261**, 12414–12419.

Chen, G.G. and Jagendorf, A.T. (1994). Chloroplast molecular chaperone-assisted refolding and reconstitution of an active multisubunit coupling factor CF1 core. Proc. Natl. Acad. Sci. USA **91**, 11497–11501.

Gatenby, A.A. (1996). Chaperonins of photosynthetic organisms. In The Chaperonins, (R.J. Ellis, ed.), pp. 65–90, Academic Press.

Gatenby, A.A. and Viitanen, P.V. (1994). Structural and functional aspects of chaperonin-mediated protein folding. Annu. Rev. Plant Physiol. Plant Mol. Biol. **45**, 469–491.

Goloubinoff, P., Christeller, J.T., Gatenby, A.A., and Lorimer, G.H. (1989). Reconstitution of active dimeric ribulose bisphosphate carboxylase from an unfolded state depends on two chaperonin proteins and Mg-ATP. Nature **342**, 884–889.

Hartman, D.J., Dougan, D., Hoogenraad, N.J., and Hoj, P.B. (1992). Heat shock proteins of barley mitochondria and chloroplasts. Identification of organellar hsp10 and 12: putative chaperonin 10 homologues. FEBS Lett. **305**, 147–150.

Landry, S.J., Zeilstra-Ryalls, J., Fayet, O., Georgopoulos, C., and Gierasch, L.M. (1993). Characterization of a functionally important mobile domain of GroES. Nature **364**, 255–258.

Viitanen, P.V., Schmidt, M., Buchner, J., Suzuki, T., Vierling, E., Dickson, R., Lorimer, G.H., Gatenby, A.A., and Soll, J. (1995). Functional characterization of the higher plant chloroplast chaperonins. J. Biol. Chem. **270**, 18158–18164.

■ Anthony A. Gatenby and Huanfeng Chen:
Central Research and Development
Experimental Station, P.O. Box 80328
DuPont
Wilmington, DE 19880-0328, USA
Tel. 1 302 695 7437
Fax. 1 302 695 1374
E-mail: gatenbaa@a1.esvax.umc.dupont.com

Mammalian Cpn60

Mammalian Cpn60 is a mitochondrial matrix protein that is a member of the CPN60/HSP60 protein family. Cpn60 and its cochaperone Cpn10 (Hsp10) constitute the chaperonin subclass of molecular chaperones, which, in mitochondria, facilitate folding of newly synthesized and imported proteins.

■ Alternative names

Hsp60 (Mizzen et al., 1989).

■ Cpn60 sequence

Amino acid sequences deduced from the nucleotide sequences of CPN60 cDNAs (Jindal et al., 1989; Peralta et al., 1990) (see also GenBank accession numbers M22882, M22383, X54793, X53585 and X53584) reveal that mammalian Cpn60 is encoded as a 573 amino acid precursor protein (p-Cpn60) from which a typical amphipathic N-terminal targeting signal is removed upon mitochondrial import. This generates a mature protein (Cpn60) of 547 amino acids with a predicted mass of c. 57.9 kDa (Peralta et al., 1993).

Members of the CPN60 family exhibit a high degree of sequence identity at the amino acid level. Thus, human Cpn60 shares 97% positional identity with rat liver Cpn60 (Peralta et al., 1990) whilst 40–50% positional identity is observed with Escherichia coli GroEL (Hemmingsen et al., 1988), chloroplast Cpn10 (Hemmingsen et al., 1988) and Saccharomyces cerevisiae Hsp60 (Reading et al., 1989) homologues. The predicted isoelectric points (pI) of various mammalian CPN60 family members are close to 6.0, whilst those of their precursors are slightly higher owing to the basic mitochondrial targeting signal (pI 6.5) (Peralta et al., 1990, 1993).

■ Cpn60 protein

Cpn60 has been isolated from a number of different mammalian sources including human lymphocytes (Waldinger et al., 1988), HeLa cells (Mizzen et al., 1989), bovine kidney (Ross et al., 1992) and porcine liver (Itoh et al., 1995). It has consistently been characterized as a mitochondrial matrix protein, although additional cellular locations have been suggested.

Cpn60 homologues display a number of common characteristics that include: induction of their synthesis by heat shock and other stress conditions (Mizzen et al., 1989); homo-oligomeric structures (chloroplast cpn60 is a hetero-oligomer, see Hemmingsen et al., 1988); a weak ATPase activity that is regulated by the cochaperone Cpn10 (a GroES homologue); and an essential but transient function in protein folding (Martinus et al., 1995).

Mammalian Cpn60 migrates as a 58 kDa monomer upon SDS–PAGE, but migrates as a typical heptamer under non-denaturing conditions (Peralta et al., 1993). The protein has been isolated as a homo-oligomeric heptamer (Viitanen et al., 1992) which is likely to associate further into a tetradecamer during its functional cycle. Cpn60 homologues from most sources, such as GroEL from E. coli, form mono-oligomeric tetradecamers (Hendrix, 1979). Whilst a three dimensional structure has yet to be reported for any eukaryotic Cpn60 homologue, it is likely to be highly similar to that of E. coli GroEL (Braig et al., 1994).

■ *CPN60* gene regulation

Cpn60 is constitutively expressed under normal cell growth conditions and is further inducible by a large number of stress conditions (Mizzen et al., 1989; Martinus et al., 1995). The mammalian chaperonin genes almost certainly are subject to modulation by a heat shock transcription factor. The genes encoding rat Cpn10 and Cpn60 are subject to regulation by a shared heat shock element (Ryan et al., 1997).

■ Function of Cpn60

Proteins synthesized within or entering mitochondria must fold in an environment with extremely high protein concentration and are therefore prone to aggregation; an undesirable side reaction that is suppressed by chaperones including Cpn60 (Gething and Sambrook, 1992; Martinus et al., 1995). Likewise, Cpn60 and Cpn10 are likely to rescue proteins that denature spontaneously within the mitochondrion (Martinus et al., 1995).

A mammalian Cpn60 expressed in E. coli was employed to refold chemically denatured ribulose-1,5-bisphosphate carboxylase in vitro (Hendrix, 1979). This process required ATP, K^+ and Cpn10. Substitution of mammalian Cpn10 with E. coli GroES abolished protein folding.

Like its prokaryotic and yeast counterparts, mammalian Cpn60 is thought to assist protein folding by binding unfolded proteins and thus preventing their aggregation. This process is thought to occur through subsequent rounds of substrate binding and release until the target protein has folded successfully without any steric 'input' from the chaperone, which merely acts as an 'Anfinsen test tube' within the cell. Neither the structural motifs in unfolded proteins that are recognised by Cpn60, nor the exact stoichiometry of ATP hydrolysis and other kinetic parameters associated with the chaperonin-mediated protein folding process are understood as yet. New mechanistic insights are likely to arise following the determination of the *E. coli* GroEL structure (Braig *et al.*, 1994).

■ References

Braig, K., Otwinowski, Z., Hedge, R., Bolsvert, D.C., Joachimiak, A., Horwich, A., and Sigler, P.B. (1994). The crystal structure of the bacterial chaperonim GroEL at 2.8 Å. Nature **371**, 578–586.

Gething, M.-J. and Sambrook, J. (1992). Protein folding in the cell. Nature **355**, 33–45.

Hemmingsen, S.M., Woolford, C., van der Vies, S.M., Tilly, K., Dennis, D.T., Georgopolous, C.P., Hendrix, R.W., and Ellis, R.J. (1988). Homologous plant and bacterial proteins chaperone oligomeric protein assembly. Nature **333**, 330–334.

Hendrix, R.W. (1979). Purification and properties of groE, a host protein involved in bacteriophage assembly. J. Mol. Biol. **129**, 375–392.

Itoh, H., Kobayashi, R., Wakui, H., Komatsuda, A., Ohtani, H., Miura, A.B., Otaka, M., Masamune, O., Andoh, H., Komaya, K., Sato, Y., and Tashima, Y. (1995). Mammalian 60-kDa stress protein (chaperonin homolog). J. Biol. Chem. **270**, 13429–13435.

Jindal, S., Dudani, A.K., Singh, B., Harley, C.B., and Gupta, R.S. (1989). Primary structure of a human mitochondrial protein homologous to the bacterial and plant chaperonins and to the 65-kilodalton mycobacterial antigen. Mol. Cell. Biol. **9**, 2279–2283.

Martinus, R.M., Ryan, M.T.R., Nayor, D.J., Herd, S.M., Hoogenraad, N.J., and Høj, P.B. (1995). Role of chaperones in the biogenesis and maintenance of the mitochondrion. FASEB J. **9**, 371–370.

Mizzen, L.A., Chang, C., Garrels, J.L., and Weich, W.J. (1989). Identification, characterisation, and purification of two mammalian stress proteins present in mitochondria, grp75, a member of the hsp70 family and hsp58, a homolog of the bacterial groEL protein. J. Biol. Chem. **264**, 20664–20675.

Peralta, D., Hartman, D.J., McIntosh, A.M., Hoogenraad, N.J., and Høj, P.B. (1990). cDNA and deduced amino acid sequence of a rat liver prehsp60 (chaperonin-60). Nucl. Acids Res. **18**, 7162.

Peralta, D., Lithgow, T., Hoogenraad, N.J., and Høj, P.B. (1993). Prechaperonin 60 and ornithine transcarbamylase share components of the import apparatus but have distinct maturation pathways in rat liver mitochondria. Eur. J. Biochem. **211**, 881–889.

Reading, D.S., Hallberg, R.L., and Myers, A.M. (1989). Characterisation of the yeast *HSP60* gene coding for a mitochondrial assembly factor. Nature **337**, 655–659.

Ross, W.R., Bertrand, W.S., and Morrison, A.R., (1992). Identification of a processed protein related to the human chaperonins (hsp 60) protein in mammalian kidney. Biochem. Biophys. Res. Commun. **185**, 683–687

Ryan, M., Herd, S.M., Sberna, G., Samuel, M., Hoogenraad, N.J., and Høy, P.B. (1997). The genes encoding mammalian chaperonin 60 and chaperonin 10 are linked head-to-head and share a bidirectional promotor. Gene In press (July 1997).

Waldinger, D., Eckerskorn, C., Lottspeich, F., and Cleve, H. (1988). Amino-acid sequence homology of a polymorphic cellular protein from human lymphocytes and chaperonins from *Escherichia coli* (groEL) and chloroplasts (Rubisco-binding protein). Biol. Chem. **369**, 1185–1189.

Viitanen, P.V., Lorimer, G.H., Seetharam, R., Gupta, R.S., Oppenheim, J., Thomas, J.O., and Cowan, N.J. (1992). Mammalian mitochondrial chaperonin 60 function as a single toroidal ring. J. Biol. Chem. **267**, 695–698.

■ Peter B. Høj:
Department of Horticulture, Viticulture and Oenology
The University of Adelaide
PMBI
Glen Osmond,
South Australia 5064, Australia
Tel. 61 8 8303 6663
Fax. 61 8 8303 7117
E-mail: phoj@waite.adelaide.edu.au

■ Nicholas J. Hoogenraad:
School of Biochemistry
La Trobe University
Dundoora
Victoria 3083, Australia
Tel. 61 3 9479 2196
Fax. 61 3 9479 2467
E-mail: N.Hoogenraad@latrobe.edu.au

■ Dadna Hartman:
Victorian Institute of Animal Science
475–485 Mickleham Rd.
Attwood
Victoria 3049, Australia
Tel. 61 3 9217 4349
Fax. 61 3 9217 4299
E-mail: hartmand@woody.as-vic.gov.au

Mammalian Cpn10

Mammalian Cpn10, the GroES homolog of mammalian mitochondria, is a member of the CPN10 chaperonin family. It is a nuclear-encoded protein that consists of seven identical c. 10 kDa subunits, organized in a single ring. Together with the mammalian mitochondrial Cpn60, its intracellular function is to assist in the folding of newly synthesized and imported mitochondrial proteins. In addition to its role as a molecular chaperone, the mammalian Cpn10 is also an extracellular growth factor, known as early pregnancy factor.

■ Alternative names

Mammalian mt-cpn10 (Lubben et al., 1990); heat shock protein 10 (HSP10), (Hartman et al., 1992); early pregnancy factor (EPF) (Cavanagh and Morton, 1994).

■ Mammalian *CPN10* gene sequence and regulation

The nucleotide sequences of mammalian *CPN10* cDNAs cloned from rat (Ryan et al., 1994), cow (Pilkington and Walker, 1993), mouse (Dickson et al., 1994) and human (Monzini et al., 1994) (GenBank accession numbers X71429, X69556, U09659 and X75821, respectively) predict proteins of 102 amino acid residues (calculated molecular mass of c.10.8 kDa). All four of these proteins exhibit limited homology to *Escherichia coli* GroES (Chandrasekhar et al., 1986) (c. 33% identity), but are virtually identical to each other. Indeed, the Cpn10 proteins of human, rat and cow differ at only one residue (Fig. 1). This extreme conservation extends to the N-terminus of the protein, which is predicted to form a positively charged amphiphilic helix (Ryan et al., 1994) and constitutes a mitochondrial targeting sequence (Jarvis et al., 1995). However, in contrast to most other mitochondrially imported proteins, the targeting sequence of the mammalian Cpn10 is not cleaved (Ryan et al., 1994). While it is currently debated whether GroES can bind adenine nucleotides (Todd et al., 1995), a putative nucleotide-binding domain has been identified in the rat Cpn10 (Hartman et al., 1993).

The levels of rat (Hartman et al., 1992) and human (Monzini et al., 1994) Cpn10 proteins are elevated under certain stress conditions, including heat shock.

■ Mammalian Cpn10 protein

Mammalian Cpn10 was first identified in bovine mitochondrial extracts (Lubben et al., 1990) by exploiting its ability to form an ATP-dependent isolatable complex

```
                      1         10        20        30        40        50
     E. coli          MNIRPLHDRVIVKRKEVETKSAGGIVLTGSAAAKSTRGEVLAVGNGRILE
     mouse            MAGQAFRKFLPLFDRVLVERSAAETVTKGGIMLPEQSQGKVLQATVVAVGSGGKGK
     rat              MAGQAFRKFLPLFDRVLVERSAAETVTKGGIMLPEQSQGKVLQATVVAVGSGGKGK
     cow              MAGQAFRKFLPLFDRVLVERSAAETVTKGGIMLPEQSQGKVLQATVVAVGSGSKGK
     human            MAGQAFRKFLPLFDRVLVERSAAETVTKGGIMLPEQSQGKVLQATVVAVGSGSKGK
                                                                              *

                                60        70        80        90       100
     E. coli          NGEVKPLDVKVGDIVIFNDGYGVKSEKIDNEE-VLI-MSESDILAIVEA
     mouse            SGEIEPVSVKVGDKVLLPE-YGG--TKVVLDDKDYFLFRDSDILGKYVD
     rat              GGEIQPVSVKVGDKVLLPE-YGG--TKVVLDDKDYFLFRDGDILGKYVD
     cow              GGEIQPVSVKVGDKVLLPE-YGG--TKVVLDDKDYFLFRDGDILGKYVD
     human            GGEIQPVSVKVGDKVLLPE-YGG--TKVVLDDKDYFLFRDGDILGKYVD
                      *   *                                     *
```

Figure 1. Comparion of the deduced amino acid sequences of CPN10 proteins. Shown are the sequences of the *E. coli* GroES protein (Chandrasekhar et al., 1986) and mammalian Cpn10 proteins from mouse (Dickson et al., 1994), rat (Ryan et al., 1994), cow (Pilkington and Walker, 1993) and human (Monzini et al., 1994). The initiator Met residues of the mammalian Cpn10s (underlined) are are lost during expression in *E. coli*. Asterisks denote the only four amino acid residues that are not identical for all four mammalian Cpn10 proteins.

with GroEL, the *E. coli* CPN60. Previous studies had shown that GroEL and GroES only interact with each other in the presence of adenine nucleotides (Chandrasekhar et al., 1986; Viitanen et al., 1990). The eukaryotic GroES homolog identified in this manner (Lubben et al., 1990) consisted of c. 10 kDa subunits, and possessed several properties that are hallmarks of GroES, including the ability to inhibit the ATPase activity of GroEL (Chandrasekhar et al., 1986; Viitanen et al., 1990) and to assist GroEL in the *in vitro* refolding of rubisco (Viitanen et al., 1990). Partial sequence analysis of the purified protein subsequently established that it was a bona fide GroES homolog (Bertsch et al., 1992). Using a different approach, Høj and co-workers (Hartman et al., 1992) identified a 10 kDa stress-inducible protein in cultured rat hepatoma cells and went on to purify it from rat liver mitochondria. This protein was also functionally compatible with GroEL and the authors reported its entire amino acid sequence (Hartman et al., 1993). The apparent molecular mass of the native rat Cpn10 protein was c. 65 kDa (Hartman et al., 1992) suggesting that, like GroES (Chandrasekhar et al., 1986), it is a single heptameric toroid.

In contrast to GroEL, mammalian Cpn60 is far more discriminating and can only function with its natural Cpn10 partner (Viitanen et al., 1992). When purified from mitochondria, the yeast (Rospert et al., 1993), cow (Lubben et al., 1990), rat (Hartman et al., 1992) and human (Cavanagh and Morton, 1994) Cpn10 proteins are blocked at their N-terminus, presumably owing to acetylation, which has been demonstrated for the rat protein (Hartman et al., 1993). Since N-terminal acetylation is thus far unique to the mitochondrial Cpn10 proteins, it was initially suggested that this modification might be necessary for interaction with the mitochondrial Cpn60s (Hartman et al., 1993). However, this hypothesis is incorrect since a non-acetylated recombinant mouse Cpn10 can still assist the mammalian Cpn60 in the facilitation of rubisco folding (Dickson et al., 1994). Thus, the unique structural features that enable mammalian Cpn10 to interact with its cognate partner remain to be determined. In this regard, it is of interest to note that all of the mammalian Cpn10s are extremely basic proteins with predicted isoelectric points greater than 9.0. In contrast, GroES and most other procaryotic Cpn10 proteins have predicted pIs that are less than 6.0.

■ Function of mammalian Cpn10

The chaperonins of *Escherichia coli* facilitate the folding of a wide variety of structurally unrelated proteins (Gatenby and Viitanen, 1994). Under 'non-permissive' conditions where spontaneous folding does not occur, the recovery of a correctly folded protein substrate from its complex with GroEL requires ATP hydrolysis and GroES (Schmidt et al., 1994). The latter is thought to coordinate the hydrolysis of ATP in the two GroEL toroids, in a manner that enhances the efficiency of protein release (Todd et al., 1994). Considering that mammalian Cpn10 can substitute for GroES in *in vitro* folding assays with GroEL (Lubben et al., 1990; Hartman et al., 1992; Dickson et al., 1994; Legname et al., 1995) it must subserve the same function as its bacterial counterpart. Accordingly, it has been demonstrated that, in addition to ATP hydrolysis, the mammalian Cpn10 is also critically required for the productive release of rubisco from the mammalian Cpn60 (Viitanen et al., 1992; Dickson et al., 1994). As already noted, GroES *cannot* substitute in this reaction.

Unexpectedly, and apart from its role as a GroES homolog, mammalian Cpn10 is also an extracellular growth factor, referred to as early pregnancy factor (EPF) (Cavanagh and Morton, 1994 and references cited therein). EPF, a potent immunosuppressant, is amongst the first substances to appear in the serum of pregnant women. Elevated EPF levels are also associated with various cancerous states. The physiological significance of the dual role of Cpn10/EPF is not yet understood.

■ References

Bertsch, U., Soll, J., Seetharam, R., and Viitanen, P.V. (1992). Identification, characterization, and DNA sequence of a functional 'double' groES-like chaperonin from chloroplasts of higher plants. Proc. Natl. Acad. Sci. USA **89**, 8696–8700.

Cavanagh, A.C. and Morton, H. (1994). The purification of early-pregnancy factor to homogeneity from human platelets and identification as chaperonin 10. Eur. J. Biochem. **222**, 551–560.

Chandrasekhar, G.N., Tilly, K., Woolford, C., Hendrix, R., and Georgopoulos, C. (1986). Purification and properties of the groES morphogenetic protein of *Escherichia coli*. J. Biol. Chem. **261**, 12414–12419.

Dickson, R., Larsen, B., Viitanen, P.V., Tormey, M.B., Geske, J., Strange, R., and Bemis, L.T. (1994). Cloning, expression, and purification of a functional nonacetylated mammalian mitochondrial chaperonin 10. J. Biol. Chem. **269**, 26858–26864.

Gatenby, A.A. and Viitanen, P.V. (1994). Structural and functional aspects of chaperonin-mediated protein folding. Annu. Rev. Plant Physiol. Plant Mol. Biol. 4b5, 469–491.

Hartman, D.J., Hoogenraad, N.J., Condron, R., and Høj, P.B. (1992). Identification of a mammalian 10-kDa heat shock protein, a mitochondrial chaperonin 10 homologue essential for assisted folding of trimeric ornithine transcarbamoylase in vitro. Proc. Natl. Acad. Sci. USA **89**, 3394–3398.

Hartman, D.J., Hoogenraad, N.J., Condron, R., and Høj, P.B. (1993). The complete primary structure of rat chaperonin 10 reveal a putative $\beta\alpha\beta$ nucleotide-binding domain with homology to p21ras. Biochim. Biophys. Acta **1164**, 219–222.

Jarvis, J.A., Ryan, M.T., Hoogenraad, N.J., Craik, D.J., and Høj, P.B. (1995). Solution structure of the acetylated, noncleavable mitochondrial targeting signal of rat chaperonin 10. J. Biol. Chem. **270**, 1323–1331.

Legname, G., Fossati, G., Gromo, G., Monzini, N., Marcucci, F., and Modena, D. (1995). Expression in *Escherichia coli*, purification and functional activity of recombinant human chaperonin 10. FEBS Lett. **361**, 211–214.

Lubben, T.H., Gatenby, A.A., Donaldson, G.K., Lorimer, G.H., and Viitanen, P.V. (1990). Identification of a groES-like chaperonin in mitochondria that facilitates protein folding. Proc. Natl. Acad. Sci. USA **87**, 7683–7687.

Monzini, N., Legname, G., Marcucci, F., Gromo, G., and Modena, D. (1994). Identification and cloning of human chaperonin 10 homologue. Biochim. Biophys. Acta **1218**, 478–480.

Pilkington, S.J. and Walker, J.E. (1993). PCR cloning of a bovine GroES homolog. J. DNA Seq. Mapping **3**, 291–295.

Ryan, M.T., Hoogenraad, N.J., and Høj, P.B. (1994). Isoloation of a cDNA clone specifying rat chaperonin 10, a stress-inducible mitochondrial matrix protein synthesised without a cleavable presequence. FEBS Lett. **337**, 152–156.

Rospert, S., Glick, B.S., Jeno, P., Schatz, G., Todd, M.J., Lorimer, G.H., and Viitanen, P.V. (1993). Identification and functional analysis of chaperonin 10, the groES homolog from yeast mitochondria. Proc. Natl. Acad. Sci. USA **90**, 10967–10971.

Schmidt, M., Buchner, J., Todd, M. J., Lorimer, G.H., and Viitanen, P.V. (1994). On the role of groES in the chaperonin-assisted folding reaction. J. Biol. Chem. **269**, 10304–10311.

Todd, M.J., Viitanen, P.V., and Lorimer, G.H. (1994). Dynamics of the chaperonin ATPase cycle: Implications for facilitated protein folding. Science **265**, 659–666.

Todd, M.J., Boudkin, O., Freire, E., and Lorimer, G.H. (1995). GroES and the chaperonin-assisted protein folding cycle: GroES has no affinity for nucleotides. FEBS Lett. **359**, 123–125.

Viitanen, P.V., Lubben, T.H., Reed J., Goloubinoff, P., O'Keefe, D.P., and Lorimer, G.H. (1990). Chaperonin-facilitated refolding of ribulose bisphosphate carboxylase and ATP hydrolysis by chaperonin 60 (groEL) are K$^+$ dependent. Biochemistry **29**, 5665–5671.

Viitanen, P.V., Lorimer, G.H., Seetharam, R., Gupta, R.S., Oppenheim, J., Thomas, J.O., and Cowan, N.J. (1992). Mammalian mitochondrial chaperonin 60 functions as a single toroidal ring. J. Biol. Chem. **267**, 695–698.

■ *Paul V. Viitanen:*
Molecular Biology Division
Central Research and Development Department
E. I. DuPont de Nemours and Company
Experimental Station
Wilmington, DE 19880-0402, USA
Tel. 1 302 695 7032
Fax. 1 302 695 4509
E-mail: ViitanPV@esvax.dnet.dupont.com

The neuroendocrine polypeptide 7B2

The 25–29 kDa 7B2 protein displays a widespread neuroendocrine-specific tissue distribution, is localized to secretory granules and is C-terminally processed to an 18–21 kDa secretory product. Intact 7B2 inhibits, while processed 7B2 can enhance the activity of the prohormone convertase PC2 (but not of the related PC1/PC3 enzyme). Transfected 7B2 facilitates the transport and maturation of transfected proPC2. In the endoplasmic reticulum, 7B2 tightly associates with proPC2 and in the trans-Golgi network/immature secretory granule it is cleaved and dissociates from the proenzyme, thus allowing proPC2 maturation. 7B2 appears to be a neuroendocrine chaperone in the secretory pathway, with PC2 as its specific physiological target.

■ Alternative names

APPG (anterior pituitary pig; Hsi *et al.*, 1982); secretogranin V (Huttner *et al.*, 1991).

■ Isolation of 7B2

The 7B2 protein was first isolated from porcine and human pituitary glands by HPLC, and the first 50 (porcine) and 77 (human) amino acid residues were determined by N-terminal amino acid sequence analysis (Hsi *et al.*, 1982; Seidah *et al.*, 1983).

■ 7B2 sequence

The nucleotide sequences of 7B2 cDNAs from human (Martens, 1988), porcine (Brayton *et al.*, 1988), *Xenopus laevis* (Martens *et al.*, 1989), mouse (Mbikay *et al.*, 1989), rat and salmon (Waldbieser *et al.*, 1991) (GenBank accession numbers, except for salmon, are Y00757, M23654, X15680/X14628, X15830, M63901, respectively) predict proteins of 207–212 amino acid residues (calculated molecular mass of c. 23 kDa) which includes a 22–26 amino acid signal peptide sequence (Fig. 1). The overall amino acid sequence identity among the 7B2 proteins of these species is 71–99% (signal peptide sequences not included), with the N-terminal 76 amino acids displaying 90–100% sequence identity and the 106–113 C-terminal amino acids 64–99% identity. The N- and C-terminal halves of 7B2 are distantly related to chaperonins-60/-10 (Braks and Martens, 1994) and members of the potato inhibitor I family (Martens *et al.*, 1994), respectively; the region of chaperonin-60 related to 7B2 (32% identity; 62% similarity) is also related to chaperonin-10 (38% identity; 62% similarity). Two 7B2 mRNAs have been identified in a number of mammalian species (as a result of alternative splicing; Paquet *et al.*, 1991a) and in *Xenopus* (two expressed genes; Braks *et al.*, 1996a). The (partial) structural organization of the 7B2 gene is known

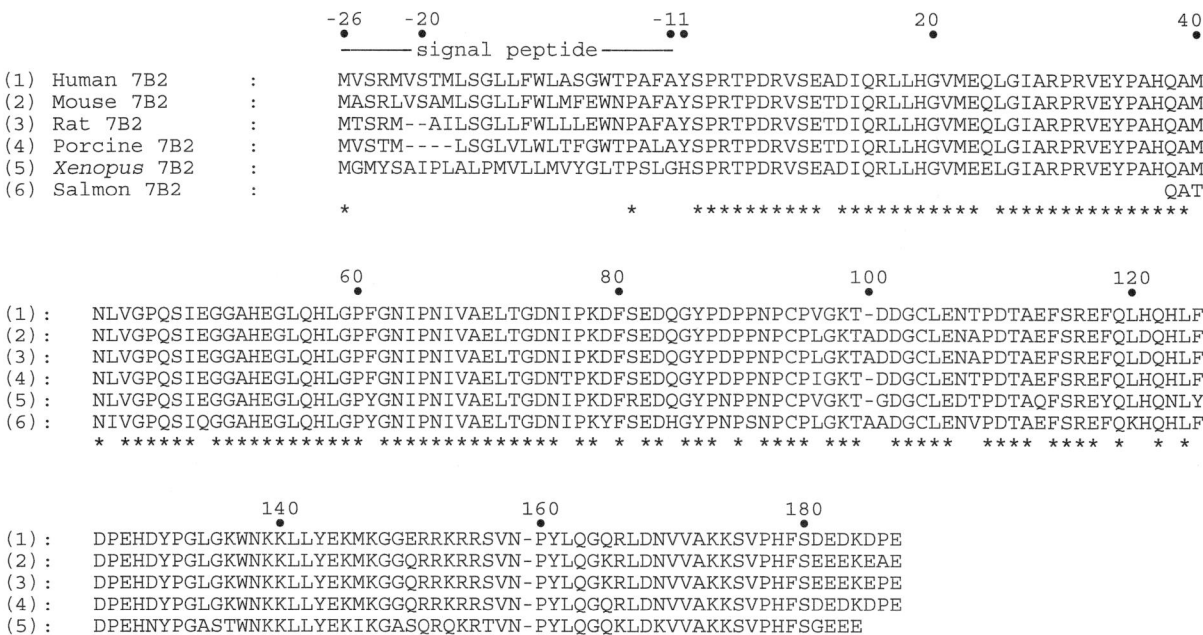

Figure 1. Comparison of the deduced amino acid sequences of 7B2 proteins. Shown are 7B2 proteins from human (Martens, 1988), mouse (Mbikay et al., 1989), rat (Waldbieser et al., 1991), porcine (Brayton et al., 1988), *Xenopus* (Martens et al., 1989) and salmon (Waldbieser et al., 1991). The one-letter amino acid notation is used. Gaps (—) have been introduced to achieve maximum identity.

in human (Mattei et al., 1990; Braks et al., 1996a; GenBank accession number X94303), rat (Waldbieser et al., 1991; accession number M63965) and *Xenopus* (Braks et al., 1996a). The 7B2 gene (locus SGNE-1 for secretory granule neuroendocrine protein-1) has been mapped to human chromosome region 15q13–14 (Roebroek et al., 1989a; Mattei et al., 1990) and mouse chromosome 2 E3-F3 (Mattei et al., 1990).

■ 7B2 protein

7B2 is a soluble, regulated secretory protein exhibiting a widespread distribution in the central nervous system and in endocrine tissues, with the highest level in the pituitary, and it is localized in dense core secretory granules (Mbikay et al., 1991 and references cited therein). Based on its expression profile, its acidic nature (pI~5) and the presence of multiple dibasic residues, 7B2 has been classified as a member of the chromogranin/secretogranin family (Huttner et al., 1991); no similarity is found between the primary structures of 7B2 and the other granins. The two (salmon), three (human, porcine and *Xenopus*) or four (mouse and rat) pairs of basic amino acids in the 7B2 sequence are potential cleavage sites for proprotein convertases (Fig. 1). In *Xenopus* intermediate pituitary, 25 kDa 7B2 is processed at the Lys^{139}–Lys^{140} site to an 18 kDa secretory product (Ayoubi et al., 1990). In the mammalian anterior pituitary, 29 kDa 7B2 is processed at Arg^{151}–Arg–Lys–Arg–Arg^{155}, possibly by the proprotein convertase furin (Paquet et al., 1994), which results in 23 kDa 7B2 (Lazure et al., 1991) and a C-terminal 31 amino acid 7B2 fragment purified from the culture media of AtT20 cells (Paquet et al., 1991b). A tridecapeptide, corresponding to the C-terminal end of 7B2 and resulting from cleavage at Lys^{173}–Lys^{174}, has been isolated from bovine adrenal medulla chromaffin vesicles. In this peptide, Ser-180 is phosphorylated (Sigafoos et al., 1993). Mouse 7B2 is sulfated prior to its processing (Paquet et al., 1994). The processed forms of porcine and human 7B2 are not glycosylated (Seidah et al., 1983); the known 7B2 sequences do not contain a potential site for N-glycosylation. From tissues 7B2 is isolated in the processed form, behaving as a dimer on gel permeation chromatography. Polyclonal antibodies have been raised against the N-terminal region (residues 23–39; Seidah et al., 1983; Zhu and Lindberg, 1995) and the C-terminal region of 7B2 (residues 156–187, Paquet et al., 1991b; residues 172–187, Zhu et al., 1996). Monoclonal antibodies against recombinant 7B2 have also been generated (Mon144, residues 64–94; Mon100/102, residues 129–144; Van Duijnhoven et al., 1991).

Biological activity of 7B2

The C-terminal 7B2 fragment (residues 175–187) induces membrane depolarization in rat supraoptic neurosecretory neurons (Senatorov et al., 1993). Recombinant, intact, but not processed, 7B2 is a potent inhibitor (half-maximal inhibition at nanomolar concentrations) of prohormone convertase PC2 enzyme activity. The PC2-related convertase PC1 (also referred to as PC3) is not inhibited by either form of 7B2 (Martens et al., 1994). A synthetic C-terminal 7B2 peptide (31-mer) represents a potent PC2 inhibitor (Lindberg et al., 1995). Moreover, intact 7B2 and the synthetic peptide can block the in vitro conversion of proPC2 to PC2 (Martens et al., 1994; Lindberg et al., 1995). 7B2 transfected into pituitary-derived AtT20 cells enhances the transport and maturation of transfected proPC2 (Zhu and Lindberg, 1995). Transfection of the processed form of 7B2 into CHO cells expressing proPC2, increases enzymatic activity (Zhu and Lindberg, 1995). In vitro, the processed form of recombinant 7B2 enhances PC2 enzyme activity (Braks and Martens, 1995).

Regulation of 7B2 synthesis and secretion

The 7B2 gene is differentially expressed in human lung cancer cells (Roebroek et al., 1989b) and hypothalami of Prader–Willi syndrome patients (Gabreëls et al., 1994). 7B2 protein levels are elevated in tissues and plasma of patients with various types of endocrine tumors, and 7B2 has been proposed as a specific neuroendocrine tumor marker (Suzuki et al., 1986).

Biological interactions

Intact 7B2 specifically interacts with the proform of PC2 in the early stages of the secretory pathway, presumably in a 1:1 ratio (Braks and Martens, 1994), and not with other proprotein convertases like PC1, furin, PACE4 and PC5 (Benjannet et al., 1995a). After 7B2 cleavage and dissociation of the N-terminal fragment in the TGN/immature secretory granule, the C-terminal fragment of 7B2 may remain associated with proPC2, thereby preventing autocatalytic conversion of the proenzyme until the appropriate site for activation in the secretory pathway is reached (Braks et al., 1996b; Zhu et al., 1996). Site-directed mutagenesis revealed that the oxyanion hole residue Asp-309 of PC2 is essential for its interaction with 7B2 (Benjannet et al., 1995b) and that a short segment within the C-terminal region of 7B2, in particluar Lys^{173}–Lys^{174}, is critical for PC2 inhibition (Van Horssen et al., 1995).

In the yeast two-hybrid system, 7B2 fused to a DNA-binding domain was found to activate transcription through two independent transcriptional activation domains (residues 8–92 and 93–145; Chaudhuri et al., 1995).

References

Ayoubi, T.A.Y., Van Duijnhoven, H.L.P., Van de Ven, W.J.M., Jenks, B.G., Roubos, E.W., and Martens, G.J.M. (1990). The neuroendocrine polypeptide 7B2 is a precursor protein. J. Biol. Chem. **265**, 15644–15647.

Benjannet, S., Savaria, D., Chrétien, M., and Seidah, N.G. (1995a). 7B2 is a specific intracellular binding protein of the prohormone convertase PC2. J. Neurochem. **64**, 2303–2311.

Benjannet, S., Lusson, J., Hamelin, J., Savaria, D., Chrétien, M., and Seidah, N.G. (1995b). Structure-function studies on the biosynthesis and bioactivity of the precursor convertase PC2 and the formation of the 7B2/PC2 complex. FEBS Lett. **362**, 151–155.

Braks, J.A.M. and Martens, G.J.M. (1994). 7B2 is a neuroendocrine chaperone that transiently interacts with prohormone convertase PC2 in the secretory pathway. Cell **78**, 263–273.

Braks, J.A.M. and Martens, G.J.M. (1995). The neuroendocrine chaperone 7B2 can enhance in vitro POMC cleavage by prohormone convertase PC2. FEBS Lett. **371**, 154–158.

Braks, J.A.M., Broers, C.A.M., Danger, J.-M.H.A., and Martens, G.J.M. (1996a). Structural organisation of the gene encoding the neuroendocrine chaperone 7B2. Eur. J. Biochem. **236**, 60–67.

Braks, J.A.M., Van Horssen, A.M. and Martens, G.J.M. (1996b). Dissociation of the complex between the neuroendocrine chaperone 7B2 and prohormone convertase PC2 is not associated with proPC2 maturation. Eur. J. Biochem. **238**, 505–510.

Brayton, K.A., Aimi, J., Qiu, H., Yazdanparast, R., Gatei, M.A., Polak, J.M., Bloom, S.R., and Dixon, J.E. (1988). Cloning, characterization, and sequencing of a porcine cDNA encoding a secreted neural and endocrine protein. DNA **7**, 713–719.

Chaudhuri, B., Huijbrechts, R.P.H., Coen, J.J., and Fürst, P. (1995). The neuroendocrine protein 7B2 contains unusually potent transcriptional activating sequences. Biochem. Biophys. Res. Commun. **216**, 1–10.

Gabreëls, B.A.Th.F., Swaab, D.F., Seidah, N.G., Van Duijnhoven, H.L.P., Martens, G.J.M., and Van Leeuwen, F.W. (1994). Differential expression of the neuroendocrine polypeptide 7B2 in hypothalami of Prader–(Labhart)–Willi syndrome patients. Brain Res. **657**, 281–293.

Hsi, K.L., Seidah, N.G., De Serres, G., and Chrétien, M. (1982). Isolation and NH_2-terminal sequence of a novel porcine pituitary polypeptide. FEBS Lett. **147**, 261–266.

Huttner, W.B., Gerdes, H.-H., and Rosa, P. (1991). The granin (chromogranin/secretogranin) family. Trends Biochem. Sci. **16**, 27–30.

Lazure, C., Benjannet, S., Seidah, N.G., and Chrétien, M. (1991). Processed forms of neuroendocrine proteins 7B2 and secretogranin II are found in porcine pituitary extracts. Int. J. Peptide Protein Res. **38**, 392–400.

Lindberg, I., Van den Hurk, W.H., Bui, C., and Batie, C.J. (1995). Enzymatic characterization of immunopurified prohormone convertase 2: potent inhibition by a 7B2 peptide fragment. Biochemistry **34**, 5486–5493.

Martens, G.J.M. (1988). Cloning and sequence analysis of human pituitary cDNA encoding the novel polypeptide 7B2. FEBS Lett. **234**, 160–164.

Martens, G.J.M., Bussemakers, M.J.G., Ayoubi, T.A.Y., and Jenks, B.G. (1989). The novel pituitary polypeptide 7B2 is a highly-conserved protein coexpressed with proopiomelanocortin. Eur. J. Biochem. **181**, 75–79.

Martens, G.J.M., Braks, J.A.M., Eib, D.W., Zhou, Y., and Lindberg, I. (1994). The neuroendocrine polypeptide 7B2 is an endogenous inhibitor of the prohormone convertase PC2. Proc. Natl. Acad. Sci. USA **91**, 5784–5787.

Mattei, M.G., Mbikay, M., Sylla, B.S., Lenoir, G., Mattei, J.F., Seidah, N.G., and Chrétien M. (1990). Assignment of the gene for neuroendocrine protein 7B2 (SGNE1 locus) to mouse chromosome region 2[E3-F3] and to human chromosome region 15q11-q15. Genomics **6**, 436–440.

Mbikay, M., Grant, S.G.N., Sirois, F., Tadros, H., Skowronski, J., Lazure, C., Seidah, N.G., Hanahan, D., and Chrétien M. (1989). cDNA sequence of neuroendocrine 7B2 expressed in beta cell tumors of transgenic mice. Int. J. Peptide Protein Res. **33**, 39–45.

Mbikay, M., Benjannet, S., Gianoulakis, C., Seidah, N.G., and Chrétien, M. (1991). The messenger RNA for pituitary protein 7B2 is widely expressed in mouse CNS and castration reduces its pituitary level. Neurochem. Int. **19**, 103–111.

Paquet, L., Lazure, C., Seidah, N.G., Chrétien, M., and Mbikay, M., (1991a). The production by alternative splicing of two mRNAs differing by one codon could be an intrinsic property of neuroendocrine protein 7B2 gene expression in man. Biochem. Biophys. Res. Commun. **147**, 156–162.

Paquet, L., Rondeau, N., Seidah, N.G., Lazure, C., Chrétien, M. and Mbikay, M., (1991b). Immunological identification and sequence characterization of a peptide derived from the processing of neuroendocrine protein 7B2. FEBS Lett. **294**, 23–26.

Paquet, L., Bergeron, F., Boudreoult, A., Seidah, N.G., Chrétien, M., Mbikay, M. and Lazure, C. (1994). The neuroendocrine precursor 7B2 is a sulfated protein proteolytically processed by an ubiquitous furin-like convertase. J. Biol. Chem. **269**, 19279–19285.

Roebroek, A.J.M., Dehaen, M.R.M., Van Bokhoven, A., Martens, G. J.M., Marynen, P., Van den Berghe, H., and Van de Ven, W.J.M. (1989a). Regional mapping of the human gene encoding the novel pituitary polypeptide 7B2 to chromosome 15q13-14 by *in situ* hybridization. Cytogenet. Cell Genet. **50**, 158–160.

Roebroek, A.J.M., Martens, G.J.M., Duits, A.J., Schalken, A.J., van Bokhoven, A., Wagenaar, S.Sc., and Van de Ven, W.J.M. (1989b). Differential expression of the gene encoding the novel pituitary polypeptide 7B2 in human lung cancer cells. Cancer Res. **49**, 4154–4158.

Seidah, N.G., Hsi, K.L., De Serres, G., Rochemont, J., Hamelin, J., Antakly, T., Cantin, M., and Chrétien, M. (1983). Isolation and NH_2-terminal sequence of a highly conserved human and porcine pituitary protein, belonging to a new superfamily. Arch. Biochem. Biophys. **255**, 525–534.

Senatorov, V.V., Yang, C.R., Marcinkiewicz, M., Chrétien, M. and Renaud, L.P. (1993). Depolarizing action of secretory granule protein 7B2 on rat supraoptic neurosecretory neurons. J. Neuroendocrinol. **5**, 533–536.

Sigafoos, J., Chestnut, W.G., Merrill, B.M., Taylor, L.C.E., Diliberto, E.R., and Viveros, O.H. (1993). Identification of a 7B2-derived tridecapeptide from bovine adrenal medulla chromaffin vesicles. Cell. Mol. Neurobiol. **13**, 271–278.

Suzuki, H., Ghatei, M.A., Williams, S.J., Uttenthal, L.O., Facer, P., Bishop, A.E., Polak, J.M., and Bloom, S.R. (1986). Production of pituitary protein 7B2 immunoreactivity by endocrine tumours and its possible diagnostic value. J. Clin. Endocrinol. Metab. **63**, 758–765.

Van Duijnhoven, J.L.P., Verschuren, M.C.M., Timmer, E.D.J., Vissers, P.M.A.M., Groeneveld, A., Ayoubi, T.A.Y., Van den Ouweland, A.M.W., and Van de Ven, W.J.M. (1991). Application of recombinant DNA technology in epitope mapping and targeting: development and characterization of a panel of monoclonal antibodies against the 7B2 neuroendocrine protein. J. Immunol. Methods **142**, 187–198.

Van Horssen, A.M., Van den Hurk, W.H., Bailyes, E.M., Hutton, J.C., Martens, G.J.M., and Lindberg, I. (1995). Identification of the region within the neuroendocrine polypeptide 7B2 responsible for the inhibition of prohormone convertase PC2. J. Biol. Chem. **270**, 14292–14296.

Waldbieser, G.C., Aimi, J., and Dixon, J.E. (1991). Cloning and characterization of the rat complementary deoxyribonucleic acid and gene encoding the neuroendocrine peptide 7B2. Endocrinology **128**, 3228–3236.

Zhu, X. and Lindberg, I. (1995). 7B2 facilitates the maturation of proPC2 in neuroendocrine cells and is required for the expression of enzymatic activity. J. Cell Biol. **129**, 1641–1650.

Zhu, X., Rouille, Y., Lamango, N.S., Steiner, D.F., and Lindberg, I. (1996). Internal cleavage of the inhibitory 7B2 carboxyl-terminal peptide by PC2: a potential mechanism for its inactivation Proc. Natl. Acad. Sci. USA **93**, 4919–4924.

■ *Gerard J.M. Martens:*
Department of Animal Physiology
University of Nijmegen
Toernooiveld
6525 ED Nijmegen, The Netherlands
Tel. 31 24 3652601
Fax. 31 24 3652714
E-mail: gmart@sci.kun.nl

7

Cytosolic Chaperonins

Cytosolic chaperonins — an overview

Cytosolic chaperonins (CCT), which are found in archaebacteria and in the eukaryotic cytosol, constitute a second family of chaperonins distinct from the CPN60 family. CCTs exhibit the structural and functional hallmarks of the eubacterial and organellar CPN60 chaperonins, namely a double ring structure, ATPase activity and ability to bind non-native proteins, but their primary structures display only weak resemblance to those of CPN60 proteins. This resemblance is confined to regions corresponding to the ATPase domain in GroEL. Archaebacterial CCT (thermosome) complexes contain two distinct subunits assembled into two stacked octameric or nonameric rings, while eukaryotic CCTs are comprised of at least eight structurally related subunits assembled into two octameric rings. The thermosome can selectively bind mesophilic proteins at temperatures where inactivation occurs, while eukaryotic CCTs apparently have a restricted range of substrates, notably actin, tubulin and their homologs. CCTs do not appear to require a cooperating GroES/CPN10-like cofactor, although unrelated cofactors play a role in CCT-assisted folding of tubulin. While the archaebacterial CCT subunits are induced markedly by heat shock, eukaryotic CCTs are not stress induced.

■ Group II cytosolic chaperonins in archaebacteria and the eukaryotic cytosol

In the cytosol of both prokaryotic and eukaryotic cells, many proteins that are newly synthesized or recovering from denatured states require folding assistance from a variety of molecular chaperones (Gething and Sambrook, 1992; Hartl, 1996). Early studies on chaperones identified a family of c. 60 kDa chaperonins (now called CPN60s) that includes GroEL of *Escherichia coli*, Hsp60 of mitochondria and the rubisco subunit binding protein of chloroplasts (Ellis and van der Vies, 1991). These chaperonins play important roles in the later stages of protein folding in the eubacterial cytosol and in the matrix of mitochondria and chloroplasts. However, for some time it appeared that members of this family were missing from the cytoplasm of eukaryotic cells. The characterization of the archaebacterial protein TF55 (Trent et al., 1991) led to the discovery of a related but distinct subfamily of cytosolic chaperonin proteins (Horwich and Willison, 1993), now termed the group II chaperonins (Willison and Kubota, 1994). In eukaryotic cells the cytosolic chaperonin complex is known as CCT (also as the TCP-1 complex, cytosolic chaperonin (c-cpn), chromobindin A or TRiC), a 960 kDa hetero-oligomeric double-torus-like complex composed of 16 subunits each of molecular weight approximately 60 kDa (Kubota et al., 1995a). TF55 and thermosome of archaebacteria are prokaryotic members of this subfamily (Phipps et al., 1991, 1993; Guagliardi et al., 1994; Knapp et al., 1994; Quaite-Randall et al., 1995). All chaperonins are thus 800–1000 kDa double-torus-like complexes consisting of 60 kDa subunits and they assist in the folding, assembly and transport of other proteins upon ATP hydrolysis. The group I chaperonins (the CPN60 family) have seven subunits per torus, whereas the group II chaperonins (CCT, TF55/thermosome) have eight or nine subunits per torus (Kubota et al., 1995a). The protein sequences of both groups of chaperonins show 15–20% amino acid identity between these groups and 30–40% identity within each group. In the group II chaperonins, each subunit species shows more than 66% identity between organisms. The proteins of both groups also contain several highly conserved motifs responsible for ATPase activity; e.g. GDGTT and VXGGG (Kim et al., 1994; Kubota et al., 1995a). One distinctive feature of CCT compared with all the other chaperonins of bacteria and endosymbiotic organelles, is that the CCT complex has at least eight different, but related, subunit species encoded by different genes (Kubota et al., 1994; 1995b).

■ The prokaryotic thermosome

Thermophilic factor 55 was originally detected as virtually the only protein synthesized in thermophilic archaebacteria under conditions of heat shock (Trent et al., 1990). The finding that prior expression of TF55 could be correlated with acquisition by the archaebacteria of thermolerance suggested a role for the protein in stabilizing cellular components against thermal inactivation. Observations that purified TF55 from two species of *Sulfolobus* and *Pyrodictium occultum* exhibited ATPase activity *in vitro* (Phipps et al., 1991; Trent et al., 1991; Knapp et al., 1994), and that *Sulfolobus* TF55 could selectively bind mesophilic proteins at temperatures where inactivation occurs (Trent et al., 1991; Guagliardi et al., 1994), provided a strong indication that TF55 functions as a molecular chaperone in the cell.

TF55 has a basal abundance in archaebacteria of at least 1–2% of soluble protein (resembling that of GroEL in the eubacterial cytosol). Studies with *P. occultum* TF55 showed that it was in fact composed of two constituent subunits (Phipps et al., 1991) and later studies showed that *Sulfolobus* TF55 also contains two distinct, closely related subunits (Knapp et al., 1994; Kagawa et al., 1995). TF55 from a number of archaebacterial species was shown to have a quaternary structure of two stacked oligomeric rings each composed of eight or nine subunits

(Phipps et al., 1991, 1993; Trent et al., 1991; Knapp et al., 1994; Marco et al., 1994b); this particle is now termed the thermosome (Phipps et al., 1991, 1993; see entry p. 211).

■ The eukaryotic CCT particle

The size of the CCT particle is estimated to be approximately 16 × 12 nm (diameter × height) (Lewis et al., 1992; Frydman et al., 1992) and averaged electron microscopic images of CCT show a torus-like structure of eightfold quasi-rotational symmetry with a central channel in top view and a cask-like structure in side view (Marco et al., 1994b; Waldmann et al., 1995). The apical portion around the central channel has been suggested to be involved in binding to substrate proteins, while the equatorial domain is probably the ATP-binding domain based upon comparisons with GroEL ATPase domains (Kim et al., 1994).

CCT is abundant in eukaryotic cytosol, at the level of $2–3 \times 10^5$ complexes per cell in mammalian cells (Lewis et al., 1992). Using monoclonal antibodies to TCP-1 (CCTα), immunolocalization studies of CCT show diffuse cytoplasmic distribution in cultured cells (Lewis et al., 1992), although one of its subunits is reported to exist also in the nucleus (Joly et al., 1994). CCT is found in early embryos and in all adult tissues in vertebrates, and is common to animals, plants, yeast and protozoa (see the entry on CCTα, p. 215) suggesting that it carries out a fundamental role in the functioning of the eukaryotic cytosol. In vitro experiments have suggested that CCT may bind membrane bound (Creutz et al., 1994) and ribosomally associated (Frydman et al., 1994) proteins, although these aspects need to be proven in vivo.

■ Function of CCT

CCT refolds denatured actin (Gao et al., 1992) and tubulin in vitro (Frydman et al., 1992; Yaffe et al., 1992) and binds these same proteins when newly synthesized in vivo (Sternlicht et al., 1993). Analysis of temperature-sensitive mutants of yeast CCT subunits shows that each of four different subunit species are involved in actin and/or tubulin organization (Ursic and Culbertson, 1991; Chen et al., 1994; Miklos et al., 1994; Ursic et al., 1994; Vinh and Drubin, 1994) and that these four and another gene (Li et al., 1994) are essential in yeast. It is also reported that a portion of CCT colocalizes with the centrosome in detergent-treated human cells (Brown et al., 1996) and that CCT shows high affinity to a middle part of the β-tubulin polypeptide (Dobrzynski et al., 1996). These results clearly indicate an important role for CCT in folding and assembly of tubulin and actin in the eukaryotic cytosol. It is also known that CCT refolds not only actin and α- and β-tubulin but also vertebrate actin-related protein and γ-tubulin (Melki et al., 1993). The binding affinity of CCT, in vitro, is higher towards denatured actin and tubulin than towards other denatured proteins like cap-binding protein, cyclin B, H-ras and c-myc (Melki and Cowan, 1994). Apart from actin, tubulin and related proteins, several other proteins are shown to be folded or bound by CCT in vitro: CCT refolds firefly luciferase (Frydman et al., 1992, 1994); CCT binds a neurofilament peptide fragment which can be released by addition of ATP (Roobol and Carden, 1993); and finally, CCT is involved in hepatitis B virus capsid assembly (Lingappa et al., 1994). These observations suggest that CCT may fold and assemble non-tubulin and non-actin cytoplasmic proteins in vivo as well.

As with all other chaperonins, CCT is an ATPase with relatively slow activity; approximately 10–20 molecules ATP hydrolysed per molecule of CCT per minute at 37°C (Martin and Creutz, 1987; Frydman et al., 1992). Classical Na$^+$/K$^+$ ATPase inhibitors (0.1 mM N-ethylmaleimide, 0.01 mM N,N-dicyclohexylcarbodiimide, 0.01 mg/ml oligomycin, 5.0 mM azide, 0.05 mM vanadate or 1.0 mM dithiothreitol) show very little or no effect, suggesting an ATPase mechanism distinct from that of the Na$^+$/K$^+$ ATPase family (Martin and Creutz, 1990). CCT requires ATP and Mg^{2+} to fold its substrate proteins (Gao et al., 1992; Yaffe et al., 1992). CCT binds and folds proteins upon ATP hydrolysis although it can bind substrate proteins without ATP. Like GroEL and Hsp60, target polypeptides of CCT undergo repetitive cycles of binding and ATP-driven release in order to fold. However, CCT differs in the specificity of substrate recognition when compared with GroEL and Hsp60 (Tian et al., 1995). It has been suggested that conformational changes in CCT are induced by ATP binding (Gao et al., 1992; Marco et al., 1994a; Hynes et al., 1995). As is the case with the other chaperonins, CCT-mediated folding of polypeptides is a slow process. The half-time of luciferase folding by CCT in vitro is estimated to be 10 minutes and that of actin in vivo to be 2–3 minutes. CCT may interact with HSP70 family proteins (Lewis et al., 1992; Kubota et al., 1994) or cooperate with cytosolic HSP70 and DnaJ homologues (Frydman et al., 1994; Frydman and Hartl, 1996). Although no GroES-like 10 kDa cochaperonin has been found in eukaryotic cytosol, four different cofactors, called cofactors A, C, D and E, have been found to facilitate tubulin folding by CCT (Gao et al., 1993, 1994; Lewis et al., 1996; Melki et al., 1996; Tian et al., 1996; see entry p. 226). These cofactors do not facilitate actin folding by CCT.

■ Composition of the CCT complex

The CCT complex has 8–10 different subunit species (Rommelaere et al., 1993; Kubota et al., 1994; 1997) in striking contrast to the group I chaperonins and the prokaryotic group II chaperonins, which contain only one or two species of subunit. Primary structures of eight constitutively expressed subunit species, CCTα, CCTβ, CCTγ, CCTδ, CCTε, CCTζ-1, CCTη and CCTθ proteins encoded by the Cct, Cct, Cct, Cct, Cct, Cct-1, Cct and Cct genes, have all been determined by cloning and sequencing of mouse Cct cDNAs (Willison et al., 1986; Kubota et al., 1994, 1995b). All the yeast homologues of these mouse genes have also been identified (Stoldt et al., 1996). Two-dimensional gel analysis of mouse testis CCT shows polypeptide spots corresponding to the eight constitutively expressed subunit

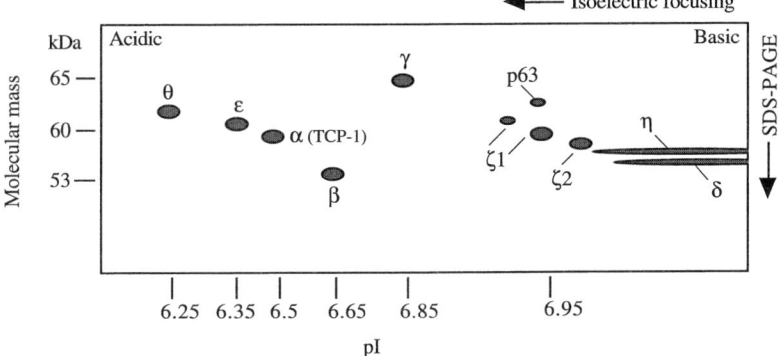

Figure 1. Mouse testis CCT subunits separated by two-dimensional gel electrophoresis. Greek letters indicate CCT subunit species and p63 is a testis-specific CCT-related polypeptide (Hynes *et al.*, 1995). ζ1 resolves into two spots probably due to post-translational modification and ζ2 is a tissue specific CCT subunit (Kubota *et al.*, 1997). This schematic figure was modified from Kubota *et al.* (1995a)

species (Kubota *et al.*, 1994, 1995b; Hynes *et al.*, 1995), one testis-specific subunit called CCTζ-2 (Kubota *et al.*, 1997) and a candidate for another testis-specific subunit called p63 (Hynes *et al.*, 1995). A panel of polyclonal antibodies against individual mouse CCT subunit species and an antibody recognizing all CCT subunits have been made (Hynes *et al.*, 1995) and other polyclonal antibodies recognizing one or a subset of the mammalian CCT subunits have been produced (Joly *et al.*, 1994; Lingappa *et al.*, 1994; Roobol *et al.*, 1995). Correspondences between these spots and the genes cloned have already been determined in the mouse (Fig. 1 and Hynes *et al.*, 1996a) and in man (Hynes *et al.*, 1996b). CCTζ-2 is a subunit that may replace CCTζ-1 in testicular germ cells (Kubota *et al.*, 1997).

■ Evolution and function of individual CCT subunit proteins

All CCT subunit proteins are approximately 57–61 kDa as calculated from their amino acid sequences and they show approximately 30% identity to one another (Kubota *et al.*, 1994, 1995b) and 40% identity to the archaebacterial chaperonin TF55 (Trent *et al.*, 1991; Kubota *et al.*, 1994, 1995b). CCT proteins also show weak similarity to the eubacterial chaperonin GroEL and the intraorganellar chaperonins Hsp60 and rubisco subunit-binding protein (Gupta 1990; Kubota *et al.*, 1994, 1995a). Each CCT protein sequence contains several chaperonin consensus motifs which have been postulated to be involved in ATPase activity (Lewis *et al.*, 1992; Rommelaere *et al.*, 1993; Kubota *et al.*, 1994, 1995a) and these motifs are contained within the equatorial ATPase domain of the *E. coli* chaperonin GroEL (Kim *et al.*, 1994).

The amino acid sequences of each subunit protein in mammalian species are very highly conserved; more than 96% of the amino acids are identical between mouse, human and other mammals. Each subunit species shows approximately 60% identity between mouse and yeast (see the entries on each subunit species, pp. 215 to 225).

These facts suggest that each subunit evolves very slowly and has a specific function as well as a common ATPase function. This specific function has been suggested to be specific binding to a particular subset of substrate proteins, or domains thereof (Kim *et al.*, 1994; Kubota *et al.*, 1994, 1995a). Another possibility for subunit-specific function might be their interactions with other molecular chaperones (Frydman *et al.*, 1994; Kubota *et al.*, 1994). All the genes encoding the eight constitutively expressed subunit species are calculated to have diverged 2×10^9 years ago and they are as old as the origin of eukaryotes (Kubota *et al.*, 1994). These genes are thought to have diverged from a homo-oligomeric chaperonin gene which was the common ancestor of the archaebacterial chaperonin TF55 and CCT subunit genes (Trent *et al.*, 1991; Lewis *et al.*, 1992; Willison and Kubota 1994). After eukaryotes and archaebacteria diverged, the eight subunit genes of CCT probably amplified and diverged almost simultaneously from the ancestral homo-oligomeric chaperonin gene (Willison and Kubota, 1994). The function of each subunit species is currently under investigation and the known characteristics of each subunit species are described in detail in the following sections for CCTα–CCTθ.

■ References

Brown, C.R., Doxsey, S.J., Hong-Brown, L.Q., Martin, R.L., and Welch, W.J. (1996). Molecular chaperones and the centrosome: a role for TCP-1 in microtuble nucleation. J. Biol. Chem. **271**, 824–832.

Chen, X., Sullivan, D.S., and Huffaker, T.C. (1994). Two yeast genes with similarity to TCP-1 are required for microtubule and actin function in vivo. Proc. Natl. Acad. Sci. USA **91**, 9111–9115.

Creutz, C.E., Liou, A., Snyder, S.L., Brownawell, A., and Willison, K. (1994). Identification of the major chromaffin granule-binding protein, chromobindin A, as the cytosolic chaperonin CCT (chaperonin containing TCP-1). J. Biol. Chem. **269**, 32035–32038.

Dobrzynski, J.K., Sternlicht, M.L., Farr, G.W., and Sternlicht, H. (1996). Newly-synthesized β-tubulin demonstrates domain-specific interactions with the cytosolic chaperonin. Biochemistry **35**, 15870–15882.

Ellis, R.J., and van der Vies, S.M. (1991). Molecular chaperones. Annu. Rev. Biochem. **60**, 321–347.

Frydman, J. and Hartl, F.U. (1996). Principles of chaperone-assisted protein folding: differences between in vitro and in vivo mechanisms. Nature **272**, 1497–1502.

Frydman, J., Nimmesgern, E., Erdjument-Bromage, H., Wall, J.S., Tempst, P., and Hartl, F.-U. (1992). Function in protein folding of TRiC, a cytosolic ring complex containing TCP-1 and structurally related subunits. EMBO J. **11**, 4767–4778.

Frydman, J., Nimmesgern, E., Ohtsuka, K., and Hartl, F.U. (1994). Folding of nascent polypeptide chain in a high molecular mass assembly with molecular chaperones. Nature **370**, 111–117.

Gao, Y., Thomas, J.O., Chow, R.L., Lee, G.-H., and Cowan, N.J. (1992). A cytoplasmic chaperonin that catalyze β-actin folding. Cell **69**, 1043–1050.

Gao, Y., Vainberg, I.E., Chow, R.I., and Cowan, N.J. (1993). Two cofactors and cytoplasmic chaperonin are required for the folding of α- and β-tubulin. Mol. Cell. Biol. **13**, 2478–2485.

Gao, Y., Melki, R., Walden, P.D., Lewis, S.A., Ampe, C., Rommelaere, H., Vandekerckhove, J., and Cowan, N.J. (1994). A novel cochaperonin that modulates the ATPase activity of cytoplasmic chaperonin. J. Cell Biol. **125**, 989–996.

Gething, M.-J. and Sambrook, J. (1992). Protein folding in the cell. Nature **355**, 33–45.

Guagliardi, A., Cerchia, L., Bartolucci, S., and Rossi, M. (1994). The chaperonin from the archaeon Sulfolobus solfataricus promotes correct refolding and prevents thermal denaturation in vitro. Protein Sci. **3**, 1436–1443.

Gupta, R.S. (1990). Sequence and structural homology between a mouse t-complex protein TCP-1 and the 'chaperonin' family of bacterial (GroEL, 60–65 kDa heat shock antigen) and eukaryotic proteins. Biochem. Int.. **20**, 833–841.

Hartl, F.-U. (1996). Molecular chaperones in cellualr protein folding. Nature **381**, 571–580.

Horwich, A.L. and Willison, K.R. (1993). Protein folding in the cell: functions of two families of molecular chaperone, hsp60 and TF55-TCPI. Phil. Trans. R. Soc. Lond. B **339**, 313–326.

Hynes, G., Kubota, H., and Willison, K.R. (1995). Antibody characterisation of two distinct conformations of the chaperonin containing TCP-1 from mouse testis. FEBS lett. **358**, 129–132.

Hynes, G., Sutton, C.W., U, S., and Willison, K.R. (1996a). Peptide mass fingerprinting of chaperonin-containing TCP-1 (CCT) and copurifying proteins. FASEB J. **10**, 137–147.

Hynes, G., Cells, J.E., Lewis, V.A., Carne, A., U, S., Lauridsen, J.B., and Willison, K.R. (1996b). Analysis of chaperonin-containing TCP-1 subunits in the human keratinocyte two-dimensional protein database: further characterisation of antibodies to individual subunits. Electrophoresis **17**, 1720–1727.

Joly, E.C., Tremblay, E., Tanguay, R.M., Wu, Y., and Bibor-Hardy, V. (1994). TRiC-P5, a novel TCP-1-related protein is located in the cytoplasm and in the nuclear matrix. J. Cell Sci. **107**, 2851–2859.

Kagawa, H.K., Osipiuk, J., Maltsev, N., Overbeek, R., Quaite-Randall, E., Joachimiak, A., and Trent, J.D. (1995). The 60 kDa heat shock proteins in the hyperthermophilic archeon Soufolobus shibatae. J. Mol. Biol. **253**, 712–725.

Kim, S., Willison, K.R., and Horwich, A.L. (1994). Cytosolic chaperonin subunits have a conserved ATPase domain but diverged polypeptide-binding domains. Trends Biochem. Sci. **19**, 543–548.

Knapp, S., Schmidt, K.I., Hebert, H., Bergman, T., Jörnvall, H., and Ladenstein, R. (1994). The molecular chaperonin TF55 from the thermophilic archaeon Sulfolobus solfataricus: a biochemical and structural characterization. J. Mol. Biol. **242**, 397–407.

Kubota, H., Hynes, G., Carne, A., Ashworth, A., and Willison, K. (1994). Identification of six Tcp-1-related genes encoding divergent subunits of the TCP-1-containing chaperonin. Curr. Biol. **4**, 89–99.

Kubota, H., Hynes, G., and Willison, K. (1995a). The chaperonin containing t-complex polypeptide 1 (TCP-1): multisubunit machinery assisting in protein folding and assembly in the eukaryotic cytosol. Eur. J. Biochem. **230**, 3–16.

Kubota, H., Hynes, G., and Willison, K. (1995b). The eighth Cct gene, Cctq, encoding the theta subunit of the cytosolic chaperonin that contains TCP-1. Gene **154**, 231–236.

Kubota, H., Hynes, G.M., Kerr, S.M., and Willison, K.R. (1997). Tissue specific subunit of the mouse cytosolic (chaperonin containing TCP-1). FEBS Lett. **402**, 53–56.

Lewis, V.A., Hynes, G.M., Zheng, D., Saibil, H., and Willison, K. (1992). T-complex polypeptide-1 is a subunit of a heteromeric particle in the eukaryotic cytosol. Nature **358**, 249–252.

Lewis, S.A., Tian, G., Vainberg, I.E., and Cowan, N.J. (1996). Chaperonin-mediated folding of actin and tubulin. J. Cell Biol. **132**, 1–4.

Li, W.-Z., Lin, P., Frydman, J., Boal, T.R., Cardillo, T.S., Richard, L.M., Toth, D., Lichtman, M.A., Hartle, F.-U., Sherman, F., and Segel, G.B. (1994). Tcp20, a subunit of the eukaryotic TRiC chaperonin from humans and yeast. J. Biol. Chem. **269**, 18616–18622.

Lingappa, J.R., Martin, R.L., Wong, M.L., Gamen, D., Welch, W.J., and Lingappa, V.R. (1994). A eukaryotic cytosolic chaperonin is associated with a high molecular weight intermediate in the assembly of hepatitis B virus capsid, a multimeric particle. J. Cell Biol. **125**, 99–111.

Marco, S., Carrascossa, J.L., and Valpuesta, J.M. (1994a). Reversible interaction of b-actin along the channel of the TCP-1 cytoplasmic chaperonin. Biophys. J. **67**, 364–368.

Marco, S., Urena, D., Carrascosa, J.L., Waldmann, T., Peters, J., Hegerl, R., Pfeifer, G., Sack, K.H., and Baumeister, W. (1994b). The molecular chaperone TF55: assessment of symmetry. FEBS Lett. **341**, 152–155.

Martin, W.H. and Creutz, C.E. (1987). Chromobindin A, a Ca^{2+} and ATP regulated chromaffin granule binding protein. J. Biol. Chem. **262**, 2803–2810.

Martin, W.H., and Creutz, C.E. (1990). Interaction of the complex scretory vesicle binding protein chromobindin A with nucleotides. J. Neurochem. **54**, 612–619.

Melki, R. and Cowan, N.J. (1994). Facilitated folding of actines and tubulins occurs via a nucleotide-dependent interaction between cytoplasmic chaperonin and distinctive folding intermediates. Mol. Cell. Biol. **14**, 2895–2904.

Melki, R., Vainberg, I.E., Chow, R.L., and Cowan, N.J. (1993). Chaperone mediated folding of verebrate actin-related protein and γ-tubulin. J. Cell Biol. **122**, 1301–1310.

Melki, R., Rommelaere, H., Leguy, R., Vanderckhove, J., and Ampe, C. (1996). Cofactor A is a molecular chaperone required for β-tubulin folding: Functional and structural characterization. Biochemistry **35**, 10422–10435.

Miklos, D., Caplan, S., Martens, D., Hynes, G., Pitluk, Z., Brown, C., Barrell, B., Horwich, A.L., and Willison, K. (1994). Primary structure and function of a second essential member of heterooligomeric TCP1 chaperonin complex of yeast, TCP1b. Proc. Natl. Acad. Sci. USA **91**, 2743–2747.

Phipps, B., Hoffmann A., Stetter, K.O., and Baumeister, W. (1991). A novel ATPase complex selectively accumulated upon heat shock is a major cellular component of thermophilic archaebacteria. EMBO J. **10**, 1711–1722.

Phipps, B.M., Typke, D., Hegerl, R., Volker, S., Hoffmann, A., Stetter, K.O., and Baumeister, W. (1993). Structure of a molecular chaperone from a thermophilic archaebacterium. Nature **361**, 475–477.

Quaite-Randall, E., Trent, J.D., Josephs, R., and Joachimiak, A. (1995). Conformation cycle of the archaeosome, a TCP-like chaperone from *Sulfolobus shibatae*. J. Biol. Chem. **270**, 28818–28823.

Rommelaere, H., van Troys, M., Gao, Y., Melki, R., Cowan, N. J., Vandekerckhove, J., and Ampe, C. (1993). Eukaryotic cytosolic chaperonin contains t-complex polypeptide 1 and seven related subunits. Proc. Natl. Acad. Sci. USA **90**, 11975–11979.

Roobol, A. and Carden, M.J. (1993). Identification of chaperonin particles in mammalian brain cytosol and t-complex polypeptide 1 as one of their components. J. Neurochem. **60**, 2327–2330.

Roobol, A., Holmes, F.E., Hayes, N.V.L., Baines, A.J., and Carden, A. J. (1995). Cytoplasmic chaperonin complex enter neurites developing in vitro and differ in subunit composition within single cells. J. Cell Sci. **108**, 1477–1488.

Sternlicht, H., Farr, G.W., Sternlicht, M.L., Driscoll, J.K., Willison, K., and Yaffe, M.B. (1993). The t-complex polypeptide 1 complex is a chaperonin for tubulin and actin in vivo. Proc. Natl. Acad. Sci. USA **90**, 9422–9426.

Stoldt, V., Radenmacher, F., Kerren, V., Ernst, J.F., Pearce, D.A., and Sherman, F. (1996). The Cct eukaryotic chaperonin subunits of *Saccharomyces cerevisiae* and other yeasts: a mini-review. Yeast **12**, 523–529.

Tian, G., Vainberg, I.E., Tap, W.D., Lewis, S.A., and Cowan, N.J. (1995). Specificity in chaperonin-mediated protein folding. Nature **375**, 250–253.

Tian, G., Huang, Y., Rommelaere, H., Vandekerckhove, J., Ampe, C., and Cowan, N.J. (1996). Pathway leading to correctly folded β-tubulin. Cell, **86**, 287–296.

Trent, J.D., Osipiuk, J., and Pinkau, T. (1990). Acquired thermotolerance and heat shock in the extremely thermophilic archaebacterium *Sulfolobus* spl. strain B12. J. Bacteriol. **172**, 1478–1484.

Trent, J.D., Nimmesgern, E., Wall, J.S., Hartl, F.-U., and Horwich, A.L. (1991). A molecular chaperone from a thermophilic archaebacterium is related to the eukaryotic protein t-complex polypeptide-1. Nature **354**, 490–493.

Ursic, D. and Culbertson, M.R. (1991). The yeast homolog to mouse *Tcp-1* affects microtubule-mediated processes. Mol. Cell. Biol. **11**, 2629–2640.

Ursic, D., Sedbrook, J.C., Himmel, K.L., and Culbertson, M.R. (1994). The essential yeast Tcp1 affects actin and microtubules. Mol. Biol. Cell **5**, 1065–1080.

Vinh, D.B.-N. and Drubin, D.G. (1994). A yeast TCP-1 like proteins is required for actin function in vivo. Proc. Natl. Acad. Sci. USA **91**, 9116–9120.

Waldmann, T., Nimmesgern, E., Nitsch, M., Peters, J., Pfeifer, G., Muller, S., Kellermann, J., Engel, A., Hartl, F.-U., and Baumeister, W. (1995). The thermosome of *Thermoplasma acidophilum* and its relationship to the eukaryotic chaperonin TRiC. Eur. J. Biochem. **227**, 848–856.

Willison, K.R., Dudley, K., and Potter, J. (1986). Molecular cloning and sequence analysis of a haploid expressed gene encoding t complex polypeptide 1. Cell **44**, 727–738.

Willison, K.R. and Kubota, H. (1994). The structure, function, and genetics of the chaperonin containing TCP-1 (CCT) in eukaryotic cytosol. In The Biology of Heat Shock Proteins and Molecular Chaperones (R.I. Morimoto, A. Tissiéres, and C. Georgopoulos, eds). Cold Spring Harbor Laboratory Press, New York, pp. 299–312.

Yaffe, M.B., Farr, G.W., Miklos, D., Horwich, A.L., Sternlicht, M.L., and Sternlicht, H. (1992). TCP1 complex is a molecular chaperone in tubulin biogenesis. Nature **358**, 245–248.

■ *Hiroshi Kubota:*
HSP Research Institute
Kyoto Research Park
17 Chudoji-minamimachi
Shimogyo-ku, Kyoto 600, Japan
Tel. 81 75 315 8656
Fax. 81 75 315 8659
E-mail: kubota@hsp.co.jp

■ *Keith Willison:*
Cancer Research Campaign Centre for Cell and Molecular Biology
Institute of Cancer Research,
Chester Beatty Laboratories
237 Fulham Road
London SW3 6JB, UK
Tel./Fax. 44 171 351 3325
E-mail: willison@icr.ac.uk

Thermosome

The thermosome, a cytosolic protein complex ubiquitously present in archaebacteria, was originally isolated from the hyperthermophilic organism *Pyrodictium occultum*. It consists of one or two polypeptide subunits that assemble into a cylindrical particle composed of two stacked octameric rings (nonameric in the case of *Sulfolobus sp.*). Structurally, the complex is the archaebacterial equivalent of the eukaryotic cytosolic chaperonin (TRiC or CCT). Functionally, it is considered to be the archaebacterial representative of the 60 kDa chaperonins.

■ Alternative names

TF55/TF56 (thermophilic factor of 55 kDa) (Trent *et al.*, 1991; GenBank accession number L34691). Archaebacterial and eukaryotic cytosolic chaperonins have been termed 'group II chaperonins' (Kubota *et al.*, 1995).

■ Thermosome sequences

Sequences are known for the two subunits of *Thermoplasma acidophilum* (GenBank accession numbers Z466494 and Z46650), *Sulfolobus shibatae* (S70432 and L34691) and *Pyrodictium occultum* [partial] (Phipps *et al.*,

1991) and for one subunit of *Pyrococcus* sp. strain KOD1 (D29672), *Methanopyrus kandleri* (Z50745) and *Methanococcus jannaschii* (1511030). They are 543–560 amino acids long (molecular mass approximately 60 kDa) and contain few aromatic residues (particularly tryptophan, which is absent entirely from five of the seven known complete sequences). The sequences are 50–70 % identical to each other and 30–40% identical to eucaryotic TRiC (Frydman *et al.*, 1992) or CCT (Kubota *et al.*, 1995) subunits, with which they can be aligned over essentially their entire length (Waldmann *et al.*, 1995b). In contrast, thermosome and TRiC/CCT subunits are only similar to GroEL-like proteins in an N- and C-terminal domain [sequence identities of approximately 30 and 22%, respectively (Waldmann *et al.*, 1995b)]. It has been proposed that the thermosome/TRiC/CCT family be called 'group II chaperonins', whereas 'group I' comprises the GroEL-like proteins (Kubota *et al.*, 1995). Three highly conserved sequences in the N-terminal domain of group I and group II sequences (including the motif GDGTT) most likely define the ATP-binding site (Kubota *et al.*, 1995). The central domain of group II subunits, which has no similarity to group I sequences, is related to a domain in the recently characterized Fab1 protein (Yamamoto *et al.*, 1995), which is essential for normal vacuole function and morphology in yeast. In a dendrogram of the group II chaperonins (Fig. 1) the archaebacterial proteins form a single branch, which is shorter than the eight branches representing the eukaryotic subunit families. This indicates that the thermosome sequences have diverged the least from the putative common ancestor of the group II chaperonins (Waldmann *et al.*, 1995b).

■ Thermosome proteins

The thermosome has a molecular mass of approximately 1 MDa (Waldmann *et al.*, 1995a). It was first purified from *Pyrodictium occultum* and has since been identified in the cytosol of many archaebacteria (Phipps *et al.*, 1991). To date, averages of electron micrographs have been obtained for the complexes from *P. brockii*, *P. occultum*, *Archaeoglobus fulgidus* (Phipps *et al.*, 1991), *T. acidophilum* (Waldmann *et al.*, 1995a), *S. shibatae* (Trent *et al.*, 1991), *S. solfataricus* (Marco *et al.*, 1994) and *M. kandleri*. (Andrä *et al.*, 1996). Their appearance is that of a barrel of c. 16 nm height and c. 16 nm diameter, in which two stacked rings enclose a large central cavity of c. 5 nm diameter. At this resolution (1–2 nm), the architecture of the thermosome appears very similar to that of TRiC and GroEL, apart from differences in the number of subunits per ring. While GroEL is sevenfold symmetric and TRiC pseudo-eightfold (Waldmann *et al.*, 1995a), the thermosome appears to be heterogeneous in that some complexes are (pseudo) eightfold symmetric [*Pyrodictium*, *Archaeoglobus*, *Thermoplasma* (Phipps *et al.*, 1991; Waldmann *et al.*, 1995a), *Methanopyrus* (Andrä *et al.*, 1996)] while some are ninefold symmetric [*Sulfolobus* (Marco *et al.*, 1994)]. Occasionally, particles with apparent ninefold symmetry have been observed in *Pyrodictium* and *Thermoplasma*, and particles with apparent eightfold symmetry in *S. shibatae*. Detailed investigation of ice-embedded complexes has revealed that this polymorphism is characteristic for the thermosome, although the subset of particles deviating from the dominant symmetry is minor (Nitsch *et al.*, 1997). For the

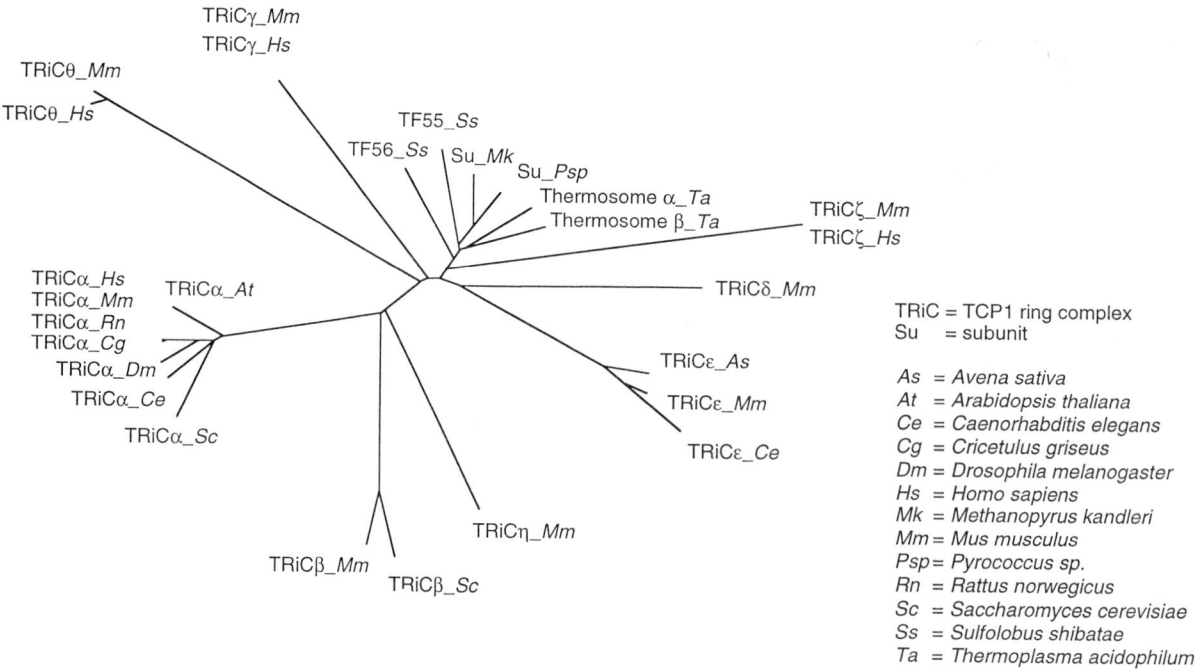

Figure 1. Dendrogram of group II chaperonins (modified from Waldmann *et al.*, 1995b).

Thermoplasma thermosome, the heterogeneity is much more pronounced when the two subunits are expressed separately (Nitsch et al., 1997).

A three-dimensional reconstruction of the *Pyrodictium* complex (Fig. 2a and b) (Phipps et al., 1993) reveals the kidney shape of the thermosome subunits. Each subunit consists of two large terminal domains, connected by a smaller central domain. These correspond to the three domains (apical, intermediate and equatorial) identified in the crystal structure of GroEL (Braig et al., 1994). The sequence similarity between thermosome subunits and GroEL can be interpreted in the light of this domain structure: the N- and C-terminal regions of sequence similarity correspond to the equatorial domain, which contains the ATP-binding site, while the central dissimilar region corresponds to the apical domain, which interacts with the denatured polypeptide substrate (Waldmann et al., 1995b).

The thermosomes of *Pyrodictium* and *Thermoplasma* are composed of equal amounts of two subunits with almost identical molecular mass (c. 60 kDa) (Waldmann et al., 1995a). For the *Thermoplasma* complex, a clear fourfold symmetry was detected in ice-embedded complexes, strongly indicating that the two rings are hetero-oligomers with the two subunits alternating (Nitsch et al., 1997). Until recently it was thought that the thermosome from *Sulfolobus* sp. consists of a single subunit, but here also a second subunit has now been identified (Knapp et al., 1994; Trent et al., 1994). In some organisms however, thermosomes probably consist of a single subunit, as in *Methanococcus jannaschii*, where the sequence of the complete genome has revealed the existence of a single thermosome gene (Bult et al., 1996).

Purified thermosomes from *Thermoplasma* and from *Sulfolobus* have been pH-dissociated and reassembled *in vitro* in a process requiring Mg^{2+} and ATP (or a non-hydrolysable ATP analogue). The reassembly did not occur in the absence of these factors (Knapp et al., 1994; Waldmann et al., 1995c). The two subunits of the *Thermoplasma* thermosome, α and β, have also been cloned and expressed in *Escherichia coli*, jointly and separately. Each subunit by itself forms thermosome-like complexes with eightfold symmetry and the two coexpressed subunits also form complexes containing both subunits (Waldmann et al., 1995c). This is in contrast to the α and β subunits of *S. solfataricus*, which have been reported to form only joint complexes (Knapp et al., 1994).

■ Gene regulation of the thermosome

The thermosome is an abundant, constitutive component of the archaebacterial cytosol. Its expression levels are strongly increased after heat shock and this increase has been found to correlate with the acquisition of thermotolerance in *Sulfolobus shibatae* (Trent et al., 1994) and *Pyrodictium occultum*. In *Pyrodictium* cells, which grow optimally at 105°C, the thermosome represents 6% of total soluble protein at 90°C, 11% at 100°C and 73% at 108°C, as judged by densitometry of SDS gels (Phipps et al., 1991).

■ Function of the thermosome

Functionally, the thermosome is considered to be the archaebacterial representative of the 60 kDa chaperonins (Horwich and Willison, 1993). *In vitro*, the thermosome exhibits a weak potassium-dependent ATPase activity and is able to bind denatured polypeptides (Phipps et al., 1991; Trent et al., 1991; Knapp et al., 1994; Waldmann et al., 1995a). Refolding of thermosome-bound polypeptides has been reported for the thermosome from *Sulfolobus solfataricus* (Guagliardi et al., 1994). These properties establish the thermosome as the archaebacterial representative of the 60 kDa family of chaperonins. Its likely function is to prevent aggregation of both newly synthesized and stress denatured polypeptides, as well as to facilitate their correct folding.

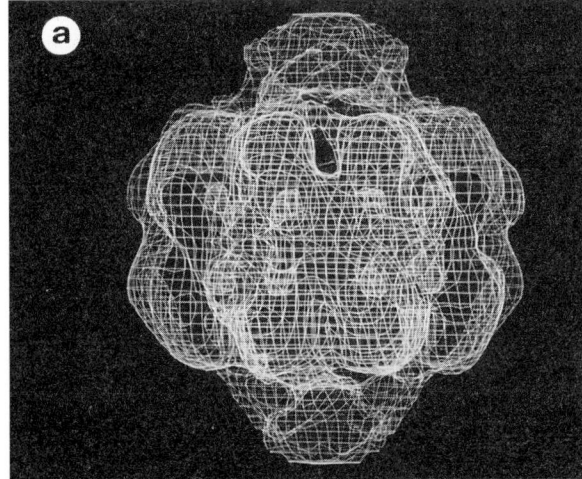
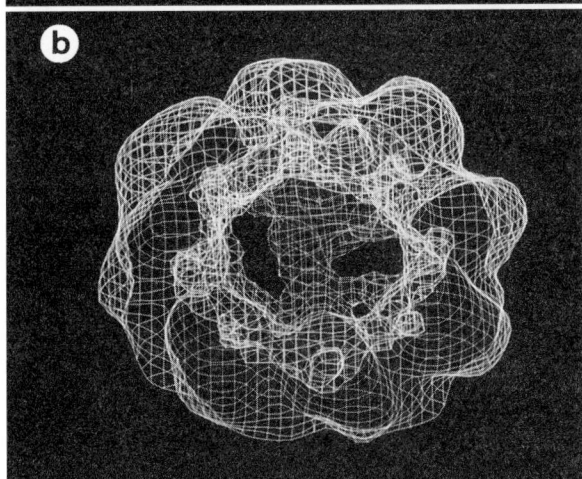

Figure 2. Network representation of a three-dimensional (3-D) model of the thermosome from *Pyrodictium occultum*. (a) side view; (b) the top part of the molecule has been cut away and the view is into the large central cavity; mean diameter of the cavity is 6.7 nm. For a detailed discussion of the 3-D reconstruction see Phipps et al. (1993).

No cofactors of the thermosome that would be comparable to GroES or the cofactors A and B of CCT (Kubota et al., 1995) are currently known. In S. solfataricus, phosphorylation of the thermosome has been observed *in vitro* and has been claimed to cause a conformational change (Knapp et al., 1994), but this change has not been quantified and its possible role in the regulation of the complex *in vivo* is as yet unclear. Post-translational modification(s) has also been proposed for the *Thermoplasma* thermosome on the basis of two-dimensional gel electrophoresis (Waldmann et al., 1995b).

■ References

Andrä, S., Frey, G., Nitsch, M., Baumeister, W., and Stetter, K.O. (1996). Purification and structural characterization of the thermosome from the hyperthermophilic archaeum *Methanopyrus kandleri*. FEBS Lett. 379, 127–131.

Braig, K., Otwinowski, Z., Hegde, R., Boisvert, D.C., Joachimiak A., Horwich, A.L., and Sigler, P.B. (1994). The crystal structure of the bacterial chaperonin GroEL at 2.8 Å. Nature 371, 578–586.

Bult, C.J., et al. (1996). Complete genome sequence of the methanogenic archaeon, *Methanococcus jannaschii*. Science 273, 1058–1073.

Frydman, J., Nimmesgern, E., Erdjument, B.H., Wall, J.S., Tempst, P., and Hartl, F.U. (1992). Function in protein folding of TRiC, a cytosolic ring complex containing TCP-1 and structurally related subunits. EMBO J. 11, 4767–4778.

Guagliardi, A., Cerchia, L., Bartolucci, S., and Rossi, M. (1994). The chaperonin from the archaeon *Sulfolobus solfataricus* promotes correct refolding and prevents thermal denaturation *in vitro*. Protein Sci. 3, 1436–1443.

Horwich, A.L. and Willison, K.R. (1993). Protein folding in the cell: functions of two families of molecular chaperone hsp 60 and TF55-TCP1. Phil. Trans. R. Soc. Lond. 339, 313–325.

Knapp, S., Schmidt, K.I., Hebert, H., Bergman, T., Jörnvall, H., and Ladenstein, R. (1994). The molecular chaperonin TF55 from the thermophilic archaeon Sulfolobus solfataricus: a biochemical and structural characterization. J. Mol. Biol. 242, 397–407.

Kubota, H., Hynes, G., and Willison, K. (1995). The chaperonin containing t-complex polypeptide 1 (TCP-1): multisubunit machinery assisting in protein folding and assembly in the eukaryotic cytosol. Eur. J. Biochem. 230, 3–16.

Marco, S., Urena, D., Carrascosa, J.L., Waldmann, T., Peters, J., Hegerl, R., Pfeifer, G., Sack, K. H., and Baumeister, W. (1994). The molecular chaperone TF55: assessment of symmetry. FEBS Lett. 341, 152–155.

Nitsch, M., Klumpp, M., Lupas, A., and Baumeister, W. (1997). The thermosome: alternating α and β-subunits within the chaperonin of the archaeon *Thermoplasma acidophilum*. J. Mol. Biol. **267**, 142–149.

Phipps, B., Hoffmann A., Stetter K.O., and Baumeister, W. (1991). A novel ATPase complex selectively accumulated upon heat shock is a major cellular component of thermophilic archaebacteria. EMBO J. 10, 1711–1722.

Phipps, B.M., Typke, D., Hegerl, R., Volker, S., Hoffmann, A., Stetter, K.O., and Baumeister, W. (1993). Structure of a molecular chaperone from a thermophilic archaebacterium. Nature 361, 475–477.

Trent, J.D., Nimmesgern, E., Wall, J.S., Hartl, F.U., and Horwich, A.L. (1991). A molecular chaperone from a thermophilic archaebacterium is related to the eukaryotic protein t-complex polypeptide-1. Nature 354, 490–493.

Trent, J.D., Gabrielsen, M., Jensen, B., Neuhard, J., and Olsen, J. (1994). Acquired thermotolerance and heat shock proteins in thermophiles from the three phylogenetic domains. J. Bacteriol. 176, 6148–6152.

Waldmann, T., Nimmesgern, E., Nitsch, M., Peters, J., Pfeifer, G., Mueller, S., Kellermann, J., Engel, A., Hartl, F.U., and Baumeister, W. (1995a). The thermosome of *Thermoplasma acidophilum* and its relationship to the eukaryotic chaperonin TRiC. Eur. J. Biochem. 227, 848–856.

Waldmann, T., Lupas, A., Kellerman, J., Peters, J., and Baumeister, W. (1995b). Primary structure of the thermosome from *Thermoplasma acidophilum*. Biol. Chem. 376, 119–126.

Waldmann, T., Nitsch, M., Klumpp, M., and Baumeister, W. (1995c). Expression of an archaeal chaperonin in *E. coli*: formation of homo- (α, β) and hetero-oligomeric ($\alpha+\beta$) thermosome complexes. FEBS Lett. 376, 67–73.

Yamamoto, A., DeWald, D.B., Boronenkov, I.V., Anderson, R.A., Emr, S.D., and Koshland D. (1995). Novel PI(4)P 5-kinase homologue Fab1p, essential for normal vacuole function and morphology in yeast. Mol. Biol. Cell 6, 525–539.

■ *Thomas Waldmann, Andrei Lupas and Wolfgang Baumeister:*
Max-Planck-Institut für Biochemie
Am Klopferspitz 18a
82152 Martinsried, Germany
Tel. 49 89 8578 2652
Fax. 49 89 8578 2641
E-mail: baumeist@alf.biochem.mpg.de

CCTα

CCTα was the first identified subunit of the CCT complex, which is a hetero-oligomeric, double-torus-shaped, 960 kDa molecular chaperone abundant in eukaryotic cytosol. The CCT complex assists in protein folding in the cytosol of eukaryotic cells in a process that is dependent upon ATP hydrolysis. CCTα is a 60.5 kDa polypeptide and is one of the eight constitutively expressed subunit species of CCT.

Alternative names

CCT has been called TCP-1 (t-complex polypeptide 1), named because the *Tcp-1* gene was found as a gene showing t-complex-encoded polymorphism in a protein product highly expressed in testis (Silver *et al.*, 1979). The t-complex is a region containing large chromosomal inversions on mouse chromosome 17 and is involved in transmission ratio distortion of t-complex-carrying mice (Silver, 1985). The yeast homologue of CCTα is called Tcp1p or Cct1p (Kubota *et al.*, 1995a).

Mutant phenotypes

Three temperature-sensitive mutants; *tcp1-1* (D_{96} to E), *tcp1-2* (G_{423} to D) and *tcp1-3* (G_{45} to S) show abnormal tubulin and actin organization in budding yeast and null alleles of the gene are not viable (Ursic and Culbertson, 1991; Ursic *et al.*, 1994). Another temperature-sensitive mutant, *tcp1a-245* (G_{48} to E), also shows an abnormal tubulin phenotype (Miklos *et al.*, 1994).

CCT α genes and sequences

The first *Cctα* cDNA (GenBank accession number M12899) was isolated as clone pB1.4 from mouse testis by differential hybridization (Dudly *et al.*, 1984) and then sequenced and identified as a t-complex-encoded gene by genetic analysis (Willison *et al.*, 1986). Its genomic DNA (accession number S467630) was cloned (Willison *et al.*, 1986) and sequenced (Kubota *et al.*, 1992) revealing that the gene consists of 12 exons. *Cctα* cDNAs of other mammals, including rat (GenBank accession number D90345; Morita *et al.*, 1991), Chinese hamster (M34665; Gupta, 1990) and human (X74801; Kirchhoff and Willison, 1990), have been cloned and the proteins encoded by these homologues show more than 96% identity to the mouse CCTα protein. The *Cctα* genes of fruit fly (M21159; Ursic and Ganetzky, 1988), nematode (U07941; Leroux and Candido, 1995), budding yeast *Saccharomyces cerevisiae* (M21160; Ursic and Culbertson, 1991) and a plant *Arabidopsis thaliana* (D11351; Mori *et al.*, 1992) have been identified and the CCTα proteins of these organisms show more than 60% identity to the mouse CCTα. The *Cctα* homologue of the protozoan *Stylonychia lemnae* has also been identified (accession number X73880, this sequence may include sequencing mistakes) (Maercker and Lipps, 1994). All CCTα proteins are approximately 550 amino acids in length and show approximately 30% identity to the other subunits of CCT (Kubota *et al.*, 1994, 1995b) and 40% identity to the archaebacterial chaperonin TF55 (Trent *et al.*, 1991). CCT proteins also show weak similarity to the eubacterial chaperonin GroEL and intraorganellar chaperonins Hsp60 and rubisco subunit-binding protein (Gupta, 1990). Each CCTα protein sequence contains several chaperonin consensus motifs which have been postulated to be involved in ATPase activity (Lewis *et al.*, 1992; Rommelaere *et al.*, 1993; Kubota *et al.*, 1994, 1995a) and these motifs are contained within sequences corresponding to the equatorial ATPase domain of the *Escherichia coli* chaperonin GroEL (Kim *et al.*, 1994).

CCTα protein

CCTα was the first identified subunit of the CCT complex (Frydman *et al.*, 1992; Gao *et al.*, 1992; Lewis *et al.*, 1992; Yaffe *et al.*, 1992). Mouse CCTα migrates as a 60 kDa spot of pI 6.5 by two-dimensional gel analysis (Kubota *et al.*, 1994; Hynes *et al.*, 1995) and makes a complex with the other seven CCT subunit species; monoclonal antibodies specific to CCTα coimmunoprecpitate all the other subunit species (Lewis *et al.*, 1992; Kubota *et al.*, 1994, 1995b). The CCT complex is approximately 960 kDa and has a double-torus-like structure (Lewis *et al.*, 1992; Marco *et al.*, 1994). The CCT complex assists in the folding of actin, tubulin and some other proteins in the eukaryotic cytosol upon ATP hydrolysis (Frydman *et al.*, 1992; Gao *et al.*, 1992; Yaffe *et al.*, 1992; Sternlicht *et al.*, 1993; Melki and Cowan 1994). The structure, chaperone function and other characteristics of the CCT complex are described in detail in the overview to this section, p. 207).

CCTα localization

Monoclonal antibodies against mouse CCTα (clones 14A, 22B, 23C, 51C, 72A, 84A and 91A) have been established (Willison *et al.*, 1989) and antibodies 14C, 84A and 91A are monospecific to CCTα (Lewis *et al.*, 1992). Immunofluoresence analysis of cultured cells with the 14C, 84A and 91A antibodies shows diffuse cytoplasmic patterns with small granule-like structures (Lewis *et al.*, 1992). Antibodies 23C and 72A react strongly with mouse

CCTα but also react with a 102 kDa coatomer protein and an unidentified 115 kDa protein (Harrison-Lavoie et al., 1993). The 23C and 72A antibodies recognize mouse, rat, hamster, dog and chicken CCTα but not monkey, human, Xenopus or yeast CCTα (Harrison-Lavoie et al., 1993). CCTα is known to enter into neurites of neuronal cells (Roobol et al., 1995).

■ Gene regulation of Cctα

The genomic DNA sequence of mouse Cctα revealed 11 introns and a CpG-rich promoter region, which has CCAAT and GC boxes but no TATA box (Kubota et al., 1992). CCTα is most likely to be expressed in all cell types and is synthesized in early stage embryos (Sanchez and Erickson, 1985) and all adult tissues subsequently. Cctα mRNA is expressed in all cells and tissues investigated (Kubota et al., 1992; Sun et al., 1995). The expression of the Cctα gene is probably up-regulated in rapidly growing cells (Kubota et al., 1992) but not by heat shock (Ursic and Culbertson, 1992; Lewis et al., 1992; Leroux and Candido, 1995).

■ References

Dudly, K., Potter, J., Lyon, M.F., and Willison, K. (1984). Analysis of male sterile mutations in the mouse using haploid stage expressed cDNA probes. Nucl. Acids Res. **12**, 4281–4293.

Frydman, J., Nimmesgern, E., Erdjument-Bromage, H., Wall, J.S., Tempst, P., and Hartl, F.-U. (1992). Function in protein folding of TRiC, a cytosolic ring complex containing TCP-1 and structurally related subunits. EMBO J. **11**, 4767–4778.

Gao, Y., Thomas, J.O., Chow, R.L., Lee, G.-H., and Cowan, N.J. (1992). A cytoplasmic chaperonin that catalyze b-actin folding. Cell **69**, 1043–1050.

Gupta, R.S. (1990). Sequence and structural homology between a mouse t-complex protein TCP-1 and the 'chaperonin' family of bacterial (GroEL, 60–65 kDa heat shock antigen) and eukaryotic proteins. Biochem. Int. **20**, 833–841.

Hynes, G., Kubota, H., and Willison, K. (1995). Antibody characterization of two distinct conformations of the chaperonin containing TCP-1 from mouse testis. FEBS. Lett. **358**, 129–132.

Harrison-Lavoie, K.J., Lewis, V.A., Hynes, G.M., and Collison, K.S. (1993). A 102 kDa subunit of a Golgi-associated particles has homology to β subunits of trimeric G proteins. EMBO J. **12**, 2847–2853.

Kim, S., Willison, K.R., and Horwich, A.L. (1994). Cytosolic chaperonin subunits have a conserved ATPase domain but diverged polypeptide-binding domains. Trends Biochem. Sci. **19**, 543–548.

Kirchhoff, C. and Willison, K.R. (1990). Nucleotide and amino-acid sequence of human testis-derived TCP1. Nucl. Acids Res. **18**, 4247.

Kubota, H., Willison, K., Ashworth, A., Nozaki, M., Miyamoto, H., Yamamoto, H., Matsushiro, A., and Morita, T. (1992). Structure and expression of the gene encoding t-complex polypeptide 1 (Tcp-1). Gene **120**, 207–215.

Kubota, H., Hynes, G., Carne, A., Ashworth, A., and Willison, K. (1994). Identification of six Tcp-1-related genes encoding divergent subunits of the TCP-1-containing chaperonin. Curr. Biol. 4, 89–99.

Kubota, H., Hynes, G., and Willison, K. (1995a). The chaperonin containing t-complex polypeptide 1 (TCP-1): multisubunit machinery assisting in protein folding and assembly in the eukaryotic cytosol. Eur. J. Biochem. **230**, 3–16.

Kubota, H., Hynes, G., and Willison, K. (1995b). The eighth Cct gene, Cctq, encoding the theta subunit of the cytosolic chaperonin that contains TCP-1. Gene **154**, 231–236.

Leroux, M.R. and Candido, E.P.M. (1995). Molecular analysis of Caenorhabditis elegans tcp-1, a gene encoding a chaperonin protein. Gene **156**, 241–246.

Lewis, V.A., Hynes, G.M., Zheng, D., Saibil, H., and Willison, K. (1992). T-complex polypeptide-1 is a subunit of a heteromeric particle in the eukaryotic cytosol. Nature **358**, 249–252.

Maercker, C. and Lipps, H.J. (1994). A macronuclear DNA molecule from the hypotrichous ciliate Stylonychia lemnae encoding a t-complex polypeptide 1-like protein. Gene **141**, 147–148.

Marco, S., Carrascossa, J.L., and Valpuesta, J.M. (1994). Reversible interaction of β-actin along the channel of the TCP-1 cytoplasmic chaperonin. Biophys. J. **67**, 364–368.

Melki, R. and Cowan, N.J. (1994). Facilitated folding of actins and tubulins occurs via a nucleotide-dependent interaction between cytoplasmic chaperonin and distinctive folding intermediates. Mol. Cell. Biol. **14**, 2895–2904.

Miklos, D., Caplan, S., Martens, D., Hynes, G., Pitluk, Z., Brown, C., Barrell, B., Horwich, A.L., and Willison, K. (1994). Primary structure and function of a second essential member of heterooligomeric TCP1 chaperonin complex of yeast, TCP1b. Proc. Natl. Acad. Sci. USA **91**, 2743–2747.

Mori, M., Murata, K., Kubota, H., Yamamoto, A., Matsushiro, A., and Morita, T. (1992). Cloning of a cDNA encoding the Tcp-1 (t complex polypeptide 1) homologue of Arabidopsis thaliana. Gene **122**, 381–382.

Morita, T., Kubota, H., Gachelin, G., Nozaki, M., and Matsushiro, A. (1991). Cloning of cDNA encoding rat TCP-1. Biochem. Biophys. Acta **1129**, 96–99.

Rommelaere, H., van Troys, M., Gao, Y., Melki, R., Cowan, N.J., Vandekerckhove, J., and Ampe, C. (1993). Eukaryotic cytosolic chaperonin contains t-complex polypeptide 1 and seven related subunits. Proc. Natl. Acad. Sci. USA **90**, 11975–11979.

Roobol, A., Holmes, F.E., Hayes, N.V.L., Baines, A.J., and Carden, A.J. (1995). Cytoplasmic chaperonin complexes enter neurites developing in vitro and differ in subunit composition within single cells. J. Cell Sci. **108**, 1477–1488.

Sanchez, E.R. and Erickson, R.P. (1985). Expression of the Tcp-1 locus of the mouse during early embryogenesis. J. Embryol. Exp. Morph. **89**, 113–122.

Silver, L.M. (1985). Mouse t haplotypes. Annu. Rev. Genet. **19**, 179–208.

Silver, L.M., Artzt, K., and Bennett, D. (1979). A major testicular cell protein specified by a mouse T/t complex gene. Cell **17**, 275–284.

Sternlicht, H., Farr, G.W., Sternlicht, M.L., Driscoll, J.K., Willison, K., and Yaffe, M.B. (1993). The t-complex polypeptide 1 complex is a chaperonin for tubulin and actin in vivo. Proc. Natl. Acad. Sci. USA **90**, 9422–9426.

Sun, H.B., Neff, A.W., Mescher, A.L., and Malacinski, G.M. (1995). Expression of the axolotl homologue of mouse chaperonin t-complex protein 1 during early development. Biochem. Biophys. Acta **1260**, 157–166.

Trent, J.D., Nimmesgern, E., Wall, J.S., Hartl, F.-U., and Horwich, A.L. (1991). A molecular chaperone from a thermophilic archaebacterium is related to the eukaryotic protein t-complex polypeptide-1. Nature **354**, 490–493.

Ursic, D. and Culbertson, M.R. (1991). The yeast homolog to mouse Tcp-1 affects microtubule-mediated processes. Mol. Cell. Biol. **11**, 2629–2640.

Ursic, D. and Culbertson, M.R. (1992). Is yeast TCP1 a chaperonin? Nature **356**, 392.

Ursic, D. and Ganetzky, B. (1988). A Drosophila melanogaster gene encodes a protein homologus to the mouse t complex polypeptide 1. Gene **68**, 267–274.

Ursic, D., Sedbrook, J.C., Himmel, K.L., and Culbertson, M.R. (1994). The essential yeast Tcp1 affects actin and microtubules. Mol. Biol. Cell **5**, 1065–1080.

Willison, K.R., Dudley, K., and Potter, J. (1986). Molecular cloning and sequence analysis of a haploid expressed gene encoding t complex polypeptide 1. Cell **44**, 727–738.

Willison, K., Lewis, V., Zuckerman, K.S., Cordell, J., Dean, C., Miller, K., Lyon, M.F., and Marsh, M. (1989). The t complex polypeptide 1 (TCP-1) is associated with the cytoplasmic aspect of Golgi membranes. Cell **57**, 621–632.

Yaffe, M.B., Farr, G.W., Miklos, D., Horwich, A.L., Sternlicht, M.L., and Sternlicht, H. (1992). TCP1 complex is a molecular chaperone in tubulin biogenesis. Nature **358**, 245–248.

■ Hiroshi Kubota:
HSP Research Institute
Kyoto Research Park
17 Chudoji-minamimachi
Shimogyo-ku, Kyoto 600, Japan
Tel. 81 75 315 8656
Fax. 81 75 315 8659
E-mail: kubota@hsp.co.jp

■ Keith Willison:
Cancer Research Campaign Centre for Cell and Molecular Biology
Institute of Cancer Research
Chester Beatty Laboratories
237 Fulham Road
London SW3 6JB, UK
Tel./Fax. 44 171 351 3325
E-mail: willison@icr.ac.uk

CCTβ

CCTβ is the β subunit of the CCT complex, which is a hetero-oligomeric, double-torus-shaped, 960 kDa molecular chaperone abundant in eukaryotic cytosol. The CCT complex assists in protein folding in the cytosol of eukaryotic cells in a process that is dependent upon ATP hydrolysis. CCTβ is the smallest (57.5 kDa as calculated from its protein sequence) subunit of the eight constitutively expressed subunit species of CCT.

■ Alternative names

The yeast homologue of CCTβ is also called Tcp1βp (Miklos et al., 1994), Bin3p (Chen et al., 1994) or Cct2p (Kubota et al., 1995a).

■ Mutant phenotype

Two temperature-sensitive mutants *tcp1β-270* and *tcp1β-326* show abnormal tubulin organization in budding yeast and disruption of the gene produces a lethal phenotype (Miklos et al., 1994). Four other temperature-sensitive mutants *bin3-1*, *bin3-2*, *bin3-3* and *bin3-4* show both abnormal tubulin and actin phenotypes (Chen et al., 1994). A temperature-sensitive mutant of CCTβ is lethal in combination with a temperature-sensitive mutant of either CCTα, CCTγ or CCTδ (Chen et al., 1994).

■ *CCTβ* genes and sequences

Mouse *Cctβ* cDNA (GenBank accession number Z31553) was isolated by hybridization using a human PCR probe produced with a mixed primer for a motif (TNDGATI) conserved between CCTα and archaebacterial chaperonin TF55 (Kubota et al., 1994). The yeast homologue of *Cctβ* (accession number X77675) was found in a yeast chromosome sequencing project (Miklos et al., 1994) and the nematode homologue of *Cctβ* (U25632) has been sequenced (Leroux and Candido, 1995). Mouse and yeast CCTβ proteins are, respectively, 535 and 543 amino acids in length and share 66% identity (Kubota et al., 1994; Miklos et al., 1994). These CCTβ proteins share approximately 30% identity with the other subunits of CCT (Kubota et al., 1994, 1995b; Miklos et al., 1994) and the archaebacterial chaperonin TF55 (Kubota et al., 1994). CCTβ proteins also show weak similarity to the eubacterial chaperonin GroEL and intraorganellar chaperonins Hsp60 and rubisco subunit-binding protein (Kubota et al., 1994, 1995a; Kim et al., 1994). The CCTβ protein sequence contains several chaperonin consensus motifs which have been postulated to be involved in ATPase activity (Kubota et al., 1994, 1995a) and these motifs are contained within sequences corresponding to the equatorial ATPase domain of the *Escherichia coli* chaperonin GroEL (Kim et al., 1994).

■ CCTβ protein

Mouse CCTβ migrates as a 53 kDa spot of pI 6.55 by two-dimensional gel analysis (Kubota et al., 1994; Hynes et al., 1995) and forms a complex with CCTα and six other subunit species; monoclonal antibodies specific to CCTα coimmunoprecipitates CCTβ and the other subunit species (Kubota et al., 1994, 1995b). Yeast CCT also makes a complex with CCTα (Miklos et al., 1994). The CCT complex is approximately 960 kDa and has a double-torus-like structure and assists in the folding of actin, tubulin and some other proteins in the eukaryotic cytosol upon ATP hydrolysis. The structure, chaperone function and other characteristics of the CCT complex are described in detail in the overview section, p. 207).

References

Chen, X., Sullivan, D.S., and Huffaker, T.C. (1994). Two yeast genes with similarity to HSP-1 are required for microtubule and actin function in vivo. Proc. Natl. Acad. Sci. USA **91**, 9111–9115.

Hynes, G., Kubota, H., and Willison, K. (1995). Antibody characterization of two distinct conformations of the chaperonin containing TCP-1 from mouse testis. FEBS. Lett. **358**, 129–132.

Kim, S., Willison, K.R., and Horwich, A.L. (1994). Cytosolic chaperonin subunits have a conserved ATPase domain but diverged polypeptide-binding domains. Trends Biochem. Sci. **19**, 543–548.

Kubota, H., Hynes, G., Carne, A., Ashworth, A., and Willison, K. (1994). Identification of six Tcp-1-related genes encoding divergent subunits of the TCP-1-containing chaperonin. Curr. Biol. 4, 89–99.

Kubota, H., Hynes, G., and Willison, K. (1995a). The chaperonin containing t-complex polypeptide 1 (TCP-1): multisubunit machinery assisting in protein folding and assembly in the eukaryotic cytosol. Eur. J. Biochem. **230**, 3–16.

Kubota, H., Hynes, G., and Willison, K. (1995b). The eighth Cct gene, Cctq, encoding the theta subunit of the cytosolic chaperonin that contains TCP-1. Gene **154**, 231–236.

Leroux, M.R. and Candido, E.P.M. (1995). Characterization of four new tcp-1-related cct genes from the nematode Caenorhabditis elegans. DNA Cell Biol. **14**, 951–960.

Miklos, D., Caplan, S., Martens, D., Hynes, G., Pitluk, Z., Brown, C., Barrell, B., Horwich, A.L., and Willison, K. (1994). Primary structure and function of a second essential member of heterooligomeric TCP1 chaperonin complex of yeast, TCP1b. Proc. Natl. Acad. Sci. USA **91**, 2743–2747.

■ *Hiroshi Kubota:*
HSP Research Institute
Kyoto Research Park
17 Chudoji-minamimachi
Shimogyo-ku, Kyoto 600, Japan
Tel. 81 75 315 8656
Fax. 81 75 315 8659
E-mail: kubota@hsp.co.jp

■ *Keith Willison:*
Cancer Research Campaign Centre for Cell and Molecular Biology
Institute of Cancer Research
Chester Beatty Laboratories
237 Fulham Road
London SW3 6JB, UK
Tel./Fax. 44 171 351 3325
E-mail: willison@icr.ac.uk

CCTγ

CCTγ is the γ subunit of the CCT complex which is a hetero-oligomeric, double-torus-shaped, 960 kDa molecular chaperone abundant in eukaryotic cytosol. The CCT complex assists in protein folding in the cytosol of eukaryotic cells in a process that is dependent upon ATP hydrolysis. Mouse CCTγ is the largest (60.6 kDa as calculated from its protein sequence) subunit of the eight constitutively expressed subunit species of CCT.

Alternative names

Mouse CCTγ is also called TRiC-P5 (Joly et al., 1994a). The yeast homologue of CCTγ is also called Bin2p (Chen et al., 1994) or Cct3p (Kubota et al., 1995a).

Mutant phenotype

A temperature-sensitive mutant called bin2-1 shows abnormal actin and tubulin organization in budding yeast and disruption of the gene creates a recessive lethal phenotype (Chen et al., 1994). A temperature-sensitive mutant of CCTγ is lethal in combination with a temperature-sensitive mutant of either CCTα, CCTβ or CCTδ (Chen et al., 1994).

CCTγ genes and sequences

Mouse Cctγ cDNA (GenBank accession number Z31556) was isolated by hybridization using a partial cDNA probe accidentally cloned from human kidney (Kubota et al., 1994) and cloned independently (accession number L20509) by screening of a cDNA library rich in nuclear matrix protein-encoding genes (Joly et al., 1994a). The exon/intron structure of the mouse Cctγ gene has been determined and is different from that of Cctα (Sevigny et al., 1995). The yeast homologue of Cctγ (accession number U09408) was cloned by complementation of a temperature-sensitive mutant gene which affects nuclear segregation (Chen et al., 1994). CCTγ from a protozoan species, Terahymena pyriformis (accession number Z34855), has also been cloned by hybridization screening for a motif (PGGG) found in microtubule-associated proteins (Soares et al., 1994). The human (X74801; Walkley et al., 1996) and Xenopus (U37062; Dunn and Mercola, 1996) homologues of Cctγ have also been cloned. Mouse, yeast and Tetrahymena CCTγ proteins are, respectively, 545, 533 and 559 amino acids in length and share approximately 60% identity with one another (Chen et al., 1994; Kubota et al., 1994; Soares et al., 1994). These CCTγ proteins share approximately 30% identity with the other subunits of CCT (Chen et al., 1994; Kubota et al., 1994, 1995b; Soares et al., 1994) and the archaebacterial chaperonin TF55 (Kubota et al., 1994). CCTγ protein also shows weak similarity to the eubacterial chaperonin GroEL and intraorganellar chaperonins Hsp60 and rubisco subunit-binding protein (Kim et al., 1994; Kubota et al., 1994, 1995a; Soares et al., 1994). The CCTγ protein sequence contains several chaperonin consensus motifs which have been postulated to be involved in ATPase activity (Kubota et al., 1994, 1995a) and the location of these motifs corresponds to the equatorial ATPase domain of the Escherichia coli chaperonin GroEL (Kim et al., 1994).

■ CCTγ protein

Mouse CCTγ migrates as a 65 kDa spot of pI 6.85 by two-dimensional gel analysis (Kubota et al., 1994; Hynes et al., 1995) and forms part of a complex with CCTα and six other subunit species; monoclonal antibodies specific to CCTα coimmunoprecipitates CCTγ and the other subunit species (Kubota et al., 1994, 1995b). The CCT complex is approximately 960 kDa and has a double-torus-like structure and assists in the folding of actin, tubulin and some other proteins in eukaryotic cytosol upon ATP hydrolysis. The structure, chaperone function and other characteristics of the CCT complex are described in detail in the overview section, p. 207).

■ CCTγ localization

CCTγ of cultured human cells distributes not only in the cytosol but also in the nucleus in association with the nuclear matrix (Joly et al., 1994b).

■ Gene regulation of CCTγ

Expression of the CCTγ gene in Tetrahymena is up-regulated during cilia recovery concomitant with tubulin gene induction and is not heat inducible (Soares et al., 1994). The human gene is also not heat inducible (Joly et al., 1994b). Like Cctα, the mouse Cctγ gene is up-regulated in testis (Walkley et al., 1996). Cctγ is highly expressed in Xenopus craninal neural crest, suggesting a novel role of CCT as a molecular chaperone (Dunn and Mercola, 1996).

■ References

Chen, X., Sullivan, D.S., and Huffaker, T.C. (1994). Two yeast genes with similarity to TCP-1 are required for microtubule and actin function in vivo. Proc. Natl. Acad. Sci. USA **91**, 9111–9115.

Dunn, M.K. and Mercola, M. (1996). Cloning and expression of Xenopus CCTγ, a chaperonin subunit developmentally regulated in neural-derived and myogenic lineages. Develop. Dynamics **205**, 387–394.

Joly, E.C., Sevigny, G., Todorov, I.T., and Bibor-Hardy, V. (1994a). cDNA encoding a novel TCP-1-related protein. Biochem. Biophys. Acta **1217**, 224–226.

Joly, E.C., Tremblay, E., Tanguay, R.M., Wu, Y., and Bibor-Hardy, V. (1994b). TRiC-P5, a novel TCP-1-related protein is located in the cytoplasm and in the nuclear matrix. J. Cell Sci. **107**, 2851–2859.

Kim, S., Willison, K.R., and Horwich, A.L. (1994). Cytosolic chaperonin subunits have a conserved ATPase domain but diverged polypeptide-binding domains. Trends Biochem. Sci. **19**, 543–548.

Kubota, H., Hynes, G., Carne, A., Ashworth, A., and Willison, K. (1994). Identification of six Tcp-1-related genes encoding divergent subunits of the TCP-1-containing chaperonin. Curr. Biol. **4**, 89–99.

Kubota, H., Hynes, G., and Willison, K. (1995a). The chaperonin containing t-complex polypeptide 1 (TCP-1): multisubunit machinery assisting in protein folding and assembly in the eukaryotic cytosol. Eur. J. Biochem. **230**, 3–16.

Kubota, H., Hynes, G., and Willison, K. (1995b). The eighth Cct gene, Cctq, encoding the theta subunit of the cytosolic chaperonin that contains TCP-1. Gene **154**, 231–236.

Sevigny, G., Lemieux, N., Steyaert, A., and Bibor-Hardy, V. (1995). Structure of the gene encoding for the mouse TRiC-P5 subunit of the cytosolic chaperonin TRiC. Genomics **31**, 107–110.

Soares, H., Penque, D., Mouta, C., and Rodrigues-Pousada, C. (1994). A Tetrahymena orthologue of the mouse chaperonin subunit CCTg and its coexpression with tubulin during cilia recovery. J. Biol. Chem. **269**, 29299–29307.

Walkley, N.A., Demaine, A.G., and Malik, A.N. (1996). Cloning, structure and mRNA expression of human Cctγ, which encodes the chaperonin subunit CCTγ. Biochem. J. **313**, 381–389.

■ *Hiroshi Kubota:*
HSP Research Institute
Kyoto Research Park
17 Chudoji-minamimachi
Shimogyo-ku, Kyoto 600, Japan
Tel. 81 75 315 8656
Fax. 81 75 315 8659
E-mail: kubota@hsp.co.jp

■ *Keith Willison:*
Cancer Research Campaign Centre for Cell and Molecular Biology
Institute of Cancer Research
Chester Beatty Laboratories
237 Fulham Road
London SW3 6JB, UK
Tel./Fax. 44 171 351 3325
E-mail: willison@icr.ac.uk

CCTδ

CCTδ is the δ subunit of the CCT complex which is a hetero-oligomeric, double-torus-shaped, 960 kDa molecular chaperone abundant in eukaryotic cytosol. The CCT complex assists in protein folding in the cytosol of eukaryotic cells in a process that is dependent upon ATP hydrolysis. Mouse CCTδ is 58.1 kDa (as calculated from its protein sequence) and is one of the eight constitutively expressed subunit species of CCT.

■ Alternative names

The yeast homologue of CCTδ is also called Anc2p (Vinh and Drubin, 1994) or Cct4p (Kubota et al., 1995a).

■ Mutant phenotype

A temperature-sensitive mutant called *anc2-1* shows abnormal actin organization in budding yeast and a null allele of the gene is a recessive lethal (Vinh and Drubin, 1994).

■ *CCTδ* genes and sequences

Mouse *Cctδ* cDNA (GenBank accession number Z31554) was isolated by hybridization using a partial nematode cDNA probe cloned by random selection (Waterston et al., 1992; Kubota et al., 1994). The complete nemtode *Cctδ* homologue cDNA (U25697) has been sequenced (Leroux and Candido, 1995). The yeast homologue of *Cctδ* (accession number Z33504) was cloned by complementation of a temperature-sensitive mutant which enhanced defects caused by actin mutations (Vinh and Drubin, 1994). The puffer fish homologue of *Cctδ* has also been cloned (D49483; Yoda et al., 1995). Mouse and yeast CCT proteins are, respectively, 539 and 559 amino acids in length and share 59% identity (Kubota et al., 1994; Vinh and Drubin, 1994). These CCTδ proteins share approximately 30% identity with the other subunits of CCT (Kubota et al., 1994, 1995b; Vinh et al., 1994) and 40% with the archaebacterial chaperonin TF55 (Kubota et al., 1994). The CCTδ protein also shows weak similarity to the eubacterial chaperonin GroEL and intraorganellar chaperonins Hsp60 and rubisco subunit-binding protein (Kim et al., 1994; Kubota et al., 1994, 1995a). The CCT protein sequence contains several chaperonin consensus motifs which have been postulated to be involved in ATPase activity (Kubota et al., 1994, 1995a) and the location of these motifs corresponds to the equatorial ATPase domain of the *Escherichia coli* chaperonin GroEL (Kim et al., 1994).

■ CCTδ protein

Mouse CCTδ migrates as a 56 kDa spot with a pI higher than 7.2 by two-dimensional gel analysis (Kubota et al., 1994; Hynes et al., 1995) and forms a complex with CCTα and six other subunit species; monoclonal antibodies specific to CCTα coimmunoprecipitate CCTδ and the other subunit species (Kubota et al., 1994, 1995b). The CCT complex is approximately 960 kDa and has a double-torus-like structure and assists in the folding of actin, tubulin and some other proteins in eukaryotic cytosol in a process that is dependent upon ATP hydrolysis. The structure, chaperone function and other characteristics of the CCT complex are described in detail in the overview section, p. 207).

■ Gene regulation of *Cctδ*

Cct mRNA is highly expressed in haematopoietic cells (Xie and Palacios, 1994).

■ References

Hynes, G., Kubota, H., and Willison, K. (1995). Antibody characterization of two distinct conformations of the chaperonin containing TCP-1 from mouse testis. FEBS. Lett. **358**, 129–132.

Kim, S., Willison, K.R., and Horwich, A.L. (1994). Cytosolic chaperonin subunits have a conserved ATPase domain but diverged polypeptide-binding domains. Trends Biochem. Sci. **19**, 543–548.

Kubota, H., Hynes, G., Carne, A., Ashworth, A., and Willison, K. (1994). Identification of six Tcp-1-related genes encoding divergent subunits of the TCP-1-containing chaperonin. Curr. Biol. **4**, 89–99.

Kubota, H., Hynes, G., and Willison, K. (1995a). The chaperonin containing t-complex polypeptide 1 (TCP-1): multisubunit machinery assisting in protein folding and assembly in the eukaryotic cytosol. Eur. J. Biochem. **230**, 3–16.

Kubota, H., Hynes, G., and Willison, K. (1995b). The eighth Cct gene, *Cctq*, encoding the theta subunit of the cytosolic chaperonin that contains TCP-1. Gene **154**, 231–236.

Leroux, M.R. and Candido, E.P.M. (1995). Characterization of four new tcp-1-related cct genes from the nematode *Caenorhabditis elegans*. DNA Cell Biol. **14**, 951–960.

Vinh, D.B.-N. and Drubin, D.G. (1994). A yeast TCP-1 like protein is required for actin function in vivo. Proc. Natl. Acad. Sci. USA **91**, 9116–9120.

Warterston, R., Martin, C., Craxton, M., Coulson, A., Hillier, L., Durbin, R., Green, P., Shownkeen, R., Halloran, N., Metzstein, M., Hawkins, T., Wilson, R., Berks, M., Du, Z., Tomas, K., Thierry-Meig, J., and Sulston, J. (1992). A survey of expressed genes in Caenorhabditis elegans. Nature Genetics **1**, 114–123.

Xie, X. and Palacios, R. (1994). Cloning and expression of a new mammalian chaperonin gene from a multipotent hematopoietic progenitor clone. Blood **84**, 2171–2174.

Yoda, T., Morita, T., Kawamatsu, K., Sueki, K., Shibata., T., and Hamano, Y. (1995). Cloning and sequencing of the chaperonin-encoding $Cct\delta$ gene from Fugu rubripes rubripes. Gene **166**, 249–253.

■ *Hiroshi Kubota:*
HSP Research Institute
Kyoto Research Park
17 Chudoji-minamimachi
Shimogyo-ku, Kyoto 600, Japan
Tel. 81 75 315 8656
Fax. 81 75 315 8659
E-mail: kubota@hsp.co.jp

■ *Keith Willison:*
Cancer Research Campaign Centre for Cell and Molecular Biology
Institute of Cancer Research
Chester Beatty Laboratories
237 Fulham Road
London SW3 6JB, UK
Tel./Fax. 44 171 351 3325
E-mail: willison@icr.ac.uk

CCTε

CCTε is the ε subunit of the CCT complex which is a hetero-oligomeric, double-torus-shaped, 960 kDa molecular chaperone abundant in eukaryotic cytosol. The CCT complex assists in protein folding in the cytosol of eukaryotic cells in a process that is dependent upon ATP hydrolysis. Mouse CCTε is 59.6 kDa (as calculated from its protein sequence) and is one of the eight constitutively expressed subunit species of CCT.

■ Alternative names

The yeast homologue of CCTε is also called Cct5p (Kubota et al., 1995a).

■ *CCTε genes and sequences*

Mouse *Cctε* cDNA (GenBank accession number Z31555) was isolated by hybridization using a human PCR probe produced with a mixed primer for a motif (TNDGATI) conserved between CCTα and the archaebacterial chaperonin TF55 (Kubota et al., 1994). The plant homologues of *Cctε* (accession number X75777) were cloned from *Avena sativa* using degenerate primers for the two consensus motifs (TITNDGA and MPKRI) of CCTα and TF55 (Ehmann et al., 1993) and *Cucumis satvus* using degenerate primers for the consensus motif (QDREIGDGTT) of CCTα (Ahnert et al., 1996). The yeast homologue of *Cctε* (accession number L37350) has also been cloned (Kim et al., 1994). A partial sequence of nematode *Cctε* homologue was first found in a random sequencing project (Warterston et al., 1992) and its complete sequence has been determined (U25698; Leroux and Candido, 1995). Mouse, *A. sativa* and yeast CCTε proteins are, respectively, 541, 535 and 551 amino acids in length and share approximately 60% identity (Ehmann et al., 1993; Kim et al., 1994; Kubota et al., 1994). The CCTε protein shares approximately 30% identity with the other subunits of CCT (Kubota et al., 1994, 1995b) and 40% identity with the archaebacterial chaperonin TF55 (Kubota et al., 1994). CCTε also shows weak similarity to the eubacterial chaperonin GroEL and intraorganellar chaperonins Hsp60 and rubisco subunit-binding protein (Kim et al., 1994; Kubota et al., 1994, 1995a). The CCTε protein sequence contains several chaperonin consensus motifs which have been postulated to be involved in ATPase activity (Kubota et al., 1994, 1995a) and the location of these motifs corresponds to the equatorial ATPase domain of the *Escherichia coli* chaperonin GroEL (Kim et al., 1994).

■ CCTε protein

Mouse CCTε migrates as a 61 kDa spot of pI 6.35 in two-dimensional gel analysis (Kubota et al., 1994; Hynes et al., 1995) and makes a complex with CCTα and six other subunit species; monoclonal antibodies specific to CCTα coimmunoprecipitate CCTε and the other subunit species (Kubota et al., 1994, 1995b). The CCT complex is approximately 960 kDa and has a double-torus-like structure and assists in the folding of actin, tubulin and some other proteins in eukaryotic cytosol upon ATP hydrolysis. The structure, chaperone function and other characteristics of the CCT complex are described in detail in the overview section, p. 207).

References

Ahnert, V., May, C., Gerke, R., and Kindl, H. (1996). Cucumber T-complex protein. Molecular cloning, bacterial expression and characterization within a 22-S cytosolic complex in cotyledons and hypocotyls. Eur. J. Biochem. **235**, 114–119 [see correction in Eur. J. Biochem. **238**, 867].

Ehmann, B., Krenz, M., Mummert, E., and Schafer, E. (1993). Two *Tcp-1*-related but highly divergent gene families exist in oat encoding proteins of assumed chaperone function. FEBS Lett. **336**, 313–316.

Hynes, G., Kubota, H., and Willison, K. (1995). Antibody characterization of two distinct conformations of the chaperonin containing TCP-1 from mouse testis. FEBS. Lett. **358**, 129–132.

Kim, S., Willison, K.R., and Horwich, A.L. (1994). Cytosolic chaperonin subunits have a conserved ATPase domain but diverged polypeptide-binding domains. Trends Biochem. Sci. **19**, 543–548.

Kubota, H., Hynes, G., Carne, A., Ashworth, A., and Willison, K. (1994). Identification of six *Tcp-1*-related genes encoding divergent subunits of the TCP-1-containing chaperonin. Curr. Biol. **4**, 89–99.

Kubota, H., Hynes, G., and Willison, K. (1995a). The chaperonin containing t-complex polypeptide 1 (TCP-1): multisubunit machinery assisting in protein folding and assembly in the eukaryotic cytosol. Eur. J. Biochem. **230**, 3–16.

Kubota, H., Hynes, G., and Willison, K. (1995b). The eighth *Cct* gene, *Cctq*, encoding the theta subunit of the cytosolic chaperonin that contains TCP-1. Gene **154**, 231–236.

Leroux, M.R. and Candido, E.P.M. (1995). Characterization of four new *tcp-1*-related *cct* genes from the nematode *Caenorhabditis elegans*. DNA Cell Biol. **14**, 951–960.

Warterston, R., Martin, C., Craxton, M., Coulson, A., Hillier, L., Durbin, R., Green, P., Shownkeen, R., Halloran, N., Metzstein, M., Hawkins, T., Wilson, R., Berks, M., Du, Z., Tomas, K., Thierry-Meig, J., and Sulston, J. (1992). A survey of expressed genes in Caenorhabditis elegans. Nature Genetics **1**, 114–123.

■ *Hiroshi Kubota:*
HSP Research Institute
Kyoto Research Park
17 Chudoji-minamimachi
Shimogyo-ku, Kyoto 600, Japan
Tel. 81 75 315 8656
Fax. 81 75 315 8659
E-mail: kubota@hsp.co.jp

■ *Keith Willison:*
Cancer Research Campaign Centre for Cell and Molecular Biology
Institute of Cancer Research
Chester Beatty Laboratories
237 Fulham Road
London SW3 6JB, UK
Tel./Fax. 44 171 351 3325
E-mail: willison@icr.ac.uk

CCTζ

CCTζ is the ζ subunit of the CCT complex which is a hetero-oligomeric, double-torus-shaped, 960 kDa molecular chaperone abundant in the eukaryotic cytosol. The CCT complex assists in protein folding in the cytosol of eukaryotic cells in a process that is dependent upon ATP hydrolysis. Mouse CCTζ1 is 58.0 kDa (as calculated from its protein sequence) and is one of the eight constitutively expressed subunit species of CCT.

■ Alternative names

The human homologue of CCTζ is also called TCP20 (Li et al., 1994). The yeast homologue of CCTζ is also called Tcp20p (Li et al., 1994) or Cct6p (Kubota et al., 1995a).

■ Mutant phenotype

A null allele of the yeast homologue of *Cctζ* gene is a recessive lethal (Li et al., 1994).

■ *CCTζ* genes and sequences

Mouse *Cctζ1* cDNA (GenBank accession number Z31557) was isolated by hybridization using a nematode partial cDNA probe cloned by random selection (Waterston et al., 1992; Kubota et al., 1994) and the human homologue of *Cctζ1* (accession number L27706) was cloned as a cDNA rescuing a mutation in an amino acid transport system (Segel et al., 1992; Li et al., 1994). The complete sequence of the nematode homologue of *Cctζ* (U13070) has been determined by the *Caenorhabdidtis elegans* genome sequencing project (Wilson et al., 1994). The yeast homologue of *Cctζ* (accession number L37350) has also been cloned (Li et al., 1994). Mouse, human and yeast CCTζ proteins are, respectively, 531, 531 and 546 amino acids in length (Kubota et al. 1994; Li et al., 1994). Mouse and human CCTζ proteins share 97% identity (Kubota et al., 1994; Li et al., 1994) and human and yeast CCtζ proteins share 57% identity. The CCtζ protein shares approximately 30% identity with the other subunits of CCT (Kubota et al., 1994, 1995b) and the archaebacterial chaperonin TF55 (Kubota et al., 1994; Li et al., 1994). CCtζ

also shows weak similarity to the eubacterial chaperonin GroEL and intraorganellar chaperonins Hsp60 and rubisco subunit-binding protein (Kim et al., 1994; Kubota et al., 1994, 1995a). The CCtζ protein sequence contains several chaperonin consensus motifs which have been postulated to be involved in ATPase activity (Kubota et al., 1994, 1995a) and the location of these motifs corresponds to the equatorial ATPase domain of the E. coli chaperonin GroEL (Kim et al., 1994).

■ CCtζ protein

Mouse CCTζ-1 migrates as two spots (62.5 kDa, pI 6.90; 62.0 kDa, pI 6.95) in two-dimensional gel analysis (Kubota et al., 1994; 1997; Hynes et al., 1995) and makes a complex with CCTα and six other subunit species; monoclonal antibodies specific to CCTα co-immunoprecipitates CCTζ and the other subunit species (Kubota et al., 1994; Kubota et al., 1995b). Recently a second member of the mouse CCTζ gene family, CCTζ-2 (Z50192), has been identified and it appears to be expressed only in testis Kubota et al., 1997). The mouse CCTζ-2 protein migrates as a 57 kDa spot of pI 7.1. A potential human homologue of CCTζ-2 (D78333) has been found although the sequence as reported may contain several sequencing errors. The CCT complex is approximately 960 kDa and has a double-torus-like structure and assists in the folding of actin, tubulin and some other proteins in the eukaryotic cytosol in a process that is dependent upon ATP hydrolysis. The structure, chaperone function and other characteristics of the CCT complex are described in detail in the overview section. p. 207).

■ References

Hynes, G., Kubota, H., and Willison, K. (1995). Antibody characterization of two distinct conformations of the chaperonin containing TCP-1 from mouse testis. FEBS. Lett. **358**, 129–132.

Kim, S., Willison, K.R., and Horwich, A.L. (1994). Cytosolic chaperonin subunits have a conserved ATPase domain but diverged polypeptide-binding domains. Trends Biochem. Sci. **19**, 543–548.

Kubota, H., Hynes, G., Carne, A., Ashworth, A., and Willison, K. (1994). Identification of six Tcp-1-related genes encoding divergent subunits of the TCP-1-containing chaperonin. Curr. Biol. **4**, 89–99.

Kubota, H., Hynes, G., and Willison, K. (1995a). The chaperonin containing t-complex polypeptide 1 (TCP-1): Multisubunit machinery assisting in protein folding and assembly in the eukaryotic cytosol. Eur. J. Biochem. **230**, 3–16.

Kubota, H., Hynes, G., and Willison, K. (1995b). The eighth Cct gene, Cctq, encoding the theta subunit of the cytosolic chaperonin that contains TCP-1. Gene **154**, 231–236.

Kubota, H., Hynes, G.M., Kerr, S.M., and Willison, K.R. (1997). Tissue-specific subunit of the mouse cytosolic chaperonin-containing TCP-1, FEBS Lett. **402**, 53–56.

Li, W.-Z., Lin, P., Frydman, J., Boal, T.R., Cardillo, T.S., Richard, L.M., Toth, D., Lichtman, M.A., Hartle, F.-U., Sherman, F., and Segel, G.B. (1994). Tcp20, a subunit of the eukaryotic TRiC chaperonin from humans and yeast. J. Biol. Chem. **269**, 18616–18622.

Ozaki, K., Kuroki, T., Hayashi, S., and Nakamura, Y. (1996). Isolation of three testis-specific genes (TSA303, TSA806, TSA903) by a differential mRNA display method. Genomics **36**, 316–319.

Segel, G.B., Boal, T.R., Cardillo, T.S., Murant, F.G., Lichtman, M.A., and Sherman, F. (1992). Isolation of a gene encoding a chaperonin-like protein by complementation of yeast amino acid transport mutants with human cDNA. Proc. Natl. Acad. Sci. USA **89**, 6060–6064.

Waterston, R., Martin, C., Craxton, M., Coulson, A., Hillier, L., Durbin, R., Green, P., Shownkeen, R., Halloran, N., Metzstein, M., Hawkins, T., Wilson, R., Berks, M., Du, Z., Tomas, K., Thierry-Meig, J., and Sulston, J. (1992). A survey of expressed genes in Caenorhabditis elegans. Nature Genetics **1**, 114–123.

Wilson, R., Ainscough, R., Anderson, K., Baynes, C., Berks, M., Bonfield, J., Burton, J., Connell, M., Copsey, T., Cooper, J., Coulson, A., Craxton, M., Dear, S., Du, Z., Durbin, R., Favello, A., Fraser, A., Fulton, L., Gardner, A., Green, P., Hawkins, T., Hillier, L., Jier, M., Johnston, L., Jones, M., Kershaw, J., Kirsten, J., Laisster, N., Lateille, P., Lightning, J., Lloyd, C., Mortomore, B., O'Calloghan, M., Parsons, J., Percy, C., Rifken, L., Roopra, A., Saunders, D., Shownkeen, R., Sims, M., Smaldon, N., Smith, A., Smith, M., Sonnhammer, E., Staden, R., Sulston, J., Thierry-Mieg, J., Thomas, K., Vaudin, K., Waterston, R., Watson, A., Weinstock, L., Wilkinson-Sproat, J., and Wohldman, P. (1994). 2.2 Mb of contiguous nucleotide sequence from chomosome III of C. elegans. Nature **368**, 32–38.

■ Hiroshi Kubota:
HSP Research Institute
Kyoto Research Park
17 Chudoji-minamimachi
Shimogyo-ku, Kyoto 600, Japan
Tel. 81 75 315 8656
Fax. 81 75 315 8659
E-mail: kubota@hsp.co.jp

■ Keith Willison:
Cancer Research Campaign Centre for Cell and Molecular Biology
Institute of Cancer Research
Chester Beatty Laboratories
237 Fulham Road
London SW3 6JB, UK
Tel./Fax. 44 171 351 3325
E-mail: willison@icr.ac.uk

CCTη

CCTη is the η subunit of the CCT complex which is a hetero-oligomeric, double-torus-shaped, 960 kDa molecular chaperone abundant in eukaryotic cytosol. The CCT complex assists in protein folding in the cytosol of eukaryotic cells in a process that is dependent upon ATP hydrolysis. Mouse CCTη is 59.6 kDa (as calculated from its protein sequence) and is one of the eight constitutively expressed subunit species of CCT.

■ Alternative names

The yeast homologue of CCTη is also called Cct7p (Kubota et al., 1995a).

■ CCTη genes and sequences

Mouse *Cctη* cDNA (GenBank accession number Z31399) was isolated by random selection from a testis cDNA library (Kubota et al., 1994). The mouse CCT protein is 544 amino acids in length and it shares approximately 30% identity with the other subunits of CCT and 40% with the archaebacterial chaperonin TF55 (Kubota et al., 1994). The CCTη protein also shows weak similarity to the eubacterial chaperonin GroEL and intraorganellar chaperonins Hsp60 and rubisco subunit-binding protein (Kim et al., 1994; Kubota et al., 1994, 1995a). The yeast homologue of *Cctη* has been found by a yeast genome sequencing project (X85021; Rasmussen, 1995). Two tetrahymena species homologues of CCTη were cloned using degenerate primer PCR (U46028 and U46030; Cryne et al., 1996). The CCTν protein sequences contain several chaperonin consensus motifs which have been postulated to be involved in ATPase activity (Kubota et al., 1994, 1995a) and the location of these motifs corresponds to the equatorial ATPase domain of the *Escherichia coli* chaperonin GroEL (Kim et al., 1994).

■ CCTη protein

Mouse CCTη migrates as a 57 kDa spot of pI 7.1 or a slightly more basic streak on two-dimensional gel analysis (Kubota et al., 1994; Hynes et al., 1995) and makes a complex with CCTα and six other subunit species; monoclonal antibodies specific to CCTα coimmunoprecpitate CCTη and the other subunit species (Kubota et al., 1994, 1995b). The CCT complex is approximately 960 kDa and has a double-torus-like structure and assists in the folding of actin, tubulin and some other proteins in eukaryotic cytosol in a process that is dependent upon ATP hydrolysis. The structure, chaperone function and other characteristics of the CCT complex are described in detail in the overview section, p. 207.

■ References

Cyrne, L., Guerreiro, P., Cardoso, A.C., Rodrigues-Pousada, G., and Soares, H. (1996). The tetrahymena chaperonin subunit CCTν gene is coexpressed with CCTγ gene during cilia biogenesis and cell sexual reproduction. FEBS. Lett. **383**, 277–283.

Hynes, G., Kubota, H., and Willison, K. (1995). Antibody characterization of two distinct conformations of the chaperonin containing TCP-1 from mouse testis. FEBS. Lett. **358**, 129–132.

Kim, S., Willison, K.R., and Horwich, A.L. (1994). Cytosolic chaperonin subunits have a conserved ATPase domain but diverged polypeptide-binding domains. Trends Biochem. Sci. **19**, 543–548.

Kubota, H., Hynes, G., Carne, A., Ashworth, A., and Willison, K. (1994). Identification of six Tcp-1-related genes encoding divergent subunits of the TCP-1-containing chaperonin. Curr. Biol. **4**, 89–99.

Kubota, H., Hynes, G., and Willison, K. (1995a). The chaperonin containing t-complex polypeptide 1 (TCP-1): multisubunit machinery assisting in protein folding and assembly in the eukaryotic cytosol. Eur. J. Biochem. **230**, 3–16.

Kubota, H., Hynes, G., and Willison, K. (1995b). The eighth Cct gene, Cctq, encoding the theta subunit of the cytosolic chaperonin that contains TCP-1. Gene **154**, 231–236.

Rasmussen, S.W. (1995). A 37.5 kb region of yeast chromosome X includes the SEM1, MEF2, GSH1 and CSD3 genes, a TCP-1-related gene, an open reading frame similar to the DAL80 gene, and a tRNA (Arg). Yeast **11**, 873–883.

■ *Hiroshi Kubota:*
HSP Research Institute
Kyoto Research Park
17 Chudoji-minamimachi
Shimogyo-ku, Kyoto 600, Japan
Tel. 81 75 315 8656
Fax. 81 75 315 8659
E-mail: kubota@hsp.co.jp

■ *Keith Willison:*
Cancer Research Campaign Centre for Cell and Molecular Biology
Institute of Cancer Research,
Chester Beatty Laboratories
237 Fulham Road
London SW3 6JB, UK
Tel./Fax. 44 171 351 3325
E-mail: willison@icr.ac.uk

CCTθ

CCTθ is the θ subunit of the CCT complex which is a hetero-oligomeric, double-torus-shaped, 960 kDa molecular chaperone abundant in eukaryotic cytosol. The CCT complex assists in protein folding in the cytosol of eukaryotic cells in a process that is dependent upon ATP hydrolysis. Mouse CCTθ is 59.6 kDa (as calculated from its protein sequence) and is one of the eight constitutively expressed subunit species of CCT.

■ Alternative names

The yeast homologue of CCTθ is also called Cct8p (Kubota et al., 1995a).

■ CCTθ genes and sequences

Mouse Cctθ cDNA (GenBank accession number Z37164) was isolated by hybridization using a human PCR probe produced with a mixed primer for a motif (TNDGATI) conserved between CCTα and the archaebacterial chaperonin TF55 (Kubota et al., 1995b). The human homologue of Cctθ (GenBank accession number D13627) was also cloned by random selection from a myeloid cell cDNA library (Nomura et al., 1994). The mouse and human CCTθ proteins are 548 amino acids in length (Nomura et al., 1994; Kubota et al., 1995b) and share 96% identity. The CCTθ protein shares approximately 25% identity with the other subunits of CCT and appears to be the most divergent CCT subunit (Nomura et al., 1994; Kubota et al., 1995b). It shares 30% identity with the archaebacterial chaperonin TF55 (Kubota et al., 1995b). CCTθ also shows weak similarity to the eubacterial chaperonin GroEL and intraorganellar chaperonins Hsp60 and rubisco subunit- binding protein (Kubota et al., 1995b). Candia albicans (U37371; Stoldt et al., 1996) and Saccharomyces cerevisiae (Z49284; D. Van, J. Perea and C. Jacq, unpublished data) homologues of Cctθ have been identified recently. The CCTθ protein sequence contains several chaperonin consensus motifs which have been postulated to be involved in ATPase activity (Kubota et al., 1995a, b) and the location of these motifs correspond to the equatorial ATPase domain of the Escherichia coli chaperonin GroEL (Kim et al., 1994).

■ CCTθ protein

Mouse CCTθ migrates as a 62 kDa spot of pI 6.25, or slightly more basic pI, on two-dimensional gel analysis (Kubota et al., 1994, 1995b; Hynes et al., 1995) and makes a complex with CCTα and six other subunit species; monoclonal antibodies specific to CCTα coimmunoprecipitate CCTθ and the other subunits (Kubota et al., 1995b). Mouse CCTθ is post-translationally modified; probably including adenylation on tyrosine-308 (Hynes et al., 1996). The CCT complex is approximately 960 kDa and has a double-torus-like structure and assists in the folding of actin, tubulin and some other proteins in eukaryotic cytosol in a process that is dependent upon ATP hydrolysis. The structure, chaperone function and other characteristics of the CCT complex are described in detail in the overview section, p. 207).

■ References

Hynes, G., Kubota, H., and Willison, K. (1995). Antibody characterization of two distinct conformations of the chaperonin containing TCP-1 from mouse testis. FEBS. Lett. **358**, 129–132.

Hynes, G., Sutton, C.W., U, S. and Willison, K. (1996). Peptide mass fingerprinting of chaperonin-containing TCP-1 (CCT) and co-purifying proteins. FASEB J. **10**, 137–147.

Kim, S., Willison, K.R., and Horwich, A.L. (1994). Cytosolic chaperonin subunits have a conserved ATPase domain but diverged polypeptide-binding domains. Trends Biochem. Sci. **19**, 543–548.

Kubota, H., Hynes, G., Carne, A., Ashworth, A., and Willison, K. (1994). Identification of six Tcp-1-related genes encoding divergent subunits of the TCP-1-containing chaperonin. Curr. Biol. **4**, 89–99.

Kubota, H., Hynes, G., and Willison, K. (1995a). The chaperonin containing t-complex polypeptide 1 (TCP-1): multisubunit machinery assisting in protein folding and assembly in the eukaryotic cytosol. Eur. J. Biochem. **230**, 3–16.

Kubota, H., Hynes, G., and Willison, K. (1995b). The eighth Cct gene, Cctq, encoding the theta subunit of the cytosolic chaperonin that contains TCP-1. Gene **154**, 231–236.

Nomura, N., Miyajima, N., Sazuka, T., Tanaka, A., Kawarabayashi, Y., Sato, S., Nagase, T., Seiki, N., Ishikawa, K.-I., and Tabata, S. (1994). Prediction of the coding sequences of unidentified human genes; I. The coding sequences of 40 new genes (KIAA0001-KIAA0040) deduced by analysis of randomly sampled cDNA clones from human immature myeloid cell line KG-1. DNA Res. **1**, 27–35.

Stoldt, V., Radenmacher, F., Kerren, V., Ernst, J.F., Pearce, D.A., and Sherman, F. (1996). The cut eukaryotic chaperonin subunits of Saccharomyces cerevisiae and other yeasts: a mini-review. Yeast **12**, 523–529.

■ Hiroshi Kubota:
HSP Research Institute
Kyoto Research Park
17 Chudoji-minamimachi
Shimogyo-ku, Kyoto 600, Japan
Tel. 81 75 315 8656
Fax. 81 75 315 8659
E-mail: kubota@hsp.co.jp

Keith Willison:
Cancer Research Campaign Centre for Cell and Molecular Biology
Institute of Cancer Research,
Chester Beatty Laboratories

237 Fulham Road
London SW3 6JB, UK
Tel./Fax. 44 171 351 3325
E-mail: willison@icr.ac.uk

Cofactors in the facilitated folding of α- and β-tubulin

The facilitated folding of α- and β-tubulin requires ATP- and GTP-dependent interaction of these target proteins with cytosolic chaperonin (CCT), followed by a cascade of ATP-independent interactions with a series of protein cofactors termed A–E. Homologs of these cofactors exist in lower eukaryotes. The function of the cofactors is unclear, but some of them probably serve to stabilize quasi-native intermediates produced as a result of one or more cycles of interaction with the chaperonin. Curiously, in spite of the fact that α- and β-tubulin are related to one another, a different but overlapping set of cofactors is required to facilitate the productive folding of these two proteins.

Alternative names

Cofactor A, Rbl2 (*S. cerevisiae*) (Archer et al., 1995); cofactor D, Cin1 (*S. cerevisiae* and *S. pombe*) (Stearns et al., 1990); cofactor E = Pac2 (*S. cerevisiae* and *S. pombe*) (GenBank accession number U16814).

Isolation of cofactors

The tubulin cofactors were originally discovered as two crude fractions obtained by ion exchange chromatography that were together required (in addition to ATP, GTP, CCT and native tubulin heterodimers, the latter included so as to provide the opportunity for stabilization by exchange of newly folded tubulin polypeptides) for the productive folding of α- and β-tubulin (Gao et al., 1993). Further purification of crude extracts of bovine testis has identified four homogeneous proteins (cofactors A, D, E and C) that participate in the β-tubulin folding pathway. The purification of cofactor A is described in Gao et al. (1994); the schemes for the purification of cofactors D, E and C are described in Tian et al. (1996).

Cofactor genes and sequences

Amino acid sequences deduced from cloned cDNAs encoding cofactors A, C, D and E have been described (Gao et al., 1994; Tian et al., 1996); GenBank accession numbers are U05333, U61234, U61233 and U61232, respectively. Cofactor A is distantly related to a region within the rod domain of myosin heavy chain, and is the homolog of Rbl2 in *S. cerevisiae* (Archer et al., 1995). Cofactor D is the homolog of Cin1 in *S. cerevisiae* (Stearns et al., 1990), while cofactor E is the homolog of Pac2 in *S. cerevisiae* and *Schizosaccharomyces pombe*. Cofactor C contains a domain that is predicted to form amphipathic α helices, but it shares no homology with any protein of known function. Cofactor E contains a region with homology within the microtubule-binding domain of CLIP-170, a microtubule-associated protein (MAP) that links endocytic vesicles to microtubules (Pierre et al., 1992). The significance of this homology in terms of cofactor function is not yet known.

Cofactor proteins and biological activities

A characteristic feature of cofactors A, D and E is that they generate β-tubulin-containing intermediates (termed lower, middle and super-middle band, respectively) that are identifiable upon non-denaturing gel electrophoresis of the products of CCT-mediated *in vitro* folding reactions containing these cofactors (Tian et al., 1996). The formation of these intermediates depends on the presence of GTP, but not on ATP or GTP hydrolysis. Polyclonal antisera have been raised against cofactors A, D and E; these have been used to identify the cofactor composition of the aforementioned intermediates. The intracellular distribution of cofactors is not known, but consistent with the sequence homology between cofactor E and CLIP-170, this cofactor (and cofactor C) co-cycle with native brain tubulin through multiple cycles of polymerization and depolymerization.

Biological regulation of cofactors

While it is clear that the tubulin cofactors play a role in the facilitated folding of α- and β-tubulin, the observation that cofactors C and E behave as MAPs raises the possibility that they also influence microtubule behavior. For example, the partitioning of cofactors C and E between free and microtubule-bound forms might serve

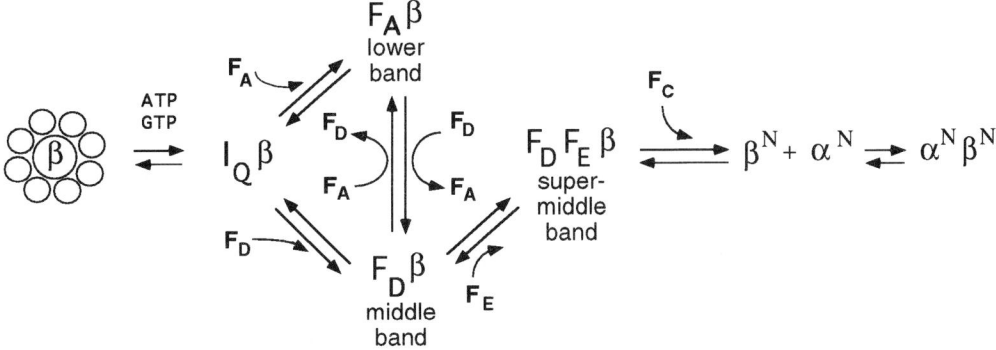

Figure 1. Pathways leading to the generation of assembly-competent β-tubulin. β: β-tubulin target protein. I$_Q$β: quasi-native β-tubulin folding intermediates generated via ATP-dependent interaction with CCT (shown as an eight subunit toroid). FA, FD, FE, FC: cofactors A, D, E and C, respectively. αN, βN: native monomeric α- and β-tubulin. From Tian et al. (1996).

to regulate the amount of these cofactors available for tubulin folding.

Genetic studies

Rbl2, the yeast homolog of cofactor A (and two other yeast non-tubulin genes) has been isolated as a result of genetic screens designed to identify genes that, when overexpressed, rescue cells from the otherwise lethal overexpression of β-tubulin (Archer et al., 1995). Rbl2 is not an essential gene, consistent with the alternate pathways leading to properly folded β-tubulin defined by biochemical experiments using purified cofactors (see below). Mutations in Cin1p (the *S. cerevisiae* homolog of cofactor D), while not lethal, result in chromosome instability and increased sensitivity of microtubules to cold and to the antimicrotubule drug benomyl (Stearns et al., 1990). Mutations in Pac2p, the homolog of cofactor E, result in defects in microtubule motor function (Hoyt, et al., 1997).

Biological interactions

The overall pathway leading to the generation of native β-tubulin has been deduced from *in vitro* β-tubulin folding reactions containing CCT and purified cofactors (Fig. 1) (Tian et al., 1996). It seems probable that the substrates recognized by cofactors A and D are quasi-native β-tubulin intermediates (termed I$_Q$: Tian et al., 1995; Lewis et al., 1996) produced as a result of one or more cycles of ATP- and GTP-dependent interaction of the target protein with CCT. The quasi-native nature of these intermediates is deduced from their relative resistance to proteolysis (compared with molecules that have not undergone ATP-dependent cycles of interaction with CCT), or, in the case of α-tubulin, the incorporation of non-exchangeably bound GTP (Tian et al., 1995). Inclusion of cofactors A, D, C and E in CCT-mediated α-tubulin folding reactions does not result in the generation of native α-tubulin (Tian et al., 1996); hence, in spite of the extensive homology shared by α- and β-tubulins, the sets of cofactors required for their facilitated folding are different, although they may be overlapping.

References

Archer, J.E., Vega, L.R., and Solomon, F. (1995). Rbl2, a yeast protein that binds to β-tubulin and participates in microtubule function in vivo. Cell **82**, 425–434.

Gao, Y., Vainberg, I.E., Chow, R.L., and Cowan, N.J. (1993). Two cofactors and cytoplasmic chaperonin are required for the folding of α- and β-tubulin. Mol. Cell. Biol. **13**, 2478–2485.

Gao, Y., Melki, R., Walden, P., Lewis, S.A., Ampe, C., Rommelaere, H., Vandekerckhove, J., and Cowan, N.J. (1994). A novel cochaperonin that modulates the ATPase activity of cytoplasmic chaperonin. J. Cell Biol. **125**, 989–996.

Hoyt, M.A., Macke, J.P., Roberts, B.T. and Geiser, J.R. (1997). Saccharomyces cerevisiae PAC2 functions with CIN1, 2 and 4 in a pathway leading to normal microtubule stability. Genetics, in press.

Lewis, S.A., Tian, G., Vainberg, I.E., and Cowan, N.J. (1996). Chaperonin-mediated folding of actin and tubulin. J. Cell Biol. **132**, 1–4.

Pierre, P., Scheel, J., Rickard, J.E., and Kreis, T. (1992). CLIP-170 links endocytic vesicles to microtubules. Cell **70**, 887–900.

Stearns, T., Hoyt, M.A., and Botstein, D. (1990). Yeast mutants sensitive to antimicrotubule drugs define threee genes that affect microtubule function. Genetics **124**, 251–262.

Tian, G., Vainberg, I.E., Tap, W.D., Lewis, S.A., and Cowan, N.J. (1995). Quasi-native chaperonin-bound intermediates in facilitated protein folding. J. Biol. Chem. **270**, 23910–23913.

Tian, G., Huang, Y., Rommelaere, H., Vandekerckhove, J., Ampe, C., and Cowan, N.J. (1996). Pathway leading to correctly folded β-tubulin. Cell **86**, 287–296.

■ Nicholas J. Cowan:
Department of Biochemistry
NYU Medical Center
550 First Avenue
New York, NY 10016, USA
Tel. 1 212 263 5809
Fax. 1 212 263 8166
E-mail: Cowann01@MCRCR6.MED.NYU.EDU

8

HSP100 Proteins (Clps)

The HSP100 family — an overview

The HSP100 proteins have a unique function in disassembling protein oligomers and aggregates. In some cases their substrates are presented to a protease for degradation. The HSP100 proteins are the most recently discovered of the heat shock protein families. Like other stress protein families, the HSP100 family has constitutive and stress inducible members, and in eukaryotic organisms distinct forms are found in mitochondria, chloroplasts, and the cytosol and/or the nucleus. Several subfamilies are distinguishable based upon the number and size of their subdomains and the presence of specific amino acid sequences.

■ Alternative names

ClpA, ClpB, ClpC, etc., component A of protease Ti. ClpP (Ti) is not a member of this family, but functions in a complex with ClpA or ClpX (Hwang et al., 1987; Katayama-Fujimura et al., 1987; Gottesman et al., 1993). HslU — a subunit of a distinct two component protease, HslVU (Chang et al., 1993).

■ HSP100 sequences

Prototypical HSP100 proteins (class 1) have two predicted ATP-binding domains which contain Walker-type A and B consensus regions (Fig. 1) similar to those of the P-type transporters and the F1-ATPase (Walker et al., 1982). In most other proteins with two ATP-binding domains, the two domains are homologous. The HSP100 family is distinguished by the very limited homology between its two domains, although each is itself highly conserved in evolution suggesting they arose by gene fusion rather than gene duplication. N-terminal, middle, and C-terminal regions flank the two ATP-binding domains; consensus sequences within these regions help to distinguish subfamilies (Table 1). In the absence of protein structure information, the boundaries of these regions are defined as positions where an abrupt change in amino acid homology occurs. Thus, the boundaries of the middle region, initially based upon comparisons of the ClpA, B, and C subfamilies (Gottesman, 1990b), have been changed to accomodate a new homology border that became apparent when four newly described subfamilies containing a single ATP-binding domain (class 2 HSP100s) were compared with the class 1 subfamilies. The size of the middle region appears to be 54 and c. 172–207 amino acids, respectively, for the A- and B-type HSP100 proteins, while the C- and D-type HSP100 proteins all have a similarly sized middle domain of c. 101–118 amino acids. The latter subfamilies are readily distinguished from each other by the duplication of the N-terminal motif in the C-type HSP100 proteins, and the absence of this region may define a third subfamily containing a middle region of this size. The *Heterosigma carterae* chloroplast genome encodes a C-type HSP100 with charged insertions of c. 90 amino acids in the N- and C-terminal regions (Schirmer et al., 1996).

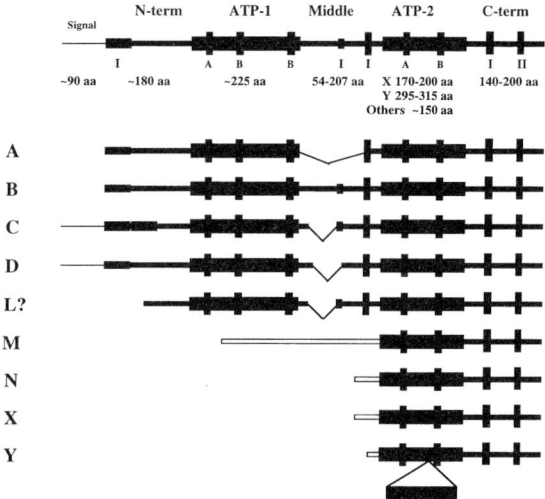

Figure 1. Cartoon of the HSP100 protein subfamilies. Areas of greater conservation are marked by the thickness of the lines. Approximate sizes of the individual regions are indicated under the region.

Class 2 HSP100s (M-, N-, X-, and Y-type) contain only one ATP-binding domain (which most closely resembles the second of the two ATP-binding domains of the longer forms) and the C-terminal domain (Fig. 1). In some cases (M-type) these two domains are fused to large N-terminal domains which have no homology to other family members. Relatedness among these N-terminal fusions may define additional subfamilies. Subfamilies of both long and short HSP100 proteins can be further distinguished on the basis of sequence characteristics (Table 1) and an insertion of c. 140 amino acids in the ATP-binding domain in the Y-type HSP100 proteins. The short M- and N-type HSP100 proteins exhibit c. 86% amino acid similarity with the second ATP-binding domain and C-terminus of the full length HSP100 proteins (A–D), but the short X- and Y-type HSP100 proteins exhibit only c. 30% similarity, suggesting separate evolutionary pathways (Schirmer et al., 1996).

Table 1. Homology regions of the HSP100 proteins

```
Amino terminus:   Signature Sequence I
        Con:  xxFTxxhxxhhxxAxxxAxxLxHxxhxxxHLLLGLh
        A:    xxxxxxxLxxxhxxAxAxAxxxRHExxTVEHLLLALh
        B:    xKFTxxxxxALAxAxxLAxxxxHxxhxPhHLAxALh
        C1:   ERFTEKAxKVIxLAQEEARRLGHNFVGTEQILLGLI
        C2:   IPFTPRAKRVLELSxEEARQLGHNYIGTEHLLLGLL
        D:    ERFTERAxRAIxxSQxEAKxLGKxxVxTxHLLLGLI

ATP-binding 1:  Walker consensus distinctions*

                        A           B1              B2
        Classical:  GX4GKT      RX$_{6-8}$h$_4$D   RX$_{6-8}$h$_4$D
        A:             "        KX$_9$h$_4$D       RX$_6$AIDLhD
        B:             "        KX$_{6-10}$h$_4$D       "
        C, D:          "        R/KX$_6$h$_4$D          "

Middle:   Signature Sequences

                    II                              III
        Con: RxxDxxxAxELRxxxxx          Con:  xWTGIPVxKh
        Bp:  RxGDLARxSELQYGxIP          A:    RIARIPxRxV
        Be:  RxxDLxxAADLRYxxIP          B:    RWTGIPVxKh
        C:   RxQDFEKAGxLRDxExx          C:    SWTGIPVxKh
                                        D:    VWSGIPVQQh

ATP-binding 2:  Walker consensus distinctions

                        A           B
        Classical    GX4GKT      RX$_{6-8}$h$_4$D
        A, B, C, D:     "        RX$_6$h$_4$D
        M, N, X:        "        RX$_{4-12}$h$_4$D
        Y:              "        RX$_{4-12}$h$_4$D

Carboxyl-terminus:   Signature Sequences

                    IV                              V
        Con:     FRPEFLNRLDEIIVFxxL     Con:     GARPLRRxI
        A:       FTPEFRNRLDxIIxFxxL     Others:  GARPLRRxI
        Others:  FRPEFLNRLDEIIVFxxL     X:       GARGLRRxI
        X, Y:    LRPEFxGRLPIxVxLxxL     Y:       GARRLRRxI
```

* the first ATP-binding domain contains two matches to the Walker B consensus

Finally, most prokaryotic HSP100 proteins (long and short) contain potential internal initiation sites which, at least in the cases of the *Escherichia coli clpA* and *clpB* genes, are used to produce shorter proteins (Park et al., 1993; Seol et al., 1994).

■ Accession numbers for HSP100 sequences

Genbank accession numbers for HSP100 sequences are: A subfamily: *Escherichia coli* M31045; B subfamily: *E. coli* M29364, *Haemophilus influenzae* HI0859, *Dichelobacter nodosus* M32229, *Synechococcus* U20646, *Saccharomyces cerevisiae* (Hsp104) M67479, *Trypanosoma brucei* M92325, *Leishmania major* Z38058, *Arabidopsis thaliana* U13949, *Glycine max* L35272, *Plasmodium berghei* U46549, *Corynebacterium glutamicum* U43536, *S. cerevisiae* (Hsp78) L16533, *Mycoplasma genitalium* U39719; C subfamily: *Odontella sinesis* — chloroplast Z67753, *Herterosigma carterae* Z25810, *Pisum sativum* L09547, *Brassica napus* X75328, *A. thaliana* ClpC (initial sequence from N. Hoffman, sequencing continued by E.C. Schirmer and S. Lindquist unpublished and partial internal sequence) Z29026, *Zea mays* T18272, *Synechococcus* U16134, *Synechocystis* PCC6803 (note this one was D6400 which was a whole cosmid, they now have a number for the specific gene, D90914), *Bacillus subtilis* U02604, *Serpulina hyodysenteriae* X73140, *Bos taurus* L34677; D subfamily: *A. thaliana* ERD1 D17582; M subfamily: *Plasmodium falciparum* X95276, *Mus musculus* U09874; N subfamily: *Brevibacterium sp* R312 M76451, *Pseudomonas aeruginosa* X77160; X subfamily: *E. coli* L18867, *Haemophilus influenzae* HI0715, *Bacillus subtilis* U18229, *Homo sapiens* R74599, *A. thaliana* F13836; Y subfamily: *Borrelia burgdorferi* U43739, *H. influenzae* HI0497, *B. subtilis* U13634. Additionally partial sequences exist for *Mycoplasma capricolum* Z33267, *Lactococcus lactis* L36907, *Pseudomonas aeruginosa* L06015, *Streptococcus pneumoniae* L20558.

Accession numbers for sequences available through Swiss-Prot are: A subfamily: *Rhodopseudomonas blastica* P05444; C subfamily: *Lycopersicon esculentum* (A) P31541, *L. esculentum* (B) P31542, *Mycobacterium leprae* P24428, *L. lactis* Q06716; N subfamily: *Methylobacterium extorquens* P30621; X subfamily: *Azotobacter vinelandii* P33683, *S. cerevisiae* P38323; Y subfamily: *E. coli* P32168, *Pasteurella haemolytica* P32180; Partial sequence for *Neurospora crassa* S28174.

Sequences not accessible from GenBank or Swiss-Prot are: C subfamily: *Porphyra purpurea* (communicated by Mike Reith); M subfamily: *H. sapiens* THC88419, THC79885 (TIGR database); X subfamily: *Synechocystis* sycslllh_85 (GenPept Updates); Y subfamily: *P. aeruginosa* S29303 (PIR database); partial sequences for *Streptococcus mutans* H35905 (PIR database), *H. sapiens* THC 95377 (TIGR database).

Note: The alignment of HSP100 sequences is available at "http://http.bsd.uchicago.edu/~hsplab/index.html".

■ HSP100 gene regulation

Most known HSP100 proteins were isolated in sequencing projects and their patterns of regulation have not yet been characterized. The three organisms in which multiple HSP100 proteins have been characterized in some detail are *Escherichia coli*, *Saccharomyces cerevisiae*, and *Arabidopsis thaliana*. In *E. coli* the ClpA protein is constitutively expressed (Katayama et al., 1988), the ClpB protein is induced by heat stress (Squires et al., 1991), the ClpX protein is induced by heat stress in the same operon as ClpP (Gottesman et al., 1993), and the ClpY gene, despite a perfect match to the heat shock promoter consensus, is only weakly induced by heat (Chuang et al., 1993). In yeast, the HSP100 protein of the nucleus and cytosol, Hsp104 (B-type), is induced by heat, ethanol, metal ions, growth on non-fermentable carbon sources, entrance into stationary phase, and sporulation (Sanchez et al., 1992). The yeast mitochondrial protein Hsp78 (B-type) is constitutively expressed but is also induced by growth on non-fermentable carbon sources and by heat shock (Leonhardt et al., 1993). Both genes have characteristic consensus elements for the heat shock transcription factor HSF and *HSP104* additionally has consensus elements for the general stress response (STREs) (Schuller et al., 1994). The nuclear and cytosolic protein of *Arabidopsis* Hsp101 (B-type) is strongly induced by heat in both roots and leaves. The mRNA for the *Arabidopsis* ClpC protein is constitutively expressed and its level is reduced by heat (Schirmer et al., 1994). The mRNA for the *ERD1* gene (D-type) is induced by desiccation stress, but not by heat stress (Kiyosue et al., 1993) and expression of the X-type protein from *Arabidopsis* has not been examined. All B-type HSP100 proteins that have been tested are induced by heat. The tested C-type HSP100 proteins are constitutively expressed, and some are further induced by heat.

■ Cellular localization of HSP100 proteins

As in other heat shock protein families, several subfamilies of HSP100 proteins exist and many organisms contain members from multiple subfamilies. Thus far in *E. coli* HSP100 proteins from four subfamilies have been identified. For those organisms whose genomes have been completely sequenced, *Haemophilus influenzae* has one each of B-, X-, and Y-types (Fleischmann et al., 1995) and *Saccharomyces cerevisiae* has two B- and one X-type HSP100 proteins. In eukaryotes some subfamilies of heat shock proteins are localized to various subcellular compartments. Thus in yeast, Hsp104 is mostly in the cytosol when expressed at normal temperatures, and a fraction of the protein is found in the nucleus after heat shock; Hsp78 is localized to mitochondria (Leonhardt et al., 1993). The *Leishmania donovani* Hsp100 is mostly in the cytosol at both normal and elevated temperatures (Hubel et al., 1995). Hsp98 from *Neurospora crassa* is reported to be concentrated in polyribosomes (Vassilev et al., 1992). The pea C-type HSP100 localizes to chloroplasts (Moore and Keegstra, 1993), and most eukaryotic C-type HSP100 proteins have putative chloroplast localization signals present within conserved 90 residue N-terminal extensions. With four separate HSP100 proteins identified to date, *Arabidopsis* should soon provide information on the tissue-specific distribution of these proteins, but so far all that is known is that ClpC and Hsp101 are expressed in leaf tissue (Schirmer et al., 1994) and Hsp101 is expressed in roots and developing seeds (E.C. Schirmer, S. Lindquist, and E. Vierling, unpublished results).

■ HSP100 protein characteristics *in vitro*

HSP100 proteins have been purified and characterized from *E. coli* (ClpA, B, X, and Y) and yeast (Hsp104). ClpA, ClpB, and Hsp104 all exhibit ATPase activity with K_m values in the range 0.2–1.1 mM (Woo et al., 1992; Maurizi et al., 1994; D.A. Parsell and S. Lindquist, unpublished data). ClpA and Hsp104 assemble into hexamers (Singh and Maurizi, 1994; Parsell et al., 1994a) in the presence of nucleotide. ClpB has been reported to form tetramers (Woo et al., 1992), but by electron microscopy appears to be a hexamer (M. Maurizi, personal communication). Mutational analyses conducted on the *E. coli* ClpA (Singh and Maurizi, 1994) and the *S. cerevisiae* Hsp104 (Parsell et al., 1991) proteins indicate that both domains are required for HSP100 biological functions. Curiously, these studies showed that the effects of individual point mutations on the biochemical activity of the protein differ for ClpA and Hsp104. Mutations in the first nucleotide-binding domain of Hsp104 interfere with ATP hydrolysis, while mutations in the second ATP-binding domain interfere with the ability of the protein to form oligomers. When similar mutations were analysed in ClpA, the converse was shown to be true: mutations in the first domain inhibit oligomerization while those in the second domain interfere with ATP hydrolysis.

The ClpA and ClpB mRNAs each contain two translation initiation sites resulting in proteins of 84 and 65 kDa (ClpA) or 93 and 79 kDa (ClpB) (Squires and Squires, 1992; Park et al., 1993; Seol et al., 1994). The product of the second ClpB initiation site (ClpB79) inhibits the protein-stimulatable ATPase activity of the larger (ClpB93) protein (Park et al., 1993). In contrast, the shorter product of the ClpA gene (ClpA65) has lower basal ATPase activity, but does not interfere with that of the larger form (ClpA84) or its stimulation by polypeptides (Seol et al., 1994); however, it does interfere with casein degradation *in vitro*.

■ HSP100 function

Three general functions have been identified for HSP100 proteins: stress tolerance, proteolysis, and regulation. Hsp104 from yeast is critical for cells to survive extreme heat (Sanchez and Lindquist, 1990), but can also protect cells from other types of stress such as exposure to ethanol. This function is conserved among the B-type HSP100 proteins since the *Arabidopsis* Hsp101 protein confers thermotolerance protection to yeast cells deleted for *HSP104* (Schirmer et al., 1994) and the *B. subtilis* ClpC also protects that organism from heat and salt stress (Kruger et al., 1994). This protection appears to be owing to the ability of these proteins to promote the solubilization of protein aggregates produced by stressful conditions. The yeast Hsp104 protein promotes the resolubilization and reactivation of aggregates of a heat-denatured luciferase test substrate *in vivo*, and Hsp104 is also responsible for the dissapearance of electron-dense material that appears in the nucleus and cytosol of yeast cells (lacking the artificial luciferase construct) under conditions of heat stress (Parsell et al., 1994b). ClpX has been shown to promote the solubilization of heat-denatured λO protein *in vitro* (Wawrzynow et al., 1995). Additionally, some chaperone functions appear to be conserved among HSP100 proteins since: (i) ClpA from *E. coli* protects luciferase from irreversible heat denaturation (Wickner et al., 1994); (ii) yeast Hsp104 expression compensates for reduced expression of Hsp70 (Sanchez et al., 1993); (iii) Hsp78 from yeast can partially substitute for Hsp70 chaperone functions in mitochondria (Schmitt et al., 1995); and (iv) the shorter ClpX protein from *E. coli* prevents aggregation of the λO protein (Wawrzynow et al., 1995).

ClpA was originally identified as a subunit of the ClpP (Ti) protease (Hwang et al., 1987; Katayama-Fujimura et al., 1987); however, ClpA by itself has no proteolytic activity. In the presence of ClpP, ClpA promotes the degradation of casein *in vitro* (hence the clp designation as 'caseinolytic protease') and of a ClpA–β-galactosidase fusion protein *in vivo* (Katayama-Fujimura et al., 1987; Gottesman et al., 1990a). ClpX has a distinct set of substrates: deletion of ClpX increases the half-life of λO protein, but not the half-life of the ClpA fusion protein *in vivo* (Gottesman et al., 1993; Wojtkowiak et al., 1993). Since HSP100 proteins do not have intrinsic proteolytic activity, their function in proteolysis may be to utilize an underlying disaggregase/chaperone function to present an unfolded substrate to the ClpP protease. This idea is supported by the *in vitro* finding that ClpA activates the plasmid P1 RepA replication initiator protein by converting RepA dimers into monomers in the absence of ClpP (Wickner et al., 1994); in the presence of ClpP, ClpA supports the degradation of the RepA protein. Additionally, ClpY seems to function with a unique protease component in supporting proteolysis (Rohrwild et al., 1996).

Further supporting a role for HSP100 proteins in regulation, ClpX promotes the binding of λO to oriλ in the absence of ClpP (Wawrzynow et al., 1995). A similar scenario applies for ClpX with bacteriophage Mu, where ClpX but not ClpP was found to be required for Mu replication (Mhammedi et al., 1994; Levchenko et al., 1995) thus separating proteolysis from the regulatory function. The Y-type HSP100 proteins may also play a role in transcriptional regulation as the *Pasteurella haemolytica* ClpY was isolated as an inhibitor of leukotoxin gene expression (Highlander et al., 1993). A regulatory role for HSP100 proteins is not restricted to class 2 members; the disassembly of RepA dimers by ClpA (Wickner et al., 1994) and the control of competence gene expression by the *B. subtilis* C-type HSP100 protein (Msadek et al., 1994) are also forms of regulation. The finding of a regulatory role for the *B. subtilis* protein in addition to its stress tolerance roles (Kruger et al., 1994) further supports the idea that a shared function in disassembling quaternary protein structures and aggregates underlies the regulatory, proteolytic, and stress functions of this diverse protein family.

References

Chuang, S.-E., Burland, V., Plunkett, G., III, Daniels, D.L., and Blattner, F.R. (1993). Sequence analysis of four new heat-shock genes constituting the hslTS/ibpAB and hslVU operons in Escherichia coli. Gene **134**, 1–6.

Fleischmann, R.D. et al. (1995). Whole-genome random sequencing and assembly of Haemophilus influenzae Rd. Science **269**, 496–512.

Gottesman, S., Clark, W.P., and Maurizi, M.R. (1990a). The ATP-dependent Clp protease of Escherichia coli: sequence of clpA and identification of a Clp-specific substrate. J. Biol. Chem. **265**, 7886–7893.

Gottesman, S., Squires, C., Pichersky, E., Carrington, M., Hobbs, M., Mattick, J.S., Dalrymple, B., Kuramitsu, H., Shiroza, T., and Foster, T. (1990b). Conservation of the regulatory subunit for the Clp ATP-dependent protease in prokaryotes and eukaryotes. Proc. Natl. Acad. Sci. USA **87**, 3513–3517.

Gottesman, S., Clark, W.P., de Crecy-Lagard, V., and Maurizi, M.R. (1993). ClpX, an alternative subunit for the ATP-dependent Clp protease of Escherichia coli: sequence and in vivo activities. J. Biol. Chem. **268**, 22618–22626.

Highlander, S.K., Wickersham, E.A., Garza, O., and Weinstock, G.M. (1993). Expression of the Pasteurella haemolytica leukotoxin in inhibited by a locus that encodes an ATP-binding cassette homolog. Infect. Immun. **61**, 3942–3951.

Hubel, A., Brandau, S., Dresel, A., and Clos, J. (1995). A member of the clpb family of stress proteins is expressed during heat shock in Leishmania spp. Mol. Biochem. Parasitol. **70**, 107–118.

Hwang, B.J., Park, W.J., Chung, C.H., and Goldberg, A.L. (1987). Escherichia coli contains a soluble ATP-dependent protease (Ti) distinct from protease La. Proc. Natl. Acad. Sci. USA **84**, 5550–5554.

Katayama, Y., Gottesman, S., Pumphery, J., Rudikoff, S., Clark, W.P., and Maurizi, M.R. (1988). The two-component, ATP-dependent Clp protease of Escherichia coli. J. Biol. Chem. **263**, 15226–15236.

Katayama-Fujimura, Y., Gottesman, S., and Maurizi, M.R. (1987). A multiple-component, ATP-dependent protease from Escherichia coli. J. Biol. Chem. **262**, 4477–4485.

Kiyosue, T., Yamaguchi-Shinozaki, K., and Shinozaki, K. (1993). Characterization of cDNA for a dehydration-inducible gene that encodes a Clp A, B-like protein in Arabidopsis thaliana L. Biochem. Biophys. Res. Commun. **196**, 1214–1220.

Kruger, E., Volker, U., and Hecker, M. (1994). Stress induction of clpC in Bacillus subtilis and its involvement in stress tolerance. J. Bacteriol. **176**, 3360–3367.

Leonhardt, S.A., Fearson, K., Danese, P.N., and Mason, T.L. (1993). HSP78 encodes a yeast mitochondrial heat shock protein in the Clp family of ATP-dependent proteases. Mol. Cell. Biol. **13**, 6304–6313.

Levchenko, I., Luo, L., and Baker, T.A. (1995). Disassembly of the Mu transposase tetramer by the ClpX chaperone. Genes Dev. **9**, 2399–2408.

Maurizi, M.R., Thompson, M.W., Singh, S.K., and Kim, S.-H. (1994). Endopeptidase Clp: ATP-dependent Clp protease from Escherichia coli. Methods Enzymol. **244**, 314–331.

Mhammedi, A.A., Pato, M., Gama, M.J., and Toussaint, A. (1994). A new component of bacteriophage Mu replicative transposition machinery: the Escherichia coli ClpX protein. Mol. Microbiol. **11**, 1109–1116.

Moore, T. and Keegstra, K. (1993). Characterization of a cDNA clone encoding a chloroplast-targeted Clp homologue. Plant Mol. Biol. **21**, 525–537.

Msadek, T., Kunst, F., and Rapoport, G. (1994). MecB of Bacillus subtilis, is a member of the ClpC ATPase family, is a pleiotropic regulator controlling competence gene expression and growth at high temperature. Proc. Natl. Acad. Sci. USA **91**, 5788–5792.

Park, S.K., Kim, K.I., Woo, K.M., Seol, J.H., Tanaka, K., Ichihara, A., Ha, D.B., and Chung, C.H. (1993). Site-directed mutagenesis of the dual translational initiation sites of the clpB gene of Escherichia coli and characterization of its gene products. J. Biol. Chem. **268**, 20170–20174.

Parsell, D.A., Sanchez, Y., Stitzel, J.D., and Lindquist, S. (1991). Hsp104 is a highly conserved protein with two essential nucleotide-binding sites. Nature **353**, 270–273.

Parsell, D.A., Kowal, A.S., and Lindquist, S. (1994a). Saccharomyces cerevisiae Hsp104 protein: purification and characterization of ATP-induced structural changes. J. Biol. Chem. **269**, 4480–4487.

Parsell, D.A., Kowal, A.S., Singer, M.A., and Lindquist, S. (1994b). Protein disaggregation mediated by heat-shock protein Hsp104. Nature **372**, 475–478.

Rohrwild, M., Coux, O., Huang, H.-C., Moerschell, R.P., Yoo, S.J., Seol, J.H., Chung, C.H., and Goldberg, A.L. (1996). HslV-HslU: a novel ATP-dependent protease complex in Escherichia coli related to the eukaryotic proteasome. Proc. Natl. Acad. Sci. USA **93**, 5808–5813.

Sanchez, Y. and Lindquist, S.L. (1990). HSP104 required for induced thermotolerance. Science **248**, 1112–1115.

Sanchez, Y., Taulien, J., Borkovich, K.A., and Lindquist, S. (1992). Hsp104 is required for tolerance to many forms of stress. EMBO J. **11**, 2357–2364.

Sanchez, Y., Parsell, D.A., Taulien, J., Vogel, J.L., Craig, E.A., and Lindquist, S. (1993). Genetic evidence for a functional relationship between Hsp104 and Hsp70. J. Bacteriol. **175**, 6484–6491.

Schirmer, E.C., Lindquist, S., and Vierling, E. (1994). An Arabidopsis heat shock protein complements a thermotolerance defect in yeast. Plant Cell **6**, 1899–1909.

Schirmer, E.C., Glover, J.R., Singer, M.A., and Lindquist, S. (1996). HSP100/Clp proteins: a common mechanism explains diverse functions. Trends Biochem. Sci. **21**, 289–296.

Schmitt, M., Neupert, W., and Langer, T. (1995). Hsp78, a Clp homologue within mitochondria, can substitute for chaperone functions of mt-hsp70. EMBO J. **14**, 3434–3444.

Schuller, C., Brewster, J.L., Alexander, M.R., Gustin, M.R., and Ruis, H. (1994). The HOG pathway controls osmotic regulation of transcription via the stress response element (STRE) of the Saccharomyces cerevisiae CTT1 gene. EMBO J. **13**, 4382–4389.

Seol, J.H., Yoo, S.J., Kim, K.I., Kang, M.-S., Ha, D.B., and Chung, C.H. (1994). The 65-kDa protein derived from the internal translational initiation site of the ClpA gene inhibits the ATP-dependent protease Ti in Escherichia coli. J. Biol. Chem. **269**, 29468–29473.

Singh, S.K. and Maurizi, M.R. (1994). Mutational analysis demonstrates different functional roles for the two ATP-binding sites in ClpAP protease from Escherichia coli. J. Biol. Chem. **269**, 29537–29545.

Squires, C. and Squires, C.L. (1992). The Clp proteins: protein regulators or molecular chaperones? J. Bacteriol. **174**, 1081–1085.

Squires, C.L., Pedersen, S., Ross, B.M., and Squires, C. (1991). ClpB is the Escherichia coli heat shock protein F84.1. J. Bacteriol. **173**, 4254–4262.

Vassilev, A.O., Plesofsky, V.N., and Brambl, R. (1992). Isolation, partial amino acid sequence, and cellular distribution of heat-shock protein hsp98 from Neurospora crassa. Biochim. Biophys. Acta **1156**, 1–6.

Walker, J.E., Saraste, M., Runswick, M.J., and Gay, N.J. (1982). Distantly related sequences in the α- and β-subunits of ATP synthase, myosin, kinases and other ATP-requiring enzymes and a common nucleotide binding fold. EMBO J. **1**, 945–951.

Wawrzynow, A., Wojtkowiak, D., Marszalek, J., Banecki, B., Jonsen, M., Graves, B., Georgopoulos, C., and Zylicz, M. (1995). The ClpX heat-shock protein of Escherichia coli, the ATP-dependent substrate specificity component of the ClpP-ClpX protease, is a novel molecular chaperone. EMBO J. **14**, 1867–1877.

Wickner, S., Gottesman, S., Skowyra, D., Hoskins, J., McKenney, K., and Maurizi, M.R. (1994). A molecular chaperone, ClpA, functions like DnaK and DnaJ. Proc. Natl. Acad. Sci. USA **91**, 12218–12222.

Wojtkowiak, D., Georgopoulos, C., and Zylicz, M. (1993). Isolation and characterization of ClpX, a new ATP-dependent specificity component of the Clp protease of Escherichia coli. J. Biol. Chem. **268**, 22609–22617.

Woo, K.M., Kim, K.I., Goldberg, A.L., Ha, D.B., and Chung, C.H. (1992). The heat-shock protein ClpB in Escherichia coli is a protein-activated ATPase. J. Biol. Chem. **267**, 20429–20434.

■ Eric C. Schirmer and Susan Lindquist:
*Department of Molecular Genetics and Howard Hughes Medical Institute
The University of Chicago
5841 S. Maryland Avenue
Room N339, MC1028
Chicago, IL 60637, USA
Tel. 1 702 773 8048/8049
Fax. 1 702 773 7254
E-mail: er93@midway.uchicago.edu and
S-Lindquist@uchicago.edu*

Escherichia coli ClpA

ClpA is a soluble ATPase originally identified in Escherichia coli *as the ATPase subunit of the ATP-dependent ClpAP protease. It is the founding member of a family of proteins (HSP100 proteins, ClpA, B, C), many of which are heat shock proteins. In vivo, ClpA participates with ClpP in the degradation of specific fusion proteins and some abnormal proteins. In vitro, in addition to participating with ClpP in ATP-dependent degradation, ClpA has chaperone activity and substrate-stimulated ATPase activity.*

■ Alternative name

Component A of protease Ti (Hwang et al., 1987, 1988).

■ ClpA gene and sequence

The *clpA* nucleotide sequence from *E. coli* (GenBank accession number M31045) predicts a protein of 758 amino acids, containing sequences that are highly conserved between ClpA and ClpB, and other HSP100 family members (Gottesman et al., 1990b, 1996; Squires and Squires, 1992; see also the overview of HSP100/Clp proteins, p. 231). The first highly conserved domain (amino acids 183–415) contains an ATP-binding site consensus; the second highly conserved domain (amino acids 421–608) contains a second consensus ATP-binding site motif (Gottesman et al., 1990b, 1996).

ClpA protein

ClpA is an 84 kDa cytoplasmic protein isolated as a mixture of monomers and dimers (Hwang et al., 1988; Katayama et al., 1988; Maurizi et al., 1994) In the presence of ATP or non-cleavable ATP analogs, ClpA assembles into a ring-shaped hexamer, as judged by centrifugation and electron microscopic examination (Kessel et al., 1995). The active hexamer has ATPase activity, can bind substrates such as RepA, and associates with a 14-mer of ClpP to form the ATP-dependent protease (Maurizi et al., 1994; Wickner et al., 1994; Kessel et al., 1995).

■ ClpA function

ClpA is the ATPase subunit of the two-component ClpAP ATP-dependent protease (Katayama-Fujimura et al., 1987). Its ATPase activity is stimulated by the addition of substrates and ClpP (Hwang et al., 1988; Katayama et al., 1988; Maurizi et al., 1994). The ClpAP protease degrades casein and other high molecular weight proteins only when coupled to ATP hydrolysis (Katayama-Fujimura et al., 1987). Protein degradation is highly processive, releasing peptide products of 5–15 amino acids in the presence of ATP (Thompson and Maurizi, 1994). ClpA alone acts as a chaperone in the ATP cleavage-dependent activation of the DNA-binding function of P1 protein RepA, restructuring it from an inactive dimer to an active monomer (Wickner et al., 1994). In vivo, ClpAP has been shown to function in the degradation of abnormal proteins, such as certain Lac fusion proteins, including β-galactosidase derivatives with abnormal N-terminal amino acids and of the *E. coli* MazE protein, which has homology to components of low copy nuclear plasmid post-segregational killing systems (Katayama-Fujimura et al., 1987; Gottesman et al., 1990a; Maurizi et al., 1990; Tobias et al., 1991; Aizenman et al., 1996).

■ clpA gene regulation

The clpA gene is in a monocistronic operon, mapping at 19 minutes on the E. coli chromosome (Gottesman et al., 1990a). ClpA, as opposed to many other members of the Clp family in E. coli, is not under σ^{32}-dependent heat shock regulation (Katayama-Fujimura et al., 1987).

■ ClpA mutagenesis studies

Site-directed mutations of the conserved lysine in the first ATP-binding site consensus sequence (K220Q) can block assembly (Singh and Maurizi, 1994). Other changes at the same site (e.g. K220R) perturb assembly only moderately; one change (K220V) blocks chaperone activity on RepA. Modifications at the conserved lysine in the second domain (e.g. K510Q) generally block degradation of proteins such as casein but allow assembly of the ClpAP protease and degradation of short proteins and peptides (Singh and Maurizi, 1994; Seol et al., 1995; Gottesman et al., 1996). Therefore, the first ATP site is necessary for assembly of the mature hexameric ClpA and also probably contributes to substrate processing. The second site is necessary for ATP-dependent protein degradation, but not for assembly or chaperone activity with RepA (Singh and Maurizi, 1994; Seol et al., 1995; Gottesman et al., 1996).

■ References

Aizenman, E., Engelbery-Kulka, H., and Glaser, G. (1996). An Escherichia coli chromosomal 'addiction module' regulated by ppGpp: a model for programmed cell death. Proc. Natl. Acad. Sci. USA **93**, 6059–6063.

Gottesman, S., Clark, W.P., and Maurizi, M.R. (1990a). The ATP-dependent Clp protease of Escherichia coli: sequence of clpA and identification of a Clp-specific substrate. J. Biol. Chem. **265**, 7886–7893.

Gottesman, S., Squires, C., Pichersky, E., Carrington, M., Hobbs, M., Mattick, J.S., Dalrymple, B., Kuramitsu, H., Shiroza, T., Foster, T., Clark, W.P., Ross, B., Squires, C.L., and Maurizi, M.R. (1990b). Conservation of the regulatory subunit for the Clp ATP-dependent protease in prokaryotes and eukaryotes. Proc. Natl. Acad. Sci. USA **87**, 3513–3517.

Gottesman, S., Wickner, S., Jubete, Y., Singh, S.K., Kessel, M., and Maurizi, M.R. (1996). Selective, energy dependent proteolysis in E. coli. Cold Spring Harbor Symp. Quant. Biol. **60**, 553–548.

Hwang, B.J., Park, W.J., Chung, C.H., and Goldberg, A.L. (1987). Escherichia coli contains a soluble ATP-dependent protease (Ti) distinct from protease La. Proc. Natl. Acad. Sci. USA **84**, 5550–5554.

Hwang, B.J., Woo, K.M., Goldberg, A.L., and Chung, C.H. (1988). Protease Ti, a new ATP-dependent protease in Escherichia coli contains protein-activated ATPase and proteolytic functions in distinct subunits. J. Biol. Chem. **263**, 8727–8734.

Katayama, Y., Gottesman, S., Pumphrey, J., Rudikoff, S., Clark, W.P., and Maurizi, M.R. (1988). The two-component ATP-dependent Clp protease of Escherichia coli: purification, cloning, and mutational analysis of the ATP-binding component. J. Biol. Chem. **263**, 15226–15236.

Katayama-Fujimura, Y., Gottesman, S., and Maurizi, M.R. (1987). A multiple-component ATP-dependent protease from Escherichia coli. J. Biol. Chem. **262**, 4477–4485.

Kessel, M., Maurizi, M.R., Kim, B., Kocsis, E., Trus, B.L., Singh, S.K., and Steven, A.C. (1995). Homology in structural organization between E. coli ClpAP protease and the eukaryotic 26S proteasome. J. Mol. Biol. **250**, 587–594.

Maurizi, M.R., Clark, W.P., Katayama, Y., Rudikoff, S., Pumphrey, J., Bowers, B., and Gottesman, S. (1990). Sequence and structure of ClpP, the proteolytic component of the ATP-dependent Clp protease of Escherichia coli. J. Biol. Chem. **265**, 12536–12445.

Maurizi, M.R., Thompson, M.W., Singh, W.K., and Kim, S.-H. (1994). Endopeptidase Clp: the ATP-dependent Clp protease from Escherichia coli. Methods Enzymol. **244**, 314–331.

Seol, J.H., Baek, S.H., Kang, M.-S., Ha, D.B., and Chung, C.H. (1995). Distinctive roles of the two ATP-binding sites in ClpA, the ATPase component of protease Ti in Escherichia coli. J. Biol. Chem. **270**, 8087–8092.

Singh, S.K. and Maurizi, M.R. (1994). Mutational analysis demonstrates different functional roles for the two ATP-binding sites in ClpAP protease from Escherichia coli. J. Biol. Chem. **269**, 29537–29545.

Squires, C. and Squires, C.L. (1992). The Clp proteins: proteolysis regulators or molecular chaperones? J. Bacteriol. **174**, 1081–1085.

Thompson, M.W. and Maurizi, M.R. (1994). Activity and specificity of Escherichia coli ClpAP protease in cleaving model peptide substrates. J. Biol. Chem. **269**, 18201–18208.

Tobias, J.W., Shrader, T.E., Rocap, G., and Varshavsky, A. (1991). The N-End rule in bacteria. Science **254**, 1374–1376.

Wickner, S., Gottesman, S., Skowyra, D., Hoskins, J., McKenney, K., and Maurizi, M.R. (1994). A molecular chaperone, ClpA, functions like DnaK and DnaJ. Proc. Natl. Acad. Sci. USA **91**, 12218–12222.

■ *Susan Gottesman, Sue Wickner, and Michael R. Maurizi:*
Bldg. 37, Rm. 2E18
Laboratories of Molecular Biology and Cell Biology
National Cancer Institute
Bethesda, MD 20892, USA
Tel. 1 301 496 3524
Fax:. 1 301 496 3875
E-mail: susang@helix.nih.gov

Escherichia coli ClpB

The *Escherichia coli* ClpB heat shock protein is a member of the HSP100 protein family. This family is apparently universal, very highly conserved, and possesses many attributes commonly associated with chaperone protein families. ClpB protects E. coli from the adverse effects of high temperatures, but is not essential for survival at normal growth temperatures. The clpB gene encodes two products, ClpB93 and ClpB79; however, the significance of this unusual property is not yet fully understood. The name given to this protein suggests its involvement in proteolysis. This property has never been demonstrated for ClpB, but does remain one of several possible mechanisms for its action.

■ Alternative names

ClpB93 has also been called F84.1 and E89 (VanBogelen et al., 1992; Fuge and Farr, 1993). ClpB79 has been called E72, F68.5, and ClpB′ (Squires et al., 1991).

■ ClpB sequence

The *Escherichia coli* ClpB heat shock protein (NCBI sequence identification number 147365) (Kitagawa et al., 1991; Squires et al., 1991) is a member of a recently discovered family (Gottesman et al., 1990; Squires and Squires, 1992), which has been referred to as the HSP100 family (Parsell et al., 1994). E. coli ClpB and other members of this family share two distinct ATP-binding regions (each about c. 200 amino acids in length) that possess very high sequence similarity. For example, E. coli and *Trypanosoma brucei* ClpBs share 66% identical and 25% similar amino acids in their ATP-binding regions. An unusual feature of HSP100 proteins is that the two ATP-binding regions are non-homologous. The two HSP100 (ClpB) regions are designated ATP-1 and ATP-2, and contain unique Prosite signatures (PS00870 and PS00871). ATP-1 and ATP-2 are delimited by somewhat less conserved leader (N-terminal), spacer, and trailer (C-terminal) sequences. Differences in size and composition of the spacer regions define three subfamilies (ClpA, ClpB, and ClpC), representatives of which have been identified in many different organisms, and which may represent functionally distinct classes (Squires and Squires, 1992). An intriguing recent development is the discovery of three half-Clp protein sequences, in bacteria and the mouse, that are homologous to the ATP-2 region of ClpB (Soubrier et al., 1992; Perier et al., 1995; Wilson et al., 1995). The E. coli clpB gene possesses a single σ^{32}-specific heat shock promoter (Kitagawa, et al., 1991, Squires, et al., 1991), and is followed by a Rho-independent transcription terminator (Squires et al., 1991).

■ *clpB* gene expression

ClpB is one of the most abundant heat shock proteins in E. coli. This property was discovered by two-dimensional gel analysis of mutant and overproducing strains (Squires et al., 1991), and by sequence analysis of major heat shock transcripts (Kitagawa et al., 1991). Following thermal induction, the E. coli clpB gene produces two related proteins, ClpB93 (856 amino acids, 93 kDa) and ClpB79 (714 amino acids, 79 kDa), which are formed by separate translational initiations on the same mRNA at codons 1 and 149 of the clpB sequence (Squires et al., 1991; Pontis et al., 1994). ClpB93 and ClpB79 are induced by a number of stress conditions in addition to heat [EtOH, Cd^{2+}, phosphate starvation, carbon starvation, treatment with nalidixic acid and the quinone ACDQ (VanBogelen et al., 1992)], and are produced in a 5:2 ratio under all physiological conditions tested to date (Squires et al., 1991).

■ ClpB proteins

ClpB93 and ClpB79 both have intrinsic ATPase activity. The activity of ClpB93 is stimulated by denatured proteins (Woo et al., 1992). However, the ATPase of ClpB79 does not respond to proteins, suggesting that the leader region (first 148 aa) of ClpB93 contains a protein recognition domain (Park et al., 1993). Gel filtration of ClpB results in its elution at c. 350 kDa, suggesting that it is normally tetrameric, and localization experiments suggest that both ClpB93 and ClpB79 are cytosolic proteins. Decreased ATPase activity of ClpB93 is observed upon the addition of ClpB79, eliciting the suggestion that ClpB79 somehow serves a negative regulatory purpose in the cell (Park et al., 1993). Another interesting property of E. coli ClpB is that it binds the alarmone, AppppA, which also binds to several other heat shock proteins, but not to ClpB's E. coli homologue, ClpA, which is synthesized constitutively at normal growth temperatures (Fuge and Farr, 1993).

■ ClpB function

The presence of ClpB protects E. coli at high temperatures, but the precise mechanism of this protection remains unknown (Squires et al., 1991). It was first thought that ClpB might be involved in the degradation of toxic heat-damaged proteins because of the role played by its *E. coli* homologue ClpA in activating the non-homologous ClpP protease. Experiments, however, indicate that ClpB does not closely resemble ClpA in its mode of action. Despite the intriguing circumstance that ClpP and ClpB are both expressed during heat shock (Kitagawa et al., 1991; Kroh and Simon, 1991; Squires et al., 1991; Squires and Squires, 1992), purified ClpB protein does not substitute for ClpA in activating the non-homologous ClpP protease, nor does ClpB form detectable complexes with ClpP (Woo et al., 1992). Thus, to date, no function in proteolysis of proteins has been detected for ClpB. The recent discovery that ClpA mediates either activation or destruction of P1 plasmid RepA protein depending on the presence of ClpP (Wickner et al., 1994) suggests that ClpB might also play a similar dual role with some as yet unidentified target protein.

■ References

Fuge, E. and Farr, S. (1993). AppppA-binding protein E89 is the *Escherichia coli* heat shock protein ClpB. J. Bacteriol. **175**, 2321–2326.

Gottesman, S., Squires, C., Pichersky, E., Carrington, M., Hobbs, M., Mattick, J., Dalrymple, B., Kuramitsu, H., Shiroza, T., Foster, T., Clark, W., Ross, B., Squires C., and Maurizi M. (1990). Conservation of the regulatory subunit for the Clp ATP-dependent protease in prokaryotes and eukaryotes. Proc. Natl. Acad. Sci. USA **87**, 3513–3517.

Kitagawa, M., Wada, C., Yoshioka S., and Yura T. (1991). Expression of ClpB, an analog of the ATP-dependent protease regulatory subunit in *Escherichia coli*, is controlled by a heat shock sigma factor (s32). J. Bacteriol. **173**, 4247–4253.

Kroh, H. and Simon L. (1991). Increased ATP-dependent proteolytic activity in Lon-deficient Escherichia coli strains lacking the DnaK protein. J. Bacteriol. **173**, 2691–2695.

Park, S., Kim, K., Woo, K., Seol, J., Tanaka, K., Ichihara, A., Ha D., and Chung C. (1993). Site-directed mutagenesis of the dual translational initiation sites in the clpB gene of *Escherichia coli* and characterization of its gene products. J. Biol. Chem. **268**, 20170–20174.

Parsell, D.A., Kowal, A.S., Singer M.A., and Lindquist S. (1994). Protein disaggregation mediated by heat-shock protein Hsp104. Nature **372**, 475–478.

Perier, F., Radeke, C.M., Raab-Graham K.F., and Vandenberg C.A. (1995). Expression of a putative ATPase suppresses the growth defect of a yeast potassium transport mutant: identification of a mammalian member of the Clp/HSP104 family. Gene **152**, 157–163.

Pontis, E., Sun, X., Jornvall, H., Krook M., and Reichard P. (1994). ClpB proteins copurify with the anaerobic *Escherichia coli* reductase. Biochem. Biophys. Res. Commun. **180**, 1222–1226.

Soubrier, F., Levy-Schil, S., Mayaux, J.F., Petre, D., Arnaud A., and Crouzet J. (1992). Cloning and primary structure of the wide-spectrum amidase from Brevibacterium sp. R312: high homology to the amiE product from Pseudomonas aeruginosa [published erratum appears in Gene 1993, **124**(2), 309] Gene **116**, 99–104.

Squires, C. and Squires C.L. (1992). The Clp Proteins: proteolysis regulators or molecular chaperones? J. Bacteriol. **174**, 1081–1085.

Squires, C.L., Pedersen, S., Ross B., and Squires C. (1991). ClpB is the *Escherichia coli* heat shock protein F84.1. J. Bacteriol. **173**, 4254–4262.

VanBogelen, R., Sankar, P., Clark, R., Bogan J., and Neidhardt F. (1992). The gene-protein database of *Escherichia coli*: Edition 5. Electrophoresis **13**, 1014–1054.

Wickner, S., Gottesman, S., Skowyra, D., Hoskins, J., McKenney K., and Maurizi M.R. (1994). A molecular chaperone, ClpA, functions like DnaK and DnaJ. Proc. Natl. Acad. Sci. USA **91**, 12218–12222.

Wilson, S. A., Williams, R.J., Pearl, L.H., and Drew, R.E. (1995). Identification of two new genes in the Pseudomonas aeruginosa amidase operon, encoding an ATPase (AmiB) and a putative integral membrane protein (AmiS). J. Biol. Chem. **270**, 18818–18824.

Woo, K., Kim, K., Goldberg, A., Ha, D., and Chung, C. (1992). The heat-shock protein ClpB in *Escherichia coli* is a protein-activated ATPase. J. Biol. Chem. **267**, 20429–20434.

■ *Catherine L. Squires and Craig Squires:*
Department of Molecular Biology and Microbiology
Tufts University School of Medicine
136 Harrison Ave.
Boston, MA 02111, USA
Tel. 1 617 636 6947
Fax. 1 617 636 0337
E-mail: csquires_rib@opal.tufts.edu

Escherichia coli ClpX

The Escherichia coli *ClpX* heat shock protein is homologous to the members of prokaryotic and eukaryotic HSP100/Clp ATPases family. ClpX was isolated as a specificity component of the ATP-dependent Clp proteases which maintains certain polypeptides in a form competent for proteolysis by the ClpP protease subunit. ClpX can act as a molecular chaperone, in the absence of ClpP, by activating the initiation proteins involved in DNA replication.

■ Alternative name

LopC (Wojtkowiak et al. 1993).

■ *clpX* gene and sequence

The *clpX* gene (GenBank accession number 223278) together with the *clpP* gene forms a heat shock-regulated operon located upstream of the *lon* gene at 10 minutes on the E. coli map (Gottesman et al., 1993). *clpX* can also be expressed independently from *clpP* using its own promotor (Yoo et al., 1994). The *clpX* gene encodes a protein of 424 amino acids (Gottesman et al., 1993). The *clpX* gene was identified following N-terminal sequence analysis of the ClpX (LopC) protein, which is the specificity component of ClpXP ATP-dependent protease (Wojtkowiak et al., 1993). That sequence, after the removal of the NH_2-terminal Met, perfectly matched the first 13 amino acids predicted from the open reading frame downstream of the termination codon for *clpP* gene (Gottesman et al., 1993). The amino acid sequence deduced from the *clpX* nucleotide sequence predicts a single ATP-binding site consensus with significant homology to members of the HSP100 family of multimeric ATPases present in prokaryotes (ClpA, B, C and Y) and eukaryotes (HSP100s) (for review see Parsell and Lindquist, 1993; Gottesman et al., 1993; Wawrzynow et al., 1996; Schirmer et al., 1996; and the overview of the HSP100/Clp family, see p. 231). Most of the members of the HSP100 family possess two ATP-binding sites. Uniquely, ClpX has a cluster of cysteine residues in the N-terminal region. The ClpX motif $CXXC(X)_{18}CXXC$, possibly a zinc binding domain, is similar to those found in the DNA-binding domains of hormone-responsive receptor proteins (Gottesman et al., 1993).

■ ClpX protein

ClpX is a 46 kDa component of a multienzymatic protease which in an ATP-dependent reaction efficiently hydrolyses the λO replication protein (Wojtkowiak et al., 1993). When purified to homogeneity and stored at –80°C in the presence of non-ionic detergent, 20% glycerol, 10 mM DTT and at relatively high ionic strength (150 mM KCl, 10 mM $MgCl_2$), ClpX does not lose its activity even after one year. However, in low salt concentrations (< 50 mM KCl) ClpX requires the presence of ClpP subunit for stabilization (Wojtkowiak et al., 1993; Wawrzynów et al., 1995).

ClpX displays a weak ATPase activity, with K_m=500 μM. The presence of λO stimulates the V_{max} of this reaction with no significant change in the affinity of ClpX for ATP (Wawrzynów et al., 1995). The presence of ATP or ATPγS stimulates the binding of ClpX to the λO substrate and to the ClpP proteolytic subunit (Wawrzynów et al., 1995). In the absence of ClpX, ClpP does not bind to λO and no proteolysis of λO was detected. Proteins that do not interact with ClpX (casein, RepA, σ^{32}) are not degraded by ClpXP protease (Wickner et al., 1994; Wawrzynów et al., 1995). However, substitution of ClpX by ClpA in ClpP-dependent protease reaction leads to a change in the proteolytic specificity such that efficient degradation of casein is observed (Wawrzynów et al., 1995). Thompson and Maurizi (1994) suggest that in the case of ClpA ATPase, the ATP hydrolysis step is not important for the processivity of protease action, but rather is involved in the initiation steps of capturing and presenting the substrates to the ClpP protease subunit. A possibility that cannot be excluded is that ClpX and ClpA are not a bona fide subunits of Clp protease. Rather, they are molecular chaperones which after presentation of the substrate to the ClpP protease leave the complex and recycle by binding to another substrate (Wojtkowiak et al., 1993; Wawrzynow et al., 1996).

In the absence of ClpP, purified ClpX protein performs chaperone activity (Wawrzynów et al., 1995). In an ATP-stimulated reaction, ClpX protects λO from aggregation and also enhances the binding of λO to the *ori*λ sequence. In the situation where λO has already aggregated, ClpX in ATP-dependent reaction could dissolve and reactivate λO. ClpA and to some extent DnaK/DnaJ/ GrpE chaperones could substitute for ClpX in this reaction (Wawrzynów et al., 1995; A. Wawrzynów and M. Zylicz, unpublished results). Moreover, the ClpA chaperone activity (but not ClpX) was shown to substitute for DnaK/DnaJ/GrpE chaperones in the activation of RepA protein for binding to *oriP1* (Wickner et al., 1994). The biological significance of these *in vitro* reconstituted reactions, i.e. activation of RepA by ClpA and λO by ClpX, are still not clear.

■ ClpX function

The *clpX* gene can be expressed alone and/or coexpressed with *clpP* in cells depending on the physiological conditions (Yoo et al., 1994). Inactivation of the *clpX* gene results in up to a tenfold stabilization of the λO protein *in vivo*. (Gottesman et al., 1993). Neither absence nor excess of ClpX significantly affects the transformation efficiency and copy number of λ plasmid, or the kinetics of λ phage growth, suggesting that ClpXP protease does not affect λDNA replication *in vivo* (Gottesman et al., 1993; Szalewska et al., 1994). However, this interpretation is complicated since in *E. coli* ClpX and ClpP analogues are present (Wojtkowiak et al., 1993; Missiakas et al., 1996; Rohrwild et al., 1996). Therefore, in certain genetic backgrounds, these analogues could bypass the requirement for the ClpX protein.

In vivo experiments show that the ClpXP protease system is also capable of degrading the bacteriophage Mu *vir* repressor (Laachouch et al., 1996), the Phd protein which is involved in the stabilization of P1 plasmid (Lehnherr and Yarmolinsky, 1995), the σ^s transcription factor involved in the starvation-mediated differentiation process (Schweder et al., 1996), and UmuD′ (Frank et al., 1996).

How ClpX recognizes its protein substrates is still not clear. Deletion analysis of MuA transposase revealed that its C-terminal sequence is important for ClpX-dependent degradation *in vitro* (Levchenko et al., 1995). In the case of another ClpX substrate, Mu *vir* repressor, a single point mutation in its seven amino acid C-terminal sequence was sufficient to decrease the degradation of the protein. After fusion of that particular sequence at the C-terminal end of CcdA and CcdB proteins, the naturally stable CcdB stayed unaffected, while CcdA, which was normally degraded by Lon protease, became a substrate for the ClpXP (Laachouch et al., 1996). These, and recent genetic experiments using σ^s mutants, suggest that the C-terminal sequence of the substrate is not the only factor required for recognition by ClpX (Schweder et al., 1996).

Bacterial ClpX can work as a molecular chaperone in the absence of ClpP. In the case of bacteriophage Mu replication, mutations in *clpX*, but not *clpP*, affect Mu replicative transposition (Mhammedi-Alaoui et al., 1994). Additionally, *in vitro* experiments show that ClpX, with the help of host factors, activates the MuA transposase in a nucleoprotein complex during strand transfer reactions. Following DNA recombination undegraded MuA transposase is removed from the DNA leading to Mu DNA replication (Levchenko et al., 1995; Nakai and Kruklitis, 1995; Kruklitis et al., 1996). The deletion of C-terminal MuA sequences inhibited the disassembly of the Mu–DNA complex (Levchenko et al., 1995). Molecular chaperone activity of members of the HSP100/Clp family was previously postulated (Squires and Squires, 1992). The yeast member of this family (Hsp104) which is required for thermotolerance is involved, *in vivo*, in disaggregation of overproduced luciferase (Parsell et al., 1994) and propagation of the yeast prion-like factor [psi+] (Chernoff et al., 1995).

■ References

Chernoff, Y.O., Lindquist, S.L., Ono B., Igne-Vechtomov, S.G., and Liebman, S.W. (1995). Role of the chaperone protein Hsp104 in propagation of yeast prion-like factor [psi+]. Science **268**, 880–883.

Frank, E.G., Ennis, D.G., Gonzales, M., Levine, A.S., and Woodgate, R. (1996). Regulation of SOS mutagenesis by proteolysis. Proc. Natl. Acad. Sci. USA **93**, 10291–10296.

Gottesman, S., Clark, W.P., de Crecy-Lagard, V., and Maurizi, M.R. (1993). ClpX, an alternative subunit for ATP-dependent Clp protease of E. coli. J. Biol. Chem. **268**, 22618–22626.

Kruklitis, R., Welty, D.J., and Nakai, H. (1996). ClpX protein of *Escherichia coli* activates bacteriophage Mu transposase in the strand transfer complex for initiation of Mu DNA synthesis. EMBO J. **15**, 935–944.

Laachouch, J.E., Desrnet, L., Geusken, V., Grimaud, R., and Toussaint, A. (1996). Bacteriophage Mu repressor as a target for the *Escherichia coli* ATP-dependent Clp protease. EMBO J. **15**, 437–444.

Lehnherr, H. and Yarmolinsky, M.B. (1995). Addiction protein Phd of plasmid prophage Pl is a substrate of the ClpXP serine protease of *Escherichia coli*. Proc. Natl. Acad. Sci. USA **91**, 3274–3277.

Levchenko, I., Luo, L., and Baker, T.A. (1995). Disassembly of the Mu transposase tetramer by the ClpX chaperone. Genes Dev. **9**, 2399–2409.

Mhammedi-Alaoui, A., Pato, M., Gama, M.-J., and Toussaint, A. (1994). A new component of bacteriophage Mu replicative transposition machinery: the *Escherichia coli* ClpX protein. Mol. Microbiol. **11**, 1109–1116.

Missiakas, D., Schwager, F., Betton, J.-M., Georgopoulos, C., and Raina, S. (1996). Identification and characterization of the HslV HslU (ClpQ ClpY) proteins involved in overall proteolysis of misfolded proteins in *Escherichia coli*. EMBO J. **15**, 6899–6909.

Nakai, H. and Kruklitis R. (1995). Disassembly of the bacteriophage Mu transposase for the initiation of Mu DNA replication. J. Biol. Chem. **270**, 19591–19598.

Parsell, D.A. and Lindquist, S. (1993). The function of heat-shock proteins in stress tolerance: degradation and reactivation of demaged proteins. Annu. Rev. Genet. **27**, 437–496.

Parsell, D.A., Kowal, A.S., Singer, M.A., and Lindquist, S. (1994). Protein disaggregation mediated by heat-shock protein Hsp104. Nature **372**, 475–478.

Rohrwild, M., Coux, O., Huang, H.-C., Moerschall, R.P., Yoo, S.J., Seol, J.H., Chung, C.H., and Goldberg, A.L. (1996). HslV HslU: a novel. ATP-dependent protease complex in *Escherichia coli* related to the eukaryotic protease. Proc. Natl. Sci. USA **93**, 5808–5813.

Schirmer, E.C., Glover, J.R., Singer, M.A., and Lindquist, S. (1996). HSP100/ Clp proteins: a common mechanism explains diverse functions. Trends Biochem. Sci. in press.

Schweder, T., Lee, K.-H., Lomovskaya, O., and Matin, A. (1996). Regulation of *Escherichia coli* starvation sigma factor (σ^s) by ClpX protease. J. Bacteriol. **178**, 470–476.

Squires, C. and Squires, C.L. (1992). The Clp proteins: proteolysis regulations or molecular chaperones? J. Bacteriol. **174**, 1081–1085.

Szalewska, A., Wegrzyn, G., and Taylor, K. (1994). Neither absence nor excess of λO initiator-digesting ClpXP protease affects λ plasmid or phage replication in *Escherichia coli*. Mol. Microbiol. **13**, 469–474.

Thompson, M.W. and Maurizi, M.R. (1994). Activity and specificity of *Escherichia coli* ClpAP protease in cleaving model peptide substrates. J. Biol. Chem. **269**, 18201–18208.

Wawrzynów, A., Wojtkowiak, D., Marszalek, J., Banecki, B., Jonsen, M., Graves, B., Georgopoulos, C., and Zylicz, M. (1995). The ClpX heat-shock protein of *Escherichia coli*, the ATP-dependent substrate specificity component of the ClpP-ClpX protease, is a novel molecular chaperone. EMBO J. **14**, 1867–1877.

Wawrzynow, A., Banecki, B., and Zylicz, M. (1996). The Clp ATPases define a novel class of molecular chaperones. Mol. Microbiol. **21**, 895–899.

Wickner, S., Gottesman, S., Skowyra, D., Hoskins, J., and McKenney, K. (1994). A molecular chaperone, ClpA, functions like DnaK and DnaJ. Proc. Natl. Acad. Sci. USA **91**, 12218–12222.

Wojtkowiak, D., Georgopoulos, C., and Zylicz, M. (1993). Isolation and characterization of ClpX, a new ATP-dependent specificity component of the Clp protease of *Escherichia coli*. J. Biol. Chem. **268**, 22609–22617.

Yoo, S.J., Seol, J.H., Kang, M.S., Ha, D.B., and Chung, C.H. (1994). *clpX* encoding an alternative ATP-binding subunit of protease Ti (Clp) can be expressed independently from *clpP* in *Escherichia coli*. Biochem. Biophys. Res. Commun. **203**(2), 798–804.

■ *Alicja Wawrzynów and Maciej Zylicz*
Department of Molecular and Cellular Biology
Division of Biophysics
University of Gdansk
Kladki 24
80-822 Gdansk, Poland
Tel. 48 58 319222
Fax. 48 58 310072
E-mail: zylicz@biotech.univ.gda.pl

Escherichia coli ClpY

ClpY is a cytoplasmic ATPase closely related to ClpX. Like ClpX and ClpA, ClpY can act in combination with a second protein to promote energy-dependent protein degradation. While ClpX and ClpA interact with ClpP, ClpY interacts with ClpQ, a protein with homology to the β subunits of proteasomes. clpQ and clpY form an operon in E. coli, and closely linked clpQ and clpY genes have also been found in Pasteurella haemolytica, Bacillus subtilis, and Haemophilus influenzae.

■ Alternative names

HslV (Chuang *et al.*, 1993); CodX (Slack *et al.*, 1995), LapA (Highlander *et al.*, 1993).

clpY gene organization and regulation

The E. coli clpY gene is the second gene of a heat shock operon; the first gene is clpQ (also called hslU), which resembles a β-type proteasome subunit (Chuang et al., 1993). The operon is at 88.9 minutes on the E. coli genetic map. Both genes show increased expression in vivo after heat shock and are transcribed in vitro by σ^{32} (Chuang et al., 1993). The clpY homologs in other bacteria are also found following a gene encoding a ClpQ-like protein (Fleischmann et al., 1995; Highlander et al., 1993; Slack et al., 1995). Accession numbers: E. coli ClpY (P31059), B. subtilis ClpY (CodX) (Z33639), P. haemolytica ClpY (LapA) (M59210).

The *clpY* gene encodes a protein of 443 amino acids with extended homology to ClpX, including a highly conserved Walker ATP-binding motif. ClpY differs from ClpX in not carrying a putative N-terminal zinc finger motif and in the addition of an extra domain of 47–67 amino acids between part A and part B of the Walker-type ATPase consensus (Gottesman et al., 1996).

■ ClpY protein

ClpY can be obtained as a homogeneous multimeric protein but can also be purified in a complex with ClpQ, provided ATP is added during all purification steps (Rohrwild *et al.*, 1996). A ClpY/ClpQ complex of about 600 kDa is observed using gel filtration columns and can be immunoprecipitated from wild-type cells using antibodies against either of the two proteins (Missiakas *et al.*, 1996; Rohrwild *et al.*, 1996). Site-directed mutagenesis of the ATPase active site of ClpY impairs the ability of the protein to form a complex with ClpQ both *in vivo* and *in vitro* (Missiakas *et al.*, 1996). ClpQ, like ClpP, forms a ring-like structure *in vivo* (Missiakas *et al.*, 1996; Kessel *et al.*, 1996; Rohrwild *et al.*, 1996) and possesses an ATP-independent protease activity for short peptides or denatured proteins such as casein. *In vitro*, ClpY did not show any interaction with ClpP, ClpA, or ClpX, and ClpQ was unable to act with ClpA or ClpX (Missiakas *et al.*, 1996).

■ ClpY function

No report of ClpY function *in vivo* has yet been published, but the organization of the genes and data from *in vitro* activity supports the notion that ClpY participates with ClpQ in energy-dependent proteolysis. By analogy with the similarly organized ClpXP protease, the ClpY ATPase subunit is likely to present substrates to the ClpQ protease subunit. Plasmids carrying the *clpQ clpY* region have a number of properties consistent with increased proteolysis. Plasmids carrying this operon have been found as multicopy suppressors of the temperature-sensitive growth of a strain carrying a mutation in the *htrC* gene. A plasmid expressing ClpQ and ClpY restores to normal levels both heat shock proteins and the rate of abnormal protein degradation, both of which are altered in the *htrC* mutant (Missiakas *et al*., 1996). In addition, plasmids carrying the *clpQ clpY* genes can suppress phenotypes of *lon* mutants, suggesting that the ClpYQ protease is capable of degrading at least some Lon substrates (W.-F. Wu and S. Gottesman, unpublished observations). These same plasmids did not suppress the phenotypes of *clpA*, *clpP*, or *clpX* mutants (W.-F. Wu and S. Gottesman, unpublished observations). Mutations in *clpY* and *clpQ* lead to a longer half-life for the high temperature sigma factor, σ^E, although it is not yet clear if this is a direct or indirect effect (Missiakas *et al*., 1996).

References

Chuang, S.-E., Burland, V., Plunkett, G., III, Daniels, D.L., and Blattner, F.R. (1993). Sequence analysis of four new heat-shock genes constituting the *hslTS/ibpAB* and *hslVU* operons in *Escherichia coli*. Gene **134**, 1–6.

Fleischmann, R.D., Adams, M.D., White, O., and Clayton, R.A. (1995). Whole-genome random sequencing and assembly of Haemophilus influenzae Rd. Science **269**, 496–512.

Gottesman, S., Wickner, S., Jubete, Y., Singh, S.K., Kessel, M., and Maurizi, M.R. (1996). Selective, energy dependent proteolysis in *E. coli*. Cold Spring Harbor Symp. Quant. Biol. **60**, 533–548.

Highlander, S.K., Wickersham, E.A., Garza, O., and Weinstock, G.M. (1993). Expression of the *Pasteurella haemolytica* leukotoxin is inhibited by a locus that encodes an ATP-binding cassette homolog. Infect. Immunol. **61**, 3942–3951.

Kessel, M., Wu, W.-F., Gottesman, S., Kocsis, E., Steven, A. and Maurizi, M.R. (1996). Six-fold rotational symmetry of ClpQ, the *E. coli* homolog of the 20 S proteasome, and its ATP-dependent activator, ClpY. FEBS Lett. **398**, 274–278.

Missiakas, D., Schwager, F., Betton, J.M., Georgopoulos, C., and Raina, S. (1996). Identification and characterization of HslV-HslU (ClpQ ClpY) proteins involved in overall proteolysis of misfolded proteins in *E. coli*. EMBO J. **15**, 6899–6909.

Rohrwild, M., Coux, O., Huang, H.-C., Moerschell, R.P., Yoo, S.J., Seol, J.H., Chung, C.H., and Goldberg, A.L. (1996). HslV–HslU: a novel ATP-dependent protease complex in *Escherichia coli* related to the eukaryotic proteasome. Proc. Nalt. Acad. Sci. USA **93**, 5808–5813.

Slack, F.J., Serror, P., Joyce, E., and Sonenshein, A.L. (1995). A gene required for nutritional repression of the *Bacillus subtilis* dipeptide permease operon. Mol. Microbiol. **15**, 689–702.

■ *Susan Gottesman and Whi-Fin Wu:*
Laboratory of Molecular Biology
National Cancer Institute
Bethesda, MD 20892-4255, USA
Tel. 1 301 496 3524
Fax. 1 301 496 3875
E-mail: susang@helix.nih.gov

■ *Dominique Missiakas:*
National de Recherche Scientifique
LIDSM-CBBM
31 Chemin J. Aiguier
13402 Marseille Cedex 20, France
Tel. 33 4 91 76 03 59
Fax. 33 4 91 71 21 24
E-mail: missaka@ibsm.cnrs-mrs.fr

Bacillus subtilis ClpC

Bacillus subtilis *ClpC, a member of the ClpC subfamily of stress response-related Clp ATPases, is implicated in several physiological functions such as cell viability under heat and salt stress conditions. This general stress protein is directly involved in proteolysis and seems to play a role in motility and cell division. Furthermore, the protein acts as pleiotropic regulator, controlling competence gene expression and degradative enzyme synthesis.*

■ Isolation of *clpC* and alternative names

The *clpC* gene of *B. subtilis* was identified by a PCR technique using degenerate primers hybridizing with the conserved ATP-binding domains (Krüger *et al*., 1994). Independently, the gene was cloned and designated as *mecB* in a genetic complementation approach (Msadek *et al*., 1994). In addition, the complete nucleotide sequence of the *clpC* locus was determined in the *B. subtilis* genome sequencing project (Ogasawara *et al*., 1994). GenBank accession numbers: X75930, U02604, D26185.

■ *clpC* gene structure

The *clpC/mecB* gene encodes a protein of 810 amino acid residues with a calculated molecular mass of 90 kDa and an isoelectric point of 7.0 (Msadek *et al*., 1994). On two-dimensional protein gels the protein migrates within the range of the values estimated (Krüger *et al*., 1994). Hydropathy profiles as well as the protein extraction procedure used for two-dimensional protein electrophoresis suggest that the ClpC protein is cytoplasmic. The ClpC/MecB protein is characterized by two regions distinguishing ClpC-type proteins (Squires and Squires, 1992): a leader sequence with a conserved tandem 32 amino acid duplication and a 61 amino acid spacer, separating the two highly conserved nucleotide-binding regions (Krüger et al., 1994; Msadek et al., 1994). Within the ClpC subfamily, ClpC of *B. subtilis* shares very strong similarity to the tomato plant CD4A protein (Gottesmann et al., 1990) (60% identity) and to the *Mycobacterium leprae* ClpC protein (Nath and Laal, 1990) (63% identity). Prokaryotic ClpC proteins have so far been identified only in Grampositive bacteria. DNA sequence analyses of the *clpC* locus indicate that ClpC is encoded by the fourth gene of a 7.2 kb operon containing six genes (Ogasawara et al., 1994).

■ *clpC* gene regulation

Under normal growth conditions the expression level of *B. subtilis clpC* is rather low. When cells are exposed to stress factors such as heat shock, ethanol, salt or the antibiotic puromycin, *clpC* mRNA increases very dramatically (more than 100-fold by heat shock and 20–50-fold by other stress conditions). Treatment with hydrogen peroxide, limitation of oxygen and exhaustion of glucose or amino acids result in a weaker induction (Krüger et al., 1994, 1996). Northern blot analysis suggests that *clpC* is part of a six gene operon cotranscribed in a polycistronic stress inducible mRNA. Transcription of this operon starts at two promoters (P_A and P_B) upstream of the first gene. P_A resembles promoters recognized by the vegetative RNA polymerase $E\sigma^A$. The other promoter (P_B) is dependent on σ^B, the general stress transcription factor in *B. subtilis* (Haldenwang, 1995). Strong induction by heat, ethanol and salt stress occurs at the σ^B-dependent promoter, whereas the vegetative promoter is only weakly induced under these conditions. In a *sigB* mutant, transcription at the σ^A promoter becomes inducible by heat and ethanol stress, completely compensating for *sigB* deficiency. Even in the absence of σ^B, this promoter switch ensures a similar rate of ClpC synthesis as in the wild-type. Oxidative stress or puromycin treatment lead to induced transcription at the σ^A promoter. Initiation of transcription seems to be the predominant target of regulation, involving a repression mechanism together with additional activation (Krüger et al., 1996).

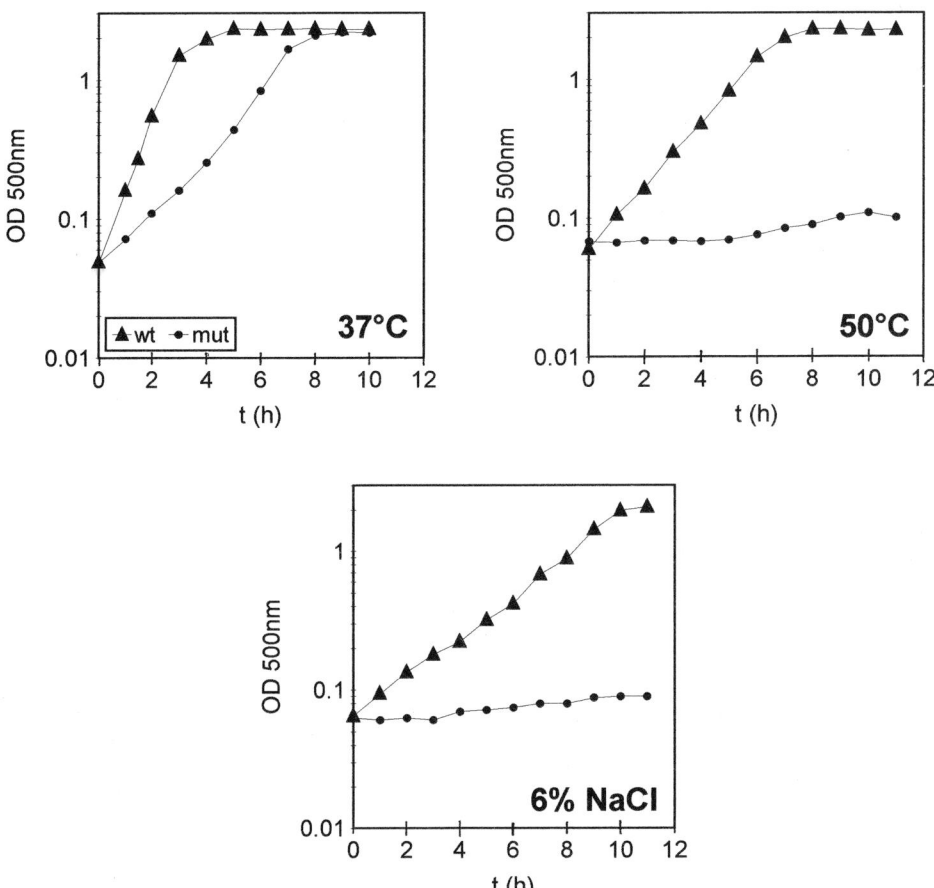

Figure 1. Growth of *B. subtilis* wild-type strain IS58 (triangles) and *clpC* mutant strain BEK1 (closed circles) at 37°C and after exposure to 50°C and 6% (w/v) NaCl in synthetic medium (adapted from Krüger et al., 1994).

Function of ClpC

Deletion or disruption of *clpC* in *B. subtilis* causes a very pleiotropic phenotype. While the HSP70 homologue DnaK does not seem to be necessary for normal growth (Schulz et al., 1995), *clpC* mutants show impaired growth in ammonium based minimal medium, even under normal conditions. However, growth can be restored by addition of certain amino acids (Krüger et al., 1994; Kunst et al., 1994). The presence of ClpC in the cell is apparently crucial for stress tolerance, because *clpC* mutants cannot grow under severe heat and salt stress conditions [50°C or 6% (w/v) NaCl, Fig. 1] (Krüger et al., 1994). This phenomenon is probably a result of a deficiency in the ability of those mutants to solubilize or degrade damaged and aggregated proteins after stress, the accumulation of which is toxic for the cell (Parsell et al., 1994). Aggregates observed by electron microscopic studies in *clpC* mutant cells exposed to 50°C heat shock or treated with puromycin may support this assumption (E. Krüger and M. Hecker, unpublished data). However, assays for detection of defects in protein degradation suggest a direct involvement of ClpC in proteolysis. Turnover rates for abnormal proteins after puromycin treatment are significantly decreased in *clpC* mutant cells in comparison to the wild-type, whereas the difference under normal conditions is less pronounced (S. Ohlmeier, E. Krüger and M. Hecker, unpublished). As shown for the *E. coli* ClpA protein (Maurizi, 1992), heterologous fusion proteins (e.g. LacZ fusion proteins) seem to be another substrate for ClpC-mediated proteolysis in *B. subtilis* (E. Krüger, F. Kunst and M. Hecker, unpublished data). In this context it is interesting to note that the gene encoding the *B. subtilis* ClpP protease is regulated in the same manner as *clpC* (U. Gerth, E. Krüger and M. Hecker, unpublished data).

ClpC was found to be involved in several stationary phase phenomena in *B. subtilis*. A *clpC* mutation triggers the formation of filamentous cells (Krüger et al., 1994) which are devoid of flagella, non-motile, more resistant against autolysins and show increased levels of intracellular flagellin (E. Krüger and M. Hecker, unpublished observations). Inappropriate expression of flagellin and autolysins results in similar phenotypes (Ordal et al., 1993) suggesting a subtle effect, possibly related to a role of ClpC in correct assembly of the flagella, or secretion of flagellin, or both. Moreover, ClpC (MecB) controls competence gene expression and degradative enzyme synthesis (Msadek et al., 1994). Originally, the *mecA* and *mecB* mutations were isolated as allowing competence gene expression in complex media (Dubnau and Roggiani, 1990). Both MecB and MecA act as negative effectors of ComK, a transcriptional activator required for the expression of late competence genes. An interaction of MecB with MecA, which shows some similarities to ClpP proteases, is thought to be necessary for negative regulation of ComK synthesis (Kong and Dubnau, 1994; Msadek et al., 1994; Turgay et al., 1997).

References

Dubnau, D. and Roggiani, M. (1990). Growth medium-independent genetic competence mutants of *Bacillus subtilis*. J. Bacteriol. **172**, 4048–4055.

Gottesman S., Squires, C., Pichersky, E., Carrington, M., Hobbs, M.,,. Mattick, J.S, Dalrymple, B., Kuramitsu, H., Shiroza, T., Foster, T., Clark, W.P, Ross, B., Squires, C.L., and Maurizi, M.R. (1990). Conservation of the regulatory subunit for the Clp ATP-dependent protease in prokaryotes and eukaryotes. Proc. Natl. Acad. Sci. USA **87**, 3513–3517.

Haldenwang, W.G. (1995). The sigma factors of *Bacillus subtilis*. Microbiol. Rev. **59**, 1–30.

Kong, L. and Dubnau, D. (1994). Regulation of competence-specific gene expression by Mec-mediated protein-protein interaction in *Bacillus subtilis*. Proc. Natl. Acad. Sci. USA **91**, 5793–5797.

Krüger, E., Msadek, T., and Hecker, M. (1996). Alternate promoters direct stress induced transcription of the *Bacillus subtilis clpC* operon. Mol. Microbiol. **20**, 713–723.

Krüger, E., Völker, U., and Hecker, M. (1994). Stress induction of *clpC* in *Bacillus subtilis* and its involvement in stress tolerance. J. Bacteriol. **176**, 3360–3367.

Kunst, F., Msadek, T., Bignon, J., and Rapoport, G. (1994). The DegS/DegU and ComP/ComA two-component systems are part of a network controlling degradative enzyme synthesis and competence in *Bacillus subtilis*. Res. Microbiol. **145**, 393–402.

Maurizi, M.R. (1992). Proteases and protein degradation in *Escherichia coli*. Experientia **48**, 178–201.

Msadek, T., Kunst, F., and Rapoport, G. (1994). MecB of *Bacillus subtilis*, a member of the ClpC ATPase family, is a pleiotropic regulator controlling competence gene expression and survival at high temperature. Proc. Natl. Acad. Sci. USA **91**, 5788–5792.

Nath, I. and Laal, S. (1990). Nucleotide sequence and deduced amino acid sequence of *Mycobacterium leprae* gene showing homology to bacterial *atp* operon. Nucl. Acids Res. **18**, 4935.

Ogasawara, N., Nakai S., and Yoshikawa, H. (1994). Systematic sequencing of the 180 kilobases region of the *Bacillus subtilis* chromosome containing the replication origin. DNA Res. **1**, 1–14.

Ordal, G.W., Marquez-Magana, L., and Chamberlin, M.J. (1993). Motility and chemotaxis. In *Bacillus subtilis and other Gram-positive bacteria: biochemistry, physiology, and molecular genetics*, (Sonenshein, A.L., Hoch, J.A., and Losick, R. eds). American Society for Microbiology, Washington, D.C., pp. 765–784.

Parsell, D.A., Kowall, A.S., Singer, M.A., and Lindquist, S. (1994). Protein disaggregation mediated by heat shock protein Hsp104. Nature **372**, 475–478.

Schulz, A., Tschaschel, B., and Schumann, W. (1995) Isolation and analysis of mutants of the *dnaK* operon of *Bacillus subtilis*. Mol. Microbiol. **15**, 421–429.

Squires, C. and Squires, C. L. (1992). The Clp proteins: proteolysis regulators or molecular chaperones? J. Bacteriol. **174**, 1081–1085.

Turgay, K., Hamoen, L.W., Venema, G., and Dubnau, D. (1997). Biochemical characterization of a molecular switch involving the heat shock protein ClpC, which controls the acitivity of ComK, the competence transcription factor of *Bacillus subtilis*. Genes Dev. **11**, 119–128.

■ *Elke Krüger and Michael Hecker:*
Institut für Mikrobiologie und Molekularbiologie
Ernst-Moritz-Arndt-Universität Greifswald
Jahnstr. 15, D-17487 Greifswald, Germany
Tel. 49 3834 864200
Fax. 49 3834 864202
E-mail: hecker@microbio7.biologie.uni-greifswald.de
and ekrueger@microbio4.biologie.uni-greifswald.de

Synechococcus sp. PCC 7942 ClpB

The Synechococcus *heat shock protein ClpB is a member of a recently discovered group of proteins designated the HSP100/Clp family. ClpB proteins are present in many different bacteria and eukaryotic organisms. The cyanobacterium* Synechococcus *sp. PCC 7942 possesses two clpB genes that code for polypeptides around 80% similar. During heat stress, the clpB1 gene is strongly induced, producing two different sized polypeptides of 78 and 92 kDa. The clpB2 gene, however, lacks the second translation initiation site that produces the two ClpB1 forms. Induction of ClpB1 is essential for developing thermotolerance in* Synechococcus *sp. PCC 7942.*

■ *Synechococcus clpB* gene and sequence

Two *clpB* genes exist in the unicellular cyanobacterium *Synechococcus* sp. PCC 7942. The first, *clpB1* (GenBank accession number U20646), consists of an uninterrupted 2649 bp open reading frame coding for a polypeptide of 883 amino acids (92 kDa). The second gene, *clpB2* (Genbank accession number U97124), is 2685 bp and codes for a predicted protein of 94 kDa. Both genes have no obvious promoter motifs for either the constitutive σ^{70} or heat shock σ^{32} sigma factors (Curtis and Martin, 1994). The predicted ClpB proteins contain the two dissimilar ATP-binding domains common to all large Clp proteins, and the long intervening spacer region characteristic of ClpB (i.e. 160–180 amino acids) (Squires and Squires, 1992; see overview of HSP100/Clp family, p. 231). ClpB1 and ClpB2 are around 80% similar to each other. Both are also similar to ClpB proteins from other prokaryotes, with 53–57% identity (72–74% similarity) to the *Escherichia coli*, *Bacteroides nodosus* and *Haemophilus influenzae* homologues. They are equally similar to ClpB from higher plants, with 52–53% identity to the *Arabidopsis thaliana* and soybean proteins (Eriksson and Clarke, 1996). As with other Clp subfamilies, the various ClpB proteins share most similarity in the two nucleotide-binding domains. Within the flanking and spacer regions, however, there exists little sequence conservation between the different ClpB proteins, which are the most variable of the Clp subfamilies.

■ Synechococcus *clpB* gene expression

Expression of the *Synechococcus clpB1* gene is relatively low under normal growth conditions, but greatly increases during thermal stress. The *clpB1* gene produces two proteins of different size, one of 92 kDa matching the predicted full length protein, the other of 78 kDa (Eriksson and Clarke, 1996). The synthesis of a truncated form of ClpB (known as ClpB′ or ClpB78) also occurs in *E. coli*, where it originates from a second translational initiation site within the single *clpB* transcript (Squires et al., 1991; Park et al., 1993). This type of additional translational start for ClpB synthesis appears to be confined to prokaryotes (Clarke, 1996). The *clpB2* gene in *Synechococcus*, however, lacks the conserved motifs for the second translational initiation site and presumably does not produce a ClpB′ form.

Besides heat shock, ClpB1 is also induced by low temperature treatments in *Synechococcus* sp. PCC 7942. Similarly, ClpB proteins from other organisms are induced during physiological changes other than high temperature. In yeast, one of the ClpB homologues (Hsp104) is induced during the transition from exponential to stationary growth phase, in the early stages of sporulation, and after exposure to ethanol and arsenite (Sanchez and Lindquist, 1990; Sanchez et al., 1992). Synthesis of ClpB in *E. coli* is also stimulated by ethanol and cadmium treatment, and periods of nutrient starvation (Squires et al., 1991).

■ *Synechococcus* ClpB1 function

ClpB proteins protect the cell from thermal damage, although the specific characteristics of this activity apparently differ somewhat between various organisms. In *E. coli*, loss of ClpB greatly reduces cell survival after a sudden and extreme heat shock, but does not affect the cells' capacity to develop thermotolerance (Squires et al., 1991); that is, resistance to a normally lethal heat treatment through preconditioning at a non-lethal high temperature. In yeast, however, Hsp104 is essential for acquired thermotolerance, but does not influence cell viability after a rapid shift to extreme temperatures (Sanchez and Lindquist, 1990). It is thought that other eukaryotic homologues of the yeast Hsp104 also function in a similar manner. Complementation studies have shown that expression of higher plant *clpB* genes in a yeast *clpB* null mutant can restore the strain's capacity to become thermotolerant (Lee et al., 1994; Schirmer et al., 1994). Although of prokaryotic origin, ClpB1 from *Synechococcus* sp. PCC 7942 appears to be functionally more similar to eukaryotic homologues. Inactivation of *clpB1* expression in *Synechococcus* reduces the cells' ability to become thermotolerant by at least fourfold, but it has no affect on cell survival during sudden, severe heat shocks.

In yeast, the two ClpB homologues apparently function as molecular chaperones during heat shock. The Hsp104 protein which localizes within the cytoplasm and nucleus facilitates the resolubilization of inactive protein aggregates that accumulate at high temperatures (Sanchez et al., 1993; Parsell et al., 1994), and it is this function which presumably is critical for developing thermotolerance. It is therefore likely that the ClpB homologues in both plants and cyanobacteria also function in the same manner since they confer thermotolerance as well. In contrast, the yeast mitochondrial ClpB protein, Hsp78, is dispensable under all growth conditions, and is not involved in the acquisition of thermotolerance (Leonhardt et al., 1993). This ClpB homologue, however, appears to function in concert with Hsp70 chaperones in preventing protein denaturation, thereby reducing the increasing likelihood of protein aggregation during heat shock (Schmitt et al., 1995).

References

Clarke, A.K. (1996). Variations on a theme: combined molecular chaperone and proteolysis functions in the Clp/HSP100 family. J. Biosci. **21**, 161–177.

Curtis, S.E. and Martin, J.A. (1994). The transcription apparatus and the regulation of transcription initiation. In The Molecular Biology of Cyanobacteria, (D.A. Bryant, ed.). Kluwer Academic Publishers, The Netherlands, pp. 613–639.

Eriksson, M.-J. and Clarke, A.K. (1996). The heat shock protein ClpB mediates the development of thermotolerance in the cyanobacterium Synechococcus sp. strain PCC 7942. J. Bacteriol. **178**, 4839–4846.

Lee, Y.-R.J., Nagao, R.T., and Key, J.L. (1994). A soybean 101-kD heat shock protein complements a yeast HSP104 deletion mutant in acquiring thermotolerance. Plant Cell **6**, 1889–1897.

Leonhardt, S.A., Fearon, K., Danese, P.N., and Mason, T.L. (1993). Hsp78 encodes a yeast mitochondrial heat shock protein in the Clp family of ATP-dependent proteases. Mol. Cell. Biol. **13**, 6304–6313.

Park, S.K., Kim, K.I., Woo, K.M., Seol, J.H., Tanaka, K., Ichihara, A., Ha, D.B., and Chung, C.H. (1993). Site-directed mutagenesis of the dual translational initiation sites of the clpB gene in Escherichia coli and characterization of its gene products. J. Biol. Chem. **268**, 20170–20174.

Parsell, D.A., Kowal, A.S., Singer, M.A., and Lindquist, S. (1994). Protein disaggregation mediated by heat shock protein Hsp104. Nature **372**, 475–478.

Sanchez, Y. and Lindquist, S.L. (1990). HSP104 required for induced thermotolerance. Science **248**, 1112–1115.

Sanchez, Y., Taulien, J., Borkovich, K.A., and Lindquist, S. (1992). Hsp104 is required for tolerance to many forms of stress. EMBO J. **11**, 2357–2364.

Sanchez, Y., Parsell, D.A., Taulien, J., Vogel, J.L., Craig, E.A., and Lindquist, S. (1993). Genetic evidence for functional relationship between Hsp104 and Hsp70. J. Bacteriol. **175**, 6484–6491.

Schirmer, E.C., Lindquist, S., and Vierling, E. (1994). An Arabidopsis heat shock protein complements a thermotolerance defect in yeast. Plant Cell **6**, 1899–1909.

Schmitt, M., Neupert, W., and Langer, T. (1995). Hsp78, a Clp homologue within mitochondria, can substitute for chaperone functions of mt-hsp70. EMBO J. **14**, 3434–3444.

Squires, C.L., Pedersen, S., Ross, B., and Squires, C. (1991). ClpB is the Escherichia coli heat shock protein F84.1. J. Bacteriol. **173**, 4254–4262.

Squires, C. and Squires, C.L. (1992). The Clp proteins: proteolysis regulators or molecular chaperones? J. Bacteriol. **174**, 1081–1085.

■ *Adrian K. Clarke and Mats-Jerry Eriksson:*
Department of Plant Physiology
University of Umeå
901 87 Umeå, Sweden
Tel. 46 90 7865209
Fax. 46 90 7866676
E-mail: adrian.clarke@plantphys.umu.se and
mats-jerry.eriksson@plantphys.umu.se

Synechococcus sp. PCC 7942 ClpC

ClpC is a member of a large, ubiquitous group of proteins known as the HSP100/Clp family which function as molecular chaperones and/or regulators of ATP-dependent proteolysis. The ClpC protein in the unicellular cyanobacterium Synechococcus sp. PCC 7942 is closely related, both structurally and functionally, to the chloroplast ClpC from higher plants. The single copy clpC gene in Synechococcus encodes a protein around 90% similar to the chloroplastic homologue. It is constitutively expressed as a monocistronic transcript, and both ClpC expression and synthesis in Synechococcus is relatively unaffected by heat shock. The content of ClpC, however, increases considerably under conditions of rapid growth, both with either increasing light intensities or CO_2 concentrations. Synthesis of ClpC in Synechococcus is essential for cell viability.

■ ClpC sequence

The Synechococcus sp. PCC 7942 ClpC (Clarke and Eriksson, 1996) is an 87 kDa polypeptide (840 amino acids) containing the two conserved ATP-binding domains characteristic of all larger Clp proteins (i.e. ClpA, B and C; see overview of HSP100/Clp proteins, p. 231). The length of the intervening region between the ATP-binding domains (i.e. 100–110 amino acids) is also representative of proteins within the ClpC subfamily, as are the two highly conserved repeat motifs located within the N-terminal region (Squires and Squires, 1992). These two motifs are situated within the first 130 amino acids and are each 32 amino acids long. This region in the cyanobacterial protein is over 95% identical to the chloroplast homologues, with two of the three differing amino acids being conserved substitutions. In contrast to the ClpA and B subfamilies, several other smaller conserved domains were also present among the different ClpC proteins, within both the central and C-terminal spacer regions.

A comparison with all other known Clp proteins shows that the *Synechococcus* sp. PCC 7942 polypeptide is most similar to the chloroplastic ClpC protein from higher plants (Clarke and Eriksson, 1996). Overall, ClpC from *Synechococcus* is around 77% identical (88% similar) to the various higher plant ClpC proteins. It shares lower homology with the other full length bacterial ClpC protein, that from *Bacillus subtilis* (61% identity), and considerably less with the chloroplast-encoded polypeptide from the chromophytic algae *Heterosigma carterae* (55% identity). Apart from the two ATP-binding domains, the cyanobacterial ClpC shares no additional sequence homology to Clp proteins from other subfamilies (Clarke and Eriksson, 1996).

ClpC is found in a wide variety of organisms. It has so far been identified in several strains of Gram-positive bacteria, such as *B. subtilis*, and in different photobionts including cyanobacteria, algae, and higher plants. In higher plants, ClpC is a nuclear-encoded protein that is post-translationally imported into the chloroplast, where it localizes within the stroma (Moore and Keegstra, 1993; Shanklin *et al.*, 1995). Using PCR with degenerate primers directed to the two ATP-binding domains, *clpC* genes homologous to that in *Synechococcus* sp. PCC 7942 have been identified and cloned from several other strains of uni- and multicellular freshwater cyanobacteria, including *Synechococcus* sp. PCC 6301, *Synechocystis* sp. PCC 6701, *Synechocystis* sp. PCC 6803, *Nostoc* sp. PCC 7120, *Calothrix* sp. PCC 7601, and *Gloeothece* sp. PCC 73107 (Clarke and Eriksson, 1996).

■ *Synechococcus clpC* gene and expression

Synechococcus sp. PCC 7942 *clpC* is a single copy gene (GenBank accession number U16134), and has many features in common with *clpC* genes from higher plants (Schirmer *et al.*, 1994; Shanklin *et al.*, 1995). The cyanobacterial homologue is constitutively expressed as a monocistronic transcript of 2.7 kb. Its expression is relatively unaffected by heat shock, as is the level of ClpC protein. As in higher plants, ClpC is essential for cell viability in *Synechococcus*. These similarities between the cyanobacterial and plastidic ClpC proteins are in strong contrast to the *B. subtilis* homologue. In *B. subtilis*, *clpC* is part of a three gene operon which is expressed only in small amounts under optimal growth conditions (Msadek *et al.*, 1994). Instead, ClpC is primarily a HSP in *B. subtilis*, while also being inducible by ethanol treatments and nutrient deprivation (Krüger *et al.*, 1994). Loss of the *B. subtilis* protein has no significant effect on normal growth, but it is essential for survival during heat stress (Msadek *et al.*, 1994).

In *Synechococcus*, the constitutive level of ClpC rises with increasing growth rates. At the optimal growth temperature (37°C), the level of ClpC more than doubles in cells grown with increasing photon irradiance (20 to 125 μmol m^{-2} s^{-1}) or higher CO_2 concentrations (ambient to 5%) (Clarke and Eriksson, 1996).

■ *Synechococcus* ClpC function

Although the precise function of ClpC is unknown at this stage, it is most likely that in photobionts ClpC has primarily evolved into a constitutive protein whose function is essential for normal growth, whereas in non-photosynthetic bacteria ClpC function remains mainly stress related. It has also been suggested that ClpC in *B. subtilis* regulates the turnover of ComK, a protein that controls competence gene expression (Msadek *et al.*, 1994). Part of the role of ClpC in photobionts is undoubtedly involved in protein turnover, via association to the proteolytic subunit, ClpP. A recent study has shown that purified chloroplast ClpP has similar peptidase activity *in vitro* as ClpP from *E. coli*, while purified chloroplast ClpC can replace ClpA in stimulating the proteolytic activity of the *E. coli* ClpP protein (Shanklin *et al.*, 1995). ClpC is also likely to function independently of ClpP as a molecular chaperone, as do members of the ClpA and ClpB subfamilies (Parsell *et al.*, 1994; Wickner *et al.*, 1994). The fact that ClpC activity is indispensable in photobionts suggests that it participates in the folding and/or degradation of metabolically important polypeptides such as those related to oxygenic photosynthesis.

■ References

Clarke, A.K. and Eriksson, M-J. (1996). The cyanobacterium *Synechococcus* sp. PCC 7942 possesses a close homologue to the chloroplast ClpC protein of higher plants. Plant Mol. Biol. **31**, 721–730.

Krüger, E., Völker, U., and Hecker, M. (1994). Stress induction of *clpC* in *Bacillus subtilis* and its involvement in stress tolerance. J. Bacteriol. **176**, 3360–3367.

Moore, T. and Keegstra, K. (1993). Characterization of a cDNA clone encoding a chloroplast targeted Clp homologue. Plant Mol. Biol. **21**, 525–537.

Msadek, T., Kunst, F., and Rapoport, G. (1994). MecB of *Bacillus subtilis*, a member of the ClpC ATPase family, is a pleiotrophic regulator controlling competence gene expression and growth at high temperature. Proc. Natl. Acad. Sci. USA **91**, 5788–5792.

Parsell, D.A., Kowal, A.S., Singer, M.A., and Lindquist, S. (1994). Protein disaggregation mediated by heat shock protein Hsp104. Nature **372**, 475–478.

Schirmer, E.C., Lindquist, S., and Vierling, E. (1994). An arabidopsis heat shock protein complements a thermotolerance defect in yeast. Plant Cell **6**, 1899–1909.

Shanklin, J., DeWitt, N.D., and Flanagan, J.M. (1995). The stroma of higher plant plastids contain ClpP and ClpC, functional homologs of *Escherichia coli* ClpP and ClpC: an archetypal two-component ATP-dependent protease. Plant Cell **7**, 1713–1722.

Squires, C. and Squires, C.L. (1992). The Clp proteins: proteolysis regulators or molecular chaperones? J. Bacteriol. **174**, 1081–1085.

Wickner, S., Gottesman, S., Skowyra, D., Hoskins, J., McKenney, K., and Maurizi, M.R. (1994). A molecular chaperone, ClpA, functions like DnaK and DnaJ. Proc. Natl. Acad. Sci. USA **91**, 12218–12222.

■ Adrian K. Clarke:
Department of Plant Physiology
University of Umeå
901 87 Umeå, Sweden
Tel. 46 90 7865209
Fax. 46 90 7866676
E-mail: adrian.clarke@plantphys.umu.se

Saccharomyces cerevisiae Hsp104

In Saccharomyces cerevisiae, Hsp104 plays an important role in helping cells survive extreme environmental stresses such as high temperatures and high concentrations of ethanol. Its function in stress tolerance is related to its ability to promote the resolubilization of protein aggregates. In addition to its role in stress tolerance, Hsp104 participates in the control of a prion-like factor known as [psi] in yeast.

■ *HSP104* gene structure and regulation

The *S. cerevisiae HSP104* gene (GenBank accession number M67479) codes for a 908 amino acid protein with predicted M_r of 102 kDa and estimated pI of 5.14. On two-dimensional gels it migrates close to the predicted values indicating that it is not subject to major post-translational modification (Parsell et al., 1991). The protein contains two domains with predicted Walker-type nucleotide-binding folds. These domains share little sequence homology other than the consensus elements themselves (Walker et al., 1982). The 5′ flanking region of the gene contains a heat shock element (HSE) 302 nucleotides before the start codon, as well as three general stress response elements (STREs) (Schuller et al., 1994) at 252, 220, and 172 nucleotides before the start. The end of the mRNA is c. 120 nucleotides beyond the stop codon (R.P. Dellavale and S. Lindquist, unpublished data). Hsp104 is expressed at a low level during log-phase growth on glucose and is weakly induced by respiratory carbon sources (mildly by galactose, more by acetate), weakly induced by heavy metals (Na arsenite, Cd, Cu), and strongly induced by EtOH, heat, stationary phase growth, and sporulation (Sanchez et al., 1992).

■ Hsp104 structural characteristics and purification

Hsp104 protein is purified from yeast by chromatography of yeast lysates through Affi-Gel Blue (Bio-Rad), DEAE, and hydroxyapatite — eluting with KCl gradients (Parsell et al., 1994a). Hsp104 modified with a 6 × His tag and expressed in bacteria is purified by chromatography on a nickel column followed by a Resource-Q column (Pharmacia). STEM analysis, sizing columns, and glutaraldehyde cross-linking experiments indicate that Hsp104 assembles into hexamers in the presence of nucleotide under conditions of low ionic strength, and at high protein concentrations (> 1 mg/ml) (Parsell et al., 1994a; E.C. Schirmer, A.S. Kowal, C. Queitsch, and S. Lindquist, unpublished data). Hsp104 hydrolyses ATP with a V_{max} of c. 1 nmol min^{-1} and a K_m of c. 5 mM under physiological conditions; however, under conditions of low ionic strength the K_m is reduced to 0.6 mM (E.C. Schirmer, C. Queitsch, A.S. Kowal, D.A. Hattendorf, D.A. Parsell and S. Lindquist, unpublished results). The ATPase activity of Hsp104 is stimulated by some peptides under a variety of conditions; however, a subset of peptides stimulate only under conditions of physiological or high salt concentrations. The best peptides stimulate ATP hydrolysis up to 13-fold (E.C. Schirmer and S. Lindquist, unpublished data). Hsp104 protein purified from bacterial expression systems behaves like Hsp104 purified from yeast in ATPase assays and glutaraldehyde cross-linking assays. A panel of peptide and monoclonal antibodies against Hsp104 have been generated: of these, an antibody designated 2-3 is particularly useful for recognizing related proteins in widely divergent organisms. Staining with antibodies indicates that Hsp104 is distributed in the cytoplasm and nucleus (J. Glover, and S. Lindquist, unpublished data).

■ Hsp104 biological activity

Hsp104 is critical for thermotolerance in yeast. While most heat shock proteins are important for growth at mildly elevated temperatures (eg. 37°C), Hsp104 is not important for growth at 25 or 37.5°C (Sanchez and Lindquist, 1990). However, cells lacking Hsp104 protein die at extreme temperatures (eg. 50°C) at 100–1000 times the rate of wild-type (WT) cells (Fig. 1) (Sanchez and Lindquist, 1990). Using a heterologous induction system which allows production of Hsp104 without the induction of other heat shock proteins (Louvion, 1993), Hsp104 could

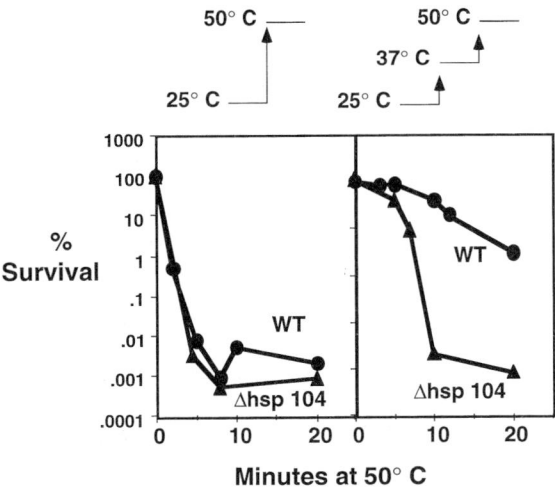

Figure 1. Cells not given a mild pre-heat shock at 37°C prior to an extreme heat stress of 50°C die rapidly. Cells in which Hsp104 has been induced by a pre-heat shock survive 1000-fold better than cells lacking Hsp104. Adapted from Sanchez and Lindquist (1990).

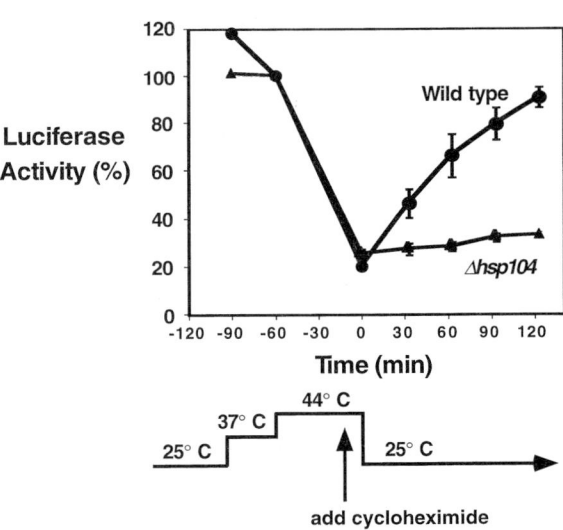

Figure 2. WT cells can reactivate a heat-denatured temperature-sensitive bacterial luciferase reporter protein, but isogenic cells deleted for Hsp104 cannot. Adapted from Parsell et al. (1994b).

provide thermotolerance in the absence of other heat shock proteins. Moreover, the effect of Hsp104 was titratable and the quantity of Hsp104 present was proportional to the number of cells surviving the severe heat shock (Lindquist and Kim, 1996). Thus Hsp104 is both critical and sufficient for thermotolerance. If other HSPs are required as cofactors, constitutively expressed versions must suffice. Hsp104 also protects cells from high concentrations of ethanol (20%) (Sanchez et al., 1992). A temperature of 50°C and 20% EtOH are extreme stresses, but the combination of milder stresses that might be more commonly encountered in the environment can also be lethal. When moderate EtOH stress (6%) is combined with moderate heat stress (40°C), yeast cells not expressing Hsp104 die, while cells expressing Hsp104 survive (Lindquist et al., 1995).

The ability to survive extreme stress appears to be related to the ability of Hsp104 to promote the solubilization and reactivation of damaged and aggregated protein. A heat inactivated temperature-sensitive luciferase protein regained activity after return to normal temperatures in yeast cells containing Hsp104, but not in yeast cells in which Hsp104 had been deleted (Fig. 2). The luciferase was soluble before heat inactivation, but after heat inactivation it disappeared from the supernatants of both wild-type (WT) and hsp104 cell lysates subjected to ultracentrifugation. After recovery, the luciferase remained pelletable in the Δhsp104 cells, but reappeared in the soluble fraction in the WT cells (Parsell et al., 1994b). Thus Hsp104 participates in the solubilization of heat damaged proteins. These conclusions are supported by scanning electron microscopic analysis of cells that were free of artificial reporter constructs: dense particles appeared in the cytosol and nucleus following heat stress in both WT and Δhsp104 cells, but disappeared upon return to normal temperatures (25°C) in

only the WT cells (Parsell et al., 1994b). In vitro reconstitution of aggregation and solubilization of luciferase with purified proteins suggests that Hsp104 cooperates with Ydj1 and Hsc70 in the formation of folding-competent aggregates and their subsequent solubilization (J. Glover, and S. Lindquist, unpublished data).

Two specific biological processes in which Hsp104 plays a role have been identified: the repair of RNA splicing after heat shock and the maintenance of the prion factor known as [psi]. When cells are subjected to a 41°C heat stress RNA splicing is disrupted. Δhsp104 cells recover splicing more slowly than WT cells containing Hsp104 (Yost and Lindquist, 1991), and purified Hsp104 restores splicing in heat treated lysates (Vogel et al., 1995). [psi] is believed to be a prion-like factor (Cox, 1994) caused by an altered conformation of the Sup35 translation-termination factor that results in a loss of translational fidelity. Hsp104 was recently isolated as a suppressor of [psi] (Chernoff et al., 1995): overexpression or underexpression of Hsp104 eliminated [psi]. Hsp104 influences the aggregation state of Sup35 (Patino, et al. 1996) and aggregation of Sup35 correlates with the appearance of the [psi] phenotype serving as a further parallel between the yeast phenomenon and mammalian prions.

■ Hsp104 mutational analysis

Proteins carrying directed mutations in the consensus (Walker A) glycine-rich loop of the first ATP-binding domain (Parsell et al., 1991) fail to hydrolyse ATP, but assemble into hexamers like WT protein. Proteins carrying equivalent mutations in the second ATP-binding domain are deficient in their ability to assemble into hexamers (Parsell et al., 1994a). This deficiency in oligomerization

Wagner, I., Arlt, H., Van Dyck, L., Langer, T., and Neupert, W. (1994). Molecular chaperones cooperate with PIM1 protease in the degradation of misfolded proteins in mitochondria. EMBO J. **13**, 5135–5145.

Wojtkowiak, D., Georgopoulos, C., and Zylicz, M. (1993). Isolation and characterization of ClpX, a new ATP-dependent specificity component of the Clp protease of *Escherichia coli*. J. Biol. Chem. **268**, 22609–22617.

■ Matthias Schmitt and Thomas Langer:
Adolf Butenandt Institut für Physiologische Chemie
Goethestr. 33
80336 München, Germany
Tel. 49 89 5996 283
Fax. 49 89 5996 270
E-mail: langer@bio.med.uni-muenchen.de

Plant Hsp101/ClpB

Plants are the only higher eukaryotes from which a close homolog to the yeast Hsp104 protein has been characterized. Although the function of plant Hsp101 remains undefined, the protein can replace the thermotolerance function of yeast Hsp104, indicating that it is also likely to be important for acquired thermotolerance in plants.

■ Plant Hsp101 genes and proteins

Full length cDNAs encoding Hsp101 have been isolated from *Arabidopsis thaliana* (Schirmer et al., 1994) and soybean (*Glycine max*) (Lee et al., 1994). The *Arabidopsis* cDNA encodes a protein of 911 amino acids with a predicted molecular weight of 101 267 Da and therefore the protein has been designated AtHsp101. The protein encoded by the open reading frame in the soybean gene is also 911 amino acids with a predicted molecular weight of 101 328 Da. The soybean protein is designated as GmHsp101. Overall AtHsp101 is approximately 94% similar and 87% identical at the amino acid level to GmHsp101. When compared with other members of the HSP100/Clp protein family, two large conserved regions are recognized (Gottesman et al., 1990), both of which contain consensus ATP-binding motifs as defined by Walker et al. (1982) (Fig. 1). These large conserved regions have been referred to as 'ATP-binding domains', although confirmation of the boundaries of these domains will require three-dimensional structure data. In both plant proteins the first conserved domain starts at amino acid 176 and is 233 amino acids long. The second conserved domain starts after a 172 residue 'spacer' at amino acid 570 and is 152 amino acids in length. Identity between the *Arabidopsis* and soybean proteins is highest in the two large consensus domains (93% identity), is high in the intervening spacer region (89% identity), and lowest in the N- and C-terminal domains (78 and 81% identity, respectively). The plant proteins are both around 50% identical and 70% similar to *E. coli* ClpB and are 44% identical and 65% similar to yeast Hsp104. These homologies, along with the length of the spacer region, indicate that AtHsp101 and GmHsp101 belong to the ClpB class of HSP100 chaperones. The multiple sequence alignment of the plant proteins to ClpB from *E. coli* and yeast Hsp104 (Fig. 1) underscores the sequence divergence outside the regions encompassing the Walker ATP-binding motifs.

Gene copy number appears to be low, with probably only one gene being present in *Arabidopsis* (Schirmer et al., 1994). Characterization of genomic clones has not been published and the map position of the *HSP101* gene in *Arabidopsis* has not yet been determined.

■ *HSP101* gene expression

The mRNAs encoding AtHsp101 and GmHsp101 are not present in vegetative tissues of plants in the absence of heat stress, but are induced to high levels with rapid kinetics at elevated temperatures (Lee et al., 1994; Schirmer et al., 1994). In fact, the AtHsp101 cDNA was isolated by differential screening of a heat shock cDNA library. The AtHsp101 protein is not detected in vegetative tissues grown under optimal conditions, but it accumulates rapidly during heat stress. Interestingly, the AtHsp101 protein is also expressed as part of the normal seed maturation program. Protein accumulation begins during the mid-maturation stage of seed development; the protein is stored in the dry seed and declines in abundance upon germination (E. Vierling, unpublished observations). This pattern of expression is similar to that observed for small HSPs and certain HSP70 isoforms in plants (DeRocher and Vierling, 1994, 1995), implicating a role for these chaperones in normal seed development.

■ Complementing activity of plant Hsp101 proteins in yeast

Both AtHsp101 and GmHsp101 will, at least partially, replace the function of Hsp104 in yeast with respect to

```
                                          40                                                  80                                                 120
At101    .MNPEKFTHK TNEtiAtAhE LAvnaGHAQF TPlHLagALI SDP.tGifPq aIsSA.GGEn aAqsaERVIN QALKkLPsQS PPPDDIPaSs sLIkViRRAQ AaQKSRGDTh lAVDQLIMgL
Gm101    .MNPEKFTHK TNEALAsAhE LAMssGHAQL TPiHLahALI SDP.nGifvl aINSAGGGEe sARavERVlN QALKkLPcQS PPPDEVPaSt nLVraiRRAQ AaQKSRGDTr lAVDQLILgi
EcClpB   .MrlDrLTnK fqlALAdAqs LALgHdnqfi ePlHLmsALl nqe.gGsvsp lltSAGinag qlRtd...IN QALnrLPQve gtggDVqpSq dLVrVlnlcd nvaqkRGDnF issElFVLaa
Sc104    mndqtqFTer altiLtlAqk LAsdHqHpQL qPiHilaAFI etPedGsvPy lqNliekGry dydlfkkVVN rnLvriPQQq PaPaEItpSy aLgkVlqdAa kiQKqqkDsF iAqDhilFaL
Con      -------T-- -------A-- LA-------- -P-H---A-- -----G---- ---------- ---------N --L---P--- --------S- -L-------- -------D-- ----------

                                         160                                                 200                                                 240
At101    LED.SQIrDl LnEvGVAtAr VKsEVEKLRG KEGKKVESAS GDTNFQALKt YGrDLvEQA. .GKLDPVIGR DEEIRRvVRI LSRRTKNNPV LIGEPGVGKT AVVEGLAQRI VkGDVPnsLt
Gm101    LED.SQIgDl LKEAGVAvAk VesEVDKLRG KEGKKVESAS GDTNFQALKt YGrDLvEQA. .GKLDPVIGR DEEIRRvVRI LSRRTKNNPV LVGEPGVGKT AVVEGLAQRI VrGDVPsnLa
EcClpb   LEsrgtVaDi LKaAGattAn ItqaIEqMRG ..GesVndqg aEdqrQALKk YtiDLtErAe QGKLDPVIGR DEEIRRtIqV LqRRTKNNPV LIGEPGVGKT AIVEGLAQRI InGEVPegLk
Sc104    FnD.SsIqqi FKEAqVdiea IKqqaleLRG ntridsrgAd tnTpLeyLsk YaiDMtEQAr QGKLDPVIGR EEEIRstIRV LaRRiKsNPc LIGEPGIGKT AIIEGvAQRI IddDVPtiLq
Con      ---------- ---------- --------RG ---------- -------L-- Y--D--E-A- -GKLDPVIGR -EEIR----- L-RR-K-NP- L-GEPG-GKT A--EG-AQRI ----VP--L-

                                         280                                                 320                                                 360
At101    dVRLisLDMG ALVAGAKYRG EFEERLKsVL KEVEDAEGKV ILFIDEIHLv LGAGKTEGSM DAANLFKPmL ARGQLRCIGA TTLEEYRKYV EKDAAFERRF QqVYVAEPSV pDTIsILRGL
Gm101    dVRLiaLDMG ALVAGAKYRG EFEERLKaVL KEVEEAEGKV ILFIDEIHLv LGAGrTEGSM DAANLFKPmL ARGQLRCIGA TTLEEYRKYV EKDAAFERRF QqVFVAEPSV vDTIsILRGL
EcClpB   grRvlaLDMG ALVAGAKYRG EFEERLKGVL nDlakqEGnV ILFIDElHtm vGAGKaDGaM DAgNMLKPaL ARGeLhCVGA TTLDEYRqYI EKDAALERRF QkVFVAEPSV eDTIaILRGL
Sc104    gakLfsLDLa ALtAGAKYkG DFEERFKGVL KEIEEsktlI VLFIDEIHMl MgnGKd.... DAANiLKPaL sRGQLkvIGA TTnnEYRsiV EKDgAFERRF QkIeVAEPSV rqTVaILRGL
Con      ------LD-- AL-AGAKY-G -FEER-K-VL ---------- -LFIDE-H-- -G-G------ DA-N--KP-L -RG-L---GA TT--EYR--- EKD-A-ERRF Q---VAEPSV --T--ILRGL

                                         400                                                 440                                                 480
At101    KEkYEGHHGV RIQDRALInA AQLSaRYITG RHLPDKAIDL VDEACANVRV QLDSqPEEiD NLERKRMQLE IELHALErEK DKAS..KARL IeVRKELDDL RDKLQPLtMK YRKEKERIDE
Gm101    KErYEGHHGV RIQDRALVmA AQLSnRYITG RHLPDKAIDL VDEACANVRV QLDSqPEEiD NLERKRMQLE VELHALEkEK DKAS..KARL VeVRKELDDL RDKLQPLmMK YRKEKERVDE
EcClpB   KErYElHHhV qItDpAiVaA AtLShRYIad RqLPDKAIDL IDEAassIRm QiDSkPEElD rLDRriiQLk lEqqALmkEs DeAS..KkRL dmlneELsDk erqyseLeee WkaEKaslsg
Sc104    qpkYEiHHGV RIlDsALVtA AQLakRYlpy RrLPDsAlDL VDisCAgVaV arDSkPEElD skERqlqliq VEikALErDe DadSttKdRL klaRqkeasL qEeLePLrqr YneEKhghEE
Con      ---YE-HH-V -I-D-A---A A-L--RY--- R-LPD-A-DL -D-------- --DS-PEE-D ---R------ -E--AL---- D--S--K-RL ---------- ------L--- ---EK-----

                                         520                                                 560                                                 600
At101    IRRLKqKrEE LMFsLQEAER RYDLARAADL RYGAIqEVEs aIaQLEGtSs EE......Nv MLTenVGPEh IAEVVsRWTG IPVTRLGqnE KErLIGLaDr LHkRVVGQNq AVNAVSeAIl
Gm101    IRRLKKKrEE LLFALQEAER RYDLARAADL RYGAIqEVEt aIqQLEG.St EE......Nl MLTetVGPEq IAEVVsRWTG IPVTRLGqnE KErLIGLgDr LHSRVVGQdq AVNAVaeAVl
EcClpB   tqtiKaelEq akiAieqArR vgDLARmsEL qYGkIpElEK qlEaatqleg kt......mr LLrnkVtdae IAEVlaRWTG IPVsRMmesE rEkLlrMEqe LHhRVIGQNe AVdAVSnAIr
Sc104    ltqaKKKlDE LenkalDAER RYDtAtAADL RYfAIpDIkK qIEkLEdqva EEerragaNs MiqnvVdsDt IsEtaaRlTG IPVkkLsesE nEkLIhMErd LsSeVVGQmd AIkAVSnAVr
Con      ----K----- -------A-R --D-A----L -Y--I----- ---------- ---------- -----V---- I-E---R-TG IPV------E -E-L------ L---V-GQ-- A--AV--A--

                                         640                                                 680                                                 720
At101    RSRAGLgRaQ QPTGSFLFLG PTGVGKTELA KALAEqLFDD EnLLVRIDMS EYMEqHSVSR LIGAPPGYVG hEEGGqLTEA VRRRPYcVIL FDEVEKAHva VFNtLLQVLD DGRLTDGQGR
Gm101    RSRAGLgRPQ QPTGSFLFLG PTGVGKTELA KALAEqLFDn EnqLVRIDMS EYMEqHSVSR LIGAPPGYVG hEEGGqLTEA VRRRPYSVVL FDEVEKAHts VFNtLLQVLD DGRLTDGQGR
EcClpB   RSRAGLadPn rPiGSFLFLG PTGVGKTELc KALAnfMFDs DeaMVRIDMS EFMEkHSVSR LVGAPPGYVG yEEGGyLTEA VRRRPYSVIL LDEVEKAHPD VFNiLLQVLD DGRLTDGQGR
Sc101    lSRsGLanPr QP.aSFLFLG lsGsGKTELA KkvAgfLFnD EdMMIRVDcS ELsEkyaVSk LlGttaGYVG yDEGGfLTnq lqykPYSVlL FDEVEKAHPD VLtvMLQmLD DGRiTsGQGk
Con      -SR-GL---- -P--SFLFLG --G-GKTEL- K--A---F-- -----R-D-S E--E---VS- L-G---GYVG --EGG-LT-- ----PY-V-L -DEVEKAH-- V----LQ-LD DGR-T-GQG-

                                         760                                                 800                                                 840
At101    TVDFRNsVII MTSNLGAEHL LaGLtGKVTM EVArDcVMrE VRkHFRPELL NRlDEIVVFD PLSHDQlRKV ArlQMKDVAv RLa..ErGvA LAVTDaALDY iLaEsYDPVY GARPiRRWme
Gm101    TVDFRNTVII MTSNLGAEHL LSGLsGKcTM qVArDrVMqE VRrqFRPELL NRlDEIVVFD PLSHDQlRKV ARlQMKDVAs RLa..EkGiA LAVTDaALDY iLsEsYDPVY GARPiRRWle
EcClpB   TVDFRNTVVI MTSNLGsD.L iqerfGeldy ahmkELVLGv VshnFRPEFi NRiDEVVVFh PLgeqhiasI AqiQLKrlyk RLe..ErGye ihIsDeALkL lsengYDPVY GARPlkRaiq
Sc104    TIDcsNcIVI MTSNLGAEfi nSqqgsKIq. EstknLVMGa VRqHFRPEFL NRissIVIFn kLSrkaihKI vdirLKEIee RFeqnDkhyk LnlTqeAkDF lakygYsddm GARPlnRliq
Con      T-D--N---I MTSNLG---- ---------- ------V--- V---FRPE-- NR----V-F- -L-------- -----K---- R--------- ------A--- -----Y---- GARP--R---

                                         880                                                 920
At101    KKVVtELskM VvREEIDENS TVYIDAG..a GDLVYRvE.s GGlVDAsTGk KSDVLIhIaN G.pkrsDAaQ AVKKMrIEEI EDDDnEEmie d*
Gm101    KKVVtELsrM lvREEIDENS TVYIDAGpng GELVYRvEKN GGvVnptTGq KSDILIqIPN GhApktDAvQ AVKKMkIEEI DDDEMEE*.. ..
EcClpB   qqIenpLaqq IlsgEl.... .........v pgkViRlEvN edRIvAVq*. .......... .......... .......... .......... ..
Sc104    neIlnkLalr IlknEIkDke TVnVv..... ......lkKg ksRdEnVpee aeEcLevlPN heAtigadtl gdddnedsme iDDDLD*... ..
Con      ------L--- ----E----- ---------- ---------- ---------- ---------- ---------- ---------- ---------- --
```

Figure 1. Amino acid sequence alignment of AtHsp101 (At101), GmHsp101 (Gm101), *E. coli* ClpB (EcClpB), and yeast Hsp104 (Sc104). Residues shown in upper case correspond to identical or conservative replacements for two or more sequences. The bottom line shows the consensus identity for all sequences. The two large conserved domains [as defined by Gottesman *et al.* (1990)] encompassing the Walker ATP-binding motifs are highlighted in bold in the consensus identity line. Residues comprising the Walker motifs are highlighted in bold and underlined within the alignment. Analysis was performed with the Wisconsin GCG package using 'Pileup' set with default parameters to align the sequences, and 'Pretty' to derive the consensus. Dots within the sequence indicate gaps introduced to maximize the alignment.

the development of thermotolerance (Lee et al., 1994; Schirmer et al., 1994). In wild-type yeast cells, but not in cells with an *HSP104* deletion, a 37°C pretreatment protects cells from subsequent treatment at 50°C. When either AtHsp101 or GmHsp101 are expressed in cells with an *HSP104* deletion, survival at 50°C is enhanced >100-fold compared to cells with the expression vector alone. The plant proteins appear to be at least tenfold less effective than yeast Hsp104, but this difference is difficult to quantify because relative levels of active Hsp104 and Hsp101 protein have not been determined. The ability of the plant proteins to substitute for other activities of Hsp104 have not yet been tested.

Foster, T., Clark, W.P., Ross, B., Squires, C.L., and Maurizi, M.R. (1990). Conservation of the regulatory subunit for the Clp ATP-dependent protease in prokaryotes and eukaryotes. Proc. Natl. Acad. Sci. USA **87**, 3513–3517.

Lee, Y.-R.J., Nagao, R.T., and Key, J.L. (1994). A soybean 101-kD heat shock protein complements a yeast HSP104 deletion mutant in acquiring thermotolerance. Plant Cell **6**, 1889–1897.

Schirmer, E.C., Lindquist, S., and Vierling, E. (1994). An Arabidopsis heat shock protein complements a thermotolerance defect in yeast. Plant Cell **6**, 1899–1909.

Walker, J.E., Saraste, M., Runswick, M.J., and Gay, N.J. (1982). Distantly related sequences in the α- and β-subunits of ATP synthase, myosin, kinases and other ATP-requiring enzymes and a common nucleotide binding fold. EMBO J. **1**, 945–951.

■ References

DeRocher, A.E. and Vierling, E. (1994). Developmental control of small heat shock protein expression during pea seed maturation. Plant J. **5**, 93–102.

DeRocher, A. and Vierling, E. (1995). Cytoplasmic HSP70 homologues of pea: differential expression in vegetative and embryonic organs. Plant Mol. Biol. **27**, 441–456.

Gottesman, S., Squires, C., Pichersky, E., Carrington, M., Hobbs, M., Mattick, J.S., Dalrymple, B., Kuramitsu, H., Shiroza, T.,

■ *Elizabeth Vierling:*
Department of Biochemistry
University of Arizona
Life Sciences South
1007 E. Lowell Street
Tucson, AZ 85721, USA
Tel. 1 520 621 1601
Fax. 1 520 621 3709
E-mail: eliz@biosci.arizona.edu

Chloroplast-localized Clp proteins

Two divergent members of the Clp family of proteins have now been identified and localized to the chloroplasts of higher plants. One of these, ClpC, is most likely involved in ATP-dependent proteolysis within plastids, analogous to the ClpA proteins from Escherichia coli. *Functional data are not yet available for the second family member, ERD1.*

■ Sequence characteristics

Complete derived amino acid sequences of higher plant ClpC proteins are available from *Pisum sativum* (Moore and Keegstra, 1993; GenBank accession number L09546) and *Lycopersicon esculentum* (Gottesman et al., 1990; P31541 and P31542), and a sequence missing less than 70 codons at the amino terminus is available from *Brassica napus* (X75328). Overall the ClpC amino acid sequences from these three species are c. 88% identical. Partial sequences are also reported as ESTs for *Arabidopsis thaliana* (Squires and Squires, 1992; Z29026) and *Zea mays* (T18272). As members of the ClpC family, the proteins have two conserved domains encompassing Walker ATP-binding motifs separated by a spacer, as defined by Gottesman et al. (1990), of approximately 75–77 amino acids (Fig. 1). The N-terminus of the isolated, mature *Pisum sativum* ClpC protein is Met-91 (Shanklin et al., 1995), indicating that the first 90 amino acids comprise the organelle targeting sequence. Significant homology between the *Pisum*, *Lycopersicon*, and *Brassica* sequences is seen only after residue 90, consistent with the location of the mature N-terminus. A cyanobacterial ClpC homologue (*Synechococcus* sp. U16134) has also been identified and is c. 76% identical to the higher plant proteins (see entry p. 247).

ERD1 has currently only been cloned from *Arabidopsis thaliana* (Kiyosue et al., 1993; D17582). Schirmer and Lindquist (this volume, p. 231) have classified ERD1 as a member of a distinct Clp family, designated ClpD. Like the plant ClpC proteins, ERD1 has a predicted chloroplast transit peptide, although the exact processing site has not been identified. *Arabidopsis* ERD1 is only c. 48% identical to *Arabidopsis* ClpC, and therefore clearly represents a different class of Clp protein.

ClpC and ERD1 have also diverged significantly from the plant ClpB homologues (see entry p. 253). Overall, ClpC and ERD1 are only 45 and 42% identical to AtHSP101, respectively (Fig. 1).

```
                    1                                                        40                                        80
PsClpC              marvlaqsLS  vpglvaghKd  sqhkgSgkSk  rsVktmcAlr  tSgLrmSgFs  ...gIRTfnh  LnTmmrpgld  Fhskvskavs
ERD1                ......mevLS  tsspltlHsh  rllsaSsssSs  h.VtsiaAss  lSsFassYLg  islsnRTiihr  FsTtptnlrr  F.........
AtHSP101            ..........  ..........  ..........  ..........  ..........  ..........  ..........  ..........
Consensus           ----------  ----------  ----------  ----------  ----------  ----------  ----------  ----------

                    81                +                                       120                                      160
PsClpC              srRaRaKrfi  prAmFERFTE  KAIkvImLAQ  EEArrLGHnf  VgTeqiLLGL  IgEgTGIaak  vLkSmGInlk  dARveV....
ERD1                pqRkRkKkftp  isAvFERFTE  rAIraIFsQ  kEAksLGkdm  VyTqHLLLGL  IaEdrd..PQ  gFlgsGitid  kAReAVwsIw
AtHSP101            ..mnpEkFTh  KtnetIatAh  ELAvnaGHaq  ftplHLagaL  IsDpTGIfPQ  aisSaG.gen  aAqsAeRvIn
Consensus           ----------  ------E-FT-  ---------I  --A---G---  -------L--  ----------  --------A-  ----------

                    161                                                    200                                         240
PsClpC              .......ek   iigrgSgfva  vEIPFtpraK  RVLEIsQEea  RqLGhnYIgs  EHLlIGLLre  gEGvAaRVLe  nLGAdp....
ERD1                deansdskQe  easstSysks  tDmPFSiStK  RVFEaAvEys  RtMdcqYIAp  EHiavGLFtv  DDGsAgRVLk  rLGAnmnllt
AtHSP101            qalkklpsQs  pp.......p  dDIPaSsSli  kVirrAQaaq  ksrGdthlAv  DqLiMGLL..  EDsqirdlLn  evGv......
Consensus           ----------  ----P-----  ----------  ---V------  ----------  ---GL-----  ----------  ---G------

                    241                                                    280                                         320
PsClpC              ...TnIrtq   VirMvGEsad  SVtat.....  ...vGSGsS  nnKtp.tLEe  YgtnLTkLAe  EGKLDPVvGR  qpqIeRVtQI
ERD1                aaAlTrlKgE  IaKdgrEpss  Sskgsfespp  sgriaGSGpg  GkKaknvLEq  FcvDlTarAs  EGliDPVIGR  EkEVqRVIQI
AtHSP101            .AtaRVKsE   VeKLrGkegk  kVe......  ......saS  Gdtnfqalkt  YGrDLveqA.  DeEIrRVVrI  LsRRTKNNPV
Consensus           ----------  ----------  ----------  ----------  -----L----  ------A---  -G--DPV-GR  -----RV--I

                    321                                                    360                                         400
PsClpC              LgRRTKNNPc  LIGEPGVGKT  AIAEGLAQRI  AnGDVPetie  gKkVItLDMG  LLIVAGtKYRG  EFEERLKkLM  eEIKqS..dd
ERD1                LcRRTKNNPi  LIGEaGVGKT  AIAEGLAisI  AeasaPgfLl  tKRImSLDiG  LLmAGAkeRG  ELEaRvtaLi  sEVKKS.GKV
AtHSP101            .GKLDPVIGR  LIGEPGVGKT  AVvEGLAQRI  vkGDVPnsLt  dvRLISLDMG  aLVAGAkYRG  EFEERLKsvL  kEVedaeGKV
Consensus           L-RRTKNNP-  -GE-GVGKT  A--EGLA--I  ------P---  ------LD-G  -L-AG-K-RG  E-E-R-----  -E--------

                    401                                                    440                                         480
PsClpC              ILFIDEVHTL  IGAGAa....  .EGaiDAANi  LKPalARGEL  QCIGATTLDE  YRKhIEKDpd  LERRFQPVkV  pEPtVDETIq
ERD1                ILFIDEVHTL  IGsGtvgrgn  kgsgLDiANL  LKPsLgRGEL  QCIasTTLDE  FRsqfEKDkA  LaRRFQPVlI  nEPSeEDaVk
AtHSP101            ILFIDEIHlv  lGAGkt....  .EGsMDAANL  FKPmLARGqL  rCIGATTLEE  YRKyVEKDaA  FERRFQqVyV  aEPSVpDTIs
Consensus           ILFIDE-H--  -G-G------  --------  -KP-L-RG-L  -CI---TTL-E  -R----EKD-  --RRFQ-V--  -EP-------

                    481                                                    520                                         560
PsClpC              ILkGLRerYE  iHHklRYTDE  ALIAAAQLSy  qYIsDRFLPD  KAIDLVDEAG  SRVRIQhaql  PEE.......  .........A
ERD1                ILIGLREKYE  aHHnckYTmE  AidAAvyLss  RYIaDRFLPD  KAIDLIDEAG  SRaRI.....  .E........  .........A
AtHSP101            ILrGLkEkYE  gHHgvRiqDr  ALInAAQLsa  RYItgRhLPD  KAIDLVDEAc  anVRVQldsq  PEEidnlerk  rmqleielhA
Consensus           IL-GL-E-YE  -HH-------  A---A--LS-  -YI--R-LPD  KAIDL-DEA-  --R-------  -E--------  ---------A
```

```
                561                                                              640
PsClpC      kElDKE.... .......... ......vrk iVKeKeEyvR nQDfekageL ........D lkAqisAliE
ERD1        FrkkKE.... .......... .....dai  cIlskppnDy wQEiktvqaM hEvvLsSRQK qD.......D gDAisdesgE
AtHSP101    LErEKDkask arlievrkel ddlrdklqpl tmKyrkEkER iDEIrrlkqk rEelMfslQe aErrydlara aDlrygAiqE
Consensus   ----K----- ---------- ---------- ---------- ---------- ---------- ---------- ---------E

                641                                                              720
PsClpC      kgkEmSkaEt etaDEgpi.. ..VtevDIqh IVSsWTGIPV dkVsADEsDR LlkMEDtLHK RIIGQDEAVq AISRAIrRaR
ERD1        lveEsSlppa agdDEpIL.. ..VGPDDIAa VaSvWsGIPV qqItADErml LMsLEDqLrg RVVGQDEAVa AISRAVkRSR
AtHSP101    vesaiaqlEg tssEEnVMlt enVGPEhIAe VVSrWTGIPV trlgqnEkER LiglaDrLHK RVVGQnqAVn AVSeAIlRSR
Consensus   ---------- ---------- ----V----I --S-W-GIPV ---------E L--------- --------------------

                721                                                              800
PsClpC      VGLKnPnRPI ASFiFsGPTG VGKsELAKAL AAyYFGSEEa MIRLDMSEFM ERHTVSKLIG SPPGYVGYtE GGQLTEAVRR
ERD1        VGLKdPdRPI AamLFCGPTG VGKTELtKAL AAnYFGSEES MIRLDMSEYM ERHTVSKLIG SPPGYVGFEE GGmLTEAIRR
AtHSP101    aGLgraqqPt gSFLFlGPTG VGKTELAKAL AeqlFddEnl LVRiDMSEYM EqHsVSrLIG aPPGYVGhEE GGQLTEAVRR
Consensus   -GL------P ------GPTG VGK-EL-KAL A---F---E- --R-DMSE-M E-H-VS-LIG -PPGYVG--E GG-LTEA-RR

                801                                                              880
PsClpC      RPYTVVLFDE IEKAHPDVFN mMLQILEDGR LTDSkGRTVD FKNtLIIMTS NVGSsvIeKG gr.rIGFDLD yDEkdsSYnr
ERD1        RPFTVVLFDE IEKAHPDIFN iLLQlFEDGh LTDSQGRrVs FKNaLIIMTS NVGSlaIaKG rhGsIGFilD dDEeaaSYtg
AtHSP101    RPYcVILFDE VEKAHvaVFN tLLQVLDDGR LTDgQGRTVD FrNsVIIMTS NlGaehIlaG ltGkVtmEva rDc.......
Consensus   RP--V-LFDE -EKAH----FN --LQ----DG- LTD--GR-V- F-N---IMTS N-G------- ---------- -D--------

                881                                                              960
PsClpC      iKsLVtEELK qYFRPEFLNR LDEmIVFRQL tKlevkEIAd iMLKEVfqRL ktkeIeLqVT ErfrDrVvdE GYnPsYGARP
ERD1        mKaLVvEELK nYFRPELLNR iDEIVIFRQL eKaQMmEIln vaIGVgLeVs EpvkELIckq GYDPaYGARP
AtHSP101    ...VmrEvr  khFRPELLNR LDEIVFdpL  shdQLrkVAr LqMKDVavRL aerGVaLaVT DaaLDyIlaE sYDPvYGARP
Consensus   ----V--E-- ---FRPE-LNR -DE---F--L ---------- --------RL ---------L-V- -------------- -Y-P-YGARP

                961                                                             1040
PsClpC      LRRaImrlLE DsMaEkmLAr EIKEGDsViV DvDsdGkvIV lngSsGtpEs lpEalSI*.. .......... ..........
ERD1        LRRtVteiVE DplSEaflAg sfKpGDTaFV vlDdtGnpsV Rtkpdsstir vTDktSIa.. .......... ..........
AtHSP101    iRRwmekkVv teLSkmvvre EIdEnsTVYI D.agaGdlVy RveSgGlvDa sTgKkSdvli hiangpkrsd aaqavkkmri
Consensus   -RR------- ---------- ---------- ---------- ---------- ---------- ---------S ---------- ----------
```

Figure 1. Amino acid sequence alignment of *Pisum sativum* ClpC (PsClpC), ERD1, and *Arabidopsis* Hsp101 (AtHsp101). Residues shown in upper case correspond to identical or conservative replacements for two or more sequences. The bottom line shows the consensus identity for all sequences. The two large conserved domains [as defined by Gottesman et al. (1990)] encompassing the Walker ATP-binding motifs are highlighted in bold in the consensus identity line. Residues comprising the Walker motifs are underlined within the alignment. The position of the mature N-terminus of PsClpC is indicated with a '+' above the alignment. Analysis was performed with the Wisconsin GCG package using 'Pileup' set with default parameters to align the sequences, and 'Pretty' to derive the consensus. Dots within the sequence indicate gaps introduced to maximize the alignment.

Localization and regulation

ClpC and ERD1 were tentatively identified as chloroplast-localized proteins based on the presence of the predicted targeting sequence in the N-terminal portion of the derived amino acid sequences. Subsequently, proteins translated from the transcribed ClpC or ERD1 cDNAs were found to be imported and proteolytically processed by isolated chloroplasts, an experimental criteria for demonstrating chloroplast localization (Moore and Keegstra, 1993; J. Froelich, unpublished results). Anti-ClpC antibodies have confirmed the chloroplast localization of ClpC in vivo (Shanklin et al., 1995).

ClpC is constitutively expressed in all tissues of *Arabidopsis* examined to date, indicating that this protein functions in both chloroplasts and non-photosynthetic plastids. ClpC accounts for c. 0.1% of total protein and is present in approximately stoichiometric levels with respect to ClpP. It colocalizes with ClpP in the stroma of plastids (Shanklin et al., 1995). In contrast to Hsp101, ClpC mRNA levels decrease during heat stress (Schirmer et al., 1994).

ERD1 expression has only been examined in leaves and whole plants. An increase in mRNA levels was observed in response to both desiccation and senescence, from a low basal level in all but the youngest unstressed leaves (Kiyosue et al., 1993; L.M. Weaver, unpublished data). Messenger RNA levels are unaffected by several other ion, hormone, and stress treatments, including heat shock (Kiyosue et al., 1993; L.M. Weaver, unpublished data).

Function of chloroplast Clp proteins

Purified mature recombinant plant ClpC (from *Pisum sativum*) was able to facilitate the degradation of casein by the *E. coli* ClpP protease in an ATP-dependent fashion. Thus, chloroplast ClpC is a functional homologue of *E. coli* ClpA rather than ClpB or ClpX (Shanklin et al., 1995). ClpC/ClpP represents the only energy dependent protease identified in the plastid to date. Interestingly the cyanobacterial homologue appears to be an essential gene in *Synechococcus* (Clarke and Eriksson, 1996), further emphasizing the important role of these proteins.

Recent data suggest chloroplast ClpC may also have a role in the import of proteins into chloroplasts. ClpC co-immunoprecipitates with complexes containing chloroplast protein precursors (Nielsen et al. 1997).

The biochemical function of ERD1 is unknown, although given its similarity to ClpC and presumed presence within senescing chloroplasts a proteolytic regulatory role would seem likely.

References

Clarke, A.K. and Eriksson, M.-J. (1996). The cyanobacterium *Synechococcus* sp. PCC 7942 possesses a close homologue to the chloroplast ClpC protein of higher plants. Plant Mol. Biol. **31**, 721–730.

Gottesman, S., Squires, C., Pichersky, E., Carrington, M., Hobbs, M., Mattick, J.S., Dalrymple, B., Kuramitsu, H., Shiroza, T., Foster, T., Clark, W.P., Ross, B., Squires, C.L., and Maurizi, M.R. (1990). Conservation of the regulatory subunit for the Clp ATP-dependent protease in prokaryotes and eukaryotes. Proc. Natl. Acad. Sci. USA **87**, 3513–3517.

Kiyosue, T., Yamaguchi-Shinozaki, K., and Shinozaki, K. (1993). Characterization of cDNA for a dehydration-inducible gene that encodes a Clp A, B-like protein in *Arabidopsis thaliana* L. Biochem. Biophys. Res. Commun. **196**, 1214–1220.

Moore, T. and Keegstra, K. (1993). Characterization of a cDNA clone encoding a chloroplast-targeted Clp homologue. Plant Mol. Biol. **21**, 525–537.

Nielsen, E., Akita, M., Davila-Aponte, J., and Keegstra, K. (1997). Stable association of chloroplastic precursors with protein translocation complexes that contain proteins from both envelope membranes and a stromal Hsp100 molecular chaperone. EMBO J. **16**, 935–946.

Schirmer, E.C., Lindquist, S., and Vierling, E. (1994). An Arabidopsis heat shock protein complements a thermotolerance defect in yeast. Plant Cell **6**, 1899–1909.

Shanklin, J., DeWitt, N.D., and Flanagan, J.M. (1995). The stroma of higher plant plastids contain ClpP and ClpC, functional homologs of *Escherichia coli* ClpP and ClpA: an archetypal two-component ATP-dependent protease. Plant Cell **7**, 1713–1722.

Squires, C. and Squires, C.L. (1992). The Clp proteins: proteolysis regulators or molecular chaperones? J. Bacteriol. **174**, 1081–1085.

Acknowledgements

I thank L.M. Weaver, J.E. Froehlich, R.M. Amasino, and A.K. Clarke for communicating unpublished data, and R. Amasino and J. Shanklin for comments on this review.

■ *Elizabeth Vierling:*
Department of Biochemistry
University of Arizona
Life Sciences South
1007 E. Lowell Street
Tucson, AZ 85721, USA
Tel. 1 520 621 1601
Fax. 1 520 621 3709
E-mail: eliz@biosci.arizona.edu

Leishmania Hsp100

The *Leishmania* ClpB *gene encodes an Hsp100 protein that is barely detectable in the insect stage of this obligatory parasite, but becomes increasingly abundant during heat stress or during* in vitro *differentiation into the pathogenic amastigote form. Like its homologue in yeast, Hsp104, Hsp100 confers stress tolerance since a Δclpb knock-out mutant of* L. major *shows increased temperature sensitivity* in vitro. *Moreover, Hsp100 contributes to the virulence of* L. major *in laboratory animals.*

■ *Leishmania*

Leishmaniae have a biphasic life-cycle. In their vectors, the sandflies of the genus *Phlebotomus,* they inhabit the digestive tract as a flagellated promastigote form. In the mammalian host the promastigotes are phagocytized by macrophages, differentiate into aflagellated amastigotes, and parasitize the macrophages of the skin (*L. major, L. tropica*), the mucous membranes (*L. mexicana*), or the viscera (*L. donovani, L. infantum*). During the transmission from a fly into a mammalian host the parasites encounter a temperature upshift of c. 10°C. This heat shock induces the synthesis of heat shock proteins and in some species can trigger the differentiation into the amastigote form.

■ *Leishmania* ClpB/*hsp100* gene and sequence

Based upon the amino acid sequences of various known members of the ClpB subfamily of HSP100 proteins (Gottesman *et al.*, 1990; Parsell *et al.*, 1991), the sequences that encode the highly conserved ATP-binding domains were amplified by using the polymerase chain reaction. The product was then employed to screen a *Leishmania major* phage library to isolate the *ClpB* gene. The deduced amino acid sequence of the *ClpB* gene product has a high degree of sequence identity in the two ATP-binding domains (> 60%) when compared to the ClpB family members from *Trypanosoma cruci, Escherichia coli* and *Saccharomyces cerevisiae* (Hübel *et al.*, 1995). However, the N-terminal and C-terminal amino acid sequences show a high degree of divergence.

■ *Leishmania* Hsp100 protein

The product of the *L. major ClpB* gene, Hsp100, has a predicted molecular mass of 97 000 daltons. The recombinant gene was expressed with a N-terminal histidine tail in the bacterial expression vector pJC45 (Clos and Brandau, 1994), purified by Ni^{2+} agarose affinity chromatography and used to immunize laying hens. Anti-Hsp100 antibodies were purified from egg yolk. These IgY antibodies specifically recognize a 97 kDa protein in extracts from heat shocked promastigotes of several *Leishmania* species (Hübel *et al.*, 1995). Biochemical fractionation studies, as well as immunogold staining experiments, suggest a predominantly cytoplasmic localization of the protein.

■ *Leishmania* ClpB/*hsp100* gene regulation

Like the other known heat shock proteins in *Leishmania*, the *ClpB* gene is transcribed constitutively without noticable heat inducibility (Brandau *et al.*, 1995). Hsp100 synthesis, however, increases eightfold during heat stress which indicates a post-transcriptional gene regulation (S. Brandau, unpublished results). This regulation is mediated by *ClpB* flanking sequences which in transient expression assays confer heat inducible expression of a CAT reporter gene (A. Hübel, unpublished results).

The synthesis of Hsp100 remains induced throughout a sustained heat stress of up to three days, during which

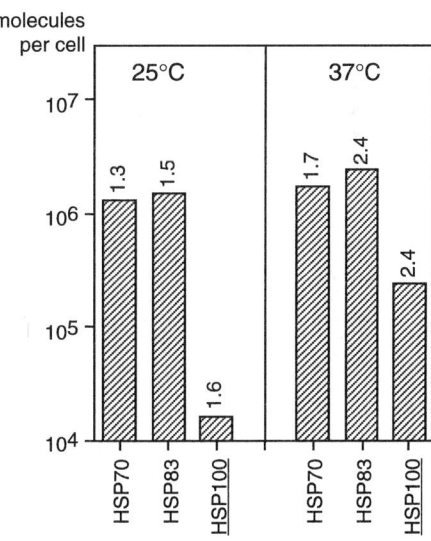

Figure 1. Number of HSP molecules per *Leishmania donovani* promastigote cell cultivated either at 25°C or at 37°C for three days. HSPs were quantified by a calibrated immunoblot analysis (Brandau *et al.*, 1995). From the total mass of each HSP the number of molecules was calculated by using Avogadro's number.

Hsp100 protein levels increase from 0.03% of the total cellular protein (1.6×10^4 molecules per cell) to c. 0.5% (2.4×10^5 molecules per cell) (Fig. 1). High concentrations of Hsp100 are also found in axenically derived amastigote stages of *L. donovani* (J. Clos, unpublished results). These findings indicate a role for Hsp100 only under conditions of heat stress.

Under all conditions the concentration of Hsp100 in *Leishmania* is one order of magnitude lower compared with Hsp70 or with Hsp83 (Fig. 1).

None of the known chemical inducers of the stress response can induce Hsp100 synthesis (S. Brandau and C. Hoyer, unpublished results). Since chemical inducers trigger the stress response in higher eukaryotes via activation of a heat shock transcription factor (HSF), the total lack of transcriptional regulation in *Leishmania* may preclude response to any stress but heat.

■ Function of *Leishmania* Hsp100

Nothing is known at present about the cellular targets and the mode of function of the *Leishmania* Hsp100.

■ Mutagenesis studies with *Leishmania* Hsp100

Δ*clpb* knock-out mutants of *L. major* display wild-type growth rates when incubated at standard cultivation temperatures (Fig. 2a). At 35°C, however, the Δ*clpb* mutant does not proliferate. This temperature still allows growth, albeit at a reduced rate, of the wild-type *L. major* (Fig. 2b). These results confirm the hypothesis that Hsp100 is required for the growth of the parasite at the upper limit of the permissive temperature range. In this it seems to play a role similar to that of the *S. cerevisiae* Hsp104 which is also required for growth under stressful conditions (Sanchez and Lindquist, 1990, Sanchez et al., 1992).

Recently, we have performed infection experiments using a susceptible inbred mouse strain (BALB/c). Preliminary results indicate a significantly reduced viability or proliferation of the Δ*clpb* mutants during the initial stages of the infection (Fig. 2c). Whether Hsp100 exerts its role by improving the parasite's tolerance towards host-specific stresses or by affecting the immune response of the mouse is the subject of an ongoing investigation.

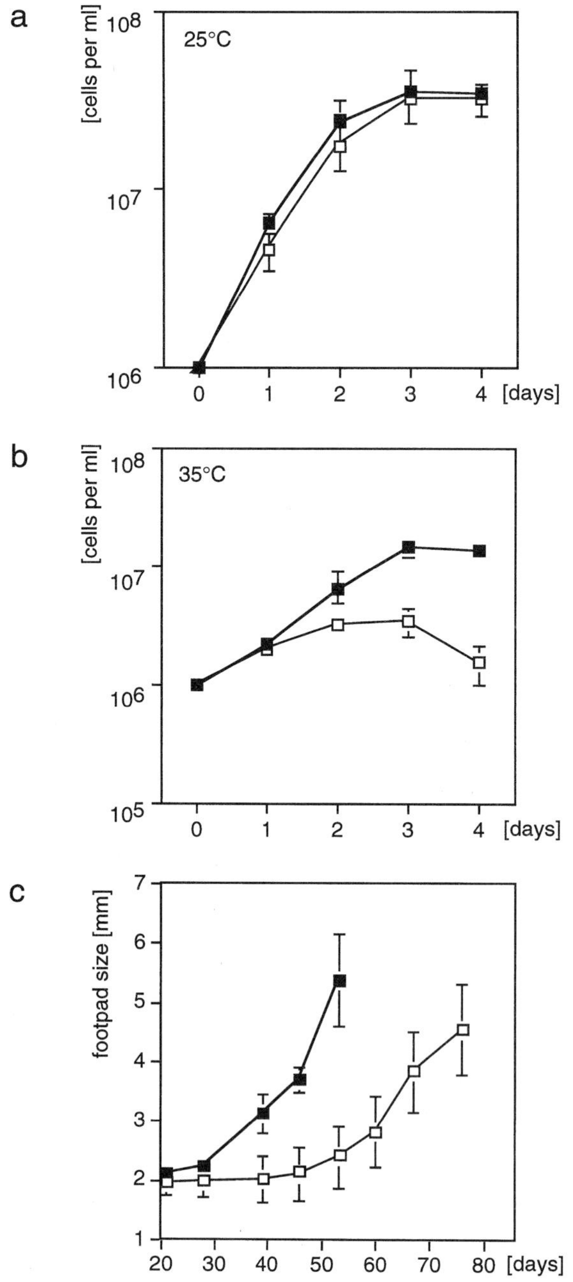

Figure 2. Phenotypic properties of the *L. major* Δ*clpb* knock-out mutant. Proliferation kinetics of *L. major* wild-type (solid squares) and *L. major* Δ*clpb* (open squares) were obtained at (a) 25°C or (b) 35°C: promastigotes of either strain were seeded at 10^6 cells per ml in supplemented M199 medium and cultivated for four days. Cells were counted daily. The data represent average counts from four individual experiments. (c) Progression of footpad lesions caused by infection of BALB/c mice with *L. major* wild-type (solid squares) or *L. major* Δ*clpb* (open squares). Stationary phase *L. major* promastigotes (2×10^7) were inoculated into footpads and lesion size was measured weekly as increase of overall footpad thickness. The data represent the mean of four infected animals.

References

Brandau, S., Dresel, A., and Clos, J. (1995). High constitutive levels of heat shock proteins in human-pathogenic parasites of the genus Leishmania. Biochem. J. **310**, 225–232.

Clos, J. and Brandau, S. (1994). pJC20 and pJC40: two high-copy-number vectors for T7 RNA polymerase-dependent expression of recombinant genes in Escherichia coli. Protein Express. Purif. **5**, 133–137

Gottesmann, S., Squires, C., Pichersky, E., Carrington, M., Hobbs, M., Mattick, J.S., Dalrymple, B., Kuramitsu, H., Shiroza, T., Foster, T., Clark, W.P., Ross, B., Squires, C.L., and Maurizi, M.R. (1990). Conservation of the regulatory subunit for the Clp ATP-dependent protease in prokaryotes and eukaryotes. Proc. Natl. Acad. Sci. USA **87**, 3513–3517.

Hübel, A., Brandau, S., Dresel, A., and Clos, J. (1995). A member of the ClpB family of stress proteins is expressed during heat shock in Leishmania sp.. Mol. Biochem. Parasitol. **70**, 107–118.

Parsell, D.A., Sanchez, Y., Stitzel, J.D., and Lindquist, S. (1991). Hsp104 is a highly conserved protein with two essential nucleotide-binding sites. Nature **353**, 270–273.

Sanchez, Y. and Lindquist, S.L. (1990). Hsp104 required for induced thermotolerance. Science **248**, 1112–1115.

Sanchez, Y., Taulien, J., Borkovich, K.A., and Lindquist, S. (1992). Hsp104 is required for tolerance to many forms of stress. EMBO J. **11**, 2357–2364.

■ *Joachim Clos, Andreas Hübel, Sven Brandau, Annette Dresel and Achim Hörauf:*
Bernhard Nocht Institute for Tropical Medicine
Bernhard Nocht St. 74
20359 Hamburg, Germany
Tel. 49 40 31182485
Fax. 49 40 31182400
Email: joachim_clos@MagicVillage.de

Malaria plastid ClpC

Malaria and related parasites have two extrachromosomal genomes, one of which is believed to belong to a vestigial plastid organelle. The 35 kb plastid-like DNA encodes around 50 different genes, only two of which have functions not involved with genetic expression. One of these two genes encodes a modified member of the HSP100/Clp family of chaperones.

■ Malaria *clp* gene and sequence

ClpC, the nucleus-encoded equivalent of Hsp100 in higher plants, is encrypted by a transposed plastid gene with an evolved leader sequence targeting its product back to the plastid organelle (Moore and Keegstra, 1993). A leaderless form of this gene is still maintained on the primitive plastid genome of the red alga *Porphyra purpurea* (M. Reith, personal communication). A form of *clp*, possibly analogous to *clpC*, is also carried on the vestigial plastid genome of malaria and related parasites (Fig. 1). This is consistent with the hypothesis that these organisms originated from a photosynthetic progenitor (possibly a dinoflagellate) carrying an algal plastid acquired by secondary endosymbiosis (Wilson *et al.*,1994). The malarial *clp* gene's modified form suggests it has a specialized function.

The open reading frame of the *clp* gene comprises an uninterrupted sequence of 765 codons specifying a putative peptide with a predicted mass of 91.2 kDa. The EMBL database accession number is X87631. Alignments of predicted amino acids show the malarial sequence resembles double nucleotide-binding forms of Clp proteins, rather than single nucleotide-binding forms such as ClpX. The garden pea ClpC and the malarial sequence have two leucine-rich blocks with similar spacing close to the N-terminus (Fig. 2). However, the malarial sequence has little other homology with the N-terminal region of conventional double nucleotide-binding forms of Clp, neither the first ATP-binding site nor the 'spacer' section being conserved. In addition, the ATPase consensus sequence ($Y-X_8-T-X_{13}-Y$) preceding the B2 region (Gottesman *et al.*, 1990) is not fully conserved in the predicted malarial peptide, T being substituted by N in the centre of the motif owing to a possible transversion (C to A) in the second position of the codon. In contrast, a relatively high level of homology is found in the second ATP-binding domain, especially in the ATP-binding regions and in the tail.

In the first of these homologous regions (aa 650–750 in Fig. 2) the predicted amino sequence of the malarial peptide and the garden pea ClpC are 66% identical. Moreover, the GC composition of the first part of the nucleotide-binding region is strikingly elevated (34%) compared with the rest of the gene, which is extremely AT-rich (88%); the vestigial genome's overall composition is 87% AT. Notably, residue 658 in the malarial sequence (Fig. 2), corresponding to Lys-620 in the nucleotide-binding site of Hsp104 in *S. cerevisiae*, is conserved. This residue is essential for oligomerization of the protein into a six-membered ring (Parsell *et al.*, 1994).

■ Transcription of the malarial *clp* gene

Transcripts of the malarial *clp* gene have been detected at a low level using RNA from forms of the parasite that infect vertebrate red blood cells (Wilson *et al.*, 1996).

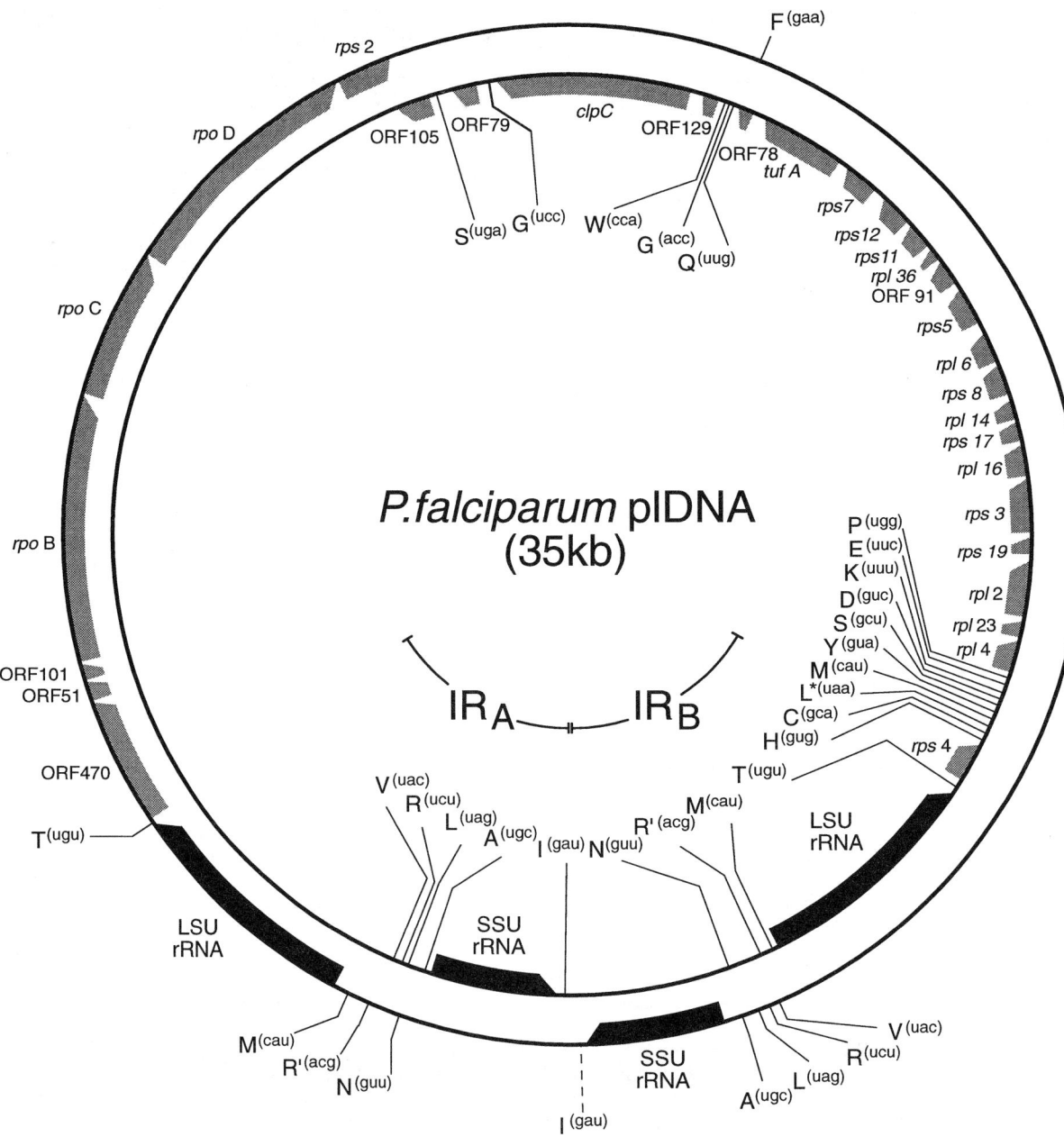

Figure 1. Gene map of plastid-like DNA from the malaria parasite *Plasmodium falciparum* showing position of *clpC*-like gene (from Wilson *et al.*, 1996).

Nothing is known about transcription in the invertebrate host, the mosquito.

■ Biological activities of malarial ClpC

Like virtually all the genes identified on the 35 kb DNA, the *clp* gene is likely to function in an organelle house-keeping capacity. However, at present, its role is unknown. Our interpretation of the origin of the plastid-like DNA is that malaria belongs to a group of organisms that acquired an algal plastid by secondary endosymbiosis. Unlike conventional plastids, such organelles are surrounded by quadruple membranes (Cavalier-Smith *et al.*, 1994) but nothing is known about the specialized organelle import system that must have evolved. Protein

```
                      10         20         30         40         50
pea.clpC     1 MARVLAQSLS VPGLVAGHKD SQHKGSGKSK RSVKTMCALR TSGLRMSGFS     50
Pfclp        1 ---------- ---------- ---------- ---------- ----------     50

                      60         70         80         90        100
pea.clpC    51 GRLTFNHLNT MMRPGLDFHS KVSKAVSSRR ARAKRFIPRA M-----FERF    100
Pfclp       51 ---------- ---------- ---------- ---------- ---MIILNNL    100

                     110        120        130        140        150
pea.clpC   101 --TEKAIKVI MLAQEEARRL GHNFVGTEQI LLGLIGEGTG IAAKVLKSMG    150
Pfclp      101 YCTKELIIIF IKSEYLALKY NNNFIMPIHL LLGLLLT-DN LCTKFLKINK    150

                     160        170        180        190        200
pea.clpC   151 INLKDARVEV EKIIGRGSGF VAVEIPFTPR AKRVLELSQE EARQLGHNYI    200
Pfclp      151 KIINNKIILS LLNKYKYNNK NIININFSN- --KVINILIK LNFNFK---I    200

                     210        220        230        240        250
pea.clpC   201 GSEHLLLGLL REGEGVAARV LENLGADPTN IRTQVIRMVG ESADSVTATV    250
Pfclp      201 NSFNLLLLLL EEKNNNKDIN YLFKYLNLNF SNLNLNNYIK TNIFSNNIRI    250

                     260        270        280        290        300
pea.clpC   251 GSGSSNNKTP TLEEYGTNLT KLAEEGKLDP VVGRQPQIER VTQILGRRTK    300
Pfclp      251 KLKEISVNLL NLNYIYNNNL NFYKQQYIQL LQILNLKIKK HI-IL-----    300

                     310        320        330        340        350
pea.clpC   301 NNPCLIGEPG VGKTAIAEGL AQRIANGDVP ETIEGKKVIT LDMGLLVAGT    350
Pfclp      301 --------EG VNDNIFIFLQ LLINNIKNKI IPIYLKYTEI WVLNDLLTYD    350

                     360        370        380        390        400
pea.clpC   351 KYRGEFEERL KKLMEEIK-Q SDDIILFIDE VHTLIGAGAA E-GAIDAANI    400
Pfclp      351 IQTLIYKILN ISKYFTNKYK LILIIKNI-E IFNLSDNINN DNNKLYYLFL    400

                     410        420        430        440        450
pea.clpC   401 LKPALARGEL QCIGATTLDE YRKHIEKDPD LERRFQPVKV PEPTVDETIQ    450
Pfclp      401 LLNKLYGYNI HIIIVTNKKE YNTYFKYNII KDSYFYKIRI KDLSILQTFL    450

                     460        470        480        490        500
pea.clpC   451 ILKGLRERYE IHHKLRYTDE ALIAAAQLSY QYISDRFLPD KAIDLVDEAG    500
Pfclp      451 IIKNNIYKYI NYYKININNY IIYELINLSK KYIKPLILPT TPLILLENSC    500

                     510        520        530        540        550
pea.clpC   501 SRVRLQHAQL PEEAKELDKE VRKIVKEKEE YVRNQDFEKA GELRDKEMDL    550
Pfclp      501 SNKYLLNNKI SYSNFNYLFT YNNNIIYNNK NNNLTIEDIK NSISN-----    550

                     560        570        580        590        600
pea.clpC   551 KAQISALIEK GKEMSKAETE TADEGPIVTE VDIQHIVSSW TGIPVDKVSA    600
Pfclp      551 ---------- ---------- ---------- ------YLNI SKTILFKDNK    600

                     610        620        630        640        650
pea.clpC   601 DESDRLLKME DTLHKRIIGQ DEAVQAISRA IRRARVGLKN PNRPIASFIF    650
Pfclp      601 LTKLNLTKLE NYLYNHIYGQ NHIFNKIIPF IKQNFIGLKN KNKPIGSWIL    650

                     660        670        680        690        700
pea.clpC   651 SGPTGVGKSE LAKALAAYYF GSEEAMIRLD MSEFMERHTV SKLIGSPPGY    700
Pfclp      651 CGPSGTGKTE LAKILSKQLF GSEKELIRFD MSEYMEKHSI SRLIGSPPGY    700

                     710        720        730        740        750
pea.clpC   701 VGYTEGGQLT EAVRRRPYTV VLFDEIEKAH PDVFNMMLQI LEDGRLTDSK    750
Pfclp      701 VGYSEGGQLT EQVYKKPNSV ILFDEIEKAH PDIYNIMLQI LDEGRLTDST    750

                     760        770        780        790        800
pea.clpC   751 GRTVDFKNTL LIMTSNVGSS VIEKGGRRIG FDLDYDEKDS SYNRIKSLVT    800
Pfclp      751 GKLIDFTHTI ILLTSNLGCP ----KNYDLY LKNKNFLSKS DLKEIEKNIK    800

                     810        820        830        840        850
pea.clpC   801 EELKQYFRPE FLNRLDEMIV FRQLTKLEVK EIADIMLKEV FQRLKTKEIE    850
Pfclp      801 ININNYFKPE LLNRLTNILI FNPLNINNLL FIFNKFINEL KIKLYLNKLN    850

                     860        870        880        890        900
pea.clpC   851 LQVTE-RFRD R-VVDEGYNP SYGARPLRRA IMRLLEDSMA EKMLAREIKE    900
Pfclp      851 IIIHINKELK YFLVKLMYNP LYGARPLKRI LELIFDKSIS DLLLTYNKHY    900

                     910        920        930        940        950
pea.clpC   901 GDSVIVDVDS D-GKVIVLNG SSGTP-ESLP EALSI..... ..........    950
Pfclp      901 FIKNKYILYY YLNKYYKLNF NIYLL..... .......... ..........    950
```

Figure 2. Alignment of nucleus-encoded ClpC from the garden pea *Pisum sativum* with the predicted malarial peptide (Pfclp) encoded on the 35 kb circle. Identical amino acids are highlighted.

remodeling by an energy-dependent chaperone activity would seem an appropriate role for the product of the malarial *clp* gene. Whether the modified Clp protein could carry out such a function is debatable since it contains only one of the two ATP-binding sites required for such activities in the yeast Hsp104 (Parsell *et al.*, 1991). However, ClpX, the 'half-Clp' equivalent, evidently can act as a chaperone (Wawrzynow *et al.*,1995).

Could the Clp protein be associated with a degradative pathway? Perhaps this is improbable because unlike the plastid genomes of higher plants, including the remnant genome of the non-photosynthetic parasitic plant *Epifagus virginiana* (Wolfe *et al.*, 1992), the 35 kb circle does not encode the proteolytic subunit ClpP. Could a heat shock role during transmission from invertebrate to vertebrate host be another reason for conservation of the *clp* gene? Up-regulation of ClpB following heat shock has been recorded for other protozoan parasites (Hubel *et al.*, 1995). There are two other possibilities. The first revolves around the suggestion by Moore and Keegstra (1993) that ClpC has a kinase-like motif AHPDVFN (around aa 730 in Fig. 2). Secondly, it has been suggested (see Squires and Squires, 1992) that Hsp110 protects ribosomal proteins or their assembly. This last context would fit with the 35 kb DNA's bias towards conservation of a gene expression system.

The EMBL accession number for a more conventional, nucleus-encoded malarial *clpB* gene is U46549.

■ References

Cavalier-Smith, T., Allsopp, M.T.E.P., and Chao, E.E. (1994). Chimeric conundrus: are nucleomorphs and chromists monophyletic or polyphyletic? Proc. Natl. Acad. Sci. USA **91**, 11368–11372.

Gottesman, S., Squires, C., Pichersky, E., Carrington, M., Hobbs, M., Mattick, J.S., Dalrymple, B., Kuramitsu, H., Shiroza, T., Foster,T., Clark, W.P., Ross, B., Squires, C.L., and Maurizi, M.R. (1990). Conservation of the regulatory subunit for the Clp ATP-dependent protease in prokaryotes and eukaryotes. Proc. Natl. Acad. Sci. USA **87**, 3513–3517.

Hubel, A., Brandau, S., Dresel, A., and Clos, J. (1995). A member of the clpb family of stress proteins is expressed during heat shock in *Leishmania* spp. Mol. Biochem. Parasitol. **70**, 107–118.

Moore, T. and Keegstra, K. (1993). Characterization of a cDNA clone encoding a chloroplast-targeted Clp homologue. Plant Mol. Biol. **21**, 525–537.

Parsell, D.A., Sanchez, Y., Stitzel, J.D., and Lindquist, S. (1991). Hsp104 is a highly conserved protein with two essential nucleotide-binding sites. Nature **353**, 270–273.

Parsell, D.A., Kowal, A.S., and Lindquist, S. (1994). *Saccharomyces cerevisiae* Hsp104 protein purification and characterization of ATP-induced structural changes. J. Biol. Chem. **269**, 4480–4487.

Squires, C. and Squires, C.L. (1992). The Clp proteins: proteolysis regulators or molecular chaperones? J. Bacteriol. **174**, 1081–1085.

Wawrzynow, A., Wojtkowiak, D., Marszalek, J., Banecki, B., Jonsen, M., Graves, B., Georgopoulos, C., and Zylicz, M. (1995). The ClpX heat-shock protein of *Escherichia coli*, the ATP-dependent substrate specificity component of the ClpP-ClpX protease, is a novel molecular chaperone. EMBO J. **14**, 1867–1877.

Wilson, R.J.M., Williamson, D.H., and Preiser, P. (1994). Malaria and other Apicomplexans: the 'plant' connection. Infect. Agents Dis. **3**, 29–37.

Wilson, R.J.M., Denny, P.W., Preiser, P.R., Rangachari, K., Roberts, K., Roy, A., Whyte, A., Strath, M., Moore, D.J., Moore, P.W. and Williamson, D.H. (1996). Complete gene map of the plastid-like DNA of the malaria parasite *Plasmodium falciparum*. J. Mol. Biol. **261**, 155–172.

Wolfe, K.H., Morden,C.W., and Palmer, J.D. (1992). Function and evolution of a minimal plastid genome from a non-photosynthetic parasitic plant. Proc. Natl. Acad. Sci. USA **89**, 10648–10652.

■ Iain Wilson:
National Institute for Medical Research
Mill Hill, London, NW7 1AA, UK
Tel. 44 181 959 3666 ext. 2177
Fax. 44 181 913 8593
E-mail: r-wilson@nimr.mrc.ac.uk

Mammalian HSP100

To date only a single representative of the HSP100 family has been cloned from a mammalian source. The gene (SKD3) was cloned as a suppressor of a potassium transport defect in yeast, suggesting a potential function for HSP100 proteins in solute transport. In addition to the gene from mouse, the human genome sequencing project has thus far contributed a number of partial cDNA sequences corresponding to HSP100s of at least two subfamilies. The prevalence of HSP100s in prokaryotes and in multiple subcellular compartments of eukaryotes suggests that many more mammalian HSP100s will be identified.

■ Alternative name

SKD3.

■ Mammalian *hsp100* genes and regulation

The *SKD3* gene (GenBank accession number U09874) encodes a 76 kDa protein composed of 677 amino acid residues (Périer et al., 1995). The C-terminal 378 amino acids align with the second nucleotide-binding and C-terminal domains of the HSP100 family (Schirmer and Lindquist, this vol. p. 231). The N-terminal domain of SKD3 contains four ankyrin-like repeats. Based on the characteristics of the primary protein sequence, the SKD3 gene product can be classified as an M-type HSP100. Northern blot analysis indicates that SKD3 is expressed in a variety of mouse tissues and is particularly abundant in testis, heart, kidney and skeletal muscle.

Partial human cDNA sequences (TIGR: THC8819, THC79885) encode an M-type HSP100 nearly identical to mouse SKD3 in the C-terminal (HSP100) domains and containing similar ankyrin repeats in the N-terminal region. Partial cDNA sequences encoding, most probably, X-type HSP100s have also been contributed (Genbank: R74599, TIGR: THC95377). An additional partial sequence containing the C-terminal HSP100 signature sequence V has been deposited (GenBank Z15857), but this clone provides too little sequence information to assign the protein positively to a particular subfamily.

Function of mammalian HSP100

No mammalian HSP100s have been functionally characterized in the biological context in which they may normally occur. SKD3 was cloned as a suppresser of a growth defect in yeast caused by mutations in genes encoding potassium transporters (Périer et al., 1995). Although there was no direct evidence at the molecular level defining a role for SKD3 in this suppression, it seems plausible that SKD3 may act as an ATP-binding component of a two-component transporter. The ankyrin repeat region could mediate the association of the nucleotide-binding domain with an integral membrane component of the transporter. The involvement of HSP100 proteins in solute transport is also suggested by the occurrence of genes encoding both an integral membrane protein (AmiS) and an N-type HSP100 ATPase (AmiB; Genbank X77160) in the amidase operon of *Pseudomonas aeruginosa* encoding genes involved in aliphatic amide utilization (Wilson et al., 1995).

References

Périer, F., Radeke, C.M., Raab-Graham, K.F., and Vandenberg, C.A. (1995). Expression of a putative ATPase suppresses the growth defect of a yeast potassium transport mutant: identification of a mammalian member of the Clp/Hsp104 family. Gene **265**, 157–163.

Wilson, S.A., Williams, R.J., Pearl, L.H., and Drew, R.E. (1995). Identification of two new genes in the *Pseudomonas aeruginosa* amidase operon, encoding an ATPase (AmiB) and a putative integral membrane protein (AmiS). J. Biol. Chem. **270**, 18818–18824.

John R. Glover:
Department of Molecular Genetics and Cell Biology
The University of Chicago
5841 S. Maryland Avenue
Room N334, MC1028
Chicago, IL 60637, USA
Tel. 1 773 702 8048
Fax. 1 773 702 7254
E-mail: jrglover@uchicago.edu

9

Small HSPs

The small heat shock protein (sHSP) family — an overview

Small HSPs (sHSPs) are a diverse class of proteins that differ from other HSP families in that only certain short sequence motifs are conserved. Characteristic features in common are their low molecular mass (15–30 kDa), their quaternary structure, ranging from 9 to 50 subunits, and their ATP-independent chaperone function. Additional functions attributed to small HSPs range from RNA storage to inhibition of actin polymerization and a contribution to the optical properties of the eye lens in the case of α-crystallin. At present it is unclear how these seemingly different functions can be integrated into a common theme. Recent evidence suggests that at least under heat shock conditions sHSPs and α-crystallin bind several non-native polypeptides per oligomeric complex. The function of sHSPs in this context is hypothesized to be to create a reservoir of non-native proteins for subsequent refolding under physiological conditions. Refolding may occur in cooperation with other, ATP-dependent chaperones.

■ Alternative names for family members

11 kDa-late embryogenesis abundant protein, 14 kDa-*Myobacterium tuberculosis* antigen, 18 kDa-*M. leprae* antigen, 25 kDa-growth-related protein, 25 kDa-inhibitor of actin polymerization, αB-crystallin, α-crystallin B-chain, AV25 protein, C14B9.1 protein, estrogen-regulated '24K' protein, estrogen receptor-related protein, Hsp16, Hsp16.3, Hsp16.9, Hsp17, Hsp17.1, Hsp17.3, Hsp17.4, Hsp17.5, Hsp17.6, Hsp17.7, Hsp17.8, Hsp17.9, Hsp18, Hsp18.1, Hsp18.2, Hsp18.5, Hsp18.6, Hsp20, Hsp21, Hsp22, Hsp22.3, Hsp22.5, Hsp22.7, Hsp23, Hsp23.5, Hsp23.9, Hsp25, Hsp26, Hsp26.6, Hsp27, Hsp30, Hsp42, IbpA, IpbB, low molecular weight/mass Hsp, IAP-25, p25, p27, pPf203.

■ Accession numbers for small HSP sequences

Acanthocheilonema viteae (h, 25) X68667; *Anas platyrhynchos* (a) U16124/L08078; *Arabidopsis thaliana* (h, 17.6) X89504, (h, 17.6) X63443, (h, 17.4) X17293, (h, 17.6) X16076, (h, 18.2) X17295, (h, 21) X54102/M94455, (h,22), U11501; *Aspergillus nidulans* (h, 30) D32070; *Brugia malayi* (h, 27) U48407; *B. pahangi* (h) X87901; *C. acetobutylicum* (h, 18) X65276/S52385; *Caenorhaibditis elegans* (a) C14B9.1/L15188/L18807, (a) F38E11.1/Z68342, (a) F38E11.2/Z68342, (a) ZK1128.7/Z47357, (h, 16) K01863, (h, 16) K01864; *C. familiaris* (h, 27) U19368; *C. longicaudatus* (h, 27) X51747; *Chenopodium rubrum* (h) X53870/S51533; *Chlamydomonas reinhardtii* (h, 22) X15053; *Citrus maxima* (h, 20) U19461; *D. carota* (h, 17.5) X53850, (h, 17.7) X53851/S48793, (h, 17.9) X53852/S48795; *D. discoideum* (h) L39778; *Dirofilaria immitis* (h, 27) U48406; *Drosophila melanogaster* (h, 22) X03888/J01098, (h, 23) X03889, (h, 26) X03890/J01099, (h, 27) X03891; *Escherichia coli* M94101/L10328; *G. gallus* (a) U26661/S53164, (h, 25) X59541; *Glycine max* (h, 17.5) M11318/M11395, (h, 17.6) M11317, (h, 17.9) X07159, (h, 18.5) X07160, (h, 22) X07188, (h, 22) X63198, (h, 22.3) U21723, (h, 22.5) U21724, (h. 23.9) U21722, (h, 26) M20363; *H. annuus* (h, 11) X59700/S41963, (h, 17.6) X59701/S41964, (h, 17.7) U46545, (h, 17.9) Z29554, (h, 18.6) U46544; *Haemophilus influenzae* U00084/L42023-HI1527; *Heliobacter pylori* (h) L23798; *Homo sapiens* (a) M28638/S45630, (h, 27) X03900; *Hordeum vulgare* (h, 17) X64560, (h, 18) X64561; *Lilium longiflorum* (h, 17.6) D21816, (h, 16.5) D21818; *Lycopersicon esculentum* (h) X56138; *M. auratus* (a) M12016/J03849; ; *Medicago sativa* (h, 18.1) X58710, (h, 18.2) X58711; *Mus musculus* (a) M63170/M73741, (h, 25) L07577/U03560-U03562; *Myobacterium leprae* (18) M22587/J03840; *M. tuberculosis* S79751; *Neurospora crassa* (h, 30) M55672/J05601; *N. tabacum* (h, 18) X70688; *O. cuniculus* (a) X95383; *Oryza sativa* (h) M80186/D12635(h, 16.9) X60820/M80938/M80939, (h, 17.8) X75616; *O. tschwawytscha* (h, 30) U19370; *O. volvulus* (h, 25) X68668/X68669; *Petunia hybrida* (h, 21) X54103; *Plasmodium crispum* (h, 17.9) X95716; *P. falciparum* (h) X15292; *P. glaucum* (h, 16.9) X94192, (h, 17) X94191, (h, 17.9) X94193; *P. nil* (h) M99429/M99430; *Picea glauca* (h, 17) L47717, (h, 17.1) L47740, (h, 18.2) L47609, (h, 23.5) L47741; *P. menziesii* (h) X92983/X92984; *Pisum sativum* (h, 17.7) M33901, (h, 17.9) M33900, (h, 18.1) M33899, (h, 21) X07187, (h, 22) X86222, (h, 22.7) M33898; *P. somniferum* (h) U08601; *R. norwegicus* (a) M55534/J05699/U04320/ S74229/ S77138/X60351-2, (h, 27) M86389/S67755; *Rana catesbeiana* (a) X87114, (h, 30) U44894; *S. albus* (h, 18) U17419; *Saccharomyces cerevisiae* (h, 26) M23871/M26942 , (h, 42) U41401; *S. lycopersicum* (h) U44386; *S. tuberosum* (h) S70186; *T. aestivum* (h) X13431, (h, 16.9) X64618/ L37065, (h, 17.3) S58279, (h, 26.6) X67328/X58280; *Xenopus laevis* (h, 30) X02511/ X02512/ X57962/ X57963/ X57964/L22211; *Zea mays* (h, 17.2) X65725, (h, 18) S59777/ X54075/X54076, (h, 26) L28712.

Note: a denotes αB-crystallin, h denotes heat shock protein and apparent or sequence-derived molecular weight if reported.

■ Sequences of small HSPs

The small HSP gene family comprises a diverse group of proteins, ranging in size from c. 15 to 40 kDa, that includes the small HSPs from eukaryotes, the α-crystallins of the vertebrate eye lens, surface antigens of *Schistosoma* and *Mycobacterium* (deJong et al., 1993), and has been extended to encompass inclusion body-associated proteins (IbpA and B) from *E. coli* (Allen et al., 1992; Jakob and Buchner, 1994). In contrast to other families of HSPs, the small HSPs are much more variable in amino acid sequence. For example, within a single plant species different small HSPs can be as little as 30% identical (Vierling, 1991). Identity between small HSPs and the related α-crystallins and bacterial surface antigens is even lower. However, all members of this family share a conserved C-terminal domain of about 100 residues (Plesofsky-Vig et al., 1992; deJong et al., 1993). As shown in Fig. 1, consensus motifs have been identified in the C-terminal domain within the N-terminal block I and the C-terminal block II (Plesofsky-Vig et al., 1992; deJong et al., 1993). It has been hypothesized that residues in this domain, particularly in block II, are necessary for the formation of oligomeric complexes, which form the native structure of these proteins.

■ Regulation of small HSPs

Because of the diversity of the small HSP family it is difficult to summarize all aspects of their regulation. However, the regulation of sHSP expression proceeds mainly by a common mechanism at the transcriptional level: expression of the protein under physiological conditions is significantly increased after different stresses owing to binding of the heat shock transcription factor to heat shock promoter elements (HSEs) (Gurley and Key, 1991). Not surprisingly, additional post-transcriptional mechanisms have also been described affecting small HSP expression (Edington and Hightower, 1990; Gotthardt et al., 1996). The αB-crystallins are also clearly stress-induced proteins (Klemenz et al., 1991; Kato et al., 1993; Boelens and deJong, 1995), although fewer studies have been performed to dissect this regulation mechanisticaly. Both the small HSPs and α-crystallins are expressed constitutively in some mammalian tissues (Klemenz et al., 1993), or are induced under other specific but non-stress conditions. For example, small HSPs are induced by estrogen (Fuqua et al., 1989) and they are developmentally regulated in a tissue-specific fashion (Pauli et al., 1989; Gernold et al., 1993; Waters et al., 1996). This diversity of regulation, at a minimum, suggests that this family of proteins participate in a wide range of cellular functions as is true of other chaperones.

At the post-translational level, several modifications of sHSPs have been detected, including phosphorylation (Welch, 1985, Voorter et al., 1986), deamidation and acylation, as well as mixed intermolecular disulfide formation, oxidation and glycation (for review see Groenen et al., 1994). Besides an age-dependent phosphorylation of α-crystallins by cAMP-dependent protein kinase (Spector et al., 1985), sHSP phosphorylation proceeds in mammals in a stress-dependent manner and is catalysed by the sHSP kinases MAPKAP kinase 2 (Stokoe et al., 1992) and 3pK (Ludwig et al., 1996), which are downstream targets of the p38 MAP kinase cascade (Rouse et al., 1994). An influence of sHSP phosphorylation on thermoresistance is not clearly defined (Knauf et al., 1994; Lavoie et al., 1995) but has been detected for its inhibitory properties in the process of actin polymerization *in vitro* (Benndorf et al., 1994). It is suggested that this latter phosphorylation-dependent property could be involved in stress-dependent reorganization of the cytoskeleton (Lavoie et al., 1993). Asparagine deamidation is a non-enzymatic modification, mainly of the α-crystallins, whereas acylation has been described for α-crystallins as well as other small HSPs. Finally, mixed disulfide formation, oxidation and glycation seem to diminish the chaperone properties of the sHSPs (Cherian and Abraham, 1995).

■ Intracellular localization of small HSPs

Small HSPs are found in the cytosol of eukaryotes, although nuclear translocation of these proteins in response to stress has been reported in several organisms (Kim et al., 1984; Arrigo et al., 1988). In prokaryotes, the proteins are cystosolic and homologues have been detected in the periplasm. In plants, small HSPs are found not only in the cytosol, but also in the endoplasmic reticulum, chloroplasts and mitochondria (Waters et al., 1996). Other eukaryotes are not known to have organelle-localized small HSPs, with the exception of

```
Block I

    (1)    ELxVxVxDxxVLxIxGxR

    (2)    EEVxVxVxxxxxLQIxGxx
           DL I I        VEV
              L          I


Block II

    (1)    FxRRFxLP-x(7-9)-VxAxLxxDGVLTVxxPK

    (2)    FxRRFxLP-x(7-9)-VxxxxxxxGVLTVxxPK
           Y KKY I         I       M SI  R
                                     L
```

Figure 1. Conserved residues within the carboxyl-terminal domain of sHSP family members. The motifs are shown as formulated by Plesofsky-Vig et al. (1992) (1) or by deJong et al. (1993) (2). Alternative common residues at several positions as indicated in the latter reference are also shown.

Drosophila which has recently been reported to have a mitochondrial small HSP (Tanguay, this volume, p. 280).

■ Structure and stability of small HSPs

The three-dimensional structure of small HSPs is unknown. CD spectroscopy suggests that the subunits are predominantly β structured (M. Ehrnsperger, M. Gaestel and J. Buchner, unpublished data). This feature is conserved between prokaryotic, plant and mammalian sHSPs. In contrast, the quaternary structure of small HSPs from different oganisms varies significantly from smaller oligomers of nine subunits (*Mycobacterium tuberculosis*, Hsp 16.3) (Chang et al., 1996) or 12 subunits (*Pisum sativum*, Hsp 18.1 and 17.7) (Lee et al., 1995), to much larger oligomers in the case of the mammalian small HSPs (up to 32 subunits) or the α-crystallins (Groenen et al., 1994). In addition, at least for some family members, the quaternary structure seems to be dynamic allowing different oligomers to exist depending on the physiological state and solvent conditions used *in vitro*. In electron micrographs sHSP complexes exhibit a globular shape with diameters typically in the range 12–18 nm. Upon stress these complexes can increase in size to several MDa.

The stability of sHSPs against chemical and thermal unfolding varies significantly between different family members (M. Ehrnsperger, S. Walke, M. Gaestel and J. Buchner, unpublished data). In some cases the thermal unfolding seems to be completely reversible.

■ Function of small HSPs

The function of small HSPs has remained unclear. sHSP proteins were assigned a wide variety of different and seemingly unrelated cellular functions (cf. Arrigo and Landry, 1994) ranging from action as a molecular chaperone (Horwitz, 1992; Jakob et al., 1993; Merck et al., 1993; Lee et al., 1995) and mediator of thermo- and chemo-resistance (Landry et al., 1989; Huot et al., 1991), to function as an inhibitor of actin polymerization (Miron et al., 1991) and elastase (Merck et al., 1993), ability to be a substrate for transglutaminase (Merck et al., 1993) and protection of RNAs or a regulator of apoptosis (Mehlen et al., 1996). Mammalian sHSPs are also targets for a unique stress-dependent signal transduction pathway which leads to a rapid phosphorylation of these proteins. Furthermore, α-crystallin, the sHSP homologue in the eye lens, has been assumed to be mainly of structural importance in context with the liquid crystal state of the lens. A clear *in vivo* function and a distinct phenotype of deletion mutants is still lacking. However, recent evidence suggests that sHSPs, like the other major HSP families, interact as molecular chaperones with unfolding proteins. In the eye lens α-crystallin seems to prevent the irreversible aggregation and subsequent cataract formation of other lenticular proteins (Horwitz, 1992). sHSPs interact with non-native proteins in an ATP-independent way. Several non-native proteins can be bound to one sHSP complex. sHSPs appear to be specialized in that they seem to be involved in stably binding unfolding proteins under stress conditions, thus creating a reservoir of non-native proteins that can be refolded to the native state after restoration of physiological conditions, potentially in cooperation with other, ATP-dependent chaperones (Ehrnsberger et al., 1997; Lee et al., 1997). How all the other functions attributed to sHSPs can be integrated into this picture is under intensive investigation.

■ References

Allen, S.P., Polazzi, J.O., Gierse, J.K., and Easton, A.M. (1992). Two novel heat shock genes encoding proteins produced in response to heterologous protein expression in *Escherichia coli*. J. Bacteriol. **174**, 6938–6947.

Arrigo, A.P. and Landry, J. (1994). Expression and function of the low-molecular-weight heat shock proteins. In *The Biology of Heat Shock Proteins and Molecular Chaperones*, (R.I. Morimoto, A. Tissiére and C. Georgopoulos, eds), pp. 335–373,.Cold Spring Habor Laboratory Press, New York.

Arrigo, A.P., Suhan, J.P., and Welch, W.J. (1988). Dynamic changes in the structure and intracellular locale of the mammalian low-molecular-weight heat shock protein. Mol. Cell. Biol. **8**, 5059–5071.

Benndorf, R., Hayess, K., Ryazantsev, S., Wieske, M., Behlke, J., and Lutsch, G. (1994). Phosphorylation and supramolecular organization of murine small heat shock protein HSP25 abolish its actin polymerization-inhibiting activity. J. Biol. Chem. **269**, 20780–20784.

Boelens, W.C. and deJong, W.W. (1995). α-crystallins, versatile stress-proteins. Mol. Biol. Rep. **21**, 75–80.

Chang, Z., Primm, T.P., Jakana, J., Lee, I.H., Serysheva, I., Chiu, W, Gilbert, H.F., and Quiocho, F.A. (1996). *Mycobacterium tuberculosis* 16-kDa antigen (Hsp16.3) functions as an oligomeric structure *in vitro* to suppress thermal aggregation. J. Biol. Chem. **271**, 7218–7223.

Cherian, M. and Abraham, E.C. (1995). Decreased molecular chaperone property of alpha-crystallins due to posttranslational modifications. Biochem. Biophys. Res. Commun. **208**, 675–679.

deJong, W.W., Leunissen, J.A., and Vooter, C.E. (1993). Evolution of the α-crystallin/small heat-shock protein family. Mol. Biol. Evol. **10**, 103–126

Edington, B.V. and Hightower, L.E. (1990). Induction of a chicken small heat shock (stress) protein: evidence of multilevel posttranscriptional regulation. Mol. Cell. Biol. **10**, 4886–4898.

Ehrnsperger, M., Gäbler, S., Gaestel, M., and Buchner, J. (1997). Binding of non-native protein to Hsp25 during heat shock creates a reservoir of folding intermediates for reactivation. EMBO J. **16**, 221–229.

Fuqua, S.A., Blum-Salingaros, M., and McGuire, W.L. (1989). Induction of the estrogen-regulated '24K' protein by heat shock. Cancer Res. **49**, 4126–4129

Gernold, M., Knauf, U., Gaestel, M., Stahl, J., and Kloetzel, P.M. (1993). Development and tissue-specific distribution of mouse small heat shock protein hsp25. Develop. Genet. **14**, 103–111.

Gotthardt, R., Neininger, A. and Gaestel, M. (1996). The anticancer drug cisplatin induces HSP25 in Ehrlich ascites tumor cells by a mechanism different from transcriptional stimulation influencing predominatly HSP25 translation. Int. J. Cancer **66**, 790–795.

Groenen, P.J., Merck, K.B., deJong, W.W., and Bloemendal, H. (1994). Structure and modifications of the junior chaperone alpha-crystallin. From lens transparency to molecular pathology. Eur. J. Biochem. **225**, 1–19.

Gurley, W.B. and Key, J.L. (1991). Transcriptional regulation of the heat-shock/response: a plant perspective. Biochemistry **30**, 1–12.

Horwitz, J. (1992). Alpha-crystallin can function as a molecular chaperone. Proc. Natl. Acad. Sci. USA **89**, 10449–10453.

Huot, J., Roy, G., Lambert, H., Chretien, P., and Landry, J. (1991). Increased survival after treatments with anticancer agents of Chinese hamster cells expressing the human Mr 27,000 heat shock protein. Cancer Res. **51**, 5245–5252.

Jakob, U. and Buchner, J. (1994). Assisting spontaneity the role of Hsp90 and small Hsps as molecular chaperones. Trends Biochem. Sci. **19**, 205–211.

Jakob, U., Gaestel, M., Engel, K., and Buchner, J. (1993). Small heat shock proteins are molecular chaperones. J. Biol. Chem. **268**, 1517–1520.

Kato, K., Goto, S., Hasegawa, K., and Inaguma, Y. (1993). Coinduction of two low-molecular-weight stress proteins, B-crystallin and HSP28, by heat or arsenite stress in human glioma cells. J. Biochem. (Tokyo) **114**, 640–647.

Kim, Y.-J., Sette, M., and Przybyla, A. (1984). Nuclear localization and phosphorylation of three 25-kilodalton rat stress proteins. Mol. Cell. Biol. **4**, 468–474.

Klemenz, R., Fröhli, E., Steiger, R.H., Schäfer, R., and Aoyama, A. (1991). B-crystallin is a small heat shock protein. Proc. Natl. Acad. Sci. USA **88**, 3652–3656.

Klemenz, R., Andres, A.-C., Fröhli, E., Schäfer, R., and Aoyama, A. (1993). Expression of the murine small heat shock proteins hsp 25 and B crystallin in the absence of stress. J. Cell Biol. **120**, 639–645.

Knauf, U., Jakob, U., Engel, K., Buchner, J., and Gaestel, M. (1994). Stress- and mitogen-induced phosphorylation of the small heat shock protein Hsp25 byMAPKAP kinase 2 is not essential for chaperone properties and cellular thermoresistance. EMBO J. **13**, 54–60.

Landry, J., Chretien, P., Lambert, H., Hickey, E., and Weber, L.A. (1989). Heat shock resistance conferred by expression of the human HSP27 gene in rodent cells. J. Cell Biol. **109**, 7–15.

Lavoie, J.N., Hickey, E., Weber, L.A., and Landry, J. (1993). Modulation of actin microfilament dynamics and fluid phase pinocytosis by phosphorylation of heat shock protein 27. J. Biol. Chem. **268**, 24210–24214.

Lavoie, J.N., Lambert, H., Hickey, E., Weber, L.A., and Landry, J. (1995). Modulation of cellular thermoresistance and actin filament stability accompanies phosphorylation-induced changes in the oligomeric structure of heat shock protein 27. Mol. Cell. Biol. **15**, 505–516.

Lee, G.J., Pokala, N., and Vierling, E. (1995). Structure and in vitro molecular chaperone activity of cytosolic small heat shock proteins from pea. J. Biol. Chem. **270**, 10432–10438.

Lee, G.J., Roseman, A.M., Saibil, H.R. and Vierling, E. (1997). A small heat shock protein stably binds heat-denatured model substrates and can maintain a substrate in a folding competent state. EMBO J. **16**, 659–671.

Ludwig, S., Engel, K., Hoffmeyer, A., Sithanandam, G., Neufeld, B., Palm, D., Gaestel, M., and Rapp, U.R. (1996). 3pK, a novel MAP kinase associated protein kinase is targeted by three MAP kinase pathways. Mol. Cell Biol. **16**, 6687–6697.

Mehlen, P., Schulze-Osthoff, K., and Arrigo, A.P. (1996). Small stress proteins as novel regulators of apoptosis: heat shock protein 27 blocks Fas/Apo 1 and staurosporine induced cell death. J. Biol. Chem. **271**, 16510–16514.

Merck, K.B., Groenen, P.J., Voorter, C.E., de Haard Hoekman, W.A., Horwitz, J., Bloemendal, H., and de Jong, W.W. (1993). Structural and functional similarities of bovine alpha crystallin and mouse small heat shock protein. A family of chaperones. J. Biol. Chem. **268**, 1046–1052.

Miron, T., Vancompernolle, K., Vandekerckhove, J., Wilchek, M., and Geiger, B. (1991). A 25-kD inhibitor of actin polymerization is a low molecular mass heat shock protein. J. Cell Biol. **114**, 255–261.

Pauli, D., Arrigo, A.P., Vazquez, J., Tonka, C.H., and Tissieres, A. (1989). Expression of the small heat shock genes during Drosophila development: comparison of the accumulation of hsp23 and hsp27 mRNAs and polypeptides. Genome **31**, 671–676.

Plesofsky-Vig, N., Vig, J., and Brambl, R. (1992). Phylogeny of the α-crystallin-related heat-shock proteins. J. Mol. Evol. **35**, 537–545.

Rouse, J., Cohen, P., Trigon, S., Morange, M., Alonso-Llamazares, A., Zamanillo, D., Hunt, T., and Nebreda, A.R. (1994). A novel kinase cascade triggered by stress and heat shock that stimulates MAPKAP kinase-2 and phosphorylation of the small heat shock proteins. Cell **78**, 1027–1037.

Spector, A., Chiesa, R., Sredy, J., and Garner, W. (1985). cAMP-dependent phosphorylation of bovine lens alpha-crystallin. Proc. Natl. Acad. Sci. USA **82**, 4712–4716.

Stokoe, D., Engel, K., Campbell, D.G., Cohen, P., and Gaestel, M. (1992). Identification of MAPKAP kinase-2 as a major enzyme responsible for the phosphorylation of the small mammalian heat shock proteins. FEBS Lett. **313**, 307–313.

Vierling, E. (1991). The roles of heat shock proteins in plants. Annu. Rev. Plant Physiol. Plant Mol. Biol. **42**, 579–620.

Voorter, C.E., Mulders, J.W., Bloemendal, H., and deJong, W.W. (1986). Some aspects of the phosphorylation of alpha-crystallin A. Eur. J. Biochem. **160**, 203–210.

Waters, E.R., Lee, G.J., and Vierling, E. (1996). Evolution, structure and function of the small heat shock proteins in plants. J. Exp. Bot. **47**, 325–338.

Welch, W.J. (1985). Phorbol ester, calcium ionophore, or serum added to quiescent rat embryo fibroblast cells all result in the elevated phosphorylation of two 28,000-dalton mammalian stress proteins. J. Biol. Chem. **260**, 3058–3062.

■ Matthias Gaestel:
AG Stressproteine
Max-Delbrück-Center for Molecular Medicine
R.-Rössle-Str. 10
13122 Berlin, Germany
Tel. 49 30 9406 3785
Fax. 49 30 9406 3798
E-mail: Gaestel@mdc-berlin.de

■ Elizabeth Vierling:
Department of Biochemistry
Life Sciences South
University of Arizona
Tucson, AZ 85721, USA
Tel. 1 520 621 1601
Fax. 1 520 621 3709
E-mail: eliz@biosci.arizona.edu

■ Johannes Buchner:
Universität Regensburg
Institut für Biophysik und Physikalische Biochemie
93040 Regensburg, Germany
Tel. 49 941 943 3039
Fax. 49 941 943 2813
E-mail: johannes.buchner@biologie.uni-regensburg.de

Escherichia coli small heat shock proteins IbpA and IbpB

Prokaryotic small heat shock proteins have been discovered as specific components of inclusion bodies after expression of recombinant proteins in Escherichia coli *and have thus been termed inclusion body proteins (IbpA and IbpB). Little is known about their function and properties. The recent identification of a homologous gene in the minimal genome of* Haemophilus influenzae *supports the view that Ibps, like their mammalian counterparts, may play an important role in intracellular protein aggregation.*

■ Alternative names

Inclusion body-associated protein A and B, IbpA=Hs1T, IbpB = Hs1S (Chuang *et al.*, 1993).

■ *ibpA* and *ibpB* genes and sequences

The genes encoding IbpA and IbpB (GenBank accession number M94104) are organized in a common operon located at about 83 minutes in the chromosome (Allen *et al.*, 1992; Chuang *et al.*, 1993). The operon is controlled by a σ^{32}-dependent promoter. It contains a 110 bp gap between the coding sequences of IbpA and IbpB and a transcription terminator downstream from IbpB.

At the amino acid level, there is 52% identity between IbpA and IbpB. IbpA and IbpB show homology to the global stress protein (GspA) of *Legionella pneumophila* (41 and 36% sequence identity, respectively; Abu-Kwaik and Engleberg, 1994), to the periplasmic molecular chaperone Caf1M from *Yersinia pestis* (Zav'yalov *et al.*, 1995), to the predicted gene product #1527 of *Haemophilus influenzae* (Fleischmann *et al.*, 1995), and weak homology to the small heat shock proteins of higher organisms (Jakob and Buchner, 1994; Zar'yalov *et al.*, 1995).

■ Ibp proteins

Although IbpA and IbpB are highly homologous in sequence, the purified proteins differ significantly in solubility, with IbpB being more soluble (P. Goloubinoff and J. Buchner unpublished data). Like their mammalian counterparts, Ibps bind non-native proteins (A. Azem and P. Goloubinoff, unpublished data).

■ Antibodies to IbpA

Rabbit polyclonal antibodies were raised against the 15 N-terminal amino acids of IbpA and have been used in Western blot analysis (Allen *et al.*, 1992).

■ Biological activities of the Ibp proteins

IbpA and IbpB are not only induced by heat shock (Chuang *et al.*, 1993), but also by high level production of heterologous proteins in *E. coli* (Allen *et al.*, 1992). Both proteins are tightly associated with intracellularly aggregated endogenous proteins (Laskowska *et al.*, 1996) and with inclusion bodies which are eventually formed during high level expression. This suggests a biological function for the Ibp proteins in intracellular protein aggregation.

■ References

Abu-Kwaik, Y. and Engleberg, N.C. (1994). Cloning and molecular characterization of a *Legionella pneumophila* gene induced by intracellular infection and by various *in vitro* stress conditions. Mol. Microbiol. **13**, 243–251.

Allen, S.P., Polazzi, J.O., Gierse, J.K., and Easton, A.M. (1992). Two novel heat shock genes encoding proteins produced in response to heterologous protein expression in *Escherichia coli*. J. Bacteriol. **174**, 6938–6947.

Chuang, S.E., Burland, V., Plunkett, G., Daniels, D.L., and Blattner, F.R. (1993). Sequences analysis of four new heat-shock genes constituting the hs1TS/ibpAB and hs1VU operons in *Escherichia coli*. Gene **134**, 1–6.

Fleischmann, R.D., Adams, M.D., White, O., Clayton, R.A., Kirkness, E.F., Kerlavage, A.R., Bult, C.J., Tomb, J.F., Dougherty, B.A., Merrick, J.M., *et al.* (1995). Whole-genome random sequencing and assembly of *Haemophilus influenzae* Rd. Science **269**, 496–512.

Jakob, U. and Buchner, J. (1994). Assisting spontaneity: the role of Hsp90 and small Hsps as molecular chaperones. Trends Biochem. Sci. **19**, 205–211.

Laskowska, E., Wawrzynow, A., and Taylor, A. (1996). IbpA and IbpB, the new heat-shock proteins, bind to endogenous *Escherichia coli* proteins aggregated intracellularly by heat shock. Biochimie **78**, 117–122.

Zav'yalov, V.P., Zav'yalova, G.A., Denesyuk, A.I., Gaestel, M., and Korpela, T. (1995). Structural and functional homology between periplasmic bacterial molecular chaperones and small heat shock proteins. FEMS Immunol. Med. Microbiol. **11**, 265–272.

■ Matthias Gaestel:
AG Stressproteine
Max-Delbrück-Center for Molecular Medicine
R.-Rössle-Str. 10
13122 Berlin, Germany
Tel. 49 30 9406 3785
Fax. 49 30 9406 3798
E-mail: gaestel@mdc-berlin.de

■ Johannes Buchner:
Institut für Biophysik und Physikalische Biochemie
Universität Regensburg
93040 Regensburg, Germany

Tel. 49 941 943 3039
Fax. 49 941 943 2813
E-mail: johannes.buchner@biologie.
uni-regensburg.de

Saccharomyces cerevisiae Hsp26

Hsp26 is one of the major stress-induced proteins of the yeast Saccharomyces cerevisiae *and is encoded by a single copy gene (*HSP26*) located on chromosome II. It shares many of the properties characteristic of small heat shock proteins (sHSPs) from higher eukaryotes including the ability to assemble into high molecular weight complexes called heat shock granules (HSG). This apparent self-assembly process requires the presence of a highly conserved hydrophobic amino acid-rich sequence GVLTXXXP.* HSP26 *gene deletion and overexpression studies have failed to reveal the cellular role of Hsp26 in stressed cells, although its ability to* inhibit actin polymerization *in vitro might suggest a role in cytoskeletal reorganization. Transcription of the* HSP26 *gene is activated by two different types of cis-acting promoter elements; heat shock elements (HSEs) and stress response elements (STREs).*

■ *HSP26* gene and sequence

The *HSP26* gene was first isolated from *Saccharomyces cerevisiae* by the Lindquist laboratory (Petko and Lindquist, 1986) from a genomic library screen using cDNA probes made from heat shock mRNAs. The sequence of the *HSP26* gene (GenBank accession number M23871) predicts a 214 amino acid protein (including the initiator Met) with a deduced molecular weight of 27 630 (Bossier *et al*., 1989; Susek and Lindquist, 1989) which is in good agreement with its molecular weight estimated from its electrophoretic mobility on SDS–PAGE (Petko and Lindquist, 1986). *HSP26* is a single copy gene located on chromosome II (Petko and Lindquist, 1986; Susek and Lindquist, 1989) that is transcribed into a 900 nucleotide transcript with a single start site 78 nucleotides 5′ to the initiator codon (Bossier *et al*., 1989; Susek and Lindquist, 1990). The calculated codon bias of 0.43 (Bossier *et al*., 1989) is consistent with its high level of expression in stressed cells.

The deduced sequence of the *S. cerevisiae* Hsp26 protein shows significant amino acid identity to sHSPs from plants and animals, particularly within the C-terminal half of the proteins. This homology is centred upon an amino acid motif Glu–Val–Leu–Thr–X–X–X–Pro that is part of a predicted hydrophobic β-sheeted region of the polypeptide. This same motif is also present in the B chain of mammalian α-crystallin, another class of proteins that assemble into highly ordered macromolecular complexes (Wistow and Piatigorsky, 1988). These observations would suggest that this motif is important for the assembly of the HSGs, a supposition recently supported by site-directed mutagenesis studies of this motif in yeast Hsp26; these studies implicate the critical residues are GV<u>L</u>TXX<u>P</u> (Rahman, 1996).

■ Isolation of Hsp26 protein

In stressed yeast cells, Hsp26 represents an abundant protein which can be readily detected on coomassie blue staining of total protein extracts fractionated on SDS–PAGE gels. Hsp26 is particularly abundant in cells grown on the surface of an agar-based medium in which it represents the major cellular protein (Rahman, 1996). The protein can be readily purified to apparent homogeneity by exploiting the tendency of the protein to assemble into the high molecular weight HSGs; such aggregates have a molecular weight of approximately 550 000 Da and appear to contain no macromolecular species other than Hsp26. The protocol for purification involves preparing extracts from stationary phase or heat shocked cells, pelleting the HSGs by differential centrifugation and then fractionating the pelleted HSGs on a 0.5–1.0M sucrose gradient (Bentley *et al*., 1992). Yeast Hsp26 expressed in *Escherichia coli* can be purified by the same strategy indicating that Hsp26 HSGs form without the requirement for a homologous chaperone (Rahman, 1996).

■ Hsp26 protein

The predicted Hsp26 amino acid sequence contains no Met residues other than the one encoded by the initiator codon (Bossier *et al*., 1989; Susek and Lindquist, 1989). The N-terminal Met residue is removed post-translationally since Hsp26 cannot be radiolabelled with [^{35}S]-Met *in vivo* (Bossier *et al*., 1989), and thus, to monitor the synthesis of Hsp26, one must resort to radiolabelling with another amino acid, e.g.[^{3}H]-Ile, or immunoblot analysis

(Petko and Lindquist, 1986). With the exception of a short stretch of amino acids at the C-terminus of the protein, Hsp26 has relatively few hydrophobic regions, with the majority of the polypeptide being either neutral or hydrophilic in nature (Rahman, 1996).

Hsp26 is predominantly localized to the cytoplasm although it does localize to the nucleus when log-phase cells grown in glucose are heat shocked; it cannot be detected in the nuclei of stationary phase cells or cells engineered to express Hsp26 in non-stressed log-phase cells (Rossi and Lindquist, 1989; Bentley, 1991). Thus the intracellular localization of Hsp26 in S. cerevisiae depends on the physiological status of the cell rather than exposure to stress per se.

The majority of Hsp26 found in stressed cells exists as high molecular weight HSGs (Rossi and Lindquist, 1989; Bentley et al., 1992). sHSPs from other species form similar aggregates, although it is unclear whether this form of the protein can be considered the 'functional' form or is simply the result of aberrant aggregation (Parsell and Lindquist, 1993). Unlike the sHSP HSGs formed in plants, yeast Hsp26 HSGs have no associated RNA species (Bentley, 1991) and appear to consist simply of approximately 20 molecules of Hsp26 assembled into a monomorphic structure (Bentley et al., 1992; Fig. 1). The yeast Hsp26 HSGs can be dissociated with urea and subsequently reassembled in vitro in a manner that does not require ATP or other cellular components (Rahman, 1996). These data suggest that Hsp26 self-assembles into HSGs in the absence of any additional assembly factors and/or chaperones.

Higher eukaryotic sHSPs are phosphorylated (e.g. human Hsp27; Arrigo and Welch, 1987) and, in keeping with this, the S. cerevisiae Hsp26 protein has several potential phosphorylation sites including five casein kinase II phosphorylation sites (S/TXXD/E) and three protein kinase C phosphorylation sites (S/TXR/K) (Rahman, 1996). Although there has been no formal demonstration of a phosphorylated form of yeast Hsp26 in vivo, high resolution two-dimensional SDS–PAGE has identified three minor isoforms of Hsp26 which may represent differentially phosphorylated forms of the protein (Bentley et al., 1992).

∎ Biological activities of Hsp26

A full understanding of the precise cellular role of the yeast Hsp26 protein has remained elusive and indeed confused by the apparent phenotypic neutrality of either an hsp26 null mutation (Petko and Lindquist, 1986) or of cells engineered to overexpress constitutively the HSP26 gene (Bentley et al., 1992; Susek and Lindquist, 1989). Such studies have also failed to implicate Hsp26 in the acquisition of thermotolerance.

Mammalian sHSPs and αB-crystallin are known to act as ATP-independent molecular chaperones in vitro (Jakob et al., 1993), and while there are no data to confirm that the yeast Hsp26 polypeptide can act as a chaperone, given the degree of amino acid conservation it shares with its mammalian homologues, this property would not be unexpected.

Another recently identified property of mammalian sHSPs is that they are able to interact with the actin filament network and inhibit actin polymerization (e.g. Lavoie et al., 1993). Intriguingly, the GVLTXXXP domain required for sHSP HSG assembly (see above) shows homology to a region within actin known to be important for the polymerization of G-actin into F-actin (Rahman et al., 1995). The yeast Hsp26 protein can also inhibit actin polymerization in vitro and this suggests a possible role for Hsp26 in the determination of cell morphogenesis, possibly acting as a stress-sensitive modulator of actin microfilament assembly (Rahman et al., 1995; Rahman, 1996).

∎ Biological regulation of HSP26

In S. cerevisiae Hsp26 is one of the major stress-induced proteins, being particularly abundant in cells subjected to physical or chemical stress or in cells during the transition from log-phase to stationary phase (McAlister et al., 1979; Kurtz et al., 1986; Petko and Lindquist, 1986). The HSP26 gene is transcribed at barely detectable levels in cells growing exponentially on glucose at growth temperatures below 30°C, but as cells begin to enter stationary phase in response to nutrient starvation so the HSP26 gene is transcriptionally activated (Susek and Lindquist, 1989, 1990). This transcriptional activation rapidly results

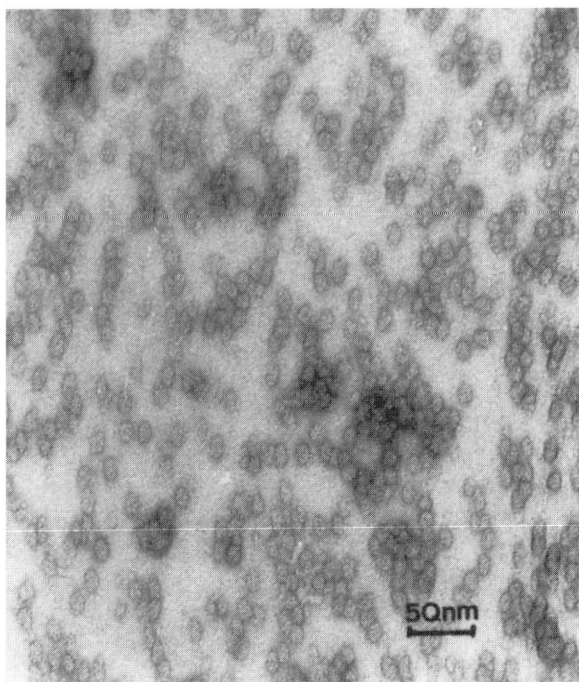

Figure 1. Hsp26-containing heat shock granules purified from S. cerevisiae and viewed at a magnification of 65 000. A 50 nm bar is shown.

in high levels of Hsp26 synthesis in cells entering the non-dividing G0 phase of the cell cycle and is coincident with the transcriptional shut-off of most other genes (Petko and Lindquist, 1986). In addition, cells artificially forced to enter the G0 stage of the cell cycle in the presence of nutrients, e.g. by growth in the presence of rapamycin, also activate transcription of the *HSP26* gene (Barbet et al., 1996). *HSP26* transcription is also activated in exponentially growing cells if they are subjected to physical (e.g. heat shock), osmotic (e.g. 0.7M NaCl) or chemical (e.g. H_2O_2) induced stresses (Petko and Lindquist, 1986; Varela et al., 1992, C. Stokes and M.F. Tuite, unpublished results) or are starved for nitrogen (Kurtz and Lindquist, 1984; Kurtz et al., 1986). There may also be a secondary level of control of expression of the *HSP26* gene at the post-transcriptional stage in heat shocked cells (Hartley et al., 1991).

The *HSP26* promoter is relatively complex in organization, containing multiple copies of the *cis*-acting elements that are required for transcriptional activation of the gene by heat shock (heat shock elements or HSEs: GAA/TTC) or by other stresses (stress response elements or STREs: CCCCT) (Bossier et al., 1989; Susek and Lindquist, 1990; Varela, 1995). The promoter may also contain elements that are required for the repression of *HSP26* transcription in non-stressed cells (Susek and Lindquist, 1990) although the location of such *cis*-acting elements remains to be defined.

■ References

Arrigo, A. and Welch, W.J. (1987). Characterisation and purification of the small 28,000 Da mammalian heat shock protein. J. Biol. Chem. **262**, 15359–15369.

Barbet, N.C., Scheider, U., Helliwell, S.B., Stansfield, I., Tuite, M.F., and Hall, M.N. (1996). TOR controls translation initiation and early G1 progression in yeast. Mol. Biol. Cell **7**, 25–42.

Bentley, N.J. (1991). PhD Thesis, University of Kent at Canterbury, UK.

Bentley, N.J., Fitch, I.T., and Tuite, M.F. (1992). The small heat-shock protein, Hsp26 of *Saccharomyces cerevisiae* assembles into a high molecular weight aggregate. Yeast **8**, 95–106.

Bossier, P., Fitch, I.T., Boucherie, H., and Tuite, M.F. (1989). Structure and expression of a yeast gene encoding a small heat shock protein Hsp26. Gene **78**, 323–330.

Hartley, A.D., Girstel, B., McCarthy, J.E.G., and Tuite, M.F. (1991). Role of the 5′mRNA leader in heat shock gene expression in yeast. Biochem. Soc. Trans. **19**, S280.

Jakob, U., Gaestel, M., Engel, K., and Buchner, J. (1993). Small heat shock proteins are molecular chaperones. J. Biol. Chem. **268**, 1517–1520.

Kurtz, S. and Lindquist, S. (1984). Changing patterns of gene expression during sporulation in yeast. Proc. Natl. Acad. Sci. USA **81**, 7323–7327.

Kurtz, S., Rossi, J., Petko, L., and Lindquist, S. (1986). An ancient developmental induction: heat shock proteins induced in sporulation and oogenesis. Science **231**, 1154–1157.

Lavoie, J.N., Hickey, E., Weber, L.A., and Landry, J. (1993) Modulation of actin microfilament dynamics and fluid phase pinocytosis by phosphorylation of the heat shock protein Hsp27. J. Biol. Chem. **268**, 24210–24214.

McAlister, L., Strausberg, S., Kulaga, A., and Finkelstein, D.B. (1979). Altered patterns of protein synthesis induced by heat shock of yeast. Curr. Genetics **1**, 63–74.

Parsell, D.A and Lindquist, S. (1993). The function of heat shock proteins in stress tolerance: degradation and reactivation of damaged proteins. Annu. Rev. Genet. **27**, 437–496.

Petko, L. and Lindquist, S. (1986). Hsp26 is not required for growth at high temperatures, nor for thermotolerance, spore development, or germination. Cell **45**, 885–894.

Rahman, D.R.J. (1996). PhD Thesis, University of Kent at Canterbury, UK.

Rahman, D.R.J., Bentley, N.J., and Tuite, M.F. (1995) The *Saccharomyces cerevisiae* small heat shock protein Hsp26 inhibits actin polymerisation. Biochem. Soc. Trans. **23**, S77.

Rossi, J.M. and Lindquist, S. (1989). The intracellular location of yeast heat-shock protein 26 varies with metabolism. J. Cell Biol. **108**, 425–439.

Susek, R.E. and Lindquist, S.L. (1989). hsp26 of *Saccharomyces cerevisiae* is related to the superfamily of small heat shock proteins but is without demonstratable function. Mol. Cell. Biol. **9**, 5265–5271.

Susek, R.E. and Lindquist, S. (1990). Transcriptional derepression of the *Saccharomyces cerevisiae* HSP26 gene during heat shock. Mol. Cell. Biol. **10**, 6362–6373.

Varela, J.C.S (1995). PhD thesis, Free University of Amsterdam, The Netherlands.

Varela, J.C.S., van Beekvelt, C.A., Planta, R.J., and Mager, W.H. (1992) Osmostress-induced changes in yeast gene expression. Mol. Microbiol. **6**, 2183–2190.

Wistow, G. and Piatigorsky, J. (1988). Lens crystallins: the evolution and expression of proteins for a highly specialised tissue. Annu. Rev. Biochem. **57**, 479–504.

■ *Mick F. Tuite, Daisy R. J. Rahman and Nicola J. Bentley:*
Research School of Biosciences
University of Kent
Canterbury, Kent CT2 7NJ, UK
Tel. 44 1227 823699
Fax. 44 1227 763912
E-mail: M.F.Tuite@ukc.ac.uk

Plant small heat shock proteins (sHSPs)

Higher plants are unusual among eukaryotes in that they synthesize numerous sHSPs in response to heat and other stresses, as well as in response to endogenous developmental signals. As seen for HSP70 and HSP100 proteins, sHSPs in plants are found localized not only in the cytosol, but also in the endoplasmic reticulum, chloroplasts and mitochondria. Plant sHSPs have only recently been shown to have chaperone activity in vitro, but their presence in multiple cellular compartments and pattern of regulation is consistent with the proposal that they act as chaperones in preventing irreversible protein denaturation and aggregation.

■ Plant sHSP gene families

Plants synthesize sHSPs belonging to five, and possibly more, nuclear-encoded gene families (Waters et al., 1996). There are over 50 plant sHSP gene sequences available in current databases. The plant sHSPs are categorized according to amino acid sequence similarity, immunological cross-reactivity and intracellular localization. The five best characterized families comprise two classes of cytosolic proteins (class I and II) and three families of organelle-localized proteins, which have been shown to be targeted to the endoplasmic reticulum, chloroplast or mitochondrion. The presence of organelle-localized sHSPs is also unique to plants; no sHSPs have been identified in organelles of other eukaryotes. [Note that unpublished data indicate Drosophila Hsp22 is a mitochondrial protein (Tanguay, this volume, p. 280)]. Within a species, amino acid sequence identity of gene family members is greater that 80%, while identity between gene families ranges from c. 30 to 50%. Between species, members of the same gene family are typically 60 to >80% identical (Vierling, 1991). Evolutionary analysis of sHSP sequences from multiple plant species has led to the hypothesis that these sHSP gene families arose by gene duplication and divergence over 150 million years ago (Waters, 1995). A putative sixth sHSP gene family was recently identified in Glycine max (soybean), and also appears to specify an organelle-targeted protein in that the cDNA encodes a signal peptide and the mRNA is translated on membrane-bound ribosomes (LaFayette et al., 1996).

Considering mature molecular weight, plant sHSPs range in size from c. 17 to 22 kDa. They contain the conserved C-terminal domain identified in other eukaryotic sHSPs and the α-crystallin proteins; however, they lack the consensus phosphorylation sites of the mammalian HSPs (Waters et al., 1996). Figure 1 shows an alignment of individual members of the five well-characterized sHSP gene families from Pisum sativum. Identity between the sequences is limited to the C-terminal domain which spans c. 100 amino acids. This domain can be further subdivided into regions termed consensus I and II, which are separated by a variable length hydrophilic region (Vierling, 1991). In consensus I, the residues Pro–X_{14}Gly–Val–Leu are found in almost all small HSPs, and a similar motif appears in the consensus II region, Pro–X_{14}–X–Val/Leu/Ile–Val/Leu/Ile. The significance of these conserved motifs is not known.

■ Expression of plant sHSPs

Unlike many other chaperones, there is no evidence that sHSPs are expressed at significant levels in plant vegetative tissues under optimal, controlled growth conditions. However, proteins from each of the five gene families accumulate rapidly in response to heat stress, and the class I and II proteins can account for over 1% of total cell protein under some conditions (DeRocher et al., 1991; Hsieh et al., 1992). In plants the typical temperature for maximum sHSP expression ranges from 38 to 45°C depending on the species. Specific sHSP family members are also expressed in response to endogenous developmental signals, documented most extensively in pollen and seed development (for review see Waters et al., 1996). Thus the chaperone function of the sHSPs appears to be confined to defined stress and developmental states, as opposed to being required for normal cellular function.

■ Protein characteristics

Similar to sHSPs from mammals (Arrigo and Landry, 1994), plant sHSPs are found in 200–300 kDa complexes in both the cytosol and organelles (Nover et al., 1983, 1989; Helm et al., 1993; Chen et al., 1994; Lenne and Douce, 1994; Jinn et al., 1995). cDNAs encoding either class I, class II or chloroplast-localized sHSPs have been expressed in E. coli and shown to assemble into discrete oligomers similar in size to those observed in vivo (Lee et al., 1995; T. Suzuki and E. Vierling, unpublished data). Assembly of the recombinant sHSPs is consistent with the interpretation that sHSP complexes are homo-oligomeric in vivo. Although the class I and II proteins both appear to be cytosolic, in vivo they are found in distinct complexes (Helm et al., 1997). Furthermore, while the recombinant proteins of both classes can be dissociated in urea or guanidine hydrochloride and will reassemble on dialysis, if class I and II proteins are combined they will not coassemble (Helm et al., 1997). These findings

```
                       1                                                              60
PsHsp18.1-I    .......... .......... .......... .......... .......... ..........
PsHsp22.7-ER   .......... .......... .......... .......... ...mslkpln mllvpfllli
PsHsp17.7-II   .......... .......... .......... .......... .......... ..........
PsHsp21.0-CP   magsvslsti aspilsgkpg ssvkstppcm asfplrrglp rlglrnvraq aggdgdnkdn
PsHsp22.0-MT   masslalkrf lssgl...ls ssflrpvass asrsf..... .....ntnam rqydqhsddr
Consensus      ---------- ---------- ---------- ---------- ---------- ----------

                       61                                                             120
PsHsp18.1-I    ......msli psffs.grrs nvf..Dpfsl dvwDPLkdf. .......... pfsnsspsas
PsHsp22.7-ER   laadfplkak asllpfidsp ntllsDlwsd rfpDPFr... .......... vleqipygve
PsHsp17.7-II   ......mdld splfntlhhi mdltDDttek nlnaPtrty. .......... ..........
PsHsp21.0-CP   svevhrvnkd dqgtaverkp rrssiDispf gllDPWspmr smrqmldtm  rifedaitip
PsHsp22.0-MT   nvdvyr.... ......hsfp rtrrdDllls dvfDPFsppr slsqvlnmvd lltdnpvl..
Consensus      ---------- ---------- -----D---- ---DPF---- ---------- ----------
                                          *         *

                       121                                                            180
PsHsp18.1-I    fprenpafvs trvDwKEtpe ahvFkaDLPG lkKEEVKVeV EDdrvlqiSg ERsvEkEdkn
PsHsp22.7-ER   khepsitlsh arvDwKEtpe ghvimvDvPG lkKDDIKIeV EEnrvlrvSg ERkkEeDkkg
PsHsp17.7-II   .vrdakamaa tpaDvKEHpn syvFmvDMPG vksgDIKVqV EDenvlliSg ERkrEeEkeg
PsHsp21.0-CP   grnigggeir vpwEiKDeeh eirMrfDMPG vsKEDVKVsV EDdvlvikSd hReenggedc
PsHsp22.0-MT   .....saasr rgwDarEted alfLrlDMPG lgKEDVKIsV Eqntltikge EgakEsEeke
Consensus      ---------- ---D-KE--- ---F--DMPG --KEDVKV-V ED--VL--S- ER--E-E---
                                *  **        ***  *       * **            *
                                                                     consensus II

                       181                                                            240
PsHsp18.1-I    dewhrveRss gkFlrrFrLP ENakmDkV.K AsMenGVLtV TVpK...Eei kkaevksIei
PsHsp22.7-ER   dhwhrveRsy gkFwrqFkLP qNvdlDsV.K AkMenGVLtl TlhKlshDki kgprmVsIve
PsHsp17.7-II   vkylkmeRri gkLmrkFvLP ENaniEaI.s AisqdGVLtV TVnKlpppep kkpktIqVkv
PsHsp21.0-CP   ....wsrksy scYdtrLkLP DNcekEkV.K AeLkdGVLyI TIpKtkiE.. rtvidVqIq.
PsHsp22.0-MT   ....ksgRrf ss...ridLP EklykidViK AeMknGVLkV TVpKmkeEer nnvinVkVd.
Consensus      -------R-- --F---F-LP EN-----V-K A-M--GVL-V TV-K---E-- -----V-I--
                                **                *  *** *  *
                                           consensus I

                       241
PsHsp18.1-I    sg........ ....
PsHsp22.7-ER   eddkpskivn delk
PsHsp17.7-II   a......... ....
PsHsp21.0-CP   .......... ....
PsHsp22.0-MT   .......... ....
Consensus      ---------- ----
```

Figure 1. Amino acid sequence alignment of *Pisum sativum* representatives from the five well-characterized sHSP gene families in plants. The consensus sequence, defined as similar residues in four of the five sequences, appears below the alignment. Those residues that are identical in all sequences are indicated with asterisks. The conserved heat shock domain comprises the C-terminal region in which consensus residues appear. Consensus regions I and II, as discussed in the text, are underlined. Gaps introduced to optimize the alignment are represented by a point (.). The N-terminal targeting signals are underlined in the chloroplast, mitochondrial and ER-localized protein sequences [note that the ER signal peptide has not been directly determined, but is derived by prediction (Helm *et al.*, 1993)]. The putative ER retention signal is also underlined. CP = chloroplast localized. MT = mitochondria localized. ER = endoplasmic reticulum localized. Amino acid sequences were aligned using 'PileUp' in the Wisconsin GCG program, with final adjustments made by hand. The consensus was derived using 'Pretty' in the GCG program. Database accession numbers for sequences shown are: PsHsp18.1-I: P19243; PsHsp22.7-ER: M33898; PsHsp17.7-II: S12720; PsHsp21.0-CP: P09886; PsHsp22.0-MT: P46254.

demonstrate that the cytosolic class I and II proteins are structurally distinct and imply that amino acid sequences outside the conserved C-terminal domain contribute to assembly of the oligomer. The evolutionary conservation of the class I and II families and their distinct structures also suggest that there are functional differences between these classes of proteins.

In mammalian systems sHSPs are phosphorylated through a MAP kinase cascade resulting in a decrease in oligomer size and potentially modifying function (Benndorf et al., 1994; Freshney et al., 1994; Kato et al., 1994; Knauf et al., 1994; Rouse et al., 1994). In contrast, plant sHSPs are not phosphorylated (Nover and Scharf, 1984; T. Suzuki and E. Vierling, unpublished data), and there is no evidence that the oligomers decrease in size under different stress conditions or during recovery.

All sHSPs, including the plant cytosolic and chloroplast proteins, will associate into insoluble 'heat shock granules' (estimated greater than 1 MDa) under certain, typically severe stress conditions (Nover, 1991; Osteryoung and Vierling, 1994). These aggregates may represent sHSPs associated with substrates that require protection from stress. Nover et al. (1989) suggest that the substrates are untranslated mRNAs, although it is equally likely that the substrates are denatured proteins. The significance of these aggregates requires further investigation.

■ Purification and structural characteristics of recombinant plant cytosolic sHSPs

Recombinant Hsp18.1 (a class I protein) and Hsp17.7 (a class II protein) from pea (*Pisum sativum*) overexpressed in *E. coli* can be purified to homogeneity by ammonium sulfate precipitation, sucrose gradient centrifugation and DEAE chromatography in the presence of 3 M urea, followed by dialysis (Lee et al., 1995). Similar steps have been used to purify class I and II proteins from wheat (*Triticum aestivum*) (E. Basha and E. Vierling, unpublished data). Sedimentation equilibrium and electron microscopy suggest that Hsp18.1 and Hsp17.7 are c. 215 kDa globular proteins composed of 12 subunits (Lee et al., 1995). Photomodification of Hsp18.1 with the hydrophobic probe bis-ANS implicates the presence of a hydrophobic surface on the sHSP, which may be involved in substrate binding (Lee et al., 1997). Hsp18.1 shows a reversible increase in bis-ANS photoincorporation with increasing temperature.

■ Chaperone activity of the plant sHSPs

Jinn et al. (1989) first obtained evidence that plant sHSPs could prevent irreversible protein aggregation at high temperatures. Recent work with recombinant sHSPs from the class I and II cytosolic families reveals that they have several activities *in vitro* consistent with a chaperone function (Lee et al., 1995). Class I and II proteins can facilitate refolding of chemically or heat denatured model substrates and prevent aggregation of proteins heated at 45°C. Stoichiometric amounts of pea Hsp18.1 (class I) and Hsp17.7 (class II) enhance the refolding of guanidine hydrochloride-denatured citrate synthase and lactate dehydrogenase, as well as increase the refolding of 38°C-treated citrate synthase at 22°C. Hsp18.1, and to a much lesser extent Hsp17.7, prevents thermal aggregation of citrate synthase (Lee et al., 1995), mitochondrial malate dehydrogenase, firefly luciferase and glyceraldehyde-3-phosphate dehydrogenase (Lee et al., 1997). Activity in these assays is nucleotide independent as has been found for the mammalian sHSPs (Jakob and Buchner, 1994). Size-exclusion HPLC indicates that Hsp18.1 prevents substrate aggregation by stably and selectively binding heat-denatured substrates to form higher molecular weight complexes. SDS–PAGE of isolated Hsp18.1/malate dehydrogenase complexes indicates that each Hsp18.1 dodecamer can bind two times its own weight in substrate, which is the equivalent of 12 malate dehydrogenase monomers. Bound substrates do not dissociate with decreased temperature, ATP or high ionic strength. However, heat-denatured firefly luciferase bound to HSP18.1 can be refolded in the presence of ATP and rabbit reticulocyte or wheat germ extracts, suggesting that bound substrates can be refolded in conjunction with ATP-dependent molecular chaperones (Lee et al., 1997).

The organelle-localized sHSPs from plants have not yet been tested for chaperone activity. Furthermore, there are no data demonstrating sHSP chaperone activity *in vivo*, and potential sHSP substrates *in vivo* remain to be identified.

■ References

Arrigo, A.-P. and Landry, J. (1994). Expression and function of the low-molecular weight heat shock proteins. In *The Biology Of Heat Shock Proteins And Molecular Chaperones*, (R. Morimoto, A. Tissieres, and C. Georgopolous, eds). Cold Spring Harbor Laboratory Press, New York, pp. 335–373.

Benndorf, R., Hayess, K., Ryazantsev, S., Wieske, M., Behlke, J., and Lutsch, G. (1994). Phosphorylation and supramolecular organization of murine small heat shock protein HSP25 abolish its actin polymerization-inhibiting activity. J. Biol. Chem. **269**, 20780–20784.

Chen, Q., Osteryoung, K., and Vierling, E. (1994). A 21-kDa chloroplast heat shock protein assembles into high molecular weight complexes in vivo and in organelle. J. Biol. Chem. **269**, 13216–13223.

DeRocher, A.E., Helm, K.W., Lauzon, L.M., and Vierling, E. (1991). Expression of a conserved family of cytoplasmic low molecular weight heat shock proteins during heat stress and recovery. Plant Physiol. **96**, 1038–1047.

Freshney, N.W., Rawlinson, L., Guesdon, F., Jones, E., Cowle, S., Hsuan, J., and Saklatvala, J. (1994). Interleukin-1 activates a novel protein kinase cascade that results in the phosphorylation of Hsp27. Cell **78**, 1039–1049.

Helm, K.W., LaFayette, P.R., Nagao, R.T., Key, J.L., and Vierling, E. (1993). Localization of small heat shock proteins to the higher plant endomembrane system. Mol. Cell. Biol. **13**, 238–247.

Helm, K.W., Lee, G.J. and Vierling, E. (1997). Expression and structure of cytosolic class II small heat shock proteins. Plant Physiol., In press.

Hsieh, M.-H., Chen, J.-T., Jinn, T.-L., Chen, Y.-M., and Lin, C.-Y. (1992). A class of soybean low molecular weight heat shock proteins. Immunological study and quantitation. Plant Physiol. **99**, 1279–1284.

Jakob, U. and Buchner, J. (1994). Assisting spontaneity: the role of Hsp90 and small Hsps as molecular chaperones. Trends Biochem. Sci. **19**, 205–211.

Jinn, T.L., Yeh, Y.C., Chen, Y.M., and Lin, C.-Y. (1989). Stabilization of soluble proteins in vitro by heat shock proteins-enriched ammonium sulfate fraction from soybean seedlings. Plant Cell Physiol. **30**, 463–469.

Jinn, T.-L., Chen, Y.-M., and Lin, C.-Y. (1995). Characterization and physiological function of class I low molecular weight heat shock protein complexes in soybean. Plant Physiol. **108**, 693–701.

Kato, K., Hasegawa, K., Goto, S., and Inaguma, Y. (1994). Dissociation as a result of phosphorylation of an aggregated form of the small stress protein, hsp27. J. Biol. Chem. **269**, 11274–11278.

Knauf, U., Jakob, U., Engel, K., Buchner, J., and Gaestel, M. (1994). Stress and mitogen-activated phosphorylation of the small heat shock protein Hsp25 by MAPKAP kinase-2 is not essential for chaperone properties and cellular thermoresistance. EMBO J. **13**, 54–60.

LaFayette, P.R., Nagao, R.T., O'Grady, K., Vierling, E., and Key, J.L. (1996). Molecular characterization of cDNAs encoding low-molecular-weight heat shock proteins of soybean. Plant Mol. Biol. **30**, 159–169.

Lee, G.J., Pokala, N., and Vierling, E. (1995). Structure and in vitro molecular chaperone activity of cytosolic small heat shock proteins from pea. J. Biol. Chem. **270**, 10432–10438.

Lee, G.J., Roseman, A.M., Saibil, H.R. and Vierling, E. (1997). A small heat shock protein stably binds heat-denatured model substrates and can maintain a substrate in a folding-competent state. EMBO J. **16**, 659–671.

Lenne, C. and Douce, R. (1994). A low molecular mass heat-shock protein is localized to higher plant mitochondria. Plant Physiol. **105**, 1255–1261.

Nover, L. (1991). *Heat Shock Response*. CRC Press, Boca Raton.

Nover, L. and Scharf, K. (1984). Synthesis, modification and structural binding of heat shock proteins in tomato cell cultures. Eur. J. Biochem. **139**, 303–313.

Nover, L., Scharf, K.-D., and Neumann, D. (1983). Formation of cytoplasmic heat shock granules in tomato cell cultures and leaves. Mol. Cell. Biol. **3**, 1648–1655.

Nover, L., Scharf, K.-D., and Neumann, D. (1989). Cytoplasmic heat shock granules are formed from precursor particles and are associated with a specific set of mRNAs. Mol. Cell. Biol. **9**, 1298–1308.

Osteryoung, K.W. and Vierling, E. (1994). Dynamics of small heat shock protein distribution within the chloroplasts of higher plants. J. Biol. Chem. **269**, 28676–28682.

Rouse, J., Cohen, P., Trigon, S., Morange, M., Alonso-Llamazares, A., Zamanillo, D., Hunt, T., and Nebreda, A.R. (1994). A novel kinase cascade triggered by stress and heat shock that stimulates MAPKAP kinase-2 and phosphorylation of the small heat shock proteins. Cell **78**, 1027–1037.

Vierling, E. (1991). The roles of heat shock proteins in plants. Annu. Rev. Plant Physiol. Plant Mol. Biol. **42**, 579–620.

Waters, E. (1995). The molecular evolution of the small heat shock proteins in plants. Genetics **141**, 785–795.

Waters, E.R., Lee, G.J., and Vierling, E. (1996). Evolution, structure and function of the small heat shock proteins in plants. J. Exp. Bot. **47**, 325–338.

■ *Elizabeth Vierling and Garrett J. Lee:*
Department of Biochemistry
University of Arizona
Life Sciences South
1007 E. Lowell Street
Tucson, AZ 85721, USA
Tel. 1 520 621 1601
Fax. 1 520 621 3709
E-mail: eliz@biosci.arizona.edu

Drosophila melanogaster small heat shock proteins

The small heat shock proteins (sHSPs) of Drosophila melanogaster are prominent among a group of proteins preferentially synthesized after exposure to high temperatures or other stress agents. The four sHSPs of D. melanogaster (22, 23, 26 and 27 kDa) are also differentially expressed during development in the absence of stress. Although it has been suggested that some of these proteins may confer thermotolerance and resistance to oxidative stress, their precise functions in the cell, either during development or under stress conditions, remain to be clearly defined.

■ Discovery of *Drosophila melanogaster* sHSPs

The specific heat shock response in *D. melanogaster* was first observed as the induction of new puffs on salivary gland polytene chromosomes following exposure to 37°C (Ashburner, 1970). This response was shown to correlate with the preferential expression of a small number of heat shock proteins and a reduction of synthesis of normal proteins (Tissières et al., 1974). ^{35}S pulse-labelling experiments in both cultured cells and salivary glands were used by Mirault et al. (1978) to identify a subset of four sHSPs of less than 30 kDa.

Drosophila melanogaster sHsp genes and sequences

The genes encoding the sHSPs are located within a 12 kb section of the 67B locus of chromosome 3L (Corces et al., 1980; Southgate et al., 1983). Three other developmentally regulated genes, which exhibit some structural similarity to the sHSPs as well as heat shock inducibility, are also located within this stretch of the chromosome (Ayme and Tissières, 1985), but their classification as HSPs is still debated. Sequence analysis of the four *sHsp* genes reveals single ORFs of 174, 186, 208 and 213 amino acids coding for proteins of 22, 23, 26 and 27 kDa (GenBank accession numbers J01098, J01100, J01094 and J01096, respectively).

Three major protein domains are shared by the sHSPs (see Fig. 1). An 80 amino acid domain homologous to the C-terminal sequence of mammalian α-crystallin (a structural protein of the lens, see entry p. 288) is found in the second half of all sHSPs (Ingolia and Craig, 1982). Attached to the C-terminus of the α-crystallin domain in all *D. melanogaster* sHSPs is a conserved region of 25 amino acids. No particular property or function inherent to this domain has been described so far. Finally, a very hydrophobic N-terminal domain of 15 amino acids is found in Hsp23, Hsp26 and Hsp27, and may play a role in protein–membrane interactions (Arrigo and Pauli, 1988).

Drosophila melanogaster sHSPs: localization and modifications

The cellular localization of *D. melanogaster* sHSPs under normal and heat shock conditions has mainly been determined by cell fractionation and immunohistochemistry or immunocytochemistry of the S3 and Kc cultured cell lines (Arrigo and Tanguay, 1991). The different sHSPs have different intracellular localizations (see Table 1). Hsp27 is nuclear in cultured cells (Beaulieu et al., 1988) but also shows a cytoplasmic locale at specific stages of oogenesis and during embryogenesis. Following heat shock or other stress conditions most of the sHSPs tend to relocalize to the perinuclear region or into the nucleus (Arrigo and Landry, 1994). Hsp22 is localized in mitochondria.

Post-translational modification in the form of phosphorylation has been demonstrated for two of the sHSPs,

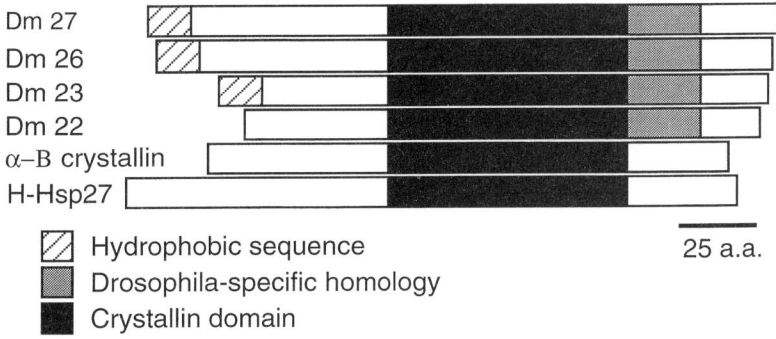

Figure 1. Structure of the sHSPs.

Table 1. Localization and tissue-specific expression of sHSPs

	Intracellular localization		Tissue-specific expression during development		
sHSP	normal	heat shock	embryonic	larval	adult
22	–	mitochondria	–	–	–
23	cytosolic	perinuclear/ nuclear	– restricted cells in the brain and midline precursor cells	–	– brain, central neuropile – CNS – fat body – gonads of young flies
26	cytosolic	perinuclear	– all cells	– epithelium – imaginal discs – proventriculus – neurocytes – spermatocytes – nurse cells	– ovarian nurse cells – spermatocytes – neurocyte of brain – CNS
27	perinuclear / nuclear	nuclear	– all cells	– CNS – gonads – imaginal discs	– germ line – CNS

Hsp26 and Hsp27 (Rollet and Best-Belpomme, 1986). Since Hsp27 is present in multiple isoforms during development (apparently related to the level of phosphorylation), and some of these isoforms are lost following heat shock (Marin et al., 1996b), a functional role for this modification seems likely. Hsp27 can be phosphorylated by a mammalian MAP kinase, MAPKAP-K2; however, the kinase responsible for the phosphorylation of Hsp27 in Drosophila has yet to be identified. A Drosophila homologue of this kinase has recently been cloned (Larochelle and Suter, 1995) but it remains to be determined if this enzyme is responsible for the phosphorylation of any of the Drosophila sHSPs.

Biological activities of D. melanogaster sHSPs

Some experiments have suggested that the expression of the sHSPs may confer thermal tolerance (Berger and Woodward, 1983; Rollet et al., 1992) and oxidative stress resistance (Mehlen et al., 1995). However, the precise cellular functions of the sHSPs, both during development or under stress conditions, remain unknown at this time.

Biological regulation of D. melanogaster sHSPs

In addition to being induced by various cellular stresses, the sHSPs show constitutive expression during development and embryogenesis (Zimmerman et al., 1983; Pauli et al., 1990; Marin et al., 1993). Expression in the absence of stress is specific for each sHSP, while induction after heat shock is coordinate and seems to occur in every cell of the organism, although some exceptions have already been identified. For example, the heat shock-induced expression of Hsp23 in the eye ommatidial unit of D. melanogaster is restricted to a specific group of cells, the cone cells (Marin et al., 1996a). Expression following heat shock is mediated by the heat shock factor (HSF) which binds directly to a specific regulator sequence, the heat shock element (HSE) (Wu et al., 1994). Different numbers of the HSE, each consisting of repeating units of five base pairs (5′–X–G–A–A–X–3′) in a head-to-tail configuration (Xiao and Lis, 1988), are found in the promoter regions of the sHsp genes and more than one copy may be necessary for full heat shock induction (Riddihough and Pelham, 1986).

Studies on the developmental expression of Hsp23 (Marin et al., 1993), Hsp26 (Zimmerman et al., 1983; Glaser et al., 1986; Marin et al., 1993) and Hsp27 (Pauli et al., 1990; Marin and Tanguay, 1996) clearly demonstrated that each sHSP displays a unique tissue- and time-specific pattern of expression (see Table 1). The developmental expression of the sHSPs can be under the control of the ecdysterone responsive element (EcRE) found in the promoter region of some of the sHSPs genes. The molting hormone, β-ecdysterone, induces transcription of the sHsp genes, both in cultured cells and in imaginal discs (Ireland and Berger, 1982). The consensus sequence for the EcRE is G/T–X–T–C–A–X–T– X–X–A/C–A/C (Luo et al., 1991). In addition to EcRE, other elements direct the tissue-specific expression of some of the sHSPs. Ovarian and spermatocyte expression elements for hsp26 and hsp27 have been mainly identified by germ line transformation studies (Cohen and Meselson, 1985; Hoffman et al., 1987; Glaser and Lis, 1990).

Mutagenesis studies with D. melanogaster sHSPs

No mutants of any of the sHSPs have been identified. However, a P-element insertion in the 3′ region of the Hsp27 promoter seems to disrupt its expression both at the RNA (Eissenberg and Elgin, 1987) and protein level (S. Michaud and R.M. Tanguay, unpublished results). However, the effects of this disruption on development or on the heat shock response have not been studied.

References

Arrigo, A.P. and Landry, J. (1994). Expression and function of the low-molecular-weight heat shock proteins. In *The Biology of Heat Shock Proteins and Molecular Chaperones* (R.I. Morimoto, A. Tissières, and C. Georgopoulos, eds), pp. 335–373. Cold Spring Harbor Laboratory Press, New York.

Arrigo, A.P. and Pauli, D. (1988). Characterization of Hsp27 and of three immunologically related polypeptides during Drosophila development. Exp. Cell. Res. **175**, 169–183.

Arrigo, A.P. and Tanguay, R.M. (1991). Expression of heat shock proteins during development in Drosophila. In *Heat Shock and Development* (L. Hightower and L. Nover, eds), pp. 106–119. Springer-Verlag, Berlin.

Ashburner, M. (1970). Patterns of puffing activity in the salivary gland chromosomes of Drosophila. V. Responses to environmental treatments. Chromosoma **31**, 356–376.

Ayme, A. and Tissières, A. (1985). Locus 67B of Drosophila melanogaster contains seven, not four, closely related heat shock genes. EMBO J. **4**, 2949–2954.

Beaulieu, J.F., Arrigo, A.P., and Tanguay, R.M. (1988). Interaction of Drosophila 27 000 Mr heat-shock protein with the nucleus of heat-shocked and ecdysone-stimulated culture cells. J. Cell Sci. **92**, 29–36.

Berger, E.M. and Woodward, M.P. (1983). Small heat shock proteins in Drosophila may confer thermal tolerance. Exp. Cell Res. **147**, 437–442.

Cohen, R.S. and Meselson, M. (1985). Separate regulatory elements for the heat-inducible and ovarian expression of the Drosophila hsp26 gene. Cell **43**, 737–746.

Corces, V., Holmgren, R., Freund, R., Morimoto, R., and Meselson, M. (1980). Four heat shock proteins of Drosophila melanogaster coded within a 12-kilobase region in chromosome subdivision 67B. Proc. Natl. Acad. Sci. USA **77**, 5390–5393.

Eissenberg, J. and Elgin, S.C.R. (1987). Hsp28stl: a P-element insertion mutation that alters the expression of a heat shock gene in Drosophila melanogaster. Genetics **115**, 333–340

Glaser, R.L. and Lis, J.T. (1990). Multiple, compensatory regulatory elements specify spermatocyte specific expression of the Drosophila melanogaster hsp26 gene. Mol. Cell. Biol. **10**, 131–137.

Glaser, R.L., Wolfner, M.F., and Lis, J.T. (1986). Spatial and temporal pattern of Hsp26 expression during normal developmental. EMBO J. **5**, 747–754.

Hoffman, E.P., Gerring, S.L. and Corces, V.G. (1987). The ovarian, ecdysterone, and heat-shock-responsive promoters of the *Drosophila melanogaster hsp*27 gene react very differently to perturbations of DNA sequences. Mol. Cell. Biol. **7**, 973–981.

Ingolia, T.D. and Craig, E. (1982). Four small *Drosophila* heat shock proteins are related to each other and to mammalian alpha-crystallin. Proc. Natl. Acad. Sci. USA **79**, 2360–2364.

Ireland, R.C. and Berger, E.M. (1982). Synthesis of low molecular weight heat shock peptides stimulated by ecdysterone in a cultured *Drosophila* cell line. Proc. Natl. Acad. Sci. USA **79**, 855–859.

Larochelle, S. and Suter, B. (1995). The *Drosophila melanogaster* homolog of the mammalian MAPK-activated protein kinase-2 (MAPKAP-2) lacks a proline-rich N terminus. Gene **163**, 209–214.

Luo, Y., Amin, J., and Voellmy, R. (1991). Ecdysterone receptor is a sequence specific transcription factor involved in the developmental regulation of heat shock genes. Mol. Cell. Biol. **11**, 3660–3675.

Marin, R., and Tanguay, R.M. (1996). Expression and stage-specific localization of the small heat shock protein Hsp27 during oogenesis in Drosophila. Chromosoma **105**, 142–149.

Marin, R., Valet, J.-P., and Tanguay, R.M. (1993). Hsp23 and Hsp26 exhibit distinct spatial and temporal patterns of constitutive expression in *Drosophila* adults. Develop. Genet. **14**, 112–118.

Marin, R., Demers, M., and Tanguay, R.M. (1996a). Cell-specific heat-shock induction of Hsp23 in the eye of *Drosophila melanogaster*. Cell Stress Chaperones **1**, 40–46.

Marin, R., Landry, J., and Tanguay, R.M. (1996b). Tissue-specific posttranslational modification of the small heat shock protein HSP27 in *Drosophila*. Exp. Cell Res. **223**, 1–8.

Mehlen, P., Preville, X., Chareyron, P., Briolay, J., Klemenz, R., and Arrigo A.P. (1995). Constitutive expression of human Hsp27, Drosophila Hsp27, or human alpha B-crystallin confers resistance to TNF- and oxidative stress-induced cytotoxicity in stably transfected murine L929 fibroblasts. J. Immunol. **154**, 363–374.

Mirault, M.E., Goldschmidt-Clermont, M., Moran, L., Arrigo, A.P., and Tissières, A. (1978). The effect of heat shock on gene expression in *Drosophila melanogaster*. Cold Spring Harbor Symp. Quant. Biol. **42**, 819–827.

Pauli, D., Tonka, C.-H., Tissières, A., and Arrigo, A.P. (1990). Tissue specific expression of the heat shock protein Hsp27 during *Drosophila melanogaster* development. J. Cell Biol. **111**, 817–828.

Riddihough, G. and Pelham, R.B. (1986). Activation of the *Drosophila hsp*27 promoter by heat shock and by ecdysone involves independent and remote regulatory sequences. EMBO J. **5**, 1653–1658.

Rollet, E. and Best-Belpomme, M. (1986). Hsp26 and 27 are phosphorylated in response to heat and ecdysterone in *Drosophila melanogaster* cells. Biochem. Biophys. Res. Commun. **163**, 301–308.

Rollet, E., Lavoie, J.N., Landry, J., and Tanguay, R.M. (1992). Expression of *Drosophila*'s 27kDa heat shock protein into rodent cells confers thermal resistance. Biochem. Biophys. Res. Commun. **185**, 116–120.

Southgate, R., Ayme, A., and Voellmy, R. (1983). Nucleotide sequence analysis of the *Drosophila* small heat shock gene cluster at locus 67B. J. Mol. Biol. **165**, 35–57.

Tissières, A., Mitchell, A.K., and Tracy, U.M. (1974). Protein synthesis in salivary glands of *Drosophila melanogaster*: relation to chromosome puffs. J. Mol. Biol. **84**, 389.

Wu, C., Clos, J., Giorgi, G., Haroun, R.I., Kim, S., Rabindran, S.K., Westwood, J.T., Wisniewski, J., and Yim, G. (1994). Structure and regulation of heat shock transcription factor. In *The Biology of Heat Shock Proteins and Molecular Chaperones* (R.I. Morimoto, A. Tissières, and C. Georgopoulos, eds), pp. 395–416. Cold Spring Harbor Laboratory Press, New York.

Xiao, H. and Lis, J.T. (1988). Germline transformation used to define key features of heat-shock response elements. Science **239**, 1139–1142.

Zimmerman, J.L., Petri, W., and Meselson, M. (1983). Accumulation of a specific subset of *D. melanogaster* heat shock mRNAs in normal development without heat shock. Cell **32**, 1161–1170.

■ *Sébastien Michaud, Denis R. Joanisse and Robert M. Tanguay:*
Laboratoire de génétique cellulaire et développementale
RSVS, Pav. CE Marchand
Université Laval
Ste-Foy, Québec G1K 7P4, Canada
Tel. 1 418 656 3339
Fax. 1 418 656 7176
E-mail: Robert.Tanguay@rsvs.ulaval.ca

Mammalian Hsp27

Hsp27 is a member of the heat shock protein family. Increased expression of Hsp27 confers cell resistance to heat shock and other stresses. Hsp27 is phosphorylated in response to stress and a variety of normal agonists and is thought to be involved in the stabilization/reorganization of the actin cytoskeleton.

■ Alternative names

Hsp25, Hsp28, mammalian small or low-molecular-weight heat shock protein (Arrigo and Landry, 1994).

■ Mammalian *hsp27* gene and sequence

The nucleotide sequences of mammalian *hsp27* genes encode proteins with highly conserved sequences. Hsp27 proteins from human (GenBank accession number X03900), dog (U19368), mouse (L07577), Chinese hamster (X51747) and rat (M86389) are more than 90% identical and share with αA- and αB-crystallin a domain of high homology thought to be responsible for the propensities of these proteins to form multimeric complexes (Merck et al., 1993b; de Jong et al., 1993). Homologous proteins are found in virtually all organisms. Human Hsp27 contains three known sites of phosphorylation found in the motif Arg–X–X–Ser at Ser-15, Ser-78 and Ser-82 (Landry et al., 1992). Chinese hamster and mouse Hsp27 are phosphorylated on two sites (Gaestel et al., 1991): the residue equivalent to human Ser-78 is replaced by Asp.

Function of mammalian Hsp27

Avian, mouse and yeast homologs of Hsp27 behave in vitro as F-actin cap-binding proteins and can inhibit actin polymerization (Miron et al., 1991; Benndorf et al., 1994; Rahman et al., 1995). In vivo, Hsp27 has an enhanced localization in membrane protrusions (lamellipodia and ruffles) of motile cells. Overexpression of Hsp27 modifies the organization of the actin microfilaments, increases pinocytosis and enhances growth factor-induced actin polymerization (Lavoie et al., 1993b). The actin cytoskeleton is stabilized by Hsp27 in cells exposed to oxidative stress or heat shock and this is thought to help cells to survive adverse conditions. Overexpression of Hsp27 has been correlated with increased resistance to a number of toxic agents (Huot et al., 1991, 1996, 1997; Lavoie et al., 1993a, 1995). Hsp27, like the closely related α-crystallins, has chaperone activity and can protect protein from denaturation in vitro (Merck et al., 1993a; Ehrnsperger et al., 1997).

Mammalian *hsp27* regulation

hsp27 is regulated both at the transcriptional and post-translational levels. The *hsp27* gene contains heat shock and estrogen-responsive elements (Hickey et al., 1986; Gaestel et al., 1993). Transcriptional activation occurs after heat shock and other stresses and results in an increase in the level of the protein by as much as tenfold in some cells. This increase in concentration probably contributes to the development of a thermotolerant state after heat shock (Landry et al., 1989, 1991). Hsp27 is the major polypeptide synthesized by certain mammalian cells in response to steroid hormones and is expressed in several estrogen-sensitive human tissues (Fuqua et al., 1989; Ciocca and Luque, 1991). Hsp27 is also phosphorylated upon cell exposures to a variety of physiological agonists such as growth/differentiation factors and cytokines, and also after potentially toxic treatments such as heat shock or oxidative stress (Arrigo and Landry, 1994). Hsp27 is phosphorylated by MAPKAP kinase-2, a serine protein kinase activated in vivo in a major signal transduction pathway involving the p38 MAP kinase (Stokoe et al., 1992; Rouse et al., 1994; Huot et al., 1995, 1997; Guay et al., 1997). Phosphorylation affects the supramolecular structure of Hsp27 and modulates its protective and actin polymerization activities (Kato et al., 1994; Benndorf et al., 1994; Lavoie et al., 1995). Hsp27 is a substrate of the serine protease myeloblastin and its function may also be regulated by hydrolysis (Spector et al., 1995).

References

Arrigo, A.P. and Landry, J. (1994). Expression and function of the low-molecular-weight heat shock proteins. In *The Biology of Heat Shock Proteins and Molecular Chaperones* (R.I. Morimoto, A. Tissières, and C. Georgopoulos, eds), pp. 335–373. Cold Spring Harbor Laboratory Press, New York

Benndorf, R., Hayess, K., Ryazantsev, S., Wieske, M., Behlke, J., and Lutsch, G. (1994). Phosphorylation and supramolecular organization of murine small heat shock protein HSP25 abolish its actin polymerization-inhibiting activity. J. Biol. Chem. **269**, 20780–20784.

Ciocca, D.R. and Luque, E.H. (1991). Immunological evidence for the identity between the hsp27 estrogen-regulated heat shock protein and the p29 estrogen receptor-associated protein in breast and endometrial cancer. Breast Cancer Res. Treat. **20**, 33–42.

de Jong, W.W., Leunissen, J.A., and Voorter, C.E. (1993). Evolution of the alpha-crystallin/small heat-shock protein family. Mol. Biol. Evol. **10**, 103–126.

Ehrnsperger, M., Gräber, S., Gaestel, M., and Buchner, J. (1997). Binding of non-native protein to Hsp25 during heat shock creates a reservoir of folding intermediates for reactivation. EMBO J. **16**, 221–229.

Fuqua, S.A., Blum-Salingaros, M., and McGuire, W.L. (1989). Induction of the estrogen-regulated '24K' protein by heat shock. Cancer Res. **49**, 4126–4129.

Gaestel, M., Schroder, W., Benndorf, R., Lippmann, C., Buchner, K., Hucho, F., Erdmann, V.A., and Bielka, H. (1991). Identification of the phosphorylation sites of the murine small heat shock protein hsp25. J. Biol. Chem. **266**, 14721–14724.

Gaestel, M., Gotthardt, R., and Muller, T. (1993). Structure and organisation of a murine gene encoding small heat-shock protein Hsp25. Gene **28**, 279–283.

Guay, J., Lambert, H., Gingras-Breton, G., Lavoie, J.N., Huot, J., and Landry, J. (1997). Regulation of actin dynamics by p38 map kinase-mediated phosphorylation of heat shock protein 27. J. Cell Sci. **110**, 357–368.

Hickey, E., Brandon, S.E., Potter, R., Stein, G., Stein, J., and Weber, L.A. (1986). Sequence and organization of genes encoding the human 27 kDa heat shock protein. Nucl. Acids Res. **14**, 4127–4145.

Huot, J., Roy, G., Lambert, H., Chrétien, P., and Landry, J. (1991). Increased survival after treatments with anticancer agents of Chinese hamster cells expressing the human Mr 27,000 heat shock protein. Cancer Res. **51**, 5245–5252.

Huot, J., Lambert, H., Lavoie, J.N., Guimond, A., Houle, F., and Landry, J. (1995). Characterization of 45-kDa/54-kDa HSP27 kinase, a stress-sensitive kinase which may activate the phosphorylation-dependent protective function of mammalian 27-kDa heat-shock protein HSP27. Eur. J. Biochem. **227**, 416–427.

Huot, J., Houle, F., Spitz, D.R., and Landry, J. (1996) HSP27 phosphorylation-mediated resistance against actin fragmentation and cell death induced by oxidative stress. Cancer Res. **56**, 273–279.

Huot, J., Houle, F. Marceau, F., and Landry, J. (1997). Oxidative stress-induced actin reorganization mediated by the p38 mitogen-activated protein kinase/heat shock protein 27 pathway in vascular endothelial cells. Circ Res. **80**, 383–390.

Kato, K., Hasegawa, K., Goto, S., and Inaguma, Y. (1994). Dissociation as a result of phosphorylation of an aggregated form of the small stress protein, hsp27. J. Biol. Chem. **269**, 11274–11278.

Landry, J., Chrétien, P., Lambert, H., Hickey, E., and Weber, L.A. (1989). Heat shock resistance conferred by expression of the human HSP27 gene in rodent cells. J. Cell Biol. **109**, 7–15.

Landry, J., Chrétien, P., Laszlo, A., and Lambert, H. (1991). Phosphorylation of HSP27 during development and decay of thermotolerance in Chinese hamster cells. J. Cell Physiol. **147**, 93–101.

Landry, J., Lambert, H., Zhou, M., Lavoie, J.N., Hickey, E., Weber, L.A., and Anderson, C.W. (1992). Human HSP27 is phosphorylated at serines 78 and 82 by heat shock and mitogen-activated kinases that recognize the same amino acid motif as S6 kinase II. J. Biol. Chem. **267**, 794–803.

Lavoie, J.N., Gingras-Breton, G., Tanguay, R.M., and Landry, J. (1993a). Induction of Chinese hamster HSP27 gene expression in mouse cells confers resistance to heat shock. HSP27 stabilization of the microfilament organization. J. Biol. Chem. **268**, 3420–3429.

Lavoie, J.N., Hickey, E., Weber, L.A., and Landry, J. (1993b). Modulation of actin microfilament dynamics and fluid phase pinocytosis by phosphorylation of heat shock protein 27. J. Biol. Chem. **268**, 24210–24214.

Lavoie, J.N., Lambert, H., Hickey, E., Weber, L.A., and Landry, J. (1995). Modulation of cellular thermoresistance and actin filament stability accompanies phosphorylation-induced changes in the oligomeric structure of heat shock protein 27. Mol. Cell. Biol. **15**, 505–516.

Merck, K.B., Groenen, P.J., Voorter, C.E., de Haard-Hoekman, W.A., Horwitz, J., Bloemendal, H., and de Jong, W.W. (1993a). Structural and functional similarities of bovine alpha-crystallin and mouse small heat-shock protein. A family of chaperones. J. Biol. Chem. **268**, 1046–1052.

Merck, K.B., Horwitz, J., Kersten, M., Overkamp, P., Gaestel, M., Bloemendal, H., and de Jong, W.W. (1993b). Comparison of the homologous carboxy-terminal domain and tail of alpha-crystallin and small heat shock protein. Mol. Biol. Rep. **18**, 209–215.

Miron, T., Vancompernolle, K., Vandekerckhove, J., Wilchek, M., and Geiger, B. (1991). A 25-kD inhibitor of actin polymerization is a low molecular mass heat shock protein. J. Cell Biol. **114**, 255–261.

Rahman, D.R., Bentley, N.J., and Tuite, M.F. (1995). The *Saccharomyces cerevisiae* small heat shock protein Hsp26 inhibits actin polymerisation. Biochem. Soc. Trans. **23**, 77S.

Rouse, J., Cohen, P., Trigon, S., Morange, M., Alonso-Llamazares, A., Zamanillo, D., Hunt, T., and Nebreda, A.R. (1994). A novel kinase cascade triggered by stress and heat shock that stimulates MAPKAP kinase-2 and phosphorylation of the small heat shock proteins. Cell **78**, 1027–1037.

Spector, N.L., Hardy, L., Ryan, C., Miller, W.H., jun., Humes, J.L., Nadler, L.M., and Luedke, E. (1995). 28-kDa mammalian heat shock protein, a novel substrate of a growth regulatory protease involved in differentiation of human leukemia cells. J. Biol. Chem. **270**, 1003–1006.

Stokoe, D., Engel, K., Campbell, D.G., Cohen, P., and Gaestel, M. (1992). Identification of MAPKAP kinase 2 as a major enzyme responsible for the phosphorylation of the small mammalian heat shock proteins. FEBS Lett. **313**, 307–313.

■ *Jacques Landry:*
Centre de recherche en cancérologie de l'Université Laval
L'Hôtel-Dieu de Québec
11, côte du Palais
Québec G1R 2J6, Canada
Tel. 1 418 691 5555
Fax. 1 418 691 5439
E-mail: jacques.landry@med.ulaval.ca

Small heat shock protein Hsp25 from mouse and rat

Hsp25 is a member of the family of mammalian small Hsps. It is highly homologous in its primary structure to the small HSPs from human (Hsp27), hamster (Hsp27) and turkey (IAP25) as well as to the α-crystallins. It shares in vitro chaperone properties with human Hsp27 and α-crystallin and confers thermoresistance when overexpressed in cellular systems. Hsp25 is rapidly phosphorylated at two serine residues in response to mitogenic signals and stress as a result of activation of the MAP kinase activated protein kinase-2. The phosphorylation of Hsp25 does not influence its in vitro chaperone and thermoresistance-mediating properties in NIH/3T3 cells, but may alter its influence on actin polymerization.

■ Alternative names

Rodent/murine low molecular weight heat shock protein (Arrigo and Landry, 1994); mouse Hsp27/28 (Zantema *et al.*, 1992); rat Hsp27/28 (Uoshima *et al.*, 1993); 25 kDa growth-related protein p25 (Benndorf *et al.*, 1988; Oesterreich *et al.*, 1990).

■ Isolation of Hsp25

Hsp25 has been isolated from Ehrlich ascites tumor cells by 'giant' two-dimensional PAGE for analytical purposes (Benndorf *et al.*, 1988). Recombinant material was obtained from *Escherichia coli* BL 21 (DE3) carrying a T7 expression plasmid for Hsp25 (Gaestel *et al.*, 1989) and purified by a two-step procedure using ion-exchange and size-exclusion chromatography (Engel et al., 1991).

■ *HSP25* gene and sequence

The protein sequence of mouse Hsp25 (GenBank accession number P14602) was first deduced from the nucleotide sequences of cDNA clones (GenBank numbers X14686, X14687; Gaestel et al., 1989). PCR analysis in mouse osteoblasts revealed two further splice variants of Hsp25 (U03561, U03562; Cooper and Uoshima, 1994). In mice, a transcribed gene (L07577, L11608) and a pseudogene of Hsp25 (L11610) have been detected using the cDNA as probe (Gaestel et al., 1993; Fröhli et al., 1993). The rat cDNA (GenBank accession number M86389) and a rat gene (S67755) have also been cloned (Uoshima et al., 1993). As for the human Hsp27 (Hickey et al., 1986), the transcribed Hsp25 genes contain HSEs and SP1 elements in the promoter and two introns. The rodent Hsp25 amino acid sequence is highly homologous to the small Hsps from human (Hsp27, Hickey et al., 1986), hamster (Hsp27; Lavoie et al., 1990) and turkey (IAP25, Miron et al., 1991) as well as to the α-crystallins (de Jong et al., 1993).

■ Hsp25 protein

Hsp25 forms large homo-oligomers. Distinct complexes of 32 subunits (Behlke et al., 1991) or 16 subunits (Ehrnsperger et al., 1997) have been detected depending on the solvent conditions used. Furthermore, the size of the oligomer increases upon incubation at temperatures above 40°C, and Hsp25 seems to be mainly β sheet structured as deduced from CD measurements M. Ehrnsperger, M. Gaestel and J. Buchner, unpublished results).

■ Antibodies against Hsp25

Polyclonal antibodies against Hsp25 have been used to identify the differentially phosphorylated isoforms of Hsp25 in immunoblotting experiments (Benndorf et al., 1988), to detect Hsp25 expression in transfected cells (Knauf et al., 1992) and to analyse Hsp25 expression in the mouse embryo (Gernold et al., 1993) and in different tissues of adult mice (Klemenz et al., 1993). Antibodies that cross-react between rodent Hsp25 and human Hsp27 have been prepared against a Hsp25/27 hybrid protein (Engel et al., 1991). Recombinant Hsp25 and polyclonal antibodies suited for immunoblotting are now commercially available from StressGen (Canada). Using various different antibodies a cytoplasmaic localization of Hsp25 could be detected. However, rat Hsp25 has also been detected in isolated nuclei of stressed cells (Kim et al., 1984).

■ Tissue-specific expression of Hsp25

In mice Hsp25 is constitutively expressed in many normal adult tissues. In the absence of stress the protein is most abundant in the eye lens, heart, stomach, colon, lung and bladder (Klemenz et al., 1993). Hsp25 is detectable also in neurones of the spinal cord and the purkinje cells (Gernold et al., 1993). During embryogenesis the relative amount of Hsp25 increases. For days 13–20 of embryogenesis, Hsp25 accumulation is predominant in the various muscle tissues, including the heart, the bladder and the back muscles (Gernold et al., 1993).

■ Post-translational modifications of Hsp25

Mouse Hsp25 is phosphorylated as a result of stress and mitogenic stimuli at serines Ser-15 and Ser-86 (Gaestel et al., 1991). The enzyme responsible for Hsp25 phosphorylation is the MAP kinase activated protein kinase-2 (Stokoe et al., 1992b). This enzyme can be activated by the MAP kinases ERK1 and ERK2 *in vitro* (Stokoe et al., 1992b) and by the mammalian HOG1 homolog p38/40 reactivating kinase *in vivo* (Freshney et al., 1994; Rouse et al., 1994). Thus Hsp25 phosphorylation is the result of activation of one or several MAP kinase cascades.

Approximately one-third of the Hsp25 purified from Ehrlich ascites tumor (EAT) cells appears to be acylated (Oesterreich et al., 1991a).

■ Regulation of Hsp25

Hsp25 expression is increased in stationary phase murine EAT cells compared with the exponentially growing cells. Heat shock and anticancer drugs can induce Hsp25 (Oesterreich et al., 1990, 1991; Bielka et al., 1994). Expression of Hsp25 in murine embryonal carcinoma and embryonic stem cells is differentiation dependent (Stahl et al., 1992). In adenovirus-transformed baby rat kidney (BRK) cells, expression of Hsp25 is inversely correlated to the oncogenicity of the cells (Zantema et al., 1989) and in these cells Hsp25 can form complexes with αB-crystallin, which dissociate as a result of heat shock (Zantema et al., 1992). Leukemia inhibitory factor/D-factor induces phosphorylation of Hsp25 in a mouse myelomonocytic leukemic cell line (Michishita et al., 1991).

■ Biological activities of Hsp25

Induction of Hsp25 provides tolerance to heat shock and chemical stress (Oesterreich et al., 1991). Ectopic overexpression of Hsp25 in EAT cells results in a thermoresistant phenotype and decreases cell proliferation (Knauf et al., 1992). Hsp25 can act as a molecular chaperone *in vitro*, preventing thermal aggregation of citrate synthase and assisting in refolding of chemically denatured β-glucosidase (Jakob et al., 1993). Upon heat shock several non-native molecules of citrate synthase stably bind to one Hsp25 oligomer. Under permissive folding conditions, citrate synthase can be released from Hsp25 and, in cooperation with the ATP-dependent chaperone Hsp70, the native state can be restored (Ehrnsperger et al. 1997). Non-phosphorylated Hsp25 purified from EAT cells inhibits *in vitro* actin polymerization, whereas recombinant or phosphorylated EAT Hsp25 is not able to influence this process (Benndorf et al., 1994).

■ Mutagenesis studies with Hsp25

The phosphorylation sites of Hsp25, serines Ser-15 and Ser-86, have been mutated to alanine (Knauf et al., 1994). The single and double phosphorylation site mutants can still confer thermoresistance to NIH/3T3 cells. A deletion mutant of Hsp25 lacking the N-terminal half of the molecule (HSP25-2Dt) can still form aggregates but does not prevent β-crystallin from thermal aggregation (Merck et al., 1993).

■ References

Arrigo, A.P. and Landry, J. (1994). Expression and function of the low-molecular-weight heat shock proteins. In *The Biology of Heat Shock Proteins and Molecular Chaperones* (R.I. Morimoto, A. Tissières, and C. Georgopoulos, eds), pp. 335–373. Cold Spring Harbor Laboratory Press, New York.

Behlke, J., Lutsch, G., Gaestel, M., and Bielka, H. (1991). Supramolecular structure of the recombinant murine small heat shock protein hsp25. FEBS Lett. **288**, 119–122.

Benndorf, R., Kraft, R., Otto, A., Stahl, J., Bohm, H., and Bielka, H. (1988). Purification of the growth-related protein p25 of the Ehrlich ascites tumor and analysis of its isoforms. Biochem. Int. **17**, 225–234.

Benndorf, R., Hayess, K., Ryazantsev, S., Wieske, M., Behlke, J., and Lutsch, G. (1994). Phosphorylation and supramolecular organisation of murine small heat shock protein Hsp25 abolish its actin polymerization-inhibiting activity. J. Biol. Chem. **269**, 20780–20784.

Bielka, H., Hoinkis, G., Oesterreich, S., Stahl, J., and Benndorf, R. (1994). Induction of the small stress protein, Hsp25, in Ehrlich ascites carcinoma cells by anticancer drugs. FEBS Lett. **343**, 165–167.

Cooper, L.F. and Uoshima, K. (1994). Differential estrogenic regulation of small M(R), heat shock protein expression in osteoblasts. J. Biol. Chem. **269**, 7869–7873.

de Jong, W.W., Leunissen, J.A.M., and Voorter, C.E.M. (1993). Evolution of the alpha-crystallin/small heat-shock protein family. Mol. Biol. Evol. **10**, 103–126.

Ehrnsperger, M., Gäbler, S., Gaestel, M., and Buchner, J. (1997). Binding of non-native protein to Hsp25 during heat shock creates a reservoir of folding intermediates for reactivation. EMBO J. **16**, 221–229.

Engel, K., Knauf, U., and Gaestel, M. (1991). Generation of antibodies against human hsp27 and murine hsp25 by immunization with a chimeric small heat shock protein. Biomed. Biochim. Acta **50**, 1065–1071.

Freshney, N.W., Rawlinson, L., Guesdon, F., Jones, E., Cowley, S., Hsuan, J., and Saklatvala, J. (1994). Interleukin-1 activates a novel protein kinase cascade that results in the phosphorylation of Hsp27. Cell **78**, 1039–1049.

Fröhli, E., Aoyama, A., and Klemenz, R. (1993). Cloning of the mouse hsp25 gene and an extremely conserved hsp25 pseudogene. Gene **128**, 273–277.

Gaestel, M., Gross, B., Benndorf, R., Strauss, M., Schunk, W.H., Kraft, R., Otto, A., Bohm, H., Stahl, J., Drabsch, H., and Bielka, H. (1989). Molecular cloning, sequencing and expression in *Escherichia coli* of the 25-kDa growth-related protein of Ehrlich ascites tumor and its homology to mammalian stress proteins. Eur. J. Biochem. **179**, 209–213.

Gaestel, M., Schroder, W., Benndorf, R., Lippmann, C., Buchner, K., Hucho, F., Erdmann, V.A., and Bielka, H. (1991). Identification of the phosphorylation sites of the murine small heat shock protein hsp25. J. Biol. Chem. **266**, 14721–14724.

Gaestel, M., Gotthardt, R., and Muller, T. (1993). Structure and organisation of a murine gene encoding small heat-shock protein hsp25. Gene **128**, 279–283.

Gernold, M., Knauf, U., Gaestel, M., Stahl, J., and Kloetzel, P.M. (1993). Development and tissue-specific distribution of mouse small heat shock protein hsp25. Develop. Genet. **14**, 103–111.

Hickey, E., Brandon, S.E., Potter, R., Stein, G., Stein, J., and Weber, L.A. (1986). Sequence and organization of genes encoding the human 27 kDa heat shock protein [published erratum appears in Nucl. Acids Res. 1986, **14**(20), 8230]. Nucl. Acids Res. **14**, 4127–4145.

Jakob, U., Gaestel, M., Engel, K., and Buchner, J. (1993). Small heat shock proteins are molecular chaperones. J. Biol. Chem. **268**, 1517–1520.

Kim, Y.-J., Sette, M., and Przybyla, A. (1984). Nuclear localization and phosphorylation of three 25-kilodalton rat stress proteins. Mol. Cell. Biol. **4**, 468–474.

Klemenz, R., Andres, A.C., Frohli, E., Schafer, R., and Aoyama, A. (1993). Expression of the murine small heat shock proteins hsp 25 and alphaB crystallin in the absence of stress. J. Cell Biol. **120**, 639–645.

Knauf, U., Bielka, H., and Gaestel, M. (1992). Over-expression of the small heat-shock protein, hsp25, inhibits growth of Ehrlich ascites tumor cells. FEBS Lett. **309**, 297–302.

Knauf, U., Jakob, U., Engel, K., Buchner, J., and Gaestel, M. (1994). Stress- and mitogen-induced phosphorylation of the small heat shock protein Hsp25 by MAPKAP Kinase 2 is not essential for chaperone properties and cellular thermoresistance. EMBO J. **13**, 54–60.

Lavoie, J., Chretien, P., and Landry, J. (1990). Sequence of the Chinese hamster small heat shock protein HSP27. Nucl. Acids Res. **18**, 1637.

Merck, K.B., Horwitz, J., Kersten, M., Overkamp, P., Gaestel, M., Bloemendal, H., and de Jong, W.W. (1993). Comparison of the homologous carboxy-terminal domain and tail of alpha-crystallin and small heat shock protein. Mol. Biol. Rep. **18**, 209–215.

Michishita, M., Satoh, M., Yamaguchi, M., Hirayoshi, K., Okuma, M., and Nagata, K. (1991). Phosphorylation of the stress protein hsp27 is an early event in murine myelomonocytic leukemic cell differentiation induced by leukemia inhibitory factor/D-factor. Biochem. Biophys. Res. Commun. **176**, 979–984.

Miron, T., Vancompernolle, K., Vandekerckhove, J., Wilchek, M., and Geiger, B. (1991). A 25-kD inhibitor of actin polymerization is a low molecular mass heat shock protein. J. Cell Biol. **114**, 255–261.

Oesterreich, S., Benndorf, R., and Bielka, H. (1990). The expression of the growth-related 25kDa protein (p25) of Ehrlich ascites tumor cells is increased by hyperthermic treatment (heat shock). Biomed. Biochim. Acta **49**, 219–226.

Oesterreich, S., Benndorf, R., Reichmann, G., and Bielka, H. (1991a). Phosphorylation and acylation of the growth-related murine small stress protein p25. In *Cellular Regulation by Protein* Phosphorylation (L.M.G. Heilmeyer, ed.), Springer Verlag, Heidelberg, pp. 489–493.

Oesterreich, S., Schunck, H., Benndorf, R., and Bielka, H. (1991b). Cisplatin induces the small heat shock protein hsp25 and thermotolerance in Ehrlich ascites tumor cells. Biochem. Biophys. Res. Commun. **180**, 243–248.

Rouse, J., Cohen, P., Trigon, S., Morange, M., Alonsollamazares, A., Zamanillo, D., Hunt, T., and Nebreda, A.R. (1994). A novel kinase cascade triggered by stress and heat shock that stimulates MAPKAP kinase-2 and phosphorylation of the small heat shock proteins. Cell **78**, 1027–1037.

Stahl, J., Wobus, A.M., Ihrig, S., Lutsch, G., and Bielka, H. (1992). The small heat shock protein hsp25 is accumulated in p19 embryonal carcinoma cells and embryonic stem cells of line BLC6 during differentiation. Differentiation **51**, 33–37.

Stokoe, D., Campbell, D.G., Nakielny, S., Hidaka, H., Leevers, S.J., Marshall, C., and Cohen, P. (1992a). MAPKAP Kinase-2 — A novel protein kinase activated by mitogen-activated protein kinase. EMBO J. **11**, 3985–3994.

Stokoe, D., Engel, K., Campbell, D.G., Cohen, P., and Gaestel, M. (1992b). Identification of MAPKAP Kinase-2 as a major enzyme responsible for the phosphorylation of the small mammalian heat shock proteins. FEBS Lett. **313**, 307–313.

Uoshima, K., Handelman, B., and Cooper, L.F. (1993). Isolation and characterization of a rat HSP-27 gene. Biochem. Biophys. Res. Commun. **197**, 1388–1395.

Zantema, A., de Jong, E., Lardenoije, R., and van der Eb, A.J. (1989). The expression of heat shock protein hsp27 and a complexed 22-kilodalton protein is inversely correlated with oncogenicity of adenovirus-transformed cells. J. Virol. **63**, 3368–3375.

Zantema, A., Verlaandevries, M., Maasdam, D., Bol, S., and van der Eb, A. (1992). Heat shock protein-27 and alphaB-Crystallin can form a complex, which dissociates by heat shock. J. Biol. Chem. **267**, 12936–12941.

■ *Matthias Gaestel:*
AG Stressproteine
Max-Delbrueck-Center for Molecular Medicine
R.-Roessle-Str. 10
13122 Berlin, Germany
Tel. 49 30 9406 3785
Fax. 49 30 9406 3798
E-mail: gaestel@mdc-berlin.de

■ *Johannes Buchner:*
Institut für Biophysik und Physikalische Biochemie
Universität Regensburg
93040 Regensburg, Germany
Tel. 49 941 943 3039
Fax. 49 941 943 2813
E-mail: johannes.buchner@biologie.uni-regensburg.de

Mammalian α-crystallins

The two types of α-crystallin subunits, αA and αB, are members of the small HSP family. Both types occur abundantly as 700–800 kDa heteropolymeric complexes in the vertebrate eye lens, while αB-crystallin is also present at significant levels in many tissues outside the lens. Like the other small Hsps, αA- and αB-crystallin prevent the aggregation of denaturing proteins and confer cellular thermotolerance. αB-Crystallin is stress inducible and implicated in the pathogenesis of various degenerative diseases.

■ Isolation of α-crystallin

α-Crystallin, comprising up to one-third of total lens protein in mammals, can be isolated in large quantities by gel filtration of aqueous lens extracts (Groenen *et al.*, 1994). Lenticular α-crystallin is composed of acidic αA and more basic αB subunits, generally in a ratio of about 3:1. These can be separated by ion exchange chromatography under denaturing conditions, or by reverse-phase HPLC. The amino acid sequences of calf lens αA and αB, determined in the early 1970s, are 57% identical. αA and αB are 173 and 175 residues in length and appear on SDS–PAGE as 20 and 22 kDa bands, respectively. Both proteins are N-terminally acetylated.

α-Crystallin gene and sequence

The αA- and αB-crystallin genes are located, in humans, on chromosomes 21 and 11, respectively (for reviews see Sax and Piatigorsky, 1994; Wistow, 1995). Both genes are single copy and no pseudogenes have been reported. Their exon–intron structures are identical, but differ from those of the related mammalian small HSPs. Two introns are present in the coding sequence. Exons 2 and 3 encode the region of homology that is characteristic for members of the small HSP family. In various mammals, alternative splicing of a 69 bp insert exon in the first intron of the αA gene results in a minor subunit, αA^{ins}. Sequences are available for the αA genes of human (GenBank accession number X14789), mouse (V00730), hamster (X02950), mole rat (Y00464) and bovine (M26142), and for the αB genes of human (M28638), hamster (M12014-5), rat (U04320) and mouse (M73741).

■ α-Crystallin protein

The 700–800 kDa α-crystallin complex has resisted crystallization, probably owing to its polydisperse, non-compact and dynamic nature (for reviews see Groenen *et al.*, 1994; Boelens and de Jong, 1995; Wistow, 1995). Electron microscopy displays globular particles of approximately 14–18 nm. Torus-like structures have sometimes been observed. α-Crystallin is predominantly a β sheet protein, with less than 10% α helical structure. There is evidence that the αA and αB subunits have a two-domain structure with a flexible C-terminal extension (e.g. Carver et al., 1993). The putative C-terminal domain of about 80 residues is conserved in the small HSP family. Proposed models for the quaternary structure range from three-layered spherical structures to micellar complexes. Some experiments indicate that the α-crystallin complex may have tetrameric building blocks (Kantorow et al., 1995). In skeletal muscle, αB may form mixed complexes with available small HSPs (Kato et al., 1994).

In the lens, up to 30% of αA and αB may occur in a phosphorylated form. In αA, Ser-122 is the major site of phosphorylation, whereas in αB the serines at positions 19, 45 and 59 can be phosphorylated. Also, the extralenticular αB is often found in a partially phosphorylated form. This phosphorylation is cAMP dependent and essentially irreversible in the lens fiber cells. In vitro, α-crystallin is capable of serine-dependent autophosphorylation (Kantorow et al., 1995). No noticeable structural or functional effects of phosphorylation of α-crystallin have yet been reported. Glycosylation of αA and αB by N-acetylglucosamine has been demonstrated, and αB is a substrate for the cross-linking enzyme transglutaminase. As for the intracellular localization, αB is normally diffusely localized in the cytoplasm in cultured cells. Under heat shock it translocates transiently to the perinuclear region. Unlike the other small Hsps, the αB complex does not dissociate under these conditions (Kato et al., 1994). Various antisera are available for αA and αB (e.g. Boelens and de Jong, 1995).

■ Biological activities of α-crystallin

α-Crystallin has chaperone-like activities in that it suppresses the non-specific aggregation of unfolding proteins in vitro, as induced by heat, UV radiation, oxidative stress or reduction of disulfide bonds (Horwitz, 1992; Raman and Rao, 1994; Farahbakhsh et al., 1995; Wang and Spector, 1995). Soluble denaturing proteins bind relatively stably to α-crystallin, up to a 1:1 monomer ratio, depending on the type of protein. Bound proteins are not clustered (Farahbakhsh et al., 1995). Each monomer in the α-crystallin aggregate can probably bind a denatured substrate. The presence of α-crystallin also facilitates, under certain conditions, the refolding of chemically denatured proteins (Horwitz, 1992).

α-Crystallin undergoes structural transitions above 30°C, resulting in an increased exposure of hydrophobic surfaces and paralleled by an enhanced chaperoning capacity (Raman and Rao, 1994; Das and Surewicz, 1995). This suggests that α-crystallin prevents aggregation of denatured proteins by providing suitably exposed hydrophobic surfaces. However, a polar environment and charged residues have also been implicated in this process, as appears from EPR analyses (Farahbakhsh et al., 1995) and site-directed mutagenesis (Smulders et al., 1995).

Lens transparency requires the maintenance of short-range order of the highly concentrated structural proteins in the lens cells. The chaperoning properties of α-crystallin are probably important for preventing the aggregation of damaged or aged proteins in the lens. Targeted disruption of the αA-crystallin gene in mice indeed indicates that αA is essential for maintaining solubility of lenticular αB in vivo (Brady et al., 1997). Outside the lens, the presence of αB has been correlated with cells of high oxidative activity. It is constitutively present at relatively high levels in heart, striated muscle, kidney and brain. In these cells it may, like the other small HSPs, provide protection against stress. Indeed, increased expression of αB enhances resistance to various forms of stress (e.g. Klemenz et al., 1991; Mehlen et al., 1995). αA-crystallin also has the capacity to confer cellular thermoresistance.

■ Biological regulation of α-crystallin

The αA gene is more specialized for lens expression than is the αB gene, which is constitutively expressed in many tissues (Bhat and Nagineni, 1989). As for the αA gene, there are remarkable differences between mouse and chicken in the proposed regulatory elements and possible transcription factors. Putative heat shock elements are present in the 5′-flanking region of the mammalian αB gene, but not in the duck αB gene, nor in the αA genes. The mammalian αB gene is indeed inducible by stress, hormones and oncogene products. It is overexpressed in various disorders, notably in neurodegenerative diseases and specific types of tumors (see, for example Iwaki et al., 1992; Pinder et al., 1994). In multiple sclerosis, αB is the immunodominant myelin autoantigen (van Noort et al., 1995). The αB-crystallin gene has two principal transcription initiation sites, which are used in a tissue preferred manner. An enhancer between the two transcription initiation sites is more active in muscle than in lens. This enhancer contains at least four cis-acting regulatory elements that are differentially utilized in various tissues. For reviews see Sax and Piatigorsky (1994) and Wistow (1995).

■ Mutagenesis studies with α-crystallin

The presence of an insert sequence in the alternative splicing product αA[ins] results in decreased protective capacity, as does the mutation D69S in αA. Truncation of the C-terminus of α-crystallin, as well as various age-related modifications also diminish the chaperone activity. The relationship (direct or indirect) between these structural changes and the altered activity is not clear. There are no disease states known involving the absence or malfunction of αA or αB (for reviews see Boelens and de Jong, 1995; Smulders et al., 1995).

■ Biological interactions of α-crystallin

αB-Crystallin is associated with various cytoskeletal proteins, in vivo as well as in vitro. In degenerative diseases it is detected in cytoplasmic inclusions with neurofilaments, glial fibrillary acidic protein and cytokeratins. In such inclusions αB-crystallin can be ubiquitinated. In heart and slow muscle, αB is localized in the Z bands, where desmin filaments occur (Bennardini et al., 1992). *In vitro*, αB binds to actin and various types of intermediate filaments (e.g. Carter et al., 1995). It also shows *in vitro* binding and consequent inhibition of elastase. Of functional interest as well, is the specific interaction of α-crystallins with lens membranes (for reviews see Boelens and De Jong, 1995).

■ References

Bennardini, F., Wrzosek, A., and Chiesi, M. (1992). αB-Crystallin in cardiac tissue. Association with actin and desmin filaments. Circ. Res. **71**, 288–294.

Bhat, S.P. and Nagineni, C.N. (1989). αB Subunit of lens-specific protein α-crystallin is present in other ocular and non-ocular tissues. Biochem. Biophys. Res. Commun. **158**, 319–325.

Boelens, W.C. and de Jong, W.W. (1995). α-Crystallins, versatile stress-proteins. Mol. Biol. Rep. **21**, 75–80.

Brady, J.P., Garland, D., Duglas-Tabor, Y., Robison, W.G., Groome, A., and Wawrousek, E.F. (1997). Targeted disruption of the mouse αA-crystallin gene induces cataract and cytoplasmic inclusion bodies containing the small heat shock protein αB-crystallin. Proc. Natl. Acad. Sci. USA **94**, 884–889.

Carter, J.M., Hutcheson, A.M., and Quinlan, R.A. (1995). *In vitro* studies on the assembly properties of the lens proteins CP49, CP115: coassembly with α-crystallin but not with vimentin. Exp. Eye Res. **60**, 181–192.

Carver, J.A., Aquilina, J.A., and Truscott, R.J. (1993). An investigation into the stability of α-crystallin by NMR spectroscopy; evidence for a two-domain structure. Biochim. Biophys. Acta **1164**, 22–28.

Das, K.P. and Surewicz, W.K. (1995). Temperature-induced exposure of hydrophobic surfaces and its effect on the chaperone activity of α-crystallin. FEBS Lett. **369**, 321–325.

Farahbakhsh, Z.T., Huang, Q.L., Ding, L.L., Altenbach, C., Steinhoff, H.J., Horwitz, J., and Hubbell, W.L. (1995). Interaction of α-crystallin with spin-labeled peptides. Biochemistry **34**, 509–516.

Groenen, P.J., Merck, K.B., de Jong, W.W., and Bloemendal, H. (1994). Structure and modifications of the junior chaperone α-crystallin. Eur. J. Biochem. **225**, 1–19.

Horwitz, J. (1992). α-Crystallin can function as a molecular chaperone. Proc. Natl. Acad. Sci. USA **89**, 10449–10453.

Iwaki, T., Wisniewski, T., Iwaki, A., Corbin, E., Tomokane, N., Tateishi, J., and Goldman, J.E. (1992). Accumulation of αB-crystallin in central nervous system glia and neurons in pathologic conditions. Am. J. Pathol. **140**, 345–356.

Kantorow, M., Horwitz, J., van Boekel, M.A., de Jong, W.W., and Piatigorsky, J. (1995). Conversion from oligomers to tetramers enhances autophosphorylation by lens αA-crystallin. J. Biol. Chem. **270**, 17215–17220.

Kato, K., Goto, S., Inaguma, Y., Hasegawa, K., Morishita, R., and Asano, T. (1994). Purification and characterization of a 20-kDa protein that is highly homologous to αB crystallin. J. Biol. Chem. **269**, 15302–15309.

Klemenz, R., Frohli, E., Steiger, R.H., Schafer, R., and Aoyama, A. (1991). αB-crystallin is a small heat shock protein. Proc. Natl. Acad. Sci. USA **88**, 3652–3656.

Mehlen, P., Preville, X., Chareyron, P., Briolay, J., Klemenz, R., and Arrigo, A.P. (1995). Constitutive expression of human hsp27, *Drosophila* hsp27, or human αB-crystallin confers resistance to TNF- and oxidative stress-induced cytotoxicity in stably transfected murine L929 fibroblasts. J. Immunol. **154**, 363–374.

Pinder, S.E., Balsitis, M., Ellis, I.O., Landon, M., Mayer, R.J., and Lowe, J. (1994). The expression of αB-crystallin in epithelial tumours: a useful tumour marker? J. Pathol. **174**, 209–215.

Raman, B. and Rao, C.M. (1994). Chaperone-like activity and quaternary structure of α-crystallin. J. Biol. Chem. **269**, 27264–27268.

Sax, C.M. and Piatigorsky, J. (1994). Expression of the α-crystallin/small heat-shock protein/molecular chaperone genes in the lens and other tissues. Adv. Enz. Rel. Areas Mol. Biol. **69**, 155–201.

Smulders, R.H.P.H., Merck, K.B., Aendekerk, J., Horwitz, J., Takemoto, L., Slingsby, C., Bloemendal, H., and de Jong, W.W. (1995). The mutation Asp69-Ser affects the chaperone-like activity of αA-crystallin. Eur. J. Biochem. **232**, 834–838.

van Noort, J.M., van Sechel, A.C., Bajramovic, J.J., el Ouagmiri, M., Polman, C.H., Lassmann, H., and Ravid, R. (1995). The small heat-shock protein αB-crystallin as candidate autoantigen in multiple sclerosis. Nature **375**, 798–801.

Wang, K.Y. and Spector, A. (1995). α-Crystallin can act as a chaperone under conditions of oxidative stress. Invest. Ophthalmol. Visual Sci. **36**, 311–321.

Wistow, G.J. (1995). Molecular biology and evolution of crystallins: gene recruitment and multifunctional proteins in the eye. R.G. Landes Company, Austin, Texas, pp. 1–167.

■ *Wilfried W. de Jong and Wilbert C. Boelens:*
Department of Biochemistry
University of Nijmegen
P.O.Box 9101
6500 HB Nijmegen, The Netherlands
Tel. 31 24 3616848/3614254
Fax. 31 24 3540525
E-mail: w.dejong@bioch.kun.nl

10

Calnexin and Calreticulin

Calnexin and calreticulin proteins — an overview

Calnexin and calreticulin represent a new class of endoplasmic reticulum (ER)-localized chaperones that exhibit a marked preference for N-linked glycoproteins. The basis for this preference is that both proteins possess a lectin-binding site with specificity for the $Glc_1Man_9GlcNAc_2$ oligosaccharide structure. Although generally not essential for cell viability, calnexin (and probably calreticulin as well) enhances the efficiency of protein folding/assembly and participates in the quality control functions of the ER.

■ Calnexin/calreticulin sequence and topology

cDNA sequences have been determined for calnexin and calreticulin from a variety of species including diverse mammals (see calreticulin and calnexin entries, pp. 299 and 304), invertebrates (e.g. GenBank accession numbers L08641 for *Schistosoma mansoni* calnexin, Hawn et al., 1993; X59589 for *Caenorhabditis elegans* calreticulin, Smith, 1992; M20565 for *Onchocerca volvulus* calreticulin, Rokeach et al., 1994), and plants (e.g. *Arabidopsis thaliana* calnexin Z18242, Huang et al., 1993; *N. tabacum* calreticulin X85382, Denecke et al., 1995; *Hordeum vulgare* calreticulin L27348, Chen et al., 1994). Calreticulin does not appear to be present in yeasts, although calnexin genes have been cloned from *Schizosaccharomyces pombe* and from *Saccharomyces cerevisiae* (see entries pp. 298 and 296). No prokaryotic homologs of either protein have been reported.

The domain structures of mammalian calnexin and calreticulin are depicted in Fig. 1 (Michalak et al., 1992; Williams, 1995). Calnexin is a type I integral membrane protein whereas calreticulin is soluble. Both are restricted to the ER by virtue of localization signals present at their C-termini. The most conserved regions between calnexin and calreticulin, and within the calnexins and calreticulins from various species, are the two segments indicated by the boxes in Fig. 1. The larger segment is characterized by two repeated motifs. Motif 1, IXDP(D/E)(A/D)XKP(E/D)DWD(D/E), which occurs four times in tandem in calnexin

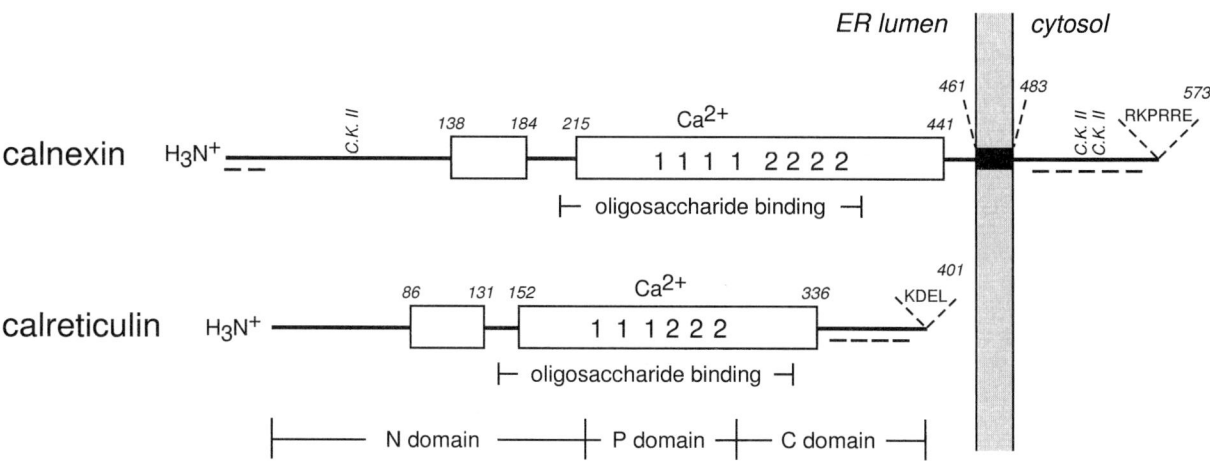

Figure 1. Domain structures of mammalian calnexin and calreticulin. Homologous regions are indicated by the open boxes, and the numbers '1' and '2' refer to repeated motifs found in all calnexins and calreticulins characterized to date. ER localization is conferred by the KDEL and RKPRRE sequences present at the C-termini of calreticulin and calnexin, respectively. Sites on calnexin that can be phosphorylated *in vitro* by casein kinase II are labelled 'C.K. II'. Segments rich in acidic amino acids are indicated by the minus signs. A high-affinity calcium-binding site as well as the oligosaccharide binding site are located in the region of the repeated motifs for both proteins. Calreticulins are typically subdivided into an N-domain that encompasses the N-terminal half of the molecule, a proline-rich P-domain that spans the next quarter of the sequence, and a C-terminal C-domain as shown.

and three times in calreticulin, is followed by motif 2, GXWXXPXIXNPXY, again repeated four times in tandem in calnexin and three times in calreticulin. The only exception to this repeat pattern is calnexin from *S. cerevisiae* in which several of the repeats, although clearly present, diverge significantly from the consensus sequences. A high affinity calcium-binding site is located in the region of the repeated motifs in both calnexin and calreticulin. Calcium binding has not been detected for the *S. cerevisiae* homolog. The lectin site in both calnexin and calreticulin has also been localized to a segment that encompasses the motif 1 and motif 2 repeats (Fig. 1, A. Vassilakos, M. Michalak, M.A. Lehrman, and D.B. Williams, manuscript submitted). The less conserved segments (indicated by the heavy lines in Fig. 1) may be highly divergent between species. For example, the stretches of negatively charged residues near the N- and C-termini of mammalian calnexins are largely absent in calnexins from *A. thaliana*, *S. mansoni*, and *S. pombe*. Likewise, the acidic segment near the C-termini of mammalian calreticulins is variable in length and, in the case of *O. volvulus* calreticulin, is replaced by basic residues. The length of the cytoplasmic domain of calnexin is also highly variable, ranging from c. 89 residues in mammals to 43 residues in *A. thaliana*, to a single residue in *S. cerevisiae*.

Related sequences

cDNAs encoding a calnexin homolog referred to as calmegin (accession number D14117, Watanabe et al., 1994) or calnexin-t (U08373, Ohsako et al., 1994) have been isolated from mouse testis. The deduced protein sequence predicts a 592 residue mature protein that is 54% identical to mouse calnexin. The highly conserved motif 1 and 2 repeats and a transmembrane segment are present. The cytoplasmic domain of the protein is c. 25 amino acids longer than that of calnexin and is less hydrophilic. Interestingly, calmegin/calnexin-t expression is restricted to specific stages of developing spermatogenic cells. mRNA and protein first appear in middle pachytene spermatocytes, last until step 15 elongated spermatids, and are absent from late spermatids and spermatozoa.

A cDNA corresponding to an alternative form of calreticulin has been isolated from bovine brain (Liu et al., 1993). The deduced amino acid sequence indicates a mature protein of 388 residues that is similar to calreticulin throughout its 318 C-terminal amino acids, but is completely divergent in its N-terminal region. Northern blot analysis of brain RNA reveals two hybridizing mRNAs, one of 3.75 kb that corresponds to the alternative form and one of 1.9 kb that corresponds to calreticulin. Two cDNAs have also been isolated from barley that encode calreticulins differing in their putative N-terminal signal sequences. The cDNAs correspond to two separate genes and elevated expression of both genes was observed following pollination and during early embryogenesis (Chen et al., 1994).

Gene regulation

Expression of calreticulin can be upregulated in response to heat, amino acid starvation, depletion of intracellular calcium, hormone stimulation, and following activation of cultured smooth muscle cells or stimulation of T lymphocytes (Booth and Koch, 1989; Opas et al., 1991; Burns et al., 1992; Hensel et al., 1994; Plakidou-Dymock and McGivan, 1994; Conway et al., 1995; see calreticulin entry, p. 304). Regulation of calnexin expression in response to various stress conditions has only been demonstrated in *S. pombe* where mRNA levels increased in response to heat, calcium depletion, β-mercaptoethanol, and 2-deoxyglucose (Jannatipour and Rokeach, 1995; Parlati et al., 1995, see entry p. 298). Little information exists concerning the regulation of expression of calnexin in multicellular organisms although one study reported that calnexin was induced to a lesser degree than calreticulin following T lymphocyte activation in rats (Clementi et al., 1994). However, calnexin expression has been shown to be developmentally regulated in *S. mansoni*. Calnexin protein increased sixfold during the transformation from cercaria to schistosomula. This stage-specific induction was accompanied by the phosphorylation of calnexin and by its unexpected expression at the surface of schistosomula (Hawn and Strand, 1994).

Chaperone function of calnexin and calreticulin

Consistent with a chaperone function for both proteins, calnexin and calreticulin have been shown to bind transiently to diverse membrane and secretory proteins during their passage through the ER (see mammalian calnexin and calreticulin entries, pp. 299 and 304). Interaction is restricted to folding or assembly intermediates and prolonged binding is observed with proteins that are unable to fold or assemble correctly. Calnexin and calreticulin associate almost exclusively with glycoproteins that possess N-linked oligosaccharides. Results from oligosaccharide analysis of bound glycoproteins, as well as *in vitro* oligosaccharide binding assays, have indicated that both proteins are lectins with selectivity for oligosaccharides carrying a single terminal glucose residue (Hebert et al., 1995; Peterson et al., 1995; Ware et al., 1995; Spiro et al., 1996). This property has been exploited to assess chaperone function by blocking the formation of monoglucosylated oligosaccharides with castanospermine and hence preventing calnexin/calreticulin binding. In these studies, the folding and subsequent assembly of several glycoproteins was impaired. Furthermore, calnexin (and probably calreticulin as well) participates in the ER retention of incompletely folded or misfolded proteins. Experiments in which normal calnexin function was lost, either through the use of a heterologous insect expression system or by removal of calnexin's ER localization signal, resulted in aberrant expression of glycoprotein assembly intermediates at distal locations in the secretory pathway (Williams, 1995).

There is some debate concerning whether calnexin and calreticulin recognize polypeptide segments of unfolded glycoproteins in addition to binding monoglucosylated oligosaccharides. Two proposed mechanisms of action involving either oligosaccharide binding exclusively or both oligosaccharide and polypeptide binding are provided in the calreticulin and mammalian calnexin entries, respectively (pp. 299 and 304; Hebert et al., 1995; Ware et al., 1995). The question also arises as to whether calnexin and calreticulin are redundant in their action or whether they possess different binding specificities. Examination of the spectrum of newly synthesized proteins that binds to each chaperone revealed substantial differences (Peterson et al., 1995; Wada et al., 1995). Different specificities are also suggested by the finding that as human class I histocompatibility molecules assemble they are passed sequentially from calnexin to calreticulin (Sadasivan et al., 1996). However, there is significant overlap in specificity since both calnexin and calreticulin have been shown to associate with influenza hemagglutinin, α-fetoprotein, transferrin, GLUT-1 glucose transporter, integrins, T cell receptor, and HIV gp160. In the case of the HIV envelope glycoprotein, gp160, a ternary complex containing both calnexin and calreticulin has been detected (Otteken and Moss, 1996). Interestingly, the pattern of proteins bound by a recombinant, membrane-anchored form of calreticulin was similar to that bound by calnexin, suggesting that the two chaperones may have essentially similar specificities but that the difference in their soluble versus membrane environments influences the proteins they encounter (Wada et al., 1995).

It is clear that the chaperone functions of calnexin and calreticulin are not essential for the survival of mammalian cells. Cells that are unable to form monoglucosylated oligosaccharides either through mutation (Kearse et al., 1994) or during prolonged treatment with castanospermine (Hammond et al., 1994; Kearse et al., 1994) remain viable even though calnexin and calreticulin binding is largely abolished. Only in S. pombe has calnexin been shown to have an essential function (Jannatipour and Rokeach, 1995; Parlati et al., 1995). Although most cells can manage without these chaperones (possibly owing to up-regulation of other ER chaperones such as BiP) (Pahl and Baeuerle, 1995), calnexin and calreticulin do play a significant role in ER quality control and in increasing the efficiency of protein folding/assembly processes. The existence of one or both of these chaperones in all eukaryotes so far examined may explain why the complex process of N-glycosylation via a preassembled, glucosylated precursor has been preserved throughout eukaryotic evolution.

■ References

Booth, C. and Koch, G.L. (1989). Perturbation of cellular calcium induces secretion of lumenal ER proteins. Cell 59, 729–737.

Burns, K., Helgason, C.D., Bleakley, R.C., and Michalak, M. (1992). Calreticulin in T lymphocytes. Identification of calreticulin in T lymphocytes and demonstration that activation of T cells correlates with increased levels of calreticulin mRNA and protein. J. Biol. Chem. **267**, 19039–19042.

Chen, F., Hayes, P.M., Mulroony, D., and Pan, A. (1994). Identification and characterization of cDNA clones encoding plant calreticulin in barley. Plant Cell **6**, 835–843.

Clementi, E., Martino, G., Grimaldi, L.M., Brambilla, E., and Meldolesi, J. (1994). Intracellular Ca^{2+} stores of T lymphocytes: changes induced by in vitro and in vivo activation. Eur. J. Immunol. **24**, 1365–1371.

Conway, E.M., Liu, L.L., Nowakowski, B., Steinermosonyi, M., Ribeiro, S.P., and Michalak, M. (1995). Heat shock-sensitive expression of calreticulin — in vitro and in vivo up-regulation. J. Biol. Chem. **270**, 17011–17016.

Denecke, J., Carlsson, L., Vidal, S., Hoglund, A.S., Ek, B., van Zeijl, M., Sinjorgo, K., and Palva, T. (1995). The tobacco homolog of mammalian calreticulin is present in protein complexes in vivo. Plant Cell **7**, 391–406.

Hammond, C., Braakman, I., and Helenius, A. (1994). Role of N-linked oligosaccharide recognition, glucose trimming, and calnexin in glycoprotein folding and quality control. Proc. Natl. Acad. Sci. USA **91**, 913–917.

Hawn, T.R. and Strand, M. (1994). Developmentally regulated localization and phosphorylation of SmIrV1, a Schistosoma mansoni antigen with similarity to calnexin. J. Biol. Chem. **269**, 20083–20089.

Hawn, T.R., Tom, T.D., and Strand, M. (1993). Molecular cloning and expression of SmIrV1, a Schistosoma mansoni antigen with similarity to calnexin, calreticulin, and OvRal1. J. Biol. Chem. **268**, 7692–7698.

Hebert, D.N., Foellmer, B., and Helenius, A. (1995). Glucose trimming and reglucosylation determine glycoprotein association with calnexin in the endoplasmic reticulum. Cell **81**, 425–433.

Hensel, G., Assmann, V., and Kern, H.F. (1994). Hormonal regulation of protein disulfide isomerase and chaperone synthesis in the rat exocrine pancreas. Eur. J. Cell Biol. **63**, 208–218.

Huang, L., Franklin, A.E., and Hoffman, N.E. (1993). Primary structure and characterization of an Arabidopsis thaliana calnexin-like protein. J. Biol. Chem. **268**, 6560–6566.

Jannatipour, M. and Rokeach, L.A. (1995). The Schizosaccharomyces pombe homologue of the chaperone calnexin is essential for viability. J. Biol. Chem. **270**, 4845–4853.

Kearse, K.P., Williams, D.B., and Singer, A. (1994). Persistence of glucose residues on core oligosaccharides prevents association of TCRa and TCRb proteins with calnexin and results specifically in accelerated degradation of nascent TCRa proteins within the endoplasmic reticulum. EMBO J. **13**, 3678–3686.

Liu, N., Fine, R.E., and Johnson, R.J. (1993). Comparison of cDNAs from bovine brain coding for two isoforms of calreticulin. Biochim. Biophys. Acta **1202**, 70–76.

Michalak, M., Milner, R.E., Burns, K., and Opas, M. (1992). Calreticulin. Biochem. J. **285**, 681–692.

Ohsako, S., Hayashi, Y., and Bunick, D. (1994). Molecular cloning and sequencing of calnexin-t: an abundant male germ cell-specific calcium-binding protein of the endoplasmic reticulum. J. Biol. Chem. **269**, 14140–14148.

Opas, M., Dziak, E., Fliegel, L., and Michalak, M. (1991). Regulation of expression and intracellular distribution of calreticulin, a major calcium binding protein of nonmuscle cells. J. Cell. Physiol. **149**, 160–171.

Otteken, A. and Moss, B. (1996). Calreticulin interacts with newly synthesized human immunodeficiency virus type 1 envelope glycoprotein, suggesting a chaperone function similar to that of calnexin. J. Biol. Chem. **271**, 97–103.

Pahl, H.L. and Baeuerle, P.A. (1995). A novel signal transduction pathway from the endoplasmic reticulum to the nucleus is mediated by transcription factor NF-kB. EMBO J. **14**, 2580–2588.

Parlati, F., Dominguez, M., Bergeron, J.J.M., and Thomas, D.Y. (1995). The calnexin homologue cnx1+ in *Schizosaccharomyces pombe* is an essential gene which can be complemented by its soluble ER domain. EMBO J. **14**, 3064–3072.

Peterson, J.R., Ora, A., Van, P.N., and Helenius, A. (1995). Transient, lectin-like association of calreticulin with folding intermediates of cellular and viral glycoproteins. Mol. Biol. Cell 6, 1173–1184.

Plakidou-Dymock, S. and McGivan, J.D. (1994). Calreticulin — a stress protein induced in the renal epithelial cell line NBL-1 by amino acid deprivation. Cell Calcium **16**, 1–8.

Rokeach, L.A., Zimmerman, P.A., and Unnasch, T.R. (1994). Epitopes of the *Onchocerca volvulus* RAL1 antigen, a member of the calreticulin family of proteins, recognized by sera from patients with onchocerciasis. Infect. Immun. **62**, 3696–3704.

Sadasivan, B., Lehner, P.J., Ortmann, B., Spies, T. and Cresswell, P. (1996). Roles for calreticulin and a novel glycoprotein, tapasin, in the interaction of MHC class I molecules with TAP. Immunity **5**, 103–114.

Smith, M.J. (1992). A *C. elegans* gene encodes a protein homologous to mammalian calreticulin. DNA Seq. **2**, 235–240.

Spiro, R.G., Zhu, Q., Bhoyroo, V., and Söling, H.-D. (1996). Definition of the lectin-like properties of the molecular chaperone, calreticulin, and demonstration of its copurification with endomannosidase from rat liver Golgi. J. Biol. Chem. **271**, 11588–11594.

Wada, I., Imai, S.-I., Kai, M., Sakane, F., and Kanoh, H. (1995). Chaperone function of calreticulin when expressed in the endoplasmic reticulum as the membrane-anchored and soluble forms. J. Biol. Chem. **270**, 20298–20304.

Ware, F., Vassilakos, A., Jackson, M., Lehrman, M.A., and Williams, D.B. (1995). The molecular chaperone calnexin binds $Glc_1Man_9GlcNAc_2$ oligosaccharide as an initial step in recognizing unfolded glycoproteins. J. Biol. Chem. **270**, 4697–4704.

Watanabe, D., Yamada, K., Nishina, Y., Tajima, Y., Koshimizu, U., Nagata, A., and Nishimune, Y. (1994). Molecular cloning of a novel Ca^{2+}-binding protein (calmegin) specifically expressed during male meiotic germ cell development. J. Biol. Chem. **269**, 7744–7749.

Williams, D.B. (1995). Calnexin: a molecular chaperone with a taste for carbohydrate. Biochem. Cell Biol. **73**, 123–132.

■ David B. Williams:
Department of Biochemistry
Medical Sciences Building
University of Toronto
Toronto, Ontario, M5S 1A8, Canada
Tel. 1 416 978 2546
Fax. 1 416 978 8548
E-mail: david.williams@utoronto.ca

Saccharomyces cerevisiae Cne1

Calnexin homologues have been found in two yeast species, Saccharomyces cerevisiae *and* Schizosaccharomyces pombe. *The* S. cerevisiae *calnexin homologue,* CNE1, *is a non-essential gene that encodes a type 1 ER integral membrane glycoprotein and functions as a constituent of the ER quality control apparatus.*

■ Alternative names

Saccharomyces cerevisiae calnexin (De Virgilio *et al.*, 1993; Parlati *et al.*, 1995).

■ *Saccharomyces cerevisiae CNE1* gene and sequence

The *CNE1* sequence (GenBank accession numbers L11012, X66470 and U12980) predicts a 502 amino acid protein, Cne1p, with a 19 amino acid N-terminal signal sequence (Parlati *et al.*, 1995). After signal peptide cleavage, the sequence predicts a protein with a molecular mass of 54 685 daltons. There are five consensus N-linked glycosylation sites. In contrast to its mammalian homologues, which encode a C-terminal transmembrane domain and a highly acidic cytosolic tail (Bergeron *et al.*, 1994), Cne1p has a C-terminal transmembrane domain and no cytosolic tail (see Fig. 1). Cne1p is 24% identical and 31% similar to canine calnexin (Wada *et al.*, 1991) and 21% identical to mouse calreticulin (Smith and Koch, 1989). Mammalian calnexins encode four KPEDWDE repeats thought to be important for calcium binding (Tjoelker *et al.*, 1994), whereas Cne1p contains only one motif between amino acids 255 and 261 (KPHDWDD).

■ Cne1 protein

Cne1p has been shown to be an integral membrane ER resident protein and migrates as a protein doublet on SDS–PAGE with an apparent molecular weight of c. 76 kDa. Upon Endo H (endoglycosidase H) digestion, the protein migrates as a single band on SDS–PAGE, with an apparent molecular weight of c. 60 kDa, indicating that all five potential N-linked glycosylation sites are used (Parlati *et al.*, 1995). Unlike its mammalian homologues, Cne1p does not appear to be a calcium binding protein. Cne1p has no known C-terminal ER retention motifs such as the HDEL or KKXX amino acid sequences.

■ *CNE1* function

CNE1 is a non-essential gene, and strains deleted for *CNE1* have no apparent phenotype (Parlati *et al.*, 1995).

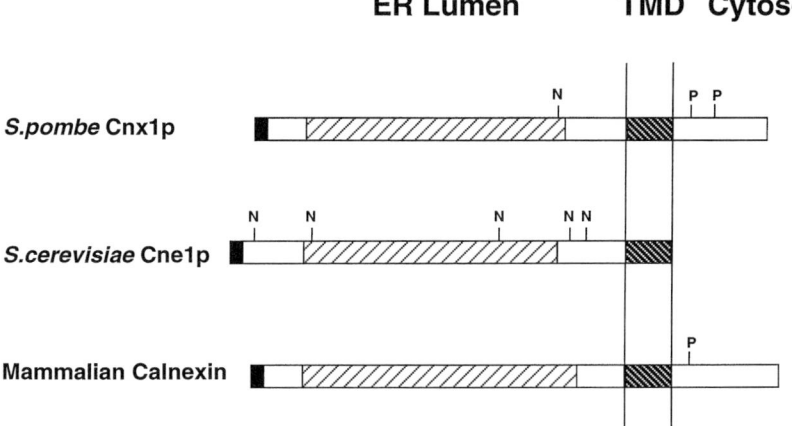

Figure 1. Topology of *S. pombe* Cnx1p, *S. cerevisiae* Cne1p and mammalian calnexin. Predicted topology of Cnx1p, Cne1p, and mammalian calnexin showing the signal sequence (■), the highly conserved central domain (▨), and transmembrane domain (▧). The predicted N-linked glycosylation sites (N) and the putative phosphorylation sites (P) on the cytoplasmic tail are shown.

However, *CNE1* is involved in the retention of mutant secretory proteins and therefore has a role in the ER quality control apparatus. Strains deleted for *CNE1* show an increase in the cell surface expression of a temperature-sensitive mutant of the α pheromone receptor Ste2-3p at non-permissive temperature (Parlati et al., 1995). There is also a 2–3-fold increase in the secretion of α_1-antitrypsin expressed in *S. cerevisiae* (Parlati et al., 1995). There seems to be no effect on the secretion or cell surface expression of non-mutant proteins such as α factor and acid phosphatase (Parlati et al., 1995).

References

Bergeron, J.J.M., Brenner, M.B., Thomas, D.Y., and Williams, D.B. (1994). Calnexin: a membrane-bound chaperone of the endoplasmic reticulum. Trends Biochem. Sci. **19**, 124–128.

De Virgilio, C., Burckert, N., Neuhaus, J.-M., Boller, T., and Wiemken, A. (1993). *CNE1*, a *Saccharomyces cerevisiae* homologue of the genes encoding mammalian calnexin and calreticulin. Yeast **9**, 185–188.

Parlati, F., Dominguez, M., Bergeron, J.J.M., and Thomas, D.Y. (1995). *Saccharomyces cerevisiae CNE1* encodes an endoplasmic reticulum (ER) membrane protein with sequence similarity to calnexin and calreticulin and functions as a constituent of the ER quality control apparatus. J. Biol. Chem. **270**, 244–253.

Smith, M.J. and Koch, G.L.E. (1989). Multiple zones in the sequence of calreticulin (CRP55, calregulin, HACBP), a major calcium binding ER/SR protein. EMBO J. **8**, 3581–3586.

Tjoelker, L.W., Seyfried, C.E., Eddy, R.L., Byers, M.G., Shows, T.B., Calderon, J., Schreiber, R.B., and Gray, P.W. (1994). Human, mouse, and rat calnexin cDNA cloning: identification of potential calcium binding motifs and gene localization to human chromosome 5. Biochemistry **33**, 3229–3236.

Wada, I., Rindress, D., Cameron, P., Ou, W.-J., Doherty, J.J., Louvard, D., Bell, A.W., Dignard, D., Thomas, D.Y., and Bergeron, J.J.M. (1991). SSRα and associated calnexin are major calcium binding proteins of the endoplasmic reticulum membrane. J. Biol. Chem. **266**, 19599–195610.

■ Francesco Parlati:
Department of Biology
Stewart Biology Building
McGill University
1205 Dr. Penfield Av.
Montreal, PQ H3A 1B1, Canada
Tel. 1 514 496 6155
Fax. 1 514 496 6213
E-mail: parlati@biotech.lan.nrc.ca

■ David Y. Thomas:
Eukaryotic Genetics Group
Biotechnology Research Institute
National Research Council
6100 Ave. Royalmount
Montreal, PQ H4P 2R2, Canada
Tel. 1 514 496 6156
Fax. 1 514 496 6213
E-mail: thomas@biotech.lan.nrc.ca

■ John J.M. Bergeron:
Department of Anatomy and Cell Biology
McGill University
3640 University Street
Montreal, PQ H3A 2B2, Canada
Tel. 1 514 398 6351
Fax. 1 514 398 5049

Schizosaccharomyces pombe Cnx1

Calnexin homologues have been found in two yeast species, Saccharomyces cerevisiae *and* Schizosaccharomyces pombe. *The* S. pombe *calnexin homologue,* cnx1+, *is an essential gene which also encodes a type 1 integral membrane calcium-binding glycoprotein. The essential* cnx1+ *function can be complemented by its ER luminal domain.*

■ Alternative names

Schizosaccharomyces pombe calnexin (Jannatipour and Rokeach, 1995; Parlati *et al.*, 1995b).

■ *Schizosaccharomyces pombe cnx1⁺* gene and sequence

The nucleotide sequence of *cnx1⁺* (GenBank accession numbers M98799 and U13389) predicts an open reading frame encoding a 560 amino acid protein (Cnx1p) with a 22 amino acid N-terminal signal sequence. After signal peptide cleavage, the sequence predicts a 538 amino acid protein with a molecular mass of 61 064 daltons. Cnx1p has one consensus N-linked glycosylation site. It is a type I integral membrane protein with a C-terminal transmembrane domain and a 49 amino acid cytosolic tail which, unlike its mammalian counterparts, is not highly acidic (Bergeron *et al.*, 1994). *S. pombe* Cnx1p is 38% identical to *Arabidopsis thaliana* calnexin (Huang *et al.*, 1993), 34% identical to canine calnexin (Wada *et al.*, 1991), 25% identical to mouse calreticulin (Smith and Koch, 1989) but only 22% identical to *S. cerevisiae* Cne1p (Parlati *et al.*, 1995a). *S. pombe* Cnx1p contains four repeats related to the motif KPEDWDE, thought to be important for calcium binding (Tjoelker *et al.*, 1994).

■ Cnx1 protein

The Cnx1 protein has been shown to be an integral membrane protein and has an abnormally low mobility on SDS–PAGE (c. 91 kDa) (Parlati *et al.*, 1995b). As with mammalian calnexin (Wada *et al.*, 1991), the low pI of Cnx1p (calculated as 4.13) is likely to be responsible for this discrepancy. After Endo H treatment, the molecular weight of Cnx1p shifts to 88 kDa, suggesting that the glycosylation site is used. Like its mammalian homologues (Bergeron *et al.*, 1994), Cnx1p is a calcium-binding protein (Parlati *et al.*, 1995b).

■ Gene regulation and function of Cnx1p

The nucleotide sequence upstream of *cnx1⁺* includes two heat shock elements consisting of three or more repeats of the sequence nGAAn in an alternating orientation, which are in the correct position to act as elements controlling *cnx1⁺* transcription. The *cnx1⁺* transcript was found to be induced approximately 1.6-fold by a transient heat shock (Parlati *et al.*, 1995b). Treatment with the glycosylation inhibitor tunicamycin or with 2-deoxyglucose had no effect, but treatment with the calcium ionophore A23187 gave a 2.6-fold increase in *cnx1⁺* transcription (Parlati *et al.*, 1995b). This suggests that Cnx1p may have a role in Ca^{2+} regulation or sequestration.

cnx1⁺ is an essential gene in *S. pombe* (Jannatipour and Rokeach., 1995; Parlati *et al.*, 1995b). Complementation of the lethal *cnx1:ura4⁺* disruption strain has shown that the C-terminal part of the molecule is not essential for viability, since amino acids 1–474 of Cnx1p could complement the lethal phenotype (Parlati *et al.*, 1995b). Deletion of the transmembrane as well as the predicted cytosolic domain led to a soluble truncated form of Cnx1p that was secreted from cells but still complemented the lethal phenotype. However, canine calnexin, mouse calreticulin and *S. cerevisiae CNE1* could not complement the lethal phenotype of a *cnx1⁺* deletion in haploid *S. pombe* cells (Parlati *et al.*, 1995b). Cnx1p also acts as a constituent of the ER quality control apparatus. When heterologously expressed in *S. pombe*, the Pl Z α_1-antitrypsin variant is efficiently secreted when the Cnx1p cytosolic tail is deleted. Otherwize, the Pl Z variant is intracellularly retained in *S. pombe* expressing full length Cnx1p (F. Parlati, unpublished results).

■ References

Bergeron, J.J.M., Brenner, M.B., Thomas, D.Y., and Williams, D.B. (1994). Calnexin: a membrane-bound chaperone of the endoplasmic reticulum. Trends Biochem. Sci. **19**, 124–128.

Huang, L., Franklin, A. E., and Hoffman, N.E. (1993). Primary structure and characterization of an *Arabidopsis thaliana* calnexin-like protein. J. Biol. Chem. **268**, 6560–6566.

Jannatipour, M. and Rokeach, L.A. (1995). The *Schizosaccharomyces pombe* homologue of the chaperone calnexin is essential for viability. J. Biol. Chem. **270**, 4845–4853.

Parlati, F., Dominguez, M.M., Bergeron, J.J.M., and Thomas, D.Y. (1995a). *Saccharomyces cerevisiae CNE1* encodes an endoplasmic reticulum (ER) membrane protein with sequence similarity to calnexin and calreticulin and functions as a constituent of the ER quality control apparatus. J. Biol. Chem. **270**, 244–253.

Parlati, F., Dignord, D., Bergeron, J.J.M., and Thomas, D.Y. (1995b). The calnexin homologue *cnx1⁺* in *Schizosaccharomyces pombe*, is an essential gene which can be complemented by its soluble ER domain. EMBO J. **14**, 3064–3072.

Smith, M.J. and Koch, G.L.E. (1989). Multiple zones in the sequence of calreticulin (CRP55, calregulin, HACBP), a major calcium binding ER/SR protein. EMBO J. **8**, 3581–3586.

Tjoelker, L.W., Seyfried, C.E., Eddy, R.L., Byers, M.G., Shows, T.B., Calderon, J., Schreiber, R.B., and Gray, P.W. (1994). Human, mouse, and rat calnexin cDNA cloning: identification of potential calcium binding motifs and gene localization to human chromosome 5. Biochemistry **33**, 3229–3236.

Wada, I., Rindress, D., Cameron, P., Ou, W.-J., Doherty, J.J., Louvard, D., Bell, A.W., Dignard, D., Thomas, D.Y., and Bergeron, J.J.M. (1991). SSRα and associated calnexin are major calcium binding proteins of the endoplasmic reticulum membrane. J. Biol. Chem. **266**, 19599–195610.

■ *Francesco Parlati*
Department of Biology
Stewart Biology Building
McGill University
1205 Dr. Penfield Av.
Montreal, PQ H3A 1B1, Canada
Tel. 1 514 496 6155
Fax. 1 514 496 6213
E-mail: parlati@biotech.lan.nrc.ca

■ *David Y. Thomas:*
Eukaryotic Genetics Group
Biotechnology Research Institute
National Research Council
6100 Ave. Royalmount
Montreal, PQ H4P 2R2, Canada
Tel. 1 514 496 6156
Fax. 1 514 496 6213
E-mail: thomas@biotech.lan.nrc.ca

■ *John J.M. Bergeron:*
Department of Anatomy and Cell Biology
McGill University
3640 University Street
Montreal, PQ H3A 2B2, Canada
Tel. 1 514 398 6351
Fax. 1 514 398 5049

Mammalian calnexin

Calnexin is a type I transmembrane protein localized to the endoplasmic reticulum (ER) of all mammalian cell types. It associates transiently with a broad range of proteins synthesized at this location and exhibits prolonged binding to proteins that are unable to fold properly or assemble into complete complexes. Calnexin facilitates the folding and subunit assembly of nascent proteins and, in many cases, regulates their export from the ER. The proteins that calnexin binds are almost exclusively N-glycosylated, reflecting the fact that calnexin utilizes a unique lectin activity specific for $Glc_1Man_9GlcNAc_2$ oligosaccharides as a component of its binding interaction.

■ Alternative names

p88 (Degen and Williams, 1991), IP90 (Hochstenbach *et al.*, 1992), or Band VII (Cala *et al.*, 1993) in early studies, but the calnexin designation has now been universally adopted.

■ Isolation of mammalian calnexin

Calnexin was identified as one of four ER proteins that are phosphorylated following incubation of canine pancreatic rough microsomes with $[\gamma^{32}P]GTP$ or $[\gamma^{32}P]ATP$ (Wada *et al.*, 1991). It was purified as a complex with three other proteins: the 35 kDa signal sequence receptor (SSR)α, 25 kDa SSRβ, and an unknown glycoprotein, gp25L. Since these proteins have not been detected as major components of calnexin complexes in other cell types, the significance of their copurification from canine pancreas remains unknown.

Calnexin was also identified on a functional basis as a protein that is cross-linked to newly synthesized major histocompatibility complex (MHC) class I heavy chains (Degen and Williams, 1991) or that coimmunoprecipitates with partially assembled forms of the T cell receptor (TCR) complex, membrane-bound immunoglobulin, and MHC class I molecules (Hochstenbach *et al.*, 1992).

■ Calnexin gene and sequence

cDNA clones have been sequenced for calnexins from dog (GenBank accession numbers X53616 and L05594; Wada *et al.*, 1991; Cala *et al.*, 1993), human (L10284, L18887, M98452; David *et al.*, 1993; Tjoelker *et al.*, 1994), mouse (L23865, L18888; Tjoelker *et al.*, 1994), and rat (L18889;

Tjoelker et al., 1994). The gene for human calnexin has been localized to the distal end of the long arm of chromosome 5, at 5q35 (Tjoelker et al., 1994). Calnexin mRNA may be alternatively spliced since two cDNAs were isolated for dog calnexin that differed only in their 5′-untranslated regions (Cala et al., 1993).

Sequence comparison reveals a high degree of identity between calnexin proteins from the above species (93–98%). The deduced amino acid sequence predicts a translation product of 591–593 amino acids with the molecular weight of the mature protein being c. 65 kDa. A type I membrane protein is predicted that can be divided into a 20 amino acid signal sequence directing the protein to the ER, a c. 461 residue ER luminal domain, a c. 22 amino acid membrane-spanning segment, and a c. 89 residue cytoplasmic domain (see Fig. 1 in the calnexin/calreticulin overview, p. 293). Clusters of acidic amino acids near the N-terminus of the protein and long acidic stretches in the cytoplasmic domain contribute to a predicted pI of c. 4.5. Basic residues at positions −3 and −5 relative to the C-terminus are involved in the localization of this protein to the ER (Rajagopalan et al., 1994). The most striking feature of the calnexin sequence is its similarity to calreticulin, the abundant 401 residue, Ca^{2+}-binding protein located in the lumen of the ER (see entry p. 304). This homology is most evident between residues 138 and 441 of calnexin and residues 86 to 336 of calreticulin. In particular, two repeated sequence motifs are highly conserved between calreticulin and the calnexins (see Fig. 1 in the overview p. 293). Mammalian calnexin does not contain any consensus sites for N-linked glycosylation, but does contain potential protein kinase C and casein kinase II phosphorylation sites.

■ Calnexin protein

Biochemical analyses have shown calnexin to be an abundant, phosphorylated, Ca^{2+}-binding protein of the ER and nuclear envelope. In many of these studies antibodies recognizing epitopes on the luminal (Hochstenbach et al., 1992; Ou et al., 1995) or cytoplasmic (Ou et al., 1993; Jackson et al., 1994) domains of calnexin have proven to be invaluable (available from StressGen, Victoria, Canada). In contrast to the 65 kDa molecular weight predicted from the cDNA sequence, calnexin migrates on SDS–PAGE as an 88–90 kDa protein, depending on the species (Degen and Williams, 1991; Wada et al., 1991; Hochstenbach et al., 1992). This anomalous behaviour is presumed to be due to the long stretches of acidic amino acids found at both the amino and carboxyl regions of the protein.

Calnexin can be phosphorylated in vitro by casein kinase II (Ou et al., 1992; Cala et al., 1993) and in vivo by growing cells in $[^{32}P]O_4$ (Capps and Zúñiga, 1994; Le et al., 1994). Ser-535 and/or Ser-545 in the cytoplasmic domain have been identified as the probable phosphorylation sites (Cala et al., 1993). In addition, Thr-74 of the intraluminal domain can be phosphorylated in vitro in the presence of detergents, and reportedly also in vivo (Cala et al., 1993).

Calnexin binds $^{45}Ca^{2+}$ and ruthenium red on overlay blots (Wada et al., 1991; Gilchrist and Pierce, 1993; Tjoelker et al., 1994) and stains blue with Stains-All (Cala et al., 1993; Gilchrist and Pierce, 1993), features that are diagnostic of Ca^{2+}-binding proteins. A subdomain analysis of calnexin has shown that calnexin residues 254–334, which contain motif 1, binds $^{45}Ca^{2+}$ with highest affinity, whereas binding to motif 2 cannot be detected (Tjoelker et al., 1994). Consistent with these findings, the P-domain of calreticulin (contains both motif 1 and 2 repeats) also contains a high affinity Ca^{2+}-binding site (K_d c. 1.6 μM) which is novel in that it does not contain an 'E–F-hand' motif. Both the acidic N- and C-terminal regions of calnexin bind $^{45}Ca^{2+}$ with moderate affinity, most likely through electrostatic interactions, although it is not known if this binding occurs in vivo. Ca^{2+} binding may be essential for calnexin function, since calcium depletion can prevent calnexin–glycoprotein complexes from forming in intact cells (Capps and Zúñiga, 1994) or can disrupt preformed complexes in vitro (Le et al., 1994). Furthermore, chelation of calcium from the purified luminal domain of calnexin results in its increased sensitivity to exogenous proteases and the formation of oligomers (Ou et al., 1995). The purified luminal domain of calnexin also binds Mg^{2+}–ATP, although no ATPase activity has been detected (Ou et al., 1995). Calnexin's lectin site has been localized to a lumenal segment (aa 204–391) which encompasses all motif 1 and motif 2 repeats and the Ca^{2+}-binding site (see Fig. 1 on p. 293), and chelation of calcium also results in loss of calnexin's oligosaccharide binding activity (A. Vassilakos, M. Michalak, M.A. Lehrman, and D.B. Williams, manuscript submitted).

Microsequencing of purified calnexin has demonstrated that the putative signal sequence is indeed cleaved (Wada et al., 1991; David et al., 1993). Furthermore, protease protection assays and in vitro phosphorylation studies using intact microsomes have confirmed the predicted topology (depicted in Fig. 1 in the overview p. 293) (Wada et al., 1991; Cala et al., 1993; Ou et al., 1995). Subcellular localization of calnexin by immunofluorescence has revealed a reticular pattern and nuclear envelope staining characteristic of the ER (Wada et al., 1991; Galvin et al., 1992; Hochstenbach et al., 1992; David et al., 1993). Immunogold electron microscopy detected intense staining of the nuclear envelope and between nuclear membranes, as well as the rough ER (Hochstenbach et al., 1992). In addition, analysis of highly purified membrane fractions has shown calnexin to be a component of the inner nuclear membrane (Gilchrist and Pierce, 1993) and the sarcoplasmic reticulum of cardiac muscle (Cala et al., 1993).

■ Involvement of calnexin in quality control and protein folding

In pulse-radiolabeled cells, upwards of 50 different newly synthesized proteins associate transiently with calnexin (Galvin et al., 1992; Hochstenbach et al., 1992; David et al., 1993; Ou et al., 1993; Kearse et al., 1994). Those proteins that have been identified represent a wide array

Table 1. Proteins that have been assessed for association with calnexin

Associate with calnexin	Do not associate with calnexin
Transmembrane proteins	
MHC class I heavy chains	CD28
Mutant class I heavy chain H-2L^{d*}	CD5
TCR partial complexes	Transferrin receptor
ϵ subunit of TCR*	
Immunoglobulin μ-heavy chain	
MHC class II α, β, I chains	
Influenza virus HA protein	
Vesicular stomatitis virus G protein	
Integrin α_6, β_1 chains	
Interferon-γ receptor	
Human immunodeficiency virus gp160	
Soluble secretory proteins	
α_1-Antitrypsin (WT and null$_{Hong\ Kong}$ variant)	Albumin*
α_1-Antichymotrypsin	
Transferrin	
Complement component C3	
apoB100	
α-Fetoprotein	
gp80 (MDCK cells)	
Thyroglobulin	
Polytopic membrane proteins	
CFTR (wt and ΔF508)	Ca^{2+}–ATPase*
P-Glycoprotein	
Mutant P-glycoprotein*	
Acetylcholine receptor α subunit	
GLUT-1 glucose transporter	

* not N-glycosylated.

of both the soluble and membrane-bound itinerant proteins of the exocytic pathway (Table 1). Initial binding of calnexin usually occurs rapidly and, in the case of influenza hemagglutinin, can occur cotranslationally (Chen et al., 1995). The $t_{1/2}$ for dissociation of calnexin is dependent on the protein it binds, varying from 2.5–5 minutes for α_1-antitrypsin, to several hours for the integrin β_1 subunit. In several instances the $t_{1/2}$ of dissociation from calnexin corresponds closely to the $t_{1/2}$ at which the protein is transported from the ER to the Golgi apparatus (Degen and Williams, 1991; Degen et al., 1992; Anderson and Cresswell, 1994; Lenter and Vestweber, 1994). Furthermore, in mutant cells that cannot form complete oligomeric complexes owing to the lack of individual subunits, the $t_{1/2}$ of dissociation is dramatically prolonged (Degen et al., 1992; David et al., 1993). This suggests that calnexin may retain subunits in an assembly competent state until they are either properly assembled or degraded. Direct demonstrations that calnexin retains proteins in the ER were provided by reconstituting calnexin function in *Drosophila melanogaster* cells (Jackson et al., 1994), by manipulating the cellular localization of calnexin by removing its C-terminal ER retention motif (Rajagopalan et al., 1994), and by preventing calnexin association with proteins using castanospermine (Hammond and Helenius, 1994). In each of these instances, interfering with calnexin function results in misfolded or unassembled protein subunits appearing inappropriately at the plasma membrane.

Consistent with a molecular chaperone function, calnexin associates preferentially with folding or assembly intermediates and its dissociation can frequently be correlated with completion of some stage of folding or assembly (Helenius, 1994; Williams, 1995). Attempts to demonstrate calnexin function more directly have generally focused on blocking calnexin binding with castanospermine (see below). Under these conditions, the formation of fully disulfide-bonded VSV-G and influenza HA proteins is dramatically impaired (Hammond and Helenius, 1994; Hebert et al., 1996), folding of MHC class I heavy chains is slowed (Tector and Salter, 1996) or impaired (Vassilakos et al., 1996), and heavy chain assembly with the β_2-microglobulin subunit is markedly reduced (Vassilakos et al., 1996). Data obtained with castanospermine must be interpreted with caution since the function of other ER proteins such as calreticulin may also be blocked and indirect effects owing to the altered oligosaccharide structure are difficult to assess. However, in the case of MHC class I biogenesis, the castanospermine results were corroborated by reconstituting calnexin function in *Drosophila melanogaster* cells and showing that calnexin facilitates heavy chain folding, reduces aggregate formation, and enhances assembly with β_2-microglobulin (Vassilakos et al., 1996). The conclusion from these studies is that calnexin associates with partially folded intermediates of nascent proteins and promotes productive folding reactions. Effects on subunit assembly are likely to be a result of improved efficiency of subunit folding. Interaction with calnexin also protects some nascent proteins from degradation in the ER (Jackson et al., 1994; Kearse et al., 1994; Hebert et al., 1996).

■ Mechanism of calnexin binding

Almost all proteins that calnexin binds are N-linked glycoproteins (Table 1) and prevention of glycosylation with tunicamycin usually blocks calnexin association (Ou et al., 1993). Calnexin binding can also be blocked by treating cells with castanospermine or deoxynojirimycin (Hammond et al., 1994). These compounds inhibit the ER processing enzymes glucosidase I and II and thereby prevent glucose removal from the $Glc_3Man_9GlcNAc_2$ oligosaccharide that is attached to nascent polypeptides. This and other data led to the suggestion that removal of the outer two glucose residues is required in order for a glycoprotein to be recognized by calnexin (Helenius, 1994; Hammond et al., 1994), a prediction borne out by the direct demonstration that calnexin functions as a lectin with specificity for the $Glc_1Man_9GlcNAc_2$ oligosaccharide (Ware et al., 1995). It has been suggested that the lectin–oligosaccharide interaction is the sole mode of calnexin binding and that calnexin functions mainly to retain glycoproteins in the ER while they fold and assemble (see model presented in the calreticulin entry, p. 304) (Helenius, 1994; Hebert et al., 1995). A lectin-only mode of binding is supported by the finding that complexes between calnexin and ribonuclease B can be dissociated following removal of oligosaccharides by digestion with endoglycosidase H (Rodan et al., 1996; Zapun et al., 1997). However, there is also substantial evidence that protein–protein interactions are important determinants of calnexin function. In contrast to the results obtained with ribonuclease B, complexes of calnexin with a variety of other soluble or membrane glycoproteins are not disrupted following complete deglycosylation with endoglycosidase H (Ware et al., 1995; Arunachalam and Cresswell, 1995; Zhang et al., 1995). In addition, several unglycosylated proteins (Table 1; Rajagoplan et al., 1994; Williams, 1995) as well as certain glycoproteins in castanospermine-treated cells (Zhang et al., 1995; van Leeuwen and Kearse, 1996) have been shown to bind to

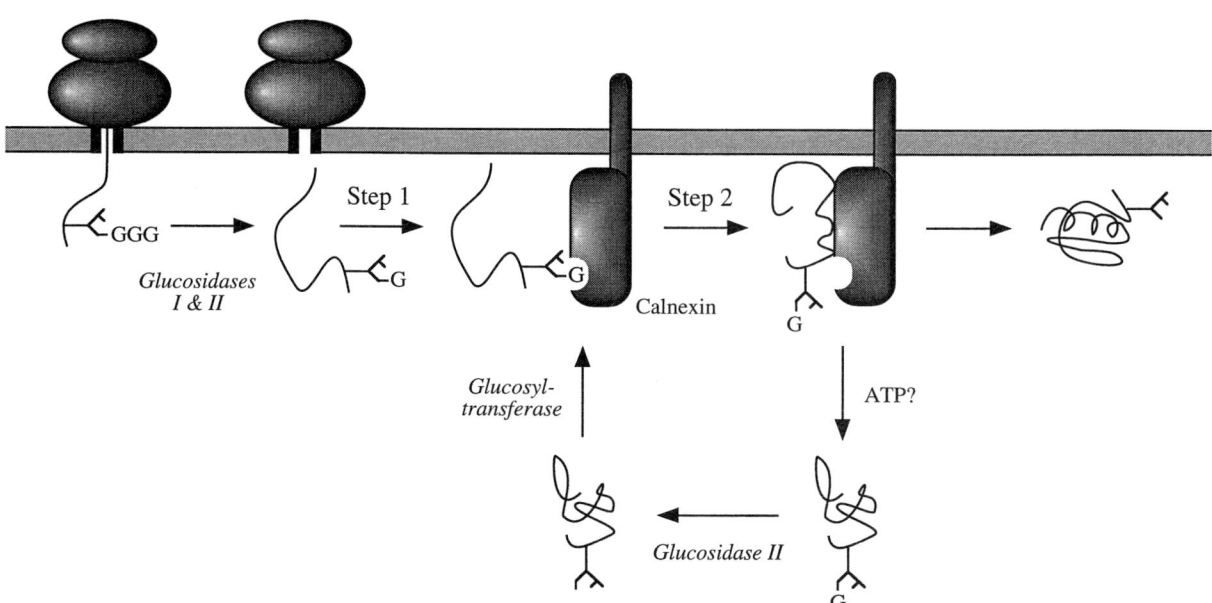

Figure 1. Two-step model for binding of calnexin to unfolded glycoproteins. Nascent polypeptide chains are glycosylated by the addition of the $Glc_5Man_9GlcNAc_2$ oligosaccharide. Removal of two glucose residues by the sequential action of glucosidases I and II reveals the monoglucosylated oligosaccharide that is recognized by calnexin (Step 1). Having been placed in proximity to calnexin by this first interaction, the unfolded polypeptide associates directly with additional sites on calnexin (Step 2). At this stage, the oligosaccharide chains are completely accessible and can be removed by digestion with endoglycosidase H. A binding cycle may take place in which the nucleotide-binding capacity of calnexin is utilized to release the glycoprotein and then re-binding occurs through the two-step process. Such a cycle would continue until it is disrupted either by burying all polypeptide sites for calnexin interaction in the folded glycoprotein or by loss of the terminal glucose residue. During the folding process, the $Glc_1Man_9GlcNAc_2$ oligosaccharide may be maintained through cycles of deglucosylation catalysed by glucosidase II and reglucosylation by another ER enzyme, UDP-glucose:glycoprotein glucosyltransferase. Since the glucosyltransferase only reglucosylates non-native glycoproteins, final release from calnexin may be facilitated by the failure of the glucosyltransferase to reglucosylate the folded glycoprotein. Adapted from Ware et al. (1995).

calnexin. Further evidence for protein–protein interactions is provided by the observations that the membrane-spanning regions of MHC class I heavy chains and calnexin are sufficiently close to be chemically cross-linked (Margolese et al., 1993) and that the associations between calnexin and glycoproteins are often detergent sensitive (Degen and Williams, 1991; Anderson and Cresswell, 1994; Le et al., 1994). A two-step model that accomodates both oligosaccharide and polypeptide modes of binding has been proposed (Fig. 1) (Ware et al., 1995). The initial step in calnexin association involves capture of glycoproteins (possibly as nascent chains) via oligosaccharide binding and then, as the unfolded glycoprotein is brought into proximity with calnexin, conventional chaperone interactions between the respective polypeptides predominate. It is not yet known whether folding occurs while the glycoprotein is associated with calnexin or during cycles of binding and release.

■ Protein folding in the absence of calnexin

In contrast to the lethal effect of calnexin disruption in *Schizosaccharomyces pombe* (see entry p. 298), calnexin is not absolutely required for viability of mammalian cells. Glucosidase I or II deficient cell lines are viable, even though in these cells $Glc_1Man_9GlcNAc_2$ is not formed and therefore glycoproteins are unable to bind to calnexin. Furthermore, a calnexin-negative human cell line has been described that shows no growth defects and exhibits no obvious impairment in assembly of MHC class I molecules (Scott and Dawson, 1995). Presumably, there is sufficient redundancy in folding factors within the ER such that in the face of chronic calnexin depletion most glycoproteins find alternative pathways for folding and assembly.

■ References

Anderson, K.S. and Cresswell, P. (1994). A role for calnexin (IP90) in the assembly of class II MHC molecules. EMBO J. **13**, 675–682.

Arunachalam, B. and Cresswell, P. (1995). Molecular requirements for the interaction of class II major histocompatibility complex molecules and invariant chain with calnexin. J. Biol. Chem. **270**, 2784–2790.

Cala, S.E., Ulbright, C., Kelley, J.S., and Jones, L.R. (1993). Purification of a 90-kDa protein (band VII) from cardiac sarcoplasmic reticulum: identification as calnexin and localization of casein kinase II phosphorylation sites. J. Biol. Chem. **268**, 2969–2975.

Capps, G.G. and Zúñiga, M.C. (1994). Class I histocompatibility molecule association with phosphorylated calnexin: implications for rates of intracellular transport. J. Biol. Chem. **269**, 11634–11639.

Chen, W., Helenius, J., Braakman, I., and Helenius, A. (1995). Cotranslational folding and calnexin binding during glycoprotein synthesis. Proc. Natl. Acad. Sci. USA **92**, 6229–6233.

David, V., Hochstenbach, F., Rajagopalan, S., and Brenner, M.B. (1993). Interaction with newly synthesized and retained proteins in the endoplasmic reticulum suggests a chaperone function for human integral membrane protein IP90 (calnexin). J. Biol. Chem. **268**, 9585–9592.

Degen, E. and Williams, D.B. (1991). Participation of a novel 88-kDa protein in the biogenesis of murine class I histocompatibility molecules. J. Cell Biol. **112**, 1099–1115.

Degen, E., Cohen-Doyle, M.F., and Williams, D.B. (1992). Efficient dissociation of the p88 chaperone from major histocompatibility complex class I molecules requires both β_2-microglobulin and peptide. J. Exp. Med. **175**, 1653–1661.

Galvin, K., Krishna, S., Ponchel, F., Frohlich, M., Cummings, D.E., Carlson, R., Wands, J. R., Isselbacher, K.J., Pillai, S., and Ozturk, M. (1992). The major histocompatibility complex class I antigen-binding protein p88 is the product of the calnexin gene. Proc. Natl. Acad. Sci. USA **89**, 8452–8456.

Gilchrist, J.S.C. and Pierce, G.N. (1993). Identification and purification of a calcium-binding protein in hepatic nuclear membranes. J. Biol. Chem. **268**, 4291–4299.

Hammond, C. and Helenius, A. (1994). Folding of VSV G protein: sequential interaction with BiP and calnexin. Science **266**, 456–458.

Hammond, C., Braakman, I., and Helenius, A. (1994). Role of N-linked oligosaccharide recognition, glucose trimming, and calnexin in glycoprotein folding and quality control. Proc. Natl. Acad. Sci. USA **91**, 913–917.

Hebert, D.N., Foellmer, B. and Helenius, A. (1995). Glucose trimming and reglucosylation determine glycoprotein association with calnexin in the endoplasmic reticulum. Cell **81**, 425–433.

Hebert, D.N., Foellmer, B. and Helenius, A. (1996). Calnexin and calreticulin promote folding, delay oligomerization and suppress degradation of influenza hemagglutinin in microsomes. EMBO J. **15**, 2961–2968.

Helenius, A. (1994). How N-linked oligosaccharides affect glycoprotein folding in the endoplasmic reticulum. Mol. Biol. Cell **5**, 253–265.

Hochstenbach, F., David, V., Watkins, S., and Brenner, M.B. (1992). Endoplasmic reticulum resident protein of 90 kilodaltons associates with the T- and B-cell antigen receptors and major histocompatibility complex antigens during their assembly. Proc. Natl. Acad. Sci. USA **89**, 4734–4738.

Jackson, M.J., Cohen-Doyle, M.F., Peterson, P., and Williams, D.B. (1994). Regulation of MHC class I transport by the molecular chaperone, calnexin (p88, IP90). Science **263**, 384–387.

Kearse, K.P., Williams, D.B., and Singer, A. (1994). Persistence of glucose residues on core oligosaccharides prevents association of TCRa and TCRb proteins with calnexin and results specifically in accelerated degradation of nascent TCRa proteins within the endoplasmic reticulum. EMBO J. **13**, 3678–3686.

Le, A., Steiner, J.L., Ferrell, G.A., Shaker, J.C., and Sifers, R.N. (1994). Association between calnexin and a secretion-incompetent variant of human α_1-antitrypsin. J. Biol. Chem. **269**, 7514–7519.

Lenter, M. and Vestweber, D. (1994). The integrin chains β_1 and α_6 associate with the chaperone calnexin prior to integrin assembly. J. Biol. Chem. **269**, 12263–12268.

Margolese, L., Waneck, G.L., Susuki, C.K., Degen, E., Flavell, R.A., and Williams, D.B. (1993). Identification of the region on the class I histocompatibility molecule that interacts with the molecular chaperone, p88 (calnexin, IP90). J. Biol. Chem. **268**, 17959–17966.

Ou, W.-J., Thomas, D.Y., Bell, A.W., and Bergeron, J.J.M. (1992). Casein kinase II phosphorylation of signal sequence receptor a and the associated membrane chaperone calnexin. J. Biol. Chem. **267**, 23789–23796.

Ou, W.-J., Cameron, P.H., Thomas, D.Y., and Bergeron, J.J.M. (1993). Association of folding intermediates of glycoproteins with calnexin during protein maturation. Nature **364**, 771–776.

Ou, W.-J., Bergeron, J.J.M., Li, Y., Kang, C.Y., and Thomas, D.Y. (1995). Conformational changes induced in the endoplasmic reticulum luminal domain of calnexin by Mg-ATP and Ca^{2+}. J. Biol. Chem. **270**, 18051–18059.

Rajagopalan, S., Xu, Y., and Brenner, M.B. (1994). Retention of unassembled components of integral membrane proteins by calnexin. Science **263**, 387–390.

Rodan, A.R., Simons, J.F., Trombetta, E.S. and Helenius, A. (1996). N-linked oligosaccharides are necessary and sufficient for RNase B binding to calnexin and calreticulin, EMBO J. **15**, 6921–6930.

Scott, J. E. and Dawson, J.R. (1995). MHC class I expression and transport in a calnexin-deficient cell line. J. Immunol. **155**, 143–148.

Tector, M. and Salter, R.D. (1995). Calnexin influences folding of human class I histocompatibility proteins but not their assembly with β_2-microglobulin. J. Biol. Chem. **270**, 19638–19642.

Tjoelker, L.W., Seyfried, C.E., Eddy, jun., R.L., Byers, M.G., Shows, T.B., Calderon, J., Schreiber, R.B., and Gray, P.W. (1994). Human, mouse, and rat calnexin cDNA cloning: identification of potential calcium binding motifs and gene localization to human chromosome 5. Biochemistry **33**, 3229–3236.

van Leeuwen, J.E.M. and Kearse, K.P. (1996). Calnexin associates exclusively with individual CD3d and T cell antigen receptor (TCR) a proteins containing incompletely trimmed glycans that are not assembled into multisubunit TCR complexes. J. Biol. Chem. **271**, 9660–9665.

Vassilakos, A., Cohen-Doyle, M.F., Peterson, P.A., Jackson, M.R., and Williams, D.B. (1996). The molecular chaperone calnexin facilitates folding and assembly of class I histocompatibility molecules. EMBO J. **15**, 1495–1506.

Wada, I., Rindress, D., Cameron, P.H., Ou, W.-J., Doherty II, J.J., Louvard, D., Bell, A.W., Dignard, D., Thomas, D.Y., and Bergeron, J.J.M. (1991). SSRa and associated calnexin are major calcium binding proteins of the endoplasmic reticulum membrane. J. Biol. Chem. **266**, 19559–19610.

Ware, F., Vassilakos, A., Jackson, M., Lehrman, M.A., and Williams, D.B. (1995). The molecular chaperone calnexin binds $Glc_1Man_9GlcNAc_2$ oligosaccharide as an initial step in recognizing unfolded glycoproteins. J. Biol. Chem. **270**, 4697–4704.

Williams, D.B. (1995). Calnexin: A molecular chaperone with a taste for carbohydrate. Biochem. Cell Biol. **73**, 123–132.

Zapun, A., Petrescu, S.M., Rudd, P.M., Dwek, R.A., Thomas, D.Y. and Bergeron, J.J.M. (1997). Conformation independent binding of monoglucosylated ribonuclease B to calnexin. Cell **88**, 29–38.

Zhang, Q., Tector, M., and Salter, R.D. (1995). Calnexin recognizes carbohydrate and protein determinants of class I major histocompatibility complex molecules. J. Biol. Chem. **270**, 3944–3948.

■ *Steven Pind and David B. Williams:*
Department of Biochemistry
Medical Sciences Building
University of Toronto
Toronto
Ontario, M5S 1A8, Canada
Tel. 1 416 978 2546
Fax. 1 416 978 8548
E-mail: david.williams@utoronto.ca
   spind@cc.umanitoba.ca

Mammalian calreticulin

Although originally characterized as a calcium-sequestering protein, increasing evidence indicates that calreticulin also functions as a lectin-like molecular chaperone for newly synthesized glycoproteins in the lumen of the endoplasmic reticulum (ER). Like calnexin, it specifically binds monoglucosylated oligosaccharides which are present as glycan maturation intermediates on proteins of non-native structure. Nevertheless, the population of substrates that bind calreticulin is only partially overlapping with the population of calnexin substrates.

■ Alternative names

CaBP3 (Van et al., 1989), CAB-63 (Waisman et al., 1985), CRP55 (Booth and Koch, 1989).

■ Calreticulin sequence

The calreticulin cDNA sequence (GenBank accession numbers M84739, human; X53363, rat; L13462, bovine) predicts a soluble protein of c. 400 amino acids, with an N-terminal signal sequence and C-terminal tetrapeptide KDEL ER retrieval signal (Michalak et al., 1992). Sequence analysis suggests that the protein can be divided into three domains: the globular N-terminus, rich in beta structure; the middle, proline-rich segment, containing two repeated motifs found also in calnexin; and the C-terminal acidic domain, shown to be important for retention of the protein in the ER (Sönnichsen et al., 1994). In some species a consensus N-linked glycosylation site is utilized (Michalak et al., 1992).

Calreticulin has high sequence similarity, clustered in five homologous domains, to the transmembrane chaperone calnexin (David et al., 1993). Two of the homologous domains are present in multiple copies in both calreticulin and calnexin (see Fig. 1 of calnexin/calreticulin overview, p. 293). Motif 1 is repeated three times in calreticulin followed by three copies of motif 2. In calnexin, four copies of motif 1 are followed by fours copies of motif 2. The significance of this matched pairing is unknown.

Calreticulin protein

Calreticulin has now been isolated from a number of species and tissues since its original discovery in 1974 (reviewed in Michalak et al., 1992). Corresponding DNA sequences are available from an even wider spectrum, including plant (Denecke et al., 1995) and invertebrate (Smith, 1992), as well as mammalian species (Michalak et al., 1992), testifying to the early evolution and strong conservation of this protein.

A highly acidic protein (pI 4.6), calreticulin migrates in SDS–PAGE as an c. 60 kDa species, although its calculated molecular weight is 46 kDa. A frictional coefficient (f/f_0) of 1.69, sedimentation constant of 3.15 and a Stokes radius of 44.2 Å are consistent with an asymmetrically shaped monomer (Waisman et al., 1985). Subcellular localization in many systems has shown calreticulin to be a soluble protein of the ER lumen, consistent with its hydrophobic signal sequence and C-terminal KDEL motif (Michalak, et al., 1992; Sönnichsen, et al., 1994).

Calreticulin binds one mole of calcium/mole of protein with high (μM) affinity and exhibits >20 low (mM) affinity calcium-binding sites, the significance of which, with respect to chaperone function, is unknown (Michalak et al., 1992).

Chaperone function of calreticulin

The transient interaction of numerous newly synthesized cellular and viral glycoproteins with calreticulin in living cells has been described (Nauseef et al., 1995; Otteken and Moss, 1996; Peterson, et al., 1995; Wada, et al., 1995; Oliver, et al., 1996; Sadasivan, et al., 1996; Van Leeuwen and Kearse, 1996a; Van Leeuwen and Kearse, 1996b). The kinetics of these interactions correlate well with early steps of folding and maturation of the substrates. Consistent with a chaperone function, an inefficiently folding mutant of the influenza hemagglutinin (HA) glycoprotein remained permanently bound to calreticulin (Peterson et al., 1995).

The association of all substrates with calreticulin depends on both ER glucosidases I and II trimming the two outermost glucose residues from core N-linked oligosaccharides present on the substrate (Peterson et al., 1995; Wada, et al., 1995). Calreticulin-bound HA was shown to have monoglucosylated glycans of the form $Glc_1Man_{7-9}GlcNAc_2$, prompting the speculation that calreticulin, like calnexin (Ware et al., 1995), is a lectin with specificity for oligosaccharides of this structure which arise as trimming intermediates during the initial steps of glycan processing (Peterson et al., 1995). Using a calreticulin–Sepharose column, Spiro et al. (1996) subsequently demonstrated a direct binding to monoglucosylated oligosaccharides

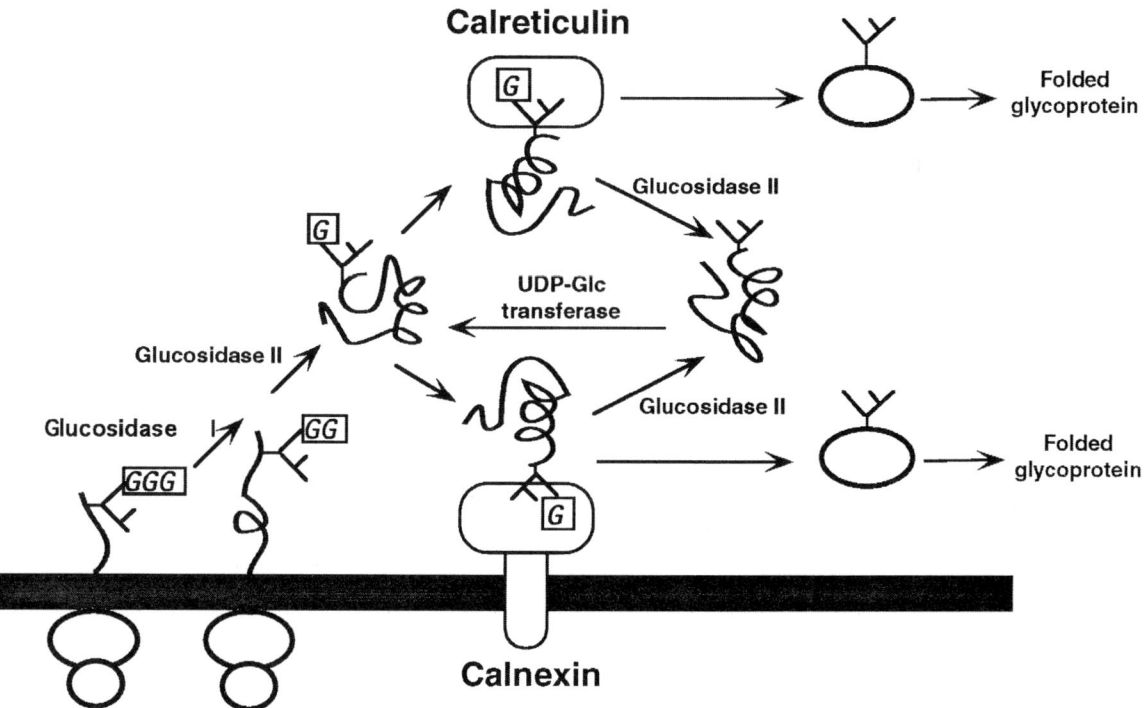

Figure 1. A model for the association and dissociation of substrates from calreticulin/calnexin, emphasizing the lectin-like nature of the binding (Helenius 1994; Hebert et al., 1995; Peterson et al., 1995). [Reproduced from *Molecular Biology of the Cell*, Vol. 6, pp. 1173–1184 (1995) with permission from the American Society for Cell Biology.]

regardless of whether the glycan was associated with a polypeptide. Calreticulin exhibited a high specificity for glycoforms containing a single glucose residue, with Glc_1Man_5 required as the minimal structure. Increasing the mannose number from Glc_1Man_5 to Glc_1Man_9 resulted in increased binding to calreticulin. For the substrate RNase, this lectin-like binding is both necessary and sufficient for calreticulin association (Rodan, et al., 1996). Nevertheless, calreticulin and calnexin exhibit some differences in both their cellular and viral substrate specificity, suggesting that, owing to steric considerations, the soluble calreticulin may have access to different glycans than the transmembrane calnexin (Otteken and Moss, 1996; Peterson et al., 1995; Van Leeuwen and Kearse, 1996b).

The following model (Fig. 1) has been proposed for the association and dissociation of substrates from calreticulin/calnexin, emphasizing the lectin-like nature of the binding (Helenius 1994; Hebert et al., 1995; Peterson et al., 1995). Nascent glycoproteins are glycosylated with the core $Glc_3Man_9GlcNAc_2$ oligosaccharide. The sequential action of glucosidases I and II trims this structure to the monoglucosylated state that is a substrate for the chaperones. Glucosidase II cleaves the final glucose residue, resulting in a $Man_9GlcNAc_2$ structure which is released from calreticulin/calnexin. UDP-glucose:glycoprotein glucosyltransferase catalyses the reglucosylation of glycoproteins having non-native structures, reforming the $Glc_1Man_{7-9}GlcNAc_2$ glycan and driving the substrate back onto the chaperones. Repetitive cycles of glycan processing and chaperone binding and release accompany substrate folding until the glycoprotein reaches its native conformation and is released for the last time by the action of glucosidase II.

The precise role of calreticulin, like calnexin, awaits further characterization. Minimally, it may play a role in the retention of misfolded or incompletely folded glycoproteins in the ER, mediated by its C-terminus. Inhibitors of the ER glucosidases, however, have been shown to perturb the maturation of several glycoproteins, suggesting that calreticulin and calnexin may play a more active role in folding substrates (Hammond and Helenius, 1994; Helenius, 1994). Furthermore, if the dissociation of HA from calreticulin/calnexin is prevented, HA folding is inhibited (Hebert et al., 1995). This suggests that cycles of binding and release are required for substrate folding, much like those proposed for HSP70- and chaperonin-assisted folding. Sadisivan et al. (1996) have proposed that the interaction of calreticulin with the MHC class I-β_2-microglobulin heterodimer iis important in presenting the complex to the TAP peptide transporter for peptide loading, implying an additional role for calreticulin in the assembly of heteroligomeric complexes. A postulated role for ATP and calcium for the function of calreticulin and calnexin (Nigam et al., 1994) remains to be shown. Calreticulin–substrate complexes, however, were not affected by the addition of ATP or calcium chelators (Wada et al., 1995; J.R. Peterson and A. Helenius, unpublished data).

Transcriptional/translational regulation of calreticulin

Cloning of a human genomic sequence revealed upstream regulatory elements shared with GRP78 (BiP), GRP94 and protein disulfide isomerase (PDI), postulated to be involved in the stress-related up-regulation of ER chaperones and foldases (McCauliffe et al., 1992). Indeed, coordinate up-regulation of calreticulin and PDI (but not BiP) at the protein level was observed in activated primary cultures of smooth muscle cells (Patton et al., 1995). Furthermore, calreticulin has been shown to be up-regulated at the transcriptional level in response to heat shock (Conway et al., 1995) and under hormonal stimulation (Hensel et al., 1994) both in cultured cells and in vivo. Amino acid starvation, as well as depletion of intracellular calcium, has also been demonstrated to result in increased levels of calreticulin protein (Booth and Koch, 1989; Plakidou-Dymock and McGivan, 1994).

References

Booth, C. and Koch, G.L. (1989). Perturbation of cellular calcium induces secretion of lumenal ER proteins. Cell **59**, 729–737.

Conway, E.M., Liu, L.L., Nowakowski, B., Steinermosonyi, M., Ribeiro, S.P., and Michalak, M. (1995). Heat shock-sensitive expression of calreticulin — in vitro and in vivo up-regulation. J. Biol. Chem. **270**, 17011–17016.

David, V., Hochstenbach, F., Rajagopalan, S., and Brenner, M.B. (1993). Interaction with newly synthesized and retained proteins in the endoplasmic reticulum suggests a chaperone function for human integral membrane protein IP90 (calnexin). J. Biol. Chem. **268**, 9585–9592.

Denecke, J., Carlsson, L.E., Vidal, S., Hoglund, A.S., Ek, B., van, Z.M., Sinjorgo, K.M., and Palva, E.T. (1995). The tobacco homologue of mammalian calreticulin is present in protein complexes in vivo. Plant Cell **7**, 391–406.

Hammond, C. and Helenius, A. (1994). Folding of VSV G protein: sequential interaction with BiP and calnexin. Science **266**, 456–458.

Hebert, D.N., Foellmer, B., and Helenius, A. (1995). Glucose trimming and reglucosylation determine glycoprotein association with calnexin in the endoplasmic reticulum. Cell **81**, 425–433.

Hebert, D.N., Foellmer, B., and Helenius, A. (1996). Calnexin and calreticulin promote folding, delay oligomerization, and suppress degradation of influenza hemagglutinin in microsomes. EMBO J. **15**, 2961–2968.

Helenius, A. (1994). How N-linked oligosaccharides affect glycoprotein folding in the endoplasmic reticulum. Mol. Biol. Cell **5**, 253–265.

Hensel, G., Assmann, V., and Kern, H.F. (1994). Hormonal regulation of protein disulfide isomerase and chaperone synthesis in the rat exocrine pancreas. Eur. J. Cell Biol. **63**, 208–218.

McCauliffe, D.P., Yang, Y.S., Wilson, J., Sontheimer, R.D., and Capra, J.D. (1992). The 5'-flanking region of the human calreticulin gene shares homology with the human GRP78, GRP94, and protein disulfide isomerase promoters. J. Biol. Chem. **267**, 2557–2562.

Michalak, M., Milner, R.E., Burns, K., and Opas, M. (1992). Calreticulin. Biochem. J. **285**, 681–692.

Nauseef, W.M., McCormick, S.J., and Clark, R.A. (1995). Calreticulin functions as a molecular chaperone in the biosynthesis of myeloperoxidase. J. Biol. Chem. **270**, 4741–4747.

Nigam, S.K., Goldberg, A.L., Ho, S., Rohde, M.F., Bush, K.T., and Sherman, M. (1994). A set of endoplasmic reticulum proteins possessing properties of molecular chaperones includes Ca(2+)-binding proteins and members of the thioredoxin superfamily. J. Biol. Chem. **269**, 1744–1749.

Oliver, J.D., Hresko, R.C., Mueckler, M., and High, S. (1996). The Glut 1 glucose transporter interacts with calnexin and calreticulin. J. Biol. Chem. **271**, 13691–13696.

Otteken, A. and Moss, B. (1996). Calreticulin interacts with newly synthesized human immunodeficiency virus type 1 envelope glycoprotein, suggesting a chaperone function similar to that of calnexin. J. Biol. Chem. **271**, 97–103.

Patton, W.F., Erdjument-Bromage, H., Marks, A.R., Tempst, P., and Taubman, M. B. (1995). Components of the protein synthesis and folding machinery are induced in vascular smooth muscle cells by hypertrophic and hyperplastic agents. J. Biol. Chem. **270**, 21404–21410.

Peterson, J.R., Ora, A., Van, P.N., and Helenius, A. (1995). Transient, lectin-like association of calreticulin with folding intermediates of cellular and viral glycoproteins. Mol. Biol. Cell **6**, 1173–1184.

Plakidou-Dymock, S. and McGivan, J.D. (1994). Calreticulin — a stress protein induced in the renal epithelial cell line NBL-1 by amino acid deprivation. Cell Calcium **16**, 1–8.

Rodan, A.R., Simons, J.F., Trombetta, E.S., and Helenius, A. (1996). N-linked oligosaccharides are necessary and sufficient for association of glycosylated forms of bovine RNase with calnexin and calreticulin. EMBO J. **15**, 6921–6930.

Sadasivan, B., Lehner, P.J., Ortmann, B., Spies, T., and Cresswell, P. (1996). Roles for calreticulin and a novel glycoprotein, tapasin, in the interaction of MHC Class I molecules with TAP. Immunity **5**, 103–114.

Smith, M.J. (1992). A *C. elegans* gene encodes a protein homologous to mammalian calreticulin. DNA Seq. **2**, 235–240.

Sönnichsen, B., Fullekrug, J., Nguyen, V.P., Diekmann, W., Robinson, D.G., and Mieskes, G. (1994). Retention and retrieval: both mechanisms cooperate to maintain calreticulin in the endoplasmic reticulum. J. Cell Sci. **107**, 2705–2717.

Spiro, R.G., Zhu, Q., Bhoyroo, V., Soling, H.D. (1996). Definition of the lectin-like properties of the molecular chaperone, calreticulin, and demonstration of its copurification with endomannosidase from rat liver Golgi. J. Biol. Chem. **271**, 11588–11594.

Van, P.N., Peter, F., and Soling, H.D. (1989). Four intracisternal calcium-binding glycoproteins from rat liver microsomes with high affinity for calcium. No indication for calsequestrin-like proteins in inositol 1,4,5-trisphosphate-sensitive calcium sequestering rat liver vesicles. J. Biol. Chem. **264**, 17494–17501.

Van Leeuwen, J.E.M., and Kearse, K.P. (1996a). The related molecular chaperones calnexin and calreticulin differentially associate with nascent T cell antigen receptor proteins within the endoplasmic reticulum. J. Biol. Chem. **271**, 25345–25349.

Van Leeuwen, J.E.M., and Kearse, K.P. (1996b). Deglucosylation of N-linked glycans is an important step in the dissociation of calreticulin-class I-TAP complexes. Proc. Natl Acad. Sci. USA **93**, 13997–14001.

Wada, I., Imai, S., Kai, M., Sakane, F., and Kanoh, H. (1995). Chaperone function of calreticulin when expressed in the endoplasmic reticulum as a the membrane-anchored and soluble forms. J. Biol. Chem. **270**, 20298–20304.

Waisman, D.M., Salimath, B.P., and Anderson, M.J. (1985). Isolation and characterization of CAB-63, a novel calcium-binding protein. J. Biol. Chem. **260**, 1652–1660.

Ware, F., Vassilakos, A., Jackson, M., Lehrman, M.A., and Williams, D.B. (1995). The molecular chaperone calnexin binds $Glc_1Man_9GlcNAc_2$ oligosaccharide as an initial step in recognizing unfolded glycoproteins. J. Biol. Chem. **270**, 4697–4704.

■ *Jeffrey R. Peterson and Ari Helenius:*
Department of Cell Biology
Yale University School of Medicine
New Haven, CT 06520, USA
Tel. 1 203 785 4313
Fax. 1 203 785 7226
E-mail: JPETERSO@biomed.med.yale.edu and ARI_HELENIUS@qm.yale.edu

11

PDI and Thioredoxin-Related Proteins

PDI and thioredoxin-related proteins — an overview

The formation of native disulfide bonds is often a vital step in the folding of secreted proteins and in the stabilization of their native structures. In contrast, disulfide reduction appears important for maintaining the structure and controlling the activity of cytosolic proteins. The thickness of this book proves Anfinsen's 1973 proposal, that many substances may assist in the protein folding process and in the formation of disulfide bonds (Anfinsen, 1973). Most if not all of the proteins that catalyse protein thiol–disulfide exchange belong to the thioredoxin super family. They all carry out variations of the reaction shown in Fig. 1. Thiol–disulfide exchange is one of the few covalent modifications that occur during protein folding. The study of a reaction that is well defined in chemical terms provides a number of advantages, and as a result the reaction mechanisms of a number of these catalysts are very well known (Gilbert, 1990; Creighton et al., 1995; Freedman, 1995; Wynn et al., 1995). The reaction starts with a nucleophilic attack by a reactive thiol group which results in the formation of an unstable mixed disulfide. This is followed by a second nucleophilic attack which results in the transfer of the disulfide between the proteins.

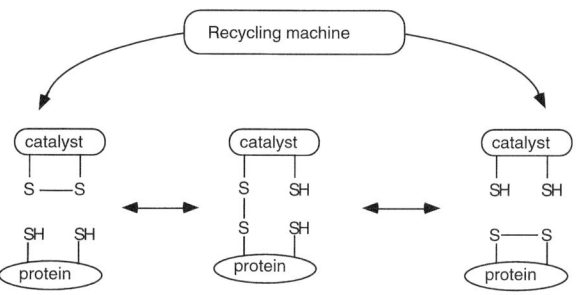

Figure 1. Catalysed disulfide exchange. The reaction proceeds through an unstable mixed disulfide intermediate. The regeneration of the catalyst is accomplished by a disulfide exchange reaction with a recycling machine. If the substrate protein has more than two SH groups, disulfide isomerization can occur by a rapid equilibrium between the mixed disulfide intermediate and the reduced catalyst.

■ Functional diversity of thioredoxin-related proteins

Strongly oxidizing catalysts such as DsbA from *Escherichia coli* tend to act as disulfide donors (Wunderlich and Glockshuber, 1993; Bardwell, 1994), while more reducing enzymes such as thioredoxin tend to remove disulfides from their substrates (Gleason and Holmgren, 1988). Protein disulfide isomerase, an endoplasmic reticulum (ER) protein of intermediate redox potential can catalyse the formation, isomerization and breakage of disulfides in a protein folding reaction (Freedman, 1995). The large number of different types of thioredoxin-like proteins found within both eukaryotes and prokaryotes is probably not just a reflection of different oxidation and isomerization requirements within specific cellular compartments, but of different oxidation and isomerization needs of individual proteins within each compartment.

In the periplasm of *E. coli* there exist at least five proteins involved in disulfide exchange. Available genetic and biochemical evidence suggests that DsbA is a rather strong oxidant, with wide substrate specificity, functioning as the immediate disulfide donor for periplasmic proteins (Bardwell, 1994). DsbB, an inner membrane protein, appears to reoxidize DsbA specifically. The formation of correct disulfide bonds in proteins with multiple cysteines requires protein disulfide isomerase activity. *In vitro* tests show that the periplasmic protein DsbC has more isomerase activity than DsbA (Zapun et al., 1995). DsbC appears to act synergistically with DsbA to catalyse correct disulfide bond formation in specific proteins that contain multiple cysteines. DsbD, an inner membrane protein, and DsbE, a periplasmic protein, appear to act in opposition to DsbA to reduce cytochromes and other periplasmically located proteins (Missiakas et al., 1995; Missiakas and Raina, 1997). Thioredoxin-like proteins that are involved in the biogenesis of cytochromes have also been found in *Bradyrhizobium japonicum* (TlpA and TlpB) and in *Rhodobacter capsulatus* (HelX). The heme moiety of c-type cytochromes and the copper center of cytochromes aa_3 are ligated to cysteine residues, and it seems likely that at least one of the functions of these proteins may be to keep these cysteines in a reduced state as a prerequisite for heme attachment (Loferer and Hennecke, 1994). Different redox proteins having very different redox potentials and recognizing different substrates can potentially function independently within the same cellular compartment as long as they do not treat each other as substrates. This appears to be true for TlpA and DsbA (Loferer et al., 1995).

The presence of multiple thiol disulfide oxidoreductases within the same cellular compartment appears to be more the rule than the exception. At least four PDI-like proteins are present in the ER of eukaryotes including conventional PDI, a minor PDI isoform, ERp61, and ERp72 (Freedman et al., 1994). Two major forms of thioredoxin exist in spinach chloroplasts: thioredoxin f, which activates fructose-1,6-bisphosphatase, and thioredoxin m, which activates NADP-dependent malate dehydrogenase (Gleason and Holmgren, 1988). Up to five cytoplasmic h-type thioredoxins are found in *Arabidopsis* (Rivera-Madrid et al., 1995). In the cytoplasm of *E. coli* there exist at least three glutaredoxins (Åslund et al., 1994). These multiple, closely related proteins within each cellular compartment may differ in substrate range, redox potential, or in expression pattern.

■ Structural similarity of thioredoxin-related proteins

High resolution structures of thioredoxin, glutaredoxin, DsbA, and the thioredoxin domain of PDI show considerable similarity to each other (Kemmink et al., 1995; Martin, 1995), and at least one domain that shows recognizable amino acid homology to the thioredoxin fold is present in nearly all of the other proteins discussed in this section. Some small proteins, like the 12 kDa thioredoxin and the 8 kDa glutaredoxin, contain just the thioredoxin fold. Others, like PDI and ERp72, contain several thioredoxin-like domains (Freedman et al., 1994). Most contain additional inferred domains, unrelated to thioredoxin, that are added on to or inserted within the thioredoxin-like domain. These additional domains may be involved in substrate interaction and membrane localization, but often have an unknown function (Freedman, 1994). They inevitably increase the size and complexity of the protein. The thioredoxin fold is minimally composed of a four-stranded β sheet flanked by three α helices (Martin, 1995). The presence of this common fold increases the probability that results obtained from studies on one protein within this family will be, if not generally applicable, at least illuminating as to the function of the other members.

■ Active site of thioredoxin-like proteins

The active site of thioredoxin-like proteins consists of a pair of cysteines in the motif CXYC that can be oxidized to form a disulfide bond. The first cysteine in the active site of the catalysts lies near the end of an α helix and is solvent exposed (Martin, 1995). It serves as the reactive nucleophile and is essential for oxidoreductase activity. In PDI, DsbA, and glutaredoxin this reactivity is enhanced by a low pK_a value of this cysteine (Nelson and Creighton, 1994; Darby and Creighton, 1995; Grauschopf et al., 1995; Kemmink et al., 1995). The redox properties of the various members of the thioredoxin family can be explained by electrostatic interactions that affect the pK_a and reactivity of the active site cysteines (Gane et al., 1995; Grauschopf et al., 1995; Takahashi and Creighton, 1996). The second cysteine is buried and reacts only with the first cysteine to form the intramolecular disulfide bond between them (Kallis and Holmgren, 1980; Wunderlich et al., 1995).

■ Redox recycling of thioredoxin-related proteins

The catalytic efficiency of these enzymes is shown by their ability to carry out thiol disulfide exchange reactions two to five orders of magnitude more rapidly than small molecule redox buffers (Gleason and Holmgren, 1988; Weissman and Kim, 1993; Creighton et al., 1995). To function as catalysts all of these enzymes need to be recycled. Their reoxidization or reduction is performed by additional components appropriate to the catalyst and its location. *In vivo*, protein disulfide isomerase may be recycled by the small molecule disulfide glutathione, while thioredoxins use thioredoxin reductase as a hydrogen donor, glutaredoxin uses glutathione, and DsbA is reoxidized by an inner membrane protein called DsbB (Hwang et al., 1992; Holmgren, 1989; Guilhot et al., 1995). Owing to their common mechanism of action, many thioredoxin-like molecules can be recycled *in vitro* by small molecule thiol compounds and proteins that they may not see *in vivo*. By adjusting the *in vitro* redox conditions, one can generally drive disulfide oxidoreductases to perform a wide spectrum of reduction, isomerization, and oxidation reactions, which may or may not have relevance *in vivo*. To draw firm conclusions about their *in vivo* function requires a more extensive knowledge of the redox conditions present within the various cellular compartments in the cell, and how mutations in the thioredoxin-like proteins alter these conditions, in addition to biochemical characterization of the purified proteins.

■ Other functions of thioredoxin-related proteins

Thioredoxin, glutaredoxin, and PDI all have a number of functions apart from their role in protein folding. Thioredoxin uses its redox activity for a wide variety of regulatory functions, and in serving as the hydrogen donor for ribonucleotide and sulfate reduction. It also has functions that are not dependent on a redox cycle in bacteriophage replication and assembly (Gleason and Holmgren, 1988). Glutaredoxin is involved in reduction of many small molecule disulfides and can substitute for thioredoxin in ribonucleotide and sulfate reduction (Gleason and Holmgren, 1988). These alternative functions of thioredoxin and glutaredoxin may, in fact, be more important *in vivo* than their roles in protein folding and stability. PDI has been demonstrated to serve as subunits of prolyl-4-hydroxylase and the triglyceride transfer protein. It apparently functions to keep its partner protein in an active and soluble state (Freedman et al.,

1994). PDI is also reported to have chaperone activity (Puig et al., 1994). These additional functions of PDI may be more of a reflection of the peptide-binding activity of PDI than of its oxidoreductase activity. Although PDI has also been proposed to have a wide variety of additional roles, many of these are doubtful and most still require verification (Freedman et al., 1994).

■ Remaining questions

Although much has been accomplished in the past few years in our understanding of these proteins, much remains to be learned.

1. Why are so many different thioredoxin-like proteins found within each cellular compartment?
2. If this is because of differing substrate specificities how are these specificities determined?
3. How do the disulfide catalysts interact with each other, with other folding catalysts, and with the chaperone machinery?
4. At exactly what point(s) on the folding pathway do the thiol disulfide catalysts interact with proteins?
5. Do they have roles in protein folding that do not involve their redox activity?
6. Are disulfide-linked folding pathways the same *in vivo* as they are *in vitro*?
7. What is the physical basis for the differences in redox properties of the disulfide bonds of the members of the thioredoxin family?
8. The equilibrium redox behavior of some model disulfide catalysts are beginning to be understood in detailed structural and biophysical terms, but why are these proteins capable of such rapid disulfide exchange kinetics?
9. What makes a protein efficient in disulfide isomerization reactions as opposed to the more straight-forward oxidation or reduction reactions?
10. What molecules serve as electron acceptors and donors for all of the newly discovered thioredoxin-like proteins?
11. Perhaps because the functional handles are so easily grasped, questions of gene regulation have so far been largely ignored.

■ References

More extensive references to the various thioredoxin-related proteins cited can be found in the various individual references listed here.

Anfinsen, C.B. (1973). Principles that govern the folding of protein chains. Science **181**, 223–230.

Åslund, F., Ehn, B., Miranda-Vizuete, A., Pueyo, C. and Holmgren, A. (1994). Two additional glutaredoxins exist in *Escherichia coli*: glutaredoxin 3 is a hydrogen donor for ribonucleotide reductase in a thioredoxin/glutaredoxin 1 double mutant. Proc. Natl. Acad. Sci. USA **91**, 9813–9817.

Bardwell, J.C.A. (1994). Building bridges: disulphide bond formation in the cell. Mol. Microbiol. **14**, 199–205.

Creighton, T.E., Zapun, A., and Darby, N.J. (1995). Mechanisms and catalysts of disulfide bond formation in proteins. Trends Biotechnol. **13**, 18–23.

Darby, N.J. and Creighton, T.E. (1995). Characterization of the active site cysteine residues of the thioredoxin-like domains of protein disulfide isomerase. Biochemistry **34**, 16770–16780.

Freedman, R.B. (1995). The formation of the disulfide bonds. Curr. Opin. Struct. Biol. **5**, 85–91.

Freedman, R.B., Hirst, T.R. and Tuite, M.F. (1994). Protein disulfide isomerase: building bridges in protein folding. Trends Biochem. Sci. **19**, 331–336.

Gane, P.J., Freedman, R.B., and Warwicker, J. (1995). A molecular model for the redox potential difference between thioredoxin an DsbA, based on electrostatics calculations. J. Mol. Biol. **249**, 376–387.

Gilbert, H.F. (1990). Molecular and cellular aspects of thiol-disulfide exchange. Adv. Enzymol. **63**, 69–172.

Gleason, F.K. and Holmgren, A. (1988). Thioredoxin and related proteins in Prokaryotes. FEMS Micro. Rev. **54**, 271–298.

Grauschopf, U., Winther, J.R., Korber, P., Zander, T., Dallinger, P., and Bardwell, J.C.A. (1995). Why is DsbA such an oxidizing disulfide catalyst? Cell **83**, 947–955.

Guilhot, C., Jander, G., Martin, N.L., and Beckwith, J. (1995). Evidence that the pathway of disulfide bond formation in *Escherichia coli* involves interactions between the cysteines of DsbB and DsbA. Proc. Natl. Acad. Sci. USA **92**, 9895–9899.

Holmgren, A. (1989). Thioredoxin and glutaredoxin systems. J. Biol. Chem. **264**, 13963–13966.

Hwang, C., Sinskey, A.J., Lodish, H.F. (1992). Oxidized redox state of glutathione in the endoplasmic reticulum. Science **257**, 1496–1502.

Kallis, G.-B. and Holmgren, A. (1980). Differential reactivity of the functional sulfhydryl groups of cysteine-32 and cysteine-35 present in the reduced form of thioredoxin from *Escherichia coli*. J. Biol. Chem. **255**, 10261–10265.

Kemmink J., Darby, N.J., Dijkstra, K., Scheek, R.M., and Creighton, T.E. (1995). Nuclear magnetic resonance characterization of the N-terminal thioredoxin-like domain of protein disulfide isomerase. Protein Sci. **4**, 2587–2593.

Loferer, H. and Hennecke, H. (1994). Protein disulfide oxidoreductases in bacteria. Trends Biochem. Sci. **19**, 169–171.

Loferer, H., Wunderlich, M., Hennecke, H., and Glockshuber, R. (1995). A bacterial thioredoxin-like protein that is exposed to the periplasm has redox properties comparable with those of cytoplasmic thioredoxins. J. Biol. Chem. **270**, 26178–26183.

Martin, J.L. (1995). Thioredoxin: a fold for all reasons. Structure **3**, 245–250.

Missiakas, D. and Raina, S. (1997). Protein folding in the bacterial periplasm. J. Bacteriol. **179**, 2465–2471.

Missiakas, D., Schwager, F., and Raina, S. (1995). Identification and characterization of a new disulfide isomerase-like protein (DsbD) in *Escherichia coli*. EMBO J. **14**, 3415–3424.

Nelson, J.W. and Creighton, T.E. (1994). Reactivity and ionization of the active site cysteine residues of DsbA, a protein required for disulfide bond formation in vivo. Biochemistry **33**, 5974–5983.

Puig, A., Lyles, M.M., Noiva, R., and Gilbert H.F. (1994). The role of the thiol/disulfide centers and peptide binding site in the and anti-chaperone activities of protein disulfide isomerase. J Biol. Chem. **269**, 19128–19135.

Rivera-Madrid, R., Mestres, D., Marinho, P., Jacquot, J-P., Decottignies, P., Miginiac-Maslow, M., and Meyer, Y., (1995). Evidence for five divergent thioredoxin *h* sequences in *Arabidopsis thaliana*. Proc. Natl. Acad. Sci. USA **92**, 5620–5624.

Takahashi, N. and Creighton, T.E. (1996). On the reactivity and ionization of the active site cysteine residues of *Escherichia coli* thioredoxin. Biochemistry **35**, 10517–10528.

Weissman, J.S. and Kim, P.S. (1993). Efficient catalysis of disulfide bond rearrangements by protein disulphide isomerase. Nature **365**, 185–188.

Wunderlich, M. and Glockshuber, R. (1993). Redox properties of protein disulfide isomerase (DsbA) from *Escherichia coli*. Protein Sci. **2**, 717–726.

Wunderlich, M., Otto, A., Maskos, K., Mücke, M., Seckler, R., and Glockshuber, R. (1995). Efficient catalysis of disulfide formation during protein folding with a single active-site cysteine. J. Mol. Biol. **247**, 28–33.

Wynn, R., Cocco, M.J., and Richards, F.M. (1995). Mixed disulfide intermediates during the reduction of disulfides by *Escherichia coli* thioredoxin. Biochemistry **34**, 11807–11813.

Zapun, A., Missiakas, D., Raina, S., and Creighton, T.E. (1995). Structural and functional characterization of DsbC, a protein involved in disulfide bond formation in *Escherichia coli*. Biochemistry **34**, 5075–5089.

■ James C. A. Bardwell:
Department of Biology
University of Michigan
Ann Arbor, MI 48109-1048, USA
Tel. 1 313 764 8028
Fax. 1 313 647 0884
E-mail: Jbardwel@biology.lsa.umich.edu

Escherichia coli thioredoxin

Thioredoxin (Trx) from Escherichia coli *is a multifunctional 12 kDa protein catalysing thiol–disulfide exchange reactions via two redox active cysteine residues (CGPC) in a unique protrusion, either forming a disulfide in the oxidized form or a dithiol in the reduced form. Thioredoxin, together with the FAD-containing enzyme thioredoxin reductase and NADPH (the Trx system), is a general protein disulfide reductase and a specific hydrogen donor for enzymes like ribonucleotide reductase.*

■ Isolation

Escherichia coli B thioredoxin was originally purified to homogeniety from a neutralized pH 5.0 supernatant by heat treatment to 85°C, followed by chromatography on DEAE–cellulose and Sephadex G-50 (Holmgren and Reichard, 1967). A source of E. coli thioredoxin is the K12 strain, SK 3981 (Lunn, et al., 1984), which contains a derivative of the plasmid pBR325 (pBHK8) into which a 3 kb *Pvu* II fragment of *E. coli* DNA containing the thioredoxin gene was inserted. Growth of SK3981 into stationary phase results in the overproduction of thioredoxin 100- to 200-fold compared to wild-type cells, or 10^6 copies of thioredoxin per cell. Thioredoxin purified from SK3981 is identical to the *E. coli* B thioredoxin (Holmgren and Björnstedt, 1995).

■ *trxA* gene and sequence

The thioredoxin gene in *E. coli* is denoted *trxA*, while the gene for thioredoxin reductase is denoted *trxB*. *trxA* is located at 84 minutes on the *E. coli* K12 genetic map (Mark, et al., 1977; Russel, 1995). The nucleotide sequence of the thioredoxin gene from the plasmid pBHK8 was determined (GenBank accession number M26133; Höög et al., 1984) and a number of mutations in the gene have been analysed (Russel, 1995; Eklund et al., 1991).

■ *Escherichia coli* thioredoxin protein

Escherichia coli thioredoxin consists of 108 amino acid residues with its active site cysteine residues in positions 32 and 35. The three-dimensional structure of *E. coli* thioredoxin (Fig. 1) has been determined to high resolution both by X-ray crystallography (Holmgren et al., 1975; Katti et al., 1990) and NMR (Jeng et al., 1994). The *E. coli* thioredoxin structure contains a twisted β sheet composed of five strands (β1–β5) flanked by four α helices (α1–α4) (Holmgren, 1985; Eklund et al., 1991). In thioredoxin-(SH)$_2$ and thioredoxin-S$_2$ the active center disulfide/dithiol is located in a loop making a unique protrusion of the three-dimensional structure between β strands (β2) and the long α2 helix (Holmgren, 1995). Thioredoxins from different species may show less than 25% amino acid sequence identity, but have an active site disulfide/dithiol with the sequence Trp–Cys–Gly–Pro–Cys conserved (Eklund et al., 1991). This is surrounded by a conservative hydrophobic surface area, consisting of Gly-33, Pro-34, Ile-75, Pro-76, Gly-92 and Ala-93 in *E. coli* thioredoxin (Eklund, et al., 1991; Holmgren, 1995). The hydrophobic surface plays a role in protein–protein interactions. Based on the structure of thioredoxin and the low pK_a value of the SH group of Cys-32 (Kallis and Holmgren, 1980), a reaction mechanism has been proposed for the reduction of a protein disulfide by thioredoxin-(SH)$_2$. Recently, a hydrogen bond between

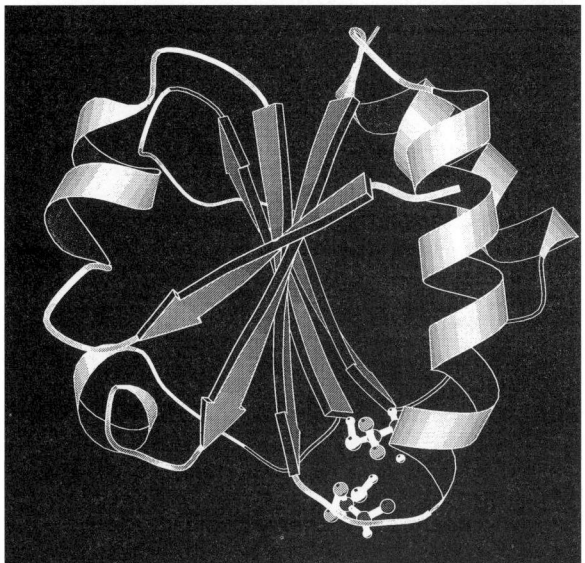

Figure 1. Structure of *E. coli* thioredoxin in reduced form from high-resolution NMR (Jeng *et al.*, 1994). The side chains of Cys-32 and Cys-35 are shown. The structure consists of a $\beta\alpha\beta\alpha\beta\alpha\beta\beta\alpha$ characteristic of the thioredoxin superfamily of proteins (Martin, 1995).

the SH group of Cys-35 and the sulfur of the Cys-32 thiolate has been proposed based on NMR data (Jeng *et al.*, 1995). The thiolate of Cys-32 makes a nucleophilic attack leading to the formation of a mixed disulfide complex linking the thioredoxin molecule and the target protein covalently. The then deprotonated Cys-35 attacks this mixed disulfide resulting in the formation of a dithiol in the target protein and thioredoxin-(SH)$_2$. Conformational changes occur in the target protein and thioredoxin during this reaction (Fig. 2). *E. coli* thioredoxin as well as thioredoxin reductase and corresponding antibodies are commercially available (IMCO Ltd, POB 21 195, S-10031 Stockholm, Sweden).

Biological activities of *E. coli* thioredoxin

Thioredoxin-S$_2$ is reduced to thioredoxin-(SH)$_2$ by thioredoxin reductase (TR) and NADPH. The thioredoxin-(SH)$_2$ form is a direct protein disulfide reductase reacting with insulin with a rate constant in excess of 5×10^4 M^{-1}s^{-1}. Thioredoxin, thioredoxin reductase and NADPH, collectively called the thioredoxin system, operate as a powerful NADPH-dependent protein disulfide reductase system (Holmgren, 1985; Holmgren and Björnstedt, 1995).

$$\text{Trx-S}_2 + \text{NADPH} + \text{H}^+ \underset{}{\overset{TR}{\rightleftharpoons}} \text{Trx-(SH)}_2 + \text{NADP}^+$$
$$\text{Trx-(SH)}_2 + \text{proteins-S}_2 \rightleftharpoons \text{Trx-S}_2 + \text{proteins-(SH)}_2$$

Apart from operating as a hydrogen donor to ribonucleotide reductase and a protein disulfide reductase, thioredoxin from *E. coli* has several other functions, including the reduction of sulfate (PAPS), methionine sulfoxide and selenium compounds. Thioredoxin-(SH)$_2$ is an essential subunit of bacteriophage T7 polymerase and is required for the assembly of the filamentous viruses f1 and M13 (Holmgren, 1985).

Mutagenesis studies with *E. coli* thioredoxin

The details of mutagenesis studies are reviewed elswhere (Russel, 1995). A Pro-34 to His mutant that changes the thioredoxin active site to one that mimics the two proposed active sites of protein disulfide isomerase (Cys–Gly–His–Cys) has been constructed. This mutant protein has a higher redox potential and reduces insulin more slowly than wild-type thioredoxin. However, it is about tenfold more efficient in disulfide formation, making it about 10% as efficient as authentic protein disulfide isomerase on a molar basis (Lundström *et al.*, 1992).

Biological interactions

Reduced thioredoxin from *E. coli* releases the binding of human α_2-macroglobulin and interleukin-1β (Borth *et al.*,

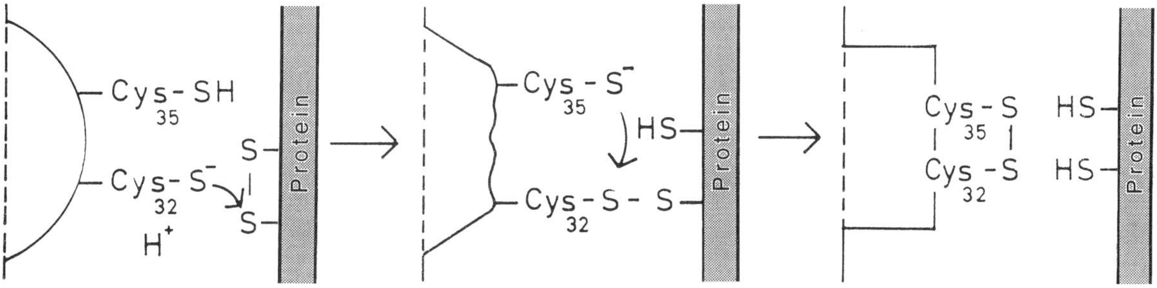

Figure 2. Mechanism of reduction of a protein disulfide by reduced thioredoxin (Holmgren, 1985).

1990). A thioredoxin gene fusion expression system provides a a new technology to produce a protein in a soluble form that is properly folded and biologically active (LaVallie, et al., 1993).

References

Borth, W., Scheer, B., Urbansky, A., Luger, T.A. and Sottrup, Jensen L. (1990). Binding of IL-1 beta to alpha-macroglobulins and release by thioredoxin. J. Immunol. **145**, 3747–3754.

Eklund, H., Gleason, F.K., and Holmgren, A. (1991). Structural and functional relations among thioredoxins of different species. Proteins **11**, 13–28.

Holmgren, A. (1985). Thioredoxin. Annu. Rev. Biochem. **54**, 237–271.

Holmgren, A. (1995). Thioredoxin structure and mechanism: conformational changes on oxidation of the active site sulfhydryls to a disulfide. Structure **3**, 239–243.

Holmgren, A. and Björnstedt, M. (1995). Thioredoxin and thioredoxin reductase. Meth. Enzymol. **252**, 199–208.

Holmgren, A. and Reichard, P. (1967). Thioredoxin 2: cleavage with cyanogen bromide. Eur. J. Biochem. **2**, 187–196.

Holmgren, A., Söderberg, B.-O., Eklund, H., and Brändén, C.-I. (1975). Three-dimensional structure of Escherichia coli thioredoxin-S_2 to 2.8 Å resolution. Proc. Natl. Acad. Sci. USA.**72**, 2305–2309.

Höög, J.O, von Bahr-Lundström, H., Josephson, S., Wallace, B.J., Kushner, S.R., Jörnvall, H., and Holmgren, A. (1984). Nucleotide sequence of the thioredoxin gene from Escherichia coli. Biosci. Rep. **4**, 917–923.

Jeng, M-F., Campbell, A.P., Begley, T., Holmgren, A., Case, D.A., Wright, P.E., and Dyson, H.J. (1994). High-resolution solution structures of oxidized and reduced Escherichia coli thioredoxin. Structure **2**, 853–868.

Jeng, M.-F., Holmgren, A., and Dyson, H.J. (1995). Proton sharing between cysteine thiols in Escherichia coli thioredoxin: implications for the mechanism of protein disulfide reduction. Biochemistry **34**, 10101–10105.

Kallis, G.-B. and Holmgren, A. (1980). Differential reactivity of the functional sulfhydryl groups of cysteine-32 and cysteine-35 present in the reduced form of thioredoxin from Escherichia coli. J. Biol. Chem. **255**, 10261–10265.

Katti, S. K., LeMaster, D.M., and Eklund, H. (1990). Crystal structure of thioredoxin from Escherichia coli at 1.68 Å resolution. J. Mol. Biol. **212**, 167–184.

LaVallie, E.R., DiBlasio, E.A., Grant, K.L., Schendel, P.F., and McCoy, J.M. (1993). A thioredoxin gene fusion expression system that circumvents inclusion body formation in the E. coli cytoplasm. Biotechnology **11**, 187–193.

Lundström, J., Krause, G., and Holmgren, A. (1992). A Pro to His mutation in active site of thioredoxin increases its disulfide-isomerase activity 10-fold. New refolding systems for reduced or randomly oxidized ribonuclease. J. Biol. Chem. **267**, 9047–9052.

Lunn, C.A., Kathju, S., Wallace, B.J., Kushner, S.R., and Pigiet, V. (1984). Amplification and purification of plasmid-encoded thioredoxin from Escherichia coli K12. J. Biol. Chem. **259**, 10469–10474.

Mark, D.F., Chase, J.W., and Richardson, C.C. (1977). Genetic mapping of trxA, a gene affecting thioredoxin in Escherichia coli K12. Mol. Gen. Genet. **155**, 145–152.

Martin, J.L. (1995). Thioredoxin: a fold for all reasons. Structure **3**, 245–250.

Russel, M. (1995). Thioredoxin genetics. Meth. Enzymol. **252**, 264–274.

■ *Hajime Nakamura, Mikael Björnstedt and Arne Holmgren:*
Medical Nobel Institute for Biochemistry
Department of Medical Biochemistry and Biophysics
Karolinska Institute
S-171 77 Stockholm, Sweden
Tel. 46 8 728 7686
Fax. 46 8 728 4716
E-mail: Arne.Holmgren@mbb.ki.se

Escherichia coli glutaredoxin

Glutaredoxin (Grx) was originally identified as a glutathione (GSH)-dependent hydrogen donor for ribonucleotide reductase in a mutant of Escherichia coli *lacking thioredoxin. In addition, glutaredoxins exhibit a general GSH-disulfide oxidoreductase activity with either protein disulfides or GSH-mixed disulfides. The glutaredoxin system comprising NADPH, GSH, glutathione reductase and glutaredoxin, thus catalyses NADPH-dependent reductions of disulfides.*

■ Isolation, gene and sequence

Purification and characterization of E. coli glutaredoxin Grx1 (see below) showed an acidic 10 kDa small protein exhibiting hydrogen donor activity with E. coli ribonucleotide reductase in the presence of 4 mM GSH and a general GSH-disulfide oxidoreductase activity with hydroxyethyl disulfide (Holmgren, 1976, 1979a, b). Grx1, as isolated, contains 85 amino acids, including a redox active disulfide within an active site sequence of –Cys–Pro–Tyr–Cys– located in positions 11 and 14 of the molecule (M_r 10 000) (Höög, et al., 1983). A clear homology was found to a protein originally known as phage T4 thioredoxin (Sjöberg and Holmgren, 1972). Since T4 thioredoxin also showed general GSH-disulfide oxidoreductase activity and similarity in three-dimensional structure (Holmgren, 1978), it is now called T4 glutaredoxin (Eklund et al., 1992). The gene for glutaredoxin (grx) was cloned (GenBank accession number M13449; Höög et al., 1986) and a deletion of the gene is viable (Russel and Holmgren, 1988). High level expression of Grx and the identification of a form with five additional N-terminal residues has been described (Björnberg and Holmgren, 1991).

Recently, two additional glutaredoxins (Grx2 and Grx3) with the conserved active site sequence of –Cys–Pro–Tyr–Cys– were identified in a mutant of E. coli lacking both thioredoxin and Grx1 (Åslund et al., 1994). Grx3 (9 kDa) has only 5% of the hydrogen donor activity determined for Grx1 with ribonucleotide reductase (Åslund et al., 1994). However, the combination of the hydrogen donor activity of Grx3 and a more than 20-fold induction of the ribonucleotide reductase levels in the double mutant is likely to be sufficient for the supply of deoxyribonucleotides in the absence of thioredoxin and Grx1, the major electron donors for ribonucleotide reductase (Åslund et al., 1994; Miranda et al., 1994). Grx2 has a larger size of 27 kDa and is inactive as a hydrogen donor for ribonucleotide reductase (Åslund et al., 1994). The amino acid sequence of Grx3 was reported recently (Åslund et al., 1996).

■ Grx1 protein

The three-dimensional structure of E. coli glutaredoxin (Grx1) has been determined in solution by high-resolution NMR spectroscopy in both the oxidized and reduced form (Sodano et al., 1991; Xia et al., 1992). Also, the solution structure of the mixed disulfide of Grx1 and GSH has been solved using a mutant of glutaredoxin, Grx-C14S (Bushweller et al., 1994). A GSH-binding site on glutaredoxin has been defined from this structure. The three-dimensional structure of glutaredoxin, consisting of four β strands and three α helices, is similar to that of thioredoxin (Holmgren et al., 1975; Martin, 1995). A common feature within the thioredoxin protein family is that the active site is located at the end of a β strand followed by an α helix. The N-terminal of the two active site cysteine residues (Cys-11 in Grx1) is exposed, whereas the other is buried. The more exposed SH group has a low pK_a value and is the nucleophile initiating the attack on a disulfide substrate.

■ Biological activities of Grx proteins

Grx1 reduces an exposed acceptor disulfide located in the C-terminal part of the R1 subunit of ribonucleotide reductase (RR), in a mechanism dependent on both of the two active site cysteines with a dithiol mechanism similar to that of thioredoxin (Bushweller et al., 1992).

$$Grx1\text{-}(SH)_2 + RR\text{-}S_2 \rightarrow Grx1\text{-}S_2 + RR(SH)_2$$

In a subsequent step the reduced acceptor disulfide transfers electrons to the active site of ribonucleotide reductase for the reduction of a ribonucleotide, and becomes oxidized to a disulfide. Grx-S_2 is reduced by two molecules of GSH, forming GSSG. Grx1 is also an electron donor for methionine sulfoxide reductase and PAPS reductase (Holmgren and Åslund, 1995).

In contrast, the activity of glutaredoxins with GSH-mixed disulfide substrates only requires the exposed Cys-11 for catalysis (Bushweller et al., 1992). Grx1 C14S and the wild-type protein synergize with protein disulfide isomerase in catalysis of native disulfide formation with GSSG (Lundström-Ljung and Holmgren, 1995). Since the bulk of GSH-disulfide reducing activity in E. coli cells is not associated with Grx1, and hence not with deoxyribonucleotide and DNA synthesis, it seems likely that Grx2 and Grx3 serve as general GSH-disulfide reductases in the cell. Since oxidative stress is accompanied by the formation of large amounts of GSH-mixed disulfides (Thomas et al., 1995), it is likely that glutaredoxins are important reductants and scavengers of such GSH-mixed disulfides.

■ References

Åslund, F., Ehn, B., Miranda-Vizuete A., Pueyo, C., and Holmgren, A. (1994). Two additional glutaredoxins exist in Escherichia coli: glutaredoxin 3 is a hydrogen donor for ribonucleotide reductase in a thioredoxin/glutaredoxin 1 double mutant. Proc. Natl. Acad. Sci. USA **91**, 9813–9817.

Åslund, F., Nordstrand, K., Berndt, K.D. Nikkola, M., Bergman, T., Ponstingl, H., Jörnvall, H., Otting, G., and Holmgren, A. (1996). Glutaredoxin-3 from E. coli: Amino acid sequence, ^{1}H and ^{15}N NMR assignments and structural analysis. J. Biol. Chem. **271**, 6736–6745.

Björnberg, O. and Holmgren, A. (1991). Characterization of homogenous recombinant glutaredoxin from Escherichia coli: purification from an inducible λ_{PL} expression system and properties of a novel elongated form. Protein Expr. Purif. **2**, 287–295.

Bushweller, J.H., Billeter, M., Holmgren, A., and Wüthrich, K. (1994). The nuclear magnetic resonance solution structure of the mixed disulfide between Escherichia coli glutaredoxin(C14S) and glutathione. J. Mol. Biol. **235**, 1585–1597.

Bushweller, J.H., Åslund, F., Wüthrich, K., and Holmgren, A. (1992). Structural and functional characterization of the mutant Escherichia coli glutaredoxin (C14→S) and its mixed disulfide with glutathione. Biochemistry **31**, 9288–9293.

Eklund, H., Ingelman, M., Söderberg, B.O., Uhlin, T., Nordlund, P., Nikkola, M., Sonnerstam, U., Joelson, T., and Petratos, K. (1992). Structure of oxidized bacteriophage T4 glutaredoxin (thioredoxin). Refinement of native and mutant proteins. J. Mol. Biol. **228**, 596–618.

Holmgren, A. (1976). Hydrogen donor system for Escherichia coli ribonucleoside-diphosphate reductase dependent upon glutathione. Proc. Natl. Acad. Sci. USA **73**, 2275–2279.

Holmgren, A. (1978). Glutathione-dependent enzyme reactions of the phage T4 ribonucleotide reductase system. J. Biol. Chem. **253**, 7424–7430.

Holmgren, A. (1979a). Glutathione-dependent synthesis of deoxyribonucleotides. Purification and characterization of glutaredoxin from Escherichia coli. J. Biol. Chem. **254**, 3664–3671.

Holmgren, A. (1979b). Glutathione-dependent synthesis of deoxyribonucleotides. Characterization of the enzymatic mechanism of Escherichia coli glutaredoxin. J. Biol. Chem. **254**, 3672–3678.

Holmgren, A., and Åslund, F. (1995). Glutaredoxin. Meth. Enzymol. **252**, 283–292.

Holmgren, A., Söderberg, B.O., Eklund, H., and Brändén, C.I. (1975). Three-dimensional structure of Escherichia coli thioredoxin-S2 to 2.8 Å resolution. Proc. Natl. Acad. Sci. USA **72**, 2305–2309.

Höög, J.O., Jörnvall, H., Holmgren, A., Carlquist, M., and Persson, M. (1983). The primary structure of Escherichia coli glutaredoxin. Distant homology with thioredoxins in a superfamily of small proteins with a redox-active cystine disulfide/cysteine dithiol. Eur. J. Biochem. **136**, 223–232.

Höög, J.-O., von Bahr-lindström, H., Jörnvall, H., and Holmgren, A. (1986). Cloning and expression of the glutaredoxin (grx) gene of Escherichia coli. Gene **43**, 13–21.

Lundström-Ljung, J. and Holmgren, A. (1995). Glutaredoxin accelerates glutathione dependent folding of reduced ribonuclease A together with protein disulfide-isomerase. J. Biol. Chem. **270**, 7822–7828.

Martin, J.L. (1995). Thioredoxin: a fold for all reasons. Structure **3**, 245–250.

Miranda-Vizuete, A., Martinez-Galisteo, E., Åslund, F., Lopez-Barea, J., Pueyo, C., and Holmgren, A. (1994). Null thioredoxin and glutaredoxin E. coli K12 mutants have no enhanced sensitivity to mutagenesis due to a new GSH-dependent hydrogen donor and high increases in ribonucleotide reductase activity. J. Biol. Chem. **269**, 16631–16637.

Russel, M. and Holmgren, A. (1988). Construction and characterization of glutaredoxin-negative mutants of Escherichia coli. Proc. Natl. Acad. Sci. USA **85**, 990–994.

Sjöberg, B.M. and Holmgren, A. (1972). Studies on the structure of T4 thioredoxin. II. Amino acid sequence of the protein and comparison with thioredoxin from Escherichia coli. J. Biol. Chem. **247**, 8063–8068.

Sodano, P., Xia, T.H., Bushweller, J.H., Björnberg, O., Holmgren, A., Billeter, M., and Wüthrich, K. (1991). Sequence-specific 1H n.m.r. assignments and determination of the three-dimensional structure of reduced Escherichia coli glutaredoxin. J. Mol. Biol. **221**, 1311–1324.

Thomas, J.A., Poland, B., and Honzatko, R. (1995). Protein sulfhydryls and their role in the antioxidant function of protein S-thiolation. Arch. Biochem. Biophys. **319**, 1–9.

Xia, T.H., Bushweller, J.H., Sodano, P., Billeter, M., Björnberg, O., Holmgren, A., and Wüthrich, K. (1992). NMR structure of oxidized Escherichia coli glutaredoxin: comparison with reduced E. coli glutaredoxin and functionally related proteins. Protein Sci. **1**, 310–321.

■ *Hajime Nakamura, Fredrik Åslund and Arne Holmgren:*
Medical Nobel Institute for Biochemistry
Department of Medical Biochemistry and Biophysics
Karolinska Institute
S-171 77 Stockholm, Sweden
Tel. 46 8 728 7686
Fax. 46 8 728 4716
E-mail: Arne.Holmgren@mbb.ki.se

Escherichia coli DsbA

DsbA is a periplasmic protein in Escherichia coli *that catalyses disulfide bond formation. It appears to be the immediate donor of disulfide bonds to secreted proteins. DsbA functions by rapidly transferring the extremely oxidizing disulfide that is present at its active site to substrate proteins as they are secreted into the periplasm.*

■ Alternative names

PpfA (Kamitani et al., 1992); IarA (Rasmussen et al., 1991; Alksne et al., 1995); E. coli protein disulfide isomerase, PDI (Wunderlich and Glockshuber, 1993a; Joly and Swartz, 1994).

■ *dsbA* mutant phenotypes

dsbA null mutants in *E. coli* have a severe and general defect in disulfide bond formation within the periplasm (reviewed in Bardwell, 1994). Thus, they have strongly decreased levels of proteins that require disulfides for stability, including alkaline and acid phosphatases, endotoxins and eukaryotic disulfide-containing proteins expressed in the periplasm of *E. coli*. *dsbA* null mutants form disulfides about two orders of magnitude more slowly than wild-type strains. Other phenotypes associated with their disulfide defect include sensitivity to DTT and benzylpenicillin, resistance to filamentous phages, poor growth on minimal media lacking cysteine and failure to assemble the p-ring of the flagella motor (for review see Bardwell, 1994). Periplasmic expression of heterologous proteins whose stability requires the absence of disulfide bonds is dependent on the absence of DsbA (Alksne et al., 1995). Most of these phenotypes are severe enough to allow for the isolation of *dsbA* mutants.

■ *dsbA* gene and sequence

The *dsbA* gene is located at 87.1 minutes on the *E. coli* chromosome. Its DNA sequence (GenBank accession number M77746) predicts a 21.1 kDa protein product preceeded by a cleavable signal sequence. It is most closely homologous to DsbA proteins found in other prokaryotes. Although the sequence shows only very limited homology to thioredoxin, the three-dimensional structure of DsbA, which is known at high resolution, shows that one of its two domains closely resembles thioredoxin (see the next entry on the DsbA structure, p. 320, and Martin et al., 1993).

DsbA protein

DsbA is a monomeric periplasmic protein that is capable of rapidly oxidizing many proteins *in vitro* and *in vivo* (Bardwell *et al.*, 1991; Bardwell and Beckwith, 1993; Wunderlich *et al.* 1993a; Zapun and Creighton, 1994). The active site in DsbA consists of two cysteines, Cys-30 and Cys-33, that can reversibly form a disulfide bond. This disulfide bond is unstable, extremely oxidizing and is very rapidly transferred to its substrate proteins (Wunderlich and Glockshuber, 1993a; Wunderlich *et al.*, 1993b; Zapun *et al.*, 1993; Zapun and Creighton, 1994). DsbA can be purified in a single chromatographic step from periplasmic extracts of overproducing strains (Zapun *et al.*, 1993).

Role of DsbA in protein folding *in vivo*

Proteins such as alkaline phosphatase that require disulfides for their stability are rapidly degraded in DsbA mutant strains (Bardwell *et al.*, 1991). Proteins that are stable in the absence of their disulfide, such as OmpA, accumulate folding intermediates in DsbA mutants. The proper folding of disulfide-containing eukaryotic proteins exported to the periplasm of *E. coli* is absolutely dependent on the presence of the *dsbA* gene. Furthermore, DsbA overexpression in conjunction with addition of thiol compounds to the media can result in substantially higher yields of properly folded eukaryotic proteins (Wunderlich and Glockshuber, 1993b).

Mutagenesis studies with DsbA

Residues in the area of the active site have been extensively mutated (Bardwell, 1994; Zapun *et al.*, 1993, 1994; Wunderlich *et al.*, 1995). The first cysteine in the active site, Cys-30, is exposed, reactive and essential for the activity of DsbA both *in vivo* and *in vitro*. The second cysteine, Cys-33, is, in contrast, buried, unreactive and can be altered by mutation with only partial disruption of activity *in vitro* and *in vivo*. Mutants that change the two residues Pro-31 and His-32, located in between the active site cysteines, have dramatic effects on the redox power of DsbA. An explanation for these effects is found in the extremely low pK_a of Cys-30, which appears to be critical in determining the oxidizing power of DsbA and the instability of the disulfide (Nelson and Creighton, 1994; Grauschopf *et al.*, 1995). The low pK_a of Cys-30 in turn appears to be the result of electrostatic interactions between Cys-30 and a relatively small number of residues, including His-32 (Gane *et al.*, 1995). Thus the extreme oxidizing power of DsbA may be determined by relatively few residues.

Biological interactions of DsbA

It is thought that DsbA is reoxidized by DsbB (see entry p. 322), allowing it to complete its catalytic cycle of protein oxidation (Bardwell *et al.*, 1993). A direct interaction between the two proteins has been shown by the isolation of a DsbA–DsbB dimer covalently linked by a disulfide bond (Guilhot *et al.*, 1995; Kishigami *et al.*, 1995). This dimer is stably formed between DsbB and a mutant form of DsbA that lacks the second cysteine at its active site. This mutant form of DsbA also allows isolation of a mixed disulfide intermediate in the reaction between DsbA and a variant of ribonuclease T1 (Frech *et al.*, 1995). In this reaction intermediate the conformational stability of DsbA is increased by 5 kJ/mol and the stability of the substrate protein is reduced by the same amount. This reciprocal effect provides evidence that DsbA binds specifically to unfolded substrate proteins.

DsbA, a prokaryotic protein disulfide isomerase?

The designation of DsbA as *E. coli* protein disulfide isomerase is somewhat confusing. Eukaryotic PDI complements DsbA null mutations and DsbA has detectable, but weak, isomerization activity under some conditions (Akiyama *et al.*, 1992; Wunderlich *et al.*,1993a; Joly and Swartz, 1994; Humphreys *et al.*, 1995). However, DsbA is much more effective in oxidizing proteins than it is in catalysing intramolecular disulfide rearrangement reactions (Zapun *et al.*, 1993, 1995; Zapun and Creighton; 1994; Darby and Creighton, 1995). These results can be reconciled if the primary *in vivo* function of DsbA is as an oxidant, while PDI can function both as an oxidant and as an isomerase. Another *E. coli* protein, DsbC (see entry p. 324), shows more isomerase activity *in vitro*, and may perform the isomerase function of PDI in prokaryotes (Zapun *et al.*, 1995).

References

Akiyama, Y., Kamitani, S., Kusukawa, N., and Ito, K. (1992). In vitro catalysis of oxidative folding of disulfide-bonded proteins by the *Escherichia coli dsbA* (*ppfA*) gene product. J. Biol. Chem. **267**, 22440–22445.

Alksne, L.E., Keeney, D., and Rasmussen, B.A. (1995). A mutation in either *dsbA* or *dsbB*, a gene encoding a component of a periplasmic disulfide bond-catalyzing system, is required for high level expression of the *Bacteroides fragilis* metallo-β-lactamase, CrrA, in *Escherichia coli*. J. Bacteriol. **177**, 462–464.

Bardwell, J.C.A. (1994). Building bridges: disulphide bond formation in the cell. Mol. Microbiol. **14**, 199–205.

Bardwell, J.C.A. and Beckwith, J. (1993). The bonds that tie: catalyzed disulfide bond formation. Cell **74**, 769–771.

Bardwell, J.C.A., McGovern, K., and Beckwith, J. (1991). Identification of a protein required for disulfide bond formation in vivo. Cell **67**, 581–589.

Bardwell, J.C.A., Lee, J.-O., Jander, G., Martin, N., Belin, D., and Beckwith, J. (1993). A pathway for disulfide bond formation in vivo. Proc. Natl. Acad. Sci. USA. **90**, 1038–1042

Darby, N.J. and Creighton, T.E. (1995). Catalytic mechanism of DsbA and its comparison with that of protein disulfide isomerase. Biochemistry **34**, 3576–3587.

Frech, C., Wunderlich, M., Glockshuber, R., and Schmid, F.X. (1995). Preferential binding of an unfolded protein to DsbA. EMBO J. **15**, 392–398.

Gane, P.J., Freedman, R.B., and Warwicker, J. (1995). A molecular model for the redox potential difference between Thioredoxin and DsbA, based on electrostatics calculations. J. Mol. Biol. **249**, 376–387.

Grauschopf, U., Winther, J., Korber, P., Zander, T., Dallinger, P., Warwicker, J., Gane, P.J., and Bardwell, J.C.A. (1995). Why is DsbA such an oxidizing disulfide catalyst? Cell **83**, 947–955.

Guilhot, C., Jander, G., Martin, N. L., and Beckwith, J. (1995). Evidence that the pathway of disulfide bond formation in *Escherichia coli* involves interactions between the cysteines of DsbB and DsbA. Proc. Natl. Acad. USA **92**, 9895–9899.

Humphreys, D.P., Weir, N., Mountain, A., and Lund, P.A. (1995). Human protein disulfide isomerase functionally complements a *dsbA* mutation and enhances the yield of pectate lyase C in *Escherichia coli*. J. Biol. Chem. **270**, 28210–28215.

Joly, J.C. and Swartz, J.R. (1994). Protein folding activities of *Escherichia coli* protein disulfide isomerase. Biochemistry **33**, 4231–4236.

Kamitani, S., Akiyama, Y., and Ito, K. (1992). Identification and characterization of an *Escherichia coli* gene required for the formation of correctly folded alkaline phosphatase, a periplasmic enzyme. EMBO J. **11**, 57–62.

Kishigami, S., Kanaya, E., Kikuchi, M., and Ito, K. (1995). DsbA-DsbB interaction through their active site cysteines: evidence from an odd cysteine mutant of DsbA. J. Biol. Chem. **270**, 17072–17074.

Martin, J.L., Bardwell, J.C.A., and Kuriyan, J. (1993). Crystal structure of DsbA protein required for disulphide bond formation in vivo. Nature **36**, 464–468.

Nelson, J.W. and Creighton, T.E. (1994). Reactivity and ionization of the active site cysteine residues of DsbA, a protein required for disulfide bond formation in vivo. Biochemistry **33**, 5974–5983.

Rasmussen, B.A. Gluzman, Y., and Tally, F.P. (1991). *Escherichia coli* chromosomal mutations that permit direct cloning of the *Bacteroides fragilis* metallo-β-lactamase gene, *ccrA*. Mol. Microbiol. **5**, 1211–1219.

Wunderlich, M. and Glockshuber, R. (1993a). Redox properties of protein disulfide isomerase (DsbA) from *Escherichia coli*. Protein Sci. **2**, 717–726.

Wunderlich, M. and Glockshuber, R. (1993b). In vivo control of redox potential during protein folding catalyzed by bacterial protein disulfide-isomerase (DsbA). J. Biol. Chem. **268**, 24547–24550.

Wunderlich, M., Otto, A., Seckler, R., and Glockshuber, R. (1993a). Bacterial protein disulfide isomerase: efficient catalysis of oxidative protein folding at acidic pH. Biochemistry **32**, 12251–12256.

Wunderlich, M., Jaenicke, R. and Glockshuber, R. (1993b). The redox properties of protein disulfide isomerase (DsbA) of *Escherichia coli* result from a tense conformation of its oxidized form. J. Mol. Biol. **233**, 559–566.

Wunderlich, M., Otto, A., Maskos, K., Mücke, M., Seckler, R., and Glockshuber, R. (1995). Efficient catalysis of disulfide formation during protein folding with a single active-site cysteine J. Mol. Biol. **247**, 28–33.

Zapun, A. and Creighton, T.E. (1994). Effects of DsbA on the disulfide folding of bovine pancreatic trypsin inhibitor and α-lactalbumin. Biochemistry **33**, 5202–5211.

Zapun, A., Bardwell, J.C.A., and Creighton, T.E. (1993). The reactive and destabilizing disulfide bond of DsbA, a protein required for protein disulfide bond formation in vivo. Biochemistry **32**, 5083–5092.

Zapun, A., Cooper, L., and Creighton, T.E. (1994). Replacement of the active-site cysteine residues of DsbA, a protein required for disulfide bond formation in vivo. Biochemistry **33**, 1907–1914.

Zapun, A., Missiakas, D., Raina, S., and Creighton, T.E. (1995). Structural and functional characterization of DsbC, a protein involved in disulfide bond formation in *Escherichia coli*. Biochemistry **34**, 5075–5089.

■ James C.A. Bardwell:
Department of Biology
University of Michigan
Ann Arbor, MI 48109-1048, USA
Tel. 1 313 764 8028
Fax. 1 313 647 0884
E-mail: Jbardwel@biology.lsa.umich.edu

Structure of DsbA

The three-dimensional structure of DsbA includes a thioredoxin domain and a helical domain. The active site disulfide of the protein is found in the thioredoxin domain and is surrounded by exposed hydrophobic residues that may be important for substrate interaction.

■ Structure

The structure of oxidized *Escherichia coli* DsbA has been determined by X-ray crystallography, originally at a resolution of 2.0 Å (Martin et al., 1993) and more recently to 1.7 Å (Guddat et al., 1977). The structure consists of two domains, a thioredoxin domain and a helical domain (Fig. 1a). Although the classical thioredoxin fold incorporates a four-stranded β sheet flanked by three helices (Martin, 1995), the thioredoxin domain of DsbA (comprising residues 1–61 and 140–189) has an additional N-terminal β strand that extends the central β sheet to five strands.

The DsbA helical domain (residues 62–139) is inserted into the thioredoxin domain and includes five helices. The first three α helices of this domain form an antiparallel three-helix bundle, while the remaining two α helices wrap around the bundle and make the connection back to the thioredoxin domain. The fifth α helix of the helical domain can also be considered an extension of the second helix of the thioredoxin domain (see Fig. 1a).

The active site disulfide of DsbA (Cys^{30}–Pro^{31}–His^{32}–Cys^{33}) is located in the thioredoxin domain. Of the two cysteines, only Cys-30 is accessible to solvent (Martin et al., 1993; Zapun et al., 1993). The residues Met-64 (from

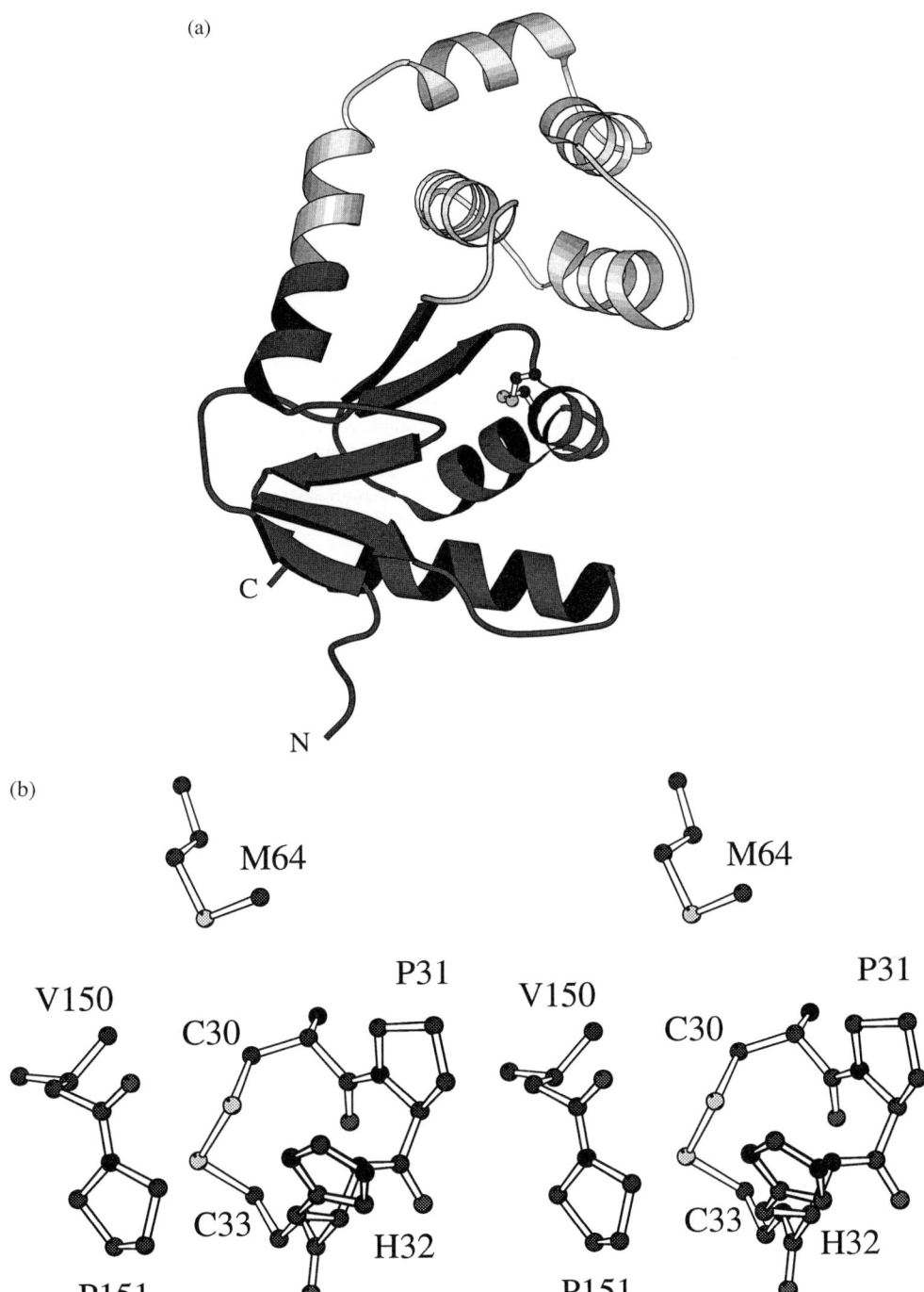

Figure 1. Structure of DsbA. (a) Schematic diagram of the backbone structure of oxidized *E. coli* DsbA. Arrows represent β strands while α helices are depicted as spirals. To highlight the two domains of the protein, the thioredoxin domain is shown in darker shading than the helical domain. The cysteines of the active site disulfide are represented as ball-and-stick atoms and the N- and C- termini are labeled. (b) Stereo view of the oxidized active site disulfide bridge (Cys^{30}–Pro^{31}–His^{32}–Cys^{33}) of DsbA, shown as a ball-and-stick model. Also shown are residues Met-64 (from the helical domain), Val-150 and Pro-151 (from the *cis*-Pro loop of the thioredoxin domain) which are in close proximity to Cys-30. The sulfur atoms from the disulfide and methionine are light grey while all other atoms are black. Figures (a) and (b) were prepared with the program MolScript (Kraulis, 1991).

the helical domain) and Pro-151 (thiroredoxin domain) make close contact with the active site disulfide (Fig. 1b). These two residues form part of an extensive hydrophobic region surrounding the active site, including a hydrophobic patch and a groove lined with hydrophobic residues (Martin et al., 1993). Such features may be important for the interaction of DsbA with unfolded or partly folded polypeptide substrate. Additionally, Pro-151, a cis-proline, is located in a loop (Arg^{148}–Gly^{149}–Val^{150}–Pro^{151}) that is highly conserved in thioredoxin-like proteins. In several structural examples from the thioredoxin family of proteins, the cis-Pro loop has been identified as making important interactions to substrate binding (Nikkola et al., 1991; Sinning et al., 1993; Bushweller et al., 1994; Qin et al., 1995).

Coordinates for the structure of oxidized E. coli DsbA have been deposited with the Brookhaven Protein Data Bank (accession codes 1dsb and 1fvk).

■ References

Bushweller, J.H., Billeter, M., Holmgren, A., and Wüthrich, K. (1994). The nuclear magnetic resonance solution structure of the mixed disulfide between Escherichia coli glutaredoxin (C14S) and glutathione. J. Mol. Biol. **235**, 1585–1597.

Guddat, L.W., Bardwell, J.C.A., Zander, T. and Martin, J.L. (1997). The uncharged surface features surrounding the active-site of E. coli DsbA are conserved and are implicated in peptide binding. Protein Science in press.

Kraulis, P. (1991). MolScript: a program to produce both detailed and schematic plots of protein structures. J. Appl. Cryst. **24**, 946–950.

Martin, J.L. (1995). Thioredoxin: a fold for all reasons. Structure **3**, 245–250.

Martin, J.L., Bardwell, J.C.A., and Kuriyan, J. (1993). Crystal structure of the DsbA protein required for disulphide bond formation in vivo. Nature **365**, 464–468.

Nikkola, M., Gleason, F.K., Saarinen, M., Joelson, T., Bjornberg, O., and Eklund, H. (1991). A putative glutathione-binding site in T4 glutaredoxin investigated by site directed mutagenesis. J. Biol. Chem. **266**, 16105–16112.

Qin, J., Clore, G.M., Kennedy, W.M.P., Huth, J.R., and Gronenborn, A.M. (1995). Solution structure of human thioredoxin in a mixed disulfide intermediate complex with its target peptide from the transcription factor NFkB. Structure **3**, 289–297.

Sinning, I., Kleywegt, G.J., Cowan, S.W., Reinemer, P., Dirr, H.W., Huber, R., Gilliland, G.L., Armstrong, R.N., Ji, X., Board, P.G., Olin, B., Mannervik, B., and Jones, T.A. (1993). Structure determination and refinement of human class alpha glutathione transferase A1-1, and a comparison with the Mu and Pi class enzymes. J. Mol. Biol. **232**, 192–212.

Zapun, A., Bardwell, J.C.A., and Creighton, T.E. (1993). The reactive and destabilizing disulfide bond of DsbA, a protein required for protein disulfide bond formation in vivo. Biochemistry **32**, 5083–5092.

■ *Luke W. Guddat and Jennifer L. Martin:*
Centre for Drug Design and Development
University of Queensland
St Lucia
Queensland 4072, Australia
Tel. 61 7 3365 4942
Fax. 61 7 3365 1990
E-mail: J.Martin@mailbox.uq.oz.au

Escherichia coli DsbB

DsbB is an inner membrane protein from Escherichia coli that is essential for disulfide bond formation in the periplasmic space. Genetic evidence strongly suggests that it is required for the reoxidation of DsbA, which in turn directly oxidizes substrate proteins.

■ Alternative names

DsbX (Belin and Boquet, 1993); IarB (Alksne et al., 1995).

■ *dsbB* mutant phenotypes

dsbB null mutants in E. coli have a severe and general defect in disulfide bond formation within the periplasm. The pleiotropic phenotypes associated with this defect are very similar to those of *dsbA* mutants (see entry p. 318). *dsbB* mutants have been isolated based on these phenotypes (Bardwell et al., 1993; Belin and Boquet, 1993; Dailey and Berg, 1993; Alksne et al., 1995) and by the DTT-resistant phenotype associated with multicopy clones of *dsbB* (Missiakas et al., 1993).

■ *dsbB* gene and localization

The *dsbB* gene is located at 26.5 minutes on the E. coli chromosome (Bardwell et al., 1993; Missiakas et al., 1993). Its sequence (GenBank accession number L03721) predicts a 20.3 kDa protein with inner membrane localization (Jander et al., 1994). Although DsbB shows no recognizable homology to thioredoxin-like proteins, the two cysteines in the first periplasmic domain form a CXXC motif, a sequence that is present in active sites of thioredoxin, DsbA and other related oxidoreductases. Protein

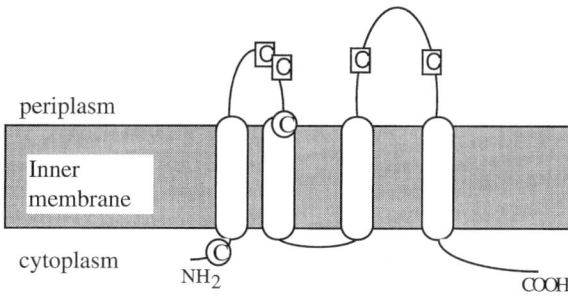

Figure 1. Membrane topology of DsbB, as determined by the alkaline phosphatase fusion approach. The cysteines essential to DsbB function (Cys-41, Cys-44, Cys-104, Cys-130) are shown in boxes. The dispensable cysteines (Cys-8 and Cys-49) and are shown in ovals.

fusions to alkaline phosphatase and biotin demonstrate that DsbB is localized to the inner membrane and has four transmembrane regions, with a topology as shown in Fig. 1 (Jander et al., 1994).

∎ *dsbB* genetic interactions

Evidence that DsbB and DsbA are involved in the same pathway for disulfide bond formation (Fig. 2) includes the severity and non-additivity of the defects associated with mutations in both proteins (Bardwell et al., 1993).

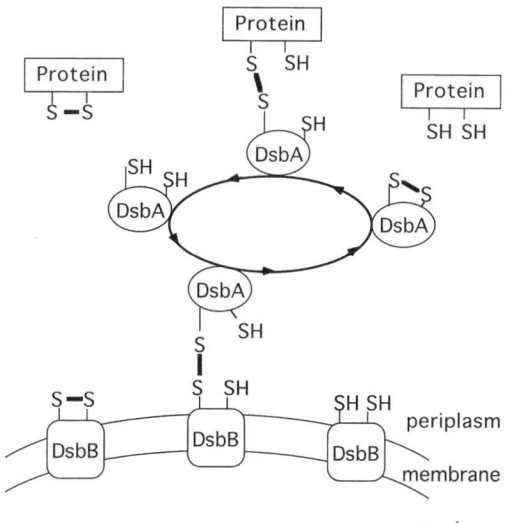

Figure 2. A pathway for disulfide bond formation in the periplasm. Proteins are secreted in a reduced state. DsbA rapidly transfers its disulfide to the folding protein. Reduced DsbA is then reoxidized by the inner membrane protein DsbB.

That DsbB is required for the reoxidation of DsbA was suggested by the observation that DsbA accumulates in a reduced form in DsbB mutants (Bardwell et al., 1993; Kishigami et al. 1995a). A direct interaction between the two proteins has been shown by the isolation of a DsbA–DsbB dimer covalently linked by a disulfide bond that links Cys-30 of DbsA with Cys-104 of DsbB (Guilhot et al., 1995 Kishigami et al., 1995b; Kishigami and Ito, 1996). It is not known what interacts with DsbB to affect its reoxidation; it could be reoxidized via the electron transport system or by an as yet uncharacterized specific oxidant.

∎ Mutagenesis studies with DsbB

All four cysteines present in the periplasmic domains of DsbB are essential for its function in reoxidizing DsbA. It is thought that a disulfide formed between two of these cysteines is transferred to DsbA resulting in its reoxidation (Jander et al., 1994; Guilhot et al., 1995; Kishigami and Ito, 1996).

∎ References

Alksne, L.E., Keeney, D., and Rasmussen, B.A. (1995). A mutation in either *dsbA* or *dsbB*, a gene encoding a component of a periplasmic disulfide bond-catalysing system, is required for high level expression of the *Bacteroides fragilis* metallo-β-lactamase, CrrA, in *Escherichia coli*. J. Bacteriol. **177**, 462–464.

Bardwell, J.C.A., Lee, J.-O., Jander, G., Martin, N., Belin, D., and Beckwith, J. (1993). A pathway for disulfide bond formation in vivo. Proc. Natl. Acad. Sci. USA **90**, 1038–1042

Belin, P. and Boquet, P.-L. (1993). Un second gène impliqué dans la formation des ponts disulfure de protéines localisées dans l'espace périplasmique de *Escherichia coli*. C. R. Acad. Sci. Paris **316**, 469–473.

Dailey, F.E. and Berg, H.C. (1993). Mutants in disulfide bond formation that disrupt flagellar assembly in *Escherichia coli*. Proc. Natl. Acad. Sci. USA **90**, 1043–1047.

Guilhot, C., Jander, G., Martin, N.L., and Beckwith, J. (1995). Evidence that the pathway of disulfide bond formation in *Escherichia coli* involves interactions between the cysteines of DsbB and DsbA. Proc. Natl. Acad. USA **92**, 9895–9899.

Jander, J., Martin, N.L., and Beckwith, J. (1994). Two cysteines in each periplasmic domain of the membrane protein DsbB are required for its function in protein disulfide bond formation. EMBO J. **13**, 5121–5127.

Kishigami, S., Akiyama, Y., and Ito, K. (1995a). Redox states of DsbA in the periplasm of *Escherichia coli*. FEBS Lett. **364**, 55–58.

Kishigami, S., Kanaya, E., Kikuchi, M., and Ito, K. (1995b). DsbA-DsbB interaction through their active site cysteines: evidence from an odd cysteine mutant of DsbA. J. Biol. Chem. **270**, 17072–17074.

Kishigami, S. and Ito, K. (1996). Roles of cysteine residues of DsbB in its activity to reoxidize DsbA, the protein disulfide bond catalyst of *Escherichia coli*. Genes Cells **1**, 201–208.

Missiakas, D., Georgopoulos, C., and Raina, S. (1993). Identification and characterization of the *Escherichia coli* gene *dsbB*, whose product is involved in the formation of disulfide bonds in vivo. Proc. Natl. Acad. Sci. USA **90**, 784–7088.

James C. A. Bardwell:
Department of Biology
University of Michigan
Ann Arbor, MI 48109-1048, USA

Tel. 1 313 764 8028
Fax. 1 313 647 0884
E-mail: J bardwel@biology.lsa.umich.edu

Escherichia coli DsbC

DsbC is a periplasmic protein originally identified as a multicopy suppressor of bacteria lacking the DsbA protein. This identification suggested that, like DsbA, DsbC belongs to the thioredoxin superfamily and is involved in the disulfide-coupled folding of secreted proteins.

■ *dsbC* gene and sequence

The sequence of the *dsbC* gene (Swiss-Prot accession number P21892) predicts a 25 kDa protein (Missiakas et al., 1994). The amino acid sequence homology between DsbC and other members of the thioredoxin superfamily is restricted to around the active site, with the following conserved motif, FXXXXCXXC. Except for this, the amino acid sequence of DsbC does not show any strong homology with other thiol–disulfide oxidoreductases from *Escherichia coli*, such as thioredoxin or DsbA, nor with the eukaryotic protein disulfide isomerase (PDI) proteins. *Escherichia coli* DsbC shares limited sequence homology with TcpG, the *Vibrio cholerae* DsbA homologue (Missiakas et al., 1994). A DsbC homologue has been described in *Erwinia chrysanthemi* (66% identical to *E. coli* DsbC; Shevchik et al., 1994).

The first 20 amino acids of DsbC are processed *in vivo* to yield a mature species of predicted M_r 23 460 Da containing 216 amino acids. The crystallographic structure of the protein is as yet still under investigation, but from the known three-dimensional structure of the 21 kDa DsbA protein (Martin et al., 1993), one could predict that just like DsbA, DsbC is also folded into two domains with its active site, FTDITCGYC, arranged into a thioredoxin fold. This thioredoxin fold is very well conserved in the DsbA structure (Martin et al., 1993), despite the absence of overall sequence homology between the two proteins. As with DsbA, the additional domain in DsbC might be more specifically involved in substrate recognition.

■ DsbC protein

DsbC was purified from periplasmic fractions and found to be a dimer (Missiakas et al., 1994; Zapun et al., 1995). The molecular weight of the monomeric form as determined by electrospray mass spectrometry was 23 458, which is in good agreement with the 23 460 Da predicted mass.

The thiol–disulfide chemistry of the active site of the protein has been investigated using techniques similar to those used for DsbA (Zapun et al., 1993, 1995; Wunderlich and Glockshuber, 1993). The disulfide bond appears to be very unstable and reactive (Zapun et al., 1995). The equilibrium constant for its formation by interchange with glutathione is 200 mM which is rather close to the value found for DsbA (80 mM). As a comparison, a 10 M equilibrium constant value has been measured for the formation of the active site of thioredoxin (Holmgren, 1981) and values up to 10^5 M can be measured for normal stabilizing bonds in folded proteins (Creighton and Goldenberg, 1984). In addition, as is the case with DsbA and thioredoxin, only the most N-terminal cysteine residue in the active site of DsbC has an accessible thiol (Zapun et al., 1995).

The unstable and reactive disulfide bond of DsbC is rapidly transferred *in vitro* to reduced BPTI (bovine pancreatic trypsin inhibitor, which contains three disulfides in the native state). Comparative analyses of the kinetic properties of the instability of both DsbA and DsbC active sites have led to the suggestion that DsbA is more efficient at transferring its disulfide bond to other proteins (Zapun et al., 1993) whereas DsbC is more efficient at catalysing disulfide rearrangements (Zapun et al., 1995). Like PDI, DsbC is also able to bypass disulfide rearrangements in quasi-native species of BPTI by incorporating directly the missing disulfide. This suggests a putative unfoldase activity for the DsbC protein. The fact that DsbC is a dimer, unlike DsbA, could somehow be a clue to this more PDI-like behavior, since the active monomeric PDI contains two redox active sites in its amino acid sequence.

There are two additional cysteine residues in the DsbC sequence. However, replacement of these two cysteines by site-directed mutagenesis did not affect the catalytic properties of the protein (Zapun et al., 1995). These two residues, cysteine-141 and cysteine-163, were found to be paired in a disulfide bond.

Biological activity of DsbC

DsbC was originally isolated because it could, when expressed from a multicopy plasmid, revert most of the phenotypic defects caused by the lack of functional DsbA protein *in vivo* (such as the lack of alkaline phosphatase activity or motility) (Missiakas et al., 1994; Shevchik et al., 1994). *In vivo*, it appears that DsbC functions rather independently of the DsbA/DsbB system (Missiakas et al., 1993; Bardwell, 1994) and that unlike DsbA, the DsbC redox active site is not recycled by the inner membrane DsbB protein (Missiakas et al., 1994; Kishigami et al., 1995). A *dsbC* null mutant impairs more severely the folding of multiple disulfide bond-containing proteins such as alkaline phosphatase and penicillin-binding protein 4 (Missiakas et al., 1994, 1995). Although these folding defects are not always as dramatic as in a *dsbA* null mutant, they are severe enough to trigger a stress response from the extracytoplasmic space (Raina et al., 1995). This stress response is reminiscent of those induced by protein misfolding in the cytoplasm (known as the heat shock response; Georgopoulos et al., 1994) or in the endoplasmic reticulum (the unfolded protein response; Gething et al., 1994).

References

Bardwell, J.C.A. (1994). Building bridges: disulphide formation in the cell. Mol. Microbiol. **14**, 199–205.

Creighton, T.E. and Goldenberg, J. (1984). Kinetic role of a meta-stable native-like two-disulfide species in the folding transition of bovine pancreatic trypsin inhibitor. J. Mol. Biol. **179**, 497–526.

Georgopoulos, G., Liberek, K., Zylicz, M., and Ang, D. (1994). Properties of the heat shock proteins of Escherichia coli and the autoregulation of the heat shock response. In *The biology of heat shock proteins and molecular chaperones*, (R.I. Morimoto, A. Tissières and C. Georgopoulos, eds). Cold Spring Harbor Laboratory Press, New York, pp. 209–249.

Gething, M.-J., Blond-Elguindi, S., Mori, K., and Sambrook, J.F. (1994). Structure, function and regulation of the endoplasmic reticulum chaperone, BiP. In *The biology of heat shock proteins and molecular chaperones*, (R.I. Morimoto, A. Tissières and C. Georgopoulos, eds). Cold Spring Harbor Laboratory Press, New York, pp. 111–135.

Holmgren, A. (1981). Thioredoxin: structure and functions. Trends Biochem. Sci. **6**, 26–29.

Kishigami, S., Kanaya, E., Kikuchi, M., and Ito, K. (1995). DsbA-DsbB interaction through their active site cysteines. J. Biol. Chem. **270**, 17072–17074.

Martin, J.L., Bardwell, J.C.A., and Kuriyan, J. (1993). Crystal structure of the DsbA protein required for disulphide bond formation *in vivo*. Nature **365**, 464–468.

Missiakas, D., Georgopoulos, C., and Raina, S. (1993). Identification and characterization of the *Escherichia coli* gene *dsbB*, whose product is involved in the formation of disulfide bonds *in vivo*. Proc. Natl. Acad. Sci. USA **90**, 7084–7088.

Missiakas, D., Georgopoulos, C. and Raina, S. (1994). The *Escherichia coli dsbC* (*xprA*) gene encodes a periplasmic protein involved in disulfide bond formation. EMBO J. **13**, 2013–2020.

Missiakas, D., Schwager, F., and Raina, S. (1995). Identification and characterization of a new disulfide-isomerase like protein (DsbD) in *Escherichia coli*. EMBO J. **14**, 3415–3424.

Raina, S., Missiakas, D., and Georgopoulos, C. (1995). The *rpoE* gene encoding the σ^E (σ^{24}) heat shock sigma factor of *Escherichia coli*. EMBO J. **14**, 1043–1055.

Shevchik, V.E., Condemine, G., and Robert-Baudouy, J. (1994). Characterization of DsbC, a peiplasmic protein of *Erwinia chrysanthemi* and *Escherichia coli* with disulfide isomerase activity. EMBO J. **13**, 2007–2012.

Wunderlich, M. and Glockshuber, R. (1993). Redox properties of protein disulfide isomerase (DsbA) from *Escherichia coli*. Protein Sci. **2**, 717–726.

Zapun, A., Bardwell, J.C.A., and Creighton, T.E. (1993). The reactive and destabilizing disulfide bond of DsbA, a protein required for protein disulfide bond formation *in vivo*. Biochemistry **32**, 5083–5092.

Zapun, A., Missiakas, D., Raina, S., and Creighton, T.E. (1995). Structural and functional characterization of DsbC, a protein involved in disulfide bond formation in *Escherichia coli*. Biochemistry **34**, 5075–5089.

Dominique Missiakas:
CNRS UPR9027
31, Chemin J. Aiguier
13402 Marseille Cedex 20, France
Tel. 33 4 91 76 03 59
Fax. 33 4 91 71 21 24
E-mail. missiaka@ibsm.cnrs-mrs.fr

André Zapun and Tom E. Creighton:
European Molecular Biology Laboratory
Meyerhofstrasse 1
D-69012 Heidelberg, Germany

Satish Raina:
Université de Genève
Centre Médical Universitaire
1, Rue Michel-Servet 1211 Genève 4, Switzerland
Tel. 41 22 702 55 11
Fax. 41 22 702 55 02
E-mail: Satish.Raina@medecine.unige.ch

Escherichia coli DsbD

The dsbD gene was identified during a search for extragenic mutations able to compensate for the lack of dsbA oxidase activity in vivo. DsbD is an inner membrane protein with a soluble 16 kDa C-terminal domain sharing 45% identity with the active site-carrying domain of eukaryotic PDI. The role of DsbD in the disulfide coupled folding of secreted proteins in vivo is likely to be related to its reductase activity.

■ Alternative names

DipZ (Crooke and Cole, 1995), CutA2 (Fong et al., 1995).

■ Isolation of *dsbD*

DsbA and DsbC periplasmic proteins of *Escherichia coli* (see entries pp. 318 and 324) are two key players involved in disulfide bond formation. DsbA has been shown to act as an oxidase (Wunderlich and Glockshuber, 1993; Zapun et al., 1993) and DsbC exhibits the dual properties of an oxidase and an isomerase (Zapun et al., 1995). These findings suggested that a reductase activity could also be required in the bacterial periplasm to facilitate disulfide bond rearrangement and some specific processes requiring reduced cysteines. Isolation of extragenic suppressors of *dsbA* null mutants led to the identification of such a putative activity encoded by a gene designated as *dsbD* (Missiakas et al., 1995). Such suppressors partly alleviated the lethal growth phenotype of *dsbA* mutants (but not of *dsbC⁻* bacteria) which had been observed when these mutant cells were grown in the presence of dithiothreitol or benzylpenicillin (Missiakas et al., 1993). However, this suppression effect requires a functional DsbC protein. This substantiates further the importance of both DsbA and DsbC proteins for *in vivo* disulfide bond formation.

The *dsbD* gene was also cloned on the basis of its requirement for cytochrome biogenesis (Crooke and Cole, 1995). DsbD could maintain the cysteine residues of the apocytochrome into a reduced state to allow proper covalent linkage with the heme.

■ *dsbD* gene and sequence

The nucleotide sequence of the *dsbD* gene (Swiss-Prot accession number P36655) predicts that it encodes a 53 kDa polypeptide of 489 amino acids with at least 6–8 potential transmembrane domains within the 37 kDa N-terminal domain (Missiakas et al., 1995). The protein is indeed located in the cytoplasmic membrane (D. Missiakas and S. Raina, unpublished data). The amino acid sequence of the last 16 kDa C-terminal domain suggests that DsbD could belong to the thioredoxin superfamily. This domain is also predicted to be highly hydrophilic and is presumed to face the periplasmic space. A global sequence search shows that this C-terminal part is most closely related to eukaryotic PDI and a score of about 40–45% sequence identity with a range of eukaryotic PDIs is obtained. There are two CXXC motifs in the DsbD sequence, but the Cys^{403}–Val–Ala–Cys^{406} redox site has been proposed to be the catalytic site. This was based on the finding that the secreted truncated C-terminal domain carrying only this motif could partly restore the phenotypic defects associated with *dsbD⁻* bacteria (Missiakas et al., 1995).

■ DsbD protein

Because of its membrane location, only a 15 kDa soluble C-terminal domain of the protein has been purified. This small domain has been subcloned in frame with a typical *E. coli* signal sequence, in order to target it to the periplasm (its suspected cellular location). The potential thiol–disulfide oxidoreductase activity of this fragment was tested *in vitro* using an assay described by Holmgren (1979) for thioredoxin. In this assay, the catalyst added accelerates the reduction between the two chains of insulin in the presence of dithiothreitol, leading to the aggregation of the B chain. The purified PDI-like domain of DsbD did accelerate this aggregation process (Missiakas et al., 1995). The redox properties of the protein are currently being investigated.

■ Biological activities of DsbD

In vivo analyses of bacteria carrying mutations only in the *dsbD* gene have shown that disulfide bond formation is partly impaired (Missiakas et al., 1995). This has been attributed to an imbalance in the ratio between oxidized and reduced forms of both DsbA and DsbC proteins. Both proteins accumulate in the oxidized state and the phenotype observed with *dsbD⁻* bacteria is probably owing to excessive oxidation processes of disulfide-cysteine containing proteins mediated by DsbA and DsbC in a random manner. These observations are in agreement with the presumed function of DsbD, i.e. that of a thiol–disulfide reductase. It is possible that DsbC isomerase activity directly depends on DsbD. Indeed, such an activity implies that the thiol group of cysteine-98 in the active site of DsbC is free and accessible.

Surprisingly, the typical σ^{24}/σ^E-dependent stress response which reflects the damaging of proteins in the extracytoplasmic space is more induced in a dsbD⁻ bacteria compared with dsbA⁻ or dsbC⁻ strains (Missiakas et al., 1995; Raina et al., 1995). Also, unlike other dsb mutants, dsbD⁻ bacteria are temperature sensitive (ts⁻) for growth at temperatures above 42°C. These two observations suggest an absolute requirement for DsbD function under stress conditions and a multiplicity of roles in the physiology of E. coli.

Biological regulation of DsbD

dsbD transcription has been shown to initiate from a unique site, located 162 nucleotides upstream of the putative GTG initiation codon; a two-gene operon dsbD orf2 is transcribed from this start site (Missiakas et al., 1995). The nucleotide sequence of this start site does not resemble those transcribed by σ^{70}, the housekeeping E. coli transcription factor. Since dsbD mutants are ts⁻ for growth, the putative regulatory effect of the temperature was examined (heat shock genes in E. coli are transcribed by specific sigma factors). Transcription of the gene was found to be decreased at 50°C but not completely turned off, as otherwise observed for housekeeping genes (Missiakas et al., 1995). The specific mechanism of transcriptional regulation of the operon remains to be elucidated.

References

Crooke, H. and Cole, J. (1995). The biogenesis of c-type cytochromes in Escherichia coli requires a membrane-bound protein, DipZ, with a protein disulphide isomerase-like domain. Mol. Microbiol. **15**, 1139–1150.

Fong, S.-T., Camakaris, J., and Lee, B.T.O. (1995). Molecular genetics of a chromosomal locus involved in copper tolerance in Escherichia coli K-12. Mol. Microbiol. **15**, 1127–1137.

Holmgren, A. (1979). Thioredoxin catalyzes the reduction of insulin disulfides by dithiothreitol and dihydrolipoamide. J. Biol. Chem. **254**, 9627–9632.

Missiakas, D., Georgopoulos, C., and Raina, S. (1993). Identification and characterization of the Escherichia coli gene dsbB, whose product is involved in the formation of disulfide bonds in vivo. Proc. Natl. Acad. Sci. USA **90**, 7084–7088.

Missiakas, D., Schwager, F., and Raina, S. (1995). Identification and characterization of a new disulfide-isomerase like protein (DsbD) in Escherichia coli. EMBO J. **14**, 3415–3424.

Raina, S., Missiakas, D., and Georgopoulos, C. (1995). The rpoE gene encoding the σ^E (σ^{24}) heat shock sigma factor of Escherichia coli. EMBO J. **14**, 1043–1055.

Wunderlich, M. and Glockshuber, R. (1993). Redox properties of protein disulfide isomerase (DsbA) from Escherichia coli. Protein Sci. **2**, 717–726.

Zapun, A., Bardwell, J.C.A., and Creighton, T.E. (1993). The reactive and destabilizing disulfide bond of DsbA, a protein required for protein disulfide bond formation in vivo. Biochemistry **32**, 5083–5092.

Zapun, A., Missiakas, D., Raina, S., and Creighton, T.E. (1995). Structural and functional characterization of DsbC, a protein involved in disulfide bond formation in Escherichia coli. Biochemistry **34**, 5075–5089.

■ *Dominique Missiakas:*
CNRS UPR9027
31, Chemin J. Aiguier
13402 Marseille Cedex 20, France
Tel. 33 4 91 76 03 59
Fax. 33 4 91 71 21 24
E-mail. missiaka@ibsm.cnrs-mrs.fr

■ *Satish Raina:*
Université de Genève
Centre Médical Universitaire
1, Rue Michel-Servet 1211 Genève 4, Switzerland
Tel. 41 22 702 55 11
Fax. 41 22 702 55 02
E-mail: Satish.Raina@medecine.unige.ch

Escherichia coli DsbE

Mutations in the dsbE *gene were isolated because they conferred hypersensitivity to changes in the redox potential of* Escherichia coli *growth medium. The wild-type gene encodes a periplasmic protein with a thiol–disulfide reductase activity .*

Isolation of *dsbE*

One of the genetic approaches that has led to the identification of genes whose products are involved in thiol–disulfide exchange reactions was based on the rationale that if such activities existed they should be inhibited in vivo by reducing agents such as dithiothreitol (DTT) (Missiakas et al., 1993). Indeed, wild-type E. coli could cope with concentrations of DTT as high as 10 mM in the growth medium, whereas some insertional mutations could be selected as a result of their increased sensitivity to low DTT concentration (Missiakas et al., 1993). Five major targets of insertional mutations were the dsbA, dsbB, dsbD, trxA and trxB genes. Some other 'DTT sensitive' loci have been re-examined more recently. One of those was mapped at 49 minutes on the E. coli chromosome and the corresponding wild-type gene was designated as dsbE (D. Missiakas and S. Raina, manuscript submitted).

■ dsbE gene and sequence

The dsbE gene (Swiss-Prot accession number P33926) encodes a 20 809 Da protein of 185 amino acids in length, which is processed to a mature species of 17 737 Da localized to the periplasmic space (D. Missiakas and S. Raina, manuscript submitted). From its amino acid sequence, DsbE appears to be more related to the cytoplasmic thioredoxin protein, especially around the presumed active site. This predicted site in DsbE reads VWATWCPTC, and in thioredoxin it has been shown to be FWAEWCGPC. As was observed with DsbC or DsbA, the DsbE protein probably carries an extra domain in addition to its thioredoxin-like moiety.

■ DsbE protein

The DsbE protein has been purified from the periplasmic fraction; it partly copurifies with the soluble periplasmic cytochromes. Cysteines-63 and -66 form the redox active site. The purified DsbE protein accelerated the aggregation of insulin in the presence of DTT as described for thioredoxin by Holmgren (1979). The protein contains no other cysteines than those of the presumed redox active site Cys^{63}–Pro–Thr–Cys^{66}. Replacement of cysteine-63 by an alanine completely abolished the ability of DsbE to reduce insulin in vitro (D. Missiakas and S. Raina, manuscript submitted).

■ Biological activity of DsbE

In vivo some of the phenotypic defects associated with dsbD mutant bacteria could be partly restored by overexpressing DsbE (Missiakas et al., 1995). For example, the stress response induced upon accumulation of unstable protein in dsbD⁻ bacteria was suppressed by introduction of a dsbE-containing plasmid. It is tempting to suggest that the major activity of DsbE is that of a reductase, since increasing this activity in a dsbD null mutant helps correct the excessive oxidation processes mediated by DsbA and DsbC. The fact that a maturation defect of c-type cytochromes is also observed in a dsbE⁻ strain (D. Missiakas and S. Raina, manuscript submitted), as is the case with dipZ (dsbD) mutants (Crooke and Cole, 1995), suggests that both proteins might be involved in some common cellular processes.

■ References

Crooke, H. and Cole, J. (1995). The biogenesis of c-type cytochromes in Escherichia coli requires a membrane-bound protein, DipZ, with a protein disulphide isomerase-like domain. Mol. Microbiol. **15**, 1139–1150.

Holmgren, A. (1979). Thioredoxin catalyzes the reduction of insulin disulfides by dithiothreitol and dihydrolipoamide. J. Biol. Chem. **254**, 9627–9632.

Missiakas, D., Georgopoulos, C., and Raina, S. (1993). Identification and characterization of the Escherichia coli gene dsbB, whose product is involved in the formation of disulfide bonds in vivo. Proc. Natl. Acad. Sci. USA **90**, 7084–7088.

Missiakas, D., Schwager, F., and Raina, S. (1995). Identification and characterization of a new disulfide-isomerase like protein (DsbD) in Escherichia coli. EMBO J. **14**, 3415–3424.

■ *Dominique Missiakas:*
CNRS UPR9027
31, Chemin J. Aiguier
13402 Marseille Cedex 20, France
Tel. 33 4 91 76 03 59
Fax. 33 4 91 71 21 24
E-mail. missiaka@ibsm.cnrs-mrs.fr

■ *Satish Raina:*
Université de Genève
Centre Médical Universitaire
1, Rue Michel-Servet 1211 Genève 4, Switzerland
Tel. 41 22 702 55 11
Fax. 41 22 702 55 02
E-mail: Satish.Raina@medecine.unige.ch

Synechococcus TxlA

The TxlA protein is present in cyanobacteria and is a member of the protein disulfide oxidoreductase family. The gene has been sequenced, but the subcellular location of the protein is not known. Based on gene inactivations, TxlA protein is involved, either directly or indirectly, in controlling the function and structure of the photosynthetic apparatus.

▪ txlA gene and sequence

The nucleotide sequence of the *txlA* gene was determined (Collier and Grossman, 1995) from a genomic clone isolated from the cyanobacterium *Synechococcus* sp. strain PCC7942 (GenBank accession number U05044). The sequence predicts a protein of 191 amino acids, with an isolelectric point of 4.2. While the N-terminal 37 amino acids and the C-terminal 44 amino acids show no homology to sequences in the GenBank database, the central sequence of 110 amino acids is similar to a number of protein disulfide oxidoreductases, including thioredoxins (Buchanan, 1991) and protein disulfide isomerases (Bardwell and Beckwith, 1993). As an example, the central region of the molecule is 29% identical and 60% similar to TrxM (thioredoxin *m* of *Synechococcus* sp. strain PCC7942) (Muller and Buchanan, 1989).

▪ Biological regulation of TxlA

The *txlA* gene is expressed under all conditions that have been tested (J. Collier and A. Grossman, unpublished data). Interestingly, during sulfur and nitrogen limitation, transcripts that originate from the promoter of a gene designated *nblA*, which encodes a protein involved in the degradation of phycobilisomes during nutrient limitation (Collier and Grossman, 1994), read through the *txlA* gene on the antisense strand. This suggests that the synthesis of the TxlA protein may be regulated during nutrient limitation.

▪ Mutagenesis studies and possible function of TxlA

While we still do not know the precise function of TxlA, it either directly or indirectly influences the structure and function of the cyanobacterial photosynthetic apparatus (Collier and Grossman, 1995). A lesion in the *txlA* gene leads to the production of excess phycobilisomes. These phycobilisomes are structurally and functionally indistinguishable from the phycobilisomes present in wild-type cells. However, the excess phycobilisomes present in one of the mutant strains, designated *txlXb*, were not able to transfer harvested light energy to the photosynthetic reaction centers. Furthermore, the growth of the *txlXb* strain, under conditions that favor the rapid growth of *Synechococcus* sp. strain PCC7942, was slower than that of wild-type cells. It is hypothesized that the changes in pigmentation in the *txlXb* mutant may be indirect and result from an underlying abnormality in cellular metabolism (Collier and Grossman, 1995).

▪ References

Bardwell, J.C.A. and Beckwith J. (1993). The bonds that tie: catalyzed disulfide bond formation. Cell **74**, 769–771.

Buchanan, B.B. (1991). Regulation of CO_2 assimilation in oxygenic photosynthesis: the ferredoxin/thioredoxin system. Arch. Biochem. Biophys. **288**, 1–9.

Collier, J.L. and Grossman, A.R. (1994). A small polypeptide triggers complete degradation of the light harvesting phycobilisomes in nutrient-deprived cyanobacteria. EMBO J. **13**, 1039–1047.

Collier, J.L. and Grossman A.R. (1995). Disruption of a gene encoding a novel thioredoxin-like protein alters the cyanobacterial photosynthetic apparatus. J. Bacteriol. **177**, 3269–3276.

Muller, E.G.D. and Buchanan, B.B. (1989). Thioredoxin is essential for photosynthetic growth: the thioredoxin m gene of Anacystis nidulans. J. Biol. Chem. **264**, 4008–4014.

▪ Arthur Grossman:
The Carnegie Institution of Washington
Department of Plant Biology
290 Panama Street
Stanford, CA 94305, USA
Tel. 1 415 325 1521 ext. 212
Fax. 1 415 325 6857
E-mail. arthur@andrew.stanford.edu

▪ Jackie Collier:
Rensselaer Polytechnic Institute
Biology 304 Materials Research Center
110 8th Street
Troy, NY 12180-3590 USA
Tel. 1 518 276 2178
Fax. 1 518 276 2344
E-mail: collij3@rpi.edu

Bacillus brevis Bdb

The Bdb protein is a protein thiol–disulfide oxidoreductase found in Bacillus brevis, a Gram-positive bacterium. It is a secreted protein bound to the cell periphery in the wild-type bacterium, and found mostly in the culture supernatant in the overproducing B. brevis. Bdb contains a well-conserved motif, Cys–X–X–Cys, in its amino acid sequence, and is enzymatically active in both disulfide bond reduction and oxidative refolding of reduced RNase A. Bdb showed significant homology to several bacterial thioredoxins, but it showed no sequence similarity to that of Escherichia coli DsbA, although it is functionally similar to DsbA.

■ *bdb* gene and sequence

bdb was cloned from *Bacillus brevis* as a gene that complements the *Escherichia coli dsbA* mutation, and encodes an 118 amino acid precursor with a signal sequence of 27 amino acids (Ishihara et al., 1995) (GenBank accession number D37936). The calculated molecular weight of the mature Bdb protein is 10 268 Da. Bdb contains, near the N-terminus, the short and characteristic amino acid sequence Cys–X–X–Cys, which is highly conserved among disulfide oxidoreductases such as eukaryotic protein disulfide isomerase and thioredoxin. The sequence motif Cys–X–X–Cys functions as the active site of the thiol–disulfide exchange of protein. The deduced amino acid sequence of Bdb showed no sequence similarity to that of *E. coli* DsbA, which is a protein disulfide oxidoreductase and essential for the disulfide bond formation of periplasmic proteins (Bardwell et al., 1991; Kamitani et al., 1992; see entry p. 318). On the other hand, Bdb showed significant sequence similarity to several different bacterial thioredoxins; for example, 23% identity with *E. coli* thioredoxin (Ishihara et al., 1995).

■ Bdb protein

A protein that cross-reacted with the antibody to Bdb was detected mainly in cell lysates, whereas negligible amounts of Bdb were found in the culture supernatant (Ishihara et al., 1995). The molecular mass of cellular Bdb was consistent with that of mature Bdb, suggesting that Bdb is localized at the periphery of the cell. The Bdb protein overproduced in *B. brevis* accumulated mostly in the culture broth, but the amount of Bdb detected in the lysate of overproduced cells was equivalent to that of non-overproducing cells.

■ Bdb function

Bdb stimulated insulin reduction, and the oxidative refolding of reduced RNase A, which was determined by RNase activity (Ishihara et al., 1995). These data indicate that Bdb of *B. brevis* is active *in vitro* in both reduction and oxidation of disulfide bonds. Bdb is active in *E. coli* cells, since the *bdb* gene complemented the *E. coli dsbA* mutation, restoring motility by means of flagella and alkaline phosphatase activity. Thus, Bdb seems to be functionally similar to *E. coli* DsbA (Ishihara et al., 1995).

Since no proteins, including cell wall proteins (Tsuboi et al., 1988), secreted by *B. brevis* have any disulfide bonds, identification of true substrate proteins for Bdb is essential to characterize the biological role of Bdb in *B. brevis*. A vector system for efficient production of heterologous proteins has been developed with *B. brevis* as the host. Using this system, several proteins with multiple disulfide bonds and the same activity as native proteins can be efficiently produced (Udaka and Yamagata, 1993). Bdb may play an important role in the disulfide bond formation of these heterologous proteins on the cell surface of *B. brevis*.

■ References

Bardwell, J.C.A., McGovern, K., and Beckwith, J. (1991). Identification of a protein required for disulfide bond formation in vivo. Cell **67**, 581–589.

Ishihara, T., Tomita, H., Hasegawa, Y., Tsukagoshi, N., Yamagata, H., and Udaka, S. (1995). Cloning and characterization of the gene for a protein thiol-disulfide oxidoreductase in *Bacillus brevis*. J. Bacteriol. **177**, 745–749.

Kamitani, S., Akiyama, Y., and Ito, K. (1992). Identification and characterization of an *Escherichia coli* gene required for the formation of correctly folded alkaline phosphatase, a periplasmic enzyme. EMBO J. **11**, 57–62.

Tsuboi, A., Uchihi, R., Adachi, T., Sasaki, T., Hayakawa, S., Yamagata, H., Tsukagoshi, N., and Udaka, S. (1988). Characterization of the gene for the hexagonally arranged surface layer protein in protein-producing *Bacillus brevis* 47: complete nucleotide sequence of the middle wall protein gene. J. Bacteriol. **170**, 935–945.

Udaka, S. and Yamagata, H. (1993). High-level secretion of heterologous proteins by *Bacillus brevis*. Meth. Enzymol. **217**, 23–33.

■ Shigezo Udaka:
Department of Brewing and Fermentation
Tokyo University of Agriculture
1-1 Sakuragaoka, Setagaya
Tokyo 156, Japan
Tel. 81 3 5477 2385
Fax. 81 3 5477 2622

Rhodobacter capsulatus HelX

HelX is a thioredoxin-related protein exposed on the outside of the cytoplasmic membrane of bacteria. It is required for the periplasmic biosynthesis of c-type cytochromes. HelX appears to be a specific thiol reducing protein (for vicinal CXXC motifs) rather than a general thioredoxin. Its in vivo substrate may be the Ccl2 protein which itself possesses a CPVC motif whereby Ccl2 subsequently reduces the CXXCH domain of apocytochromes c for the thioether ligation to heme.

■ Alternative names

TlpB, CycY, CcmG (Thony-Meyer et al., 1995).

■ *helX* gene and sequence analysis

The *helX* gene of the photosynthetic bacterium *Rhodobacter capsulatus* (Beckman and Kranz, 1993; GenBank accession number M96013) encodes a protein of 176 amino acids with a traditional signal sequence that is related to the class of disulfide oxidoreductases called thioredoxins. Proteins of this class contain a conserved active site (FWAXWCXPCR). The functions of members of the disulfide oxidoreductase family of proteins in bacteria have recently been reviewed (Loferer and Hennecke, 1994). DsbA is a general periplasmic disulfide oxidoreductase, the function of which is to catalyse the formation of disulfide bonds in proteins that are secreted through the cytoplasmic membrane (also reviewed by Bardwell, 1994; Creighton et al. 1995; see entry p. 318). TrxA is the classic cytoplasmic thioredoxin involved in general thiol reduction reactions inside the cell.

■ Mutagenesis studies with *helX*

The *helX* gene, when deleted from the chromosome of *Rhodobacter capsulatus* yields a strain that is unable to make c-type cytochromes (Beckman and Kranz, 1993). Cytochromes that are c-type contain heme with vinyl groups that are covalently bonded to the protein by two thioether linkages (i.e. to CXXCH of the polypeptide where histidine acts as an axial ligand to the Fe of heme). This assembly is proposed to occur outside the cytoplasmic membrane in bacteria and both the cysteine residues and the heme are required to be in the reduced state for ligation.

Mutation of the active site cysteine residues of HelX demonstrated that both are required for function (E.M. Monika, B.S. Goldman, D.L. Beckman and R.G. Kranz, manuscript submitted).

■ Topological and immunological studies with HelX

Beckman and Kranz (1993) showed that the HelX protein is secreted outside the cytoplasmic membrane in both *R. capsulatus* and *Escherichia coli*. This was proven by using *helX* gene fusions to the topological reporters *lacZ* and *phoA*. Recently, HelX has been overproduced in *E. coli* and purified from the periplasmic shock fraction (E.M. Monika, B.S. Goldman, D.L. Beckman and R.G. Kranz, manuscript submitted). This purified preparation contains 70% unprocessed (i.e. signal not cleaved) and 30% processed HelX. Using antibodies to HelX it appears that the major form in wild-type *R. capsulatus* is the unprocessed form whereby HelX is loosely associated with the outer surface of the cytoplasmic membrane. Such a location is consistent with the view that other components required for cytochrome c biogenesis (e.g. HelABCD and Ccl12) are all integrally associated with the cytoplasmic membrane, with putative active sites exposed to the outside (Kranz, 1989; Beckman et al., 1992).

■ HelX biochemical activities and functions

At least four biochemical functions could be envisioned for HelX. (1) A general thioredoxin (like TrxA) for proteins in the periplasm, including apocytochromes c. (2) A general thioredoxin for the reduction of heme iron which is a prerequisite for ligation. (3) A specific thioredoxin for apocytochromes c cysteines (i.e. CXXCH). (4) A specific thioredoxin for the Ccl2 protein which contains CXXC and may directly reduce apocytochromes c.

These four possibilities have been addressed (Kranz et al., 1996). The purified HelX protein does not act as a general thioredoxin, at least by the criterion that it does not reduce the disulfide of insulin (as does TrxA). Moreover, HelX does not reduce oxidized heme as detected spectrophotometrically in solution with DTT as the reductant for HelX. However, apocytochrome c peptide (containing CXXCH) and Ccl2 both oxidize the cysteines of HelX at micromolar concentrations. Additionally, oxidized Ccl2 is partially reduced by HelX at these same concentrations. Thus, HelX is envisioned as a specific thioredoxin protein, probably for Ccl2, with Ccl2 subsequently delivering and reducing apocytochromes c to reduced hemin. It is proposed that this later step, heme presentation, is carried out by the Ccl1 protein using a conserved putative heme-binding, tryptophan-rich domain (i.e. WWD domain; Beckman et al., 1992).

References

Bardwell, J.C. (1994) Building bridges: disulphide bond formation in the cell. Mol. Microbiol. **14**, 199–205.

Beckman, D.L. and Kranz, R.G. (1993). Cytochromes c biogenesis in a photosynthetic bacterium requires a periplasmic thioredoxin-like protein. Proc. Natl. Acad. Sci. USA **90**, 2179–2183.

Beckman, D.L., Trawick, D.R., and Kranz, R.G. (1992). Bacterial cytochromes c biogenesis. Genes Dev. **6**, 268–283.

Creighton, T.E., Zapun, A., and Darby, N. (1995). Mechanisms and catalysts of disulphide bonde formation in proteins. Trends Biotechnol. **13**, 18–23.

Kranz, R.G. (1989). Isolation of mutants and genes involved in cytochromes c biosynthesis in *Rhodobacter capsulatus*. J. Bacteriol. **171**, 456–464.

Loferer, H. and Hennecke, H. (1994). Protein disulfide oxidoreductases in bacteria. Trends Biochem. Sci. **19**, 169–171.

Thony-Meyer, L., Fischer, F., Kunzler, P., Ritz, D., and Hennecke, H. (1995). Escherichia coli genes required for cytochrome c maturation. J. Bacteriol. **177**, 4321–4326.

■ Robert G. Kranz:
Biology Department
Washington University
One Brookings Dr.
St. Louis, MO 63130, USA
Tel. 1 314 935 4278
Fax. 1 314 935 4432
E-mail: kranz@wustlb.wustl.edu

Bradyrhizobium japonicum TlpA

The thioredoxin-like protein TlpA was detected by analysing a respiration-deficient transposon mutant of Bradyrhizobium japonicum, a Gram-negative soil bacterium that undergoes root nodule symbiosis with soybean plants. TlpA is anchored to the cytoplasmic membrane at its N-terminus, leaving its thioredoxin-like domain exposed to the periplasm. The phenotype of a transposon mutant strain indicates that TlpA is involved in the biogenesis of the terminal oxidase cytochrome aa_3 and in the development of functional endosymbiosis. Biophysical studies on a purified soluble form suggest that TlpA functions as a disulfide reductase in the periplasm.

■ Isolation of the *tlpA* gene

A random Tn5 mutagenesis of *Bradyrhizobium japonicum* strain 110 was performed and the mutant bank screened for respiration-deficient strains using the electron donor TMPD (N,N,N',N'-tetramethyl-p-phenylenediamine). Three independent Tn5 insertions, which resulted in defective TMPD oxidation, were located in the *tlpA* gene (Loferer et al., 1993).

■ TlpA sequence

The *tlpA* gene was mapped at 54 minutes on the *Bradyrhizobium japonicum* chromosome (Kündig et al., 1993). Its nucleotide sequence is 666 base pairs in length (GenEMBL accession number Z23140) and encodes a protein of 221 amino acid residues (Loferer et al., 1993). A hydrophobic stretch of 22 amino acids (from position 14 to 35) is present at the N-terminal region of TlpA. The amino acids preceding this hydrophobic domain exhibit a bias for positively charged side chains suggesting a cytoplasmic location of the N-terminus. The remaining bulk portion of the polypeptide shows extensive homology with the thioredoxin family of proteins (32% overall identity to thioredoxin from *Escherichia coli*), including the active site sequence WCVPC (positions 106–110) (Eklund et al., 1991; Loferer et al., 1993). However, a stretch of 28 amino acids (positions 150–177) with no detectable homology to any known protein sequence is inserted into the thioredoxin-like domain of TlpA (Fig. 1). TlpA contains cysteine residues not only in its active site but also at positions 45 and 190 in the linear sequence (Loferer et al., 1993).

■ TlpA protein

TlpA is a 23 kDa member of the thioredoxin family. Experiments using translational fusions to β-galactosidase and alkaline phosphatase reporters demonstrate that TlpA is anchored to the cytoplasmic membrane at its N-terminal hydrophobic segment, whereas the thioredoxin-like domain is exposed to the periplasmic space (Loferer et al., 1993). A soluble form of TlpA (TlpA$_{sol}$), which lacks its membrane anchor, was produced using two different expression systems in *E. coli* and was consequently purified to homogeneity. First, TlpA$_{sol}$ was affinity

```
                         +    ++
BjTlpA      MLDTKPSATRRIPLVIATVAVGGLAGFAALYGLGLSRAPTGDPACRAAVATAQKIAPLAH      60
BjTlpB              MSEQSTSANPQQRRTFLMVLPLIAFIGLALLFWFRLGSGDPSRIPS              46
RcHelX                        MAKPLMFLPLLVMAGFVGAGYFAMQQND                  28

                                           ——β₂——        ——α₂——
BjTlpA      GEVAALTMASAPLKLPDLAFEDADGKPKKLSDFRGKTLLVNLWATWCVPCRKEMPALDEL   120
EcTrxA                  SDKIIHLTDDSFDTDVLKADGAILVDFWAEWCGPCKMIAPILDEI    45
RsTrxA                  STVPVTDATFDTEVRKSDVPVVVDFWAEWCGPCRQIGPALEEL    43
RrTrxA                  MKQVSDASFEEDVLKADGPVxVDFWAEWCGPCRQxAPALEEL     42
CnTrx-c2                SATIVNTTDENFQADVLDAETPVLVDFWAGWCAPCKAIAPVLEEL   45
BjTlpB      ALIGRPAPQTALPPLEGLQADNVQVPGLDPAAFKGKVSLVNVWASWCVPCHDEAPLLTEL   106
RcHelX      PNAMPTALAGKEAPAVRLEPLGAEAPFTDADLRDGKIKLVNFWASWCAPCRVEHPNLIGL    88
SpTrxm                  VQDVNDSGWKEFVLQSSEPSMVDFWAPWCGPCKLIAPVIDEL     42
HsTrx               MVKQIESKTAFQEALDAAGDKLVVVDFSATWCGPCKMINPFFHSL       45
BovPDI-A1   H₂N-19aa-GAPDEEDHVLVLHKGNFDEALAAHKYLLVEFYAPWCGHCKALAPEYAKA    68
BovPDI-A2   -QELPDDWDKQPVKVLVGKNFEEVAFDEKKNVFVEFYAPWCGHCKQLAPIWDKL        412
                         D  F          V  FWA WCGPCK  AP L EL

                 ————
BjTlpA      QGKLSGPNFEVVAINIDTRDPEKPKTFLKEANLTRLGYFNDQKAKVFQDLKAIGRALGMP   180
EcTrxA      ADEY-QGKLTVAKLNIDQNPGTAPKYGIR---------------------------GIP    76
RsTrxA      SKEY-AGKVKIVKVNVDENPxSPAMLGVR---------------------------GIP    74
RrTrxA      ATAL-GDKVTVAKINIDENPQTPSKYGVR---------------------------GIP    73
CnTrx-c2    SNEY-AGKVKIVKVDVTSCEDTAVKYNIR---------------------------NIP    76
BjTlpB      GKDKRFQLVGINYKDAADNARRFLGRYGNPFGRVGVDANGRASIEWGVY--------GVP   158
RcHelX      KQDGIEIMGVNWKDTPDQAQGFLAEMGSPYTRLGA-DPGNKMGLDWGVA--------GVP   139
SpTrxm      AKEY-SGKIAVTKLNTDEAPGIATQYNIR---------------------------SIP    73
HsTrx       SEKY-SNVIFLEVDVDDCQDVASECEVK----------------------------CTP    75
BovPDI-A1   AGKLKAEGSEIRLAKVDATEESDLAQQYGVR--------------------------GYP   102
BovPDI-A2   GETYKDHEN-IVIAKMDSTANEVEAVKVH---------------------------SFP   443
                Y       N D                                    G P

BjTlpA      TSVLVDPQGCEIATIAGPAEWASEDALKLIRAATGKAAAAL                      221
EcTrxA      TL-LLFKNGEVAATKVGALSKGQLKEFLDANLA                              108
RsTrxA      AL-FLFKNGQVVSNKVGAAPKAALATWIASAL                               105
RrTrxA      TL-MIFKDGQVAATKIGALPKTKLFEWVEASV                               104
CnTrx-c2    AL-LMFKDGEVVAQQVGAAPRSKLAAFIDQNI                               107
BjTlpB      GVPETFVVGREGTIVYKLVGPITDNLRSVLLPQMEKALK                        198
RcHelX      ---ETFVVDGAGRILTRIAGPLTEDVITKKIDPLLAGTAD                       176
SpTrxm      TV-LFFKNGERKESIIGDVSKYQL                                        96
HsTrx       TF-QFFKKGQKVGEFSGANKEKLEATINELV                                105
BovPDI-A1   TI-KFFKNGDTASPKEYTA-240aa-PDI-A2-                              120
BovPDI-A2   TL-KFFPASADRTVIDYN-50aa-COOH                                   460
             T  F   G      G
```

Figure 1. Alignment of TlpA with thioredoxins (Trx) and thioredoxin-homologous domains of protein disulfide isomerase (PDI). Thioredoxin sequences are from *E. coli* (Ec), *Rhodobacter sphaeroides* (Rs), *Rhodospirillum rubrum* (Rr), *Corynebacterium nephredii* (Cn), spinach (Sp), and human (Hs). BovPDI-A1/A2 are the two thioredoxin-homologous domains of bovine PDI. All of these sequences were taken from the NBRF protein data base. BjTlpB is a Trx-like protein with 26% amino acid sequence identity to *E. coli* TrxA. The *tlpB* gene is located in a cluster of *B. japonicum* genes important for the biogenesis of *c*-type cytochromes. RcHelX from *Rhodobacter capsulatus* shows 39% amino acid sequence identity with BjTlpB, and *helX* is also located in a cluster of genes essential for cytochrome *c* biogenesis. Conserved amino acid residues are printed in bold, with an overall consensus shown in the bottom line. The N-terminal transmembrane domain of TlpA and the signal peptides of HelX and TlpB are underlined. The positively charged amino acids at the N terminus of TlpA are marked by (+) symbols. β-Sheet two and α-helix two, which are strictly conserved in all thioredoxins, are overscored. Note: the name BjTlpB has recently been changed to BjCycY.

purified as a fusion to the maltose-binding protein, cleaved with the protease factor Xa and subsequently isolated by anion exchange chromatography (Loferer and Hennecke, 1994b). A second approach involved the export of soluble TlpA to the periplasm using the OmpA signal peptide and a purification scheme of anion exchange, cation exchange and hydrophobic interaction chromatography (Loferer et al., 1995). A polyclonal antiserum against TlpA$_{sol}$ was raised (Loferer and Hennecke, 1994b).

Purified TlpA$_{sol}$ was obtained as a monomer and exhibited protein thiol–disulfide oxidoreductase activity in vitro on heterologous substrates (Loferer and Hennecke, 1994b). SDS–PAGE under reducing or non-reducing conditions and sequencing of CNBr-generated peptides of oxidized or reduced TlpA$_{sol}$ revealed that the two additional cysteines, 45 and 190, form a structural disulfide bond in vivo and in vitro (Loferer and Hennecke, 1994b). Comparison of the CD spectra of oxidized and fully reduced TlpA$_{sol}$ suggest that the structural disulfide bond formed by Cys-45 and Cys-190 is not essential for the maintenance of TlpA's native conformation. Furthermore, near-UV CD spectra of TlpA$_{sol}$ indicate a high α helical content of the protein (Loferer et al., 1995). Using fluorescence quenching of conserved tryptophan residues by the active site disulfide, a redox titration of the active site cysteines was performed with glutathione as a redox partner. In this experiment, TlpA's active site turned out to be strongly reducing, having a midpoint redox potential of −213 mV (Loferer et al., 1995).

■ TlpA function

TlpA is one of several protein thiol–disulfide oxidoreductases that have been discovered recently (see Loferer and Hennecke, 1994a for a review). Phenotypical analyses of a mutant strain carrying a Tn5 insertion in the tlpA gene indicates that TlpA is essential for establishment of a functional root nodule endosymbiosis, as well as for the biogenesis of the terminal oxidase cytochrome aa_3 (Loferer et al., 1993). Like many other bacteria (Saraste et al., 1991), Bradyrhizobium japonicum expresses a branched respiratory chain containing several terminal oxidases, which are used under different growth conditions (Bott et al., 1992). Cytochrome aa_3, which is functional under free-living aerobic conditions, consists of three subunits, two of which carry the redox active prosthetic groups, namely subunit II for the binuclear copper center Cu_A and subunit I for hemes a, a_3 and copper B (Saraste, 1990). In a tlpA mutant strain, cytochrome aa_3 is spectroscopically undetectable, whereas cytochromes b and c are at wild-type levels (Loferer et al., 1993). However, subunit I can be immunologically detected in membranes, demonstrating that TlpA affects cytochrome aa_3 maturation post-translationally (Loferer et al., 1993). Subunit I of cytochrome aa_3 is unlikely to be a direct target of TlpA, since it contains only one cysteine residue, which is located in one of the transmembrane domains. Given the redox properties and the periplasmic location of TlpA's active site, the current working model for TlpA function in aa_3 biogenesis is that it keeps the cysteine residues in subunit II reduced, which is a prerequisite for Cu_A ligation. Such a reducing function is conceivable if one considers the oxidative conditions in the periplasmic space. However, this hypothesis has still to be tested experimentally. A defect in cytochrome aa_3 assembly, however, is no explanation for the symbiotic phenotype of a tlpA mutant, since this terminal oxidase is not functionally expressed under these conditions. This implies that TlpA is also involved in other cellular processes independent of cytochrome aa_3 maturation.

It should be noted that additional thioredoxin-like proteins have been discovered recently that are involved in the biogenesis of c-type cytochromes (Loferer and Hennecke, 1994a). In this type of cytochrome, the heme moiety is covalently linked to the protein via two adjacent cysteine residues. The proteins are CycY from Bradyrhizobium japonicum (formerly ORF132; Ramseier et al., 1991), HelX from Rhodobacter capsulatus (Beckman and Kranz, 1993, see entry p. 331) and DipZ (DsbD) from Escherichia coli (Crooke and Cole, 1995, see entry p. 326). HelX has been described as a periplasmic protein, whereas DipZ and CycY (Fabianek et al., 1997) are located to the cytoplasmic membrane. For these proteins it is also speculated that they keep cysteines of apocytochrome c reduced within the periplasm as a prerequisite for heme ligation. However, biochemical proof is currently lacking.

■ References

Beckman, D.L. and Kranz, R.G. (1993). Cytochrome c biogenesis in a photosynthetic bacterium requires a periplasmic thioredoxin-like protein. Proc. Natl. Acad. Sci. USA **90**, 2179–2183.

Bott, M., Preisig, O., and Hennecke, H. (1992). Genes for a second terminal oxidase in Bradyrhizobium japonicum. Arch. Microbiol. **158**, 335–343.

Crooke, H. and Cole, J. (1995). The biogenesis of c-type cytochromes in Escherichia coli requires a membrane-bound protein, DipZ, with a protein disulfide isomerase-like domain. Mol. Microbiol. **15**, 1139–1150.

Eklund, H., Gleason, F.K., and Holmgren, A. (1991). Structural and functional relations among thioredoxins of different species. Proteins **11**, 13–28.

Fabianek, R.A., Huber-Wunderlich, M., Glockshuber, R., Künzler, P., Hennecke, H., and Thöny-Meyer, L. (1997). Characterization of the Bradyrhizobium japonicum CycY protein, a membrane-anchored periplasmic thioredoxin that may play a role as a reductant in the biogenesis of c-type cytochromes. J. Biol. Chem. **272**, 4467–4473.

Kündig, C., Hennecke, H., and Göttfert, M. (1993). Correlated physical and genetic map of the Bradyrhizobium japonicum 110 genome. J. Bacteriol. **175**, 613–622.

Loferer, H. and Hennecke, H. (1994a). Protein disulfide oxidoreductases in bacteria. Trends Biochem. Sci. **19**, 169–171.

Loferer, H. and Hennecke, H. (1994b). Expression, purification and functional properties of Bradyrhizobium japonicum TlpA, a thioredoxin-like protein. Eur. J. Biochem. **223**, 339–344.

Loferer, H., Bott, M., and Hennecke, H. (1993). Bradyrhizobium japonicum TlpA, a novel membrane-anchored thioredoxin-like protein involved in the biogenesis of cytochrome aa_3 and development of symbiosis. EMBO J. **12**, 3373–3383.

Loferer, H., Wunderlich, M., Hennecke, H., and Glockshuber, R. (1995). A bacterial thioredoxin-like protein that is exposed to the periplasm has redox properties comparable with those of cytoplasmic thioredoxins. J. Biol. Chem. **270**, 26178–26183.

Ramseier, T., Winteler, H., and Hennecke, H. (1991). Discovery and sequence analysis of bacterial genes involved in the biogenesis of c-type cytochromes. J. Biol. Chem. **266**, 7793–7803.

Saraste, M. (1990). Structural features of cytochrome oxidase. Q. Rev. Biophys. **23**, 331–366.

Saraste, M., Holm, L., Lemieux, L., Lübben, M., and van der Oost, J. (1991). The happy family of cytochrome oxidases. Biochem. Soc. Trans. **19**, 608–612.

■ Hannes Loferer:
Geneva Biomedical Research Institute
Glaxo Wellcome Research and Development S.A.
14, chemin des Aulx
CH-1228 Plan-les-Ouates
Geneva, Switzerland
Tel. 41 22 706 9608
Fax. 41 22 794 6965
E-mail: jhl39437@ggr.co.uk

Saccharomyces cerevisiae protein disulfide isomerase (PDI)

Protein disulfide isomerase (PDI) is an enzyme involved in the catalysis of disulfide bond formation in the lumen of the endoplasmic reticulum (ER). In the yeast *Saccharomyces cerevisiae* PDI is encoded by the PDI1 gene and shows many of the structural features of PDIs from higher eukaryotes, including two copies of the thioredoxin-like active site sequence, WCGHCK, and a C-terminal ER retention signal. One unusual feature of yeast PDI is that it is N-glycosylated, although this post-translational modification is not required for disulfide isomerase activity or for ER retention. Disruption of the PDI1 gene results in cell inviability, but this is not owing to the loss of the disulfide isomerase activity, thereby bringing into question the essential cellular role played by this key protein folding catalyst.

■ Alternative names

Thioredoxin-related glycoprotein (TRG) (Gunther et al., 1991); glycosylation site binding protein (GSBP) (LaMantia et al., 1991).

■ Protein disulfide isomerases

Protein disulfide isomerase (PDI; EC 5.3.4.1) is the archetypal member of a family of sequence-related proteins located in the lumen of the endoplasmic reticulum (ER) of eukaryotic cells (Freedman et al., 1994). These proteins contain one or more copies of the –CGHC– motif which is related to the thioredoxin active site sequence –WCGPCK–. The role of PDI in the ER lumen is to catalyse thiol–disulfide interchange and is therefore one of the key protein folding catalysts in the secretory pathway, ensuring that native disulfide bond formation in secretory and cell-surface proteins takes place efficiently (Freedman, 1991). The presence of PDI in *Saccharomyces cerevisiae* was first demonstrated biochemically in 1990 (Mizunaga et al., 1990) with the gene encoding PDI (*PDI1*) being cloned and sequenced the following year by at least five different laboratories (Farquhar et al., 1991; Gunther et al., 1991; LaMantia et al., 1991; Scherens et al., 1991; Tachikawa et al., 1991).

■ Isolation of the *S. cerevisiae* PDI protein

In mammalian tissues, PDI is an abundant protein representing up to 0.8% of the total cellular proteins (Freedman et al., 1994). In contrast, yeast PDI represents less than 0.05% of the total soluble protein fraction (Mizunaga et al., 1990). Since PDI is located solely within the membrane-bound ER, it can be enriched for by preparing microsomal fractions, although its activity can only be detected in such fractions after the integrity of the ER membrane is destroyed by, for example, sonication or the detergent Triton X-100. While feasible for mammalian and plant tissue samples, there is as yet no effective method available for obtaining pure microsomes in high yields from *S. cerevisiae*. For a detailed description of a protocol to purify yeast PDI from total cell extracts, see Mizunaga et al. (1990).

■ *PDI1* gene and sequence

The gene encoding PDI was isolated from *S. cerevisiae* and sequenced by five different laboratories (Farquhar et al., 1991; Gunther et al., 1991; LaMantia et al., 1991; Scherens et al., 1991; Tachikawa et al., 1991). The gene (designated *PDI1*) is located on chromosome III adjacent to the *GLK1* gene (encoding glucokinase). The *PDI1* gene sequences reported by Farquhar et al. and Gunther et al.

predict a 530 amino acid protein, while the sequences reported by the other three groups (LaMantia et al., 1991; Scherens et al., 1991; Tachikawa et al., 1991) predict a protein eight amino acids shorter (i.e. 522 amino acids) lacking the sequence EADAEAEA in the acidic C-terminal region of the protein. It is not clear whether this discrepancy in deduced PDI sequence is due to a natural gene polymorphism or to a DNA sequencing error. For the purposes of this article, the 530 amino acid sequence will be used (EMBL accession number X54535).

The predicted sequence of the yeast PDI protein identifies a number of key functional and/or structural regions of the protein and these are summarized in Fig. 1. The mature PDI sequence is preceeded by a hydrophobic region (aa 1–22) with the characteristics of a secretory signal sequence and with a predicted signal peptidase cleavage site following the Ala residue at position 22. Direct experimental confirmation of this cleavage site has not been reported since the N-terminus of the mature protein appears to be blocked (Mizunaga et al., 1990). The mature protein sequence (aa 23–530) contains two copies of the thioredoxin-like active site sequence –WCGHCK– (aa 60–65 and aa 405–410). These two –WCGHCK– motifs lie within the so-called a and a' domains, respectively, large internal duplications of approximately 100 amino acids that are also present in mammalian and avian PDI proteins (Freedman et al., 1994). There are five N-linked glycosylation sites (Asn–X–Ser/Thr) at amino acids 82, 117, 155, 174 and 425. Between amino acids 252 and 277 lies a sequence resembling the peptide-binding site present in the C-terminus of mammalian PDI (LaMantia and Lennarz, 1993), although it has yet to be demonstrated that peptides do bind this sequence in yeast PDI. Finally, the C-terminus of PDI, like its mammalian homologues, is very acidic and ends with the tetrapeptide sequence –His–Asp–Glu–Leu (-HDEL) which is present at the C-termini of other proteins known to be retained in the lumen of the ER in yeast (Pelham, 1990).

■ PDI protein

When purified from S. cerevisiae, PDI was found to have an apparent molecular mass of 70 kDa (Mizunaga et al.,1990), while the gene sequence predicted a molecular mass of 59 kDa (Farquhar et al., 1991; Gunther et al., 1991). This discrepancy is a result of the occupation of the five N-linked glycosylation sites (Fig. 1) by core oligosaccharide (Mizunaga et al., 1990). Treatment of PDI with endoglycosidase H or growth of cells in the presence of tunicamycin (an inhibitor of N-linked core glycosylation) results in a decrease in the apparent molecular mass of PDI to the expected value of 59–60 kDa (Mizunaga et al., 1990; Tachibana and Stevens, 1992). Intriguingly, the mammalian and avian PDIs are not glycosylated, while deglycosylated yeast PDI has full in vitro isomerase activity (LaMantia and Lennarz, 1993) and is correctly localized to the ER (Natalia, 1994). The question of whether this additional post-translational modification has any functional significance remains to be answered. Yeast PDI is glycosylated by a Sec59p-dependent pathway confirming that it is core glycosylated in the ER (Natalia, 1994).

A variety of studies, including subcellular fractionation (Gunther et al., 1991) and whole-cell immunofluoresence analysis (Natalia, 1994), have confirmed that PDI is localized to the ER of S. cerevisiae and is present in both the rough and smooth membrane vesicle fractions (Natalia, 1994). In vitro translation studies with PDI1 mRNA show that fully glycosylated PDI is only synthesized in a mammalian cell-free translation system if ER membrane vesicles are added, thus confirming that PDI is translocated into, and core glycosylated in, the lumen of the

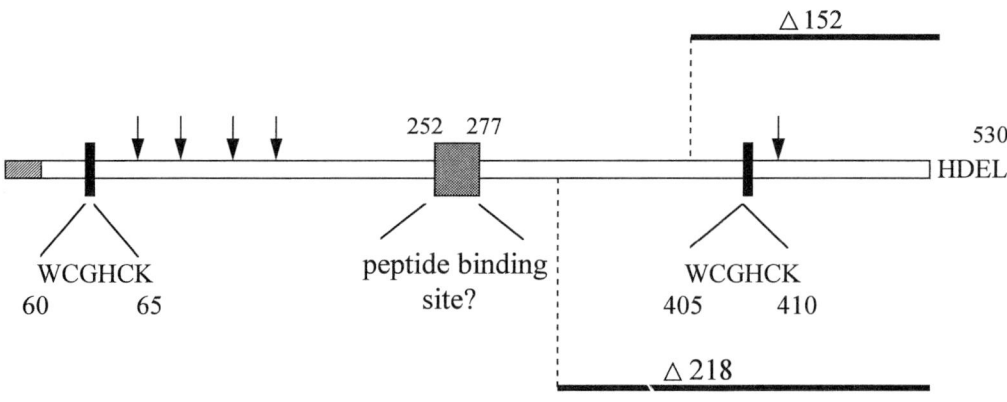

Figure 1. The structural features of yeast protein disulfide isomerase (PDI). The major structural features shown are: the N-terminal signal sequence (cross hatches), the five N-linked glycosylation sites (arrows), the two thioredoxin-like active sites (WCGHCK), the C-terminal HDEL ER retention signal and the putative peptide-binding site (aa 252–277). Also shown are the extent of two C-terminal deletions (Δ152 and Δ218) described by LaMantia and Lennarz (1993), which define the essential region of PDI.

ER (Gunther et al., 1991). The retention of PDI in the ER is mediated by an interaction between its C-terminal HDEL sequence and the *ERD2*-encoded receptor protein Erd2p (Natalia, 1994). Erd2p, which retrieves luminal ER proteins from the post-ER compartments (Semenza et al., 1990), is required for the retention of all soluble proteins of the yeast ER. PDI is also secreted by *erd1* strains which are defective in ER retention (LaMantia and Lennarz, 1993), and when the *PDI1* gene is overexpressed in an otherwise wild-type strain (Natalia, 1994). The latter result suggests that the ER retention system for PDI (and other ER proteins) in yeast is readily saturable.

Like its mammalian counterpart, yeast PDI has a native molecular mass of 140 kDa indicating that it exists as a homodimer. PDI is also a very acidic protein (containing 14% Asp, 13.8% Glu residues) with a pI of 4.02 (Mizunaga et al., 1990). The majority of the acidic residues are located in the C-terminus of the protein, although it should be noted that this region can be deleted without loss of the essential function of PDI (LaMantia and Lennarz, 1993; see below).

■ Biological activities of yeast PDI

Studies with purified mammalian PDI have demonstrated that PDI catalyses thiol–disulfide interchanges that can lead to the net formation, the net breakage or the net rearrangement of protein disulfide bonds, depending on the redox conditions and the nature of the (protein) substrate (reviewed in Freedman et al., 1994). While no analogous biochemical studies have yet been carried out with purified yeast PDI, studies using total yeast cell lysates or recombinant PDI (expressed in *E. coli*) suggest that the yeast protein has similar catalytic properties to the mammalian protein (LaMantia and Lennarz 1993; Natalia, 1994).

The active site cysteine residues of PDI have different enzymatic roles. Mutational analysis of the PDI active sites –CXXC– indicate that the first cysteine residue in each copy of the active site is necessary for isomerisation whereas the second cysteine is necessary for disulphide bond formation (Laboissiere et al., 1995; Walker et al., 1996).

Reports by several laboratories that disruption of the *PDI1* gene resulted in loss of cell viability (Farquhar et al., 1991; Gunther et al., 1991; LaMantia et al., 1991; Scherens et al., 1991; Tachikawa et al., 1991) would suggest that the enzyme is essential. However, the subsequent studies of Laboissiere et al., (1995), who showed that a *pdi1* null mutant could not be complemented by expression of rat PDI containing the active site mutation–SXXC -, indicates that the essential role of yeast PDI is to act as a catalyst of the isomeration of existing disulphide bonds. Interestingly, Chivers et al. (1996) have demonstrated that the gene encoding *E. coli* thioredoxin (*trx*, see entry p. 314) cannot complement *S. cerevisiae* *PDI1* null mutants unless the Trx sequence is mutated to have a CXXC motif with a disulfide bond of high reduction potential and a thiol group of low pK_a. Thus an enzymatic thiolate appears to be necessary.

Up to 152 amino acids can be deleted from the C-terminus of PDI (see Fig. 1) without loss of viability, although a mutant protein lacking 218 amino acids from the C-terminus will not support viability even though the truncated PDI shows isomerase activity *in vitro* (LaMantia and Lennarz, 1993). Thus, there is a region between the putative peptide-binding site and the second active site that is important for the essential function of PDI in yeast. What that function is remains to be defined.

A number of viable strains of yeast with no, or reduced, PDI isomerase activity have been constructed. In the active site mutants reported by LaMantia and Lennarz (1993), a defect in the disulfide bond formation and transport of carboxypeptidase Y, a disulfide-bonded vacuolar protein, was detected. This secretory defect is also seen in a strain in which expression of the *PDI1* gene was down-regulated using the *GAL1* promoter (Tachibana and Stevens, 1992). A viable strain, which only expresses the first 116 N-terminal amino acids of PDI (Natalia, 1994; Luz, 1994), shows a defect in the secretion of yeast killer toxin, a heterodimeric protein that is secreted by strains of *S. cerevisiae* containing the M double-stranded RNA mycovirus and which has interchain disulfide bonds (Dunn et al., 1995). The toxin-secretion defect can be eliminated by reintroducing a single copy of the full length *PDI1* gene (Natalia, 1994; Dunn et al., 1995) thus confirming that secretion of a native disulfide-bonded protein from *S. cerevisiae* requires the presence of PDI with two active sites.

Mammalian PDI is also found as a component of prolyl-4-hydroxylase and triacylglycerol transfer protein (Freedman et al., 1994). Whether yeast PDI also functions as a subunit of more complex enzyme systems is not known.

■ Biological regulation of yeast PDI

PDI1 mRNA has a codon bias index of 0.60 (Farquhar et al., 1991) which is consistent with the encoded protein being moderately to highly expressed, although this is not borne out by biochemical studies (Mizunaga et al., 1990). Like other genes encoding ER-resident proteins (e.g. BiP encoded by the *KAR2* gene and Eug1p encoded by the *EUG1* gene), the transcription of the *PDI1* gene is regulated by the unfolded protein-response pathway (Shamu et al., 1994). The accumulation of unfolded proteins in the ER results in the activation of this pathway via a transmembrane kinase (Ire1p) and a *cis*-acting DNA sequence (GXXXGCGXXTC) located in the promoter regions of these genes. Thus, growth of yeast in tunicamycin, which causes cells to accumulate unfolded proteins in the ER, results in an increase in the steady-state levels of PDI (Cox et al., 1993).

■ Biological and genetic interactions of yeast PDI

At least three genes encoding PDI-related proteins have been discovered in *S. cerevisiae*. The first described was *EUG1* encoding a 65 kDa glycoprotein (designated Eug1p) located in the ER and showing 43% amino acid identity to yeast PDI (Tachibana and Stevens, 1992; see entry p. 339). It has a hydrophobic N-terminal signal sequence, two PDI-like active sites (–WCLHSQ– and –WCIHSK–), and a C-terminal HDEL ER retrieval signal. Importantly, when overexpressed, it will complement the lethality of a *PDI1* gene disruption, yet the Eug1p protein does not itself appear to have disulfide isomerase activity (Webster, 1995). The *MPD1* and *MPD2* genes were identified as multicopy suppressors of an otherwise lethal *PDI1* gene disruption (Tachikawa et al., 1995). *MPD2* corresponds to an ORF (O0941) reported on chromosome XV (Zumstein et al., 1995). The encoded Mpd1p glycoprotein has a predicted molecular mass of 31 kDa, an N-terminal hydrophobic signal sequence, one PDI-like active site (WCGHCK) and a C-terminal HDEL motif (Tachikawa et al., 1995). The purified protein has disulfide isomerase activity (H. Tachikawa, unpublished data). The *MPD2* gene encodes a polypeptide of 277 amino acids whose structural features resemble those of Mpd1p (H. Tachikawa, unpublished data).

The existence of three other putative ER proteins with structural and/or functional identity to yeast PDI, but which may not themselves be important for the catalysis of disulfide bond formation, further raises the possibility that PDI may be involved either directly or indirectly in a number of other cellular processes in *S. cerevisiae*.

■ References

Chivers, P.T., Laboissière, M.C.A., and Raines, R.T. (1996). The CXXC motif: imperatives for the formation of native disulfide bonds in the cell. EMBO J. **15**, 2659–2667.

Cox, J.S., Shamu, C.E., and Walter, P. (1993). Transcriptional induction of genes encoding endoplasmic reticulum resident proteins requires a transmembrane protein kinase. Cell **73**, 1197–1206.

Dunn. A., Luz, J.M., Natalia, D., Gamble, J.A., Freedman, R.B., and Tuite, M.F. (1995). Protein disulphide isomerase (PDI) is required for the secretion of a native disulphide-bonded protein from *S. cerevisiae*. Biochem. Soc. Trans. **23**, 78S.

Farquhar, R., Honey, N., Murant, S.J., Bossier, P., SChultz, L., Montgomery, D., Ellis, R.W., Freedman, R.B., and Tuite, M.F. (1991). Protein disulfide isomerase is essential for viability in *Saccharomyces cerevisiae*. Gene **108**, 81-89.

Freedman, R.B. (1991). Protein disulfide isomerase — an enzyme that catalyses protein folding in the test tube and in the cell. In *Conformations and forces in protein folding*, (E.T. Nall and K.A. Dill, eds), Vol. **16**, pp. 204–216. AAAS.

Freedman, R.B., Hirst, T.R., and Tuite, M.F. (1994) Protein disulphide isomerase: building bridges in protein folding. Trends Biochem. Sci. **19**, 331–336.

Gunther, R., Brauer, C., Janetzky, B., Forster, H.-H., Ehbrecht, E.-M., Lehle, L., and Kuntzel, H. (1991) The *Saccharomyces cerevisiae TRG1* gene is essential for growth and encodes a lumenal endoplasmic reticulum glycoprotein involved in the maturation of vacuolar carboxypeptidase. J. Biol. Chem. **266**, 24557–24563.

Laboissiere, M.C.A., Sturley, S.L., and Raines, R.T. (1995). The essential function of protein-disulfide isomerase is to unscramble non-native disulfide bonds. J. Biol. Chem. **270**, 28006–28009.

LaMantia, M.L. and Lennarz, W.J. (1993). The essential function of yeast protein disulfide isomerase does not reside in its isomerase activity. Cell **74**, 899–908.

LaMantia, M.L., Miura, T., Tachikawa, H., Kaplan, H.A., Lennarz, W.J., and Mizunaga, T. (1991). Glycosylation site binding protein and protein disulfide isomerase are identical and essential for cell viability in yeast. Proc. Natl. Acad. Sci. USA **88**, 4453–4457.

Luz, J.M.C.G.A. (1994). PhD Thesis, University of Kent, UK.

Mizunaga, T., Katakura, Y., Miura, T., and Marutama, Y. (1990) Purification and characterisation of yeast protein disulphide isomerase. J. Biochem. **108**, 846–851.

Natalia, D. (1994). PhD Thesis, University of Kent, UK.

Pelham, H.R.B. (1990). The retention signal for soluble proteins of the endoplasmic reticulum. Trends Biochem. Sci. **15**, 483–486.

Scherens, B., Dubois, E. and Messenguy, F. (1991). Determination of the sequence of the yeast *YCL313* gene localized on chromosome III: homology with the protein disulphide isomerase (PDI) gene product of other organisms. Yeast **7**, 185–193.

Semenza. J.C., Hardwick, K.G., Dean, N., and Pelham, H.R.B. (1990). *ERD2*, a yeast gene required for the receptor-mediated retrieval of luminal ER proteins in the secretory pathway. Cell **61**, 1349–1357.

Shamu, C.E., Cox, J.S., and Walter, P. (1994). The unfolded-protein-response pathway in yeast. Trends Cell Biol. **4**, 56–60.

Tachibana, C. and Stevens, T.H. (1992). The yeast *EUG1* gene encodes an endoplasmic reticulum protein that is functionally related to protein disulfide isomerase. Mol. Cell. Biol. **12**, 4601–4611.

Tachikawa, H., Miura, T., Katakura, Y., and Mizunaga, T. (1991). Molecular structure of a yeast gene, *PDI1*, encoding protein disulfide isomerase that is essential for cell growth. J. Biochem. **110**, 306–313.

Tachikawa, H., Takeuchi, Y., Funahashi, W., Miura, T., Gao, X-D., Fujimoto, D., Mizunaga, T., and Onodera, K. (1995). Isolation and characterisation of a yeast gene, *MPD1*, the overexpression of which suppresses the inviability caused by protein disulfide isomerase. FEBS Lett. **369**, 212–216.

Walker, K.W., Lyles, M.M., and Gilber, H.F. (1996). Catalysis of oxidative protein folding by mutants of protein disulfide isomerase with a single active-site cysteine. Biochemistry **35**, 1972–1980.

Webster, P.A., Ph.D. Thesis, University of Kent.

Zumstein, E., Pearson, B. M., Kalogeropoulos, A., and Schweizer, M. (1995). A 29.425 kb segment on the left arm of yeast chromosome XV contains more than twice as many unknown as known open reading frames. Yeast **11**, 975–986.

■ *Mick F. Tuite and Angela Dunn:*
Research School of Biosciences
University of Kent
Canterbury
Kent CT2 7NJ, UK
Tel. 44 1227 823699
Fax. 44 1227 763912
E-mail: M.F.Tuite@ukc.ac.uk

Saccharomyces cerevisiae Eug1

The EUG1 gene of Saccharomyces cerevisiae encodes a 65 kDa glycoprotein that is functionally related to yeast PDI, the product of the PDI1 gene. The Eug1 protein (Eug1p) contains two copies of the PDI active site sequence (CGHC) in which a serine substitutes for the second cysteine at both sites in Eug1p (CLHS and CIHS). Eug1p, like PDI, is an HDEL-containing endoplasmic reticulum (ER) luminal protein. However, whereas the PDI1 gene is required for yeast cell viability, eug1Δ cells exhibit no observable phenotypes. Overexpression of Eug1p allows yeast cells lacking PDI to survive, indicating that Eug1p is functionally related to PDI and may interact with newly translocated proteins in the ER lumen.

■ Mutant phenotype

EUG1 (ER protein unnecessary for growth under standard laboratory conditions) was named for its observed mutant phenotype (Tachibana and Stevens, 1992). Overproduction or deletion of the EUG1 gene appears to have no effect on cell viability or function. Yeast cells deleted for EUG1 grow with wild-type rates on rich or minimal media at temperatures from 18 to 40°C and can grow anaerobically. Sporulation is not affected in homozygous eug1Δ cells, nor is the transport of vacuolar and secreted proteins through the secretory pathway.

■ EUG1 gene and sequence

The EUG1 gene (GenBank accession number M84796) contains a single open reading frame predicted to encode a 517 amino acid protein containing an N-terminal hydrophobic signal sequence (Tachibana and Stevens, 1992). Eug1p exhibits 43% amino acid identity to yeast PDI (see entry p. 335) and contains two highly similar PDI-like active site sequences. The Eug1p active site sequences, CLHS and CIHS, differ from the two identical sequences in PDI, CGHC, in that a serine is found in place of the second cysteine in each (Fig. 1). The Eug1p sequence also reveals two internal cysteine residues (non-active site) that are identical in position with two internal cysteines present in yeast PDI (Farquhar et al., 1991; Gunther et al., 1991; LaMantia et al., 1991; Scherens et al., 1991; Tachikawa et al., 1991). This is particularly intriguing, as in vitro studies performed with yeast PDI suggest a possible functional role for these internal cysteines (LaMantia and Lennarz, 1993). Interestingly, the conserved positioning of these cysteines in Eug1p and yeast PDI differs substantially from the two conserved internal cysteines found in mammalian PDIs. Eug1p also possesses the C-terminal ER retention signal, HDEL, as well as five potential sites for Asn-linked glycosylation. The upstream region of EUG1 contains an 'unfolded protein-response element' (Mori et al., 1992), consistent with the observation that EUG1 mRNA is greatly elevated under conditions that lead to elevated levels of unfolded proteins (Tachibana and Stevens, 1992). Genes containing the unfolded protein-response element are induced under conditions that result in the accumulation of unfolded proteins in the ER (Mori et al., 1992; Tachibana and Stevens, 1992; Kohno et al., 1993).

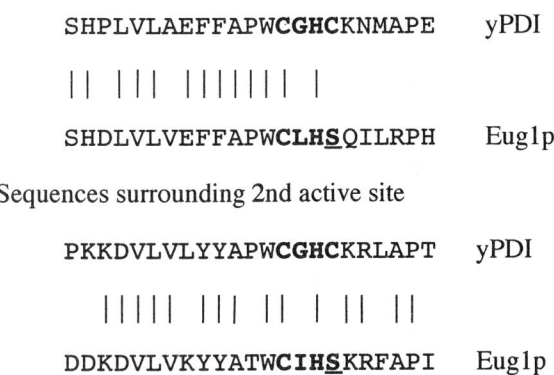

Figure 1. Comparison of active site sequences within yeast PDI and Eug1p.

■ Eug1 protein

Eug1p is a glycoprotein with an apparent molecular mass of 65 kDa, and localizes to the ER based on indirect immunofluorescence (Tachibana and Stevens, 1992). Removal of N-linked glycosylation with Endo H reduces Eug1p to an apparent molecular mass of about 55 kDa. In yeast cells that fail to retain ER luminal proteins owing to a defect in the HDEL receptor (erd2), Eug1p is secreted and the extracellular Eug1p is more heavily glycosylated than the ER-retained Eug1p (Tachibana, 1991). While it is unknown whether Eug1p has disulfide isomerase activity, rat PDI with both active sites altered to CGHS (as in Eug1p) can functionally replace yeast PDI (Laboissiere et al., 1995). PDI with two SCHS active sites (rather than CGHS active sites) cannot functionally replace yeast PDI. PDI with two CGHS active sites efficiently catalyses the shuffling of disulfide bonds ('shufflease' activity), but is incapable of catalysing the oxidation of dithiols or the reduction of disulfide bonds (Laboissiere et al., 1995). Consistent with the CGHS PDI activity, and based on the ability of elevated levels of Eug1p to compensate for loss of PDI in yeast, Eug1p is predicted to exhibit shufflease activity.

Eug1p appears to be a 'peptide-binding protein', as it is capable of binding and cross-linking to the photoactivatable peptides originally used to identify yeast PDI (LaMantia et al., 1991; M. LaMantia, unpublished data). Interestingly, the recently identified peptide-binding domain within mammalian PDI (Noiva et al., 1993) exhibits maximal sequence similarity with a region towards the center of the yeast PDI. The corresponding region in Eug1p maintains this high sequence conservation (Fig. 2). Like the ER protein Kar2p (BiP), the level of Eug1p is elevated approximately tenfold in response to conditions such as treatment with the drug tunicamycin or a mutation in the SEC18 gene that result in accumulation of unglycosylated proteins in the ER or delayed protein transport from the ER, respectively. This increase in Eug1p production is the result of an increase in transcription (Tachibana and Stevens, 1992).

■ Overproduction of Eug1 protein

PDI1 is an essential gene in yeast (Farquhar et al., 1991; Gunther et al., 1991; LaMantia et al., 1991; Scherens et al., 1991; Tachikawa et al., 1991). The lethality resulting from the pdi1Δ allele in yeast can be suppressed by the overexpression of EUG1 (Tachibana and Stevens, 1992). Haploid pdi1Δ cells that contain PDI1 on a plasmid are able to survive loss of the PDI1 plasmid if these cells also contain EUG1 on a 2 μm multicopy plasmid. Similarly, haploid spores carrying the pdi1Δ mutation are able to germinate and grow if they are overexpressing Eug1p. Moreover, the cells rescued by Eug1p overexpression grow nearly as well as wild-type cells. Western analysis of the rescued pdi1Δ cells indicates that Eug1p is being synthesized at very high levels. However, depletion of PDI resulted in delay of movement of CPY out of the ER regardless of the level of production of Eug1p (Tachibana and Stevens, 1992).

■ References

Farquhar, R., Honey, N., Murant, S.J., Bossier, P., Schultz, D., Montgomery, D., Ellis, R.W., Freedman, R.B., and Tuite, M.F. (1991). Protein disulfide isomerase is essential for viability in Saccharomyces cerevisiae. Gene **108**, 81–89.

Günther, R., Brauer, C., Janetzky, B., Forster, H.-H., Ehbrecht, E.-M., Lehle, L., and Küntzel, H. (1991). The Saccharomyces cerevisiae TRG1 gene is essential for growth and encodes a lumenal endoplasmic reticulum glycoprotein involved in the maturation of vacuolar carboxypeptidase. J. Biol. Chem. **266**, 24557–24563.

Kohno, K., Normington, K., Sambrook, J., Gething, M.-J., and Mori, K. (1993). The promoter region of the yeast KAR2 (BiP) gene contains a regulatory domain that responds to the presence of unfolded proteins in the endoplasmic reticulum. Mol. Cell. Biol. **13**, 877–890.

Laboissiere, M.C., Sturley, S.L., and Raines, R.T. (1995). The essential function of protein-disulfide isomerase is to unscramble non-native disulfide bonds. J. Biol. Chem. **270**, 28006–28009.

LaMantia, M.L. and Lennarz, W.J. (1993). The essential function of yeast protein disulfide isomerase does not reside in its isomerase activity. Cell **74**, 899–908.

LaMantia, M., Miura, T., Tachikawa, H., Kaplan, H.A., Lennarz, W.J., and Mizunaga, T. (1991). Glycosylation site binding protein and protein disulfide isomerase are identical and essential for cell viability in yeast. Proc. Natl. Acad. Sci. USA **88**, 4453–4457.

Mori, K., Sant, A., Kohno, K., Normington, K. Gething, M.-J., and Sambrook, J.F. (1992). A 22 bp cis-acting element is necessary and sufficient for the induction of the yeast KAR2 (BiP) gene by unfolded proteins. EMBO J. **11**, 2583–2593.

Noiva, R., Freedman, R., and Lennarz, W.J. (1993). Peptide binding to protein disulfide isomerase occurs at a site distinct from the active sites. J. Biol. Chem. **268**, 19210–19217.

Scherens, B., Dubois, E., and Messenguy, F. (1991). Determination of the sequence of the yeast YCL313 gene localized on chromosome III: homology with the protein disulfide

```
FLESGGQDGAG-DNDDLDLEEALEPDI       mPDI peptide binding fragment

::|||      |     ||: :|||    |

YVESGLPLGYLFYNDEEELEE-YKPLF       yPDI maximal alignment

YVESGLPLGYLFYNDEEELEEYKPLF        yPDI (fragment above)

|: |  |||  |  ||    |||||:|   ||

YISSNLPLAYFFYTSEEELEDYTDLF        yEug1p corresponding region
```

Figure 2. Maximal sequence alignment between mammalian PDI (mPDI) peptide binding site, yeast PDI and Eug1p. The vertical bars (|) indicate identical residues and the colons (:) indicate conservative amino acid substitutions.

isomerase (PDI) gene product of other organisms. Yeast **7**, 185–193.

Tachibana, C. (1991). An analysis of the protein disulfide isomerase homologs encoded by the *PDI1* and *EUG1* genes of *Saccharomyces cerevisiae*. Ph.D. Thesis, University of Oregon, Eugene, Oregon.

Tachibana, C. and Stevens, T.H. (1992). The yeast *EUG1* gene encodes an endoplasmic reticulum protein that is functionally related to protein disulfide isomerase. Mol. Cell. Biol. **12**, 4601–4611.

Tachikawa, H., Miura, T., Katakura, Y., and Mizunaga, T. (1991). Molecular structure of a yeast gene, *PDI1*, encoding protein disulfide isomerase that is essential for cell growth. J. Biochem. **110**, 306–313.

■ Tom H. Stevens:
Institute of Molecular Biology
University of Oregon
Eugene, OR 97403-1229, USA
Tel. 1 541 346 5884
Fax. 1 541 346 4854
E-mail: stevens@molbio.uoregon.edu

Saccharomyces cerevisiae Mpd1p and Mpd2p

Mpd1p and Mpd2p are protein disulfide isomerase (PDI) related proteins found in Saccharomyces cerevisiae. Both of them have a single disulfide isomerase active site-like sequence (Mpd1p has APWCGHCK and Mpd2p has TSWCQHCK), and appear to have disulfide isomerase activity. They contain putative signal sequences at their N-termini and endoplasmic reticulum (ER) retention signals (HDEL) at their C-termini. Overproduction of either protein can suppress lethality caused by PDI depletion, indicating that they can substitute for the essential function of yeast PDI.

■ *MPD1* and *MPD2* genes and sequences

The *MPD1* and *MPD2* genes were identified as multicopy suppressors of a *pdi1* deletion mutant (Tachikawa et al., 1995). Unlike the *PDI1* gene (see entry on *S. cerevisiae* PDI, p. 335), neither *MPD1* nor *MPD2* is essential for viability. The *MPD1* gene encodes a polypeptide of 318 amino acids (GenBank accession number D34633), which is a member of the PDI family. While PDI has two disulfide isomerase active site sequences and mammalian PDI related proteins have two or three of those sites (Freedman et al., 1994), Mpd1p retains only one active site sequence, APWCGHCK. The amino acid sequence around the active site is similar to that found in the first active (called 'thioredoxin-like') domain of rat PDI isozyme Q-2 (Srivastava et al., 1993) (43% in 93 amino acids), and in the second thioredoxin-like domain of yeast PDI (27% in 97 amino acids). Mpd1p also has an N-terminal putative signal sequence, a C-terminal ER retention signal, HDEL (Pelham et al., 1990), and two N-linked glycosylation sites (NXT or NXS). Thus Mpd1p is proposed to be a glycosylated ER protein.

The *MPD2* gene encodes a polypeptide of 277 amino acids (GenBank accession number D34634). The protein structure of Mpd2p resembles Mpd1p. Mpd2p is another member of the PDI family, and retains a single TSWCQHCK sequence, which is similar to, but distinct from the disulfide isomerase active site sequence of PDI and Mpd1p (H. Tachikawa, Y. Takeuchi, W. Funahashi, X.D. Gao, R. Nishihara, H. Nakanishi, T. Mizunaga and D. Fujimoto, unpublished data). It also has an N-terminal putative signal sequence, a C-terminal ER retention signal and an N-linked glycosylation site. The amino acid sequence around the active site is again similar to that found in the second thioredoxin-like domain of rat PDI isozyme Q-2 (34% identity in 76 amino acids), in the first thioredoxin-like domain of yeast PDI (23% identity in 105 amino acids), and in the thioredoxin-like domain of Mpd1p (26% identity in 115 amino acids).

■ Mpd1 and Mpd2 proteins

Mpd1p was detected by immunoblot with anti-Mpd1p antibody in a yeast lysate only when it was overproduced from a multicopy *MPD1* plasmid (Tachikawa et al., 1995). The protein could not be detected in the wild-type cell extract, indicating that the expression level of *MPD1* is quite low under normal growth conditions. The apparent molecular mass of Mpd1p was 36 kDa, which was shifted to 31 kDa by endoglycosidase H treatment, indicating that Mpd1p has core-type carbohydrates. This result is consistent with the idea that Mpd1p resides in the ER.

Mpd2p could be detected by immunoblotting with anti-Mpd2p antibody in the wild-type cell lysate as a 32 kDa protein with 3 kDa equivalent of N-linked carbohydrates (H. Tachikawa, Y. Takeuchi, W. Funahashi, X.D. Gao, R. Nishihara, H. Nakanishi, T. Mizunaga and D. Fujimoto, unpublished data). Hence, it is also proposed to be an ER protein.

Function of Mpd1p and Mpd2p

Mpd1p has disulfide isomerase activity; glutathione S-transferase-Mpd1p fusion protein, expressed in *E. coli* and purified using glutathione-Sepharose, can reactivate ribonuclease A with scrambled disulfide bonds (H. Tachikawa, Y. Takeuchi, W. Funahashi, X.D. Gao, R. Nishihara, H. Nakanishi, T. Mizunaga and D. Fujimoto, unpublished data). It is not yet clear whether Mpd2p has disulfide isomerase activity.

Although neither *MPD1* nor *MPD2* is essential for viability, overexpression of either one rescues the lethality of *PDI1* deletion and partially restores the transport defect of vacuolar protein carboxypeptidase Y in this mutant. So, they can substitute for the essential function of yeast PDI. Another yeast gene *EUG1* (see entry p. 339) also can suppress the lethality of *PDI1* deletion when overexpressed (Tachibana and Stevens, 1992). Further studies are required to clarify the functional relationship and substrate specificity amongst PDI, Mpd1p, Mpd2p and Eug1p in disulfide bond formation and folding of secretory proteins in the yeast ER.

References

Freedman, R.B., Hirst, T.R., and Tuite, M.F. (1994). Protein disulphide isomerase: building bridges in protein folding. Trends Biochem. Sci. **19**, 331–336

Pelham, H.R.B. (1990). The retention signal for soluble proteins of the endoplasmic reticulum. Trends Biochem. Sci. **15**, 483–486

Srivastava, S.P., Fuchs, J.A., and Holzman, J.L. (1993). The reported cDNA sequence for phospholipase Cα encodes protein disulfide isomerase isozyme Q-2 and not phospholipase-C. Biochem. Biophys. Res. Commun. **193**, 971–978

Tachibana, C. and Stevens, T.H. (1992). The yeast *EUG1* gene encodes an endoplasmic reticulum protein that is functionally related to protein disulfide isomerase. Mol. Cell. Biol. **12**, 4601–4611

Tachikawa, H., Takeuchi, Y., Funahashi, W., Miura, T., Gao, X.D., Fujimoto, D., Mizunaga, T., and Onodera, K. (1995) Isolation and characterization of a yeast gene, *MPD1*, the overexpression of which suppresses inviability caused by protein disulfide isomerase depletion. FEBS Lett. **369**, 212–216

Hiroyuki Tachikawa:
Department of Applied Biological Science
Tokyo University of Agriculture and Technology
3-5-8 Saiwai-cho, Fuchu
Tokyo 183, Japan
Tel. and Fax. 81 423 67 5703
E-mail: tachi@cc.tuat.ac.jp

Takemitsu Mizunaga:
Keisen College
1436 Sannomiya, Isehara City
Kanagawa 259-11, Japan
Tel. 81 463 95 1010
Fax. 81 463 96 6219

BS2: *Trypanosoma brucei* protein disulfide isomerase homologue

BS2 is a putative homologue of protein disulfide isomerase (PDI) from the parasitic protozoan Trypanosoma brucei. *It was identified as a developmentally regulated gene in the bloodstream stage of the parasitic life-cycle. Its identity is based solely on deduced amino acid homology and the gene product has not been characterized.*

Trypanosoma brucei

African trypanosomes of the *T. brucei* species (Order Kinetoplastida) are parasitic protozoa that are the causative agents of African trypanosomiasis (African sleeping sickness) in humans and Nagana in cattle (Vickerman et al., 1993). Analyses of 18S rRNA sequence homologies indicate that these organisms comprise one of the most ancient of all eukaryotic lineages, being as distant phylogenetically from yeast as they are from humans (Sogin et al., 1989). They are digenetic parasites living alternatively in the bloodstream of the mammalian host and the midgut and mouthparts of the insect vector, the tsetse fly. Complex differentiation processes occur at each stage of the life-cycle.

BS2 gene and sequence

The cloned *BS2* gene sequence (GenBank accession number J02865), derived from a genomic library, is a single contiguous open reading frame of 1494 nucleotides (Hsu et al., 1989). Like all known trypanosomal genes it has no introns. The deduced amino acid sequence indicates a protein of 498 residues (55.5 kDa) including a putative N-terminal signal sequence, 12 potential N-linked glycosylation sites, two internal thioredoxin-like domains and the C-terminal tetrapeptide Lys–Gln–Asp–Leu (KQDL).

The deduced amino acid sequence of the BS2 gene has significant similarity to rat protein disulfide isomerase (PDI) (Edman et al., 1985); amino acid identity is most striking in and around the thioredoxin-like domains. However, equivalent levels of similarity were also found to a rat gene reported to be form I phosphatidylinositol-specific phospholipase C (PIPLC) (Bennett et al., 1988) creating some uncertainty as to the likely identity of BS2. Subsequently, it has been definitively shown that the PIPLC was misidentified and is in fact a protein disulfide isomerase isoenzyme called ERp61 (Srivastava et al., 1991; Mazzarella et al., 1994). Thus it is likely that the BS2 gene is a trypanosomal homologue of PDI.

■ BS2 protein

Nothing is known about the actual BS2 gene product. Attempts to make antibody to recombinant protein have failed (J.D. Bangs and J.C. Boothroyd, unpublished observations) and no direct analysis of the protein has been performed. However, the C-terminal tetrapeptide, KQDL, has been shown to be a functional ER localization signal in trypanosomes (see TbBiP entry, p. 41), consistent with the putative PDI function of the BS2 gene product.

■ *BS2* gene regulation

By Northern analysis the BS2 message is about 15-fold more abundant in the bloodstream stage than in the procyclic insect stage (Hsu et al., 1989).

■ References

Bennett, C.F., Balcarek, J.M., Varrichio, A., and Crooke, S.T. (1988). Molecular cloning and complete amino acid sequence of form I phosphoinositde-specific phospholipase C. Nature **334**, 268–270.

Edman, J.C., Ellis, L., Blacher, R.W., Roth, R.A., and Rutter, W.J. (1985). Sequence of protein disulfide isomerase and implications of its relationship to thioredoxin. Nature **317**, 267–270.

Hsu, M.P., Muhich, M.L., and Boothroyd, J.C. (1989). A developmentally regulated gene of trypanosomes encodes a homologue of rat protein-disulfide isomerase and phosphoinositol-phospholipase C. Biochemistry **28**, 6440–6446.

Mazzarella, R.A., Marcus, N., Haugejorden, S.M., Balcarek, J.M., Baldassare, J.J., Roy, B., Li, L., Lee, A.S., and Green, M. (1994). ERp61 is GRP58, a stress-inducible lumenal endoplasmic reticulum protein, but is devoid of phosphatidylinositide-specific phospholipase C activity. Arch. Biochem. Biophys. **308**, 454–460.

Sogin, M.L., Grunderson, J.H., Elwood, H.J., Alonoso, R.A., and Peattie, D.A. (1989). Phylogenetic meaning of the kingdom concept: an unusual ribosomal RNA from *Giardia lamblia*. Science **243**, 75–77.

Srivastava, S.P., Chen, N.Q., Liu, Y.X., and Holtzman, J.L. (1991). Purification and characterization of a new isozyme of thiol:protein disulfide oxidoreductase from rat hepatic microsomes. J. Biol. Chem. **266**, 20337–20344.

Vickerman, K., Myler, P.J., and Stuart, K.D. (1993). African Trypanosomiasis. In *Immunology and Molecular Biology of Parasitic Infections*, (K.S. Warren, ed.). Blackwell Scientific Publications, Boston, pp. 170–212.

■ *James D. Bangs:*
Department of Medical Microbiology and Immunology
University of Wisconsin–Madison School of Medicine
1300 University Avenue
Madison, WI 53706, USA
Tel. 1 608 262 3110
Fax. 1 608 262 8418
E-mail: bangs@macc.wisc.edu

Mammalian thioredoxin

Mammalian thioredoxin is a small (M_r 12 000) multifunctional protein with a redox active disulfide/dithiol in the conserved active site sequence: –Trp–Cys–Gly–Pro–Cys– . Mammalian thioredoxin may be secreted by cells and has a large variety of biological activities.

■ Isolation

Mammalian thioredoxin has been purified from several sources including rat Novikoff hepatoma, calf liver, rat liver, rabbit bone marrow and human placenta (Holmgren, 1985; Holmgren and Björnstedt, 1995). A protein that was purified from the conditioned medium of human T cell lymphotropic virus type-I (HTLV-I)-transformed T cells and Epstein–Barr virus-transformed B cells, has been identified as human thioredoxin [then called adult T cell leukemia-derived factor (ADF) (Tagaya et al., 1989; Wollman et al., 1988; Wakasugi et al., 1990)].

■ Thioredoxin gene and sequence

The nucleotide sequences of thioredoxin cDNA from mammalian species (GenBank accession numbers JO3882, JO4026, X14878, X77585, D21855-21859) and amino acid sequence determined by tandem mass spectrometry show a protein of 104 or 105 amino acids that includes the conserved active site sequence –Trp–Cys–Gly–Pro–Cys–. Genomic sequences of human and murine thioredoxin gene have been reported (GenBank accession numbers X54539-54541, X70286-70288, D21855-21859) (Wollman et al., 1988; Tagaya et al., 1989; Tonissen and Wells, 1991; Kaghad et al., 1994; Matsui et al., 1995).

■ Thioredoxin protein

Thioredoxin is ubiquitously distributed in mammalian cells (Holmgren and Björnstedt, 1995). Thioredoxin is secreted by malignant and normal cells through a leaderless secretory pathway (Tagaya et al., 1989; Rosén et al., 1995; Rubartelli et al., 1995). Expression of thioredoxin is induced by a variety of stresses including mitogens, phorbol myristate acetate, virus infection, UV exposure and hydrogen peroxide (Yodoi and Uchiyama, 1992). As with the Escherichia coli thioredoxin system, oxidized mammalian thioredoxin is reduced by thioredoxin reductase and NADPH. Thioredoxin, thioredoxin reductase and NADPH, collectively called the thioredoxin system, operate as a powerful NADPH-dependent protein disulfide reductase system. In contrast to the E. coli thioredoxin reductase, mammalian thiredoxin reductase is not specific to the homologous thioredoxin and many other substances have been reported to be substrates for mammalian thioredoxin reductase, e.g. organic and inorganic selenium compounds (Björnstedt et al., 1995a; Holmgren and Björnstedt, 1995) and hydroperoxides (Björnstedt et al., 1995b). Mammalian thioredoxins have, in addition to the active site disulfide, two (three in human) additional structural half-cystine residues. Oxidation of these half-cystine residues to intramolecular or intermolecular disulfides leads to aggregation of the protein and loss of its activity as a protein disulfide reductase (Ren et al., 1993). The three-dimensional structure of human thioredoxin is quite similar to that of E. coli (Forman et al., 1991).

Protein disulfide isomerase (PDI, see entry p. 348) contains two domains with strong sequence homology to thioredoxin. PDI is a substrate for mammalian thioredoxin reductase and has thioredoxin-like activity. Two additional proteins from the endoplasmic reticulum (CaBP1 and CaBP2, see entries pp. 354 and 353) also contain thioredoxin domains and they share catalytic properties with PDI in thiol–disulfide interchange reactions during protein folding (Lundström-Ljung et al., 1995). A set of endoplasmic reticulum proteins show properties of molecular chaperones including calcium-binding proteins and members of the thioredoxin superfamily (Nigam et al., 1994).

■ Biological activities of thioredoxin

With the exception of rabbit bone marrow, mammalian thioredoxins serve as hydrogen donors to their homologous ribonucleotide reductase (Holmgren, 1985). In addition, mammalian thioredoxins have a variety of biological functions such as operating as general protein disulfide reductases, regulation of the glucocorticoid receptor, growth promotion, cytoprotection against oxidative stress, redox regulation of transcription factors and reduction of selenium compounds (Holmgren, 1985; Yodoi and Uchiyama, 1992; Björnstedt et al., 1995; Holmgren and Björnstedt, 1995). Human thioredoxin is an efficient electron donor to human plasma glutathione peroxidase and can thereby exert a protective effect against hydroperoxides in human plasma which is an environment almost free from glutathione (Björnstedt et al., 1994). Truncated forms of human thioredoxin show the activity of eosinophil cytotoxicity enhancing factor (Silberstein et al., 1993). Thioredoxin is a serum component of 'early pregnancy factor' (Tonissen et al., 1993).

■ Mutagenesis studies with mammalian thioredoxin

C32S lacks cytoprotective activity against oxidative stress (Nakamura et al., 1994). C32S, C35S and C32S/C35S lack growth promoting activity (Oblong et al., 1994). C72S avoids the dimerization of the molecules and inactivation through oxidation (Ren et al., 1993) and lacks the activity of early pregnancy factor (Tonissen et al., 1993).

■ Biological interactions

Human thioredoxin is responsible for activating the DNA-binding properties of cellular transcription factors, including NF-κB and AP-1. DNA-binding of NF-κB is activated by reduction of a disulfide bond involving Cys-62 of the p50 subunit (Matthews et al., 1992). In addition, the complex is stabilized by numerous hydrogen bonding, electrostatic and hydrophobic interactions which involve residues 57–85 of the NF-κB peptide and confer substrate specificity (Qin et al., 1995). Activation of the AP-1 DNA-binding activity is mediated via a protein Ref-1. This protein is reduced by thioredoxin and activates AP-1 DNA binding by reduction of conserved cysteine residues.

■ References

Björnstedt, M., Xue, J., Huang, W., Åkesson, B., and Holmgren, A. (1994). The thioredoxin and glutaredoxin systems are efficient electron donors to human plasma glutathione peroxidase. J. Biol. Chem. **269**, 29382–29384.

Björnstedt, M., Kumar, S. and Holmgren, A. (1995a). Selenite and selenodigultathione: reactions with thioredoxin systems. Meth. Enzymol. **252**, 209–219.

Björnstedt, M., Hamberg, M., Kumar, S., Xue, J., and Holmgren, A. (1995b). Human thioredoxin reductase directly reduces lipid hydroperoxides by NADPH and selenocystine strongly stimulates the reaction via catalytically generated selenols. J. Biol. Chem. **270**, 11761–11764.

Forman, Kay Jd, Clore, G.M., Wingfield, P.T., and Gronenborn, A.M. (1991). High-resolution three-dimensional structure of reduced recombinant human thioredoxin in solution. Biochemistry **30**, 2685–98.

Holmgren, A. (1985). Thioredoxin. Annu. Rev. Biochem. **54**, 237–271.

Holmgren, A. and Björnstedt, M. (1995). Thioredoxin and thioredoxin reductase. Meth. Enzymol. **252**, 199–208.

Kaghad, M., Dessarps, F., Jacquemin, Sablon H., Caput, D., Fradelizi, D., and Wollman, E.E. (1994). Genomic cloning of human thioredoxin-encoding gene: mapping of the transcription start point and analysis of the promoter. Gene **140**, 273–278.

Lundström-Ljung, J., Birnbach, U., Rupp, K., Söling, H.D., and Holmgren, A. (1995). Two resident ER-proteins, CaBP1 and CaBP2, with thioredoxin domains, are substrates for thioredoxin reductase: comparison with protein disulfide isomerase. FEBS Lett. **357**, 305–308.

Matsui, M., Taniguchi, Y., Hirota, K., Taketo, M., and Yodoi, J. (1995). Structure of the mouse thioredoxin-encoding gene and its processed pseudogene. Gene **152**, 165–171.

Matthews, J.R., Wakasugi, N., Virelizier, J.L., Yodoi, J., and Hay, R.T. (1992). Thioredoxin regulates the DNA binding activity of NF-kappa B by reduction of a disulphide bond involving cysteine 62. Nucl. Acids Res. **20**, 3821–3830.

Nakamura, H., Matsuda, M., Furuke, K., Kitaoka, Y., Jwata, S., Toda, K., Inamoto, T., Yamaoka, Y., Ozawa, K. and Yodoi, J. (1994). Adult T cell leukemia-derived factor/human thioredoxin protects endothelial F-2 cell injury caused by activated neutrophils or hydrogen peroxide (published erratum appears in Immunol. Lett. 1994, 42(3), 213). Immunol. Lett. **42**, 75–80.

Nigam, S.K., Goldberg, A.L., Ho, S., Rohde, M.F., Bush, K.T., and Sherman, MYu (1994). A set of endoplasmic reticulum proteins possessing properties of molecular chaperones includes Ca(2+)-binding proteins and members of the thioredoxin superfamily. J. Biol. Chem. **269**, 1744–1749.

Oblong, J.E., Berggren, M., Gasdaska, P.Y., and Powis, G. (1994). Site-directed mutagenesis of active site cysteines in human thioredoxin produces competitive inhibitors of human thioredoxin reductase and elimination of mitogenic properties of thioredoxin. J. Biol. Chem. **269**, 11714–11720.

Qin, J., Clore, G.M., Kennedy, W.M.P., Huth, J.R., and Gronenborn, A.M. (1995). Solution structure of human thioredoxin in a mixed disulfide intermediate complex with its target peptide from the transcription factor NFκB. Structure **3**, 289–297.

Ren, X., Björnstedt, M., Shen, B., Ericson, M.L., and Holmgren, A. (1993). Mutagenesis of structural half-cystine residues in human thioredoxin and effects on the regulation of activity by selenodiglutathione. Biochemistry **32**, 9701–9708.

Rosén, A., Lundman, P., Carlsson, M., Bhavani, K., Shinivasa, B.R., Kjellström, G., Nilssen, K. and Holmgren, A. (1995). A CD4+ T cell line-secreted factor, growth promoting for normal and leukemic B cells, identified as thioredoxin. Int. Immunol. **7**, 625–633.

Rubartelli, A., Bonifaci, N., and Sitia, R. (1995). High rates of thioredoxin secretion correlate with growth arrest in hepatoma cells. Cancer Res. **55**, 675–680.

Silberstein, D.S., McDonough, S., Minkoff, M.S., and Balcewicz, Sablinska Mk (1993). Human eosinophil cytotoxicity-enhancing factor. Eosinophil-stimulating and dithiol reductase activities of biosynthetic (recombinant) species with COOH-terminal deletions. J. Biol. Chem. **268**, 9138–9142.

Tagaya, Y., Maeda, Y., Mitsui, A., Kondo, N., Matsui, H. Hamuro, Y., Brown, N., Arai, K., Yokota, T., Wakazugi, H. and Yodoi, J. (1989). ATL-derived factor (ADF), an IL-2 receptor/Tac inducer homologous to thioredoxin; possible involvement of dithiol-reduction in the IL-2 receptor induction (published erratum appears in EMBO J. 1994, **13**(9),2244). EMBO J. **8**, 757–764.

Tonissen, K.F. and Wells, J.R. (1991). Isolation and characterization of human thioredoxin-encoding genes. Gene **102**, 221–228.

Tonissen, K., Wells, J., Cock, I., Perkins, A., Orozco, C., and Clarke, F. (1993). Site-directed mutagenesis of human thioredoxin. Identification of cysteine 74 as critical to its function in the 'early pregnancy factor' system. J. Biol. Chem. **268**, 22485–22489.

Wakasugi, N., Tagaya, Y., Wakasugi, H., Mitsui, A., Maeda, M., Yodoi, J., and Tursz, T. (1990). Adult T-cell leukemia-derived factor/thioredoxin, produced by both human T-lymphotropic virus type I- and Epstein-Barr virus-transformed lymphocytes, acts as an autocrine growth factor and synergizes with interleukin 1 and interleukin 2. Proc. Natl. Acad. Sci. USA **87**, 8282–8286.

Wollman, E.E., d'Auriol, L., Rimsky, L., Shaw, A., Jacquot, J.P., Wingfield, P., Graber, P., Dessarps, F., Robin, P., Galibert, F., Fradelizi, D. (1988). Cloning and expression of a cDNA for human thioredoxin. J. Biol. Chem. **263**, 15506–15512.

Yodoi, J. and Uchiyama, T. (1992). Diseases associated with HTLV-I: virus, IL-2 receptor dysregulation and redox regulation. Immunol. Today **13**, 405–411.

■ *Hajime Nakamura, Mikael Björnstedt and Arne Holmgren:*
Medical Nobel Institute for Biochemistry
Department of Medical Biochemistry and Biophysics
Karolinska Institute
S-171 77 Stockholm, Sweden
Tel. 46 8 728 7686
Fax. 46 8 728 4716
E-mail: Arne.Holmgren@mbb.ki.se

Mammalian glutaredoxin

Glutaredoxin (Grx) from mammalian cells is a 12 kDa protein originally identified as a hydrogen donor for ribonucleotide reductase in calf thymus. The protein contains a redox active disulfide/dithiol in its active site and catalyses GSH–disulfide oxidoreductions and is the mammalian equivalent of Escherichia coli *glutaredoxin.*

■ Alternative names

Glutathione-homocystine transhydrogenase (Racker, 1955); thioltransferase (Askelöf et al., 1974).

■ Isolation, gene and sequence

Mammalian glutaredoxin was first purified to homogeneity from calf thymus (Luthman et al., 1979; Luthman and Holmgren, 1982), which later enabled determination of the primary structure (Klintrot et al., 1984) demonstrating an active site with the sequence Cys–Pro–Tyr–Cys, identical to that of *Escherichia coli* glutaredoxin. Calf thymus Grx showed a general GSH–disulfide oxidoreductase activity with hydroxyethyldisulfide in a coupled system with NADPH, GSH and glutathione reductase. Similar types of GSH–disulfide oxidoreductase activity with small disulfides had previously been observed in bovine liver (Racker, 1955) and rat liver (Eriksson and Mannervik, 1970) referred to as glutathione–homocystine disulfide reductase and thioltransferase (Askelöf et al., 1974), respectively. Since pure thioltransferase originally was suggested to contain 10% carbohydrate it was assumed to be different from glutaredoxin (Luthman and Holmgren, 1982). A pI shift method improved the yields in the purification of mammalian thioltransferase (Gan and Wells, 1987a) resulting in homogenous preparations. The amino acid sequence determination of thioltransferase clearly demonstrated its identity to the sequence of glutaredoxin with no carbohydrate (Wells et al., 1993). Amino acid sequences of mammalian glutaredoxins have been obtained from pig liver, calf liver, rabbit bone marrow and human placenta and red blood cells (Klintrot et al., 1984; Gan and Wells, 1987b; Hopper et al., 1989; Papayannopoulos et al., 1989; Papov et al., 1994; Padilla et al., 1995). A glutaredoxin with strong amino acid sequence homology to mammalian glutaredoxins has been identified in the vaccinia virus particle (Ahn and Moss, 1992). Complete nucleotide sequences of mammalian glutaredoxin cDNAs were obtained in pig and human (GenBank accession numbers D21238, X76648) (Yang et al., 1989; Fernando et al., 1994; Padilla et al., 1995). High-level expression of a mammalian glutaredoxin in *E. coli* was reported for pig glutaredoxin (Yang and Wells, 1990) and human glutaredoxin (Padilla et al., 1995).

■ Grx protein and biological activities

Glutaredoxin (Grx) is a 12 kDa protein with two additional structural SH groups which are sensitive to oxidation and may regulate activity (Holmgren and Åslund, 1995). Grx is widely distributed in mammalian cells (Rozell et al., 1993). Glutaredoxins from calf thymus and human placenta have hydrogen donor activity in the ribonucleotide reductase assay in the presence of GSH, glutathione reductase and NADPH (Luthman et al., 1979; Padilla et al., 1995). This activity is species specific and may be compared to that observed with the mammalian thioredoxin system (Padilla et al., 1995). In contrast, a glutaredoxin from rabbit bone marrow was reported to lack activity with a partially purified bone marrow ribonucleotide reductase (Hopper and Tomko, 1986). Glutaredoxins have a GSH-binding site, and a specificity for reducing glutathionyl mixed disulfides. The activity with GSH mixed disulfides has been shown to be dependent on only the presence of the N-terminal of the active site half-cystines (Yang and Wells, 1991). Glutaredoxin exhibits activity in catalysing reduction of dehydroascorbic acid (DHA) with GSH in the presence of NADPH and glutathione reductase (Wells et al., 1990).

Reduction of GSH mixed disulfides of proteins in oxidative environments such as red blood cells (Gravina and Mieyal, 1993) is one function of glutaredoxin. In addition, under oxidative stress, glutaredoxin may be involved in removing GSH mixed disulfides (Thomas et al., 1995). *Escherichia coli* glutaredoxin (Grx1) can stimulate native disulfide formation in proteins such as ribonuclease A in a GSH/GSSG redox buffer together with protein disulfide isomerase. Whether glutaredoxin is present in the endoplasmic reticulum remains to be determined.

The most important biological function exerted by glutaredoxin is as GSH protein disulfide reductase. Since glutaredoxin is found in many non-dividing cells (Martínez-Galisteo et al., 1995), its main function is as a general reductant to maintain protein SH groups in a reduced state and catalyse GSH-dependent reactions.

■ References

Ahn, B.Y. and Moss, B. (1992). Glutaredoxin homolog encoded by vaccinia virus is a virion-associated enzyme with thioltransferase and dehydroascorbate reductase activities. Proc. Natl. Acad. Sci. USA **89**, 7060–7064.

Askelöf, P., Axelsson, K., Eriksson, S., and Mannervik, B. (1974). Mechanism of action of enzymes catalyzing thiol-disulfide interchange. Thioltransferases rather than transhydrogenases. FEBS Lett.s **38**, 263–267.

Ericksson, S.A. and Mannervik, B. (1970). The reduction of the L-cystine-glutathione mixed disulfide in rat liver. Invovement of an enzyme catalysing thiol-disulfide interchange. FEBS Lett. **7**, 26–28.

Fernando M.R., Suminoto, H., Nanri, H., Kawabata, S., Iwanaga, S., Minakami, S., Fukumaki, Y., Takeshige K. (1994). Cloning and sequencing of the cDNA encoding human glutaredoxin. Biochim. Biophys. Acta **1218**, 229–231.

Gan, Z.-R. and Wells, W.W. (1987a). Preparation of homologous pig liver thioltransferase by a thiol: disulfide mediated pI shift. Anal. Biochem. **162**, 265–273.

Gan, Z-R. and Wells, W.W. (1987b). The primary structure of pig liver thioltransferase. J. Biol. Chem. **262**, 6699–6703.

Gravina, S.A. and Mieyal, J.J. (1993). Thioltransferase is a specific glutathionyl mixed disulfide oxidoreductase. Biochemistry **32**, 3368–3376.

Holmgren, A. and Åslund, F. (1995). Glutaredoxin. Meth. Enzymol. **252**, 283–292.

Hopper, S. and Tomko, M.A. (1986). Thioredoxin and a glutaredoxin-like protein from rabbit bone marrow. In *Thioredoxin and Glutaredoxin Systems*, (A. Holmgren, C.I. Brändén, H. Jörnvall and B.M. Sjöberg, eds). Raven Press, New York, pp. 103–110.

Hopper, S., Johnson, R.S., Vath, J. E. and Biemann, K. (1989). Glutaredoxin from rabbit bone marrow. Purification, characterization, and amino acid sequence determined by tandem mass spectrometry. J. Biol. Chem. **264**, 20438–20447.

Klintrot, I.M., Höög, J.O., Jörnvall, H., Holmgren, A. and Luthman, M. (1984). The primary structure of calf thymus glutaredoxin. Homology with the corresponding Escherichia coli protein but elongation at both ends and with an additional half-cystine/cysteine pair. Eur. J. Biochem. **144**, 417–423.

Luthman, M. and Holmgren, A. (1982). Glutaredoxin from calf thymus. Purification to homogeneity. J. Biol. Chem. **257**, 6686–6690.

Luthman, M., Eriksson, S., Holmgren, A. and Thelander, L. (1979). Glutathione-dependent hydrogen donor system for calf thymus ribonucleoside-diphosphate reductase. Proc. Natl. Acad. Sci. USA **76**, 2158–2162.

Martínez-Galisteo, E., Padilla, C.A., Holmgren, A. and Bárcena, J.A. (1995). Characterization of mammalian thioredoxin reductase, thioredoxin and glutaredoxin by immunochemical methods. Comp. Biochem. Physiol. **111B**, 17–25.

Padilla, C.A., Martínez-Galisteo, E., Bárcena, J.A., Spyrou, G. and Holmgren, A. (1995). Purification from placenta, amino acid sequence, structure comparisons and cDNA cloning of human glutaredoxin. Eur. J. Biochem. **227**, 27–34.

Papayannopoulos, I.A., Gan, Z.R., Wells, W.W. and Biemann, K. (1989). A revised sequence of calf thymus glutaredoxin. Biochem. Biophys. Res. Commun. **159**, 1448–1454.

Papov, V.V., Gravina, S.A., Mieyal, J.J. and Biemann, K. (1994). The primary structure and properties of thioltransferase (glutaredoxin) from human red blood cells. Protein Sci. **3**, 428–434.

Racker, E. (1955). Glutathione-homocystine transhydrogenase. J. Biol. Chem. **217**, 867–874.

Rozell, B., Bárcena, J.A., Martínez-Galisteo, E., Padilla, C.A. and Holmgren, A. (1993). Immunochemical characterization and tissue distribution of glutaredoxin (thioltransferase) from calf. Eur. J. Cell Biol. **62**, 314–323.

Thomas, J.A., Poland, B. and Honzatko, R. (1995). Protein sulfhydryls and their role in the antioxidant function of protein S-thiolation. Arch. Biochem. Biophys. **319**, 1–9.

Wells, W.W., Xu, D.P., Yang, Y. and Rocque, P.A. (1990). Mammalian thioltransferase (glutaredoxin) and protein disulfide isomerase have dehydroascorbate reductase activity. J. Biol. Chem. **265**, 15361–15364.

Wells, W.W., Yang, Y., Deits, T.L. and Gan, Z.R. (1993). Thioltransferases. Adv. Enzymol. Relat. Areas Mol. Biol. **66**, 149–201.

Yang, Y., Gan, Z.-R. and Wells, W.W. (1989). Cloning and sequencing the cDNA encoding pig liver thioltransferase. Gene **83**, 339–346.

Yang, Y. and Wells, W.W. (1990). High-level expression of pig liver thioltransferase (glutaredoxin) in Escherichia coli. J. Biol. Chem. **265**, 589–593.

Yang, Y. and Wells, W.W. (1991). Identification and characterization of the functional amino-acids at the active center of pig liver thioltransferase by site-directed mutagenesis. J. Biol. Chem. **266**, 12759–12765.

Hajime Nakamura, Fredrik Åslund, Mikael Björnstedt and Arne Holmgren:
Medical Nobel Institute for Biochemistry
Department of Medical Biochemistry and Biophysics
Karolinska Institute
S-171 77 Stockholm, Sweden
Tel. 46 8 728 7686
Fax. 46 8 728 4716
E-mail: Arne.Holmgren@mbb.ki.se

Mammalian protein disulfide isomerases

Protein disulfide isomerase (PDI), a major soluble protein of the ER lumen, catalyses protein folding associated with the formation of native disulfide bonds. Its interactions in vivo with nascent proteins have been detected by cross-linking, and its catalytic properties in vitro have been defined in some detail. PDI contains two active sites which are functionally and structurally homologous to that of thioredoxin and are responsible for its catalysis of thiol–disulfide chemistry. PDI also functions as a component of other ER luminal enzymes, a property that may be associated with its ability to interact as a chaperone with partly folded protein intermediates.

■ Isolation of PDI

PDI is readily purified from mammalian liver, or from other highly secretory tissues such as pancreas or placenta. [Here and in subsequent paragraphs where no specific citation is given see Freedman and Tuite (1994), Freedman et al., (1994, 1995) and Freedman (1995) for pre-1994 references]. Purifications from bovine, rat and human tissues have been reported following basically similar procedures exploiting the enzyme's abundance, acidic pI and thermostability. A current purification of PDI from bovine liver, and its derivation from earlier procedures, is described elsewhere (Freedman et al., 1995).

■ Alternative names

The enzyme catalyses various thiol–disulfide interchange processes and hence has previously been named as glutathione–insulin transhydrogenase (GIT), and more generally as thiol–protein–disulfide oxidoreductase. On the basis of putative affinity labelling reactions, it has also been described as a thyroid hormone-binding protein and a glycosylation site-binding protein, but these are not physiological activities (Freedman et al., 1994). PDI is, however, found as a subunit of two other enzymes (see below) and the polypeptide is appropriately named in such contexts as the prolyl-4-hydroxylase β subunit or as the microsomal triglyceride transfer protein small subunit.

■ Mammalian *PDI* gene and sequence

cDNA sequences encoding PDI have been cloned from a wide variety of organisms and show a high degree of similarity. They encode proteins of approximately 500 residues. Multiple alignment of higher eukaryote PDI sequences and cladistic analysis shows the expected clustering of vertebrate sequences and of higher plant sequences, with fungal PDIs forming a distinct branch (Freedman et al., 1994). Human PDI (GenBank accession number X05130) is 86% identical to chicken PDI (X13110), 43% to *Caenorhabditis elegans* PDI (Z37139), 36% to alfalfa PDI (A41440) and 31% to yeast PDI (X54535).

The sequences of the human and chicken PDI genes have been defined. The human gene (GenBank accession numbers J04049 and J04050), is located in the long arm of human chromosome 17, at band 17q25; it is transcribed as a 16.5 kb species, with the coding sequence divided into 11 exons. The chick gene (GenBank accession number X06768), is similar in intron/exon organization but has shorter introns so that it is only about 9 kb in size. Promoter analysis of the human PDI gene reveals the presence of 11 promoter elements in the 5´-flanking region from nucleotide −630, including six CCAAT boxes between −108 and −378 which appear to show some functional redundancy.

The sequence of the PDI polypeptide has been analysed as a clue to its structural organization. Internal sequence homologies and homologies to thioredoxin were recognized when the first PDI sequence was determined. The sequence databases also contain other PDI-like sequences which clearly represent related proteins with ER targeting and retention signals, but in most cases the proteins corresponding to these sequences have not been purified or expressed. Some of these sequences were originally cloned mistakenly and given misleading names (e.g. as a deoxycytidine kinase, or as a PI-specific phospholipase C). Others have been characterized to some extent as ER-specific proteins (ERps), as glucose-regulated proteins (GRPs) or as calcium-binding proteins (CaBPs). One group of sequences corresponds to a family of PDI isoforms, sometimes termed ERp60 or ERp61 (see entry p. 351) or GRP58 (Mazzarella et al., 1994). These sequences are similar to PDI in length, and in the presence of two thioredoxin-like sequences in the same relative positions as in PDI. A second group of sequences, known as P5 or CaBP1, is similar in length to PDI but has its two thioredoxin-like sequences in close proximity in the N-terminal half of the protein (see entry p. 354). A third family, known as Erp72 or CaBP2 (see entry p. 353), contains three thioredoxin-like sequences and resembles PDI with the addition of an N-terminal thioredoxin-like domain. The limited functional work on these proteins suggests that they share redox and catalytic properties with PDI (Lundstrom-Ljung et al., 1995).

■ Mammalian PDI protein

PDI is a soluble acidic protein which is an abundant component of the lumen of the endoplasmic reticulum (e.g. in dog pancreas microsomes), from which it can be removed by alkaline or detergent washes. The polypeptide has a relative molecular mass of approximately 56 kDa, dependent on species, and runs on SDS–PAGE with a M_r of 58–59 kDa against conventional standards, and in native gel-filtration with an M_r of 110 kDa, indicating that the native state is predominantly a homodimer. The mammalian protein, unlike yeast PDI (see entry p. 335), has no sites for N-glycosylation, and is not glycosylated. In the protein as isolated, the cysteine residues of the two active sites are in the oxidized state, each forming a vicinal disulfide within the sequence ...WCGHCK....

The complete three-dimensional structure of PDI is not known. Domain models were derived from sequence information (above) and have been tested by determination of sites of sensitivity to proteolysis (H.C. Hawkins and R.B. Freedman, unpublished data) and by the expression and characterization of putative domains (Darby et al., 1996). Data from both sources indicate the existence of four main structural domains (a-b-b'-a') with boundaries around residues 114–116, 217–230 and 329–352; it is not clear whether the acidic C-terminal region from residue 460 onwards is structured. The a and a' domains show high sequence identity and close homology to thioredoxin; they have been expressed and functionally characterised (see below) and the structure of the a domain, determined to high resolution by NMR, resembles that of thioredoxin (Kemmink et al., 1996). The b and b' domains show lower identity to each other and no obvious homology to thioredoxin and yet recent work suggests that they also have structures based on the thioredoxin fold (Kemmink et al., 1997).

■ Catalytic activities *in vitro* of mammalian PDI

Purified PDI catalyses thiol–disulfide interchange reactions within proteins or between proteins and low molecular weight thiol compounds, leading to net protein disulfide reduction, formation or isomerization. It demonstrates a very broad protein substrate specificity (see e.g. Freedman, 1995). Among low molecular weight thiols, there is some evidence that it shows a preference for glutathione (Darby et al., 1994), the major physiological thiol in the lumen of the ER. The most informative studies on the catalytic action of PDI have employed reduced protein substrates with a defined folding pathway, such as BPTI or specific intermediates in the pathway of reoxidation of BPTI. Other studies have used reduced bovine ribonuclease and ribonuclease T1 derivatized to have each of its four Cys residues in a mixed disulfide with glutathione (Ruoppolo et al., 1996). Recently, peptides with two Cys residues have been used to study in detail the process of net protein disulfide formation catalysed by PDI (Darby et al., 1994; Ruddock et al., 1996).

PDI catalyses by up to 1000-fold protein disulfide isomerizations in which the reactions are limited by conformational constraints in the substrate protein (Weissmann and Kim, 1993). The mechanism of catalysis of net formation of protein disulfides in the presence of a low molecular weight disulfide is mainly by a series of thiol–disulfide interchanges, via mixed disulfide intermediates,

P(SH)2 + RSSR ⇔ AE P(SH)SSR + RSH ⇔ AE P(SS) + 2RSH

but may also involve a contribution from direct oxidation by PDI itself, with reduced PDI subsequently reoxidized by RSSR (Darby et al., 1994).

Detailed studies on the reactions between PDI and alkylating agents and disulfides have clarified the unusual properties of the dithiol–disulfide active sites which are responsible for its catalytic activity in thiol–disulfide interchange reactions. Data on the whole PDI molecule have now been backed up by more easily interpretable results on the isolated recombinant a and a' domains (Darby and Creighton, 1995b). The N-terminal Cys residue in each active site (WCGHCK) is highly acidic and a good attacking nucleophile at physiological pH; however, the mixed disulfides that it forms, with reagents such as GSH or with protein substrate Cys residues, are extremely unstable and reactive owing to the presence of the vicinal Cys residue (WCGHCK) in the active site. Furthermore, the active site dithiol–disulfide redox couples are unusually oxidizing with standard redox potentials of approximately –110 to –170 mV (i.e. the disulfide form is unusually unstable) (Lundstrom and Holmgren, 1993). These features rationalize the catalytic reactivity of PDI but they cannot themselves be fully explained in the absence of a high-resolution structure. Preliminary insights can be gained from a comparative study of the high-resolution structures of homologous proteins (Gane et al., 1995).

The action of PDI in catalysing disulfide formation and isomerization associated with protein folding sugggests that it may have folding or 'chaperone' activities in addition to activity in thiol–disulfide chemistry. In some systems it has been shown to have classical 'chaperone' activity *in vitro*, suppressing aggregation during refolding of non-disulfide bonded proteins (Cai et al., 1994).

■ Biological activities and interactions of mammalian PDI

The catalytic activities of PDI and its subcellular location suggest that it is involved in catalysing the formation of native disulfide bonds during protein biosynthesis. This is supported by the results of depletion/reconstitution studies with dog pancreas microsomes in *in vitro* translation experiments and by the phenotype of yeast in which the *PDI1* gene has been disrupted (see entry on *S. cerevisiae* PDI, p. 335). The direct interaction of PDI with nascent proteins in mammalian cells has also been demonstrated by cross-linking (Persson and Pettersson, 1991; Klappa et al., 1995).

Disulfide formation pathways *in vivo* have been determined for a few eukaryotic proteins such as influenza haemagglutinin and chorionic gonadotropin subunit. Comparison of the rates and pathways of such processes within whole cells, cotranslationally and post-translationally within isolated microsomes, and in well-defined *in vitro* refolding systems, confirms the essential role of PDI in facilitating rapid formation of native disulfides (Freedman, 1995).

PDI has been isolated as a component of two multi-component enzymes, prolyl-4-hydroxylase, an α_2–β_2 tetramer, and microsomal triglyceride transfer protein, an α–β dimer, and the evidence suggests that it plays an essential role in these systems. Studies on the denaturation and renaturation of these complexes *in vitro*, and on their coexpression *in vivo* in insect cells, strongly suggest that the major role of the PDI polypeptide in these complexes is to stabilize the other subunits in a soluble active conformation (Vuori *et al*., 1992). This role may be comparable to that of some molecular chaperones which form stable complexes with their protein substrates, such as HSP90 proteins (see HSP90 section, this volume, pp. 231–264).

Mutagenesis studies with mammalian PDI

Significant results have been derived from mutagenesis of one or more of the Cys residues in the active site WCGHCK sequences. Mutation, separately, of the more N-terminal Cys residues in each of these sequences inactivated each site and showed that the two active sites were independent and each contributed 50% to the overall PDI activity of the protein (Vuori *et al*., 1992); it has since been shown that the detailed kinetic properties of the two sites are distinct (Lyles and Gilbert, 1994). Comparison of the effects of mutating the more C-terminal Cys residue in each site with the more conventional mutation of the N-terminal residue suggests that mutation of either residue inactivates the site for catalysis of net protein disulfide formation or reduction, but that the active site sequence, WCGHSK, is fully active in catalysing disulfide isomerizations (Laboissiere *et al*., 1995).

Mutations of these active site residues have no effect on the ability of PDI to coassemble with other polypeptides to form functional prolyl-4-hydroxylase and microsomal triglyceride transfer protein, indicating that the essential thiol–disulfide active site of PDI is not involved in its function within these complexes (Vuori *et al*., 1992; Lamberg *et al*., 1996).

References

Cai, H., Wang, C.-C. and Tsou, C.-L. (1994). Chaperone-like activity of protein disulfide isomerase in the refolding of a protein with no disulfide bonds. J. Biol. Chem. **269**, 24550–24552.

Darby, N.J. and Creighton, T.E. (1995a). Functional properties of the individual thioredoxin-like domains of protein disulfide isomerase. Biochemistry **34**, 11725–11735.

Darby, N.J. and Creighton, T.E. (1995b). Characterization of the active site cysteine residues of the thioredoxin-like domains of protein disulfide isomerase. Biochemistry **34**, 16770–16780.

Darby, N.J., Freedman, R.B., and Creighton, T.E. (1994). Dissecting the mechanism of protein disulfide isomerase: catalysis of disulfide bond formation in a model peptide. Biochemistry **33**, 7937–7947.

Darby, N.J., Kemmink, J. and Creighton, T.E. (1996). Identifying and characterizing a structural domain of protein disulfide isomerase. Biochemistry **35**, 10517–10528.

Freedman, R.B. (1995). The formation of protein disulphide bonds. Curr. Opin. Struc. Biol. **5**, 85–91.

Freedman, R.B. and Tuite, M.F. (1994). Protein disulphide-isomerase (PDI). In *Guidebook to the Secretory Pathway*, pp. 83–86, ed. J. Rothblatt, P. Novick and T. Stevens, Sambrook and Tooze at OUP.

Freedman, R.B., Hirst, T.R., and Tuite, M.F. (1994). Protein disulphide isomerase: building bridges in protein folding Trends Biochem. Sci. **19**, 331–336.

Freedman, R.B., Hawkins, H.C., and McLaughlin, S.H. (1995). Protein disulfide-isomerase. Meth. Enzymol. **251**, 397–406.

Gane, P.J., Freedman, R.B., and Warwicker, J. (1995). A molecular model for the redox potential difference between thioredoxin and DsbA based on electrostatics calculations J. Mol. Biol. **249**, 376–387.

Kemmink, J., Darby, N.J., Dijkstra, K., Nilges, M. and Creighton, T.E. (1996). Structure determination of the N-terminal thioredoxin-like domain of protein disulfide isomerase using multidimensional heteronuclear ^{13}C/^{15}N NMR spectroscopy. Biochemistry **35**, 7684–7691.

Kemmink, J., Darby, N.J., Dijkstra, K., Nilges, M. and Creighton, T.E. (1997). The folding catalyst protein disulfide isomerase is constructed of active and inactive thioredoxin modules. Current Biology **7**, 239–245.

Klappa, P., Freedman, R.B., and Zimmermann, R. (1995). Protein disulphide isomerase and a lumenal cyclophilin-type peptidyl prolyl cis-trans isomerase are in transient contact with secretory proteins during late stages of translocation. Eur. J. Biochem. **232**, 755–764.

Laboissiere, M.C.A., Sturley, S.L., and Raines, R.T. (1995). The essential function of protein-disulfide isomerase is to unscramble non-native disulfide bonds. J. Biol. Chem. **270**, 28006–28009.

Lamberg, A., Jauhiainen, M., Metso, J., Enholm, C., Shoulders, C., Scott, J., Pihlajaniemi, T., and Kivirikko, K.I. (1996). The role of protein disulfide isomerase in the microsomal triacylglycerol transfer protein does not reside in its isomerase activity. Biochem. J. **315**, 533–536.

Lundstrom, J. and Holmgren, A. (1993). Determination of the reduction-oxidation potential of the thioredoxin-like domains of protein disulfide-isomerase from the equilibrium with glutathione and thioredoxin. Biochemistry **32**, 6649–6655.

Lundstrom-Ljung, J., Birnbach, U., Rupp, K., Soling, H.-D., and Holmgren, A. (1995). Two resident ER-proteins, CaBP1 and CaBP2, with thioredoxin domains, are substrates for thioredoxin reductase: comparison with protein disulfide isomerase. FEBS Lett. **357**, 305–308.

Lyles, M.M. and Gilbert, H.F. (1994). Mutations in the thioredoxin sites of protein disulfide isomerase reveal functional nonequivalence of the N- and C-terminal domains J. Biol. Chem. **269**, 30946–30952.

Mazzarella, R.A., Marcus, N., Haugejorden, S.M., Balcarek, J.M., Baldassare, J.J., Roy, B., Li, L.-J., Lee, A.S., and Green, M. (1994). ERp61 is GRP58, a stress-inducible luminal endoplasmic reticulum protein, but is devoid of phosphatidylinositide-specific phospholipase C activity. Arch. Biochem. Biophys. **308**, 454–460.

Persson, R. and Pettersson, R.F. (1991). Formation and intracellular transport of a heterodimeric viral spike protein complex. J. Cell Biol. **112**, 257–266.

Ruddock, L.W., Hirst, T.R., and Freedman, R.B. (1996). pH dependence of the dithiol oxidizing activity of DsbA and protein disulphide-isomerase: studies with a novel simple peptide substrate. Biochem. J. **315**, 1001–1005.

Ruoppolo, M., Freedman, R.B., Pucci, P. and Marino, G. (1996). The glutathione-dependent pathways of refolding of ribonuclease T$_1$ by oxidation and disulfide isomerization: catalysis by protein disulfide-isomerase. Biochemistry **35**, 13636–13646.

Vuori, K., Pihlajaniemi, T., Myllylä, R., and Kivirikko, K.I. (1992). Site-directed mutagenesis of human protein disulphide isomerase: effect on the assembly, activity and endoplasmic reticulum retention of human prolyl-4-hydroxylase in *Spodoptera frugiperda* insect cells. EMBO J. **11**, 4213–4217.

Weissmann, J.S. and Kim, P. (1993). Efficient catalysis of disulphide bond rearrangements by protein disulphide isomerase. Nature **365**, 185–188.

■ Robert B. Freedman:
Department of Biosciences
University of Kent
Canterbury
Kent, CT2 7NJ, UK
Tel. 44 1227 823226
Fax. 44 1227 763912
E-mail: r.b.freedman@ukc.ac.uk

Mammalian ERp61

ERp61 is a luminal ER stress protein and member of the PDI protein family. Its biological activities and physiological role have not been well defined.

■ Alternative names

ERp61 has also been identified as Grp58 (Lee, 1981), HIP-70 (Mobbs *et al*., 1990), Q-2 (Srivastava *et al*., 1991), a thiol protease of 60 kDa (Urade *et al*., 1992), a 54 kDa carnitine medium/long chain acyltransferase (Murthy and Pande, 1993), a protein covalently associated with a metabolite of halothane (Martin *et al*., 1991) and, incorrectly, as form I phosphoinositide-specific phospholipase C (Bennett *et al*., 1988).

■ ERp61 gene and sequence

The sequence of an ERp61 cDNA from the rat (Bennett *et al*., 1988; GenBank accession number X13255) encodes a protein with a predicted molecular weight of 56 559 Da. This precursor protein includes an N-terminal signal sequence of 24 amino acids, which is cleaved to form the mature protein. There are no consensus sequences for glycosylation (Bennett *et al*., 1988), consistent with the finding that the size of the mature protein is not altered by endoglycosidase H treatment (Lewis *et al*., 1986). The protein has two thioredoxin/PDI regions containing CGHC motifs (Bennett *et al*., 1988). Hydropathy analysis indicates that the mature protein has one hydrophobic domain near the N-terminus. Overall, the protein is hydrophilic with five consecutive lysines near the C-terminus and several repeating patterns of acidic amino acid residues. It terminates with the ER retrieval sequence QEDL (Bennett *et al*., 1988) and removal of these amino acids results in secretion of ERp61 (Mazzarella *et al*., 1994). Several isoforms of ERp61 have been identified after separation on two-dimensional gels (Mobbs *et al*., 1990). One group has identified eight different forms of ERp61 (Urade *et al*., 1992) but the differences between the species are not defined.

■ Localization of ERp61

Several studies have localized this protein to the ER. The protein has been purified from the rough endoplasmic reticulum (RER) (Lewis *et al*., 1986; Urade *et al*., 1992). Biosynthetic sorting studies and subcellular fractionation studies demonstrate that it is a resident ER protein (Lewis *et al*., 1986). Immunocytochemical staining has shown colocalization of ERp61 with two resident ER proteins, PDI (see entry p. 348) and Hsp47, a collagen-binding heat shock protein (Kozaki *et al*., 1994; see entry p. 465). ERp61 is a luminal ER protein which lacks cytoplasmic domains, as shown by its resistance, in ER microsomes, to proteases in the absence of detergents (Lewis *et al*., 1986).

■ Function of ERp61

Although ERp61 is a member of the PDI family, it is not clear that it functions as a disulfide isomerase *in vivo*. Activities of ERp61 in the glutathione insulin transhydrogenase assay are one-third to one-fifth of those of

PDI (Srivastava et al., 1991; Kozaki et al., 1994). Additionally, ERp61 is reported to have thiol protease activity which can cleave the ER proteins PDI and calreticulin (Urade et al., 1992). Autocatalytic degradation can occur but the optimal conditions are non-physiological, requiring 4°C in the presence of 3 M urea and 0.1 βM -mercaptoethanol. One study showed that ERp61 in liver is covalently labeled by the trifluoroacetylchloride metabolite of halothane (Martin et al., 1991). Another group found carnitine medium/long chain acyltransferase (CPT) activity in purified preparations of ERp61 (Murthy and Pande, 1993). Furthermore, membrane fractions isolated from cells transfected with murine ERp61 cDNA showed significant increases in CPT activity and immunoreactive protein for ERp61 over controls (Murthy and Pande, 1994). Additionally, in vitro transcription/ translation of the ERp61 cDNA yielded a product with CPT activity. However, the physiological role for ERp61 as a CPT is speculative. While the function(s) of ERp61 is puzzling, several groups have shown definitively that it is not a phosphoinositol phospholipase C (Martin et al., 1991; Srivastava et al., 1991; Urade et al., 1992; Mazzarella et al., 1994) as originally reported (Bennett et al., 1988).

■ Regulation of ERp61

Levels of ERp61 protein increase in response to stress such as depletion of glucose (Lee, 1981). ERp61 protein and mRNA levels increase in a mutant K12 hamster cell line cultured at non-permissive temperatures in which glycosylation is blocked (Mazzarella et al., 1994). Hormonal treatments may also regulate ERp61. Uterine ERp61 mRNA levels increase with estrogen treatment (Kaplitt et al., 1993). ERp61 can undergo a post-translational modification of unknown nature which shifts its isoelectric point to that of a more basic isoform. This occurs in the ventromedial hypothalamus in response to estrogen and in the pituitary in response to estrogen and luteinizing hormone-releasing hormone (Mobbs et al., 1990).

■ References

Bennett, C.F., Balcarek, J.M., Varrichio, A., and Crooke, S.T. (1988). Molecular cloning and complete amino-acid sequence of form-I phosphoinositide-specific phospholipase C. Nature **234**, 268–270.
Kaplitt, M.G., Kleopoulos, S.P., Pfaff, D.W., and Mobbs, C.V. (1993). Estrogen increases HIP-70/PLC-α messenger ribonucleic acid in the rat uterus and hypothalamus. Endocrinology **133**, 99–103.
Kozaki, K., Miyaishi, O., Asai, N., Iida, K., Sakata, K., Hayashi, M., Nishida,T., Matsuyama, M., Shimizu, S., Kaneda, T., and Saga, S. (1994). Tissue distribution of ERp61 and association of its increased expression with IgG production in hybridoma cells. Exptl. Cell Res. **213**, 348–358.
Lee, A. (1981). The accumulation of three specific proteins related to glucose-regulated proteins in a temperature-sensitive hamster mutant cell line K12. J. Cell. Physiol. **106**, 119–125.
Lewis, M.J., Mazzarella, R.A., and Green, M. (1986). Structure and assembly of the endoplasmic reticulum: biosynthesis and intracellular sorting of ERp61, ERp59, and ERp49, three protein components of murine endoplasmic reticulum. Arch. Biochem. Biophys. **245**, 389–403.
Martin, J.L., Pumford, N.R., LaRosa, A.C., Martin, B.M., Gonzaga, H.M.S., Beaven, M. A., and Pohl, L.R. (1991). A metabolite of halothane covalently binds to an endoplasmic reticulum protein that is highly homologous to phosphatidylinositol-specific phospholipase C-α but has no activity. Biochem. Biophys. Res. Commun. **178**, 679–685.
Mazzarella, R.A., Marcus, N., Haugejorden, S.M., Balcarek, J.M., Baldassare, J.J., Roy, B., Li, L., Lee, A.S., and Green, M. (1994). ERp61 is GRP58, a stress-inducible luminal endoplasmic reticulum protein, but is devoid of phosphatidylinositide-specific phospholipase C activity. Arch. Biochem. Biophys. **308**, 454–460.
Mobbs, C.V., Fink, G., and Pfaff, D.W. (1990). HIP-70: a protein induced by estrogen in the brain and LH-RH in the pituitary. Science **247**, 1477–1479.
Murthy, M.S.R. and Pande, S.V. (1993). Carnitine medium/long chain acyltransferase of microsomes seems to be the previously cloned ~ 54 kDa protein of unknown function. Mol. Cell. Biochem. **122**, 133–138.
Murthy, M.S.R. and Pande, S.V. (1994). A stress-regulated protein, GRP58, a member of thioredoxin superfamily, is a carnitine palmitoyltransferase isoenzyme. Biochem. J. **304**, 31–34.
Srivastava, S.P., Chen, N.Q., Liu,Y.X., and Holtzman, J.L. (1991). Purification and characterization of a new isozyme of thiol:protein-disulfide oxidoreductase from rat hepatic microsomes. J. Biol. Chem. **266**, 20337–20344.
Urade, R., Nasu, M., Moriyama, T., Wada, K., and Kito, M. (1992). Protein degradation by the phosphoinositide-specific phospholipase C-α family from rat liver endoplasmic reticulum. J. Biol. Chem. **267**, 15152–15159.

■ *Nancy Marcus and Michael Green:*
Department of Molecular Microbiology and Immunology
Saint Louis University School of Medicine
1402 S. Grand Boulevard
St. Louis, MO 63104, USA
Tel. 1 314 577 8445
Fax. 1 314 773 3403
E-mail: greenmi@wpogate.slu.edu

Mammalian ERp72

ERp72 is an abundant, soluble, ER-luminal protein with sequence homology to thioredoxin-like or PDI-like proteins. The role of this protein in the ER is not yet understood, although it may play a role in protein folding or protein degradation.

■ Alternative name

CaBP2 (Nguyen Van *et al.*, 1993).

■ ERp72 gene and sequence

ERp72 DNA/protein sequences have been reported from several sources [EMBL/SWISS-PROT accession numbers J05016/P13667 (human), S59335/P38659 (rat), J05186/P15841 (mouse), L15188/P34329 (*Caenorhabditis elegans*)]. The DNA sequences are *c.* 1.9 kb long and encode an N-terminal signal sequence of around 20 amino acids that is cleaved to generate a highly conserved *c.* 625 amino acids protein (Mazzarella *et al.*, 1990; Nguyen Van *et al.*, 1993). The human protein has a molecular mass of 72 932 Da, and shares 95% homology with both rat and mouse proteins, and 69% homology with *C. elegans* ERp72. Shortly into the sequence (aa 39–55) is an acidic cluster, followed by three copies of the highly conserved CGHC tetrapeptide, also known as the thioredoxin box (trx) motif. The C-terminal tetrapeptide KEEL (Lys–Glu–Glu–Leu) is responsible for the retrieval of the protein to the ER from later compartments of the secretory pathway. ERp72 also shares primary sequence homology with other PDI-like proteins (e.g. PDI, CaBP1; see entries p. 348 and p. 354) around the trx tetrapeptides.

■ ERp72 protein

ERp72, like Grp94 and BiP (see entries p. 161 and p. 59), is stress inducible (Kozutsumi *et al.*, 1988; Dorner *et al.*, 1990; Nguyen Van *et al.*, 1993). ERp72 is known to possess both low and high affinity Ca^{2+}-binding sites (Nguyen Van *et al.*, 1989). The rat protein has a redox potential of greater than –260 mV (Lundstrom-Ljung *et al.*, 1995), possesses significant disulfide isomerase activity (Nguyen Van *et al.*, 1993; Rupp *et al.*, 1994), and is able to catalyse renaturation of denatured anti-creatine kinase (CK)–Fab fragments and RNase AIII (Rupp *et al.*, 1994).

■ ERp72 function

Despite structural differences, both mammalian PDI and mouse ERp72 can complement PDI-deficient yeast (Günther *et al.*, 1993). Recent work on rat PDI (Laboissiere *et al.*, 1995), in which the N-terminal cysteine of the trx motif was replaced by a serine residue, showed that the alleviation of the lethal PDI deletion phenotype is dependent on the presence of this reactive cysteine, suggesting that the ERp72 complementation effect may be due to the catalytic disulfide isomerase activity of the trx motif. Indeed, it could be shown that the N-terminal cysteines in the trx boxes in rat ERp72 are required for yeast PDI deficiency complementation (B. Kramer, S. Bonitz, N. Pöhlmann, D. Ferrari and H.D. Söling, manuscript in preparation). Nigam *et al.* (1994) have shown that ERp72 can bind reversibly to denatured proteins covalently coupled to Sepharose/agarose columns. Possible interactions between ERp72 and other ER proteins such as BiP and Grp94 during *in vitro* refolding of denatured thyroglobulin have been indicated by Kuznetsov *et al.* (1994). It has further been reported that ERp72 may possess a Ca^{2+}-dependent proteolytic activity (Urade *et al.*, 1993).

■ References

Dorner, A.J., Wasley, L.C., Raney, P., Haugejorden, S., Green, M., and Kaufman, R.J. (1990). The stress response in chinese hamster ovary cells. Regulation of ERp72 and protein disulfide isomerase expression and secretion. J. Biol. Chem. **265**, 22029–22034.

Günther, R., Srinivasan, M., Haugejorden, S., Green, M., Ehrbrecht, I.M., and Küntzel, H. (1993). Functional replacement of the *Saccharomyces cerevisiae* Trg1/Pdi1 protein by members of the mammalian protein disulfide isomerase family. J. Biol. Chem. **268**, 7728–7732.

Kozutsumi, Y., Segal, M., Normington, K., Gething, M.J., and Sambrook, J. (1988). The presence of malfolded proteins in the endoplasmic reticulum signals the induction of glucose-regulated proteins. Nature **332**, 462–464.

Kuznetsov, G., Chen, L.B., and Nigam, S.K. (1994). Several endoplasmic reticulum stress proteins, including ERp72, interact with thyroglobulin during its maturation. J. Biol. Chem. **269**, 22990–22995.

Laboissiere, M.C.A., Sturley, S.L., and Raines, R.T. (1995). The essential function of protein-disulfide isomerase is to unscramble non-native disulfide bonds. J. Biol. Chem. **270**, 28006–28009.

Lundström-Ljung, J., Birnbach, U., Rupp, K., Söling, H.D., and Holmgren, A. (1995). Two resident ER-proteins, CaBP1 and CaBP2, with thioredoxin domains, are substrates for thioredoxin reductase: comparison with protein disulfide isomerase. FEBS Lett. **357**, 305–308.

Mazzarella, R.A., Srinivasan, M., Haugejorden, S.M., and Green, M. (1990). ERp72, an abundant luminal endoplasmic reticulum protein, contains three copies of the active site sequences of protein disulfide isomerase. J. Biol. Chem. **265**, 1094–1101.

Nguyen Van, P., Peter, F., and Söling, H.D. (1989). Four intracisternal calcium-binding glycoproteins from rat liver microsomes with high affinity for calcium. J. Biol. Chem. **264**, 17494–17501.

Nguyen Van, P., Rupp, K., Lampen, A., and Söling, H.D. (1993). CaBP2 is a rat homolog of ERp72 with protein disulfide isomerase activity. Eur. J. Biochem. **213**, 789–795.

Nigam, S.K., Goldberg, A.L., Ho, S., Rohde, M.E., Bush, K.T., and Sherman, M.Y. (1994). A set of endoplasmic reticulum proteins possessing properties of molecular chaperones includes Ca^{2+}-binding proteins and members of the thioredoxin superfamily. J. Biol. Chem. **269**, 1744–1749.

Rupp, K., Birnbach, U., Lundström, J., Nguyen Van, P., and Söling, H.D. (1994). Effects of CaBP2, the rat analog of ERp72, and of CaBP1 on the refolding of denatured reduced proteins. J. Biol. Chem. **269**, 2501–2507.

Urade, R., Takenaka, Y., and Kito, M. (1993). Protein degradation by ERp72 from rat and mouse liver endoplasmic reticulum. J. Biol. Chem. **268**, 22004–22009.

■ *David Ferrari and Hans-Dieter Söling:*
Department of Clinical Biochemistry
University of Göttingen
Robert-Koch-Str. 40
370 75 Göttingen, Germany
Tel. 49 551 396389
Fax. 49 551 392953
E-mail: dferrari@med.uni-goettingen.de

Mammalian CaBP1

CaBP1 (calcium binding protein 1) is a 49 kDa luminal protein of the endoplasmic reticulum with high affinity binding sites for calcium. It contains two thioredoxin-like domains and catalyses, similarly to PDI, the refolding of denatured proteins in vitro.

■ Alternative name

P5 (Chaudhuri et al., 1992).

■ CaBP1 gene and sequence

The nucleotide sequences of cDNAs encoding rat liver CaBP1 (EMBL X79328; Füllekrug et al., 1994) and the homologous P5 protein (95 % identity) from CHO cells (EMBL X62678; Chaudhuri et al., 1992) reveal two internal repeats of a thioredoxin-like domain including the putative catalytic site CGHC (75% internal homology, 65% homology to *Escherichia coli* thioredoxin). A highly acidic region (residues 416–436) is followed by the C-terminal tetrapeptide sequence KDEL which is necessary for retrieval by the KDEL receptor (Pelham, 1989). Figure 1 describes the locations of these domains and motifs in the CaBP1 sequence.

Other members of the thioredoxin-like protein family in the endoplasmic reticulum of mammalian cells include protein disulfide isomerase (PDI, reviewed by Freedman, 1989; see entry p. 348), ERp72/CaBP2 (Mazzarella et al., 1990; Nguyen Van et al., 1993; see entry p. 353) and ERp61 (Bennett et al., 1988; see entry p. 351).

The sequence of human CaBP1 has been reported recently (Hayano and Kikuchi, 1995). The database of expressed sequence tags (EPFL, dbest 11.11.1995) contains several cDNAs highly homologous to residues 267–440 of CaBP1 (the protein minus signal sequence and thioredoxin-like domains), as well as various human cDNAs (accession numbers T92500, R06994, Z205999,

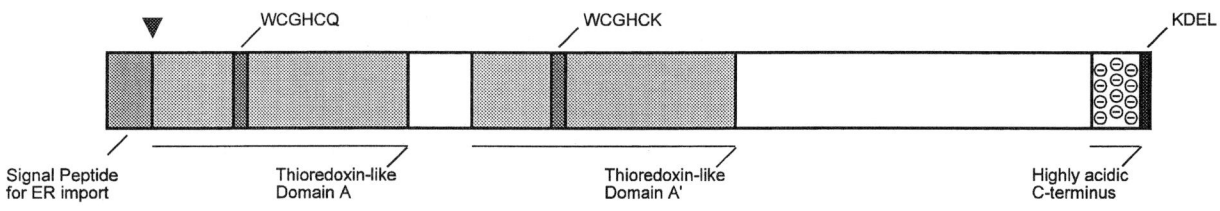

Figure 1. The locations of functional domains and motifs in the CaBP1 sequence.

R71140, Z30232, etc.). There are also entries for mouse (Z31321) and rice (D23362) sequences.

■ CaBP1 protein

CaBP1 was purified from rat liver microsomes on the basis of its calcium-binding properties (Nguyen Van et al., 1989). The mature protein is lacking the N-terminal signal sequence and contains 421 amino acids (46 kDa) but displays an apparent molecular mass of 52–54 kDa by SDS–PAGE. It is not glycosylated and is observed in a monomeric state by non-denaturing gel analysis. CaBP1 from RH35 cells is not a glucose-regulated protein like BiP or ERp72/CaBP2 (U. Birnbach and H.D. Söling, unpublished data).

■ CaBP1 localization

CaBP1 carries the C-terminal ER localization/sorting signal KDEL but was initially reported to be a resident protein of the ER–Golgi intermediate compartment (Peter et al., 1992; Schweizer et al., 1993). However, a detailed study has shown that CaBP1 localizes to the endoplasmic reticulum (Füllekrug et al., 1994) and these data are now generally accepted. The amount of CaBP1 present in the intermediate compartment or Golgi is below the detection limit when using subcellular fractionation or immunofluorescence, even after high level overexpression. However, by immuno-EM, CaBP1 and other KDEL proteins like PDI have been shown to label the intermediate compartment and the first cisternae of the Golgi stack (Griffiths et al., 1994).

■ CaBP1 function

Several protein disulfide isomerase (PDI)-like activities have been reported for CaBP1. It catalyses the refolding of denatured reduced model proteins (Rupp et al., 1994) and the reduction of insulin (Füllekrug et al., 1994). Like PDI and CaBP2/ERp72, it is also a substrate for thioredoxin reductase (Lundström-Ljung et al., 1995).

CaBP1 is able to rescue the lethal phenotype of a PDI gene disruption in *Saccharomyces cerevisiae* and is therefore assumed to replace PDI functionally (B. Kramer, S. Bonitz, N. Pöhlmann, D. Ferrari and H.D. Söling, manuscript in preparation).

The ATP-dependent binding of CaBP1 to a denatured protein column is indicative of an activity as a molecular chaperone (Nigam et al., 1994).

The P5 protein homologous to CaBP1 was induced in a hydroxyurea-resistant hamster cell line but is considered to be coamplified with ornithine decarboxylase and not functionally related to hydroxyurea resistance (Chaudhuri et al., 1992).

■ References

Bennett, C.F., Balcarek, J.M., Varrichio, A., and Crooke, S.T. (1988). Molecular cloning and complete amino-acid sequence of form-I phosphoinositide-specific phospholipase C. Nature **334**, 268–270.

Chaudhuri, M.M., Tonin, P.N., Lewis, W.H., and Srinivasan, P.R. (1992). The gene for a novel protein, a member of the protein disulfide isomerase/form I phosphoinositide-specific phospholipase c family, is amplified in hydroxyurea-resistant cells. Biochem. J. **281**, 645–650.

Freedman, R.B. (1989). Protein disulfide isomerase: multiple roles in the modification of nascent secretory proteins. Cell **57**, 1069–1072.

Hayano, T. and Kikuchi, M. (1995). Cloning and sequencing of the cDNA encoding human P5. Gene **164**, 377–378.

Mazzarella, R., Srinivasan, M., Haugejorden, S., and Green, M. (1990). ERp72, an abundant luminal endoplasmic reticulum protein, contains three copies of the active site sequences of protein disulfide isomerase. J. Biol. Chem. **265**, 1094–1101.

Nguyen Van, P., Peter, F., and Söling, H.D. (1989). Four intracisternal calcium-binding glycoproteins from rat liver microsomes with high affinity for calcium. No indication for calsequestrin-like proteins in inositol 1,4,5-trisphosphate-sensitive calcium sequestering rat liver vesicles. J. Biol. Chem. **264**, 17494–17501.

Nguyen Van, P., Rupp, K., Lampen, A., and Söling, H.D. (1993). CaBP2 is a rat homolog of ERp72 with protein disulfide isomerase activity. Eur. J. Biochem. **213**, 789–795.

Pelham, H.R.B. (1989). Control of protein exit from the endoplasmic reticulum. Annu. Rev. Cell Biol. **5**, 1–23.

Peter, F., Nguyen Van, P., and Söling, H.D. (1992). Different sorting of Lys-Asp-Glu-Leu proteins in rat liver. J. Biol. Chem. **267**, 10631–10637.

Rupp, K., Birnbach, U., Lundström, J., Nguyen Van, P., and Söling, H.D. (1994). Effects of CaBP2, the rat analog of ERp72, and of CaBP1 on the refolding of denatured reduced proteins. J. Biol. Chem. **269**, 2501–2507.

Schweizer, A., Peter, F., Nguyen Van, P., Söling, H.D., and Hauri, H.P. (1993). A luminal calcium-binding protein with a KDEL endoplasmic reticulum retention motif in the ER-Golgi intermediate compartment. Eur. J. Cell Biol. **60**, 366–370.

■ *Joachim Fuellekrug:*
Cell Biology Programme
European Molecular Biology Laboratory
Meyerhofstr. 1
D-69112 Heidelberg, Germany
Tel. 49 6221 387408
Fax. 49 6221 387512
E-mail: fuellekr@embl-heidelberg.de

■ *Hans-Dieter Söling:*
Department of Clinical Biochemistry
University of Goettingen
Robert-Koch-Str. 40
D-37075 Goettingen, Germany
Tel. 49 551 6389
Fax. 49 551 2953
E-mail: hsoelin@gwdg.de

12

Peptidyl-Prolyl Isomerases (PPIases)

Peptidyl–prolyl isomerases — an overview of the cyclophilin, FKBP and parvulin families

Peptidyl–prolyl isomerases catalyse the interconversion between cis and trans forms of the peptide bond preceding proline residues in proteins. Three distinct families of prolyl isomerases have been identified: the cyclophilins, the FKBPs, and the parvulins. Cyclophilins and FKBPs have attracted attention as the cellular receptors for the immunosuppressants cyclosporin A, FK506, and rapamycin, but this role in drug action is distinct from enzymatic activity. Indeed, it is the prolyl isomerase–drug complexes that are the active in vivo agents. In general, members of all three prolyl isomerase families are highly conserved, abundant, and expressed in multiple intracellular compartments, suggesting that these enzymes play a critical role in cell physiology. Because prolyl isomerases can accelerate slow protein refolding steps in vitro, it has been speculated that this is what they do in vivo. However, most prolyl isomerases are not required for viability and thus can not play an essential role in protein folding. A number of proteins physically associated with different prolyl isomerases have been identified; with a few exceptions, the physiological relevance of these associations remains to be established. Thus, although a multitude of prolyl isomerases has been identified and their role in immunosuppressant action elucidated, much remains to be learned concerning their normal biological functions.

■ Overview

Because the peptide bond has partial double bond character, like all appropriately substituted double bonds it can exist in two isomers: *cis* and *trans*. For every amino acid except proline, there is a profound steric clash between adjacent side chains for the *cis* isomer but not for the *trans* isomer. In the case of proline, because the side chain is a secondary amide bound back to the main chain, this steric conflict is roughly equal for the two isomers. If one looks at the peptide bond in proteins of known structures, for all residues except proline the bond is invariably *trans*, but for those preceding proline residues in proteins of known structures, roughly one in twelve is in the unusual *cis* isomer. It is generally thought, but not yet proven, that the ribosome stereospecifically synthesizes the peptide bond in the *trans* configuration. These unusual *cis* peptidyl–prolyl bonds must then arise co- or post-translationally. This isomerization reaction can occur spontaneously at reasonable rates. However, several distinct enzymes have been identified that catalyse *trans* to *cis*, and *cis* to *trans* peptidyl–prolyl isomerization. Possible roles for such enzymes could include protein folding, protein refolding following heat shock, protein assembly and disassembly, and protein trafficking in the cell.

Three distinct families of peptidyl–prolyl isomerases are known: the cyclophilins, the FKBPs, and the parvulins (for reviews see Heitman et al., 1992; Schmid et al., 1993; Fischer, 1994). Although marked sequence identity and structural similarity define each individual family, remarkably, there is no primary sequence identity between cyclophilins, FKBPs, and parvulins, and there is no tertiary structural similarity between cyclophilins and FKBPs. Either it is very easy to make a binding pocket with prolyl isomerase activity or nature has gone to the trouble to evolve three completely different families of such enzymes.

Members of each peptidyl–prolyl isomerase family are conserved from bacteria (Table 1) and unicellular eukaryotes (Table 2) to humans (Table 3). Moreover, multiple members of each family are expressed in different intracellular compartments. For example, the yeast *Saccharomyces cerevisiae* expresses twelve different prolyl isomerases, including seven cyclophilins, four FKBPs, and one parvulin (Table 2). Three cyclophilins are cytoplasmic in yeast (Cpr1, Cpr6, Cpr7), three are in the secretory pathway (Cpr2, Cpr4, Cpr5), and one is mitochondrial (Cpr3). In most cases, mammalian homologs of each of these enzymes have been identified. Because these enzymes are so highly conserved and can catalyse protein refolding *in vitro*, it was initially thought that they might play a critical essential role in protein folding *in vivo*. However, none of the cyclophilins or FKBPs is essential for cell viability in yeast or in any other microorganism in which the genes have been deleted. In fact, remarkably few mutant phenotypes have been associated with mutations in specific prolyl isomerases. Initially it was thought that other prolyl isomerases might have overlapping functions; however, the construction of multiply mutant strains lacking several prolyl isomerases has not yet revealed any functional overlap. Thus, neither the cyclophilins nor the FKBPs can play an essential role in protein folding under standard conditions.

The only prolyl isomerase in yeast that has been found to be essential is the one member of the parvulin family, Ess1. It is not yet clear that the essential function of Ess1 is its prolyl isomerase activity. Interestingly, the essential function of Ess1 can be provided by either the human (Pin1) or the *Drosophila* (dodo) homolog, suggesting that the essential function of Ess1 has been conserved. However, although Ess1 is essential in yeast, and either

Table 1. Bacterial peptidyl–prolyl isomerases

Gene (synonym)	Molecular weight (kDa)	Mammalian homolog	Localization	Characteristics/ mutant phenotypes	References
Cyclophilins					
CypA (RotA)	18	CypB	Periplasm		Liu and Walsh (1990), Hayano et al. (1991)
CypB (RotB)	18	CypA	Cytoplasm		Hayano et al. (1991)
FKBPs					
MIP	24		Cell surface	• *Legionella* virulence factor	Cianciotto et al. (1989), Cianciotto and Fields (1992)
FkpA	12		Periplasm	• mutation induces stress response	Missiakas et al. (1996)
trigger factor (SecI)	48		Cytoplasm	• chaperone found in ribosome and GroEL complexes	Lill et al. (1988), Guthrie and Wickner (1990)
Parvulins					
parvulin (PpiC)	10	Pin1	Cytoplasm		Rudd et al. (1995)
SurA			Periplasm	• involved in folding outer membrane proteins	Tormo et al. (1990), Lazar and Kolter (1996), Missiakas et al. (1996)

Table 2. Yeast peptidyl–prolyl isomerases

Gene (synonym)	Molecular weight (kDa)	Mammalian homolog	Localization	Characteristics/ mutant phenotypes	References
Cyclophilins					
CPR1 (CYP1, CPH1)	17	CypA	Cytoplasm	• primary receptor for CsA • induced by heat shock	Haendler et al. (1989)
CPR2 (CYP2)	20	CypB	Secretory pathway	• induced by heat shock	Koser et al. (1990), Sykes et al. (1993)
CPR3 (CYP3)	20	CypD	Mitochondria	• Δcpr3 has growth defect on lactate at 37°C • accelerates Su9-DHFR refolding	Davis et al. (1992), McLaughlin et al. (1992), Matouschek et al. (1995), Rospert et al. (1996)
CPR4 (SCC3)	33	CypC	Secretory pathway ER (HDEL)		Franco et al. (1991)
CPR5	23				Frigerio and Pelham (1993)
CPR6	45	Cyp40	Cytoplasm	• interacts with Hsp82	Chang and Lindquist (1994)
CPR7	45	Cyp40	Cytoplasm	• CPR6 homolog	Duina et al., (1996a, b)
FKBPs					
FPR1 (FKB1)	12	FKBP12	Cytoplasm	• primary receptor for FK506 and rapamycin • Δfpr1 exhibits slow growth	Heitman et al. (1991a, b), Koltin et al. (1991), Wiederrecht et al. (1991)
FPR2 (FKB2)	12.5	FKBP13	Secretory pathway	• induced by heat shock and tunicamycin	Nielsen et al. (1992), Partaledis and Berlin (1993)
FPR3 (NPI46)	70	FKBP25	Nucleolus	• phosphorylated • Ptp1 substrate	Shan et al. (1994), Benton et al. (1994), Manning-Krieg et al. (1994), Wilson et al. (1995)
FPR4	60	FKBP25	Nucleolus	• FPR3 homolog	
Parvulin					
ESS1 (PTF1)	19	Pin1	Nucleus	• essential gene, mitotic regulator	Hanes et al. (1989), Hani et al. (1995), Lu et al. (1996)

Table 3. Mammalian peptidyl–prolyl isomerases

Gene (synonym)	Molecular weight (kDa)	Localization	Characteristics/ mutant phenotypes	References
Cyclophilins				
CypA	18	Cytoplasm	• primary receptor for CsA • interacts with HIV gag, Hsp90, YY1	Handschumacher et al. (1984), Liu et al. (1991), Nadeau et al. (1993), Luban et al. (1993), Yang et al. (1995)
CypB	23	Secretory pathway	• in complex with Hsp47 and procollagen • interaction with CAML (Ca^{2+} transport) • interacts with HIV gag	Bergsma et al. (1991), Hasel et al. (1991), Price et al. (1991), Bram and Crabtree (1994), Smith et al. (1995) Friedman and Weissman (1991), Friedman et al. (1993)
CypC	23	Secretory pathway	• interacts with 70 kDa glycoprotein (CAP)	
CypD	20	Mitochondria	• involved in regulation of mitochondria permeability transition pore	Connern and Halestrap (1992), Nicolli et al. (1996)
Cyp40	40	Cytoplasm	• Hsp90, Hsp70 binding, component of inactive steroid receptor complex • binds calmodulin	Kieffer et al. (1992, 1993), Ratajczek et al. (1993, 1995), Hoffman and Handschumacher (1995)
FKBPs				
FKBP12	12	Cytoplasm	• primary receptor for FK506 and rapamycin • found in ryanodine and IP3 receptor complexes • interacts with YY1	Liu et al. (1991), Jayaraman et al. (1992), Wiederrech et al. (1992), Timerman et al. (1993), Cameron et al. (1995a, b)
FKBP12.6 FKBP13	12.6 13	Cytoplasm ER	• induced by tunicamycin	Bush et al. (1994), Jin et al. (1991)
FKBP25	25	Nucleus and cytoplasm	• interacts with nucleolin, casein kinase II	Gala et al. (1992), Riviere et al. (1993)
FKBP52 (Hsp56, FKBP59, FKBP51)	52	Cytoplasm and nucleus	• Hsp90, Hsp70 binding, component of inactive steroid receptor complex • binds calmodulin, ATP, GTP	Tai et al. (1992)
FKBP54	54	Cytoplasm	• Hsp90, Hsp70 binding, component of inactive steroid receptor complex	Smith et al. (1993)
Parvulins				
Pin1	19	nucleus	• mitotic regulator, interacts with NIMA kinase, has WW domain	Lu et al. (1996)

human Pin1 or fly dodo can satisfy this function, the dodo enzyme is not essential in *Drosophila*. Once the structure of the Ess1, Pin1, or dodo enzyme has been solved and active site residues identified, it should be possible to isolate inactive mutants to test whether or not prolyl isomerase activity is required for the essential function in yeast.

Thus, despite a remarkable degree of evolutionary conservation, a wealth of biochemical and structural information, a well-defined enzymatic activity capable of folding proteins *in vitro*, and potent active site inhibitors, the normal cellular functions of the prolyl isomerases are only beginning to emerge. We now discuss each protein family in turn to consider some of what is known about specific family members.

■ Cyclophilins

The first cyclophilin to be identified was cyclophilin A, an 18 kDa cytoplasmic protein whose peptidyl–prolyl isomerase activity is potently inhibited by the immunosuppressant cyclosporin A (Fischer et al., 1989; Handschumacher et al., 1984; Takahashi et al., 1989). Cyclophilin A is highly conserved from yeast to humans, with 65% sequence identity over roughly one billion years of evolution (Haendler et al., 1989). Yeast mutants lacking cyclophilin A are viable (Heitman et al., 1991a), indicating that cyclosporin does not kill yeast by inhibiting cyclophilin function. That both cyclophilin A and cyclosporin A are required for toxicity strongly supports a model in which the cyclophilin A–cyclosporin A (CsA) complex is the active agent (Tropschug et al., 1989; Breuder et al., 1994). One target of this complex has been identified as calcineurin, a calcium-calmodulin regulated serine/threonine-specific phosphatase, which is highly conserved from yeast to humans (Cyert et al., 1991; Liu et al., 1991). Calcineurin is essential for viability in CsA-sensitive yeast strains (Parent et al., 1993; Breuder et al., 1994), and mutations that confer CsA resistance in yeast are single amino acid substitutions in the calcineurin A catalytic subunit that reduce binding by the cyclophilin A–CsA complex (Cardenas et al., 1995). Similarly, overexpression of cyclophilin A in mammalian cells increases CsA sensitivity, suggesting that this protein mediates CsA action in both yeast and human (Bram et al., 1993).

Although the role of cyclophilin A in CsA action has been well established, the normal cellular functions have been elusive. Yeast cells lacking cyclophilin A are viable and only two subtle phenotypes have been described. First, mutants lacking the protein are subtly more sensitive to extreme heat shock (Sykes et al., 1993). Second, the ability of a mutant form of calcineurin A to promote recovery from pheromone arrest is compromised in cyclophilin A mutants, suggesting a normal interaction between the two enzymes (Cardenas et al., 1994).

One very interesting finding was the discovery that human cyclophilin A physically interacts with the HIV-1 capsid protein, and that cyclophilin A is a normal constituent of HIV-1 virions (Luban et al., 1993; reviewed in Cullen and Heitman, 1994; Franke and Luban, 1995). The cyclophilin A-binding target has been identified as a short conserved peptide encompassing proline-90 of the capsid protein (Franke et al., 1994). The NMR structure of the HIV-1 capsid protein has recently been determined, revealing that the peptidyl–prolyl bond preceding proline-90 is a mixture of isomers: roughly 86% *trans* and 14% *cis*, which sounds like an ideal binding target for cyclophilin A (Gitti et al., 1996). CsA and non-immunosuppressive analogs perturb the cyclophilin A–capsid protein complex, and HIV-1 virions produced from cells cultured in the presence of CsA are defective for some early step in infection, possibly viral disassembly (Thali et al., 1994). HIV-1 mutants resistant to CsA have been described and result from amino acid substitutions adjacent to proline-90 (Aberham et al., 1996). Curiously, these CsA-resistant isolates are also CsA dependent, suggesting that cyclophilin A may now interfere with some step in the viral life cycle. Interestingly, cyclophilin A does not interact with the capsid proteins from closely related retroviruses, such as SIV-1. This is clearly a fertile area for further research.

Several cyclophilins are localized to the secretory pathway in both yeast and human (see Table 3). The yeast secretory cyclophilins are Cpr2, Cpr4 (Scc3), and Cpr5 (also known as cyclophilin D). None of these proteins are essential, and their functions are not yet known. Mutants lacking Cpr2 are, like cyclophilin A mutants, moderately more sensitive to extreme heat shock, suggesting a role in survival under stress conditions (Sykes et al., 1993). Human cyclophilins B and C are also localized to the secretory pathway. Incubation of cultured cells with CsA has subtle effects on the folding of secretory or membrane proteins, such as transferrin, collagen, and ligand-gated ion channels, which may reflect a role for these cyclophilins in protein folding or trafficking during secretion (Lodish and Kong, 1991; Steinmann et al., 1991; Helekar et al., 1994; Klappa et al., 1995). Cyclophilin B localization in the ER is perturbed by CsA, suggesting that cyclophilin–protein interactions may normally retain the protein in the cell (Price et al., 1994). In addition, cyclophilin C has been shown to associate with a 70 kDa glycoprotein, termed CAP for cyclophilin-associated protein, whose function is as yet unknown (Friedman et al., 1993). The best understood secretory pathway cyclophilin is the *Drosophila* ninaA protein, which is a cyclophilin homolog with a C-terminal transmembrane domain (Schneuwly et al., 1989; Shieh et al., 1989). ninaA is required for the proper localization of two of the four fly rhodopsins, and correspondingly *ninaA* mutants accumulate rhodopsin in the ER and have visual system defects (Colley et al., 1991; Stamnes et al., 1991). Interestingly, ninaA forms a very stable complex with rhodopsin, and *ninaA* mutants exhibit haploinsufficiency, suggesting that ninaA acts stoichiometrically rather than catalytically (Baker et al., 1994). These findings suggest that ninaA may act as a chaperone rather than as a folding enzyme. Further studies will be required to determine if ninaA has prolyl isomerase activity and, if so, whether or not this activity is required for rhodopsin biogenesis.

Yeast and humans also express a mitochondrial cyclophilin, known as Cpr3 in yeast or cyclophilin D in mammals. Yeast mutants lacking mitochondrial cyclophilin are unable to utilize lactate as a carbon source at 37°C, indicative of a role for this enzyme in mitochondrial physiology at elevated temperature (Davis et al., 1992). In both *Neurospora crassa* and *S. cerevisiae*, refolding of the model protein DHFR is delayed in mitochondria exposed to CsA or missing the mitochondrial cyclophilin, indicating that this protein indeed plays a role in refolding proteins following mitochondrial import (Matouschek et al., 1995; Rassow et al., 1995). It is not yet known whether the prolyl isomerase activity of mitochondrial cyclophilin is required for this action. Notably, the structure of mammalian DHFR is known and all of the peptidyl–prolyl bonds are in the *trans* configuration (Volz et al., 1982). Moreover, kinetic studies suggest that slow DHFR refolding steps are not likely to be attributable to prolyl isomerization (Jennings et al., 1993). Thus, it is also not yet clear whether DHFR requires prolyl isomerization during folding.

Finally, yeast and mammalian cells express a larger conserved cytoplasmic cyclophilin, known as cyclophilin 40 based on its molecular weight of 40 kDa (Kieffer et al., 1992). Yeast cells express two cyclophilin 40 homologs, known as Cpr6 (Chang and Lindquist, 1994) and Cpr7 (Duina et al., 1996a, b). Most interestingly, cyclophilin 40 family members are found in complexes with the heat shock proteins Hsc70 and Hsp90 via a direct interaction with Hsp90. In mammalian cells, this large complex is found in association with members of the steroid receptor family (Kieffer et al., 1993; Ratajczak et al., 1993). Although the role of cyclophilin 40, if any, in steroid receptor function remains to be established, it is intriguing that this complex of proteins is the target for multiple immunosuppressive drugs, including steroids, CsA, 15-deoxyspergualin (which binds Hsc70), and FK506 (described below). The function of cyclophilin 40 proteins in yeast is not yet known; however, the Cpr6 protein is found in complex with Hsp90 in *S. cerevisiae* (Chang and Lindquist, 1994), and the cyclophilin 40 homolog in *Schizosaccharomyces pombe* was isolated as a suppressor of a cell cycle mutant strain (Weisman et al., 1996).

■FKBPs

The first FKBP to be identified was the small 12 kDa cytoplasmic FKBP12 protein (Harding et al., 1989; Siekierka et al., 1989). FKBP12 is a peptidyl–prolyl isomerase whose activity is inhibited by either FK506 or rapamycin. FKBP12 is conserved from yeast to humans (Heitman et al., 1991a, b); the two proteins share 54% overall sequence identity and have virtually superimposable X-ray crystal structures (Van Duyne et al., 1991; Rotonda et al., 1993). As with cyclophilin A, yeast cells lacking FKBP12 are viable. Hence the toxic effects of FK506 and rapamycin are not mediated by inhibition of an essential FKBP12 activity. Instead, as with cyclophilin A–CsA, the FKBP12–FK506 and FKBP12–rapamycin complexes are the active *in vivo* agents (Heitman et al., 1991a, b; Koltin et al., 1991; Foor et al., 1992; Breuder et al., 1994). The target of FKBP12–FK506 is calcineurin (Liu et al., 1991), and the X-ray structure of the FKBP12–FK506–calcineurin AB complex has been solved (Griffith et al., 1995; Kissinger et al., 1995). The target of the FKBP12– rapamycin complex was first identified in yeast as the Tor1 and Tor2 kinase homologs (Heitman et al., 1991b; Cafferkey et al., 1993; Kunz et al., 1993; Helliwell et al., 1994; Cardenas and Heitman, 1995; Lorenz and Heitman, 1995). A mammalian homolog of the yeast Tor proteins has been identified (Brown et al., 1994; Sabatini et al., 1994) and found to have a domain with function conserved from yeast to humans (Alarcon et al., 1996).

Yeast mutants lacking FKBP12 are viable. Two subtle phenotypes have been described. First, FKBP12 mutant cells have a growth defect (Heitman et al., 1991a, b). Second, FKBP12 mutants recover from pheromone arrest more rapidly than isogenic wild-type cells, suggesting that FKBP12 may normally inhibit calcineurin function *in vivo* (Cardenas et al., 1994). Mammalian cells express two closely related forms of FKBP12, FKBP12 and FKBP12.6, which may represent redundant homologs or specialized forms of the same original protein. In mammalian cells, FKBP12 has been shown to be a subunit of two distinct, large calcium channels, the ryanodine receptor and the IP3 receptor (Jayaraman et al., 1992; Timerman et al., 1993; Cameron et al., 1995a, b). Reconstitution of the ryanodine receptor in artificial bilayers yields poor channel activity, whereas addition of purified FKBP12 restores the normal gating properties of the channel (Brillantes et al., 1994). Interestingly, FKBP12 mutants lacking prolyl isomerase activity fully support ryanodine receptor channel activity *in vitro* (Timerman et al., 1995). In addition, FKBP12 has been found to target calcineurin to the ryanodine receptor in the absence of any immunosuppressive ligand (Cameron et al., 1995a, b) and ligand-independent interactions have been previously described in yeast (Cardenas et al., 1994). FKBP12 also interacts with the type I TGFβ receptor (Wang et al., 1994). Finally, we have recently shown that FKBP12 is required for function of the multidrug resistance (MDR) drug efflux pump (Hemenway and Heitman, 1996). These observations suggest that FK506 may take advantage of pre-existing FKBP12–calcineurin interactions, that FKBP12 may serve as a targeting subunit for calcineurin, and that FKBP12 may play a broad role in regulating the function of large membrane proteins.

Yeast and human cells also express a conserved ER-localized form of FKBP, known as FKBP13 (Fpr2/Fkb2 in yeast) (Jin et al., 1991; Nielsen et al., 1992). Yeast mutants lacking FKBP13 are viable. The FKBP13 gene is induced by misfolded proteins in the lumen of the ER, and the promoter contains a consensus unfolded protein response element (UPRE) (Partaledis and Berlin, 1993). This suggests that, like the secretory cyclophilins, FKBP13 may play a role in protein folding in the ER, but this has not yet been established, and proteins that might interact with FKBP13 are as yet unknown. Yeast FKBP13 mutants might lack a phenotype if one or more of the secretory cyclophilins were to have overlapping functions, but experiments to address this have not been reported.

Yeast and human cells also express a conserved nuclear/nucleolar FKBP, known as FKBP25 in mammals and FKBP70/Fpr3/Npi46 in yeast cells. Mammalian FKBP25 is a 25 kDa nuclear protein found in physical association with both casein kinase II and the nucleolar protein nucleolin (Jin and Burakoff, 1993). In contrast, the yeast homolog is much larger, c. 70 kDa, and contains an FKBP domain fused to a nucleolin domain (Benton et al., 1994; Manning-Krieg et al., 1994; Shan et al., 1994). Although the yeast protein has been found to bind to nuclear targeting signals (NLSs), is nucleolar, and can be tyrosine phosphorylated (Wilson et al., 1995), the protein is non-essential and no phenotypes have been found in mutant strains. A role in ribosome assembly, based on nucleolar localization, has been suggested but as yet no evidence has been reported.

Mammalian cells express at least two and possibly more FKBPs in the 50 to 60 kDa size range. There is no yeast homolog of these proteins in the completed yeast genome, suggesting that these forms are mammalian specific. As with cyclophilin 40, FKBP52 and FKBP54 are found in association with HSPs and steroid receptor complexes (Callebaut et al., 1992; Peattie et al., 1992; Yem et al., 1992). These forms of FKBP have in fact two, or possibly three, FKBP domains within the protein. At least one FKBP active site is unoccupied in the complex, as FKBP52–steroid receptor complexes are retained on FK506 affinity resin. The role of these larger FKBPs is as yet unknown.

Finally, there are two examples of what may be prokaryotic-specific FKBPs. The first is the MIP protein, named after its role in potentiating macrophage infection by Legionella pneumophila and Chlamydia trachomitis (Engleberg et al., 1989; Lundemose et al., 1991). MIP is expressed on the bacterial cell surface, has an FKBP conserved domain with prolyl isomerase activity sensitive to FK506–rapamycin (Fischer et al., 1992), and may function by inhibiting phagosome–lysosome fusion in the infected cell. A second, distinct prokaryotic FKBP has recently been identified as a trigger factor (Callebaut and Mornon, 1995; Stroller et al., 1995), originally identified via its role in export across the plasma membrane of some non-traditional secretory proteins. The trigger factor FKBP is one of the most divergent FKBPs yet identified, and is not inhibited by FK506 or rapamycin. The protein is associated with the ribosome and has been suggested to play a role in folding of nascent chains. Alternatively, we suggest that trigger factor might act on the ribosome to promote ribosomal conformational changes during protein synthesis.

∎ Parvulins

As if two completely distinct families of prolyl isomerases were not bizarre enough, recent studies have defined a third distinct family of prolyl isomerases, named the parvulins. The protein was first purified from bacterial cells based on its ability to catalyse cis to trans isomerization of a tetrapeptide substrate (Rahfield et al., 1994).

Thus far parvulin family members have also been identified in bacteria, yeast, Drosophila, and humans. The parvulins share no sequence identity with either cyclophilins or FKBPs and, correspondingly, parvulins are not sensitive to either CsA or FK506. Intriguingly, the yeast parvulin, Ess1, is essential (Hanes et al., 1989). Although Ess1 has not yet been shown to have prolyl isomerase activity, the corresponding human protein Pin1 is an active prolyl isomerase. While one could suggest that yeast mutants lacking cyclophilins and FKBPs are viable because of the presence of Ess1, it seems unlikely that a single, presumably cytoplasmic, parvulin could substitute for missing prolyl isomerases in other cellular compartments. Thus, it remains to be established that Ess1 has prolyl isomerase activity and to test whether this is its essential function in vivo.

Interestingly, the human Pin1 protein and the fly dodo protein can both substitute for the yeast Ess1 protein (Lu et al., 1996; Maleszka et al., 1996). A mutant form of the human Pin1 protein altered in three conserved C-terminal residues lacks prolyl isomerase activity in vitro and fails to complement a yeast ess1 mutant strain in vivo (Lu et al., 1996). Although it has been suggested, based on this finding, that the prolyl isomerase activity is the essential in vivo function, these mutations might alter some other feature of Pin1 to result in loss of function. In this regard, Pin1 was originally identified in a two-hybrid screen for proteins that would interact with the nimA protein kinase mitotic regulator (Lu et al., 1996).

∎ Conclusions

While progress has been made in identifying many prolyl isomerases, much remains to be learned about their functions. The structures of both cyclophilins and FKBPs have been determined at high resolution and active site residues have been identified. In only one case have prolyl isomerase active site mutants been tested for function; remarkably, these FKBP12 mutant proteins were fully functional in restoring ryanodine receptor function. It remains to be tested whether prolyl isomerase activity is the essential in vivo function of any of these proteins.

A collection of proteins that physically interact with prolyl isomerases have been identified but the biological relevance of most of these interactions remains to be established. Most prolyl isomerases are not essential for life, challenging the view that these enzymes play a broad, general, essential role in protein folding. Why then might these proteins be so highly conserved? Perhaps an analogous example might be histones, which are also highly conserved because they exist in complex with a small number of conserved partner proteins. Similarly perhaps, the prolyl isomerases do not interact with many other proteins, but with only a few conserved partners. In this view, cyclophilins, FKBPs, and parvulins might be highly specialized binding proteins, akin to SH2, SH3, or pleckstrin homology domains, with prolyl isomerase activity a later addition or a fortuitous feature of a hydrophobic-binding pocket.

References

Aberham, C., Weber, S., and Phares, W. (1996). Spontaneous mutations in the human immunodeficiency virus type 1 *gag* gene that affect viral replication in the presence of cyclosporins. J. Virol. **70**, 3536–3544.

Alarcon, C.M., Cardenas, M.E., and Heitman, J. (1996). Mammalian RAFT1 kinase domain provides rapamycin-sensitive TOR function in yeast. Genes Dev. **10**, 279–288.

Baker, E.K., Colley, N.J., and Zuker, C.S. (1994). The cyclophilin homolog NinaA functions as a chaperone, forming a stable complex *in vivo* with its protein target rhodopsin. EMBO J. **13**, 4886–4895.

Benton, B.M., Zang, J.H., and Thorner, J. (1994). A novel FK506- and rapamycin-binding protein (FPR3 gene product) in the yeast Saccharomyces cerevisiae is a proline rotamase localized to the nucleolus. J. Cell Biol. **127**, 623–639.

Bergsma, D.J., Eder, C., Gross, M., Kersten, H., Sylvester, D., Appelbaum, E., Cusimano, D., Livi, G.P., McLaughlin, M.M., Kasyan, K., Porter, T.G., Silverman, C., Dunnington, D., Hand, A., Prichett, W.P, Bossard, M.J., Brant, M., and Levy, M.A. (1991). The cyclophilin multigene family of peptidyl-prolyl isomerases. Characterization of three separate human isoforms. J. Biol. Chem. **266**, 23204–23214.

Bram, R.J. and Crabtree, G.R. (1994). Calcium signalling in T cells stimulated by a cyclophilin B-binding protein. Nature **371**, 355–358.

Bram, R.J., Hung, D.T., Martin, P.K., Schreiber, S.L., and Crabtree, G.R. (1993). Identification of the immunophilins capable of mediating inhibition of signal transduction by cyclosporin A and FK506: roles of calcineurin binding and cellular location. Mol. Cell. Biol. **13**, 4760–4769.

Breuder, T., Hemenway, C.S., Movva, N.R., Cardenas, M.E., and Heitman, J. (1994). Calcineurin is essential in cyclosporin A- and FK506-sensitive yeast strains. Proc. Natl. Acad. Sci. USA **91**, 5372–5376.

Brillantes, A.-M.B., Ondrias, K., Scott, A., Kobrinsky, E., Ondriasova, E., Moschella, M.C., Jayaraman, T., Landers, M., Ehrlich, B.E., and Marks, A.R. (1994). Stabilization of calcium release channel (ryanodine receptor) function by FK506-binding protein. Cell **77**, 513–523.

Brown, E.J., Albers, M.W., Shin, T.B., Ichikawa, K., Keith, C.T., Lane, W.S., and Schreiber, S.L. (1994). A mammalian protein targeted by G1-arresting rapamycin-receptor complex. Nature **369**, 756–759.

Bush, K.T., Hendrickson, B.A., and Nigam, S.K. (1994). Induction of the FK506-binding protein, FKBP13, under conditions which misfold proteins in the endoplasmic reticulum. Biochem. J. **303**, 705–708.

Cafferkey, R., Young, P.R., McLaughlin, M.M., Bergsma, D.J., Koltin, Y., Sathe, G.M., Faucette, L., Eng, W.-K., Johnson, R.K., and Livi, G.P. (1993). Dominant missense mutations in a novel yeast protein related to mammalian phosphatidylinositol 3-kinase and VPS34 abrogate rapamycin cytotoxicity. Mol. Cell. Biol. **13**, 6012–6023.

Callebaut, I., and Mornon, J.-P. (1995). Trigger factor, one of the *Escherichia coli* chaperone proteins, is an original member of the FKBP family. FEBS Lett. **374**, 211–215.

Callebaut, I., Renoir, J.-M., Lebeau, M.-C., Massol, N., Burny, A., Baulieu, E.-E., and Mornon, J.-P. (1992). An immunophilin that binds Mr 90 000 heat shock protein: main structural features of a mammalian p59 protein. Proc. Natl. Acad. Sci. USA **89**, 6270–6274.

Cameron, A.M., Steiner, J.P., Roskams, A.J., Ali, S.M., Ronnett, G.V., and Snyder, S.H. (1995a). Calcineurin associated with the inositol 1,4,5-trisphophate receptor-FKBP12 complex modulates Ca^{2+} flux. Cell **83**, 463–472.

Cameron, A.M., Steiner, J.P., Sabatini, D., Kaplin, A.I., Walensky, L.D., and Snyder, S.H. (1995b). Immunophilin FK506 binding protein associated with inositol 1,4,5-trisphosphate receptor modulates calcium flux. Proc. Natl. Acad. Sci. USA **92**, 1784–1788.

Cardenas, M.E. and Heitman, J. (1995). FKBP12-rapamycin target TOR2 is a vacuolar protein with an associated phosphatidylinositol-4 kinase activity. EMBO J. **14**, 5892–5907.

Cardenas, M.E., Hemenway, C., Muir, R.S., Ye, R., Fiorentino, D., and Heitman, J. (1994). Immunophilins interact with calcineurin in the absence of exogenous immunosuppressive ligands. EMBO J. **13**, 5944–5957.

Cardenas, M.E., Muir, R.S., Breuder, T., and Heitman, J. (1995). Targets of immunophilin-immunosuppressant complexes are distinct highly conserved regions of calcineurin A. EMBO J. **14**, 2772–2783.

Chang, H.J. and Lindquist, S. (1994). Conservation of Hsp90 macromolecular complexes in *Saccharomyces cerevisiae*. J. Biol. Chem. **269**, 24983–24988.

Cianciotto, N.P. and Fields, B.S. (1992). *Legionella pneumophila mip* gene potentiates intracellular infection of protozoa and human macrophages. Proc. Natl. Acad. Sci. USA **89**, 5188–5191.

Cianciotto, N.P., Eisenstein, B.I., Mody, C.H., Toews, G.B., and Engleberg, N.C. (1989). A *Legionella pneumophila* gene encoding a species-specific surface protein potentiates initiation of intracellular infection. Infect. Immun. **57**, 1255–1262.

Colley, N.J., Baker, E.K., Stamnes, M.A., and Zuker, C.S. (1991). The cyclophilin homolog ninaA is required in the secretory pathway. Cell **67**, 255–263.

Connern, C.P. and Halestrap, A.P. (1992). Purification and N-terminal sequencing of peptidyl-prolyl cis-trans-isomerase from rat liver mitochondrial matrix reveals the existence of a distinct mitochondrial cyclophilin. Biochem J. **284**, 381–385.

Cullen, B.R. and Heitman, J. (1994). Chaperoning a pathogen. Nature **372**, 319–320.

Cyert, M.S., Kunisawa, R., Kaim, D., and Thorner, J. (1991). Yeast has homologs (*CNA1* and *CNA2* gene products) of mammalian calcineurin, a calmodulin-regulated phosphoprotein phosphatase. Proc. Natl. Acad. Sci. USA. **88**, 7376–7380.

Davis, E.S., Becker, A., Heitman, J., Hall, M.N., and Brennan, M.B. (1992). A yeast cyclophilin gene essential for lactate metabolism at high temperature. Proc. Natl. Acad. Sci. USA **89**, 11169–11173.

Duina, A.A., Marsh, J.A., and Gaber, R.F. (1996a). Identification of two CyP-40-like cyclophilins in *Saccharomyces cerevisiae*, one of which is required for normal growth. Yeast **12**, 943–949.

Duina, A.A., Chang, H.C., Marsh, J.A., Lindquist, S., and Gaber, R.F. (1996b). A cyclophilin function in Hsp90-dependent signal transduction. Science **274**, 1613–1614.

Engleberg, N.C., Carter, C., Weber, D.R., Cianciotto, N.P., and Eisenstein, B. I. (1989). DNA sequence of *mip*, a *Legionella pneumophila* gene associated with macrophage infectivity. Infect. Immun. **57**, 1263–1270.

Fischer, G. (1994). Peptidyl-prolyl *cis/trans* isomerases and their effectors. Angew. Chem. Int. Ed. Engl. **33**, 1415–1436.

Fischer, G., Wittmann-Liebold, B., Lang, K., Kiefhaber, T., and Schmid, F.X. (1989). Cyclophilin and peptidyl-prolyl cis-trans isomerase are probably identical proteins. Nature **337**, 476–478.

Fischer, G., Bang, H., Ludwig, B., Mann, K., and Hacker, J. (1992). Mip protein of *Legionella pneumophila* exhibits peptidyl-prolyl-*cis/trans* isomerase. Mol. Microbiol. **6**, 1375–1383.

Foor, F., Parent, S.A., Morin, N., Dahl, A.M., Ramadan, N., Chrebet, G., Bostian, K.A., and Nielsen, J.B. (1992). Calcineurin mediates inhibition by FK506 and cyclosporin of recovery from a-factor arrest in yeast. Nature **360**, 682–684.

Franco, L., Jiménez, A., Demolder, J., Molemans, F., Fiers, W., and Contreras, R. (1991). The nucleotide sequence of a third cyclophilin-homologous gene from *Saccharomyces cerevisiae*. Yeast **7**, 971–979.

Franke, E.K. and Luban, J. (1995). Cyclophilin and Gag in HIV-1 replication and pathogenesis. In *Cell Activation and Apoptosis in HIV Infection*, (J.-M. Andrieu and W. Lu, eds). Plenum Press, New York, pp. 217–228.

Franke, E.K., Yuan, H.E.H., and Luban, J. (1994). Specific incorporation of cyclophilin A into HIV-1 virions. Nature **372**, 359–362.

Friedman, J. and Weissman, I. (1991). Two cytoplasmic candidates for immunophilin action are revealed by affinity for a new cyclophilin: one in the presence and one in the absence of CsA. Cell **66**, 799–806.

Friedman, J., Trahey, M., and Weissman, I. (1993). Cloning and characterization of cyclophilin C-associated protein: A candidate natural cellular ligand for cyclophilin C. Proc. Natl. Acad. Sci. USA **90**, 6815–6819.

Frigerio, G. and Pelham, H.R.B. (1993). A *Saccharomyces cerevisiae* cyclophilin resident in the endoplasmic reticulum. J. Mol. Biol. **233**, 183–188.

Galat, A., Lane, W.S., Standaert, R.F., and Schreiber, S.L. (1992). A rapamycin-selective 25-kDa immunophilin. Biochemistry **31**, 2427–2434.

Gitti, R.K., Lee, B.M., Walker, J., Summers, M.F., Yoo, S., and Sundquist, W.I. (1996). Structure of the amino-terminal core domain of the HIV-1 capsid protein. Science **273**, 231–235.

Griffith, J.P., Kim, J.L., Kim, E.E., Sintchak, M.D., Thomson, J.A., Fitzgibbon, M.J., Felming, M.A., Caron, P.R., Hsiao, K., and Navia, M.A. (1995). X-ray structure of calcineurin inhibited by the immunophilin-immunosuppressant FKBP12-FK506 complex. Cell **82**, 507–522.

Guthrie, B. and Wickner, W. (1990). Trigger factor depletion or overproduction causes defective cell division but does not block protein export. J. Bacteriol. **172**, 5555–5562.

Haendler, B., Keller, R., Hiestand, P.C., Kocher, H.P., Wegmann, G., and Movva, N.R. (1989). Yeast cyclophilin: isolation and characterization of the protein, cDNA and gene. Gene **83**, 39–46.

Handschumacher, R.E., Harding, M.W., Rice, J., and Drugge, R.J. (1984). Cyclophilin: a specific cytosolic binding protein for cyclosporin A. Science **226**, 544–547.

Hanes, S.D., Shank, P.R., and Bostian, K.A. (1989). Sequence and mutational analysis of ESS1, a gene essential for growth in *Saccharomyces cerevisiae*. Yeast **5**, 55–72.

Hani, J., Stumpf, G., and Domdey, H. (1995). PTF1 encodes an essential protein in *Saccharomyces cerevisiae*, which shows strong homology with a new putative family of PPIases. FEBS Lett. **365**, 198–202.

Harding, M.W., Galat, A., Uehling, D.E., and Schreiber, S.L. (1989). A receptor for the immunosuppressant FK506 is a cis-trans peptidyl-prolyl isomerase. Nature **341**, 758–760.

Hasel, K.W., Glass, J.R., Godbout, M., and Sutcliffe, J.G. (1991). An endoplasmic reticulum-specific cyclophilin. Mol. Cell. Biol. **11**, 3484–3491.

Hayano, T., Takahashi, N., Kato, S., Maki, N., and Suzuki, M. (1991). Two distinct forms of peptidylprolyl-*cis-trans*-isomerase are expressed separately in periplasmic and cytoplasmic compartments of *Escherichia coli* cells. Biochemistry **30**, 3041–3048.

Heitman, J., Movva, N.R., Hiestand, P.C., and Hall, M.N. (1991a). FK506-binding protein proline rotamase is a target for the immunosuppressive agent FK506 in *Saccharomyces cerevisiae*. Proc. Natl. Acad. Sci. USA **88**, 1948–1952.

Heitman, J., Movva, N.R., and Hall, M.N. (1991b). Targets for cell cycle arrest by the immunosuppressant rapamycin in yeast. Science **253**, 905–909.

Heitman, J., Movva, N.R., and Hall, M.N. (1992). Proline isomerases at the crossroads of protein folding, signal transduction, and immunosuppression. New Biologist **4**, 448–460.

Helekar, S.A., Char, D., Neff, S., and Patrick, J. (1994). Prolyl isomerase requirement for the expression of functional homo-oligomeric ligand-gated ion channels. Neuron **12**, 179–189.

Helliwell, S.B., Wagner, P., Kunz, J., Deuter-Reinhard, M., Henriquez, R., and Hall, M.N. (1994). TOR1 and TOR2 are structurally and functionally similar but not identical phosphatidylinositol kinase homologues in yeast. Mol. Biol. Cell **5**, 105–118.

Hemenway, C.S. and Heitman, J. (1996). Immunosuppressant target protein FKBP12 is required for P-glycoprotein function in yeast. J. Biol. Chem. **270**, 18527–18534.

Hoffman, K. and Handschumacher, R.E. (1995). Cyclophilin-40: evidence for a dimeric complex with hsp90. Biochem. J. **307**, 5–8.

Jayaraman, T., Brillantes, A.-M., Timerman, A.P., Fleischer, S., Erdjument-Bromage, H., Tempst, P., and Marks, A.R. (1992). FK506 binding protein associated with the calcium release channel (ryanodine receptor). J. Biol. Chem. **267**, 9474–9477.

Jennings, P.A., Finn, B.E., Jones, B.E., and Matthews, C.R. (1993). A reexamination of the folding mechanism of dihydrofolate reductase from *Escherichia coli*: verification and refinement of a four-channel model. Biochemistry **32**, 3783–3789.

Jin, Y.J. and Burakoff, S.J. (1993). The 25-kDa FK506-binding protein is localized in the nucleus and associates with casein kinase II and nucleolin. Proc. Natl. Acad. Sci. USA **90**, 7769–7773.

Jin, Y.J., Albers, M.W., Lane, W.S., Bierer, B.E., Schreiber, S.L., and Burakoff, S.J. (1991). Molecular cloning of a membrane-associated human FK506- and rapamycin-binding protein, FKBP-13. Proc. Natl. Acad. Sci. USA **88**, 6677–6681.

Kieffer, L.J., Thalhammer, T., and Handschumacher, R.E. (1992). Isolation and characterization of a 40-kDa cyclophilin-related protein. J. Biol. Chem. **267**, 5503–5507.

Kieffer, L.J., Seng, T.W., Li, W., Osterman, D.G., Handschumacher, R.E., and Bayney, R.M. (1993). Cyclophilin-40, a protein with homology to the P59 component of the steroid receptor complex. J. Biol. Chem. **268**, 12303–12310.

Kissinger, C.R., Parge, H.E., Knighton, D.R., Lewis, C.T., Pelletier, L.A., Tempczyk, A., Kalish, V.J., Tucker, K.D., Showalter, R.E., Moomaw, E.W., Gastinel, L.N., Habuka, N., Chen, X., Maldonado, F., Barker, J.E., Bacquet, R., and Villafranca, J.E. (1995). Crystal structures of human calcineurin and the human FKBP12-FK506-calcineurin complex. Nature **378**, 641–644.

Klappa, P., Freedman, R.B., and Zimmermann, R. (1995). Protein disulphide isomerase and a lumenal cyclophilin-type peptidyl prolyl *cis-trans* isomerase are in transient contact with secretory proteins during late stages of translation. Eur. J. Biochem. **232**, 755–764.

Koltin, Y., Faucette, L., Bergsma, D.J., Levy, M.A., Cafferkey, R., Koser, P.L., Johnson, R.K., and Livi, G.P. (1991). Rapamycin sensitivity in *Saccharomyces cerevisiae* is mediated by a peptidyl-prolyl cis-trans isomerase related to human FK506-binding protein. Mol. Cell. Biol. **11**, 1718–1723.

Koser, P.L., Sylvester, D., Livi, G.P., and Bergsma, D.J. (1990). A second cyclophilin-related gene in *Saccharomyces cerevisiae*. Nucl. Acids Res. **18**, 1643.

Kunz, J., Henriquez, R., Schneider, U., Deuter-Reinhard, M., Movva, N.R., and Hall, M.N. (1993). Target of rapamycin in yeast, TOR2, is an essential phosphatidylinositol kinase homolog required for G_1 progression. Cell **73**, 585–596.

Lazar, S. and Kolter, R. (1996). SurA assists the folding of *Escherichia coli* outer membrane proteins. J. Bacteriol. **178**, 1770–1773.

Lill, R., Crooke, E., Guthrie, B., and Wickner, W. (1988). The 'trigger factor cycle' includes ribosomes, presecretory proteins, and the plasma membrane. Cell **54**, 1013–1018.

Liu, J. and Walsh, C.T. (1990). Peptidyl-prolyl *cis-trans*-isomerase from *Escherichia coli*: a periplasmic homolog of cyclophilin that is not inhibited by cyclosporin A. Proc. Natl. Acad. Sci. USA **87**, 4028–4032.

Liu, J., Farmer, J.D., Lane, W.S., Friedman, J., Weissman, I., and Schreiber, S.L. (1991). Calcineurin is a common target of cyclophilin-cyclosporin A and FKBP-FK506 complexes. Cell **66**, 807–815.

Lodish, H.F. and Kong, N. (1991). Cyclosporin A inhibits an initial step in folding of transferrin within the endoplasmic reticulum. J. Biol. Chem. **266**, 14835–14838.

Lorenz, M.C. and Heitman, J. (1995). TOR mutations confer rapamycin resistance by preventing interaction with FKBP12-rapamycin. J. Biol. Chem. **270**, 27531–27537.

Lu, K.P., Hanes, S.D., and Hunter, T. (1996). A human peptidyl-prolyl isomerase essential for regulation of mitosis. Nature **380**, 544–547.

Luban, J., Bossolt, K.L., Franke, E.K., Kalpana, G.V., and Goff, S.P. (1993). Human immunodeficiency virus type 1 gag protein binds to cyclophilins A and B. Cell **73**, 1067–1078.

Lundemose, A.G., Birkelund, S., Fey, S.J., Larsen, P.M., and Christiansen, G. (1991). *Chlamydia trachomatis* contains a protein similar to the *Legionella pneumophila mip* gene product. Mol. Microbiol. **5**, 109–115.

Maleszka, R., Hanes, S.D., Hackett, R.L., Couet, H.G.D., and Miklos, G.L.G. (1996). The *Drosophila melanogaster* dodo (dod) gene, conserved in humans, is functionally interchangeable with the ESS1 cell division gene of *Saccharomyces cerevisiae*. Proc. Natl. Acad. Sci. USA **93**, 447–451.

Manning-Krieg, U.C., Henriquez, R., Cammas, F., Graff, P., Gaveriaux, S., and Movva, N.R. (1994). Purification of FKBP-70, a novel immunophilin from *Saccharomyces cerevisiae*, and cloning of its structural gene, FPR3. FEBS Lett. **352**, 98–103.

Matouschek, A., Rospert, S., Schmid, K., Glick, B.S., and Schatz, G. (1995). Cyclophilin catalyzes protein folding in yeast mitochondria. Proc. Natl. Acad. Sci. USA **92**, 6319–6323.

McLaughlin, M.M., Bossard, M.J., Koser, P.L., Cafferkey, R., Morris, R.A., Miles, L.M., Strickler, J., Bergsma, D.J., Levy, M.A., and Livi, G.P. (1992). The yeast cyclophilin multigene family: purification, cloning and characterization of a new isoform. Gene **111**, 85–92.

Missiakas, D., Betton, J.-M., and Raina, S. (1996). New components of protein folding in the extra-cytoplasmic compartments of Eschericia coli: SurA, FkpA and Skp/OmpH. Mol. Microbiol. **21**, 871–884.

Nadeau, K., Das, A., and Walsh, C.T. (1993). Hsp90 chaperonins possess ATPase activity and bind heat shock transcription factors and peptidyl prolyl isomerases. J. Biol. Chem. **268**, 1479–1487.

Nicolli, A., Basso, E., Petronilli, V., Wenger, R.M., and Bernardi, P. (1996). Interactions of cyclophilin with the mitochondrial inner membrane and regulation of the permeability transition pore, and cycloporin A-sensitive channel. J. Biol. Chem. **271**, 2185–2192.

Nielsen, J.B., Foor, F., Siekierka, J.J., Hsu, M.-J., Ramadan, N., Morin, N., Shafiee, A., Dahl, A. M., Brizuela, L., Chrebet, G., Bostian, K.A., and Parent, S.A. (1992). Yeast FKBP-13 is a membrane-associated FK506-binding protein encoded by the nonessential gene *FKB2*. Proc. Natl. Acad. Sci. USA **89**, 7471–7475.

Parent, S.A., Nielsen, J.B., Morin, N., Chrebet, G., Ramadan, N., Dahl, A.M., Hsu, M.-J., Bostian, K.A., and Foor, F. (1993). Calcineurin-dependent growth of an FK506- and CsA-hypersensitive mutant of *Saccharomyces cerevisiae*. J. Gen. Microbiol. **139**, 2973–2984.

Partaledis, J.A. and Berlin, V. (1993). The *FKB2* gene of *Saccharomyces cerevisiae*, encoding the immunosuppressant-binding protein FKBP-13, is regulated in response to accumulation of unfolded proteins in the endoplasmic reticulum. Proc. Natl. Acad. Sci. USA **90**, 5450–5454.

Peattie, D.A., Harding, M.W., Fleming, M.A., DeCenzo, M.T., Lipke, J.A., Livingston, D.J., and Benasutti, M. (1992). Expression and characterization of human FKBP52, an immunophilin that associates with the 90-kDa heat shock protein and is a component of steroid receptor complexes. Proc. Natl. Acad. Sci. USA **89**, 10974–10978.

Price, E.R., Zydowsky, L.D., Jin, M., Baker, C.H., McKeon, F.D., and Walsh, C.T. (1991). Human cyclophilin B: a second cyclophilin gene encodes a peptidyl-prolyl isomerase with a signal sequence. Proc. Natl. Acad. Sci. USA **88**, 1903–1907.

Price, E.R., Jin, M., Lim, D., Pati, S., Walsh, C.T., and McKeon, F.D. (1994). Cyclophilin B trafficking through the secretory pathway is altered by binding of cyclosporin A. Proc. Natl. Acad. Sci. USA **91**, 3931–3935.

Rahfield, J.-U., Rucknagel, K.P., Schelbert, B., Ludwig, B., Hacker, J., Mann, K., and Fischer, G. (1994). Confirmation of the existence of a third family among peptidyl-prolyl *cis/trans* isomerase. Amino acid sequence and recombinant production of parvulin. FEBS Lett. **352**, 180–184.

Rassow, J., Mohrs, K., Koidl, S., Barthelmess, I.B., Pfanner, N., and Tropschug, M. (1995). Cyclophilin 20 is involved in mitochondrial protein folding in cooperation with molecular chaperones hsp70 and hsp60. Mol. Cell. Biol. **15**, 2654–2662.

Ratajczak, T., Carrello, A., Mark, P.J., Warner, B.J., Simpson, R.J., Moritz, R.L., and House, A.K. (1993). The cyclophilin component of the unactivated estrogen receptor contains a tetratricopeptide repeat domain and shares identity with p59 (FKBP59). J. Biol. Chem. **268**, 13187–13192.

Riviere, S., Menez, A., and Galat, A. (1993). On the localization of FKBP25 in T-lymphocytes. FEBS Lett. **315**, 247–251.

Rospert, S., Looser, R., Dubaquie, Y., Matouschek, A., Glick, B.S., and Schatz, G. (1996). Hsp60-independent protein folding in the matrix of yeast mitochondria. EMBO J. **15**, 764–774.

Rotonda, J., Burbaum, J.J., Chan, H.K., Marcy, A.I., and Becker, J.W. (1993). Improved calcineurin inhibition by yeast FBKP12-drug complexes. J. Biol. Chem. **268**, 7607–7609.

Rudd, K.E., Sofia, H.J., Koonin, E.V., G. Plunkett, I., Larzar, S., and Rouviere, P.E. (1995). A new family of peptidyl-prolyl isomerases. Trends Biochem. Sci. **20**, 12–14.

Sabatini, D.M., Erdjument-Bromage, H., Lui, M., Tempst, P., and Snyder, S.H. (1994). RAFT1: a mammalian protein that binds to FKBP12 in a rapamycin-dependent fashion and is homologous to yeast TORs. Cell **78**, 35–43.

Schmid, F.X., Mayr, L.M., Mucke, M., and Schonbrunner, E.R. (1993). Prolyl isomerases: role in protein folding. Adv. Protein Chem. **44**, 25–66.

Schneuwly, S., Shortridge, R.D., Larrivee, D.C., Ono, T., Ozaki, M., and Pak, W.L. (1989). *Drosophila ninaA* gene encodes an eye-specific cyclophilin (cyclosporine A binding protein). Proc. Natl. Acad. Sci. USA **86**, 5390–5394.

Shan, X., Xue, Z., and Melese, T. (1994). Yeast NPI46 encodes a novel prolyl *cis-trans* isomerase that is located in the nucleolus. J. Cell Biol. **126**, 853–862.

Shieh, B.-H., Stamnes, M.A., Seavello, S., Harris, G.L., and Zuker, C.S. (1989). The *ninaA* gene required for visual transduction in *Drosophila* encodes a homologue of cyclosporin A-binding protein. Nature **338**, 67–70.

Siekierka, J.J., Hung, S.H.Y., Poe, M., Lin, C.S., and Sigal, N.H. (1989). A cytosolic binding protein for the immunosuppressant FK506 has peptidyl-prolyl isomerase activity but is distinct from cyclophilin. Nature **341**, 755–757.

Smith, D.F., Albers, M.W., Schreiber, S.L., Leach, K.L., and Deibel, M.R. (1993). FKBP54, a novel FK506-binding protein in avian progesterone receptor complexes and HeLa extracts. J. Biol. Chem. **268**, 24270–24273.

Smith, T., Ferreira, L.R., Hebert, C., Norris, K., and Sauk, J.J. (1995). Hsp47 and cyclophilin B traverse the endoplasmic reticulum with procollagen into pre-golgi intermediate vesicles. J. Biol. Chem. **270**, 18323–18328.

Stamnes, M.A., Shieh, B.-H., Chuman, L., Harris, G.L., and Zuker, C.S. (1991). The cyclophilin homolog ninaA is a tissue-specific integral membrane protein required for the proper synthesis of a subset of Drosophila rhodopsins. Cell **65**, 219–227.

Steinmann, B., Bruckner, P., and Superti-Furga, A. (1991). Cyclosporin A slows collagen triple-helix formation in vivo: indirect evidence for a physiologic role of peptidyl-prolyl cis-trans-isomerase. J. Biol. Chem. **266**, 1299–1303.

Stroller, G., Rucknagel, K.P., Nierhaus, K.H., Schmid, F.X., Fischer, G., and Rahfeld, J.-U. (1995). A ribosome-associated peptidyl-prolyl *cis/trans* isomerase identified as the trigger factor. EMBO J. **14**, 4939–4948.

Sykes, K., Gething, M.-J., and Sambrook, J. (1993). Proline isomerases function during heat shock. Proc. Natl. Acad. Sci. USA **90**, 5853–5857.

Tai, P.-K. K., Albers, M.W., Chang, H., Faber, L.E., and Schreiber, S.L. (1992). Association of a 59-kilodalton immunophilin with the glucocorticoid receptor complex. Science **256**, 1315–1318.

Takahashi, N., Hayano, T., and Suzuki, M. (1989). Peptidyl-prolyl cis-trans isomerase is the cyclosporin A-binding protein cyclophilin. Nature **337**, 473–475.

Thali, M., Bukovsky, A., Kondo, E., Rosenwirth, B., Walsh, C.T., Sodroski, J., and Gottlinger, H.G. (1994). Functional association of cyclophilin A with HIV-1 virions. Nature **372**, 363–365.

Timerman, A.P., Ogunbumni, E., Freund, E., Wiederrecht, G., Marks, A.R., and Fleischer, S. (1993). The calcium release channel of sarcoplasmic reticulum is modulated by FK-506-binding protein. J. Biol. Chem. **268**, 22992–22999.

Timerman, A.P., Wiederrecht, G., Marcy, A., and Fleischer, S. (1995). Characterization of an exchange reaction between soluble FKBP-12 and the FKBP-ryanodine receptor complex. J. Biol. Chem. **270**, 2451–2459.

Tormo, A., Almiron, M., and Kolter, R. (1990). *surA*, an *Escherichica coli* gene essential for survival in stationary phase. J. Bacteriol. **172**, 4339–4347.

Tropschug, M., Barthelmess, I.B., and Neupert, W. (1989). Sensitivity to cyclosporin A is mediated by cyclophilin in *Neurospora crassa* and *Saccharomyces cerevisiae*. Nature **342**, 953–955.

Van Duyne, G.D., Standaert, R.F., Karplus, P.A., Schreiber, S.L., and Clardy, J. (1991). Atomic structure of FKBP-FK506, an immunophilin-immunosuppressant complex. Science **252**, 839–842.

Volz, K.W., Matthews, D.A., Alden, R.A., Freer, S.T., Hansch, C., Kaufman, B.T., and Kraut, J. (1982). Crystal structure of avian dihydrofolate reductase containing phenyltriazine and NADPH. J. Biol. Chem. **257**, 2528–2536.

Wang, T., Donahoe, P.K., and Zervos, A.S. (1994). Specific interaction of type 1 receptors of the TGF-β family with the immunophilin FKBP-12. Science **265**, 674–676.

Weisman, R., Creanor, J., and Fantes, P. (1996). A multicopy suppressor of a cell cycle defect in *S. pombe* encodes a heat shock-inducible 40 kDa cyclophilin-like protein. EMBO J. **15**, 447–456.

Wiederrecht, G., Brizuela, L., Elliston, K., Sigal, N.H., and Siekierka, J.J. (1991). *FKB1* encodes a nonessential FK506-binding protein in *Saccharomyces cerevisiae* and contains regions suggesting homology to the cyclophilins. Proc. Natl. Acad. Sci. USA **88**, 1029–1033.

Wiederrecht, G., Hung, S., Chan, H.K., Marcy, A., Martin, M., Calaycay, J., Boulton, D., Sigal, N., Kincaid, R.L., and Siekierka, J.J. (1992). Characterization of high molecular weight FK-506 binding activities reveals a novel FK-506-binding protein as well as a protein complex. J. Biol. Chem. **267**, 21753–21760.

Wilson, L.K., Benton, B.M., Zhou, S., Thorner, J., and Martin, G.S. (1995). The yeast immunophilin Fpr3 is a physiological substrate of the tryosine-specific phosphoprotein phosphatase Ptp1. J. Biol. Chem. **270**, 25185–25193.

Yang, W.M., Inouye, C.J., and Seto, E. (1995). Cyclophilin A and FKBP12 interact with YY1 and alter its transcriptional activity. J. Biol. Chem. **270**, 15187–15193.

Yem, A.W., Tomasselli, A.G., Heinrikson, R.L., Zurcher-Neely, H., Ruff, V.A., Johnson, R.A., and Deibel, M.R., Jr. (1992). The Hsp56 component of steroid receptor complexes binds to immobilized FK506 and shows homology to FKBP-12 and FKBP-13. J. Biol. Chem. **267**, 2868–2871.

■ *Kara Dolinski[1] and Joseph Heitman:[2]*
[1]Department of Genetics, [2]Departments of Genetics and Pharmacology
Howard Hughes Medical Institute
Duke University Medical Center
322 CARL Bldg, Research Drive
Durham, NC 27710, USA
Tel. 1 919 684 2809/2824
Fax. 1 919 684 5458
E-mail: kd2@acpub.duke.edu and
    heitm001@mc.duke.edu

12a

Cyclophilin PPIases

Escherichia coli cyclophilins

Although the peptidyl–prolyl cis-trans isomerase (PPIase) activities of most cyclophilins originating from eukaryotic cells are inhibited by the immunosuppressive drug cyclosporin A (CsA), the activity of Escherichia coli cyclophilins is essentially not sensitive to CsA. No definitive evidence for the physiological role of cyclophilins was identified in E. coli cells; however, they are believed to be involved in protein folding in the cell.

■ Alternative names

Peptidyl prolyl *cis–trans* isomerase (Hayano *et al.*, 1991), rotamase (Liu and Walsh, 1990; Rosen *et al.*, 1990), *cis–trans* peptidyl prolyl isomerase (Harding *et al.*, 1989), cyclosporin A-binding protein (Takahashi *et al.*, 1989), EC: 5.2.1.8.

■ Isolation of *E. coli* cyclophilins

Because no binding activity for cyclosporin A could be detected in *E. coli* cells, *E. coli* cyclophilins were isolated on the basis of their PPIase activity (Hayano *et al.*, 1991). In *E. coli* cells, two forms of cyclophilin have so far been identified. One is present in the cytosol and another in the periplasm of the cell. Oligonucleotides that were synthesized based on peptide sequences obtained from the proteolytic fragments of the cyclophilins were used to isolate the corresponding genes from a genomic DNA library of *E. coli* HB101 cells (GenBank accession numbers M55429 for the periplasmic form and M55430 for the cytosolic form).

■ *Escherichia coli* cyclophilin sequences

The cytosolic cyclophilin of *E. coli* consists of 164 amino acid residues with a molecular weight of 18 184 Da. Cytosolic cyclophilin lacks a hydrophobic amino acid stretch that could serve as a signal sequence or a transmembrane domain (Hayano *et al.*, 1991). Periplasmic cyclophilin is encoded as a 190 amino acid precursor with a signal sequence of 24 amino acids (Liu and Walsh, 1990). The sequence of the periplasmic form is identical with that of the hypothetical protein encoded by the ORF190 gene adjacent to the *fic* gene that is involved in the cell filamentation in *E. coli* induced by cyclic AMP (Kawamukai *et al.*, 1989). The initiation codon of the *fic* gene starts 262 bp downstream of the termination codon of the periplasmic cyclophilin gene. The calculated molecular weight of the mature periplasmic protein is 18 077 Da. The amino acid sequences of the *E. coli* cyclophilins are about 50% identical with each other, and they display c. 25% sequence identity with other cyclophilins from eukaryotic species, such as mammal, insect, fungi and yeast. Although the degree of homology shared among all cyclophilin molecules is much lower than that shared between eukaryotic cyclophilins, two highly homologous regions are present in the middle of the sequences in many species including *E. coli*; one is located around residues 30–70 and the other around residues 90–120 of the cyclophilin sequences. The *E. coli* cyclophilins do not share any sequence homology with FK506-binding protein, FkpA (see entry p. 000) or PpiC (see entry p. 000), the members of the other two PPIase families (FKBPs and parvulins) in *E. coli* (Maki *et al.*, 1990; Tran *et al.*, 1990; Wulfing *et al.*, 1994; Rudd *et al.*, 1995).

Biological activities and function of cyclophilins in *E. coli*

Cyclophilins catalyse the slow *cis–trans* isomerization of Xaa–Pro peptide bonds in oligopeptides (Lang et al., 1987). The catalytic activity originally detected in porcine kidney was sulfhydryl dependent and heat sensitive, and was abolished by incubation with trypsin but not chymotrypsin (Fischer et al., 1984). *Escherichia coli* cyclophilins are also trypsin sensitive, but are heat stable. The PPIase activity of the cytosolic form is affected by sulfhydryl modifying reagents, however, that of the periplasmic form is not, because it contains no cysteine residues (Hayano et al., 1991). *Escherichia coli* cyclophilins exhibit significantly (roughly 100-fold) reduced affinity for CsA (Schonbrunner et al., 1991). When assayed with the synthetic peptide succinyl–Ala–Ala–Pro–Phe–4-nitroanilide (which is cleaved by chymotrypsin only when the Ala–Pro bond is in the trans conformation), the k_{cat}/K_m value of *E. coli* cytosolic cyclophilin was 1.9×10^{-6} $M^{-1}s^{-1}$. This value is very similar to those of the other cyclophilins, such as those from mammals, fungi and yeast, which vary overall by no more than a factor of three (Schonbrunner et al., 1991). All of the cyclophilins tested so far catalyse the slow refolding reactions of RNase T1, whose slow refolding kinetics are determined by the isomerization of two prolines, Pro-39 and Pro-55, both of which are in *cis* configurations in the native protein. However, the pattern of refolding of RNase T1 by *E. coli* cytosolic cyclophilin differs somewhat from that by cyclophilins of other species. *Escherichia coli* cyclophilin accelerates efficiently the slow *trans* to *cis* isomerization of both prolines in denatured RNase T1. On the other hand, cyclophilins from other species catalyse only the isomerization of Pro-55. *Escherichia coli* cyclophilin therefore seems to be able to catalyse prolyl isomerization in sterically more restricted regions of a protein chain, or has different steric requirements in general (Schonbrunner et al., 1991). Since all eukaryotic cyclophilins have strong affinity for cyclosporin A at the catalytic site of PPIase, the fact that *E. coli* cyclophilins have relatively poor affinities for cyclosporin A may be related to their ability to access Pro-39 during refolding of RNase T1. The X-ray structural analysis of *E. coli* cytosolic cyclophilin suggested that the presence of an extra polar (Gln) residue that sticks out over the active site, prevents the access of CsA to the hydrophobic pocket of the cyclophilin molecule (Konno et al., 1996).

Mutagenesis of *E. coli* cyclophilin

The *E. coli* cytosolic cyclophilin lacks the single highly conserved tryptophan residue of eukaryotic cyclophilins. Mutation of the natural Phe-112 to Trp-112 enhances *E. coli* rotamase susceptibility to inhibition by cyclosporin A by 23-fold (Liu et al., 1991). The corresponding Trp-121 to Phe-121 mutation in human cyclophilin A decreases the sensitivity of the protein to cyclophilin A by 75-fold, confirming the importance of the indole side chain as a major determinant in immunosuppressant drug recognition.

References

Fischer, G., Bang, H., and Mech, C. (1984). Nachweis einer exzymkatalyse fur die cis-trans-isomerisierung der peptidbindung in prolinhaltigen peptiden. Biomed. Biochim. Acta **43**, 1101–1111.

Harding, M.W., Galat, A., Uehling, D.E., and Schreiber, S.L. (1989). A receptor for the immunosuppressant FK506 is a cis-trans peptidyl prolyl isomerase. Nature **341**, 758–760.

Hayano, T., Takahashi, N., Kato, S., Maki, N., and Suzuki, M. (1991). Two distinct forms of peptidyl prolyl cis-trans isomerase are expressed separately in periplasmic and cytoplasmic compartments of Escherichia coli cells. Biochemistry **30**, 3041–3048.

Kawamukai, M., Matsuda, H., Fujii, W., Utsumi, R., and Komano, T. (1989). Nucleotide sequence of fic and fic-1 genes involved in cell filamentation induced by cyclic AMP in Escherichia coli. J. Bacteriol. **171**, 4525–4529.

Konno, M., Ito, M., Hayano. T., and Takahashi, N. (1996). The substrate-binding site in Escherichia coli cyclophilin A preferably recognizes a cis-proline-isomer or a largely distorted trans-isomer. J. Mol. Biol. **256**, 897–908.

Lang, K., Schmid, F.X., and Fischer, G. (1987). Catalysis of protein folding by prolyl isomerase. Nature **329**, 268–270.

Liu, J. and Walsh, C.T. (1990). Peptidyl prolyl *cis-trans* isomerase from *Escherichia coli*:: a periplasmic homolog of cyclophilin that is not inhibited by cyclosporin A. Proc. Natl. Acad. Sci. USA **87**, 4028–4032.

Liu, J., Chen, C.-M., and Walsh, C.T. (1991). Human and *Escherichia coli* cyclophilins: sensitivity to inhibition by the immunosuppressant cyclosporin A correlate with a specific tryptophan residue. Biochemistry **30**, 2306–2310.

Maki, N., Sekiguchi, F., Nishimaki, J., Miwa, K., Hayano, T., Takahashi, N., and Suzuki, M. (1990). Complementary DNA encoding the human T-cell FK506-binding protein, a peptidyl prolyl cis-trans isomerase distinct from cyclophilin. Proc. Natl. Acad. Sci. USA **87**, 5440–5443.

Rosen, M.K., Standaert, R.F., Galat, A., Nakatsuka, M., and Schreiber, S.L. (1990). Inhibition of FKBP rotamase activity by immunosuppressant FK506: twisted amide surrogate. Science **248**, 863–866.

Rudd, K.E., Sofia, H.J., Koonin, E.V., Plunkett, G., Lazar, S., and Rouviere, P.E. (1995). A new family of peptidyl-prolyl isomerase. Trends Biochem. Sci. **20**, 12–14.

Schonbrunner, E.R., Mayer, S., Tropschug, M., Fischer, G., Takahashi, N., and Schmid, F.X. (1991). Catalysis of protein folding by cyclophilins from different species. J. Biol. Chem. **266**, 3630–3635.

Takahashi, N., Hayano, T., and Suzuki, M. (1989). Peptidyl prolyl cis-trans isomerase is the cyclosporin A-binding protein cyclophilin. Nature **337**, 473–475.

Tran, P.V., Bannor, T.A., Doktor, S.Z., and Nichols, B.P. (1990). Chromosomal organization and expression of Escherichia coli pabA. J. Bacteriol. **172**, 397–410.

Wulfing, C., Lombardero, J., and Pluckthum, A. (1994). An Escherichia coli protein consisting of a domain homologous to FK506-binding protein (FKBP) and a new metal binding motif. J. Biol. Chem. **269**, 2895–2901.

■ *Nobuhiro Takahashi:*
Tonen Coporation
Corporate Research and Development Laboratory
1-3-1 Nishi-tsurugaoka
Ohi-machi, Iruma-gun
Saitama 356, Japan

Tel. 81 492 66 8371
Fax. 81 492 66 8359
E-mail: LDM03454@niftyserve.or.jp

Saccharomyces cerevisiae cyclophilin A/Cpr1/Cyp1

Cyclophilin A (CypA) is the most abundant cyclophilin in Saccharomyces cerevisiae. It is a cytosolic protein which, like the other family members, catalyses the cis–trans isomerization of peptidyl–prolyl bonds. Cyclosporin A (CsA) blocks the enzymatic activity of all cyclophilins but its toxic effects are mediated through binding to CypA and inhibition of yeast calcineurin.

■ Alternative names

CPH (Haendler et al., 1989); Cpr1 (Heitman et al., 1991); Cyp1 (Koser et al., 1991).

■ Isolation

CypA was isolated from a *S. cerevisiae* cell-free lysate using dye-ligand chromatography, isoelectric focusing and gel filtration (Haendler et al., 1989).

■ *CYP1* gene and sequence

The cDNA sequence predicts a protein of 162 amino acids (Haendler et al., 1989; Dietmeier and Tropschug, 1990; GenBank accession number X17505). Comparison with the protein sequence of other soluble cyclophilins from yeast or higher eukaryotes shows the existence of a highly conserved central region and of specific N- and C-terminal parts.

The *S. cerevisiae CYP1* gene has no intron and contains three putative TATA boxes in the 5′ promoter region and a transcription termination sequence in the 3′ region (Haendler et al., 1989). It maps to chromosone IV (Koser et al., 1991).

■ Cyclophilin A protein

No structural data are available on *S. cerevisiae* CypA but owing to the high sequence similarity a configuration comparable to that of mammalian CypA is likely (Cardenas et al., 1995a).

Polyclonal antibodies against *Escherichia coli*-expressed recombinant CypA detect the native protein in *S. cerevisiae* extracts (Zydowski et al., 1992).

■ Biological activities of cyclophilin A

CypA purified from *S. cerevisiae* and the recombinant form purified from *E. coli* possess peptidyl–prolyl isomerase activity (Haendler et al., 1989; Zydowski et al., 1992; Cardenas et al., 1995a). CsA inhibits this activity (Zydowski et al., 1992). Genomic disruption of CypA is compatible with life in *S. cerevisiae* (Tropschug et al., 1989; Heitman et al., 1991; Koser et al., 1991; Tanida et al., 1991). An increased sensitivity to high temperatures was, however, observed (Sykes et al., 1993). Absence of CypA or mutation to a form that does not bind to CsA renders previously sensitive *S. cerevisiae* strains resistant to CsA, indicating that CypA is the mediator of the toxic effects of the drug (Tropschug et al., 1989; Koser et al., 1991; Breuder et al., 1994; Cardenas et al., 1995a, b).

Saccharomyces cerevisiae CypA complexed to CsA binds to the protein phosphatase calcineurin (Foor et al., 1992; Cardenas et al., 1995b; Hemenway et al., 1995). It also interacts with HSP90 (Nadeau et al., 1993).

Biological regulation of cyclophilin A

Protein activity. The isomerase activity of *S. cerevisiae* cyclophilins is blocked by CsA.
Gene regulation. An increased expression of CypA (and CypB) has been observed after heat shock (Sykes et al., 1993).

■ Mutagenesis studies with cyclophilin A

The Gly^{70}Ser, His^{90}Tyr and Gly^{102}Ala mutants of CypA confer CsA resistance to a *S. cerevisiae* strain that was previously CsA sensitive (Cardenas et al., 1995a).

References

Breuder, T., Hemenway, C.S. Movva, N.R., Cardenas, M.E., and Heitman, J. (1994). Calcineurin is essential in Cyclosporin A- and FK506-sensitive yeast strains. Proc. Natl. Acad. Sci. USA **91**, 5372–5376.

Cardenas, M.E., Lim, E., and Heitman, J. (1995a). Mutations that perturb cyclophilin A ligand binding pocket confer cyclosporin A resistance in Saccharomyces cerevisiae. J. Biol. Chem. **270**, 20997–21002.

Cardenas, M.E., Muir, R.S., Breuder, T., and Heitman, J. (1995b). Targets of immunophilin-immunosuppressant complexes are distinct highly conserved regions of calcineurin A. EMBO J. **14**, 2772–2783.

Dietmeier, K. and Tropschug, M. (1990). Nucleotide sequence of a full-length cDNA coding for cyclophilin of Saccharomyces cerevisiae. Nucl. Acids Res. **18**, 373.

Foor, F., Parent, S.A., Morin, N., Dahl, A.M., Ramadan, N., Chrebet, G., Bostian K.A., and Nielsen, J.B. (1992). Calcineurin mediates inhibition by FK506 and cyclosporin of recovery from -factor arrest in yeast. Nature **360**, 682–684.

Haendler, B., Keller, R., Hiestand, P.C., Kocher, H.P., Wegmann, G., and Movva, N.R. (1989). Yeast cyclophilin: isolation and characterization of the protein, cDNA and gene. Gene **83**, 39–46.

Heitman, J., Movva, N.R., Hiestand, P.C., and Hall, M.N. (1991). FK506-binding protein rotamase is a target for the immunosuppressive agent FK506 in *Saccharomyces cerevisiae*. Proc. Natl. Acad. Sci. USA **88**, 1948–1952.

Hemenway, C.S., Dolinski, K., Cardenas, M.E., Hiller, M.A., Jones, E.W., and Heitman, J. (1995). vph6 mutants of Saccharomyces cerevisiae require calcineurin for growth and are defective in vacuolar H^+-ATPase assembly. Genetics **141**, 833–844.

Koser, P.L., Bergsma, D.J., Cafferkey, R., Eng, W.-K., McLauglin, M.M., Ferrara, A., Silverman, C., Kasyan, K., Bossard, M.J., Johnson, R.K., Porter, T.G., Levy, M.A., and Livi, G.P. (1991). The CYP2 gene of Saccharomyces cerevisiae encodes a cyclosporin A-sensitive peptidyl-prolyl cis-trans isomerase with an N-terminal signal sequence. Gene **108**, 73–80.

Nadeau, K., Das, A., and Walsh, C.T. (1993). Hsp90 chaperonins possess ATPase activity and bind heat shock transcription factors and peptidyl prolyl isomerases. J. Biol. Chem. **268**, 1479–1487.

Sykes, K., Gething, M.J., and Sambrook, J. (1993). Proline isomerases function during heat shock. Proc. Natl. Acad. Sci. USA **90**, 5853–5857.

Tanida, I., Yanagida, M., Maki, N., Yagi, S., Namiyama, F., Kobayashi, T., Hayano, T., Takahashi, N., and Suzuki, M. (1991). Yeast cyclophilin-related gene encodes a nonessential second peptidyl-prolyl cis-trans isomerase associated with the secretory pathway. Transplant. Proc. **23**, 2856–2861.

Tropschug, M., Barthelmess, I.B., and Neupert, W. (1989). Sensitivity to cyclosporin A is mediated by cyclophilin in Neurospora crassa and Saccharomyces cerevisiae. Nature **342**, 953–955.

Zydowsky, L.D., Ho, S.I., Baker, C.H., McIntyre, K., and Walsh, C.T. (1992). Overexpression, purification, and characterization of yeast cyclophilins A and B. Protein Sci. **1**, 961–969.

■ *Bernard Haendler:*
Research Laboratories of Schering AG
13342 Berlin, Germany
Tel. 49 30 468 12669
Fax. 49 30 468 16707
E-mail: Bernard.Haendler@Schering.DE

Saccharomyces cerevisiae cyclophilin B/Cpr2/Cyp2

Cyclophilin B belongs to the family of yeast cyclophilin-related peptidyl–prolyl cis–trans isomerases (PPIases), enzymes that catalyse the isomerization of prolyl–peptide bonds in vitro. The protein was originally identified as the deduced translation product of a gene cloned as a homolog of a human cyclophilin A-encoding cDNA. Yeast cyclophilin B is directed to the ER/secretory pathway by a 20 amino acid hydrophobic signal sequence, where it may function to catalyse the folding of secretory proteins. Cyclophilin B is also involved in the heat shock response. Although purified recombinant cyclophilin B protein possesses cyclosporin A (CsA)-sensitive PPIase activity in vitro, cyclophilin B does not play a role in CsA cytotoxicity in yeast, which is mediated instead by cytosolic cyclophilin A.

■ Alternative names

Cyp2 (Koser et al., 1991; McLaughlin et al., 1992), yCyPB (Zydowsky et al., 1992), Cpr2 (Davis et al., 1992), CRG (Tanida et al., 1991).

■ *CYP2* gene isolation and sequence

Cyclophilin B is encoded by the *CYP2* gene (EMBL accession number X51497), which was isolated from a *Saccharomyces cerevisiae* genomic library using a human

T cell cyclophilin (cyclophilin A) cDNA as a probe (Koser et al., 1990). CYP2, also known as CPR2 (Davis et al., 1992) and CRG (Tanida et al., 1991), maps to yeast chromosome VIII (Koser et al., 1991).

The 1042 bp CYP2 nucleotide sequence contains a 615 bp open reading frame predicting a protein of 205 amino acids (Koser et al., 1990). The 5′-untranslated region (UTR) contains two putative (albeit imperfect) TATAA elements, and the 3′-UTR contains a near perfect TAG...TAGT...TTT consensus sequence for transcription termination (Zaret and Sherman, 1982).

■ Cyclophilin B protein sequence

The predicted cyclophilin B protein shows significant overall homolgy to other yeast cyclophilins [e.g. 56% identity to the yeast cytosolic cyclophilin A (Haendler et al., 1989; see entry p. 372)] as well as to cyclophilins from other organisms [e.g. 52% identity to human cytosolic cyclophilin A (Haendler et al., 1987; see entry p. 386)]. The protein sequence includes a hydrophobic, 34 residue N-terminal extension (relative to yeast cyclophilin A); part of this N-terminal extension is processed, presumably in the ER, suggesting that it serves as a signal sequence (see below). Unlike yeast cyclophilin D (Frigerio and Pelham, 1993; see entry p. 380), cyclophilin B does not contain a C-terminal HDEL tetrapeptide sequence shown to mediate retention of soluble proteins within the lumen of the ER (Pelham, 1990). The tryptophan residue (Trp-155), which is required for CsA binding (Bossard et al., 1991; Liu et al., 1991), is conserved in cyclophilin B.

Humans possess a secreted form of cyclophilin which is a counterpart to yeast cyclophilin B; in fact, it is more similar to yeast cyclophilin B than it is to human cyclophilin A, suggesting that the yeast and human cyclophilin B proteins were derived from a common ancestral gene. The human protein, which has been purified from milk, also contains an ER-directed signal sequence and lacks the canonical C-terminal ER retention signal; human cyclophilin B has been localized to subcellular compartments of the secretory pathway (Hasel et al., 1991; Price et al., 1991; Spik et al., 1991; Arber et al., 1992). There is 50% overall amino acid sequence identity between human and yeast cyclophilin B proteins, with 75% identity within a core 77 amino acid region comprising part of the CsA-binding and PPIase activity domain (Friedman and Weissman, 1991; Koser et al., 1991; Price et al., 1991).

In addition to cyclophilins A and B, at least three other S. cerevisiae cyclophilin proteins have been identified to date, each containing an N-terminal signal sequence for subcellular compartmentalization [i.e. mitochondrial cyclophilin C (Cpr3) (Davis et al., 1992; McLaughlin et al., 1992; Matouschek et al., 1995; see entry p. 378), ER-localized cyclophilin D (Cpr5) which contains a C-terminal HDEL sequence (Frigerio and Pelham, 1993), and an uncharacterized integral membrane protein, Scc3p (Cpr4), which contains a luminal cyclophilin-related domain and is associated with the secretory pathway (Franco et al., 1991; see entry p. 381)].

■ Cyclophilin B protein

Polyclonal antisera raised to cyclophilin B have failed to detect any intracellular protein in yeast cells, although a faint protein signal could be detected in highly concentrated culture media, suggesting that the protein is secreted (Zydowsky et al., 1992). The low level of protein correlates with low levels of CYP2 mRNA (Zydowsky et al., 1992). When engineered for overexpression, cyclophilin B is secreted from yeast cells as a mature protein lacking the first 20 amino acid residues deduced from the gene sequence (Tanida et al., 1991). While overexpression may have resulted in protein mis-sorting, these data suggest that the first 20 amino acids serve as a signal for targeting the protein to the ER. The precise location of cyclophilin B has yet to be determined.

Attempts to express the full length protein in Escherichia coli have been unsuccessful; however, truncated forms lacking in the first 20 (Zydowsky et al., 1992) or the first 34 (Koser et al., 1991) amino acids have been expressed in E. coli and purified to homogeneity. The purified truncated proteins can be easily visualized by Western blotting, e.g. using either a protein-specific antiserum (Zydowsky et al., 1992) or a cross-reacting antiserum raised to yeast mitochondrial cyclophilin (McLaughlin et al., 1992).

■ Biological activity of cyclophilin B

The purified recombinant truncated forms of cyclophilin B, as well as the secreted mature form present in yeast culture medium, exhibit PPIase activity which is inhibitable by the anti-fungal and immunosuppressive drug CsA (Koser et al., 1991; Tanida et al., 1991; Zydowsky et al., 1992). Use of the chromogenic substrate succinyl–Ala–Ala–Pro–Phe–4-nitroanilide in a chymotrypsin-coupled PPIase assay, in which the cis–trans isomerization of the Ala–Pro bond is monitored at 10°C (Fischer et al., 1984), has yielded k_{cat}/K_m values of 1.1×10^7 M^{-1} s^{-1} and 5.77×10^6 M^{-1} s^{-1} for the E. coli-derived 34 and 20 amino acid-truncated proteins, respectively (Koser et al., 1991; Zydowsky et al., 1992). No data are available on substrate specificity. CsA binds recombinant cyclophilin B and inhibits its PPIase activity with an IC_{50} of c. 100 nM, a value 2.5-fold less than that observed for yeast cyclophilin A (Zydowsky et al., 1992).

Cyclophilin B is not essential for viability under normal conditions, since cells harboring a CYP2 gene disruption exhibit no obvious unconditional growth defects (Koser et al., 1991; Tanida et al., 1991). Although cyclophilin B possesses PPIase activity in vitro, it does not mediate CsA cytotoxicity in yeast. Instead, CsA's growth inhibitory effect is mediated by cytosolic cyclophilin A: genomic disruption of the cyclophilin A-encoding gene (CYP1) results in a CsA-resistant phenotype (Breuder et al., 1994), whereas disruption of CYP2 has no effect on CsA sensitivity (Koser et al., 1991; Tanida et al., 1991). This lack of drug sensitivity in whole cells is apparently a result of the subcellular compartmentalization of cyclophilin B.

A precise role for yeast cyclophilin B *in vitro* has not yet been determined, but it is speculated to catalyse the folding, and perhaps enhance the stability, of secretory polypeptides. The fact that cyclophilin B has been detected extracellularly suggests that it may act to facilitate protein transport and stability through the secretory pathway. However, as cyclophilin B-deficient cells do not exhibit any growth defects, the spontaneous rate of peptidyl–prolyl bond conversion for cyclophilin B substrates *in vitro* may be adequate to support normal growth. Alternatively, additional PPIases [possibly cyclophilin D (Cpr5) or the putative transmembrane cyclophilin-related protein (Cpr4), both predicted to be associated with the secretory pathway] can compensate for the function of cyclophilin B. Since *CYP2* (as well as *CYP1*) mRNA is induced following heat shock, and gene disruption results in increased sensitivity to elevated temperatures, cyclophilin B may be involved in the heat shock response, perhaps functioning to refold or stabilize secretory proteins essential for recovery following exposure to stressful growth conditions (Sykes *et al.*, 1993).

References

Arber, S., Frause, K.-H., and Caroni, P. (1992). S-cyclophilin is retained intracellulary via a unique COOH-terminal sequence and colocalizes with the calcium storage protein calreticulin. J. Cell Biol. **116**, 113–125.

Bossard, M.J., Koser, P.L., Brandt, M., Bergsma, D.J., and Levy, M.A. (1991). A single Trp121 to Ala121 mutation in human cyclophilin alters cyclosporin A affinity and peptidyl-prolyl isomerase activity. Biochem. Biophys. Res. Commun. **176**, 1142–1148.

Breuder, T., Hemenway, C.S., Movva, N.R., Cardenas, M.E., and Heitman, J. (1994). Calcineurin is essential in Cyclosporin A- and FK506-sensitive yeast strains. Proc. Natl. Acad. Sci. USA **91**, 5372–5376.

Davis, E.S., Becker, A., Heitman, J., Hall, M.N., and Brennan, M.B. (1992). A yeast cyclophilin gene essential for lactate metabolism at high temperature. Proc. Natl. Acad. Sci. USA **89**, 1169–1173.

Fischer, G., Bang, H., and Mech, C. (1984). Detection of enzyme catalysis for *cis-trans* isomerization of peptide bonds using proline containing peptides as substrates. Biomed. Biochim. Acta **43**, 1101–1112.

Franco, L., Jimenez, A., Demolder, J., Molemans, F., Fiers, W., and Contrera, R. (1991). The nucleotide sequence of a third cyclophilin-homologous gene from *Saccharomyces cerevisiae*. Yeast **7**, 971–979.

Friedman, J. and Weissman, I. (1991). Two cytoplasmic candidates for immunophilin action are revealed by affinity for a new cyclophilin: one in the presence and one in the absence of CsA. Cell **66**, 799–806.

Frigerio, G. and Pelham, H.R.B. (1993). A *Saccharomyces cerevisiae* cyclophilin resisdent in the endoplasmic reticulum. J. Mol. Biol. **233**, 183–188.

Haendler, B., Hofer-Warinek, R., and Hofer, E. (1987). Complementary DNA for T-cell cyclophilin. EMBO J. **6**, 947–950.

Haendler, B., Keller, R., Hiestand, P.C., Kocher, H.P., Wegmann, G., and Movva, N.R. (1989). Yeast cyclophilin: isolation and characterization of the protein, cDNA and gene. Gene **83**, 39–46.

Hasel, K.W., Glass, J.R., Godbout, M., and Sutcliffe, J.G. (1991). An endoplasmic reticulum specific cyclophilin. Mol. Cell. Biol. **11**, 3484–3491.

Koser, P.L., Sylvester, D., Livi, G.P., and Bergsma, D.J. (1990). A second cyclophilin-related gene in *Saccharomyces cerevisiae*. Nucl. Acids Res. **18**, 1643.

Koser, P.L., Bergsma, D.J., Cafferkey, R., Eng, W.K., McLaughlin, M.M., Ferrara, A., Silverman, C., Kasyan, K., Bossard, M.J., Johnson, R.K., Porter, T.G., Levy, M.A., and Livi, G.P. (1991). The *CYP2* gene of *Saccharomyces cerevisiae* encodes a cyclosporin A-sensitive peptidyl-prolyl *cis-trans* isomerase with an N-terminal signal sequence. Gene **108**, 73–80.

Liu, J., Chen, C.-M., and Walsh, C.T. (1991). Human and *E. coli* cyclophilins: sensitivity to inhibition by the immunosuppressant cyclosporin A correlates with a specific tryptophan residue. Biochemistry **30**, 2306–2310.

McLaughlin, M.M., Bossard, M.J., Koser, P.L., Cafferkey, R., Morris, R.A., Miles, L.M., Strickler, J., Bergsma, D.J., Levy, M.A., and Livi, G.P. (1992). The yeast cyclophilin multigene family: purification, cloning and characterization of a new isoform. Gene **111**, 85–92.

Matouschek, A., Rospert, S., Schmid, K., Glick, B.S., and Schatz, G. (1995). Cyclophilin catalyzes protein folding in yeast mitochondria. Proc. Natl. Acad. Sci. USA **92**, 6319–6323.

Pelham, H.R.B. (1990). The retention signal for soluble proteins of the endoplasmic reticulum, Trends Biol. Sci. **15**, 483–486.

Price, E.R., Zydowsky, L.D., Jin, M., Baker, C.H., McKeon, F.D., and Walsh, C.T. (1991). Human cyclophilin B: a second cyclophilin gene encodes a peptidyl-prolyl isomerase with a signal sequence. Proc. Natl. Acad. Sci. USA **88**, 1903–1907.

Spik, G., Haendler, B., Delmas, O., Mariller, C., Chamoux, M., Maes, P., Tartar, A., Montreuil, J., Stedman, K., Kocher, H.P., Keller, R., Hiestand, P.C., and Movva, N.R. (1991). A novel secreted cyclophilin-like protein (SCYLP). J. Biol. Chem. **266**, 10735–10738.

Sykes, K., Gething, M.-J., and Sambrook, J. (1993). Proline isomerases function during heat shock. Proc. Natl. Acad. Sci. USA **90**, 5853–5857.

Tanida, I., Yanagida, M., Maki, N., Yagi, S., Namiyama, F., Kobayashi, T., Hayano, T., Takahashi, N., and Suzuki, M. (1991). Yeast cyclophilin-related gene encodes a nonessential second peptidyl-prolyl *cis-trans* isomerase associated with the secretory pathway. Transplant. Proc. **23**, 2856–2861.

Zaret, K.S. and Sherman, F. (1982). DNA sequence required for efficient transcription termination in yeast. Cell **28**, 563–573.

Zydowsky, L.D., Ho, S.I., Baker, C.H., McIntyre, K., and Walsh, C.T. (1992). Overexpression, purification, and characterization of yeast cyclophilins A and B. Protein Sci. **1**, 961–969.

■ *Robert Cafferkey,[1] Katie Freeman[2] Rodica Stan[2] and George P. Livi:[2]*
[1]*Department of Molecular Diagnostics,* [2]*Department of Comparative Genetics, UE0548*
SmithKline Beecham Pharmaceuticals
P.O. Box 1539
King of Prussia, PA 19406, USA
Tel. 1 610 270 6614/7535/7709/7717 (R.C./K.F./R.S./G.P.L.)
Fax. 1 610 270 5093/5093/5093/7962 (R.C./K.F./R.S./G.P.L.)
E-mail: RobertCafferkey-1@sbphrd.com,
Katie_B_Freeman@sbphrd.com
Rodica_Stan-1@sbphrd.com and
George_P_Livi@sbphrd.com

Saccharomyces cerevisiae Cpr6

Cpr6 (cyclophilin 6) was identified as a 45 kDa, Hsp90-associated protein in Saccharomyces cerevisiae. *It is constitutively expressed and is highly induced upon heat shock. The primary sequence of Cpr6 is highly homologous to mammalian cyclophlin 40 (Cyp40), a component of estrogen receptor (ER)::Hsp90 complexes.*

■ Alternative names

The yeast 45 kDa cyclophilin homolog (Chang and Lindquist, 1994).

■ Isolation of Cpr6

Cpr6 was isolated in a search for proteins that bind to Hsp90 in *S. cerevisiae* (Chang and Lindquist, 1994). Nickel-affinity chromatography was employed to isolate histidine-tagged Hsp90 as well as Hsp90-associated proteins (Chang and Lindquist, 1994). Histidine-tagged Hsp90 possessed wild-type function *in vivo*, and, hence, the associated protein complexes are expected to be functionally relevant. Cpr6, a 45 kDa protein, was one of four proteins identified in this manner (Chang and Lindquist, 1994).

■ *CPR6* gene and sequence

The *CPR6* gene, located on chromosome XII at position L8167.23 (GenBank: U48867; PIR: S48567), was cloned in the yeast genome sequencing project. This gene encodes a protein of 371 amino acids with a predicted molecular weight of 42 kDa. The primary amino acid sequence of Cpr6 shares 44% identity with mammalian cyclophilin 40, each of which has two characteristic domains. The N-terminal region (aa 1–175) of Cpr6 is homologous to the CsA binding/PPIase domain of Cyp18, an 18 kDa cytosolic cyclophilin (Kieffer *et al.*, 1993; Ratajczak.et *al.*, 1993; Duina *et al.*, 1996a), while the C-terminal region (aa 175–371) of Cpr6 shares high homology with the three TPR (tetratricopeptide repeat) motifs of FKBP59 (Kieffer *et al.*, 1993; Ratajczak.et *al.*, 1993; Duina *et al.*, 1996a). These C-terminal TPR motifs of Cpr6 are sufficient for Hsp90 binding (Duina *et al.*, 1996b).

■ Cpr6 protein and biological activities

Immunoblotting reveals that Cpr6 is moderately abundant and highly induced upon heat shock (H.-C.J. Chang and S. Lindquist, unpublished data). Cpr6 appears as two major spots in the two-dimensional gel electrophoresis of total yeast cellular proteins, suggesting that the protein experiences post-translational modifications (H.-C.J. Chang and S. Lindquist, unpublished data). These results have been confirmed with two different peptide antibodies to amino acids 156–171 and 356–371 of Cpr6.

A search of the complete genomic sequence of *S. cerevisiae* reveals that Cpr7 is a Cpr6 homolog, with both comprising the Cyp40 subfamily in yeast. Genetic analysis of *cpr6* mutations reveals that it is not essential for normal growth (J. Heitman, unpublished data; Duina *et al.*, 1996b). Unlike *cpr7* mutations, *cpr6* mutations have no effects on the growth of cells carrying an Hsp90 ($\Delta hsc82$) mutation (Duina *et al.*, 1996b). Also, D*cpr6* $\Delta cpr7$ confers similar phenotypes as $\Delta cpr7$ alone (Duina *et al.*, 1996a). Thus, the functional relationships between these two Cyp40 homologs remain unknown.

Recently, Cpr6 was also identified in a yeast two-hybrid screen for proteins that interact with Rpd3, a yeast homolog of histone deacetylase (Duina *et al.*, 1996a; Taunton *et al.*, 1996). Although physical interactions between Cpr6 and Rpd3 were clearly demonstrated, *cpr6* null mutation does not affect Rpd3-related phenotypes tested to date (Duina *et al.*, 1996a). The functions of Cpr6 in Rpd3-related pathways remain unexamined.

■ References

Chang, H-C.J. and Lindquist, S. (1994). Conservation of Hsp90 macromolecular complexes in *Saccharomyces cerevisiae*. J. Biol. Chem. **269**, 24983–24988.

Duina, A.A., Marsh, J.A., and Gaber, R.F. (1996a). Identification of two Cyp-40-like cyclophilins in *Saccharomyces cerevisiae*, one of which is required for normal growth. Yeast **12**, 943–952.

Duina, A.A., Chang, H.C.J, Marsh, J.A., Lindquist, S., and Gaber, R.F. (1996b). A cyclophilin function in Hsp90-mediated signal transduction. Science **274**, 1713–1715.

Kieffer, L.J., Seng, T.W., Osterman, D.G., Handshumacher, R.E., and Bayney, R.M. (1993). Cyclophilin-40, a protein with homology to the p59 component of the steroid receptor complex. J. Biol. Chem. **268**, 12303–12310.

Ratajczak, T., Carrello, A., Mark, P.J., Warner, B.J., Simpson, R.J., Moritz, R.L., and House A.K. (1993). The cyclophilin component of the unactivated estrogen receptor contains a tetratricopeptide repeat domain and shares identity with p59 (FKBP59). J. Biol. Chem. **268**, 13187–13192.

Taunton J., Hassig, C.A., and Schreiber, S.L. (1996) A mammalian histone deacetylase related to the yeast transcriptional regulator Rpd3p. Science **272**, 408–411.

Hui-Chen Jane Chang and Susan Lindquist:
Department of Molecular Genetics and Cell Biology
The University of Chicago
5841 S. Maryland Ave.
Room N339, MC 1028
Chicago, IL 60637, USA
Tel. 1 773 702 8049
Fax. 1 773 702 7254
E-mail: jhchang@midway.uchicago.edu and
S-Lindquist@uchicago.edu

Saccharomyces cerevisiae Cpr7

Cpr7 (Cyclophilin 7) was identified as a 45 kDa, Rpd3-associated protein in Saccharomyces cerevisiae. The primary seqence of Cpr7 is highly homologous to yeast Cpr6, with both comprising the cyclophilin 40 (Cyp40) subfamily, a component of estrogen receptor– Hsp90 complexes. Cpr7 associates with Hsp90 and exhibits strong genetic interactions with Hsp90. Its functions are required for the full activation of specific Hsp90 target proteins such as glucocorticoid receptors (GR) and p60$^{v\text{-}src}$.

Alternative names

j1585p (Zagulski *et al.*, 1995).

Isolation of Cpr7

Cpr7 was isolated in a two-hybrid screen for proteins that interact with Rpd3, a yeast homolog of histone deacetylase (Duina *et al.*, 1996a; Taunton *et al.*, 1996). CPR7 was also identified in the yeast genome sequencing project (Zagulski *et al.*, 1995).

CPR7 gene and sequence

The *CPR7* gene (GenBank accession number U48868), located on chromosome X, encodes a protein of 393 amino acids with a predicted molecular weight of 45 kDa. The primary amino acid sequence of Cpr7 shares 38% amino acid identity with Cpr6, a Cyp40 homolog (Duina *et al.*, 1996a). A search of the complete genomic sequence of *S. cerevisiae* indicates that Cpr6 and Cpr7 are the only two cyclophilins in the Cyp40 subfamily.

Cpr7 shares 24% amino acid identity with Cyp40, across both of its characteristic domains. The N-terminal region (aa 1–193) of Cpr7 is homologous to the CsA binding/PPIase domain of Cyp18, an 18 kDa cytosolic cyclophilin (Duina *et al.*, 1996a; Kieffer *et al.*, 1993; Ratajczak *et al.*, 1993), while the C-terminal region (aa 194–393) of Cpr7 shares high homology with the three TPR (tetratricopeptide repeat) motifs of FKBP59 (Duina *et al.*, 1996a; Kieffer *et al.*, 1993; Ratajczak *et al.*, 1993).

Cpr7 protein and biological activities

Several Cpr7 features remain to be examined thoroughly. Although Cpr7 was identified originally by its interaction with Rpd3 (Duina *et al.*, 1996), a *cpr7* null mutant does not affect Rpd3-related phenotypes tested to date (Duina *et al.*, 1996a). Moreover, although Cpr7 is a Cpr6 homolog in *S. cerevisiae*, Δ *cpr6* Δ*cpr7* exhibits no synthetic enhancement of phenotypes tested (Duina *et al.*, 1996a, b). The functional relationship between Cpr6 and Cpr7 remains unknown.

Since Cpr7 belongs to the subfamily of Cyp40, a component of the estrogen receptor::Hsp90 complex (Kieffer *et al.*, 1993; Ratajczak *et al.*, 1993), investigation of the interaction between Cpr7 and Hsp90 was undertaken. In vitro, GST–Cpr7 fusion proteins bind to Hsp90 when mixed with yeast lysate; binding occurs within the Cpr7 carboxy-terminal domain (Duina *et al.*, 1996b). Thus, Cpr7 is a part of the Hsp90 large heteromeric chaperone complex (Chang and Lundquist, 1994).

A *cpr7* null mutation confers a slow growth phenotype (Duina *et al.*, 1996a; K. Dolinski and J. Heitman, unpublished data); in combination with the *hsc82* null mutation, *cpr7* exhibits a much reduced growth rate (Duina *et al.*, 1996b). Thus, *cpr7* has genetic interactions with *hsc82* (Duina *et al.*, 1996b), suggesting strongly that Cpr7 plays a role in Hsp90 complexes.

To determine the role of Cpr7 in Hsp90 complexes, the activities of two specific Hsp90 targets were examined in *cpr7* mutants. The activity of rat glucocorticoid receptor (GR) and of avian pp60$^{v\text{-}src}$ tyrosine kinase depends greatly on Hsp90 (Nathan and Lindquist, 1995). Null mutations in *cpr7* decrease their activity significantly (Duina *et al.*, 1996b), indicating that Cpr7 potentiates the activation of these Hsp90-associated targets (Duina *et al.*, 1996b). Cpr7 exhibits concerted functions with Hsp90 and modulates Hsp90-mediated signal transduction.

■ References

Chang, H-C.J. and Lindquist, S. (1994). Conservation of Hsp90 macromolecular complexes in *Saccharomyces cerevisiae*. J. Biol. Chem. **269**, 24983–24988.

Duina, A.A., Marsh, J.A., and Gaber, R.F. (1996a). Identification of two Cyp-40-like cyclophilins in *Saccharomyces cerevisiae*, one of which is required for normal growth. Yeast **12**, 943–952.

Duina, A.A., Chang, H.-C.J, Marsh, J.A., Lindquist, S., and Gaber, R. F. (1996b). A cyclophilin function in Hsp90-mediated signal transduction, Science **274**, 1713–1715.

Kieffer, L.J., Seng, T.W., Osterman, D.G., Handshumacher, R.E., and Bayney, R.M. (1993). Cyclophilin-40, a protein with homology to the p59 component of the steroid receptor complex. J. Biol. Chem. **268**, 12303–12310.

Nathan, D.F. and Lindquist, S. (1995). Mutational analysis of Hsp90 function: interactions with a steroid receptor and a protein kinase. Mol. Cell. Biol. **15**, 3917–3925.

Ratajczak, T., Carrello, A., Mark, P.J., Warner, B.J., Simpson, R.J., Moritz, R.L., and House A.K. (1993). The cyclophilin component of the unactivated estrogen receptor contains a tetratricopeptide repeat domain and shares identity with p59 (FKBP59). J. Biol. Chem. **268**, 13187–13192.

Taunton J., Hassig, C.A., and Schreiber, S.L. (1996). A mammalian histone deacetylase related to the yeast transcriptional regulator Rpd3p. Science **272**, 408–411.

Zagulski, M., Babinska, B., Gromadka, R., Migdalski, A., Rytka, J., Sulicka, J., and Herbert, C.J. (1995). The sequence of 24.3 kb from chromosome(reveals five complete open reading frames, all of which correspond to new genes, and a tandem insertion of a Ty1 transposon. Yeast **11**, 1179–1186.

■ *Hui-Chen Jane Chang and Susan Lindquist:*
Department of Molecular Genetics and Cell Biology
The University of Chicago
5841 S. Maryland Ave.
Room N339, MC 1028
Chicago, IL 60637, USA
Tel. 1 773 702 8049
Fax. 1 773 702 7254
E-mail: jhchang@midway.uchicago.edu and
S-Lindquist@uchicago.edu

Saccharomyces cerevisiae mitochondrial cyclophilin, Cpr3

The gene for yeast mitochondrial cyclophilin (CPR3) is nuclear; its protein product is transported to the matrix of mitochondria. Disruption of the gene causes yeast to grow slowly at elevated temperatures on media containing the non-fermentable L-lactate as the carbon source. Yeast mitochondrial cyclophilin catalyses the folding of a newly imported protein in the mitochondrial matrix.

■ Alternative names

CYP3 (McLaughlin *et al.*, 1992), cyclophilin C (see Cafferkey *et al.* p. 373).

■ Isolation

The yeast mitochondrial cyclophilin gene was isolated as *CPR3* from a *S. cerevisiae* genomic library by hybridization with a rat cyclophilin cDNA probe (Davis *et al.*, 1992). Independently, yeast mitochondrial cyclophilin protein was purified from yeast extracts following cyclosporin A-sensitive peptidyl–prolyl *cis–trans* isomerase activity (McLaughlin *et al.*, 1992). The corresponding gene (*CYP3*) was then isolated from *S. cerevisiae* genomic libraries by hybridization with a probe derived from the N-terminal protein sequence.

■ *CPR3* gene and sequence

CPR3 (GenBank accession numbers: X56962, S50399, M84758) maps to chromosome XIII (McLaughlin *et al.*, 1992). It contains two putative TATAA elements at positions −58 and −84 in the 5′-untranslated region, and a consensus sequence for termination of transcription in the 3′-untranslated region (McLaughlin *et al.*, 1992).

Cpr3 protein

CPR3 codes for a soluble protein with a calculated relative molecular weight of 19 918 Da. The protein has a 20 amino acid N-terminal targeting sequence which is cleaved (McLaughlin *et al.*, 1992). The protein has 62% amino acid identity with rat cyclophilin, 70% identity with *S. cerevisiae* cyclophilin A (Cpr1/Cyp1, see entry p. 372) and 48% identity with *S. cerevisiae* cyclophilin B (Cpr2/Cyp2, see entry p. 373). Using anti-Cpr3p antibodies raised against peptides, differential centrifugation and mitochondrial fractionation show that Cpr3p is localized to the mitochondrial matrix (M.B. Brennan, personal communication). Cpr3p has been purified and shows *in vitro* peptidyl–prolyl *cis–trans* isomerase activity that is inhibited by cyclosporin A binding (McLaughlin *et al.*, 1992; M.B. Brennan, personal communication; K. Dolinski and J. Heitman, unpublished data).

Biological function of Cpr3p

Gene disruptions have shown that neither *CPR3* nor any combination of *CPR1*, *CPR2*, *CPR3* and the FK506-binding peptidyl–prolyl isomerase *FPR1* are essential for vegetative growth, mating or sporulation at 30°C (Davis *et al.*, 1992). Growth at 37°C on ethanol/glycerol is not affected, and on pyruvate is only slightly affected, indicating that *CPR3* is not essential for respiration or general mitochondrial function (Davis *et al.*, 1992; McLaughlin *et al.*, 1992). However, disruption of *CPR3* does prevent growth on media with L-lactate as the carbon source at 37°C (Davis *et al.*, 1992). This defect is recessive.

Experiments with mitochondria purified from *S. cerevisiae* and from *Neurospora crassa* have shown that Cpr3p catalyses the folding of a newly imported precursor protein (Matouschek *et al.*, 1995; Rassow *et al.*, 1995). Folding of dihydrofolate reductase in the matrix of mitochondria after import is slowed down by a factor of 2–3 by the addition of micromolar concentrations of cyclosporin A or by the disruption of *CPR3*. Cpr3p appears not to be part of a stable protein complex and to be forming only a transient enzyme–substrate complex. Inhibition of Cpr3p with CsA affects the interaction of dihydrofolate reductase with the mitochondrial chaperones Ssc1p and Hsp60 (Rassow *et al.*, 1995; Rospert *et al.*, 1996). Cpr3p does not catalyse the folding of all newly imported proteins in mitochondria and it is not clear what, if any, functions it has other than catalysis of protein folding.

The inner membrane of mammalian mitochondria contains a cyclosporin A sensitive unspecific channel, frequently called the permeability transition pore. Rat mitochondrial cyclophilin appears to bind to and regulate this channel (Connern and Halestrap, 1994; Nicolli *et al.*, 1996). The enzymatic properties of rat mitochondrial cyclophilin have been analysed in detail (Connern and Halestrap, 1992).

Biological and genetic interactions

Searches for multicopy suppressors of the growth defect at 37°C on L-lactate medium of *cpr3::HIS3* disruptant strains yielded one novel gene, *JEN1*, which weakly rescues the growth defect (M.B. Brennan, personal communication). *JEN1* appears to encode a lactate transporter. A screen for genomic single copy chromosomal suppressors yielded a gene, *JEN2*, of unknown function (M.B. Brennan, personal communication).

To find proteins stably interacting with Cpr3p a C-terminally hexa-histidine-tagged Cpr3p was constructed. Although various conditions were tried no proteins could be copurified with the tagged Cpr3p expressed into mitochondria (Matouschek *et al.*, 1995).

References

Connern, C.P. and Halestrap, A.P. (1992). Purification and N-terminal sequencing of peptidyl-prolyl cis-trans isomerase from rat liver mitochondrial matrix reveals the existence of a distinct mitochondrial cyclophilin. Biochem. J. **284**, 382–384.

Connern, C.P. and Halestrap, A.P. (1994). Recruitment of mitochondrial cyclophilin to the mitochondrial inner membrane under conditions of oxidative stress that enhance the opening of a calcium sensitive non-specific channel. Biochem. J. **302**, 321–324.

Davis, E.S., Becker, A., Heitman, J., Hall, M.N., and Brennan, M.B. (1992). A yeast cyclophilin gene essential for lactate metabolism at high temperature. Proc. Natl. Acad. Sci. USA **89**, 11169–11173.

Matouschek, A., Rospert, S., Schmid, K., Glick, B.S., and Schatz, B. (1995). Cyclophilin catalyzes protein folding in yeast. Proc. Natl. Acad. Sci. USA **92**, 6319–6323.

McLaughlin, M.M., Bossard, M.J., Koser, P.L., Cafferkey, R., Morris, R.A., Miles, L.M., Strickler, J., Bergsma, D.J., Levy, M. A., and Livi, G.P. (1992). The yeast cyclophilin multigene family: purification, cloning and characterization of a new isoform. Gene **111**, 85–92.

Nicolli, A., Basso, E., Petronilli, V., Wenger, R.M., and Bernardi, P. (1996). Interactions of cyclophilin with the mitochondrial inner membrane and regulation of the permeability transition pore, a cyclosporin A-sensitive channel. J. Biol. Chem. **271**, 2185–2192.

Rassow, J., Mohrs, K., Koidl, S., Barthelmess, I.B., Pfanner, N., and Tropschug, M. (1995). Cyclophilin 20 is involved in mitochondrial protein folding in cooperation with molecular chaperones hsp70 and hsp60. Mol. Cell. Biol. **15**, 2654–2662.

Rospert, S., Looser, R., Dubaquié, Y., Matouschek, A., Glick, B.S., and Schatz, G. (1996). Hsp60-independent protein folding in the matrix of yeast mitochondria. EMBO J. **15**, 764–774.

■ Andreas Matouschek:
Department of Biochemistry, Molecular Biology and Cell Biology
Northwestern University
2153 Sheridan Road
Evanston, IL 60208-3500, USA
Tel. 1 847 467 3570
Fax. 1 847 467 1380
E-mail: matouschek@nwu.edu

Saccharomyces cerevisiae CypD/Cpr5

The CYPD gene of Saccharomyces cerevisiae encodes a cyclophilin found in the lumen of the endoplasmic reticulum (ER). Cyclophilins are cyclosporin A-sensitive peptidyl–prolyl isomerases.

■ Alternative name

Cpr5.

■ *CYPD* sequence

The genomic *CYPD* sequence is part of cosmid 9740 from chromosome IV (GenBank accession number U28374; Johnston *et al.*, unpublished entry). The open reading frame of a previously isolated cDNA (EMBL accession number X73142; Frigerio and Pelham, 1993), predicts a protein of 225 amino acids with a 22 residue N-terminal hydrophobic signal sequence and the C-terminal ER retrieval signal HDEL (Pelham, 1990). The 162 amino acid cyclophilin core domain defined with regard to cytoplasmic cyclophilins has up to 62% amino acid identity with other *S. cerevisiae* cyclophilins. Sequence similarity is highest with a second cyclophilin found in the secretory pathway followed by the mitochondrial and the cytoplasmic form.

■ CypD protein

Epitope-tagged CypD protein was found in the endoplasmic reticulum by immunofluorescence (Frigerio and Pelham, 1993).

Mutagenesis studies with *CYPD*

Deletion of the *CYPD* gene has no effect on the growth of haploid cells under standard conditions (Frigerio and Pelham, 1993).

■ Biological activity of CypD protein

Cyclophilins are found in a number of compartments within the cell. They are thought to be involved in folding of newly synthesized proteins as well as in signaling pathways that rely on conformational changes of key proteins (Gething and Sambrook, 1992; Rutherford and Zuker, 1994). A *S. cerevisiae* cyclophilin has been shown to be required for efficient refolding of a marker protein upon translocation into the mitochondrial matrix (Matouschek *et al.*, 1995; Rassow *et al.*, 1995). The function of the CypD protein itself has not been defined.

■ References

Frigerio, G. and Pelham, H.R.B. (1993). A Saccharomyces cerevisiae cyclophilin resident in the endoplasmic reticulum. J. Mol. Biol. **233**, 183–188.

Gething, M.-J. and Sambrook, J. (1992). Protein folding in the cell. Nature **355**, 33–44.

Matouschek, A., Rospert, S., Schmid, K., Glick, G.S., and Schatz, G. (1995). Cyclophilin catalyzes protein folding in mitochondria. Proc. Natl. Acad. Sci. USA **92**, 6319–6323

Pelham, H.R.B. (1990). The retention signal for soluble proteins of the endoplasmic reticulum. Trends Biol. Sci. **15**, 483–486.

Rassow, J., Mohrs, K., Koidl, S., Barthelmess, I.B., Pfanner, N., and Tropschug, M. (1995). Cyclophilin 20 is involved in mitochondrial protein folding in cooperation with molecular chaperones hsp70 and hsp60. Mol. Cell. Biol. **15**, 2654–2662.

Rutherford, S.L. and Zuker, C.S. (1994). Protein folding and the regulation of signaling pathways. Cell **79**, 1129–1132.

■ *Gabriella Frigerio:*
Sanger Centre
Hinxton
Cambridge, CB10 1SA, UK
Tel. 44 1223 494954
Fax. 44 1223 494919
E-mail: gcf@sanger.ac.uk

Saccharomyces cerevisiae Scc3/Cpr4

Scc3 is an integral membrane protein associated with the secretory pathway in Saccharomyces cerevisiae. *It has a luminal domain related to cyclophilins.*

■ Alternative name

Cpr4.

■ SCC3 gene and sequence

The genomic sequence of the *SCC3* gene (Accession numbers: EMBL: X59720; PIR: S26658; HSSP: PO5092; PROSITE: PS00170) was originally identified during the sequencing of chromosome III in the overlap of the insert of phages PM5307 and PM6589 (Franco *et al.*, 1991). *SCC3* corresponds to the third cyclophilin gene from *S. cerevisiae*. It is present as a single copy in the genome. Analysis of the 5′ non-coding region of *SCC3* shows a TATA box at −135, a putative CAAA (at −78) and a CAPyACA (at −64) promotor motif. No upstream activating sequences are detected, which is reflected in Northern analysis which indicates that *SCC3* renders an abundant 1.2 kb transcript during exponential growth. The 3′ non-coding region of *SCC3* shows a potential sequence (TTTTTATA) at +1054 for polyadenylation.

■ Scc3 protein

The nucleotide sequence of a 1558 bp DNA fragment contains an open reading frame of 954 nucleotides which encodes a putative protein of 318 amino acids, M_r 35 800 Da, and an isoelectric point of 6.9. The amino acid sequence displays a high similarity to those of known and presumptive cyclophilins. The deduced amino acid sequence of Scc3p includes two highly hydrophobic domains at its N- and C-termini. The former is probably a signal peptide of 20 amino acids which presumably directs the protein to the ER where it could serve as a peptidyl–prolyl *cis–trans* isomerase folding enzyme. The sequence of the C-terminus is compatible with function as a transmembrane helix domain of 17 amino acids. These findings suggest that Scc3 is membrane anchored.

■ Biological activities of Scc3p

Cyclophilins are a highly conserved family of proteins. They have peptidyl–prolyl *cis–trans* isomerase activity and, as such, have been implicated in catalysing rate-limiting steps in protein folding which has been demonstrated by catalysing antibody folding *in vitro* (Lilie *et al.*, 1993). They have a wide phylogenetic distribution spanning the entire animal and plant kingdoms, from bacteria and fungi to cabbage, humans and flies. Cyclophilins have been found in all tissue types examined and distinct isoforms are residents of many intracellular compartments, including the cytosol (Haendler *et al.*, 1989), mitochondria (Davis *et al.*, 1992), endoplasmic reticulum (Frigerio and Pelham, 1993), chloroplasts (Kunz and Hall, 1993) and nucleus (Galat, 1993). The ubiquitous and highly conserved nature of cyclophilins suggests that they play a fundamental role in cellular metabolism.

The deduced amino acid sequence of Scc3 shows a high degree of similarity to those of other known or presumptive cyclophilins. The only cyclophilin with intriguing similarities of structure to that of Scc3 is ninaA from *Drosophila melanogaster*, which is a transmembrane protein that seems to be implicated in the correct folding and/or intercalation of rhodopsin in the endoplasmic reticulum of the fly photoreceptors (Baker *et al.*, 1994; see entry p. 384). By analogy with the ninaA protein, which is required for proper assembly of a subset of *D. melanogaster* rhodopsins, there might be multispanning integral membrane proteins that are dependent on the membrane-anchored yeast cyclophilins.

■ References

Baker, E.K., Colley, N.J., and Zuker, C.S. (1994). The cyclophilin homolog ninaA functions as a chaperone, forming a stable complex in vivo with its protein target rhodopsin. EMBO J. **13**, 4886–4895.

Davis, E.S., Becker, A., Heitman, J., Hall, M.N., and Brennan, M.B. (1992). A yeast cyclophilin gene essential for lactate metabolism at high temperature. Proc. Natl. Acad. Sci. USA **89**, 11169–11173.

Franco, L., Jiménez, A., Demolder, J., Molemans, F., Fiers, W., and Contreras, R. (1991). The nucleotide sequence of a third cyclophilin-homologous gene from *Saccharomyces cerevisiae*. Yeast **7**, 971–979.

Frigerio, G. and Pelham, H.R.B. (1993). A *Saccharomyces cerevisiae* cyclophilin resident in the endoplasmic reticulum. J. Mol. Biol. **233**, 183–188.

Galat, A. (1993). Peptidylproline cis-trans-isomerases: immunophilins. Eur. J. Biochem. **216**, 689–707.

Haendler, B., Keller, R., Hiestand, P.C. Kocher, H.P., Wegmann, G., and Movva, N.R. (1989). Yeast cyclophilin: isolation and characterization of the proteins. Gene **83**, 39–46.

Kunz, J. and Hall, M.N. (1993). Cyclosporin A, FK506 and rapamycin: more than just immunosuppresion. Trends Biochem. Sci. **18**, 334–338.

Lilie, H., Lang, K., Rudolph, R., and Buchner. J. (1993). Prolyl isomerases catalyze antibody folding in vitro. Protein Sci. **9**, 1490–1496.

■ Roland Contreras
Universiteit Gent
Laboratorium Moleculaire Biologie
K. L. Ledeganckstraat 35
B-9000 Gent, Belgium
Tel. 32 9 264 5136
Fax. 32 9 264 53 48
E-mail: Roland@lmb1.rug.ac.be

■ Jan Demolder
Flanders Interuniversity Institute for Biotechnology
Rijvisschestraat 118, box1
B-9052 Zwijnaarde
Tel. 32 9 2446611
Fax. 32 9 2446610
E-mail: Jan.Demolder@vib.be

Neurospora crassa cyclophilins

Neurospora crassa *contains at least four different cyclophilins (Cyps): Cyp45 is located in the cytosol; Cyp21 is synthesized with a signal sequence and translocated into the ER. Cyp20, which is responsible for the sensitivity of N. crassa towards CsA, is located in the cytosol and mitochondria; both identical proteins are the product of a single nuclear gene,* csr-1. *Cyp20 is involved in protein folding in the mitochondrial matrix in cooperation with molecular chaperones Hsp70 and Hsp60. A novel Cyp18 was identified only recently.*

■ Peptidyl–prolyl isomerases in *N. crassa*

Fungi like *Neurospora crassa* or *Saccharomyces cerevisiae* are ideal organisms to study the cellular roles of the different peptidyl–prolyl *cis–trans* isomerases (PPIases) and the molecular mechanisms of action of immunosuppressants like CsA, FK506 or rapamycin (reviewed by Kunz and Hall, 1993). These drugs were originally isolated as antifungal compounds. Three families of PPIases have been described. Two of these, the cyclophilins (Cyps) and FKBPs, have been known for some time (see reviews by Schreiber, 1992; Schmid, 1993; Fischer, 1994; Galat and Metcalfe, 1995; see also the overview in this volume, p. 359). Parvulins constitute a more recently defined family (Rahfeld *et al.*, 1994; see overview p. 365). To date only members of the Cyp (this entry) and FKBP (see entry p. 416) families have been defined in *N. crassa*.

■ Alternative names

Cyclosporin-binding proteins; peptidyl–prolyl *cis–trans* Isomerases (PPIases; EC no. 5.2.1.8.); rotamase; immunophilin (Schreiber, 1992).

■ *Neurospora crassa* cyclophilins

Abundant cytosolic and mitochondrial Cyp20 proteins were isolated using standard biochemical methods (Tropschug *et al.*, 1988). Novel Cyps were identified by CsA-affinity chromatography, SDS–PAGE and named according to their apparent molecular weights (Cyp45 = cyclophilin with an apparent *M*r of 45 kDa). For determination of the relative abundance of individual Cyps, detergent extracts of *N. crassa* were run over CsA columns, and then bound Cyps were released and separated on SDS–PAGE gels. Coomassie-stained gels were analysed by laser densitometry; see Table 1.

■ *Neurospora crassa* cyclophilin proteins and their biological activities

■ Cyp45

Nucleotide sequence analysis of a cDNA encoding Cyp45 revealed an N-terminal PPIase domain and a C-terminal domain homologous to the C-terminus of human and bovine Cyp40 (B. Solscheid and M. Tropschug, manuscript

Table 1. Relative abundance of cyclophilins

Protein	Location	Rel. abundance (%)	Gene
Cyp45	Cytosol	5	nd
Cyp21	ER	5	nd
Cyp20	Cytosol and mitochondria	85	csr-1
Cyp18	nd	5	nd

nd = not determined.

in preparation). Whether Cyp45 (like Cyp40 in higher eukaryotes; see Pratt, 1993) interacts with components of hormone receptors, is not yet known.

■ Cyp21

This cyclophilin is synthesized with a signal sequence of 27 amino acids, which is cleaved upon entry into the ER (H. Schneider, F.X. Schmid, R. Zimmerman and M. Tropschug, manuscript in preparation). Mature Cyp21 is active as a PPIase and is CsA inhibitable. The C-terminal amino acids are HVEL, which might be a *N. crassa* ER retention signal (Pelham *et al.*, 1988).

■ Cyp20

This most abundant cyclophilin has a dual location in *N. crassa*. Identical Cyp20 proteins are located in both the cytosol and mitochondria (Tropschug *et al.*, 1988). The mitochondrial protein is synthesized with a presequence of 44 amino acids, which is cleaved in two steps upon entry in the mitochondrial matrix. Mutants in the gene *csr-1*, which encodes both cytosolic and mitochondrial Cyp20 (Tropschug, 1990; Rassow *et al.*, 1995) have lost the proteins in both compartments; these mutants are CsA resistant (Tropschug *et al.*, 1989), proving that Cyp20 is the receptor for CsA in *N. crassa*. These mutants have been used to answer the question whether cyclophilins, besides their long known *in vitro* role in protein folding (Schönbrunner *et al.*, 1991; Gething and Sambrook, 1992), also act as *in vivo* folding enzymes. Intra-mitochondrial folding of imported DHFR was monitored in wild-type mitochondria treated with CsA (blocking Cyp20 activity) and in *csr-1* mitochondria, which have lost Cyp20 function. In both cases, folding but not import of DHFR was delayed; folding intermediates accumulated bound to the molecular chaperones (Hartl *et al.*, 1994; Stuart *et al.*, 1994) mtHsp70 and Hsp60 (Rassow *et al.*, 1995; reviewed by Schmid, 1995). Similar results have been obtained regarding the role of a mitochondrial cyclophilin (Davis *et al.*, 1992) in yeast (Matouschek *et al.*, 1995). These results support and extend former data on the *in vivo* role of cyclophilins (Bächinger, 1987; Lodish and Kong, 1991; Steinman *et al.*, 1991; Sykes *et al.*, 1993; Kruse *et al.*, 1995). The structural gene for Cyp20 (*csr-1*) has been cloned and sequenced (EMBL accession number X17692). It contains four introns, the longest being in the 5´-untranslated region (Tropschug, 1990).

■ Cyp18

This protein has only recently been identified by binding to a CsA-affinity column (M. Tropschug, unpublished data).

References

Bächinger, H.P. (1987). The influence of peptidyl-prolyl *cis-trans* isomerase on the *in vitro* folding of type III collagen. J. Biol. Chem. **262**, 17144–17148.

Davis, E.S., Becker, A., Heitman, J., Hall, M.N., and Brennan, M.B. (1992). A yeast cyclophilin gene essential for lactate metabolism at high temperature. Proc. Natl. Acad. Sci. USA **89**, 11169–11173.

Fischer, G. (1994). Peptidyl-prolyl *cis/trans* isomerases and their effectors. Angew. Chem. Int. Ed. Engl. **33**, 1415–1436.

Galat, A. and Metcalfe, S.M. (1995). Peptidylproline *cis/trans* isomerases. Prog. Biophys. Mol. Biol. **63**, 67–118.

Gething, M.-J. and Sambrook, J. (1992). Protein folding in the cell. Nature **355**, 33–45.

Hartl, F.-U., Hlodan, R., and Langer, T. (1994). Molecular chaperones in protein folding: the art of avoiding sticky situations. Trends Biochem. Sci. **19**, 20–25.

Kruse, M., Brunke, M., Escher, A., Szalay, A.A., Tropschug, M., and Zimmermann, R. (1995). Enzyme assembly after *de novo* synthesis in rabbit reticulocyte lysate involves molecular chaperones and immunophilins. J. Biol. Chem. **270**, 2588–2594.

Kunz, J. and Hall, M.N. (1993). Cyclosporin A, FK506 and rapamycin: more than just immunosuppression. Trends Biochem. Sci. **18**, 334–338.

Lodish, H.F. and Kong, N. (1991). Cyclosporin A inhibits an initial step in folding of transferrin within the endoplasmic reticulum. J. Biol. Chem. **266**, 14835–14838.

Matouschek, A., Rospert, S., Schmid, K., Glick, B.S., and Schatz, G. (1995). Cyclophilin catalyzes protein folding in yeast mitochondria. Proc. Natl. Acad. Sci. USA **92**, 6319–6323.

Pelham, H.R.B., Hardwick, K.G., and Lewis, M.J. (1988). Sorting of soluble ER proteins in yeast. EMBO J. **7**, 1757–1762.

Pratt, W.B. (1993). The role of heat shock proteins in regulating the function, folding and trafficking of the glucocorticoid receptor. J. Biol. Chem. **268**, 21455–21458.

Rahfeld, J.-U., Rucknagel, K.P., Schelbert, B., Ludwig, B., Hacker, J., Mann, K., and Fischer, G. (1994). Confirmation of the existence of a third family among peptidyl-prolyl *cis/trans* isomerases: amino acid sequence and recombinant production of parvulin. FEBS Lett. **352**, 180–184.

Rassow, J., Mohrs, K., Koidl, S., Barthelmess, I.B., Pfanner, N., and Tropschug, M. (1995). Cyclophilin 20 is involved in mitochondrial protein folding in cooperation with molecular chaperones Hsp70 and Hsp60. Mol. Cell. Biol. **15**, 2654–2662.

Schmid, F.X. (1993). Prolyl isomerases: enzymatic catalysis of slow protein folding reactions. Annu. Rev. Biophys. Biomol. Struct. **22**, 123–143.

Schmid, F.X. (1995). Protein folding: prolyl isomerases join the fold. Curr. Biol. **5**, 993–994

Schönbrunner, E.R., Mayer, S., Tropschug, M., Fischer, G., Takahashi, N., and Schmid, F.X. (1991). Catalysis of protein folding by cyclophilins from different species. J. Biol. Chem. **266**, 3630–3635.

Schreiber, S.L. (1992). Immunophilin-sensitive protein phosphatase action in cell signalling pathways. Cell **70**, 365–368.

Steinman, B., Bruckner, P., and Superti-Furga, A. (1991). Cyclosporin A slows collagen triple-helix formation *in vivo*: indirect evidence for a physiological role of peptidyl-prolyl *cis-trans* isomerase. J. Biol. Chem. **266**, 1299–1303.

Stuart, M.A., Cyr, D.M., Craig, E.A., and Neupert, W. (1994). Mitochondrial molecular chaperones: their role in protein translocation. Trends Biochem. Sci. **19**, 87–92.

Sykes, K., Gething, M.-J., and Sambrook, J. (1993). Proline isomerases function during heat shock. Proc. Natl. Acad. Sci. USA **90**, 5853–5857.

Tropschug, M. (1990). Nucleotide sequence of the gene coding for cyclophilin/peptidyl-prolyl *cis-trans* isomerase of *Neurospora crassa*. Nucl. Acids Res. **18**, 190.

Tropschug, M., Barthelmess, I.B., and Neupert, W. (1989). Sensitivity to cyclosporin A is mediated by cyclophilin in

Neurospora crassa and Saccharomyces cerevisiae. Nature **342**, 953–955.

Tropschug, M., Nicholson, D.W., Hartl, F.-U., Köhler, H., Pfanner, N., Wachter, E. and Neupert, W. (1988). Cyclosporin A-binding protein (cyclophilin) of Neurospora crassa: one gene codes for both the cytosolic and mitochondrial forms. J. Biol. Chem. **263**, 14433–14440.

■ Maximilian Tropschug:
Institut für Biochemie und Molekularbiologie der Albert-Ludwigs-Universität,
Hermann-Herder-Str. 7,
D-79104 Freiburg i. Br., Germany
Tel. 49 761 203 5244
Fax. 49 761 203 5253
E-mail: tropschu@sun2.ruf.uni-freiburg.de

Drosophila melanogaster ninaA

Drosophila ninaA is a photoreceptor cell-specific cyclophilin required for the maturation of Rh1 rhodopsin in the endoplasmic reticulum (ER). In the absence of ninaA, immature, core-glycosylated Rh1 opsin accumulates in the ER and is eventually degraded, causing the dramatic reduction of rhodopsin observed in ninaA mutants. The high substrate specificity of ninaA, together with the extensive genetic analysis possible in Drosophila, has made this an ideal system with which to pursue a comprehensive understanding of cyclophilin function in vivo. While it is well established that cyclophilins can accelerate protein folding in vitro, the study of ninaA provides the first strong evidence that, in vivo, cyclophilin functions as a critical component of the protein biosynthetic machinery.

■ Isolation and sequence of the *Drosophila ninaA* gene

ninaA mutants were originally isolated based on their visual defect by William Pak and co-workers (Pak 1979). Mutant flies can have as much as a 100-fold reduction in Rh1 rhodopsin levels, leading to a dramatic loss of light sensitivity and loss of the prolonged depolarizing after potential (nina = neither inactivation nor after potential) (Larivee et al. 1981; Stephenson et al. 1983); this is a physiological assay that is sensitive to relative rhodopsin levels. Analysis of rhodopsin gene expression in ninaA mutants demonstrated that, despite the massive loss of rhodopsin protein, RNA levels were normal, indicating a post-translational defect (Zuker et al. 1985). The ninaA gene was cloned by a subtractive hybridization approach using eye-specific sequences (Shieh et al. 1989). Confirmation of the identity of the gene was obtained by cytogenetic matching and transformation rescue of the mutant phenotype with the cloned sequence (Shieh et al. 1989; Stamnes et al. 1991). The gene was also cloned by a chromosomal walk (Schneuwly et al. 1989). The genomic, cDNA, and protein sequences have been deposited with GenBank (accession number M22851). ninaA maps between 24D1-2;24E2 on the second chromosome of Drosophila and encodes a 900 bp transcript expressed in the R1–6 photoreceptor cells and light-sensing ocelli. Expression begins late in pupal life and continues through adulthood.

■ ninaA protein

ninaA is a 237 amino acid protein containing an N-terminal targeting sequence and a hydrophobic C-terminal tail (Stamnes et al., 1991). The remainder of the protein contains the core cyclophilin domain which is 42% identical to mammalian CypA isoforms. ninaA is expressed specifically in photoreceptor cells and, within photoreceptor cells, is found in the ER and secretory pathway (Colley et al. 1991). Protease protection experiments combined with in vitro translations in the presence and absence of dog pancreas microsomes have demonstrated that ninaA is a type I integral membrane protein with a small C-terminus extending cytoplasmically and the entire rest of the protein located within the lumen of the ER and intracellular transport vesicles. The ninaA protein is N-linked glycosylated (Asn-68), but glycosylation is not required for its function (Stamnes et al. 1991). Antibodies against a variety of epitopes in ninaA have been generated and used in subcellular localization studies. Immunoelectron microscopy reveals ninaA colocalized with Rh1 throughout the secretory pathway (Colley et al. 1991).

■ Function of ninaA

ninaA mutants have defective visual physiology owing to their dramatically reduced levels of Rh1 rhodopsin. A unique feature of ninaA is its exceptional substrate specificity: it only affects rhodopsin, and within the rhodopsin family of proteins is only required for the biogenesis of the two visible rhodopsin forms, Rh1 and Rh2, which are expressed in the R1–6 photoreceptor cells and ocelli, respectively (Stamnes et al. 1991). The more distantly related UV-sensitive rhodopsins, Rh3 and Rh4 (expressed in non-overlapping subsets of R7 cells), do not require ninaA function. The genetic requirement for ninaA by Rh1 and Rh2, but not by Rh3 and Rh4, was directly demonstrated by generating transgenic flies expressing each of the different opsins within the same cellular context (R1–R6 photoreceptor cells) and then testing for ninaA dependency (Stamnes et al. 1991). These studies showed that Rh3 and Rh4 function normally in the ninaA mutant and therefore do not require ninaA, and also, since visual transduction is functional in these transgenic animals, demonstrated that no other component of the phototransduction cascade requires ninaA function.

Cell biological studies revealed that in ninaA mutants Rh1 accumulates in an Endo H-sensitive, high molecular weight form indicative of blocked passage from the ER to the Golgi (Colley et al. 1991). Furthermore, ninaA photoreceptors display a dramatic overproliferation of ER membranes in R1–6 cells; these are so extensive that they may fill the entire cytoplasm (Colley et al. 1991). This ER phenotype is characteristic of many cell types accumulating improperly folded proteins in the ER. In ninaA mutants the ER accumulation depends on both the loss of ninaA function and on the presence of misfolded Rh1; ninaA;ninaE double mutants no longer display an ER phenotype (ninaE is the structural gene for the Rh1 rhodopsin). Eventually, the immature Rh1 trapped in the ER is degraded, leading to the dramatic loss of Rh1 rhodopsin characteristic of ninaA mutants.

■ Mutagenesis studies with ninaA

A near-saturation mutagenesis study has provided significant insight into the structure/function relationships important for cyclophilin activity within a natural cellular context (Ondek et al. 1992). Functionally important regions in the ninaA protein were identified through the isolation and characterization of novel ninaA alleles. To this end over 700 000 individually mutagenized chromosomes were screened and 70 independent ninaA mutants were isolated and characterized. Since this screen was based on a rhodopsin-dependent phenotype it was unbiased in its requirement for a defined biochemical activity of ninaA. Interestingly, most of the mutations mapped to or near the barrel face of cyclophilin that is involved in CsA and prolyl isomerase substrate binding. This result constitutes the strongest argument to date that the functionally relevant region of the molecule, in vivo, overlaps with the peptidyl–prolyl substrate-binding site. Additional mutations identified in this screen define other functionally important regions of the ninaA protein; the hydrophobic transmembrane region is important for ninaA's interaction with Rh1 opsin (see below) and the conserved C-terminal tail, which extends into the cytoplasm, may regulate the localization of the protein in the cell.

■ Biological and genetic interactions of ninaA

Genetic studies designed to manipulate the relative levels of ninaA and Rh1 opsin reveal a quantitative requirement for ninaA during Rh1 rhodopsin biogenesis (Baker et al. 1994). Furthermore, by using transgenic flies expressing histidine-tagged Rh1 or histidine-tagged ninaA in conjunction with rapid Ni^{2+} affinity purification, Baker et al. (1994) went on to show that, in vivo, ninaA forms a specific and highly stable protein complex with Rh1, but not with Rh3 or Rh4 opsins (Baker et al. 1994). The examination of various ninaA mutants using this assay further established the functional link between ninaA activity and its tight physical association with Rh1; in the majority of cases, the severity of the mutant defect correlated with the ability of the mutant ninaA protein to associate with the Rh1 opsin. The few exceptions were telling, and suggest that additional factors may be involved in ninaA function. For example, two mutations in the C-terminal transmembrane anchoring domain of the protein cause strong phenotypic defects, yet they retain their ability to associate with Rh1 rhodopsin. Also, the C-terminal cytosolic residues are also essential for the physical interaction between ninaA and Rh1. These results provide the first evidence for a biologically relevant, physical interaction between a cyclophilin and its cellular target, and together these results suggest that this class of cyclophilin may function as a chaperone.

■ Other Drosophila cyclophilins

In addition to the ninaA cyclophilin, Drosophila express numerous other cyclophilin isoforms of a more general nature. To date, these include CYP-1 (Stamnes et al. 1991), which is the fly homolog of the abundant and ubiquitous mammalian CypA isoform, and two additional cyclophilin family members, a CypB homolog found in the secretory pathway (DroCyp-B; B. Ondek, unpublished data) and CYP-2, which shares over 90% amino acid identity with CYP-1 over the cyclophilin core domain (D. Schultz, unpublished data).

■ References

Baker, E.K., Colley, N.J., and Zuker, C.S. (1994). The cyclophilin homolog NinaA functions as a chaperone, forming a stable complex in vivo with its protein target rhodopsin. EMBO J. **13**, 4886–4895.

Colley, N.J., Baker, E.K., Stamnes, M.A., and Zuker C.S. (1991). The cyclophilin homolog ninaA is required in the secretory pathway. Cell **67**, 255–263.

Larivee, D.C., Conrad, S., Stephenson, R.S., and Pak, W.L. (1981). Mutation that selectively affects rhodopsin concentration in the peripheral photoreceptors of *Drosophila* melanogaster. J. Gen. Physiol. **78**, 521–545.

Ondek, B., Hardy, R.W., Baker, E.K., Stamnes, M.A., Shieh, B.-H., and Zuker, C.S. (1992). Genetic dissection of cyclophilin function. J. Biol. Chem. **267**, 16460–16466.

Pak, W.L. (1979). *Study of Photoreceptor Function Using Drosophila Mutants.* Elsevier North-Holland, Inc., New York.

Schneuwly, S., Shortridge, R.D., Larrivee, D.C., Ono, T., Ozaki, M., and Pak, W.L. (1989). *Drosophila ninaA* gene encodes an eye-specific cyclophilin (cyclosporin A binding protein). Proc. Natl. Acad. Sci. USA **86**, 5390–5394.

Shieh, B.-H., Stamnes, M.A., Seavello, S., Harris, G.L., and Zuker, C.S. (1989). The *ninaA* gene required for visual transduction in *Drosophila* encodes a homologue of cyclosporin A-binding protein. Nature **338**, 67–70.

Stamnes, M.A., Shieh, B.-H., Chuman, L., Harris, G.L., and Zuker, C.S. (1991). The cyclophilin homolog ninaA is a tissue-specific integral membrane protein required for the proper synthesis of a subset of *Drosophila* rhodopsins. Cell **65**, 219–227.

Stephenson, R.S., O'Tousa, J., Scavarda, N.J., Randall, L.L., and Pak, W.L. (1983). *Drosophila* mutants with reduced rhodopsin content. In *Biology of Photoreceptors*, (D. Cosens and D. Vince-Price eds). Cambridge University Press, Cambridge, pp. 477–501.

Zuker, C.S., Cowman, A.F. and Rubin, G.M. (1985). Isolation and structure of a rhodopsin gene from D. melanogaster. Cell **40**, 851–858.

■ *Charles S. Zuker:*
Howard Hughes Medical Institute and Departments of Biology and Neuroscience
University of California, San Diego
9500 Gilman Drive, CMM 355
La Jolla, CA 92093-0649, USA
Tel. 1 619 534 5528
Fax. 1 619 534 8510
E-mail: charles@flyeye.ucsd.edu

■ *Suzanne L. Rutherford:*
Howard Hughes Medical Institute
The University of Chicago
5841 S. Maryland Avenue, MC1028
AMB-101
Chicago, IL 60637, USA
Tel. 1 773 702 0868
Fax. 1 773 702 7254
E-mail: srutherf@midway.uchicago.edu

Mammalian cyclophilin A

Cyclophilin A (CypA) was discovered as the major cytosolic ligand for the immunosuppressive drug cyclosporin A (CsA). It is abundantly expressed in all tissues and cell types tested. It has high sequence similarity to the secreted and/or endoplasmic reticulum-resident forms CypB and -C, to the mitochondrial form CypD and is more distantly related to another cytosolic form named Cyp40. The cyclophilins belong, together with the structurally unrelated FK506-binding proteins and parvulins, to the larger family of peptidyl–prolyl cis–trans isomerases, or rotamases, that facilitate the folding and trafficking of target proteins.

■ Alternative names

sp18 (Sherry *et al.*, 1992), CyP-18 (Galat, 1993), NUC18 (Montague *et al.*, 1994).

■ Isolation

CypA was originally isolated from calf thymus as the major ligand for CsA using molecular filtration, isoelectric focusing, phenylsepharose chromatography and weak cation exchange HPLC (Handschumacher *et al.*, 1984).

■ *CYPA* gene and sequence

The human cDNA was isolated from a T cell library and predicts a protein sequence of 165 amino acids (Haendler *et al.*, 1987; GenBank accession number Y00052). Comparison with the protein sequence of other soluble cyclophilins shows the existence of a highly conserved central region and of specific N- and C-terminal parts.

Cyclophilin cDNAs from phylogenetically distantly related organisms, such as yeasts, bacteria and plants, have been isolated and their sequence was found to be highly conserved between species (see Galat, 1993 and GenBank accession numbers therein).

The human *CYPA* gene is organized in five exons and four introns (Haendler and Hofer, 1990; GenBank accession number X52851) and maps to chromosome 7p11.2-p13 (Willenbrink *et al.*, 1995). Several, most probably inactive, pseudogenes have been identified and four of them have been mapped to chromosomes 3, 10, 14 and 18.

Cyclophilin A protein

Cyclophilin A forms an eight-stranded antiparallel β barrel closed on each side by an amphipathic helix (Ke, 1992; Spitzfaden et al., 1994). Aromatic and hydrophobic residues form a compact hydrophobic core inside the barrel.

CypA is essentially cytosolic but nuclear localization has also been reported (Le Hir et al., 1995). Anti-peptide antibodies able to discriminate between CypA and the closely related CypB have been generated (Allain et al., 1995).

Biological activities of cyclophilin A

Cyclophilins catalyse the *cis–trans* isomerization of peptidyl–prolyl bonds, a rate-limiting step in protein folding (Fischer et al., 1989; Takahashi et al., 1989; Gething and Sambrook, 1992). CsA blocks this activity. Roles as a nuclease in apoptosis (Montague et al., 1994) and as a chemoattractant for leukocytes (Sherry et al., 1992; Xu et al., 1992) have also been reported.

CypA complexed to CsA binds to and inhibits the Ca^{2+}/calmodulin-dependent protein phosphatase, calcineurin (Liu et al., 1991), thereby preventing the dephosphorylation and nuclear translocation of NF-AT, a transcription factor involved in IL-2 gene activation (McCaffrey et al., 1993). Furthermore, CypA interacts with Hsp90 (Nadeau et al., 1993), the p55 gag protein of HIV (Luban et al., 1993) and the YY1 transcription factor (Yang et al., 1995).

Biological regulation of cyclophilin A

Protein activity: The isomerase activity of cyclophilins is blocked by CsA.

Gene regulation: CypA is synthesized constitutively and at fairly constant high levels. A higher expression has been observed in tumours (Koletsky et al., 1986) and in lymphocytes treated with concanavalin A (Richards et al., 1992).

Mutagenesis studies with cyclophilin A

The Trp[121]Ala mutant of CypA has only low affinity for CsA and impaired enzymatic activity (Bossard et al., 1991). The four Cys residues are not essential for CsA binding or for catalytic activity (Liu et al., 1990). Mutation of Arg-69 lessens the inhibitory effect of the CypA–CsA complex on calcineurin but not the isomerase activity (Etzkorn et al., 1994). Conversely, mutation of Arg-54, Phe-60 or His-126 dramatically reduces the isomerase activity without affecting the calcineurin inhibitory activity in the presence of CsA (Zydowsky et al., 1992).

References

Allain, F., Boutillon, C., Mariller, C., and Spik, G. (1995). Selective assay for CyPA and CyPB in human blood using highly specific anti-peptide antibodies. J. Immunol. Meth. **178**, 113–120.

Bossard, M.J., Koser, P.L., Brandt, M., Bergsma, D.J., and Levy, M.A. (1991). A single Trp121 to Ala121 mutation in human cyclophilin alters cyclosporin A affinity and peptidyl-prolyl isomerase activity. Biochem. Biophys. Res. Commun. **176**, 1142–1148.

Etzkorn, F.A., Chang, Z.Y., Stolz, L.A., and Walsh, C.T. (1994). Cyclophilin residues that affect noncompetitive inhibition of the protein serine phosphatase activity of calcineurin by the cyclophilin/cyclosporin A complex. Biochemistry **33**, 2380–2388.

Fischer, G., Wittmann-Liebold, B., Lang, K., Kiefhaber, T., and Schmid, F.X. (1989). Cyclophilin and peptidyl-prolyl cis-trans isomerase are probably identical proteins. Nature **337**, 476–478.

Galat, A. (1993). Peptidyl proline cis-trans-isomerases: immunophilins. Eur. J. Biochem. **216**, 689–707.

Gething, M.-J. and Sambrook, J. (1992). Protein folding in the cell. Nature **355**, 33–45.

Haendler, B. and Hofer, E. (1990). Characterization of the human cyclophilin gene and of related processed pseudogenes. Eur. J. Biochem. **190**, 477–482.

Haendler, B., Hofer-Warbinek, R., and Hofer, E. (1987). Complementary DNA for human T-cell cyclophilin. EMBO J. **6**, 947–950.

Handschumacher, R.E., Harding, M.W., Rice, J., and Drugge, R.J. (1984). Cyclophilin: a specific cytosolic binding protein for cyclosporin A. Science **226**, 544–547.

Ke, H. (1992). Similarities and differences between human cyclophilin A and other beta-barrel structures. Structural refinement at 1.63 Å resolution. J. Mol. Biol. **228**, 539–550.

Koletsky, A.J., Harding, M.W., and Handschumacher, R.E. (1986). Cyclophilin: distribution and variant properties in normal and neoplastic tissues. J. Immunol. **137**, 1054–1059.

Le Hir, M., Su, Q., Weber, L., Woerly, G., Granelli-Piperno, A., and Ryffel, B. (1995). In situ detection of cyclosporin A: evidence for nuclear localization of cyclosporine and cyclophilins. Lab. Invest. **73**, 727–733.

Liu, J., Albers, M.W., Chen, C.-M., Schreiber, S.L., and Walsh, C.T. (1990). Cloning, expression, and purification of human cyclophilin in Escherichia coli and assessment of the catalytic role of cysteines by site-directed mutagenesis. Proc. Natl. Acad. Sci. USA **87**, 2304–2308.

Liu, J., Farmer, J.D., Lane, W.S., Friedman, J., Weissman, I., and Schreiber, S.L. (1991). Calcineurin is a common target of cyclophilin-cyclosporin A and FKBP-FK506 complexes. Cell **66**, 807–815.

Luban, J., Bossolt, K.L., Franke, E.K., Kalpana, G.V., and Goff, S.P. (1993). Human immunodeficiency virus type 1 gag protein binds to cyclophilins A and B. Cell **73**, 1067–1078.

McCaffrey, P.G., Perrino, B.A., Soderling, T.R. and Rao, A. (1993). NF-ATp, a T lymphocyte DNA-binding protein that is a target for calcineurin and immunosuppressive drugs. J. Biol. Chem. **268**, 3747–3752.

Montague, J.W., Gaido, M.L., Frye, C., and Cidlowski, J.A. (1994). A calcium- dependent nuclease from apoptotic rat thymocytes is homologous with cyclophilin. Recombinant cyclophilins A, B, and C have nuclease activity. J. Biol. Chem. **269**, 18877–18880.

Nadeau, K., Das, A., and Walsh, C.T. (1993). Hsp90 chaperonins possess ATPase activity and bind heat shock transcription factors and peptidyl prolyl isomerases. J. Biol. Chem. **268**, 1479–1487.

Richards, F.M., Milner, J., and Metcalfe, S. (1992). Inhibition of the serine/ threonine protein phosphatases PP1 and PP2A in lymphocytes: effect on mRNA levels for interleukin-2, IL-2R alpha, krox-24, p53, hsc70 and cyclophilin. Immunology **76**, 642–647.

Sherry, B., Yarlett, N., Strupp, A., and Cerami, A. (1992). Identification of cyclophilin as a proinflammatory secretory product of lipopolysaccharide- activated macrophages. Proc. Natl. Acad. Sci. USA **89**, 3511–3515.

Spitzfaden, C., Braun, W., Wider, G., Widmer, H., and Wuthrich, K. (1994). Determination of the NMR solution structure of the cyclophilin A-cyclosporin A complex. J. Biomol. NMR **4**, 463–482.

Takahashi, N., Hayano, T., and Suzuki, M. (1989). Peptidyl-prolyl cis-trans isomerase is the cyclosporin A-binding protein cyclophilin. Nature **337**, 473–475.

Willenbrink, W., Halaschek, J., Schuffenhauer, S., Kunz, J., and Steinkasserer, A. (1995). Cyclophilin A, the major intracellular receptor for the immunosuppressant cyclosporin A, maps to chromosome 7p11.2-p13: four pseudogenes map to chromosomes 3, 10, 14, and 18. Genomics **28**, 101–104.

Xu, Q., Leiva, M.C., Fischkoff, S.A., Handschumacher, R.E., and Lyttle, C.R. (1992). Leukocyte chemotactic activity of cyclophilin. J. Biol. Chem. **267**, 11968–11971.

Yang, W.M., Inouye, C.J., and Seto, E. (1995). Cyclophilin A and FKBP12 interact with YY1 and alter its transcriptional activity. J. Biol. Chem. **270**, 15187–15193.

Zydowsky, L.D., Etzkorn, F.A., Chang, H.Y., Ferguson, S.B., Stolz, L.A., Ho, S.I., and Walsh, C.T. (1992). Active site mutants of human cyclophilin A separate peptidyl-prolyl isomerase activity from cyclosporin A binding and calcineurin inhibition. Protein Sci. **1**, 1092–1099.

■ *Bernard Haendler:*
Research Laboratories of Schering AG
13342 Berlin, Germany
Tel. 49 30 468 12669
Fax. 49 30 468 16707
E-mail: Bernard.Haendler@Schering.DE

Vertebrate cyclophilin B

Like the prototypic cyclophilin A (CypA), cyclophilin B (CypB) is a peptidyl–proline isomerase (PPIase) which can be inhibited by binding to the immunosuppressant cyclosporin A (CsA). Both cyclophilins have a similar primary sequence and tertiary structure. CypB, like CypA, is highly conserved throughout evolution, and ubiquitously expressed in tissues at high levels. CypB is distinguished from CypA by N- and C-terminal extensions. The N-terminal extension contains a signal sequence that directs CypB to the secretory pathway and the cell surface. Additionally, CypB is secreted.

■ Isolation of cyclophilin B

Cyclophilin B was identified by low stringency hybridization screens of cDNA libraries using the cyclophilin A sequence as a probe (Bergsma et al., 1991; Hasel et al., 1991; Price et al., 1991). Cyclophilin B was also identified as a protein in human breast milk (Spik et al., 1991) and in the media of COS cells transfected with a chicken cDNA library (Caroni et al., 1991). The mouse CypB homologue was found by purification of a membrane-associated protein induced during differentiation (Schumacher et al., 1991). Finally, a related protein (and perhaps homologue) was identified in rats depleted of sodium (Iwai and Inagami, 1990). The chromosomal location of the human CypB homologue has been mapped to chromosome 15 (Peddada et al., 1992).

■ Cyclophilin B homologues

The CypB homologues are listed in Table 1 with their GenBank accession numbers.

Table 1. Cyclophilin B homologues

Species	Alternative name	Accession	Reference
Human	cyclophilin B (hCypB)	M60857	(Price et al., 1991)
	secreted cyclophilin-like protein (SCYLP)	M63573	(Spik et al., 1991)
	hCPH2	M60457	(Hasel et al., 1991)
	hCyp2	M80254	(Bergsma et al., 1991)
Cow	cyclophilin B (bCypB)	D14073	(Carrello et al., unpublished data)
Mouse	cyclophilin-S1 (mCyp-S1)	X58990	(Schumacher et al., 1991)
	mCPH2	M60456	(Hasel et al., 1991)
Chicken	S-cyclophilin	M63553	(Caroni et al., 1991)

Cyclophilin B protein

The CypB cDNA encodes a 208 amino acid basic polypeptide (pI = 9.3) with a predicted molecular weight of 22.7 kDa. The CypB protein sequence is 65% identical to that of CypA. Like CypA, CypB is a peptidyl–proline isomerase, but functions less efficiently than CypA in in vitro PPIase assays using an artificial peptide substrate (Price et al., 1991) and has a different peptide substrate specificity compared with CypA (Bergsma et al., 1991). CypB also binds CsA which inhibits CypB's PPIase activity. Although the NMR and X-ray crystal structures clearly show the structural conservation of the CsA binding pocket in CypB (Neri et al., 1991; Mikol et al., 1994), CypB has a lower affinity for CsA relative to CypA (Bergsma et al., 1991; Price et al., 1991; Spik et al., 1991).

The CypB protein is phylogenetically conserved, as evidenced by the 86% sequence identity shared between the human and chicken homologues. CypB is abundant and expressed in all tissues, albeit at lower levels compared with CypA (Hasel et al., 1991; Price et al., 1991; Schumacher et al., 1991). The 25 amino acid hydrophobic N-terminal sequence of CypB is essential for targeting to the secretory pathway (Schumacher et al., 1994). CypB is proteolytically processed to remove its signal sequence, yielding a protein with an approximate molecular weight of 20.3 kD (Schumacher et al., 1991; Spik et al., 1991). It has been argued that the C-terminal extension restricts CypB to a part of the endoplasmic reticulum (ER) called the calciosome (Arber et al., 1992). Two lines of evidence argue against exclusive localization to this compartment. First, transiently transfected human CypB in baby hamster kidney cells distributes throughout the secretory pathway, on the cell surface, and is secreted into the media (Price et al., 1994). Second, both the human and the chick CypB homologue were isolated as secreted products (Caroni et al., 1991; Spik et al., 1991).

Biological activities of cyclophilin B

Convincing evidence for the identification of true physiological target(s) of CypB has been lacking. Recently, there has emerged some intriguing evidence for various candidates. Because collagen is proline-rich and secreted, the collagen molecule is a logical candidate. CsA slows the folding of procollagen in chick embryo tendon fibroblasts (Steinmann et al., 1991). Furthermore, a complex of Hsp47, CypB, and procollagen I has been demonstrated. CsA pretreatment reduces the amount of CypB bound to the procollagen and reduces procollagen secretion (Smith et al., 1995). Another strategy to identify a CypB target used CypB in a two-hybrid screen (Bram and Crabtree, 1994). A CypB-specific binding protein, CAML, was identified. Overexpression of this membrane protein, CAML, increases calcium concentration and facilitates T cell activation. Whether CypB interacts with CAML from the lumen of the secretory pathway remains to be demonstrated.

Biological or genetic interactions

In two cases, both CypA and CypB strongly interact with the same protein in vitro, but in all likelihood it is the cyclophilin's subcellular localization that dictates which interaction is truly physiologically relevant. In a two-hybrid screen, the HIV gag protein was shown to interact with both CypA and CypB (Luban et al., 1993), yet only CypA has been found within the HIV virions that assemble in the cytoplasm (Franke et al., 1994; Thali et al., 1994). Similarly, CypB, like CypA, when complexed to CsA will bind and inhibit the phosphatase calcineurin. The CypB–CsA complex was found to be a 2–5-fold better inhibitor of calcineurin's phosphatase activity compared with CypA (Swanson et al., 1992). This high affinity in vitro interaction was exploited in mapping the immunophilin/immunosuppressant binding site on calcineurin (Milan et al., 1994). But since CypB resides within the secretory compartment, it most likely does not interact with the cytoplasmic calcineurin protein. Oddly, overexpression of not only CypA, but also CypB increases T cell sensitivity to CsA (Bram, et al., 1993). Clearly, the workings of CypB in the cell are not fully understood at this time.

References

Arber, S., Krause, K.H., and Caroni, P. (1992). S-cyclophilin is retained intracellularly via a unique COOH-terminal sequence and colocalizes with the calcium storage protein calreticulin. J. Cell. Biol. **116**, 113–125.

Bergsma, D.J., Eder, C., Gross, M., Kersten, H., Sylvester, D., Appelbaum, E., Cusimano, D., Livi, G.P., McLaughlin, M.M., Kasyan, K., and et al. (1991). The cyclophilin multigene family of peptidyl-prolyl isomerases. Characterization of three separate human isoforms. J. Biol. Chem. **266**, 23204–23214.

Bram, R.J. and Crabtree, G.R. (1994). Calcium signalling in T cells stimulated by a cyclophilin B-binding protein. Nature **371**, 355–358.

Bram, R.J., Hung, D.T., Martin, P.K., Schreiber, S.L., and Crabtree, G.R. (1993). Identification of the immunophilins capable of mediating inhibition of signal transduction by cyclosporin A and FK506: roles of calcineurin binding and cellular location. Mol. Cell. Biol. **13**, 4760–4769.

Caroni, P., Rothenfluh, A., McGlynn, E., and Schneider, C. (1991). S-cyclophilin. New member of the cyclophilin family associated with the secretory pathway. J. Biol. Chem. **266**, 10739–10742.

Franke, E.K., Yuan, H.E., and Luban, J. (1994). Specific incorporation of cyclophilin A into HIV-1 virions. Nature **372**, 359–362.

Hasel, K.W., Glass, J.R., Godbout, M., and Sutcliffe, J.G. (1991). An endoplasmic reticulum-specific cyclophilin. Mol. Cell. Biol. **11**, 3484–3491.

Iwai, N. and Inagami, T. (1990). Molecular cloning of a complementary DNA to rat cyclophilin-like protein mRNA. Kidney Int. **37**, 1460–1465.

Luban, J., Bossolt, K.L., Franke, E.K., Kalpana, G.V., and Goff, S.P. (1993). Human immunodeficiency virus type 1 Gag protein binds to cyclophilins A and B. Cell **73**, 1067–1078.

Mikol, V., Kallen, J., and Walkinshaw, M.D. (1994). X-ray structure of a cyclophilin B/cyclosporin complex: comparison with cyclophilin A and delineation of its calcineurin-binding domain. Proc. Natl. Acad. Sci. USA **91**, 5183–5186.

Milan, D., Griffith, J., Su, M., Price, E.R., and McKeon, F. (1994). The latch region of calcineurin B is involved in both immunosuppressant-immunophilin complex docking and phosphatase activation. Cell **79**, 437–447.

Neri, P., Gemmecker, G., Zydowsky, L.D., Walsh, C.T., and Fesik, S.W. (1991). NMR studies of [U-13C]cyclosporin A bound to human cyclophilin B. FEBS Lett. **290**, 195–199.

Peddada, L.B., McPherson, J.D., Law, R., Wasmuth, J.J., Youderian, P., and Deans, R.J. (1992). Somatic cell mapping of the human cyclophilin B gene (PPIB) to chromosome 15. Cytogenet. Cell. Genet. **60**, 219–221.

Price, E.R., Zydowsky, L.D., Jin, M.J., Baker, C.H., McKeon, F.D., and Walsh, C.T. (1991). Human cyclophilin B: a second cyclophilin gene encodes a peptidyl-prolyl isomerase with a signal sequence. Proc. Natl. Acad. Sci. USA **88**, 1903–1907.

Price, E.R., Jin, M., Lim, D., Pati, S., Walsh, C.T., and McKeon, F.D. (1994). Cyclophilin B trafficking through the secretory pathway is altered by binding of cyclosporin A. Proc. Natl. Acad. Sci. USA **91**, 3931–3935.

Schumacher, A., Schroter, H., Multhaup, G., and Nordheim, A. (1991). Murine cyclophilin-S1: a variant peptidyl-prolyl isomerase with a putative signal sequence expressed in differentiating F9 cells. Biochim. Biophys. Acta **1129**, 13–22.

Schumacher, A., Westermann, B., Osborn, M., and Nordheim, A. (1994). The N-terminal signal peptide of the murine cyclophilin mCyP-S1 is required in vivo for ER localization. Eur. J. Cell Biol. **63**, 182–191.

Smith, T., Ferreira, L.R., Hebert, C., Norris, K., and Sauk, J.J. (1995). Hsp47 and Cyclophilin B traverse the endoplasmic reticulum with procollagen into pre-Golgi intermediate vesicles — A role for Hsp47 and cyclophilin B in the export of procollagen from the endoplasmic reticulum. J. Biol. Chem. **270**, 18323–18328.

Spik, G., Haendler, B., Delmas, O., Mariller, C., Chamoux, M., Maes, P., Tartar, A., Montreuil, J., Stedman, K., Kocher, H.P., and et al. (1991). A novel secreted cyclophilin-like protein (SCYLP). J. Biol. Chem. **266**, 10735–10738.

Steinmann, B., Bruckner, P., and Superti-Furga, A. (1991). Cyclosporin A slows collagen triple-helix formation in vivo: indirect evidence for a physiologic role of peptidyl-prolyl cis-trans-isomerase. J. Biol. Chem. **266**, 1299–1303.

Swanson, S.K., Born, T., Zydowsky, L.D., Cho, H., Chang, H.Y., Walsh, C.T., and Rusnak, F. (1992). Cyclosporin-mediated inhibition of bovine calcineurin by cyclophilins A and B. Proc. Natl. Acad. Sci. USA **89**, 3741–3745.

Thali, M., Bukovsky, A., Kondo, E., Rosenwirth, B., Walsh, C.T., Sodroski, J., and Gottlinger, H.G. (1994). Functional association of cyclophilin A with HIV-1 virions. Nature **372**, 363–365.

■ *E. Roydon Price and Frank McKeon:*
Cell Biology
Harvard Medical School
Boston, MA. 02115, USA
Tel. 1 617 432 0327
Fax. 1 617 432 1144
E-mail: eprice@warren.med.harvard.edu

Mammalian cyclophilin C

Cyclophilin C (CypC) is a member of a family of cyclosporin A (CsA)-binding proteins. Cyclophilins have peptidyl–prolyl isomerase (PPIase) activity and may play a role in the folding and chaperoning of proteins. The isomerase activity is inhibited by CsA binding. CypC is unique among the mammalian cyclophilins in its restricted tissue distribution, its unique pattern of expression in the kidney and its ability to interact with CypCAP (cyclophilin C-associated protein).

■ Isolation of cyclophilin C

CypC was originally identified as an IL-1-induced transcript from a murine bone marrow stromal cDNA library (GenBank accession number M74227; Friedman and Weissman, 1991). The human cDNA was subsequently isolated from a kidney cDNA library (GenBank accession number S71018; Schneider et al., 1994). A complex of CypC and the immunosuppressant drug CsA was shown to bind to and inhibit the calcium and calmodulin-regulated serine/threonine phosphatase activity of calcineurin (Friedman and Weissman, 1991; Liu et al., 1991). In T cells, this block of calcineurin activity results in an inhibition of IL-2 transcription which is a key element in the immunosuppressive effect of CsA (Clipstone and Crabtree, 1992; O'Keefe et al., 1992). Although CsA bound CypA, -B, and -C can complex with calcineurin in vitro, only CypA and -B are in the correct cellular compartments to interact with calcineurin in vivo (Bram et al., 1993).

■ Cyclophilin C sequence

The murine cDNA sequence predicts a protein of 212 amino acids with a predicted pI of 7.37 and molecular weight of c. 23 kDa (Friedman and Weissman, 1991). The human cDNA encodes a protein that is 90% identical at the amino acid level to the mouse homologue (Schneider et al., 1994). CypC shares a core region of homology with other cyclophilins and is most closely related to human CypB (77% identity over 165 amino acids) (Friedman and Weissman, 1991). Relative to CypA, the major cytosolic cyclophilins CypB and -C are flanked by N- and C-terminal extensions which, in part, direct these proteins to their unique subcellular compartments (Arber et al., 1992; Bram et al., 1993).

■ Cyclophilin C protein

The structure of CypC is an eight-stranded anti-parallel barrel with a hydrophobic core closed on each end by a helix (Ke et al., 1993). While CypC shares this structure with human CypA it differs from CypA in the position of three loops, as well having N- and C-terminal extensions that protrude from the barrel. These differences most likely account for the differences in cellular function, localization, and protein interaction of these molecules (Ke et al., 1993; Bram et al., 1993).

A number of monoclonal antibodies specific for mouse CypC are available (Friedman et al., 1994). Using indirect immunofluorescence and surface labeling, CypC has been found in the ER, Golgi, and at the cell surface (M. Trahey and I.L. Weissman, unpublished data). The mechanism of cell surface attachment is currently unknown although a glycosylphosphatidylinositol (GPI) linkage has been ruled out (M. Trahey and I. L. Weissman, unpublished data).

CypC differs from CypA and -B because it has a more restricted tissue distribution (Friedman et al., 1994). For example, by in situ hybridization, CypA and -B were shown to be expressed in thymus, spleen, and lymph nodes whereas CypC was not detected in these tissues (Friedman et al., 1994). CypC also differs from A and B in its unique pattern of expression in the kidney. By indirect immunofluorescence and in situ hybridization of kidney sections (Friedman et al., 1994) and by RT–PCR of dissected kidney tubules (Otsuka et al., 1994), CypC was found to be expressed at high levels in the proximal convoluted and proximal straight tubules, regions of the kidney that are sensitive to CsA toxicity.

■ Biological activities of cyclophilin C

Like other members of the cyclophilin family, CypC has PPIase activity that is inhibited by CsA (Friedman and Weissman, 1991). Cyclophilins have been shown to catalyse the cis–trans isomerization of peptidyl–prolyl bonds, a step that can be rate limiting in the folding of some proteins (Fischer et al., 1989; Takahashi et al., 1989). This enzyme activity, the localization to the secretory pathway, and its ability to associate with at least two proteins (see below) suggest a role for CypC in protein folding/chaperoning. However, this has not been demonstrated directly.

Cyp A, -B, and -C can act as Ca^{2+}/Mg^{2+}-dependent nucleases in vitro, raising the possibility that a cyclophilin may be involved in programmed cell death or apoptosis (Montague et al., 1994; 1997).

■ Biological interactions of cyclophilin C

CypC, but not CypA or -B, associates with a secreted glycoprotein, CypCAP (cyclophilin C-associated protein) and this association is inhibited by CsA (Friedman and Weissman, 1991; Friedman et al., 1993). CypCAP is the mouse homologue of the human Mac-2-binding protein/90K tumour-associated antigen (M. Trahey and I. L. Weissman, unpublished data) a protein found elevated in the serum of patients with breast and ovarian cancer (Scambia et al., 1988; Iacobelli et al., 1994) and AIDS (Natoli et al., 1991).

Although CypA, -B, and -C can associate in vitro with the gag protein of HIV-1, only CypA is found incorporated into HIV-1 virions and required for infectious virion formation (Franke et al., 1994).

■ References

Arber, S., Krause, K.-H., and Caroni, P. (1992). s-Cyclophilin is retained intracellularly via a unique COOH-terminal sequence and colocalizes with the calcium storage protein calreticulin. J. Cell Biol. **116**, 113–125.

Bram, R.J., Hung, D.T., Martin, P.K., Schreiber, S.L., and Crabtree, G.R. (1993). Identification of the immunophilins capable of mediating inhibition of signal transduction by cyclosporin A and FK506: roles of calcineurin binding and cellular location. Mol. Cell. Biol. **13**, 4760–4769.

Clipstone, N.A. and Crabtree, G.R. (1992). Identification of calcineurin as a key signalling enzyme in T-lymphocyte activation. Nature **357**, 695–697.

Fischer, G., Wittmann-Liebold, B., Lang, K., Kiefhaber, T., and Schmid, F.X. (1989). Cyclophilin and peptidyl-prolyl cis-trans isomerase are probably identical proteins. Nature **337**, 476–478

Franke, E.K., Yuan, H.E.H., and Luban, J. (1994). Specific incorporation of cyclophilin A into HIV-1 virions. Nature **372**, 359–362.

Friedman, J. and Weissman, I. (1991). Two cytoplasmic candidates for immunophilin action are revealed by affinity for a new cyclophilin: one in the presence and one in the absence of CsA. Cell **66**, 799–806

Friedman, J., Trahey, M., and Weissman, I. (1993). Cloning and characterization of cyclophilin C-associated protein: a candidate natural cellular ligand for cyclophilin C. Proc. Natl. Acad. Sci. USA **90**, 6815–6819.

Friedman, J., Weissman, I., Friedman, J., and Alpert, S. (1994). An analysis of the expression of cyclophilin C reveals tissue restriction and an intriguing pattern in the mouse kidney. Am. J. Pathol. **144**, 1247–1256.

Iacobelli, S., Sismondi, P., Giai, M., D'Egidio, M., Tinari, N., Amatetti, C., DiStefano, P., and Natoli, C.(1994). Prognostic value of a novel circulating serum 90K antigen in breast cancer. Br. J. Cancer **69**, 172–176.

Ke, H., Zhao, Y., Luo,F., Weissman, I., and Friedman, J. (1993). Crystal structure of murine cyclophilin C complexed with the immunosuppressive drug cyclosporin A. Proc. Natl. Acad. Sci. USA **90**, 11850–11854 .

Liu, J., Farmer, J.D., Jr., Lane, W.S., Friedman, J., Weissman, I., and Schreiber, S. L. (1991). Calcineurin is a common target of cyclophilin-cyclosporin A and FKBP-FK506 complexes. Cell **66**, 807–815.

Montague, J.W., Gaido, M.L., Frye, C. and Cidlowski, J.A. (1994). A calcium-dependent nuclease from apoptotic rat thymocytes is homologous with cyclophilin. J. Biol. Chem. **269**, 18877–18880.

Montague, J.W., Hughes, F.M. Jr., and Cidlowski, J.A. (1997). Native recombinant cyclophilins A, B and C degrade DNA independently of peptidylprolyl cis-trans-isomerase activity. J. Biol. Chem. **272**, 6677–6684.

Natoli, C., Iacobelli, S., and Ghinelli, F. (1991). Unusually high level of a tumour-associated antigen in the serum of human immunodeficiency virus-seropositive individuals. J. Infect. Dis. **164**, 616–617.

O'Keefe, S.J., Tamura, J., Kincaid, R.L., Tocci, M.J., and O'Neill, E.A. (1992). FK-506- and CsA-sensitive activation of the interleukin-2 promoter by calcineurin. Nature **357**, 692–694.

Otsuka, M., Terada, Y., Yang, T., Nonoguchi, H., Tomita, K., and Marumo, F. (1994). Localization of cyclophilin A and cyclophilin C mRNA in murine kidney using RT-PCR. Kidney Int. **45**, 1340–1345

Scambia, G., Benedetti Panici, P., Baiocchi, G., Perrone, L., Iacobelli, S., and Mancuso, S. (1988). Measurement of a monoclonal antibody defined antigen (90K) in the sera of patients with ovarian cancer. Anticancer Res. **8**, 761–764.

Schneider, H., Charara, N., Schmitz, R., Wehrli, S., Mikol, V., Zurini, M.G.M., Quesniaux, V.F.J., and Movva, N.R.(1994). Human cyclophilin C: primary structure, tissue distribution, and determination of binding specificity for cyclosporins. Biochemistry **33**, 8218–8224.

Takahashi, N., Hayano, T., and Suzuki, M. (1989). Peptidyl-prolyl cis-trans isomerase is the cyclosporin A binding protein. Nature **337**, 473–475.

■ *Meg Trahey and Irving L. Weissman:*
Stanford University
Department of Pathology
Stanford University
Stanford, CA 94305, USA
Tel. 1 415 723 7389
Fax. 1 415 498 6255
E-mail: traheym@selway.umt.edu

Mammalian cyclophilin D

Cyclophilin D is a member of the highly conserved family of cyclosporin A (CsA)-sensitive peptidyl–prolyl cis–trans isomerases (PPIases). The human protein was orginally deduced from a cDNA isolated by cross-hybridization with a human cyclophilin A-coding sequence. This cDNA predicts a 207 amino acid, c. 22 kDa protein, which, like the ER-localized cyclophilins B and C, contains an N-terminal hydrophobic extension used to direct subcellular localization. Cyclophilin D mRNA is ubiquitously expressed but at much lower levels relative to the other cyclophilins. Mammalian cyclophilin D is localized to mitochondria where it appears likely that it catalyses the folding of imported proteins. Cyclophilin D may also be involved in regulating the function of the CsA-sensitive mitochondrial inner membrane ion channel.

■ Alternative names

For human cyclophilin D: hCyP3 (Bergsma et al., 1991), Cyp-D (Schneider et al., 1994).

■ cDNA isolation, sequence and mRNA expression

Human cyclophilin D is encoded by a cDNA (GenBank accession number M80254) cloned from a Jurkat T cell cDNA library using the cyclophilin A-coding sequence as the probe (Bergsma et al., 1991). The 841 bp nucleotide sequence contains a 621 bp open reading frame predicting a protein of 207 amino acids. Southern blot analysis of human genomic DNA probed with the cyclophilin D cDNA reveals a band pattern consistent with the existence of a single copy gene (Bergsma et al., 1991). Nothern blot analysis of Jurkat T cell mRNA reveals a band of c. 2 kb, indicating that the cDNA clone lacks an appreciable amount of 3′-untranslated sequence. Published tissue distribution analyses of cyclophilin A–D mRNAs have indicated that, whereas each isoform is ubiquitously expressed, cyclophilin D mRNA levels are relatively quite low in all human cell and tissue types examined, i.e. cyclophilin A mRNA is approximately 5–10, >10 and 10–50 times more abundant than the mRNAs for cyclophilins B, C and D, respectively (Bergsma et al., 1991; Schneider et al., 1994).

■ Cyclophilin D protein sequence

The deduced human cyclophilin D protein shows a significant degree of overall sequence homology to other cyclophilin family members. The protein includes a hydrophobic, 43 amino acid N-terminal extension (relative to cytoplasmic cyclophilin A) that appears to serve as a signal sequence to target the protein to a subcellular organelle, most likely mitochondria (see below; Bergsma et al., 1991). In fact, this putative signal sequence contains several basic residues not present in the signal sequences of the ER-localized cyclophilins B and C that are characteristic of leader sequences used to target proteins to mitochondria (Gavel and von Heijne, 1990). In the regions of sequence overlap, and excluding the N-terminal signal sequences, human cyclophilin D shares 74, 65 and 53% identity with human cyclophilins A (Harding et al., 1986), B (Price et al., 1991) and C (Schneider et al., 1994), respectively.

Cyclophilin D protein

Western blot analysis of whole-cell extracts using cyclophilin D-specific antisera has detected a protein of c. 18 kDa, signficantly smaller than the size of the protein predicted from its cDNA (Bergsma et al., 1991). This size is consistent with post-translational processing and removal of the N-terminal signal sequence, and further supports its role in targeting the protein to a subcellular location. Cyclophilin D has been shown to be associated with a particulate fraction of Jurkat T cells containing membranes and organelles (Bergsma et al., 1991). Furthermore, mitochondrial matrix cyclophilins from rat liver (18.6 kDa; Connern and Halestrap, 1992) and bovine heart (19 kDa; Inoue et al., 1993) have been purified, partially sequenced and found to be highly homologous to human cyclophilin D. For example, alignment of the N-terminal amino acid sequence of native mature rat liver mitochondrial cyclophilin (determined for 19 residues) shows 15 identities to a region of human cyclophilin D starting at amino acid 41 (see Fig. 1 below). This suggests that the first 40 amino acids of the human protein probably delineate the signal sequence (Connern and Halestrap, 1992; Kay, 1992). Taken together, these data strongly suggest that the cloned human cyclophilin D is a mitochondrial cyclophilin.

Biological activity of cyclophilin D

Expression of full length (22 kDa) human cyclophilin D protein has been achieved in Escherichia coli (Bergsma et al., 1991); the protein required solubilization prior to purification. A truncated, more soluble 18 kDa form of human cyclophilin D lacking the 43 N-terminal amino acids has also been expressed and purified (Bergsma et al., 1991). Each of the purified proteins exhibits PPIase activity that is inhibitable by CsA (with complete inhibition at 500 nM). Using the tetrapeptide chromogenic substrate succinyl–Ala–Ala–Pro–Phe–4-nitroanilide, truncated cyclophilin D was found to possess a specific activity comparable to human recombinant cyclophilin A (23 versus 22 $\mu M^{-1}s^{-1}$). However, CsA is a weaker inhibitor of human cyclophilin D (K_i of 8 ± 3 nM) than of human cyclophilin A (K_i of 1.6 ± 0.4 nM), consistent with the protein's lower affinity for the drug. The CsA inhibition constants compare favorably between human recombinant 22 kDa cyclophilin D and both the mature 19 kDa bovine heart mitochondrial cyclophilin (K_i of 6.9 ± 1.4 nM) (Inoue et al., 1993) and mature 18.6 kDa rat mitochondrial cyclophilin (K_i of 3.6 ± 0.6 nM) (Connern and Halestrap, 1992).

Genetic studies of multiple cyclophilin gene homologs in yeast (Koser et al., 1991; McLaughlin et al., 1992), combined with biochemical studies of the available human recombinant cyclophilins using synthetic CsA analogs (Bergsma et al., 1991; Schneider et al., 1994), suggest that cyclophilin D is not the primary cellular target of the drug. Given that the cyclophilin PPIases evolved to carry out specific cellular functions unrelated to their role as mediators of the action of CsA, what then is the likely cellular function of cyclophilin D? The mitochondrial cyclophilins from Saccharomyces cerevisiae (Davis et al., 1992; McLaughlin et al., 1992) and Neurospora crassa (Tropschug et al., 1988) catalyse protein folding in the mitochondria and appear to act in concert with the molecular chaperones Hsp60 and Hsp70 (Matouschek et al., 1995; Rassow et al., 1995). CsA has been shown to affect directly the rate of folding of imported mitochondrial preproteins in these systems, rather than block protein translocation across the membrane (Matouschek et al., 1995; Rassow et al., 1995). Thus, mammalian cyclophilin D is predicted to play a similar role in mitochondrial protein folding.

Finally, cyclophilin D may also be involved in regulating the function of the mitochondrial inner membrane ion channel, or permeability transition pore, which appears to regulate Ca^{2+} homeostasis. In rat liver mitochondria, this membrane channel is inhibited by nanomolar concentrations of CsA and is found to associate with the homologous mitochondrial matrix cyclophilin (Nicolli et al., 1996).

References

Bergsma, D.J., Eder, C., Gross, M., Kersten, H., Sylvester, D., Appelbaum, E., Cusimano, D., Livi, G.P., McLaughlin, M.M., Kasyan, K., Poter, T.G., Silverman, C., Dunnington, D., Hand, A., Prichett, W.P., Bossard, M.J., Brandt, M., and Levy, M.A. (1991). The cyclophilin multigene family of peptidyl-prolyl isomerases. Characterization of three separate human isoforms. J. Biol. Chem. **266**, 23204–23214.

Connern, C.P. and Halestrap, A.P. (1992). Purification and N-terminal sequencing of peptidyl-prolyl cis-trans-isomerase from rat liver mitochondrial matrix reveals the existence of a distinct mitochondrial cyclophilin. Biochem. J. **284**, 381–385.

Davis, E.S., Becker, A., Heitman, J., Hall, M.N., and Brennan M.B. (1992). A yeast cyclophilin gene essential for lactate metabolism at high temperature. Proc. Natl. Acad. Sci. USA **89**, 11169–11173.

Gavel, Y. and von Heijne, G. (1990). Cleavage-site motifs in mitochondrial targeting peptides. Protein Eng. **4**, 33–37.

Harding, M., Handschumacher, R.E., and Speicher, D.W. (1986). Isolation and amino acid sequence of cyclophilin. J. Biol. Chem. **261**, 8547–8555.

```
hCyP3  MLALRCGSRWLGLLSVPRSVPLRLPAARACSKGSGDPSSSSSSGNPLVYLDVDANGKPL
                                              *** ******** * * **
rat mitochondrial cyclophilin                 SSSQNPLVYLDVGADGQPL
```

Figure 1. Comparison of the N-terminal sequences of hCyP3 and rat mitochondrial cyclophilin.

Inoue, T., Yoshida, Y., Isaka, Y., and Tagawa, K. (1993). Isolation of mitochondrial cyclophilin from bovine heart. Biochem. Biophys. Res. Commun. **190**, 857–863.

Kay, J.E. (1992). Mitochondrial cyclophilins. Biochem. J. **288**, 1074–1075.

Koser, P.L., Bergsma, D.J., Cafferkey, R., Eng, W.K., McLaughlin, M.M., Ferrara, A., Silverman, C., Kasyan, K., Bossard, M.J., Johnson, R.K., Porter, T.G., Levy, M.A., and Livi, G.P. (1991). The CYP2 gene of Saccharomyces cerevisiae encodes a cyclosporin A-sensitive peptidyl-prolyl cis-trans isomerase with an N-terminal signal sequence. Gene **108**, 73–80.

McLaughlin, M.M., Bossard, M.J., Koser, P.L., Cafferkey, R., Morris, R.A., Miles, L.M., Strickler, J., Bergsma, D.J., Levy, M.A., and Livi, G.P. (1992). The yeast cyclophilin multigene family: purification, cloning and characterization of a new isoform. Gene **111**, 85–92.

Matouschek, A., Rospert, S., Schmid, K., Glick, B.S., and Schatz, G. (1995). Cyclophilin catalyzes protein folding in yeast mitochondria. Proc. Natl. Acad. Sci. USA. **92**, 6319–6323.

Nicolli, A., Basso, E., Petronelli, V., Wenger, R.M., and Bernardi, P. (1996). Interactions of cyclophilin with the mitochondrial inner membrane and regulation of the permeability transition pore, a cyclosporin A-sensitive channel. J. Biol. Chem. **271**, 2185–2192.

Price, E.R., Zydowsky, L.D., Jin, M., Baker, C.H., McKeon, F.D., and Walsh, C.T. (1991). Human cyclophilin B: a second cyclophilin gene encodes a peptidyl-prolyl isomerase with a signal sequence. Proc. Natl. Acad. Sci. USA **88**, 1903–1907.

Rassow, J., Mohrs, K., Koidl, S., Barthelmess, I.B., Pfanner, N., and Tropschug, M. (1995). Cyclophilin 20 is involved in mitochondrial protein folding in cooperation with molecular chaperones Hsp70 and Hsp60. Mol. Cell. Biol. **15**, 2654–2662.

Schneider, H., Charara, N., Schmitz, R., Wehrli, S., Mikol, V., Zurini, M.G.M., Quesniaux, V.F.J., and Movva, N.R. (1994). Human cyclophilin C: primary structure, tissue distribution, and determination of binding specificity for cyclosporins. Biochemistry **33**, 8218–8224.

Tropschug, M., Nicholson, D.W., Hartl, F.-U., Köhler, H., Pfanner, N., Wachter, E., and Neupert, W. (1988). Cyclosporin A-binding protein (cyclophilin) of Neurospora crassa. One gene codes for both the cytosolic and the mitochondrial forms. J. Biol. Chem. **263**, 14433–14440.

■ *George P. Livi[1] and Derk J. Bergsma:[2]*
[1]Department of Comparative Genetics,
[2]Department of Molecular Genetics, UE0548
SmithKline Beecham Pharmaceuticals
P.O. Box 1539
King of Prussia, PA 19406, USA
Tel. 1 610 270 7717/7610 (G. P. L./ D. J. B.)
Fax. 1 610 270 7962
E-mail: George_P_Livi@sbphrd.com and Derk_J_Bergsma@sbphrd.com

Mammalian cyclophilin 40 (Cyp40)

Cyp40 is a member of the abundant cyclophilin protein family. It is a peptidyl–prolyl cis–trans isomerase and binds and is inhibited by the immunosuppressive drug cyclosporin A (CsA). Cyp40 is a component of the unactivated steroid receptor complex, an association that is mediated by heat shock protein 90 (Hsp90), which is also found to exist in dimeric complexes with Cyp40.

■ Alternative names

Estrogen receptor binding cyclophilin (ERBC; Ratajczak et al., 1993).

■ Cyclophilin 40 sequence

The open reading frame in a human pancreatic islet cell Cyp40 cDNA clone (Kieffer et al., 1993; GenBank accession number L11667) is 1110 nucleotides in length. Alignments of the deduced amino acid sequence reveal two structural domains (Fig. 1). The N-terminal region (amino acid residues 9–185) shares about 60% identity with Cyp18, the cyclophilin prototype, and includes the postulated CsA binding and isomerase activity sites. The C-terminal domain of the protein has no similarities to any known Cyp, but amino acids 220–369 are about 30% identical to the C-terminus of FKBP59 (FK506 binding protein 59), a 59 kDa immunophilin of the FK506 binding

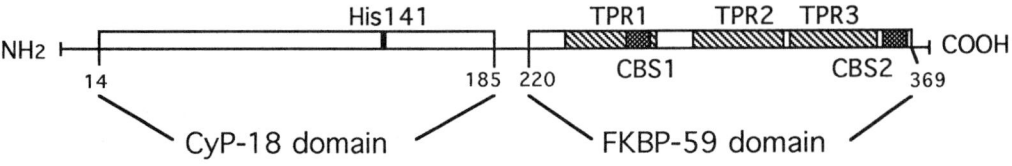

Figure 1. Linear structure of Cyp40 protein. The locations of the Cyp18 domain, which includes the isomerase-active site and the CsA-binding site with His-141, and the FKBP59 domain, which binds to Hsp90 and includes three TPR domains (TPR1-3) and two calmodulin-binding sites (CBS1-2), are shown.

class (Fretz et al., 1991). The Hsp90-binding region, a TPR domain and two calmodulin-binding sites are located in this domain.

■ Cyclophilin 40 protein and mutagenesis studies

The protein consists of a single peptide chain of 370 amino acids with a predicted molecular mass of 40.6 kDa (not including the initial methionine). Isoelectric focusing, done with the bovine counterpart, indicated two isoforms with pI values of 5.3 and 5.5. Cyp40 is enzymatically active and accelerates the *cis–trans* interconversion of proline peptide bonds. This activity is inhibited by CsA with an IC_{50} of 1.8 µM, which is about 60-fold higher than for Cyp18. It has been shown, by site-directed mutagenesis, that this weak affinity for CsA is caused by His-141 that replaces a Trp shown to be critical for CsA binding and highly conserved in other Cyp proteins that have high affinity for CsA (Hoffmann et al., 1995).

Truncation studies have shown that the two domains of Cyp40 are structurally and functionally independently organized (Hoffmann et al., 1995). The two domains of Cyp40, when independently expressed, retain their specific activities compared with the intact Cyp40. The N-terminal Cyp18-like domain is an active isomerase and inhibitable by CsA whereas the C-terminal domain binds Hsp90. In the intact Cyp40, CsA binding does not affect the affinity for Hsp90 and, conversely, Hsp90 has no apparent effects on isomerase activity.

■ Cyclophilin 40 antibodies

Polyclonal rabbit antibodies were raised against the intact Cyp40 protein and against two Cyp40 sequence-derived peptides. The antibodies are commercially available (Affinity Bioreagents, Colorado) and are effective in western blots, immunofluorescence and immunoprecipitation.

■ Biological features of cyclophilin 40

Cyp40 was originally isolated by the means of CsA-affinity chromatography (Kieffer et al., 1992) from calf brain homogenates. Northern and immuno analysis indicate a widespread tissue distribution and a nucleolar association within the cell (J.K. Owens-Grillo, personnel communication). Although there is no evidence for a bacterial Cyp40 homolog, a yeast counterpart with a M_r of 45 kDa has recently been identified (Chang et al., 1994; see entry p. 376).

Cyclophilins have been discovered as receptors for the immunosuppressive drug CsA (Handschumacher et al., 1984). The drug–cyclophilin complex interacts with and inhibits the phosphatase calcineurin (Liu et al., 1991), subsequently preventing IL-2 gene expression and T cell proliferation. Because of the weak affinity for CsA, Cyp40 is apparently not involved in immunosuppression. However, several biological interactions have been found and characterized for Cyp40. It is a subcomponent of steroid receptors (Ratajczak et al., 1993), it forms a dimeric complex with Hsp90 (Hoffmann and Handschumacher, 1995) and it binds calmodulin in a Ca^{2+}-dependent fashion (Ratajczak et al., 1995). The presence of Cyp40 in steroid receptor complexes is well established. It binds exclusively to the non-DNA-binding form of the receptors and may be involved in receptor activation and/or trafficking (Johnson and Toft, 1994; Renoir et al., 1995). Cyp40 has also been found to form dimeric complexes with Hsp90, which is by itself a component of steroid receptors and may mediate Cyp40's presence there (Owens-Grillo et al., 1995). However, the function of the immunophilin–Hsp complex does not relate exclusively to steroid receptors, but has a more common role in many cell types where it may function to alter the conformation of selected proteins through either a catalytic or stoichiometric mechanism.

■ References

Chang, H.C.J. and Lindquist, S. (1994). Conversation of hsp90 macromolecular complexes in *Saccharomyces cerevisiae*. J. Biol. Chem. **269**, 24983–24988.

Fretz, H., Albers, M.W., Galat, A., Standaert, R.F., Lane, W.S., Burakoff, S.J., Bierer, B.E., and Schreiber, S.L. (1991). Rapamycin and FK506 binding proteins (immunophilins). J. Am. Chem. Soc. **113**, 1409–1411.

Handschumacher, R.E., Harding, M.W., Rice, J., Drugge, R.J., and Speicher, D.W. (1984). Cyclophilin: A specific cytosolic binding protein for cyclosporin A. Science **226**, 544–546.

Hoffmann, K. and Handschumacher, R.E. (1995). Cyclophilin-40: evidence for a dimeric complex with hsp90. Biochem. J. **307**, 5–8.

Hoffmann, K., Kakalis, L.T., Anderson, K.S., Armitage, I.M., and Handschumacher, R.E. (1995). Human cyclophilin-40: expression and the effect of the H141W mutation on catalysis and cyclosporin A binding. Eur. J. Biochem. **299**, 188–193.

Johnson, J.L. and Toft, D.O. (1994). A novel chaperone complex for steroid receptors involving heat shock proteins, immunophilins, and p23. J. Biol. Chem. **269**, 24989–24993.

Kieffer, L., Thalhammer, T., and Handschumacher, R.E. (1992). Isolation and characterization of a 40 kDa cyclophilin-related protein. J. Biol. Chem. **267**, 5503–5507.

Kieffer, L., Seng, T.W., Li, W., Osterman, D.G., Handschumacher, R.E. and Bayney, R.M. (1993). Cyclophilin-40, a protein with homology to the p59 component of the steroid receptor complex. J. Biol. Chem. **268**, 12303–12310.

Liu, J., Farmer, J.D., Lane, W.S., Friedman, J., Weissman, I., and Schreiber, S.L. (1991). Calcineurin is a common target of cyclophilin-cyclosporin A and FKBP-FK506 complexes. Cell **66**, 807–815.

Owens-Grillo, J.K., Hoffmann, K., Hutchinson, K.A., Yem, A.W., Deibel, M.R., Handschumacher, R.E., and Pratt, W.B. (1995). The cyclosporin A-binding immunophilin CyP-40 and the FK506-binding immunophilin hsp56 (FKBP52) bind to a common site on hsp90 and exist in independent cytosolic heterocomplexes with the untransformed glucocorticoid receptor. J. Biol. Chem. **270**, 20479–20484.

Ratajczak, T., Carrello, A., Mark, P.J., Warner, B.J., Simpson, R.J., Moritz, R.L., and House, A.K. (1993). The cyclophilin component of the unactivated estrogen receptor contains a tetratricopeptide repeat domain and shares identity with p59 (FKBP59). J. Biol. Chem. **268**, 13187–13192.

Ratajczak, T., Carrello, A., and Minchin R.F. (1995). Biochemical and calmodulin binding properties of estrogen receptor binding cyclophilin expressed in *Escherichia coli*, Biochem. Biophys. Res. Commun. **209** (1), 117–125.

Renoir, J.-M., Mercier-Bodard, C., Hoffmann, K., Le Bihan, S., Ning, Y.-M., Sanchez, E.R., Handschumacher, R.E., and Baulieu, E.-E. (1995). Cyclosporin A potentiates the dexamethasone-induced mouse mammary tumor virus-chloramphenicol acetyltransferase activity in LMCAT cells: a possible role for different heat shock protein-binding immunophilins in glucocorticosteroid receptor-mediated gene expression. Proc. Natl. Acad. Sci. USA **92**, 4977–4981.

■ *Kai Hoffmann and Robert E. Handschumacher:*
Department of Pharmacology
Yale University School of Medicine
333 Cedar Street
New Haven, CT 06520, USA
Tel. 1 203 785 4385
Fax. 1 203 785 7670
E-mail: HandschuRE@maspo2.mas.yale.edu

12b

FK506-Binding Proteins (FKBPs)

Three-dimensional structure of FKBPs

The techniques of X-ray diffraction and NMR spectroscopy have defined the structures of FKBP12 in both the free and drug-complexed forms.

■ FK506-binding proteins

The mammalian cytosolic FKBP12 protein was discovered through its ability to bind the immunosuppressive agents FK506 (tacrolimus) and rapamycin (sirolimus) (Harding et al., 1989; Siekierka et al., 1989). The FKBP family, for FK506 binding proteins, is growing rapidly, and the best characterized members are the mammalian proteins FKBP12, FKBP13 (Jin et al., 1991), FKBP25 (Galat et al., 1992), and FKBP59 (Tai et al., 1992). Since the immunosuppressive effects of FK506 and rapamycin are moderated by FKBP12, it has received by far the most scrutiny. FKBP12 is a peptidyl–prolyl isomerase (PPIase), and it catalyses isomerization via a 'distortion' mechanism (Rosen et al., 1990; Fischer et al., 1993). However, the involvement of FKBP12 in the immunosuppressive effects of FK506 and rapamycin is not associated with the inhibition of its PPIase activity (Schreiber, 1991). Immunosuppression is a gain, not a loss of function phenomenon and involves inhibition of calcineurin by the FKBP12–FK506 complex (Liu et al., 1991) and of FRAP (also TOR1, TOR2, and RAFT) by the FKBP12–rapamycin complex (Cafferkey et al., 1993; Kunz et al., 1993; Brown et al., 1994; Sabers et al., 1995).

■ Three-dimensional structure of FKBP12 and complexes

High-resolution X-ray crystal structures of FKBP12–FK506 (1.4 Å) (Van Duyne et al., 1991a), FKBP12–rapamycin (1.7 Å) (Van Duyne et al., 1991b), and FKBP12 (2.3 Å) (Wilson et al., 1995) have all been determined. A solution NMR structure has been reported for FKBP12 (Michnick et al., 1991; Moore et al., 1991). All of these studies show the overall fold seen in Fig. 1a. FKBP12 has a five-stranded antiparallel ß sheet wrapping around a short a helix. The protein core is composed exclusively of hydrophobic residues, and many of them are highly conserved among FKBPs found in different organisms. The FK506-binding pocket is a deep hydrophobic cavity between the α helix and the interior wall of the ß sheet. The binding pocket, which is also the catalytic pocket, is lined with completely conserved aromatic residues. A tryptophan (Trp-59) forms the bottom of the pocket, and phenylalanines (Phe-46 and Phe-99) and tyrosines (Tyr-26 and Tyr-82) form the sides. The pocket is flanked by three loops (Fig. 1a): a bulge in ß5 (the 40S loop), a 20 residue loop connecting ß2 to ß3 (the 80S loop), and the loop connecting ß5 to the α helix (the 50S loop). Each loop appears to have a different function. The 40S loop, which extends away from the binding pocket, has a nuclear localization site in FKBP25 (Galat et al., 1992), and therefore may help localize the protein. The 50S loop functions as a recognition site through interactions resembling antiparallel ß strands with the ligands (Fig. 1b) (Clardy, 1995), and the 80S loop mediates protein–protein interactions in the inhibitory complexes of FKBP12–FK506 and FKBP12–rapamycin (Rosen et al., 1993; Yang et al., 1993). Atomic details of the interactions between FKBP12–FK506–calcineurin and FKBP–rapamycin–FRAP are provided by crystal structures of these triple complexes

(Griffith et al., 1995; Kissinger et al., 1995; Choi et al., 1996). In both structures, FK506 and rapamycin utilize their distinctive effector domains to interact with two different targets, calcineurin or FRAP.

■ Implications for PPIase activity

The high-resolution structural studies provide some insights into the mechanism of PPIase activity. Since there are no nucleophiles in the active site of FKBP12, addition-elimination mechanisms are not possible. The generally accepted view has FK506 and rapamycin mimic a peptide substrate where the pipecolinyl ring serves as a proline analog and the twisted α-keto amide group mimics the PPIase transition state (Rosen et al., 1990; Van Duyne et al., 1993). The hydrophobicity of the pocket appears to play a general role in catalysis and several hydrogen bonds and polar interactions have been suggested to play important specific roles (Fischer et al., 1993). Perhaps the most unusual suggestion is that the twisted carbonyl group is stabilized by interactions with the edges of three conserved aromatic residues (Van Duyne et al., 1993). Unlike cyclophilin, which forms complexes with a variety of peptides, FKBPs have not been reported to bind ligands composed exclusively of amino acids.

■ Other FKBPs

Other members of the FKBP family have been characterized. All FKBP family members contain FKBP12 homologous domains with similar PPIase activity; however, they differ in cellular localization and their ability to form inhibitory complexes with calcineurin and FRAP. High-resolution crystal structures for FKBP13 (Schultz et al., 1994) and the FKBP12 homologous domain of FKBP25 (Liang et al., 1995) are available, and they have conformations and binding pockets essentially identical to that of FKBP12. However, they have major differences in the loop regions.

■ References

Brown, E.J., Albers, M.W., Shin, T.B., Ichikawa, K., Keith, C.T., Lane, W.S., and Schreiber, S.L. (1994). A mammalian protein targeted by G1-arresting rapamycin-receptor complex. Nature **369**, 756–758.

Cafferkey, R., Young, P.R., Mclaughlin, M.M., Bergsma, D.J., Koltin, Y., Sathe, G.M., Faucette, L., Eng, W.K., and Livi, G.P. (1993). Dominant missense mutations in a novel yeast protein related to mammalian phosphatidylinositol 3-kinase and VPS34 abrogate rapamycin cytotoxicity. Mol. Cell. Biol. **13**, 6012–6023.

Choi, J., Chen, J., Schreiber, S.L., and Clardy, J. (1996). Rapamycin's role in binding FKBP12 and FRAP. Science **273**, 239–242.

Clardy, J. (1995). The chemistry of signal transduction. Proc. Natl. Acad. Sci. USA **92**, 56–61.

Fischer, S., Michnick, S.W., and Karplus, M. (1993). A mechanism for rotamase catalysis by the FK506 binding protein. Biochemistry **32**, 13830–13837.

Galat, A., Lane, W.S., Standaert, R.F., and Schreiber, S.L. (1992). A rapamycin-selective, 25-kDa immunophilin. Biochemistry **31**, 2427–2434.

Griffith, J.P., Kim, J.L., Kim, E.E., Sintchak, M.D., Thomson, J.A., Fitzgibbon, M.J., Fleming, M.A., Caron, P.R., Hsiao, K., and Navia, M.A. (1995). X-ray structure of calcineurin inhibited by the immunophilin-immunosuppressant FKBP12-FK506 complex. Cell **82**, 507–522.

Harding, M.W., Galat, A., Uehling, D.E., and Schreiber, S.L. (1989). A receptor for the immunosuppressant FK506 is a cis-trans peptidyl-prolyl isomerase. Nature **341**, 758–760.

Jin, Y.-J., Albers, M.W., Lane, W.S., Bierer, B.E., Schreiber, S.L., and Burakoff, S.J. (1991). Molecular cloning of a membrane-associated human FK506- and rapamycin-binding protein, FKBP13. Proc. Natl. Acad. Sci. USA **88**, 6677–6681.

Kissinger, C.R., Parge, H.E., Knighton, D.R., Lewis, C.T., Pelletier, L.A., Tempczyk, A., Kalish, V.J., Tucker, K.D., Showalter, R.E., Moomaw, E.W., Gastinel, L.N., Habuka, N., Chen, X., Maldonado, F., Barker, J.E., Bacquet, R., and Villafrance, J.E. (1995). Crystal structures of human calcineurin and the human FKBP12-FK506-calcineurin complex. Nature **378**, 641–644.

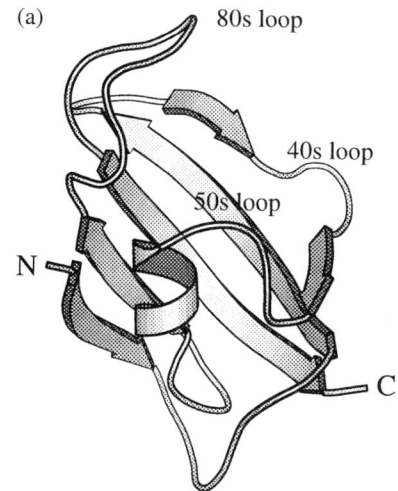

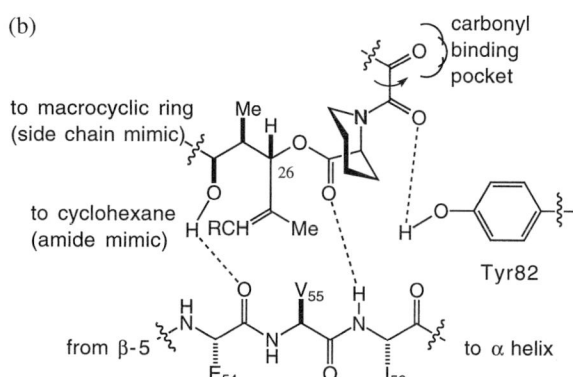

Figure 1. Three-dimensional structure of FKBP12. (a) Ribbon diagram of FKBP12. (b) Interactions between the 50S loop of FKBP12 and the macrocyclic ring of immunosuppressants.

Kunz, J., Henriquez, R., Schneider, U., Deuter-Reinhard, M., Movva, N.R., and Hall, M.N. (1993). Target of rapamycin in yeast TOR2 is an essential phosphatidylinositol kinase homolog required for G-1 progression. Cell **73**, 585–596.

Liang, J., Hung, D., Schreiber, S.L., and Clardy, J. (1996). The atomic structure of the 25 kDa FK506 binding protein complexed with rapamycin. J. Am. Chem. Soc. **118**, 1231–1232.

Liu, J., Farmer, J.D., Lane, W.S., Friedman, J., Weissman, I., and Schreiber, S.L. (1991). Calcineurin is a common target of cyclophilin-cyclosporin A and FKBP-FK506 complexes. Cell **66**, 807–815.

Michnick, S.W., Rosen, M.K., Wandless, T.J., Karplus, M., and Schreiber, S.L. (1991). Solution structure of FKBP, a rotamase enzyme and receptor for FK506 and rapamycin. Science **251**, 836–839.

Moore, J.M., Pettie, D.A., Fitzgibbon, M.J., and Thomson, J.A. (1991). Solution structure of the major binding protein for the immunosuppressant FK506. Nature **351**, 248–250.

Rosen, M.K., Standaert, R.F., Galat, A., Nakatuska, M., and Schreiber, S.L. (1990). Inhibition of FKBP rotamase activity by immunosuppressant FK506: twisted amide surrogate. Science **248**, 863–866.

Rosen, M.K., Yang, D., Martin, P.K., and Schreiber, S.L. (1993). Activation of an inactive immunophilin by mutagenesis. J. Am. Chem. Soc. **115**, 821–822.

Sabers, C.J., Martin, M.M., Brunn, G.J., Williams, J.M., Dumont, F.J., Wiederrecht, G., and Abraham, R.T. (1995). Isolation of a protein target of the FKBP12-rapamycin complex in mammalian cells. J. Biol. Chem. **270**, 815–822.

Schreiber, S.L. (1991). Chemistry and biology of the immunophilins and their immunosuppressive ligands. Science **251**, 283–287.

Schultz, L.W., Martin, P.K., Liang, J., Schreiber, S.L., and Clardy, J. (1994). Atomic structure of the immunophilin FKBP13-FK506 complex: insights into the composite binding surface for calcineurin. J. Am. Chem. Soc. **116**, 3129–3130.

Siekierka, J.J., Hung, S.H.Y., Poe, M., Lin, C.S., and Sigal, N.H. (1989). A cytosolic binding protein for the immunosuppressant FK506 has peptidyl-prolyl isomerase activity but is distinct from cyclophilin. Nature **341**, 755–757.

Tai, P.K.K., Albers, J.W., Chang, H., Faber, L.E., and Schreiber, S. (1992). Association of a 59-kilodalton immunophilin with the glucocorticoid receptor complex. Science **256**, 1315–1318.

Van Duyne, G.D., Standaert, R.F., Karplus, P.A., Schreiber, S.L., and Clardy, J. (1991a). Atomic structure of FKBP-FK506, an immunophilin-immunosuppressant complex. Science **251**, 839–842.

Van Duyne, G.D., Standaert, R.F., Schreiber, S.L., and Clardy, J. (1991b). Atomic structure of the rapamycin human immunophilin FKBP-12 complex. J. Am. Chem. Soc. **113**, 7433–7434.

Van Duyne, G.D., Standaert, R.F., Karplus, P.A., Schreiber, S.L., and Clardy, J. (1993). Atomic structures of the human immunophilin FKBP-12 complexes with FK506 and rapamycin. J. Mol. Biol. **229**, 105–124.

Wilson, K.P., Yamashita, M.M., Sintchak, M.D., Rotstein, S.H., Murcko, M.A., boger, J., Thomson, J.A., Fitzgibbon, M.J., Black, J.R., and Navia, M.A. (1995). Comparative X-ray structures of the major binding protein for the immunosuppressant FK506 (tacrolimus) in unliganded form and in complex with FK506 and rapamycin. Acta Cryst. **D51**, 511–521.

Yang, D., Rosen, M.K., and Schreiber, S.L. (1993). A composite FKBP12-FK506 surface that contacts calcineurin. J. Am. Chem. Soc. **115**, 819–820.

■ *Jun Liang and Jon Clardy:*
Department of Chemistry—Baker Laboratory
Cornell University
Ithaca, NY 14853-1301, USA
Tel. 1 607 255 7685
Fax. 1 607 255 1253
E-mail: jcc12@cornell.edu and jl40@cornell.edu

Mip

The group of the Mip (macrophage infectivity potentiator) proteins should be thought of as a subfamily of the FK506-binding proteins (FKBPs). Proteins that are combined in this group share a C-terminal region homologous to FKBP12, and an N-terminal extension of about 100 amino acid residues. Almost exclusively, the members are of prokaryotic origin, deriving from pathogenic facultative or obligate intracellular bacteria. The peptidyl–prolyl cis–trans isomerase (PPIase) activity is inhibited by FK506. The Mip protein of Legionella pneumophila is the archetype of the subfamily.

■ Alternative names

Legionella pneumophila FKBP25mem (Fischer, 1994), Lpn25, lpFKBP-25 (Kay, 1996), LpMip (Mo et al., 1995).

■ Further members of the Mip family

Legionella micdadei Mip (Bangsborg et al., 1991; accession number PIR2: A43596), (*L. micdadei* FKBP25,

LmFKBP25), *Legionella lonbeachae* Mip (accession number EMBL: X83036), *Chlamydia trachomatis* Mip (Lundemose et al., 1991; accession number PIR2: S28639), (*C. trachomatis* FKBP27, CtFKBP27, CtMip, Ctr27), *Chlamydia psittaci* Mip (accession number EMBL: L39892), *Coxiella burnetti* Mip (Mo et al., 1995; accession number PATCHX: U14170), (*C. burnetti* FKBP25, CbMip), *Escherichia coli* Mip (Horne and Young, 1995; accession number PATCHX: U18997 KK), (*E. coli* FKBP26, FkpA; see following entry p. 402), *Trypanosoma cruzi* Mip (Moro et al., 1995; accession number EMBL: X69655), (*T. cruzi* FKBP, TcMip), *Haemophilus influenzae* Mip (accession number GenBank: 42023).

■ Mip sequence

The *mip* gene from *Legionella pneumophila* Philadelphia I (accession: PIR3: S22665) encodes a protein of 233 amino acids in length. N-terminal sequencing of the mature protein indicated a signal sequence cleaved off at position 20 of the precursor protein chain (Fischer et al., 1992). The proteolytic specificity points to the signal peptidase I as the processing protease. The authentic protein in the mature state lacks any post-translational modification as evaluated by determining the molecular mass of 22 842 dalton using electrospray mass spectrometry (Schmidt et al., 1994). The Mip protein is composed of two domain-like regions. The C-terminal part of the protein is similar (29.6% identity) to the eukaryotic FKBP12 (see entry p. 420). A comparison of the 30 highly conserved amino acid residues, which were considered to form the consensus sequence of FKBPs (Trandinh et al., 1992), yields 67% identity. The N-terminal region comprises as much as 106 amino acid residues of the mature protein. The function of the latter is unclear and no motifs have been identified as yet. Secondary structure prediction for this region proposed that α helices are likely to be the predominant structures.

Among pathogenic and non-pathogenic *Legionella pneumophila* strains the Mip protein sequence is strongly conserved. Only a single amino acid exchange was found by examination of five strains (Ser-115/Ala for the pathogenic *L. pneumophila* Corby and the non-pathogenic isolate *L. pneumophila* U21S6). However, this variant does not have altered PPIase properties of the respective protein (Ludwig et al., 1994).

Until now, the *Chlamydia trachomatis* Mip [73% identity with the FKBP consensus sequence (Trandinh et al., 1992)] is the only member of the Mip subfamily of the FKBPs that had been shown to be modified post-translationally. *Chlamydia trachomatis* Mip contains a recognition site for signal peptidase II and is post-translationally modified by palmitic acid to become a lipoprotein in the mature state (Lundemose et al., 1993).

■ Mip protein

Legionella pneumophila causes severe pneumonia in humans, termed Legionnaires' disease (McDade et al., 1977). The Gram-negative faculative bacterium is able to survive and multiply in monocytes like lung macrophages, leading to tissue damage and quite often to lethal consequences. To initiate the identification of putative virulence factors of the *L. pneumophila* strains, several surface antigens of the bacterium were cloned and expressed in *E. coli* (Engleberg et al., 1984). A 24 kDa protein was identified on the cell surface of the host cells and chosen for further investigation after identification using anti-*Legionella* polyclonal antobodies. A *L. pneumophila* mutant strain defective in the expression of that protein was constructed. The mutant strain was phenotypically characterized by an 80-fold decrease of its ability to infect U937 cells and human alveolar macrophages. The phenomenon was abolished by re-introduction of an inact 24 kDa gene. This indicates a strong relation between the 24 kDa protein and the full infectivity of the *Legionella pneumophila*, which is the reason the gene is termed *mip* (macrophage infectivity potentiator) (Cianciotto et al., 1989; Cianciotto and Fields, 1992). The gene encoding this protein was sequenced. The deduced amino acid sequence contained 233 amino acid residues. At that time no similarity was found between the amino acid sequence of the *mip* gene product and other known bacterial proteins (Cianciotto et al., 1989). In 1990 the DNA sequences of various eukaryotic FKBPs were published (Standaert et al., 1990; Tropschug et al., 1990) and then the putative FKBP-like nature of Mip became obvious by sequence comparison (Tropschug et al., 1990). The *mip* gene of *L. pneumophila* Philadelphia I was cloned and expressed in *E. coli*. The recombinant protein was purified and the enzymatic properties of the PPIase were investigated (Fischer et al., 1992). The authentic Mip proteins were isolated from pathogenic and non-pathogenic *L. pneumophila* strains. A comparison of the recombinant with the authentic protein did not show any differences (Ludwig et al., 1994).

The molecular mass of the native proteins was determined by size exclusion chromatography to be 62 kDa. This indicates that the Mip protein is a dimer in solution (Schmidt et al., 1994). This conclusion was verified by cross-linking both Mip protein in solution and on the surface of the bacteria. Calculations based on small-angle X-ray solution scattering also indicated dimerization (Schmidt et al., 1995). The latter method allowed the authors to hypothesize that the contact region is located in the N-terminal part of the molecule.

Among the FKBPs the activity-related homo-oligomerization of Mip seems to be a unique property. In addition, it was found that the related *E. coli* FKBP22, *E. coli* FKBP26, and two Mip-like FKBPs from *Serratia marescens* are also dimeric enzymes (J. Rahfeld and G. Fischer, unpublished data).

The cationic character of human FKBP12 is retained in Mip because an isoelectric point of 9.8 was determined.

The close relationship between *L. pneumophila* Mip and the eukaryotic FKBPs, like human FKBP12, was confirmed by investigation of its PPIase properties (Fischer et al., 1992). The pattern of subsite specificity using peptide substrates (Suc–Ala–Xaa–Pro–Phe–4NA) resembles that of human FKBP12. The value of the second-order rate constant calculated for the dimeric protein is 2.2×10^6 M^{-1}s^{-1} (substrate Suc–Ala–Leu–Pro–Phe–4NA) (Schmidt et al., 1994). The PPIase activity is inhibited by FK506 but is not affected by CsA (Fischer et al., 1992). The value of the inhibition constant K_i is 211 nM. In comparison to human FKBP12, the K_i value of Legionella Mip for FK506 is increased more than 500-fold. However, considering the K_i of E. coli FKBP26 of 16.2 nM (J.-U. Rahfeld, unpublished data), insufficient inhibition by FK506 does not seem to be a general occurrence within the bacterial FKBP subfamily of Mip proteins.

Using site-directed mutagenesis a number of L. pneumophila FKBP25 mutant proteins were expressed in E. coli and the protein variants were purified to homogeneity. Replacements of the amino acid residues Asp-142L and Phe-202A were carried out. The values of the specificity constant k_{cat}/K_m decreased to 6% and 2.5% of the wild-type protein activity, respectively. The sensitivity toward FK506 was altered in the Asp-142L and the Phe-202A variants of the Mip protein to K_i values of 47 and 4 mM, respectively (Ludwig et al., 1994; B. Schmidt, unpublished results).

■ Mip function

The Mip protein is the first and only verified virulence factor of the genus Legionella. The Mip sequences are highly conserved amongst virulent and non-virulent strains of L. pneumophila and other Legionella species (Cianciotto et al., 1990).

The mechanism of how the Mip protein affects the pathogenic consequences of infection is unknown. Infection by L. pneumophila of monocytes like alveolar macrophages is characterized by inhibition of formation of the phagolysosome, suppression of the oxidative burst, and lysis of the host cell. The localisation of the protein loosely attached to the outer membrane and its secretion strongly suggest that the Mip protein interacts with the host cell. To investigate the role of the catalytic activity of L. pneumophila Mip in the poorly understood infection mechanism, Mip-negative L. pneumophila mutants were complemented with genes encoding wild-type and site-directed mutagenized Mip protein variants. The intracellular survival of L. pneumophila mutants was measured after introducing the appropriate genes. The mutants transfected with genes that encode for Mip proteins with strongly reduced PPIase activity fully restored intracellular survival that could not be distinguished from the wild-type organism (Wintermeyer et al., 1995). However, experiments involving transfection of enzymatically inactive Mip variants are still missing. There is still a lack of evidence for intracellular substrates or binding proteins in the bacterium itself as well as for these types of interaction with proteins of the host cells.

■ Mip regulation

The gene of the Mip protein mip is not localized within the two genetic loci dot (defect in organelle trafficking) (Berger and Isberg, 1993) and icm (intracellular multiplication) (Marre et al., 1992), already known to be involved in pathogenic mechanisms. These loci were identified by transcomplementation analysis of L. pneumophila mutants that were reduced in their ability to multiply in macrophages.

Little is known concerning the regulation of the mip gene. The nucleotide sequences immediately upstream of the mip gene of L. pneumophila and L. micdadei are very conserved. Inverted repeats represents a factor-independent transcription termination signal downstream of the genes (Bangsborg et al., 1991).

■ References

Bangsborg, J.M., Cianiotto, N.P., and Hinderson, P. (1991). Nucleotide sequence analysis of the Legionella micdadei mip gene, encoding a 30-kDa analog of the Legionella pneumophila Mip protein. Infect. Immun. **59**, 3836–3840.

Berger, K.H. and Isberg, R.R. (1993). Two distinct defects in intracellular growth complemented by a single genetic locus in Legionella pneumopnila. Mol. Microbiol. **7**, 7–19.

Cianciotto, N.P. and Fields, B.S. (1992). Legionella pneumophila mip gene potentiates intracellular infection of protozoa and human macrophages. Proc. Nalt. Acad. Sci. USA **89**, 5188–5191.

Cianciotto, N.P., Eisenstein, B.I., Mody, C.H., Toews, G.B., and Engleberg, N.C. (1989). A Legionella pneumophila gene encoding a species-specific surface protein potentiates initiation of intracellular infection. Infect. Immun. **57**, 1255–1262.

Cianciotto, N.P., Bangsborg, J.M., Eisenstein, B.I., and Engleberg, N.C. (1990). Identification of mip-like genes in the genus Legionella. Infect. Immun. **58**, 2912–2918.

Engleberg, N.C., Pearlman, E., and Eisenstein, B.I. (1984). Legionella pneumophila surface antigens cloned and expressed in Escherichia coli are translocated to the host cell surface and interact with specific anti-Legionella antibodies. J. Bacteriol. **160**, 199–203.

Fischer, G. (1994). Peptidyl-prolyl cis/trans isomerases and their effectors. Angew. Chem. Int. Ed. Engl. **33**, 1415–1436.

Fischer, G., Bang, H., Ludwig, B., Mann, K., and Hacker, J. (1992). Mip protein of Legionella pneumophila exhibits peptidyl prolyl cis/trans isomerase (PPIase) activity. Mol. Microbiol. **6**, 1375–1383.

Horne, S.M. and Young, K.D. (1995). Escherichia coli and other species of the Enterobacteriaceae encode a protein similar to the family of Mip-like FK506-binding proteins. Arch. Microbiol. **163**, 357–365.

Kay, J.E. (1996). Structure-function relationships in FK506-binding protein (FKBP) family of peptidylprolyl cis/trans isomerases. Biochem. J. **314**, 316–385.

Ludwig, B., Rahfeld, J., Schmidt, B., Mann, K., Wintermeyer, E., Fischer, G., and Hacker, J. (1994). Characterization of Mip proteins of Legionella pneumophila. FEMS Lett. **118**, 23–30.

Lundemose, A.G., Birkelund, S., Fey, S.J., Larson, P., and Christiansen, G. (1991). Chlamydia trachomatis contains a protein similar to the Legionella pneumophila gene product. Mol. Microbiol. **5**, 109–115.

Lundmose, A.G., Rouch, D.A., Penn, C.W., and Pearce, J.H. (1993). The Chlamydia trachomatis Mip-like protein is a lipoprotein. J. Bacteriol. **175**, 3669–3671.

Marre, A., Blander, S.J., Horwitz, M.A., and Shuman, H.A. (1992). Identification of a *Legionella pneumophila* locus required for intracellular multiplication in human macrophages. Proc. Natl. Acad. Sci. USA **89**, 9607–9611.

McDade, J.E., Shepard, C.C., Fraser, D.W., Tsai, T.R., Redus, M.A., and Dowdle, W.R. (1977). Legionnaires' disease: isolation of a bacterium and demonstration of its role in other respiratory disease. N. Engl. J. Med. **297**, 1197–1203.

Mo, Y.Y., Cianciotto, N.P., and Mallavia, L.P. (1995). Molecular cloning of a *Coxiella burnetti* gene encoding a macrophage infectivity potentiator (Mip) analogue. Microbiology **11**, 2861–2871.

Moro, A., Ruiz-Cabello, F., Fernadez-Cano, A., Stock, R.P., and Gronzalez, A. (1995). Secretion by *Trypanosoma cruzi* of peptidyl prolyl *cis/trans* isomerase involved in cell infection. EMBO J. **14**, 2483–2490.

Schmidt, B., Rahfeld, J., Schierhorn, A., Ludwig, B., Hacker, J., and Fischer, G. (1994). A homodimer represents an active species of the peptidyl prolyl *cis/trans* isomerase FKBP25mem from *Legionella pneumophila*. FEBS Lett. **352**, 185–190.

Schmidt, B., König, S., Svergun, D., Volkov, V., Fischer, G., and Koch, M.H.J. (1995). Small-angle X-ray solution scattering study on the dimerization of the FKBP25mem from *Legionella pneumophila*. FEBS Lett. **372**, 169–172.

Standaert, R.F., Galat, A., Verdine, G.L., and Schreiber, S.L. (1990). Molecular cloning and overexpression of the human FK506-binding protein. FKBP. Nature **346**, 671–674.

Trandinh, C.C., Pao, G.M., and Saier, M.H. (1992). Structural and evolutionary relationship among the immunophilins: two ubiquitous families of peptidyl prolyl *cis/trans* isomerases. FASEB J. **6**, 3410–3420.

Tropschug, M., Wachter, E., Mayer, S., Schonbrunner, E.R., and Schmidt, F.X. (1990). Isolation and sequence of an FK506-binding protein from *N. crassa* which catalyses protein folding. Nature **342**, 953–955.

Wintermeyer, E., Ludwig, B., Steiner, M., Schmidt, B., Fischer, G., and Hacker, J. (1995). Influence of site specifically altered Mip proteins on intracellular survival of *Legionella pneumophila* in eukaryotic cells. Infect. Immun. **63**, 4576–4683.

■ *Gunter Fischer and Jens-U. Rahfeld:*
Max-Planck-Research
Unit-Enzymology of Protein Folding
Kurt-Mothes-Str. 3
D-06120 Halle, Germany
Tel. 49 345 552 2801
Fax. 49 345 551 1972
E-mail: fischer@cis.biochemtech.uni-halle.de

Escherichia coli FkpA

FkpA is a periplasmic protein originally identified as a multicopy suppressor of bacteria accumulating unfolded proteins in the periplasm. From its deduced amino acid sequence and its in vitro function, FkpA belongs to the family of FK506-binding proteins (FKBPs) peptidyl–prolyl cis–trans isomerases (PPIases), also called rotamases.

■ Isolation

In *Escherichia coli*, misfolded proteins accumulating in the cell envelope, that is the periplasm or the outer membrane, induce a stress response that is controlled at the transcriptional level by the Eσ^E polymerase (Mecsas *et al.*, 1993; Raina *et al.*, 1995; Missiakas *et al.*, 1996a). Unfolding of proteins is quite pronounced in strains lacking major periplasmic catalysts such as the Dsb proteins, which are involved in the formation of disulfide bonds of exported proteins (Bardwell, 1994; see entries on the Dsb proteins, pp. 318–328). Misfolded proteins accumulate even more in a double *dsbC htrA* mutant triggering a more significant Eσ^E-dependent response (Missiakas *et al.*, 1996b). HtrA is a periplasmic protease (Lipinska *et al.*, 1988; Strauch and Beckwith, 1988) and DsbC is a thiol–disulfide exchange enzyme capable of reshuffling wrongly paired disulfides, which implies an unfoldase activity for this enzyme (Missiakas *et al.*, 1994; Zapun *et al.*, 1995). Genetic evidence has demonstrated that the increased Eσ^E-dependent response constitutively induced by slow folding species accumulating in the double *dsbC htrA* mutant can be reduced by overexpressing either SurA or FkpA (Missiakas *et al.*, 1996b). Both these proteins are PPIases belonging to two different classes, SurA to the parvulin family (Rudd *et al.*, 1995; see entry p. 436) and FkpA to the FKBP family (Missiakas *et al.*, 1996b).

■ *fkpA* gene and sequence

The *fkpA* gene maps in the region corresponding to the 75 minute area of the *E. coli* chromosome. There are two sources available for its sequence, one from the genome sequencing project (GenBank accession number U18997) and another obtained by Horne and Young (1995; accession number L28082). Sequence analysis of the *fkpA* gene predicts the presence of a 270 amino acid long open reading frame. The predicted amino acid sequence is very similar to the eukaryotic FKBPs, a class of well-characterized

PPIases which have been shown to be inhibited by the macrolide FK506 (Fischer and Schmid, 1990). Interestingly, the *fkpA* gene is located immediately downstream of the *slyD* gene and was sequenced because of this particular location (Horne and Young, 1995). SlyD is a cytoplasmic host protein required for bacteriopahge φX174 lysis and its sequence shows homology to the FKBPs as well (Roof *et al.*, 1994). Finally, transcription of the *fkpA* gene is regulated by the Eσ^E polymerase (A. Mathey-Dupraz, C. Dartigalongue, D. Missiakas and S. Raina, manuscript submitted).

Sequence analysis of the FkpA protein predicts two domains. Clearly, the C-terminal part is very homologous to the eukaryotic FKBP12 (see entry p. 420), being about 50% identical. The sequence of the N-terminal half of the protein does not show any clear homology with other proteins and contains many charged residues. This domain, although of unknown function, is also conserved among the other bacterial homologues of FkpA.

FkpA homologues are present in many pathogenic bacteria such as *Legionella pneumophila*. The *Legionella* FkpA homologue, designated Mip for macrophage infectivity potentiator, has PPIase activity which is thought to be crucial for virulence (Engleberg *et al.*, 1989; Fischer *et al.*, 1992; see entry p. 399).

FkpA protein

In vivo, the precursor form of FkpA can be chased into a mature species consistent with the presence of a predicted cleavable signal sequence after amino acid 25 (Missiakas *et al.*, 1996b). This mature species fractionates with the periplasmic proteins confirming its periplasmic location and arguing against a previously thought membrane localization. The putative rotamase activity of FkpA was assayed *in vitro* using the model substrate succinyl–Ala–Ala–Pro–Phe–4-nitroanilide. The estimated k_{cat}/K_m value is 90 mM^{-1}s^{-1}, a rate comparable with that of the FkpA homologue MipA from *L. pneumophila* (Fischer *et al.*, 1992). This finding was important since the genetic isolation of the *fkpA* gene was based on the ability of its product to prevent the accumulation of unfolded periplasmic proteins. FkpA is the first example of an FKBP-like rotamase localized in the periplasm. Interestingly, there are at least two other PPIases in the periplasm, RotA (Liu and Walsh, 1990) and SurA (Lazar and Kolter, 1996; Missiakas *et al.*, 1996b). RotA belongs to the cyclophilin subfamily of PPIases whereas SurA belongs to a third class of PPIases (parvulins) conserved again in both prokaryotes and eukaryotes (Rudd *et al.*, 1995).

Biological activity of FkpA

fkpA is not an essential gene (Horne and Young, 1995; Missiakas *et al.*, 1996b). Probably, the presence of multiple rotamases in the periplasm can compensate for the loss of one of them. However, in an *fkpA* null mutant, an increased activity of the σ^E regulon is observed (Missiakas *et al.*, 1996b) and transcription of the *fkpA* gene itself is induced by protein misfolding since it is regulated by Eσ^E (A. Mathey-Dupraz, C. Dartigalongue, D. Missiakas and S. Raina, manuscript submitted). This is an additional proof that FkpA plays an important role in protein folding. Nevertheless, it appears that the three periplasmic PPIases, FkpA, SurA and RotA, have different substrate specificities. SurA participates in the folding of subunits of outer membrane proteins (Lazar and Kolter, 1996; Missiakas *et al.*, 1996b). Although both *surA* and *fkpA* mutants exhibit phenotypes of bacteria with leaky membranes, only *surA* mutants present a drastic imbalance in outer membrane protein content (Missiakas *et al.*, 1996b). Also, the fact that, unlike *fkpA* mutants, *rotA* mutants by themselves do not induce a significant σ^E-dependent response argues that the two catalysts do not share many substrates in common.

References

Bardwell, J.C.A. (1994). Building bridges: disulphide formation in the cell. Mol. Microbiol. **14**, 199–205.

Engleberg, N.C., Carter, C., Weber, D.R., Cianciotto, N.P., and Eisenstein, B.I. (1989). DNA sequence of *mip*, a *Legionella pneumophila* gene associated with macrophage infectivity. Infect. Immun. **57**, 1263–1270.

Fischer, G. and Schmid, F.X. (1990). The mechanism of protein folding. Implications of *in vitro* refolding models for *de novo* protein folding and translocation in the cell. Biochemistry **29**, 2205–2212.

Fischer, G., Bang, H., Ludwig, B., Mann, K., and Hacker, J. (1992). Mip protein of *Legionella pneumophila* exhibits peptidyl-prolyl-*cis/trans* isomerase (PPIase) activity. Mol. Microbiol. **6**, 1375–1383.

Horne, S.M. and Young, K.D. (1995). *Escherichia coli* and other species of Enterobacteriaceae encode a protein similar to the family of Mip-like FK506-binding proteins. Arch. Microbiol. **163**, 357–365.

Lazar, S. and Kolter, R. (1996). SurA assists the folding of *Escherichia coli* outer membrane proteins. J. Bacteriol. **178**, 1770–1773

Lipinska, B., Sharma, S., and Georgopoulos, C. (1988). Sequence analysis and regulation of the *htrA* gene of *Escherichia coli*: a sigma 32-independent mechanism of heat-inducible transcription. Nucl. Acids Res. **16**, 10053–10067.

Liu, J. and Walsh, C.T. (1990). Peptidyl-prolyl *cis-trans* isomerase from *Escherichia coli*: a periplasmic homolog of cyclophilin that is not inhibited by cyclosporin A. Proc. Natl. Acad. Sci. USA **87**, 4028–4032.

Mecsas, J., Rouvière, P.E., Erickson, J.W., Donohue, T.J., and Gross, C.A. (1993). The activity of E, an *Escherichia coli* heat-inducible sigma factor, is modulated by expression of outer membrane proteins. Genes Dev. **7**, 2618–2628.

Missiakas, D., Georgopoulos, C. and Raina, S. (1994). The *Escherichia coli dsbC (xprA)* gene encodes a periplasmic protein involved in disulfide bond formation. EMBO J. **13**, 2013–2020.

Missiakas, D., Raina, S., and Georgopoulos, C. (1996a) The heat shock system. In *Regulation of gene expression in Escherichia coli*, (E.C.C. Lin and S.A. Lynch, eds). R.G. Landes Company, Austin TX. pp. 481–501.

Missiakas, D., Betton, J.-M., and Raina, S. (1996b). New components of protein folding in the extra-cytoplasmic compartments of *Escherichia coli*: SurA, FkpA and Skp/OmpH. Mol. Microbiol. **21**, 871–884.

Raina, S., Missiakas, D., and Georgopoulos, C. (1995). The *rpoE* gene encoding the E (24) heat shock sigma factor of *Escherichia coli*. EMBO J. **14**, 1043–1055.

Roof, W.D., Horne, S.M., Young, K.D., and Young, R. (1994). *slyD*, a host gene required for X174 lysis, is related to the FK506-binding protein family of peptidyl-prolyl *cis-trans* isomerases. J. Biol. Chem. **269**, 2902–2910.

Rudd, K.E., Sofia, H.J., Koonin, E.V., Plunkett, G., Lazar, S., and Rouvière, P.E. (1995). New family of peptidyl-prolyl isomerases. Trends Biochem. Sci. **20**, 12–14.

Strauch, K.L. and Beckwith, J. (1988). An *Escherichia coli* mutation preventing degradation of abnormal periplasmic proteins. Proc. Natl. Acad. Sci. USA **85**, 1576–1580.

Zapun, A., Missiakas, D., Raina, S., and Creighton, T.E. (1995). Structural and functional characterization of DsbC, a protein involved in disulfide bond formation in *Escherichia coli*. Biochemistry **34**, 5075–5089.

■ Satish Raina:
Université de Genève
Centre Médical Universitaire
1, Rue Michel-Servet 1211 Genève 4, Switzerland
Tel. 41 22 702 55 11
Fax. 41 22 702 55 02
E-mail: Satish.Raina@medecine.unige.ch

■ Dominique Missiakas:
CNRS UPR9027
31, Ch. J. Aiguier
13402 Marseille Cedex 20, France
Tel. 33 491 76 0359
Fax. 33 491 71 2124
E-mail: missiaka@ibsm.cnrs-mrs.fr

Escherichia coli trigger factor

In Escherichia coli *cells trigger factor (TF) is a 42 kDa molecular chaperone which also has peptidyl–prolyl cis–trans isomerase (PPIase) activity that is not inhibited by the immunosuppressants cyclosporin A and FK506. Thus, trigger factor is not an immunophilin. Its protein sequence shares only 18.2 and 28.3% overall identity with the conserved region of the amino acid sequences of cyclophilins and FKBPs, respectively. However, the enzymatic properties of trigger factor resemble those of the FKBPs. Trigger factor is found in* E. coli *as a free protein, in association with the 50S ribosome or in complexes with the molecular chaperone, GroEL. It was first discovered by its ability to bind to secreted proteins and facilitate their translocation into isolated membrane vesicles. When associated with GroEL, trigger factor can stimulate GroEL binding to many unfolded proteins in vitro. Formation of the TF–GroEL–substrate complex is also the rate-limiting step in the degradation of certain polypeptide substrates. In the ribosome, trigger factor is bound to nascent polypeptide chains. The relationship between the cochaperone function and PPIase activity is unclear at present.*

■ Alternative name

SecI (Tai *et al.*, 1992). SecI was proposed to have a regulatory function in protein translocation across bacterial cytoplasmic membranes (Tai *et al.*, 1992).

■ Trigger factor gene and sequence

Trigger factor was described originally as an abundant cytosolic protein in *E. coli* (Guthrie and Wickner, 1990). The *E. coli tig* gene (GenBank accession number M34066) encodes a protein (PIR accession number A36129) consisting of 432 amino acid residues. The *tig* gene is located at a map position of about 9.9 minutes on the *E. coli* genome (Rudd, 1992). The three genes located immediately clockwise of *tig* are *clpP–clpX–lon* (Gottesman *et al.*, 1993). Interestingly, trigger factor functions with the ClpP protease in intracellular protein degradation. *clpX* (Gottesman *et al.*, 1993) and *lon* genes (Chin *et al.*, 1988) both encode subunits of major cellular proteases. However, unlike *tig*, these three genes are induced by heat shock (Goff *et al.*, 1984; Kroh and Simon, 1990; Gottesman *et al.*, 1993). *bolA* is located immediately counter-clockwise of *tig*, however, *bolA* is expressed during entry into stationary phase (Aldea *et al.*, 1989) when TF levels are unchanged (Kandror and Goldberg, 1997). Recently *tig* genes were identified in *Haemophilus influenzae* (Fleischmann *et al.*, 1995) and *Campylobacter jejuni* (Griffiths *et al.*, 1995).

Comparison of the sequence of trigger factor with known peptidyl–prolyl *cis–trans* isomerases (PPIases) reveals extensive similarities with the FKBP class of PPIases, especially in the region of the active site. The isoleucine in position 56 of human FKBP12, which hydrogen bonds to FK506 (Michnick *et al.*, 1991; Van Duyne *et al.*, 1991), is conserved in both the various homologs of FKBP and the trigger factor proteins of *E. coli* (Callebaut and Mornon, 1995), *H. influenzae* (Fleischmann *et al.*, 1995), *C. jejuni* (Griffiths *et al.*, 1995) and *Mycoplasma genitalium* (Petersen, 1992). Within a 50 residue region surrounding Ile-56 of human FKBP12 or Ile-196 in *E. coli* trigger factor, these PPIases are 30% identical and 66% similar. The residues in the region of FKBP that are in the binding pocket or involved in hydrogen binding to FK506 are largely conserved or identical in trigger factor and other FKBPs (Callebaut and Mornon, 1995). However, trigger factor does not appear to be similar to either cyclophilins or parvulins, the other two classes of PPIases (see overview p. 359).

■ Trigger factor protein

The molecular weight of trigger factor, deduced from its nucleotide sequence, is 48 023 Da and its calculated pI is 4.5. However, on SDS–PAGE, the protein has an apparent molecular mass of about 55 kDa and a pI of about 5.2 (Kandror et al., 1995; Stoller et al., 1995). Determination of the molecular mass by electrospray mass spectrometry yielded a value of 48 228 Da (Stoller et al., 1995). These disparities indicate a probable, as yet unidentified, post-translational modification of trigger factor. The N-terminus of the protein, by microsequencing, matches the deduced sequence, including the initiator methionine residue (Kandror et al., 1995).

A purification method was first developed by Wickner and colleagues (Crooke et al., 1988), based in part on the ability of trigger factor to bind to the secreted polypeptide, proOmpA. After affinity chromatography using immobilized proOmpA protein, the eluted material was further purified by ion-exchange chromatography on S-sepharose Fast Flow cation exchange resin and MonoQ anion exchange FPLC (Crooke et al., 1988). This three-step procedure yields a 150-fold purification with a 54% yield. Alternative traditional chromatographic approaches for rapid isolation have been described more recently (Stoller et al., 1995). Trigger factor can be readily isolated from ribosomes by a LiCl salt wash followed by column chromatography (Lill et al., 1988; Stoller et al., 1995).

The trigger factor protein appears to function as a cochaperone in complex with GroEL, in which one trigger factor molecule is present per GroEL 14-mer (Kandror et al., 1995). This complex of trigger factor and GroEL can be immunoprecipitated with antibodies to either trigger factor or GroEL (Kandror et al., 1995). These complexes can be isolated by affinity columns using unfolded proteins such as fetuin or CRAG as ligands (Kandror et al., 1995).

■ Biological activity of trigger factor

The enzymatic activity of trigger factor was discovered in the course of identification of a ribosome-associated PPIase. Screening was performed with Suc–Ala–Phe–Pro–Phe–4-nitroanilide as a substrate. PPIase activity was found in fractions of crude 70S ribosomes, coupled 70S ribosomes, tightly coupled 70S ribosomes and 50S ribosomal subunits from E. coli. PPIase activity was not detected in samples of 30S ribosomal subunits (Stoller et al., 1995). Purification and N-terminal sequencing of the homogenous protein showed its identity with the known trigger factor.

A specificity constant k_{cat}/K_M of 7.4×10^5 $M^{-1}s^{-1}$ was determined by measurement with the substrate Suc–Ala–Phe–Pro–Phe–4-nitroanilide in the standard PPIase assay (Stoller et al., 1995). Comparative experiments with 10 different oligopeptide substrates show that trigger factor resembles FKBP in the subsite specificity concerning the P_1 position of the substrate. However, the PPIase activity of trigger factor is not affected by 100 µM concentrations of FK506 or cyclosporin A. Therefore, trigger factor should not be considered as an immunophilin.

The effect of trigger factor on the slow refolding reaction of a RNase T1 variant (Mücke and Schmid, 1994) was investigated. Trigger factor is extremely efficient in catalysis of the Pro-39-limited folding of the RNase T1 variant RCM Ser^{54}Gly/Pro^{55}Asn owing to its catalytic effect on the trans to cis isomerization of Pro-39 (Stoller et al., 1995). As much as a 12-fold higher concentration of the cytoplasmic E. coli cyclophilin is necessary to accomplish the same acceleration factor.

■ Function of trigger factor

Trigger factor was originally discovered as one of a few proteins that are important in translocation of proOmpA across the cytoplasmic membrane in a process called 'trigger factor cycle' (Crooke and Wickner, 1987; Lill et al., 1988; Lecker et al., 1989). Like the chaperones GroEL and SecB, trigger factor binds to the proOmpA and maintains it in a form competent for translocation in vitro (Lecker et al., 1989). The protein was purified using its ability to bind proOmpA, a precursor of the outer membrane protein A (OmpA) of E. coli (Crooke and Wickner, 1987). The protein was termed 'trigger factor', suggesting that it triggers the folding of proOmpA into a membrane assembly-competent form after denaturation with urea. However, it was later found that the trigger factor is not necessary for translocation of the proOmpA. Large reductions in the cellular content of trigger factor did not reduce the proOmpA translocation process in vivo (Guthrie and Wickner, 1990).

Trigger factor was also shown to be a component of the large ribosomal particle (Lill et al., 1988; Stoller et al., 1995) and it was suggested that trigger factor may bind to nascent peptide chains and assist protein translocation across membranes. Recent studies utilizing cross-linking agents have shown that trigger factor binds to many nascent polypeptide chains on E. coli ribosomes (Valent et al., 1995; Hesterkamp et al., 1996). Trigger factor then may play an important role in assisting the proper folding of nascent polypeptides, especially in light of its ability to stimulate the binding of GroEL to unfolded proteins (Kandror et al., 1995; Kandror et al., 1997). A small fraction of the trigger factor in E. coli is present in complexes with GroEL, and increasing the intracellular level of trigger factor enhances markedly the ability of GroEL to bind to a variety of unfolded proteins (Kandror et al., 1997). On the other hand, decreasing trigger factor content greatly reduces this binding. The trigger factor–GroEL complexes that were bound to a denatured protein (such as fetuin), could be released from the ligand upon addition of ATP. However, after release, the trigger factor and GroEL complex remained associated and could not be separated by gel filtration (Kandror et al., 1997).

Trigger factor appears to regulate the ability of GroEL to bind to denatured proteins by a different mechanism to that for GroEL phosphorylation. The phosphorylation of GroEL, which occurs during heat shock, also promotes its binding to unfolded proteins (Sherman and Goldberg, 1992, 1994). Phosphorylated GroEL binds to some proteins that trigger factor–GroEL complexes do not bind (Kandror et al., 1997). Furthermore, phosphorylation promotes substrate binding at high temperatures, while trigger factor enhances binding at low temperatures (Kandror et al., 1997).

Trigger factor is a rate-limiting factor in the rapid degradation of the fusion protein CRAG and the formation of trigger factor–GroEL–CRAG complexes is a critical step for CRAG degradation (Kandror et al., 1995). In vivo, large fragments of CRAG which are short-lived intermediates in proteolysis can be found in association with trigger factor–GroEL complexes (Kandror et al., in preparation). GroES is also necessary for CRAG breakdown and the generation of these intermediates (Kandror et al., 1994). Apparently, the ability of GroEL and GroES to release polypeptides in an unfolded form facilitates proteolytic attack on CRAG (O.A. Kandror, M.Y. Sherman and A.L. Goldberg, in preparation). In addition to favoring proteolysis, the ability of trigger factor to enhance GroEL's capacity to bind to unfolded proteins is likely to serve other functions in the cell. By stimulating GroEL's binding, trigger factor probably promotes the translocation or folding of certain polypeptides such as newly synthesized chains emerging from the ribosome, or perhaps stimulates the refolding of some damaged proteins

It is noteworthy that trigger factor, which has the capacity to bind strongly to unfolded proteins, also has PPIase activity which can catalyse the rate-limiting step for refolding of many proteins. Purified trigger factor catalyses prolyl isomerization in some model peptides and proteins such as RNase T1 (Stoller et al., 1995). Perhaps trigger factor's PPIase activity is responsible for the ability to stimulate the binding of unfolded polypeptides to GroEL through isomerization of a critical proline residue in the substrate. Trigger factor might facilitate the polypeptide's entry into GroEL's central cavity. Alternatively, this PPIase activity may function synergistically with the chaperone function of GroEL to enhance the folding of proteins as they emerge from the GroEL–trigger factor complex.

■ Regulation of trigger factor

Trigger factor, which corresponds to B50.3 in the two-dimensional map of E. coli proteins (Guthrie and Wickner, 1990) is an abundant cell protein during balanced cell growth (Pedersen et al., 1978). Unlike most proteins involved in polypeptide folding or degradation, trigger factor is not a heat shock protein. In fact, TF content increases progressively as growth temperature decreases from 42°C to 16°C and even rises in cells stored at 4°C. Upon temperature downshift or exposure to chloramphenicol, TF synthesis is induced like that of cold shock proteins (Kandror and Goldberg, 1997). At low temperatures, the effects of trigger factro on GroEL binding and proteolysis are much greater than at 37°C.

At 37°C, cell growth rates are unaffected by increased or reduced expression of trigger factor, although both over- and under-production of trigger factor inexplicably leads to a filamentous phenotype (Guthrie and Wickner, 1990; Tai et al., 1992). The filamentation of trigger factor overproducing cells is suppressed by coexpression of the essential cell division gene ftsZ (Guthrie and Wickner, 1990). Heterologous expression of the Campylobacter tig gene in E. coli also leads to the same phenotype of E. coli cells (Griffiths et al., 1995). Therefore, it has been suggested that trigger factor plays an important role in cell division.

Curiously, for cells grown in minimal medium, changes in trigger factor content have much smaller effects on GroEL binding and protein degradation than in rich medium (O. Kandror, unpublished result). In contrast to 20°C or 37°C, trigger factor levels do have marked effects on cell viability at low temperature. At 4°C, E. coli colonies slowly lose viability; reduced trigger factor content dramatically decreases the cell's ability to survive, while trigger factor overproduction makes the cell less sensitive to storage in the cold (Kandror and Goldberg 1997). Cells producing large amounts of the abnormal protein CRAG also require normal levels of trigger factor for viability (Kandror et al., 1995). Trigger factor thus appears to play a major role under conditions where cells may accumulate abnormally folded polypeptides. Interestingly, at 50°C, where E. coli die rapidly, TF overproduction enhanced this process, while a reduction in TF content prolonged viability. Thus, at high temperatures, where heat shock proteins are induced and appear protective, TF is repressed and appears harmful (Kandror and Goldberg, 1997).

■ Mutagenesis studies

The PPIase properties of trigger factor are not affected by the removal of 47 or 59 N-terminal amino acid residues, respectively (Stoller et al., 1995).

■ References

Aldea, M., Garrido, T., Hernandez-Chico, C., Vicente, M., and Kushner, S.R. (1989). Induction of a growth-phase-dependent promoter triggers transcription of bolA, an Escherichia coli morphogene. EMBO J. **8**, 3923–3931.

Callebaut, I. and Mornon, J.P. (1995). Trigger factor, one of the Escherichia coli chaperone proteins, is an original member of the FKBP family. FEBS Lett. **347**, 211–215.

Chin, D.T., Goff, S.A., Webster, T., Smith, T., and Goldberg, A.L. (1988). Sequence of the lon gene in Escherichia coli. A heat-shock gene which encodes the ATP-dependent protease La. J. Biol. Chem. **263**, 11718–11728.

Crooke, E. and Wickner, W. (1987). Trigger factor: a soluble protein that folds pro-OmpA into a membrane-assembly-competent form. Proc. Natl. Acad. Sci. USA **84**, 5216–5220.

Crooke, E., Guthrie, B., Lecker, S., Lill, R., and Wickner, W. (1988). ProOmpA is stabilized for membrane translocation by either purified *E. coli* trigger factor or canine signal recognition particle. Cell **54**, 1003–1011.

Fleischmann, R.D., Adams, M.D., White, O., Clayton, R.A., Kirkness, E.F., Kerlavage, A.R., Bult, C.J., Tomb, J.F., Dougherty, B.A., Merrick, J.M., et al. (1995). Whole-genome random sequencing and assembly of *Haemophilus influenzae* Rd [see comments]. Science **269**, 496–512.

Goff, S.A., Casson, L.P., and Goldberg, A.L. (1984). Heat shock regulatory gene htpR influences rates of protein degradation and expression of the lon gene in *Escherichia coli*. Proc. Natl. Acad. Sci. USA **81**, 6647–6651.

Gottesman, S., Clark, W.P., de Crecy-Lagard, V., and Maurizi, M.R. (1993). ClpX, an alternative subunit for the ATP-dependent Clp protease of *Escherichia coli*. Sequence and *in vivo* activities. J. Biol. Chem. **268**, 22618–22626.

Griffiths, P.L., Park, R.W.A., and Connerton, I.F. (1995). The gene for *Campylobacter* trigger factor: evidence for multiple transcription start sites and protein products. Microbiology **141**, 1359–1367.

Guthrie, B. and Wickner, W. (1990). Trigger factor depletion or overproduction causes defective cell division but does not block protein export. J. Bacteriol. **172**, 5555–5562.

Hesterkamp, T., Hauser, S., Lütcke, H. and Bukau, B. (1996). *Escherichia coli* trigger factor is a prolyl isomerase that associates with nascent polypeptide chains. Proc. Natl. Acad. Sci. USA **93**, 4437–4441.

Kandror, O.A., Goldberg, A.L. (1997). Trigger factor is induced upon cold shock and enhances viability of *E. coli* at low temperatures. Proc. Natl. Acad. Sci. USA. **94**, 4978–4981.

Kandror, O., Busconi, L., Sherman, M., and Goldberg, A.L. (1994). Rapid degradation of an abnormal protein in *Escherichia coli* involves the chaperones GroEL and GroES. J. Biol. Chem. **269**, 23575–23582.

Kandror, O., Sherman, M., Rhode, M., and Goldberg, A.L. (1995). Trigger factor is involved in GroEL-dependent protein degradation in *E. coli* and promotes binding of GroEL to unfolded proteins. EMBO J. **14**, 6021–6027.

Kandror, O.A., Sherman, M.Y., Moerschell, R., Goldberg, A.L. (1997) Trigger factor associates with GroEL in vivo and promotes its binding to certain polypeptides. J. Biol. Chem. **272**, 1730–1734.

Kroh, H.E. and Simon, L.D. (1990). The ClpP component of Clp protease is the sigma 32-dependent heat shock protein F21.5. J. Bacteriol. **172**, 6026–6034.

Lecker, S., Lill, R., Ziegelhoffer, T., Georgopoulos, C., Bassford, P.J., Jr., Kumamoto, C.A., and Wickner, W. (1989). Three pure chaperone proteins of *Escherichia coli*–SecB, trigger factor and GroEL–form soluble complexes with precursor proteins *in vitro*. EMBO J. **8**, 2703–2709.

Lill, R., Crooke, E., Guthrie, B., and Wickner, W. (1988). The "trigger factor cycle" includes ribosomes, presecretory proteins, and the plasma membrane. Cell **54**, 1013–1018.

Michnick, S.W., Rosen, M.K., Wandless, T.J., Karplus, M., and Schreiber, S.L. (1991). Solution structure of FKBP, a rotamase enzyme and receptor for FK506 and rapamycin. Science **252**, 836–839.

Mücke, M. and Schmid, F.X. (1994). Folding mechanism of ribonuclease T1 in the absence of the disulfide bonds. Biochemistry **33**, 14608–14619.

Pedersen, S., Bloch, P.L., Reeh, S., and Neidhardt, F.C. (1978). Patterns of protein synthesis in *E. coli*: a catalog of the amount of 140 individual proteins at different growth rates. Cell **14**, 179–190.

Petersen, S.N. (1992). Characterization and analysis of the Mycoplasma genitalium genome. Ph.D. Thesis University of North Carolina Medical School, Chapel Hill.

Rudd, K.E. (1992). *A Short Course in Bacterial Genetics: A Laboratory Handbook for Escherichia coli and Related Bacteria*. Cold Spring Harbor Laboratory Press, New York.

Sherman, M.Y. and Goldberg, A.L. (1992). Heat shock in *Escherichia coli* alters the protein-binding properties of the chaperonin groEL by inducing its phosphorylation. Nature **357**, 167–169.

Sherman, M. and Goldberg, A.L. (1994). Heat shock — induced phosphorylation of GroEL alters its binding and dissociation from unfolded proteins. J. Biol. Chem. **269**, 31479–31483.

Stoller, G., Rücknagel, K.P., Niehaus, K.H., Schmid, F.X., Fischer, G., and Rahfeld, J.-U. (1995). A ribosome-associated peptidyl-prolyl *cis/trans* isomerase identified as the trigger factor. EMBO J. **14**, 4939–4948.

Tai, P.C., Lian, J., Yu, N.-J., Fandl, J.,Xu, H., and Vidugiriene, J. (1992). On protein translocation across cytoplasmic membranes. Antonie Leeuwenhoek **61**, 105–109.

Valent, Q.A., Kendall, D.A. High, S., Kuster, R., Oudega, B., and Luirink, J. (1995). Early events in preprotein recognition in *E. coli*: interaction of SRP and trigger factor with nascent polypeptides. EMBO J. **14**, 5494–5505.

Van Duyne, G.D., Standaert, R.F., Karplus, P.A., Schreiber, S.L., and Clardy, J. (1991). Atomic structure of FKBP-FK506, an immunophilin-immunosuppressant complex. Science **252**, 839–842.

■ *Richard Moerschell, Olga Kandror and Alfred L. Goldberg:*
Harvard Medical School
Deptartment of Cell Biology
240 Longwood Ave.
Boston, MA 02115, USA
Tel. 1 617 432 1855
Fax. 1 617 232 0173
E-mail: agoldber@bcmp.med.harvard.edu

■ *Gunter Fischer and Jens-U. Rahfeld:*
Max-Planck-Research,
Unit-Enzymology of Protein Folding
Kurt-Mothes-Str. 3
D-06120 Halle, Germany
Tel. 49 345 552 2801
Fax. 49 345 551 1972
E-mail: fischer@cis.biochemtech.uni-halle.de

Saccharomyces cerevisiae FKBP12

FKBP12 is a prolyl isomerase originally identified from T cell and thymic extracts as a binding protein for the immunosuppressants FK506 and rapamycin. The yeast Saccharomyces cerevisiae *expresses a homolog of FKBP12 that is markedly similar in sequence and structure to human FKBP12 and mediates FK506 and rapamycin actions in yeast. Yeast mutants lacking FKBP12 are viable with only subtle phenotypes, and thus the biological functions of FKBP12 remain largely unknown.*

■ Alternative names

Fkb1 protein (Wiederrecht et al., 1991), Rbp1 protein (Koltin et al., 1991).

■ FKBP12 sequence

The yeast FKBP12 protein was purified by FK506-affinity chromatography and sequenced (Heitman et al., 1991a; Koltin et al., 1991; Wiederrecht et al., 1991). The gene encoding FKBP12, *FPR1* (Genbank accession number M60877), was cloned by screening yeast genomic libraries with a PCR product obtained with degenerate oligonucleotide primers based on protein sequence (Heitman et al., 1991a; Wiederrecht et al., 1991). The *FPR1* gene was also isolated by complementation of an *fpr1* mutation (Koltin et al., 1991). The *FPR1* gene encodes a 114 residue protein with a molecular weight of 12 kDa. The protein sequences of yeast and human FKBP12 are 54% identical (Heitman et al., 1991a).

■ FKBP12 protein and function

Yeast FKBP12 is an abundantly expressed cytoplasmic protein that exhibits peptidyl–prolyl isomerase activity *in vitro* that is competitively inhibited by FK506 or rapamycin (Heitman et al., 1991a, b; Koltin et al., 1991; Wiederrecht et al., 1991). In contrast to several other yeast prolyl isomerases, expression of FKBP12 is not induced by heat shock or tunicamycin (Partaledis and Berlin, 1993; Sykes et al., 1993; Cardenas et al., 1994). As is the case with mammalian FKBP12, yeast FKBP12 forms a complex with FK506 that inhibits calcineurin, a serine/threonine calcium/calmodulin-dependent phosphatase. Calcineurin plays an important role in signal transduction and ion homeostasis in several organisms. In yeast, calcineurin is required for recovery from pheromone-induced cell cycle arrest, and calcineurin mutation or inhibition by FKBP12–FK506 prevents recovery (Cyert et al., 1991; Foor et al., 1992; Cardenas et al., 1994). Calcineurin mutations or inhibition by FKBP12–FK506 also render yeast cells sensitive to cation stress (Nakamura et al., 1993; Breuder et al., 1994; Hemenway et al., 1995). Lastly, calcineurin is required for expression of one of the subunits of yeast 1,3ß-D-glucan synthase (Fks2 protein), and mutants lacking the homologous Fks1 protein subunit are sensitive to FKBP12-mediated FK506 toxicity (Douglas et al., 1994; Eng et al., 1994; Garrett-Engele et al., 1995; Mazur et al., 1995).

FKBP12 is also the target of rapamycin in yeast (Heitman et al., 1991b; Koltin et al., 1991). The immunosuppressive and antifungal effects of the FKBP12–rapamycin complex are mediated through a different pathway than inhibition of calcineurin. Genetic and biochemical studies in yeast identified TOR1 and TOR2 as the targets of FKBP12–rapamycin (Heitman et al., 1991b). The TOR proteins share homology with phosphatidylinositol-3 and –4 kinases (PI3 and PI4) and protein kinases (Cafferkey et al., 1993; Kunz et al., 1993; Helliwell et al., 1994), and TOR2 has been shown to be tightly associated with a PI4 kinase activity (Cardenas and Heitman, 1995). A mammalian FKBP12–rapamycin-binding protein shares marked identity with the yeast TOR proteins (Brown et al., 1994; Chiu et al., 1994; Sabatini et al., 1994) and also has an associated PI4 kinase activity (Sabatini et al., 1995). The TOR proteins are likely components of a novel signal transduction cascade responsible for cellular proliferation in response to growth factors and possibly nutrients.

■ Structure of FKBP12

The X-ray crystal structures of FK506 in complex with both yeast and human FKBP12 have been solved to high resolution (Michnick et al., 1991; Van Duyne et al., 1991; Rotonda et al., 1993). The structures of yeast and human FKBP12 are remarkably similar (Fig. 1). The overall architecture of both proteins is a five-stranded antiparallel ß sheet (Fig. 1). Residues comprising the hydrophobic ligand binding/active site are virtually identical between the two proteins (Tyr-33, Phe-43, Ile-63, Trp-66, Tyr-89, Ile-98, Phe-106 in yeast FKBP12) with two minor substitutions: Glu-54 and His-87 in human FKBP12 are replaced by glutamine and phenylalanine in yeast FKBP12, which most likely makes an already quite hydrophobic pocket more hydrophobic. A conserved tyrosine in this pocket (Tyr-89 in yeast FKBP12) is required for rapamycin binding (Koser et al., 1993). An FKBP12 surface residue (Arg-49 in yeast) is conserved and required for binding of either yeast or human FKBP12–FK506 complexes to calcineurin (Aldape et al., 1992; Yang et al., 1993; Cardenas et al., 1994).

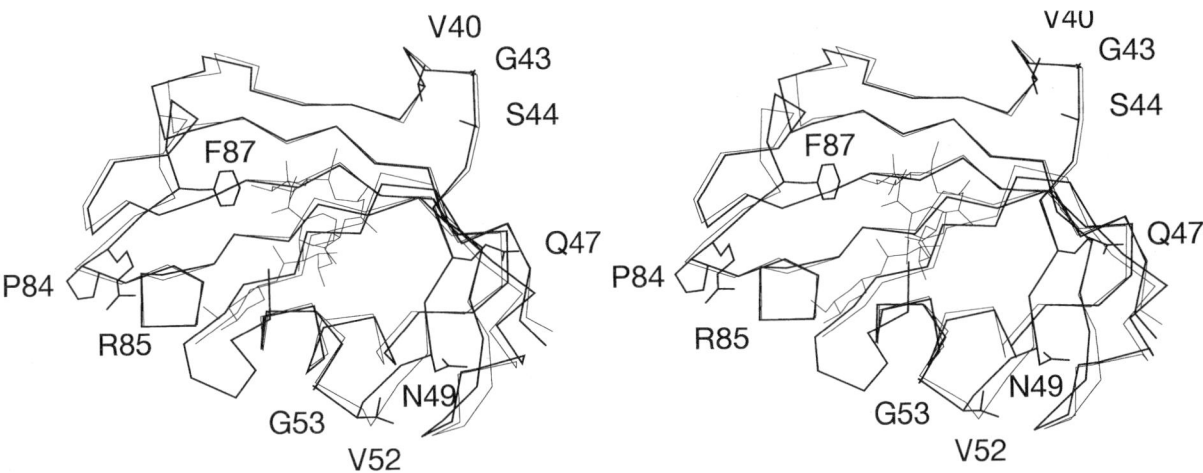

Figure 1. The structure of FKBP12 has been conserved from yeast to human. Stereo view of the α carbon trace of yeast FKBP12 (heavy lines) superimposed on that of human FKBP12 (fine lines). Bound FK506 is depicted in the center of the figure. Side chains of residues on the effector face of FKBP12 that differ between the yeast and human proteins are depicted for yeast FKBP12. Adapted from Rotonda *et al.* (1993) with permission, and kindly provided by Joseph Becker.

Studies of an FK506 analog have revealed an interesting difference between yeast and human FKBP12. L-685,818, a non-immunosuppressive FK506 analog, does not inhibit bovine calcineurin when complexed with human FKBP12, but does inhibit it when complexed with yeast FKBP12 (Rotonda *et al.*, 1993). Sequence differences between yeast and human FKBP12 most likely overcome the decreased calcineurin affinity that results from a hydrophilic hydroxyl group introduced on the effector surface of L-685,818. This, or similar FK506 analogs, may prove useful as novel antifungal agents.

■ FKBP12 mutants

Yeast mutants lacking FKBP12 are viable and FK506 and rapamycin resistant, clearly demonstrating that the effects of these drugs are not mediated via FKBP12 inhibition, but rather as FKBP12–drug complexes. However, these studies reveal little about the normal cellular roles of FKBP12 in the absence of FK506 or rapamycin. Thus far, a large number of possible phenotypes have been excluded for FKBP12 mutant strains, including temperature sensitivity, or defects in heat shock, thermotolerance, mating, sporulation, germination, Ty transposition, production of pheromones and barrier protease, growth on different carbon sources or in the absence of essential nutrients (Heitman *et al.*, 1991a; Cardenas *et al.*, 1994). Yeast mutants lacking FKBP12 do grow more slowly than isogenic wild-type strains, with a doubling time increased by 15–30% (Heitman *et al.*, 1991a). In addition, FKBP12 mutations promote recovery from α factor-induced arrest, suggesting that FKBP12 might normally inhibit calcineurin function. Consistent with this, yeast FKBP12 can weakly bind calcineurin in the absence of FK506, both *in vitro*, by affinity chromatography, and *in vivo*, in the yeast two-hybrid system (Cardenas *et al.*, 1994).

The phenotypes of FKBP12 mutant strains might be subtle if other proteins were to have overlapping functions. However, yeast mutants lacking FKBP12 and other prolyl isomerases, including cyclophilins A, B, and D (Cpr1, Cpr2, and Cpr3 in yeast) (Heitman *et al.*, 1991a; Davis *et al.*, 1992), or FKBP13 (Fpr2) and FKBP70 (Fpr3) (Nielsen *et al.*, 1992; Heitman *et al.*, 1993; Benton *et al.*, 1994; Manning-Krieg *et al.*, 1994; Lorenz *et al.*, 1995) are all viable. These findings suggest little or no functional overlap between different FKBPs or cyclophilins in yeast. Further studies will be required to uncover the *in vivo* functions of FKBP12 that underlie the extreme conservation of a non-essential protein over the billion years of evolution that separate yeast and humans.

■ References

Aldape, R.A., Futer, O., DeCenzo, M.T., Jarrett, B.P., Murcko, M.A., and Livingston, D.J. (1992). Charged surface residues of FKBP12 participate in formation of the FKBP12-FK506-calcineurin complex. J. Biol. Chem. **267**, 16029–16032.

Benton, B.M., Zang, J., and Thorner, J. (1994). A novel FK506-and rapamycin-binding protein (*FPR3* gene product) in the yeast *Saccharomyces cerevisiae* is a proline rotamase localized to the nucleolus. J. Cell Biol. **127**, 623–639.

Breuder, T., Hemenway, C.S., Movva, N.R., Cardenas, M.E., and Heitman, J. (1994). Calcineurin is essential in cyclosporin A- and FK506-sensitive yeast strains. Proc. Natl. Acad. Sci. USA **91**, 5372–5376.

Brown, E.J., Albers, M.W., Shin, T.B., Ichikawa, K., Keith, C.T., Lane, W.S., and Schreiber, S.L. (1994). A mammalian protein targeted by G1-arresting rapamycin-receptor complex. Nature **369**, 756–759.

Cafferkey, R., Young, P.R., McLaughlin, M.M., Bergsma, D.J., Koltin, Y., Sathe, G.M., Faucette, L., Eng, W.-K., Johnson, R.K., and Livi, G.P. (1993). Dominant missense mutations in a novel yeast protein related to mammalian phosphatidylinositol 3-kinase and VPS34 abrogate rapamycin cytotoxicity. Mol. Cell. Biol. **13**, 6012–6023.

Cardenas, M.E. and Heitman, J. (1995). FKBP12-rapamycin target TOR2 is a vacuolar protein with an associated phosphatidylinositol-4 kinase activity. EMBO J. **14**, 5892–5907.

Cardenas, M., Hemenway, C., Muir, R.S., Ye, R., Fiorentino, D., and Heitman, J. (1994). Immunophilins interact with calcineurin in the absence of exogenous immunosuppressive ligands. EMBO J. **13**, 5944–5957.

Chiu, M.I., Katz, H., and Berlin, V. (1994). RAPT1, a mammalian homolog of yeast Tor, interacts with the FBKP12/rapamycin complex. Proc. Natl. Acad. Sci. USA **91**, 12574–12578.

Cyert, M.S., Kunisawa, R., Kaim, D., and Thorner, J. (1991). Yeast has homologs (CNA1 and CNA2 gene products) of mammalian calcineurin, a calmodulin-regulated phosphoprotein phosphatase. Proc. Natl. Acad. Sci. USA **88**, 7376–7380.

Davis, E.S., Becker, A., Heitman, J., Hall, M.N., and Brennan, M.B. (1992). A yeast cyclophilin gene essential for lactate metabolism at high temperature. Proc. Natl. Acad. Sci. USA **89**, 11169–11173.

Douglas, C.M., Foor, F., Marrinan, J.A., Morin, N., Nielsen, J.B., Dahl, A.M., Mazur, P., Baginsky, W., Li, W., El-Sherbeini, M., Clemas, J.A., Mandala, S.M., Frommer, B.R., and Kurtz, M.B. (1994). The Saccharomyces cerevisiae FKS1 (ETG1) gene encodes an integral membrane protein which is a subunit of 1,3-β-D-glucan synthase. Proc. Natl. Acad. Sci. USA **91**, 12907–12911.

Eng, W.-K., Faucette, L., McLaughlin, M.M., Cafferkey, R., Koltin, Y., Morris, R.A., Young, P.R., Johnson, R.K., and Livi, G.P. (1994). The yeast FKS1 gene encodes a novel membrane protein, mutations in which confer FK506 and cyclosporin A hypersensitivity and calcineurin-dependent growth. Gene **151**, 61–71.

Foor, F., Parent, S.A., Morin, N., Dahl, A.M., Ramadan, N., Chrebet, G., Bostian, K.A., and Nielsen, J.B. (1992). Calcineurin mediates inhibition by FK506 and cyclosporin of recovery from α-factor arrest in yeast. Nature **360**, 682–684.

Garrett-Engele, P., Moilanen, B., and Cyert, M.S. (1995). Calcineurin, the Ca^{2+}/calmodulin dependent protein phosphatase, is essential in yeast mutants with cell integrity defects and in mutants that lack a functional vacuolar H^+-ATPase. Mol. Cell. Biol. **15**, 4103–4114.

Heitman, J., Movva, N.R., Hiestand, P.C., and Hall, M.N. (1991a). FK506-binding protein proline rotamase is a target for the immunosuppressive agent FK506 in Saccharomyces cerevisiae. Proc. Natl. Acad. Sci. USA **88**, 1948–1952.

Heitman, J., Movva, N.R., and Hall, M.N. (1991b). Targets for cell cycle arrest by the immunosuppressant rapamycin in yeast. Science **253**, 905–909.

Heitman, J., Koller, A., Kunz, J., Henriquez, R., Schmidt, A., Movva, N.R., and Hall, M.N. (1993). The immunosuppressant FK506 inhibits amino acid import in Saccharomyces cerevisiae. Mol. Cell. Biol. **13**, 5010–5019.

Helliwell, S.B., Wagner, P., Kunz, J., Deuter-Reinhard, M., Henriquez, R., and Hall, M.N. (1994). TOR1 and TOR2 are structurally and functionally similar but not identical phosphatidylinositol kinase homologues in yeast. Mol. Biol. Cell **5**, 105–118.

Hemenway, C.S., Dolinski, K., Cardenas, M.E., Hiller, M.A., Jones, E.W., and Heitman, J. (1995). vph6 mutants of Saccharomyces cerevisiae require calcineurin for growth and are defective in vacuolar H^+-ATPase assembly. Genetics **141**, 833–844.

Koltin, Y., Faucette, L., Bergsma, D.J., Levy, M.A., Cafferkey, R., Koser, P.L., Johnson, R.K., and Livi, G.P. (1991). Rapamycin sensitivity in Saccharomyces cerevisiae is mediated by a peptidyl-prolyl cis-trans isomerase related to human FK506-binding protein. Mol. Cell. Biol. **11**, 1718–1723.

Koser, P.L., Eng, W.-K., Bossard, M.J., McLaughlin, M.M., Cafferkey, R., Sathe, G.M., Faucette, L., Levy, M.A., Johnson, R.K., Bergsma, D.J., and Livi, G.P. (1993). The tyrosine[89] residue of yeast FKBP12 is required for rapamycin binding. Gene **129**, 159–165.

Kunz, J., Henriquez, R., Schneider, U., Deuter-Reinhard, M., Movva, N.R., and Hall, M.N. (1993). Target of rapamycin in yeast, TOR2, is an essential phosphatidylinositol kinase homolog required for G1 progression. Cell **73**, 585–596.

Lorenz, M.C., Muir, R.S., Lim, E., McElver, J., Weber, S.C., and Heitman, J. (1995). Gene disruption with PCR products in Saccharomyces cerevisiae. Gene **158**, 113–117.

Manning-Krieg, U.C., Henriquez, R., Cammas, F., Graff, P., Gaveriaux, S., and Movva, N.R. (1994). Purification of FKBP-70, a novel immunophilin from Saccharomyces cerevisiae, and cloning of its structural gene, FPR3. FEBS Lett. **352**, 98–103.

Mazur, P., Morin, N., Baginsky, W., El-Sherbeini, M., Clemas, J.A., Nielsen, J.B., and Foor, F. (1995). Differential expression and function of two homologous subunits of yeast 1,3-β-D-glucan synthase. Mol. Cell. Biol. **15**, 5671–5681.

Michnick, S.W., Rosen, M.K., Wandless, T.J., Karplus, M., and Schreiber, S.L. (1991). Solution structure of FKBP, a rotamase enzyme and receptor for FK506 and rapamycin. Science **252**, 836–839.

Nakamura, T., Liu, Y., Hirata, D., Namba, H., Harada, S.-i., Hirokawa, T., and Miyakawa, T. (1993). Protein phosphatase type 2B (calcineurin)-mediated, FK506-sensitive regulation of intracellular ions in yeast is an important determinant for adaptation to high salt stress conditions. EMBO J. **12**, 4063–4071.

Nielsen, J.B., Foor, F., Siekierka, J.J., Hsu, M.-J., Ramadan, N., Morin, N., Shafiee, A., Dahl, A. M., Brizuela, L., Chrebet, G., Bostian, K.A., and Parent, S.A. (1992). Yeast FKBP-13 is a membrane-associated FK506-binding protein encoded by the nonessential gene FKB2. Proc. Natl. Acad. Sci. USA **89**, 7471–7475.

Partaledis, J.A. and Berlin, V. (1993). The FKB2 gene of Saccharomyces cerevisiae, encoding the immunosuppressant-binding protein FKBP-13, is regulated in response to accumulation of unfolded proteins in the endoplasmic reticulum. Proc. Natl. Acad. Sci. USA **90**, 5450–5454.

Rotonda, J., Burbaum, J.J., Chan, H.K., Marcy, A.I., and Becker, J.W. (1993). Improved calcineurin inhibition by yeast FBKP12-drug complexes. J. Biol. Chem. **268**, 7607–7609.

Sabatini, D.M., Erdjument-Bromage, H., Lui, M., Tempst, P., and Snyder, S.H. (1994). RAFT1: a mammalian protein that binds to FKBP12 in a rapamycin-dependent fashion and is homologous to yeast TORs. Cell **78**, 35–43.

Sabatini, D.M., Pierchala, B.A., Barrow, R.K., Schell, M.J., and Snyder, S.H. (1995). The rapamycin and FKBP12 target (RAFT) displays phosphatidylinositol 4-kinase activity. J. Biol. Chem. **270**, 20875–20878.

Sykes, K., Gething, M.-J., and Sambrook, J. (1993). Proline isomerases function during heat shock. Proc. Natl. Acad. Sci. USA **90**, 5853–5857.

Van Duyne, G.D., Standaert, R.F., Karplus, P.A., Schreiber, S.L., and Clardy, J. (1991). Atomic structure of FKBP-FK506, an immunophilin-immunosuppressant complex. Science **252**, 839–842.

Wiederrecht, G., Brizuela, L., Elliston, K., Sigal, N.H., and Siekierka, J.J. (1991). FKB1 encodes a nonessential FK506-binding protein in *Saccharomyces cerevisiae* and contains regions suggesting homology to the cyclophilins. Proc. Natl. Acad. Sci. USA **88**, 1029–1033.

Yang, D., Rosen, M.K., and Schreiber, S.L. (1993). A composite FKBP12-FK506 surface that contacts calcineurin. J. Am. Chem. Soc. **115**, 819–820.

■ *Kara Dolinski[1] and Joseph Heitman:[2]*
[1]*Department of Genetics,* [2]*Departments of Genetics and Pharmacology*
Howard Hughes Medical Institute
Duke University Medical Center
322 CARL Bldg, Research Drive
Durham, NC 27710 USA
Tel. 1 919 684 2809/2824
Fax. 1 919 684 5458
E-mail: kd2@acpub.duke.edu and heitm001@mc.duke.edu

Saccharomyces cerevisiae FKBP13

Yeast FKBP13 (yFKBP13) belongs to a conserved, ubiquitous family of FK506-binding proteins (FKBPs) that bind the immunosuppressive drugs FK506 and rapamycin and possess peptidyl–prolyl cis–trans isomerase (PPIase) activity. yFKBP13 is most similar to its mammalian homolog FKBP13 located in the lumen of the endoplasmic reticulum (ER). Expression of both proteins is induced under conditions leading to accumulation of unglycosylated, unfolded proteins in the ER, suggesting yFKBP13 facilitates protein folding or trafficking in this organelle.

■ Alternative names

Fkb2 (Nielsen *et al.*, 1992; Partaledis *et al.*, 1992); Fpr2 (Kunz and Hall, 1993).

■ *FKB2* gene and sequence

The yeast *FKB2* gene encodes a single open reading frame of 135 amino acids (GenBank accession numbers M90767, M90646; Nielsen *et al.*, 1992; Partaledis *et al.*, 1992). At its start are two consecutive ATG initiator codons. Depending on which start site is used, there is a signal peptide of 16 or 17 amino acids, which is presumably removed by processing. The mature protein is 118 amino acids with a predicted molecular weight of 12 446 daltons. Among the FKBPs, it is most homologous to human FKBP13 (hFKBP13) (Jin *et al.*, 1991), sharing 64% identity in the core regions of these proteins. Its identity with the major human cytosolic FKBP (hFKBP12) in the same region is 51%. The amino acid residues of hFKBP12 and yFKBP12, whose side chains interact with FK506, are conserved (Van Duyne *et al.*, 1991; Rotonda *et al.*, 1993). Human FKBP13 has an ER retention signal at its C-terminus (RTEL) and is membrane associated (Jin *et al.*, 1991). Although yFKBP13 lacks a putative ER retention signal at its C-terminus, the protein is membrane-associated and several experiments suggest it resides in this organelle (see below). Yeast and human FKBP13 share two cysteine residues in position for a possible disulfide bridge (Jin *et al.*, 1991; Nielsen *et al.*, 1992). The *FKB2* gene is located on the right arm of chromosome IV (Partaledis *et al.*, 1992) immediately downstream of the *EUG1* gene, which encodes an ER protein with homology to protein disulfide isomerase (Tachibana and Stevens, 1992).

■ yFKBP13 protein

Yeast FKBP13 was purified by virtue of its affinity for FK506. We identified the protein by its affinity for ^{3}H-FK506 in yeast *fkb1* mutants lacking the major cytosolic FK506-binding protein (Fkb1p or yFKBP12, see entry p. 408) (Nielsen *et al.*, 1992). The purification procedure employed standard chromatographic methods, including a final step on an FK506-affinity resin. Binding activity was measured with a rapid filtration assay optimized for the somewhat greater hydrophobicity of yFKBP13 than yFKBP12. Berlin and colleagues (Partaledis *et al.*, 1992) identified yFKBP13 as a protein that bound to and could be eluted from an FK506-affinity resin. Protein extracts from wild-type cells were adsorbed to the affinity resin, washed and eluted with FK506. Yeast FKBP13 and FKBP12 were resolved by SDS–PAGE. yFKBP12 migrates at 12 kDa, while yFKBP13 migrates anomalously at 15 kDa. Microsequencing of the purified yFKBP13s provided sufficient N-terminal sequences to design PCR primers that amplified DNA encoding the N-termini (Nielsen *et al.*, 1992; Partaledis *et al.*, 1992). These DNAs were used as probes to identify clones containing the *FKB2* gene encoding yFKBP13.

Yeast FKBP13 binds FK506 and rapamycin with high affinity (Nielsen et al., 1992). In competitive displacement assays with [^{3}H]-dihydro-FK506, the K_d values for FK506 and rapamycin are 18 and 11 nM, respectively. These affinities are only slightly lower than that of yFKBP12 for both drugs. The protein is also an active PPIase (Nielsen et al., 1992), with turnover comparable to that of yFKBP12. It shows a marked preference for leucine or phenylalanine preceding proline in the model peptide substrate, probably reflecting activity on more hydrophobic substrates, or in a more hydrophobic environment than cytosolic yFKBP12. FK506 and rapamycin inhibit the PPIase activity with K_i values of 8.3 and 0.7 nM, respectively, closely resembling their affinities with FKBP12 homologs.

In contrast to the FK506–yFKBP12 complex (Foor et al., 1992; Parent et al., 1993), the FK506–yFKBP13 complex does not inhibit calcineurin in vitro (Nielsen et al., 1992), the target of the immunosuppressive activities of the FK506–FKBP12 and CsA–cyclophilin complexes in mammalian cells (Liu et al., 1991). These results are consistent with in vivo genetic experiments that reveal that yeast FKBP12 mediates the inhibitory action of FK506 on several calcineurin-dependent processes, including mating factor recovery (Foor et al., 1992), growth of calcineurin-dependent yeast mutants (Parent et al., 1993; Eng et al., 1994), expression of several calcineurin-responsive genes (Mazur et al., 1995; Cunningham and Fink, 1996) and ion homeostasis (Nakamura et al., 1993). Exchange of hFKBP13 residues Pro-97 and Lys-98 with their hFKBP12 counterparts (Gly-89 and Ile-90) confers calcineurin affinity to FKBP13 (Rosen et al., 1993). The high degrees of homology between the pairs of yeast and human-binding proteins suggest that similar changes in yFKBP13 would also confer calcineurin affinity.

The mammalian homolog of yFKBP13 is localized to the lumen of the ER (Nigam et al., 1993) and induced under conditions that misfold proteins in this organelle (Bush et al., 1994), suggesting FKBP13 plays a role in folding proteins in the ER. Although yFKBP13 has not been localized to the ER by immunofluorescence, its sedimentation profile in sucrose gradients is very similar to that of the ER HSP70 family member Kar2p (J. Nielsen, unpublished observations), banding over a range of densities characteristic of ER membranes. Optimal extraction of yFKBP13 from resident membrane fractions requires use of a non-ionic detergent during isolation (Nielsen et al., 1992).

■ Gene regulation of *FKB2*

Several experiments analysing expression of yFKBP13 indicate that the protein plays a role in the ER (Partaledis and Berlin, 1993). *FKB2* RNA is expressed constitutively and is induced in response to increases in unglycosylated, unfolded proteins in tunicamycin-treated cells or in the temperature-sensitive *sec53-6* mutant grown at the nonpermissive temperature. Tunicamycin blocks core glycosylation of nascent polypeptides in the ER, leading to the accumulation of unfolded proteins in this organelle and induction of several eukaryotic ER chaperones, including yeast homologs of mammalian BiP (Kar2p) (Normington et al., 1989; Rose et al., 1989) and protein disulfide isomerase (Eug1p) (Tachibana and Stevens, 1992). *SEC53* encodes phosphomannomutase and mutations in this gene block core glycosylation of proteins in the ER (Kepes and Schekman, 1988). This *sec53* phenotype is specific in that several *sec* mutations affecting different steps of the secretory pathway have little effect on *FKB2* RNA levels.

The promoters of the yeast *EUG1*, *FKB2* and *KAR2* genes share a conserved unfolded protein response element (UPRE) that mediates transcription induction in response to the presence of unfolded ER proteins (Mori et al., 1992; Tachibana and Stevens, 1992; Partaledis and Berlin, 1993). The *FKB2* UPR is a classic yeast upstream activation sequence that functions in a heterologous promoter. As predicted from the analysis of steady-state *FKB2* RNAs, this UPR element responds to tunicamycin treatment, as well as to the *sec53-6* mutation. Like many chaperone genes, including *KAR2* (Normington et al., 1989; Rose et al., 1989), *FKB2* is induced by heat shock (Partaledis and Berlin, 1993). Although the *FKB2* promoter contains a putative heat shock element (HSE) upstream of the UPR site, the role of this HSE is unclear. Heterologous promoters containing this site are not heat shock inducible. Interestingly, expression of the yeast genes *CYP1* and *CYP2*, encoding cytosolic and secretory cyclophilins, is also induced by heat shock and the promoter of the *CYP1* gene contains a HSE that functions in a heterologous promoter (Sykes et al., 1993). Cyclophilins bind the immunosuppressive drug CsA and, like the FKBPs, they possess PPIase activity.

■ Mutagenesis studies with yFKBP13

Mutants lacking yFKBP13 possess several wild-type properties (Nielsen et al., 1992) and no discernible phenotype has been identified. Strains lacking this protein are viable in the presence and absence of several other FKBPs and cyclophilins (Nielsen et al., 1992; Kunz and Hall, 1993). The mutants exhibit wild-type growth rates on several carbon sources, they are not heat sensitive for growth on standard rich medium and they possess wild-type sensitivities to FK506 and rapamycin.

■ yFKBP13 function

Although the precise cellular role(s) of this protein is unknown, the biochemical and molecular genetic data described above indicate that yFKBP13 plays a role in the ER, either as a chaperone and/or by assisting in the folding of proteins via its PPIase activity. The non-essential nature of this protein and the finding that knock-out mutations of the *FKB2* gene do not significantly affect growth rate (Nielsen et al., 1992) suggest that yFKBP13 does not have a global role in protein trafficking and/or folding and that it may function to fold one or a few non-essential proteins (Partaledis and Berlin, 1993). Yeast FKBP13 function may also be redundant in the cell. Insight into the role of this PPIase is likely to emerge from genetic and biochemical experiments identifying proteins that interact with FKBP13. The yeast systems described here provide an ideal model for these studies.

References

Bush, K.T., Hendrickson, B.A., and Nigam, S.K. (1994). Induction of the FK506-binding protein, FKBP13, under conditions which misfold proteins in the endoplasmic reticulum. Biochem. J. **303**, 705–708.

Cunningham, K.W. and Fink, G.R. (1996). Calcineurin inhibits VCX1-dependent H+/Ca2+ exchange and induces Ca2+ ATPases in Saccharomyces cerevisiae. Mol. Cell. Biol. **16**, 2226–2237.

Eng, W.-K., Faucette, L., McLaughlin, M.M., Cafferkey, R., Koltin, Y., Morris, R.A., Young, P.R., Johnson, R.K., and Livi, G.P. (1994). The yeast FKS1 gene encodes a novel membrane protein, mutations in which confer FK506 and cyclosporin A hypersensitivity and calcineurin-dependent growth. Gene **151**, 61–71.

Foor, F., Parent, S.A., Morin, N., Dahl, A.M., Ramadan, N., Chrebet, G., Bostian, K.A., and Nielsen, J.B. (1992). Calcineurin mediates inhibition by FK506 and cyclosporin of recovery from a-factor arrest in yeast. Nature **360**, 682–684.

Jin, Y.-J., Albers, M.W., Lane, W.S., Bierer, B.E., Schreiber, S.L., and Burakoff, S.J. (1991). Molecular cloning of a membrane-associated human FK506- and rapamycin-binding protein, FKBP-13. Proc. Natl. Acad. Sci. USA **88**, 6677–6681.

Kepes, F. and Schekman, R. (1988). The yeast SEC53 gene encodes phosphomannomutase. J. Biol. Chem. **263**, 9155–9161.

Kunz, J. and Hall, M.N. (1993). Cyclosporin A, FK506 and rapamycin: more than just immunosuppression. Trends Biochem. Sci. **18**, 334–338.

Liu, J., Farmer, J.D., Jr., Lane, W.S., Friedman, J., Weissman, I., and Schreiber, S.L. (1991). Calcineurin is a common target of cyclophilin-cyclosporin A and FKBP-FK506 complexes. Cell **66**, 807–815.

Mazur, P., Morin, N., Baginsky, W., El-Sherbeini, M., Clemas, J.A., Nielsen, J.B., and Foor, F. (1995). Differential expression and function of two homologous subunits of yeast 1,3-b-D-glucan synthase. Mol. Cell. Biol. **15**, 5671–5681.

Mori, K., Sant, A., Kohno, K., Normington, K., Gething, M.-J., and Sambrook, J.F. (1992). A 22 bp cis-acting element is necessary and sufficient for the induction of the yeast KAR2 (BiP) gene by unfolded proteins. EMBO J. **11**, 2583–2593.

Nakamura, T., Liu, Y., Hirata, D., Namba, H., Harada, S., Hirokawa, T., and Miyakawa, T. (1993). Protein phosphatase type 2B (calcineurin)-mediated, FK506-sensitive regulation of intracellular ions in yeast is an important determinant for adaptation to high salt stress conditions. EMBO J. **12**, 4063–4071.

Nielsen, J.B, Foor, F., Siekierka, J.J., Hsu, M.-J., Ramadan, N., Morin, N., Shafiee, A., Dahl, A.M., Brizuela, L., Chrebet, G., Bostian, K.A., and Parent, S.A. (1992). Yeast FKBP-13 is a membrane-associated FK506-binding protein encoded by the nonessential gene FKB2. Proc. Natl. Acad. Sci. USA **89**, 7471–7475.

Nigam, S.K., Jin, Y.-J., Jin, M.-J., Bush, K.T., Bierer, B.E., and Burakoff, S.J. (1993). Localization of the FK506-binding protein, FKBP 13, to the lumen of the endoplasmic reticulum. Biochem. J. **294**, 511–515.

Normington, K., Kohno, K., Kozutsumi, Y., Gething, M.-J., and Sambrook, J. (1989). S. cerevisiae encodes an essential protein homologous in sequence and function to mammalian BiP. Cell **57**, 1223–1236.

Parent, S.A., Nielsen, J.B., Morin, N., Chrebet, G., Ramadan, N., Dahl, A.M., Hsu, M.-J., Bostian, K.A., and Foor, F. (1993). Calcineurin-dependent growth of an FK506- and CsA-hypersensitive mutant of Saccharomyces cerevisiae. J. Gen. Microbiol. **139**, 2973–2984.

Partaledis, J.A. and Berlin, V. (1993). The FKB2 gene of Saccharomyces cerevisiae, encoding the immunosuppressant-binding protein FKBP-13, is regulated in response to accumulation of unfolded proteins in the endoplasmic reticulum. Proc. Natl. Acad. Sci. USA **90**, 5450–5454.

Partaledis, J.A., Fleming, M.A., Harding, M.W., and Berlin, V. (1992). Saccharomyces cerevisiae contains a homolog of human FKBP-13, a membrane-associated FK506/rapamycin binding protein. Yeast **8**, 673–680.

Rose, M.D., Misra, L.M., and Vogel, J.P. (1989). KAR2, a karyogamy gene, is the yeast homolog of the mammalian BiP/GRP78 gene. Cell **57**, 1211–1221.

Rosen, M.K., Yang, D., Martin, P.K., and Schreiber, S.L. (1993). Activation of an inactive immunophilin by mutagenesis. J. Am. Chem. Soc. **115**, 821–822.

Rotonda, J., Burbaum, J.J., Chan, H.K., Marcy, A.I., and Becker, J.W. (1993). Improved calcineurin inhibition by yeast FKBP12-drug complexes. J. Biol. Chem. **268**, 7607–7609.

Sykes, K., Gething, M.-J., and Sambrook, J. (1993). Proline isomerases function during heat shock. Proc. Natl. Acad. Sci. USA **90**, 5853–5857.

Tachibana, C. and Stevens, T.H. (1992). The yeast EUG1 gene encodes an endoplasmic reticulum protein that is functionally related to protein disulfide isomerase. Mol. Cell. Biol. **12**, 4601–4611.

Van Duyne, G.D., Standaert, R.F., Karplus, P.A., Schreiber, S.L., and Clardy, J. (1991). Atomic structure of FKBP-FK506, an immunophilin-immunosuppressant complex. Science **252**, 839–842.

■ *Stephen A. Parent and Jennifer B. Nielsen:*
Merck Research Laboratories
P.O. Box 2000
RY 80Y-300/210
Rahway, NJ 07065, USA
Tel. 1 908 594 5787/6799
Fax. 1 908 594 5878/1399
E-mail: steve_parent@merck.com and
jennifer_nielsen_kahn@merck.com

Saccharomyces cerevisiae Npi46p

Saccharomyces cerevisiae *Npi46p is a prolyl cis–trans isomerase that recognizes nuclear localization sequences (NLS) in vitro. It belongs to the FKBP family, proteins that form a complex with the immunosuppressant drugs FK506 and rapamycin. The C-terminal domain of Npi46p possesses the isomerase activity, whereas the N-terminal domain is capable of binding specifically to nuclear localization signals (NLS). Yeast cells lacking Npi46p have no observable phenotype. The cellular function of Npi46p is unknown.*

■ Alternative names

Fpr3 (Benton *et al.*, 1994); FKBP70 (Manning-Krieg *et al.*, 1994).

■ *NPI46* gene and sequence

NPI46 was isolated by screening a λZAP yeast genomic expression library, using antiserum raised against yeast NLS-binding proteins (Shan *et al.*, 1994). It encodes a protein of 411 amino acid residues with a molecular mass of 46 541 Da (Genbank accession number X79379). It was also identified as Fpr3, a protein that binds to FK520, a FK506-related immunosuppressant agent (Benton *et al.*, 1994; Manning-Krieg *et al.*, 1994). The C-terminal domain of Npi46p shares 45% identity with both human and yeast FKBP12. The N-terminal domain is highly charged, contains a bipartite NLS, and a number of acidic stretches with serine residues. The serine residues form potential casein kinase II phosphorylation sites. The structure of the N-terminal domain is strikingly similar to a previously identified group of nucleolar NLS-binding proteins, mammalian nucleolin, Nopp140 and B23 (No38), as well as yeast Nsr1p (Fig. 1).

■ Npi46p function

Npi46p is the first known proline isomerase to be found in the nucleolus. It binds specifically to a wild-type H2B NLS, but not to a mutant H2B NLS that is incompetent for nuclear localization (Shan *et al.*, 1994). It can be readily purified as a GST–Npi46 fusion protein from *Escherichia coli*. The purified fusion protein retains the prolyl isomerase activity, when assayed *in vitro* with either a succinyl–Ala–Leu–Pro–Phe–*p*-nitroanilide or a succinyl–Ala–Ala–Pro–Phe–*p*-nitroaniliide peptide as substrate (Shan *et al.*, 1994). Unpublished results from our laboratory showed that the N-terminal acidic domain is the domain that binds to the NLS. When fused to the IgG-binding domain of protein A, the N-terminal domain can also target the fusion protein to the nucleolus of yeast (X. Shan and T. Melese, unpublished results). Different functional domains of several nucleolar proteins, when fused to non-nucleolar proteins, direct the fusion proteins to the nucleolus (Peculis and Gall, 1992; Creancier *et al.*, 1993; Schmidt-Zachmann and Nigg, 1993; Yan and Melese, 1993; Girard *et al.*, 1994). Thus, nucleolar proteins most likely accumulate in the nucleolus by binding to resident proteins or nucleic acids (Melese and Xue, 1995). The N-terminal domain of Npi46p is most likely

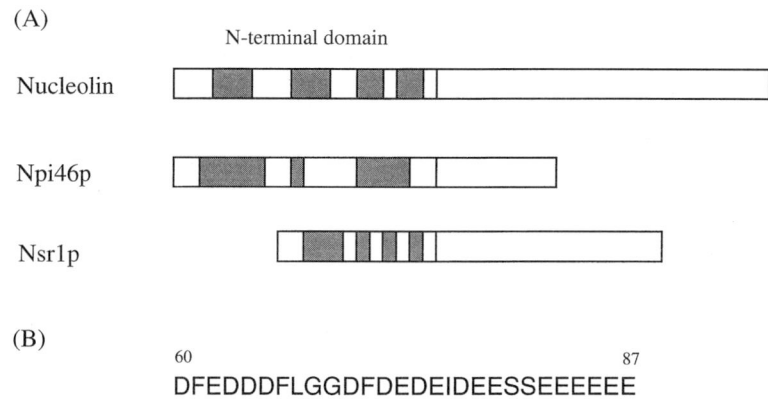

Figure 1. (A) Structural features of the N-terminal domain of nucleolin, Npi46p, and Nsr1p. The shaded regions represent stretches of acidic amino acid residues. (B) The amino acid sequence of the first acidic stretch of Npi46p (from residue 60 to residue 87 of Npi46p).

involved in these types of interactions, while the C-terminal domain possesses proline isomerase activity (Benton et al., 1994).

The significance of the NLS-binding ability of Npi46p is not yet understood. However, a number of nucleolar proteins have been shown to bind to NLSs specifically, and several of them have a role in ribosome biogenesis (Lapeyre et al., 1987; Lee et al., 1991; Xue et al., 1993; Meier and Blobel, 1992; Shan et al., 1994). Thus, the NLS-binding domain of these proteins was proposed to interact with NLSs of ribosome proteins (Xue and Melese, 1994). A number of ribosomal proteins contain multiple NLS-like sequences that distinguish them from other nuclear proteins with only one. The finding that a ribosomal protein, S22, copurifies with FKBP70 (Npi46p) on a FK520-affinity column (Manning-Krieg et al., 1994) is consistent with the idea that Npi46p interacts with certain ribosomal proteins. The isomerase activity of Npi46p could play a role in catalysing the folding or unfolding of ribosomal proteins, which might be associated with the assembly of preribosome particles. However, more definitive evidence, such as a direct demonstration of an interaction between ribosomal proteins and Npi46p, is needed to confirm this hypothesis.

■ Mutations in *NPI46* and post-translational modification of Npi46p

NPI46 is not essential for cell growth. No detectable phenotype is observed in a strain carrying a Npi46 null allele (Benton et al., 1994; Shan et al., 1994). A yeast strain that lacks all three FKBPs, Fpr1, Fpr2, and Fpr3/Npi46p, is also viable (Manning-Krieg et al., 1994), suggesting that the role of these proteins might be to increase the rate of a spontaneous process. Overexpression of the C-terminal isomerase domain of Fpr3p/Npi46p under the yeast *GAL1* promoter, in a *fpr1* strain that is rapamycin resistant, restores rapamycin sensitivity (see FKBP12 entry, p. 408, for a discussion on FK506 and rapamycin sensitivity). Overexpression of Fpr3p or the N-terminal domain of Fpr3p is toxic to both wild-type and *fpr1* mutant yeast strains (Benton et al., 1994). However, we were unable to corroborate this toxic effect when either Npi46p/Fpr3p, or a fusion protein consisting of the N-terminal domain of Npi46p fused to the IgG-binding domain of protein A, was expressed from the yeast *GAL1* promoter (X. Shan and T. Melese, unpublished results). By immunoblot, both the native protein and the fusion protein were highly expressed. A possibility is that the discrepancy between our results and those of Benton and co-workers is owing to differences in strain background.

Npi46p contains a number of sites that are phosphorylated by casein kinase II (Benton et al., 1994; Shan et al., 1994). This feature is shared by all nucleolar NLS-binding proteins. Phosphorylation at the casein kinase II sites may serve to enhance the binding affinity between these proteins and their physiological substrates. Npi46p is also phosphorylated on tyrosine-184 by an unknown protein kinase (Benton et al., 1994; Wilson et al., 1995). This phosphorylation site is the only known physiological substrate of the tyrosine-specific phosphoprotein phosphatase Ptp1 (Wilson et al., 1995). However, the functional significance of this site is not clear. Identification of the kinase that phosphorylates Npi46p at tyrosine-184 might shed light on the cellular role of Npi46p.

■ References

Benton, B. M., Zang, J., and Thorner, J. (1994). A novel FK506- and rapamycin-binding protein (*FPR3* gene product) in the yeast *Saccharomyces cerevisiae* is a proline rotamase localized to the nucleolus. J. Cell Biol. **127**, 623–639.

Creancier, L., Prats, H., Zanibellato, C., Amalric, F., and Bugler, B. (1993). Determination of the functional domains involved in nucleolar targeting of nucleolin. Mol. Biol. Cell **4**, 1239–1250.

Girard, J.P., Bagni, C., Caizergues-Ferrer, M., Amalric, F., and Lapeyre, B. (1994). Identification of a segment of the small nucleolar ribonucleoprotein-associated protein GAR1 that is sufficient for nuclear accumulation. J. Biol. Chem. **269**, 18499–18506.

Lapeyre, B., Bourbon, H., and Amalric, F. (1987). Nucleolin, the major nucleolar protein of growing eukaryotic cells: an unusual protein structure revealed by the nucleotide sequence. Proc. Natl. Acad. Sci. USA **86**, 8808–8812.

Lee, W-C., Xue, Z., and Melese, T. (1991). The *NSR1* gene encodes a protein that specifically binds nuclear localization sequences and has two RNA recognition motifs. J. Cell Biol. **113**, 1–12.

Manning-Krieg, U.C., Henriquez, R., Cammas, F., Graff, P., Gaveriaux, S., and Movva, N.R. (1994). Purification of FKBP-70, a novel immunophilin from *Saccharomyces cerevisiae*, and cloning of its structural gene, *FPR3*. FEBS Lett. **352**, 98–103.

Meier, U.T. and Blobel, G. (1992). Nopp140 shuttles on tracks between nucleus and cytoplasm. Cell. **70**, 127–138.

Melese T. and Xue, Z. (1995). The nucleolus: an organelle formed by the act of building a ribosome. Curr. Opin. Cell Biol. **7**, 319–324.

Peculis, B.A. and Gall, J.G. (1992). Localization of the nucleolar protein NO38 in amphibian oocytes. J. Cell Biol. **116**, 1–14.

Schmidt-Zachmann, M.S. and Nigg, E.A. (1993). Protein localization to the nucleolus: a search for targeting domains in nucleolin. J. Cell Sci. **105**, 799–806.

Shan, X., Xue, Z., and Melese, T. (1994). Yeast *NPI46* encodes a novel prolyl *cis-trans* isomerase that is located in the nucleolus. J. Cell Biol. **126**, 853–862.

Wilson, L.K., Benton, B.M., Zhou, S., Thorner, J., and Martin, G.S. (1995). The yeast immunophilin Fpr3 is a physiological substrate of the tyrosine-specific phosphoprotein phosphatase Ptp1. J. Biol. Chem. **270**, 25185–25193.

Xue, Z. and Melese, T. (1994). Nucleolar proteins that bind NLSs: a role in nuclear import or ribosome biogenesis? Trends Cell Biol. **4**, 414–417.

Xue, Z., Shan, X., Lapeyre, B., and Melese, T. (1993). The amino terminus of mammalian nucleolin specifically recognizes SV-40 T-antigen type nuclear localization sequences. Eur. J. Cell Biol. **62**, 13–21.

Yan, C. and Melese, T. (1993). Multiple regions of *NSR1* are sufficient for accumulation of a fusion protein within the nucleolus. J. Cell Biol. **123**, 1081–1091.

■ *Xiaoyin Shan and Teri Melese:*
Department of Biological Sciences
702 Fairchild
Columbia University

New York, NY 10027, USA
Tel. 1 212 854 5443
Fax. 1 212 865 8246
E-mail: shan@cubsps.bio.columbia.edu and
teri@cubsps.bio.columbia.edu

■ Zhixiong Xue:
DuPont Central Research and Development
P.O. Box 80402
Wilmington, DE 19880-0402, USA
Tel. 1 302 695 9465
Fax. 1 302 695 8480
E-mail: xuez@esvax.dnet.dupont.com

Neurospora crassa FKBPs

Neurospora crassa contains at least two different FKBPs (binding proteins for FK506 and rapamycin). FKBP22 is located in the secretory pathway. FKBP13, which is responsible for N. crassa's sensitivity towards FK506, is located in the cytosol and mitochondria; both identical proteins are the product of a single nuclear gene.

■ Alternative names

FK506-binding proteins; rapamycin-binding proteins; peptidyl–prolyl *cis–trans* isomerases (PPIase, EC no. 5.2.1.8.); rotamases, immunophilins (Schreiber, 1992).

■ Identification and isolation of FKBPs in *Neurospora crassa*

Neurospora crassa is sensitive against the immunosuppressants FK506 and rapamycin, which originally were isolated as antifungal agents. This led to the search for receptor proteins for these immunsuppressants. In 1990, the abundant human (Standaert *et al.*, 1990) and *N. crassa* cytosolic FKBPs (Tropschug *et al.*, 1990) were cloned (see separate entry on *N. crassa* Cyps, p. 382). FKBPs are found in every prokaryotic and eukaryotic cell and in all subcellular localizations, like cytosol, endoplasmic reticulum (ER) or nucleus (reviewed by Schreiber, 1992; Kunz and Hall, 1993; Schmid, 1993; Fischer, 1994; Galat and Metcalfe, 1995).

Cytosolic and mitochondrial FKBP13 were isolated from *N. crassa* using standard biochemical methods (Tropschug *et al.*, 1990). A novel FKBP22 was identified by ascomycin-affinity chromatography, SDS–PAGE and named according to its molecular weight of 22 kDa (M. Tropschug, unpublished work). For determination of the relative abundance of the FKBPs, detergent extracts of *N. crassa* were run over ascomycin columns and then the bound FKBPs released and separated on SDS gels. Coomassie-stained gels were analysed by laser densitometry (see Table 1).

■ *Neurospora crassa* FKBP proteins and their biological activities

■ FKBP22

cDNA cloning of FKBP22 revealed an N-terminal PPIase (FK506-binding) domain (B. Solscheid, G. Fischer, F.X. Schmid and M. Tropschug, manuscript in preparation). FKBP22 is synthesized as a precursor with a signal sequence of 20 amino acids, which is cleaved upon entry in the ER. Mature FKBP22 is active as a PPIase in a chymotrypsin-coupled peptide assay; its substrate specificity (Harrison and Stein, 1990) resembles that of typical FKBPs (G. Fischer, personal communication). FKBP22 is also active as a protein folding enzyme in assays of *in vitro* refolding of RNase T1 (F.X. Schmid, personal communication; B. Solscheid, G. Fischer, F.X. Schmid and M. Tropschug, manuscript in preparation).

■ FKBP13

This protein was originally isolated from a cytosolic fraction (Tropschug *et al.*, 1990). Later it was found that an

Table 1. Analyses of FKBP22 and FKBP13

Protein	Location	Rel. abundance (%)	Gene
FKBP22	ER	10	nd
FKBP13	Cytosol and mitochondria	90	*fkr-2* ?

nd = not determined.

identical protein can be identified in the mitochondrial matrix (C. Knühl and M. Tropschug, manuscript in preparation). Therefore, FKBP13 has a dual location in *N. crassa*. Both proteins are the product of a single nuclear gene, probably *fkr-2* (see below). The mitochondrial protein is synthesized as a precursor with a long presequence of 56 amino acids, which is cleaved in two steps upon entry into the mitochondrial matrix. To our knowledge, this is the first mitochondrial FKBP described.

Mutants in the gene *fkr-2* lead to loss of FKBP13 in both the cytosol and mitochondria; these mutants are FK506 resistant (Barthelmess and Tropschug, 1993), proving that FKBP13 is the receptor for FK506 in *N. crassa*. In *fkr-2* mitochondria, which have lost FKBP13, folding of imported DHFR is slower than in wild-type mitochondria, suggesting a similar role for FKBP13 in intramitochondrial protein folding as for Cyp20 (Rassow et al., 1995). Why *N. crassa* uses the same strategy (one gene, two locations) for two abundant PPIases, Cyp20 (Tropschug et al., 1988) and FKBP13, is unknown.

■ References

Barthelmess, I.B. and Tropschug, M. (1993). FK506-binding protein of *Neurospora crassa* (NcFKBP) mediates sensitivity to the immunosuppressant FK506; resistant mutants identify two loci. Curr. Genet. **23**, 54–58.

Fischer, G. (1994). Peptidyl-prolyl cis/trans isomerases and their effectors. Angew. Chem. Int. Ed. Engl. **33**, 1415–1436.

Galat, A. and Metcalfe, S.M. (1995). Peptidylproline cis/trans isomerases. Prog. Biophys. Mol. Biol. **63**, 67–118.

Harrison, R.K. and Stein, R.L. (1990). Substrate specificities of the peptidyl-prolyl cis-trans isomerase activities of cyclophilin and FK506 binding protein: evidence for the existence of a family of distinct enzymes. Biochemistry **29**, 3813–3816.

Kunz, J. and Hall, M.N. (1993). Cyclosporin A, FK506 and rapamycin: more than just immunosuppression. Trends Biochem. Sci. **18**, 334–338.

Rassow, J., Mohrs, K., Koidl, S., Barthelmess, I.B., Pfanner, N., and Tropschug, M. (1995). Cyclophilin20 is involved in mitochondrial protein folding in cooperation with molecular chaperones Hsp70 and Hsp60. Mol. Cell. Biol. **15**, 2654–2662.

Schmid, F.X. (1993). Prolyl isomerases: enzymatic catalysis of slow protein folding reactions. Annu. Rev. Biophys. Biomol. Struct. **22**, 123–143.

Schreiber, S.L. (1992). Immunophilin-sensitive protein phosphatase action in cell signalling pathways. Cell **70**, 365–368.

Standaert, R.F., Galat, A., Verdine, G.L., and Schreiber, S.L. (1990). Molecular cloning and overexpression of the human FK506-binding protein FKBP. Nature **346**, 671–674.

Tropschug, M., Nicholson, D.W., Hartl, F.-U., Köhler, H., Pfanner, N., Wachter, E., and Neupert, W. (1988). Cyclosporin A-binding protein (cyclophilin) of *Neurospora crassa*: one gene codes for both the cytosolic and mitochondrial forms. J. Biol. Chem. **263**, 14433–14440.

Tropschug, M., Wachter, E., Mayer, S., Schönbrunner, E.R., and Schmid, F.X. (1990). Isolation and sequence of an FK506-binding protein from *N. crassa* which catalyses protein folding. Nature **346**, 674–677.

■ Maximilian Tropschug:
Institut für Biochemie und Molekularbiologie
der Albert-Ludwigs-Universität,
Hermann-Herder-Str. 7
D-79104 Freiburg i. Br., Germany
Tel. 49 761 203 5244
Fax. 49 761 203 5253
Email: tropschu@sun2.ruf.uni-freiburg.de

Plant FKBP73

FKBP73 of wheat is a PPIase (peptidyl–prolyl cis–trans isomerase) member of the FKBP (FK506-binding protein) family, being most homologous to mammalian FKBP52. Uniquely, the deduced amino acid sequence contains three FKBP12-like domains, a calmodulin-binding site and a putative TPR (tetratricopeptide) motif. The wheat FKBP73 is abundant in highly dividing tissues and an additional isoform of 71 kDa is associated with mature cells.

■ FKBP73 sequence

The amino acid sequence deduced from the nucleotide sequence of a cDNA clone isolated from a wheat root tip cDNA library (EMBL accession number X86903) reveals an open reading frame of 559 amino acids with calculated molecular mass of 62 kDa and with electrophoresis mobility of 73 kDa on SDS–PAGE (Blecher et al., 1996). The N-terminal amino acid sequence does not reveal similarity to mitochondrial or chloroplast targeting sequences. The protein sequence contains three FKBP12-like domains (Standaert et al., 1990; Van Duyne et al., 1991) starting from the first methionine of the open reading frame: amino acids 42–148 (domain I), amino acids 158–265 (domain II) and amino acids 275–384 (domain III). The FKPB12-like domains are followed by a putative TPR motif (amino acids 404–521) and a calmodulin-binding site (amino acids 530–546). Wheat FKBP73 thus belongs to the FKBP gene family (for reviews see Trandinh et al., 1992; Galat, 1993; Fruman et al., 1994; introduction to this section, p. 359). Its amino acid sequence displays a strong similarity with the mammalian FKBP59 (Yem et al.,

1992; see entry on FKBP52, p. 430), known to consist of two FKBP12-like domains followed by a putative TPR motif and a calmodulin-binding site (Callebaut et al., 1992). The sequence contains fourteen putative phosporylation sites belonging to five groups; myosin I heavy chain kinase (KXXS*X), cGMP-dependent protein kinase (XS*RX), proline-dependent protein kinase (XS*PR), casein kinase I (XSXXS*X) and casein kinase II (XS*XXEX).

Chromosomal mapping of the *FKBP1* locus shows that it is located on the centromeric region of the short arm of the wheat group 7 chromosomes.

■ Wheat FKBP73 protein

The cDNA encoding wheat FKBP73 was expressed in *E. coli* and the recombinant wheat FKBP73 protein exhibited PPIase activity determined by the acceleration of the isomerization of a synthetic peptide (Fischer et al., 1984; Harrison and Stein, 1990). The immunosupprevive drugs FK506 and rapamycin were found to be potent inhibitors of this PPIase activity (about 90% inhibition on addition of 0.66 μM FK506 or 1.48 μM rapamycin).

The recombinant wheat FKBP73 was retained on a calmodulin column in the presence of Ca^{2+} and was eluted by the addition of EGTA.

■ Expression of plant FKBP73 proteins

Using polyclonal antibodies against the wheat FKBP73, the abundance of FKBP73 and its homologs was examined. The antibodies recognized a 73 kDa protein in barley and corn (monocots) and a less abundant higher molecular weight protein in pea and beans (dicots).

In wheat, a differential display of 73 and 71 kDa isoforms was detected in the wheat tissues in correlation with their cellular age. The 73 kDa isoform was found predominately in young meristematic tissues, whereas the 71 kDa isoform was found predominately in older tissues.

■ Other plant FKBPs

By using FK506-affinity columns, FK506-binding proteins of 55, 25, 18 and 12 kDa were identified in *Vicia faba* extracts (Luan et al., 1994). The FKBP13 localized in chloroplasts was regulated by light, and was not expressed in etiolated tissues or in roots. The FKBP18 protein was associated with *Vicia faba* mitochondria. These results are consistent with PPIase activity inhibited by rapamycin in pea and wheat mitochondria (Breiman et al., 1992).

■ References

Blecher, O., Erel, N., Callebaut, I., Aviezer, K., and Breiman, A. (1996). A novel plant Peptidyl-prolyl-*cis-trans* isomerase (PPIase): cDNA cloning, structural analysis, enzymatic activity and expression. Plant Mol. Biol. **32**, 493–504.

Breiman, A., Fawcett, T.W., Ghirardi, M.I., and Mattoo, A.K. (1992). Plant organelles contain distinct peptidylprolyl cis, trans-isomerases. J. Biol. Chem. **267**, 21293–21296.

Callebaut, I., Renoir, J.M., Lebeau, M.C., Massol, N., Burny, A., Baulieu, E.E., and Mornon, J.P. (1992). An immunophilin that binds Mr 90,000 heat shock protein: main structural features of a mammalian p59 protein. Proc. Natl. Acad. Sci. USA **89**, 6270–6274.

Fischer, G., Bang, H., Berger, E., and Schellenberger, A. (1984). Conformational specificity of chymotrypsin toward proline containing substrates. Biochim. Biophys. Acta **791**, 87–97.

Fruman, D.A., Burakoff, S.J., and Bierer, B.E. (1994). Immunophilins in protein folding and immunosuppression. FASEB J. **8**, 391–400.

Galat, A. (1993). Peptidyl prolyl cis trans Isomerases: immunophilins. Eur. J. Biochem. **216**, 689–707.

Harrison, R.K. and Stein, R. (1990). Substrate specificities of the peptidyl prolyl cis trans isomerase activity of cyclophilin and FK506 binding protein: evidence for existence of family of distinct enzymes. Biochemistry **29**, 3813–3816.

Luan, S., Albers, M.W., and Schreiber, S.L. (1994). Light-regulated, tissue-specific immunophilins in higher plant. Proc. Natl. Acad. Sci. USA **91**, 984–988.

Standaert, R.F., Galat, A., Verdine, G.L., and Schreiber, S.L., (1990). Molecular cloning and overexpression of the human FK506-binding protein FKBP. Nature **346**, 671–674.

Trandinh, C.C., Pao, G.M., and Saier, M.H. (1992). Structural and evolutionary relationships among the immunophilins: two ubiquitous families of peptidyl prolyl cis trans isomerases. FASEB J. **6**, 3410–3420.

Van Duyne, G.D., Standaert, R.F., Karplus, P.A., Schreiber, S.L. and Clardy, J. (1991). Atomic structure of FKBP-FK506, an immuniphilin-immunosuppressant complex. Science **252**, 839–842.

Yem, A.W., Tomasselli, A.G., Heinrikson, R.L., Zurcher-Neely, H., Ruff, V.A., Johnson, R.A., and Deibel, M.R. (1992). The Hsp56 component of steroid receptor complexes binds to immobilized FK506 and shows homology to FKBP-12 and FKBP-13. J. Biol. Chem. **267**, 2868–2871.

■ *Oshra Blecher and Adina Breiman:*
Department of Botany
George S. Wise Faculty of Life Sciences
Tel Aviv University
Tel Aviv 69978, Israel
Tel. 972 3 640 9377
Fax. 972 3 640 9380
E-mail: adina@ccsg.tau.ac.il

Drosophila FKBP39

Drosophila FKBP39 is the first member of the FK506-binding proteins found in this species. FKBP39 shows the highest similarity to a subfamily of these proteins that comprises one yeast and another insect protein. These FKBPs are characterized by a highly charged N-terminal domain followed by a C-terminal domain that binds the immunosuppressive drug FK506.

■ Drosophila FKBP39 sequence

The cDNA isolated from a library of immune-induced adult flies (GenBank accession number Z46894) shows an open reading frame which codes for a protein with a bipartite domain structure (Theopold et al., 1995): the N-terminus contains two stretches of amino acids with acidic residues, separated by highly basic stretches (Fig. 1). Part of the second basic stretch comprises two overlapping, possible nuclear localization signals. The C-terminal domain shows high similarity to proteins that bind the immunosuppressive drug FK506 (Gething and Sambrook, 1992; Kay, 1996; see overview, p. 359). Proteins with a similar domain structure have been described in yeast (FKBP70, encoded by the gene NPI46, FPR3; Benton et al., 1994; Manning-Krieg et al., 1994; Shan et al., 1994; see entry p. 414), and in Spodoptera frugiperda (FKBP46; Alnemri et al., 1994). Other features shared between the members of this new subfamily of FKBPs include a high pI of the FK506-binding domain (9.9 in the case of Drosophila FKBP39, cf. Alnemri et al., 1994) and a higher sequence conservation of the FKBP domain between members of the subfamily compared with other FKBPs: Drosophila FKBP39 is 51% identical to its possible yeast homologue FKBP70, compared with 48% and 46% identity to yeast FKBP12 and FKBP13, respectively.

■ Gene regulation of Drosophila FKBP39

Drosophila FKBP39 is expressed throughout development with the highest expression in embryos. The mRNA could also be detected in the hemocyte cell line mbn-2. The antibacterial defense response in these cells seems to be unaffected by FK506. In situ hybridizations specific for Drosophila FKBP39 revealed a strong signal in the ovaries. This expression pattern does not suggest a highly specialized function as for the other Drosophila immunophilin, the ninaA gene product (Schieh et al., 1989). A second FKBP sequence has been isolated from Drosophila (GenBank accession number Z49079) which indicates that Drosophila FKBP39, as in other organisms, is a member of a family of proteins that share the FK506-binding domain. The product of the yeast FPR3 gene is located exclusively in the nucleolus (Benton et al., 1994; Shan et al., 1994). It has been isolated by its affinity for nuclear localization signals (Shan et al., 1994). The lepidopteran FKBP46 appears to associate with a nuclear kinase (Alnemri et al., 1994) and may be involved in the execution of apoptosis (Ahmad et al., 1996).

■ References

Ahmad, M., Srinivasula, S.M., Wang, L., Litwack, G., Fernandes-Alnemri, T., and Alnemri, E.S. (1996). *Spodoptera frugiperda* caspase-1, a novel insect death protease that cleaves the nuclear immunophilin FKPB46, is the target of the baculovirus antiapoptotic protein p35. J. Biol. Chem. **272**, 1421–1424.

Alnemri, E.S., Fernandes-Alnemri, T., Pomerenke, K., Robertson, N.M., Dudley, K., DuBois, G.C., and Litwack, G. (1994). FKBP46, a novel Sf9 insect cell nuclear immunophilin that forms a protein-kinase complex. J. Biol. Chem. **269**, 30828–30834.

Benton, B.M., Zang, J.-H., and Thorner, J. (1994). A novel FK506- and rapamycin-binding protein (*FPR3* gene product) in the yeast *Saccharomyces cerevisiae* is a proline rotamase localized to the nucleolus. J. Cell Biol. **127**, 3623–3639.

Gething, M.-J. and Sambrook, J. (1992). Protein folding in the cell. Nature **355**, 33–45.

Kay, J.E. (1996). Structure-function relationships in the FK506-binding protein (FKBP) family of peptidylprolyl cis-trans isomerases. Biochem. J. **314**, 361–385.

Manning-Krieg, U., Henríquez, R., Cammas, P., Gavériaux, S., and Movva, N.R. (1994). Purification of FKBP-70, a novel immunophilin from *Saccharomyces cerevisiae*, and cloning of its structural gene, *FPR3*. FEBS Lett. **352**, 98–103.

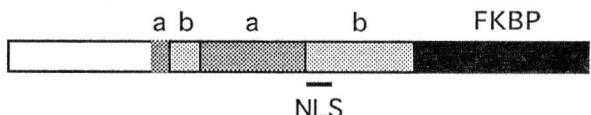

Figure 1. Schematic representation of Drosophila FKBP39, showing basic (b) and acidic (a) stretches, the FKBP homologous domain and two adjacent nuclear localization signals (NLS).

Schieh, B.-H., Stamnes, M.A., Seavello, S., Harris, G.L., and Zuker, C.S. (1989). The *ninaA* gene required for visual transduction in *Drosophila* encodes a homologue of cyclosporin A-binding protein. Nature **338**, 67–70.

Shan, X., Zhixiong, X., and Mélèse, T. (1994). Yeast *NPI46* encodes a novel prolyl cis-trans isomerase that is located in the nucleolous. J. Biol. Chem. **126**, 4853–4862.

Theopold, U., DalZotto, L., and Hultmark, D. (1995). FKBP39, a *Drosophila* member of a family of proteins that bind the immunosuppressive drug FK506. Gene **156**, 247–251.

■ Ulrich Theopold:
Department of Crop Protection
University of Adelaide
Glen Osmond
South Australia 5064, Australia
Tel. 61 8 8303 6565
Fax. 61 8 8379 4095
E-mail: utheopold@waite.adelaide.edu.au

■ Dan Hultmark:
Department of Developmental Biology
Stockholm University
S-10691 Stockholm, Sweden
Tel. 46 8 164 153
Fax. 46 8 152 350
E-mail: dan@molbio.su.se

Mammalian FKBP12

FKBP12 is a member of a class of proteins that bind the immunosuppressive agents FK506 and rapamycin (RAP) and possess peptidyl–prolyl cis–trans isomerase activity (PPIase). The FKBP12–FK506 complex acts as a specific and potent inhibitor of the activity of protein phosphatase 2B (calcineurin), which is the basis of the immunosuppressive activity of FK506 (Liu et al., 1991). The FKBP12–RAP complex does not interact with calcineurin, but rather with a novel PI3-kinase homolog, RAFT1–FRAP. The immunosuppressive mechanism of action of RAP is currently not fully understood.

■ FKBP12 gene and sequence

The amino acid sequence deduced from a single cDNA clone isolated from a phytohaemagglutinin (PHA)-stimulated human T-cell library (GenBank accession number M34539; Maki et al., 1990) reveals a protein of 108 amino acids and a molecular weight of 11 951 Da in agreement with the molecular mass estimated by SDS–PAGE (Siekierka et al., 1989b). No significant homology was observed at that time with other sequences in the GenBank and EMBL databases. Hydropathy analysis did not reveal any significant hydrophobic stretches indicative of a signal peptide or transmembrane region and is consistent with the cytoplasmic localization of FKBP12. A physical map of the human FKBP12 gene has been obtained and the gene has been mapped to chromosome 20 (DiLella and Craig, 1991).

■ FKBP12 protein

FKBP12 is a heat stable, highly abundant (c. 0.2–0.4%) and ubiquitous cytoplasmic protein (Siekierka et al., 1990). FKBP12 binds the immunosuppressant drugs FK506 and RAP with high affinity (K_d ~ 1 nM). FKBP12 exhibits PPIase activity which is inhibitable by FK506 and RAP (K_i ~ 1 nM). The crystal and solution structure for FKBP12 has been solved (Van Duyne et al., 1991).

■ FKBP12 function

FKBP12 was first identified as the molecular target for the immunosuppressive agents FK506 and RAP (Harding et al., 1989; Siekierka et al., 1989a, b). The FKBP12–FK506 complex binds with high affinity to the 'latch region' of the B subunit of calcineurin and non-competitively inhibits its catalytic activity (Milan et al., 1994). The FKBP12–RAP complex binds to a distinct target termed RAFT1, or FRAP, which exhibits significant homology to several PI3 kinases (Brown et al., 1994; Sabatini et al., 1994). FKBP12–RAP does not inhibit RAFT1/FRAP kinase activity, indicating that the immunosuppressive mechanism of action of RAP is more complex (Sabatini et al., 1995). The crystal structures for the FKBP12–FK506–calcineurin and FKBP12–rapamycin–FRAP complexes have been solved (Griffith et al., 1995; Choi et al., 1996).

It is unlikely that FKBP12 plays a direct role in regulating calcineurin or RAFT1/FRAP activity, since interaction with these proteins does not occur in the absence of the relevant immunosuppressive ligand. In the absence of FK506 and RAP, FKBP12 is found complexed with the ryanodine receptor calcium release channel in muscle sarcoplasmic reticulum and appears to modulate the activity of this channel (Timerman et al., 1993). FKBP12 PPIase activity is required for this function, whereas it is not for FKBP1–FK506-mediated inhibition of calcineurin activity (Wiederrecht et al., 1992). This function is likely to be physiologically relevant. FKBP12 has also been found complexed to the inositol-1,4,5-triphosphate receptor (IP3R) which shares homology with the ryanodine receptor (Cameron et al., 1995). In addition, complex formation between FKBP12 and the TGFβ-I receptor and transcription factor YY1 has been reported (Wang et al., 1994; Yang et al., 1995). The physiological relevance of these latter interactions is unknown.

■ References

Brown, E.J., Albers, M.W., Shin, T.B., Ichikawa, K., Keith, C.T., Lane, W.S., and Schreiber, S.L. (1994). A mammalian protein targeted by G1 arresting rapamycin receptor complex. Nature **369**, 756–758.

Cameron, A.M., Steiner, J.P., Sabatini, D.M., Kaplin, A.I., Walensky, L.D., and Snyder, S.H. (1995). Calcineurin associated with the inositol 1,4,5-triphosphate receptor-FKBP12 complex modulates Ca+2 flux. Proc. Natl. Acad. Sci. USA **92**, 1784–1788.

Choi, J., Chen, J., Schreiber, S.L., and Clardy, J. (1996). Structure of the FKBP-rapamycin complex interacting with the binding domain of human FRAP. Science **273**, 239–242.

DiLella, A.G. and Craig, R.J. (1991). Exon organization of the human FKBP-12 gene: correlation with structural and functional protein domains. Biochemistry **30**, 8512–8517.

Griffith, J.P., Kim, J.L., Eunice, E., Sintchak, M.D., Thomson, J.A., Fitzgibbon, M.J., Fleming, M.A., Caron, P.R., Hsiao, K., and Navia, M.A. (1995). X-ray structure of calcineurin inhibited by the immunophilin-immunosuppressant FKBP12-FK-506 complex. Cell **82**, 507–522.

Harding, M.W., Galat, A., Uehling, D.E., and Schreiber, S.L. (1989). A receptor for the immunosuppressant FK-506 is a cis-trans peptidyl-prolyl isomerase. Nature **341**, 758–760.

Liu, J., Farmer, J.D., Lane, W.S., Friedman, J., Weissman, I., and Schreiber, S.L. (1991). Calcineurin is a common target of cyclophilin-cyclosporin A and FKBP-FK-506 complexes. Cell **66**, 807–815.

Maki, N., Sekiguchi, F., Nishimaki, J., Miwa, K., Hayano, T., Takahashi, N., and Suzuki, M. (1990). Complementary DNA encoding the human T-cell FK506-binding protein, a peptidylprolyl cis-trans isomerase distinct from cyclophilin. Proc. Natl. Acad. Sci. USA **87**, 5440–5443.

Milan, D., Griffith, J., Su, M., Price, E.R., and McKeon, F. (1994). The latch region of calcineurin B is involved in both immunosuppressant-immunophilin complex docking and phosphatase activation. Cell **79**, 437–447.

Sabatini, D.M., Erdjument-Bromage, H., Lui, M., Tempst, P., and Snyder, S.H. (1994). RAFT1: a mammalian protein that binds in a rapamycin-dependent fashion and is homologous to yeast TORs. Cell **78**, 35–43.

Sabatini, D.M., Pierchala, B.A., Barrow, R.K., Schell, M.J., and Snyder, S.H. (1995). The rapamycin and FKBP12 target (RAFT) displays phosphatidylinositol 4-kinase activity. J. Biol. Chem. **270**, 20875–20878.

Siekierka, J.J., Staruch, M.J., Hung, S.H., and Sigal, N.H. (1989a). FK-506, a potent novel immunosuppressive agent, binds to a cytosolic protein which is distinct from the cyclosporin A-binding protein, cyclophilin. J. Immunol. **143**, 718–726.

Siekierka, J.J., Hung, S.H., Poe, M., Lin, C.S., and Sigal, N.H. (1989b). A cytosolic binding protein for the immunosuppressant FK-506 has peptidyl-prolyl isomerase activity but is distinct from cyclophilin. Nature **341**, 755–757.

Siekierka, J.J., Wiederrecht, G., Greulich, H., Boulton, D., Hung, S.H.Y., Cryan, J., Hodges, P.J., and Sigal, N.H. (1990). The cytosolic binding protein for the immunosuppressant FK-506 is both a ubiquitous and highly conserved peptidyl-prolyl cis-trans isomerase. J. Biol. Chem. **265**, 21011–21015.

Timerman, A.P., Ogunbumni, E., Freund, E., Wiederrecht, G., Marks, A.R., and Fleischer, S. (1993). The calcium release channel of sarcoplasmic reticulum is modulated by FK-506-binding protein. J. Biol. Chem. **268**, 22992–22999.

Van Duyne, G.D., Standaert, R.F., Karplus, P.A., Schreiber, S.L., and Clardy, J. (1991). Atomic structure of FKBP-FK-506, an immunophilin-immunosuppressant complex. Science **252**, 839–842.

Wang, T., Donahoe, P.K., and Zervos, A.S. (1994). Specific interaction of type I receptors of the TGF-β family with the immunophilin FKBP-12. Science **265**, 674–676.

Wiederrecht, G., Hung, S., Chan, K.H., Marcy, A., Martin, M., Calaycay, J., Boulton, D., Sigal, N., Kincaid, R.L., and Siekierka, J.J. (1992). Characterization of high molecular weight FK-506 binding activities reveals a novel FK-506-binding protein as well as a protein complex. J. Biol. Chem. **267**, 21753–21760.

Yang, W-M., Inouye, C.J., and Seto, E. (1995). Cyclophilin A and FKBP12 interact with YY1 and alter its transcriptional activity. J. Biol. Chem. **270**, 15187–15193.

■ *John J. Siekierka:*
The R.W. Johnson Pharmaceutical Research Institute
Route 202 South
Raritan, NJ 08869, USA
Tel. 1 908 704 4599
Fax. 1 908 526 7118
E-mail: jsiekierka@prius.jnj.com

Mammalian FKBP12.6

Among the known mammalian members of the family of FK506-binding proteins, FKBP12.6 is the one most closely related to FKBP12. The affinities of FK506 and rapamycin for FKBP12.6 are equivalent to their affinities for FKBP12. Importantly, the FKBP12.6–FK506 complex is as potent an inhibitor of calcineurin as the FKBP12–FK506 complex, indicating that FKBP12.6 may be associated with some of the immunosuppressive or toxic side effects of FK506 therapy. The physiological function of FKBP12.6 is distinct from that of FKBP12. Whereas FKBP12 binds to and regulates the ryanodine receptor (RyR-1) of skeletal muscle sarcoplasmic reticulum, FKBP12.6 binds to and regulates the ryanodine receptor isoform (RyR-2) of cardiac muscle.

■ Alternative names

OTK4 (Arakawa et al., 1994); FKBP-C (cardiac) (Timerman et al., 1994).

■ FKBP12.6 cDNA and amino acid sequence

The bovine FKBP12.6 protein was originally purified from brain and completely sequenced (PIR accession number A53924; Sewell et al., 1994). The human cDNA (GenBank accession numbers D38037 and L37086) was cloned by both random sequencing (Arakawa et al., 1994) and by use of amino acid sequence information obtained from the bovine protein (Lam et al., 1995). The human cDNA, isolated from a brain library, is 881 nucleotides in length. The FKBP12.6 transcript is expressed in most tissues surveyed, with the highest steady-state levels of mRNA observed in brain and thymus. FKBP12.6 transcript levels are significantly lower than those observed for FKBP12. The 324 bp ORF encodes an 11.6 kDa protein of 108 amino acids, identical in length to the 11.8 kDa FKBP12 protein (Maki et al., 1990). An alternatively spliced form of the FKBP12.6 cDNA encodes an 80 amino acid protein with a predicted molecular weight of 8.8 kDa (Arakawa et al., 1994). To date, no FKBP12.6 homologs have been found in other species and nothing is known about its transcriptional or translational regulation.

Human and bovine FKBP12.6 are identical in amino acid sequence and there are only 18 amino acid differences between human FKBP12 and FKBP12.6. The calculated molecular weight, 11.6 kDa, of FKBP12.6 is 200 Da lower than that of FKBP12. The most significant amino acid difference between FKBP12 and FKBP12.6 is the substitution of a phenylalanine residue in FKBP12.6 for a tryptophan residue at position 59 in FKBP12. The tryptophan residue is one of the most highly conserved amino acids among all of the FKBP family members and forms the base of the cavity into which FK506 and rapamycin insert (Van Duyne et al., 1991).

■ FKBP12.6 protein

The FKBP12.6 protein has been characterized from two sources. Using FK506 binding as an assay, bovine FKBP12.6 has been purified to homogeneity using a six-column protocol that includes cation exchange, hydrophobic interaction, size exclusion, and affinity chromatography (Sewell et al., 1994). Human FKBP12.6 has been purified to homogeneity from protein expressed in Escherichia coli (Lam et al., 1995). FKBP12.6 expressed in E. coli localizes to the periplasm and large amounts can be quickly purified to homogeneity by cracking the periplasm and applying the extract to a cation exchange (CM) HPLC column (Lam et al., 1995). Despite being slightly lower in molecular weight than FKBP12, the migration of purified FKBP12.6 on SDS–PAGE gels is slightly retarded relative to that of FKBP12. Among the known FKBP family members, FKBP12.6 is, at 85% amino acid sequence identity, the closest relative of FKBP12 and many of the antibodies developed against FKBP12 will also bind to FKBP12.6 (Sewell et al., 1994). The high amino acid similarity suggests that FKBP12.6 has a tertiary structure quite similar to that of FKBP12.

■ Biological activities of FKBP12.6

Like FKBP12, FKBP12.6 is an enzyme that catalyses the cis to trans isomerization of peptidyl–prolyl bonds in tetrapeptides of the general structure N-succinyl–Ala–Xaa–cis-Pro-Phe–p-nitroanilide (Sewell et al., 1994) and is termed a peptidyl–prolyl isomerase (PPIase). However, the catalytic activity of FKBP12.6 is about twofold lower than that of FKBP12. Like FKBP12, FKBP12.6 prefers substrates in which the amino acid immediately preceding the proline is a hydrophobic amino acid such as Leu, Ile, or Phe (Lam et al., 1995). The isomerase activity of FKBP12.6 has not, to date, been demonstrated upon any naturally occurring substrates.

The best characterized activities of FKBP12.6 occur in the presence of the immunosuppressive drugs FK506 and rapamycin (RAP). FKBP12.6 binds FK506 (K_d=0.55 nM) with an affinity almost equivalent to that of FKBP12 (K_d=0.4 nM) (Siekierka et al., 1989). The binding of FK506 inhibits the PPIase activity of FKBP12.6 in a competitive manner. Like FKBP12, FKBP12.6 also binds rapamycin (RAP). RAP is a twofold more potent inhibitor of PPIase activity than FK506, suggesting that it binds with twice the affinity to the active site. RAP's greater potency, relative to FK506, as a PPIase inhibitor has also been observed with FKBP12 (Bierer et al., 1990).

■ Biological and pharmacological interactions

FKBP12.6 and FKBP12 are, at present, the only mammalian FKBP family members that, in the presence of FK506, bind to and become potent inhibitors of the protein phosphatase calcineurin (CaN). The inhibition of CaN is associated with both the immunosuppressive and toxic side effects of FK506 (Dumont et al., 1992). The abilities of the FKBP12.6–FK506 and FKBP12–FK506 complexes to inhibit CaN phosphatase activity in in vitro assays are identical (Sewell et al., 1994; Lam et al., 1995). Moreover, FKBP12.6 and FKBP12 are equipotent at mediating the FK506 sensitivity of a Jurkat T cell line (Lam et al., 1995), suggesting that, where present, FKBP12.6 can mediate the immunosuppressive action and may be responsible for mediating some of the toxic side effects of FK506. FKBP12.6 and FKBP12 are also, at present, the only mammalian FKBP family members that, in the presence of rapamycin, are capable of binding to the 288 kDa protein termed the mammalian target of rapamycin (mTOR) (Lam et al., 1995). The affinity of the FKBP12.6–RAP complex for mTOR has not been quantified. Qualitatively, it appears to bind mTOR less well than the FKBP12–RAP complex. However, it seems likely that FKBP12.6 is also capable of mediating the immunosuppressive and toxic side effects of rapamycin. When considering the immunosuppressive and toxic side effects of FK506 and rapamycin, it is important to recognize that FKBP12.6 is much less abundant in most tissues than FKBP12, thereby suggesting that FKBP12 is likely to be responsible for mediating the majority of FK506's effects.

Like FKBP12, the physiological function of FKBP12.6 is unrelated to its pharmacological role. In the absence of drug, the best characterized function of FKBP12 is its association with and role in modulating the activity of the skeletal muscle ryanodine receptor isoform-1 (RyR-1), an intracellular Ca^{2+} release channel (CRC) required for excitation-contraction coupling (Jayaraman et al., 1992; Timerman et al., 1993). The CRC isolated from the sarcoplasmic reticulum of skeletal muscle is a tetramer of 565 kDa RyR-1 protomers, with each protomer associated with a single FKBP12 protein (Timerman et al., 1993). Thus, the overall stoichiometry of the skeletal muscle CRC can be represented by the formula $(RyR-1)_4(FKBP12)_4$. FKBP12 insures cooperativity among the RyR-1 protomers and stabilizes the closed conformation of the channel (Timerman et al., 1993; Brillantes et al., 1994; Mayrleitner et al., 1994). In the absence of FKBP12, the channel flickers between subconductance states, while in the presence of FKBP12 the channel opens to the full conductance state (Brillantes et al., 1994). The CRC in cardiac muscle (RyR-2) also contains four 565 kDa ryanodine protomers 64% identical to RyR-1. However, unlike the skeletal muscle CRC, the cardiac CRC is associated with four FKBP12.6 molecules (Timerman et al., 1996; Lam et al., 1995) and the stoichiometry of the complete cardiac CRC is $(RyR-2)_4(FKBP12.6)_4$. It seems likely that the function of FKBP12.6 on the cardiac CRC will be analogous to the function of FKBP12 on the skeletal muscle CRC.

■ References

Arakawa, H., Nagase, H., Hayashi, N., Fujiwara, T., Ogawa, M., Shin, S., and Nakamura, Y. (1994). Molecular cloning and expression of a novel human gene that is highly homologous to human FK506-binding protein 12kDa (hFKBP-12) and characterization of two alternatively spliced transcripts. Biochem. Biophys. Res. Commun. **200**, 836–843.

Bierer, B., Mattila, P., Standaert, R., Herzenberg, L., Burakoff, S., Crabtree, G., and Schreiber, S. (1990). Two distinct signal transmission pathways in T lymphocytes are inhibited by complexes formed between an immunophilin and either FK506 or rapamycin. Proc. Natl. Acad. Sci. USA **87**, 9231–9235.

Brillantes, A.-M., Ondrias, K., Scott, A., Kobrinsky, E., Ondriasova, E., Moschella, M., Jayaraman, T., Landers, M., Ehrlich, B., and Marks, A. (1994). Stabilization of calcium release channel (ryanodine receptor) function by FK506-binding protein. Cell **77**, 513–523.

Dumont, F., Staruch, M., Koprak, S., Siekierka, J., Lin, C., Harrison, R., Sewell, T., Kindt, V., Beattie, T., Wyvratt, M., and Sigal, N. (1992). The immunosuppressive and toxic effects of FK-506 are mechanistically related: pharmacology of a novel antagonist of FK-506 and rapamycin. J. Exp. Med. **176**, 751–760.

Jayaraman, T., Brillantes, A.-M., Timerman, A., Fleischer, S., Erdjument-Bromage, H., Tempst, P., and Marks, A. (1992). FK506 binding protein associated with the calcium release channel (Ryanodine receptor). J. Biol. Chem. **267**, 9474–9477.

Lam, E., Martin, M.M., Timerman, A.P., Sabers, C., Fleischer, S., Lukas, T., Abraham, R.T., O'Keefe, S.J., O'Neill, E.A., and Wiederrecht, G.J. (1995). A novel FK506 binding protein can mediate the immunosuppressive effects of FK506 and is associated with the cardiac ryanodine receptor. J. Biol. Chem. **270**, 26511–26522.

Maki, N., Sekiguchi, F., Nishimaki, J., Miwa, K., Hayano, T., Takahashi, N., and Suzuki, M. (1990). Complementary DNA encoding the humanT-cell FK506-binding protein, a peptidylprolyl cis-trans isomerase distinct from cyclophilin. Proc. Natl. Acad. Sci. USA **87**, 5440–5443.

Mayrleitner, M., Timerman, A., Wiederrecht, G., and Fleischer, S. (1994). The calcium release channel of sarcoplasmic reticulum is modulated by FK-506 binding protein: effect of FKBP-12 on single channel activity of the skeletal muscle ryanodine receptor. Cell Calcium **15**, 99–108.

Sewell, T., Lam, E., Martin, M., Leszyk, J., Weidner, J., Calaycay, J., Griffin, P., Williams, H., Hung, S., Cryan, J., Sigal, N., and Wiederrecht, G. (1994). Inhibition of calcineurin by a novel FK-506 binding protein. J. Biol. Chem. **269**, 21094–21102.

Siekierka, J.J., Hung, S.H.Y., Poe, M., Lin, C. S., and Sigal, N.H. (1989). A cytosolic binding protein for the immunosuppressant FK-506 has peptidyl-prolyl isomerase activity but is distinct from cyclophilin. Nature **341**, 755–757.

Timerman, A., Ogunbumni, E., Freund, E., Wiederrecht, G., Marks, A., and Fleischer, S. (1993). The calcium release channel of sarcoplasmic reticulum is modulated by FK-506-binding protein. J. Biol. Chem. **268**, 22992–22999.

Timerman, A., Jayaraman, T., Wiederrecht, G., Onoue, H., Marks, A., and Fleischer, S. (1994). The ryanodine receptor from canine heart sarcoplasmic reticulum is associated with a novel FK-506 binding protein. Biochem. Biophys. Res. Commun. **198**, 701–706.

Timerman, A.P., Onoue, H., Xin, H.-B., Barg, S., Copello, J., Wiederrecht, G., and Fleischer, S. (1996). Selective binding of FKBP12.6 by the cardiac ryanodine receptor. J. Biol. Chem. **271**, 20385–20391.

Van Duyne, G., Standaert, R., Karplus, P., Schreiber, S., and Clardy, J. (1991). Atomic structure of FKBP-FK506, an immunophilin-immunosuppressant complex. Science **252**, 839–842.

■ Gregory J. Wiederrecht:
Merck Research Laboratories
P.O. Box 2000; Mail Code R80M-260B
Rahway, NJ 07065-0900, USA
Tel. 1 908 594 6576
Fax. 1 908 594 7140
E-mail: greg_wiederrecht@merck.com

Mammalian FKBP25

FKBP25 was isolated to homogeneity from bovine brain. It was the second FKBP for which peptidyl–prolyl cis–trans (PPIase) activity was established. The immunosuppressive drug rapamycin binds to mammalian FKBP25 with a high inhibitory constant, $K_i = 0.9$ nM, while the FKBP25–FK506 complex has a much weaker K_i of 160 nM. Mammalian FKBP25 comprises two domains, the N-terminal domain whose function remains unknown and the C-terminal domain which binds the immunosuppressive drugs. In the C-terminal domain of FKBP25 there exists a nuclear translocation signal (NLS) but the protein has been localized both in the cytosolic and nuclear fractions of human T cells. FKBP25 binds strongly to cation exchangers, including double-stranded DNA.

■ Alternative names

The prokaryotic homologs of mammalian FKBP25 are named macrophage infectivity potentiator (Mip) (Engleberg et al., 1989; Bangsborg et al., 1991; see entry p. 399).

■ Isolation

FKBP25 was isolated to homogeneity from bovine brain and other organs using two different procedures (Galat et al., 1992). Brain tissues were extracted with (1) phosphate or (2) Tris buffer, each containing 0.3–0.5 M NaCl and 0.5% Triton X-100. After dialysis the solubilized proteins were passed through a DEAE–cellulose (in Tris buffer, method 1) or were trapped on a CM–cellulose column (in phosphate buffer, method 2). The first fractions from DEAE (1) or the proteins eluted with high salt from CM (2) were fractionated on a chromatofocusing (PBE118) gel or were resolved on a preparative flat-bed isoelectrofocusing (ampholites from 11 to 7) gel. Final purification of FKBP25 was carried out on MonoS (Pharmacia).

■ FKBP25 sequence

The N-terminus of bovine FKBP25 is blocked. Almost the entire sequence (216 aa of 224) of bovine FKBP25 was established by Edman degradation by W. Lane at Harvard University (PIR accession number A40050). Two years later, cDNAs encoding human FKBP25 were cloned and sequenced (GenBank accession number M90309). The sequences established by Hung and Schreiber (1992) (PIR accession number JQ1522) and Wiederrecht et al. (1992) (PIR accession number JT0605) are identical, but that given by Jin et al. (1992) (Genbank accession number M90820; PIR A42774) has one amino acid difference (Ala-181/Thr-181). The entire sequence of bovine FKBP25 was established from cDNA (GenBank accession number M95123).

■ FKBP25 protein and function

Peptidyl–prolyl cis–trans isomerases have recently been reviewed in detail (Galat and Metcalfe, 1995; see also the overview p. 359). FKBP25 is a member of the mammalian family of FK506-binding proteins, which currently comprises six different classes of proteins, namely FKBP12, FKBP13, FKBP25, FKBP38, FKBP52 and FKBP65 (see Fig. 1). Human FKBP25 is a very basic protein (pI=9.7) and its molecular mass established from cDNA is 25 177 daltons. Both mammalian and prokaryotic FKBP25s contain a large fraction of charged residues but only the mammalian FKBP25s are highly hydrophilic proteins. Owing to its high pI, bovine FKBP25 migrates as a 30 kDa protein on a 12% SDS–PAGE. FKBP25 has a different substrate specificity from FKBP12, namely it isomerases Suc–Ala–Ala–Pro–Phe–p-NA 35% faster than Suc–Ala–Ala–Leu–Phe–p-NA which is the preferred substrate of FKBP12 (Harrison and Stein, 1990). With the preferred substrate it has $k_{cat}/K_m = 0.8 \times 10^6$ $M^{-1}s^{-1}$ (Galat et al., 1992; Galat, 1996). FKBP25 binds the immunosuppressive drugs FK506

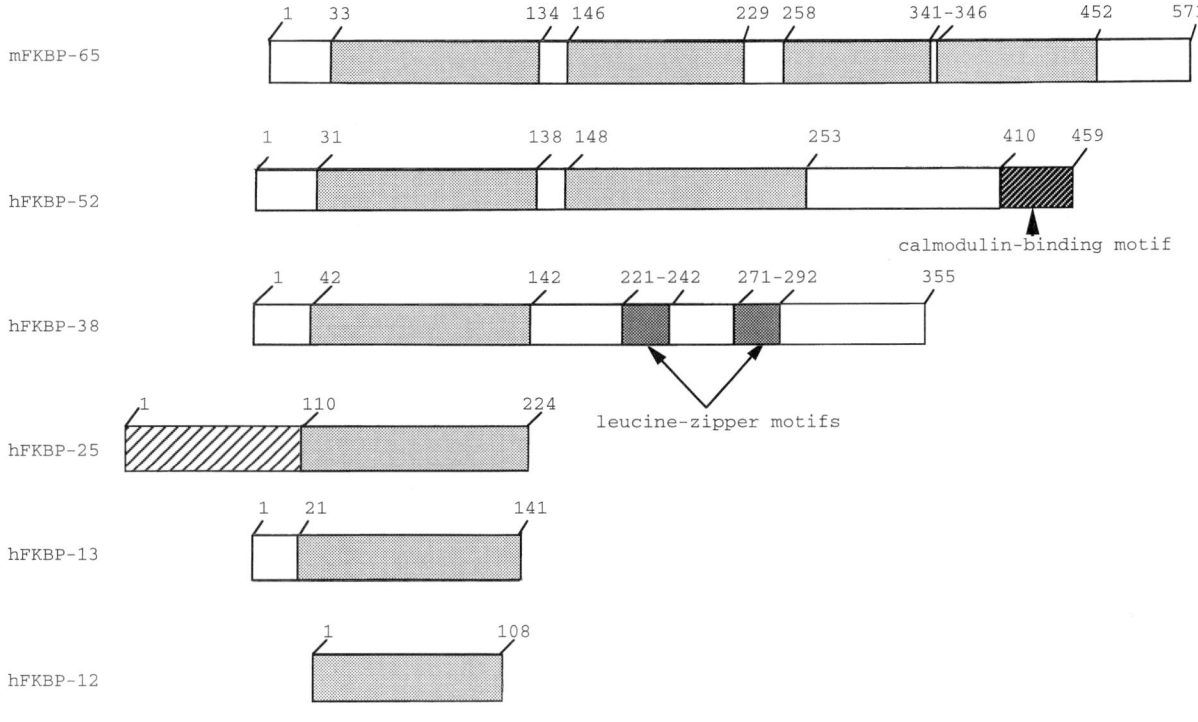

Figure 1. Global comparison of six sequences of the FKBP family of proteins. Murine mFKBP65, GenBank: L07063 (Coss *et al.*, 1995) contains a cleavable 33 amino acid presequence; four FK506–rapamycin-binding domains are denoted by grey rectangles, M_m=61.003 kDa, estimated pI=5.0, hydrophobicity index H_i=33.0% (data for processed protein). Human hFKBP52, GenBank: M88279 (Peattie *et al.*, 1992), two FK506-binding domain are present in the N-terminal half of the protein, three tetratricopeptide repeat motifs (TPR) occur at amino acids 273–306, 322–355, 356–389, and a calmodulin-binding domain in the C-terminal half; hFKBP38 (Lam *et al.*, 1995), two consensus leucine zipper repeats (denoted by deep black rectangles) are at aa 221–242 and aa 271–292, three imperfect (34 amino acid) TPR-motifs are in the C-terminal domain, M_m=38.341 kDa, estimated pI=7.6, H_i=42.3%. hFKBP25, GenBank: M90309 (Hung and Schreiber, 1992); hFKBP13, GenBank: M65128, contains a cleavable 21 amino acid presequence (Jin *et al.*, 1991); hFKBP12, GenBank: M34539 contains cleavable methionine at its N-terminus (Standaert *et al.*, 1990).

or rapamycin, but the inhibitory constants of these molecules are different, namely K_i=0.9 nM (rapamycin) and K_i=160 nM (FK506) (Galat *et al.*, 1992). The inhibitory constants (K_i) of these drugs and FKBP12 are 0.2 and 0.4 nM, respectively (Bierer *et al.*, 1990). Neither of the complexes FKBP25–(immunosuppressive drug) inhibit the phosphatase activity of calcineurin. Fractionation of proteins from human lymphocytes showed that FKBP25 is associated with the insoluble fractions (Jin and Burakoff, 1993; Riviere *et al.*, 1993). However, a proportion of the protein also exists in the cytosolic fraction as detected on immunoblots using polyclonal anti-FKBP25 antibodies (Riviere *et al.*, 1993). In the rapamycin-binding domain of FKBP25 the motif KK(X)$_7$KK(X)$_{26}$KKKK is present, which is a typical feature of NLS. Immunoblotting showed that FKBP25 is associated with the insoluble fraction and thus it may be reasoned that the protein is localized in the nucleus of human and murine T cells. Mammalian FKBP25 binds strongly to many cation exchangers including double-stranded DNA–cellulose matrix and thus one may speculate that the protein may interact *in vivo* with DNA. However, the prokaryotic homologs of FKBP25 are localized in the outer membrane of bacteria, whereas a part of the total cellular content of FKBP25 is in the cytosolic fraction of T lymphocytes. Perhaps *in situ* localization of FKBP25 may show if FKBP25 is a nuclear protein in mammalian cells.

■ Structure of FKBP25

The X-ray structure of FKBP25 remains to be elucidated but certain model structures have been proposed (Galat *et al.*, 1992; Riviere *et al.*, 1993). FKBP25 is composed of two domains, the N-terminal domain whose function remains unknown and the C-terminal domain which

```
hFKBP-25                                         1-MAAAVPQRAWT-VEQLR-SEQLPKKDIIKFLQEHGSDSFLA-EHKLLGNIKNVAKTANKDHLV  (60)
                                                   :::        :::   : ::: ::: :  :     :      :     : :::  :
lpFKBP-25                                1-MKMKL-VTAAVMGLAMSTAMAATDATSLATDKDKLSYSIGADLGKNFKNQGIDVNPEAMAKGMQDAMSGAQLALTEQQMKDV  (81)
                                           :::::  :::::::::::                ::::::::::::::  ::::  ::::::: ::    ::::::
lmFKBP-25                       1-MKMRLVAAAMGLAMSTTIAATATTDATTSAPGTSLTTDTEKLSYSIGADLGKNFKKQGIEISPAAMAKGLQDMSGGQLLTDDQMKDV  (90)
                                                         :: ::   : ::::::::::::::::::::::  :   ::::::  :           :::::
ctFKBP-25                       1-MKNILSWMLMFAVALPIVGCDNGGSQTSATEKSMVEDSALTDNQKLSRTFGHLLARQLSRTEDFSLDLVEVIKGMQSEIDGQ---SA-PLTDT  (90)

Consensus                                                                          T                    L



hFKBP-12                                                                                         1-MG-VQVET-ISPGDGRTFPKRGQTCVVH--YTGMLEDGKKFD-----SS  (40)
                                                                                                   ::  ::::  ::: ::  :::: :: ::  :::: :::  ::
lpFKBP-25      TAYNHLFETKRF-KGTESISKVSEQVKNVKLNEDKPKETKSEETLDEG-PPKYTKSVLKKGDKTNFPKKGDVVHCWYTGTLQDGTVFDTNIQTSA (152)
                                                                 :         ::::   :      :::   ::  ::  ::::  :::   :::::  TE-
lpFKBP-25      ---LNKFQKDLMAKRTAEFNKKADENK--VKG-EAFLTENKNKPGVVVL-PSGLQYKVINSGNGVK-PGKSDTVTVEYTGRLIDGTVFDS---TE- (165)
                  :::::::::::: :: :: :: :::  :::  ::::: :: ::::::: :::::: ::: : ::  :       ::::: ::: :::::
lmFKBP-25      ---LNKFQKDLMMKRSAEFNKKAEENK--SKG-EAFLNENKSKEGVVSL-PSGLQYNILERGDGAK-PTKDDVVTVEYTGKLIDGQVFDS---TE- (174)
                  :  :   : ::                 :::   :     :  : :                                       :::::: SE-
chFKBP-25      ----E-YEKQMAEVQKASFEAKCSENLASA--EEFLKENKEKAGVIELEPNKLQYRVVKEGTGRV-LSGKPTALLHYTGSFIDGKVFDS---SE- (173)

Consensus               E         E K                   G                                       YTG    DG   FD



hFKBP-12       R-DRN-KPFKFMLGKQEVIRGWEEGVAQMSVGQRAKLTISPDYAYGATGHPGI-IPPHATLVFDVELLKLE*                      (108)
                :     ::  :::::  :: :     :: :  ::  :                :     : :: :    :::
hFKBP-25       KKKKNAKPLSFKVGVGKVIRGWDEALLAMSKGEKARLEIEPEWAYGKKGQPDAKIPPNAKLTFEVELVDID*                     (224)
                       :: ::     ::          :::             :::  :         ::::::::::::
lpFKBP-25      KT---GKPATFQVS--QVIPGWTEALQLMPAGSTWEIYVPSGLAYGPRS-VGGPIGPNETLIFKIHLISVKKSS*                  (233)
                ::  ::::: ::  :::::::::::::::::::::: :::::::::: ::::::::::::::::::::::::
lmFKBP-25      KT---GKPATFKVS--QVIPGWTEALQLMPAGSTWEVYIPSNLAYGPRS-VGGPIGPNETLIFKIHLISVKKSDA*                 (243)
                :    :: :: ::  ::::::: ::  :::::: :   :: :::: :  :  ::: :: ::  :: ::: ::
chFKBP-25      KN---KEPILLPLT--KVIPGFSQGMQGMKEGEVRVLYIHPDLAYG--T--AGQLPPNSLLIFEVKLIEAND-DNVSVTE*             (243)

Consensus      K      P          VI G       M  G              AYG         PN  L F   L
```

Figure 2. Sequence alignment of the FKBP25 family of proteins. hFKBP12, GenBank: M34539 (Standaert *et al.*, 1990); hFKBP25, GenBank: M90309 (Hung and Schreiber, 1992); lpFKBP25, FKBP25 from *Legionella pneumphilia*, PIR: A30591 (Engleberg *et al.*, 1989); lmFKBP25, FKBP25 from *Legionella micdadei*, PIR: A43596 (Bangsborg *et al.*, 1991); ctFKBP25, FKBP25 from *Chlamydia trachomatis*, EMBL: X66126. The highest overall sequence similarity exists between hFKBP25 and lpFKBP25 (Galat *et al.*, 1992). The C-terminal domains of hFKBP25 and lpFKBP25 have good sequence correlation, whereas the N-terminal domains show poor correlation. Although both proteins have very basic pIs (about 9.5–10.0) they differ in hydrophobicity, namely $H_i=17.9\%$ (very hydrophilic) for hFKBP25 and $H_i=36.9\%$ (moderately hydrophobic) for lpFKBP25.

binds to FK506 or rapamycin (see Fig. 1). Sequence analyses suggest that a helix–loop–helix motif may exist in the N-terminal part of mammalian FKBP25. Figure 2 shows a sequence alignment of human FKBP12 and FKBP25 with four prokaryotic homologs of FKBP25. There is a significant sequence similarity between the C-terminal part of FKBP25 and FKBP12, but the former has an additional seven amino acids inserted into the loop between the two parts of the third strand of sheet. Also, in the C-terminal part of FKBP25 there exists the NLS signal sequence, which is absent in FKBP12. Several amino acid substitutions in the C-terminal part of FKBP25 vs. FKBP12 create different interaction networks between each protein and the immunosuppressive drugs. Both FKBP25 and its complexes with FK506 or rapamycin bind well to double-stranded DNA–cellulose matrix, while FKBP12 does not bind to DNA (Riviere et al., 1993). The cellular function and biological significance of FKBP25–(immunosuppressive drugs) complexes remain to be established.

References

Bangsborg, J.M., Cianciotto, N.P., and Hindersson, P. (1991). Nucleotide sequence analysis of the *Legionella micdadei* mip gene, encoding a 30-kilodalton analog of the *Legionella pneumophilia* Mip protein. Infect. Immun. **59**, 3836–3840.

Bierer, B.E., Mattila, P.S., Standaert, R.F., Herzenberg, L.A., Burakoff, S.J., Crabtree, G., and Schreiber, S.L. (1990). Two distinct signal transmission pathways are inhibited in T lymphocytes by complexes formed between an immunophilin and either FK506 or rapamycin. Proc. Natl. Acad. Sci. USA **87**, 9231–9235.

Coss, M.C., Winterstein, D., Sowder, R.C., and Simek, S.L. (1995). Molecular cloning, DNA sequence analysis, and biochemical characterization of a novel 65-kDa FK506-binding protein (FKBP65). J. Biol. Chem. **270**, 29336–29341.

Engleberg, N.C., Carter, C., Weber, D.R., Cianciotto, N.P., and Eisenstein, B.I. (1989). DNA sequence of mip, a *Legionella pneumophila* gene associated with macrophage infectivity. Infect. Immun. **57**, 1263–1270.

Galat, A. (1996). A large-scale processing of kinetic data files with derivation of the inhibitory constant Ki: an application to proline isomerases. Comp. Chem. **20**, 279–281.

Galat, A. and Metcalfe, S.M. (1995). Peptidylproline cis/tras isomerase. Progr. Biophys. Mol. Biol. **63**, 67–118.

Galat, A., Lane, W.S., Standaert, R.F., and Schreiber, S.L. (1992). A rapamycin-selective 25-kDa immunophilin. Biochemistry **31**, 2427–2434.

Harrison, R.K. and Stein, R.L. (1990). Substrate specificities of the peptidyl prolyl cis-trans isomerase activities of cyclophilin and FK-506 binding protein: evidence for the existence of a family of distinct enzymes. Biochemistry **29**, 3813–3816.

Hung, D.T. and Schreiber, S.L. (1992). cDNA cloning of a human 25 kDa FK506 and rapamycin binding protein. Biochem. Biophys. Res. Commun. **184**, 733–738.

Jin, Y.J. and Burakoff, S.J. (1993). The 25 kDa FK506 binding protein is localised in the nucleus and associates with casein kinase II and nucleolin. Proc. Natl. Acad. Sci. USA **90**, 7769–7773.

Jin, Y.-J., Albers, M.W., Lane, W.S., Bierer, B.E., Schreiber, S.L., and Burakoff, S.J. (1991). Molecular cloning of a membrane-associated human FK506- and rapamycin-binding protein FKBP-13. Proc. Natl. Acad. Sci. USA **88**, 6677–6681.

Jin, Y.-J., Burakoff, S.J., and Bierer, B.E. (1992). Molecular cloning of a 25-kDa high affinity rapamycin binding protein, FKBP25. J. Biol. Chem. **267**, 10942–10945.

Lam, E., Martin, M., and Wiederrecht, G. (1995). Isolation of a cDNA encoding a novel human FK506-binding protein homolog containing leucine zipper and tetratricopeptide repeat motifs. Gene **160**, 297–302.

Peattie, D.A., Harding, M.W., Fleming, M.A., DeCenzo, M.T., Lippke, J.A., Livingston, D.J., and Benasutti, M. (1992). Expression and characterization of human FKBP52, an immunophilin that associates with the 90-kDa heat shock protein and is a component of steroid receptor complex. Proc. Natl. Acad. Sci. USA **89**, 10974–10978.

Rivière, S., Menez, A., and Galat, A. (1993). On the localization of FKBP25 in T-lymphocytes. FEBS Lett. **315**, 247–251.

Standaert, R.F., Galat, A., Verdine, G.L., and Schreiber, S.L. (1990). Molecular cloning and overexpression of the human FK506-binding protein, FKBP. Nature **346**, 671–674.

Wiederrecht, G., Martin, M., Sigal N.H., and Siekierka J.J. (1992). Isolation of a human cDNA encoding a 25 kDa FK-506 and rapamycin binding protein. Biochem. Biophys. Res. Commun. **185**, 298–303.

Andrzej Galat:
DIEP/DSV, Bat. 152
CE-Saclay
91191 Gif-sur-Yvette, France
Tel. 33 1 69083040
Fax. 33 1 69089137

Mammalian FKBP51

FKBP51 is a member of the FK506-binding family of immunophilins. It shares about 60% amino acid sequence identity with FKBP52, and also has peptidyl–prolyl cis–trans isomerase activity. Like FKBP52, FKBP51 is an Hsp90-binding protein that has been found as a component in steroid receptor complexes. FKBP51 is expressed in a wide range of tissues and has been shown to be up-regulated during glucocorticoid-induced apoptosis of thymocytes and during adipocyte differentiation. Little is currently known about specific functions for FKBP51.

■ Alternative names

p54 (Smith et al., 1990), FKBP54 (Smith et al., 1993b).

■ Isolation of FKBP51

FKBP51 was originally noted as a component of native chicken progesterone receptor complexes (Smith et al., 1990), but cDNAs for mouse FKBP51 were independently isolated in two laboratories. The gene for FKBP51 was one of several identified in the WEHI-7TG murine thymoma cell line whose expression was increased during glucocorticoid-induced apoptosis (Baughman et al., 1991). A cDNA for FKBP51 was subsequently sequenced and shown to be a novel FKBP (Baughman et al., 1995). In a separate study (Yeh et al., 1995), a cDNA for FKBP51 was cloned based on the accumulation of the mRNA during the clonal expansion phase of 3T3-L1 cell growth. A cDNA sequence for human FKBP51 has also recently been obtained (Nair et al., 1997).

■ *fkbp51* gene and sequence

FKBP51 is a recent addition to the multiprotein family of FK506-binding immunophilins (reviewed in Kay, 1996). Two cDNA sequences for murine FKBP51 have been submitted to GenBank: accession numbers U16959 (Baughman et al., 1995) and U36220 (Yeh et al., 1995). A cDNA sequence for most of the ORF of human FKBP51 has also been submitted: accession number U42031. From deduced amino acid sequences, there is 100% identity between the two mouse sequences and about 90% identity between human and mouse. FKBP51 shares about 60% amino acid sequence identity with mammalian FKBP52.

■ FKBP51 protein

The cDNAs for FKBP51 encode a protein with a predicted size of approximately 51 000 Da. Similarly to FKBP52, FKBP51 has an N-terminal peptidyl–prolyl isomerase (PPIase) domain that is also the binding site for FK506. In the C-terminal two-thirds of the protein there is a region of tetratricopeptide repeats, which, in FKBP52 and several other proteins, have been shown to be responsible for Hsp90 binding.

■ Biological regulation of FKBP51

Despite one report that FKBP51 expression is restricted to thymocytes (Baughman et al., 1995), other studies have shown a broad pattern of tissue expression for FKBP51 protein (Smith et al., 1993a) and mRNA (Yeh et al., 1995; Nair et al., 1997). FKBP51 protein in rabbit reticulocyte lysate is approximately one-fifth and one-tenth the levels for FKBP52 and Cyp40, respectively (Nair et al., 1997). Expression of FKBP51 is up-regulated during glucocorticoid-induced apoptosis of mouse thymocytes (Baughman et al., 1995) and during the clonal expansion phase of 3T3-L1 cell growth, so its expression can be sensitive, in a tissue-specific manner, to growth and differentiation factors. The promoter region for the *fkbp51* gene has not yet been examined for specific regulatory elements.

■ Biological interactions

FKBP51 is an Hsp90-binding protein (Smith et al., 1993c; Nair et al., 1997) and through its interaction with Hsp90 is a component of progesterone receptor complexes. In vitro assembly studies (Nair et al., 1996) have also suggested that FKBP51 may be a component of Hsp90-containing complexes with estrogen receptor, certain tyrosine kinases, the arylhydrocarbon receptor, and the heat shock transcription factor HSF1. In the case of progesterone receptor complexes, FKBP51 is restricted to functionally mature receptor complexes also containing Hsp90 and p23 (Smith et al., 1995). Immunophilin interactions with steroid receptor complexes appear to be interchangeable and competitive (Owens-Grillo et al., 1995; Ratajczak and Carrello, 1996). Recent evidence with cell-free assembly reactions suggests that progesterone receptors (PR) have a preference for FKBP51 over FKBP52, even though FKBP52 and FKBP51 compete equally for Hsp90 binding in a purified system (Nair et al., 1997). The preferential interaction of FKBP51 with PR complexes relates to an earlier observation (Smith et al., 1990; Smith et al., 1993a) that FKBP51 binding to PR complexes is more sensitive to progesterone than FKBP52–PR binding.

Thus, while the immunophilins enter steroid receptor complexes through their association with Hsp90, the particular immunophilin present may influence Hsp90 binding to the receptor or the immunophilin may contact both Hsp90 and the receptor. It is not yet clear what relationship these interactions with progesterone receptor have on the regulation and function of receptor activity. One suggestion (reviewed by Pratt, 1993) has been that the immunophilins serve, through Hsp90, to tether targets such as steroid receptors to cytoskeletal elements for directed transport, but this model remains controversial. Although FKBP51, like the other steroid receptor-associated immunophilins FKBP52 and Cyp40, has PPIase activity, it has not been shown that Hsp90 or steroid receptors are substrates for this enzymatic activity.

■ References

Baughman, G., Harrigan, M.T., Campbell, N.F., Nurrish, S.J., and Bourgeois, S. (1991). Genes newly identified as regulated by glucocorticoids in murine thymocytes. Mol. Endocrinol. **5**, 637–644.

Baughman, G., Wiederrecht G.J., Campbell, N.F., Martin, M.M. and Bourgeois, S. (1995). FKBP51, a novel T-cell-specific immunophilin capable of calcineurin inhibition. Mol. Cell. Biol. **15**, 4395–4402.

Kay, J.E. (1996). Structure-function relationships in the FK506-binding protein (FKBP) family of peptidylprolyl *cis-trans* isomerases. Biochem. J. **314**, 361–385.

Nair, S.C., Toran, E.J., Rimerman, R.A., Hjermstad, S., Smithgall, T.E., and Smith, D.F. (1996). A pathway of multi-chaperone interactions common to diverse regulatory proteins: estrogen receptor, Fes tyrosine kinase, heat shock transcription factor HSF1, and the arylhydrocarbon receptor. Cell Stress Chap., **1**, 237–250.

Nair, S.C., Rimerman, R.A., Toran, E.J., Chen S., Prapapanich, V., Butts R.N., and Smith, D.F. (1997). Molecular cloning of human FKBP51 and comparisons of immunophilin interactions with hsp90 and progesterone receptor. Mol. Cell Biol. **17**, 594–603.

Owens-Grillo, J.K., Hoffman, K., Hutchison, K.A., Yem, A.W., Deibel, M.R., Jr., Handschumacher, R.E., and Pratt, W.B. (1995). The cyclosporin A-binding immunophilin CyP-40 and the FK506-binding immunophilin hsp56 bind to a common site on hsp90 and exist in independent cytosolic heterocomplexes with the untransformed glucocorticoid receptor. J. Biol. Chem. **270**, 20479–20484.

Pratt, W.B. (1993). The role of the heat shock proteins in regulating the function, folding, and trafficking of the glucocorticoid receptor. J. Biol. Chem. **268**, 21455–21458.

Ratajczak, T. and Carrello, A. (1996). Cyclophilin 40 (CyP-40), mapping of its hsp90 binding domain and evidence that FKBP52 competes with CyP-40 for hsp90 binding. J. Biol. Chem. **271**, 2961–2965.

Smith, D.F., Faber, L.E., and Toft, D.O. (1990). Purification of unactivated progesterone receptor and identification of novel receptor-associated protein. J. Biol. Chem. **265**, 3996–4003.

Smith, D.F., Baggenstoss, B.A.,Marion, T.N., and Rimerman, R.A.. (1993a). Two FKBP-related proteins are components of progesterone receptor complexes. J. Biol. Chem. **268**, 18365–18371.

Smith, D.F., Albers, M.W., Schreiber, S.L., Leach, K.L., and Deibel M.R. (1993b). FKBP54, a novel FK506-binding protein in avian progesterone receptor complexes and HeLa extracts. J. Biol. Chem. **268**, 24270–24273.

Smith, D.F., Sullivan, W.P., Marion, T.N., Zaitsu, K., Madden, B.J., McCormick, D.J., and Toft, D.O. (1993c). Identification of a 60-kilodalton stress-related protein, p60, which interacts with hsp90 and hsp70. Mol. Cell. Biol. **13**, 869–876.

Smith, D.F., Whitesell, L., Nair, S.C., Chen, S., Prapapanich, V., and Rimerman, R.A. (1995). Progesterone receptor structure and function altered by geldanamycin, an hsp90-binding agent. Mol. Cell. Biol. **15**, 6804–6812.

Yeh, W.-C., Li, T.-K., Bierer, B.E., and McKinght, S.L. (1995). Identification and characterization of an immunophilin expressed during the clonal expansion pahse of adipocyte differentiation. Proc. Natl. Acad. Sci. USA **92**, 11081–11085.

■ *David F. Smith:*
Department of Pharmacology
University of Nebraska Medical Center
Omaha, NE 68198-6260, USA
Tel. 1 402 559 8604
Fax. 1 402 559 7495
E-mail: dfsmith@mail.unmc.edu

Mammalian FKBP52

FKBP52 was originally isolated as a protein in association with progesterone receptor complexes and was called p59. It was characterized as an FKBP through its amino acid sequence and its ability to bind FK506. The most characteristic feature of this protein is its ability to bind to Hsp90 and its biological function remains unknown.

■ Alternative names

FKBP59 (Tai et al., 1992), FKBP51 (Wiederrecht et al., 1992), Hsp56 (Sanchez, 1990), p59 (Tai et al., 1986), p50 (Smith et al., 1993), HBI (HSP-binding immunophilin) (Chambraud et al., 1993).

■ FKBP52 gene and sequence

Nucleotide sequences encoding FKBP52 proteins from the rabbit (Lebeau et al., 1992; GenBank accession number M84474) and the human (Peattie et al., 1992; M88279) reveal an acidic protein of 459 amino acid residues with a molecular weight of 51 810 Da (human). A domain with homology to FKBP12 exists at the N-terminus. Structural modeling suggests the presence of two additional domains having lesser homology with FKBP12 and a calmodulin-binding domain at the C-terminus (Callebaut et al., 1992). The C-terminal half of FKBP52 shows some sequence homology with cyclophilin 40 (Ratajczak et al., 1993). This region contains three tetratricopeptide (TPR) domains, 34 residue repeats predicted to contain common structural features including short amphipathic helices. The TPR regions appear to be required for the binding of FKBP52 to Hsp90 (Radanyi et al., 1994). It should be noted that the sequence of FKBP52 is distinct from that of FKBP54 (FKBP51), which also binds to steroid receptor complexes (Smith et al., 1993; see entry p. 428).

■ FKBP52 protein

FKBP52 is quite abundant and ubiquitous; it has been observed in several tissues and species. It is localized predominantly in the cell nucleus (Gasc et al., 1990), but it also exists in the cytoplasm where it appears to associate with microtubules (Czar et al., 1994b). FKBP52 has been expressed and purified from Sf9 insect cells (Alnemri et al., 1993) and from bacteria (Chambraud et al., 1993) where several mutant forms were also expressed. Size estimates for native FKBP52 suggest that it is a dimer (Wiederrecht et al., 1992). Several isoforms of unknown origin have been observed upon two-dimensional electrophoresis (Sanchez et al., 1990). FKBP52 has been reported to be up-regulated by heat shock, thus, the alternative name of Hsp56 (Sanchez, 1990).

■ FKBP52 function

The biological function of FKBP52 remains unknown. It has been studied mainly because of its association with inactive complexes of steroid receptors (Tai and Faber, 1985; Tai et al., 1986, 1992; Renoir et al., 1990; Smith et al., 1993) where it appears to be bound to Hsp90 (Renoir et al., 1990; Sanchez et al., 1990). Its possible role in the folding or trafficking of the receptor has been suggested. The majority of FKBP52 in cell extracts appears to exist in complexes. Immune isolation of FKBP52 results in the coisolation of Hsp90 and Hsp70, and the binding of purified FKBP52 to Hsp90 has been demonstrated (Radanyi et al., 1994; Czar et al., 1994a). Competition between FKBP52 and cyclophilin 40 for Hsp90 binding indicates that these proteins bind at the same site on Hsp90 (Owens-Grillo et al., 1995). FKBP52 has also been shown to bind ATP and GTP, but the significance of this binding is unknown (Alnemri et al., 1993; Bihan et al., 1993).

FKBP52 binds FK506 and rapamycin (Tai et al., 1992; Wiederrecht et al., 1992; Yem et al., 1992). It contains peptidyl–prolyl *cis–trans* isomerase activity (Peattie et al., 1992; Wiederrecht et al., 1992; Chambraud et al., 1993), and it has been reported to be a weak inhibitor of calcineurin phosphatase activity (Lam et al., 1995; Abraham and Wiederrecht, 1996).

■ References

Abraham, R.T. and Wiederrecht, G.T. (1996). Immunopharmacology of rapamycin. Annu. Rev. Immunol. **14**, 483–510.

Alnemri, E.S., Fernandes-Alnemri, T., Nelki, D.S., Dudley, K., DuBois, G.C., and Litwack, G. (1993). Overexpression, characterization, and purification of a recombinant mouse immunophilin FKBP-52 and identification of an associated phosphoprotein. Proc. Natl. Acad. Sci. USA **90**, 6839–6843.

Bihan, S.L., Renoir, J.-M., Radanyi, C., Chambraud, B., Joulin, V., Catelli, M.-G., and Baulieu, E.-E. (1993). The mammalian heat shock protein binding immunophilin (p59/HBI) is an ATP and GTP binding protein. Biochem. Biophys. Res. Commun. **195**, 600–607.

Callebaut, I., Renoir, J.-M., Lebeau, M.-C., Massol, N., Burny, A., Baulieu, E.-E., and Mornon, J.-P. (1992). An immunophilin that binds Mr 90,000 heat shock protein: Main structural features of a mammalian p59 protein. Proc. Natl. Acad. Sci. USA **89**, 6270–6274.

Chambraud, B., Rouviere-Fourmy, N., Radany, C., Hsiao, K., Peattie, D.A., Livingston, D.J., and Baulieu, E.-E. (1993). Overexpression of p59-HBI (FKBP59), full length and domains, and characterization of PPIase activity. Biochem. Biophys. Res. Commun. **196**, 160–166.

Czar, M.J., Owens-Grillo, J.K., Dittmar, K.D., Hutchison, K.A., Zacharek, A.M., Leach, K.L., Deibel, Jr., M.R., and Pratt, W.B. (1994a). Characterization of the protein-protein interactions determining the heat shock protein (hsp90-hsp70-hsp56) heterocomplex. J. Biol. Chem. **269**, 11155–11161.

Czar, M.J., Owens-Grillo, J.K., Yem, A.W., Leach, K.L., Deibel, Jr., M.R., Welsh, M.J., and Pratt, W.B. (1994b). The hsp56 immunophilin component of untransformed steroid receptor complexes is localized both to microtubules in the cytoplasm and to the same nonrandom regions within the nucleus as the steroid receptor. Mol. Endocrinol. **8**, 1731–1741.

Gasc, J.-M., Renoir, J.-M., Faber, L.E., Delahaye, F., and Baulieu, E.-E. (1990). Nuclear localization of two steroid receptor-associated proteins, hsp90 and p59. Exp. Cell Res. **186**, 362–367.

Lam, E., Martin, M.M., Timmerman, A.P., Sabers, C., Fleicher, S., Lukas, T., Abraham, R.T., O'Keefe, S.J., O'Neill, E.A., and Wiederrecht, G.J. (1995). A novel FK506 binding protein can mediate the immunosuppressive effects of FK506 and is associated with the cardiac ryanodine receptor. J. Biol. Chem. **270**, 26511–26522.

Lebeau, M.-C., Massol, N., Herrick, J., Faber, L.E., Renoir, J.-M., Radanyi, C., and Baulieu, E.-E. (1992). P59, an hsp 90-binding protein. J. Biol. Chem. **267**, 4281–4284.

Owens-Grillo, J.K., Hoffmann, K., Hutchison, K.A., Yem, A.W., Deibel, Jr., M.R., Handschumacher, R.E., and Pratt, W.B. (1995). The cyclosporin A-binding immunophilin CyP-40 and the FK506-binding immunophilin hsp56 bind to a common site on hsp90 and exist in independent cytosolic heterocomplexes with the untransformed glucocorticoid receptor. J. Biol. Chem. **270**, 20479–20484.

Peattie, D.A., Harding, M.W., Fleming, M.A., DeCenzo, M.T., Lippke, J.A., Livingston, D.J., and Benasutti, M. (1992). Expression and characterization of human FKBP52, an immunophilin that associates with the 90-kDa heat shock protein and is a component of steroid receptor complexes. Proc. Natl. Acad. Sci. USA **89**, 10974–10978.

Radanyi, C., Chambraud, B., and Baulieu, E.-E. (1994). The ability of the immunophilin FKBP59-HBI to interact with the 90-kDa heat shock protein is encoded by its tetratricopeptide repeat domain. Proc. Natl. Acad. Sci. USA **91**, 11197–11201.

Ratajczak, T., Carrello, A., Mark, P.J., Warner, B.J., Simpson, R.J., Moritz, R.L., and House, A.K. (1993). The cyclophilin component of the unactivated estrogen receptor contains a tetratricopeptide repeat domain and shares identity with p59 (FKBP59). J. Biol. Chem. **268**, 13187–13192.

Renoir, J.-M., Radanyi, C., Faber, L.E., Baulieu, E.-E. (1990). The non-DNA-binding heterooligomeric form of mammalian steroid hormone receptors contains a hsp90-bound 59-kilodalton protein. J. Biol. Chem. **265**, 10740–10745.

Sanchez, E.R. (1990). Hsp56: a novel heat shock protein associated with untransformed steroid receptor complexes. J. Biol. Chem. **265**, 22067–22070.

Sanchez, E.R., Faber, L.E., Henzel, W.J., and Pratt, W.B. (1990). The 56-59-kilodalton protein identified in untransformed steroid receptor complexes is a unique protein that exists in cytosol in a complex with both the 70- and 90-kilodalton heat shock proteins. Biochemistry **29**, 5145–5152.

Smith, D.F., Baggenstoss, B.A., Marion, T.N., and Rimerman, R.A. (1993). Two FKBP-related proteins are associated with progesterone receptor complexes. J. Biol. Chem. **268**, 18365–18371.

Tai, P.-K.K. and Faber, L.E. (1985). Isolation of dissimilar components of the 8.5S nonactivated uterine progestin receptor. Can. J. Biochem. Cell Biol. **63**, 41–49.

Tai, P.-K.K., Maeda, Y., Nakao, K., Wakim, N.G., Duhring, J.L., and Faber, L.E. (1986). A 59-kilodalton protein associated with progestin, estrogen, androgen, and glucocorticoid receptors. Biochemistry **25**, 5269–5275.

Tai, P.-K.K., Albers, M.W., Chang, H., Faber, L.E., and Schreiber, S.L. (1992). Association of a 59-kilodalton immunophilin with the glucocorticoid receptor complex. Science **256**, 1315–1318.

Wiederrecht, G., Hung, S., Chan, H.K., Marcy, A., Martin, M., Calaycay, J., Boulton, D., Sigal, N., Kincaid, R.L., and Siekierka, J.J. (1992). Characterization of high molecular weight FK-506 binding activities reveals a novel FK-506-binding protein complex. J. Biol. Chem. **267**, 21753–21760.

Yem, A.W., Tomasselli, A.G., Heinrikson, R.L., Zurcher-Neely, H., Ruff, V.A., Johnson, R.A., and Deibel, Jr., M.R. (1992). The hsp56 component of steroid receptor complexes binds to immobilized FK506 and shows homology to FKBP-12 and FKBP-13. J. Biol. Chem. **267**, 2868–2871.

■ *David O. Toft:*
Department of Biochemistry and Molecular Biology
Mayo Clinic
Rochester, MN 55905, USA
Tel. 1 507 284 8401
Fax. 1 507 284 2053
E-mail: toft@mayo.edu

■ *Lee E. Faber:*
Department of Physiology
Medical College of Ohio
Toledo, OH 43699, USA
Tel. 1 419 381 4584
Fax. 1 419 381 3124
E-mail: lfaber@gemini.mco.edu

Mammalian FKBP13

The ER luminal protein FKBP13 is a member of a growing family of proteins, the FKBPs (for FK506-binding proteins), many of which bind the fungal metabolite and immunosuppressant drug, FK506. The size of the FKBP family, the distribution of the members into multiple subcellular compartments, and the degree of conservation even in lower organisms, all suggest a fundamental role in cellular function.

■ Isolation

FKBP13 was originally isolated from bovine thymus by use of a rapamycin affinity matrix (Jin et al., 1991).

■ FKBP13 gene and sequence

A cDNA clone isolated from a human colon carcinoma cell library (GenBank accession number M65128) encodes a protein of 141 amino acids and a molecular weight of 13 200 Da in agreement with the molecular mass estimated by SDS–PAGE (Jin et al., 1991). Sequence analysis revealed 51% nucleotide homology and 43% amino acid homology to FKBP12 (Jin et al., 1991; see entry p. 420). There is a particularly strong homology with the PPIase active and drug-binding site of FKBP12, suggesting both similarity of higher structure and molecular mechanism. However, the binding affinity of FK506 for FKBP13 (K_i=55 nM) is roughly 50-fold less than for FKBP12, suggesting that it may not be an *in vivo* target of the drug (Bram et al., 1993). The gene for a yeast homolog, FKB2, is 57% identical to its human counterpart (Partaledis et al., 1992; Partaledis and Berlin 1993; see entry p. 411).

■ FKBP13 protein

Cell fractionation studies and immunolocalization studies of an epitope-tagged construct suggest that FKBP13 is a luminal protein of the endoplasmic reticulum (Nigam et al., 1993). Although FKBP13 does not possess a classical C-terminal tetrapeptide ER retention signal (KDEL), the RTEL tetrapeptide at its C-terminus may serve an equivalent function; however, this remains to be proven.

■ Function and regulation of FKBP13

FKBP13 is a member of a growing family of proteins, the FKBPs (for FK506-Binding Proteins), many of which bind the fungal metabolite and immunosuppressant drug, FK506 (Fruman et al., 1994). Other members of the family have molecular masses of 12, 25, and 52, and are localized to the cytosol and nucleus. The FKBP family is part of a larger group of immunosuppressant drug-binding proteins, the immunophilins, which include the cyclosporin-binding proteins, the cyclophilins (see overview of PPIase families, p. 359). These proteins are broadly conserved across species, including yeast. The size of the FKBP family, the distribution of the members into multiple subcellular compartments, and the degree of conservation even in lower organisms, all suggest a fundamental role in cellular function.

In general, FKBPs also serve as high affinity receptors for a related drug, rapamycin. *In vitro*, these drugs inhibit the peptidyl–prolyl isomerase activity (PPIase, also refered to as rotamase) of these proteins, which is the ability to catalyse the interconversion of *cis–trans* rotamers of peptidyl–prolyl bonds. PPIase activity appears to be important for the refolding of at least some denatured proteins (Fischer and Bang, 1985).

Perhaps the best understood member of the group is FKBP12. The crystal structure of FKBP12, resolved to 1.7 Å, reveals a large pocket that forms the PPIase active site (Michnick et al., 1991). The drug appears to bind at this active site (Van Duyne et al., 1991). Upon binding FK506, FKBP12 complexes with the protein phosphatase calcineurin A, thereby inhibiting its activity. In T cells, this blocks a key signalling pathway necessary for the immune response (Liu, 1993).

Support for the notion that immunophilins might serve a chaperone function came from studies of the *Drosophila* cyclophilin homolog, ninaA, which is believed to play a key role in movement of rhodospin though the secretory pathway in photoreceptor cells (Stamnes et al., 1991; Ondek et al., 1992). When FKBP13 was shown to be an ER luminal protein (Nigam et al., 1993), it was hypothesized that, particularly given its potential PPIase activity, it might participate in the folding of secretory proteins transiting the ER. ER-enriched subcellular fractions have been shown to possess PPIase activity (Bose and Freedman, 1994). Although this has been attributed to the presence of an ER cyclophilin, it is possible that some fraction of the PPIase activity may be contributed by FKBP13.

Indirect evidence, from mammalian cells as well as yeast, suggest a chaperone function for FKBP13 within the ER lumen. When cultured cells are treated with tunicamycin or calcium ionophores, an ER-specific stress response is elicited. The major proteins synthesized appear to be proteins thought to function as ER chaperones, including BiP, Grp94, and Erp72, presumably because the stresses cause protein misfolding in the ER (Cox et al., 1993; Mori et al., 1993). FKBP13 message is also increased up to fivefold under these conditions, though to a greater extent by ionophores compared with tunicamycin (Bush et al., 1994). Moreover, the 5′ flanking region of the *fkbp13* gene contains sequences similar to those found in equivalent regions of the genes for BiP and Grp94, including a 37 base pair region with c. 50% identity with the unfolded protein response element (UPRE) of the BiP gene (Bush et al., 1994). Similar results have been obtained with the yeast gene, *FKB2* (Partaledis and Berlin, 1993). Inhibition of glycosylation, either with tunicamycin or in mutant stains, results in an increase in mRNA levels of *FKB2*. There is similarity of its 5′ flanking region with the UPRE of *KAR2*, the yeast BiP homolog (Partaledis and Berlin, 1993). Furthermore, a 21 base pair sequence in this region appears to mediate the transcriptional response (Partaledis and Berlin, 1993). Nevertheless, a direct association of FKBP13 or the FKB2 gene product with secretory proteins transiting the ER has yet to be demonstrated.

■ References

Bose, S. and Freedman, R.B. (1994). Peptidyl prolyl *cis-trans*-isomerase activity associated with the lumen of the endoplasmic reticulum. Biochem. J. **300**, 865–870.

Bram, R.J., Hung, D.T., Martin, P.K., Schreiber, S.L., and Crabtree, G.R. (1993). Identification of the immunophilins capable of mediating inhibition of signal transduction by cyclosporin A and FK506: roles of calcineurin binding and cellular location. Mol. Cell. Biol. **13**, 4760–4769.

Bush, K.T., Hendrickson, B.A., and Nigam, S.K. (1994). Induction of the FK506-binding protein, FKBP 13, under conditions which misfold proteins in the endoplasmic reticulum. Biochem. J. **303**, 705–708.

Cox, J.S., Shamu, C.E., and Walter, P. (1993). Transcriptional induction of genes encoding endoplasmic reticulum resident proteins requires a transmembrane protein kinase. Cell **73**, 1197–1206.

Fischer, G. and Bang, H. (1985). The refolding of urea-denatured ribonuclease A is catalysed by peptidyl-prolyl *cis-trans* isomerase. Biophys. Acta. **828**, 39–42.

Fruman, G., Burakoff, S.J., and Bierer, B.E. (1994). Immunophilins in protein folding and immunosuppression. FASEB J. **8**, 391–400.

Jin, Y.-J., Albers, M.W., Lane, W.S., Bierer, B.E., Schreiber, S.L., and Burakoff, S.J. (1991). Molecular cloning of a membrane-associated human FK506- and rapamycin-binding protein, FKBP-13. Proc. Natl. Acad. Sci. USA **88**, 6677–6681.

Liu, J. (1993). FK506 and cicloporin: molecular probes for studying intracellular signal transduction. Trends Pharmacol. Sci. **14**, 182–188.

Michnick, S.W., Rosen, M.K., Wandless, T.J., Karplus, M.K., and Schreiber, S.L. (1991). Solution structure of FKBP, a rotamase enzyme and receptor for FK506 and rapamycin. Science **252**, 836–839.

Mori, K., Ma, W., Gething, M.-J., and Sambrook, J. (1993). A transmembrane protein with a CDC2$^+$/CDC28 — related kinase activity is required for signalling from the ER to the nucleus. Cell **74**, 743–756.

Nigam, S.K., Jin, Y.-J, Jin, M,-J, Bierer, B.E., Bush, K.T., and Burakoff, S.J. (1993). Localization of the FK506-binding protein, FKBP 13, to the lumen of the endoplasmic reticulum. **294**, 511–514.

Ondek, B., Hardy, R.W., Baker, E.K., Stamnes, M.A., Shieh, B.H., and Zuker, C.S. (1992). Genetic dissection of cyclophilin function: Saturation mutagenesis of the *Drosophila* cyclophilin homolog ninaA. J. Biol. Chem. **267**, 16460–16466.

Partaledis, J.A. and Berlin, V. (1993). The *FKBP* gene of *Saccharomyces cerevisiae*, encoding the immunosuppressant-binding protein FKBP-13, is regulated in response to accumulation of unfolded proteins in the endoplasmic reticulum. Proc. Natl. Acad. Sci. USA **90**, 5450–5454.

Partaledis, J.A., Fleming, M.A., Harding, M.W., and Berlin, V. (1992). *Saccharomyces cerevisiae* contains a homolog of human FKBP-13, a membrane-associated FK506/rapamycin binding protein. Yeast **8**, 673–680. [published erratum appears in Yeast **8**, 815 (1992)].

Stamnes, M.N. Shieh, B., Chuman, L., Harris, G.L., and Zuker, C.S. (1991). The cyclophilin homolog ninaA is a tissue specific integral membrane protein required for the proper synthesis of a subset of *Drosphila* rhodopsins. Cell **65**, 214–227.

Van Duyne, G.D., Standaert, R.F., Karplus, P.A, Schreiber, S.L., and Clardy, J. (1991). Atomic structure of FKBP-FK506, an immunophilin-immunosuppressant complex. Science **252**, 839–842.

■ *Sanjay K. Nigam:*
Brigham and Woman's Hospital
and Harvard Medical School
75 Francis Street
Boston, MA 02115, USA
Tel. 1 617 525 8880
Fax. 1 617 525 8881
E-mail: sknigam@bics.bwh.harvard.edu

■ *Steven J. Burakoff:*
Dana Farber Cancer Institute and Harvard Medical School
Boston, MA 02115, USA

12c

Parvulin PPIases

Escherichia coli parvulin

Parvulin (latin: parvulus, very small) is the first member of a new family of PPIases, sharing no sequence homology in their amino acid sequence to the cyclophilins or FKBPs. The enzyme was purified from Escherichia coli cells by monitoring the enzymatic activity with the protease-coupled PPIase assay. The mature enzyme consists only of 92 amino acid residues. To our knowledge parvulin represents the smallest known enzyme being enzymatically active as a monomer.

■ Alternative name

PpiC (Rudd *et al.*, 1995).

■ Parvulin sequence

The amino acid sequence of parvulin was determined by Edman degradation (Rahfeld *et al.*, 1994a). The corresponding gene was identified by translation from the amino acid sequence using the TFASTA (Rahfeld *et al.*, 1994a) or TBLASTN (Rudd *et al.*, 1995) program. It was localized in the *E. coli* genome between 84.5 and 86.5 minutes, a previously unidentified open reading frame between the *E. coli ilvC* and *rep* genes in the ECOUW85U sequence (Daniels *et al.*, 1992). This gene encodes a protein consisting of 93 amino acid residues (PIR accession number S45525). However, the N-terminal Met residue is cleaved off in the mature protein. Database searches with the protein sequence of parvulin showed highly significant similarity with a number of prokaryotic and eukaryotic protein sequences (Rahfeld *et al.*, 1994a; Hani *et al.*, 1995; Rudd *et al.*, 1995). It is not yet known whether all these homologues possess PPIase activity. Multiple alignment of the protein sequences showed the conservation of distinctive amino acid residues within the 12 members of the putative PPIase family, the twofold repeated motif, HI(L)L(V),(X)n, H(A;E)V(I;L)I(L); n=73–97, is generally noticed especially.

Several of the prokaryotic homologues were well characterized. PrsA (Kontinen *et al.*, 1993), PrtM (Vos *et al.*, 1989), and NifM (Dean *et al.*, 1993) were thought to be involved in maturation or transport of specific proteins. In addition, SurA (Tormo *et al.*, 1990; see entry p. 436) was described as necessary for the survival of *E. coli* in stationary phase. An essential gene, *PTF1*, in *Saccharomyces cerevisiae* (also called *ESS1*; Hanes *et al.*, 1989; see entry p. 438) was identified encoding a putative 3'-end processing or transcription termination factor of pre-mRNAs (Hani *et al.*, 1995). Recently, additional family members have been characterized in *Drosophila melanogaster* (dodo; Maleszka *et al.*, 1996; see entry p. 440) and mammalian cells (Pin1, Lu *et al.*, 1996; see entry p. 443).

■ Parvulin protein

Parvulin was purified from an *E. coli* cell extract using its enzymatic property to accelerate the *cis* to *trans* interconversion of the oligopeptide Suc–Ala–Phe–Pro–Phe–4-nitroanilide. The molecular mass of the protein was 10 101 Da as determined by electrospray mass spectrometry. Thus, the protein contains no post-translational modification. From the amino acid composition it is deduced that parvulin is very hydrophilic (28% charged residues). The pI value was determined by isoelectric focusing as >9.5, corresponding to the theoretical value of 9.65 derived from the protein sequence. Application of the Chou–Fasman–Rose algorithm to the sequence of parvulin predicted the protein to consist predominantly of helices and only few β-sheets. This was qualitatively confirmed by analysing the far UV–CD spectrum.

The purified parvulin showed a specificity constant k_{cat}/K_m of 1.69×10^7 $M^{-1}s^{-1}$ for the substrate Suc–Ala–Leu–Pro–Phe–4-nitroanilide, using the standard PPIase assay (Fischer et al., 1984). The enzyme shows the typical pattern of substrate specificity of FKBPs. It prefers amino acid residues with hydrophobic side chains in the P_1-position of the peptide substrates (Rahfeld et al., 1994b). The remarkable activity toward substrates having an Ala residue in this position contrasts to known FKBPs.

The enzyme was not inhibited by cyclosporin A and the value of the IC_{50} for FK506 is about 40 μM (Rahfeld et al., 1994b).

■ Parvulin function

To determine the location of parvulin in E. coli, cells were fractionated into periplasmic, membrane, and cytosolic fractions (J.-U. Rahfeld, unpublished results). The fractions were examined by immunoblotting. The experiments indicated a cytosolic localization of the protein. Parvulin represented less than 1% of the total PPIase activity of the cells measured with the protease-coupled assay. Homologous overexpression of the protein in E. coli does not lead to any phenotype. Low molecular weight effectors or binding proteins of the parvulin have not been found yet. At present, the cellular function of the protein remains unknown.

In in vivo refolding assays using denatured RNase T1 variants with one or two cis prolines, parvulin has similar activity to cytosolic cyclophilin and FKPB12 (F.X. Schmid, personal communication).

■ References

Daniels, D., Plunkett, G.,III., Burland, V.D., and Blattner, F.R. (1992). Analysis of the Escherichia coli sequence of the region from 84.5 to 86.5 minutes. Science **257**, 771–778.

Dean, D.R., Bolin, J.T., and Zheng, L. (1993). Nitrogenase metalloclusters: structures, organization, and synthesis. J. Bacteriol. **175**, 6737–6744.

Fischer, G., Bang, H., and Mech, C. (1984). Determination of enzymatic catalysis for the cis/trans-isomerization of peptide binding in proline-containing peptides. Biomed. Biochim. Acta **43**, 1101–1111.

Hanes, S.D., Shank, P.R., and Bostian, K.A. (1989). Sequence and mutational analysis of ESS1, a gene essential for growth in Saccharomyces cerevisiae. Yeast **5**, 55–72.

Hani, J., Stumpf, G., and Domdey, H. (1995). PTF1 encodes an essential protein in Saccharomyces cerevisiae, which shows strong homology with a new putative family of PPIases. FEBS Lett. **365**, 198–202.

Kontinen, V.P. and Sarvas, M. (1993). The PrsA lipoprotein is essential for protein secretion in Bacillus subtilis and sets a limit for high-level secretion. Mol. Microbiol. **8**, 727–737.

Lu, K.P., Hanes, S.D., and Hunter, T. (1996). A human peptidyl-prolyl isomerase essential for regulation of mitosis. Nature **380**, 544–547.

Maleszka, R., Hanes, S.D., Hackett, R.L., De Couet, H.G., and Gabor Miklos, G.L. (1996). The Drosophila melanogaster dodo (dod) gene, conserved in humans, is functionally interchangeable with the ESS1 cell division gene of Saccharomyces cerevisiae. Proc. Natl. Acad. Sci. USA **93**, 447–451.

Rahfeld, J.-U., Rücknagel, K.P., Schelbert, B., Ludwig, B., Hacker, J., Mann, K., and Fischer, G. (1994a). Confirmation of the existence of a third family among peptidyl-prolyl cis/trans isomerases. Amino acid sequence and recombinant production of parvulin. FEBS Lett. **352**, 180–184.

Rahfeld, J.-U., Schierhorn, A., Mann, K., and Fischer, G. (1994b). A novel peptidyl-prolyl cis/trans isomerase from Escherichia coli. FEBS Lett. **343**, 65–69.

Rudd, K.E., Sofia, H.J., Koonin, E.V., Plunkett III, G., Lazar, S., and Rouviere, P.E. (1995). A new family of peptidyl-prolyl isomerases. Trends Biochem. Sci. **20**, 12–14.

Tormo, A., Almirón, M., and Kolter, R. (1990). SurA, an Escherichia coli gene essential for survival in stationary phase. J. Bacteriol. **172**, 4339–4347.

Vos, P., van Asseldonk, M., van Jeveren, F., Siezen, R., Simons, G., and de Vos, W.M. (1989). A maturation protein is essential for production of active forms of Lactococcus lactis SK11 serine proteinase located in or secreted from the cell envelope. J. Bacteriol. **171**, 2795–2802.

■ *Gunter Fischer and Jens-U. Rahfeld:*
Max-Planck-Research
Unit-Enzymology of Protein Folding
Kurt-Mothes-Str. 3
D-06120 Halle, Germany
Tel. 49 345 552 2801
Fax. 49 345 551 1972
E-mail: fischer@cis.biochemtech.uni-halle.de

Escherichia coli SurA

SurA is a periplasmic protein with two tandem domains that exhibit similarity to the entire sequence of parvulin, the founding member of a family of peptidyl–prolyl isomerases (PPIases). Purified SurA has low peptidyl–prolyl isomerase activity with synthetic substrates. Cells lacking SurA are defective in assembly and production of some outer membrane porins.

■ Isolation

SurA was originally identified in a search for *Escherichia coli* genes essential for survival in stationary phase. A Tn10 insertion in *surA* leads to a pronounced loss in the viability of stationary phase cells; hence the name *surA* (Tormo et al., 1990). *surA* has also been identified because insertions in this gene lead to a sevenfold increase in σ^E activity (Rouvière and Gross, 1996). σ^E senses and responds to extracytoplasmic stress (Mecsas et al., 1993; Raina et al., 1995; Rouvière and Gross, 1996). In addition, a gene that has been called *ostA* may be *surA*. Mutations in *ostA* lead to tolerance for n-hexane (Aono et al., 1994a, b). This gene is very close to *surA* and may be coincident with it.

■ *surA* gene and sequence

surA encodes a protein of 430 amino acids, with a predicted molecular weight of 47 kDa. A signal sequence is found at the extreme N-terminus, with a predicted cleavage site at Ala-20, leaving a mature protein of 410 amino acids. N-terminal analysis of the periplasmic protein indicates the sequence Ala–Pro–Glu–Val–Val as expected for this cleavage position (Rouvière and Gross, 1996). The C-terminus of SurA contains two domains with similarity to parvulin, a 93 amino acid PPIase (Rahfeld et al., 1994a, b; Rudd et al., 1995; Lazar and Kolter, 1996; Rouvière and Gross, 1996; see previous entry, p. 434). Sequence alignments (Devereux et al., 1984) indicate that the domain I (encompassing aa 150–252 of mature SurA) exhibits 36% identity and 55% similarity to parvulin, whereas domain II (encompassing aa 265–364 of mature SurA) exhibits 36% identity and 51% similarity to parvulin (L. Connolly, unpublished results). [Accession numbers: Swiss-Prot P21202; SURA_ECOLI; Gen Bank D10483 (Yura et al., 1992)].

■ SurA protein

The signal sequence of SurA predicts a periplasmic location and, indeed, both the native protein and a hexahistidine C-terminally tagged SurA have been purified from the periplasm of osmotically shocked cells (Lazar and Kolter, 1996; Rouvière and Gross, 1996). The PPIase activity of SurA has been demonstrated with a variety of artificial substrates. As expected, this activity is not inhibited by either cyclosporin A or FK506. However, the PPIase activity of SurA is low; about 1% of that exhibited by parvulin itself. The low activity is not owing to inhibition by the N-terminal region of SurA, since a construct containing only the C-terminal portion of SurA has the same activity as the intact protein. Each domain with similarity to parvulin has been tested for PPIase activity. Domain I lacks PPIase activity, whereas domain II exhibits activity comparable to that of the intact protein (Missiakas et al., 1996; Rouvière and Gross, 1996).

■ Biological activities of SurA

Cells lacking SurA have a defective cell envelope as evidenced by the fact that SurA colonies lysed on plates and exhibited increased sensitivity to a variety of chemicals whose entrance to the cell is normally blocked by the outer membrane (Rouvière and Gross, 1996). In addition, the outer membrane protein (OMP) profile indicated that several OMPs are missing or reduced in *surA*$^-$ cells (Rouvière and Gross, 1996). Analysis of the assembly of the LamB porin indicated that *surA*$^-$ cells are specifically defective in converting unfolded monomer to a more rapidly migrating species. Trimer-specific antibodies recognize this species, suggesting that it possesses aspects of the barrel structure characteristic of the trimeric porin (Rouvière and Gross, 1996). The PPIase activity, a chaperone function of SurA, or both may be required for this conversion event (Rouvière and Gross, 1996). In addition, both OmpF and OmpA have been shown to exhibit folding defects in the absence of SurA (Lazar and Kolter, 1996). None of the four periplasmic proteins examined exhibited folding defects in the absence of SurA (Lazar and Kolter, 1996). SurA is also part of the feedback loop that activates σ^E, an alternative sigma factor that coordinates the response to extracytoplasmic stress. Cells lacking surA are highly induced for σ^E (Missiakas et al., 1996, Rouviere and Gross, 1996). Conversely, overexpression of SurA downregulates normal σ^E induction in response to defects in lipopolysaccharide caused by mutations in *htrM* (*rfaD*) or to the accumulation of unfolded proteins (Missiakas et al., 1996).

■ Biological regulation of *surA*

The *surA* gene is located at about 1 minute on the *E. coli* chromosome and is contained in Kohara phage #104, #105, and #106 (Tormo et al., 1990). *surA* is part of a complex operon, whose downstream members include *pdxA* (synthesis of vitamin B6), *ksgA* (kasugamycin sensitivity), and *apaG* and *apaH* (ApppA hydrolase) (Roa et al., 1989; Tormo et al., 1990). Insertions in *surA* are polar only on *pdxA*, since the operon contains internal promoters driving expression of the other genes. It has been verified that the *surA*$^-$ phenotypes described above are solely a result of *surA*, and not caused by polarity on downstream genes. Nothing is known about the regulation of *surA* expression.

■ Mutagenesis studies

Cells lacking SurA die more rapidly during stationary phase, are induced for σ^E, and are defective in maturation of selected outer membrane porins and for the barrier functions of the cell envelope (Tormo et al., 1990; Lazar and Kolter, 1996; Rouvière and Gross, 1996). In addition, *surA*$^-$ cells are defective in expression of some OMPs.

■ References

Aono, R., Negishi, T., Aibe, K., Inoue, A., and Horikoshi, K. (1994a). Mapping of organic solvent tolerance gene *ostA* in *Escherichia coli* K-12. Biosci. Biotechnol. Biochem. **58**, 1231–5.

Aono, R., Negishi, T., and Nakajima, H. (1994b). Cloning of organic solvent tolerance gene *ostA* that determines n-hexane tolerance level in Escherichia coli. Appl. Envrion. Microbiol. **60**, 4624–66.

Devereux, J., Haeberli, P., and Smithies, O. (1984). A comprehensive set of sequence analysis programs for the VAX. Nucl. Acids Res. **12**, 387–95.

Lazar, S.W. and Kolter, R. (1996). SurA assists the folding of *E. coli* outer membrane proteins. J. Bacteriol. **178**, 1770–3.

Mecsas, J., Rouvière, P.E., Erickson, J.W., Donohue, T.J., and Gross, C.A. (1993). The activity of σ^E, an *Escherichia coli* heat-inducible sigma-factor, is modulated by expression of outer membrane proteins. Genes Dev. **7**, 2618–28.

Missiakas, D., Betton, J.-M., and Raina, S. (1996). New components of protein folding in extracytoplasmic compartments of *Escherichia coli* SurA, FkpA and Skp/OmpH. Mol. Micro. **21**, 871–884.

Rahfeld, J.U., Rucknagel, K.P., Schelbert, B., Ludwig, B., Hacker, J., Mann, K., and Fischer, G. (1994a). Confirmation of the existence of a third family among peptidyl-prolyl cis/trans isomerases. Amino acid sequence and recombinant production of parvulin. FEBS Lett. **352**, 180–4.

Rahfeld, J.U., Schierhorn, A., Mann, K., and Fischer, G. (1994b). A novel peptidyl-prolyl cis/trans isomerase from *Escherichia coli*. FEBS Lett. **343**, 65–9.

Raina, S., Missiakas, D., and Georgopoulos, C. (1995). The rpoE gene encoding the σE ($\sigma 24$) heat shock sigma factor of *Escherichia coli*. EMBO J. **14**, 1043–55.

Roa, B.B., Connolly, D.M., and Winkler, M.E. (1989). Overlap between *pdxA* and *ksgA* in the complex *pdxA-ksgA-apaG-apaH* operon of Escherichia coli K-12. J. Bacteriol. **171**, 4767–77.

Rouvière, P.E. and Gross, C.A. (1996). SurA, a periplasmic protein with peptidyl-prolyl isomerase activity, participates in the assembly of outer membrane porins. Genes Dev. **10**, 1–43.

Rouvière, P.E., De Las Penas, A., Mecsas, J., Lu, C.Z., Rudd, K.E., and Gross, C.A. (1995). rpoE, the gene encoding the second heat-shock sigma factor, σE, in Escherichia coli. EMBO J. **14**, 1032–42.

Rudd, K.E., Sofia, H.J., Koonin, E.V., Plunkett, G.R., Lazar, S., and Rouviere, P.E. (1995). A new family of peptidyl-prolyl isomerases. Trends Biochem. Sci. **20**, 12–14.

Tormo, A., Almiron, M., and Kolter, R. (1990). *surA*, an *Escherichia coli* gene essential for survival in stationary phase. J. Bacteriol. **172**, 4339–47.

Yura, T., Mori, H., Nagai, H., Nagata, T., Ishihama, A., Fujita, N., Isono, K., Mizobuchi, K., and Nakata, A. (1992). Systematic sequencing of the *Escherichia coli* genome: analysis of the 0-2.4 min region. Nucl. Acids Res. **20**, 3305–8.

■ *Carol A. Gross[1] and Lynn Connolly:[2]*
[1]*Departments of Stomatology and Microbiology and*
[2]*Departments of Biochemistry and Biophysics*
University of California at San Francisco
513 Parnassus Ave
San Francisco, CA 94143, USA
Tel. 1 415 476 4161/1493
Fax. 1 415 476 4204
E-mail: cgross@cgl.ucsf.edu and lcon@itsa.ucsf.edu

■ *Pierre Rouvière:*
Environmental Biotechnology
Central Research and Development
DuPont Company
Wilmington, DE 19880-0328, USA
Tel. 1 302 695 1782
Fax. 1 302 695 1829
E-mail: rouviepe@al.esvax.umc.dupont.com

Saccharomyces cerevisiae Ess1

Ess1 is a small protein (19.2 kDa) composed of two distinct modules, a WW domain proposed to mediate protein–protein interactions and a PPIase domain that isomerizes peptidyl–prolyl bonds. Ess1 is essential for growth in Saccharomyces cerevisiae; depletion of Ess1 causes mitotic arrest and nuclear fragmentation. Ess1 is highly conserved among eukaryotic cells (see Drosophila dodo and human Pin1 entries, pp. 440 and 443) and might serve as a universal mitotic regulator by down-regulating G2/M-specific cell cycle kinases to allow cells to exit mitosis.

Alternative name

Ptf1 (Hani et al. 1995).

Isolation

The *ESS1* gene was originally isolated by cross-hybridization with the simian sarcoma virus oncogene v-*sis*, known to promote the G1/S transition. However, Ess1 protein bears little similarity to p28^{sis}. *ESS1* mapped to the right arm of chromosome 10 and was not allelic to any known yeast mutation (Hanes et al., 1989). *PTF1* was identified in a screen for pre-mRNA processing mutants defective in 3′-end formation (Hani et al., 1995).

ESS1 gene and sequence

ESS1 encodes a highly charged 170 amino acid protein that is 45% identical to the *Drosophila melanogaster dodo* gene product (Maleszka et al., 1996) and 45% identical to the human *PIN1* gene product (Lu et al., 1996). (For alignments see entries on human Pin1 p. 443 and dodo p. 440.) At its N-terminus, Ess1 contains a 38 amino acid sequence known as the WW domain that is found in diverse proteins from organisms ranging from yeast to humans (Bork and Sudol, 1994; Staub and Rotin, 1996). WW domains are characterized by two invariant tryptophans, and like SH3 domains, are protein-interaction modules that bind proline-rich peptides (Sudol, 1996). At its C-terminus Ess1 is similar to a new family of PPIases (peptidyl–prolyl *cis*–*trans* isomerases) called parvulins (Rahfeld, et al., 1994; Rudd et al., 1995). Parvulins are thought to be structurally distinct from the well-studied cyclophilin and FKBP families of PPIases (Fruman et al., 1994). (Accession numbers: Ess1, Swiss-Prot: P22696, PIR: S07867; *ESS1*, Yeast Genome Project: YJR017c; *PTF1*, EMBL: X85972.)

Ess1 protein

Ess1 is not well characterized biochemically. It has not been shown to have PPIase activity, perhaps as a result of sensitivity to the chymotrypsin used in the PPIase assay (Hani et al., 1995). However, the human homolog, Pin1, does have PPIase activity that, like parvulin, is not inhibited by cyclosporin or FK506 (Lu et al., 1996). The intracellular localization of Ess1 is not known; however, Pin1 is localized to the nucleus in HeLa cells. Pedigree analysis in yeast indicates that the Ess1 protein segregates equally to mother and daughter cells at mitosis (Hanes, 1988). Antibodies to Ess1 are not yet available.

Biological activities of Ess1

Ess1 is a mitotic regulator. Its function in yeast can be replaced by *Drosophila* dodo or human Pin1 (Table 1). Given its similarity to known PPIases, it is likely that Ess1 works by catalysing the isomerization of peptidyl–prolyl bonds that may control the folding, association or activity of cell cycle regulatory proteins. Clues to the biological role of Ess1 come from analysis of an *ess1*–amber mutant (Hanes et al., 1989) and from galactose shut-off experiments (Lu et al., 1996). *ess1*–amber mutants arrest with a multibudded phenotype suggesting a defect in late stages of cell division such as cytokinesis or cell separa-

Table 1. *Drosophila* dodo and human *PIN1* complement *ess1* mutant yeast[a]

Plasmid	Total Tetrads	Viable spores per tetrad				
		0	1	2	3	4
Vector	50	1	6	41	2	0
dodo	61	0	5	30	4	22
dodo-rev	32	1	2	29	0	0
PIN1	83	5	6	46	6	20
ESS1	29	0	0	6	6	17

[a]Yeast strain MGG3/pSHU (*ess1::URA3/ESS1*) was transformed with episomal plasmids that express the indicated gene. Cells were induced to undergo sporulation and tetrads dissected. Viable spores were scored for growth using tetrads that showed proper segregation of independent markers. Plasmid *dodo*-rev carries the *dodo* cDNA in reverse orientation relative to the promoter. (Data adapted from Maleszka et al., 1996; Lu et al., 1996.)

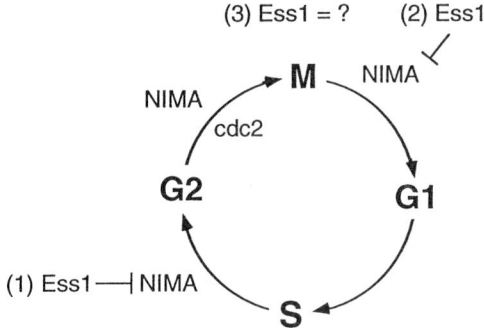

(1) Ess1 keeps NIMA activity low until G2/M

(2) Ess1 lowers NIMA activity allowing exit from M

(3) Ess1 has role in mitosis unrelated to NIMA

Figure 1. Models For Ess1/dodo/Pin1 Function

tion. However, this phenotype is weakly penetrant perhaps owing to leakiness of the SUP4-3 tRNA suppressor used in these experiments. In contrast, depletion experiments using GAL1–PIN1 or GAL1–ESS1 constructs in an ess1::URA3 null background show a uniform terminal phenotype of mitotic arrest. DAPI staining reveals that nuclear division appears to proceed to completion, albeit more slowly than normal. This is followed by a dramatic fragmentation of nuclei, reminiscent of apoptosis in higher eukaryotic cells. Studies with NIMA kinase and human Pin1 in HeLa cells and yeast suggest possible models (Fig. 1), one of which is that Ess1/Pin1 acts as a negative regulator of NIMA kinase to allow cells to exit mitosis (Lu et al., 1996).

■ Biological regulation of *ESS1*

ESS1 mRNA is moderately abundant (c. 0.01% of total mRNA) in growing cells but is absent in cells entering stationary phase. Steady-state levels of *ESS1* mRNA are similar in all stages of the cell cycle (Hanes, 1988). However, it is not known whether Ess1 protein levels or PPIase activity vary during the cell cycle. *ESS1* mRNA is not induced by moderate (37°C) heat shock.

■ Mutagenesis studies

Point mutagenesis of the Ess1 WW domain gave the surprising result that the signature tryptophans in the WW domain can be individually replaced by phenylalanine or doubly replaced by phenylalanine or valine without effect; all four mutants complement an ess1::URA3 null mutant (Table 2). Similar experiments with Drosophila dodo show that individual alanine substitutions are tolerated, but a double mutant (dodo W11A, W34A) no longer complemented yeast ess1 mutant cells,

demonstrating the importance of the WW domain *in vivo* (see also entry on *Drosophila* dodo this volume, p. 440). In a random mutagenesis screen, we isolated a large number of missense mutants that abolish Ess1 function, many of which are conditional (R. Hackett and S. Hanes, unpublished data). The mutations are found in both the WW and PPIase domains and tend to map to residues highly conserved between Ess1, Dodo and Pin1. In addition, a C-terminal deletion or a triple alanine substitution in Pin1 (G155A, H157A, I159A) abolishes PPIase activity *in vitro* and renders Pin1 incapable of complementing *ess1* null mutants (Lu et al., 1996). Analogous mutations in Ess1 might be expected to behave similarly. Overexpression of wild-type *ESS1* does not lead to a growth defect or other obvious phenotypes.

■ Biological or genetic interactions

Thus far, no biological partner proteins of Ess1 are known. The human homolog, Pin1, was originally identified by its interaction with *Aspergillus nidulans* NIMA kinase in a two-hybrid assay and was shown to interact *in vitro* with NIMA (Lu et al., 1996). A two-hybrid screen with Ess1 and genetic suppression studies are underway.

■ References

Bork, P. and Sudol, M. (1994). The WW domain: a signalling site in dystrophin? Trends Biochem. Sci. **19**, 531–533.

Fruman, D.A., Burakoff, S.J., and Bierer, B.E. (1994). Immunophilins in protein folding and immunosuppression. FASEB J. **8**, 391–400.

Hanes, S.D. (1988). Isolation, sequence and mutational analysis of *ESS1*, a gene essential for growth in *Saccharomyces cerevisiae*. Ph. D. Thesis, Brown University, USA.

Table 2. *In vivo* assay of WW domain mutants[a]

Plasmid	Total Tetrads	Viable spores per tetrad				
		0	1	2	3	4
Vector	30	0	2	28	0	0
ESS1	29	0	0	6	6	17
ESS1 W15F	61	1	3	21	5	31
ESS1 W38F	70	0	2	33	4	31
ESS1 W15F, W38F	58	0	4	21	5	28
ESS1 W15V, W38V	51	0	6	20	10	15
dodo	61	0	5	30	4	22
dodo W34F	45	1	4	18	5	17
dodo W11A	42	2	1	9	9	21
dodo W34A	43	0	4	17	11	11
dodo W11A, W34A	88	0	10	74	4	0

[a] Plasmids encoding the indicated mutant proteins were transformed into diploid strain MGG3/pSH-U (ess1::URA3/ESS1), cells were induced to sporulate and tetrads dissected. Growth was scored after 3 days (R. Maleszka, R. Hackett, K. Ryan, and S. Hanes, unpublished data).

Hanes, S.D., Shank, P.R., and Bostian, K.A. (1989). Sequence and mutational analysis of *ESS1*, a gene essential for growth in *Saccharomyces cerevisiae*. Yeast **5**, 55–72.

Hani, J., Stumpf, G., and Domdey, H. (1995). *PTF1* encodes an essential protein in *Saccharomyces cerevisiae*, which shows strong homology with a new putative family of PPIases. FEBS Lett. **365**, 198–202.

Lu, K.P, Hanes, S. D., and Hunter, T. (1996). A human peptidyl-prolyl isomerase essential for regulation of mitosis. Nature **380**, 544–547.

Maleszka, R., Hanes, S.D., Hackett, R.L., De Couet, H.G., and Miklos, G.L.G. (1996). The *Drosophila melanogaster* dodo (*dod*) gene, conserved in humans, is functionally interchangeable with the *ESS1* cell division gene of *Saccharomyces cerevisiae*. Proc. Natl. Acad. Sci. USA **93**, 447–451.

Rahfeld, J.U., Rücknagel, K.P., Schelbert, B., Ludwig, B., Hacker, J., Mann, K., and Fischer, G. (1994). Confirmation of the existence of a third family among peptidyl-prolyl *cis*/*trans* isomerases. Amino acid sequence and recombinant production of parvulin. FEBS Lett. **352**, 180–184.

Rudd, K.E., Sofia, H.J., Koonin, E.V., Plunkett, G.R., Lazar, S., and Rouviere, P.E. (1995). A new family of peptidyl-prolyl isomerases. Trends Biochem. Sci. **20**, 12–14.

Staub O. and Rotin D. (1996). WW domains. Struct. Curr. Biol. **4**, 495–499.

Sudol, M. (1996). The WW module competes with the SH3 domain? Trends Biochem. Sci. **21**, 161–163.

■ Steven D. Hanes:
Wadsworth Center
New York State Department of Health and
Department of Biomedical Sciences
State University of New York
Albany, NY 12203, USA
Tel. 1 518 473 4213
Fax. 1 518 474 3181
E-mail: steven.hanes@wadsworth.org

Drosophila melanogaster dodo

The Drosophila melanogaster dodo *gene encodes a modular, evolutionarily conserved protein that, so far, has been found only in eukaryotes. It has a WW domain involved in protein–protein interactions; a peptidyl–prolyl cis–trans isomerase (PPIase) domain belonging to a recently described third family of PPIases involved in protein folding and unfolding; a nuclear localization motif: and, finally, a predicted surface exposed amphipathic helix that could provide a docking surface for other proteins that contain coiled-coil structures. The deletion of* dodo *from the genome does not lead to drastic consequences on viability. The* dodo *gene is functionally interchangeable with the essential yeast gene* ESS1, *which is required for progression through mitosis.*

■ Gene isolation

The *dodo* gene (*dod*) was identified in the functional analysis of the *flightless* (*fli*) region at the base of the X chromosome in *Drosophila* (Maleszka *et al.*, 1996; see Fig. 1). The genomic analysis was followed by the sequencing of a number of *dodo* cDNAs from larval and pupal libraries that revealed the existence of two classes of *dodo* transcripts differing in the lengths of their 3'-untranslated regions. The longer transcript is 1013 bp in length.

■ Mapping and sequencing of *dodo*

The genetic and molecular map of the *flightless* region is shown in Fig. 1. The putative 5'-end of the *dodo* transcription unit is 279 bp from the 3'-end of the *flightless* transcript, whereas the 3'-ends of the *dodo* and *penguin* transcripts overlap by 353 bp. The gene has two introns (111 and 369 bp, respectively), the first of which is not always removed. Judging by standard molecular techniques *dodo* is single copy in the fly genome (Maleszka *et al.*, 1996). The GenBank accession number for *dodo* is U35140.

■ Characteristics of the dodo protein

The predicted product of the *dodo* gene is a protein of 166 amino acids (18.3 kD). It is 45% identical to the product of the yeast *ESS1* gene (Ess1). The dodo protein contains two, well-characterized modules; a WW domain involved in protein–protein interactions (Bork and Sudol, 1994) and a peptidyl–prolyl *cis–trans* isomerase (PPIase) domain belonging to a recently described third family of PPIases involved in protein folding and unfolding (Rahfeld *et al.*, 1994; Rudd *et al.*, 1995; see Fig. 2). This family is structurally unrelated to the previously described PPIases, namely the cyclophilins and FK506-binding proteins. In addition, dodo contains a nuclear localization motif (Lu *et al.*, 1996; underlined in Fig. 2) and, finally, a predicted surface exposed amphipathic helix (R. Maleszka, A. Lupas, S.D. Hanes and G.L.G. Miklos, *et al.*,

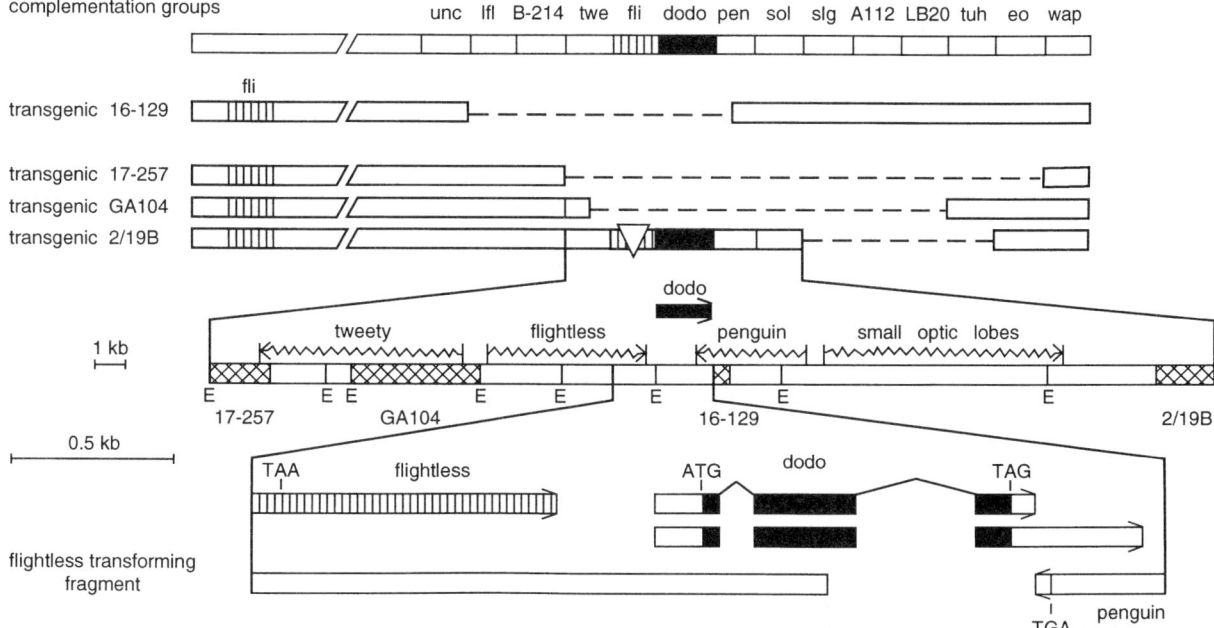

Figure 1. (Top) Genetic and molecular map of the *flightless* (*fli*) region: genetic complementation groups and characteristics of the rearranged chromosomes. (Middle) The five complete transcription units, the *Eco*RI restriction sites, and the extent of the four chromosomal breakpoints (cross-hatched boxes). (Bottom) Intron/exon structure of the *dodo* transcription unit and the positions of the termination codons in the *flightless*, *dodo* and *penguin* genes. The most proximal extent of the *flightless* transforming fragment is as shown (after Maleszka *et al.* 1996).

unpublished results; indicated with hhhh in Fig. 2), which is likely to be involved in binding to a cell cycle serine/threonine kinase (Lu *et al.*, 1996).

The multimodular structure of dodo and its yeast and human homologues is shown in Fig. 2. The alignment also includes a prototypical member of the third family of PPIases, the bacterial protein parvulin (Rahfeld *et al.*, 1994; Rudd *et al.*, 1995; see entry p. 434). Although parvulin shares the three conserved motifs (denoted as I, II and III in Fig. 2) with fly dodo, yeast Ess1 and human Pin1, the alignment reveals that the PPIase domain in the parvulin family is interrupted by a segment of 17 amino acids that is not present in the bacterial enzyme. This region, containing a conserved loop, may represent an evolutionary addition to an ancestral prokaryotic PPIase domain that created a structural motif that allowed for a gain of function. The proteins belonging to the dodo family contain a conserved motif R/EVR/HCL/SHLLVKH–SRRP that has been proposed to act as a nuclear localization signal for the human Pin1 (Lu *et al.*, 1996). This motif evolved as a fusion of the first PPIase motif and the N-terminal region of the insertion that interrupts the PPIase domain in various dodo proteins. Thus, dodo and its homologues are likely to be restricted to the nucleus and interact with nuclear protein partners. Indeed, the Pin1 protein localizes almost exclusively to the nucleus, within a substrate of the nuclear scaffold, called the nuclear speckle (Lu *et al.*, 1996) that breaks down during mitosis.

Furthermore, the genomic insertions that gave rise to the dodo family during its evolutionary history allowed the emergence of an extended, surface-exposed amphipathic helix (Maleszka, A. Lupas, S.D. Hanes and G.L.G. Miklos, unpublished results; see Fig. 2) that could dock with the coiled-coil region of another protein. In accord with this notion, the Pin1 protein has been shown to interact with the specific region of the fungal serine/threonine kinase called NIMA (Lu *et al.*, 1996) that has a potential to adopt a coiled-coil conformation (Fry and Nigg, 1995).

■ Interchangeability between organisms

By transgenic shuttling between *Drosophila* and *Saccharomyces* we demonstrated that the fly protein is the functional homologue of the yeast Ess1 protein (Maleszka *et al.*, 1996), so-called because it is essential (ESS) for vegetative growth in yeast (Hanes *et al.*, 1989; see entry p. 438). The human *pin1* gene also functionally substitutes for the yeast *ESS1* gene (Lu *et al.*, 1996), although not as efficiently as the product of the fly gene. By these criteria, the fly *dodo*, yeast *ESS1* and human *pin1* genes are homologues. In yeast, *ESS1* is required for progression through mitosis; yeast cells depleted of Ess1 (or the human *pin1* product) arrest in mitosis and

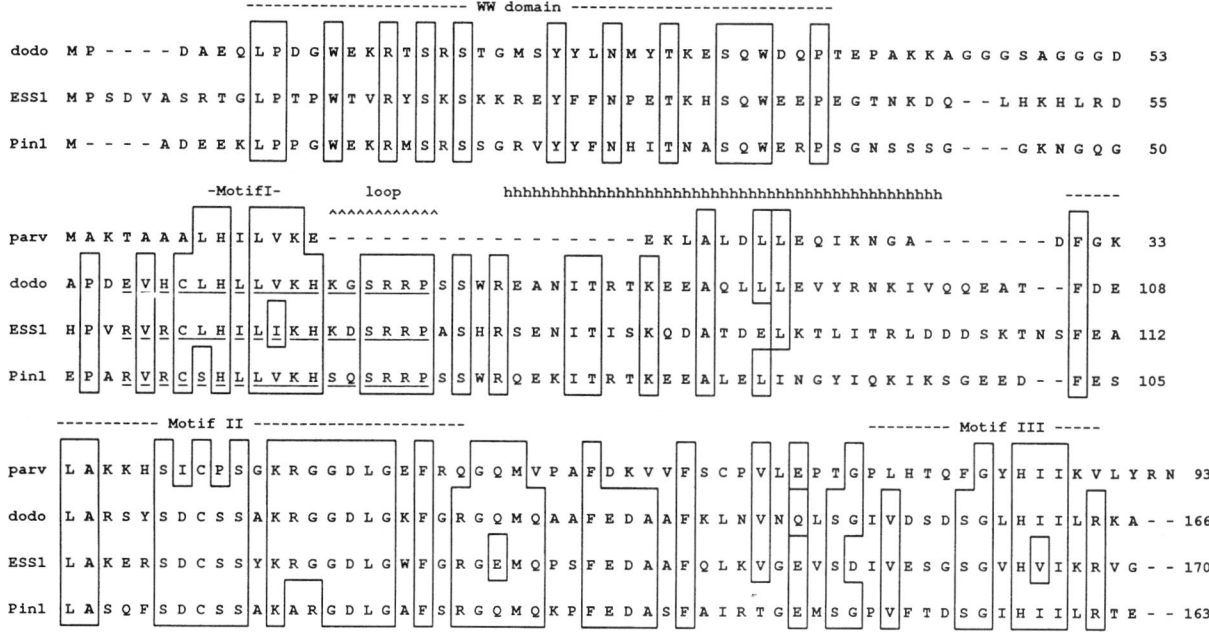

Figure 2. Alignment of parvulin, dodo, Ess1 and Pin1. Identical residues in at least three sequences are boxed. The WW domain and three conserved motifs in the PPIase domain are labelled above the sequence. The conserved loop and the extended amphipathic α helix are indicated by <^><^> and hhhh respectively. Underlined residues form a consensus nuclear localization signal. The corresponding database codes are: parvulin (Swiss-Prot P39159), dodo (GenBank U35140), Ess1 (Swiss-Prot P22696) and Pin1 (GenBank U49070).

undergo nuclear fragmentation and cell death (Lu et al., 1996; R.L. Hackett and S.D. Hanes, unpublished).

■ Mutagenesis and transgenic analyses

In contrast to Ess1, dodo is not essential. Transgenic and deficiency analysis revealed that in the region containing the *dodo* gene only the *flightless* gene is vital to the organism. The simultaneous absence of *tweety*, *dodo* and *penguin* (see Fig. 1) allows the overriding majority of individuals to develop to adulthood and to fly normally, suggesting that *dodo* is neither essential for development, nor is it required for the integrated neuromuscular activities needed for flight control (Maleszka et al., 1996). Since *dodo* appears not to be a member of a multigene family its absence is compensated for either by other unrelated proteins (Maleszka et al., 1996) or by degenerate networks (Edelman, 1993). The function of the human *pin1* gene has thus far been examined only in tissue culture, where expression of antisense constructs causes mitotic arrests phenotypes in about 50–60% of transfected HeLa cells (Lu et al., 1996).

The significance of individual modules has been demonstrated by mutating the PPIase domain in the human Pin1 protein (Lu et al., 1996) and by a double replacement of tryptophan residues in the WW domain of the fly dodo (See previous entry p. 438).

■ References

Bork, P. and Sudol, M. (1994). The WW domain: a signalling site for dystrophin? Trends Biochem. Sci. **19**, 531–533.

Edelman, G.M. (1993). A golden age for adhesion. Cell Adhes. Commun. **1**, 1–7.

Fry, A.M. and Nigg, E.A. (1995). The NIMA kinase joins forces with Cdc2. Curr. Biol. **5**, 1122–1125.

Hanes, S.D., Shank, P.R., and Bostian, K.A. (1989). Sequence and mutational analysis of *ESS1*, a gene essential for growth in *Saccharomyces cerevisiae*. Yeast **5**, 55–72.

Lu, K.P., Hanes, S.D., and Hunter, T. (1996). A human peptidyl-prolyl isomerase essential for regulation of mitosis. Nature **380**, 544–547.

Maleszka, R., Hanes, S.D., Hackett, R.L., De Couet, H.G., and Gabor Miklos, G.L. (1996). The *Drosophila melanogaster* dodo (*dod*) gene, conserved in humans, is functionally interchangeable with the *ESS1* cell division gene of *Saccharomyces cerevisiae*. Proc. Natl. Acad. Sci. USA **93**, 447–451.

Rahfeld, J.-U., Rücknagel, K.P., Schelbert, B., Ludwig, B., Hacker, J., Mann, K., and Fischer, G. (1994). Confirmation of the existence of a third family among peptidyl-prolyl *cis/trans* isomerase. Amino acid sequence and recombinant production of parvulin. FEBS Lett. **352**, 180–184.

Rudd, K.E., Sofia, H.J., Koonin, E.V., Plunkett, III, G., Lazar, S., and Rouviere, P.E. (1995). A new family of peptidyl-prolyl isomerases. Trends Biochem. Sci. **20**, 12–13.

■ R. Maleszka:
Visual Sciences, Research School of Biological Sciences
The Australian National University
Canberra ACT 0200, Australia
Tel. 61 6 249 0451
Fax. 61 6 249 3784
E-mail: maleszka@rsbs-central.anu.edu.au

■ G.L. Gabor Miklos:
The Neurosciences Institute
10640 John Jay Hopkins Drive
San Diego, CA 92121, USA
Tel. 1 619 626 2000
Fax. 1 619 626 2099
E-mail: miklos@nsi.edu

Mammalian Pin1

Pin1, a human protein interacting with the mitotic NIMA kinase isolated from *Aspergillus nidulans*, contains a catalytic domain characteristic of the highly conserved third family of peptidyl–prolyl *cis*–*trans* isomerases that are distinct from either the cyclophilins or the FK506-binding proteins. Pin1 is a widely expressed 18 kDa protein localized in nuclear speckles which also contain spliceosomes. Pin1 is structurally and functionally related to Ess1/Ptf1, an essential protein in budding yeast, and to the *Drosophila* dodo protein. The biological function of this protein includes an essential role in regulation of mitotic progression, via direct interaction with a subset of mitosis-specific phosphoproteins, including NIMA and CDC25.

■ Isolation

Pin1 was originally isolated as a protein that interacts with the *Aspergillus* NIMA protein kinase and suppresses its mitosis-promoting activity in a yeast two-hybrid screen (Lu *et al*., 1996).

■ Pin1 protein and homologues

Pin1 contains 163 amino acids that display about 40% identity either to Ess1/Ptf1 from *Saccharomyces cerevisiae* (Hanes *et al*., 1989; Hani *et al*., 1995; see entry p. 438) or dodo from *Drosophila melanogaster* (Maleszka *et al*., 1996; see entry p. 440). All three proteins contain an N-terminal WW domain, a C-terminal peptidyl–prolyl *cis*–*trans* isomerase (PPIase) domain and a putative bipartite nuclear localization signal located at the beginning of the PPIase domain (Fig. 1). The WW domain is a 38–39 amino acid motif that is characterized by two invariant tryptophans and other conserved residues. This motif is found in a number of unrelated proteins and has been shown to bind short Pro-rich sequences (Sudol *et al*., 1995; Chan *et al*., 1996). The C-terminal two-thirds of Pin1 contains motifs that are characteristic of the third family of PPIases, with parvulin (see entry p. 434) being the prototype, but share no similarity to either the cyclophilins or the FK506-binding proteins (Rahfeld *et al*., 1994; Rudd *et al*., 1995). Furthermore, based on a standard chymotrypsin-coupled spectrophotometric PPIase assay, Pin1 catalyses *cis*–*trans* isomerization of peptidyl–prolyl peptide bonds *in vitro* and this activity is not inhibited by either cyclosporin A or an FK506 derivative (Lu *et al*., 1996). Thus, Pin1 represents a human member of the new family of PPIases that are insensitive to the immunosuppressive drugs.

■ *pin1* gene

cDNAs and/or genomic sequences for *pin1* and its homologues have been cloned and sequenced from *Homo sapiens* (GenBank accession number U49070) (Lu *et al*., 1996), *Drosophila melanogaster* (U35140) (Maleszka *et al*., 1996) and *Saccharomyces cerevisiae* (X85972) (Hanes *et al*., 1989; Hani *et al*., 1995).

■ Purification of Pin1

Although the Pin1 protein has not yet been purified in the native form from human tissue or cells, the recombinant Pin1 protein can be easily expressed in *E. coli* as a histidine-tagged fusion protein and purified in large quantities using Ni^{2+}–NTA–agarose column (Lu *et al*., 1996). Purified Pin1 has PPIase activity (Lu *et al*., 1996) and has been used to generate a crystal structure (Ranganathan *et al*., 1997).

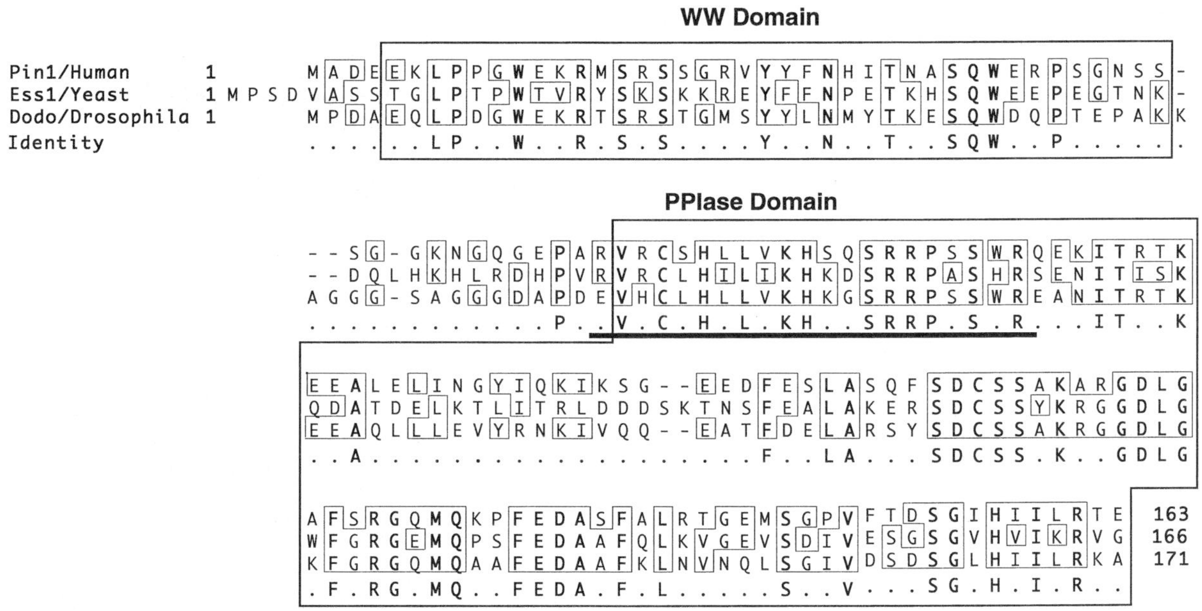

WW Domain

```
Pin1/Human      1        M A D E E K L P P G W E K R M S R S S G R V Y Y F N H I T N A S Q W E R P S G N S S -
Ess1/Yeast      1 M P S D V A S S T G L P T P W T V R Y S K S K K R E Y F F N P E T K H S Q W E E P E G T N K -
Dodo/Drosophila 1        M P D A E Q L P D G W E K R T S R S T G M S Y Y L N M Y T K E S Q W D Q P T E P A K K
Identity                 . . . . . . L P . . W . . R . S . S . . . . . Y . . N . . T . . S Q W . . P . . . . . .
```

PPIase Domain

```
- - S G - G K N G Q G E P A R V R C S H L L V K H S Q S R R P S S W R Q E K I T R T K
- - D Q L H K H L R D H P V R V R C L H I L I K H K D S R R P A S H R S E N I T I S K
A G G G - S A G G G D A P D E V H C L H L L V K H K G S R R P S S W R E A N I T R T K
. . . . . . . . . . . . P . . V . C . H . L . K H . . S R R P . S . R . . . I T . . K

E E A L E L I N G Y I Q K I K S G - - E D F E S L A S Q F S D C S S A K A R G D L G
Q D A T D E L K T L I T R L D D D S K T N S F E A L A K E R S D C S S Y K R G G D L G
E E A Q L L L E V Y R N K I V Q Q - - E A T F D E L A R S Y S D C S S A K R G G D L G
. . A . . . . . . . . . . . . . . . . . . . . . . F . . L A . . . S D C S S . K . . G D L G

A F S R G Q M Q K P F E D A S F A L R T G E M S G P V F T D S G I H I I L R T E   163
W F G R G E M Q P S F E D A A F Q L K V G E V S D I V E S G S G V H V I K R V G   166
K F G R G Q M Q A A F E D A A F K L N V N Q L S G I V D S D S G L H I I L R K A   171
. F . R G . M Q . . F E D A . F . L . . . . . . S . . V . . . S G . H . I . R . .
```

Figure 1. Sequence comparison and domain structure of human Pin1, budding yeast Ess1 and *Drosophila* dodo proteins.

■ Antibodies against Pin1

Rabbit polyclonal antisera against the C-terminal peptide (residues 147–163) of Pin1 have been produced, which can be used for immunoprecipitation and immunoblotting (Lu et al., 1996).

■ Biological activity of Pin1

Pin1 has PPIase activity which can be assayed based on isomer-specific cleavage of the synthetic peptide *N*-succinyl–Ala–Ala–Pro–Phe–*p*-nitroanilide by chymotrypsin (Lu et al., 1996). Site-directed mutagenesis reveals that the PPIase domain of Pin1 is required for its ability to perform the essential function of Ess1 in the budding yeast (Lu et al., 1996). Physiological targets for Pin1 have not been identified. It is possible that the main function of Pin1 is to interact with target proteins via a specific proline residue to transmit a signal, rather than to catalyse isomerization. In this regard, Pin1 directly associates with the NIMA protein interaction domain (NID) and attenuates the mitosis-promoting activity of NIMA when coexpressed *in vivo* (Lu et al., 1996).

■ Function of Pin1

In budding yeast, human *pin1* can functionally substitute for the *ESS1* gene (Lu et al., 1996), whose disruption is lethal (Hanes et al., 1989; Hani et al., 1995). However, PPIase-negative *pin1* mutants generated either by a C-terminal truncation beyond Ser-114 or Ala substitutions of three highly conserved residues in the catalytic core (Gly-155/Ala, His-157/Ala, Ile-159/Ala) fail to rescue the *ess1* null mutation (Lu et al., 1996). Furthermore, depletion of Pin1 from GAL–*PIN1*-dependent yeast cells results in a concentration-dependent mitotic arrest phenotype; cells accumulate in mitosis with 2*n* DNA content and a large bud, with the nucleus stuck in the neck between mother and bud, and eventually undergo nuclear fragmentation (Lu et al., 1996). Thus, Pin1/Ess1 plays an essential role in mitotic progression in yeast. Neither the cyclophilins nor the FK506-binding proteins have been found to be essential for yeast cell viability (Fischer, 1994). In contrast, Pin1/Ess1 represents a PPIase that is required for cell growth in yeast and HeLa cells (Lu et al., 1996), although dodo apparently is not essential in flies (Maleszka et al., 1996).

In human HeLa cells, exogenously expressed epitope-tagged Pin1 is almost exclusively nuclear, concentrating at nuclear speckles that are identical to those detected using anti-SC35 splicing factor monoclonal antibody (Fig. 2; Lu et al., 1996). When cells enter mitosis, Pin1 colocalizes with the condensed chromatin (Fig. 2). Thus, Pin1 is associated with the nuclear speckle, which is a key component of the nuclear scaffold or matrix and undergoes disassembly in mitosis. Furthermore, whereas overexpression of Pin1 induces a G2 arrest in HeLa cells, depletion of Pin1 using antisense RNA expression results in mitotic events including cell rounding, nuclear lamin disassembly, chromatin condensation and eventual nuclear fragmentation (Lu et al., 1996). These results indicate that Pin1 is also a mitotic regulator in HeLa cells.

The above phenotypes associated with Pin1 are reciprocal to those resulting from manipulation of the NIMA kinase (Lu and Means, 1994; Lu and Hunter, 1995a, b).

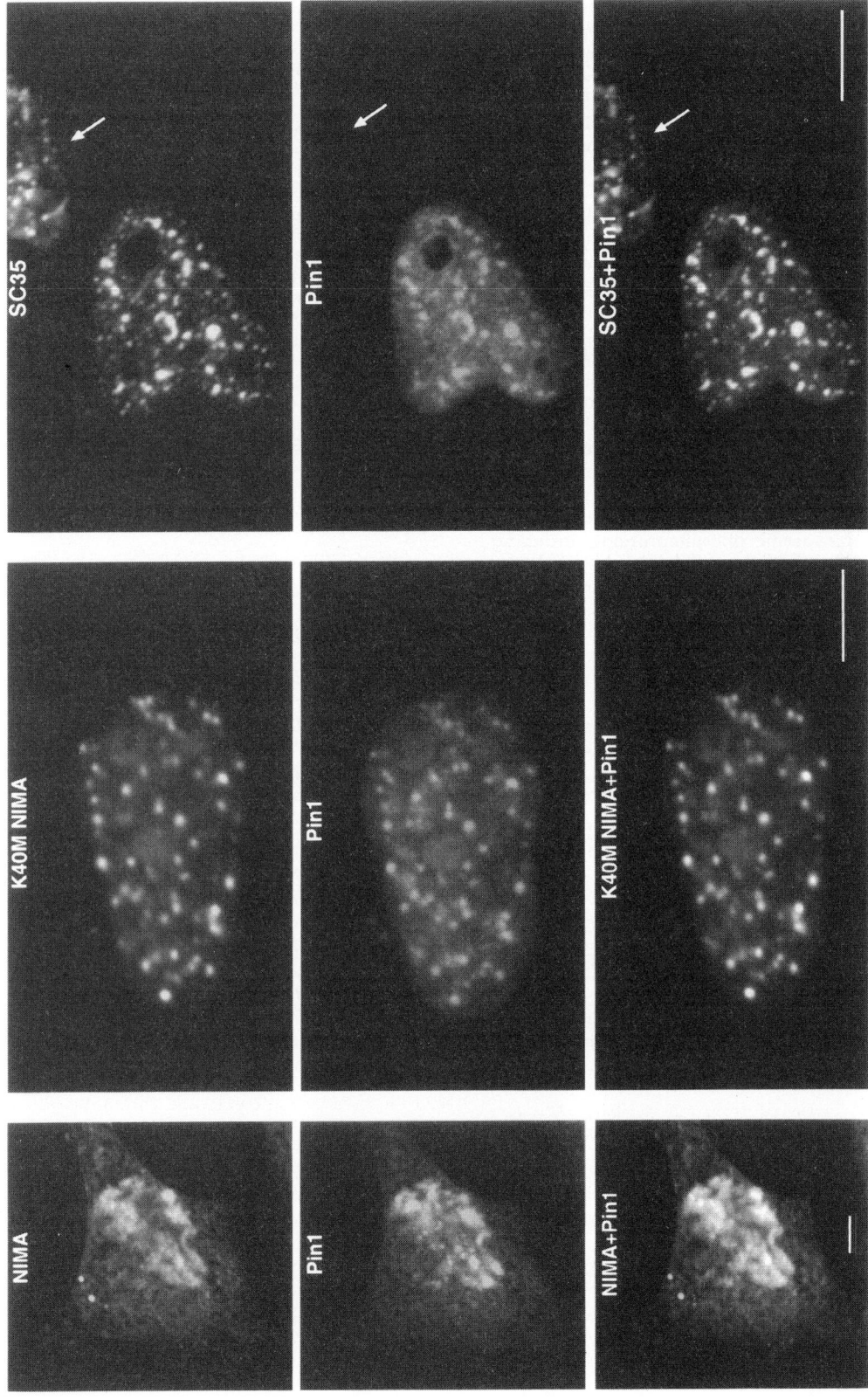

Figure 2. Immunolocalization of the epitope-tagged Pin1 and its colocalization with NIMA in nuclear speckles detected by the splicing factor SC35 antibody during the interphase and in the condensed chromatin at mitosis, in HeLa cells. Bar, 1 m. Modified from Lu *et al.* (1996).

Expression of dominant-negative NIMA induces G2 arrest, but expression of the wild-type kinase promotes premature entry into mitosis in HeLa cells (Lu and Means, 1994; Lu and Hunter, 1995a). This NIMA mitosis-promoting activity is attenuated by coexpression of Pin1 (Lu et al., 1996). Furthermore, although Pin1 is neither a substrate nor an inhibitor of NIMA kinase, it interacts with the C-terminal regulatory domain of NIMA both in vitro and in vivo. Specifically, it binds with the NID domain of NIMA (Lu et al., 1996), which is essential for NIMA in vivo function and shows significant similarity to other protein-interacting domains (Lu and Means, 1994; Lu and Hunter, 1995a). Moreover, in HeLa cells Pin1 colocalizes with NIMA in the nuclear speckles during interphase and in the condensed chromatin at mitosis (Fig. 2; Lu et al., 1996). These results indicate that Pin1 binds the NID domain of NIMA and somehow inhibits its mitotic function, thereby regulating progression through mitosis.

Recent determination of the Pin1 crystal structure (Ranganathan et al., 1997) and biochemical identification of Pin1 target proteins (M. Shen, T. Stukenberg, M. Kirschner and K.P. Lu, manuscript submitted) demonstrate that Pin1 is a general mitotic inhibitor which sequesters mitosis-specific phosphoproteins, including NIMA and CDC25, likely through interacting with the common mitotic phosphorylation motif phosphoser/thr-Pro.

■ References

Chan, D.C., Bedford, M.T., and Leder, P. (1996). Formin binding proteins bear WWP/WW domains that bind proline-rich peptides and functionally resemble SH3 domains. EMBO J. **15**, 1045–1054.

Fischer, G. (1994). Peptidyl-prolyl cis/trans isomerases. Angew. Chem. Intl. Ed. Engl. **63**, 67–118.

Hanes, S.D., Shank, P.R., and Bostian, K.A. (1989). Sequence and mutational analysis of ESS1, a gene essential for growth in Saccharomyces cerevisiae. Yeast **5**, 55–72.

Hani, J., Stumpf, G., and Domdey, H. (1995). PTF1 encodes an essential protein in Saccharomyces cerevisiae, which shows strong homology with a new putative family of PPIases. FEBS Lett. **365**, 198–202.

Lu, K.P. and Hunter, T. (1995a). Evidence for a NIMA-like mitotic pathway in vertebrate cells. Cell **81**, 413–424.

Lu, K.P. and Hunter, T. (1995b). The NIMA kinase: A mitotic regulator in Aspergillus nidulans and vertebrate cells. In Progress in Cell Cycle Research, (L. Meijer, S. Guidet and H.Y.L. Tung, eds), pp. 187–205. Plenum Press, New York.

Lu, K.P. and Means, A.R. (1994). Expression of the noncatalytic domain of the NIMA kinase causes a G2 arrest in Aspergillus nidulans. EMBO J. **13**, 2103–2113.

Lu, K.P., Hanes, S.D., and Hunter, T. (1996). A human peptidyl-prolyl isomerase essential for regulation of mitosis. Nature **380**, 544–547.

Maleszka, R., Hanes, S.D., Hackett, R.L., de, C.H., and Miklos, G.L. (1996). The Drosophila melanogaster dodo (dod) gene, conserved in humans, is functionally interchangeable with the ESS1 cell division gene of Saccharomyces cerevisiae. Proc. Natl. Acad. Sci. USA **93**, 447–451.

Rahfeld, J.U., Schierhorn, A., Mann, K., and Fischer, G. (1994). A novel peptidyl-prolyl cis/trans isomerase from Escherichia coli. FEBS Lett. **343**, 65–69.

Ranganathan, R., Lu, K.P., Hunter, T. and Noel, J. (1997). Structural and functional analysis of the Pin1 mitotic peptidylprolyl isomerase suggests substrate recognition is phosphorylation dependent. Cell (in press).

Rudd, K.E., Sofia, H.J., Koonin, E.V., Plunkett, III, G., Lazar, S. and Rouviere, P.E. (1995). A new family of peptidyl-prolyl isomerase. Trends Biochem. Sci. **20**, 12–14.

Sudol, M., Chen, H.I., Bougeret, C., Einbond, A., and Bork, P. (1995). Characterization of a novel protein-binding module: the WW domain. FEBS Lett. **369**, 67–71.

■ *Kun Ping Lu*
Cancer Biology Program, HIM 1047
Beth Israel Deaconess Medical Center
Harvard Medical School
330 Brookline Avenue
Boston, MA 02215, USA
Tel. 1 617 667 4143
Fax. 1 617 667 0610
E-mail: kLu@bidmc.harvard.edu

■ *Tony Hunter:*
Molecular Biology and Virology Laboratory
The Salk Institute
10010 North Torrey Pines Road
La Jolla, CA 92037, USA
Tel. 1 619 453 4100 ext. 1387
Fax. 1 619 457 4765
E-mail: Tony_Hunter@Salk.edu

13

Individual Chaperones

Escherichia coli SecB

SecB is a molecular chaperone in Escherichia coli that was originally identified in genetic studies as a factor that facilitated the export of a number of proteins destined for either the periplasmic space or the outer membrane. SecB binds the precursor forms of the secretory proteins and blocks aggregation and/or folding to maintain them in a state that is competent for translocation through the cytoplasmic membrane. There is no obvious consensus in sequence or in structure among the polypeptides that bind SecB, thus it is unclear what features are recognized. The binding is selective for non-native structure and is characterized by high affinity.

■ Identification of SecB

Escherichia coli SecB was originally identified in genetic studies as a factor that facilitated the export of a number of proteins destined for either the periplasmic space or the outer membrane (Kumamoto and Beckwith, 1983).

■ *secB* gene and sequence

Genes with sequence similarity to *secB* have been identified in *Buchnera aphidicola* (GenBank accession number M90644) and *Haemophilus influenzae* (GenBank accession number L42023). Both bacteria are members of the gamma subdivision of Proteobacteria, which also includes *E. coli*. The deduced sequence of the protein encoded by the *Buchnera* gene is 47% identical (Lai and Baumann, 1992) to that of the *E. coli* SecB protein (Kumamoto and Nault, 1989) and the deduced sequence of the protein from *Haemophilus* is 56% identical to the *E. coli* protein (Fleischmann et al., 1995).

■ SecB protein

SecB can be readily purified from strains of *E. coli* harboring plasmids that contain the *secB* gene under regulated strong promoters (Randall et al., 1996). The purification involves passage over a Q-Sepharose ion-exchange column followed by chromatography using a molecular sieve resin, Sephacryl S-300, and finally a Mono-Q ion-exchange column. The extinction coefficient for SecB at 280 nm is 0.69 ml mg^{-1}cm^{-1}. The molecular mass of SecB calculated from the deduced amino acid sequence is 17 278 Da. Analysis of the denatured purified protein by electrospray mass spectrometry (Smith et al., 1996) reveals two species, one of mass 17 146 Da, which most likely represents the polypeptide from which the N-terminal methionine has been removed, and a larger species with a mass of 17 189 Da, which is likely to be lacking the methionine and acetylated at the N-terminus. Mass spectrometry of native SecB indicates that it is a tetramer of molecular mass 68 610 Da. The protein is acidic with a pI determined to be 3.95–4.1 (Weiss et al., 1988). Analyses by circular dichroism (CD) show that SecB has a high content of β sheet conformation (Breukink et al., 1992; Fasman et al., 1995). SecB has proven to be difficult to crystallize. Crystals have been grown but they diffract to only 8 Å resolution (Vrielink et al., 1995).

■ Biological activity of SecB

SecB is a cytosolic bacterial chaperone for which no eukaryotic homologue has yet been found. Among the bacterial chaperones it appears to be the only one dedicated to protein export since it is the only one that has affinity for SecA, the peripheral-membrane ATPase of the export apparatus (Hartl et al., 1990). The affinity of SecA for SecB is modulated by the presence of the membrane and also by the binding of precursor proteins to SecB. SecB differs from other molecular chaperones in that ATP seems to have no effect on binding or release of ligand (Hardy and Randall, 1991). However, it should be remembered that SecB interacts with SecA, which does hydrolyse ATP when it binds precursor at the membrane translocation apparatus (Hartl et al., 1990). Whether binding or hydrolysis of ATP is necessary for the transfer of precursor from SecB to SecA is not known.

The distinguishing characteristic of all polypeptides that interact with SecB during export is the possession of a leader sequence (for a review see Randall and Hardy, 1995). However, SecB does not recognize the leader directly. Rather, the leader plays an indirect role by retarding the folding of the polypeptide, allowing SecB to bind before the polypeptide can fold. The binding is selective for non-native structure and is characterized by high affinity. Folding can be a very rapid process, thus the rate at which SecB binds its non-native ligand must be high. The final distribution of polypeptides between the folded state and that complexed to SecB is determined by a partitioning that is dependent on the rate of folding relative to the rate of binding to the chaperone, i.e. a kinetic partitioning.

■ Interaction of SecB with its ligands

A model to account for the high selectivity for binding of non-native proteins over proteins in their native states has been proposed based on studies of SecB with short peptide ligands (Randall, 1992). The model invokes the ordered binding of portions of the ligand to two types of interacting sites: one binding extended flexible regions and the other being hydrophobic. Detailed analyses of the region of ligands that occupy the binding sites on SecB have been completed for maltose-binding protein (Topping and Randall, 1994), for galactose-binding protein (Khisty et al., 1995), and for oligopeptide-binding protein (Smith et al., 1997). In all cases the binding frame is large, spanning minimally 150–170 residues. Studies of species of SecA that have the C-terminal 66 or 70 amino acid residues deleted indicate that this region is involved in binding to SecB (Rajapandi and Oliver, 1994; Breukink et al., 1995).

■ Mutational analyses of SecB

Both classical genetic approaches involving random chemical mutagenesis followed by screening for defects in export, as well as site-directed mutagenesis have led to the identification of residues in SecB that are crucial to its function (Gannon and Kumamoto, 1993; Kimsey et al., 1995). Mutations at the alternating positions Phe-74, Cys-76, Val-78 or Gln-80 compromise the ability of SecB to bind precursor maltose-binding protein in vivo. In contrast, mutations at residues Leu-75 or Glu-77 enhance the binding of SecB to substrate proteins in vivo and/or in vitro. Mutations affecting the acidic residues Asp-20 and Glu-24 confer a phenotype that resembles the phenotype of Leu-75 or Glu-77 mutations.

■ References

Breukink, E., Kusters, R., and de Kruijff, B. (1992). In vitro studies on the folding characteristics of the Escherichia coli precursor protein prePhoE. Eur. J. Biochem. **208**, 419–425.

Breukink, E., Nouwen, N., van Raalte, A., Mizushima, S., Tommassen, J., and de Kruijff, B. (1995). The C-terminus of SecA is involved in both lipid binding and SecB binding. J. Biol. Chem. **270**, 7902–7907.

Fasman, G.D., Park, K., and Randall, L.L. (1995). Chaperone SecB: conformational changes demonstrated by circular dichroism. J. Protein Sci. **14**, 595–600.

Fleischmann, R.D. et al. (1995). Whole-genome random sequencing and assembly of Haemophilus influenzae Rd. Science **269**, 496–512.

Gannon, P.M. and Kumamoto, C.A. (1993). Mutations of the molecular chaperone protein SecB which alter the interaction between SecB and maltose-binding protein. J. Biol. Chem. **268**, 1590–1595.

Hardy, S.J.S. and Randall, L.L. (1991). A kinetic partitioning model of selective binding of nonnative proteins by the bacterial chaperone SecB. Science **251**, 439–443.

Hartl, F.-U., Lecker, S., Schiebel, E., Hendrick, J.P., and Wickner, W. (1990). The binding cascade of SecB to SecA to SecY/E mediates preprotein targeting to the E. coli plasma membrane. Cell **63**, 269–279.

Khisty, V.J., Munske, G.R., and Randall, L.L. (1995). Mapping of the binding frame for the chaperone SecB within a natural ligand, galactose-binding protein. J. Biol. Chem. **270**, 25920–25927.

Kimsey, H.H., Dagarag, M.D., and Kumamoto, C.A. (1995). Diverse effects of mutation on the activity of the Escherichia coli export chaperone SecB. J. Biol. Chem. **270**, 22831–22835..

Kumamoto, C.A. and Beckwith, J. (1983). Mutations in a new gene, secB, cause defective protein localization in Escherichia coli. J. Bacteriol. **154**, 253–260.

Kumamoto, C.A. and Nault, A.K. (1989). Characterization of the Escherichia coli protein export gene secB. Gene **75**, 167–175.

Lai, C.-Y. and Baumann, P. (1992). Sequence analysis of a DNA fragment from Buchnera aphidicola (an endosymbiont of aphids) containing genes homologous to dnaG, rpoD, cysE, and secB. Gene **119**, 113–118.

Rajapandi, T. and Oliver, D. (1994). Carboxy-terminal region of Escherichia coli SecA ATPase is important to promote its protein translocation activity in vivo. Biochem. Biophys. Res. Commun. **200**, 1477–1483.

Randall, L.L. (1992). Peptide binding by chaperone SecB: implications for recognition of nonnative structure. Science **257**, 241–245.

Randall, L.L. and Hardy, S.J.S. (1995). High selectivity with low specificity; how SecB has solved the paradox of chaperone binding. Trends Biochem. Sci. **20**, 65–69.

Randall, L.L., Topping, T.B., Smith, V.F., Diamond, D., and Hardy, S.J.S. (1996). SecB, a chaperone from E. coli. Meth. Enzymol. in press.

Smith, V.F., Schwartz, B.L., Randall, L.L., and Smith, R.D. (1996). Electrospray mass spectrometric investigations of SecB. Protein Sci. **5**, 488–494..

Smith, V.F., Hardy, S.J.S., and Randall, L.L. (1997). Determination of the binding frame of the chaperone SecB within the physiological ligand oligopeptide-binding protein. Prot. Sci. In press.

Topping, T.B. and Randall, L.L. (1994). Determination of the binding frame in a physiological ligand for the chaperone SecB. Prot. Sci. **3**, 730–736.

Vrielink, A., Beamer, L., Le, T., and Eisenberg, D. (1995). Crystallization of the chaperone SecB. Protein Sci. **4**, 1651–1653.

Weiss, J.B., Ray, P.H., and Bassford, P.J., Jr. (1988). Purified SecB of Escherichia coli retards folding and promotes membrane translocation of the maltose-binding protein in vitro. Proc. Natl. Acad. Sci. USA **85**, 8978–8982.

■ Linda L. Randall:
Department of Biochemistry and Biophysics
Washington State University
Pullman, WA 99164-4660, USA
Tel. 1 509 335 6398
Fax. 1 509 335 9688

Escherichia coli FtsH

FtsH is a membrane-bound ATPase found in the cytoplasmic membrane of prokaryotic cells. The cytoplasmic domain of FtsH contains an ATPase segment that is homologous to the corresponding domains of 'AAA family' members as well as a segment with a zinc protease signature. FtsH has proteolytic activity towards a selected set of unstable proteins. It has been postulated that FtsH has chaperone-like activities by which it assists in proper assembly of some membrane proteins.

■ Alternative name

HflB (Herman et al., 1993).

■ Identification of FtsH

Escherichia coli FtsH was identified genetically in several independent studies. The temperature-sensitive *ftsH1* mutant was originally described as filamentation temperature-sensitive (Santos and Almeida, 1975). However, this mutant proved to carry a second mutation in a cell division gene (*ftsI*), the *ftsH1* mutation being primarily responsible for the temperature-sensitive cell growth phenotype (Begg et al., 1992). A classical mutation, *hflB29*, that causes high-frequency lysogenization of phage λ by stabilizing the CII protein (Banuett et al., 1986), has been localized within *ftsH* (Herman et al., 1993). The *std* (stop transfer defective) alleles of *ftsH* were isolated in screens in which mutations that allowed an unusual mode of protein export were sought; such mutations allow translocation of the mature sequence of alkaline phosphatase (PhoA) that had been attached to the last cytoplasmic domain of the SecY protein (Akiyama et al., 1994a). Finally, a major class of mutations that stabilize otherwise rapidly eliminated forms of SecY missing its partner (SecE) affect *ftsH* (Kihara et al., 1995). In *Bacillus subtilis* and *Lactococcus lactis*, *ftsH* mutants are sensitive to salts (Deurling et al., 1995; Nilsson et al., 1994). Recently it was reported that mutations in YTA10 (*AFG3*) and YTA12 (*RCA1*) that encode mitochondrial FtsH homologs in *Saccharomyces cerevisiae* cause defective assembly as well as degradation of membrane proteins (Arlt et al., 1996; Suzuki et al., 1997).

■ *ftsH* gene and sequence

The nucleotide sequence of *E. coli ftsH* (GenBank accession number M83138) predicts a product of 644 amino acids (Tomoyasu et al., 1993b). An alternative product using a TTG initiation codon resulting in three additional N-terminal amino acids was also proposed (accession number U01376). Homologs have been reported in *B. subtilis* (D26185), *L. lactis* (X69123) and *Haemophilus influenzae* (L42023). The N-terminal region of FtsH contains two hydrophobic stretches that serve as transmembrane segments (Tomoyasu et al., 1993a). The cytoplasmic domain contains one set of ATP-binding consensus motifs at residues 192–199 (motif A) and a second at residues 241–251 (motif B) (Tomuyasu et al., 1993b), as well as a zinc metalloprotease motif (HEXXH) at residues 414–418 (Tomoyasu et al., 1995). The central region (residues 153–361) of the cytoplasmic domain shares homology with members of the AAA family (Tomoyasu et al., 1993b) [AAA stands for ATPases associated with diverse cellular activities (Kunau et al., 1993; Confalonieri and Duguent, 1995)]. Among its members in eukaryotic cells, some in mitochondria and chloroplasts show close similarity to FtsH (Confalonieri and Duguet, 1995).

■ FtsH protein

FtsH is an integral cytoplasmic membrane protein as shown by cell fractionation and immunoelectron microscopy (Tomoyasu et al., 1993a). Topology analysis using alkaline phosphatase fusions indicates that both termini of FtsH face the cytoplasm, while it is anchored to the membrane with two N-terminally located transmembrane segments (Tomoyasu et al., 1993a). A preparation of FtsH protein purified from overproducing cells has an ATPase activity (specific activity 230 nmol min^{-1}mg^{-1}) and catalyses an ATP-dependent degradation of the heat shock sigma factor, σ^{32} (Tomoyasu et al., 1995). SecY (Akiyama et al., 1996a) and λ CII (Kihara et al., 1997) degradation activity was also demonstrated using a purified system. FtsH in wild-type cells is in a large complex, as shown by gel filtration after solubilization with a non-ionic detergent (Akiyama and Ito, 1995). Immunoprecipitation, histidine-tagging, and cross-linking experiments indicate that the FtsH complex contains more than one molecule of FtsH (Akiyama and Ito, 1995) and at least a pair of two other proteins, HflC and HflK, which appear to modulate FtsH function (Kihara et al., 1996, 1997).

■ Biological activities and genetic interactions of FtsH

FtsH is essential for viability of *E. coli* (Begg et al., 1992; Akiyama et al., 1994a). It is required for proteolytic degradation of some unstable proteins that include both soluble proteins such as σ^{32} (Herman et al., 1995; Tomoyasu et al., 1995) and λCII (Herman et al., 1993; Kihara et al., 1997), and membrane proteins including uncomplexed forms of SecY (Kihara et al., 1995) and the F_o subunit α of the proton ATPase (Akiyama et al., 1996b). Thus, FtsH is a key proteolytic element that governs transcriptional control over stress response and the lysis/lysogeny decision, as well as quality control of the membrane. FtsH is the only protease that is known to be essential for the life of prokaryotic organisms, and may be involved in proteolytic control of diverse biological processes.

In addition, this membrane-bound ATPase might be more than just a protease (Fig. 1). A loss of its function leads to the stop transfer defective behavior of at least a SecY–PhoA fusion protein, and this phenomenon is not a consequence of stabilization of this particular protein (Akiyama et al., 1994a; Kihara et al., 1995). FtsH mutants are partially defective in export of normally secreted proteins (Akiyama et al., 1994a, b). These phenotypes are differentially suppressible by overproduction of some other chaperones (Shirai et al., 1996). It is tempting to speculate that its interaction with substrate proteins sometimes leads to protein assembly/maturation rather than to degradation. In apparent agreement with this notion, FtsH can bind to denatured PhoA without degrading it (Y. Akiyama, A. Kihara, M. Ehrman and K. Ito, unpublished results).

■ Transcriptional regulation of *ftsH*

The *E. coli ftsH* gene constitutes an operon with the upstream *ftsJ* gene (Herman et al., 1995). Expression of the *ftsJ–ftsH* operon is directed by two promoters, one constitutive and the other subject to heat shock regulation by σ^{32} (Herman et al., 1995). The *B. subtilis* FtsH protein is also induced by temperature upshift as well as by osmotic shock (Deurling et al., 1995).

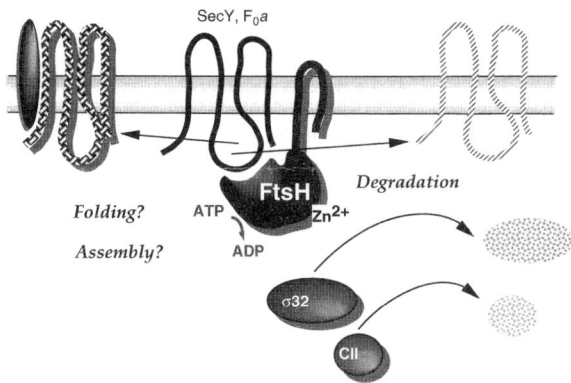

Figure 1. A schematic representation of the FtsH protein and its functions.

■ Mutagenesis studies with FtsH

Alterations of the conserved lysine in the ATPase motif A inactivate FtsH and cause a dominant negative phenotype (Akiyama et al., 1994b). Moreover, expression of the N-terminally located and membrane-associated segment alone causes a dominant interference (Akiyama and Ito, 1995; Herman et al., 1995). This region is required for FtsH–FtsH cross-linking and coimmunoprecipitation (Akiyama and Ito, 1995).

■ References

Akiyama, Y., Yoshisa, T., and Ito, K. (1995). FtsH, a membrane-bound ATPase, forms a complex in the cytoplasmic membrane of Escherichia coli. J. Biol. Chem., J. Biol Chem. **270**, 23485–23490.

Akiyama, Y., Ogura T., and Ito, K. (1994a). Involvement of FtsH in protein assembly into and through the membrane. I. Mutations that reduce retention efficiency of a cytoplasmic reporter. J. Biol. Chem. **269**, 5218–5224.

Akiyama, Y., Shirai, Y., and Ito, K. (1994b). Involvement of FtsH in protein assembly into and through the membrane. II. Dominant mutations affecting FtsH functions. J. Biol. Chem. **269**, 5225–5229.

Akiyama, Y., Kihara, A., Tokuda, H., and Ito, K. (1996a). FtsH (HflB) is an ATP-dependent protease selectively acting on SecY and some other membrane proteins. J. Biol. Chem. **271**, 31196–31201.

Akiyama, Y., Kihara, A., and Ito, K. (1996b). Subunit a of proton ATPase F_0 sector is a substrate of the FtsH protease in Escherichia coli. FEBS Lett. **399**, 26–28.

Arlt, H., Tauer, R., Feldmann, H., Neupert, W., Langer, T. (1996). The YTA10-12 complex, an AAA protease with chaperone-like activity in the inner membrane of mitochondria. Cell. **85**, 875–885.

Banuett, F., Hoyt, M.A., McFarlane, L., Echols, H., and Herskowitz, I. (1986). hflB, a new Escherichia coli locus regulating lysogeny and the level of bacteriophage lambda cII protein. J. Mol. Biol. **187**, 213–244.

Begg, K.J., Tomoyasu, T., Donachie, W.D., Khattar, M., Niki, H., Yamanaka, K., Hiraga, S., and Ogura, T. (1992). Escherichia coli mutant Y16 is a double mutant carrying thermosensitive ftsH and ftsI mutations. J. Bacteriol. **174**, 2416–2417.

Confalonieri, F. and Duguet, M. (1995). A 200-amino acid ATPase module in search of a basic function. BioEssays **17**, 639–650.

Deurling, E., Paeslack, B., and Schumann, W. (1995). The ftsH gene of Bacillus subtilis is transiently induced after osmotic and temperature upshift. J. Bacteriol. **177**, 4105–4112.

Herman, C., Ogura, T., Tomoyasu, T., Hiraga, S., Akiyama, Y., Ito, K., Thomas, R., D'Ari, R., and Bouloc, P. (1993). Cell growth and λ phage development controlled by the same essential Escherichia coli gene, ftsH/hflB. Proc. Natl. Acad. Sci. USA **90**, 10861–10865.

Herman, C., Thévenet, D., D'Ari, R., and Bouloc, P. (1995). Degradation of σ^{32}, the heat shock regulator in Escherichia coli, is governed by HflB. Proc. Natl. Acad Sci. USA **92**, 3516–3520.

Kihara, A., Akiyama, Y., and Ito, K. (1995). FtsH is required for proteolytic elimination of uncomplexed forms of SecY, an essential protein translocase subunit. Proc. Natl. Acad. Sci. USA **92**, 4532–4536.

Kihara, A., Akiyama, Y., and Ito, K. (1996). A protease complex in the Escherichia coli plasma membrane: HflKC (HflA) forms a complex with FtsH (HflB), regulating its proteolytic activity against SecY. EMBO J. **15**, 6122–6131.

Kihara, A., Akiyama, Y., and Ito, K. (1997). Host regulation of lysogenic decision in bacteriophage λ: transmembrane modulation of FtsH (HflB), the cII degrading protease, by HflKC (HflA) Proc. Natl. Acad. Sci. USA. **94**, in press.

Kunau, W.H., Beyer, A., Franken, T., Götte, K., Marzioch, M., Saidowsky, J., Skaletz-Rorowski, A., and Wiebel, F.F. (1993). Two complementary approaches to study peroxisome biogenesis in Saccharomyces cerevisiae: Forward and reverse genetics. Biochimie **75**, 209–224.

Nilsson, D., Lauridsen, A.A., Tomoyasu, T., and Ogura, T. (1994). A Lactococcus lactis gene encodes a membrane protein with putative ATPase activity that is homologous to the essential Escherichia coli ftsH gene product. Microbiology **140**, 2601–2610.

Santos, D. and Almedia, D.F. (1975). Isolation and characterization of a new temperature-sensitive cell division mutant of Escherichia coli K-12. J. Bacteriol. **142**, 1502–1507.

Shirai, Y., Akiyama, Y., and Ito, K., (1996). Suppression of ftsH mutant phenotypes by overproduction of molecular chaperones. J. Bacteriol. **198**, 1141–1145.

Suzuki, C.K., Rep, M., van Dijl, J.M., Suda, K., Grivell, L.A., and Schatz, G. (1997). ATP-dependent proteases that also chaperone protein biogenesis. Trends Biochem. Sci. **22**, 118–123.

Tomoyasu, T., Yamanaka, K., Murata, K., Suzaki, T., Bouloc, P., Kato, A., Niki, H., Hiraga, S., and Ogura, T. (1993a). Topology and subcellular localization of FtsH protein in Escherichia coli. J. Bacteriol. **175**, 1352–1357.

Tomoyasu, T., Yuki, T., Morimura, S., Mori, H., Yamanaka, K., Niki, H., Hiraga, S., and Ogura, T. (1993b). The Escherichia coli FtsH protein is a prokaryotic member of a protein family of putative ATPase involved in membrane functions, cell cycle control, and gene expression. J. Bacteriol. **175**, 1344–1351.

Tomoyasu, T., Gamer, J., Bukau, B., Kanemori, M., Mori, H., Rutman, A.J., Oppenheim, A.B., Yura, T., Yamanaka, K., Niki, H., Hiraga, S., and Ogura, T. (1995). Escherichia coli FtsH is a membrane-bound, ATP-dependent protease which degrades the heat-shock transcription factor s32. EMBO J. **14**, 2551–2560.

■ *Yoshinori Akiyama and K. Ito:*
Department of Cell Biology
Institute for Virus Research
Kyoto University
Kyoto 606-01, Japan
Tel. 81 75 751 4050/4015
Fax. 81 75 761 5626/5699
E-mail: yakiyama@virus.kyoto-u.ac.jp and kito@virus.kyoto-u.ac.jp

■ *Teru Ogura:*
Department of Molecular Cell Biology
Institute of Molecular Embryology and Genetics
Kumamoto University School of Medicine
Kumamoto 862, Japan
Tel. 81 96 373 5336
Fax. 81 96 371 2408
E-mail: ogura@gpo.kumamoto-u.ac.jp

Mammalian MSF

MSF (mitochondrial import stimulation factor) was isolated as a N-ethyl maleimide (NEM)-sensitive cytosolic factor required for in vitro protein import into mitochondria. MSF catalyses unfolding of precursor proteins, keeps them in an import-competent conformation, targets the unfolded proteins to mitochondria, and eventually mediates ATP-dependent mitochodrial protein import.

■ Isolation of MSF

Mitochondrial precursor proteins synthesized in the wheat germ lysate system are in an aggregated state and, thus, are either not imported or poorly imported into mitochondria *in vitro*, whereas those synthesized in the reticulocyte lysate are efficiently imported (Murakami et al., 1988). MSF was biochemically identified in a rat liver cytosol fraction as a factor that is required for the *in vitro* import of the wheat germ-synthesized precursor protein into mitochondria (Hachiya et al., 1993). The factor was purified 10 000-fold from rat liver cytosol using a COXIV presequence–β-galactosidase fusion protein-conjugated resin as the affinity matrix and several conventional chromatographic steps. Purified MSF stimulates mitochondrial import of all of proteins so far examined, suggesting that it generally recognizes mitochondrial precursor proteins.

■ MSF gene and sequence

The two subunits of the purified MSF protein were sequenced and found to be members of the collectively named '14-3-3 protein family' (Alam et al., 1994). The 32 kDa subunit is most similar to the isoform of 14-3-3 protein (GenBank accession number D30739), while the 30 kDa subunit is most similar to the ζ isoform (GenBank accession number D30740).

MSF protein

The purified MSF was composed of nearly stoichiometric amounts of two polypeptides with molecular weights of 30 and 32 kDa, which could not be separated under nondenaturing conditions. Import stimulation activity in cytoplasm was completely inhibited by polyclonal antibodies against either the recombinant 32 kDa subunit or the recombinant 30 kDa subunit (Alam et al., 1994).

MSF maintains the unfolded conformation of urea-denatured precursor proteins. MSF also depolymerizes both wheat germ-synthesized precursor proteins and aggregated recombinant precursor proteins purified from *Escherichia coli*. (Hachiya et al., 1993, 1994). The depolymerization process is dependent on ATP hydrolysis and accompanied by conformational changes in the precursor proteins (Hachiya et al., 1993).

MSF exhibits strong ATPase activity in the presence of the mitochondrial precursor proteins or synthetic peptides containing a functional mitochondrial presequence (Hachiya et al., 1993; Komiya et al., 1994). The precursor protein-induced ATPase activity of MSF is also dependent on the aggregated states of the effector proteins; the import-competent, unfolded precursor protein only induced a low activity, whereas aggregated, import-incompetent precursor induced a higher ATPase activity. MSF therefore recognizes not only the primary sequences of the target proteins but also their conformation (Hachiya et al., 1994). The precursor protein-induced ATPase activity is inhibited by mitochondrial outer membrane (Hachiya et al., 1994). The component of the yeast outer mitochondrial membrane responsible for the inhibition of ATPase activity has been identified as the Mas70p–Mas37p complex which has been identified as a mitochondrial import receptor (Hachiya et al., 1995).

NEM treatment inhibits the mitochondrial import stimulation activity of MSF, leaving the depolymerization function intact. The NEM-treated MSF retains the precursor-induced ATPase activity, but the activity is no longer inhibited by mitochondrial outer membrane. The treatment of MSF with NEM inhibits binding of the MSF–precursor complex to the receptor on the outer mitochondrial membrane (Hachiya et al., 1993, 1994).

Biological function of MSF

MSF recognizes the presequence of precursor proteins and maintains import-competent conformation of nascent precursor protein by forming a stable complex with it. MSF can also restore the import-competent conformation of aggregated and import-incompetent precursor molecules in an ATP-dependent fashion. The unfolded precursor–MSF complex docks on to the Mas70p–Mas37p complex in yeast (Hachiya et al., 1995), MSF is then released from the docking complex depending on ATP hydrolysis, and the precursor protein is transferred to the Mas20p–Mas22p complex, and finally delivered to the import channel in the outer mitochondrial membrane (Hachiya et al., 1995; Komiya and Mihara, 1996).

Hsc70, a general molecular chaperone, has also been reported to mediate mitochondrial import (Deshaies et al., 1988). In Hsc70-mediated import, the precursor protein directly interacts with the Mas20p–Mas22p complex, and is then imported into mitochondria (Komiya et al., 1996; Mihara and Omura, 1996).

References

Alam, R., Hachiya, N., Sakaguchi, M., Kawabata, S., Iwanaga, S., Kitajima, M., Mihara, K., and Omura, T. (1994). cDNA cloning and characterization of mitochondrial import stimulation factor (MSF) purified from rat liver cytosol. J. Biochem. **166**, 416–425.

Deshaies, R.J., Koch, B.D., Werner-Washburne, M., Craig, A.E., and Schekman, R. (1988). A subfamily of stress proteins facilitates translocation of secretory and mitochondrial precursor polypeptides. Nature **332**, 800–805.

Hachiya, N., Alam, R., Sakasegawa, Y., Sakaguchi, M., Mihara, K., and Omura, T. (1993). A mitochondrial import factor purified from rat liver cytosol is an ATP-dependent conformational modulator for precursor proteins. EMBO J. **12**, 1579–1586.

Hachiya, N., Komiya, T., Alam, R., Iwahashi, J., Sakaguchi, M., Omura, T., and Mihara, K. (1994). MSF, a novel cytoplasmic chaperone which functions in precursor targeting in mitochondria. EMBO J. **13**, 5146–5154.

Hachiya, N., Mihara, K., Suda, K., Horst,M., Schatz, G., and Lithgow, T. (1995). Reconstitution of the initial steps of mitochondrial protein import. Nature **376**, 705–709.

Komiya, T., and Mihara, K. (1996). Protein import into mammalian mitochondria: Characterization of the intermediates along the import pathway of the precursor into the matrix. J. Biol. Chem., **271**, 22105–22110.

Komiya, T., Hachiya, N., Sakaguchi, M., Omura, T., and Mihara, K. (1994). Recognition of mitochondria-targeting signals by a cytosolic import stimulation factor, MSF. J. Biol. Chem. **269**, 30893–30897.

Komiya, T., Sakaguchi, M., and Mihara, K. (1996). Cytoplasmic chaperones determine the targeting pathway of precursor proteins to mitochondria. EMBO J. **15**, 399–407.

Mihara, K., and Omura, T. (1996). Cytoplasmic chaperones in precursor targeting to mitochondria: the role of MSF and hsp70. Trends Cell Biol. **6**, 65–68.

Murakami, H., Pain, D., and Blobel, G. (1988). 70-kD heat shock-related protein is one of at least two distinct cytosolic factors stimulating protein import into mitochondria. J. Cell Biol. **107**, 2051–2057.

■ *Katsuyoshi Mihara and Masao Sakaguchi:*
Graduate School of Medical Science
Kyushu University
Higashiku
Fukuoka 812-12, Japan
Tel. 81 92 642 6176
Fax. 81 92 642 6183
E-mail: mihara@cell.med.kyushu-u.ac.jp

Mammalian Hip

The Hsc70-interacting protein, Hip, participates in the regulation of the 70 kDa heat shock cognate Hsc70 in the eukaryotic cytosol. Hip binds to the ATPase domain of Hsc70. Complex formation depends on activation of the Hsc70–ATPase by Hsp40, a eukaryotic DnaJ homologue. Hip functions as a regulator of Hsc70 during the stabilization of newly translated polypeptide chains and during the conformational regulation of signalling molecules known to interact with Hsc70 and Hsp90.

■ Alternative name

p48 (Smith et al., 1995; Prapapanich et al., 1996a).

■ *hip* gene and sequence

The nucleotide sequence of the *hip* cDNA of rat predicts a 368 amino acid protein of 41.3 kDa (GenBank accession number X82021). The protein migrates with reduced mobility in SDS–PAGE resulting in an apparent mass of c. 50 kDa. Hip forms a homo-oligomer of c. 160 kDa probably containing four subunits. Rat Hip is 90.5% identical to p48 of the human Hsp90–progesterone receptor complex (accession number U28918) and 36 % identical to a so-called heat shock-related protein of *Plasmodium berghei* (accession numbers L04508 and L21710), revealing conservation of the Hip protein during eukaryotic evolution.

Hip contains three consecutively-arranged tetratricopeptide repeat (TPR) domains (Fig. 1) which are found in proteins of diverse biological activity and are supposed to mediate protein–protein interactions (Lamb et al., 1995; Prapapanich et al., 1996b; Frydman and Höhfeld, 1997; Irmer and Höhfeld, 1997). Another internally repeated sequence motif, the tetrapeptide GGMP, is shared with cytosolic Hsp70 proteins (Boorstein et al., 1994). The GGMP repeats probably form a flexible, rather hydrophobic loop of as yet unknown function. The C-terminus of Hip resembles that of the p60/Sti1 protein (see entries pp. 458 and 459) which is part of a chaperone complex in the eukaryotic cytosol that includes Hsp70, Hsp90 and immunophilins (Nicolet and Craig, 1989; Smith et al., 1993; Chang and Lindquist, 1994; see review p. 518). Remarkably, Sti1p also belongs to the TPR protein family (Fig. 1).

■ Hip function

The cytosolic Hip protein binds to the ATPase domain of constitutively expressed Hsc70. This interaction provided the molecular basis for the identification of Hip in a screen for Hsc70-interacting proteins (Höhfeld et al., 1995). Similar binding characteristics have been observed for *Escherichia coli* GrpE which functions as a nucleotide exchange factor during the ATPase cycle of the bacterial HSP70 homologue DnaK (Liberek et al., 1991; Buchberger et al., 1994; Szabo et al., 1994). Hip, however, is neither structurally nor functionally related to GrpE. In contrast to the ADP-releasing activity of GrpE on DnaK, Hip stabilizes the ADP-bound state of Hsc70 (Höhfeld et al., 1995). Notably, the ADP-bound form of Hsc70 interacts most stably with unfolded polypeptide substrates (Palleros et al., 1991; Schmid et al., 1994). Hip thus promotes Hsc70–substrate interactions.

In the presence of ATP, the eukaroytic DnaJ homologue Hsp40 (see entry p. 121) is required for efficient

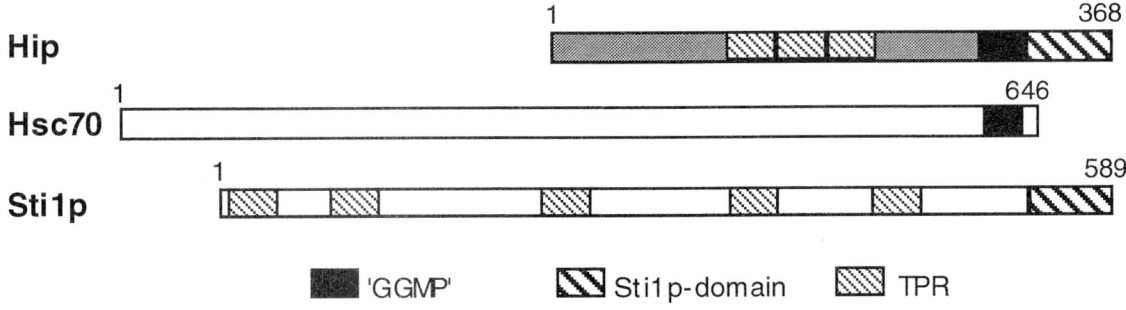

Figure 1. Hip, a member of the tetratricopeptide repeat (TPR) protein family, has a distinct domain structure sharing domains with Hsc70 and the TPR protein Sti1p, which is a partner protein of Hsc70 and Hsp90 in the eukaryotic cytosol. The Hsc70-related sequence comprises multiple repeats of the tetrapeptide GGMP.

formation of Hip–Hsc70 complexes (Höhfeld et al., 1995). Hsp40 stimulates the ATP hydrolytic activity of Hsc70 and thus drives Hsc70 into the ADP-bound conformation recognized by Hip.

Hip cooperates with Hsc70 and Hsp40 in the stabilization of newly synthesized polypeptide chains emerging from the ribosome and their subsequent transfer on to the TRiC chaperonin (Frydman et al., 1994; Höhfeld et al., 1995). Hip is also involved in the Hsc70/Hsp40-mediated assembly of complexes between Hsp90 and signalling molecules, such as the transforming tyrosine kinase pp60$^{v\text{-}src}$ (Caplan et al., 1995; Höhfeld et al., 1995; Kimura et al., 1995; Smith et al., 1995; Prapapanich et al., 1996). The homo-oligomeric structure of Hip (probably a pentamer) appears to be of importance in these reactions. Under in vitro conditions at least two Hsc70 molecules associate with the Hip oligomer. Hip may thus act as a scaffolding protein in arranging multiple Hsc70 molecules in close proximity to the unfolded polypeptide substrate, and may initiate the assembly of a multifunctional chaperone complex.

References

Boorstein, W., Ziegelhoffer, T. and Craig, E.A. (1994). Molecular evolution of the HSP70 multigene family. J. Mol. Evol. **38**, 1–17.

Buchberger, A., Schröder, H., Büttner, M., Valencia, A., and Bukau, B. (1994). A conserved loop in the ATPase domain of the DnaK chaperone is essential for stable binding of GrpE. Nat. Struct. Biol. **1**, 95–101.

Caplan, A., Langley, E., Wilson, E.M., and Vidal, J. (1995). Hormone-dependent transactivation by the human androgen receptor is regulated by a dnaJ protein. J. Biol. Chem. **270**, 5251–5257.

Chang, J.H.-C. and Lindquist, S. (1994). Conservation of Hsp90 macromolecular complexes in Saccharomyces cerevisiae. J. Biol. Chem. **269**, 24983–24988.

Frydman, J., and Höhfeld, J. (1997). Chaperones get in touch: the Hip-Hop connection. Trends Biochem. Sci. **22**, 87–92.

Frydman, J., Nimmesgern, E., Ohtsuka, K., and Hartl, F.-U. (1994). Folding of nascent polypeptide chains in a high molecular mass assembly with molecular chaperones. Nature **370**, 111–117.

Höhfeld, J., Minami, Y. and Hartl, F.-U. (1995). Hip, a new cochaperone involved in the eukaryotic Hsc70/Hsp40 reaction cycle. Cell **83**, 589–598.

Irmer, H., and Höhfeld, J. (1997). Characterization of functional domains of the eukaryotic co-chaperone Hip. J. Biol. Chem. **272**, 2230–2235.

Kimura, Y., Yahara, I. and Lindquist, S. (1995). Role of the protein chaperone YDJ1 in establishing Hsp90-mediated signal transduction pathways. Science **268**, 1362–1365.

Lamb, J.R., Tugendreich, S. and Hieter, P. (1995). Tetratrico peptide repeat interactions: to TPR or not to TPR? Trends Biochem. Sci. **20**, 257–259.

Liberek, K., Marszalek, J., Ang, D., Georgopoulos, C. and Zylicz, M. (1991). Escherichia coli DnaJ and GrpE heat shock proteins jointly stimulate ATPase activity of DnaK. Proc. Natl. Acad. Sci. USA **88**, 2874–2878.

Nicolet, C.M. and Craig, E.A. (1989). Isolation and characterization of STI1, a stress-inducible gene from Saccharomyces cerevisiae. Mol. Cell. Biol. **9**, 3638–3646.

Palleros, D.R., Welch, W.J., and Fink, A.L. (1991). Interaction of hsp70 with unfolded proteins: effects of temperature and nucleotides on the kinetics of binding. Proc. Natl. Acad. Sci. USA **88**, 5719–5723.

Prapapanich, V., Chen, S., Nair, S.C., Rimerman, R.A., and Smith, D.F. (1996a). Molecular cloning of human p48, a transient component of progesterone receptor complexes and an Hsp70-binding protein. Mol. Endocrinol. **10**, 420–431.

Prapapanich, V., Chen, S., Toran, E.J., Rimerman, R.A., and Smith, D.F. (1996b). Mutational analysis of the hsp70-interacting protein Hip. Mol. Cell. Biol. **16**, 6200–6207.

Schmid, D., Baici, A., Gehring, H., and Christen, P. (1994). Kinetics of molecular chaperone action. Science **263**, 971–973.

Smith, D.F., Sullivan, W.P., Marion, T.N., Zaitsu, K., Madden, B., McCormick, D.J., and Toft, D.O. (1993). Identification of a 60-kilodalton stress-related protein, p60, which interacts with hsp90 and hsp70. Mol. Cell. Biol. **13**, 869–876.

Smith, D.F., Whitesell, L., Nair, S.C., Chen, S., Prapapanich, V., and Rimerman, R.A. (1995). Progesterone receptor structure and function altered by geldanamycin, an hsp90-binding agent. Mol. Cell. Biol. **15**, 6804–6812.

Szabo, A., Langer, T., Schröder, H., Flanagan, J., Bukau, B., and Hartl, F.U. (1994). The ATP hydrolysis-dependent reaction cycle of the Escherichia coli Hsp70 system — DnaK, DnaJ, and GrpE. Proc. Natl. Acad. Sci. USA **91**, 10345–10349.

■ *Jörg Höhfeld:*
Zentrum für Molekulare Biologie, ZMBH
Universität Heidelberg
Im Neuenheimer Feld 282
D-69120 Heidelberg, Germany
Tel. 49 6221 54 6837
Fax. 49 6221 54 5891
E-mail: j-hoehfeld@sun0.urz.uni-heidelberg.de

Vertebrate p23

p23 is a Hsp90-binding protein that is a component of mature steroid receptor complexes. p23 forms a complex with Hsp90 and immunophilins that is disrupted by the Hsp90-binding agent geldanamycin. p23, or the p23 complex, is necessary for the formation of mature steroid receptor complexes, and may be involved in several diverse signal transduction pathways.

■ Isolation

p23 was found to coimmunoprecipitate with inactive chick progesterone receptor complexes, along with Hsp90, p60, and the immunophilins FKBP52, FKBP54 and Cyp40 (Smith et al., 1990; Smith and Toft, 1993; for review see p. 518). Monoclonal and peptide-directed polyclonal antibodies were used to isolate a chicken cDNA from an expression library (Johnson et al., 1994).

■ p23 gene and sequence

The human and chicken cDNAs (GenBank accession numbers L24804 and L24898, respectively) encode proteins of 160 amino acids that are 96.3% identical. p23 is a hydrophilic protein with an observed pI of 4.5–5.2, and contains an aspartic acid-rich C-terminus. Other than candidate phosphorylation sites, no common structural motifs are present (Johnson et al., 1994). p23-related sequences in *Saccharomyces cerevisiae* and *Schizosaccharomyces pombe* have been reported (GenBank accession numbers Z28117 and L41166, respectively).

■ p23 protein

Monoclonal antibodies raised against recombinant p23 that recognize p23 from a variety of species have been described. These antibodies specifically immunoprecipitate p23 in complex with Hsp90 and other proteins. p23 is a predominantly cytosolic protein expressed at similar levels in a variety of chicken and mammalian tissues. p23 is not heat inducible (Johnson et al., 1994).

■ Biological interactions of p23

p23, along with Hsp90, FKBP52, FKBP54 (also called FKBP51), and cyclophilin 40 is a component of the mature progesterone receptor capable of high-affinity hormone binding. Upon hormone binding, p23 and the other associated proteins dissociate from the receptor (Smith et al., 1990). p23 can bind directly to Hsp90 in the presence of ATP and molybdate (Johnson et al., 1996). In rabbit reticulocyte lysate, p23 forms a complex with Hsp90, FKBP52, FKBP54, and Cyp40 in an ATP-dependent manner (Johnson and Toft, 1994). The formation of this p23 complex is disrupted by the Hsp90-binding agent geldanamycin (Whitesell et al., 1994; Johnson and Toft, 1995; Smith et al., 1995). p23 and/or formation of the p23 complex is necessary for the reconstitution of steroid receptor complexes with high-affinity steroid binding (Johnson and Toft, 1994; Hutchinson et al., 1995; Smith et al., 1995). In addition, p23 is able to bind some unfolded proteins, indicating it has some properties of molecular chaperones (Bose et al., 1996; Freeman et al., 1996). Recent evidence suggests that the p23 complex may also be involved in hepatitis B virus replication (Hu and Seeger, 1996), suggesting diverse roles for p23 within the cell.

■ References

Bose, B., Weikl, T., Bügl, H., and Buchner, J. (1996). Chaperone function of Hsp90-associated proteins. Science. **274**, 1715–1717.

Freeman, B.C., Toft, D.O. and Morimoto, R.I. (1996). Molecular chaperone machines: chaperone activities of the cyclophilin Cyp40 and the steroid aporeceptor-associated protein p23. Science **274**, 1718–1720.

Hu, J. and Seeger, C. (1996). Hsp90 is required for the activity of a hepatitis B virus reverse transcriptase. Proc. Natl. Acad. Sci. USA **93**, 1060–1064.

Hutchinson, K.A., Stancato, L.F., Owens-Grillo, J.K., Johnson, J.L., Krishna, P., Toft, D.O., and Pratt, W.B. (1995). The 23-kDa acidic protein in reticulocyte lysate is the weakly bound component of the hsp foldosome that is required for assembly of the glucocorticoid receptor into a functional heterocomplex with hsp90. J. Biol. Chem. **270**, 18841–18847.

Johnson, J.L. and Toft, D.O. (1994). A novel chaperone complex for steroid receptors involving heat shock proteins, immunophilins, and p23. J. Biol. Chem. **269**, 24989–24993.

Johnson, J.L. and Toft, D.O. (1995). Binding of p23 and hsp90 during assembly with the progesterone receptor. Mol. Endocrinol. **9**, 670–678.

Johnson, J.L., Beito, T.G., Krco, C.J., and Toft, D.O. (1994). Characterization of a novel 23-kilodalton protein of unactive progesterone receptor complexes. Mol. Cell. Biol. **14**, 1956–1963.

Johnson, J., Corbisier, R., Stensgard, B., and Toft, D. (1996). The involvement of p23, hsp90, and immunophilins in the assembly of progesterone receptor complexes. J. Steroid Biochem. Mol. Biol. **56**, 31–37.

Smith, D.F. and Toft, D.O. (1993). Steroid receptors and their associated proteins. Mol. Endocrinol. **7**, 4–11.

Smith, D.F., Faber, L.E., and Toft, D.O. (1990). Purification of unactive progesterone receptor and identification of novel receptor-associated proteins. J. Biol. Chem. **265**, 3996–4003.

Smith, D.F., Whitesell, L., Nair, S.C., Chen, S., Prapapanich, V., and Rimerman, R.A. (1995). Progesterone receptor structure and function altered by geldanamycin, an hsp90-binding agent. Mol. Cell. Biol. **15**, 6804–6812.

Whitesell, L., Mimnaugh, E.G., De Costa, B., Myers, C.E., and Neckers, L.M. (1994). Inhibition of heat shock protein hsp90-pp60v-src heteroprotein complex formation by benzoquinone ansamycins: essential role for stress proteins in oncogenic transformation. Proc. Natl. Acad. Sci. USA **91**, 8324–8328.

■ Jill L. Johnson:
Department of Biomolecular Chemistry
University of Wisconsin
Madison, WI 53706, USA
Tel. 1 608 262 1358
Fax. 1 608 262 5253
E-mail. jjohns39@facstaff.wisc.edu

Vertebrate p60

p60 binds transiently to progesterone receptor complexes during assembly of mature progesterone complexes in vitro. p60 coprecipitates with Hsp70 and Hsp90 in a variety of tissues. Complexes between purified p60, Hsp70 and Hsp90 form readily, and it appears that p60 binds independently to Hsp70 and Hsp90, potentially mediating the interaction between Hsp90 and Hsp70.

■ Alternative names

IEF SSP 3521 (Honoré et al., 1992).

■ Isolation of p60

p60 was first observed to be up-regulated twofold in SV40-transformed human fibroblasts and also up-regulated by stress (Honoré et al., 1992). Using peptides generated from purified protein, p60 was further identified as the 60 kDa protein that copurifies with immunoprecipitated avian Hsp90 (Smith et al., 1993).

■ p60 gene and sequence

The human cDNA (GenBank accession number M86752) reveals that p60 shares approximately 42% amino acid identity with the *Saccharomyces cerevisiae* protein Sti1 (Nicolet and Craig, 1989). The homology between the human p60 and isolated peptide fragments from avian p60 is 70%. p60 also shows limited homology to the C-terminus of the Hsp70-interacting protein, Hip (Höhfeld et al., 1995). p60 contains multiple tetratricopeptide repeats (Honoré et al., 1992), a protein motif that is postulated to be important in protein–protein interactions (Lamb et al., 1995).

■ p60 protein

A monoclonal antibody that recognizes a protein of approximately 60 kDa in a wide range of chicken tissues, as well as in *Xenopus* liver and various mammalian cell and tissue extracts has been described (Smith et al., 1993). This antibody specifically immunoprecipitates p60 in complex with Hsp90 and Hsp70 from a variety of tissues.

■ Biological interactions of p60

p60 and Hsp70 are the major proteins that copurify with Hsp90 from a variety of tissues (Smith et al., 1993). Purified p60 binds independently to either Hsp90 or Hsp70. A complex between purified Hsp90, Hsp70 and p60 forms in the absence of ATP or other factors. No complex formation between purified Hsp90 and Hsp70 is observed in the absence of p60, suggesting p60 may mediate interactions between Hsp90 and Hsp70 (Chen et al., 1996; Johnson et al., 1996).

Hsp90, Hsp70 and p60 are associated with steroid receptors isolated from cytosolic tissue extracts, as well as those steroid receptor complexes that are reconstituted *in vitro* (Smith and Toft, 1993). During progesterone receptor assembly in rabbit reticulocyte lysate, the receptor first binds Hsp70, Hip, p60 and Hsp90, followed by the mature form of the receptor characterized by the presence of p23 and immunophilins (Smith, 1993). The intermediate form of the receptor containing Hip and p60, which is unable to bind hormone, accumulates under conditions of limiting ATP (Smith et al., 1992) or when lysate is treated with the Hsp90-binding drug geldanamycin (Smith et al., 1995).

■ References

Chen, S., Prapapanich, V., Rimerman, R.A., Honoré, B., and Smith, D.F. (1996). Interactions of p60, a mediator of progesterone receptor assembly, with heat shock proteins Hsp90 and Hsp70. Mol. Endocrinol. **10**, 682–693.

Höhfeld, J., Minami, Y., and Hartl, F.-U. (1995). Hip, a novel cochaperone involved in the eukaryotic Hsc70/Hsp40 reaction cycle. Cell **83**, 589–598.

Honoré, B., Leffers, H., Madsen, P., Rasmussen, H.H., Vandekerckhove, J. and Celis, J.E. (1992). Molecular cloning and expression of a transformation-sensitive human protein containing the TPR motif and sharing identity to the stress-inducible yeast protein STI1. J. Biol. Chem. **267**, 8485–8491.

Johnson, J., Corbisier, R., Stensgard, B., and Toft, D. (1996). The involvement of p23, hsp90, and immunophilins in the assembly of progesterone receptor complexes. J. Steroid Biochem. Mol. Biol. **56**, 31–37.

Lamb, J.R., Tugendreich, S., and Hieter, P. (1995). Tetratrico peptide repeat interactions:to TPR or not to TPR? Trends Biochem. Sci. **20**, 257–259.

Nicolet, C.M. and Craig, E.A. (1989). Isolation and characterization of STI1, a stress-inducible gene from Saccharomyces cerevisiae. Mol. Cell. Biol. **9**, 3638–3646.

Smith, D.F. (1993). Dynamics of heat shock protein 90-progesterone receptor binding and the disactivation loop model for steroid receptor complexes. Mol. Endocrinol. **7**, 1418–1429.

Smith, D.F. and Toft, D.O. (1993). Steroid receptors and their associated proteins. Mol. Endocrinol. **7**, 4–11.

Smith, D.F., Stensgard, B.A., Welch, W.J., and Toft, D.O. (1992). Assembly of progesterone receptor with heat shock proteins and receptor activation are ATP mediated events. J. Biol. Chem. **267**, 1350–1356.

Smith, D.F., Sullivan, W.P., Marion, T.N., Zaitsu, K., Madden, B., McCormick, D.J., and Toft, D.O. (1993). Identification of a 60-kilodalton stress-related protein, p60, which interacts with hsp90 and hsp70. Mol. Cell. Biol. **13**, 869–876.

Smith, D.F., Whitesell, L., Nair, S.C., Chen, S., Prapapanich, V., and Rimerman, R.A. (1995). Progesterone receptor structure and function altered by geldanamycin, an hsp90-binding agent. Mol. Cell. Biol. **15**, 6804–6812.

■ *Jill L. Johnson:*
Department of Biomolecular Chemistry
University of Wisconsin
Madison, WI 53706, USA
Tel. 1 608 262 1358
Fax. 1 608 262 5253
E-mail. jjohns39@facstaff.wisc.edu

Saccharomyces cerevisiae Sti1

The Saccharomyces cerevisiae *protein Sti1 is the apparent homolog of the vertebrate p60.* STI1 *is a non-essential, heat-inducible gene that is required for wild-type growth at extreme temperatures. Sti1 is important in Hsp90 signalling pathways, and may affect Hsp70 function by both interacting directly with members of the SSA family of cytosolic Hsp70s and mediating Hsp70 expression.*

■ Isolation

STI1 was isolated as a transactivator of the promoter of SSA4, one of the cytosolic hsp70 genes in Saccharomyces cerevisiae. STI1 mRNA levels increase approximately tenfold upon heat induction. Cells carrying a disruption of the STI1 gene grow like wild-type cells at 30°C, but show impaired growth at higher and lower temperatures (Nicolet and Craig, 1989).

■ STI1 gene and sequence

The STI1 gene (GenBank accession number M28486) encodes a protein that migrates as a 73–75 kDa protein with multiple isoforms. The Sti1 protein shares approximately 42% amino acid identity with human p60 (Honoré et al., 1992; Smith et al., 1993), and, like p60, contains multiple TPR repeats (Boguski et al., 1990), a protein motif that is postulated to be important in protein–protein interactions (Lamb et al., 1995).

■ Biological or genetic interactions of Sti1

Purified p60, the homolog of Sti1, binds directly to Hsp90 and Hsp70 (Johnson et al., 1996), and transiently associates with steroid receptors during in vitro assembly of mature receptor complexes (Smith, 1993). The homologous complex of Hsp82, Hsp70 (Ssa family) and Sti1 proteins has been detected in S. cerevisiae (Chang and Lindquist, 1994). Sti1 is important in Hsp90 signalling pathways in vivo, as STI1 has genetic interactions with HSP82 and deletion of STI1 reduces the activity of heterologous Hsp90 target proteins expressed in S. cerevisiae (Chang et al., 1997).

Overexpression of STI1 is able to transactivate the SSA4 promoter, and strains carrying mutations in both STI1 and SSA genes display growth defects more severe than either parental strain (Nicolet and Craig, 1989), suggesting Sti1 and members of the Ssa family contribute to similar cellular processes.

■ References

Boguski, M.S., Sikorski, R.S., Heiter, P., and Goebl, M. (1990). Expanding family. Nature **346**, 114.

Chang, H.J. and Lindquist, S. (1994). Conservation of hsp90 macromolecular complexes in Saccharomyces cerevisiae. J. Biol. Chem. **269**, 24983–24988.

Chang, H.J., Nathan, D.F. and Lindquist, S. (1997). In vivo analysis of the Hsp90 cochaperone Sti1 (p60). Mol. Cell. Biol. **17**, 318–325.

Honoré, B., Leffers, H., Madsen, P., Rasmussen, H.H., Vandekerckhove, J., and Celis, J.E. (1992). Molecular cloning and expression of a transformation-sensitive human protein containing the TPR motif and sharing identity to the stress-inducible yeast protein STI1. J. Biol. Chem. **267**, 8485–8491.

Johnson, J., Corbisier, R., Stensgard, B., and Toft, D. (1996). The involvement of p23, hsp90, and immunophilins in the assembly of progesterone receptor complexes. J. Steroid Biochem. Mol. Biol. **56**, 31–37.

Lamb, J.R., Tugendreich, S., and Hieter, P. (1995). Tetratrico peptide repeat interactions:to TPR or not to TPR? Trends Biochem. Sci. **20**, 257–259.

Nicolet, C.M. and Craig, E.A. (1989). Isolation and characterization of STI1, a stress-inducible gene from *Saccharomyces cerevisiae*. Mol. Cell. Biol. **9**, 3638–3646.

Smith, D.F. (1993). Dynamics of heat shock protein 90-progesterone receptor binding and the disactivation loop model for steroid receptor complexes. Mol. Endocrinol. **7**, 1418–1429.

Smith, D.F., Sullivan, W.P., Marion, T.N., Zaitsu, K., Madden, B., McCormick, D.J., and Toft, D.O. (1993). Identification of a 60-kilodalton stress-related protein, p60, which interacts with hsp90 and hsp70. Mol. Cell. Biol. **13**, 869–876.

■ *Jill L. Johnson:*
Department of Biomolecular Chemistry
University of Wisconsin
Madison, WI 53706, USA
Tel. 1 608 262 1358
Fax. 1 608 262 5253
E-mail. jjohns39@facstaff.wisc.edu

14

Protein-Specific Chaperones

Escherichia coli PapD

PapD is a periplasmic chaperone which functions to facilitate assembly of P pili in uropathogenic *Escherichia coli*. The pilus contains six different structural proteins which are organized into a thick rod-like fiber which is connected to a thinner fibrillum. PapD forms stable periplasmic complexes with each subunit, preventing aggregation and degradation. The subunits are then delivered to outer membrane assembly sites where the chaperone dissociates and the subunits are incorporated into the pilus. PapD has an immunoglobulin-like fold which is thought to provide a template for the proper folding of subunits as they emerge from the cytoplasmic membrane. PapD belongs to a large family of periplasmic chaperones which are each dedicated to the assembly of distinct cell surface structures in Gram-negative bacteria.

Homologs of PapD

PapD is the prototype member of the immunoglobulin-like superfamily of periplasmic chaperones (Hultgren et al., 1993). Most members of the family have been identified only by sequence or genetic analysis. The PapD superfamily can be divided into two subclasses according to conserved structural differences. These two subclasses assemble organelles with distinctly different architectures (Hung et al., 1996). PapD is the best characterized chaperone, but studies have been performed with other members of the family [FimC, type I pili of *E. coli* (Jones et al., 1993); FaeE, K88 pili of *E. coli* (Mol et al., 1994); and FanE, K99 pili of *E. coli* (Bakker et al., 1991)].

Identification of PapD

The genes encoding the P pilus (the *pap* operon) were cloned from clinical isolates based on their ability to confer hemagglutination of red blood cells on *E. coli* lab strains (Tennent et al., 1990). Genetic and biochemical analyses of the clones and mutant derivatives revealed that *papD* encodes a 28.5 kDa periplasmic protein (Lindberg et al., 1989). PapD is essential for piliation and for the stability of pilus subunits in the periplasm.

papD gene and sequence

papD is part of the *pap* operon (GenBank accession number X61239; Marklund et al., 1992). The pilus subunits (PapA, PapE, PapF, PapG, PapH and PapK) and the usher protein, PapC, are also encoded within this operon. Expression of the *pap* operon is subject to complex regulation (van der Woude et al., 1992).

PapD protein

Cell fractionation studies demonstrated that PapD is a periplasmic protein (Lindberg et al., 1989). PapD is routinely purified from periplasmic extracts of an overexpressing *E. coli* strain by cation-exchange chromatography (Lindberg et al., 1989). The isoelectric point of PapD is 9.3, making isoelectric focusing and acidic native gel electrophoresis useful techniques for studying PapD and its complexes with subunits (Striker et al., 1994). Polyclonal antisera against PapD, chaperone–subunit complexes, and individual pilus subunits are available (Slonim et al., 1992; Striker et al., 1994).

Structure and mutagenesis studies

The three-dimensional structure of PapD has been solved and refined to a 2.0 Å resolution (Holmgren and Brändén, 1989; D. Ogg, unpublished results). PapD has two globular domains which each have a β barrel structure similar to the structural motif known as the immunoglobulin fold (Williams and Barclay, 1988). Alignment of PapD homologs revealed several amino acids in the cleft between the two domains, which were invariant but did not appear to be structurally important. Site-directed mutagenesis revealed that mutations at two such invariant residues (Arg-8 and Lys-112) severely affected PapD function (Slonim et al., 1992). Mutations at these sites abolish pilus assembly and render PapD unable to bind to pilus subunits, as evidenced by both the absence of periplasmic complexes with subunits and the instability of the subunits (Slonim et al., 1992; Kuehn et al., 1993).

Biological activities of PapD

PapD function can be assessed in several ways (Kuehn et al., 1994), including measuring hemagglutination of red blood cells by P piliated bacteria, purification of pili and characterization of pilus components, and analysis of chaperone–subunit complexes in periplasmic extracts or in pure form.

PapD forms complexes with each of the pilus subunits. In these complexes, the subunits are protected from the alternative fate of degradation by the DegP protease. Expression of several subunits in the absence of PapD is toxic in a *degP* strain. The subunits appear to be trapped in the cytoplasmic membrane suggesting that PapD may facilitate import into the periplasm and possibly assist in subunit folding. Indeed, recent work has shown that PapG expressed in spheroplasts remains associated with the cytoplasmic membrane, but becomes released into the medium upon addition of PapD (C.H. Jones, P.N. Danese, T.J. Silhavy and S.J. Hultgren, unpublished results). Periplasmic complexes between PapD and PapA (Striker et al., 1994), PapE (Lindberg et al., 1989), PapF (C.H. Jones and S.J. Hultgren, unpublished results), PapK (Striker et al., 1994), and PapG (Hultgren et al., 1989) have been detected and purified to homogeneity using standard chromatographic techniques. The PapD–PapG complex is routinely purified using affinity chromatography taking advantage of the affinity of the PapG adhesin for its ligand Galα(1-4)Gal (Hultgren et al., 1989). Subunits appear to be in a native-like state when bound to PapD since the PapG adhesin is capable of binding Galα (1–4)Gal when complexed with PapD (Hultgren et al., 1989).

PapD targets the subunits to the other component of the pilus assembly machinery, the outer membrane usher protein PapC. In vitro assays have demonstrated binding of chaperone–subunit complexes to PapC (Dodson et al., 1993). PapD alone does not appear to bind the usher; however experiments in another pilus assembly system strongly suggests that the chaperone does have specific interactions with the usher protein (Klemm et al., 1995).

■ Biological interactions of PapD

PapD binds to ('caps') a conserved C-terminal motif present in pilus subunits via a β zippering interaction. The C-terminus of each subunit has a conserved penultimate tyrosine and a pattern of alternating hydrophobic residues. The cocrystallization of PapD with the C-terminal PapG peptide revealed that the peptide lies in an extended conformation making backbone hydrogen bonds with the exposed edge of a β sheet of PapD. In addition, the peptide is anchored in the cleft via hydrogen bonds with the invariant Arg-8 and Lys-112 of PapD (Kuehn et al., 1993). Mutagenesis studies show that this interaction is crucial for binding of subunits to PapD (Kuehn et al., 1993). A second site of interaction between PapD and PapG is being elucidated. A region of PapG that binds PapD has been identified by fusion protein and peptide-binding studies (Xu et al., 1995). The region of PapD that participates in this interaction is not yet known.

■ References

Bakker, D., Vader, C.E., Roosendaal, B., Mooi, F.R., Oudega, B., and De Graaf, F.K. (1991). Structure and function of periplasmic chaperone-like proteins involved in the biosynthesis of K88 and K99 fimbriae in enterotoxigenic Escherichia coli. Mol. Microbiol. **5**, 875–886.

Dodson, K.W., Jacob-Dubuisson, F., Striker, R.T., and Hultgren, S.J. (1993). Outer membrane PapC usher discriminately recognizes periplasmic chaperone-pilus subunit complexes. Proc. Natl. Acad. Sci. USA **90**, 3670–3674.

Holmgren, A. and Brändén, C. (1989). Crystal structure of chaperone protein PapD reveals an immunoglobulin fold. Nature **342**, 248–251.

Hultgren, S.J., Lindberg, F., Magnusson, G., Kihlberg, J., Tennent, J.M., and Normark, S. (1989). The PapG adhesin of uropathogenic Escherichia coli contains separate regions for receptor binding and for the incorporation into the pilus. Proc. Natl. Acad. Sci. USA **86**, 4357–4361.

Hultgren, S.J., Abraham, S.N., Caparon, M., Falk, P., St. Geme, III, J.W., and Normark, S. (1993). Pilus and non-pilus bacterial adhesins: assembly and function in cell recognition. Cell **73**, 887–901.

Hung, D.L., Knight, S.D., Woods, R.M., Pinkner, J.S., and Hultgren, S.J. (1996) Molecular basis of two subfamilies of immunoglobulin-like caperones. EMBO J. **15**, 3792–3805.

Jones, C.H., Pinkner, J.S., Nicholes, A.V., Slonim, L.N., Abraham, S.N., and Hultgren, S.J. (1993). FimC is a periplasmic PapD-like chaperone that directs assembly of type 1 pili in bacteria. Proc. Natl. Acad. Sci. USA **90**, 8397–8401.

Klemm, P., Jul Jorgensen, B., Kreft, B., and Christiansen, G. (1995). The export systems of Type 1 and F1C fimbriae are interchangeable but work in parental pairs. J. Bacteriol. **177**, 621–627.

Kuehn, M.J., Ogg, D.J., Kihlberg, J., Slonim, L.N., Flemmer, K., Bergfors, T., and Hultgren, S.J. (1993). Structural basis of pilus subunit recognition by the PapD chaperone. Science **262**, 1234–1241.

Kuehn, M.J., Jacob-Duboisson, F., Dodson, K., Slonim, L., Striker, R., and Hultgren, S.J. (1994). Genetic, biochemical, and structural studies of biogenesis of adhesive pili in bacteria. Meth. Enzymol. **236**, 282–306.

Lindberg, F., Tennent, J.M., Hultgren, S.J., Lund, B., and Normark, S. (1989). PapD, a periplasmic transport protein in P-pilus biogenesis. J. Bacteriol. **171**, 6052–6058.

Marklund, B.I., Tennent, J.M., Garcia, E., Hamers, A., Baga, M., Lindberg, F., Gaastra, W., and Normark, S. (1992). Horizontal gene transfer of the Escherichia coli pap and prs pili operons as a mechanism for the development of tissue-specific adhesive properties. Molecular Microbiology **6**, 2225–2242.

Mol, O., Vischers, R.W., de Graaf, F.K., and Oudega, B. (1994). Escherichia coli periplasmic chaperone FaeE is a homodimer and the chaperone-K88 subunit complex is a heterotrimer. Mol. Microbiol. **11**, 391–402.

Slonim, L.N., Pinkner, J.S., Branden, C.I., and Hultgren, S.J. (1992). Interactive surface in the PapD chaperone cleft is conserved in pilus chaperone superfamily and essential in subunit recognition and assembly. EMBO J. **11**, 4747–4756.

Striker, R., Jacob-Dubuisson, F., Frieden, C., and Hultgren, S.J. (1994). Stable fiber forming and non-fiber forming chaperone-subunit complexes in pilus biogenesis. J. Biol. Chem **269**, 12233–12239.

Tennent, J.M., Hultgren, S., Forsman, K., Goransson, M., Marklund, B.I., Uhlin, B.E., and Normark, S. (1990). Genetics of adhesin expression in Escherichia coli. In The Bacteria; Molecular Basis of Pathogenesis, (B.H. Iglewski and V.C. Clark, eds). Academic Press, New York, pp. 79–110.

van der Woude, M.W., Braaten, B.A., and Low, D.A. (1992). Evidence of global regulatory control of pilus expression in Escherichia coli by Lrp and DNA methylation: model building based on analysis of pap. Mol. Microbiol. **6**, 2429–2435.

Williams, A.F. and Barclay, A.N. (1988). The immunoglobulin superfamily-domains for cell surface recognition. Annu. Rev. Immunol. **6**, 381–405.

Xu, Z., Jones, C.H., Haslam, D., Pinkner, J.S., Dodson, K., Kihlberg, J., and Hultgren, S.J. (1995). Molecular dissection of PapD interaction with PapG reveals two chaperone binding sites. Mol. Microbiol. **16**, 1011–1020.

Mary-Jane Lombardo, David G. Thanassi and Scott J. Hultgren:
Department of Molecular Microbiology
Washington University School of Medicine
St. Louis, MO 63110, USA
Tel. 1 314 362 6788
Fax. 1 314 362 1998
E-mail: hultgren@borcim.wustl.edu

Vertebrate Hsp47

Hsp47 is a stress-inducible protein located in the lumen of the endoplasmic reticulum (ER), and belongs to the serpin (serine protease inhibitor) superfamily. Hsp47 is assumed to be a substrate-specific molecular chaperone which, in the ER, binds with different affinities to procollagen subunits as well as triple helix-formed procollagen. In addition to the demonstrated interaction of Hsp47 with collagen, the level of expression of Hsp47 in cultured cells and in tissues is always closely correlated with that of collagen.

Alternative names

Colligin (Kurkinen et al., 1984) and J6 protein (Wang and Gudas, 1990) in mice, and gp46 in rats (Cates et al., 1987) and in humans (Clarke and Sanwal, 1992).

Isolation of Hsp47

Hsp47 was originally identified as a gelatin-binding protein from chick embryo fibroblasts (Nagata and Yamada, 1986). Therefore, Hsp47 can be easily purified by a single step of affinity chromatography on gelatin–Sepharose or collagen–Sepharose following cell lysis with 1% Nonidet P-40 (Kurkinen et al., 1984; Nagata and Yamada, 1986).

hsp47 genes and sequence

cDNAs encoding Hsp47 and its homologues have been cloned from chick (GenBank accession number X57157; Hirayoshi et al., 1991), mouse (J05609; Wang and Gudas, 1990; X60676, Takechi et al., 1992), rat (M69246; Clarke et al., 1991) and human (X61598; Clarke and Sanwal, 1992). The 3278 bp chick cDNA encodes a 15 amino acid signal peptide and a mature protein of 390 amino acids. Hsp47 has an RDEL (Arg–Asp–Glu–Leu) tetrapeptide sequence at the C-terminus which acts as an ER retrieval signal (Hirayoshi et al., 1991). Hsp47 and the J6 protein belong to the serpin (serine protease inhibitor) superfamily, sharing 31% sequence identity with protein C inhibitor and 10–30% identity with other serpin family members. The reaction center common to the serpin family proteins is well conserved in Hsp47, and conserved secondary structures in the serpin family are also observed in Hsp47. The sequence identity is 80% between chick and mouse Hsp47, 98.6% between the human and rat proteins, and 80% between the human and chick proteins. Analysis of a genomic clone encoding mouse Hsp47/J6 (Wang, 1992; Hosokawa et al., 1993) showed that the mouse hsp47 gene consists of six exons separated by five introns (Hosokawa et al., 1993). The promoter region contains a TATA box, four Sp1-binding sites and one AP-1-binding site. A complete heat shock element (HSE) is located between nucleotides –61 and –79.

Hsp47 protein and biological activities

Hsp47 is a basic (pI=9.0) glycoprotein composed of a single 47 kDa polypeptide (Nagata et al., 1986) located in the lumen of the ER (Saga et al., 1987). The protein contains two asparagine-linked oligosaccharides, both of which are high mannose type, and is phosphorylated mainly in transformed cells (Nagata and Yamada, 1986). Hsp47 binds to various procollagens in the ER and is dissociated from procollagen in the cis-Golgi network (Satoh et al., 1996). Hsp47 is assumed to contribute to the processing, secretion, or triple helix formation of procollagen, and to inhibit the aggregation of nascent polypeptides of procollagen before they form triple helices (Satoh et al., 1996). When cells are heat shocked or treated with α, α'-dipyridyl, an inhibitor of triple helix formation of collagen, Hsp47 remains bound to the abnormal collagen molecules for extended periods, preventing their secretion from the cells (Nakai et al., 1992). Hsp47 is therefore assumed to have a molecular chaperone-like function which is specific for collagen (Nagata, 1996). Jain et al. (1994) have shown that colligin (Hsp47) inhibits procollagen degradation in the ER.

Hsp47 binds to types I–V collagens and gelatin (denatured types I and III collagen) in vitro, and to single subunits as well as the triple helix form of type I procollagen in vivo (Nagata and Yamada, 1986; Natsume et al., 1994; Satoh et al., 1996). Binding of Hsp47 to collagen is pH sensitive: dissociation occurs below pH 6.3 (Saga et al., 1987). The dissociation constants for types I–V collagens are 10^{-6} to 10^{-7} M; these relatively low dissociation constants result from the rapid dissociation rate constant ($k_{diss} > 10^{-2}$ s^{-1}) and high association rate constant (Natsume et al., 1994).

■ Biological regulation of Hsp47

Expression of Hsp47 is induced by cellular stresses including heat shock (Nagata et al., 1986). This regulation is at the transcriptional level (Hirayoshi et al., 1991). The synthesis of Hsp47 is down-regulated following transformation of fibroblasts by Rous sarcoma virus (Nagata and Yamada, 1986) or by simian virus 40 (Nakai et al., 1990), and up-regulated during the differentiation of the F9 mouse teratocarcinoma cell line following treatment with retinoic acid or retinoic acid plus dibutyryl cyclic AMP (Kurkinen et al., 1984; Wang and Gudas, 1990; Takechi et al. 1992). The synthesis of collagen is also down-regulated and up-regulated during transformation of fibroblasts and differentiation of F9 cells, respectively. Both Hsp47 and types I and III collagen are dramatically induced in the liver during the progression of experimental liver fibrosis caused by administration of carbon tetrachloride into rats (Masuda et al., 1994). Moreover, Hsp47 is expressed in tissues that express high amounts of various collagens such as developing murine femurs and molars (Shroff et al., 1993) and chicken chondrocytes (Kambe et al., 1994). Thus, the synthesis of Hsp47 is always closely correlated with that of collagen.

Three cDNAs which differ only in their 5'-non-coding regions have been isolated from mouse cells (Wang and Gudas, 1990; Hosokawa et al., 1993). The *hsp47* gene is alternatively spliced at this 5'-non-coding region only after heat shock; MH47b is the major transcript and MH47c the minor one under normal conditions, but MH47a becomes the major one following heat shock concomitant with the decrease in MH47c (Takechi et al., 1994). The J6 gene has been reported to be transactivated by the GATA-4 or related transcription factor, which is activated by retinoic acid during F9 cell differentiation (Wang, 1994; Bielinska and Wilson, 1995).

■ References

Bielinska, M. and Wilson, D.B. (1995). Regulation of J6 protein gene expression by transcription factor GATA-4. Biochem. J. **307**, 183–189.

Cates, G.A., Nandan, D., Brickenden, A.M., and Sanwal, B.D. (1987). Differentiation defective mutants of skeletal myoblasts altered in a gelatin-binding glycoprotein. Biochem. Cell. Biol. **65**, 767–775.

Clarke, E.P. and Sanwal, B.D. (1992). Cloning of a human collagen-binding protein, and its homology with rat gp46, chick hsp47 and mouse J6 proteins. Biochim. Biophys. Acta **1129**, 246–248.

Clarke, E.P., Cates, G.A., Ball, E.H., and Sanwal, B.D. (1991). A collagen-binding protein in the endoplasmic reticulum of myoblasts exhibits relationship with serine protease inhibitors. J. Biol. Chem. **266**, 17230–17235.

Hirayoshi, K., Kudo, H., Takechi, H., Nakai, A., Iwamatsu, A., Yamada, K.M., and Nagata, K. (1991). HSP47, a tissue-specific, transformation-sensitive, collagen-binding heat shock protein of chicken embryo fibroblasts. Mol. Cell. Biol. **11**, 4036–4044.

Hosokawa, N., Takechi, H., Yokota, S., Hirayoshi, K., and Nagata, K. (1993). Structure of the gene encoding the mouse 47-kDa heat-shock protein (HSP47). Gene **126**, 187–193.

Jain, N., Brickenden, A., Ball, E.H., and Sanwal, B.D. (1994). Inhibition of procollagen I degradation by colligin: a collagen-binding serpin. Arch. Biochem. Biophys. **314**, 23–30.

Kambe, K., Yamamoto, A., Yoshimori, T., Hirayoshi, K., Ogawa, R., and Tashiro, Y. (1994). Preferential localization of heat shock protein 47 in dilated endoplasmic reticulum of chicken chondrocytes. J. Histochem. Cytochem. **42**, 833–841.

Kurkinen, M., Taylor, A., Garrels, J.I., and Hogan, B.L.M. (1984). Cell surface-associated proteins which bind native type I collagen or gelatin. J. Biol. Chem. **259**, 5915–5922.

Masuda, H., Fukumoto, M., Hirayoshi, K., and Nagata, K. (1994). Coexpression of the collagen-binding stress protein HSP47 gene and the 1(I) and 1(III) collagen genes in carbon tetrachloride-induced rat liver fibrosis. J. Clin. Invest. **94**, 2481–2488.

Nagata, K. (1996). HSP47: a collagen-specific molecular chaperone. Trends Biochem. Sci. **21**, 23–26.

Nagata, K. and Yamada, K.M. (1986). Phosphorylation and transformation sensitivity of a major collagen-binding protein of fibroblasts. J. Biol. Chem. **261**, 7531–7536.

Nagata, K., Saga, S., and Yamada, K.M. (1986). A major collagen-binding protein of chick embryo fibroblasts is a novel heat shock protein. J. Cell Biol. **103**, 223–229.

Nakai, A., Hirayoshi, K., and Nagata, K. (1990). Transformation of BALB/3T3 cells by simian virus 40 causes a decreased synthesis of a collagen-binding heat-shock protein (hsp47). J. Biol. Chem. **265**, 992–999.

Nakai, A., Satoh, M., Hirayoshi, K., and Nagata, K. (1992). Involvement of the stress protein HSP47 in procollagen processing in the endoplasmic reticulum. J. Cell Biol. **117**, 903–914.

Natsume, T., Koide, T., Yokota, S., Hirayoshi, K., and Nagata, K. (1994). Interactions between collagen-binding stress protein HSP47 and collagen. J. Biol. Chem. **269**, 31224–31228.

Saga, S., Nagata, K., Chen, W.-T., and Yamada, K. (1987). pH-dependent function, purification, and intracellular location of a major collagen-binding glycoprotein. J. Cell Biol. **105**, 517–527.

Satoh, M., Hirayoshi, K., Yokota, S., Hosokawa, N., and Nagata, K. (1996). Intracellular interaction of collagen-specific stress protein HSP47 with newly synthesized procollagen. J. Cell. Biol. **133**, 469–483.

Shroff, B., Smith, T., Norris, K., Pileggi, R., and Sauk, J.J. (1993). HSP47 is localized to regions of type I collagen production in developing murine femurs and molars. Connect. Tissue Res. **29**, 273–286.

Takechi, H., Hirayoshi, K., Nakai, A., Kudo, H., Saga, S., and Nagata, K. (1992). Molecular cloning of a mouse 47-kDa heat-shock protein (HSP47), a collagen-binding stress protein, and its expression during the differentiation of F9 teratocarcinoma cells. Eur. J. Biochem. **206**, 323–329.

Takechi, H., Hosokawa, N., Hirayoshi, K., and Nagata, K. (1994). Alternative 5' splice site selection induced by heat shock. Mol. Cell. Biol. 1b4, 567–575.

Wang, S.-Y. (1992). Structure of the gene and its retinoic acid-regulatory region for murine J6 serpin. An F9 teratocarcinoma cell retinoic acid-inducible protein. J. Biol. Chem. **267**, 15362–15366.

Wang, S.-Y. (1994). A retinoic acid-inducible GATA-binding protein binds to the regulatory region of J6 serpin gene. J. Biol. Chem. **269**, 607–613.

Wang, S.-Y. and Gudas, L.J. (1990). A retinoic acid-inducible mRNA from F9 teratocarcinoma cells encodes a novel protease inhibitor homologue. J. Biol. Chem. **265**, 15818–15822.

■ Kazuhiro Nagata:
Department of Cell Biology, Chest Disease Research Institute
Kyoto University
Sakyo-ku, Kyoto 606-01, Japan
Tel. 81 75 751 3848
Fax. 81 75 751 4645
E-mail: nagata@chest.kyoto-u.ac.jp

15

Intramolecular Chaperones

Subtilisin

Subtilisin E is a non-specific alkaline serine protease produced from Bacillus subtilis. *Homologous enzymes, including subtilisin BPN' and subtilisin Carlsberg, exist in different species of* Bacillus. *They are all produced as precursors, called pre-pro-subtilisins. The pre-region (signal peptide) is required for protein secretion, while the pro-region (also called the propeptide or prosequence) located between the signal peptide and the mature protein, functions as an intramolecular chaperone (IMC) for folding of the mature domain. Importantly, the IMC is not a part of the mature protein and is not required for the stability of the protease.*

■ Alternative names

Subtilases and subtilopeptidase (Siezen et al., 1991).

■ Subtilisin sequence

The complete amino acid sequence of subtilisin E was deduced from the nucleotide sequences from chromosomal DNA isolated from *B. subtilis 168* (Genbank accession number K01988; Ikemura et al., 1987) and is homologous with the sequences of subtilisin BPN' (*B. amyloliquefaciens*; Stahl and Ferrari, 1984; accession number X00165) and Carlsberg (*B. licheniformis* and *B. pumilis*; Jacobs et al., 1983; accession number X03341). The protease is produced in the sporulation phase of *Bacilli*, and the late expression of the gene is mainly controlled at the level of transcription (Wong et al., 1986). The protein sequences of pre-pro-subtilisins are made up of 381 amino acid residues, 29 of which correspond to the signal peptide and 77 to the IMC (Inouye, 1991). The mature active proteins are made up of 275 amino acid residues and are devoid of the signal peptide as well as their IMCs. The molecular weights of the mature proteins vary between 28 and 30 kDa (Wong et al., 1986; Shinde and Inouye, 1994).

■ Subtilisin proteins

Subtilisin E is a non-specific alkaline serine protease produced from *Bacillus subtilis* (Wong et al., 1986). Subtilisins are all produced as precursors, called pre-pro-subtilisins (Powers et al., 1986). The pre-region (signal peptide) is required for protein secretion, while the pro-region (also called propeptides or prosequences) located between the signal peptide and the mature protein, functions as an intramolecular chaperone (IMC) for folding of the mature domain (Ikemura et al., 1987). Importantly, the IMC is not a part of the mature protein and is not required for the stability of the protease (Ikemura and Inouye, 1988). The catalytic residues of the mature protease are Asp-32, His-64 and Ser-221 (Wright et al., 1969). Maturation of the precursor leads to the formation of the active enzyme. The signal peptide is removed by signal peptidase while the IMC is cleaved by an autoprocessing mechanism that may be either an intramolecular or intermolecular type (Shinde and Inouye, 1995a). The catalytic triads of the mature protein are responsible for the autoprocessing of the propeptide, and mutations within these residues prevent maturation (Ikemura et al., 1987; Ikemura and Inouye, 1988). The mature protease domain consists of eight α helices, five β sheets and two calcium-binding sites (Wright et al., 1969).

The IMC domain, which is important for efficient folding of the subtilisin domain (Zhu et al., 1989), is unusually biased in its distribution of highly charged residues (36% charged) when compared with the mature protease domain. A comparison of the primary sequences of various propeptides within the subtilisin family of proteins has revealed the existence of two conserved motifs. The corresponding motifs from subtilisin E are motif 1 (KKYIVGFKQT) and motif 2 (AAAATLDEKAVKELK-KDPSVAVYEED), and they have been suggested to play a crucial role in protein folding. These motifs contain short hydrophobic stretches of residues flanked by charged residues (Shinde and Inouye, 1993). A recently published crystal structure of the IMC-subtilisin complex will provide clues for understanding IMC-mediated protein folding (Bryan et al., 1995).

■ Subtilisin importance and functions

Subtilisin is mainly a scavenging protease produced during the sporulation phase (Wong et al., 1986). Since it is a non-specific protease, it was used in the detergent industry as a scavenger enzyme and in the optical industry for cleaning lenses. The prohormone convertase family of proteins are homologous to subtilisin and belong to the subtilase family of proteins (Barr, 1991).

The IMC for subtilisin is essential for the folding of the protease domain. It helps in the maturation pathway by catalysing the folding reaction, preventing protein aggregation and delaying the enzymatic activity of the subtilisin by acting as a temporary competitive inhibitor (Shinde and Inouye, 1995b). Stable complexes of active site mutants of subtilisins and their IMC chaperones have been isolated, characterized and crystallized (Eder et al., 1993; Bryan et al., 1995; Shinde and Inouye, 1995a, b). Subtilisin is a model system for studying IMCs. Similar

IMCs have now been shown to exist in a number of proteins and are not exclusive to the protease family (Shinde et al., 1995).

References

Barr, P.J. (1991). Mammalian subtilisins: the long-sought dibasic processing endoprotease. Cell, **66**, 1–3.

Bryan, P., Wang, L., Hoskins, J., Ruvinov, S., Strausberg, S., Alexander, P., Almog, O., Gilliland, G., and Gallagher, T. (1995). Catalysis of a protein folding reaction: mechanistic implications of the 2.0 Å structure of the subtilisin-prodomain complex. Biochemistry, **34**, 10310–10318.

Eder, J., Rheinnecker, M., and Fersht, A. (1993) Folding of subtilisin BPN': role of the pro-sequence. J. Mol. Biol., **223**, 293–304.

Ikemura, H. and Inouye, M. (1988). In vitro processing of pro-subtilisin produced in Escherichia coli. J. Biol. Chem., **263**, 12959–12963.

Ikemura, H., Takagi, M., and Inouye, M. (1987). Requirements of pro-sequence for the production of active subtilisin E in Escherichia coli. J. Biol. Chem., **262**, 7859–7864.

Inouye, M. (1991). Intramolecular chaperone: the role of the pro-peptide in protein folding. Enzymes, **45**, 314–321.

Jacobs, M., Elisson, M., Uhlen, M., and Flock, J.J. (1983). Cloning sequencing and expression of Subtilisin Carlsberg from Bacillus licheniformis. Nucl. Acids Res., **13**, 8913–8922.

Powers, S.D., Adams, R.M., and Wells, J.A. (1986). Secretion and autoproteolytic maturation of subtilisin. Proc. Natl. Acad. Sci. USA, **83**, 3096–3100.

Shinde, U. and Inouye, M. (1993). Intramolecular chaperones and protein folding. Trends Biochem. Sci., **18**, 442–446.

Shinde, U. and Inouye, M. (1994). The structural and functional organization of intramolecular chaperones: the N-terminal propeptides which mediate protein folding. J. Biochem., **115**, 629–636.

Shinde, U. and Inouye, M. (1995a). Folding mediated by an intramolecular chaperone: Autoprocessing pathway of the precursor resolved via a 'substrate assisted catalysis' mechanism. J. Mol. Biol., **247**, 390–395.

Shinde, U. and Inouye, M. (1995b). Folding pathway mediated by an intramolecular chaperone: characterization of the structural changes in pro-subtilisin E coincident with autoprocessing. J. Mol. Biol., **252**, 25–30.

Shinde, U., Li, Y., and Inouye, M. (1995). Propeptide mediated protein folding: Intramolecular Chaperones. In Intramolecular Chaperones and Protein Folding, (U. Shinde and M. Inouye, eds), pp. 1–9. R.G. Landes & Co., Austin, Texas.

Siezen, R.J., de Vos, W.M., Leunissen, J.A.M., and Dijkstra, B.W. (1991). Homology modelling and and protein engineering strategy of subtilases, the family of subtilisin like serine proteases. Prot. Eng., **4**, 719–737.

Stahl, M.L. and Ferarri, E. (1984). Replacement of Bacillus subtilis subtilisin structural gene with an in vitro-derived deletion mutation. J. Bacteriol., **158**, 411–418.

Wong, S-L., Price C.W., Goldfarb, D.S., and Doi, R.H. (1986). The subtilisin E gene of Bacillus subtilis is transcribed from a sigma 37 promoter in vivo. Proc. Natl. Acad. Sci. USA, **81**, 1184–1188.

Wright, C.S., Adlen, R.A., and Kraut, J. (1969). The structure of Subtilisin BPN' at 2.5 Å. Nature, **221**, 235–238.

Zhu, X., Ohta, Y., Jordan, F., and Inouye, M. (1989). Pro-sequence of subtilisin can guide the refolding of denatured subtilisin in an intermolecular process. Nature, **339**, 483–484.

■ *Ujwal Shinde and Masayori Inouye:*
Department of Biochemistry
Robert Wood Johnson Medical School
675 Hoes Lane
Piscataway, NJ 08854, USA
Tel. 1 908 235 4115
Fax. 1 908 235 4559
E-mail. shinde @rwja.umdnj.edu and inouye @rwja.umdnj.edu

Carboxypeptidase Y and its propeptide

Carboxypeptidase Y (CPY) is a resident of the yeast vacuole (equivalent to the lysosome in mammals). It is synthesized as a zymogen (procarboxypeptidase Y; proCPY) which contains a 91 amino acid N-terminal propeptide. The propeptide is essential for folding of the protein in vivo. In vitro, CPY is unable to refold in the absence of high concentrations of salt, while proCPY refolds efficiently. Under high salt conditions CPY refolds with a low efficiency, and the isolated propeptide is able to stimulate folding of the enzyme through low-affinity interactions.

■ Alternative name

(pro)carboxypeptidase yscY (Knop et al., 1993).

■ Gene and protein structure

ProCPY is encoded by the PRC1 gene in Saccharomyces cerevisiae and consists of a propeptide of 91 amino acid residues and an enzyme region of 421 residues (Valls et al., 1987; GenBank accession number M15482). CPY belongs to a group of enzymes, termed serine carboxypeptidases (Remington and Breddam, 1994), that utilize a mechanism similar to that of the subtilisin and trypsin families of proteases, but have a much lower pH optimum. The three-dimensional structure of CPY has been determined (Endrizzi et al., 1994; pdb 1YSC). CPY is modified at four positions with high mannose N-linked glycosyl residues, which together comprise 10–14% of the molecular mass.

Biosynthesis of CPY

CPY traverses the early parts of the secretory pathway to reach the vacuole. It is targeted into the endoplasmic reticulum by a 20 amino acid N-terminal signal sequence (reviewed in van den Hazel et al., 1996). Here, folding takes place in conjunction with formation of five intramolecular disulfide bonds and core glycosylation of four Asn–Xaa–Thr sites (Holst et al., 1996). This pro-form of CPY, of about 67 kDa, is termed p1-proCPY or just p1-CPY. In the Golgi apparatus, α1-6 and α1-3 mannose modification of the glycosyl core gives rise to the p2 form of proCPY of 69 kDa. In a late Golgi compartment, sorting to the vacuole is mediated in an active process by a receptor encoded by the VPS10 gene (Marcusson et al., 1994). Final processing takes place in the vacuole, where the propeptide is proteolytically removed by proteinase A and B to yield the active enzyme of 63 kDa (Sørensen et al., 1994). No specific biological function has been ascribed to CPY, and $\Delta prc1$ strains show no growth phenotypes.

CPY activity assays

The activity of CPY can be detected on yeast colonies using N-acetyl-DL-phenylalanyl-β-naphthylester as substrate and employing a subsequent chromogenic reaction (Jones, 1991). In liquid samples CPY activity can be quantitated using the chromogenic substrate furyl–acryloyl–phenylalanyl–phenylalanine (Remington and Breddam, 1994).

Purification of CPY

Both CPY and proCPY (the p2 form) can conveniently be purified from the culture supernatant of yeast strains overproducing the proteins from high copy plasmids. In the case of CPY, this is best done in a vps1 mutant strain background (Nielsen et al., 1990), where correct processing of the proCPY takes place extracellularly. Subsequent purification is easily carried out using DEAE–ion exchange or affinity chromatography (Johansen et al., 1976). ProCPY can be prepared by similar means, using a vps$^+$ strain which is deleted for the vacuolar proteases proteinase A ($\Delta pep4$) and proteinase B ($\Delta prb1$; Winther and Sørensen, 1991; Sørensen et al., 1994). The isolated propeptide (termed CYPR) has been purified after production in Escherichia coli (Sørensen et al., 1993) using the T7 RNA polymerase/promoter system.

Functions of the propeptide

The CPY propeptide performs several important functions. A short segment of the most N-terminal part of the propeptide, containing the sequence $Q_{24}RPL_{27}$, contains the information necessary for correct vacuolar sorting (Valls et al., 1990). Certain mutations in this sequence (e.g. $Q_{24*ar}K_{24}$) will result in an inability to interact with the Vps10p receptor (Marcusson et al., 1994) and missorting of 85–95% of the proCPY to the culture medium.

The propeptide maintains proCPY in an inactive state, having less than 0.05% of the enzyme activity (Winther and Sørensen, 1991). Limited proteolysis of proCPY with yeast proteinase A in vitro has shown that the C-terminal one-third of the propeptide contains an element sufficient for inhibition of the activity (Sørensen et al., 1994). ProCPY is efficiently activated in vitro by addition of proteinase K (Winther and Sørensen, 1991). The propeptide appears to inhibit activity, not by sterically hindering access to the active site, but rather by perturbing the active site structure (Sørensen and Winther, 1994).

Role of the propeptide in folding of CPY

The propeptide of CPY is required for efficient folding of the protein in vitro. Thus, denaturation in 6 M guanidinium chloride and subsequent dilution of the denaturant to 0.1 M results in efficient renaturation of proCPY, whereas less than 0.5% of mature CPY subjected to the same treatment is recovered in a folded, enzymatically active state (Winther and Sørensen, 1991). In vivo analysis has shown that large deletions in the propeptide result in complete loss of CPY activity and accumulation and/or degradation of the ER (p1) form of the truncated protein (Ramos et al., 1994; Holst et al., 1996). This supports the conclusion that the propeptide is essential for folding in vivo also. While deletions covering the C-terminal one-third of the propeptide has this severe phenotype, there is a surprising tolerance towards smaller deletions in the N-terminal two-thirds of the propeptide (Ramos et al., 1994). The C-terminal part is probably more specifically involved in formation of zymogen structure, whereas the N-terminal part appears to be transiently involved in the folding process.

Further in vitro analysis has shed more light on the possible mechanism involved in this 'internal chaperone' activity (Winther et al., 1994). As discussed above, there is an absolute requirement for the propeptide for refolding in low salt buffers, while 5–10% of the activity can be recovered by renaturation in high salt buffers (e.g. 0.9 M ammonium sulfate). Furthermore, it is possible to prepare a stable folding intermediate by removal of the denaturing agent on a desalting column. The folding intermediate (I-CPY) can be induced to fold by addition of 0.9 M ammonium sulfate and will yield about 5% activity. Under these conditions the refolding efficiency can be further enhanced by the addition of the isolated propeptide (CYPR). Thus, in a typical experiment, refolding yield will increase from 5% without CYPR, to 10% with one molar equivalent, and 25% with 10 molar equivalents of CYPR. Importantly, CYPR does not inhibit CPY activity; indeed, neither CPY nor I-CPY have proven to interact strongly with CYPR. Analysis of the effect of CYPR on the refolding kinetics (measured by CPY activity) showed that CYPR did not increase the overall rate of folding but only the yield, suggesting that it reduced the significance of unproductive side reactions. Furthermore, the folding kinetics were consistent with the interpretation that CYPR interacts with a second folding intermediate (I'-CPY), which is only present in the salt-induced refolding reaction. The yield of I-CPY refolding in the absence of CYPR is inversely dependent on the concentration. Thus, in the I-CPY/CYPR system, yield may be increased by binding of CYPR to I'-CPY which is otherwise prone to aggregation. Such a mode of action is reminiscent of that of molecular chaperones. The analogy to heat shock proteins is further corroborated by the observation that CYPR significantly stabilizes CPY towards thermoinactivation, even though there is no evidence for a direct interaction with native CPY.

CYPR has been analysed by various means (NMR, fluorescence, circular dichroism) and is found to have significant amount of secondary structure but a dynamic tertiary structure (Sørensen et al., 1993). Theoretical predictions suggest that CYPR has a high probability for forming amphipathic helix structures. Such structures would, potentially, be able to interact with and solubilize a wide range of structures in the folding reaction.

The CPY propeptide shares remarkably little homology with other serine protease propeptides. In the case of CPY from Candida albicans, for instance, the identity in the enzyme parts approach 75%, while it is less than 20% in the propeptide overall (Mukhtar et al., 1992). CYPR does not stabilize unrelated proteins towards thermal denaturation (J.R. Winther, unpublished observations) and thus appears to be somewhat dedicated in its action. On the other hand, the deletion analysis has shown that there are probably few specific requirements to sequence conservation in the N-terminal two-thirds of the propeptide. These structure–function considerations pose interesting questions as to mechanism and specificity (see also the entry on proteinase A and its propeptide, p. 475).

References

Endrizzi, J.A., Breddam, K., and Remington, S.J. (1994). 2.8-Å structure of yeast serine carboxypeptidase. Biochemistry **33**, 11106–11120.

Holst, B., Bruun, A.W., Kielland-Brandt, M.C., and Winther, J.R. (1996). Competition between folding and glycosylation in the endoplasmic reticulum. EMBO J. **15**, 3538–3546.

Johansen, J.T., Breddam, K., and Ottesen, M. (1976). Isolation of carboxypeptidase Y by affinity chromatography. Carlsberg Res. Commun. **41**, 1–14.

Jones, E.W. (1991). Tackling the protease problem in Sacchromyces cerevisiae. Meth. Enzymol. **194**, 428–453.

Knop, M., Schiffer, H.H., Rupp, S., and Wolf, D.H. (1993). Vacuolar/lysosomal proteolysis: proteases, substrates, machanisms. Curr. Biol. **5**, 990–996.

Marcusson, E.G., Horazdovsky, B.F., Cereghino, J.L., Gharakhanian, E., and Emr, S.D. (1994). The sorting receptor for yeast vacuolar carboxypeptidase Y is encoded by the VPS10 gene. Cell **77**, 579–586.

Mukhtar, M., Logan, D.A., and Käufer, N.F. (1992). The carboxypeptidase Y-encoding gene from Candida albicans and its transcription during yeast-to-hyphae conversion. Gene **121**, 173–177.

Nielsen, T.L., Holmberg, S., and Petersen, J.G.L. (1990). Regulated overproduction and secretion of yeast carboxypeptidase Y. Appl. Microbiol. Biotechnol. **33**, 307–312.

Ramos, C., Winther, J.R., and Kielland-Brandt, M.C. (1994). Requirement of the propeptide for in vivo formation of active yeast carboxypeptidase Y.J. Biol. Chem. **269**, 7006–7012.

Remington, S.J. and Breddam, K. (1994). Carboxypeptidases C and D. Meth. Enzymol. **244**, 231–248.

Sørensen, S.O. and Winther, J.R. (1994). Active-site residues of procarboxypeptidase Y are accessible to chemical modification. Biochem. Biophys. Acta **1205**, 289–293.

Sørensen, P., Winther, J.R., Kaarsholm, N.C., and Poulsen, F.M. (1993). The pro region required for folding of carboxypeptidase Y is a partially folded domain with little regular structural core. Biochemistry **32**, 12160–12166.

Sørensen, S.O., Hazel, H.B.v., Kielland-Brandt, M.C., and Winther, J.R. (1994). pH-dependent processing of yeast carboxypeptidase Y by proteinase A in vivo and in vitro. Eur. J. Biochem. **220**, 19–27.

Valls, L.A., Hunter, C.P., Rothman, J.H., and Stevens, T.H. (1987). Protein sorting in yeast: the localization determinant of yeast vacuolar carboxypeptidase Y resides in the propeptide. Cell **48**, 887–897.

Valls, L.A., Winther, J.R., and Stevens, T.H. (1990). Yeast carboxypeptidase Y vacuolar targeting signal is defined by four propeptide amino acids. J. Cell Biol. **111**, 361–368.

van den Hazel, H.B., Winther, J.R., and Kielland-Brandt, M.C. (1996). Review: biosynthesis and function of yeast vacuolar proteases. Yeast **12**, 1–16.

Winther, J.R. and Sørensen, P. (1991). Propeptide of carboxypeptidase Y provides a chaperone-like function as well as inhibition of the enzymatic activity. Proc. Natl. Acad. Sci. USA. **88**, 9330–9334.

Winther, J.R., Sørensen, P., and Kielland-Brandt, M.C. (1994). Refolding of a carboxypeptidase Y folding intermediate in vitro by low-affinity binding of the proregion. J. Biol. Chem. **269**, 22007–22013.

■ Jakob R. Winther:
Department of Yeast Genetics
Carlsberg Laboratory
Gamle Carlsberg Vej 10
DK-2500 Copenhagen Valby, Denmark
Tel. 45 3327 5282
Fax. 45 3327 4765
E-mail: JRW@CRC.DK

Proteinase A and its propeptide

The aspartic protease proteinase A (PrA) plays a central role in the yeast vacuole (equivalent to the lysosome in mammals) as initiator of zymogen activation. PrA is itself synthesized as a zymogen (proPrA), which is activated by the autoproteolytic removal of the 54 residue N-terminal propeptide. The propeptide is essential for folding in vivo and propeptide-deleted mutant forms are accumulated in the endoplasmic reticulum and subsequently degraded. The propeptide shows a low requirement for sequence conservation and does not require covalent linkage to the mature part for function.

■ Alternative names

proteinase yscA, (Knop et al., 1993), saccharopepsin.

■ Gene and protein structure

ProPrA is encoded by the *PEP4* gene in *Saccharomyces cerevisiae* with a 22 residue signal peptide, a 54 residue propeptide and an enzyme part of 329 residues (Ammerer et al., 1986; Woolford et al., 1986; GenBank accession numbers M13358 and M13632). The signal sequence fulfils the conventional requirements (hydrophobicity, etc.) for directing the protein across the endoplasmic reticulum membrane, and the mature part of the protein is very similar to that of other aspartic proteases (e.g. 46% sequence identity to human cathepsin D). The propeptide, however, is not homologous to those of even closely related aspartic proteases (Ammerer et al., 1986; Jarai et al., 1995). PrA is modified with N-linked glycosylation at two positions in the enzyme part. Of these, one site is not always used, frequently giving rise to doublet bands on SDS–PAGE (Pedersen and Biedermann, 1993; Sørensen et al., 1994).

■ Biogenesis of PrA

ProPrA is transported to the vacuole via the secretory pathway (reviewed in van der Hazel et al., 1996). Thus, upon translocation into the ER, the signal sequence is cleaved off and core glycosylation takes place. During subsequent transport through the Golgi stacks, glycosyl residues are modified, and proPrA is actively sorted from secreted proteins and transported to the vacuole (Klionsky et al., 1988). In the vacuole, removal of the propeptide takes place yielding an active enzyme. In the vacuole PrA activity is responsible for proteolytic maturation of a large number of hydrolase zymogens. By analogy to other aspartic proteases, it has been hypothesized that activation is induced by the low pH environment of the vacuole (Ammerer et al., 1986). However, disruption of the vacuolar H^+–ATPase does not lead to proPrA accumulation (Yamashiro et al., 1990; Sørensen et al., 1994). Recent *in vitro* experiments have, however, shown that moderately high salt concentrations (thought to be present in the vacuole) may also trigger activation of proPrA, even at neutral pH (van den Hazel et al., 1997). The autoproteolytic cleavage takes place at a position upstream of the fully processed mature PrA. The product of this cleavage is active but final processing is dependent on proteinase B, another yeast endoprotease (Hirsch et al., 1992; van den Hazel et al., 1992; Woolford et al., 1993).

■ Purification of PrA

Overexpression of *PEP4* leads to secretion of large amounts of proPrA into the culture supernatant (Rothman et al., 1986). This can be conveniently utilized for the purification of both PrA and proPrA. Under normal growth conditions proPrA activation takes place extracellularly yielding correctly processed active PrA, which can be easily purified by ion-exchange chromatography (Sørensen et al., 1994). Similarly, if pepstatin (an inhibitor of aspartic proteases) is added to the growth medium, proPrA can be purified from the culture supernatant of a strain that is deleted for proteinase B ($\Delta prb1$) but overexpresses *PEP4* (Wolff et al., 1996).

■ PrA activity assay

On plates, PrA activity is most conveniently detected indirectly using an overlay assay for carboxypeptidase Y activity, using N-acetyl-DL-phenylalanyl-β-naphthylester as substrate (Jones, 1991). Since activation of procarboxypeptidase Y is dependent on PrA activity in a cascade-like manner, this gives a very sensitive, but not very quantitative assay. In extracts, PrA can be measured using acid-denatured hemoglobin (Jones, 1991) or internally quenched fluorescent peptides (van den Hazel et al., 1993).

■ Functions of the PrA propeptide

One of the most obvious functions of the PrA propeptide is to maintain the zymogen in its inactive state, thus preventing proteolytic activity from being unleashed prior to arrival in the vacuole. It has, furthermore, been shown that the propeptide contains information able to confer sorting of the normally secreted enzyme invertase to the vacuole (Klionsky et al., 1988). It appears, however, that this signal is not the only vacuolar sorting signal in native proPrA (Westphal et al., 1996).

■ Role of propeptide in folding of PrA

Deletion of the PrA propeptide sequence results in accumulation of the glycosylated polypeptide (PrAΔpro) in the ER and its slow degradation (van der Hazel et al., 1993). It is known from other systems that the 'quality control' system of the ER ensures that only correctly folded proteins are allowed to exit the ER. Furthermore, while PrA is normally a very stable enzyme, no activity can be detected in cells producing PrAΔpro. It should be noted that PrA and PrAΔpro have the same primary sequence. It thus appears that the only consistent interpretation is that in vivo folding is compromised by deletion of the propeptide (van den Hazel et al., 1993). Interestingly, the propeptide does not require covalent linkage to the enzyme region for function. Thus, production of equimolar amounts of the propeptide and PrAΔpro gives rise to PrA activity, correctly localized to the vacuole (van den Hazel et al., 1993). The affinity between the propeptide and the enzyme region is probably fairly weak and of a transient nature since the level of activity recovered under these conditions is only of the order of 3%. The observation that overproduction of the propeptide relative to the enzyme region results in an increase in activity recovered (up to 8%) further supports this notion (van den Hazel et al., 1993).

It is frequently found that the propeptides of even closely related enzymes show little or no homology. The requirement for sequence conservation of the propeptide has been investigated by substitution of large parts of the propeptide by random sequence. In a genetic approach, parts of PEP4 encoding either of the two halves of propeptide were substituted with randomly generated synthetic DNA (van den Hazel et al., 1995). The resulting mutant plasmids were then screened for PrA activity using the CPY plate staining assay described above. A surprisingly high number of active clones (c. 1%) were identified, giving rise to up to 66% of the wild-type activity. Apart from a specific lysine residue (Lys-53), which also seems to be conserved in the propeptides of closely related aspartic proteases, no structural similarity was found between the random inserts. This strongly supports the notion that the requirements for sequence conservation of propeptides, at least in aspartic proteases, are not stringent. It is also consistent with the idea that, similarly to the carboxypeptidase Y propeptide (see entry p. 472), a main prerequisite for propeptide function in folding is the ability to interact with an array of folding intermediates. These putative interactions would be transient and not highly specific, and their main purpose would be to increase solubility of the folding intermediates.

■ References

Ammerer, G., Hunter, C.P., Rothman, J.H., Saari, G.C., Valls, L.A., and Stevens, T.H. (1986). PEP4 gene of Saccharomyces cerevisiae encodes proteinase A, a vacuolar enzyme required for processing of vacuolar precursors. Mol. Cell. Biol. **6**, 2490–2499.

Hirsch, H.H., Schiffer, H.H., and Wolf, D.H. (1992). Biogenesis of the yeast vacuole (lysosome). Proteinase yscB contributes molecularly and kinetically to vacuolar hydrolase-precursor maturation. Eur. J. Biochem. **207**, 867–876.

Jarai, G., van den Hombergh, H., and Buxton, F.P. (1995). Cloning and characterization of the pepE gene of Aspergillus niger encoding a new aspartic protease and regulation of pepE and pepC. Gene **145**, 171–178.

Jones, E.W. (1991). Tackling the protease problem in Sacchromyces cerevisiae. Meth. Enzymol. **194**, 428–453.

Klionsky, D.J., Banta, L.M., and Emr, S.D. (1988). Intracellular sorting and processing of a yeast vacuolar hydrolase: proteinase A propeptide contains vacuolar targeting information. Mol. Cell. Biol. **8**, 2105–2116.

Knop, M., Schiffer, H.H., Rupp, S., and Wolf, D.H. (1993). Vacuolar/lysosomal proteolysis: proteases, substrates, mechanisms. Curr. Biol. **5**, 990–996.

Pedersen, J. and Biedermann, K. (1993). Characterization of proteinase A glycoforms from recombinant Saccharomyces cerevisiae. Biotechnol. Appl. Biochem. **18**, 377–388.

Rothman, J.H., Hunter, C.P., Valls, L.A., and Stevens, T.H. (1986). Overproduction-induced mislocalization of a yeast vacuolar protein allows isolation of its structural gene. Proc. Natl. Acad. Sci. USA **83**, 3248–3252.

Sørensen, S.O., van den Hazel, H.B., Kielland-Brandt, M.C., and Winther, J.R. (1994). pH-dependent processing of yeast carboxypeptidase Y by proteinase A in vivo and in vitro. Eur. J. Biochem. **220**, 19–27.

van den Hazel, H.B., Kielland-Brandt, M.C., and Winther, J.R. (1992). Autoactivation of proteinase A initiates activation of yeast vacuolar zymogens. Eur. J. Biochem. **207**, 277–283.

van den Hazel, H.B., Kielland-Brandt, M.C., and Winther, J.R. (1993). The propeptide is required for in vivo formation of stable active yeast proteinase A and can function even when not covalently linked to the mature region. J. Biol. Chem. **268**, 18002–18007.

van den Hazel, H.B., Kielland-Brandt, M.C., and Winther, J.R. (1995). Random substitution of large parts of the propeptide of yeast proteinase A. J. Biol. Chem **270**, 8602–8609.

van den Hazel, H.B., Winther, J.R., and Kielland-Brandt, M.C. (1996). Review: biosynthesis and function of yeast vacuolar proteases. Yeast **12**, 1–16.

van den Hazel, H.B., Wolff, A.M., Kielland-Brandt, M.C., and Winther, J.R. (1997). Mechanism and ion dependence of in vitro autoactivation of yeast proteinase A: possible implications for compartmentalized activation in vivo. Biochem. J. In press.

Westphal, V., Marcusson, E.G., Winther, J.R., Emr, S.D., and van den Hazel, H.B. (1996). Multiple pathways for vacuolar sorting of yeast proteinase A. J. Biol. Chem. **271**, 11865–11870.

Wolff, A.M., Din, N., and Petersen, J.G.L. (1996) Vacuolar and extracellular maturation of Saccharomyces cerevisiae proteinase A. Yeast **12**, 823–832.

Woolford, C.A., Daniels, L.B., Park, F.J., Jones, E.W., Arsdell, J.N.v., and Innis, M.A. (1986). The PEP4 gene encodes an aspartyl protease implicated in the posttranslational regulation of Saccharomyces cerevisiae vacuolar hydrolases. Mol. Cell. Biol. **6**, 2500–2510.

Woolford, C.A., Noble, J.A., Garman, J.D., Tam, M.F., Innis, M.A., and Jones, E.W. (1993). Phenotypic analysis of proteinase A mutants. J. Biol. Chem. **268**, 8990–8998.

Yamashiro, C.T., Kane, P.M., Wolczyk, D.F., Preston, R.A., and Stevens, T.H. (1990). Role of vacuolar acidification in protein sorting and zymogen activation: A genetic analysis of the yeast vacuolar proton-translocating ATPase. Mol. Cell. Biol. **10**, 3737–3749.

■ *Jakob R. Winther and H. Bart van den Hazel:*
Department of Yeast Genetics
Carlsberg Laboratory
Gamle Carlsberg Vej 10
DK-2500 Copenhagen Valby, Denmark
Tel. 45 3327 5282
Fax.45 3327 4765
E-mail: JRW@CRC.DK

16

Molecular Chaperone Machines

The role of molecular chaperones in DNA replication

Studies on the involvement of molecular chaperones in DNA replication have been limited to a few replication systems, belonging primarily to the prokaryotic world. The insights gained from these studies have substantially contributed to our detailed understanding of the eukaryotic DNA replication process as well. The finding that molecular chaperones can activate some initiation proteins before DNA synthesis, has led to the more general suggestion that molecular chaperones can influence the DNA-binding activity of transcriptional factors involved in cell regulatory systems. Studies on molecular chaperones, especially with respect to their role in DNA replication, have also suggested a synergistic and cooperative action for the DnaK/DnaJ/GrpE chaperone machine. This particular system is now a paradigm for our understanding of fundamental biochemical processes, such as protein folding, translocation, selective proteolysis and autoregulation of the heat shock response. It is likely that the recent studies on the influence of the Clp ATPase family of molecular chaperones on DNA replication will help to define finally the nature of the signals that help a chaperone to 'decide' between the refolding or the degradation of a misfolded protein.

The role of the DnaK/DnaJ/GrpE molecular chaperones has been described in detail for a few DNA replication systems, namely, λ, mini-F, mini-P1 and oriC plasmid DNA replication. Only in the case of the λ DNA replication system has it been shown that the DnaK/DnaJ/GrpE chaperone machine is directly involved in the initiation process. In all other cases, the DnaK/DnaJ/GrpE chaperones activate the initiation proteins but do not participate during the final DNA replication initiation event. Recently, studies on the bacteriophage Mu DNA replication have suggested that this system could be another example of the direct involvement of molecular chaperones (in this case Clp ATPase) in the initiation process (Levchenko et al., 1995; Kruklitis et al., 1996).

In this review we will focus on the molecular mechanisms of chaperone action. Further details on similar subjects can be found in other reviews (Georgopoulos et al., 1990, 1994; Baker and Wickner, 1992; Learn et al., 1993; Zylicz, 1993; Chattoraj, 1995; Skowyra et al., 1995; Taylor and Wegrzyn, 1995; Wawrzynow et al., 1996; Hartl, 1996).

■ Bacteriophage λ DNA replication

During infection of *Escherichia coli* with λ bacteriophage the double-stranded bacteriophage DNA is injected into the cell, then, following its circularization and supercoiling, DNA replication is initiated at a single site (oriλ) and proceeds bidirectionally, according to the theta mode (Fig. 1). Following a few rounds of theta mode replication a rolling circle mechanism of replication ensues (sigma mode, Fig. 1; for a review, see Learn et al., 1993). The early theta mode for λ DNA replication could be mimicked by a plasmid DNA system containing the minimal elements required for λ DNA replication, namely, oriλ sequence, the genes encoding λO and λP replication proteins and a promoter sequence. The latter is required not only for expression of λO and λP, but also for the transcriptional activation of the oriλ sequence, an event that regulates the frequency of λ DNA replication (for reviews see Learn et al., 1993; Taylor and Wegrzyn, 1995). *In vitro* reconstitution of the λ plasmid DNA replication system using purified proteins has allowed the identification of intermediate reactions leading to the initiation of λ DNA replication (Alfano and McMacken, 1989a; Zylicz et al., 1989). The λO dimers bind to the oriλ sequence at four inverted repeats (IR) forming a large nucleosome structure, called the O-some (Echols, 1990). The formation of this structure changes the topology of the AT-rich region located downstream of the λO binding sites. Transcription, which proceeds through the O-some structure, makes the λO accessible to the ClpXP protease (Wojtkowiak et al., 1993; A. Wawrzynów and M. Zylicz, unpublished results). This suggests that the O-some structure could not be inherited in the next round of λ DNA replication. The λO protein, especially at low salt concentrations or high temperature, has a tendency to aggregate. The ClpX chaperone, in the absence of the ClpP proteolytic subunit can protect the λO protein from aggregation and also dissociate the previously aggregated λO protein, thus enhancing the specific binding of λO to the oriλ sequence. The ClpA and DnaK/DnaJ/GrpE chaperones could partially substitute for the ClpX in this reaction (Wawrzynów et al., 1995a; A. Wawrzynów and M. Zylicz, unpublished results). The biological role of the ClpX-dependent enhancement of the oriλ–λO complex formation is still not clear.

During the initial steps of λ DNA replication, two λP–DnaB complexes (each composed of a dimer of λP and a hexamer of DnaB) interact with the λO-some structure to form an oriλ–λO-(λP–DnaB)$_2$ preprimosomal complex (Fig. 1, step 1; Alfano and McMacken, 1989a; Zylicz et al., 1989). Transcriptional activation, initiated at the p$_R$ pro-

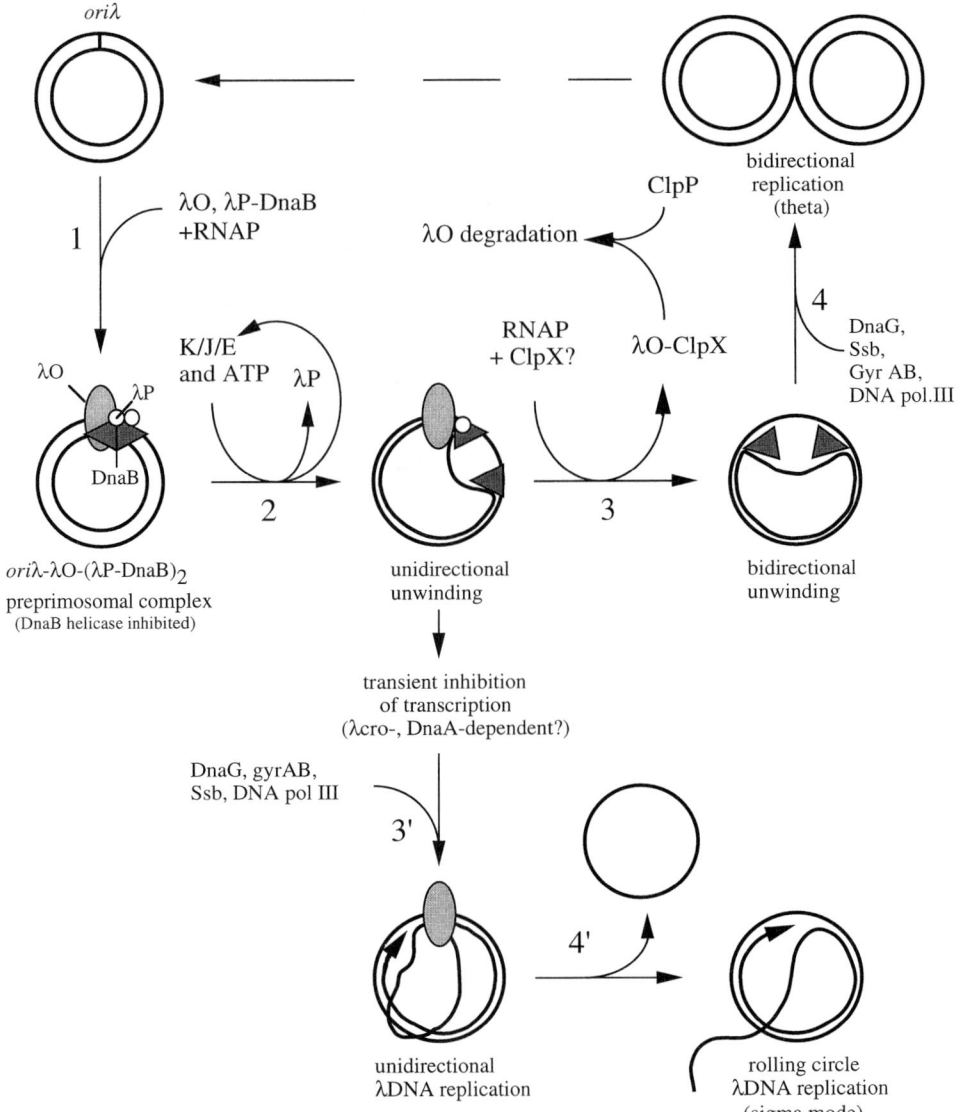

Figure 1. A hypothetical scheme for the role of molecular chaperones and RNA polymerase transcription in the initiation of λ DNA replication. (1) Transcriptional activation is required for the proper assembly of the preprimosomal structure [oriλ–λO–(λP–DnaB)$_2$]. (2) The DnaK/DnaJ/GrpE chaperone machine, in an ATP-dependent reaction, releases a fraction of the λP protein, thus triggering the unidirectional unwinding of double-stranded DNA, catalysed by the DnaB helicase. (3) Transcription, when transversing the oriλ sequence dissociates the λO-some (and remaining λP) from the DNA complex, thus allowing DnaB helicase to unwind double-stranded DNA in both directions. (4) In the presence of the single-stranded DNA-binding proteins (Ssb) and gyrase (GyrAB), the DnaB-dependent unwinding of double-stranded DNA proceeds, thus allowing the DnaG primase to synthesize RNA primers, which are subsequently extended into DNA by the DNA polymerase III. (3´) The transient inhibition of transcription (and/or ClpX chaperone activity) blocks the bidirectional unwinding of the double-stranded DNA at the oriλ sequence, resulting in unidirectional λ DNA replication. The transcriptional activation could be modulated in the presence of the λCro repressor and the *E. coli* DnaA (see text for details). (4´) The collision of a DnaB helicase/DNA polymerase III replication fork complex with an undissociated O-some structure, following the completion of a round of unidirectional DNA replication may be the key event that triggers the initiation of a rolling-circle mode of DNA replication (see text for more details). (K/J/E): DnaK/DnaJ/GrpE molecular chaperones; (RNAP): *E. coli* RNA polymerase; the exact stoichiometry of the proteins and the protein subunits is not shown.

moter, is required for the correct assembly of the preprimosomal complex (Yamamoto et al., 1987; Mensa-Wilmot et al., 1989). The λP is required for bringing the DnaB helicase into the neighbourhood of the O-some structure. The unusual stability of this preprimosomal complex is due to the fact that λP protein interacts with the single-stranded DNA (Learn et al., 1997). The weak λO–λP interaction was also detected (for a review see Zylicz, 1993). The process of transcription, which in the absence of ClpXP protease 'bypasses' the O-some structure, in the case of the oriλ–λO–(λP–DnaB)$_2$ preprimosome, terminates at the oriλ sequence (Learn et al., 1993) and makes λO inaccessible to the ClpXP protease (A. Wawrzynów and M. Zylicz, unpublished results). Genetic experiments support these findings (Wegrzyn et al., 1995). The DnaB helicase in the preprimosomal complex is not active, and its activation requires molecular chaperones (Fig. 1, step 2 and Fig. 2). The DnaJ chaperone binds to the preprimosome and, in addition, stabilizes the λO–λP complex by interacting primarily with the λP and DnaB helicase protein (Fig. 2, step 2a; Alfano and McMacken, 1989a; Zylicz et al., 1989; Zylicz, 1993). The presence of DnaJ in the preprimosomal complex facilitates the binding of the DnaK chaperone to this complex (Fig. 2, step 2b). The DnaK chaperone, depending on whether ADP or ATP is bound, possesses a different conformation and affinity for its protein substrates (Wawrzynów and Zylicz, 1995; Wawrzynów et al., 1995b; Banecki and Zylicz, 1996). After ATP hydrolysis, DnaK is converted to the so-called DnaK*–ADP conformation, which can interact only transiently with λP. The presence of DnaJ induces conformational changes in DnaK, thus leading to the formation of a more stable complex between the DnaK–ADP form and the preprimosomal complex. Subsequently, the GrpE-dependent ADP/ATP exchange in the oriλ–[λO–(λP–DnaJ–DnaB)$_2$]–DnaK–ADP complex, followed by ATP hydrolysis, changes the conformation of DnaK chaperone back to the DnaK*–ADP form which possesses a low affinity for the λP protein (Wawrzynów and Zylicz, 1995; Wawrzynów et al., 1995b; Banecki and Zylicz, 1996). It has been shown previously that the presence of GrpE and DnaJ greatly stimulates DnaK's ATPase activity (Liberek et al., 1991; Jordon and McMacken, 1995). The net result of these reactions is the selective release of DnaK, and λP, from the preprimosomal complex (Fig. 2, step 2c). On the contrary, most of the λO present in the O-some structure is not released from the DNA during the DnaK/DnaJ/GrpE chaperone machine action (Liberek et al., 1988; Alfano and McMacken, 1989b; Dodson et al., 1989).

The release of λP from the DnaB helicase leads to unidirectional unwinding of DNA near the oriλ sequence (Fig. 2, step 2c; Dodson et al., 1989; Learn et al., 1993). Significant unwinding of supercoiled DNA requires the additional presence of the DNA gyrase and Ssb proteins (Learn et al., 1993). Release of λP from the DnaB complex is the crucial event leading to the initiation of λ DNA replication. A mutant of the λP protein, called λPπ, binds so weakly to the DnaB helicase that the involvement of molecular chaperones is not required for its dissociation from DnaB (Konieczny and Marszalek, 1995). This observation explains the previous classical genetic experiments which showed that λ DNA replication could proceed in the dnaK, dnaJ and grpE mutant backgrounds, provided the λP gene was mutated (λPπ mutation; Georgopoulos et al., 1990).

The release of λP from the preprimosomal complex is not complete (Zylicz, 1993), and thus only the hexameric DnaB helicase which has dissociated from λP may be involved in the unidirectional unwinding of λ DNA. The remaining protein complex at the oriλ sequence may still possess some λP protein (Fig. 2). Amongst all in vitro isolated replication intermediates, this is the best candidate for a replication complex that is inherited by one of the daughter λ DNA circles. Genetic experiments suggest that at least during amino acid starvation in a relA genetic mutant background, the once-assembled λO-containing replication complex could be inherited, and used by one of two daughter λ DNA molecules during the second round of λ DNA replication (Taylor and Wegrzyn, 1995). The possibility that during amino acid starvation λ plasmid DNA replication proceeds unidirectionally, has not been tested.

In the absence of transcription from p_R the reconstituted λ DNA replication system leads to unidirectional DNA replication. The addition of purified RNA polymerase together with rNTPs, switches a fraction (up to 30%; Learn et al., 1993) of the unidirectional replicated molecules to bidirectional DNA replication (Fig. 1, step 3). McMacken and colleagues postulated that the assembled O-some structure could cause a physical barrier for the passage of DnaB helicase through it. Thus, the role of the transcriptional activation could be to rearrange the O-some structure and probably release additional λP, thus allowing the DnaB helicase to unwind DNA in the right to left direction on the genetic map (Learn et al., 1993).

We propose that the presence of the ClpXP protease (or the ClpX chaperone alone) may lead to a more efficient transition from uni- to bidirectional λ DNA replication (Wojtkowiak et al., 1993). For example, in the presence of ClpXP, the reassociation of λO protein into another O-some structure will almost certainly be inhibited, since λO will be efficiently degraded (Fig. 1, step 3). The deletion of the clpX gene only slightly influences λ phage growth, leading to the interpretation that ClpXP protease degrades only the surplus synthesized λO protein and does not affect λ DNA replication (Szalewska et al., 1994). It has been shown for some chaperones, for example DnaK and GroEL (Ziemienowicz et al., 1993), DnaK and SecB (Wild et al., 1992), DnaJ and CbpA (Ueguchi et al., 1994), DnaK/DnaJ/GrpE and ClpA (Wickner et al., 1994), that inactivation of any of these may be compensated by the other chaperone in the indicated pairs. From this point of view it is important to stress that some of ClpX's chaperone functions could be performed by either the ClpA protein or the DnaK/DnaJ/GrpE chaperone machine (Wawrzynów et al., 1995a; A. Wawrzynów and M. Zylicz, unpublished results). Additionally, a ClpX-close homologue has been

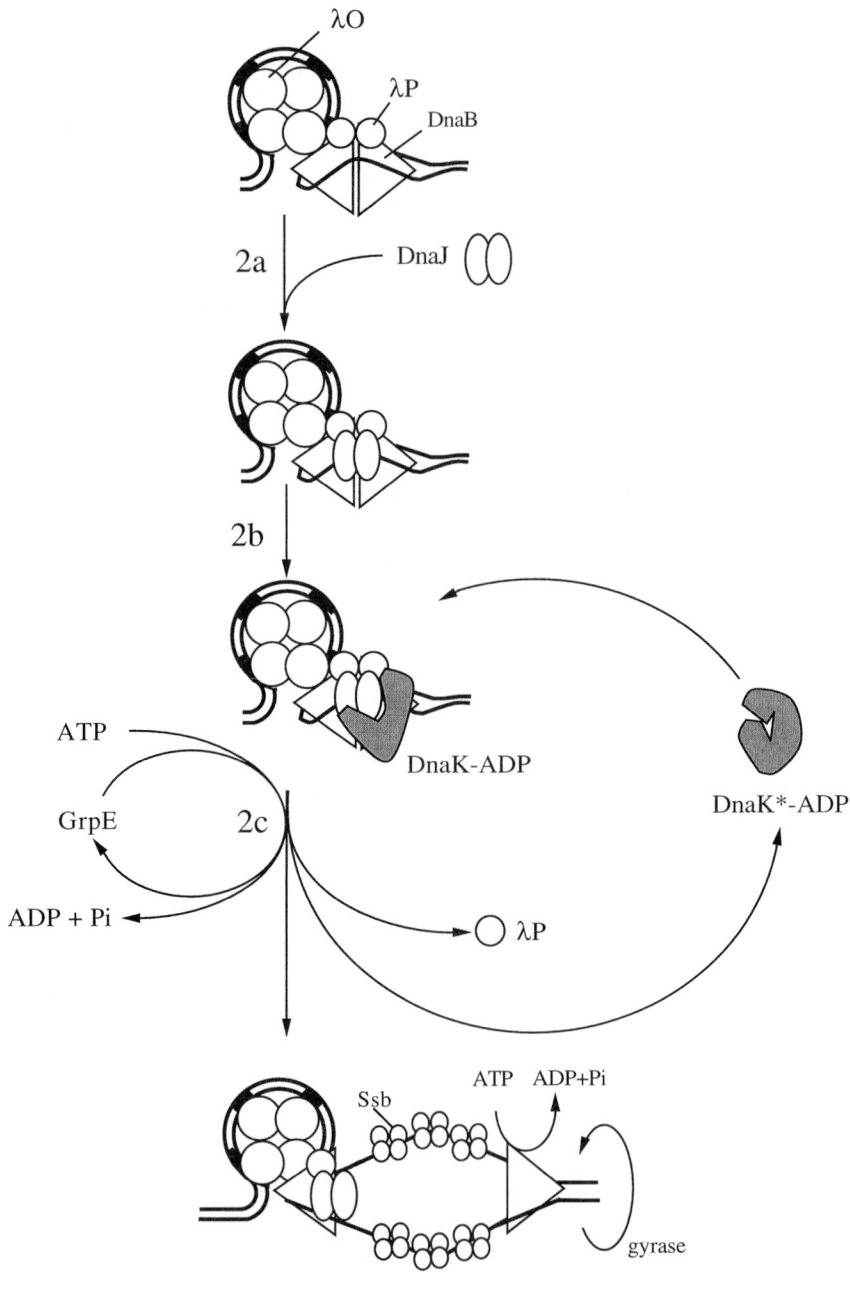

Figure 2. DnaK/DnaJ/GrpE chaperone-dependent activation of the DnaB helicase in the λ preprimosomal complex. The DnaK/DnaJ/GrpE-dependent activation of preprimosomal complex (reaction 2; Fig. 1) could be described by the following intermediate steps: (2a) DnaJ chaperone binds to λP and DnaB proteins thus attracting (2b) the DnaK chaperone to the preprimosomal structure. The presence of DnaJ changes the conformation of DnaK in such a way that more stable preprimosomal–DnaK–ADP complex could be formed. (2c) Following the GrpE-dependent ADP/ATP exchange and ATP hydrolysis, a portion of λP is released from the preprimosomal complex and DnaK is converted back to the DnaK*–ADP conformation. These molecules of DnaB helicase, which are no longer associated with λP, proceed to unwind double-stranded DNA. Filled boxes; oriλ iterons, which are recognition sites for λO. The (*) symbol represents the DnaK*–ADP conformation reached only after ATP hydrolysis, and not as a result of the simple preincubation of DnaK with ADP. The stoichiometries of the proteins and the protein subunits are not represented.

identified recently in *E. coli* bacteria (Missiakas *et al.*, 1996). These findings leave the problem of the putative involvement of ClpX chaperone in λ DNA replication still unsolved.

McMacken and colleagues have proposed that at the end of one round of unidirectional λ DNA replication (Fig 1, step 3′), when the replication fork travels around the circular DNA and reaches the *ori*λ sequence, a strand separation reaction could occur and replication may switch from the theta to the rolling circle (sigma) form (Fig. 1, step 4′; Learn *et al.*, 1993). According to this hypothesis, the switch between the theta and sigma modes of λ DNA replication could be correlated with the inhibition of transcriptional activation. We propose that the ClpX molecular chaperone could also modulate this reaction. For example, the transient inhibition of ClpX activity could favour unidirectional λDNA replication, and as a consequence initiate the switch to the rolling circle replication mode. Recent findings, namely that the elimination of *dnaA* function decreases the transcriptional activation reaction (Taylor and Wegrzyn, 1995), have lead to the hypothesis that DnaA, by modulating a transcriptional activation reaction, is required for the proper assembly of the replication complex leading to bidirectional theta λ DNA replication. In wild-type cells the utilization of DnaA by rapidly replicating λ phage DNA may inhibit transcriptional activation and thus favour the switch from bidirectional theta to the unidirectional theta replication mode and later to the sigma mode (Taylor and Wegrzyn, 1995).

■ Bacteriophage Mu DNA replication

Bacteriophage Mu employs a transposition mechanism to replicate its DNA. The bacteriophage-encoded Mu transposase (MuA) monomers bind to the multiple DNA sites at each end of the Mu DNA. Following DNA supercoiling, in the presence of the DNA-binding proteins HU and IHF, the Mu DNA ends are brought together resulting in the assembly of a complex containing the MuA tetramer. Such stable, synaptic protein–DNA structures, which are retained after strand transfer reaction (ends of the Mu DNA are cleaved and joined to the target DNA), are substrates for the initiation of Mu DNA replication. In the presence of the DnaB–DnaC complex, DnaG primase, DNA gyrase, DNA polymerase III and additional host initiation factors, Mu DNA synthesis is initiated (Nakai and Kruklitis, 1995). One of these host initiation factors was identified as the ClpX molecular chaperone. The energy of ATP hydrolysis is utilized by ClpX to induce the conformational changes in the MuA, thus triggering (in the presence of (an)other, still unidentified, host factor(s)) the release of MuA from the stable MuA tetramer–DNA complex (Nakai and Kruklitis, 1995; Levchenko *et al.*, 1995; Kruklitis *et al.*, 1996). The analogy between the DnaK/DnaJ/GrpE chaperone-dependent release of λP from λ prepimosomal complex and the ClpX-dependent release of MuA from the DNA complex is striking. The released MuA is not degraded or irreversibly denatured and is able to perform another round of recombination *in vitro* (Levchenko *et al.*, 1995; Kruklitis *et al.*, 1996). In support of the involvement of ClpX, but not ClpP, in Mu DNA replication, it was shown that the *clpX* gene product is required for the lytic growth of bacteriophage Mu. The kinetics of the block in bacteriophage growth after induction of a *clpX*-deleted lysogen suggest that ClpX functions after the strand transfer reaction but before the onset of extensive DNA replication (Mhammedi-Alaoui *et al.*, 1994).

■ Mini-F plasmid DNA replication

The mini-F plasmid carries the basic replicon of the *E. coli* sex factor (for review, see Baker and Wickner, 1992; Chattoraj, 1995). As with λ DNA replication, mini-F plasmid contains special sequences to which plasmid-encoded initiation RepE protein binds during the initiation of DNA replication (Fig. 3A). The RepE could work as an initiator protein during initiation of mini-F DNA replication, and act also as a repressor by binding to the iteron sequence that overlaps with the operator of σ^{32}-dependent RepE promoter (Fig. 3A; for a review, see Chattoraj, 1995). The σ^{32} contribution in mini-F DNA replication is not limited to the synthesis of the RepE replication protein but also stimulates the synthesis of the DnaK, DnaJ and GrpE molecular chaperones, which, in turn, activate the RepE protein. Although the same set of chaperones participate in λ and mini-F replication, their role is quite different (Kawasaki *et al.*, 1990). The mini-F origin contains four direct repeats (DR) to which the RepE monomer can bind. In contrast, the operator sequence of the *repE* gene contains the same sequence elements but arranged in an inverted repeat (IR) mode. The dimers of RepE bind to these IR elements and act as a repressor of the *repE* gene (Fig. 3A; Ishiai *et al.*, 1994). The DnaK/DnaJ/GrpE molecular chaperone machine activates RepE for binding to the direct repeat sequence, thus promoting the initiation of DNA replication. The mechanism by which the DnaK/DnaJ/GrpE molecular chaperones activate RepE for binding to the DR elements is not clear. The DnaK/DnaJ/GrpE chaperones are also thought to help in the folding of the nascent RepE polypeptide (Chattoraj, 1995). In this model, the fraction of properly folded RepE would increase and the distribution between monomeric (initiation mode) and dimeric (repressor mode) forms of RepE would depend on the affinities of the active monomers for each other (Fig. 3A). Alternatively, the DnaK/DnaJ/GrpE chaperone machine could activate RepE by dissociating the already formed RepE dimers, thereby increasing the concentration of monomeric RepE protein which can then, in turn, bind to the DR sequence (Fig. 3A; by analogy to the mini-P1 system, see Wickner *et al.*, 1991). Of course, both modes of chaperone action are not mutually exclusive.

■ Mini-P1 plasmid DNA replication

The mini-P1 replicon has a similar organization when compared with the mini-F system (Fig.3B). The P1 plasmid

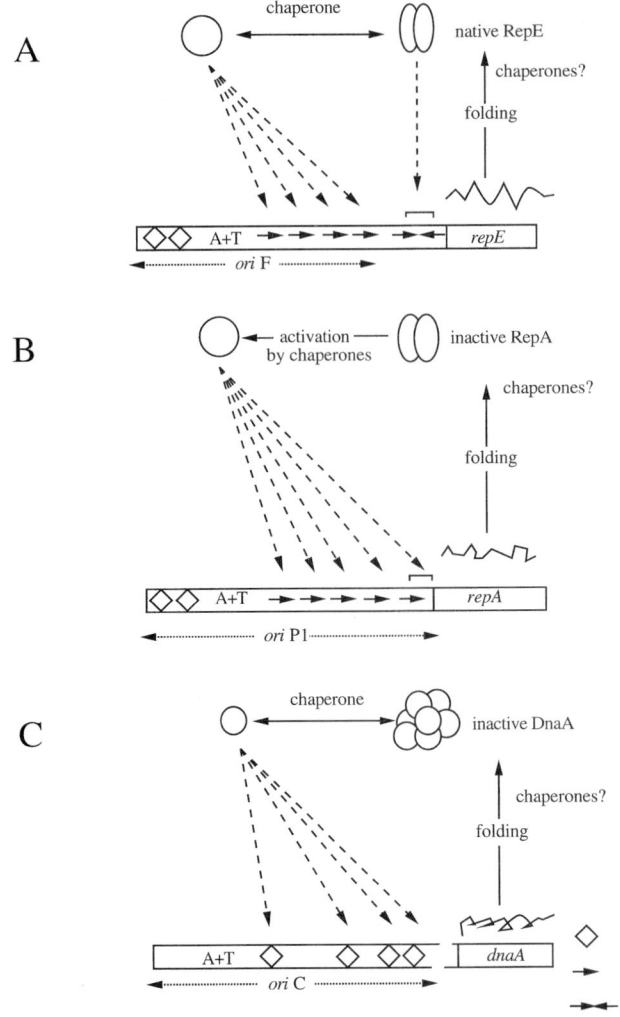

Figure 3. Chaperone-dependent activation of the replication proteins for binding to various *ori* DNA sequences. (A) mini-F; (B) mini-P1 and (C) *ori*C. See text for details.

has only the direct repeat (DR) sequences that are used to modulate both initiation of P1 replication and repression of the *repA* gene. The σ^{70}-dependent promoter overlaps the DR sequences (Fig. 3B; Sozhamannan and Chattoraj, 1993). Both *in vivo* and *in vitro* initiation of P1 DNA replication are dependent on the functional presence of the DnaK, DnaJ and GrpE chaperones (Tilly and Yarmolinsky, 1989; Wickner et al., 1992). It was proposed that the mechanism of activation of RepA for binding to the DR sequences involves dissociation of RepA dimers. This was based on the fact that inactive RepA was found as a dimer and was activated by either the DnaK, DnaJ and GrpE chaperone machine (or by refolding after urea treatment) to an active, monomeric form (Wickner et al., 1991, 1994). However, the results of others suggest that monomerization of RepA may not be sufficient for activation, since diluted solutions of RepA (where the protein is essentially monomeric) still require chaperones for activation (DasGupta et al., 1993). This last result suggests that the activation reaction may require the DnaK, DnaJ and GrpE-dependent refolding of inactive RepA monomers (Fig. 3B; Chattoraj, 1995). Both genetic and *in vitro* experiments suggest that after RepA binds to the DR sequence the presence of chaperones is not obligatory for any of the subsequent steps leading to the initiation of P1 DNA replication (Wickner et al., 1992).

It is interesting to note here that the biological role of the DnaK, DnaJ and GrpE chaperones in the RepA activation reaction and the subsequent binding to *ori*P1 DNA could be substituted not only by the presence of 6M urea but also by ClpA and ATP (Wickner et al., 1994). This result suggests that, as in the case of ClpX (Wawrzynów et al., 1995a), ClpA can perform a chaperone action in the absence of its usual ClpP catalytic protease partner.

*ori*C DNA replication

A direct role for molecular chaperones in chromosomal replication has been suggested by the isolation of the *dnaK*111 mutation, which prevents the specific *ori*C-dependent initiation of DNA replication (Sakakibara, 1988). Moreover, it has been shown *in vitro* that the DnaA initiation protein, which binds to the *ori*C DNA sequence thus triggering the DNA initiation process, aggregates easily, thereby losing its DNA-binding activity. However, in the presence of DnaK and ATP, such DnaA aggregates are dissociated, thus allowing the initiation of *ori*C DNA replication (Fig. 3C; Hwang *et al.*, 1990).

DnaK chaperone, in the presence of GrpE can also 'activate' the mutant DnaA5 and DnaA46 proteins (Hupp and Kaguni, 1993). At least in the case of DnaA46, the DnaK-dependent activation reaction could be accomplished by substituting DnaK with the GroEL chaperone (Fayet *et al.*, 1986; J. Marszalek, unpublished results).

In vitro work on *ori*C DNA replication has demonstrated that molecular chaperones may activate the DnaA initiation protein for binding to the *ori*C sequence, but they are not necessary for any of the subsequent steps leading to the *ori*C-dependent initiation of DNA synthesis (Kaguni and Kornberg, 1984). Interestingly, a consequence of chaperone-dependent activation of DnaA is the down-regulation of the *rpoH* gene (which encodes the σ^{32} subunit of RNA polymerase), thus down-regulating the *dnaK–dnaJ* and *grpE* gene expression. Binding of DnaA to the *dnaA*-boxes present in the *rpoH* gene promoter region inhibits its expression. Thus, this regulatory control circuit can result in the homeostatic regulation of both *dnaA* and heat shock gene expression.

Future prospects

Up to now, no one has seriously addressed the question of the possible involvement of the Hsp70/Hsp40/Hsp24 chaperone machine (eukaryotic DnaK, DnaJ and GrpE homologues) in eukaryotic DNA replication. Eukaryotic HSP70 proteins have been identified in all cellular compartments, namely the cytoplasm, nucleus, endoplasmic reticulum, mitochondria and chloroplasts. Most cells carry both heat-inducible and constitutive forms of HSP70 (for review, see Georgopoulos and Welch, 1993). Recently, eukaryotic DnaJ and GrpE homologues have been identified in mitochondria and DnaJ homologues in the cytosol and endoplasmic reticulum (for review see Georgopoulos *et al.*, 1994). The involvement of the Hsp70/Hsp40/Hsp24 chaperone machine in mitochondrial DNA replication has not been directly tested. Supporting the idea of the involvement of molecular chaperones in mitochondrial DNA replication, genetic experiments show that the disruption of mitochondrial *MDJ1* gene (DnaJ family) resulted in the loss of the mitochondrial DNA (Rowley *et al.*, 1994).

Based upon experiments on the replication of DNA using a cell-free extract prepared from *Xenopus* eggs, R. Laskey and J. Blow have proposed a model for the activation of eukaryotic DNA replication. Such activation depends on the presence of a licensing factor and occurs once in the cell cycle. Recently, some elements of such a licensing factor have been identified (Chong *et al.*, 1995; Madine *et al.*, 1995). Licensing factor, activated during the G1 to S transition, binds or modifies chromatin in such a way that DNA replication can be initiated. These licensing factors are subsequently inactivated during the process of DNA replication, so future initiation of DNA replication cycles cannot occur until the DNA is relicensed by passing through mitosis.

Stillman (1994) has pointed out that the λP protein in the λ DNA replication, plays a role analogous to the licensing factor in eukaryotic DNA replication. Extending this hypothesis it is possible that, as in λ DNA replication, the molecular chaperone machine could participate in the inactivation of the licensing factor, following the initiation of eukaryotic DNA replication. In such a reaction, Hsp40 could be a specificity factor which enables Hsc70 to form a transient complex with the licensing factor. Following ATP hydrolysis catalysed by Hsc70, the licensing factor could dissociate from its complex with the chromatin structure, thus inhibiting the reinitiation reaction. In agreement with this, in human cells the transcription of the *hsp70* genes is induced at the onset of the S phase, and Hsc70 migrates to the nucleus (Milarski and Morimoto, 1986). Besides the putative role of the HSP70 machine in modulation of licensing factor activity, molecular chaperones could be involved in maintaining the active form of enzymes required for DNA replication. It has been shown recently that calf thymus Hsc70, in the presence of DnaJ homologue, protects and reactivates DNA polymerase α and ϵ (Ziemienowicz *et al.*, 1995).

Acknowledgements

We thank Professor Costa Georgopoulos for discussions, critical comments and reading of this manuscript. We are grateful to Professor H. Nakai for sharing his unpublished results with us. We also thank Drs J. Marszalek, K. Liberek, B. Banecki, A. Blaszczak and G. Wegrzyn for reading and commenting on this manuscript. This work was supported by a grant from the Polish State Committee for Scientific Research.

References

Alfano, C. and McMacken, R. (1989a). Ordered assembly of nucleoprotein structures at the bacteriophage λ replication origin during the initiation of DNA replication. J. Biol. Chem. **264**, 10699–10708.

Alfano, C. and McMacken, R. (1989b). Heat shock protein-mediated disassembly of nucleoprotein structures is required for initiation of bacteriophage λDNA replication. J. Biol. Chem. **264**, 10709–10718.

Baker, T.A. and Wickner, S.H. (1992). Genetics and enzymology of DNA replication in *Escherichia coli*. Annu. Rev. Genet. **26**, 447–477.

Banecki, B., and Zylicz, M. (1996). Real time kinetics of the DnaK/DnaJ/GrpE molecular chaperone machine action. J. Biol. Chem. **271**, 6137–6143.

Chattoraj, D.K. (1995). Role of molecular chaperones in the initiation of plasmid DNA replication. Genet. Eng. (NY) **17**, 81–98.

Chong, J.P.J., Mahbubani, H.M., Khoo, C.-Y. and Blow, J.J. (1995). Purification of MCM-containing complex as a component of the DNA replication licensing system. Nature, **375**, 418–421.

DasGupta, S., Mukhopadhyay, G., Papp, P.P., Lewis, M.S. and Chattoraj, D.K. (1993). Activation of DNA binding by the monomeric form of P1 replication initiator RepA by heat shock proteins DnaJ and DnaK. J. Mol. Biol. **232**, 23–34.

Dodson, M., McMacken, R. and Echols, H. (1989). Specialized nucleoprotein structures at the origin of replication of bacteriopghage λ: protein association and disassociation reactions responsible for localized initiation of replication. J. Biol. Chem. **264**, 10719–10725.

Echols, H. (1990). Nucleoprotein structures initiating DNA replication, transcription, and site-specific recombination. J. Biol. Chem. **265**, 14697–14700.

Fayet, O.T., Louarn, J.-M. and Georgopoulos, C. (1986). Suppression of the Escherichia coli dnaA46 mutation by amplification of the groES and groEL genes. Mol. Gen. Genet. **202**, 435–445.

Georgopoulos, C. and Welch, W.J. (1993). Role of the major heat shock proteins as molecular chaperones. Annu. Rev. Cell Biol. **9**, 601–634.

Georgopoulos, C., Ang. D., Liberek, K. and Zylicz, M. (1990). Properties of the Escherichia coli heat shock proteins and their role in bacteriophage λ growth. In Stress protein in biology and medicine (ed. Morimoto, R.I., Tissieres, A. and Georgopoulos, C.). Cold Spring Harbor Laboratory Press, pp. 191–221.

Georgopoulos, C., Liberek, K., Zylicz, M. and Ang, D. (1994). Properties of the heat shock proteins of Escherichia coli and the autoregulation of the heat shock response. In The biology of the heat shock proteins and molecular chaperones (ed. Morimoto, R.I., Tissieres, A. and Georgopoulos, C.). Cold Spring Harbor Laboratory Press, pp. 209–249.

Hartl, F.U. (1996). Molecular chaperones in cellular protein folding. Nature **381**, 571–580.

Hupp, T.R. and Kaguni, J.M. (1993). Activation of mutant forms of DnaA protein of Escherichia coli by DnaK and GrpE proteins occurs prior to DNA replication. J. Biol. Chem. **268**, 13143–13150.

Hwang, D.S., Crooke, E. and Kornberg, A. (1990). Aggregated DnaA protein is dissociated and activated for DNA replication by phospholipase or DnaK protein. J. Biol Chem. **265**, 19244–19248.

Ishiai, M., Wada, C., Kawasaki, Y. and Yura, T. (1994). Replication initiator protein RepE of mini-F plasmid: functional differentiation between monomers (initiator) and dimers (autogenous repressor). Proc. Natl. Acad. Sci. USA **91**, 3839–3843.

Jordan, R. and McMacken, R. (1995). Modulation of the ATPase activity of the molecular chaperone DnaK by peptides and the DnaJ and GrpE heat shock proteins. J. Biol. Chem. **270**, 4563–4569.

Kaguni, J.M. and Kornberg, A. (1984). Replication initiated at the origin (oriC) of the E. coli chromosome reconstituted with purified enzymes. Cell **38**, 183–190.

Kawasaki, Y., Wada, C. and Yura, T. (1990). Role of Escherichia coli heat shock proteins DnaK, DnaJ and GrpE in mini-F plasmid replication. Mol. Gen. Genet. **220**, 277–282.

Konieczny, I. and Marszalek, J. (1995). The requirement for molecular chaperones in lambda DNA replication is reduced by the mutation in lambda P gene which weakens the interaction between lambda P protein and DnaB helicase. J. Biol. Chem. **270**, 9792–9799.

Kruklitis, R., Welty, D.J. and Nakai, H. (1996). ClpX protein of Escherichia coli activates Mu transposase in the strand transfer complex for initiation of Mu DNA synthesis. EMBO J. **15**, 935–944.

Learn, B., Karzai, A.W. and McMacken, R. (1993). Transcription stimulates the establishment of bidirectional lDNA replication in vitro. Cold Spring Harbor symposia on quantitative biology, vol. LVIII. Cold Spring Harbor Laboratory Press, pp. 389–402.

Learn, B.A., Um, S.-J., Huang, L., and McMacken, R. (1997). Cryptic single-stranded-DNA binding activities of the phage λP and Escherichia coli DnaC replication proteins facilitate the transfer of E. coli DnaB helicase onto DNA. Proc. Natl. Acad. Sci. USA **94**, 1154–1159.

Levchenko, I,. Luo, L. and Baker, T. A. (1995). Disassembly of the Mu transposase tetramer by the ClpX chaperone. Genes Dev. **9**, 2399–2408.

Liberek, K., Georgopoulos, C. and Zylicz, M. (1988). Role of the Escherichia coli DnaK and DnaJ heat shock proteins in the initiation of bacteriophage lDNA replication. Proc. Natl. Acad. Sci. USA **85**, 6632–6636.

Liberek, K., Marszalek, J., Ang, D., Georgopoulos, C. and Zylicz, M. (1991). Escherichia coli DnaJ and GrpE heat shock proteins jointly stimulate ATPase activity of DnaK. Proc. Natl. Acad. Sci. USA **88**, 2874–2878.

Madine, M.A., Khoo, C.-Y., Mills, A.D. and Laskey, R.A. (1995). MCM3 complex required for cell cycle regulation of DNA replication in vertebrate cells. Nature **375**, 421–424.

Mensa-Wilmot, K., Carrol, K. and McMacken, R. (1989). Transcriptional activation of bacteriophage λDNA replication in vitro. Regulatory of histone-like protein HU of Escherichia coli. EMBO J. **8**, 2393–2398.

Mhammedi-Alaoui, A., Pato, M., Gama, M.-J. and Toussaint, A. (1994). A new component of bacteriophage Mu replicative transposition machinery: the Escherichia coli ClpX protein. Mol. Microbiol. **11**, 1109–1116.

Milarski, K.L. and Morimoto, R.I. (1986). Expression of human Hsp70 during the synthetic phase of cell cycle. Proc. Natl. Acad. Sci. USA **83**, 9517–9521.

Missiakas, D., Schwager, F., Betton, J.-M., Georgopoulos, C., and Raina, S. (1996). Identification and characterisation of the HsIV HsIU (ClpQ ClpY) proteins involved in overall proteolysis of misfolded proteins in Escherichia coli. EMBO J. **15**, 6899–6909.

Nakai, H. and Kruklitis, R. (1995). Disassembly of the bacteriophage Mu transposase for the initiation of Mu DNA replication. J. Biol. Chem. **270**, 19591–19598.

Rowley, N., Prip-Buus, C., Westermann, B., Brown, C., Schwartz, E., Barrell, B. and Neupert, W. (1994). Mdj1, a novel chaperone of the DnaJ family is involved in mitochondrial biogenesis and protein folding. Cell **77**, 249–259.

Sakakibara, Y. (1988). The dnaK gene of Eschericha coli functions in initiation of chromosome replication. J. Bacteriol. **170**, 972–979.

Skowyra, D., McKenney, K. and Wickner, S.H. (1995). Function of molecular chaperones in bacteriophage and plasmid DNA replication. Semin. Virol. **6**, 43–51.

Sozhamannan, S. and Chattoraj, D.K. (1993). Heat shock proteins DnaJ, DnaK and GrpE stimulate P1 plasmid replication by promoting initiator protein binding to the origin. J. Bacteriol. **175**, 35463555.

Stillman, B. (1994). Initiation of chromosomal DNA replication in Eucaryots: lessons from lambda. J. Biol. Chem. **269**, 7047–7050.

Szalewska, A., Wegrzyn, G. and Taylor, K. (1994). Neither absence nor excess of λO initiator-digesting ClpXP protease affects λ plasmid or phage replication in Escherichia coli. Mol. Microbiol. **13**, 469–474.

Taylor, K. and Wegrzyn, G. (1995). Replication of coliphage lambda DNA. FEMS Microbiol. Rev. **17**, 109–119.

Tilly, K. and Yarmolinsky, M. (1989). Participation of *Escherichia coli* heat shock proteins DnaJ, DnaK and GrpE in P1 plasmid replication. J. Bacteriol. **171**, 6025–6029.

Ueguchi, C., Kakeda, M., Yamada, H. and Mizuno, T. (1994). An analogue of the DnaJ molecular chaperone in *Escherichia coli*. Proc. Natl. Acad. Sci. USA **91**, 1054–1058.

Wawrzynów, A. and Zylicz, M. (1995). Divergent effect of ATP on the binding of the DnaK and DnaJ chaperones to each other, or to their various native and denatured protein substrates. J. Biol. Chem. **270**, 19300–19306.

Wawrzynów, A., Wojtkowiak, D., Marszalek, J., Banecki, B., Jonsen, M., Graves, B., Georgopoulos, C. and Zylicz, M. (1995a). The ClpX heat-shock protein of *Escherichia coli*, the ATP-dependent substrate specificity component of ClpP-ClpX protease, is a novel molecular chaperone. EMBO J. **14**, 1867–1877.

Wawrzynów, A., Banecki, B., Wall, D., Liberek, K., Georgopoulos, C. and Zylicz, M. (1995b). ATP hydrolysis is required for the DnaJ-dependent activation of DnaK chaperone for binding to both native and denatured protein substrates. J. Biol. Chem. **270**, 19307–19311.

Wawrzynów, A., Banecki. B., and Zylicz, M. (1996). The ClpATPases define a novel class of molecular chaperones. Mol. Microbiol. **21**, 895–899.

Wegrzyn, A., Wegrzyn, G. and Taylor, K. (1995). Protection of coliphage λO initiator protein from proteolysis in the assembly of the replication complex *in vivo*. Virology **207**, 179–184.

Wickner, S., Hoskins, J. and McKenney, K. (1991). Function of DnaJ and DnaK as chaperones in origin-specific DNA binding by RepA. Nature **350**, 165–167.

Wickner, S., Skowyra, D., Hoskins, J., McKenney, K. (1992). DnaJ, DnaK and GrpE heat shock proteins are required in *ori*P1 DNA replication solely at the RepA monomerization step Proc. Natl. Acad. Sci. USA **89**, 10345–10349.

Wickner, S., Gottesman, S., Skowyra, D., Hoskins, J., McKenney, K. and Maurizi, M. (1994). A molecular chaperone, ClpA, functions like DnaK and DnaJ. Proc. Natl. Acad. Sci. USA **91**, 12218–12222.

Wild, J., Altman, E., Yura, T. and Gross, C.A. (1992). DnaK and DnaJ heat shock proteins participate in protein export in *Escherichia coli*. Genes Dev. **6**, 1165–1172.

Wojtkowiak, D., Georgopoulos, C. and Zylicz, M. (1993). Isolation and characterization of ClpX, new ATP-dependent specificity component of the Clp protease of *Escherichia coli*. J. Biol. Chem. **268**, 22609–22617.

Yamamoto, T., McIntyre, J., Sell., S.M., Georgopoulos, C., Skowyra, D. and Zylicz, M. (1987). Enzymology of the pre-priming steps in ldvDNA replication. J. Biol. Chem. **262**, 7996–7999.

Ziemienowicz, A., Skowyra, D., Zeilstra-Ryalls, J., Fayet, O., Georgopoulos, C. and Zylicz, M. (1993). Both the *Escherichia coli* chaperone systems, GroEL/GroES and DnaK/DnaJ/GrpE can reactivate heat-treated RNA polymerase. J. Biol. Chem. **268**, 25425–25431.

Ziemienowicz, A., Zylicz, M., Floth, C., and Hubscher, U. (1995). Calf thymus Hsc70 protein protects and reactivates prokaryotic and eukaryotic enzymes. J. Biol. Chem. **270**, 15479–15484.

Zylicz, M. (1993). The *Escherichia coli* chaperones involved in DNA replication. Phil. Trans. R. Soc. London. B **339**, 271–278.

Zylicz, M., Ang, D., Liberek, K. and Georgopoulos, C. (1989). Initiation of lDNA replication with purified host- and bactriophage-encoded proteins: role of the DnaK, DnaJ and GrpE heat shock proteins. EMBO J. **8**, 1601–1608.

■ *Alicja Wawrzynów and Maciej Zylicz:*
Division of Biophysics
Department of Molecular and Cellular Biology
University of Gdansk
80-822 Gdansk, Poland
Tel. 48 58 319 222
Fax. 48 58 310 072
E-mail: wawrzynow@biotech.univ.gda.pl

Ribosome-associated chaperones and protein synthesis: molecular machines catalysing protein targeting, folding and assembly

The process of polypeptide translation, intracellular targeting and folding is a highly complex biochemical reaction. During translation, several events occur while the polypeptide is present as a ribosome-bound nascent chain that determine its ultimate fate in the cell. First, misfolding and aggregation of nascent chains are prevented. Secondly, the correct cellular location of the emerging nascent chain is determined based on the presence of targeting sequences such as an N-terminal signal sequence. Targeting to the correct cellular location may require translocation across intracellular membranes such as those of the endoplasmic reticulum (ER) and mitochondria. In order for this to occur, nascent polypeptides must be maintained in an unfolded state. Finally, polypeptides must be folded to their correct native state or be assembled into homo- or hetero-oligomeric complexes.

In recent years a number of cellular factors have been identified that play important roles in the regulation of folding and translocation of nascent chains. Many of these factors are molecular chaperones, including members of the HSP70, DnaJ (HSP40) and chaperonin families. As the elongating polypeptide chain emerges from the ribosome, other factors, such as the signal recognition particle (SRP), also begin to interact with it in order to effect specific targeting decisions. Finally, the nascent polypeptide chain may undergo an ordered series of interactions with molecular chaperones that mediate folding and prevent aggregation reactions. In addition to their roles in folding, it has become apparent that molecular chaperones also have roles in the initiation and elongation of translation itself, suggesting the possibility of feedback mechanisms between the folding machinery and the translational apparatus. In this review, we will summarize our current knowledge of the role molecular chaperones and ribosome-bound factors play in the translation, targeting and folding of nascent polypeptides emerging from the ribosome.

■ Regulation of translation by ribosome-bound molecular chaperones

Two reports have highlighted the role of molecular chaperones belonging to the DnaJ family (Zhong and Arndt, 1993) and the HSP70 family (Nelson et al., 1992) in the regulation of protein translation. In both cases, the evidence for a role for these molecular chaperones in translation has come from genetic studies in Saccharomyces cerevisiae; the biochemical basis of their roles in the translation process has yet to be firmly established.

■ The DnaJ homologue Sis1p plays a role in translation initiation

DnaJ proteins have been demonstrated to be involved in a variety of cellular processes (Cyr et al., 1994; Hartl, 1996). In addition to their function as molecular chaperones, DnaJ proteins generally act as regulators of the ATPase activity and peptide-binding activity of the HSP70s. In the yeast Saccharomyces cerevisiae, SIS1 is an essential gene that encodes a DnaJ homologue that was originally identified as a high-copy suppressor of mutations in the SIT4 gene that encodes a serine/threonine kinase (Luke et al., 1991). The protein product of SIS1, Sis1p, is localized to the cytosol and shares several features with those members of the DnaJ family known to be molecular chaperones. Sis1p contains the absolutely conserved J-domain present in all DnaJ proteins and known to be necessary for interaction with HSP70 (Wall et al., 1994; Szabo et al., 1996). Sis1p also contains a glycine/phenylalanine-rich sequence and the less well conserved C-terminal domain, but, like the human cytosolic Hdj1 (HSP40) (Ohtsuka, 1993; Frydman et al., 1994) lacks the zinc finger domain found in the yeast cytosolic DnaJ homologue Ydj1p (Caplan and Douglas, 1991; Atencio and Yaffe, 1992) and in Escherichia coli DnaJ (Szabo et al., 1996). Sis1p also contains a region of GGMP repeats, similar to sequences located close to the peptide-binding site of several cytosolic HSP70s (Boorstein et al., 1994) and also found in the Hsc70-interacting protein (HIP), a cochaperone of the eukaryotic HSP70 system (Hohfeld et al., 1995).

Sis1p is associated mainly with free 40S ribosome subunits and small polysomes, but may also be associated with the 60S ribosomal subunit. Temperature-sensitive mutants of SIS1 have a phenotype suggesting a defect in the initiation of translation (Zhong and Arndt, 1993). At the non-permissive temperature, sis1 strains rapidly accumulate inactive 80S ribosomes and have decreased amounts of polysomes. A similar phenotype is observed in mutants of GCD1 and PRT1, probable components of eukaryotic initiation factor 2B (Cigan et al., 1991) and eukaryotic initiation factor 3 (Keierleber et al., 1986), respectively. Biochemical characterization of the 80S ribosomes that accumulate in the sis1 background revealed that they are either 80S preinitiation complexes or 80S couples (inactive 80S ribosomes not bound to mRNA), indicating that initiation is greatly reduced. Interestingly, alteration of the structure or levels of 60S ribosomes by mutation of SOS1, a yeast homologue of the L35 ribosomal subunit, or deletion of SPB2, which encodes ribosomal protein L46, suppresses the temperature-sensitive phenotype of a sis1 strain. Both the temperature-sensitive block in initiation that results in the accumulation of 80S ribosomes and the decrease in polysome levels are suppressed by mutation in SOS1. This same mutation in the 60S ribosomal subunit suppresses mutations in PAB1, the yeast poly(A)-binding protein that is required for the normal initiation of translation (Sachs and Davis, 1990). Consistent with the notion that Sis1p and Pab1p function in the same, or in different but interacting, initiation steps, overexpression of Sis1p inhibits the growth rate of a pab1 strain.

At present there is no evidence that the Sis1p-dependent event(s) involved in translation initiation requires the activity of cytoplasmic HSP70 family members. Sis1p has been reported to interact directly with the HSP70 Ssa1p to stimulate its ATPase activity (Cyr et al., 1994), but a function of Ssa1p in translation initiation has not been demonstrated. Yeast contain four members of the Ssa family and two members of the Ssb family (Craig et al., 1994) and very little is known about which yeast DnaJ homologues interact functionally with these HSP70s. The Ssa and Ssb families of HSP70 proteins are functionally distinct with respect to their regulation by the DnaJ homologue, Ydj1p. ATP hydrolysis by the Ssa family is efficiently stimulated by Ydj1p, whereas Ydj1p has little effect on ATP hydrolysis by members of the Ssb family (Cyr et al., 1992; Cyr and Douglas, 1994). In addition, only the Ssa family, and not the Ssb family, can cooperate with Ydj1p to prevent protein aggregation in vitro (Cyr, 1995). Members of the Ssb family are involved in translation elongation (see below), but there are no data regarding the ability of Sis1p to function with Ssb1p/2p in this process. It is possible that the presence of Sis1p on 40S subunits and polysomes may play a role in targeting of Ssb1p/2p to the ribosome. However, this view is probably over simplistic as the regulation of expression of these genes is different. SIS1 is heat inducible (Luke et al., 1991) whereas expression of the

Ssb family is down-regulated following heat shock (Werner-Washburne et al., 1989). In addition, recent work also indicates that, in common with a number of heat shock proteins in yeast, regulation of transcription of SIS1 is regulated by Ssa1p/2p function (Zhong et al., 1996). Clearly, further work is necessary to elucidate the role, if any, of HSP70s in cooperating with Sis1p both in translation initiation and in general.

■ The Ssb family of HSP70s plays a role in translation elongation

Genetic studies have revealed a role for the Ssb family of HSP70 proteins in the elongation step of translation (Nelson et al., 1992; Craig et al., 1994). Sucrose density gradient centrifugation indicates that the bulk of the Ssb1p/2p in yeast is associated with translating ribosomes, including both monsomes and polysomes. Treatment of polysomes with RNAase to destroy the mRNA linking the individual ribosomes results in collapse of the polysomes into a single 80S monosome peak. There is a concomitant change in the distribution of Ssb1p/2p suggesting a direct association of Ssb1p/2p with ribosomes. Furthermore, there is no association of Ssb1p/2p with monosomes obtained from cell lysates in which azide treatment has been used to deplete ATP levels and thereby prevent reinitiation. Puromycin treatment of yeast cells to release bound nascent chains does not affect the polysome distribution in sucrose gradients but dramatically reduces the amount of Ssb1p/2p migrating with the polysomes. This effect is consistent with the notion that Ssb1p/2p bind directly to the nascent chain as it emerges from the ribosome, as has been demonstrated for HSP70 proteins in mammalian systems (Beckmann et al., 1990; Frydman et al., 1994). Direct binding of Ssb1p/2p to the ribosome in such a way that nascent chain release disrupts binding is also consistent with this observation.

Support for a direct role of Ssb1p/2p in the translation process was provided by an analysis of Ssb deletion mutants lacking both members of this HSP70 family (Nelson et al., 1992). ssb1 ssb2 cells demonstrate hypersensitivity to a variety of drugs that act as inhibitors of protein synthesis, including paromomycin, hygromycin B and G418, all of which are aminoglycosides and act on the 40S ribosomal subunit to inhibit polypeptide chain elongation. In addition, these cells exhibit sensitivity to verrucarin A, a member of the trichothecene family of antibiotics that act primarily on the 60S ribosomal subunit and block polypeptide bond formation. Conversely, the ssb1 ssb2 cells are not hypersensitive to cycloheximide or anisomycin which inhibit peptide bond formation. Further evidence for a direct link between Ssb1p/2p and the translation process was provided by a multicopy suppressor screen of the ssb1 ssb2 strain. Following transformation, selection for rapid growth at a semi-permissive temperature for this strain resulted in the isolation of one gene, in addition to SSB1 and SSB2 themselves. This gene, HSB1 (HSP70 subfamily B suppressor) was a previously unidentified gene having significant sequence homology with the translation elongation factor EF-1, a highly conserved protein found in all eukaryotes. Overall identity between Hbs1p and EF-1 was 33% with similarity extending over the entire length of EF-α1 and including the three highly conserved sequence blocks found in many GTP-binding proteins. Thus, Hbs1p may be functionally related to EF-α1 and directly involved in translation, although its role is completely unknown. Whether Ssb proteins act in translation to prevent nascent chains from misfolding and thereby inhibiting the translation process owing to aggregation, or whether they are involved in a feedback mechanism interacting directly with elongation factors is unclear. How Ssb1p/2p are specifically recruited to the ribosome, and whether a specific DnaJ homologue or a ribosomal component are necessary for this effect remains to be determined.

■ Nascent chain recognition and targeting

As nascent chains begin to emerge into the eukaryotic cytosol, two complexes engage the nascent chain, namely the nascent polypeptide-associated complex (NAC) and the signal recognition particle (SRP). Although neither of these complexes have been demonstrated to be molecular chaperones, they are critical in effecting targeting decisions and in this section of this review we will briefly discuss their roles.

■ Nascent polypeptide-associated complex (NAC)

NAC is the first cytosolic factor to be encountered by a nascent polypeptide as it emerges from the ribosome (Wiedmann et al., 1994; Wickner, 1995). Genes encoding the subunits of this complex have been identified in all eukaryotes examined; however, no homologue of NAC has been identified in prokaryotes. NAC was identified by using truncated nascent chains carrying a photoactivable cross-linker to detect proteins in close proximity to the nascent chain (Wiedmann et al., 1994). Several cross-linked polypeptides were detected, including SRP54 (see below) and a 33 kDa polypeptide. This 33 kDa polypeptide was shown to be one component of a heterodimeric complex termed NAC. NAC consists of a 23 kDa subunit (NAC, corresponding to the 33 kDa cross-linked polypeptide) and an 18 kDa subunit (NAC). NAC can be cross-linked to regions of nascent polypeptides as close as 17 residues and up to 100 residues from the peptidyl transferase centre and protects them from protease digestion (Wang et al., 1995). NAC may therefore represent part of the 'ribosome tunnel' through which a nascent polypeptide has been proposed to traverse before emerging into the cytosol (Wiedmann et al., 1994).

NAC is abundant in the cytosol but associates only with ribosome-bound nascent chains and not with released chains. A major function of NAC appears to be the regulation of cotranslational translocation into the ER. Depletion of NAC from ribosome-bound nascent chains *in vitro* results in loss of the fidelity of SRP-mediated translocation across the ER membrane (Wiedmann *et al.*, 1994; Lauring *et al.*, 1995). In the absence of NAC, nascent polypeptides lacking signal sequences can be targeted to the ER membrane and translocated, ableit at lower efficiency than signal sequence-containing polypeptides. Readdition of cytosol or purified NAC to the NAC-depeleted ribosome–nascent chain complexes restores the fidelity of SRP-mediated translocation. Furthermore, in the absence of SRP, NAC prevents all ribosome–nascent chain complexes from binding to the ER membrane (Lauring *et al.*, 1995). NAC appears to have the ability to recognize essentially all regions of nascent polypeptides, but has only a low affinity for signal sequences (Wiedmann *et al.*, 1994). Binding of NAC to nascent polypeptides lacking signal sequences may thus prevent non-specific binding by SRP and also block a site on the ribosome necessary for interaction with the ER membrane.

Many questions remain regarding the function of NAC. Cross-linking of nascent chains to both subunits of NAC has been demonstrated (Wiedmann *et al.*, 1994), but the contribution of the individual subunits in forming a binding site(s) has yet to be determined. Indeed, it is not clear that NAC binds directly to nascent chains or is simply in close proximity to them. Whether NAC acts as a molecular chaperone and plays a role in the folding of nascent chains destined for a cytosolic location also remains to be analysed. Given that NAC is one of the first cytosolic factors to recognize a nascent chain, NAC may be able to act as a recruitment factor for the cytosolic chaperone machinery such as HSP70 and DnaJ proteins (Hartl, 1996).

■ The signal recognition particle (SRP)

Nascent chains destined for either secretion to the cell surface or residence in the ER must be translocated across the ER membrane. In mammalian cells, synthesis of such proteins occurs on ribosomes that are bound to the rough ER membrane and protein translocation and integration occur in a cotranslational manner. In contrast, synthesis of all other proteins is thought to occur on ribosomes that are free in the cytosol. The signal recognition particle (SRP), a cytoplasmic ribonucleoprotein complex, and its receptor in the ER membrane, SR (also called docking protein), are required for targeting of nascent chains bearing signal sequences to the ER membrane. The scope of this review is insufficient to describe the complex molecular details of this process and we will only summarize this process briefly. For a more detailed overview of this subject the reader is referred to a recent review (Walter and Johnson, 1994).

In mammalian cells, SRP consists of a 7S RNA molecule (SRP RNA) and six distinct polypeptides with molecular masses of 9, 14, 19, 54, 68 and 72 kDa. The SRP RNA forms the structural backbone on to which the SRP proteins assemble. Only SRP54 has identifiable sequence homology to other known proteins. SRP54 contains a central GTPase domain which is most closely related to a GTPase domain in the SRα subunit but only distantly related to other known GTPases. The GTPase domain is flanked on its N-terminus by the N-domain and on its C-terminus by the M-domain, which contains numerous methionine residues and is connected to the rest of SRP54 by a highly flexible hinge region. The site of signal sequence recognition has been mapped to a 6 kDa fragment located on the C-terminal portion of the M-domain. Most of the methionine residues in the M-domain have been predicted to lie on one face of a group of highly amphipathic helices that may form a signal sequence binding groove (Bernstein *et al.*, 1989; Hann *et al.*, 1989). ER signal sequences generally consist of a core of 8–12 hydrophobic amino acids that are presumed to form an α helix (von Heijne, 1985), but their amino acid sequences are not conserved and thus each must have a different shape. Methionine side chains are more flexible than other comparably hydrophobic residues which would enable them to provide a hydrophobic environment with sufficient plasticity to allow signal sequence binding despite their heterogeneity in amino acid sequence.

SRP has a low affinity for ribosomes that are not engaged in translation but its affinity is increased by three to four orders of magnitude when a signal sequence is expressed and exposed outside the ribosome as part of a nascent chain. SRP probably cycles between ribosome-bound and -free states, scanning the nascent polypeptide for signal sequences. Cells possess about one SRP for every ten ribosomes, suggesting that SRP does not remain bound to any given ribosome waiting for a signal sequence to emerge. SRP may transiently associate with the ribosome during each elongation cycle and remains bound only if a signal sequence associates with it. If a signal sequence is not detected, it may rapidly dissociate and move to another ribosome. The specificity of this scanning function would be facilitated by NAC covering those nascent chains that do not contain a signal sequence.

Following recognition of the signal sequence by SRP, translation elongation is arrested. This function is associated with the SRP9 and SRP14 polypeptides, which form a complex with one another and are bound to the opposite end of the SRP RNA to those regions associated with SRP54. Thus, SRP has one end associated with signal sequence recognition and the other end involved in elongation arrest. Elongation arrest is not essential for protein translocation *in vitro*, but most signal sequence-containing polypeptides lose their ability to be translocated if elongation proceeds too far. Therefore, translation arrest probably helps to maintain the fidelity of translocation *in vivo*.

Following elongation arrest, the complex of the ribosome, the nascent chain with its signal sequence and the SRP interact with the SR, a heterodimeric integral membrane protein composed of SRα and SRβ. This leads to release of SRP concurrent with the formation of the ribosome–translocon junction and delivery of the signal sequence into the aqueous environment of the translocon, the translocation machinery in the ER membrane. Translational arrest is also released and the growing polypeptide chain is now transferred across the ER membrane. Targeting to the ER membrane involves multiple

GTPases in SRP54, SRα and SRβ which function as molecular switches to provide unidirectionality to this complex process. Recent work indicates that a ribosomal component is also involved and acts to promote binding of GTP to SRP54, concomitantly switching SRP into its high affinity state for SR (Bacher et al., 1996). SRP and the SR act catalytically in this process to direct the ribosome to the ER membrane. They are then released from the ribosome and do not form part of the ribosome–translocon junction. For a more detailed overview of the components of the translocon and the translocation process itself, the reader is referred to a recent review (Rapoport et al., 1996). Roles played by molecular chaperones during post-translational protein translocation into the ER and into mitochondria are reviewed elsewhere in this volume (pp. 499 and 506).

■ Homologues of SRP and SRα exist in *Escherichia coli*

Translocation of proteins into the periplasmic space of bacteria has long been considered to be an evolutionarily equivalent process to that of translocation into the ER of eukaryotic cells. Unlike eukaryotes, in which this process occurs predominantly cotranslationally, the majority of exported proteins in *E. coli* utilize a pathway of post-translational translocation mediated by the cytoplasmic chaperone SecB, which recognizes mostly internal regions of the precursor polypeptide chain (Hardy and Randall, 1991, 1993; Randall, 1992). However, evidence that the two processes are indeed evolutionarily related has been provided by the discovery of homologues of eukaryotic SRP RNA, SRP54 and SR in *E. coli*. The known components of this system consist of a 4.5S RNA and the proteins Ffh (fifty-four homologue, also known as p48) and FtsY. Ffh was identified as a protein of unknown function that had a highly similar sequence and domain structure to mammalian SRP54 (Bernstein et al., 1989; Romisch et al., 1989). FtsY was identified as an SRα homologue based on sequence similarity that extends through the N-domain and GTPase domains of SRα and SRP54 (Bernstein et al., 1989; Romisch et al., 1989). The N-terminal portion of FtsY does not have any sequence similarity to the N-terminal region of SRα that mediates interaction with SRβ. No homologue of the membrane-bound SR subunit has yet been identified in *E. coli*, but FtsY itself has been reported to be associated peripherally with the cytoplasmic membrane (Lurink et al., 1994). Considerable evidence indicates that *E. coli* 4.5S RNA and Ffh are functionally homologous to the eukaryotic SRP. 4.5S RNA and Ffh form a ribonucleoprotein complex that specifically recognizes nascent polypeptides carrying signal sequences (Ribes et al., 1990). Ffh can be specifically cross-linked to signal sequences in crude extracts and it can replace mammalian SRP54 functionally in signal sequence recognition when it is assembled with mammalian SRP RNA and proteins into a chimeric particle (Bernstein et al., 1993). Depletion of Ffh *in vivo* leads to defects in translocation (Ribes et al., 1990; Phillips and Silhavy, 1992) and depletion of FtsY leads to an accumulation of presecretory proteins (Lurink et al., 1994). In a manner strikingly similar to the mammalian system, *E. coli* SRP and FtsY form a complex *in vitro* in the presence of non-hydrolysable guanine analogues, and GTP hydrolysis is stimulated significantly when they are combined (Miller et al., 1994). Thus, bacteria have a signal recognition and protein targeting apparatus that functionally resembles the SRP/SR system of eukaryotic cells. The functional relationships, if any, between *E. coli* SRP, FtsY, SecB and the components of the prokaryotic translocation apparatus (see Rapoport et al., 1996) remain to be elucidated.

■ Interaction of nascent chains with chaperones on the pathway to folding

Nascent chains bound to the ribosome cannot reach their final native state as they are topologically restricted and lack the complete polypeptide information necessary for their folding. Most proteins probably fold in a domain-wise fashion, thus, until a complete polypeptide domain has been synthesized, a nascent chain will not assume a stably folded conformation. During translation, nascent polypeptides expose hydrophobic residues to the cellular milieu that would otherwise be buried in the hydrophobic core of the native polypeptide. Given that protein concentrations in the cytosol may be as high as 340 mg/ml, nascent polypeptides are therefore at great risk of making incorrect interactions with other cellular components, leading to aggregation reactions rather than folding. Complicating this problem further, translating ribosomes are organized into polysomes and the resulting localized concentration of unfolded nascent chains would be higher than that of the averaged ribosome concentration (approximately 35 μM) (Hartl, 1996). Cells must therefore maintain a significant population of unfolded polypeptide chains in a folding or translocation-competent state as part of their normal cellular metabolism.

In recent years, it has become clear that molecular chaperones engage the nascent chain very early in the translation process. Furthermore, the vectorial nature of protein translation apparently represents a fundamentally different folding process to that observed in *in vitro* refolding experiments where complete unfolded polypeptides refold in dilute solutions containing chaperones. In these experiments, folding is a stochastic process with free partitioning of folding intermediates to those chaperone systems able to bind them. Our current paradigm of chaperone-mediated protein folding in the context of translation is of a coupled process in which the polypeptide is transferred along a pathway of defined chaperone interactions, which prevent aggregation reactions and by controlled binding and release allow the polypeptide to reach its final native state. In the remaining part of this chapter, we will briefly review our current knowledge of chaperone–nascent chain interactions that regulate folding of polypeptides destined to reside in the cytosol.

Chaperone–nascent chain interactions in the prokaryotic cytosol

DnaJ/DnaK

In vitro studies on the ability of the purified *E. coli* chaperones DnaJ, DnaK (HSP70), its nucleotide exchange factor GrpE and the chaperonin GroEL/GroES to cooperate in the refolding of denatured proteins provided important insights into the sequence of polypeptide–chaperone interactions that may occur during translation. Using unfolded bovine rhodanese as a model, DnaK and DnaJ were shown to cooperate with one another to stabilize rhodanese in an intermediate conformation lacking ordered tertiary structure which subsequently could be transferred to GroEL in a GrpE-dependent fashion for folding to the native state (Langer *et al.*, 1992). Studies using heterologous and homologous *in vitro* translation systems have subsequently demonstrated that DnaJ and DnaK are associated with nascent polypeptides (Hendrick *et al.*, 1993; Gaitanaris *et al.*, 1994; Kudlicki *et al.*, 1995; Vysokanov *et al.*, 1995). In the case of DnaJ, cotranslational binding to nascent chains resulted in folding arrest as long as DnaK and GrpE were not present, and prevented post-translational translocation into microsomes and post-translational import into mitochondria of several precursor proteins (Hendrick *et al.*, 1993). These studies point to a general ability of DnaJ proteins to bind to the folding intermediates formed by nascent chains during translation. Analogous to the data from *in vitro* refolding studies, efficient folding of rhodanese upon cell-free translation can be dependent on all five components, in that DnaJ and DnaK must bind the polypeptide before its release from the ribosome in order for efficient transfer to GroEL to occur (Kudlicki *et al.*, 1994, J. Hendrick, unpublished observations).

Although DnaJ and DnaK can be shown to be associated with nascent chains during *in vitro* translation experiments, the generality of their role in the folding of newly synthesized polypeptides *in vivo* remains to be clarified. Null mutants of either *dnaJ* or *dnaK* are viable at 30°C although they grow slowly (Paek and Walker, 1987; Sell *et al.*, 1990), indicating that neither molecular chaperone alone is essential. In *E. coli* that are null mutants of the *rpoH* gene and thus lack σ^{32}, the sigma factor necessary for the transcription of heat shock protein genes, expression of both the DnaK/DnaJ/GrpE and the GroEL/GroES systems *in vivo* is necessary to prevent the aggregation of a large fraction of newly synthesized polypeptides under normal growth conditions (Gragerov *et al.*, 1992). This indicates that loss of both chaperone systems probably has synergistic effects on protein folding. However, at present, little is known about what fraction of all nascent chains *in vivo* must bind DnaJ and DnaK as a prerequisite for their correct folding under normal conditions.

Temperature-sensitive mutations in *dnaK* result in ribosome assembly defects (Alix and Guerin, 1993) indicating that DnaK may also play a role in ribosomal assembly. Furthermore, several reports indicate that DnaJ and DnaK may be directly associated with the ribosome even in the absence of a nascent polypeptide (Gaitanaris *et al.*, 1994; Vysokanov *et al.*, 1995; J. Hendrick, unpublished observations). Whether DnaJ and DnaK play analogous roles in translation regulation in *E. coli* to those played by Sis1p and Ssb1p/2p in *S. cerevisiae* is unknown.

Trigger factor

Recent reports have shown that trigger factor (TF), a 48 kDa protein originally implicated as a chaperone involved in translocation and secretion in *E. coli* (Crooke and Wickner, 1987) is also bound to the ribosome and to nascent chains (Stoller *et al.*, 1995; Valent *et al.*, 1995; Hesterkamp *et al.*, 1996). TF is a member of the FKBP family of peptidyl–prolyl *cis–trans* isomerases (PPIase) (Callebaut and Mornon, 1995), although unlike other PPIases of this family, it is not inhibited in its activity by either FK506 or cyclosporin (Stoller *et al.*, 1995). Salt stripping and puromycin treatment abolished cross-linking of TF with nascent chains, thus TF binds polypeptides only in the context of the ribosome (Stoller *et al.*, 1995; Hesterkamp *et al.*, 1996). TF may bind to the nascent chain prior to the interaction of DnaJ and DnaK (Valent *et al.*, 1995; Hesterkamp *et al.*, 1996). Whether it plays a role in post-translational folding processes is unclear, although it has been reported to be in association with GroEL *in vivo* (Kandror *et al.*, 1995).

GroEL/GroES

Direct binding of GroEL to nascent chains has not been reported. The concentration of GroEL in the *E. coli* cytosol is much lower than the concentration of ribosomes (c. 4 μM GroEL versus c. 35 μM ribosomes) (Ellis and Hartl, 1996), thus it seems unlikely that all proteins require GroEL for their folding. Consistent with this notion, under normal growth conditions the folding of only c. 30% of all cytosolic protein species was affected in a GroEL mutant strain (Horwich *et al.*, 1993). Thus, interaction with GroEL most probably occurs post-translationally and may be restricted to those proteins which are highly aggregation sensitive.

Chaperone–nascent chain interactions in the eukaryotic cytosol

Evidence for a coupled chaperone-mediated protein folding pathway for nascent polypeptides has come from studies of the eukaryotic cytosol using firefly luciferase translated in reticulocyte lysate as a model (Frydman *et al.*, 1994). Following unsynchronized translation of full length luciferase and puromycin treatment of the isolated ribosome–nascent chain complexes, most of the released luciferase polypeptides were found in a high molecular weight complex of approximately 1200 kDa containing Hsc70, Hdj1 (Hsp40) and the chaperonin TRiC (CCT). When the same experiment was performed using a truncated luciferase chain consisting of only the

N-terminal 77 residues of luciferase, nascent luciferase chains were found exclusively in an approximately 200 kDa complex containing Hsc70 but lacking TRiC. In contrast, similar studies using a truncated luciferase chain lacking only the C-terminal 11 amino acid residues revealed that most of the nascent polypeptides were in the 1200 kDa complex containing TRiC. These studies clearly indicated a chain-length dependence of nascent chain–chaperone interactions.

Depletion of Hsc70 from the reticulocyte lysate significantly reduced the binding of luciferase nascent chains to TRiC in an unsynchronized translation mixture, whereas association of nascent chains with Hdj1 was not affected. Depletion of Hdj1 from the lysate decreased the interaction of nascent chains with both Hsc70 and with TRiC, highlighting the role of this DnaJ homologue in loading the nascent chains on to Hsc70. In contrast, depletion of TRiC from the lysate had no effect on the binding of nascent chains to either Hsc70 or Hdj1. Translation of full length luciferase in Hsc70-depleted reticulocyte lysate resulted in a 70% reduction in the final specific activity of luciferase. Readdition of Hsc70, cotranslationally but not post-translationally, restored luciferase folding. Removal of TRiC also resulted in a decrease in the final specific activity of translated luciferase in a manner that could be partially restored if TRiC was added post-translationally. These results imply the presence of an ordered sequence of chaperone– nascent chain interactions in which Hsc70, in cooperation with Hdj1, interacts first with nascent polypeptides emerging from the ribosome. As the polypeptide grows in length, there is subsequent ATP-dependent transfer of the polypeptide to the chaperonin TRiC. Transfer to TRiC may also occur concomitantly with the folding of proteolytically stable domains, as was observed for luciferase (Frydman et al., 1994).

Further analysis of this chaperone-mediated folding pathway has highlighted the difference between folding of newly synthesized polypeptides *in vivo* versus chaperone-mediated refolding of chemically denatured polypeptides *in vitro*. The results of these latter studies have suggested that the main purpose of molecular chaperones is to unfold incorrectly folded or kinetically trapped intermediates to prepare them for another folding trial in the bulk solution (Todd et al., 1994; Weissman et al., 1994). In this model, folding would be governed by the kinetic partitioning of non-native intermediates between different chaperone systems. In contrast, synthesis of polypeptides *in vivo* is a vectorial process and it is not clear that interaction of such polypeptides with chaperones mimics that which occurs during refolding of chemically denatured polypeptides *in vitro*.

This question has recently been addressed by examining the folding of luciferase and actin in reticulocyte lysate following either chemical denaturation or translation (Frydman and Hartl, 1996). Chemically denatured luciferase and actin diluted into the reticulocyte lysate formed complexes with Hsc70, Hdj1 and TRiC. Complex formation with TRiC was ATP dependent, suggesting an initial stabilization of the unfolded polypeptides by Hsc70. Two forms of GroEL able to bind polypeptides but not to release them were used as 'traps' (T-GroEL) (Weissman et al., 1994; Mayhew et al., 1996) for folding intermediates released into the bulk solution. During refolding of chemically denatured luciferase or actin, both polypeptides were efficiently captured and their folding inhibited by the added T-GroELs, even when the polypeptides were first bound to the cytosolic chaperones in the reticulocyte lysate. Thus, refolding of denatured polypeptides in reticulocyte lysate does involve release into the bulk solution of non-native intermediates that require further interaction with chaperone components to complete their folding, consistent with a kinetic partitioning model of chaperone-mediated folding.

A distinctly different form of chaperone interaction was observed when folding of synchronized translation products was followed. When full length actin and luciferase were translated in the presence of T-GroEL, no significant inhibition of either the rate or yield of folded products was observed, indicating a fundamental difference to the refolding of the chemically denatured polypeptides. Notably, ribosome-bound actin nascent chains stabilized by the translation inhibitor cycloheximide were exclusively bound to TRiC and not to T-GroEL. In contrast, release of these incomplete chains from the ribosome with puromycin allowed them to bind to T-GroEL. The ability of TRiC, but not T-GroEL, to interact with the ribosome–nascent chain complex suggested that TRiC is specifically recruited to the ribosome–nascent chain complex. As observed for luciferase (Frydman et al., 1994), interaction of synchronously translated truncated actin chains with various chaperone components in reticulocyte lysate also showed a chain length dependence. A short chain composed of the N-terminal 133 residues was able to interact with Hsc70 but not with TRiC, whereas a longer chain composed of 220 residues interacted with both chaperones. Thus, during translation, chaperones can be recruited into the folding process in a chain length-dependent process, providing the elongating polypeptide chain with a sequestered environment in which it is protected from the cellular milieu. It is believed that this chaperone pathway plays a role in the folding of at least a subset of aggregation-sensitive cytosolic polypeptides, such as actin and tubulin.

■ Future aspects

Molecular chaperones play a variety of roles in translation and folding of newly synthesized polypeptides, as summarized in Fig. 1. A major question to be investigated is how general is the requirement for chaperones in the folding of nascent polypeptides. At present, most of our knowledge of chaperone function comes from *in vitro* studies or from the analysis of cell-free translation reactions. Little is known about their actual role *in vivo*. It is apparent that both prokaryotes and eukaryotes share a general cytosolic chaperone machinery consisting of DnaJ and DnaK (or their eukaryotic homologues) that can interact with nascent polypeptides during translation and may subsequently transfer the polypeptide to a

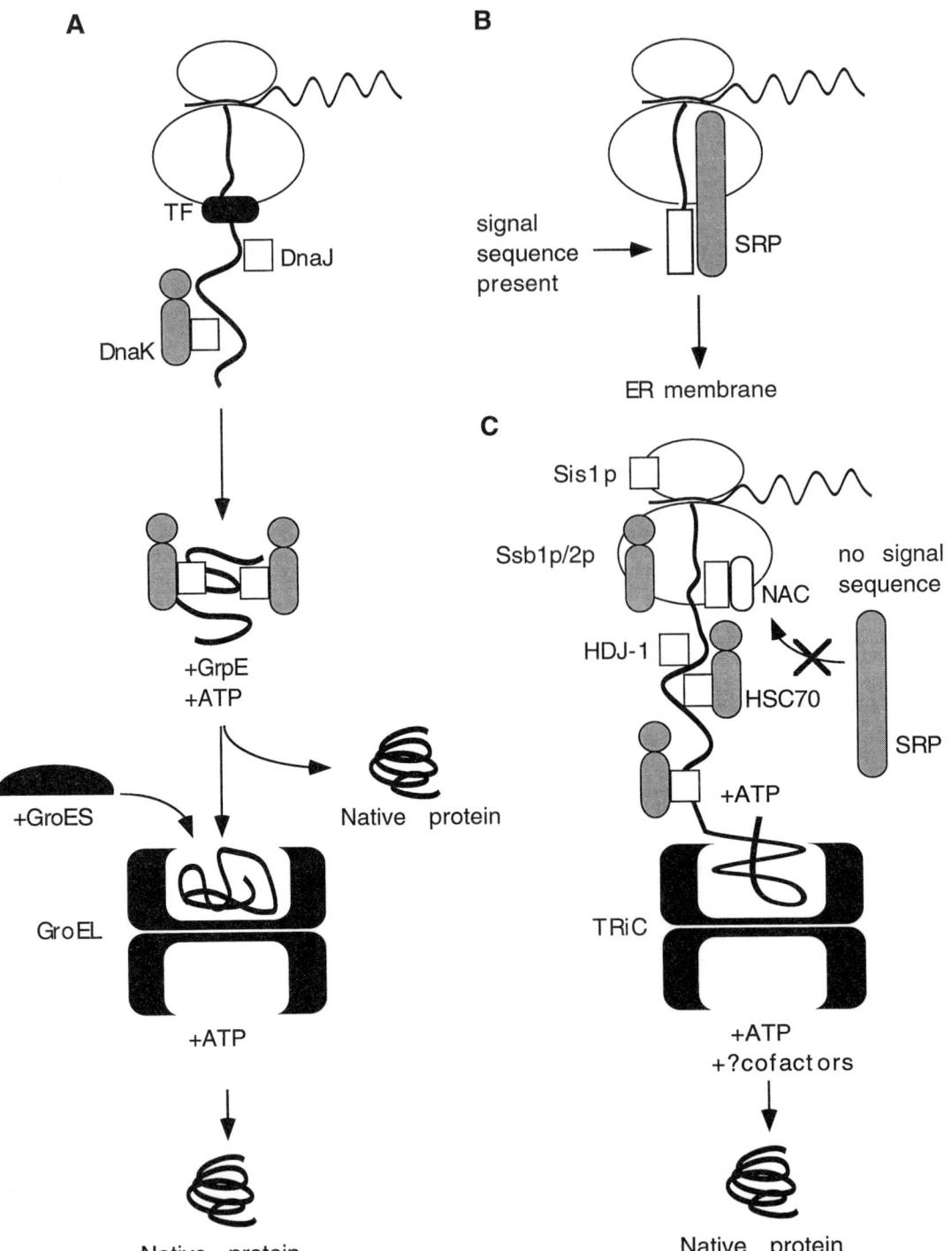

Figure 1. Multiplicity of chaperone interactions in protein translation and folding of newly synthesized polypeptides. (A) In prokaryotes, nascent chains emerging from the ribosome interact with trigger factor (TF) and DnaJ. It is possible that some polypeptides require TF specifically for their folding. Binding of DnaJ targets the nascent chain for interaction with DnaK, the HSP70 homologue of *E. coli*. Nascent chains may be released from the ribosome in association with DnaJ and DnaK and some may be able to reach the native state after ATP- and GrpE-mediated release from this chaperone system. Alternatively, transfer of the newly synthesized polypeptide to the chaperonin GroEL, probably as a partially folded intermediate, occurs post-translationally in an ATP- and GrpE-dependent process. Folding to the final native state or to a state committed to reach native occurs in an ATP-dependent process in the sequestered environment of the chaperonin cavity provided by GroEL and its cofactor, GroES. (B) In eukaryotes, signal sequences emerging from the ribosome interact with SRP. This initiates a targeting cycle in which SRP interacts with its ER receptor, SR, resulting in insertion of the signal peptide into the aqueous environment of the translocon in the ER membrane. The exact role of the SRP in prokaryotes remains to be established and is therefore not shown. (C) In eukaryotes, genetic studies in yeast indicate a role for the DnaJ homologue Sis1p in translation initiation, and a role for the Ssb family of HSP70 proteins in translation elongation. Whether these two components form a functional team remains to be determined. In mammalian cells, nascent chains destined for the cytosol first encounter the heterodimeric complex NAC, which prevents non-specific binding of SRP. Subsequently, interaction of the nascent polypeptide with Hdj1, a DnaJ homologue, recruits Hsc70. In an ATP- and chain length-dependent process, which can occur cotranslationally, some nascent polypeptides are transferred to the chaperonin TRiC for final folding to the native state. It should be noted that the pathways shown in (A) and (C) have model character and may be utilized only by specific subsets of newly-synthesized polypeptides.

chaperonin. However, the spectrum of components that may potentially interact with nascent polypeptides is much greater in eukaryotes and remains to be completely defined. For example, recent release of the complete *S. cerevisiae* genome has revealed the presence of several previously unknown DnaJ homologues and it is at present not defined what role, if any, these play in translation and folding. Other factors, such as HIP, may interact with particular subsets of nascent chains, for example, tyrosine kinases such as pp60^{v-src} (Hohfeld et al., 1995) and may act in targeting nascent chains bound to Hsc70 to other cytosolic chaperone systems such as Hsp90.

Clearly, several parallel pathways of chaperone-mediated protein folding may exist. A major problem that remains to be addressed is how the machinery of protein folding and that of protein degradation are functionally related. As is the case for folding and translocation, unfolded polypeptides have to be maintained in a non-aggregated state to allow degradation to occur (Hayes and Dice, 1996). Recent observations made in *in vitro* translation systems suggest that the chaperone machinery gives nascent polypeptides a chance to fold while simultaneously protecting them from being degraded (Frydman and Hartl, 1996). However, polypeptide chains that are unable to fold partition to the proteolysis machinery, such as the proteasome. It is possible that in eukaryotic cells Hsp90, in cooperation with Hsc70 and other factors, has an important role as a quality control system in regulating the flow of substrates into refolding and degradation pathways (Schneider et al., 1996). Hsp90 seems to act predominantly by a post-translational mechanism and would therefore have access to newly synthesized polypeptides once they have attempted to fold in conjunction with other chaperones. It will be interesting to see whether the components of the folding and degradation pathways are indeed organized in a heirarchical manner, and to understand the cross-talk between them.

References

Alix, J.H. and Guerin, M.F. (1993). Mutant DnaK chaperones cause ribosome assembly defects in *Escherichia coli*. Proc. Natl. Acad. Sci. USA **90**, 9725–9729.

Atencio, D.P. and Yaffe, M.P. (1992). MAS5, a yeast homolog of DnaJ involved in mitochondrial protein import. Mol. Cell. Biol. **12**, 283–291.

Bacher, G., Lutcke, H., Jungnickel, B., Rapoport, T.A. and Dobberstein, B. (1996). Regulation by the ribosome of the GTPase of the signal-recognition particle during protein targeting. Nature **381**, 248–251.

Beckmann, R.P., Mizzen, L.E. and Welch, W.J. (1990). Interaction of Hsp70 with newly synthesized proteins: implications for protein folding and assembly. Science **248**, 850–854.

Bernstein, H.D., Poritz, M.A., Strub, K., Hoben, P.J., Brenner, S. and Walter, P. (1989). Model for signal sequence recognition from amino-acid sequence of 54K subunit of signal recognition particle. Nature **340**, 482–486.

Bernstein, H.D., Zopf, D., Freymann, D.M. and Walter, P. (1993). Functional substitution of the signal recognition particle 54-kDa subunit by its *Escherichia coli* homolog. Proc. Natl. Acad. Sci. USA **90**, 5229–5233.

Boorstein, W.R., Ziegelhoffer, T. and Craig, E.A. (1994). Molecular evolution of the HSP70 multigene family. J. Mol. Evol. **38**, 1–17.

Callebaut, I. and Mornon, J.P. (1995). Trigger factor, one of the *Escherichia coli* chaperone proteins, is an original member of the FKBP family. FEBS Lett. **374**, 211–215.

Caplan, A.J. and Douglas, M.G. (1991). Characterization of YDJ1: a yeast homologue of the bacterial dnaJ protein. J. Cell Biol. **114**, 609–621.

Cigan, A.M., Foiani, M., Hannig, E.M. and Hinnebusch, A.G. (1991). Complex formation by positive and negative translational regulators of GCN4. Mol. Cell. Biol. **11**, 3217–3228.

Craig, E.A., Baxter, B.K., Becker, J., Halladay, J. and Ziegelhoffer, T. (1994). Cytosolic hsp70s of *Saccharomyces cerevisiae*: roles in protein synthesis, protein translocation, proteolysis and regulation. In *The Biology of Heat Shock Proteins and Molecular Chaperones* (ed. Morimoto, R.I., Tissieres, A. and

Georgopoulos, C.). Cold Spring Harbour Laboratory Press, pp. 31–52.

Crooke, E. and Wickner, W. (1987). Trigger factor: a soluble protein that folds pro-OmpA into a membrane-assembly-competent form. Proc. Natl. Acad. Sci. USA **84**, 5216–5220.

Cyr, D.M. (1995). Cooperation of the molecular chaperone Ydj1 with specific Hsp70 homologs to suppress protein aggregation. FEBS Lett **359**, 129–132.

Cyr, D.M. and Douglas, M.G. (1994). Differential regulation of Hsp70 subfamilies by the eukaryotic DnaJ homologue YDJ1. J. Biol. Chem. **269**, 9798–9804.

Cyr, D.M., Lu, X. and Douglas, M.G. (1992). Regulation of Hsp70 function by a eukaryotic DnaJ homolog. J. Biol. Chem. **267**, 20927–20931.

Cyr, D.M., Langer, T. and Douglas, M.G. (1994). DnaJ-like proteins: molecular chaperones and specific regulators of Hsp70. Trends Biochem. Sci. **19**, 176–181.

Ellis, R.J. and Hartl, F.U. (1996). Protein folding in the cell: competing models of chaperonin function. FASEB J. **10**, 20–26

Frydman, J. and Hartl, F.U. (1996). Principles of chaperone-assisted protein folding: differences between in vitro and in vivo mechanisms. Science **272**, 1497–1502.

Frydman, J., Nimmesgern, E., Ohtsuka, K. and Hartl, F.U. (1994). Folding of nascent polypeptide chains in a high molecular mass assembly with molecular chaperones. Nature **370**, 111–117.

Gaitanaris, G.A., Vysokanov, A., Hung, S.C., Gottesman, M.E. and Gragerov, A. (1994). Successive action of *Escherichia coli* chaperones in vivo. Mol. Microbiol. **14**, 861–869.

Gragerov, A., Nudler, E., Komissarova, N., Gaitanaris, G.A., Gottesman, M.E. and Nikiforov, V. (1992). Cooperation of GroEL/GroES and DnaK/DnaJ heat shock proteins in preventing protein misfolding in *Escherichia coli*. Proc. Natl. Acad. Sci. USA **89**, 10341–10344

Hann, B.C., Poritz, M.A. and Walter, P. (1989). *Saccharomyces cerevisiae* and *Schizosaccharomyces pombe* contain a homologue to the 54-kD subunit of the signal recognition particle that in *S. cerevisiae* is essential for growth. J. Cell Biol. **109**, 3223–3230.

Hardy, S.J. and Randall, L.L. (1991). A kinetic partitioning model of selective binding of nonnative proteins by the bacterial chaperone SecB. Science **251**, 439–443.

Hardy, S.J. and Randall, L.L. (1993). Recognition of ligands by SecB, a molecular chaperone involved in bacterial protein export. Phil. Trans. R. Soc. Lond. B. Biol. Sci. **339**, 343–352.

Hartl, F.U. (1996). Molecular chaperones in cellular protein folding. Nature **381**, 571–579.

Hayes, S.A and Dice, J.F. (1996). Roles of molecular chaperones in protein degradation. J. Cell Biol. **132**, 255–258.

Hendrick, J.P., Langer, T., Davis, T.A., Hartl, F.U. and Wiedmann, M. (1993). Control of folding and membrane translocation by binding of the chaperone DnaJ to nascent polypeptides. Proc. Natl. Acad. Sci. USA **90**, 10216–10220.

Hesterkamp, T., Hauser, S., Lutcke, H. and Bukau, B. (1996). *Escherichia coli* trigger factor is a prolyl isomerase that associates with nascent polypeptide chains. Proc. Natl. Acad. Sci. USA **93**, 4437–4441.

Hohfeld, J., Minami, Y. and Hartl, F.U. (1995). Hip, a novel cochaperone involved in the eukaryotic Hsc70/Hsp40 reaction cycle. Cell **83**, 589–598.

Horwich, A.L., Low, K.B., Fenton, W.A., Hirshfield, I.N. and Furtak, K. (1993). Folding in vivo of bacterial cytoplasmic proteins: role of GroEL. Cell **74**, 909–917.

Kandror, O., Sherman, M., Rhode, M. and Goldberg, A.L. (1995). Trigger factor is involved in GroEL-dependent protein degradation in *Escherichia coli* and promotes binding of GroEL to unfolded proteins. EMBO J. **14**, 6021–6027.

Keierleber, C., Wittekind, M., Qin, S.L. and McLaughlin, C.S. (1986). Isolation and characterization of PRT1, a gene required for the initiation of protein biosynthesis in *Saccharomyces cerevisiae*. Mol. Cell. Biol. **6**, 4419–4424.

Kudlicki, W., Odom, O.W., Kramer, G. and Hardesty, B. (1994). Chaperone-dependent folding and activation of ribosome-bound nascent rhodanese. Analysis by fluorescence. J. Mol. Biol. **244**, 319–331.

Kudlicki, W., Odom, O.W., Kramer, G., Hardesty, B., Merrill, G.A. and Horowitz, P.M. (1995). The importance of the N-terminal segment for DnaJ-mediated folding while bound to ribosomes as peptidyl-tRNA. J. Biol. Chem. **270**, 10650–10657.

Langer, T., Lu, C., Echols, H., Flanagan, J., Hayer, M.K. and Hartl, F.U. (1992). Successive action of DnaK, DnaJ and GroEL along the pathway of chaperone-mediated protein folding. Nature **356**, 683–689.

Lauring, B., Kreibich, G., and Wiedmann, M. (1995). Nascent polypeptide-associated complex protein prevents mistargeting of nascent chains to the endoplasmic reticulum. Proc. Natl. Acad. Sci. USA **92**, 5411–5415

Luke, M.M., Sutton, A. and Arndt, K.T. (1991). Characterization of SIS1, a *Saccharomyces cerevisiae* homologue of bacterial dnaJ proteins. J. Cell Biol. **114**, 623–638.

Lurink, J., ten Hagen-Jongman, C.M., van der Weijden, C.C., Oudega, B., High, S., Dobberstein, B. and Kusters, R. (1994). An alternative protein targeting pathway in *Escherichia coli*: studies on the role of FtsY. EMBO J. **13**, 2289–2296

Mayhew, M., da Silva, A.C., Martin, J., Erdjument-Bromage, H., Tempst, P. and Hartl, F.U. (1996). Protein folding in the central cavity of the GroEL-GroES chaperonin complex. Nature **379**, 420–426.

Miller, J.D., Bernstein, H.D. and Walter, P. (1994). Interaction of *E. coli* Ffh/4.5S ribonucleoprotein and FtsY mimics that of mammalian signal recognition particle and its receptor. Nature **367**, 657–659.

Nelson, R.J., Ziegelhoffer, T., Nicolet, C., Werner-Washburne, M. and Craig, E. A. (1992). The translation machinery and 70 kd heat shock protein cooperate in protein synthesis. Cell **71**, 97–105.

Ohtsuka, K. (1993). Cloning of a cDNA for heat-shock protein hsp40, a human homologue of bacterial DnaJ. Biochem. Biophys. Res. Commun. **197**, 235–240.

Paek, K.H. and Walker, G.C. (1987). *Escherichia coli* dnaK null mutants are inviable at high temperature. J. Bacteriol. **169**, 283–290.

Phillips, G.J. and Silhavy, T.J. (1992). The E. coli Ffh gene is necessary for viability and efficient protein export . Nature **359**, 744–746.

Randall, L.L. (1992). Peptide binding by chaperone SecB: implications for recognition of nonnative structure. Science **257**, 241–245.

Rapoport, T.A., Rolls, M.M. and Jungnickel, B. (1996). Approaching the mechanism of protein transport across the ER membrane. Curr. Opin. Cell Biol. **8**, 499–504.

Ribes, V., Romisch, K., Giner, A., Dobberstein, B. and Tollervey, D. (1990). E. coli 4.5S RNA is part of a ribonucleoprotein particle that has properties related to signal recognition particle. Cell **63**, 591–600.

Romisch, K., Webb, J., Herz, J., Prehn, S., Frank, R., Vingron, M. and Dobberstein, B. (1989). Homology of 54K protein of signal-recognition particle, docking protein and two E. coli proteins with putative GTP-binding domains. Nature **340**, 478–482.

Sachs, A.B. and Davis, R.W. (1990). Translation initiation and ribosomal biogenesis: involvement of a putative rRNA helicase and RPL46. Science **247**, 1077–1079.

Schneider, C., Sepp-Lorenzino, L., Nimmesgern, E., Ouathek, O., Danishefsky, S., Rosen., N and Hartl, F.U. Pharmacologic shifting of a balance between protein refolding and degradation mediated by Hsp90. Proc. Natl. Acad. Sci. USA **93**, 14536–14541.

Sell, S.M., Eisen, C., Ang, D., Zylicz, M. and Georgopoulos, C. (1990). Isolation and characterization of dnaJ null mutants of *Escherichia coli*. J. Bacteriol. **172**, 4827–4835.

Stoller, G., Rucknagel, K.P., Nierhaus, K. H., Schmid, F.X., Fischer, G. and Rahfeld, J.U. (1995). A ribosome-associated peptidyl-prolyl cis/trans isomerase identified as the trigger factor. EMBO J. **14**, 4939–4948.

Szabo, A., Korszun, R., Hartl, F.U. and Flanagan, J. (1996). A zinc finger-like domain of the molecular chaperone DnaJ is involved in binding to denatured protein substrates. EMBO J. **15**, 408–417.

Todd, M.J., Viitanen, P.V. and Lorimer, G.H. (1994). Dynamics of the chaperonin ATPase cycle: implications for facilitated protein folding. Science **265**, 659–666.

Valent, Q.A., Kendall, D.A., High, S., Kusters, R., Oudega, B. and Luirink, J. (1995). Early events in preprotein recognition in *E. coli*: interaction of SRP and trigger factor with nascent polypeptides. EMBO J. **14**, 5494–5505.

von Heijne, G. (1985). Signal sequences. The limits of variation. J. Mol. Biol. **184**, 99–105.

Vysokanov, A.V., Gaitanaris, G.A., Vysokanov, A., Hung, S.C., Gottesman, M.E. and Gragerov, A. (1995). Synthesis of chloramphenicol acetyltransferase in a coupled transcription-translation *in vitro* system lacking the chaperones DnaK and DnaJ. FEBS Lett. **375**, 211–214.

Wall, D., Zylixz, M. and Georgopoulos, C. (1994). The NH2-terminal 108 amino acids of the *Escherichia coli* DnaJ protein stimulate the ATPase activity of DnaK and are sufficient for lambda replication. J. Biol. Chem. **269**, 5446–5451.

Walter, P. and Johnson, A.E. (1994). Signal sequence recognition and protein targeting to the endoplasmic reticulum membrane. Annu. Rev. Cell. Biol. **10**, 87–119.

Wang, S., Sakai, H. and Wiedmann, M. (1995). NAC covers ribosome-associated nascent chains thereby forming a protective environment for regions of nascent chains just emerging from the peptidyl transferase center. J. Cell Biol. **130**, 519–528.

Weissman, J.S., Kashi, Y., Fenton, W.A. and Horwich, A.L. (1994). GroEL-mediated protein folding proceeds by multiple rounds of binding and release of nonnative forms. Cell **78**, 693–702.

Werner-Washburne, M., Becker, J., Kosic-Smithers, J. and Craig, E.A. (1989). Yeast Hsp70 RNA levels vary in response to the physiological status of the cell. J. Bacteriol. **171**, 2680–2688.

Wickner, W. (1995). The nascent-polypeptide-associated complex: having a "NAC" for fidelity in translocation. Proc. Natl. Acad. Sci. USA **92**, 9433–9434.

Wiedmann, B., Sakai, H., Davis, T.A. and Wiedmann, M. (1994). A protein complex required for signal-sequence-specific sorting and translocation. Nature **370**, 434–440.

Zhong, T. and Arndt, K.T. (1993). The yeast SIS1 protein, a DnaJ homolog, is required for the initiation of translation. Cell **73**, 1175–1186.

Zhong, T., Luke, M.M. and Arndt, K.T. (1996). Transcriptional regulation of the yeast DnaJ homologue SIS1. J. Biol. Chem. **271**, 1349–1356.

■ *Damian McColl*
Department of Biochemistry and Biophysics
University of California San Francisco
513 Parnassus Avenue
San Francisco, CA 94143-0448, U.S.A.

■ *F. Ulrich Hartl*
Max-Planck-Institute for Biochemistry
Department of Cellular Biochemistry
Am Klopferspitz 18a
D-82152 Martinsried, Germany
Tel. 49-89-8578 2244/2233
Fax. 49-89-8578 2211
E-mail: uhartl@biochem.mpg.de

Functions of molecular chaperone proteins in the biogenesis of mitochondria

Molecular chaperone proteins bind non-native protein structures and stabilize them against aggregation. By this means, they affect multiple cellular processes, ranging from folding of newly synthesized polypeptides emerging from ribosomes, cellular protein trafficking, translocation across intracellular membranes, the reactivation of heat-denatured proteins and proteolytic breakdown. This is exemplified by mitochondria whose biogenesis depends on the activity of various chaperone systems. In the cytosol, molecular chaperone proteins maintain nuclear-encoded preproteins in a conformation allowing subsequent membrane translocation. The mitochondrial Hsp70 machinery mediates the vectorial translocation of nuclear encoded preproteins across the mitochondrial membranes. In the mitochondrial matrix space, the Hsp70 and Hsp60 chaperone systems ensure proper folding of newly imported proteins and the proteolytic breakdown of misfolded polypeptides. In addition to these 'housekeeping' functions, chaperone activity is required to maintain mitochondrial functions during heat stress. Thus, studying mitochondrial biogenesis, a wide range of chaperone functions have been identified and their modes of action were subsequently characterized at the molecular level. This review will focus on the multiple roles of cytosolic and mitochondrial chaperones in various processes of mitochondrial biogenesis.

The role of cytosolic factors in mitochondrial protein import

Protein import into mitochondria can occur post-translationally. The conformation assumed by preproteins before association with mitochondria seems to be an intrinsic property of the specific preprotein. Some preproteins may fold to a native-like structure that is unfolded upon membrane translocation, while others may be prevented from premature folding. Various cytosolic factors have been identified that stabilize non-native preproteins against aggregation and, in addition, contribute to efficient mitochondrial targeting (Fig. 1).

Cytosolic HSP70 proteins and the DnaJ homologue, Ydj1p

Cytosolic HSP70 proteins were the first proteins proposed to assist in maintaining a translocation competent state of mitochondrial preproteins (Chirico et al., 1988; Deshaies et al., 1988). A yeast strain lacking SSA1, SSA2 and SSA4 can be rescued by overexpression of Ssa1p from a galactose-regulated promotor. Genetic depletion of Ssa1p resulted in the accumulation of a mitochondrial precursor, $pF_1\beta$ (β subunit of the F_1–ATPase), in the cytosol suggesting the requirement of Ssa1p for efficient protein import into mitochondria (Deshaies et al., 1988). A direct interaction of cytosolic Hsc70 with the precursor form of Tom70, an integral component of the mitochondrial outer membrane has been demonstrated by coimmunoprecipitation (Schlossmann and Neupert, 1995). Additional evidence for a role of the cytosolic HSP70 system in mitochondrial protein import came from the analysis of a temperature-sensitive yeast mutant of the cytosolic DnaJ homologue, Ydj1p (Caplan et al., 1992; see entry p. 102). A number of mitochondrial precursor proteins were observed to accumulate in the cytosol of this mutant strain under restrictive conditions. Ydj1p exerts chaperone activity on its own and triggers the release of Ssa1p-bound substrate proteins in vitro (Cyr et al., 1992; Cyr, 1995). It is conceivable that Ydj1p facilitates mitochondrial protein import in intact cells in a similar way.

Presequence binding factor (PBF)

Although import of denatured, chemically pure fusion proteins into isolated mitochondria can be performed in the absence of cytosolic factors (Becker et al., 1992), mitochondrial protein import in intact cells may be stimulated by cytosolic proteins that specifically recognize the mitochondrial targeting sequence. Such a factor, termed PBF (for presequence binding factor), has been identified in rabbit reticulocyte lysate. PBF stimulates the import of pre-ornithine transcarbamoylase (pOTC) into rat liver mitochondria (Murakami and Mori, 1990). Apparently, PBF does not act as a molecular chaperone protein in the cytosol; however, it seems to cooperate with purified cytosolic Hsc70 (Murakami and Mori, 1990). PBF is thought to confer additional targeting information to a preprotein–Hsc70 complex and might thereby ensure efficient import of certain precursor proteins into mitochondria (Murakami et al., 1992).

Mitochondrial import stimulation factor (MSF)

A heterodimeric complex has been purified from rat liver cytosol and rabbit reticulocyte lysate that stimulates mitochondrial protein import in vitro and was therefore termed MSF (for mitochondrial import stimulation factor) (Ono and Tuboi, 1988; Hachiya et al., 1993; see entry p. 453). MSF belongs to the 14-3-3 protein family, ubiquitous, acidic proteins, to which a role in signal transduction has been assigned (for review see, Aitken et al., 1992). Purified MSF binds the presequence of the adrenodoxin preprotein and displays substrate-dependent ATPase activity. Interestingly, MSF also mediates dissolution of aggregated preadrenodoxin in an ATP-dependent manner. Therefore, MSF may represent a novel chaperone for mitochondrial precursor proteins with a dual function: on the one hand, it confers translocation competence to mitochondrial preproteins; on the other hand, it recognizes presequences and may thus target precursor proteins to mitochondria (Hachiya et al., 1994). Recently, the Tom37/Tom70 receptor complex in the mitochondrial

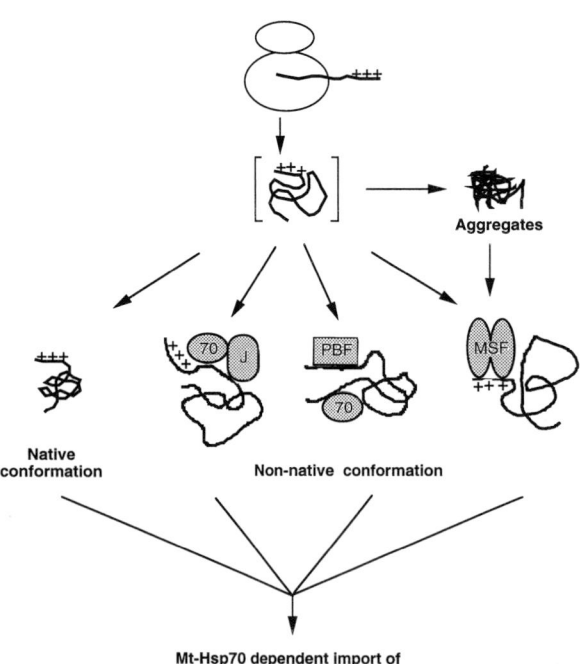

Figure 1. Maintenance of translocation competence of mitochondrial preproteins in the cytosol. See text for details. (70) HSP70; (J) Ydj1p; (MSF) mitochondrial import stimulation factor; (PBF) presequence binding factor; (mt-Hsp70) mitochondrial Hsp70.

outer membrane has been identified as a binding site for a preprotein associated with MSF (Hachiya et al., 1995). Modulation of the ATPase activity of MSF upon binding to the Tom37/ Tom70 complex has been proposed to trigger MSF release, allowing the passage of the preprotein to components of the receptor complex acting at later steps of translocation (Hachiya et al., 1995). In yeast, two homologues of mammalian MSF, the BMH1 and BMH2 gene products, have been identified (van Heusden et al., 1995). So far, evidence for a role for these proteins in mitochondrial protein import in vivo is still awaited.

■ Components of the outer membrane translocation complex contribute to preprotein unfolding

As preproteins cross the mitochondrial membranes in an extended conformation those cytosolic preproteins that contain folded structures have to undergo unfolding before or during membrane passage. Investigations on protein import into isolated outer membrane vesicles revealed that import components of the outer membrane translocation machinery contribute to precursor unfolding (Mayer et al., 1995; Lill and Neupert, 1996). The observed unfolding activity is ATP independent and depends on the interaction of the mitochondrial presequence with a binding site at the inner surface of the mitochondrial outer membrane, termed trans-site. After spontaneous unfolding at the outer surface of mitochondria, and penetration into the outer membrane translocation channel, the association with the trans-site apparently prevents refolding of preprotein domains immediately following the presequence. By this means, the unfolding activity of the outer membrane import apparatus can precede the action of the matrix-localized mitochondrial Hsp70 system in unfolding of the preproteins at the mitochondrial surface.

■ Chaperone-driven membrane translocation of mitochondrial preproteins

Protein import into mitochondria requires an electrochemical gradient across the mitochondrial inner membrane and an ATP-dependent proteinaceous translocation machinery at its inner surface (Pfanner et al., 1994; Ryan and Jensen, 1995). This machinery is composed of the matrix-located Hsp70 chaperone (mt-Hsp70; see entry on Ssc1p, p. 30), Tim44, a peripheral component of the translocation pore in the inner membrane, and Mge1p (see entry p. 142), the yeast homologue of E. coli GrpE (see entry p. 137). All these proteins are required for viability, reflecting their essential functions in mitochondrial biogenesis.

The role of mt-Hsp70 in mitochondrial protein import was first recognized in yeast by the analysis of temperature-sensitive mutant forms of mt-Hsp70 (Kang et al., 1990; Gambill et al., 1993). Under restrictive conditions, nuclear-encoded preproteins were observed to accumulate as translocation intermediates spanning the outer and inner membrane of mitochondria. mt-Hsp70 binds preproteins emerging from the import channel into the matrix and thereby confers unidirectionality on the otherwise reversible translocation process (Neupert et al., 1990). Import intermediates exposing less than 20 residues of the mitochondrial targeting sequence to the matrix space could be found in association with mt-Hsp70 (Ungermann et al., 1994).

Efficient binding of mt-Hsp70 to translocation intermediates requires its ATP-dependent interaction with Tim44 (Kronidou et al., 1994; Rassow et al., 1994; Schneider et al., 1994). Tim44 has been identified by a genetic screen as an essential protein of the inner membrane import machinery (Maarse et al., 1992). The import component is peripherally associated with the inner membrane and recruits a fraction of the otherwise soluble mt-Hsp70 to the membrane. Translocating polypeptide chains are stabilized in the matrix space by the mt-Hsp70–Tim44 complex. In addition to mt-Hsp70, Tim44 by itself seems to exert a chaperone-like activity (Schneider et al., 1994). In the presence of hydrolysable ATP, dissociation of mt-Hsp70 from Tim44 has been observed, a process stimulated by unfolded polypeptides (Schneider et al., 1994).

Mge1p represents the third component of the translocation machinery in the mitochondrial matrix. Upon cellular depletion of Mge1p, or in yeast strains carrying conditional mutant alleles of MGE1, unprocessed mitochondrial preproteins accumulate in the cytosol indicating an import defect (Laloraya et al., 1994; Westermann et al., 1995). Mge1p acts in concert with mt-Hsp70 during membrane translocation as demonstrated by several observations. First, similarly to its prokaryotic homologue, Mge1p is bound to mt-Hsp70 in an ATP-dependent manner (Bolliger et al., 1994). Secondly, translocating polypeptide chains can be simultaneously cross-linked to mt-Hsp70 and Mge1p (Voos et al., 1994). Thirdly, conditional mutations in Mge1p result in a reduced binding of mt-Hsp70 to translocating polypeptides (Laloraya et al., 1995; Westermann et al., 1995). Finally, the mt-Hsp70–Tim44 complex is more sensitive to ATP-induced dissociation in mitochondria carrying a temperature-sensitive mutant allele of MGE1 than that in wild-type mitochondria (Westermann et al., 1995). These results suggest that Mge1p acts as a nucleotide exchange factor on mt-Hsp70, similar to the highly homologous GrpE protein of E. coli. By this means, Mge1 modulates the ATP-dependent complex formation of mt-Hsp70 with Tim44 and translocating polypeptides. By analogy to the DnaK/DnaJ/GrpE chaperone machinery of E. coli, and based on a weak homology of Tim44 to DnaJ-like proteins, it has been hypothesized that Tim44 can fulfil a function in the ATPase cycle of mt-Hsp70 similar to that of E. coli DnaJ (Rassow et al., 1994). This might explain the surprising finding that the mitochondrial DnaJ homologue, Mdj1p, is not required for protein import into mitochondria (Rowley et al., 1994; see entry p. 106).

The Brownian ratchet model for membrane translocation of mitochondrial preproteins

These findings lead to a model for mitochondrial protein import according to which the membrane translocation of preproteins is driven by Brownian motion (Neupert et al., 1990; Schneider et al., 1994). Translocating polypeptides can oscillate in the proteinaceous import channel. The translocation machinery in the matrix prevents a retrograde movement of preproteins out of the translocation channel and thereby confers unidirectionality on the translocation process (Ungermann et al., 1994). The assembly of the mt-Hsp70–Tim44 complex and its binding to translocating preproteins is regulated in an ATP-dependent manner. Examination of the present understanding of nucleotide-modulated HSP70 activity with our knowledge on mitochondrial protein import leads to the following scenario (Fig. 2): mt-Hsp70 loaded with ATP associates with Tim44 (step 1). Binding of mt-Hsp70 to translocating polypeptide chains stimulates ATP hydrolysis resulting in dissociation of mt-Hsp70 from Tim44 (step 2). As a consequence, the polypeptide chain can slide in the import channel driven by Brownian motion. mt-Hsp70, which is supposed to bind preproteins with high affinity when in the ADP-bound conformation, prevents retrograde movement of the polypeptide. Subsequently, Tim44 and another ATP-loaded mt-Hsp70 bind to a more C-terminal segment initiating another cycle of translocation (step 3). Mge1p acts as a nucleotide exchange factor in this process by recycling mt-Hsp70 to the ATP-bound form which is competent to associate productively with Tim44 (Westermann et al., 1995).

According to the Brownian ratchet model, mt-Hsp70 fulfils a dual function during the translocation process. In concert with the membrane potential across the mitochondrial inner membrane, the mt-Hsp70–Tim44 complex initially stabilizes the mitochondrial presequence in the matrix, thus triggering early steps of translocation, and then mediates transmembrane movement of the rest of the preprotein. At later stages of the import process, mt-Hsp70 may be able to bind translocating polypeptides independently of Tim44. This Tim44-independent association of mt-Hsp70 with translocating chains could contribute to preventing their backsliding. Experimental evidence for two preprotein-bound populations of mt-Hsp70, Tim44 associated and non-associated, has recently been obtained (Ungermann et al., 1996).

The mitochondrial ClpB homologue Hsp78

Further support for the Brownian ratchet model, comes from the analysis of the mitochondrial homologue of E. coli ClpB, Hsp78 (Leonhardt et al., 1993; see entry p. 251), which exerts chaperone activity in the mitochondrial matrix space (Schmitt et al., 1995). Whereas disruption of HSP78 does not affect cell growth, yeast strains carrying both a conditional mutant of mt-Hsp70 and a deletion of HSP78, show a synthetic petite phenotype indicating functional overlap between mt-Hsp70 and Hsp78 (Moczko et al., 1995; Schmitt et al., 1995). Overexpression of Hsp78 rescues the import deficiency of mitochondria containing a conditional mutant form of mt-Hsp70 under permissive conditions (Schmitt et al., 1995). Similarly to mt-Hsp70, Hsp78 binds to translocating polypeptides; however, association of Hsp78 with Tim44 cannot be detected. In the presence of a conditional mutant form of mt-Hsp70, Hsp78 seems to substitute for those mt-Hsp70 molecules that bind independently of Tim44 to the translocating preprotein (Schmitt et al., 1995; see Fig. 2, step 3). By this means, Hsp78 may prevent the retrograde movement of polypeptides when the mt-Hsp70–Tim44 complex is dissociated, and could facilitate protein translocation without associating with Tim44.

Matrix-located chaperone systems facilitate folding and assembly

After import into the matrix space and cleavage of the targeting sequence, nuclear-encoded mitochondrial proteins have to fold into their native conformation at their site of function. Folding in the mitochondrial matrix space is assisted by the Hsp70 and Hsp60 chaperone systems and by a folding catalyst in mitochondria, Cyp20p (Cpr3p, see entry p. 378).

The mitochondrial Hsp70 system

Studies in yeast revealed a role for mt-Hsp70 in folding processes, in addition to its function in protein translocation. A conditional mutant of mt-Hsp70, Ssc1-2p, has

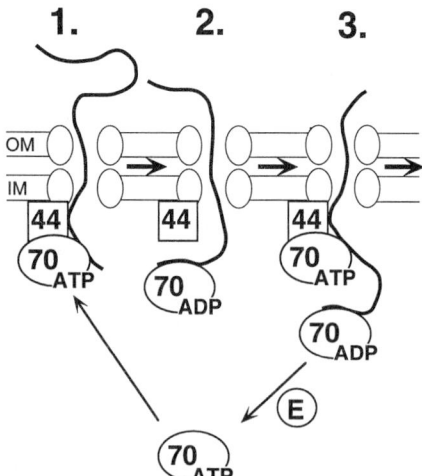

Figure 2. The Brownian ratchet model for the mt-Hsp70-driven membrane translocation of mitochondrial preproteins. (OM) mitochondrial outer membrane; (IM) mitochondrial inner membrane; (44) Tim44; (70) mt-Hsp70; (E) Mge1p.

been identified that irreversibly binds to polypeptides, most likely because of an impaired ATPase activity (Kang et al., 1990; Gambill et al., 1993). The block of mitochondrial protein import under restrictive conditions can be circumvented by artificially unfolding preproteins *in vitro*. However, although completely imported under these conditions, folding of these proteins is impaired in *ssc1-2* mitochondria (Kang et al., 1990). mt-Hsp70 also assists the assembly of certain protein complexes in the mitochondrial inner membrane. Integration of the mitochondrially encoded subunits 6 and 9 into the F_0–ATPase complex is impaired in the absence of functional mt-Hsp70 (Herrmann et al., 1994). Strikingly, mt-Hsp70 apparently does not assist folding of these ATPase subunits, but rather ensures their correct assembly into a functional complex.

The chaperone function of mt-Hsp70 during protein folding is regulated by two components, Mdj1p and Mge1p. In mitochondria lacking Mdj1p, folding of newly imported proteins is impaired (Rowley et al., 1994). Mdj1p is required for efficient binding of polypeptides to mt-Hsp70 (Wagner et al., 1994; Prip-Buus et al., 1996). By analogy to the highly homologous *E. coli* DnaJ protein, Mdj1p may exert chaperone activity on its own by preventing aggregation. This may explain the surprising finding that lack of Mdj1p leads to enhanced aggregation of a matrix-located tester protein at elevated temperature, whereas inactivation of mt-Hsp70 or Mge1p does not affect its solubility under these conditions (Prip-Buus et al., 1996).

A deficiency in the folding of newly imported polypeptides has also been demonstrated in mitochondria containing a conditional mutant form of Mge1p (Westermann et al., 1995). Thus, all three components of the mitochondrial Hsp70 system are required to ensure efficient protein folding in the matrix space, whereas only mt-Hsp70 and Mge1p, together with Tim44, suffice for the import process.

■ The mitochondrial Hsp60 system

mt-Hsp60, highly homologous to *E. coli* GroEL, is a central player of the mitochondrial folding apparatus. In yeast, mt-Hsp60 has been identified by a genetic screen for mutants lacking the enzymatically active form of imported ornithine transcarbamoylase (Cheng et al., 1989; Reading et al., 1989; Johnson et al., 1989; see entry p. 189). Folding and assembly of both imported and mitochondrially encoded proteins is impaired upon depletion of functional mt-Hsp60 (Cheng et al., 1989; Hallberg et al., 1993; Prasad et al., 1990; Horwich et al., 1992). mt-Hsp60 acts at the level of polypeptide chain folding, rather than affecting the assembly of multisubunit complexes (Ostermann et al., 1989; Zheng et al., 1993; Saijo et al., 1994). At elevated temperatures, mt-Hsp60 maintains imported DHFR in an enzymatically active conformation and prevents aggregation of thermolabile mitochondrial proteins (Martin et al., 1992).

ATP-dependent polypeptide folding by mt-Hsp60 is regulated by mt-Hsp10, a homologue of *E. coli* GroES (Lubben et al., 1990; Hartman et al., 1992a, b; Rospert et al., 1993a, b; Höhfeld and Hartl, 1994; see entry p. 191). In yeast, mt-Hsp10 is essential for viability, as is mt-Hsp60 (Rospert et al., 1993a; Höhfeld and Hartl, 1994). Folding of newly imported proteins was found to be impaired in a conditional yeast mutant of mt-Hsp10 under restrictive conditions (Höhfeld and Hartl, 1994). The temperature-sensitive mutation maps to a conserved domain of mt-Hsp10 that corresponds to a previously identified conserved loop region of *E. coli* GroES required for association with GroEL (Landry et al., 1993). Consistently, the point mutation in mt-Hsp10 results in a lower binding affinity to mt-Hsp60, thereby most likely causing the observed deficiency in the folding of newly imported proteins (Höhfeld and Hartl, 1994).

■ Sequential action of the mt-Hsp70 and mt-Hsp60 chaperone machineries

The folding of newly imported mitochondrial preproteins depends on the presence of functional mt-Hsp70 and mt-Hsp60 machineries in the matrix (Fig. 3). In contrast to mt-Hsp70, mt-Hsp60 function is not required for the membrane translocation of mitochondrial preproteins

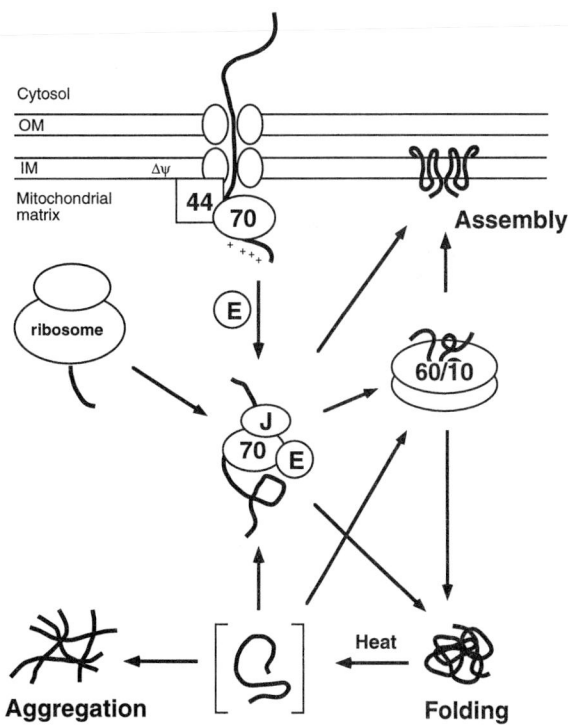

Figure 3. The role of matrix-located chaperones in protein folding and assembly. See text for details. (OM) mitochondrial outer membrane; (IM) mitochondrial inner membrane; (44) Tim44; (70) mt-Hsp70; (J) Mdj1p; (E) Mge1p; (60/10) mt-Hsp60 and mt-Hsp10.

(Cheng et al., 1989; Kang et al., 1990). This suggests a sequential interaction of mt-Hsp70 and mt-Hsp60 with newly imported polypeptides. In fact, the precursors of β-MPP and medium-chain acyl-CoA dehydrogenase were demonstrated to be associated successively with mt-Hsp70 and mt-Hsp60 (Manning-Krieg et al., 1991; Saijo et al., 1994). Furthermore, newly imported mt-Hsp60 was also found in a transient complex with mt-Hsp70 prior to its assembly (Cheng et al., 1990; Hallberg et al., 1993).

ATP hydrolysis promotes substrate release from mt-Hsp70 and binding to mt-Hsp60. It is conceivable that release of polypeptides from mt-Hsp70 is regulated by Mdj1p and Mge1p, as the corresponding E. coli homologues are required for the transfer of folding intermediates from HSP70 (DnaK) to HSP60 (GroEL) chaperones in vitro (Langer et al., 1992). The sequential interaction of molecular chaperones seems to be directed by their binding specificity and by the particular role of mt-Hsp70 in the import process (Kang et al., 1990; Langer et al., 1992; Schneider et al., 1994). Whereas completely unfolded polypeptides are recognized by HSP70 proteins, HSP60 chaperones are known to stabilize compact folding intermediates with secondary, but disordered tertiary structure (Martin et al., 1991; Langer et al., 1992). However, although the general importance of this folding pathway has been established, variations of the general scheme are likely. In particular, the requirement of specific chaperone systems for protein folding may depend on the preprotein and its specific properties.

■ The mitochondrial cyclophilin Cyp20p

The yeast cyclophilin Cyp20p, encoded by the *CPR3* gene, has recently been identified as a further component of the mitochondrial folding machinery (Matouschek et al., 1995; Rassow et al., 1995; see entry p. 378 on Cpr3p). The folding of newly imported DHFR has been demonstrated to be accelerated in the presence of Cyp20p, which presumably exerts peptidyl–prolyl *cis–trans* isomerase activity in the matrix (Matouschek et al., 1995; Rassow et al., 1995). In the absence of Cyp20p, DHFR remained associated for a prolonged time with mt-Hsp60 (Rassow et al., 1995). Thus, Cyp20p seems to affect folding of DHFR at a rather late stage. However, as deletion of *CPR3* causes only a weak growth phenotype, Cyp20p seems to play only an accessory role during protein folding within mitochondria. In any case, it will be interesting to characterize further the functional interplay of this folding catalyst with the mitochondrial Hsp70 and Hsp60 system under physiological conditions.

■ Acknowledgements

We are grateful to B. Westermann for support in the preparation of the figures. The work in the authors' laboratory was supported by grants of the Deutsche Forschungsgemeinschaft (SFB184; Majority Program "Molekulare Zellbiologie der Hitzestressantwort") and by the Genzentrum München.

■ References

Aitken, A., Collinge, D.B., van Heusden, B.P.H., Isobe, T., Roseboom, P.H., Rosenfeld, G. and Soll, J. (1992). 14-3-3 proteins: a highly conserved, widespread family of eukaryotic proteins. Trends Biochem. Sci. **17**, 498–501.

Becker, K., Guiard, B., Rassow, J., Söllner, T. and Pfanner, N. (1992). Targeting of a chemically pure preprotein to mitochondria does not require the addition of a cytosolic signal recognition factor. J. Biol. Chem. **267**, 5637–5643.

Bolliger, L., Deloche, O., Glick, B.S., Georgopoulos, C., Jenö, P., Kronidou, N., Horst, M., Morishima, N. and Schatz, G. (1994). A mitochondrial homolog of bacterial GrpE interacts with mitochondrial hsp70 and is essential for viability. EMBO J. **13**, 1988–2006.

Caplan, A., Cyr, D.M. and Douglas, M.G. (1992). YDJ1p facilitates polypeptide translocation across different intracellular membranes by a conserved mechanism. Cell **71**, 1143–1155.

Cheng, M.Y., Hartl, F.-U., Martin, J., Pollock, R.A., Kalousek, F., Neupert, W., Hallberg, E.M., Hallberg, R.L. and Horwich, A.L. (1989). Mitochondrial heat-shock protein hsp60 is essential for assembly of proteins imported into yeast mitochondria. Nature **337**, 620–625.

Cheng, M.Y., Hartl, F.-U. and Horwich, A.L. (1990). The mitochondrial chaperonin hsp60 is required for its own assembly. Nature **348**, 455–458.

Chirico, W., Waters, M. and Blobel, G. (1988). 70K heat-shock related proteins stimulated protein translocation into microsomes. Nature **332**, 805–810.

Cyr, D.M. (1995). Cooperation of the molecular chaperone Ydj1p with specific Hsp70 homologs to suppress protein aggregation. FEBS Lett. **359**, 129–132.

Cyr, D.M., Lu, X. and Douglas, M.G. (1992). Regulation of eukaryotic hsp70 function by a dnaJ homolog. J. Biol. Chem. **267**, 20927–20931.

Deshaies, R.J., Koch, B.D., Werner-Washburne, M., Craig, E.A. and Schekman, R. (1988). A subfamily of stress proteins facilitates translocation of secretory and mitochondrial precursor polypeptides. Nature **332**, 800–805.

Gambill, B.D., Voos, W., Kang, P.J., Miao, B., Langer, T., Craig, E.A. and Pfanner, N. (1993). A dual role for mitochondrial heat shock protein 70 in membrane translocation of preproteins. J. Cell Biol. **123**, 109–117.

Hachiya, N., Alam, R., Sakasegawa, Y., Sakaguchi, M., Mihara, K. and Omura, T. (1993). A mitochondrial import factor purified from rat liver cytosol is an ATP-dependent conformational modulator for precursor proteins. EMBO J. **12**, 1579–1586.

Hachiya, N., Komiya, T., Alam, R., Iwahashi, J., Sakaguchi, M., Omura, T. and Mihara, K. (1994). MSF, a novel cytoplasmic chaperone which functions in precursor targeting to mitochondria. EMBO J. **13**, 5146–5154.

Hachiya, N., Mihara, K., Suda, K., Horst, M., Schatz, G. and Lithgow, T. (1995). Recognition of the initial steps of mitochondrial protein import. Nature **376**, 705–709.

Hallberg, E.M., Shu, Y. and Hallberg, R.L. (1993). Loss of mitochondrial hsp60 function: nonequivalent effects on matrix-targeted and intermembrane-targeted proteins. Mol. Cell. Biol. **13**, 3050–3057.

Hartman, D.J., Hoogenrad, N.J. and Hoj, P.B. (1992a). Heat shock proteins of barley mitochondria and chloroplast. Identification of organellar hsp10 and 12: putative chaperonin 10 homologues. FEBS Lett. **305**, 147–150.

Hartman, D.J., Hoogenraad, N.J., Condron, R. and Hoj, P.B. (1992b). Identification of a mammalian 10-kDa heat shock protein, a mitochondrial chaperonin 10 homologue essential for assisted folding of trimeric ornithine transcarbamoylase in vitro. Proc. Natl. Acad. Sci. USA **89**, 3394–3398.

Herrmann, J.M., Stuart, R.A., Craig, E.A. and Neupert, W. (1994). Mitochondrial heat shock protein 70, a molecular chaperone for proteins encoded by mitochondrial DNA. J. Cell Biol. **127**, 893–902.

Höhfeld, J. and Hartl, F.-U. (1994). Requirement of the chaperonin cofactor HSP10 for protein folding and sorting in yeast mitochondria. J. Cell Biol. **126**, 305–315.

Horwich, A., Caplan, S., Wall, J.S. and Hartl, F.-U. (1992). Chaperonin-mediated protein folding. In *Membrane Biogenesis and Protein Targeting* (ed. Neupert, W. and Lill, R.). Elsevier, Amsterdam, pp. 329–337.

Johnson, R.B., Fearon, K., Mason, T. and Jindal, S. (1989). Cloning and characterization of the yeast chaperonin HSP60 gene. Gene **84**, 295–302.

Kang, P.-J., Ostermann, J., Shilling, J., Neupert, W., Craig, E.A. and Pfanner, N. (1990). Requirement for hsp70 in the mitochondrial matrix for translocation and folding of precursor proteins. Nature **348**, 137–143.

Kronidou, N.G., Oppliner W., Bolliger, L., Hannavy, K., Glick, B.S., Schatz, G. and Horst, M. (1994). Dynamic interaction between Isp45 and mitochondrial hsp70 in the protein import system of the yeast mitochondrial inner membrane. Proc. Natl. Acad. Sci. USA **91**, 12818–12822.

Laloraya, S., Gambill, B.D. and Craig, E.A. (1994). A role for a eukaryotic GrpE-related protein, Mge1p, in protein translocation. Proc. Natl. Acad. Sci. USA **91**, 6481–6485.

Laloraya, S., Dekker, P.J.T., Voos, W., Craig, E.A. and Pfanner, N. (1995). Mitochondrial GrpE modulates the function of matrix Hsp70 in translocation and maturation of preproteins. Mol. Cell. Biol. **15**, 7098–7105.

Landry, S., Zeilstra-Ryalls, J., Fayet, O., Georgopoulos, C., Gierasch, L.M. (1993). Characterization of a functionally important mobile domain of GroES. Nature **364**, 255–258.

Langer, T., Lu, C., Echols, H., Flanagan, J., Hayer, M.K. and Hartl, F.-U. (1992). Successive action of DnaK, DnaJ and GroEL along the pathway of chaperone-mediated protein folding. Nature **356**, 683–689.

Leonhardt, S.A., Fearon, K., Danese, P.N. and Mason, T.L. (1993). HSP78 encodes a yeast mitochondrial heat shock protein in the Clp family of ATP-dependent proteases. Mol. Cell. Biol. **13**, 6304–6313.

Lill, R. and Neupert, W. (1996). Mechanisms of protein import across the mitochondrial outer membrane. Trends Cell Biol. **6**, 56–61.

Lubben, T.H., Gatenby, A.A., Donaldson, G.K.,Lorimer, G.H. and Viitanen, P.V. (1990). Identification of a groES-like chaperonin in mitochondria that facilitates protein folding. Proc. Natl. Acad. Sci. USA **87**, 7683–7687.

Maarse, A.C., Blom, J., Grivell, L.A. and Meijer, M. (1992). MPI1, an essential gene encoding a mitochondrial membrane protein, is possibly involved in protein import into yeast mitochondria. EMBO J. **11**, 3619–3628.

Manning-Krieg, U.C., Scherer, P.E. and Schatz, G. (1991). Sequential action of mitochondrial chaperones in protein import into the matrix. EMBO J. **10**, 3273–3280.

Martin, J., Langer, T., Boteva, R., Schramel, A., Horwich, A.L. and Hartl, F.-U. (1991). Chaperonin-mediated protein folding at the surface of groEL through a 'molten globule'-like intermediate. Nature **352**, 36–42.

Martin, J., Horwich, A.L. and Hartl, F.U. (1992). Prevention of protein denaturation under heat stress by the chaperonin Hsp60. Science **258**, 995–998.

Matouschek, A., Rospert, S., Schmid, K., Glick, B.S. and Schatz, G. (1995). Cyclophilin catalyzes protein folding in yeast mitochondria. Proc. Natl. Acad. Sci. USA **92**, 6319–6323.

Mayer, A., Neupert, W. and Lill, R. (1995). Mitochondrial protein import: reversible binding of the presequence at the trans side of the outer membrane drives partial translocation and unfolding. Cell **80**, 127–137.

Moczko, M., Schönfisch, B., Voos, W., Pfanner, N. and Rassow, J. (1995). The mitochondrial ClpB homolog Hsp78 cooperates with matrix Hsp70 in maintenance of mitochondrial function. J. Mol. Biol. **254**, 538–543.

Murakami, K. and Mori, M. (1990). Purified presequence binding factor (PBF) forms an import-competent complex with a purified mitochondrial precursor protein. EMBO J. **9**, 3201–3208.

Murakami, K., Tanase, S., Morino, Y. and Mori, M. (1992). Presequence binding factor-dependent and -independent import of proteins into mitochondria. J. Biol. Chem. **267**, 13119–13122.

Neupert, W., Hartl, F.-U., Craig, E.A. and Pfanner, N. (1990). How do polypeptides cross the mitochondrial membranes? Cell **63**, 447–450.

Ono, H. and Tuboi, S. (1988). The cytosolic factor required for import of precursors of mitochondrial proteins into mitochondria. J. Biol. Chem. **263**, 3188–3193.

Ostermann, J., Horwich, A.L., Neupert, W. and Hartl, F.-U. (1989). Protein folding in mitochondria requires complex formation with hsp60 and ATP hydrolysis. Nature **341**, 125–130.

Pfanner, N., Craig, E.A. and Meijer, M. (1994). The protein import machinery of the inner mitochondrial membrane. Trends Biochem. Sci. **19**, 368–372.

Prasad, T.K., Hack, E. and Hallberg, R.L. (1990). Function of the maize mitochondrial chaperonin hsp60: specific association between hsp60 and newly synthesized F_1-ATPase alpha subunits. Mol. Cell. Biol. **10**, 3979–3986.

Prip-Buus, C., Westermann, B., Schmitt, M., Langer, T., Neupert, W. and Schwarz, E. (1996). Role of the mitochondrial DnaJ homologue, Mdj1p, in the prevention of heat-induced protein aggregation. FEBS Lett. **380**, 142–146.

Rassow, J., Maarse, A.C., Krainer, E., Kübrich, M., Müller, H., Meijer, M., Craig, E.A. and Pfanner, N. (1994). Mitochondrial protein import: biochemical and genetic evidence for interaction of matrix hsp70 and the inner membrane protein Mim44. J. Cell Biol. **127**, 1547–1556.

Rassow, J., Mohrs, K., Koidl, S., Barthelmess, I.B., Pfanner, N. and Tropschug, M. (1995). Cyclophilin 20 is involved in mtiochondrial protein folding in cooperation with molecular chaperones Hsp70 and Hsp60. Mol. Cell. Biol. **15**, 2654–2662.

Reading, D.S., Hallberg, R.L. and Myers, A.M. (1989). Characterization of the yeast HSP60 gene coding for a mitochondrial assembly factor. Nature **337**, 655–659.

Rospert, S., Junne, T., Glick, B.S. and Schatz, G. (1993a). Cloning and disruption of the gene encoding yeast mitochondrial chaperonin 10, the homolog of *E. coli* groES. FEBS Lett. **335**, 358–360.

Rospert, S., Glick, B.S., Jenö, P., Schatz, G., Todd, M.J., Lorimer, G.H. and Viitanen, P.V. (1993b). Identification and functional analysis of chaperonin 10, the groES homolog from yeast mitochondria. Proc. Natl. Acad. Sci. USA **90**, 10967–10971.

Rowley, N., Prip-Buus, C., Westermann, B., Brown, C., Schwarz, E., Barrell, B. and Neupert, W. (1994). Mdj1p, a novel chaperone of the DnaJ family, is involved in mitochondrial biogenesis and protein folding. Cell **77**, 249–259.

Ryan, K.R. and Jensen, R.E. (1995). Protein translocation across mitochondrial membranes: what a long, strange trip it is. Cell **83**, 517–519.

Saijo, T., Welch, W.J. and Tanaka, K. (1994). Intramitochondrial Folding and Assembly of Medium-chain Acyl-CoA Dehydrogenase (MCAD). J. Biol. Chem. **269**, 4401–4408.

Schlossmann, J. and Neupert, W. (1995). Assembly of the preprotein receptor MOM72/MAS70 into the protein import complex of the outer membrane of mitochondria. J. Biol. Chem. **270**, 27116–27121.

Schmitt, M., Neupert, W. and Langer, T. (1995). Hsp78, a Clp homologue within mitochondria, can substitute for chaperone functions of mt-hsp70. EMBO J. **14**, 3434–3444.

Schneider, H.-C., Berthold, J., Bauer, M.F., Dietmeier, K., Guiard, B., Brunner, M. and Neupert,W. (1994). Mitochondrial Hsp70/MIM44 complex facilitates protein import. Nature **371**, 768–774.

Ungermann, C., Neupert, W. and Cyr, D.M. (1994). The role of Hsp70 in conferring unidirectionality on protein translocation into mitochondria. Science **266**, 1250–1253.

Ungermann, C., Guiard, B., Neupert, W. and Cyr, D.M. (1996). The DY- and Hsp70-MIM44-dependent reaction cycle driving early steps of protein import into mitochondria. EMBO J. **15**, 735–744.

van Heusden, G.P.H., Griffiths, D.J.F., Ford, J.C., Chin-a-Woeng, T.F.C., Schrader, P.A.T., Carr, A.M. and Steensma, H.Y. (1995). The 14-3-3 proteins encoded by the BMH1 and BMH2 genes are essential in the yeast Saccharomyces cerevisiae and can be replaced by a plant homologue. Eur. J. Biochem. **229**, 45–53.

Voos, W., Gambill, B.D., Laloraya, S., Ang, D., Craig, E.A. and Pfanner, N. (1994). Mitochondrial GrpE is present in a complex with hsp70 and preproteins in transit across membranes. Mol. Cell. Biol. **14**, 6627–6634.

Wagner, I., Arlt, H., van Dyck, L., Langer, T. and Neupert, W. (1994). Molecular chaperones cooperate with PIM1 protease in the degradation of misfolded proteins. EMBO J. **13**, 5135–5145.

Westermann, B., Prip-Buus, C., Neupert, W. and Schwarz, E. (1995). The role of the GrpE homologue, Mge1p, in mediating protein import and protein folding in mitochondria. EMBO J. **14**, 3452–3460.

Zheng, X., Rosenberg, L.E., Kalousek, F. and Fenton, W.A. (1993). GroEL, GroES, and ATP-dependent folding and spontaneous assembly of ornithine transcarbamylase. J. Biol. Chem. **268**, 7489–7493.

■ *Thomas Langer, Walter Neupert and Elisabeth Schwarz:*
Adolf Butenandt Institut für Physiologische Chemie der Universität München
Goethestr. 33
80336 München, Germany
Tel. 49 89 5996 283
Fax. 49 89 5996 270
E-mail: Langer@bio.med.uni-muenchen.de

Protein translocation into the endoplasmic reticulum

Translocation of precursors into the endoplasmic reticulum (ER) may occur either cotranslationally, where the polypeptide is threaded through a proteinaceous pore as it spools off the ribosome; or post-translationally, where synthesis is completed before the polypeptide engages the pore. Recent advances have begun to form a coherent picture of the means by which proteins cross a lipid bilayer. In this review we examine progress towards defining the mechanism of protein translocation into the ER, with particular emphasis on the involvement of HSP70 proteins and their DnaJ-related partners in translocation in Saccharomyces cerevisiae.

Passage of secretory proteins across the membrane of the ER is their first obligate step into the secretory pathway as they proceed to their ultimate destination in the cell. Proteins enter the ER lumen by way of a proteinaceous pore that acts as a gateway across the lipid bilayer. Movement of secretory precursor proteins through this gateway is mediated by the orchestrated action of a number of proteins that support distinct phases of the translocation process: docking of the secretory precursor at the membrane; traversing the pore in the lipid bilayer; and exiting on the luminal side. The precursor engages the docking machinery by the signal sequence, an N-terminal extension of soluble secretory proteins that specifically interacts with a receptor protein to direct the precursor to the site of translocation. After this targeting event, the precursor moves across the membrane and into the lumen, where the signal sequence is cleaved by a specific processing protease. Transmembrane proteins, in contrast to soluble secretory proteins, do not proceed into the lumen but probably move laterally into the bilayer to attain their correct position (Martoglio et al., 1995). The signal sequence of such membrane proteins is generally located in an internal segment of the protein and is not cleaved off after the protein is directed to the membrane (for review see von Heijne, 1990).

Translocation of precursors may occur either cotranslationally, where the polypeptide is threaded through the pore as it spools off the ribosome; or post-translationally, where synthesis is completed before the polypeptide engages the pore. Both co- and post-translational pathways must meet the requirement that the precursor polypeptide be maintained in an extended/unfolded form that is able to pass through the confines of a membrane pore (for review see Verner and Schatz, 1988). In the case of cotranslational translocation, the problem is simplified by the fact that stretches of polypeptide emerging from the ribosome are immediately channeled into the pore. In post-translational translocation, however, polypeptides have, by definition, already been disengaged from the ribosome and so must find another means of preventing premature folding. This is accomplished by the intervention of the 70 kDa heat shock (HSP70) proteins, chaperone molecules that often work in conjunction with DnaJ-like (HSP40) proteins to modulate polypeptide folding (for reviews see Gething and Sambrook, 1992; Hartl, 1996).

■ Cotranslational translocation

The co- and post-translational pathways use both common and distinct machinery to effect the passage of a polypeptide across the ER membrane. Although the translocation pore is central to both pathways, the cytosolic accessory factors required for each pathway are unique. A pivotal component of the cotranslational path is the ribonucleoprotein complex known as the signal recognition particle (SRP), which recognizes the signal sequence of secretory proteins as the nascent chain emerges from the ribosome (for reviews see Nunnari and Walter, 1992; Rapoport, 1992; Walter and Johnson, 1994). In mammalian cells, an additional checkpoint is provided by NAC (nascent polypeptide-associated complex), a heterodimeric protein that prevents proteins lacking signal sequences from binding SRP (Wiedmann et al., 1994). When the signal sequence of a nascent secretory precursor is bound by SRP, two events occur: First, translation of the nascent polypeptide is arrested. This temporary block in protein synthesis prevents premature folding of the polypeptide chain while SRP carries out its second function, that of targeting the ribosome–nascent chain complex to the translocation site. Targeting to the ER membrane is accomplished by interaction of the SRP–ribosome–nascent chain complex with the SRP receptor (SR); an association that allows SRP to link the cytosolic machinery of translation to the membranous machinery of translocation. Upon interaction of SRP with SR, the signal sequence binds to the translocation machinery at the ER membrane, and the translational block is released. Cotranslational translocation proceeds upon GTP hydrolysis by SRP and SR (Powers and Walter, 1995), which are then free to participate in another targeting event. SRP-mediated translocation accounts for the bulk of translocation in mammalian cells, in which the cotranslational pathway is used almost exclusively (for review see Walter and Johnson, 1994). Homologs of the components of SRP have been identified in Saccharomyces cerevisiae, but none are essential for viability (for review see Nunnari and Walter, 1992; Brown et al., 1994). Saccharomyces cerevisiae provides secretory proteins with a mechanistic choice — some follow the path of SRP-dependent, cotranslational translocation, but many enter the ER via an SRP-independent post-translational translocation pathway.

■ Post-translational translocation

Post-translational translocation presents the cell with a unique challenge: proteins must be in an extended/unfolded conformation in order to pass through a pore (for review see Verner and Schatz, 1988), yet post-translational translocation occurs after the completion of protein synthesis, at which point proteins would presumably have had ample time to fold. To resolve this problem, the cell must provide a means of preventing polypeptides from folding prematurely and thus maintain their import competence. Two early studies of post-translational translocation into the mammalian ER first suggested that HSP70 molecules might fulfil this role (Wiech et al., 1987; Zimmermann et al., 1988). The ability of HSP70 proteins to bind a range of polypeptide substrates makes them ideally suited to carry out this cytosolic chaperoning function. The preferred binding motif for HSP70s is a short stretch of amino acids enriched in hydrophobic and aromatic residues that is often flanked by basic residues, while acidic residues are only rarely tolerated (Blond-Elguindi et al., 1993; Gragerov et al., 1994; Takenaka et al., 1995; Rüdiger et al., 1996). Stable association with a substrate molecule is favored when an HSP70 is in the ADP-bound state (Palleros et al., 1994; Schmid et al., 1994). The substrate-binding event stimulates ADP–ATP exchange on the HSP70 (Sadis and Hightower, 1992), and the subsequent binding of ATP causes a conformational change in the HSP70 molecule that results in release of the bound substrate (for review see Hightower et al., 1994). Hydrolysis of the bound ATP then returns HSP70 to its stable polypeptide-binding mode, and the cycle can begin again. This scenario suggests a means by which the cytosolic HSP70s could prepare a secretory precursor for post-translational translocation: as the precursor protein is translated, HSP70 molecules could bind to and 'coat' the nascent polypeptide emerging from the ribosome, preventing premature folding. Nucleotide-dependent binding and dissociation of the HSP70s would then maintain import competence of the precursor polypeptide until it was able to engage the translocation machinery at the ER membrane.

■ Cytosolic factors
■ Ssa1p/Ssa2p

The Ssa protein family in the yeast cytosol consists of four HSP70s — two that are constitutively expressed (Ssa1p/Ssa2p) as well as two that are synthesized only under conditions of cellular stress (Ssa3p/Ssa4) (for review see Craig et al., 1994 and entry p. 26). Ssa1p and Ssa2p have been implicated in the process of protein translocation by several independent approaches. Deshaies et al. (1988) demonstrated that depletion of Ssa1p (placed under regulatable control) in an *ssa1 ssa2 ssa4* deletion strain resulted in the cytosolic accumulation of untranslocated precursor proteins. The authors also showed that the precursor protein pre-pro-alpha factor (ppaF), translated *in vitro* in a wheat germ lysate, was efficiently translocated into yeast ER microsomal vesicles only in the presence of Ssa1p and a second cytosolic factor. Two distinct activities were also found in the analysis of a translocation-stimulating fraction by Chirico et al. (1988), who demonstrated that the two stimulating activities comprised an NEM-insensitive fraction containing Ssa1p and Ssa2p, plus a second fraction containing an unidentified NEM-sensitive component. A substantial clue to the mechanism of Ssa1p/Ssa2p action came from the observation that urea could substitute for HSP70 function. Precursor proteins denatured in urea no longer required Ssa1p/Ssa2p for translocation competence (Chirico et al., 1988), suggesting that the primary role of the cytosolic HSP70s in the translocation reaction was to maintain the precursor protein in a partially unfolded, import-competent state.

■ Ydj1p

Although the identity of the NEM-sensitive component of the stimulatory cytosolic fraction is not yet known, a likely candidate is the cytosolic DnaJ (HSP40) homolog Ydj1p (see entry p. 102). DnaJ homologs are often coupled with HSP70 partners, from the archetypical DnaJ–DnaK pair in *Escherichia coli* to similar pairings in yeast and higher eukaryotes (for review see Cyr et al., 1994). In *E. coli*, DnaJ is able to stimulate the ATPase activity of its HSP70 cohort DnaK (Liberek et al., 1991) — a reaction reproduced by Ydj1p and Ssa1p (Cyr et al., 1992). Ydj1p containing a temperature-sensitive (ts) mutation (*ydj1-151*p) is defective for stimulation of the ATPase activity of Ssa1p *in vitro*, and a *ydj1-151* strain accumulates the secretory precursor ppaF at the restrictive temperature *in vivo* (Caplan et al., 1992a), suggesting that regulation of Ssa1p function by Ydj1p is essential for the translocation process. The C-terminus of Ydj1p contains a consensus site for lipid modification, creating a pool of Ydj1p that is peripherally attached to the ER membrane via a farnesyl moiety (Caplan et al., 1992b). A mutant allele that is not farnesylated displays a temperature-sensitive translocation defect similar to the *ydj1-151* allele (Caplan et al., 1992a), implying that at the restrictive temperature membrane attachment is essential for Ydj1p function in protein trafficking. The association of Ydj1p with the membrane could allow it to serve as a beacon to direct an Ssa1p-coated precursor protein to the membrane site of the translocation apparatus. Upon reaching the translocation site, the precursor could be released from Ssa1p in a Ydj1p-mediated reaction (Cyr et al., 1992), allowing it to engage the docking machinery and to be subsequently translocated across the lipid bilayer. Ydj1p may also have a more direct role in binding the precursor protein and presenting it to the translocation machinery, as some evidence suggests that DnaJ can itself serve as a chaperone (Langer et al., 1992).

■ Membrane machinery

Passage of a secretory precursor across the ER membrane may be modeled as a multistep process, consisting of: (i) docking of the precursor at the translocation site; (ii) insertion into the membrane-spanning translocation pore; and (iii) transit through the pore and release into the ER lumen (see Fig. 1). Genetic and biochemical analyses of post-translational translocation have defined a set of membrane proteins that constitutes the minimum translocon: Sec62p, Sec71p, Sec72p, Sec63p, and the Sec61p complex; plus a stimulatory luminal factor, BiP, which is the major ER-located member of the HSP70 family (Rothblatt et al., 1989; Deshaies et al., 1991; Panzner et al., 1995). A variety of experimental approaches over the last several years have begun to assign functions to these components, and to clarify how they cooperate to effect the transfer of a polypeptide from the cytosol into the ER lumen.

■ Sec62p, Sec71p, Sec72p

Sanz and Meyer (1989) have shown that translocation is divided by energy requirements into two steps: ATP-independent association of the precursor with the membrane, followed by ATP-dependent passage through the membrane. On the basis of these energetics, the integral membrane protein Sec62p (Deshaies and Schekman, 1990) appears to be involved early in the translocation process. Müsch et al. (1992) found that a ribosome-tethered molecule of the precursor protein ppaF will cross-link to Sec62p in the absence of ATP, but in the presence of ATP the tethered precursor cross-links primarily to Sec61p (the translocation pore). This suggests that the Sec62p–precursor interaction is dissolved in an ATP-dependent reaction that results in transfer of the precursor to the pore. Further evidence for an early role for Sec62p comes from the observation that a translocation-defective allele of *sec62* (*sec62-1*) (Deshaies and Schekman, 1989) impairs the ability of ppaF to interact with Sec61p (Sanders et al., 1992).

Sec71p and Sec72p (previously known as Sec66p and Sec67p) (Green et al., 1992) are implicated early in translocation by both genetic and biochemical evidence. Although the *SEC72* gene is not essential, and *SEC71* is essential only at elevated temperatures, deletion or mutation of either gene impairs the translocation of a subset of precursor proteins (Feldheim et al., 1993; Kurihara and Silver, 1993; Fang and Green, 1994; Feldheim and Schekman, 1994). Feldheim and Schekman (1994) used chimeric proteins to demonstrate that the translocation defect manifested in Δ*sec72* is associated with the signal sequence portion of the secretory precursor rather than with the mature region. This observation, in conjunction with the fact that Δsec71 and Δsec72 strains show translocation defects for several of the same secretory precursors, suggests that Sec71p and Sec72p are involved in some facet of signal sequence recognition. Sec62p, Sec71p, and Sec72p comprise a translocon subcomplex that targets precursor proteins to the membrane site of translocation. Binding of a secretory precursor to this receptor complex depends on the presence of an intact signal sequence and occurs only in the absence of ATP: in the presence of ATP, the precursor is released in a reaction mediated by the luminal HSP70 BiP (Lyman and Schekman, 1997).

Sec62p, Sec71p, and Sec72p seem to function primarily in post-translational translocation rather than in cotranslational translocation. Mutation or deletion of sec62, sec71, or sec72 (Rothblatt et al., 1989; Feldheim et al., 1993; Kurihara and Silver, 1993; Fang and Green, 1994; Feldheim and Schekman, 1994) has a much more significant effect on the maturation of ppaF and pre-pro-carboxypeptidase Y, precursors that are able to translocate post-translationally (Chirico et al., 1988; Deshaies et al., 1988), than on that of invertase, which translocates chiefly cotranslationally (Brodsky et al., 1995). In addition, Ng et al. (1996) recently isolated a mutant allele of sec62 that is defective for the translocation of SRP-independent (i.e. post-translational) substrates yet does not affect the translocation of SRP-dependent (i.e. cotranslational) substrates. The function of Sec62p, Sec71p, and Sec72p as a signal sequence receptor complex in post-translational translocation is most likely paralleled in cotranslational translocation by the binding of the signal sequence to SRP and the subsequent interaction of SRP with its receptor (SR) at the ER membrane (for review see Walter and Johnson, 1994).

■ Sec61p

The central component of the yeast translocon is Sec61p, the translocation pore. Sec61p is a hydrophobic integral membrane protein of 52 kDa with ten membrane-spanning domains (Stirling et al., 1992; Wilkinson et al., 1996). Sec61p is associated with two smaller polypeptides of 14 and 8 kDa (Sbh1p and Sss1p) in a heterotrimeric complex that is the homologous counterpart of the mammalian Sec61p complex, in which Sec61α, β-, and -γ correspond to the yeast Sec61p, Sbh1p, and Sss1p, respectively (Panzner et al., 1995). Sec61p (Sec61α) is most likely the key pore-forming component. Experimental evidence has demonstrated that as secretory precursors traverse the membrane, their primary interaction is with Sec61p. In yeast, a precursor protein sterically blocked in the process of crossing the lipid bilayer can be chemically cross-linked to Sec61p (Sanders et al., 1992), and in the mammalian ER Sec61α is the major cross-linking partner of nascent chains emerging from the ribosome (Mothes et al., 1994). In mammalian cells Sec61α is also associated with translating ribosomes, suggesting that the interaction serves to 'clamp' the ribosome over the pore, forming a conduit through which the emerging polypeptide may pass during cotranslational translocation (Görlich et al., 1992; Görlich and Rapoport, 1993; Kalies et al., 1994). In yeast, the Sec61p complex exists in two distinct populations: one in which the heterotrimeric Sec61p complex is associated with membrane-bound ribosomes, and the other in which the heterotrimeric complex is associated with Sec62p, Sec63p, Sec71p, and Sec72p (Panzner et al., 1995). The authors speculate that this partitioning of the Sec61p complex could reflect a functional segregation of the pore proteins between post-translational and cotranslational pathways: an intriguing possibility that remains to be tested experimentally.

■ Sec63p, BiP (Kar2p)

Sec63p is an integral membrane protein containing three transmembrane segments that define cytoplasmic and luminal domains. Within a 120 amino acid loop protruding into the lumen is a 70 amino acid region that is 43% identical to the 'DnaJ-domain', the most highly conserved region of the DnaJ family of proteins (Sadler et al., 1989; Feldheim et al., 1992; for review see Caplan et al., 1993). The recurring evolutionary motif of HSP70–DnaJ pairings suggests that one of the functions of the J-domain in Sec63p may be to provide a means of linking the translocation machinery of the ER membrane to a luminal translocation factor — the HSP70 homolog Kar2p (BiP).

Sec63p and BiP have been both genetically and biochemically implicated in translocation. sec63-1, a ts mutation in sec63 that maps to the conserved J-domain, displays translocation defects both in vitro and in vivo (Rothblatt et al., 1989). A group of ts alleles of kar2 are defective for translocation in vitro (Sanders et al., 1992) and accumulate untranslocated precursors in vivo (Vogel et al., 1990); a phenotype that is reproduced by the cellular depletion of BiP (Vogel et al., 1990; Nguyen et al., 1991). The notion that BiP and Sec63p act together in translocation is supported by several lines of evidence. Scidmore et al. (1993) noted that certain alleles of kar2 and sec63 demonstrate allele-specific synthetic lethality. Synthetic lethality is a phenomenon in which the phenotype of a pairwise combination of two mutant alleles is more severe than that of either allele alone, often indicating that the gene products interact or act at a common step in a biological process (Kaiser and Schekman, 1990). Brodsky and Schekman (1993) provided evidence for a direct physical interaction between BiP and Sec63p: a complex containing BiP, Sec63p, Sec71p, and Sec72p can be isolated from yeast microsomal membranes and copurifies over several column steps. However, if the complex is purified from the sec63-1 strain, in which a conserved residue of the J-domain has been mutated (A179T) (Nelson et al., 1993), BiP is released from the complex. The ability of the sec63-1 mutation to destabilize interaction of BiP with the translocation complex implies that BiP interacts with Sec63p via the lumenal J-domain and that the sec63-1 mutation uncouples BiP from Sec63p. The isolated J-domain of Sec63p is sufficient to mediate stable ATP-dependent interaction with BiP, and is also able to stimulate the ATPase activity of BiP (Corsi and Schekman, 1997). ATP appears to regulate the association of BiP with Sec63p. Brodsky and Schekman (1993) showed that the presence of the non-hydrolysable analog ATPγS releases BiP from the Sec63p–Sec71p–Sec72p complex, and that non-hydrolysable analogs will not support interaction of the J-domain of Sec63p with BiP (Corsi and Schekman, 1997).

■ Mechanism of translocation: BiP and Sec63 function

An array of cytosolic, membranous, and luminal factors must cooperate in order to orchestrate the movement of a secretory protein across the ER membrane. As a prelude to post-translational translocation, the cytosolic HSP70s Ssa1p/Ssa2p most likely bind the completed precursor polypeptide and suspend its folding so that it may retain an extended, import-competent conformation. Interaction of Ssa1p/Ssa2p with the DnaJ homolog Ydj1p may simultaneously localize the precursor to the site of translocation and induce its release from the HSP70s. The secretory precursor may then dock at the ER membrane (Fig. 1, step 1) in an ATP-independent process that involves binding of the precursor to a translocon subcomplex comprising Sec62p, Sec71p, and Sec72p (Lyman and Schekman, 1997). This is followed by the ATP-dependent release of the precursor and its transfer into the pore (Müsch et al., 1992; Lyman and Schekman, 1997) in a reaction mediated by BiP and Sec63p (Lyman and Schekman, 1997) (Fig. 1, step 2). In the final step of translocation (Fig. 1, step 3), BiP and Sec63p cooperate to complete the passage of the precursor through the pore and to deposit it in the ER lumen (Sanders et al., 1992; Lyman and Schekman, 1995).

Both the HSP70, BiP, and its DnaJ cohort, Sec63p, have previously been implicated early in the translocation process. Sanders et al. (1992) characterized a mutant allele of kar2 (kar2-159) that appeared to be defective primarily for an early step of translocation, as the mutation largely prevented the formation of a complex between the precursor and the pore. The authors also showed that the sec63-1 J-domain mutation decreased the ability of a precursor protein to associate with the pore, implying that the sec63-1 allele may impair translocation at the same step as kar2-159; and that one of the

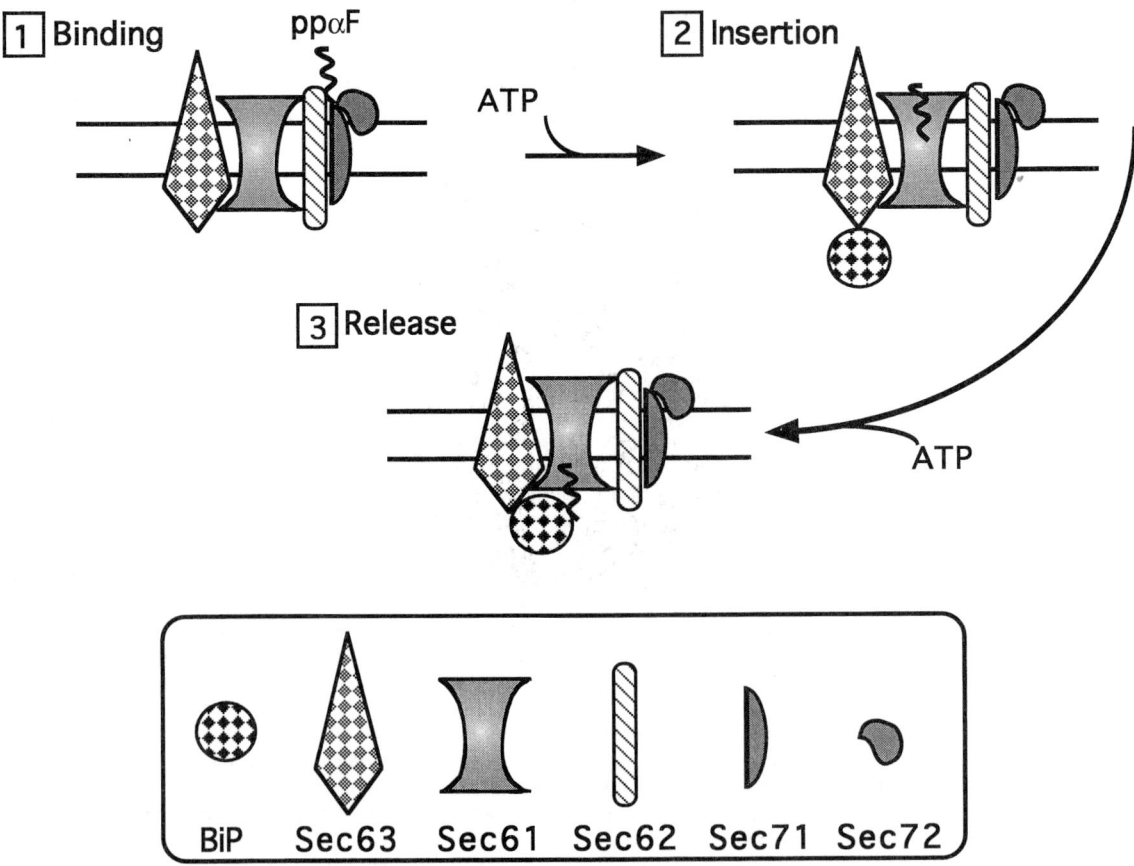

Figure 1. Model of post-translational translocation across the ER membrane (see text for details).

roles of Sec63p may be to act with BiP at this transfer step. Recent data have provided a concrete basis for these observations: After the precursor has engaged the translocon, BiP effects the release of the receptor-bound precursor in an ATP-dependent reaction that requires interaction between BiP and Sec63p (Lyman and Schekman, 1997).

BiP and Sec63p have also been implicated in the final step of translocation (Fig. 1, step 3), release of the precursor from the pore into the lumen. BiP binds to a precursor protein blocked in its passage across the membrane (Sanders et al., 1992), suggesting that it interacts with transiting polypeptides as they emerge from the pore. Compelling evidence for a role of BiP late in the translocation process comes from characterization of a mutant allele of kar2 (kar2-203) that does not allow the precursor to exit from the pore (Sanders et al., 1992), resulting in the formation of a 'stalled' precursor–pore complex. Lyman and Schekman (1995) have expanded on this observation by demonstrating that the sec63-1 allele, which uncouples BiP from its interaction with Sec63p, results in a phenotype identical to that of kar2-203 — the precursor protein is 'stalled' in the pore, unable to transit into the lumen. This suggests that Sec63p and BiP also act together in the final phase of translocation.

These observations converge on the conclusion that BiP and Sec63p must interact at two distinct points in the translocation cycle (Fig. 1, steps 2 and 3) in order to provide the impetus for the precursor to enter the pore, traverse it, and exit into the lumen. How might BiP and Sec63p accomplish this feat? The topology of Sec63p provides some clues — with its large cytosolic tail, three transmembrane domains, and lumenal J-domain (Feldheim et al., 1992), Sec63p is ideally suited to link lumenal BiP function to that of the membrane-bound translocation machinery. The observation that Sec63p exists in a complex with Sec61p, Sec62p, Sec71p, and Sec72p (Brodsky and Schekman, 1993; Deshaies et al., 1991) raises the possibility that Sec63p could serve as a mediator through which BiP could act or be acted upon by the other components of the translocon, through a series of allosteric conformational changes [as suggested in the basic tenets of the model presented by Brodsky and Schekman, (1994)].

Such communication between BiP, Sec63p, and the translocon components could have a bipartite function: earlier, in releasing the precursor from the receptor complex and possibly also in 'activating' the pore to receive the precursor (Fig. 1, step 2); and later, in facilitating the final phase of precursor transfer across the membrane. The early role of BiP may encompass both its demonstrated function in precursor release and a putative 'pore-gating' function proposed by Crowley et al. (1994). In this elegant study, the authors demonstrated that the mammalian translocation pore opens to the lumen only when the span of the nascent chain emerging from the ribosome reaches a minimum critical length of c. 70 residues, and suggested that a luminal translocation factor such as BiP may 'plug' the pore until the advancing secretory polypeptide somehow triggers its release. It is also possible that, rather than acting as a passive 'plug', BiP might actively induce the pore to open. In such a scenario, the cytosolic domain of Sec63p might serve as a sensor to alert BiP to the binding of a precursor protein at the signal sequence receptor complex, upon which an ATP-mediated conformational change originating from BiP in the lumen could simultaneously open the pore and release the bound precursor to allow it to access the pore. Alternatively, as suggested by Brodsky and Schekman (1994), one role of BiP might be to 're-prime' the translocation apparatus to ready it for a second round of import.

In the latter part of the translocation reaction (Fig. 1, step 3), BiP cooperates with Sec63p in driving the transfer of the polypeptide across the membrane and into the lumen. The two primary models for HSP70 function in this reaction, as reviewed by Glick (1995) and Brodsky (1996), propose that HSP70s work either as a Brownian ratchet or as molecular motors. The ratchet model supposes that the random thermal oscillations of Brownian motion cause the transiting polypeptide to slide back and forth in the translocation channel. Binding of a luminal HSP70 such as BiP to the end of the polypeptide flickering in and out of the pore could skew the kinetic equilibrium by trapping a portion of the polypeptide chain on the luminal side. As more of the polypeptide chain 'flickers' into the lumen, successive binding of other HSP70 molecules would then increasingly prevent retrograde movement and ultimately deposit the entire precursor into the lumen. The function of Sec63p in this model would be primarily to place BiP in proximity to the translocation machinery in order to increase its local concentration around the translocation site. In contrast, the molecular motor model proposes that an HSP70 molecule actively reels the transiting polypeptide into the lumen. In this scenario, HSP70s generate a pulling force by being anchored in the ADP-bound form to a membrane component (i.e. Sec63p) while simultaneously binding a portion of the emerging precursor protein. Polypeptide binding stimulates ADP–ATP exchange on HSP70 (Sadis and Hightower, 1992), and the resultant conformational change induced by ATP binding (Palleros et al., 1993) would allow force to be transmitted to the transiting polypeptide, pulling it into the lumen. In the ATP-bound form, the HSP70 would dissociate from the membrane anchor, and ATP hydrolysis by the HSP70 would allow it to rebind the membrane component and engage the polypeptide chain in another cycle of binding. The role of Sec63p in this model is a dynamic one — in addition to serving as an anchor for BiP, Sec63p is likely to regulate the association of BiP with the translocation machinery by stimulating the hydrolysis of bound ATP, thus governing the conversion between 'active' and 'inactive' motor states.

Since there are HSP70s placed on either side of the ER membrane (Ssa1p/Ssa2p and BiP), both of these models of luminal HSP70 action must invoke some means of preventing the precursor from being caught in a 'tug of war' between the opposing HSP70s. Brodsky et al. (1993) have shown that BiP and Ssa1p will not function interchangeably on opposite sides of the membrane, suggesting that despite their significant homology (63% identity), each molecule carries out unique functions. It is likely that the need for the HSP70s to interact with their respective DnaJ partners forms at least part of the basis for this discrimination. Just as Sec63p appears to be involved with BiP in receiving the transiting polypeptide, Ydj1p may be required for the regulated release of the precursor on the cytosolic side of the membrane. It is also possible that BiP and Sec63p differ in the tenacity of their binding, such that the coordinated forward movement of the precursor through the pore might be accomplished by gradation of binding affinity that allows a stronger BiP–precursor interaction to 'overcome' a weaker association of the precursor with Ssa1p/Ssa2p.

The common feature of both both the Brownian ratchet and the molecular motor models is that luminal HSP70 provides the energetic asymmetry necessary to prevent precursor backsliding and to deposit the transiting polypeptide into the lumen. Surprisingly, a yeast reconstituted system that lacks BiP and ATP still supports a low level of apparently faithful translocation, but addition of BiP and ATP stimulates translocation several fold to a final level comparable with that achieved with native ER microsomal vesicles (Panzner et al., 1995). The authors suggest that the BiP-independent translocation represents a basal level of 'passive' transport, or that there is enough energy 'stored' in the system to support a single round of translocation, implying, perhaps, that the directionality provided by BiP is an important, but not absolutely essential, feature of translocation. Luminal modifications such as glycosylation might also play a role in preventing retrograde precursor movement (Ooi and Weiss, 1992). Regardless, it is clear that efficient translocation in yeast requires both BiP and ATP. The role of BiP in mammalian translocation, however, is unclear. Although one report states that luminal proteins are required for the completion of translocation (Nicchitta and Blobel, 1993), a purified reconstituted mammalian system does not require BiP (Görlich and Rapoport, 1993). The apparent mechanistic variance between yeast and mammalian translocation might be traced to a choice of life-style: whereas yeast use both co- and post-translational pathways, translocation in mammalian cells is almost exclusively cotranslational (for review see Walter and Johnson, 1994). In a cotranslational reaction, the ribosome might provide a 'pushing' force as it spools nascent polypeptide into the pore, obviating the need for BiP to supply a 'pulling' force. However, Brodsky et al. (1995) recently demonstrated that mutant alleles of *kar2* and *sec63* are defective for cotranslational translocation in yeast as well as for post-translational translocation, suggesting that the roles of BiP and Sec63p may be more complex than currently supposed.

Another facet of BiP function lies not in its connection to the machinery of translocation, but in post-translocational events: since proteins must be partially unfolded in order to cross the membrane, they must be folded again upon entry into the lumen. HSP70s are thought to promote correct folding of their substrates by the binding and regulated, ATP-dependent release of exposed hydrophobic surfaces, thus preventing non-specific hydrophobic aggregation (for review see Hartl, 1996). Several studies have implicated BiP in the normal folding pathways of secretory proteins (Ng et al., 1989; Hammond and Helenius, 1994; Melnick et al., 1994; Simons et al., 1995), lending credence to the prospect of HSP70-mediated folding in the ER. Recently, Craven et al. (1996) identified a second HSP70 molecule (Lhs1p, also called Ssi1p; see entry p. 36) in the ER lumen. Although *LHS1* is not an essential gene, synthetic lethality with *kar2* alleles suggests functional overlap betweeen the two HSP70s; a notion supported by evidence implying a role for Lhs1p in protein folding and by translocation defects in an *lhs1* deletion strain. The ER lumen contains a second DnaJ-like molecule as well: the DnaJ homolog Scj1p (Blumberg and Silver, 1991) genetically interacts with BiP (Schlenstedt et al., 1995) and may cooperate with BiP and/or Lhs1p in the folding process, providing one more link in the chain of HSP70–DnaJ interactions essential for the journey of a secretory protein from cytosol to lumen.

▌Acknowledgements

We would like to thank Jeff Brodsky, Ann Corsi, and Rien Pilon for valuable comments and discussions in the course of preparing this review.

▌References

Blond-Elguindi, S., Cwirla, S.E., Dower, W.J., Lipshutz, R.J., Sprang, S.R., Sambrook, J.F. and Gething, M.H. (1993). Affinity panning of a library of peptides displayed on bacteriophages reveals the binding specificity of BiP. Cell **75**, 717–728.

Blumberg, H. and Silver, P.A. (1991). A homologue of the bacterial heat shock gene *DnaJ* that alters protein sorting in yeast. Nature **349**, 627–629.

Brodsky, J.L. (1996). Post-translational translocation: not all hsc70s are created equal. Trends Biochem. Sci. **21**, 122–126.

Brodsky, J.L. and Schekman, R. (1993). A Sec63p-BiP complex from yeast is required for protein translocation in a reconstituted proteoliposome. J. Cell Biol. **123**, 1355–1363.

Brodsky, J.L. and Schekman, R. (1994). In *The biology of heat shock proteins and molecular chaperones* (eds Morimoto, R.I., Tissieres, A. and Georgopoulos, C.). Cold Spring Harbor Laboratory Press, New York, pp. 85–109.

Brodsky, J.L., Hamamoto, S., Feldheim, D. and Schekman, R. (1993). Reconstitution of protein translocation from solubilized yeast membranes reveals topologically distinct roles for BiP and cytosolic hsc70. J. Cell Biol. **120**, 95–102.

Brodsky, J.L., Goeckeler, J. and Schekman, R. (1995). BiP and Sec63p are required for both co- and post-translational protein translocation into the yeast endoplasmic reticulum. Proc. Natl. Acad. Sci. **92**, 9643-9646.

Brown, J.D., Hann, B.C., Medzihradszky, K.F., Niwa, M., Burlingame, A.L. and Walter, P. (1994). Subunits of the *Saccharomyces cerevisiae* signal recognition particle required for its functional expression. EMBO J. **13**, 4390–4400.

Caplan, A.J., Cyr, D.M. and Douglas, M.G. (1992a). YDJ1p facilitates polypeptide translocation across different intracellular membranes by a conserved mechanism. Cell **71**, 1143–1155.

Caplan, A.J., Tsai, J., Casey, P.J. and Douglas, M.G. (1992b). Farnesylation of Ydj1p is required for function at elevated growth temperatures in *Saccharomyces cerevisiae*. J. Biol. Chem. **267**, 18890–18895.

Caplan, A.J., Cyr, D.M. and Douglas, M.G. (1993). Eukaryotic homologues of *Escherichia coli* DnaJ: A diverse protein family that functions with hsp70 stress proteins. Mol. Biol. Cell **4**, 555–563.

Chirico, W.J., Waters, M.G. and Blobel, G. (1988). 70K heat shock related proteins stimulate protein translocation into microsomes. Nature **332**, 805–810.

Corsi, A.K., and Schekman, R. (1997). The lumenal domain of Sec63p stimulates the ATPase activity of BiP and mediates BiP recruitment to the translocon in *Saccharomyces cerevisiae*. J. Cell Biol. In press.

Craig, E.A., Baxter, B.K., Becker, J., Halladay, J. and Ziegelhoffer, T. (1994). In *The biology of heat shock proteins and molecular chaperones* (eds Morimoto, R.I., Tissieres, A. and Georgopoulos, C.). Cold Spring Harbor Laboratory Press, New York, pp. 31–52.

Craven, R.A., Egerton, M. and Stirling, C.J. (1996). A novel Hsp70 of the yeast ER lumen is required for the efficient translocation of a number of protein precursors. EMBO J. **15**, 2640–2650.

Crowley, K.S., Liao, S., Worrell, V.E., Reinhart, G.D. and Johnson, A.E. (1994). Secretory proteins move through the endoplasmic reticulum membrane via an aqueous, gated pore. Cell **78**, 461–471.

Cyr, D.M., Lu, X. and Douglas, M.G. (1992). Regulation of hsp70 function by a eukaryotic DnaJ homolog. J. Biol. Chem. **267**, 20927–20931.

Cyr, D.M., Langer, T. and Douglas, M.G. (1994). DnaJ-like proteins: molecular chaperones and specific regulators of hsp70. Trends Biochem. Sci. **19**, 176–181.

Deshaies, R.J. and Schekman, R.W. (1989). SEC62 encodes a putative membrane protein required for protein translocation into the yeast endoplasmic reticulum. J. Cell Biol. **109**, 2653–2664.

Deshaies, R.J. and Schekman, R. (1990). Structural and functional dissection of Sec62p, a membrane-bound component of the yeast endoplasmic reticulum protein import machinery. Mol. Cell. Biol. **10**, 6024–6035.

Deshaies, R.J., Koch, B.D., Werner-Washburne, M., Craig, E.A. and Schekman, R. (1988). A subfamily of stress proteins facilitates translocation of secretory and mitochondrial precursor polypeptides. Nature **332**, 800–805.

Deshaies, R.J., Sanders, S.L., Feldheim, D.A. and Schekman, R. (1991). Assembly of the yeast Sec proteins involved in translocation into the endoplasmic reticulum into a membrane-bound multisubunit complex. Nature **349**, 806–808.

Fang, H. and Green, N. (1994). Nonlethal *sec71-1* and *sec72-1* mutations eliminate proteins associated with the Sec63-BiP complex from *S. cerevisiae*. Mol. Biol. Cell **5**, 933–942.

Feldheim, D. and Schekman, R. (1994). Sec72p contributes to the selective recognition of signal peptides by the secretory polypeptide translocation complex. J. Cell Biol. **126**, 935–943.

Feldheim, D., Rothblatt, J. and Schekman, R. (1992). Topology and functional domains of Sec63p, an endoplasmic reticulum membrane protein required for secretory protein translocation. Mol. Cell. Biol. **12**, 3288–3296.

Feldheim, D., Yoshimura, K., Admon, A. and Schekman, R. (1993). Structural and functional characterization of Sec66p, a new subunit of the polypeptide translocation apparatus in the yeast endoplasmic reticulum. Mol. Biol. Cell **4**, 931–939.

Gething, M. and Sambrook, J. (1992). Protein folding in the cell. Nature **355**, 33–45.

Glick, B.S. (1995). Can hsp70 proteins act as force-generating motors? Cell **80**, 11–14.

Görlich, D. and Rapoport, T.A. (1993). Protein translocation into proteoliposomes reconstituted from purified components of the endoplasmic reticulum membrane. Cell **75**, 615–630.

Görlich, D., Prehn, S., Hartmann, E., Kalies, K.-U. and Rapoport, T.A. (1992). A mammalian homolog of Sec61p and SecYp is associated with ribosomes and nascent polypeptides during translocation. Cell **71**, 489–503.

Gragerov, A., Zeng, L., Zhao, X., Burkholder, W. and Gottesman, M.E. (1994). Specificity of DnaK-peptide binding. J. Mol. Biol. **235**, 848–854.

Green, N., Fang, H. and Walter, P. (1992). Mutations in three novel complementation groups inhibit membrane protein insertion into and soluble protein translocation across the endoplasmic reticulum membrane of *Saccharomyces cerevisiae*. J. Cell Biol. **116**, 597–604.

Hammond, C. and Helenius, A. (1994). Folding of VSV G protein: sequential interaction with BiP and calnexin. Science **266**, 456–458.

Hartl, F.U. (1996). Molecular chaperones in protein folding. Nature **381**, 571–579.

Hightower, L.E., Sadis, S.E. and Takenaka, I.M. (1994). In *The biology of heat shock proteins and molecular chaperones* (eds Morimoto, R.I., Tissieres, A. and Georgopoulos, C.). Cold Spring Harbor Laboratory Press, New York, pp. 179–207.

Kaiser, C. and Schekman, R. (1990). Distinct sets of SEC genes govern transport vesicle formation and fusion early in the secretory pathway. Cell **61**, 723–733.

Kalies, K.-U., Görlich, D. and Rapoport, T.A. (1994). Binding of ribosomes to the rough endoplasmic reticulum mediated by the Sec61p-complex. J. Cell Biol. **126**, 925–934.

Kurihara, T. and Silver, P. (1993). Suppression of a *sec63* mutation identifies a novel component of the yeast endoplasmic reticulum translocation apparatus. Mol. Biol. Cell **4**, 919–930.

Langer, T., Lu, C., Echols, H., Flanagan, J., Hayer, M.K. and Hartl, F.U. (1992). Successive action of DnaK, DnaJ, and GroEL along the pathway of chaperone-mediated protein folding. Nature **356**, 683–689.

Liberek, K., Marszalek, J., Ang, D., Georgopoulos, C. and Zylicz, M. (1991). *Escherichia coli* DnaJ and GrpE heat shock proteins jointly stimulate the ATPase activity of DnaK. Proc. Natl. Acad. Sci. USA **88**, 2874–2878.

Lyman, S.K. and Schekman, R. (1995). Interaction between BiP and Sec63p is required for the completion of protein translocation into the ER of *Saccharomyces cerevisiae*. J. Cell Biol. **131**, 1163–1171.

Lyman, S.K. and Schekman, R. (1997). Binding of secretory precursor polypeptides to a translocon subcomplex is regulated by BiP. Cell **88**, 85–96.

Martoglio, B., Hofmann, M.W., Brunner, J. and Dobberstein, B. (1995). The protein-conducting channel in the membrane of the endoplasmic reticulum is open laterally toward the lipid bilayer. Cell **81**, 207–214.

Melnick, J., Dul, J.L. and Argon, Y. (1994). Sequential interaction of the chaperones BiP and GRP94 with immunoglobulin chains in the endoplasmic reticulum. Nature **370**, 373–375.

Mothes, W., Prehn, S. and Rapoport, T.A. (1994). Systematic probing of the environment of a translocating secretory protein during translocation through the ER membrane. EMBO J. **13**, 3973–3982.

Müsch, A., Wiedmann, M. and Rapoport, T.A. (1992). Yeast Sec proteins interact with polypeptides traversing the endoplasmic reticulum membrane. Cell **69**, 343–352.

Nelson, M.K., Kurihara, T. and Silver, P.A. (1993). Extragenic suppressors of mutations in the cytoplasmic C terminus of

SEC63 define five genes in Saccharomyces cerevisiae. Genetics **134**, 159–173.

Ng, D.T.W., Randall, R.E. and Lamb, R.A. (1989). Intracellular maturation and transport of the SV5 type II glycoprotein hemagglutinin-neuramidase: specific and transient association with GRP78-BiP in the endoplasmic reticulum and extensive internalization from the cell surface. J. Cell Biol. **109**, 3273–3289.

Ng, D.T.W., Brown, J.D. and Walter, P. (1996). Signal sequences specifiy the targeting route to the endoplasmic reticulum membrane. J. Cell Biol. **134**, 269–278.

Nguyen, T.H., Law, D.T.S. and Williams, D.B. (1991). Binding protein BiP is required for translocation of secretory proteins into the endoplasmic reticulum in *Saccharomyces cerevisiae*. Proc. Natl. Acad. Sci. USA **88**, 1565–1569.

Nicchitta, C.V. and Blobel, G. (1993). Lumenal proteins of the mammalian endoplasmic reticulum are required to complete protein translocation. Cell **73**, 989–998.

Nunnari, J. and Walter, P. (1992). Protein targeting to and translocation across the membrane of the endoplasmic reticulum. Curr. Opin. Cell Biol. **4**, 573–580.

Ooi, C.E. and Weiss, J. (1992). Bidirectional movement of a nascent polypeptide across microsomal membranes reveals requirements for vectorial translocation of proteins. Cell **71**, 87–96.

Palleros, D.R., Reid, K.L., Shi, L., Welch, W. and Fink, A.L. (1993). ATP-induced protein-Hsp70 complex dissociation requires K+ but not ATP hydrolysis. Nature **365**, 664–666.

Palleros, D.R., Shi, L., Reid, K.L. and Fink, A.L. (1994). Hsp70-protein complexes: Complex stability and conformation of bound substrate protein. J. Biol. Chem. **269**, 13107–13114.

Panzner, S., Dreier, L., Hartmann, E., Kostka, S. and Rapoport, T.A. (1995). Posttranslational protein transport in yeast reconstituted with a purified complex of Sec proteins and Kar2p. Cell **81**, 561–570.

Powers, T. and Walter, P. (1995). Reciprocal stimulation of GTP hydrolysis by two directly interacting GTPases. Science **269**, 1422–1423.

Rapoport, T.A. (1992). Transport of proteins across the endoplasmic reticulum membrane. Science **258**, 931–936.

Rothblatt, J.A., Deshaies, R.J., Sanders, S.L., Daum, G. and Schekman, R. (1989). Multiple genes are required for proper insertion of secretory proteins into the endoplasmic reticulum in yeast. J. Cell Biol. **109**, 2641–2652.

Sadis, S. and Hightower, L.E. (1992). Unfolded proteins stimulate molecular chaperone Hsc70 ATPase by accelerating ADP/ATP exchange. Biochemistry **31**, 9406–9412.

Sadler, I., Chiang, A., Kurihara, T., Rothblatt, J., Way, J. and Silver, P. (1989). A yeast gene important for protein assembly into the endoplasmic reticulum and the nucleus has homology to DnaJ, an *Escherichia coli* heat shock protein. J. Cell Biol. **109**, 2665–2675.

Sanders, S.L., Whitfield, K.M., Vogel, J.P., Rose, M.D. and Schekman, R.W. (1992). Sec61p and BiP directly facilitate polypeptide translocation into the ER. Cell **69**, 353–365.

Sanz, P. and Meyer, D.I. (1989). Secretion in yeast: preprotein binding to a membrane receptor and ATP-dependent translocation are sequential and separable events in vitro. J. Cell Biol. **108**, 2101–2106.

Schlenstedt, G., Harris, S., Risse, B., Lill, R. and Silver, P.A. (1995). A yeast DnaJ homologue, Scj1p, can function in the endoplasmic reticulum with BiP/Kar2p via a conserved domain that specifices interaction with hsp70s. J. Cell Biol. **129**, 979–988.

Schmid, D., Baici, A., Gehring, H. and Christen, P. (1994). Kinetics of molecular chaperone action. Science **263**, 971–973.

Scidmore, M.A., Okamura, H.H. and Rose, M.D. (1993). Genetic interactions between *KAR2* and *SEC63*, encoding eukaryotic homologues of DnaK and DnaJ in the endoplasmic reticulum. Mol. Biol. Cell **4**, 1145–1159.

Simons, J.F., Ferro-Novick, S., Rose, M.D. and Helenius, A. (1995). BiP/Kar2p serves as a molecular chaperone during carboxypeptidase Y folding in yeast. J. Cell Biol. **130**, 41–49.

Stirling, C.J., Rothblatt, J., Hosobuchi, M., Deshaies, R. and Schekman, R. (1992). Protein translocation mutants defective in the insertion of integral membrane proteins into the endoplasmic reticulum. Mol. Biol. Cell **3**, 129–142.

Takenaka, I.M., Leung, S., McAndrew, S.J., Brown, J.P. and Hightower, L.E. (1995). Hsc70-binding peptides selected from a phage display peptide library that resemble organellar targeting sequences. J. Biol. Chem. **270**, 19839–19844.

Verner, K. and Schatz, G. (1988). Protein translocation across membranes. Science **241**, 1307–1313.

Vogel, J.P., Misra, L.M. and Rose, M.D. (1990). Loss of BiP/GRP78 function blocks translocation of secretory proteins in yeast. J. Cell Biol. **110**, 1885–1895.

von Heijne, G. (1990). The signal peptide. J. Membr. Biol. **115**, 195–201.

Walter, P. and Johnson, A.E. (1994). Signal sequence recognition and protein targeting to the endoplasmic reticulum membrane. Annu. Rev. Cell Biol. **10**, 87–119.

Wiech, H., Sagstetter, M., Muller, G. and Zimmermann, R. (1987). The ATP-requiring step in assembly of M13 procoat protein into microsomes is related to preservation of translocation competence of the precursor protein. EMBO J. **6**, 1011–1016.

Wiedmann, B., Sakai, H., Davis, T.A. and Wiedmann, M. (1994). A protein complex required for signal-sequence-specific sorting and translocation. Nature **370**, 434–440.

Wilkinson, B.M., Critchley, A.J., and Stirling, C.J. (1996). Determination of the transmembrane topology of yeast Sec61p, an essential component of the endoplasmic reticulum translocation complex. J. Biol. Chem. **271**, 25590–25597.

Zimmermann, R., Sagstetter, M., Lewis, M.J. and Pelham, H.R.B. (1988). Seventy-kilodalton heat shock preins and an additional component from reticulocyte lysate stimulate import of M13 procoat into microsomes. EMBO J. **7**, 2875–2880.

■ *Susan K. Lyman and Randy Schekman:*
Department of Molecular and Cell Biology
Howard Hughes Medical Institute
University of California, Berkeley
Berkeley, CA 94720, USA
Tel. 1 510 642 5686
Fax. 1 510 642 7846
E-mail: Schekman@mendel.berkeley.edu

Quality control in the endoplasmic reticulum

The endoplasmic reticulum (ER) is unique among the protein folding compartments in the eukaryotic cell in producing proteins mainly for secretion and for residence in other intracellular compartments. In mammalian cells, the vast majority of these proteins are cotranslationally translocated into the ER lumen or inserted into the ER membrane. Folding occurs co- and post-translationally, and export from the ER occurs only when the proteins have acquired a properly folded and oligomerized form. For different proteins, the rates and efficiencies of export vary depending on the kinetics and completeness of the folding and assembly process. Misfolded proteins, unassembled subunits and incompletely assembled oligomers are, as a rule, retained in the ER and eventually degraded. The conformation-specific sorting underlying the selective transport from the ER, observed for both soluble and membrane proteins, is called quality control (Hurtley and Helenius, 1989) or architectural editing (Klausner, 1989).

The reasons for quality control are at least threefold. First, it makes sense to keep newly synthesized proteins in the ER until folding is completed because the ER contains a rich variety of chaperones and folding enzymes at high concentrations. The ER also has an ionic and redox milieu optimized for folding and disulfide formation. In the extracellular space, or other final compartments of residence, no folding machinery is available.

Secondly, quality control prevents defective or incompletely folded molecules from being deployed, thus diminishing the potentially harmful effects that these might have. Consider, for example, the damage that could be caused if defective ion channels were to be delivered to the plasma membrane in large numbers. We are learning that for many polypeptide chains folding is not an efficient process, and that the generation of defective and misassembled proteins is in fact a common occurrence, not merely a theoretical possibility (Hurtley and Helenius, 1989). Furthermore, the subunits of hetero-oligomers are often produced in non-stochiometric proportions, which creates a need for retention of incomplete complexes and excess subunits (Klausner, 1989).

Finally, quality control provides an opportunity for post-translational regulation of protein expression. Examples of this include the differentially regulated secretion of immunoglobulins in B cells (Sitia *et al.*, 1990), and the ligand-induced secretion of retinol-binding protein from hepatocytes (Ronne *et al.*, 1983). In some unusual cases, ER retention can, in addition, be used for storage of protein products (Levanony *et al.*, 1992).

■ Quality control is mediated by chaperones

For quality control to work, the ER must have the ability to distinguish between folded and unfolded proteins of widely different structure and function. Considering the number of proteins produced and all of their different folding intermediates, the number of different structures exposed in the ER is immense. It is clear that there can be no common single molecular signal in all these proteins. The principles involved in separating folded from unfolded proteins must therefore be based on general properties that differ between them.

While the details of the underlying recognition criteria remain vague, it is apparent that the ER takes advantage of the ability of molecular chaperones to recognize and associate with incompletely folded proteins (Hammond and Helenius, 1995; Pelham, 1995). In addition to their role in folding, some of these clearly participate in keeping incompletely folded and misfolded proteins in the ER. The mechanisms involve *retention* in the ER as well as *retrieval* of incompletely folded and misfolded proteins that have escaped to the Golgi complex.

Chaperones and folding factors such as BiP, Grp94, Grp170, calnexin and calreticulin associate with many different types of incompletely folded proteins (Ou *et al.*, 1993; Hammond and Helenius, 1994; Melnick *et al.*, 1994; Ortmann *et al.*, 1994; Otsu *et al.*, 1994; Suh *et al.*, 1994; Peterson *et al.*, 1995). The molecular features that they recognize in the substrate proteins include hydrophobic surface patches, flexible polypeptide loops or termini and partially trimmed N-linked oligosaccharides.

The evidence that some of these interactions lead to retention is convincingly demonstrated for some chaperones. BiP, for example, is responsible for the retention of unassembled IgG heavy and light chains (Hendershot *et al.*, 1987; Ma *et al.*, 1990), and some secretory proteins, such as tissue plasminogen activator and von Willebrand factor, whose retention correlates with the level of BiP expression (Dorner *et al.*, 1992). In BiP mutants of yeast, carboxypeptidase Y fails to fold, and is confined to the ER in a BiP-bound form (Simons *et al.*, 1995). Results from a folding mutant of vesicular stomatitis virus G protein, moreover, have implicated BiP in a back-up retrieval mechanism for misfolded proteins from *cis*-Golgi to the ER (Hammond and Helenius, 1994).

Calnexin and the closely related calreticulin, provide another well-studied example. These ER chaperones bind to glycoproteins that contain core glycans from which the two outermost glucose residues have been removed

($Glu_1Man_{9-7}GlcNAc_2$) (Hammond et al., 1994; Hebert et al., 1995; Ora and Helenius, 1995; Peterson et al., 1995; Ware et al., 1995). As discussed elsewhere in this book, calnexin and calreticulin (see entries pp. 299 and 304) are part of a complex retention/quality control system which also includes glucosidase I and II, and UDP-glucose glycoprotein glucosyltransferase. Glucosidases I and II remove the first and the second glucose residues from N-linked core oligosaccharides, respectively, allowing the protein to bind to calnexin and calreticulin. Glucosidase II also removes the third glucose residue thereby causing the release of the substrate from calnexin and calreticulin (Hebert et al., 1995; Rodan et al., 1996). However, from a quality control point of view, the glucosyltransferase is the key player. This enzyme restores the calnexin/calreticulin substrate ($Glu_1Man_{7-9}GlcNAc_2$) from already fully glucose-trimmed glycoproteins ($Man_{7-9}GlcNAc_2$) by adding back one glucose residue. Remarkably, it only recognizes conformationally immature proteins as substrates, thereby constituting the checkpoint that maturing proteins have to pass before exiting the ER (Sousa et al., 1992; Hebert et al., 1995).

That calnexin has the capacity to retain proteins in the ER was first demonstrated for a well-known substrate, the MHC class I β_2-microglobulin heterodimer (Jackson et al., 1994). Expression of the mouse class I subunits in Drosophila melanogaster cells, together with canine calnexin, resulted in a 2–5-fold slower transport from the ER to the Golgi, compared with control cells not expressing calnexin. Similarly, faster secretion of α1-antitrypsin and cell surface expression of a mutant α factor receptor was observed in a yeast strain in which the calnexin-encoding gene, CNE1, had been disrupted (Parlati et al., 1995). Finally, if substrate glycoproteins are locked in the monoglucosylated form, their folding and exit from the ER is delayed (Hebert et al., 1995; Labriola et al., 1995).

Interestingly, although the primary structures of many ER chaperones are highly conserved between different species, they seem to differ in their substrate specificities. Human immunoglobulin heavy chains expressed in D. melanogaster, for example, interact with the endogenous BiP during folding and assembly with light chains. However, if expressed in the absence of light chains, the heavy chains are not retained in the ER bound to BiP as they would be in mammalian cells, but are secreted to the extracellular space (Kirkpatrick et al., 1995). Thus, although the Drosophila BiP is able to chaperone the folding and assembly of the immunoglobulins, its ability to exert quality control of a foreign polypeptide is hampered.

The ER matrix

The chaperones themselves are also kept in the ER by retention, mediated by interactions with other ER proteins, and by retrieval from the Golgi complex.

Most of the major ER-resident proteins have KDEL or dilysine retrieval signals at their C-terminus (Pelham, 1995). This means that if they escape from the ER to the Golgi complex, they are returned by recycling vesicles. However, removal of the KDEL sequence does not lead to rapid secretion of luminal chaperones, indicating additional interactions that prevent their transport (Pelham, 1995). The most likely interactions are via the acidic calcium-binding domains that most of these proteins possess. When this acidic sequence, together with the KDEL sequence, was removed from calreticulin, transport from the ER was greatly accelerated (Sonnichsen et al., 1994). Depletion of calcium from living cells also causes secretion of some luminal proteins (Booth and Koch, 1989).

Based on these data and on cross-linking studies in live cells, it has been speculated that the luminal ER proteins, most of which are chaperones, associate with each other to form a non-covalently associated network stabilized by calcium and perhaps ATP (Hammond and Helenius, 1995; Tatu and Helenius, 1997). This network, we think, spans the lumen of the ER and connects with the membranes, serving as a retention matrix for incompletely folded proteins. Depending on the affinities that a given newly synthesized protein has for the various chaperones in the network, its diffusion within the ER will be more or less restricted. When the protein reaches a fully folded form it is no longer associated with the components of the matrix, and can freely diffuse to export sites in the ER.

Chaperone-independent ER retention

Besides chaperone-mediated quality control, other mechanisms for keeping proteins in the ER are likely. The mere aggregation of misfolded proteins into covalently or non-covalently cross-linked aggregates may confine misfolded products to the ER simply because of size (Hurtley and Helenius, 1989). Such aggregates are often observed when folding is compromised. In other cases, a classical ER retention signal may be temporarily exposed in a protein product. This is exemplified by the α subunit of the high-affinity receptor for immunoglobulin E, which is retained in the ER by virtue of a C-terminal dilysine motif. Upon assembly with the γ subunit the retention motif becomes covered allowing transport (Letourneur et al., 1995). Moreover, studies on immunoglobulin secretion have shown that exposure of free sulfhydryl groups may result in retention (Sitia et al., 1990).

ER degradation

An integral part of ER quality control is the disposal of retained misfolded or unassembled proteins. An increasing body of data indicates that the degradation of retained proteins takes place within the cytoplasm and involves the ubiquitin-proteasome pathway (Ward et al., 1995; Jensen et al., 1995; Qu et al., 1996; Hiller et al., 1996; Wiertz et al. 1996). Cytoplasmic domains of transmembrane proteins may be accessed directly by the proteasome (Ward et al., 1995; Jensen et al., 1995), whereas transmembrane and lumenal domains, and soluble proteins are translocated back to the cytoplasm by a process that appears to involve the sec61 translocation machinery (Wiertz et al. 1996). It is unclear how proteins destined for degradation are distinguished from folding intermediates and resident proteins. Some misfolded proteins are degraded very rapidly with half-times of 10–60 minutes, whereas others are stable for up to 8 hours (Klausner and Sitia, 1990). In some cases degradation is prevented by oligomeric assembly, as for TCR α and β subunits after association with the ϵ^* and γ chains (Wileman et al., 1990). In this and other similar cases, it is thought that oligomerization changes the conformation of the proteins, thereby making parts that mediate degradation inaccessible (Shin et al., 1993). The intrinsic ability of molecular chaperones to recognize incompletely folded proteins make them likely to also have a role in targeting polypeptides for degradation (Knittler et al. 1995).

Finally, it should be emphasized that although most proteins are restricted to the ER until fully folded or degraded, some proteins do not fold completely or oligomerize until reaching more distal locations such as the ER–Golgi intermediate compartment or the Golgi complex (Musil and Goodenough, 1993; Tatu et al., 1995; Wagner, 1990). For these, quality control might occur at later sites of the secretory pathway; the best known example being the retrieval from the cis-Golgi back to the ER (Hammond and Helenius, 1995; Pelham, 1995).

■ References

Booth, C. and Koch, L.E. (1989) Perturbation of cellular calcium induces secretion of luminal ER proteins. Cell **59**, 729–737.

Dorner, A.J., Wasley, L.C., and Kaufman, R. J. (1992) Overexpression of GRP78 mitigates stress induction of glucose regulated proteins and blocks secretion of selective proteins in Chinese hamster cells. EMBO J. **11**, 1563–1571.

Hammond, C. and Helenius, A. (1994) Folding of VSV G protein involves sequential interaction with Bip/GRP78 and calnexin. Science **266**, 456–458.

Hammond, C. and Helenius, A. (1995) Quality control in the secretory pathway. Curr. Op. Cell Biol. **7**, 523–529.

Hammond, C., Braakman, I., and Helenius, A. (1994) Role of N-linked oligosaccharides, glucose trimming and calnexin during glycoprotein folding in the endoplasmic reticulum. Proc. Natl. Acad. Sci. USA **91**, 913–917.

Hebert, D.N., Foellmer, B., and Helenius, A. (1995) Glucose trimming and reglucosylation determines glycoprotein association with calnexin. Cell **81**, 425–433.

Hendershot, L., Bole, D., Köhler, G., and Kearney, J. F. (1987) Assembly and secretion of heavy chains that do not associate post-translationally with immunoglobulin heavy chain binding protein. J. Cell Biol. **104**, 761–767.

Hiller, M.M., Finger, A., Schweiger, M. and Wolf, D.H. (1996). ER degradation of a misfolded lumenal protein by the cytosolic ubiquitin-proteasome pathway. Science **273**, 1725–1728.

Hurtley, S.M. and Helenius, A. (1989) Protein oligomerization in the endoplasmic reticulum. Annu. Rev. Cell Biol. **5**, 277–307.

Jackson, M. R., Cohen-Doyle, M. F., Peterson, P.A., and Williams, D.B. (1994) Regulation of MHC Class I transport by the molecular chaperone, calnexin (p88, IP90). Science **263**, 384–387.

Jensen, T.J., Loo, M.A., Pind, S., Williams, D.B., Goldberg, A.L., and Riordan, J.R. (1995). Multiple proteolytic systems, including the proteasome, contribute to CFTR processing. Cell **83**, 129–135.

Kirkpatrick, R.B., Ganguly, S., Angelichio, M., Griego, S., Shatzman, A., Silverman, C., and Rosenberg, M. (1995) Heavy chain dimers as well as complete antibodies are efficiently formed and secreted from Drosophila via a BiP-mediated pathway. J. Biol. Chem. **270**, 19800–19805.

Klausner, R.D. (1989) Architectural editing: determining the fate of newly synthesized membrane proteins. The New Biologist **1**, 3–8.

Klausner, R.D. and Sitia, R. (1990) Protein degradation in the endoplasmic reticulum. Cell **62**, 611–614.

Knittler, M.R., Dirks, S.D., and Haas, I.G. (1995). Molecular chaperones involved in protein degradation in the endoplasmic reticulum: Quantitative interaction of the heat shock cognate protein BiP with partially folded immunoglobulin light chains that are degraded in the endoplasmic reticulum. Proc. Natl. Acad. Sci. USA **92**, 1764–1768.

Labriola, C., Cazzulo, J.J., and Parodi, A. (1995) Retention of glucose units added by the UDP-GLC:glycoprotein glycosyltransferase delays exit of glycoproteins from the endoplasmic reticulum. J. Cell Biol. **130**, 771–779.

Letourneur, L., Hennecke, S., Demolliere, C., and Cosson, P. (1995) Steric masking of a dilysine endoplasmic reticulum retention motif during assembly of the human high affinity receptor for immunoglobulin E.J. Cell Biol. **129**, 971–978.

Levanony, H., Rubin, R., Altschuler, Y., and Galili, G. (1992) Evidence for a novel route of wheat storage proteins to vacuoles. J. Cell Biol. **119**, 1117–1128.

Ma, J., Kearney, J., and Hendershot, L.M. (1990) Association of transport-defective light chains with immunoglobulin heavy chain binding protein. Mol. Immunol. **27**, 623–630.

Melnick, J., Dul, J.L., and Argon, Y. (1994) Sequential interaction of the chaperones BiP and GRP94 with immunoglobulin chains in the endoplasmic reticulum. Nature **370**, 373–375.

Musil, L.S. and Goodenough, D.A. (1993) Multisubunit assembly of an integral plasma membrane channel protein, gap junction connexin43, occurs after exit from the ER. Cell **74**, 1065–1077.

Ora, A. and Helenius, A. (1995) Calnexin fails to associate with substrate proteins in glucosidase-deficient cell lines. J. Biol. Chem. 270, 26060–26062.

Ortmann, B., Androlewicz, M.J., and Cresswell, P. (1994) MHC class I/b2-microglobulin complexes associate with TAP transporters before peptide binding. Nature **368**, 864–867.

Otsu, M., Omura, F., Yoshimori, T., and Kikuchi, M. (1994) Protein disulfide isomerase associates with misfolded human lysozyme in vivo. J. Biol. Chem. **269**, 6874–6877.

Ou, W.-J., Cameron, P. H., Thomas, D.Y., and Bergeron, J. J. M. (1993) Association of folding intermediates of glycoproteins with calnexin during protein maturation. Nature **364**, 771–776.

Parlati, F., Dominguez, M., Bergeron, J.J. M., and Thomas, D.Y. (1995) Sacharomyces cerevisiae CNE1 encodes an endoplasmic reticulum (ER) membrane protein with sequence similarity to calnexin and calreticulin and functions as a constituent of the ER quality control apparatus. J. Biol. Chem. **270**, 244–253.

Pelham, H.R.B. (1995) Sorting and retrieval between the endoplasmic reticulum and the Golgi apparatus. Curr. Opin. Cell Biol. **7**, 530–535.

Peterson, J.R., Ora, A., Nguyen Van, P., and Helenius, A. (1995) Transient, lectin-like association of calreticulin with folding intermediates of cellular and viral glycoproteins. Mol. Biol. Cell **6**, 1173–1184.

Qu, D., Teckman, J.H., Omura, S., and Perlmutter, D.H. (1996). Degradation of a mutant secretory protein, α_1-antitrypsin Z, in the endoplasmic reticulum requires proteasome activity. J. Biol. Chem. **271**, 22791–22795.

Rodan, A.R., Simons, J.F., Trombetta, E.S., and Helenius, A. (1996). N-linked oligosaccharides are necessary and sufficient for association of RNase B with calnexin and calreticulin. EMBO J. **15**, 6921–6930.

Ronne, H., Ocklind, C., Wiman, K., Rask, L., Obrink, B., and Peterson, P. (1983) Ligand dependent regulation of intracellular protein transport: effect of vitamin A on the secretion of retinol-binding protein. J. Cell Biol. **96**, 907–910.

Shin, J., Lee, S., and Strominger, J.L. (1993) Translocation of TCRα chains into the lumen of the endoplasmic reticulum and their degradation. Science **259**, 1901–1904.

Simons, J.F., Ferro-Novick, S., Rose, M.D., and Helenius, A. (1995) BiP/Kar2p functions as a molecular chaperone during folding of carboxypeptidase Y in yeast. J. Cell Biol. **130**, 41–49.

Sitia, R., Neuberger, M., Alberini, C., Bet, P., Fra, A., Valetti, C., Williams, G., and Milstein, C. (1990) Developmental regulation of IgM secretion: the role of the carboxy-terminal cysteine. Cell **60**, 781–790.

Sonnichsen, B., Fullekrug, J., Nguyen, V.P., Diekmann, W., Robinson, D.G., and Mieskes, G. (1994) Retention and retrieval: both mechanisms cooperate to maintain calreticulin in the endoplasmic reticulum. J. Cell Sci. **107**, 2705–2717.

Sousa, M.C., Ferrero-Garcia, M.A., and Parodi, A.J. (1992) Recognition of the oligosaccharide and protein moieties of glucoproteins by the UDP-Glc:glycoprotein glucosyltransferase. Biochemistry **31**, 97–105.

Suh, W.-K., Cohen-Doyle, M. F., Fruh, K., Wang, K., Peterson, P., and Williams, D. B. (1994) Interaction of MHC class I molecules with the trasnporter associated with antigen processing. Science **264**, 1322–1326.

Tatu, U., Hammond, C., and Helenius, A. (1995) Folding and oligomerization of influenza hemagglutinin in the ER and the intermediate compartment. EMBO J. **14**, 1340–1348.

Tatu, U., and Helenius, A. (1997). Interactions between newly synthesized glycoproteins, calnexin and a network of resident chaperones in the endoplasmic reticulum. J. Cell Biol. In press.

Wagner, D.D. (1990) The cell biology of von Willebrand factor. Annu. Rev. Cell Biol. **6**, 217–246.

Ward, C.L., Omura, S., and Kopito, R.R. (1995). Degradation of CFTR by the ubiquitin-proteasome pathway. Cell **83**, 121–127.

Ware, F.E., Vassilakos, A., Peterson, P.A., Jackson, M.R., Lehrman, M.A., and Williams, D.B. (1995) The molecular chaperone calnexin binds $Glc_1Man_9GlcNAc_2$ oligosaccharides as an initial step in recognizing unfolded glycoproteins. J. Biol. Chem. **270**, 4697–4704.

Wiertz, E.J.H.J., Tortorella, D., Bogyo, M., Yu, J., Mothes, W., Jones, T.R., Rapoport, T.A., and Ploegh, H.L. (1996). Sec61-mediated transfer of a membrane protein from the endoplasmic reticulum to the proteasome for destruction. Nature **284**, 432–438.

Wileman, T., Carson, G.R., Concino, M., Ahmed, A., and Terhorst, C. (1990). The gamma and epsilon subunits of the CD3 complex inhibit pre-Golgi degradation of newly synthesized T cell antigen receptors. J. Cell Biol. **110**, 973–986.

■ *Jan Fredrik Simons and Ari Helenius:*
Department of Cell Biology
Yale University School of Medicine
333 Cedar Street
New Haven, CT 06520-8002, USA
Tel. 1 203 785 4303
Fax. 1 203 737 1756
E-mail: Ari Helenius@qmail.yale.edu

The pathway of assembly of the progesterone receptor

Progesterone and other steroid receptors exist in a heteromeric complex prior to binding hormone. Each of the receptor-associated proteins is a component of the molecular chaperone machinery. Biochemical studies have identified eight chaperone components participating in the assembly of progesterone receptor (PR) complexes that are competent for hormone binding. The assembly pathway involving these components is described here.

■ Individual pathway components

The PR assembly pathway minimally involves the following chaperone components: Hsp90, Hsc70, Hip, Hop (p60), p23, FKBP52, FKBP51, and Cyp40 (Smith *et al.*, 1995 and references therein; also see the individual entries in this volume for these and other chaperones mentioned in this review).

There is also genetic evidence from steroid receptor expression in yeast cells that Ydj1, a cytosolic DnaJ homolog, is required for proper function of steroid receptors (Caplan *et al.*, 1995; Kimura *et al.*, 1995). Although a mammalian DnaJ homolog has not been directly observed in native or reconstituted steroid receptor complexes, it may participate in a more transient or less stable manner than other components, or may be required for interactions that prepare other components for PR assembly.

■ Pre-existing complexes

Each of the PR assembly components can be isolated in a complex with Hsp90. Separate complexes that have been identified are Hsp90–Hop–Hsc70–Hip (Prapapanich *et al.*, 1996) and I–Hsp90–p23 (Johnson and Toft, 1994), where I represents any one of the three receptor-associated immunophilins FKBP52, FKBP51, or Cyp40. As illustrated in Fig. 1 (side-assembly A), Hip binds the ATPase domain

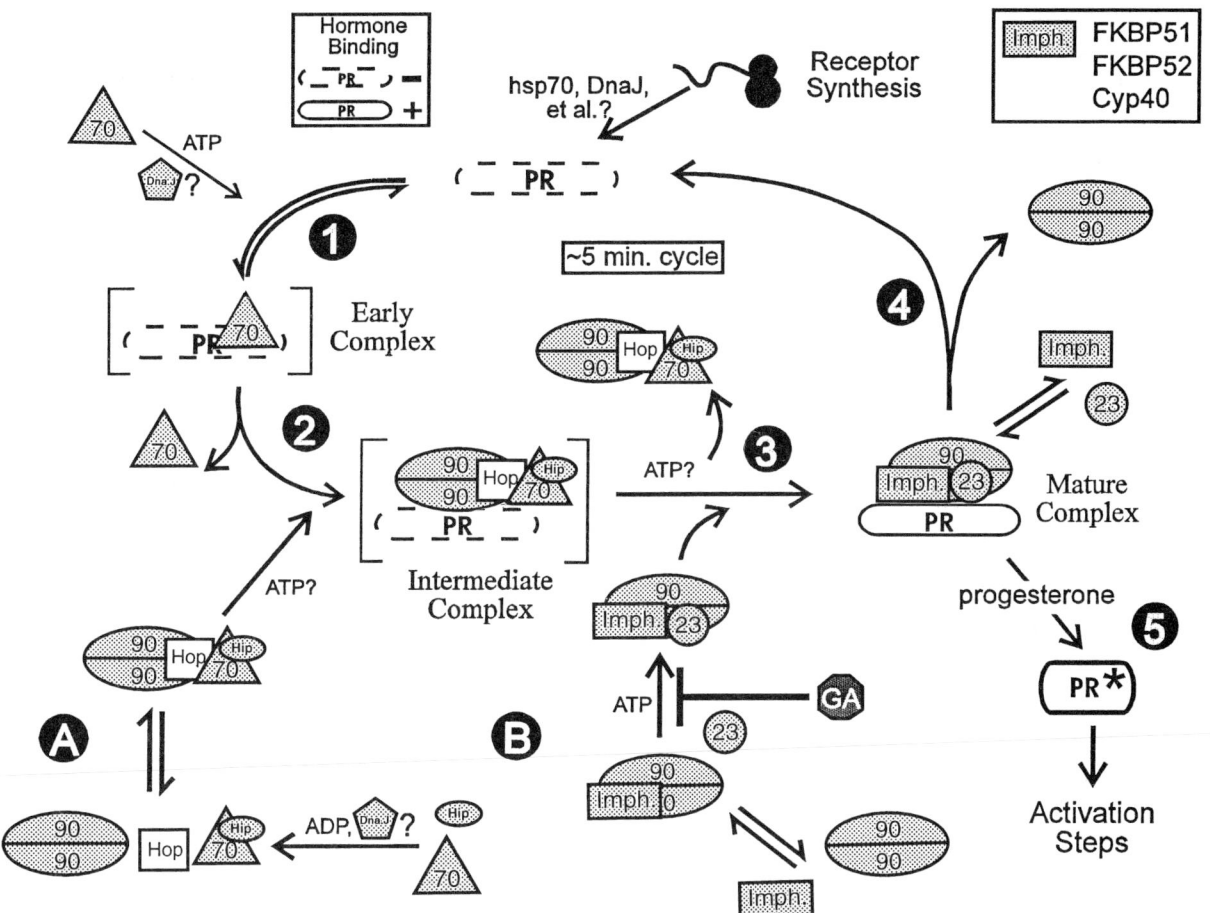

Figure 1. Pathway for assembly of progesterone receptor complexes. Functionally mature progesterone receptor complexes are formed through an ordered, dynamic pathway of interactions with multiple components of the molecular chaperone machinery. A detailed description of the proposed pathway is presented in the text.

of Hsc70 (Höhfeld et al., 1995) in a manner that is enhanced by ADP (Höhfeld et al., 1995; Prapapanich et al., 1996), but this binding may need to be assisted by a DnaJ homolog or other factors (Höhfeld et al., 1995). Hop independently and concomitantly associates with Hsp90 and Hsc70 in a manner that is neither nucleotide nor temperature dependent (Chen et al., 1996).

Interactions between Hsp90 and immunophilins and p23 are also illustrated in Fig. 1 (side-assembly B). Immunophilins directly bind Hsp90 through their tetratricopeptide repeat (TPR) regions and compete with one another for binding to Hsp90 (Owens-Grillo et al., 1995; Ratajczak and Carrello, 1996). Binding of Cyp40 and FKBP52 to Hsp90 does not require nucleotides, but FKBP51 binding to Hsp90 is enhanced in the presence of ADP (Nair et al., 1997). Binding of p23 to Hsp90 requires ATP (Johnson and Toft, 1995) and is disrupted by the Hsp90-binding drug geldanamycin (Johnson and Toft, 1995; Smith et al., 1995). See Fig. 1, side-assembly B for an illustration of these interactions.

■ Dynamic assembly pathway

As illustrated in Fig. 1, an interaction pathway of chaperone components with PR has been proposed from cell-free assembly studies in rabbit reticulocyte lysate (RL). Importantly, PR must be properly associated with Hsp90 to bind progesterone with high affinity at physiological temperatures (Smith, 1993). A similar observation was first made for glucocorticoid receptor (GR) by Pratt and his colleagues (Scherrer et al., 1990). However, Hsp90 does not bind PR as a free protein or homodimer, but instead is assembled with PR in a progressive manner involving several chaperone components. Each of these components can be found in pre-existing complexes (side-assembly pathways A and B) that may be the active components during assembly of PR complexes. The following discussion follows the numbered steps in the proposed assembly pathway.

1. The first assembly step detected is the binding of Hsc70 to PR (early complex). Evidence for this comes

from assembly time-courses showing that Hsc70 rapidly associates with PR (Smith, 1993). Hsc70 is required to facilitate later assembly stages, since subsequent steps are blocked by any of several approaches that inhibit Hsc70 function: (a) depleting ATP, Mg^{2+}, or K^+ (Hutchison et al., 1992b; Smith et al., 1992); (b) including an antibody against Hsc70 (Smith et al., 1992); or directly depleting Hsc70 from RL (Hutchison et al., 1993). At this early stage, Hsc70 appears to bind PR as a misfolded protein substrate (Chen et al., 1996).

2. In the next assembly stage (intermediate complex), Hsp90, Hip and Hop all appear in conjunction with Hsc70 (Smith, 1993; Smith et al., 1995). It has not yet been demonstrated that Hip is required for assembly of intermediate complexes, but it has been shown (Chen et al., 1996) that an antibody against Hop can inhibit further assembly steps while enhancing recovery of early complexes. Also, Pratt and his colleagues have recently shown (Dittmar et al., 1996) that Hop is required in a purified system to establish functional assembly of GR complexes. Both Hsc70 and Hsp90 in intermediate complexes behave in manners distinct from their interactions in early and mature PR complexes, respectively. Hsc70 association in the intermediate complex is sensitive to ionic strength, but not in early complexes (Chen et al., 1996). Hsp90 association in intermediate complexes is not stabilized by molybdate, in contrast to the effect of molybdate on Hsp90 binding in mature complexes (Chen et al., 1996). Thus, it is not yet clear which of these components is directly interacting with PR in intermediate complexes. Also, it is has not yet been resolved whether Hsc70 in the early complex is carried over into intermediate complexes, or whether it is displaced by pre-existing Hsp90–Hop–Hsc70–Hip complexes.

3. In order for PR to bind progesterone with high affinity, it must progress from the intermediate complex to the mature form containing p23 and one of the immunophilins. In the presence of geldanamycin, PR assembly is arrested at the intermediate stage, presumably owing to the drug's ability to block p23 binding to Hsp90. Dittmar et al. (1996) showed that p23 is required for final assembly of functionally mature GR complexes, but it is not clear that immunophilins are necessary to achieve the proper hormone-binding conformation. As with Hsc70 during the transition from early to intermediate complexes, it has not been resolved whether Hsp90 in mature complexes is carried over from intermediate assemblies or enters de novo as a pre-existing complex with p23.

4. Mature PR complexes are not stable under cell-free assembly conditions (Smith, 1993), nor do they appear to be stable in vivo (Smith et al., 1995). The immunophilins and p23 exchange on the mature complex with half-lives around 1 minute (Nair et al., 1997; D.F. Smith, unpublished observations) while Hsp90 dissociates from PR with a half-life of approximately 5 minutes. Following dissociation of Hsp90, it appears that PR re-enters the assembly pathway at step 1. The mature complex is more stable than early or intermediate complexes; therefore, at steady state most PR exists in mature complexes that are competent for hormone binding, but there is a constant cycling of receptor through assembly stages.

5. Hormone binding to unactivated receptor complexes leads to dissociation of chaperone components from the receptor; however, hormone binding does not directly promote dissociation of Hsp90 (Smith, 1993). Instead, Hsp90 dissociates at its normal rate, but the hormone-bound receptor is no longer recognized by Hsc70, thereby freeing the liganded PR from further chaperone interactions. PR then proceeds through a series of activation steps, including homodimerization, hyperphosphorylation, DNA binding, and establishing protein–protein interactions with elements of the transcriptional machinery, that lead to changes in target gene activity.

■ General implications of PR assembly pathway

Hsp90 interactions have been noted with members from several classes of regulatory or signal-transducing proteins. Amongst these are the steroid receptors, diverse transcription factors unrelated to steroid receptors, and several protein tyrosine and serine/threonine kinases. In the best studied cases other than steroid receptors, heterocomplexes of Src tyrosine kinase (Hutchison et al., 1992a) and Raf serine/threonine kinase (Stancato et al., 1993) contain several components in addition to Hsp90 that are common to the PR assembly pathway. Four target proteins known to bind Hsp90 and representing different evolutionary families were recently compared for chaperone interactions in the cell-free system used for PR and GR assembly studies (Nair et al., 1996). For every PR assembly component tested, each was found associating with estrogen receptor, the Fes tyrosine kinase, the heat shock transcription factor HSF1, and the arylhydrocarbon receptor. The only qualitative distinction in the formed complexes was the exclusive presence in the kinase complex of p50, a protein known to appear concomitantly with Hsp90 in several kinase complexes (reviewed by Brugge, 1986). From these comparisons it appears that the multichaperone pathway described for PR assembly may be one that various signalling proteins have co-opted for regulatory purposes.

■ References

Brugge, J.S. (1986). Interaction of the Rous sarcoma virus protein pp60src with the cellular proteins pp50 and pp90. Curr. Top. Microbiol. Immunol. **12**, 1–22.

Caplan, A.J., Langley, E., Wilson, E.M. and Vidal, J. (1995). Hormone-dependent transactivation by the human androgen receptor is regulated by a dnaJ protein. J. Biol. Chem. **270**, 5251–5257.

Chen, S., Prapapanich, V., Rimerman, R.A., HonorÇ, B. and Smith, D.F. (1996). Interactions of p60, a mediators of progesterone receptor assembly, with heat shock proteins Hsp90 and Hsp70. Mol. Endocrinol. **10**, 682–693.

Dittmar, K.D., Hutchison, K.A., Owens-Grillo, J.K. and Pratt, W.B. (1996). Reconstitution of the steroid recpetor-hsp90 heterocomplex assembly system of rabbit reticulocyte lysate. J. Biol. Chem. **271**, 12833–12839.

Höhfeld, J., Minami, Y. and Hartl, F.-U. (1995). Hip, a new cochaperone involved in the eukaryotic hsc70/hsp40 reaction cycle. Cell **83**, 589–598.

Hutchison, K.A., Brott, B.K., De Leon, J.H., Perdew, G.H., Jove, R. and Pratt, W.B. (1992a). Reconstitution of the multiprotein complex of pp60^{src}, hsp90, and p50 in a cell-free system. J. Biol. Chem. **267**, 2902–2908.

Hutchison, K.A., Czar, M.J., Scherrer, L.C. and Pratt, W.B. (1992b). Monovalent cation selectivity for ATP-dependent association of the glucocorticoid receptor with hsp70 and hsp90. J. Biol. Chem. **267**, 14047–14053.

Hutchison, K.A., Dittmar, K.D., Czar, M.J. and Pratt, W.B. (1993). Proof that hsp70 is required for assembly of the glucocorticoid receptor into a heterocomplex with hsp90. J. Biol. Chem. **269**, 5043–5049.

Johnson, J.L. and Toft, D.O. (1994). A novel chaperone complex for steroid receptors involving heat shock protein, immunophilins and p23. J. Biol. Chem. **269**, 24989–24993.

Johnson, J.L. and Toft, D.O. (1995). Binding of p23 and hsp90 during assembly with the progesterone receptor. Mol. Endocrinol. **9**, 670–678.

Kimura, Y., Yahura, I. and Lindquist, S. (1995). Role of the protein chaperone YDJ1 in establishing hsp90-mediated signal trandsduction pathways. Science **268**, 1362–1365.

Nair, S.C., Toran, E.J., Rimerman, R.A., Hjermstad, S., Smithgall, T.E. and Smith, D.F. (1996). A pathway of multi-chaperone interactions common to diverse regulatory proteins: estrogen receptor, Fes tyrosine kinase, heat shock transcription factor HSF1, and the arylhydrocarbon receptor. Cell Stress Chap. **1**, 237–250.

Nair, S.C., Rimerman, R.A., Toran, E.J., Chen S., Prapapanich, V., Butts R.N. and Smith, D.F. (1997). Molecular cloning of human FKBP51 and comparisons of immunophilin interactions with hsp90 and progesterone receptor. Mol. Cell Biol. **17**, 594–603.

Owens-Grillo, J.K., Hoffman, K., Hutchison, K.A., Yem, A.W., Deibel, M.R., Jr., Handschumacher, R.E. and Pratt, W.B. (1995). The cyclosporin A-binding immunophilin CyP-40 and the FK506-binding immunophilin hsp56 bind to a common site on hsp90 and exist in independent cytosolic heterocomplexes with the untransformed glucocorticoid receptor. J. Biol. Chem. **270**, 20479–20484.

Prapapanich, V., Chen, S., Nair, S.C., Rimerman, R.A. and Smith, D.F. (1996). Molecular cloning of human p48, a transient component of progesterone recpetor complexes and an Hsp70-binding protein. Mol. Endocrinol. **10**, 420–431.

Ratajczak, T., and Carrello, A. (1996). Cyclophilin 40 (CyP-40), mapping of its hsp90 binding domain and evidence that FKBP52 competes with CyP-40 for hsp90 binding. J. Biol. Chem. **271**, 2961–2965.

Scherrer, L.C., Dalman, F.C., Massa, E., Meshinchi, S. and Pratt, W. B. (1990). Structural and functional reconstitution of the glucocorticoid receptor-hsp90 complex. J. Biol. Chem. **265**, 21397–21400.

Smith, D.F. (1993). Dynamics of heat shock protein 90-progesterone receptor binding and the disactivation loop model for steroid receptor complexes. Mol. Endocrinol. **7**, 1418–1429.

Smith, D.F., Stensgard, B.A., Welch, W.J., and Toft, D.O. (1992). Assembly of progesterone receptor with heat shock proteins and receptor activation are ATP mediated events. J. Biol. Chem. **267**, 1350–1356.

Smith, D.F., Whitesell, L., Nair, S.C., Chen, S., Prapapanich, V. and Rimerman, R.A. (1995). Progesterone receptor structure and function altered by geldanamycin, an hsp90-binding agent. Mol. Cell. Biol. **15**, 6804–6812.

Stancato, L.F., Chow, Y.-H., Hutchison, K.A., Perdew, G.H., Jove, R. and Pratt, W.B. (1993). Raf exists in a native heterocomplex with hsp90 and p50 that can be reconstituted in a cell-free system. J. Biol. Chem. **268**, 21711–21716.

■ *David F. Smith:*
Department of Pharmacology
University of Nebraska Medical Center
Omaha, NE 68198-6260, USA
Tel. 1 402 559 8604
Fax. 1 402 559 7495
E-mail: dfsmith@mail.unmc.edu

17

Cellular Regulation of Chaperone Activity

The heat shock response in *Escherichia coli*

The conserved heat shock (stress) response allows cells to adapt to environmental and metabolic changes and to survive stress conditions (Morimoto et al., 1994). It is induced by a large variety of stress conditions including physicochemical factors such as heat shock, metabolically harmful substances and complex metabolic processes. In Escherichia coli, the heat shock response to temperature upshift from 30 to 42°C consists of the rapid, up to 20-fold, induction of synthesis of more than 20 heat shock proteins (HSPs), followed by an adaptation period where the rate of HSP synthesis decreases to reach a new steady state level (Gross et al., 1990; Bukau, 1993; Georgopoulos et al., 1994). Major HSPs are molecular chaperones and proteases (Georgopoulos et al., 1994), indicating that their induced synthesis during the heat shock response is an adaptation to changes in the folding status of cellular proteins. HSPs include the DnaK and GroE chaperone systems formed by DnaK, DnaJ and GrpE, and GroEL and GroES, respectively. They constitute the two major chaperone systems of E. coli as judged by their abundance [15–20% of total protein at 46°C (Neidhardt and VanBogelen, 1987)] and importance for cell survival (Georgopoulos et al., 1994).

■ Positive regulation of the *Escherichia coli* heat shock response by σ^{32}

The E. coli heat shock response is positively controlled at the transcriptional level by the product of the *rpoH* gene, the heat shock promoter-specific σ^{32} subunit of RNA polymerase (Grossman et al., 1984; Landick et al., 1984; Cowing et al., 1985) (Fig. 1). σ^{32} is required for induced expression as well as uninduced basal expression of heat shock genes (Zhou et al., 1988). Some heat shock genes, including the *groES groEL* operon, have additional σ^{70}-dependent promoters contributing to the basal levels of their expression (Zhou et al., 1988). The cellular concentration of σ^{32} is very low under steady-state conditions [10–30 copies per cell at 30°C (Craig and Gross, 1991)] and is limiting for heat shock gene transcription (Grossman et al., 1987). The heat shock response is induced as a consequence of a rapid increase in σ^{32} levels (Straus et al., 1987) and, possibly, stimulation of σ^{32} activity. The shut-off of the response occurs as a consequence of a decline in σ^{32} levels and inhibition of σ^{32} activity (Straus et al., 1987, 1989; VanBogelen et al., 1987). As described below, this mode of regulation of σ^{32} is fast and therefore allows E. coli cells to respond rapidly to sudden stress.

One mechanism mediating stress-dependent changes in σ^{32} levels affects the translation of *rpoH* mRNA (Grossman et al., 1987; Straus et al., 1987; Nagai et al., 1991). *rpoH* translation is repressed at steady-state conditions and is rapidly derepressed upon shift of the cells from 30 to 42°C. At 2–4 minutes after temperature upshift, a 12-fold higher translation level compared with the pre-heat shock condition is reached. Then, translation becomes increasingly repressed during the shut-off phase of the heat shock response, to reach a new steady-state level. *rpoH* translation control is mediated through distinct mechanisms involving three *cis*-acting elements of *rpoH* coding sequence, termed regions A, B and C (Nagai et al., 1991, 1994). Region A is a positive regulatory element comprising the initiation codon and the downstream 20 nucleotides. It is required for *rpoH* translation at steady-state conditions as well as after heat shock. Region B is a negative regulatory element located within nucleotides 110 and 247. It is required for translation repression at steady-state conditions and derepression during the induction phase of the heat shock response. Extensive base pairing between regions A and B has been predicted and was further substantiated by mutational analysis (Nagai et al., 1991; Yura et al., 1993). The formation of this secondary structure is hypothesized to mediate repression of *rpoH* translation at steady-state conditions, e.g. by preventing translation initiation, as well as thermal induction of translation by a mechanism that remains to be elucidated. Region C is a negative regulatory element located within nucleotides 364 and 433 of the *rpoH* coding sequence. It is involved in repression of *rpoH* translation during the shut-off phase of the heat shock response (Nagai et al., 1994) (Fig.2). Interestingly, two frameshift mutations introduced into the *rpoH* coding sequence to alter specifically the amino acid sequence of region C of σ^{32} (residues 122 to 144), result in defects in translational repression during the shut-off phase of the heat shock response (Nagai et al., 1994). This raises the intriguing possibility that region C acts at the protein level to mediate translational repression of *rpoH*.

A second mechanism mediating stress-dependent changes in σ^{32} levels affects stability of σ^{32} (Fig.1). During steady-state growth, σ^{32} has an extremely short half-life of less than 1 minute (Straus et al., 1987; Tomoyasu et al., 1995). Upon temperature upshift from 30 to 42°C, σ^{32} becomes transiently stabilized at least eightfold until the beginning of the shut-off phase of the heat shock response (Straus et al., 1987). The regulatory region C of σ^{32} has been shown genetically to be required for the efficient degradation of σ^{32} (Nagai et al., 1994). One protease responsible for σ^{32} degradation is the ATP-dependent metalloprotease, FtsH (Herman et al., 1995; Tomoyasu et al., 1995). FtsH is an integral cytoplasmic membrane protein, with the active site located in a

Negative modulation and stress sensing by the DnaK chaperone system

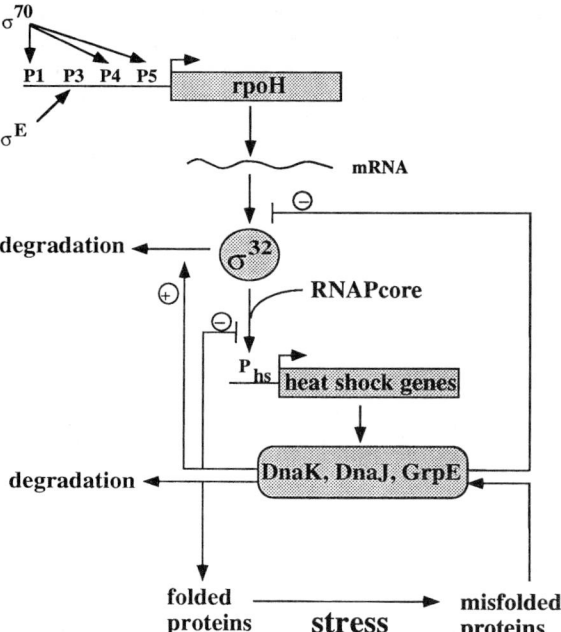

Figure 1. The *E. coli* heat shock regulon. Positive regulator is the product of the *rpoH* gene, σ^{32}, which directs RNA polymerase (RNAP core) to the promoters of heat shock genes. Transcription of *rpoH* involves at least four promoters, one of which utilizes the alternative σ^{24} factor. The cellular concentration of σ^{32} is very low and limiting for heat shock gene transcription. The heat shock response is induced as the result of an increase in σ^{32} levels, primarily owing to increases in translation of *rpoH* mRNA and stabilization of σ^{32}, and, possibly, derepression of σ^{32} activity. Adaptation is achieved through a decrease in σ^{32} levels and repression of σ^{32} activity. The DnaK, DnaJ and GrpE heat shock proteins act as negative modulators by mediating repression of *rpoH* mRNA translation during the shut-off phase of the heat shock response, efficient degradation of σ^{32} and repression of σ^{32} activity.

The DnaK chaperone system negatively modulates transcription of heat shock genes as first indicated by the regulatory defects of *dnaK*, *dnaJ* and *grpE* mutants (Gross et al., 1990; Bukau, 1993; Yura et al., 1993; Georgopoulos et al., 1994) (Fig. 1). It mediates the stress-dependent inactivation and degradation of σ^{32} (Straus et al., 1989, 1990) and repression of *rpoH* translation during the shut-off phase of the heat shock response (Grossman et al., 1987; Straus et al., 1990). Genetic data indicate that the chaperone-dependent controls of σ^{32} stability and *rpoH* translation require region C of σ^{32} (Nagai et al., 1994).

These modulatory activities result, at least in part, from reversible association of the DnaK chaperone system with σ^{32} (Gamer et al., 1992, 1996; Liberek et al., 1992; Liberek and Georgopoulos, 1993). DnaK and the cochaperone DnaJ bind independently to free σ^{32} with binding constants (K_D) of 5 μM and 20 nM, respectively (Gamer et al., 1996). ATP decreases the stablity of DnaK–σ^{32} complexes. Further analyses (Gamer et al., 1996; Liberek et al., 1992; Liberek and Georgopoulos, 1993; Liberek et al., 1995) led to the proposal of an ATP-controlled σ^{32} binding/release cycle (Gamer et al., 1996) in which DnaJ mediates the efficient binding of DnaK to σ^{32} in the presence of ATP. DnaJ rapidly associates with σ^{32} and then interacts with DnaK–ATP, thereby stimulating the hydrolysis of DnaK-bound ATP and allowing the efficient formation of DnaK–DnaJ–σ^{32} complexes containing ADP. GrpE associates with DnaK–ADP bound to σ^{32}, thereby stimulating nucleotide release and, upon rebinding of ATP, subsequent dissociation of σ^{32}. This cycle of σ^{32} binding and release is similar to the cycle proposed to operate in other chaperone activities of the DnaK system (Szabo et al., 1994) (see entry p. 22), implying that σ^{32} acts as regular substrate for the DnaK chaperone system. DnaK and DnaJ cooperatively inhibit σ^{32} activity in heat shock gene transcription, and GrpE partially reverts this inhibition (Gamer et al., 1996). This reversible inhibition of σ^{32} activity through transient association of DnaK and DnaJ is proposed to constitute a homeostatic σ^{32} activity control system operating *in vivo* (Gamer et al., 1996) (Fig. 2). It remains unclear how these interactions affect the stability of σ^{32} *in vivo*.

A recent study identified two high affinity DnaK binding sites centrally and peripherally to region C (McCarty et al., 1996). The sequence of region C including these sites is highly conserved among σ^{32} homologs of eubacteria but is missing in other σ factors. Based on these results it is proposed that binding of DnaK to region C is central to a conserved, chaperone-dependent regulatory mechanism allowing transduction of information on the stress status of the cell to the heat shock gene transcription machinery.

Homeostatic regulation models (Georgopoulos et al., 1990; Craig and Gross, 1991; Bukau, 1993) propose that induction of a heat shock response after stress treatment relies on the sequestration of the DnaK system through binding to misfolded proteins accumulating during stress

cytosolic domain which shares high sequence homology with the conserved family of AAA proteins (Tomoyasu et al., 1995; see entry p. 451).

Another level of negative regulation of σ^{32} involves its activity. This is indicated by the finding that artificial overproduction of σ^{32} at 30°C results in only transient induction of HSP synthesis (VanBogelen et al., 1987). Furthermore, a downshift of cells from 42 to 30°C causes rapid and efficient repression of heat shock gene expression by inhibition of σ^{32} activity (Straus et al., 1989). This mechanism is likely to be important for the rapid establishment of appropriate HSP levels after recovery of cells from stress treatment, and the maintenance of homeostasis of heat shock gene expression.

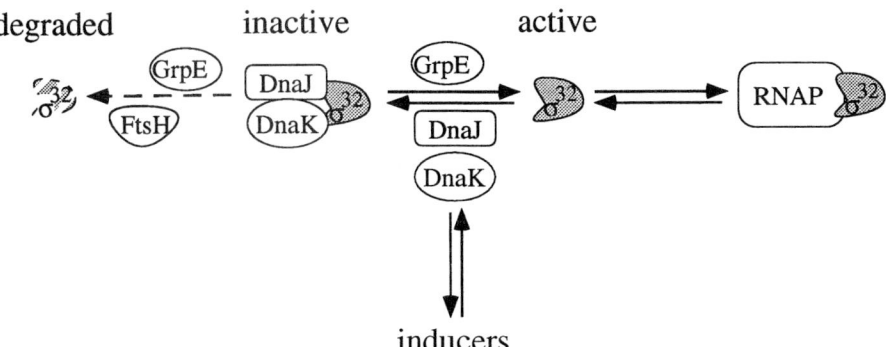

Figure 2. Model of the homeostatic control of σ^{32} activity and stability by the DnaK chaperone system. σ^{32} is in equilibrium between an active form, which can assemble with RNAP to mediate transcription of heat shock genes, and an inactive form, which is bound to DnaJ and DnaK. Inactive σ^{32} can be activated by ATP-mediated dissociation from DnaK and DnaJ, a process that is stimulated by the nucleotide exchange factor GrpE. σ^{32} can be eliminated through DnaK-, DnaJ-, GrpE-dependent degradation by FtsH and other cellular proteases. During induction of the heat shock response, the equilibrium is shifted towards active σ^{32} through sequestering of the DnaK system by binding to misfolded proteins. Upon protein damage repair, the DnaK system becomes available for binding to σ^{32} and homeostasis is re-established. Elements of this model, in particular the sequestering of chaperones through binding to misfolded proteins, have been described earlier (Craig and Gross, 1991; Bukau, 1993).

(Fig. 2). HSP-mediated refolding or degradation of misfolded proteins ameliorates the inducing signal and frees the DnaK system to shut off the heat shock response. Experimental support for these models is provided by analysis of inducers of the *E. coli* heat shock response. A common feature of inducers is their potential to generate misfolded proteins. For instance, the production of heterologous proteins and mutant proteins, and of thermolabile proteins such as firefly luciferase at non-permissive temperature, induces a heat shock response in *E. coli* (Goff and Goldberg, 1985; Ito et al., 1986; Parsell and Sauer, 1989). The DnaK system is active in refolding or degradation of misfolded proteins (Straus et al., 1988; Gaitanaris et al., 1990; Skowyra et al., 1990; Sherman and Goldberg, 1992; Schröder et al., 1993), including firefly luciferase inactivated by a heat shock to 42°C (Schröder et al., 1993). To verify further the proposed chaperone sequestration models it will be crucial to obtain evidence for the assumption that the DnaK chaperone system is limiting for heat shock gene regulation *in vivo*. It should be mentioned that sequestration of other regulatory components, such as the FtsH protease, might also contribute to the homeostatic control of heat shock gene expression in *E. coli*.

■ References

Bukau, B. (1993). Regulation of the *E. coli* heat shock response. Mol. Microbiol. **9**, 671–680.

Cowing, D.W., Bardwell, J.C., Craig, E.A., Woolford, C., Hendrix, R.W., and Gross, C.A. (1985). Consensus sequence for Escherichia coli heat shock gene promoters. Proc. Natl. Acad. Sci. USA **82**, 2679–2683.

Craig, E.A. and Gross, C.A. (1991). Is hsp70 the cellular thermometer? Trends Biochem. Sci. **16**, 135–140.

Gaitanaris, G.A., Rubock, A.G.P., Silverstein, S.J., and Gottesman, M.E. (1990). Renaturation of denatured l repressor requires heat shock proteins. Cell **61**, 1013–1020.

Gamer, J., Bujard, H., and Bukau, B. (1992). Physical interaction between heat shock proteins DnaK, DnaJ, GrpE and the bacterial heat shock transcriptional factor σ^{32}. Cell **69**, 833–842.

Gamer, J., Multhaup, G., Tomoyasu, T., McCarty, J. S., Rüdiger, S., Schönfeld, H.-J., Schirra, C., Bujard, H., and Bukau, B. (1996). A cycle of binding and release of the DnaK, DnaJ and GrpE chaperones regulates activity of the *E. coli* heat shock transcription factor σ^{32}. EMBO J. **15**, 607–617.

Georgopoulos, C., Ang, D., Liberek, K., and Zylicz, M. (1990). Properties of the *Escherichia coli* heat shock proteins and their role in bacteriophage λ growth. In *Stress proteins in biology and medicine* (ed. R. Morimoto, A. Tissieres and C. Georgopoulos). Cold Spring Harbor Laboratory Press, New York, pp. 191–222.

Georgopoulos, C., Liberek, K., Zylicz, M., and Ang, D. (1994). Properties of the heat shock proteins of *Escherichia coli* and the autoregulation of the heat shock response. In *The biology of heat shock proteins and molecular chaperones* (ed. R.I. Morimoto, A. Tissières and C. Georgopoulos). Cold Spring Harbor Laboratory Press, New York, pp. 209–250.

Goff, S.A., and Goldberg, A.L. (1985). Production of abnormal proteins in *E. coli* stimulates transcription of *lon* and other heat shock genes. Cell **4**, 587–595.

Gross, C.A., Straus, D.B., and Erickson, J.W. (1990). The function and regulation of heat shock proteins in *Escherichia coli*. In *Stress proteins in biology and medicine* (ed. R. Morimoto, A. Tissieres and C. Georgopoulos). Cold Spring Harbor Laboratory Press, New York, pp. 167–190.

Grossman, A.D., Erickson, J.W., and Gross, C.A. (1984). The *htpR* gene product of *E. coli* is a sigma factor for heat-shock promotors. Cell **38**, 383–390.

Grossman, A.D., Straus, D., Walter, W.A., and Gross, C.A. (1987). σ^{32} synthesis can regulate the synthesis of heat shock proteins in Escherichia coli. Genes Dev. **1**, 179–184.

Herman, C., Thévenet, D., D'Ari, R., and Bouloc, P. (1995). Degradation of σ^{32}, the heat shock regulator in *Escherichia coli*, is governed by HflB. Proc. Natl. Acad. Sci. USA **92**, 3516–3520.

Ito, K., Akiyama, Y., Yura, T., and Shiba, K. (1986). Diverse effects of the MalE-LacZ hybrid protein on Escherichia coli cell physiology. J. Bacteriol. **167**, 201–204.

Landick, R., Vaughn, V., Lau, E.T., VanBogelen, R.A., Erickson, J.W., and Neidhardt, F.C. (1984). Nucleotide sequence of the heat shock regulatory gene of *E. coli* suggests its protein product may be a transcription factor. Cell **38**, 175–182.

Liberek, K. and Georgopoulos, C. (1993). Autoregulation of the *Escherichia coli* heat shock response by the DnaK and DnaJ heat shock proteins. Proc. Natl. Acad. Sci. USA **90**, 11019–11023.

Liberek, K., Galitski, T. P., Zylicz, M., and Georgopoulos, C. (1992). The DnaK chaperone modulates the heat shock response of Escherichia coli by binding to the σ^{32} transcription factor. Proc. Natl. Acad. Sci. USA **89**, 3516–3520.

Liberek, K., Wall, D., and Georgopoulos, C. (1995). The DnaJ chaperone catalytically activates the DnaK chaperone to preferentially bind the σ^{32} heat shock transcriptional regulator. Proc. Natl. Acad. Sci. USA **92**, 6224–6228.

McCarty, J. S., Rüdiger, S., Schönfeld, H.-J., Schneider-Mergener, J., Nakahigashi, K., Yura, T., and Bukau, B. (1996). Regulatory region C of the *E.coli* heat shock transcription factor, σ^{32}, constitutes a DnaK binding site and is conserved among eubacteria. J. Mol. Biol. **256**, 829–837.

Morimoto, R.I., Tissières, A., and Georgopoulos, C. (ed.) (1994). *The biology of heat shock proteins and molecular chaperones*. Cold Spring Harbor Laboratory Press, New York.

Nagai, H., Yuzawa, H., and Yura, T. (1991). Interplay of two cis-acting mRNA regions in translational control of σ^{32} synthesis during the heat shock response of *Escherichia coli*. Proc. Natl. Acad. Sci. USA **88**, 10515–10519.

Nagai, H., Yuzawa, H., Kanemori, M., and Yura, T. (1994). A distinct segment of the σ^{32} polypeptide is involved in DnaK-mediated negative control of the heat shock response in *Escherichia coli*. Proc. Natl. Acad. Sci. USA **91**, 10280–10284.

Neidhardt, F.C. and VanBogelen, R.A. (1987). Heat shock response. In Escherichia coli and Salmonella typhimurium: cellular and molecular biology, Vol. 2, (ed. F.C. Neidhardt). American Society for Microbiology, Washington, D.C., pp. 1334–1345.

Parsell, D. and Sauer, R.T. (1989). Induction of a heat shock-like response by unfolded protein in Escherichia coli: dependence on protein level not protein degradation. Genes Dev. **3**, 1226–1232.

Schröder, H., Langer, T., Hartl, F.-U., and Bukau, B. (1993). DnaK, DnaJ, GrpE form a cellular chaperone machinery capable of repairing heat-induced protein damage. EMBO J. **12**, 4137–4144.

Sherman, M. and Goldberg, A.L. (1992). Involvement of the chaperonin DnaK in the rapid degradation of a mutant protein in *Escherichia coli*. EMBO J. **11**, 71–77.

Skowyra, D., Georgopoulos, C., and Zylicz, M. (1990). The E. coli *dnaK* gene product, the hsp70 homolog, can reactivate heat-inactivated RNA polymerase in an ATP hydrolysis-dependent manner. Cell **62**, 939–944.

Straus, D.B., Walter, W.A., and Gross, C.A. (1987). The heat shock response of *E. coli* is regulated by changes in the concentration of σ^{32}. Nature **329**, 348–350.

Straus, D.B., Walter, W.A., and Gross, C.A. (1988). *Escherichia coli* heat shock gene mutants are defective in proteolysis. Genes Dev. **2**, 1851–1858.

Straus, D.B., Walter, W.A., and Gross, C.A. (1989). The activity of σ^{32} is reduced under conditions of excess heat shock protein production in *Escherichia coli*. Genes Dev. **3**, 2003–2010.

Straus, D., Walter, W., and Gross, C. (1990). DnaK, DnaJ, and GrpE heat shock proteins negatively regulate heat shock gene expression by controlling the synthesis and stability of σ^{32}. Genes Dev. **4**, 2202–2209.

Szabo, A., Langer, T., Schröder, H., Flanagan, J., Bukau, B., and Hartl, F. U. (1994). The ATP hydrolysis-dependent reaction cycle of the *Escherichia coli* Hsp70 system-DnaK, DnaJ and GrpE. Proc. Natl. Acad. Sci. USA **91**, 10345–10349.

Tomoyasu, T., Gamer, J., Bukau, B., Kanemori, M., Mori, H., Rutman, A.J., Oppenheim, A.B., Yura, T., Yamanaka, K., Niki, H., Hiraga, S., and Ogura, T. (1995). *Escherichia coli* FtsH is a membrane-bound, ATP-dependent protease which degrades the heat-shock transcription factor σ^{32}. EMBO J. **14**, 2551–2560.

VanBogelen, R.A., Acton, M.A., and Neidhardt, F.C. (1987). Induction of the heat shock regulon does not produce thermotolerance in *Escherichia coli*. Genes Dev. **1**, 525–531.

Yura, T., Nagai, H., and Mori, H. (1993). Regulation of the heat-shock response in bacteria. Annu. Rev. Microbiol. **47**, 321–350.

Zhou, Y.-N., Kusukawa, N., Erickson, J. W., Gross, C.A., and Yura, T. (1988). Isolation of characterization of *Escherichia coli* mutants that lack the heat shock sigma factor σ^{32}. J. Bacteriol. **170**, 3640–3649.

■ *Bernd Bukau:*
Institut für Biochemie und Molekularbiologie
Universität Freiburg
Hermann Herder Str. 7
D-79104 Freiburg, Germany
Tel. 49 761 203 5221
Fax. 49 761 203 5257
E-mail: bukau@sun2.ruf.uni-Freiburg.de

The periplasmic unfolded protein response in *Escherichia coli*

Misfolding of proteins in Escherichia coli *triggers two stress responses. Accumulation of misfolded proteins in the cytoplasm induces the 'classical' heat shock regulon controlled by the Eσ^{32} polymerase. In contrast, misfolding of proteins in the extracytoplasm induces the newly described Eσ^E-dependent regulon. Such an intercompartmental organization implies the presence of signal-transducing molecules located in the periplasm and in the inner membrane. Eσ^E, by itself, is the master regulator of the heat shock response since it ensures, at high temperatures, the transcription of the* rpoH *gene encoding σ^{32}.*

■ Stimuli inducing the σ^E regulon

Although biochemical observations have long supported the existence of an additional sigma factor in *E. coli*, designated either as σ^E or σ^{24} (Erickson and Gross, 1989; Wang and Kaguni, 1989), the isolation of the *rpoE* gene which encodes this factor occurred only recently (Raina et al., 1995; Rouvière et al., 1995). Initial attempts to isolate the *rpoE* gene were based on the assumption that multicopy expression of *rpoE* should lead to increased β-galactosidase activity as driven from the *rpoH*P3–*lacZ* transcriptional fusion. The *rpoH* gene encodes the σ^{32} factor and its P3 promoter is controlled by the Eσ^E polymerase. Unexpectedly, this multicopy approach revealed that overexpression of some outer membrane proteins (OMPs) could induce this new regulon (Mecsas et al., 1993; Rouvière et al., 1995). It was proposed that such an overexpression leads to imbalances in the ratio of OMPs and that only those that crossed the peptidoglycan barrier could induce the transcriptional activity reflected by the *rpoH*P3–*lacZ* fusion. Also, inactivation of OmpR, a positive regulator of *ompC* and *ompF* expression, was found to down-regulate the activity of the σ^E regulon (Mecsas et al., 1993). This was initially explained by an imbalance in the ratio of OMPs occurring in the *ompR* mutant strains. Since then a better understanding of the regulation of *rpoE* has been achieved. In fact, most of the transported proteins encountering alterations, whether destined for the periplasm, outer membrane or extracellular medium, are detected/sensed by the cell, thereby prompting a σ^E-dependent response (Missiakas et al., 1995a, 1996; Raina et al., 1995). For example, altering the solubility of two periplasmic proteins, L-asparaginase II (AnsB) or the maltose carrier protein (MalE) leads to an increased transcription from *rpoH*P3, *htrA* or *rpoE*P2 promoters, which are all transcribed by the Eσ^E polymerase (Table 1). In these cases, AnsB is expressed as a C-terminal truncated polypeptide and MalE as an allele forming inclusion bodies in the periplasm (Missiakas et al., 1996).

Interestingly, σ^E belongs to a subfamily of sigma factors designated as extracytoplasmic factors (ECF), which are devoted to the regulation of extracytoplasmic activities (Lonetto et al., 1994). In *Pseudomonas aeruginosa* for example, a σ^E homologue is involved in the transcriptional regulation of genes participating in alginate synthesis (Deretic et al., 1995).

■ Folding catalysts in the extracytoplasmic compartments

Since the increased activity of the σ^E regulon reflected (at least partly) misfolding occurring in the periplasm, this measure was taken as a tool for search for folding catalysts in the extracytoplasm of *E. coli* (Missiakas et al., 1996). Mapping of mutations in loci encoding various Dsb proteins was cogent proof of the proposed model (Fig. 1). The Dsb proteins belong to a newly described class of periplasmic folding enzymes homologous to the eukaryotic protein disulfide isomerase (Bardwell, 1994; see also the entries pp. 318–328). Mutations in various *dsb* genes, such as *dsbA*, *dsbC* or *dsbD*, have already been shown to trigger a σ^E-dependent response (Missiakas et al., 1995a; Raina et al., 1995).

In addition, four other mutants were isolated and mapped to the *htrM*, *surA*, *fkpA* and *skp* genes (Raina et al., 1995; Missiakas et al., 1996). Mutations in *htrM* and *skp* strictly affect the folding and oligomerization of OMPs. *htrM* mutants affect the nature of the lipopolysaccharide (LPS) molecule. Skp is thought to act as an exchange factor of the LPS molecule associated with protrimers of OMPs, thereby triggering their insertion in the outer membrane (Chen and Henning, 1996; Missiakas et al., 1996). It was assumed for a long time that LPS is required for the proper folding of OMPs (Sen and Nikaido, 1991). It could well be that the LPS content of the membrane serves as an early sensor for detecting events occurring at the furthest envelope of the cell. In fact, the biogenesis of LPS is partly controlled by the Eσ^E polymerase (S. Raina and D. Missiakas, unpublished observations).

The SurA and FkpA proteins were found to belong to a second class of folding catalyst known as the peptidyl–prolyl *cis*–*trans* isomerases (PPIase; Fischer and Schmid, 1990; see overview p. 359). SurA appears to play an impor-

Table 1. The σ^E regulon. This figure summarizes our knowledges of the genes transcribed by the $E\sigma^E$ polymerase

Name protein/gene	M.W. (kDa)	Location	Function	References
HtrA (DegP)/htrA (degP)	50	Periplasm	Protease	Lipinska et al., 1988; Strauch et al., 1988
σ^E (σ^{24})/rpoE	21.6	Cytoplasm	Sigma factor	Raina et al., 1995; Rouvière et al., 1995
σ^{32}/rpoH	33	Cytoplasm	Sigma factor	See entry p. 000
RseA/rseA	24.3	Inner membrane	Negatively regulates σ^E activity	Raina et al., 1995; Rouvière et al., 1995; Missiakas et al., 1995b
RseB/rseB	35.8	Periplasm	Binds RseA	Missiakas et al., 1995b; unpublished observations
RseC/rseC	17	Inner membrane	Positive modulator of σ^E	Missiakas et al., 1995b; unpublished observations
RfaX/rfaX	24.5	Inner membrane	LPS biogenesis	Unpublished observations
RfaY/rfaY	14	Outer membrane	LPS biogenesis	Unpublished observations
FkpA	27	Periplasm	Peptidyl prolyl isomerase	Unpublished observations

tant role in the folding of monomers of the LamB OMP (Lazar and Kolter, 1996; see entry p. 436). However, its function could be broader, since overexpression of SurA was found to alleviate efficiently the σ^E response induced by aggregated MalE mutant or truncated AnsB (Missiakas et al., 1996). PPIases are usually distinguished according to their sensitivity to two types of inhibitors: cyclosporin A or the FK506 macrolide (Fischer and Schmid, 1990). FkpA belongs to the FK506 binding protein family (see entry p. 402), whereas SurA belongs to a new class of PPIase, which also seem to be present in both prokaryotes as well as eukaryotes (Rahfeld et al., 1994; Rudd et al., 1995).

■ Transduction pathways for signaling the presence of altered proteins

Transcription of the rpoE gene is an autoregulated process. Under conditions altering the cell envelope, when the demand in σ^E is maximal, transcription of the rpoE gene is controlled by the Eσ^E enzyme from its

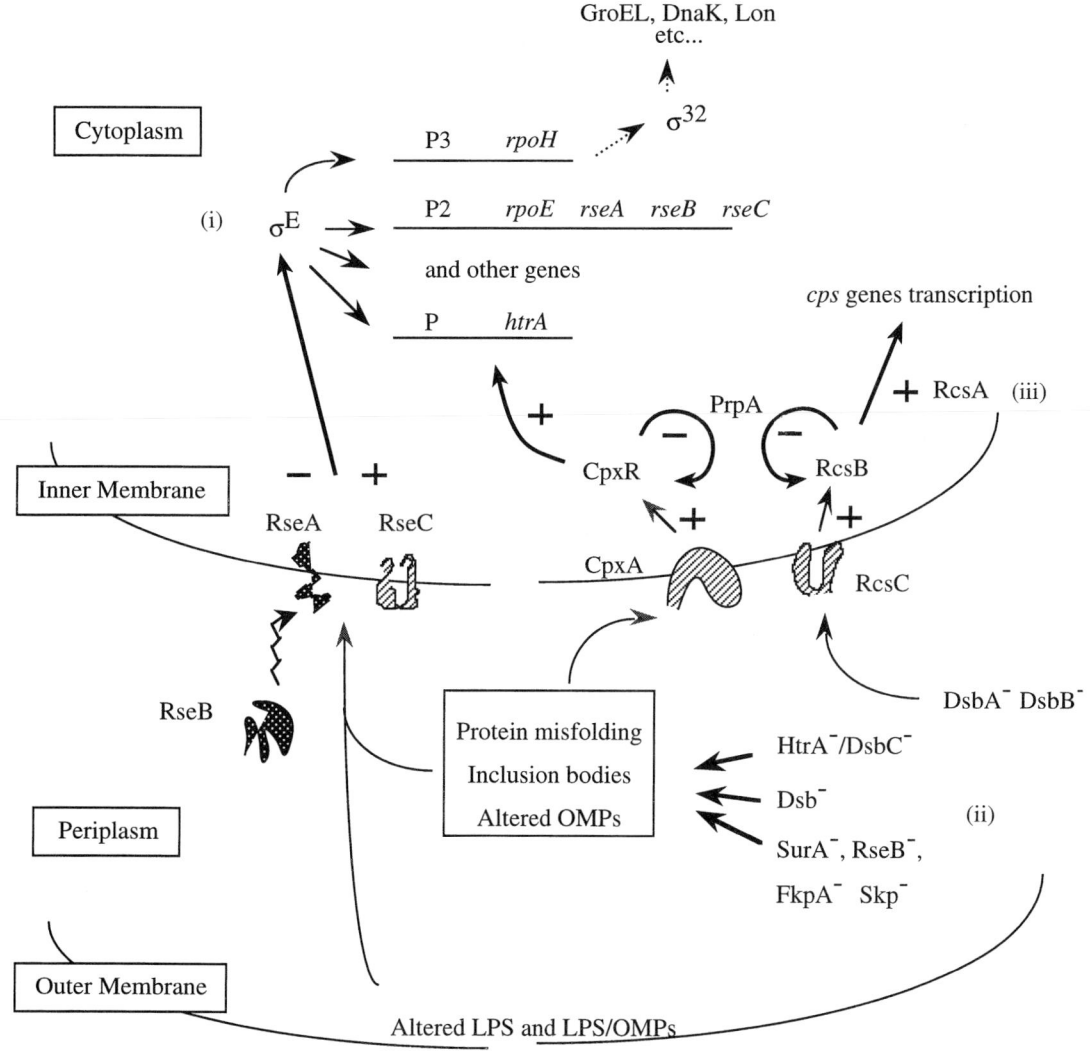

Figure 1. Model of transducing pathways signaling protein misfolding to the various transcriptional machineries. The three known pathways responding to protein misfolding in the extra-cytoplasm are shown. (i) The signal transduction pathway controlling the global σ^E-dependent response. (ii) The pathway fine tuning transcription from the htrA promoter. It is not known whether other genes are also controlled by the CpxA CpxR PrpA proteins. (iii) This last pathway leads to the transcriptional induction of the cps genes. A negative control is provided by the phosphatase activity of PrpA. It should be noted that the cps genes are not part of the σ^E regulon. Nevertheless, synthesis of the capsule, which provides an additional barrier to the cell environment, may be a way to prevent further misfolding of periplasmic proteins.

rpoEP2 promoter. In addition, genetic evidence shows that various mutations can be obtained which either block the σ^E response or lead to a constitutive σ^E activity regardless of the presence of aberrantly folded polypeptides in the cell envelope. Mutations in the rseA or rseB genes are mostly responsible for a constitutive induction of the σ^E regulon, implying that their products act as negative modulators (Fig. 1). Both genes are located just downstream of the rpoE gene and are indeed cotranscribed with rpoE. Topological analysis of the RseA protein distinguishes one transmembrane domain with an N-terminal domain facing the cytoplasm. Overexpression of this domain, although toxic in a high copy vector, leads to the down-regulation of the σ^E regulon. Addition of purified RseA in a reconstituted runoff transcription system has a negative effect on the ability of Eσ^E to transcribe the htrA promoter (Missiakas et al., 1997; De Las Peñas et al., 1997). In vivo, this negative activity of RseA is modulated by two other factors, RseB and RseC. Genetic and biochemical evidence suggest that RseB is a periplasmic 'sensor' protein, favoring the interaction between RseA and σ^E. What and how signals are detected in the first place is still a matter for speculation. Mutations in a third gene, rseC, have the opposite effect to mutations in rseA. RseC is another inner membrane protein, capable of displacing the RseA–σ^E complex, thereby ensuring a positive feedback control on σ^E activity (Fig. 1; D. Missiakas and S. Raina, unpublished observations).

A second system was found to signal protein misfolding to only one gene of the σ^E regulon, that is the htrA gene, which encodes a periplasmic protease (Table 1 and Fig. 1). Trans-acting mutations affecting htrA transcription were isolated in the cpxR cpxA operon (Danese et al., 1995; Raina et al., 1995) and in the prpA and prpB genes (Missiakas and Raina, 1997). The deduced amino acid sequence of the cloned cpxR gene suggests that the protein it encodes is the cognate regulator for the membrane sensor CpxA, a putative histidine kinase (Weber and Silverman, 1988; Dong et al., 1993). Biochemical and genetic evidence show that the prpA and prpB genes encode two phosphoprotein phosphatases. PrpA has both serine/threonine and tyrosine phosphatase activities and can modulate in the cell the phosphorylation status of various proteins including proteins of two-component systems (Missiakas and Raina, 1997). It is not yet clear how the input signal is transmitted to and by CpxA. It is interesting that in yeast, a transmembrane protein kinase activity is also required for signaling protein misfolding occurring in the endoplasmic reticulum (Cox et al., 1993; Mori et al., 1993). In E. coli, it appears that it is a cascade of phosphorylation and dephosphorylation processes that is responsible for fine-tuning htrA transcription upon accumulation of misfolded proteins in the periplasm (Fig. 1). PrpA may also play a key role in modulating overall the heat shock response, since: (i) transcription of prpA is heat shock regulated; (ii) null mutations in prpA lead to accumulation of many phosphoproteins, some of which are major chaperones; and (iii) the phosphatase activity of PrpA is stimulated at higher temperatures.

Finally, it has also been observed that misfolding of proteins occurring in dsbA or dsbB mutants not only induces a σ^E response, but also capsular polysaccharide synthesis. This is reflected in vivo by a highly mucoid phenotype (Missiakas and Raina, 1996). Mucoidy results from the production of colanic acid capsular polysaccharide which is promoted by the cps gene products. Synthesis of the capsule is adaptive in response, and under certain external stresses provides a good mechanism of defense by strengthening the cell envelope (Gottesman and Stout, 1991). Under normal conditions, capsular polysaccharide biogenesis is shut off since RcsA, a positive regulator, is rapidly degraded by the Lon protease. RcsB, a second regulator of cps transcription, is part of the two-component signaling system RcsB RcsC, and is activated by phosphorylation, in an RcsC-dependent manner (Gottesman and Stout, 1991). Hence, the RcsB RcsC two-component system is also induced upon misfolding of proteins in the extracytoplasm. Interestingly, overexpression of the PrpA phosphatase in dsbA or dsbB backgrounds leads to a reduced transcription of the cps genes by a factor of two- to threefold, almost suppressing the mucoidy phenotype of the dsbA or dsbB mutant cells in a lon-independent manner (Missiakas and Raina, 1997). Protein phosphatases in E. coli such as PrpA appear to be negative regulators of kinase-controlled transcriptional processes, by providing a rapid feedback mechanism (Fig. 1).

■ The importance of Eσ^E transcriptional activity in extreme conditions

As mentioned earlier, the rpoHP3 and rpoEP2 promoters are under the control of the Eσ^E holoenzyme. The usage pattern of rpoHP3 increases with temperature, so that at 50°C, P3 constitutes the sole operating transcription system of the rpoH gene (Erickson and Gross, 1989). At such extreme temperatures, Eσ^{70}-directed transcription from the other rpoH promoters ceases, whereas Eσ^E transcription continues unabated. Continuous transcription of the rpoH gene at high temperatures is very important to ensure unabated transcription of the classical heat shock genes encoding the heat shock proteins, such as the GroEL/GroES, DnaK/DnaJ/GrpE chaperones and the Lon and Clp protease systems. These proteins are very important for maintaining, in a folding- or degradation-competent state, a population of aggregation-prone polypeptides accumulating with elevated temperatures.

■ References

Bardwell, J.C.A. (1994). Building bridges: disulphide formation in the cell. Mol. Microbiol. **14**, 199–205.

Chen, R. and Henning, U. (1996). A periplasmic protein (Skp) of Escherichia coli selectively binds a class of outer membrane proteins. Mol. Microbiol. **19**, 1287–1294.

Cox, J.S., Shamu, C.E. and Walter, P. (1993). Trancriptional induction of genes encoding endoplasmic reticulum resident proteins requires a transmembrane protein kinase. Cell **73**, 1197–1206.

Danese, P., Snyder, W.B., Cosma, C., Davis, L. and Silhavy, T. (1995). The Cpx two-component signal transduction pathway of *Escherichia coli* regulates transcription of the gene specifying the stress-inducible periplasmic protease. Genes Dev. **9**, 387–398.

De Las Peñas, A., Connolly, L., and Gross, C.A. (1997). The σ^E-mediated response to extra cytoplasmic stress in *Escherichia coli* is transduced by RseA and RseB, two negative regulators of σ^E. Mol. Microbiol. **24**, 373–385.

Deretic, V., Schurr, M.J. and Yu, H. (1995). *Pseudomonas aeruginosa*, mucoidy and the chronic infection phenotype in cystic fibrosis. Trends Microbiol. **3**, 351–356.

Dong, J., Iuchi, S., Kwan, S.H., Lu, Z. and Lin, E.C. (1993). The deduced amino-acid sequence of the cloned *cpxR* gene suggests the protein is the cognate regulator for the membrane sensor, CpxA, in a two-component signal transduction system of *Escherichia coli*. Gene **136**, 227–230.

Erickson, J.W. and Gross, C.A. (1989). Identification of the σ^E subunit of RNA polymerase: a second alternate σ factor involved in high-temperature gene expression. Genes Dev. **3**, 1462–1471.

Fischer, G. and Schmid, F.X. (1990). The mechanism of protein folding. Implications of *in vitro* refolding models for *de novo* protein folding and translocation in the cell. Biochemistry **29**, 2205–2212.

Gottesman, S. and Stout, V. (1991). Regulation of capsular polysaccharide synthesis in *Escherichia coli* K12. Mol. Microbiol. **5**, 1599–1606.

Lazar, S. and Kolter, R. (1996). SurA assists the folding of *Escherichia coli* outer membrane proteins. J. Bacteriol. **178**, 1770–1773.

Lipinska, B., Sharma, S. and Georgopoulos, C. (1988). Sequence analysis and regulation of the *htrA* gene of *Escherichia coli*: a sigma 32-independent mechanism of heat-inducible transcription. Nucl. Acids Res. **16**, 10053–10067.

Lonetto, M., Brown, K.L., Rudd, K.E. and Buttner, M.J. (1994). Analysis of the *Streptomyces coelicolor* sigmaE gene reveals the existence of a subfamily of Eubacterial RNA polymerase sigma factors involved in the regulation of extracytoplasmic functions. Proc. Natl. Acad. Sci. USA **91**, 7573–7577.

Mecsas, J., Rouvière, P.E., Erickson, J.W., Donohue, T.J. and Gross, C.A. (1993). The activity of σ^E, an *Escherichia coli* heat-inducible sigma factor, is modulated by expression of outer membrane proteins. Genes Dev. **7**, 2618–2628.

Missiakas, D. and Raina, S. (1997). Signal transduction pathways in response to protein misfolding in the extra-cytoplasmic compartments of *E. coli*: Role of two phosphoprotein phosphatases. EMBO J., **16**, 1670–1685.

Missiakas, D., Schwager, F. and Raina, S. (1995a) Identification and characterization of a new disulfide isomerase-like protein (DsbD) in *Escherichia coli*. EMBO J. **14**, 3415–3424.

Missiakas, D., Raina, S. and Georgopoulos, C. (1995b) The heat shock system. In *Regulation of gene expression in Escherichia coli* (Lin, E.C.C. and Lynch, S.A. ed.). R.G. Landes Company, Austin TX pp. 481–501.

Missiakas, D., Betton, J.-M. and Raina, S. (1996). New components of protein folding in the extra-cytoplasmic compartments of *Escherichia coli*: SurA, FkpA and Skp/OmpH. Mol. Microbiol., **21**, 871–884.

Missiakas, D., Mayer, M.P., Lemaire, M., Georgopoulos, C., and Raina, S. (1997). Modulation of the *Escherichia coli* σ^E (RpoE) heat-shock transcription-factor activity by the RseA, RseB and RseC proteins. Mol. Microbiol. **24**, 355–371.

Mori, K., Ma, W., Gething, M.-J. and Sambrook, J. (1993). A transmembrane protein with a cdc2+/CDC28-related kinase activity is required for signaling from the ER to the nucleus. Cell **74**, 743–756.

Rahfeld, J.-U., Schierhorn, A., Mann, K. and Fischer, G. (1994) A novel peptidyl-prolyl *cis/trans* isomerase from *Escherichia coli*. FEBS Lett. **343**, 65–69.

Raina, S., Missiakas, D. and Georgopoulos, C. (1995). The *rpoE* gene encoding the σ^E (σ^{24}) heat shock sigma factor of *Escherichia coli*. EMBO J. **14**, 1043–1055.

Rouvière, P., de las Penas, A., Mecsas, J., Lu, C.Z., Rudd, K.E. and Gross, C.A. (1995). *rpoE*, the gene encoding the second heat-shock sigma factor, σ^E, in *Escherichia coli*. EMBO J. **14**, 1032–1042.

Rudd, K.E., Sofia, H.J., Koonin, E.V., Plunkett, G., Lazar, S. and Rouvière, P.E. (1995). New family of peptidyl-prolyl isomerases. Trends Biochem. Sci. **20**, 12–14.

Sen, K. and Nikaido, H. (1991). Lipopolysaccharide structure required for *in vitro* trimerization of *Escherichia coli* OmpF porin. J. Bacteriol. **173**, 926–928.

Strauch, K.L. and Beckwith, J. (1988). An *Escherichia coli* mutation preventing degradation of abnormal periplasmic proteins. Proc. Natl. Acad. Sci. USA **85**, 1576–1580.

Wang, Q. and Kaguni, J.M. (1989). A novel sigma factor is involved in expression of the *rpoH* gene of *Escherichia coli*. J. Bacteriol. **171**, 4248–4253.

Weber, R.F. and Silverman, P.J. (1988). The Cpx proteins of *Escherichia coli* K12. Structure of the CpxA polypeptide as an inner membrane component. J. Mol. Biol. **203**, 467–476.

■ *Satish Raina:*
Université de Genève
Centre Médical Universitaire
1, Rue Michel-Servet
1211 Genève 4, Switzerland
Tel. 41 22 702 55 11
Fax. 41 22 702 55 02
E-mail: Satish.Raina@medecine.unige.ch

■ *Dominique Missiakas:*
CNRS UPR9027
31, Ch. J. Aiguier
13402 Marseille, Cedex 20, France
Tel. 33 4 91 76 03 59
Fax. 30 4 91 71 21 24
E-mail: missiaka@ibsm.cnrs.mrs.fr

Transcriptional regulation of eukaryotic heat shock genes

Common to all organisms is an essential, highly conserved and exquisitely regulated cellular response to suboptimal physiological and environmental conditions that affect cell growth, maintenance of cellular activities, and development. The activation of heat shock genes is a highly regulated response to heat shock, oxidative stress, heavy metals, various toxic chemicals, bacterial and viral infection, and exposure to a number of acute and chronic disease states which results in the elevated synthesis of a family of stress-induced or heat shock proteins (Lindquist and Craig, 1988; Morimoto et al., 1990, 1994). This affords the cell a protective mechanism against acute exposure to stress which, if unchecked, leads to irreversible cell damage and cell death. Heat shock proteins, many of which have properties as molecular chaperones, have essential roles in the process of protein biosynthesis, specifically in the synthesis, folding, degradation, transport, and translocation of proteins (reviewed in Gething and Sambrook, 1992; Georgopoulos and Welch, 1993; Hendrix and Hartl, 1993; Craig et al., 1993; Morimoto et al., 1994). During cell stress, heat shock proteins and molecular chaperones diminish the deleterious effects of protein damage by limiting or preventing the appearance of misfolded proteins. Studies from many laboratories have contributed to our general understanding of molecular chaperones as essential cellular proteins that associate with nascent chains and unfolded proteins to ensure the proper folding or translocation of proteins (Bochkareva and Girshovich, 1988; Beckmann et al., 1990; Langer et al., 1992).

An understanding of the heat shock response requires an appreciation of the myriad of conditions that lead to the elevated expression of heat shock proteins. Initial studies on the heat shock response in *Drosophila* identified conditions such as exposure to elevated temperature, salicylate, dinitrophenol, ethanol, and anoxia, all of which induced the appearance of specific chromosomal puffs that corresponded to heat shock gene loci (Ritossa, 1962; Ashburner, 1970; Tissieres et al., 1974; Lindquist-McKenzie et al., 1975). Subsequent studies carried out in plant and animal systems have revealed the universal expression of heat shock proteins in response to an ever growing list of conditions (Fig. 1) that fall into three general classes: (1) environmental stresses such as exposure to heat shock, amino acid analogues, drugs and toxic chemicals, and heavy metals; (2) pathophysiology and disease states including oxidative stress, fever, inflammation, myocardial and neuronal degenerative diseases; (3) non-stress conditions including the cell cycle, growth factors, serum stimulation, development, differentiation and activation by certain oncogenes. The diversity of these conditions has posed numerous questions on how cells detect 'stress' at the molecular level.

■ Inducible transcription by a family of heat shock factors

A well-studied aspect of the heat shock response is stress-inducible transcription. Upon exposure of cells to elevated temperatures, heat shock transcription factor (HSF) is activated resulting in the rapid increase in transcription of genes encoding heat shock proteins (Lindquist and Craig, 1988; Lis and Wu, 1993; Morimoto, 1993; Wu, 1995). Initial studies to examine the chromatin structure in the 5' flanking region of the *Drosophila* HSP70 and HSP90 genes led to the identification of constitutive and inducible protein factor-binding sites (Wu, 1980; 1984). HSF binds with a high degree of specificity to the heat shock element (HSE), a series of pentameric units arranged as inverted adjacent arrays of the sequence 5'-AGAAn-3' which is located in heat shock gene promoters (Amin et al., 1988; Xiao and Lis, 1988; Perisic et al., 1989; Xiao et al., 1991; Kroeger and Morimoto, 1994). For example, in the mammalian HSP70 and HSP90 promoters, the HSE is comprised of five and six pentameric units, respectively, positioned in close proximity to the site of transcription initiation and independent of adjacent basal promoter elements (Greene and Kingston, 1990; Williams and Morimoto, 1990).

Yeast HSF binds constitutively to DNA, yet heat shock is required to activate stress-responsive transcription (Sorger and Nelson, 1988). In contrast, *Drosophila*, plant, and vertebrate HSFs are negatively regulated for DNA binding and require heat shock for both acquisition of DNA binding and transcriptional activity (Kingston et al., 1987; Zimarino and Wu, 1987, 1990; Larson et al., 1988; Mosser et al., 1988; Westwood et al., 1991; Westwood and Wu, 1993; Nover, 1994). The DNA-binding properties of the latent *Drosophila* and mammalian HSFs can be activated *in vitro* following exposure to a variety of conditions, including *in vitro* heat shock, non-ionic detergents, low pH, or chaotropes that affect protein conformation (Larson et al., 1988; Mosser et al., 1990). The biochemical characterization of HSF and the subsequent cloning of the corresponding gene was initially accomplished in *Saccharomyces cerevisiae* and *D. melanogaster* (Sorger et al., 1987; Wu et al., 1987;

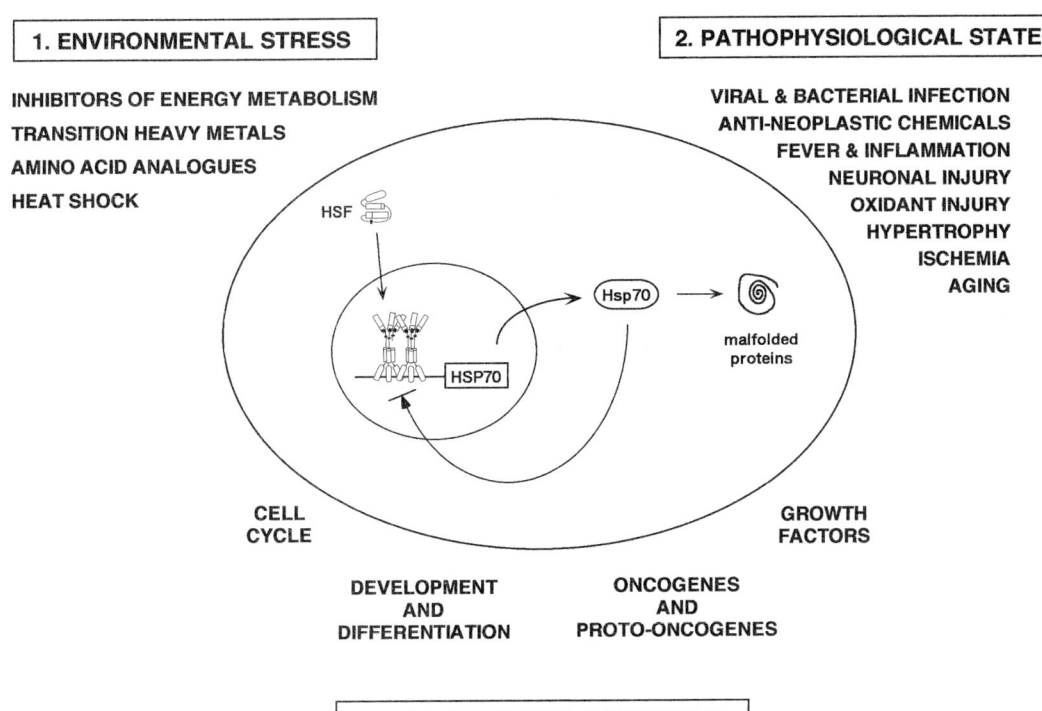

Figure 1. Conditions that result in the induction of heat shock gene expression in eukaryotes. Representation of three general classes of conditions known to result in the elevated expression of stress proteins including: (1) environmental and physiological stress; (2) pathophysiological states including conditions of disease; and (3) non-stressful conditions such as cell growth and development. Each condition acts on the cell as illustrated in the figure, and, in the case of environmental stress and certain pathophysiological states, leads to the activation of heat shock gene expression and the synthesis of heat shock proteins.

Weiderrecht et al., 1988). Subsequently, the cloning of HSF genes from larger eukaryotes uncovered the unexpected observation of a HSF multigene family. At least three HSFs have been isolated from the human, mouse, chicken, and tomato genomes (Scharf et al., 1990; Rabindran et al., 1991; Sarge et al., 1991; Schuetz et al., 1991; Nakai and Morimoto, 1993; Nakai et al., 1997). The cloned HSFs vary in size, from 301 amino acids (aa) for tomato HSF24, 512 aa for tomato HSF8, 503 aa and 529 aa for mouse and human HSF1, 517 aa and 536 aa for mouse and human HSF2, 691 aa for *Drosophila* HSF and 833 aa for *S. cerevisiae* HSF (Scharf et al., 1990; Rabindran et al., 1991; Sarge et al., 1991; Schuetz et al., 1991; Nakai and Morimoto, 1993). Within a species, members of the HSF family (i.e. mouse HSF1-HSF2 or chicken HSF1-HSF2-HSF3) are c. 40% related in amino acid sequence; this is primarily owing to sequence identity within the DNA-binding and oligomerization domains (Clos et al., 1990; Scharf et al., 1990; Rabindran et al., 1991; Sarge et al., 1991; Schuetz et al., 1991; Nakai and Morimoto, 1993; Nakai et al., 1997). Interspecies comparisons, i.e. between human, mouse, and chicken HSF1, reveals a high level of sequence identity (85–95%). Among the vertebrates HSFs, there appears to be a common ancestral progenitor from which the contemporary HSFs have evolved (Nakai and Morimoto, 1993).

The functional domains of the HSFs (Fig. 2) have been identified initially by alignment of the derived amino acid sequences, and subsequently by functional assays on deletion and point mutants. Comparison of the HSFs from yeasts, *Drosophila*, tomato, chicken, mouse, and human identified a conserved c. 100 amino acid DNA-binding domain of the winged helix–loop–helix motif located towards the amino terminus (Damberger et al., 1994; Harrison et al., 1994; Vuister et al., 1994a, b). The high degree of conservation in the DNA-binding domain is not surprising as each HSF binds to the highly conserved HSE motif. Adjacent to the DNA-binding domain is a conserved hydrophobic heptad repeat region of approximately 100 residues which is involved in oligomerization of HSF (Persic et al., 1989; Sorger and Nelson, 1989; Clos et al., 1990; Nieto-Sotelo et al., 1990; Peteranderl and Nelson, 1992; Rabindran et al., 1993). The HSFs of *S. pombe*, *Drosophila*, and larger eukaryotes contain an additional array of hydrophobic heptads positioned at the carboxyl terminus which has been suggested to have a role in the negative regulation of DNA binding through intramolecular interactions with the heptad repeat domain adjacent to the DNA-binding domain (Rabindran et al., 1993). A detailed analysis of the transcriptional activation domains of the HSF from *S. cerevisiae* and *Kluveromyces lactis* have identified separate regions required for full transcriptional activity (Nieto-Sotelo et al., 1990; Jakobsen and Pelham, 1991; Chen et al., 1993). The transcriptional activation domains of *Drosophila* and mammalian HSF1 have been positioned through the use of deletion mutants and GAL4 fusion proteins to the extreme carboxyl terminus under the negative regulation of a *cis*-element positioned in the vicinity of the DNA-binding domain (Shi et al., 1995; Zuo et al., 1995; Newton et al., 1996).

What is the role for a family of HSFs in the transcriptional regulation of heat shock genes? One possibility is that larger organisms may require multiple HSFs to provide specialized responses to the diverse developmental and environmental cues and insults that they may be exposed to during life. Consistent with this speculation, vertebrate HSF1 and HSF2, though structurally related, are distinct by a variety of regulatory and functional criteria. For example, HSF1 is the stress-responsive transcription factor whereas HSF2 is activated in response to signals during early development and differentiation (Theodorakis et al., 1989; Sistonen et al., 1992; 1994; Sarge et al., 1993, 1994). Activation of HSF1 is a complex multistep process which involves oligomerization from an inert monomer to active trimer, acquisition of DNA-binding ability, stress-induced phosphorylation, and nuclear localization (Jurivich et al., 1992; Sistonen et al., 1992, 1994; Baler et al., 1993; Sarge et al., 1993; Cotto et al., 1996). Acquisition of DNA binding precedes inducible phosphorylation; furthermore, inducible phosphorylation of HSF1 is stress dependent such that heat shock, heavy metals, and arachidonate treatment results in the fully phosphorylated state, whereas exposure to amino acid analogues or salicylate results in the activation of the DNA-binding competent non-inducibly phosphorylated state (Sarge et al., 1993; Jurivich et al., 1994; Cotto et al., 1996). In contrast, HSF2 is not post-translationally modified by heat shock or other stresses (Sistonen et al., 1992, 1994; Sarge et al., 1993). HSF2 DNA

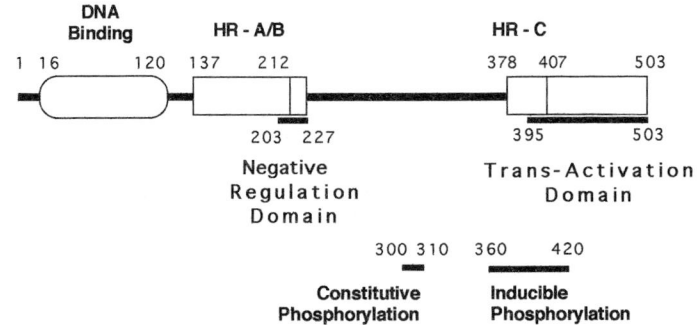

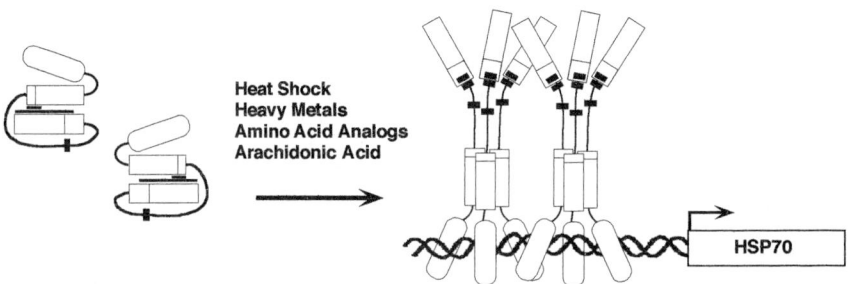

Figure 2. Functional domains of heat shock factors. The organizational structure of mouse HSF1 is indicated with the DNA-binding domain designated at the amino terminus. Adjacent to the DNA-binding domain are the conserved hydrophobic heptad repeats involved in oligomerization of HSF1 and the negative regulatory domain. At the carboxyl terminus is an additional hydrophobic heptad repeat adjacent to the transcription activation domain. The positions of sites of constitutive and inducible serine phosphorylation are indicated. Depicted below is a representation of the conversion of the inert monomeric form of HSF1 to the transcriptionally active trimer.

binding is activated, for example, following exposure of human K562 erythroleukemia cells to hemin (Singh and Yu, 1984; Theodorakis et al., 1989; Sistonen et al., 1992). The acquisition of HSF2 DNA-binding activity is accompanied by conversion from an inert dimer to a DNA-binding competent trimer; the DNA-binding properties of HSF1 and HSF2 are similar but not identical (Kroeger et al., 1993; Sistonen et al., 1994). Constitutive HSF2 DNA-binding activity has been observed in embryonal carcinoma cells (Morange et al., 1984; Mezger et al., 1989, 1994; Murphy et al., 1994). Another observation which supports a role for HSF2 during development is the elevated expression of HSF2 expression and HSF2 DNA-binding activity during spermatogenesis (Sarge et al., 1991, 1994). By comparison to HSF1 and HSF2, much less is known about HSF3, a member of the HSF gene family that is expressed ubiquitously in avian tissues and is also negatively regulated for DNA-binding activity (Nakai and Morimoto, 1993). HSF3 exhibits the unique properties of being activated, together with HSF1, only in chicken erythroblastic cells (Nakai et al. 1995). The co-activation of HSF1 and HSF3 could enhance the ability of the cell to regulate tightly the heat shock response in a cell type-specific manner (Nakai et al., 1995).

■ Regulatory aspects of activity: the HSF cycle and a proposed role for heat shock proteins in autoregulation of the heat shock response

A unique feature of the heat shock response is the rapid kinetics of activation, the magnitude of inducibility, and reversibility of the response. The attenuation of the transcriptional response occurs gradually in cells continuously exposed to intermediate heat shock temperatures or immediately upon temperature downshift, whereas, following exposure to extreme heat shock temperatures (>43°C), the heat shock response persists at high levels (Abravaya et al., 1991a). Comparison of HSF1 DNA-binding properties using in vivo genomic footprinting on the human HSP70 promoter reveals that the HSE is unoccupied in cells at control temperatures and fully occupied upon heat shock (Abravaya et al.,1991a, b; Sistonen et al., 1994). During attenuation, HSF1 releases from the HSE, and thereafter HSF1 DNA-binding activity is no longer detected. The in vivo equilibrium dissociation rate for HSF1 has a half-life of approximately 10 minutes, whereas the dissociation rate of the HSF–HSE complex formed in vitro is greater than 100 minutes. The disparity of these results suggests that there are additional components involved in the release of the activated form of HSF from DNA. The reversible modulation of HSF1 DNA-binding properties and the oligomeric state is not an intrinsic property of the HSF protein as the recombinant Drosophila, chicken, mouse, and human HSFs expressed in Escherichia coli are purified as native trimers exhibiting constitutive DNA-binding activity (Clos et al., 1990; Kroeger et al., 1993; Nakai and Morimoto, 1993; Sarge et al., 1993).

It has long been speculated from studies in Drosophila and yeast that heat shock proteins function in an autoregulatory loop to modulate the intensity and duration of the heat shock response. Although the sensor of cell stress has not be unequivocally identified, most models for regulation of the heat shock response have proposed that the appearance of misfolded proteins induced during heat shock and other forms of stress, sequesters HSP70, thus resulting in the activation of HSF1 (Morimoto et al., 1990; Abravaya et al., 1992; Baler et al., 1992; Morimoto, 1993). Additional support for the autoregulatory hypothesis comes from experimental evidence that links the activation of the heat shock response to increased levels of denatured and misfolded proteins (Ananthan et al., 1986; Baler et al., 1992). Exposure to inhibitors of protein synthesis blocks the activation of the heat shock response by interfering with induction of HSF1; these results suggest that the proper synthesis and folding of nascent polypeptides represents a critical target for detection of misfolded proteins (Mosser et al., 1988; Amici et al., 1992; Baler et al., 1992). Taken together, these results support the hypothesis that the appearance and accumulation of misfolded polypeptides are directly involved in the pathway of stress detection and response (DiDomenico et al., 1982). Molecular chaperones such as members of the HSP70 family are attractive candidates in the autoregulation of the heat shock response, as they have a primary role to associate with proteins to prevent their aggregation and to facilitate protein folding (Gething and Sambrook, 1992; Freeman et al., 1995; Freeman and Morimoto, 1996). Although there is a substantial evidence to support an involvement of molecular chaperones in the regulation of HSF1 activation, it is uncertain whether this occurs through direct or indirect effects. HSF1 trimers become associated with HSP70 during attenuation; additionally HSF1 attenuation is more rapid in cell lines expressing high levels of HSP70 (Abravaya et al., 1992; Mosser et al., 1993; Rabindran et al., 1994). If indeed it is the balance or ratio of free HSP70 that is either directly or indirectly involved in maintaining the non-DNA-binding form of HSF1 or in the attenuation of HSF1, it might be expected that overexpression of HSF1 would be either constitutively active or more readily activated. Cells overexpressing HSF1 express high levels of a nuclear localized DNA-binding competent factor independent of exogenous stress (Sarge et al., 1993). These results reveal that heat shock is not an obligatory step in HSF1 activation and are consistent with the hypothesis that there is a critical balance between HSF1 and its negative regulatory molecules to maintain HSF1 in either the non-DNA-binding or DNA-binding state.

To summarize, many of the current observations are consistent with a model for the regulation of HSF1 DNA-binding activity which is schematically represented in Fig. 3. Under non-stressful conditions, HSF1 is synthesized and maintained in a non-DNA-binding monomer through intramolecular interactions, perhaps influenced by constitutive phosphorylation of the factor. Heat shock and other stresses result in the appearance of misfolded and aggregated proteins, which creates a large pool of new

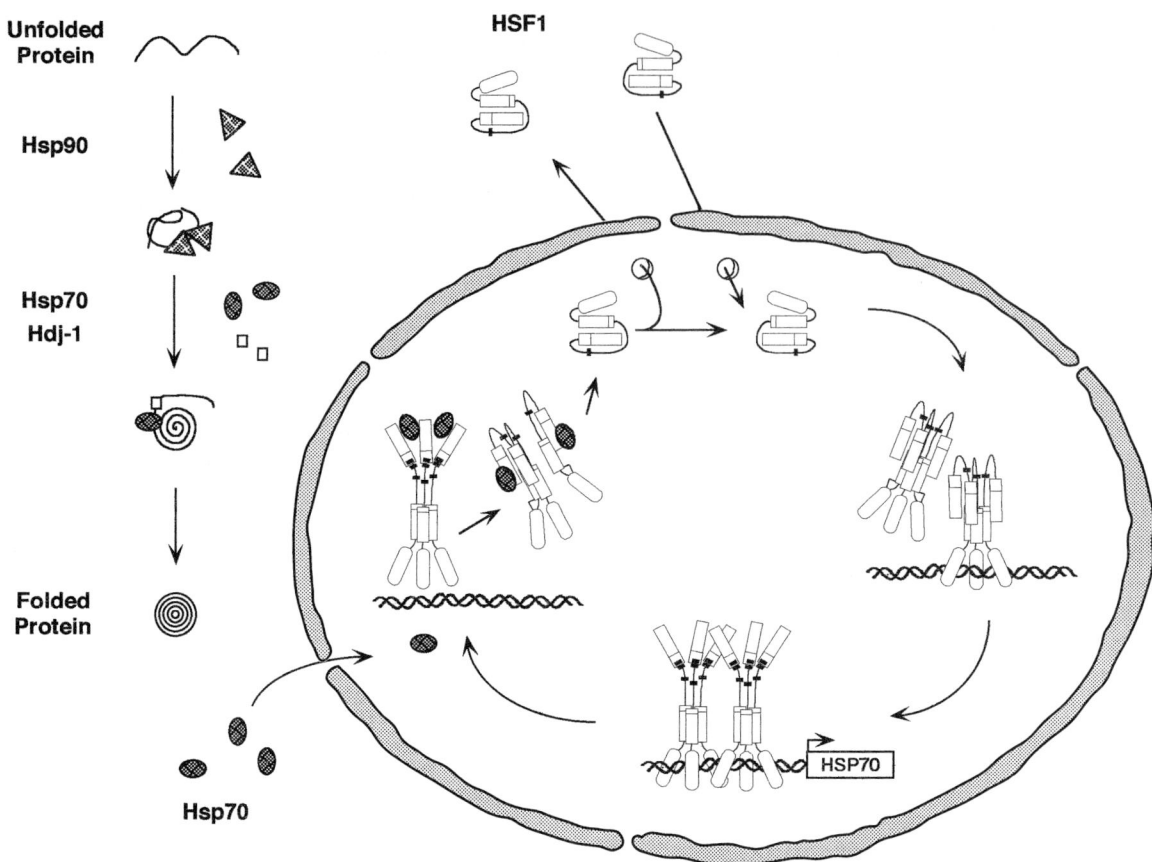

Figure 3. HSF cycle: a model of HSF1 regulation. In the unstressed cell, HSF1 is maintained in a monomeric, non-DNA-binding form. Upon heat shock or other forms of stress, HSF1 translocates and relocalizes within the nucleus and assembles into a trimer; HSF1 trimers bind to the heat shock element, in heat shock gene promoters and undergoes inducible serine phosphorylation. Transcriptional activation of the heat shock genes leads to increased levels of Hsp70 and to formation of an HSF1–Hsp70 complex. During attenuation of the heat shock transcriptional response, HSF1 dissociates from the DNA and is eventually converted to non-DNA-binding monomers.

protein substrates that compete with HSF1 for association with HSP70. Thus, heat shock and other stresses initiate the events that remove the negative regulatory influence on HSF1 DNA-binding activity. The conversion of HSF1 monomers to trimers requires a significant change in the conformation of HSF1 which may involve other activities to facilitate or stabilize the DNA-binding trimer. HSF1 also undergoes a stress-dependent inducible serine phosphorylation and acquires transcriptional activity. The activated state of HSF1 results in elevated transcription of genes encoding heat shock proteins and molecular chaperones leading to the prolonged synthesis and accumulation of heat shock proteins. The interaction of HSF1 with HSP70 detected during attenuation is proposed to have an effect on the activity of HSF1. Ultimately, these events may be important in the regulation of HSF1 transcriptional activity and/or conversion of the active protein to the control state.

Acknowledgements

The studies from our laboratory were supported by grants from the National Institutes of General Medicine. The author extends appreciation to members of the laboratory who contributed to these studies.

References

Abravaya, K., Phillips, B., and Morimoto, R.I. (1991a). Attenuation of the heat shock response in HeLa cells is mediated by the release of bound heat shock transcription factor and is modulated by changes in growth and in heat shock temperatures. Genes Dev. **5**, 2117–2127.

Abravaya, K., Phillips, B., and Morimoto, R.I. (1991b). Heat shock-induced interactions of heat shock transcription factor and the human hsp70 promoter examined by in vivo footprinting. Mol. Cell. Biol. **11**, 586–592.

Abravaya, K., Myers, M.P., Murphy, S.P., and Morimoto, R.I. (1992). The human heat shock protein hsp70 interacts with HSF, the transcription factor that regulates heat shock gene transcription. Genes Dev. **6**, 1153–1164.

Amici, C., Sistonen, L., Santoro, M.G., and Morimoto, R.I. (1992). Antiproliferative prostaglandin's activate heat shock transcription factor. Proc. Natl. Acad. Sci. USA **89**, 6227–6231.

Amin, J., Ananthan, J., and Voellmy, R. (1988). Key features of heat shock regulatory elements. Mol. Cell. Biol. **8**, 3761–3769.

Ananthan, J., Goldberg, A.L., and Voellmy, R. (1986). Abnormal proteins serve as eukaryotic stress signals and trigger the activation of heat shock genes. Science **232**, 522–524.

Ashburner, M. (1970). Pattern of puffing activity in the salivary gland chromosomes of Drosophila. V. Response to environmental treatments. Chromosoma **31**, 356–376.

Baler, R., Welch, J., and Voellmy, R. (1992). Heat shock gene regulation by nascent polypeptides and denatured proteins: hsp70 as a potential autoregulatory factor. J. Cell Biol. **117**, 1151–1159.

Baler, R., Dahl, G., and Voellmy, R. (1993). Activation of human heat shock genes is accompanied by oligomerization, modification, and rapid translocation of heat shock transcription factor HSF1. Mol. Cell. Biol. **13**, 2486–2496.

Beckmann, R.P., Mizzen, L.A., and Welch, W.J. (1990). Interaction of Hsp70 with newly synthesized proteins: Implications for protein folding and assembly. Science **248**, 850–854.

Bochkareva, E.S. and Girshovich, A.S. (1992). A newly synthesized protein interacts with GroES on the surface of chaperonin GroEL. J. Biol. Chem. **267**, 25672–25675.

Chen, Y., Barlev, N.A., Westergaard, O., and Jakobsen., B.K. (1993). Identification of the C-terminal activator domain in yeast heat shock factor: independent control of transient and sustained transcriptional activity. EMBO J. **12**, 5007–5018.

Clos, J., Westwood, J.T., Becker, P.B., Wilson, S., Lambert, K., and Wu, C. (1990). Molecular cloning and expression of a hexameric Drosophila heat shock factor subject to negative regulation. Cell **63**, 1085–1097.

Cotto, J.J., Kline, M., and Morimoto, R.I. (1996). Activation of heat shock factor 1 DNA binding precedes stress-induced serine phosphorylation. J. Biol Chem. **271**, 3355–3358.

Craig, E.A., Gambill, B.D., and Nelson, R.J. (1993). Heat shock proteins: Molecular chaperones of protein biogenesis. Microbiol. Rev. **57**, 402–414.

DiDomenico, B.J., Bugaisky, G.E., and Lindquist, S. (1982). The heat shock response is regulated at both the transcriptional and posttranscriptional levels. Cell **31**, 593–603.

Damberger, F.F., Pelton, J.G., Harrison, C.J., Nelson, H.C.M., and Wemmer, D.E. (1994). Solution structure of the DNA-binding domain of the heat shock transcription factor determined by multidimensional heteronuclear magnetic resonance spectroscopy. Protein Sci. **3**, 1806–1821.

Freeman, B.C. and Morimoto, R.I. (1996) The human cystolic molecular chaperones hsp90, hsp70 (hsc70) and hdj-1 have distinct roles in recognition of a non-native protein and protein refolding. EMBO J. **15**, 2969–2979.

Freeman, B., Meyers, M., Schumacher, R., and Morimoto, R.I. (1995). Identification of a regulatory motif in Hsp70 that affects ATPase activity, substrate binding, and interaction with HDJ-1. EMBO J. **14**, 2281–2292.

Gething, M.-J. and Sambrook, J. (1992). Protein folding in the cell. Nature **355**, 33–45.

Georgopoulos, C., and Welch, W.J. (1993). Role of major heat shock proteins as molecular chaperones. Annu. Rev. Cell Biol. **9**, 601–635.

Greene, J.M. and Kingston, R.E. (1990). TATA-dependent and TATA-independent function of the basal and heat shock elements of a human hsp70 promoter. Mol. Cell. Biol. **10**, 1319–1328.

Harrison, C.J., Bohm, A.A., and Nelson, H.C.M. (1994). Crystal structure of the DNA binding domain of the heat shock transcription factor. Science **263**, 224–227.

Hendrix, J.P., and Hartl, F.-U. (1993). Molecular chaperone functions of heat-shock proteins. Ann. Rev. Biochem. **62**, 349–384.

Jakobsen, B.K. and Pelham, H.R. (1991). A conserved heptapeptide restrains the activity of the yeast heat shock transcription factor. EMBO J. **10**, 369–375.

Jurivich, D.A., Sistonen, L., Kroes, R.A., and Morimoto, R.I. (1992). Effect of sodium salicylate on the human heat shock response. Science **255**, 1243–1245.

Jurivich, D.A., Sistonen, L., Sarge, K.D., and Morimoto, R.I. (1994). Arachidonate is a potent modulator of human heat shock gene transcription. Proc. Natl. Acad. Sci. USA **91**, 2280–2284.

Kingston, R.E., Schuetz, T.J., and Larin, Z. (1987). Heat-inducible human factor that binds to a human hsp70 promoter. Mol. Cell. Biol. **7**, 1530–1534.

Kroeger, P.E. and Morimoto, R.I. (1994). Selection of new HSF1 and HSF2 DNA binding sites reveals differences in trimer cooperativity. Mol. Cell. Biol. **14**, 7592–7603.

Kroeger, P.E., Sarge, K.D, and Morimoto, R.I. (1993). Mouse heat shock transcription factors 1 and 2 prefer a trimeric binding site but interact differently with the HSP70 heat shock element. Mol. Cell. Biol. **13**, 3370–3383.

Langer, T., Lu, C., Echols, H., Flanagan, J., Hayer, M.K., and Hartl. F.-U. (1992). Successive action of DnaK, DnaJ and GroEL along the pathway of chaperone-mediated protein folding. Nature **356**, 683–689.

Larson, J.S., Schuetz, T.J., and Kingston, R.E. (1988). Activation in vitro of sequence-specific DNA binding by a human regulatory factor. Nature **335**, 372–375.

Lindquist, S. and Craig, E.A. (1988). The heat shock proteins. Annu. Rev. Genet. **22**, 631–677.

Lindquist-McKenzie, S.L., Henikoff, S., and Meselon, M. (1975). Localization of RNA from heat-induced polysomes at puff sites in Drosophila melanogaster . Proc. Natl. Acad. Sci. USA **72**, 1117–1121.

Lis, J. and Wu, C. (1993). Protein traffic on the heat shock promoter, parking, stalling and trucking along. Cell **74**, 1–4.

Mezger, V., Bensaude, O., and Morange, M. (1989). Unusual levels of heat shock element-binding activity in embryonal carcinoma cells. Mol. Cell. Biol. **9**, 3888–3896.

Mezger, V., Rallu, M., Morimoto, R.I., Morange, M., and Renard, J.-P. (1994). Rapid communication heat shock factor 2-like activity in mouse blastocysts. Develop. Biol. **166**, 819–822.

Morange, M., Diu, A., Bensaude, O., and Babinet, C. (1984). Altered expression of heat shock proteins in embryonal carcinoma and mouse early embryonic cells. Mol. Cell. Biol. **4**, 730–735.

Morimoto, R.I. (1993). Cells in stress: transcriptional activation of heat shock genes. Science **259**, 1409–1410.

Morimoto, R.I., Tissieres, A., and Georgopoulos, C. (1990). The stress response, function of the proteins, and perspectives. In *Stress proteins in biology and medicine* (Morimoto, R.I., Tissieres, A., and Georgopoulos, C., ed.). Cold Spring Harbor Laboratory Press, New York, pp. 1–36.

Morimoto, R.I., Tissieres, A., and Georgopoulos, C. (1994). The biology of heat shock proteins and molecular chaperones. In *Stress proteins in biology and medicine* (Morimoto, R.I., Tissieres, A., and Georgopoulos, C., ed.). Cold Spring Harbor Laboratory Press, New York.

Mosser, D.D., Theodorakis, N.G., and Morimoto, R.I. (1988). Coordinate changes in heat shock element-binding activity and hsp70 gene transcription rates in human cells. Mol. Cell. Biol. **8**, 4736–4744.

Mosser, D.D., Kotzbauer, P.T., Sarge, K.D., and Morimoto, R.I. (1990). In vitro activation of heat shock transcription factor DNA-binding by calcium and biochemical conditions that affect protein conformation. Proc. Natl. Acad. Sci. USA **87**, 3748–3752.

Mosser, D.D., Duchaine, J., and Massie, B. (1993). The DNA-binding activity of the human heat shock transcription factor is regulated in vivo by hsp70. Mol. Cell. Biol. **13**, 5427–5438.

Murphy, S.P., Gorzowski, J.J., Sarge, K.D., and Morimoto, R.I. (1994). Characterization of constitutive HSF2 DNA-binding activity in mouse embryonal carcinoma cells. Mol. Cell. Biol. **14**, 5309–5317.

Nakai, A. and Morimoto, R.I. (1993). Characterization of a novel chicken heat shock transcription factor, HSF3, suggests a new regulatory pathway. Mol. Cell. Biol. **13**, 1983–1997.

Nakai, A., Kawazoe, Y., Tanabe, M., Nagata, K., and Morimoto, R. I. (1995). The DNA-binding properties of two heat shock factors, HSF1, and HSF3, are induced in an avian erythroblast cell line HD6. Mol. Cell. Biol. **15**, 5268–5278.

Nakai, A., Tanabe, M., Kawazoe, Y., Inazawa, J., Morimoto, R. I. and Nagata, K. (1997). HSF4, a new member of the human heat shock factor gene family which lacks properties of a transcriptional activator. Mol. Cell Biol. **17**, 469–481.

Newton, E.M., Knauf, U., Green, M., and Kingston, R.E. (1996). The regulatory domain of human heat shock factor 1 is sufficient to sense heat stress. Mol. Cell. Biol. **16**, 839–846.

Nieto-Sotelo, J., Wiederrecht, G., Okuda, A., and Parker, C. S. (1990). The yeast heat shock transcription factor contains a transcriptional activation domain whose activity is repressed under nonshock conditions. Cell **62**, 807–817.

Nover, L. (1994). The heat stress reponse as part of the plant stress network. In *NATO-ASI Series on biochemical and cellular mechanisms of stress tolerance in plants* (Cherry J. ed.). Springer Verlag, Berlin, pp. 3–45.

Perisic, O., Xiao, H., and Lis, J.T. (1989). Stable binding of Drosophila heat shock factor to head-to-head and tail-to-tail repeats of a conserved 5 bp recognition unit. Cell **59**, 797–806.

Peteranderl, R. and Nelson, H.C.M. (1992). Trimerization of the heat shock transcription factor by a triple-stranded a-helical coiled-coil. Biochemistry **31**, 12272–12276.

Rabindran, S.K., Giorgi, G., Clos, J., and Wu, C. (1991). Molecular cloning and expression of a human heat shock factor, HSF1. Proc. Natl. Acad. Sci. USA **88**, 6906–6910.

Rabindran, S.K., Haroun, R.I., Clos, J., Wisniewski, J., and Wu, C. (1993). Regulation of heat shock factor trimer formation: role of a conserved leucine zipper. Science **259**, 230–234.

Rabindran, S.K., Wisniewski, J., Li, L., Li, G.C., and Wu, C. (1994). Interaction between heat shock factor and HSP 70 is insufficient to suppress induction of DNA-binding activity in vivo. Mol. Cell. Biol. **14**, 6552–6560.

Ritossa, F.M. (1962). A new puffing pattern induced by a temperature shock and DNP. Drosophilia: Experientia **18**, 571–573.

Sarge, K.D., Zimarino, V., Holm, K., Wu, C., and Morimoto, R.I. (1991). Cloning and characterization of two mouse heat shock factors with distinct inducible and constitutive DNA-binding ability. Genes Dev. **5**, 1902–1911.

Sarge, K.D., Murphy, S.P., and Morimoto, R.I. (1993). Activation of heat shock gene transcription by HSF1 involves oligomerization, acquisition of DNA binding activity, and nuclear localization and can occur in the absence of stress. Mol. Cell. Biol. **13**, 1392–1407.

Sarge, K.D., Park-Sarge, O.-K., Kirby, J.D., Mayo, K.E., and Morimoto, R.I. (1994). Regulated expression of heat shock factor 2 in mouse testis: potential role as a regulator of hsp gene expression during spermatogenesis. Biol. Reprod. **50**, 334–1343.

Scharf, K.-D., Rose, S., Zott, W., Schoff, F., and Nover, L. (1990). Three tomato genes code for heat stress transcription factors with a remarkable degree of homology to the DNA-binding domain of the yeast HSF. EMBO J. **9**, 4495–4501.

Schuetz, T.J., Gallo, G.J., Sheldon, L., Tempst, P., and Kingston, R.E. (1991). Isolation of a cDNA for HSF2: evidence for two heat shock factor genes in humans. Proc. Natl. Acad. Sci. USA **88**, 6910–6915.

Shi, Y., Kroeger, P.E., and Morimoto, R.I. (1995). The carboxyl-terminal transcription domain of heat shock factor 1 is negatively regulated and stress responsive. Mol. Cell. Biol. **15**, 4309–4318.

Singh, M.K. and Yu, J. (1984). Accumulation of a heat shock-like protein during differentiation of human erythroid cell line K562. Nature **309**, 631–633.

Sistonen, L., Sarge, K.D., Phillips, B., Abravaya, K., and Morimoto, R.I. (1992). Activation of heat shock factor 2 during hemin-induced differentiation of human erythroleukemia cells. Mol. Cell. Biol. **12**, 4104–4111.

Sistonen, L., Sarge, K.D., and Morimoto, R.I. (1994). Human heat shock factors 1 and 2 are differentially activated and can synergistically induce HSP70 gene transcription. Mol. Cell. Biol. **14**, 2087–2099.

Sorger, P.K. and Nelson, H.C.M. (1988). Yeast heat shock factor is an essential DNA-binding protein that exhibits temperature-dependent phosphorylation. Cell **54**, 855–864.

Sorger, P.K. and Nelson, H.C.M. (1989). Trimerization of a yeast transcriptional activator via a coiled-coil motif. Cell **59**, 807–813.

Sorger, P.l., Lewis, M.J., and Pelham, H.R.B. (1987). Heat factor is regulated differently in yeast and HeLa cells. Nature **329**, 81–84.

Theodorakis, N.G., Zand, D.J., Kotzbauer, P.T., Williams, G.T., and Morimoto, R.I. (1989). Hemin-induced transcriptional activation of the hsp70 gene during erythroid maturation in K562 cells is due to a heat shock factor-mediated stress response. Mol. Cell. Biol. **9**, 3166–3173.

Tissieres, A., Mitchell, K.H., and Tracy, V.M. (1974). Protein synthesis in salivary glands of *Drosophila melanogaster*: Relation to chromosome puffs. J. Mol. Biol. **84**, 389–398.

Vuister, G.W., Kim, S.J., Wu, C., and Bax, A. (1994a). NMR evidence for similarities between the DNA binding regions of Drosophila melanogaster heat shock factor and the helix-turn-helix and HNF-3/forkhead families of transcription factors. Biochemistry **33**, 10–16.

Vuister, G.W., Kim, S.J., Orosz, A., Marquardt, J., Wu, C., and Bax, A. (1994b). Solution structure of the DNA binding domain of Drosophila heat shock transcription factor. Nature Struct. Biol. **1**, 605–614.

Westwood, J.T. and Wu, C. (1993). Activation of Drosophila heat shock factor: conformational change associated with a monomer-to-trimer transition. Mol. Cell. Biol. **13**, 3481–3486.

Westwood, J.T., Clos, J., and Wu, C. (1991). Stress-induced oligomerization and chromosomal relocalization of heat-shock factor. Nature **353**, 822–827.

Wiederrecht, G., Seto, D., and Parker, C.S. (1988). Isolation of the gene encoding the S. cerevisiae heat shock transcription factor. Cell **54**, 841–853.

Williams, G.T. and Morimoto, R.I. (1990). Maximal stress-induced transcription from the human hsp70 promoter requires interactions with the basal promoter elements independent of rotational alignment. Mol. Cell. Biol. **10**, 3125–3136.

Wu, C. (1980). The 5′ ends of *Drosphila* heat shock genes in chromatin are hypersensitive to DNase1. Nature **286**, 854–860.

Wu, C. (1984). Two protein-binding sites in chromatin implicated in the activation of heat-shock genes. Nature **309**, 229–234.

Wu. C. (1995). Heat stress transcription factors. Annu. Rev. Cell Biol. **11**, 441–469.

Wu, C., Wilson, S., Walker, B., David, I., Paisley, T., Zimarino, V., and Ueda, H. (1987). Purification and properties of *Drosophila* heat shock activator protein. Science **238**, 1247–1253.

Xiao, H. and Lis, J.T. (1988). Germline transformation used to define key features of the heat shock response element. Science **239**, 1139–1142.

Xiao, H., Perisic, O., and Lis, T.J. (1991). Cooperative binding of Drosophila heat shock factor to arrays of a conserved 5 bp unit. Cell **64**, 585–593.

Zimarino, V. and Wu, C. (1987). Induction of sequence-specific binding of Drosophila heat shock activator proteins without protein synthesis. Nature **327**, 727–730.

Zimarino, V., Tsai , C., and Wu, C (1990). Complex modes of heat shock factor activation. Mol. Cell. Biol. **10**, 752–759.

Zuo, J., Rungger, D., and Voellmy, R. (1995). Multiple layers of regulation of human heat shock transcription factor1. Mol. Cell. Biol. **15**, 4319–4330.

■ *Richard I. Morimoto:*
Department of Biochemistry, Molecular Biology and Cell Biology
Northwestern University
2153 Sheridan Road
Evanston, IL 60208, USA
Tel. 1 847 491 3340
Fax. 1 847 491 4461
E-mail: morimoto@casbah.acns.nwu.edu

Signalling of the unfolded protein response from the endoplasmic reticulum to the nucleus

In eukaryotic cells, the accumulation of misfolded polypeptides in the ER lumen results in the transcriptional up-regulation of many of the genes that encode ER-resident chaperones and folding catalysts. This cellular response is called the 'unfolded protein response' (UPR). The transduction of the UPR signal to its final destination in the nucleus must proceed through a number of obligatory stages, when it is considered that even though their membrane surfaces are contiguous, the ER and the nucleus are distinct membrane bounded organelles. These stages include: the detection of the presence of unfolded proteins in the ER lumen, the transduction of this signal across the ER membrane, the transmission of this signal to the transcriptional machinery in the nucleus, and finally the activation of transcription of genes encoding UPR-regulated ER-resident proteins. In recent years a number of the components of the UPR signalling pathway in yeast have been identified and promoter elements involved in the transcriptional control of UPR-regulated genes have been defined in both yeast and mammalian cells.

■ Secretory protein biosynthesis and the UPR response

The outer surface of the endoplasmic reticulum (ER) is a major site of protein synthesis in eukaryotic cells. New polypeptide chains synthesized on membrane-bound polyribosomes are transported across the ER membrane and into the ER lumen where protein folding and post-translational modifications such as N-linked glycosylation and disulfide-linkage formation occur (reviewed by Rothblatt, 1994a, b). Within the lumen a variety of highly conserved resident ER proteins assist the newly translocated polypeptides to fold into their correct tertiary and quaternary structures. These resident proteins include both molecular chaperones that recognize and stabilize partially folded intermediates during polypeptide folding and assembly, and enzymes that catalyse rate-determining steps in folding, such as protein disulfide isomerase and peptidyl–prolyl isomerases (see reviews by Gething and Sambrook, 1992; Hartl, 1996; see earlier entries in this volume).

Eukaryotic secretory proteins that are unable to fold correctly are retained in the ER (reviewed by Rose and Doms, 1988). Studies in mammalian cells have shown that these misfolded polypeptides form stable complexes with particular molecular chaperones (for review see Gething and Sambrook, 1992). In both yeast and mammalian cells, the accumulation of misfolded polypeptides in the ER lumen results in the transcriptional up-regulation of many of the genes encoding ER-resident chaperones and

Table 1. Molecular chaperones and protein folding factors of the ER that are induced by the unfolded protein response[a]

Protein family	Function	Mammalian	Yeast
HSP70	Molecular chaperone	BiP/Grp78 (1)	BiP/Kar2p (2)
			Lhs1p (3)
HSP90	Molecular chaperone	Grp94 (4)	
HSP40	Molecular chaperone	Scj1p (5)	
PDI	Protein disulfide isomerase	PDI (2)	Pdi1p (2)
	(and homologues)	Erp61 (2)	Eug1p (6)
		Erp72 (2)	
FKBP	Peptidyl–prolyl isomerase	FKB13 (7)	Fkb2p (7)

[a] Numbers in parentheses refer to the following references: (1) Kozutsumi et al., 1988 (2) Normington et al., 1989; (3) Craven et al., 1996; (4) Lee, 1992; (5) Schlenstedt et al., 1995; (6) Tachibana and Stevens, 1992; (7) Partaledis and Berlin, 1993.

folding catalysts (see Table 1 and reviews by McMillan et al., 1994; Shamu et al., 1994). This cellular response is called the 'unfolded protein response' (UPR): when unfolded proteins accumulate in the ER the cell responds by increasing its ability to fold proteins, thus alleviating the increased demand on the existing cellular protein folding machinery. In yeast, the ER chaperone BiP (Kar2p) and other members of the HSP70 family have been shown to respond in a compartment-specific manner to the presence of unfolded proteins (Normington et al., 1989). Thus the synthesis of BiP is induced only by the accumulation of secretory precursors within the ER whereas synthesis of cytosolic HSP70 protein(s) increases in response to the accumulation of unfolded precursors in the cytoplasm. The UPR in both yeast and mammals is regulated at the level of transcription (Kozutsumi et al., 1988; Normington et al., 1989; Mori et al., 1992; Kohno et al., 1993), so a pathway must exist whereby the presence of unfolded proteins in the ER is detected and then communicated to the cellular transcription machinery in the nucleus.

The UPR can be induced experimentally using various treatments, all of which interfere with the protein folding process (reviewed by McMillan et al., 1994). Treatments include the prevention of protein glycosylation by starving cells of glucose (Shiu et al., 1977), by treating cells with tunicamycin (Kozutsumi et al., 1988; Normington et al., 1989; Rose et al., 1989), or by expressing in yeast cells a mutant form of the SEC53 gene product, which is required for N-linked glycosylation (Normington et al., 1989; Mori et al., 1992, 1993). The UPR can also be induced by raising the concentration of secretory protein precursors in the ER by incubating sec mutants at non-permissive temperatures (Normington et al., 1989; Rose et al., 1989), by preventing disulfide bond formation using β-mercaptoethanol (Cox et al., 1993), or by overexpressing mutant secretory proteins that cannot fold properly (Kozutsumi et al., 1988; Tokunaga et al., 1992). In mammalian cells, the UPR is also induced by depleting the ER of calcium by the use of calcium ionophores A23187 and thapsigargin (reviewed by Lee, 1987, 1992).

The UPR is best understood in the yeast Saccharomyces cerevisiae, where a number of components of the UPR signal transduction pathway have been identified (Cox et al., 1993; Mori et al., 1993, 1996; Cox and Walter, 1996; Sidrauski et al., 1996). Although no mammalian genes encoding components of the signalling pathway have yet been cloned, the upstream elements controlling the transcription of the UPR-regulated grp78 gene have been defined in detail (reviewed by Lee, 1992; Roy and Lee, 1995; Li et al., 1997).

■ Detection of the presence of unfolded proteins in the ER lumen

Most studies of UPR regulation and control have focused on the chaperone BiP, the major ER-located member of the HSP70 protein family. BiP is the best characterized ER luminal chaperone (see entries pp. 33 and 59), and is encoded by the KAR2 gene in S. cerevisiae (Normington et al., 1989; Rose et al., 1989) and by the grp78 gene in mammalian cells (Shiu et al., 1977). These genes are transcribed constitutively at a basal level, and up-regulated under conditions of UPR stress (Lee, 1987, 1992; Normington et al., 1989). Studies in yeast demonstrate that the KAR2 gene is essential for cell survival under conditions of stress that cause the accumulation of unfolded proteins in the ER (Normington et al., 1989).

Evidence from studies in both yeast and mammalian cells suggests that the proximal signal for the induction of UPRE-controlled genes is not an increase in the concentration of unfolded or unassembled proteins or of BiP–protein complexes in the ER, but rather the consequent decrease in the concentration of free BiP (reviewed in Kohno et al., 1993, McMillan et al., 1994; Shamu et al., 1994). Accumulation in the mammalian ER of unfolded polypeptides that do not bind BiP did not elicit the UPR (Graham et al., 1990; Ng et al., 1992), while the overexpression of functional BiP in yeast cells was found to inhibit transcriptional induction of UPR-regulated genes, suggesting that the amount of BiP in the ER directly modulates initiation of the UPR pathway (Kohno et al., 1993). Furthermore, expression of BiP antisense mRNA in mammalian cells was able to down-regulate UPR induction levels in response to stress (Li et al., 1992), and expression of a mutant version of yeast BiP (Kar2p) that is not retained in the ER was sufficient to induce transcriptional upregulation of the *KAR2* gene (Hardwick et al., 1990). Both experiments imply that a reduction in the concentration of BiP is sufficient to induce the UPR.

■ What is the sensor that monitors the concentration of free BiP in the ER lumen?

Genetic screens in *Saccharomyces cerevisiae* for mutants that are unable to signal in response to the presence of unfolded proteins in the ER identified the *ERN1* (*IRE1*) gene (Cox et al., 1993; Mori et al., 1993). *ERN1* encodes a 1115 amino acid transmembrane protein (Ern1p/Ire1p) that is almost certainly localized to the ER membrane (Mori et al., 1993) and is composed of a glycosylated luminal N-terminal domain and a cytoplasmic C-terminal domain, separated by a 16 amino acid hydrophobic transmembrane spanning region (Mori et al., 1993). The hydrophilic N-terminal domain of Ern1p is unique in sequence, having no significant sequence homology to any other known protein. Because of its localization, it is presumed to be the ligand-binding domain of the molecule responsible for monitoring events in the ER lumen. The ligand generated by the presence of unfolded proteins that binds to this domain is not yet known, but, as discussed above, may be BiP itself. Residues 673–980 in the C-terminal domain display significant similarity to the catalytic domains of serine/threonine protein kinases (Cox et al., 1993; Mori et al., 1993). The 12 conserved subdomains characteristic of all known protein kinases are present in the Ern1p sequence, and 11 of the 12 invariant residues are perfectly conserved.

The *ERN1* gene is not essential for vegetative growth, but is absolutely necessary for survival under conditions that cause UPR stress (Cox et al., 1993; Mori et al., 1993). *ERN1* is a single copy gene that is constitutively expressed at a very low basal level and is not induced as part of the UPR (Mori et al., 1993), although its transcription is up-regulated at low glucose concentrations (K. Briggs, D.-H. Seog and M. J. Gething, unpublished data). *Ern1* mutants are capable of inducing transcription in response to other stimuli such as heat shock, indicating that they are not generally defective for stress responses (Mori et al., 1993; Shamu et al., 1994). Interestingly, yeast cells carrying mutations in the *ERN1/IRE1* gene are also auxotrophic for inositol (Nikawa and Yamashita, 1992; Cox et al., 1993; Mori et al., 1993). It has thus been postulated that regulation of ER-resident protein synthesis as a result of the UPR may be coupled to ER membrane biogenesis and phospholipid synthesis (Cox et al., 1993; Shamu et al., 1994), in which free inositol levels are known to play a central regulatory role (White et al., 1991). The ER membrane may coordinately expand to accommodate an increase in the volume of ER-resident proteins (Nunnari and Walter, 1996).

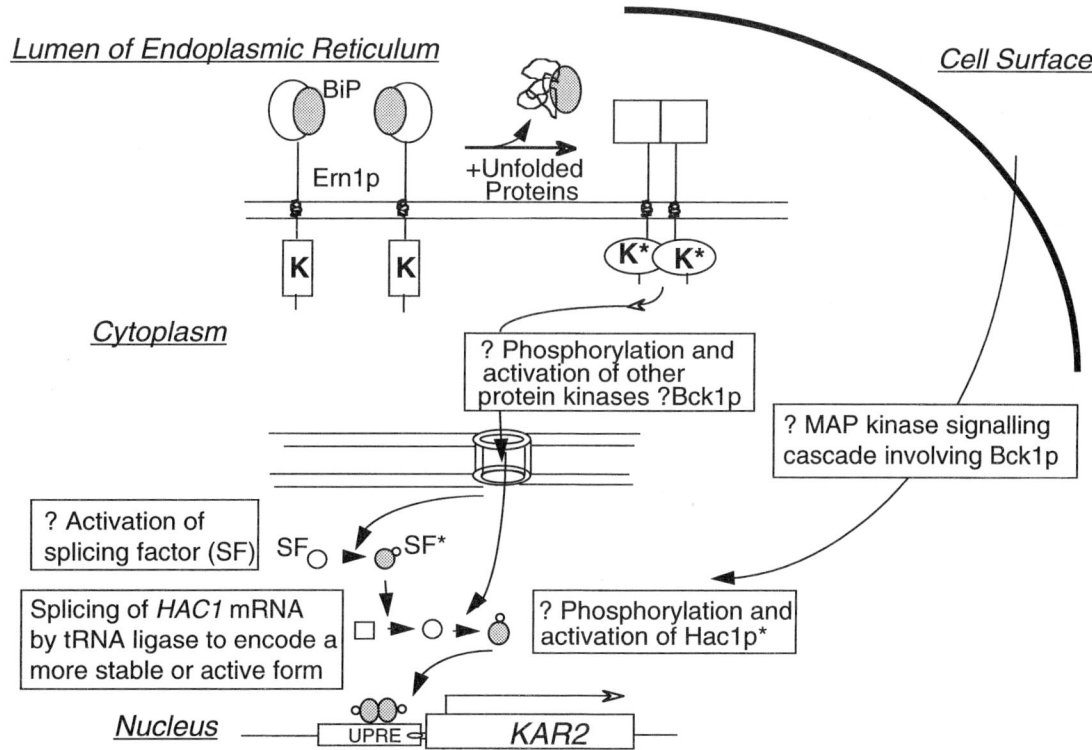

Figure 1. A model for the unfolded protein response pathway in yeast. See text for details.

Transduction of the UPR signal across the ER membrane

Characterization of the Ern1 transmembrane receptor protein kinase identifies in one molecule at least two of the essential components of the UPR pathway, the luminal sensor, and the mechanism for transducing the signal across the ER membrane. Ern1p was the first transmembrane kinase identified in yeast and shows structural similarity to class 1 growth factor receptors of higher eukaryotes (Mori et al., 1993; Shamu et al., 1994). In a manner similar to these, Ern1p undergoes ligand-directed oligomerization in response to unfolded protein accumulation in the ER lumen (Shamu and Walter, 1996; Welihinda and Kaufman, 1996). The ability to form oligomers is suggested to reside primarily in the C-terminal tail (Welihinda and Kaufman, 1996). The most likely model for Ern1p action involves chaperones like BiP acting as ligands and binding directly to Ern1p, thereby preventing oligomerization. Upon the accumulation of unfolded proteins, BiP may dissociate from the receptor, allowing oligomerization to occur (Fig. 1) (reviewed by McMillan et al., 1994; Shamu et al., 1994; Shamu and Walter, 1996; Welihinda and Kaufman, 1996).

The analysis of point mutations in the kinase domain of Ern1p, particularly of the conserved lysine-702 in the putative catalytic site, has demonstrated that the kinase activity of Ern1p is essential for transduction of the UPR across the ER membrane (Mori et al., 1993). The oligomerization of Ern1p molecules upon detection of the UPR signal results in activation of the kinase domain of Ern1p, and trans-autophosphorylation on serine and possibly threonine residues also occurs (Shamu and Walter, 1996; Welihinda and Kaufman, 1996). The activation of the Ern1p kinase domain then leads directly or indirectly to transmission of the UPR signal to the cellular transcriptional machinery in the nucleus (Fig. 1). The C-terminal tail region of Ern1p is not required for kinase activity (Welihinda and Kaufman, 1996) but may contain phosphorylation sites or may bind to other proteins responsible for transmitting the UPR signal to the nucleus (Welihinda and Kaufman, 1996). Interestingly, the C-terminal tail has been reported (Cox and Walter, 1996) to show significant homology to a mammalian RNA endonuclease which also has a protein kinase domain (Bork and Sander, 1993).

No putative homologue of Ern1p has been reported to date in mammalian cells, so the events that occur concerning the detection and transmission of the UPR signal in the mammalian ER remain unknown.

Transmission of the UPR signal through the cytoplasm to the nucleus

In S. cerevisiae, the UPR pathway between the ER and the nucleus may involve direct activation of a transcription factor by Ern1p, or may involve a signal transduction cascade comprised of other protein kinases. Cross-talk between the UPR pathway and other stress-induced signal transduction pathways activated in yeast cells is also possible. MAP (mitogen activated protein) kinase modules composed of MEKK, MEK, and MAPK enzymes play a major role in stress signal transduction in yeast (reviewed by Ruis and Schuller, 1995). Such modules are components of the Hog1 pathway (which responds to high osmolarity stress), the Pkc1/Mpk1 pathway (responding to cell wall integrity and low osmolarity stress), and the pseudohyphal growth pathway that is triggered by nitrogen starvation in diploid cells (Herskowitz, 1995; Levin and Errede, 1995; Ruis and Schuller, 1995). The gene encoding Bck1, the MEKK enzyme of the Pkc1/Mpk1 pathway, has recently been shown to suppress a defect in UPR signalling when present on a multicopy vector (L. Helfenbaum and M. J. Gething, unpublished data). This pathway has recently been reported to be required for activation of synthesis of cell wall constituents for bud growth (Drgonová et al., 1996; Kamada et al., 1996). It remains to be established whether Bck1 and other members of the Pkc1/Mpk1 module play a direct role in transmitting the UPR signal from activated Ern1p to the nucleus, or whether they feed signals into the pathway to up-regulate synthesis of ER chaperones and folding catalysts in response to demands for additional plasma membrane components.

Protein phosphorylation events have also been implicated in the transmission of the UPR signal in mammalian cells. Stress-induced transcription of grp78 is almost completely inhibited by a specific inhibitor of tyrosine kinases (Cao et al., 1995), shows some sensitivity to serine/threonine kinase inhibitors (Resendez et al., 1986; Price and Calderwood, 1992), and is mildly enhanced by an inhibitor of serine/threonine phosphatases (Price et al., 1992).

Activation of the transcription of genes regulated by the UPR

Transcription of the yeast KAR2 (BiP) gene is regulated by three independent cis-acting elements: (i) a functional heat shock element (HSE) containing four repeats of the 5 bp modular units defined by [aGAAccTTCtgGAAatTTCa] (Lis et al., 1990); (ii) a pyrimidine-rich region which contributes to the high level of basal expression of the KAR2 gene and is similar in sequence to the consensus element for binding of the mammalian transcription factor Sp1 [which regulates basal transcription of a human cytosolic hsp70 gene (Morgan, 1989)]; and (iii) a 22 bp element (UPRE) which is required for the induction of BiP mRNA by unfolded proteins (Mori et al., 1992; Kohno et al., 1993). The UPRE shows significant homology to the consensus sequence found in the promoters of the mammalian glucose-regulated genes (Resendez et al., 1988; Chang et al., 1989). Closely related sequences are also present in the promoter regions of other unfolded protein-responsive yeast genes including EUG1, which encodes an ER-resident protein related to protein disulfide isomerase (Tachibana and Stevens, 1992), FKB2, which encodes the ER-located member of the FKB family of peptidyl–prolyl isomerases (Partaledis and Berlin, 1993), and *SCJ1*, which encodes the ER homologue of DnaJ (Blumberg and Silver, 1991). Internal deletions of 10 bp segments from the HSE and from the UPRE independently abolished the heat shock response and the unfolded protein response, while a 10 bp deletion of the Sp1-like element significantly decreased the basal level of expression of the *KAR2* gene (Mori *et al.*, 1992).

Transplantation of the 22 bp UPR element into the heterologous *CYC1* promoter causes the promoter to respond to the presence of unfolded proteins in the ER (Mori *et al.*, 1992). Mutational analysis has revealed that the UPRE contains an imperfect palindrome with a spacer of one nucleotide (Mori *et al.*, 1996). Point mutations that improve the palindrome significantly increase the degree of inducibility of the promoter in response to UPR stress (Mori *et al.*, 1996).

Stress induction of both *grp78* and *grp94* in mammalian cells is also mediated at the transcriptional level (Chang *et al.*, 1989). However, in contrast to the relatively simple system in yeast cells where the 22 bp UPRE is both necessary and sufficient for UPR-stimulated transcriptional activation, the mammalian system involves a much more complicated transcriptional regulatory system. The promoters of the *grp78* and *grp94* genes contain *cis*-acting regulatory elements required for high basal level expression and for induction by misfolded proteins, glycosylation block, or the calcium ionophore A23187 (Wooden *et al.*, 1991; Lee, 1992). These include:

1. A 36 bp region called the 'grp core' which is highly conserved among many mammalian UPR-regulated genes (Chang *et al.*, 1989; Liu and Lee, 1991; Wooden *et al.*, 1991; Li *et al.*, 1993, 1994). A portion of the grp core is *c.* 60% identical to the yeast UPRE (Mori *et al.*, 1992; Kohno *et al.*, 1993) and is required for UPR stress-induced transcription of the *grp78* gene (Liu and Lee, 1991; Wooden *et al.*, 1991; Li *et al.*, 1993, 1994). Footprinting studies with the grp core revealed specific changes in factor occupancy after stress induction (Li *et al.*, 1994). However, the factors involved were not identified. A protein p70CORE (*c.* 70 kDa in monomeric form; Li *et al.*, 1994) binds to the grp core and has recently been shown to be identical to YY1, a member of the GLI zinc finger family of transcription factors (Li *et al.*, 1997).

2. A series of CCAAT- or CCAAT-like motifs, the most proximal of which to the TATAA sequence is required to mediate the effects of the upstream regulatory elements (Wooden *et al.*, 1991). A protein complex called C1F (C1 binding factor) was shown to occupy this region constitutively (Li *et al.*, 1994). The highly conserved heteromeric CCAAT binding factor (CBF) was identified as a major component of C1F in human cells (Roy and Lee, 1995). The binding of this factor was stable at low Ca^{2+} concentrations, but the off-rate was increased by high Ca^{2+} concentrations (Roy and Lee, 1995). No putative CCAAT-like motifs appear to be present in the yeast *KAR2* promoter (Kohno *et al.*, 1993).

3. A cAMP response element (CRE) which functions as a major basal level regulatory element and is also necessary to maintain high promoter activity under stress-induced conditions (Alexandre *et al.*, 1991).

Although the 3′ half of the grp78 core is homologous to the UPRE of yeast, neither it or the CCAAT-like motif mentioned above are sufficient to mediate stress induction on their own. Both of these elements appear to show functional redundancy when deleted, owing to the apparent duplication of these sequences further upstream in the promoter (Liu and Lee, 1991; Roy and Lee, 1995). However, either element can, if duplicated, confer stress inducibility on a heterologous promoter (Li *et al.*, 1993).

Recently, the gene encoding the mammalian protein CHOP [C/EBP homologous protein, previously called GADD153 (Bartlett *et al.*, 1992)] was shown to be strongly up-regulated by the presence of unfolded proteins in the ER lumen (Wang and Ron, 1996). CHOP is involved in the regulation of members of the C/EBP (CCAAT/enhancer binding protein) family of transcription factors (Ron and Habener, 1992; Ubeda *et al.*, 1996) and also possesses a stress-inducible transactivation domain (Ubeda *et al.*, 1996). Whether CHOP is a member of the UPR pathway in mammalian cells, or a downstream target of the pathway remains to be determined.

■ Hac1p is the transcription factor that binds the yeast UPRE

The gene encoding the transcription factor that binds the yeast UPRE has recently been identified by a one-hybrid screen (Mori *et al.*, 1996), and by multicopy suppression of UPR signalling defects in cells containing either a deletion of the *ERN1* gene (Cox and Walter, 1996) or a point-mutated UPRE controlling a *lacZ* reporter gene (L. Helfenbaum and M. J. Gething, unpublished data). The gene, *HAC1*, encodes a transcription factor with a basic–leucine zipper motif which is an essential component of the UPR pathway (Cox and Walter, 1996; Mori *et al.*, 1996). Cells lacking *HAC1*, like those lacking *ERN1* (Mori *et al.*, 1993), are sensitive to ER stress (Mori *et al.*, 1996). Hac1p was shown to bind specifically to the UPRE *in vitro*, but only after induction by UPR stress (Cox and Walter, 1996; Mori *et al.*, 1996). UPR stress regulation of *HAC1* was shown to be mediated at the level of splicing by a novel splicing pathway that is independent of the spliceosome (Kawahara *et al.*, 1996; Sidrauski *et al.*, 1996). In this reaction, an unknown RNA endonuclease [possibly the C-terminal tail of Ern1p (Sidrauski *et al.*, 1996)] cleaves a 252nt intron out of *HAC1* mRNA, then the mRNA halves are ligated together by yeast tRNA ligase (Sidrauski *et al.*, 1996; see below). This UPR-induced splicing is strictly dependent on the presence of a functional *ERN1* gene (Kawahara *et al.*, 1996; Sidrauski *et al.*, 1996). Removal of the intron from *HAC1* mRNA upon stress induction results in the replacement of the C-terminal 10 amino acids with a sequence of 18 different amino acids. Cox and Walter (1996) suggest that uninduced *HAC1* (*HAC1*u) mRNA is translated and is able to act as a functional transcription factor on its own, but that the resulting Hac1pu protein is degraded as rapidly as it is made by an ubiquitin-dependent proteolysis pathway stimulated by the C-terminal tail of the Hac1pu protein. The new C-terminal tail on induced Hac1p (Hac1pi) is proposed to prevent this degradation and increase stability of Hac1p, enabling it to exist long enough to up-regulate and sustain the UPR. An alternative mechanism has been proposed by Kawahara et al. (1997), who found that the unspliced Hac1 protein is not efficiently translated, and that the new C-terminal 18 residues provide an additional transcriptional activation domain to the DNA-binding protein.

Interestingly, when pre-spliced (induced) *HAC1* mRNA was expressed in cells deleted for *HAC1*, treatment with tunicamycin to cause UPR stress was found to induce the UPR by a further 1.5 times in a strictly *ERN1*-dependent manner (Cox and Walter, 1996). It was suggested that Ern1p could render Hac1pi more active by phosphorylating the protein directly, or more indirectly by modifying protein products involved in regulating the synthesis, stability, or function of Hac1p (Cox and Walter, 1996). Hac1p may be phosphorylated by another stress-activated kinase (such as Bck1 or another member of the Pkc1/Mpk1 MAP kinase pathway, see above) which may feed into the UPR pathway downstream of *ERN1*.

Yeast tRNA ligase is required for splicing of *HAC1* mRNA and transcription of UPR-regulated genes

The *RLG1* gene encoding tRNA ligase was identified as a component of the UPR pathway by genetic screens for: (i) mutations that are lethal when combined with a *kar2-ΔHDEL* mutation (Sidrauski et al., 1996), or (ii) mutations that abolish UPR signalling to a UPRE-*lacZ* reporter (D. R. McMillan, J. Sambrook and M.J. Gething, unpublished data). tRNA ligase is a well-characterized RNA-processing enzyme required for the splicing of small introns from a subset of yeast pre-tRNAs (Greer et al., 1983; Westaway et al., 1988). It is a multifunctional enzyme possessing three distinct activities arranged in separate domains on a single polypeptide chain: a polynucleotide kinase, a cyclic phosphodiesterase, and adenylate synthetase (Apostol et al., 1991; Greer et al., 1983; Phizicky et al., 1986; Xu et al., 1990). These activities catalyse a series of sequential reactions that together result in the joining of tRNA halves that have been generated by a separate tRNA endonuclease (Peebles et al., 1983). The *rlg1* mutants that are defective in UPR signalling have single amino acid substitutions within the adenylate synthetase domain (His-148 to Tyr, Sidrauski et al., 1996, or Leu-68 to Ser, D.R. McMillan, unpublished data) and are severely defective in processing of the *HAC1* mRNA. The His-148/Tyr substitution did not affect the pre-tRNA splicing activity of the tRNA ligase (Sidrauski et al., 1996), while two previously characterized *rlg1*^{ts} mutants (Phizicky et al., 1992) were found to be only partially defective in UPR signalling (D.R. McMillan, unpublished data). Yeast strains bearing *sen2* mutations that affect the tRNA endonuclease (Ho et al., 1990) appear wild-type for UPR signalling (D.R. McMillan, unpublished data). The basis for the differential effects of the different *rlg* mutations on *HAC1* pre-mRNA splicing and pre-tRNA splicing remains to be determined, as does whether the *HAC1* pre-mRNA is cleaved by another endonuclease [perhaps the Ern1p C-terminal tail (Sidrauski et al., 1996)], distinct from tRNA endonuclease.

Is the UPR in mammals and yeast regulated by a conserved pathway?

The UPR pathways in mammalian and yeast cells share some common features, suggesting evolutionary conservation and reflecting the importance of this response for ensuring the survival of a cell under conditions of stress. The UPR in both types of organisms is induced by similar stimuli and with similar kinetics, suggesting similar modes of regulation. Both organisms appear to use a decrease in the concentration of uncomplexed BiP molecules as a signal to initiate the UPR, and both responses apparently depend on the activity of one or more protein kinases to transmit the UPR signal to the nucleus. Finally, the UPR in both mammalian and yeast cells is regulated at the level of transcription via the highly conserved UPRE present upstream of all UPR-regulated genes.

However, some evolutionary divergence between the yeast and mammalian systems is also apparent. The promoter of the yeast BiP gene (*KAR2*) contains a functional HSE and is responsive to the effects of heat shock (primarily the accumulation of unfolded polypeptides in the cytosol), whereas the mammalian *grp78* gene is not inducible by heat shock. Similarly, the involvement of the CCAAT-like motifs upstream of the mammalian *grp78* gene in UPR regulation is not clear, since these motifs do not appear to be present upstream of the yeast BiP gene. While the UPRE in yeast genes is necessary and sufficient for UPR regulation of gene expression, the UPRE in mammalian genes appears to be only one component of a complex system of transcriptional control of UPR-regulated genes that acts through a series of functionally redundant elements.

The parallels between the UPRs in yeast and mammals, however, are strong enough to suggest that information gained through studies of the UPR pathway in one organism, should provide significant insights into the components and regulation of the UPR pathway in the other.

References

Alexandre, S., Nakaki, T., Vanhamme, L., and Lee, A.S. (1991). A binding site for the cyclic adenosine 3′,5′-monophosphate-response element-binding protein as a regulatory element in the grp78 promoter. Mol. Endocrinol. **5**, 1862–1872.

Apostol, B.L., Westaway, S.K., Abelson, J., and Greer, C.L. (1991). Deletion analysis of a multifunctional yeast tRNA ligase polypeptide. J. Biol. Chem. **266**, 7445–7455.

Bartlett, J., Luethy, J., Carlson, S., Sollott, S., and Holbrook, N. (1992). Calcium ionphore A23187 induces expression of the growth arrest and DNA damage-inducible CCAAT/enhancer-binding protein (C/EBP)-related gene, gadd153. J. Biol. Chem. **267**, 20465–20470.

Blumberg, H. and Silver, P.A. (1991). A homologue of the bacterial heat-shock gene DnaJ that alters protein sorting in yeast. Nature **349**, 627–630.

Bork, P. and Sander, C. (1993). A hybrid protein kinase-RNase in an interferon-induced pathway? FEBS Lett. **334**, 149–152.

Cao, X., Zhou, Y., and Lee, A.S (1995). Requirement of tyrosine- and serine/threonine kinases in the transcriptional activation of the mammalian grp78/BiP promoter by thapsigargin. J. Biol. Chem. **270**, 494–502.

Chang, S.C., Erwin, A.E., and Lee, A.S. (1989). Glucose-regulated protein (GRP94 and GRP78) genes share common regulatory domains and are coordinately regulated by common trans-acting factors. Mol. Cell. Biol. **9**, 2153–2162.

Cox, J.S. and Walter, P. (1996). A novel mechanism for regulating activity of a transcription factor that controls the unfolded protein response. Cell **87**, 391–404 (1996).

Cox, J.S., Shamu, C.E., and Walter, P. (1993). Transcriptional induction of genes encoding endoplasmic reticulum resident proteins requires a transmembrane protein kinase. Cell **73**, 1197–1206.

Craven, R.A., Egerton, M., and Stirling, C.J. (1996). A novel Hsp70 of the yeast ER lumen is required for the efficient translocation of a number of protein precursors. EMBO J. **15**, 2640–2650.

Drgonová, J., Drgon, T., Tanaka, K., Kollár, R., Chen, G.-C., Ford, R.A., Chan, C.S.M., Takai, Y., and Cabib, E. (1996). Rho1p, a yeast protein at the interface between cell polarization and morphogenesis. Science **272**, 277–279.

Gething, M.J. and Sambrook, J.F. (1992). Protein folding in the cell. Nature **355**, 33–45.

Graham, K.S., Lee, A., and Sifers, R.N. (1990). Accumulation of the insoluble PiZ variant of human α1-antitrypsin within the hepatic endoplasmic reticulum does not elevate the steady-state level of grp78/BiP. J. Biol. Chem. **265**, 20463–20468.

Greer, C.L., Peebles, C.L., Gegenheimer, P., and Abelson, J. (1983). Mechanism of action of a yeast RNA ligase in tRNA splicing. Cell **32**, 537–546.

Hardwick, K.G., Lewis, J., Semenza, N., Dean, N., and Pelham, H.R.B. (1990). *EDR1*, a yeast gene required for the retention of luminal endoplasmic reticulum proteins, affects glycoprotein processing in the Golgi apparatus. EMBO J. **6**, 620–630.

Hartl, F.U. (1996). Molecular chaperones in cellular protein folding. Nature **381**, 571–580.

Herskowitz, I. (1995). MAP kinase pathways in yeast: for mating and more. Cell **80**, 187–197.

Ho, C.K., Rauhut, R., Vijayraghavan, U., and Abelson, J. (1990). Accumulation of pre-tRNA splicing '2/3' intermediates in a *Saccharomyces cerevisiae* mutant. EMBO J. **9**, 1245–1252.

Kamada, Y., Qadota, H., Python, C.P., Anraku, Y., Ohya, Y., and Levin, D.E. (1996). Activation of yeast protein kinase C by Rho1 GTPase. J. Biol. Chem. **271**, 9193–9196.

Kawahara, T., Yanagi, H., Yura, T., and Mori, K. (1997). ER stress-induced mRNA splicing permits synthesis of transcription factor Hac1p that activates the unfolded protein response. Mol. Biol. Cell. In press.

Kohno, K., Normington, K., Sambrook, J., Gething, M.J., and Mori, K. (1993). The promoter region of the yeast KAR2 (BiP) gene contains a regulatory domain that responds to the presence of unfolded proteins in the endoplasmic reticulum. Mol. Cell. Biol. **13**, 877–890.

Kozutsumi, Y., Segal, M., Normington, K., Gething, M.J., and Sambrook, J. (1988). The presence of malfolded proteins in the endoplasmic reticulum signals the induction of glucose regulated proteins. Nature **332**, 462–464.

Lee, A.S. (1987). Coordinated regulation of a set of genes by glucose and calcium ionophores in mammalian cells. Trends Biochem. Sci. **12**, 20–23.

Lee, A.S. (1992). Mammalian stress response: induction of the glucose-regulated protein family. Curr. Opin. Cell Biol. **4**, 267–273.

Levin, D.E. and Errede, B. (1995). The proliferation of MAP kinase signaling pathways in yeast. Curr. Opin. Cell Biol. **7**, 197–202.

Li, L.-J., Li, V., Ferrario, A., Rucker, N., Liu, E.S., Wong, S., Gomer, C.J., and Lee, A.S. (1992). Establishment of a chinese hamster ovary cell line that expresses grp78 antisense transcripts and suppresses A23187 induction of both GRP78 and GRP94. J. Cell. Physiol. **153**, 575–582.

Li, W.W., Alexandre, S., Cao, X., and Lee, A.S. (1993). Transactivation of the grp78 promoter by Ca^{2+} depletion. J. Biol. Chem. **268**, 12003–12009.

Li, W.W., Sistonen, L., Morimoto, R.I., and Lee, A.S. (1994). Stress induction of the mammalian GRP78/BiP protein gene: in vivo genomic footprinting and identification of p70CORE from human nuclear extract as a DNA-binding component specific to the stress regulatory element. Mol. Cell. Biol. **14**, 5533–5546.

Li, W.W., Hsiung, Y., Zhou, Y., Roy, B., and Lee, A.S. (1997). Induction of the mammalian GRP78/BiP gene by Ca^{2+} depletion and formation of aberrant proteins: Activation of the conserved stress-inducible *grp* core promoter by the human nuclear factor YY1. Mol. Cell. Biol. **17**, 54–60.

Lis, J.T., Xiao, H., and Perisic, O. (1990). Modular units of heat shock regulatory regions: structure and function. In *Stress Proteins in Biology and Medicine* (R.I. Morimoto, A. Tissiéres, and C. Georgopoulos, ed.). Cold Spring Harbor Laboratory Press, New York, pp. 411–428.

Liu, E.S. and Lee, A.S. (1991). Common sets of nuclear factors binding to the conserved promoter sequence motif of two coordinately regulated ER protein genes, GRP78 and GRP94. Nucl. Acids Res. **19**, 5425–5431.

McMillan, D.R., Gething, M.-J., and Sambrook, J. (1994). The cellular response to unfolded proteins: intercompartmental signaling. Curr. Biol. **5**, 540–545.

Morgan, W.D. (1989). Transcription factor Sp1 binds to and activates a human hsp70 gene promoter. Mol. Cell. Biol. **9**, 4099–4104.

Mori, K., Sant, A., Kohno, K., Normington, K., Gething, M.J., and Sambrook, J.F. (1992). A 22-bp cis-acting element is necessary and sufficient for the induction of the yeast KAR2 (BiP) gene by unfolded proteins. EMBO J. **11**, 2583–2593.

Mori, K., Ma, W., Gething, M.J., and Sambrook, J. (1993). A transmembrane protein with a cdc2+/CDC28-related kinase activity is required for signaling from the ER to the nucleus. Cell **74**, 743–756.

Mori, K., Kawahara, T., Yoshida, H., Yanagi, H., and Yura, T. (1996). Signaling from the ER to the nucleus: transcription factor with a basic-leucine zipper motif is required for the unfolded protein-response pathway. Genes Cells **1**, 803–817.

Ng, D.T.W., Watowich, S.S., and Lamb, R.A. (1992). Analysis in vivo of GRP78-BiP/substrate interactions and their role in induction of the GRP78-BiP gene. Mol. Biol. Cell **3**, 143–155.

Nikawa, J. and Yamashita, S. (1992). IRE1 encodes a putative protein kinase containing a membrane-spanning domain and is required for inositol phototrophy in Saccharomyces cerevisiae. Mol. Microbiol. **6**, 1441–1446.

Normington, K., Kohno, K., Kozutsumi, Y., Gething, M.J., and Sambrook, J. (1989). S. cerevisiae encodes an essential protein homologous in sequence and function to mammalian BiP. Cell **57**, 1223–1236.

Nunnari, J. and Walter, P. (1996). Regulation of organelle biogenesis. Cell **84**, 389–394.

Partaledis, J.A. and Berlin, V. (1993). The FKB2 gene of Saccharomyces cerevisiae, encoding the immunosuppressant-binding protein FKBP-13, is regulated in response to accumulation of unfolded proteins in the endoplasmic reticulum. Proc. Natl. Acad. Sci. USA **90**, 5450–5454.

Peebles, C.L., Gegenheimer, P., and Abelson, J. (1983). Precise excision of intervening sequences from precursor transfer RNA by a membrane-associated endonuclease. Cell **32**, 525–536.

Phizicky, E.M., Schwartz, R.C., and Abelson, J. (1986). Saccharomyces cerevisiae tRNA ligase. Purification of the protein and isolation of the structural gene. J. Biol. Chem. **261**, 2978–2986.

Phizicky, E.M., Consaul, S.A., Nehrke, K.W., and Abelson, J. (1992). Yeast tRNA ligase mutants are noonviable and accumulate tRNA splicing intermediates. J. Biol. Chem. **267**, 4577–4582.

Price, B.D. and Calderwood, S.K. (1992). Gadd45 and Gadd153 messenger RNA levels are increased during hypoxia and after exposure of cells to agents which elevate the levels of the glucose-regulated proteins. Cancer Res. **52**, 3814–3817.

Price, B.D., Mannheim-Rodman, L.A. and Calderwood, S.K. (1992). Brefeldin A, thapsigargin and AlF$_4$ stimulate the accumulation of GRP78 mRNA in a cycloheximide-dependent manner, whilst induction by hypoxia is independent of protein synthesis. J. Cell Physiol. **152**, 545–552.

Resendez, E., Jr., Ting, J., Kim, K.S., Wooden, S.K., and Lee, A.S. (1986). Calcium ionophore A23187 as a regulator of gene expression in mammalian cells. J. Cell Biol. **103**, 2145–2152.

Resendez, E., Jr., Wooden, S.K., and Lee, A.S. (1988). Identification of highly conserved regulatory domains and protein-binding sites in the promoters of the rat and human genes encoding the stress-inducible 78-kilodalton glucose-regulated protein. Mol. Cell. Biol. **8**, 4579–4584.

Ron, D. and Habener, J.F. (1992). CHOP, a novel developmentally regulated nuclear protein that dimerizes with transcription factors C/EBP and LAP and functions as a dominant negative inhibitor of gene transcription. Genes Dev. **6**, 439–453.

Rose, J.K. and Doms, R.W. (1988). Regulation of protein export from the endoplasmic reticulum. Annu. Rev. Cell Biol. **4**, 257–288.

Rose, M.D., Misra, L.M., and Vogel, J.P. (1989). KAR2, a karyogamy gene, is the yeast homologue of mammalian BiP/GRP78. Cell **57**, 1211–1221.

Rothblatt, J. (1994a). Protein translocation and maturation in the yeast ER. In *Guidebook to the Secretory Pathway* (J. Rothblatt, P. Novick, and T.H. Stevens, ed). Sambrook & Tooze at Oxford University Press, Oxford, pp. 23–26.

Rothblatt, J. (1994b). Protein translocation and maturation in the mammalian ER. In *Guidebook to the Secretory Pathway* (J. Rothblatt, P. Novick, and T.H. Stevens, ed). Sambrook & Tooze at Oxford University Press, Oxford, pp. 65–67.

Roy, B. and Lee, A.S. (1995). Transduction of calcium stress through interaction of the human transcription factor CBF with the proximal CCAAT regulatory element of the *grp78*/BiP promoter. Mol. Cell. Biol. **15**, 2263–2274.

Ruis, H. and Schüller, C. (1995). Stress signalling in yeast. BioEssays **17**, 959–965.

Schlenstedt, G., Harris, S., Risse, B., Lill, R., and Silver, P.A. (1995). A yeast DnaJ homologue, Scj1p, can function in the endoplasmic reticulum with BiP/Kar2p via a conserved domain that specifies interactions with Hsp70s. J. Cell Biol. **129**, 979–988.

Shamu, C.E. and Walter, P. (1996). Oligomerization and phosphorylation of the Ire1p kinase during intracellular signalling from the endoplasmic reticulum to the nucleus. EMBO J. **15**, 3028–3039.

Shamu, C.E., Cox, J.S., and Walter, P. (1994). The unfolded-protein-response pathway in yeast. Trends Cell Biol. **4**, 56–60.

Shiu, R.P.C., Pouyssegur, J., and Pastan, I. (1977). Glucose depletion accounts for the induction of two transformation-senstive proteins in Rous sarcoma virus-transformed chick embryo fibroblasts. Proc. Natl. Acad. Sci. USA **74**, 3840–3844.

Sidrauski, C., Cos, J.S., and Walter, P. (1996). tRNA ligase is required for regulated mRNA splicing in the unfolded protein response. Cell **87**, 405–413.

Tachibana, C. and Stevens, T.H. (1992). The yeast *EUG1* gene encodes an endoplasmic reticulum protein that is functionally related to protein disulfide isomerase. Mol. Cell. Biol. **12**, 4601–4611.

Tokunaga, M., Kawamura, A., and Kohno, K. (1992). Purification and characterization of BiP/Kar2 protein from Saccharomyces cerevisiae. J. Biol. Chem. **267**, 17553–17559.

Ubeda, M., Wang, X.-Z., Zinszner, H., Wu, I., Harbener, J.F., and Ron, D. (1996). Stress-induced binding of the transcription factor CHOP to a novel DNA control element. Mol. Cell. Biol. **16**, 1479–1489.

Wang, X.-Z and Ron, D. (1996). Stress-induced phosphorylation and activation of the transcription factor CHOP (GADD153) by p38 MAP kinase. Science **272**, 1347–1349.

Welihinda, A.A. and Kaufman, R.J. (1996). The unfolded protein response pathway in *Saccharomyces cerevisiae*. J. Biol. Chem **271**, 18181–18187.

Westaway, S.K., Phizicky, E.M., and Abelson, J. (1988). Structure and function of the yeast ERNA ligase gene. J. Biol. Chem. **263**, 3171–3176.

White, M.J., Lopes, J.M., and Henry, S.A. (1991). Inositol metabolism in yeasts. Adv. Microb. Physiol. **32**, 1–51.

Wooden, S.K., Li, L.J., Navarro, D., Qadri, I., Pereira, L., and Lee, A.S. (1991). Transactivation of the grp78 promoter by malfolded proteins, glycosylation block, and calcium ionophore is mediated through a proximal region containing a CCAAT motif which interacts with CTF/N-I. Mol. Cell. Biol. **11**, 5612–5623.

Xu, Q., Teplow, D., Lee, T.D., and Abelson, J. (1990). Domain structure in yeast tRNA ligase. Biochemistry **29**, 6132–6138.

■ *Carolyn McNees and Mary-Jane Gething:*
Department of Biochemistry and Molecular Biology
University of Melbourne
Parkville
Victoria 3052, Australia
Tel. 61 3 9344 5948
Fax. 61 3 9347 9109
E-mail: gething@ariel.ucs.unimelb.edu.au

Index

7B2 201
25 kDa growth-related protein p25, see Hsp25 285

ADF, adult T cell leukaemia-derived factor, see thioredoxin 343
adult T cell leukaemia-derived factor, ADF, see thioredoxin 343
α-tubulin, folding of 226
αA-crystallin 288
αB-crystallin 288
Anc2p, see CCTδ 220
Anj1 112
APG-1, see Osp94 84
APG-2, see HSP110/SSE overview 73
APPG, see 7B2 201
auxillin 124
AV25 protein, see sHSP overview 269

B25.3, see GrpE 137
b-70, see plant BiP 38
Bacillus brevis Bdb 330
Bacillus subtilis ClpC 243
bacteriophage T4 Gp31 protein 185
band VII, see calnexin 299
Bdb 330
β-tubulin, folding of 226
Bin2p, see CCTγ 218
Bin3p, see CCTβ 217
BiP, mammalian 59
  plant 38
  of *S. cerevisiae*, see Kar2p 33
  of *S. pombe* 37
  of *Trypanosoma brucei* 41
Bradyrhizobium japonicum TlpA 332
BS2, *T. brucei* protein disulfide isomerase homologue 342

C14B9.1 protein, see sHSP overview 269
C3H strain specific antigen, see Pbp74 65
  see also mammalian mitochondrial Hsp70 67
C62.5, see HtpG 151
CAB-63, see calreticulin 304
CaBP1 354
CaBP2, see ERp72 353
CABP3, see calreticulin 304
calnexin, mammalian 299
  of *S. cerevisiae*, see Cne1 296

  of *S. pombe*, see Cnx1 298
calnexin proteins, overview 293
calreticulin, mammalian 304
  overview 293
carboxypeptidase Y 472
carboxypeptidase yscY 472
CbpA 98
CBP140, see Grp170 85
CCCS1, see cysteine string proteins 115
CcmG 331
CCTs, see cytosolic chaperonins 207
CCTα 215
CCTβ 217
CCTγ 218
CCTδ 220
CCTϵ 221
CCTζ 222
CCTη 224
CCTθ 225
Cct1p, see CCTα 215
Cct2p, see CCTβ 217
Cct3p, see CCTγ 218
Cct4p, see CCTδ 220
Cct5p, see CCTϵ 221
Cct6p, see CCTζ 222
Cct7p, see CCTη 224
Cct8p, see CCTθ 225
Cer1p, see Ssi1p 36
chaperonin 10, see CPN10 167
  see also GroES 183
  see also Hsp10 191,199
chaperonin 60, see CPN60 167, 197
  see also GroEL 178
  see also Hsp60 189
Chl chaperonin 60, see plastid Cpn60 192
Chl chaperonin 10, see plastid Cpn21 194
chloroplast chaperonin 60, see plastid Cpn60 192
chloroplast-localized Clp proteins 255
chromobindin A, see cytosolic chaperonins 207
Cin1, see cofactor D 226
clathrin uncoating ATPase, see Hsc70 53
Clp proteins, see overview of

  HSP100/Clp proteins 231
Clp proteins, chloroplast localized 255
ClpA 236
ClpB, of *E. coli* 238
  of *Leishmania*, see Hsp100 259
  malaria plasmid 264
  plant, see Hsp101 253
  of *Synechococcus* sp. PCC 7942 246
ClpB79 238
ClpB93 238
ClpC, of *B. subtilis* 243
  malaria plasmid 261
  plant 255
  of *Synechococcus* sp. PCC 7942 247
ClpP, see ClpA 236
ClpX 240
ClpY 242
Cne1 296
Cnx1 298
cochaperonin, see CPN10 167
CodX, see ClpY 242
cofactor A 226
cofactor D 226
cofactor E 226
colligin, see Hsp47 465
component A of protease Ti, see ClpA 236
CPN10 protein family, overview 167
Cpn10, mammalian 199
  plastid, see Cpn21 194
  of *S. cerevisiae*, see Hsp10 191
Cpn21, plastid 194
Cpn24, see plastid Cpn21 194
Cpn60, mammalian 197
  plastid 192
  of *S. cerevisiae*, see Hsp60 189
CPN60 protein family, overview 167
CPH, see cyclophilin A of *S. cerevisiae* 372
CPH2, see vertebrate cyclophilin B 388
Cpr1, see cyclophilin A of *S. cerevisiae* 372
Cpr2, see cyclophilin B of *S. cerevisiae* 373
Cpr3 378

Cpr4, see Scc3 381
Cpr5, see CypD 380
Cpr6 376
Cpr7 377
CRG, see cyclophilin B of S. cerevisiae 373
CRP55, see calreticulin 304
CSA, see Pbp74 65
    see also mammalian mitochondrial Hsp70 67
CutA2, see DsbD 326
cyclophilins, E. coli 370
    N. crassa 382
cyclophilin 6, see Cpr6 376
cyclophilin 7, see Cpr7 377
cyclophilin 40 394
cyclophilin A, mammalian 386
    of S. cerevisiae 372
cyclophilin B, of S. cerevisiae 373
    vertebrate 388
cyclophilin C, mammalian 390
    of S. cerevisiae, see Cpr3 378
cyclophilin D, mammalian 392
    of S. cerevisiae, see CypD 380
cyclophilin PPIases, overview 359
cyclophilin S1, see vertebrate cyclophilin B 388
cyclophilins, overview 359
cyclosporin A-binding protein, see E. coli cyclophilin 370
CycY, of R. capsulatus 331
cysteine string proteins 115
Cyp1, see cyclophilin A of S. cerevisiae 372
Cyp2, of S. cerevisiae, see cyclophilin B 373
    vertebrate, see cyclophilin B 388
Cyp3, mammalian, see cyclophilin D 392
    of S. cerevisiae, see Cpr3 378
Cyp-18, mammalian, see cyclophilin A 386
Cyp18, of N. crassa 382
Cyp20, of N. crassa 382
Cyp21, of N. crassa 382
Cyp40, see mammalian cyclophilin 40 394
Cyp48, of N. crassa 382
CypA, mammalian, see cyclophilin A 386
    of S. cerevisiae, see cyclophilin A 372
CypB, vertebrate, see cyclophilin B 388
CypC, see mammalian cyclophilin C 390
CypD, see mammalian cyclophilin D 392
    of S. cerevisiae 380

Cyp-S1, see vertebrate cyclophilin B 388
cytosolic chaperonins, overview 207

DipZ, see DsbD 326
DjlA, see RcsG 100
DNA replication, role of molecular chaperones 481
DnaJ-like proteins from leek 113
DnaJ-related (HSP40) proteins, overview 89
DnaJ 95
DnaK 22
DnaK, structure of substrate binding domain 18
dodo 440
DroE1p 141
Drosophila melanogaster chaperones and folding catalysts
    dodo 440
    DroE1p 141
    FKBP39 419
    Hsc3p 45
    Hsc4p 42
    Hsp83 154
    Hsp82, see Hsp83 154
    Hsp90, see Hsp83 154
    ninaA 384
    small heat shock proteins 280
    Tid56 117
DsbA 318
    3-dimensional structure 321
DsbB 322
DsbC 324
DsbD 326
DsbE 327
DsbX, see DsbB 322

E72, see ClpB 238
E89, see ClpB 238
early pregnancy factor, see Cpn10 199
Ens1p, see Ssc1 30
endoplasmic reticulum chaperones and folding catalysts
    mammalian BiP 59
    mammalian CaBP1 354
    mammalian calnexin 299
    mammalian calreticulin 304
    mammalian ERp61 351
    mammalian ERp72 353
    mammalian FKBP13 432
    mammalian Grp94 161
    mammalian Grp170 85
    mammalian PDI 348
    N. crassa Cyp21 382
    N. crassa FKBP22 416

    plant BiP 38
    D. melanogaster Hsc3p 45
    D. melanogaster ninaA 384
    S. cerevisiae CYPD/Cpr5 378
    S. cerevisiae Eug1p 339
    S. cerevisiae FKBP13 411
    S. cerevisiae Kar2p 33
    S. cerevisiae Mpd1p and Mpd2p 341
    S. cerevisiae PDI 335
    S. cerevisiae Sec63p 108
    S. cerevisiae Ssi1p 36
    S. cerevisiae Ssj1p 110
    S. pombe BiP 37
endoplasmic reticulum, quality control and molecular chaperones 515
endoplasmic reticulum, translocation and molecular chaperones 506
endoplasmin, see Grp94 161
EPF, see Cpn10 199
E. coli PDI, see DsbA 318
E. coli protein disulfide isomerase, see DsbA 318
estrogen receptor-binding cyclophilin, see Cyp40 394
estrogen receptor-related protein, see sHSP overview 269
estrogen-regulated '24K' protein, see sHSP overview 269
ERBC, see Cyp40 394
ERD1 255
Ern1 545
ERp61 351
ERp72 353
ERp99, see Grp94 161
Eug1 339
Escherichia coli chaperones and folding catalysts
    CbpA 98
    ClpA 236
    ClpB 238
    ClpX 240
    ClpY 242
    cyclophilins 370
    DnaK 22
    DnaJ 95
    DsbA 318
    DsbB 322
    DsbC 324
    DsbD 326
    DsbE 327
    FkpA 402
    FtsH 451
    glutaredoxin 316
    GroEL 173
    GroES 179, 183
    GrpE 137

HtpG 151
Hsc66 25
Hsp90, see HtpG 151
IbpA and IbpB 273
PapD 463
parvulin 434
RcsG 100
SecB 449
SurA 436
thioredoxin 314
trigger factor 404
Ess1 438

F68.5, see ClpB 238
F84.1, see ClpB 238
FaeE, see PapD 463
FanE, see PapD 463
FimC, see PapD 463
fish HSP70 proteins 47
FK506-binding proteins (FKBPs), overview 359
FKBP PPIases, overview 359
FKBPs, overview 359
　3-dimensional structure 397
　of N. crassa 416
FKBP12, mammalian 420
　of S. cerevisiae 408
FKBP12.6 422
FKBP13, mammalian 432
　of N. crassa 416
　of S. cerevisiae 411
FKBP22, of N. crassa 416
FKBP25, mammalian 424
FKBP39, of D. melanogaster 419
FKBP51, mammalian 428
　see also FKBP52 430
FKBP52, mammalian 430
FKBP54, see FKBP51 428
FKBP59, see FKBP52 430
FKBP70, see Npi46p 414
FKBP73, plant 417
FKBP-C (cardiac), see FKBP12.6 422
Fkb1, see FKBP12 of S. cerevisiae 408
Fkb2, see FKBP13 of S. cerevisiae 411
FkpA 402
Fpr2, see FKBP13 of S. cerevisiae 411
Fpr3, see Npi46p 414
FtsH 451

glutaredoxin, mammalian 346
　of E. coli 316
glutathione-homocystine transhydrogenase, see glutaredoxin 346
glutathione-insulin transhydrogenase, see PDI 348

glycosylation site binding protein, see S. cerevisiae PDI 335
Gp31, of T4 bacteriophage 185
gp48, see Hsp47 465
gp96, see Grp94 161
GroEL, of E. coli, structure and function 173
　of T. thermophilus 187
GroES, of E. coli 183
　structure 179
GroES, of T. thermophilus 187
GroPC, see DnaK 22
GroP-like protein, see GrpE 137
group I chaperonins, see CPN60 167
group II chaperonins, see cytosolic chaperonins 207
Grp58, see ERp61 351
Grp75, see Pbp74 65
　see also mammalian mitochondrial Hsp70 67
Grp78, see mammalian BiP 59
Grp94 161
Grp170 85
GrpE 137
GrpE family of proteins, overview 133
GrpEp, see Mge1 139
Grx1, see E. coli glutaredoxin 316
Grx2, see E. coli glutaredoxin 316
Grx3, see E. coli glutaredoxin 316

Hac1p 545
HBI, see FKBP52 430
Hdj1, see Hsp40 121
Hdj2 123
heat shock genes: transcriptional regulation in eukaryotes 534
heat shock genes: transcriptional regulation in E. coli 525
heat shock response in eukaryotes 534
heat shock response in E. coli 525
HelX 331
HflB, see FtsH 451
Hip 455
HIP-70, see ERp61 351
Hop (Hsc70/Hsp90 organizing protein) 56
human chaperones and folding catalysts
　Hdj2 123
　hsp70 genes 49
　neurone-specific Hsj1 proteins 126
Hs1t, see IbpA 273
Hs1s, see IbpB 273
Hsc3p 45
Hsc4p 45
Hsc66 25

Hsc70, three dimensional structure 13
　of D. melanogaster, see Hsc4p 42
　mammalian 53
　of S. cerevisiae, see Ssa proteins 26
Hsc72 of D. melanogaster, see Hsc3p 42
Hsc73, mammalian, see Hsc70 53
Hsc82, of S. cerevisiae, see Hsp90 152
hscA, see Hsc66 25
HSDJ, see Hdj2 123
HSF, see heat shock genes: transcriptional regulation in eukaryotes 535
Hsj1 126
Hsj1 proteins, neurone-specific 126
HslV, see ClpY 242
Hsp10, mammalian, see Cpn10 199
　plant, see plastid Cpn21 194
　of S. cerevisiae 191
Hsp22, of D. melanogaster 280
Hsp23, of D. melanogaster 280
Hsp25, mammalian 285
　from mouse and rat 285
　see also Hsp27 283
Hsp26, of D. melanogaster 280
　of S. cerevisiae 274
Hsp27, mammalian 283
　of D. melanogaster 280
Hsp28, mammalian, see Hsp27 283
HSP40 (DnaJ-related) protein family, overview 89
Hsp40, mammalian 121
Hsp47, vertebrate 465
Hsp56, see FKBP52 430
Hsp60, mammalian, see Cpn60 197
　plant, see plastid Cpn60 192
　of S. cerevisiae 189
Hsp70, mammalian 53
hsp70 genes of mice and man 49
HSP70 protein family, overview 3
　phylogenetic tree 4
　sequence comparisons 6
　sequence accession numbers 11
Hsp70h 82
Hsp70RY, see Hsp70h 82
Hsp72, mammalian, see Hsp70 53
Hsp78 251
Hsp82
　of D. melanogaster, see Hsp83 154
　of S. cerevisiae, see Hsp90 152
Hsp83, of D. melanogaster 154
Hsp84/86, mammalian, see Hsp90 158
Hsp85, mammalian, see Hsp90 158
Hsp88, of N. crassa 77
Hsp89, mammalian, see Hsp90 158

Hsp90, of *D. melanogaster*, see Hsp83 154
  mammalian 158
  plant 156
  of *S. cerevisiae* 152
HSP90 protein family, overview 147
HSP100 protein (Clp) family, overview 231
Hsp100, of *Leishmania* 259
  mammalian 264
Hsp101/ClpB, plant 253
Hsp104, of *S. cerevisiae* 249
Hsp105, mammalian, see Hsp110 81
Hsp107, mammalian, see Hsp110 81
Hsp108, mammalian, see Grp94 161
HSP110/SSE protein family, overview 73
Hsp110, mammalian 81
Hsp112, see Hsp110 81
HSP-binding immunophilin (HBI), see FKBP52 430
HtpG 151
Hup (Hsc70 unbinding protein) 56

IAP-25, see sHSP overview 269
IarA, see DsbA 318
IarB, see DsbB 322
IbpA 273
IbpB 273
IEF SSP 3251, see p60 458
immunoglobulin heavy chain binding protein
  see BiP, mammalian 59
inclusion body-associated protein A, see IbpA 273
inclusion body-associated protein B, see IbpB 273
IP90, see calnexin 299

j1585p, see Cpr7 377
J6 protein, see Hsp47 465

Kar2p 33

LapA, see ClpY 242
Ldj1, of leek 113
Ldj2, of leek 113
Legionella pneumophila Mip 399
Legionella pneumophila FKBP25 mem, see Mip 399
Leishmania Hsp100 259
Lhs1p, of *S. cerevisiae*, see Ssi1p 36
LopC, see ClpX 240
low molecular mass Hsp, see sHSP overview 269
low molecular weight Hsp, see sHSP

overview 269
LpFKBP-25, see Mip 399
LpMip, see Mip 399
Lpn25, see Mip 399
luminal binding protein, see plant BiP 38

macrophage infectivity potentiator, see Mip 399
malaria plastid ClpB 264
malaria plastid ClpC 261
Mas5, see Ydj1 102
mammalian chaperones and folding catalysts
  7B2 201
  auxillin 124
  BiP 59
  CaBP1 354
  calnexin 299
  calreticulin 304
  Cpn10 199
  Cpn60 197
  α-crystallins 288
  cyclophilin 40 394
  cyclophilin A 386
  cyclophilin B 388
  cyclophilin C 390
  cyclophilin D 392
  cytosolic chaperonins 207, 215-28
  FKBP12 420
  FKBP12.6 422
  FKBP13 432
  FKBP25 424
  FKBP51 428
  FKBP52 430
  ERp61 351
  ERp72 353
  glutaredoxin 346
  Grp94 161
  Grp170 85
  Hdj2 123
  Hip 455
  Hsc70 13, 53
  Hsj1 126
  Hsp25 285
  Hsp27 283
  Hsp40 121
  Hsp47 465
  Hsp70 53
  Hsp70h 82
  Hsp90 158
  Hsp100 264
  Hsp110 81
  MSF 453
  Osp94 84
  p23 457
  p60 458
  Pbp74 65
  PDI 348

  Pin1 443
  Protein disulfide isomerase 348
  Prp73 58
  mitochondrial GrpE 142
  mitochondrial Hsp70 67
  mt-Hsp70 67
  Mtj1 129
  STCH 69
  thioredoxin 343
Mdj1p, of *S. cerevisiae* 106
microsomal triglyceride transfer protein small subunit, see PDI 348
Mip 399
mitochondrial biogenesis and molecular chaperones 499
mitochondrial chaperones and folding catalysts
  Cpn10, mammalian 199
  Cpn60, mammalian 197
  Cpr3p, of *S. cerevisiae* 378
  Cyclophilin D, mammalian 392
  DroE1, of *D. melanogaster* 141
  GrpE, mammalian mitochondrial 142
  Hsp70, mammalian mitochondrial 67
  Hsp10, of *S. cerevisiae* 191
  Hsp60, of *S. cerevisiae* 189
  Hsp70, of *S. cerevisiae*, see Ssc1 30
  Mdj1p, of *S. cerevisiae* 106
  Mge1, of *S. cerevisiae* 139
  Pbp74, mammalian 65
  Ssc1, of *S. cerevisiae* 30
  Ssh1p, of *S. cerevisiae* 32
mitochondrial cyclophilin of *S. cerevisiae*, see Cpr3 378
mitochondrial Hsp70, mammalian 67
  of *S. cerevisiae*, see Ssc1 30
mitochondrial protein import and molecular chaperones 499
mitochondrial import stimulation factor, see MSF 453
mortalin, see mammalian mitochondrial Hsp70 67
mouse *hsp70* genes 49
Mpd1 341
Mpd2 341
MSF 453
Msi3, see Sse1 76
Mt-GrpE, mammalian 142
  of *S. cerevisiae*, see Mge1 139
Mt-Hsp10, see mammalian Cpn10 199
  of *S. cerevisiae*, see Hsp10 191
Mt-Hsp60, see mammalian Cpn60 197
  of *S. cerevisiae*, see Hsp60 189
Mt-Hsp70, see mammalian mitochondrial Hsp70 67

Mt-Hsp75, see mammalian mitochondrial Hsp70 67
Mtj1 129
murine *hsp70* genes 49

Neurospora crassa chaperones and folding catalysts
   cyclophilins 382
   FKBPs 416
   Hsp88 77
neuroendocrine peptide 7B2 201
ninaA 384
Npi46p 414
NPL1, see Sec63p 108
NUC18, see mammalian cyclophilin A 386

Omega conotoxin receptor, see cysteine string proteins 115
Osp94 84
OstA, see SurA 436
OTK4, see FKBP12.6 422

P5, see CaBP1 354
p23 457
p25, see sHSP overview 269
p27, see sHSP overview 269
p88, see calnexin 299
p48, see Hip 455
p50, see FKBP52 430
p54, see FKBP51 428
p59, see FKBP52 430
p60 458
P71, see mammalian mitochondrial Hsp70 67
Pac2, see cofactor E 226
PapD 463
parvulin 434
parvulin PPIases, overview 359
PDI and thioredoxin-related proteins, overview 311
   homologue in *T. brucei*, BS2 342
   mammalian 348
   of *S. cerevisiae* 335
PEP4, see proteinase A 475
peptidyl-prolyl isomerases, overview 359
periplasmic unfolded protein response in *E. coli* 529
Pin1 443
plant chaperones and folding catalysts
   Anj1 112
   BiP 38
   chloroplast-localized Clp proteins 255
   FKBP73 417

Ldj1 and Ldj2, from leek 113
   Hsp90 159
   Hsp101/ClpB 253
   plastid Cpn21 194
   plastid Cpn60 192
   sHSPs 277
   small heat shock proteins (sHSPs) 277
plastid Cpn21 194
plastid Cpn60 192
PpfA, see DsbA 318
pPf203, see sHSP overview 269
PPIases, see peptidyl prolyl isomerases, overview 359
   cyclophilins, overview 359
   FKBPs, overview 359
   parvulins, overview 359
PpiC, see parvulin 434
PRC1, see carboxypeptidase Y 472
procarboxypeptidase Y 472
procarboxypeptidase yscY 472
progesterone receptor, pathway of assembly 518
prolyl-4-hydroxylase β subunit, see PDI 348
protein disulfide isomerase, homologue in *T. brucei*, BS2 342
   mammalian 348
   of *S. cerevisiae* 335
proteinase A 475
proteinase yscA, see proteinase A 475
protein synthesis and molecular chaperones 489
protein translocation and molecular chaperones 506
Prp73 58
   see also Hsc70 53
Ptf1, see Ess1 438
PTL1, see Sec63p 108

Q-2, see ERp61 351
quality control in the endoplasmic reticulum 515

Rbl2, see cofactor A 226
Rbp1, see FKBP12 408
RcsG 100
Rhodobacter capsulatus HelX 331
ribosome associated chaperones 489
rotamase, see *E. coli* cyclophilins 370
Rubisco large subunit binding protein, see plastid Cpn60 192
Rubisco subunit binding protein, see plastid Cpn60 192

saccharopepsin, see proteinase A 475

sea urchin egg receptor for sperm 78
S-cyclophilin, see cyclophilin B 388
Scc3 381
Scj1p 110
SCYLP, see cyclophilin B 388
SecB 449
SecI, see trigger factor 404
Sec63p 108
secreted cyclophilin-like protein, see cyclophilin B 388
secretogranin V, see 7B2 201
sHSP family, overview 269
sHSPs, plant 277
   sequence alignments 278
   of *D. melanogaster* 280
Sis1 104
SKD3, see mammalian Hsp100 264
small heat shock protein family, overview 269
small heat shock proteins, plant 277
   of *D. melanogaster* 280
   sequence alignments 278
sp18, see mammalian cyclophilin A 386
Ssa proteins of *S. cerevisiae* 26
   Ssa1 26
   Ssa2 26
   Ssa3 26
   Ssa4 26
Ssb proteins of *S. cerevisiae* 29
   Ssb1 29
   Ssb2 29
Ssc1 30
Ssd1, see Kar2p 33
Sse proteins of *S. cerevisiae* 76
   Sse1 76
   Sse2 76
Ssh1p 32
Ssi1p 36
SurA 436
Sti1 459
subtilases, see subtilisin 471
subtilopeptidases, see subtilisin 471
subtilisin 471
Synechococcus sp. PCC 7942 chaperones and folding catalysts
   ClpB 246
   ClpC 247
   TxlA 329

T4 Gp31 185
TbBiP 41
TCP-1, see CCTα 215
TCP-1 complex, see cytosolic chaperonins 207
Tcp1p, of *S. cerevisiae*, see CCTα 215
Tcp1bp, of *S. cerevisiae*, see CCTβ 217
TCP20, human, see CCTζ 222

Tcp20p, of *S. cerevisiae*, *see* CCTζ 222
Tcsp, *see* cysteine string proteins 115
TF55/56, *see* thermosome 211
thermophilic factor of 55 kDa, *see* TF55/56 211
thermosome 211
Thermus thermophilus GroEL and GroES 187
thiol protease of 60 kDa, *see* ERp61 351
thiol protein-disulfide oxidoreductase, *see* PDI 348
thioltransferase, *see* glutaredoxin 346
thioredoxin, *E. coli* 314
  mammalian 343

thioredoxin-related glycoproteins, *see* PDI 335
thioredoxin-related proteins, overview 311
thyroid hormone binding protein, *see* PDI 348
Tid56, of *D. melanogaster* 117
TlpA, of *B. japonicum* 332
TlpB, of *R. capsulatus* 331
trigger factor 404
trxA, *see* thioredoxin 314
Trypanosoma brucei BiP 41
TRiC, *see* cytosolic chaperonins 207
TRiC-P5, *see* CCTα 218
TxlA 329

unfolded protein responses, in the endoplasmic reticulum 541
  in the *E. coli* periplasm 529
  in eukaryotic cytoplasm 534
UPR, unfolded protein response 541

yCyPB, *see* cyclophilin B of *S. cerevisiae* 373
Ydj1 102
yeast 45 kDa cyclophilin homolog, *see* Cpr6 376
Yge1, *see* Mge1 139

zuotin 105